Steve McClung

03923

D0915445

Third Edition

Principles of Field Crop Production

John H. Martin

Courtesy Professor of Agronomic Crop Science, Oregon State University, formerly Research Agronomist, Agricultural Research Service, United States Department of Agriculture

Warren H. Leonard

Late Professor of Agronomy, Colorado State University

David L. Stamp

Associate Professor, Agronomy Department, Texas Technological University

Macmillan Publishing Co., Inc.
NEW YORK

Collier Macmillan Publishers
LONDON

Copyright © 1976, Macmillan Publishing Co., Inc.

Printed in the United States of America

All rights reserved. No part of this book may be reproduced or transmitted in any form or by any means, electronic or mechanical, including photocopying, recording, or any information storage and retrieval system, without permission in writing from the Publisher.

Earlier editions, by John H. Martin and Warren H. Leonard, copyright 1949 and copyright © 1967 by Macmillan Publishing Co., Inc.

Macmillan Publishing Co., Inc.
866 Third Avenue, New York, New York 10022

Collier Macmillan Canada, Ltd.

Library of Congress Cataloging in Publication Data

Martin, John Holmes, (date)
Principles of field crop production.

Includes bibliographies and index.
1. Field crops—United States. I. Leonard, Warren H., (date) joint author. II. Stamp, David L., joint author. III. Title.
SB187.U6M3 1975 633'.00973 74-5722
ISBN 0-02-376720-0 (Hardbound)
ISBN 0-02-979500-1 (International Edition)

PRINTING 1011 YEAR 23456

Preface to the Third Edition

The extensive developments in crop research, improvement, and production methods in recent years justify a revised edition. More space is devoted to scientific principles and to the impact of cropping practices on the environment. Crop hybrids, cultivars, and pesticides are so numerous, so frequently changed, and so often merely of local interest that they are generally omitted. Many new books including symposium monographs that cover specific crops and cropping practices are now available. These are listed among the literature citations at the end of each chapter.

Much of the numerical data is expressed in metric figures, in accordance with the policy of the American Society of Agronomy. This will assist many foreign readers. The metric system may eventually be adopted in the United States.

We regret that Doctor Leonard's death in 1966 limits the improvements that might have been achieved in this revision.

Acknowledgments

The authors are deeply indebted to their associates as well as to others who have assisted in the preparation of the third edition and to Carol Ann Tornow for her assistance in typing and editing of the manuscript.

J. H. M.
D. L. S.

Preface to the Second Edition

Numerous advances in crop science, new cropping practices, new machines, improved crop varieties or hybrids, new agricultural chemicals, and shifts in the importance of many crops in different regions necessitated a revision of the first edition that would describe modern methods of crop production.

Less attention is given to varieties of field crops in this new edition. These must be treated as a local crops laboratory problem, because varieties differ from state to state as well as from year to year, especially with the current rapid adoption of new varieties and hybrids.

Acknowledgments

The authors are deeply indebted to their associates as well as to others who have assisted in the preparation of the second edition of this book. Among those whose help is gratefully acknowledged because they read portions of the manuscript, or supplied subject material, are: W. K. Bailey, W. M. Bruce, B. S. Crandall, J. C. Culbertson, C. Gordon, A. A. Hansen, C. H. Hanson, M. A. Hein, Paul Henson, H. W. Johnson, Thomas Kerr, F. A. Loeffel, J. E. McMurtrey, Jr., Walter Scholl, C. S. Slater, C. E. Steinbauer, Dewey Stewart, I. E. Stokes, and W. J. Zaumeyer.

Unless credited elsewhere, illustrations were provided by the United States Department of Agriculture.

J. H. M.
W. H. L.

Preface to the First Edition

This book presents some of the facts and fundamental principles essential to an understanding of field crop production in the United States. It is designed as a college text for a general course in field crops, especially for agricultural students who take only one course in crop production. It also should serve as a reference to those concerned with crop production. The subject matter probably is more advanced than that in other general crops books, but if so it should help raise the level of field crop instruction. The book is longer than can usually be covered in a one-semester course, but it offers a choice of subject matter to meet different institutional and local requirements. Some knowledge of botany and chemistry is desirable but not essential to an understanding of the material presented. Since the subject matter is of national scope it should be supplemented with lectures or assigned readings on local varieties and cropping practices.

* * *

The references chosen are among those that seem to be pertinent to the subjects discussed, but many other references would be as applicable. It is impossible to cite all the worthwhile published articles. Foreign references are omitted for the most part because they are not accessible to, or usable by, large undergraduate classes.

The major crops are grouped into chapters in accordance with their botanical relationships. This should help avoid confusion concerning crop plant structure and behavior. It is realized fully that the arrangement, selection, and presentation of topics and references might be better. Suggestions for improvement and reorganization of the subject matter will be very welcome.

J. H. M.
W. H. L.

Contents

Principles of

Field Crop Production

PART ONE

GENERAL PRINCIPLES OF CROP PRODUCTION

Chapter 1

The Art and Science of Crop Production

CROP PRODUCTION AS AN ART

Primitive man lived on wild game, leaves, roots, seeds, berries, and fruits.[31] * As the population increased, the food supply was not always sufficiently stable or plentiful to supply his needs. Crop production began at least 9000 years ago when domestication of plants became essential to supplement natural supplies in certain localities. The art of crop production is older than civilization, and its essential features have remained almost unchanged since the dawn of history. These features include (1) gathering and preserving the seed of the desired crop plants, (2) destroying other kinds of vegetation growing on the land, (3) stirring the soil to form a seedbed, (4) planting when the season and weather are right as shown by past experience, (5) destroying weeds, (6) protecting the crop from natural enemies, and (7) gathering, processing, and storing the products.

Farm machines merely speed the hand of man in doing these things or enable him to do the work better.

According to the story of the Creation, man originally was expected to subsist solely upon horticultural and animal foods. Only after he had tasted the fruit of the Tree of Knowledge did he consume field-crop products. Adam was banished from the Garden of Eden with the following warning: "Thorns also and thistles shall it bring forth to thee; and thou shalt eat the herb of the field. In the sweat of thy face shalt thou eat bread." It seems that the next grower of field crops clashed with a sheep-herder and was driven to still poorer lands with these words: "When thou tillest the ground, it shall not henceforth yield unto thee her strength." With this burden the problems of the crop grower have continued through the ages. Today 95 per cent of the world's tilled land is devoted to field crops.

The early husbandman cultivated a limited number of crops, the

* Superscript numbers indicate numbered references at the end of chapters.

cereals being among the first to be grown in most parts of the world. The same crop often was produced continuously on a field until low yields necessitated a shift to new land. This temporary abandonment of seemingly partly worn-out land has been almost universal in the history of agriculture. This is still common in parts of Africa, but it is also a highly effective practice in growing tobacco in southern Maryland. A modification of this practice was the introduction of bare fallow every two or three years. The primitive husbandman removed by hand the destructive insects in his fields, and appeased the gods or practiced mystic rites to drive away the evil spirits he believed to be the cause of plant diseases. With advancing civilization materials such as sulfur, brine, ashes, white-wash, soap, and vinegar were applied to plants to suppress diseases or insects.

Cultivated plants are a product of human achievement and discovery which has enabled man to provide his food and fiber needs with progressively less labor.

The first successful domestication of plants by man has recently been suggested[20] to have occurred in Thailand in Neolithic times. Remnants of rice and broadbeans or soybeans from fully 10,000 years ago were recently discovered. Emmer and barley specimens dating about 6750 B.C. were recovered at the Jarma site in Iran. Harlan[24] lists both the Middle East and Near East as centers and noncenters of agricultural origin.

Much of the record of early agriculture comes from the writings of Greek and Roman scholars such as Herodotus about 500 B.C. and Pliny about A.D. 50. Hieroglyphs of harvest scenes, and remains of plants and seeds in ancient tombs show an Egyptian agriculture as early as 5000–3400 B.C. with the cereal grains emmer and barley of major significance. This successful agricultural economy enabled men to construct pyramids and beautiful tombs and develop fine arts.

Romans of the first century A.D. intertilled crops with iron hand knives. Wood writes in 1629 (New England Prospect) in great detail about the skillful use of "clamme shell hooes," used in maize fields to control weeds. Intertillage with animal power was advocated in England in the 17th century.

The value of lime, marl, manures, and green manures for the maintenance of soil productivity was recognized 2,000 years ago. Books on agriculture written by the Romans (Pliny, Varro, and Columella) of about the 1st century A.D. describe the growing of common crops including wheat, barley, clover, and alfalfa by procedures very similar to those in use today except that more of the work was done by hand and the farm implements then used were crude.[13] However, in the experimental nursery plots of present-day agronomists, as well as in thousands of home gardens and on the small farms of many lands, one sees crops being grown and harvested by hand methods almost identical with those followed by the slaves in the Nile Valley in the time of the pharaohs 6,000 years ago.

The old art of crop production still predominates in farm practice throughout the world. Plant pathologists and entomologists have found ways to control plant diseases and insect pests more effectively. Chemists and agronomists have found supplements for the manure and ashes formerly used for fertilizers. Rotations perhaps are slightly improved. Many new crop hybrids and varieties (cultivars) have been developed. The control of weeds with herbicides was realized in the 20th century.

Improved cultural methods doubtless followed observations made by primitive farmers. They found better crops in spots where manure, ashes, or broken limestone had been dropped, or where weeds were not allowed to grow, or where the soil was dark, deep, or well watered, or where one crop followed certain other crops. Observations or empirical trials quickly revealed roughly the most favorable time, place, and manner of planting and cultivating various crops. These ideas were handed down through the generations. Observation, the only means of acquiring new knowledge until the 19th century, continued to enrich the fund of crop lore. Eventually, the exchange of ideas, observations, and experiences, through agricultural societies and rural papers and magazines, spread the knowledge of crops.

CROP PRODUCTION AS A SCIENCE

Agronomy is the branch of agriculture that treats of the principles and practice of crop production and field management. The term was derived from two Greek words, *agros* (field) and *nomos* (to manage). Scientific research in agronomy may be said to have begun with the establishment of the first experiment station by J. B. Boussingault in Alsace in 1834 and was given further impetus by Gilbert and Lawes who established the famous research facility at Rothamsted, England, in 1843 to study fertilizer use. It was 1870 before such tests were undertaken in the United States at the land grant agricultural colleges. However long before these landmark efforts were launched, many empirical tests had previously established numerous facts about crops and soils.

Agronomy has been a distinct and recognized branch of agricultural sciences only since about 1900. The American Society of Agronomy was organized in 1908. Agronomy had its origins largely in the sciences of botany, chemistry, and physics. Botanical writings describing crop production began with the ancient Greeks.

It is to Theophrastus of Eresus, who lived in about 300 B.C., to whom we have given the name father of botany. This student of Aristotle's listed plant differences which distinguish between monocots and dicots, gymnosperms and angiosperms and described germination and development and annual growth rings. He also mentioned "dusting" (pollination) of date palms which first demonstrated sexuality in plants. Botanical writings continued by herbalists in monasteries and medical practitioners were concerned chiefly with the use of plants for medicinal purposes. It was not until the 12th century that the interest in plants evolved to modern systematic botany, and later to other plant sciences. Chemistry had its origin in ancient mystic alchemy and in the work of men who compounded medicines. Lavoisier, often called the father of chemistry, did his work in the second half of the 18th century. The application of chemistry to agriculture dates from the publication of the book by Sir Humphry Davy entitled *Essentials of Agricultural Chemistry* in 1813. Physics arose from ancient philosophies. Agricultural engineering is largely applied physics.

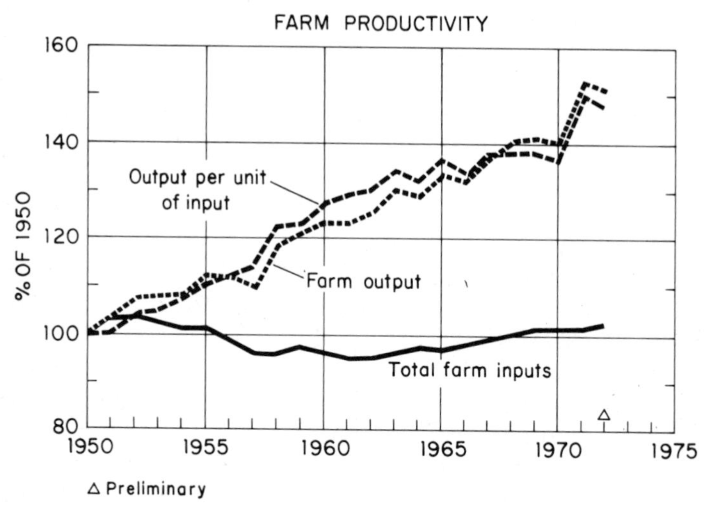

Figure 1–1. Total farm inputs have remained relatively constant since 1950 even with reduced total acreages and increased output.

The science of agronomy was built up by coordination of knowledge derived from the natural and biological sciences with the written records of observations and empirical trials, and later of controlled experiments that dealt with crop production.

Better crop production follows the use of new facts, improved machines, and pesticides, and the growing of superior cultivars. Informed farmers in advanced communities quickly adopt these improvements. This is evident from the worldwide acceptance of hybrid cultivars of corn, sorghum and sugarbeets, of dwarfed types of rice, wheat, grain sorghum and sunflowers, and of farm mechanization, fertilizers and chemical pesticides. This results in sharp increases in the output per farm worker (Figure 1–1): Improvements in crop production from research and invention often arrive after long and painstaking trials.

POPULATION AND FOOD SUPPLY

Crop culture will always be an important industry because crop products are essential to the existence of man. It has been stated that a man who goes without food for 24 hours will quarrel; one who is denied food for 48 hours will steal; and one who is without food for 72 hours will fight. Thus, the difference between peace and anarchy in most countries is a matter of only a few days without food.

Malthusian Theory

Aristotle and Plato agreed that the population of a civilized community should be kept within bounds. The problem of sufficient food for a population that continues to increase in a world of limited land area was raised again by Thomas R. Malthus in 1798.[30]

He argued that man could increase his subsistence only in arithmetic progression, whereas his numbers tended to increase in geometric progression, i.e., by the compound-interest law. Acre yields might be doubled once or even twice, but there is a limit beyond which increases are impossible. The basic check on population increase is the maximum limit set by the food supply. Population was formerly contained within that limit by war, famine, pestilence, and premature mortality. This will again occur unless the less-developed populations reduce their birth rate and increase their farm productivity. Worldwide adoption of organic farming would immediately reduce the population far below its present level by starvation.

The human population at the beginning of the Christian Era has been estimated roughly at about 250 million people.[17] The aggregate world population doubled to about 500 million by 1650. It doubled again to 1,000 million in only 200 years, or by 1850. The United Nations estimate of the world population 100 years later in 1950 was 2,406 million people. It reached 4 billion in 1973.

Farm output has exceeded population growth in the United States and other advanced countries (Figure 1–2). This resulted from increased production and low birth rates.

The Malthusian Theory has suffered setbacks since 1798.[10] Sometimes food production has outrun population growth in a few countries, following political schemes to inflate farm commodity prices. Despite this, one-third to one-half of the world population suffers at times from malnutrition, hunger, or both.[2]

In 1965 the world of roughly 3.3 billion people was increasing at a rate of 125 a minute or 65 million per year. The population merely

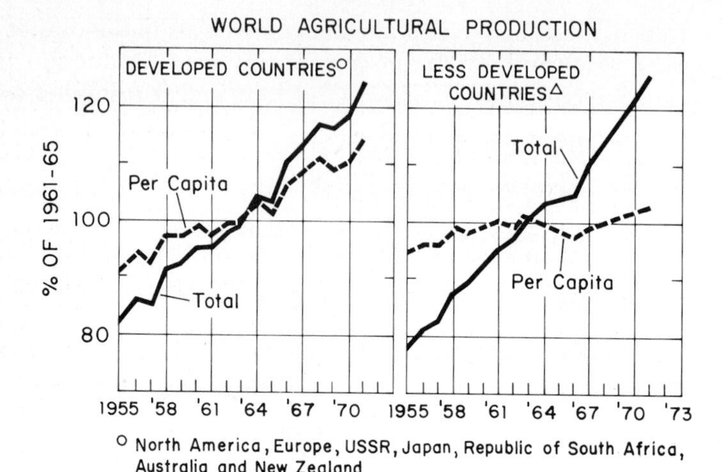

Figure 1–2. Farm output closely approximates the total population growth in developed countries. [Courtesy of USDA]

doubled from 0 A.D. to 1650 A.D., but at currently increasing rates may double in 40 years or less. The median estimate of the United Nations predicts a world population of 6,280 million people by the year 2000. Areas of the world experiencing the most rapid population increases are Latin America, Asia, Africa, and Oceania. Oceania is the only area of the four capable of excess food production on a relatively predictable basis. FAO reported that in 1962, 44 per cent of the world population was in developing countries, where economic growth is slow and a large majority of the people are illiterate and poverty stricken.

Acre yields of grain crops have been doubled or tripled in the United States since 1900 but there is a limit beyond which increases are impossible. Additional land for increasing production is limited. While the area of potentially arable land is three times the area harvested in any one year (The World Food Problem, White House, May, 1967), more than half of this untilled land (about 4 billion acres) lies in the tropics. Such land usually requires clearing, drainage, heavy fertilization, new plant varieties, and pest control for successful crop production.

In Asia there is little additional potentially arable land except by irrigation development. It is estimated that in southern Asia over 200 million additional acres could be irrigated.

Crop yields in Asia currently are being increased by improved short-stem varieties of rice and wheats, and by improved maize and sorghum hybrids with expanded geographic adaptation.

Cereal grains are a suitable common denominator for food pro-

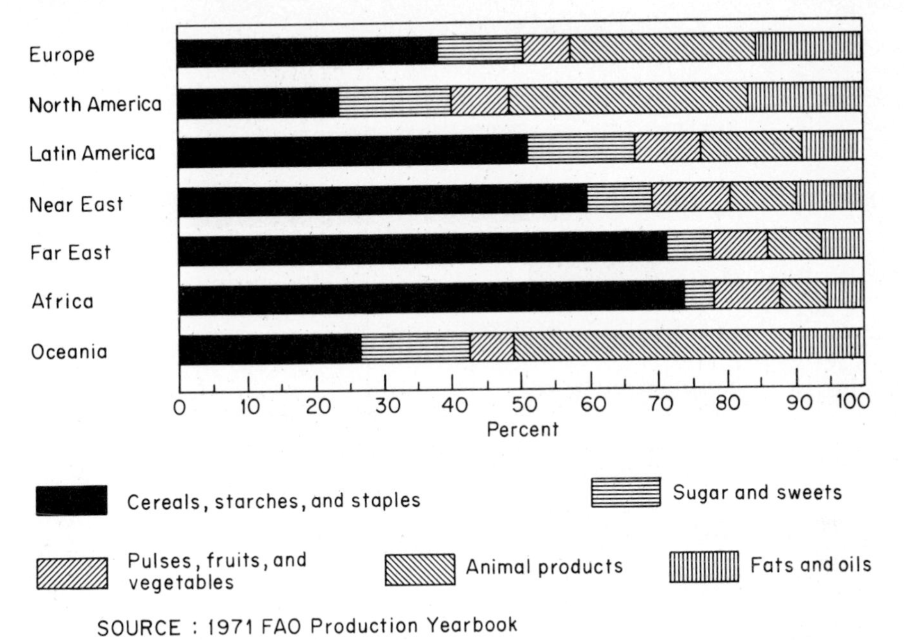

SOURCE : 1971 FAO Production Yearbook

Figure 1–3. Starchy foods from cereals, roots, and tubers supplied more than 50 per cent of the food-calorie intake in the less advanced parts of the world.

duction, consumption, and trade. On a calorie basis, cereals account for an average of 53 per cent of human food in the world by direct consumption. Probably another 20 per cent of human food comes indirectly from cereals in the form of meat, dairy products, and eggs.[12] Diets of the mass of the people in the underdeveloped countries remain appallingly inadequate, primarily due to shortages of animal proteins (Figure 1–3). The nutritional quality of the diet is usually considered unsatisfactory when more than 80 per cent of the calories are derived from cereals, starchy roots, and sugar. For the world as a whole, food supplies for adequate nutrition will need to be greatly expanded by the turn of the century.[1, 2]

Means for Increased Food Production

Agronomists, as well as all other agriculturists, are confronted with the problem of providing food for a world population that continues to grow at an accelerated rate. Underdeveloped countries of Asia, Africa, and Latin America are now in a deficit position with respect to food production for their own populations.[12] World food production can be augmented by expansion of the cultivated land area or by increased yields on present agricultural land (Figure 1–4). Aside from conventional agricultural means, food supplies also may be increased by the synthesis of foods as well as by the culture of certain lower plant forms such as yeast or chorella.

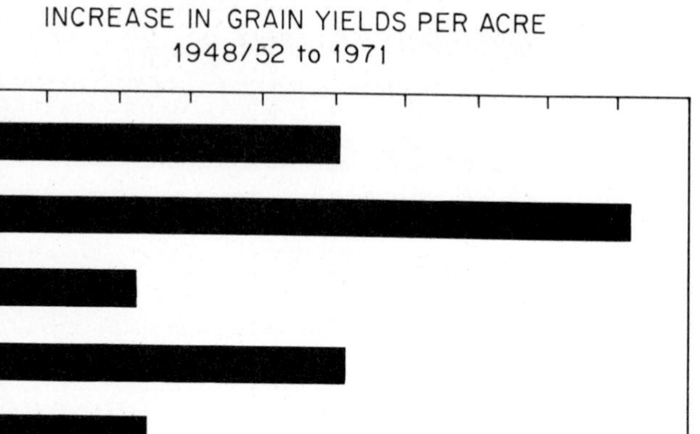

Figure 1–4. Grain yields increased in proportion to technological inputs.

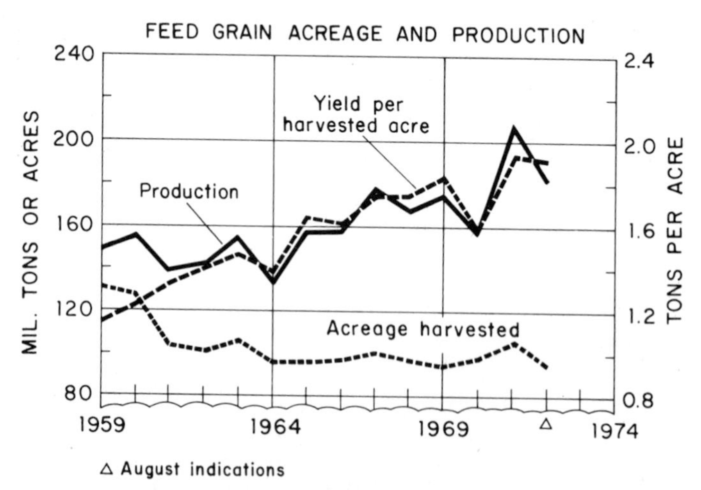

Figure 1–5. Increased technology enable production to increase as acres in production decreased. [Courtesy of USDA]

TABLE 1–1. World Consumption of N, P_2O_5, and K_2O by Geographic Location
100 Metric Tons

	N			P_2O_5			K_2O		
	1948–49 1952–53	*1970–71*	*% Increase*	*1948–49 1952–53*	*1970–71*	*% Increase*	*1948–49 1952–53*	*1970–71*	*% Increase*
World	43,086	316,077	634	60,514	206,423	240	45,027	165,380	267
Europe	18,999	96,748	409	25,527	81,456	219	25,334	74,846	195
North America	12,075	74,765	519	21,343	59,100	178	13,032	39,929	206
Latin America	1,160	14,073	1113	616	3,464	462	549	6,905	1157
Near East	938	8,003	753	178	1,838	933	51	371	627
Far East	6,172	40,187	551	2,500	12,829	413	1,626	12,382	661
Africa	325	4,752	1362	1,344	6,856	410	277	2,342	746
Oceania	174	1,629	836	4,691	10,160	116	154	1,954	1168

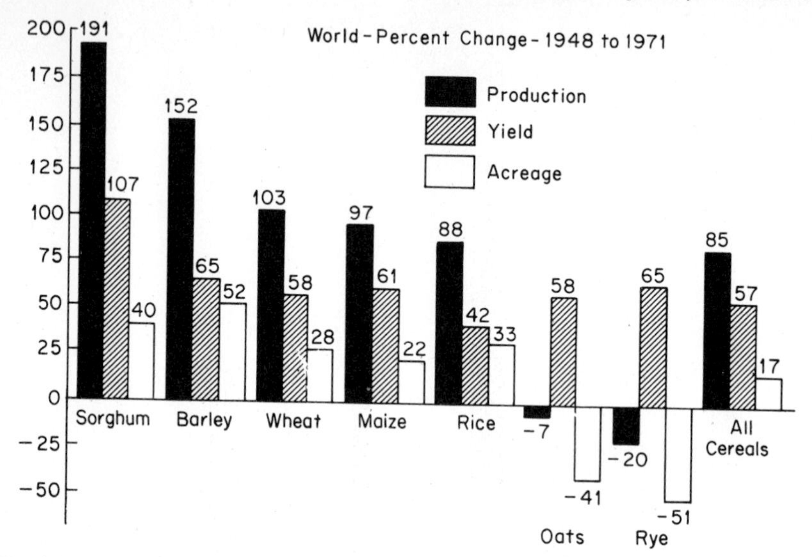

Figure 1–6. Changes in world production and yield of cereal crops from 1948 to 1971.

In traditional agriculture, new land was plowed out by the peasant or farmer for increased food production. This simple means of expansion of arable land is seldom possible today. Many countries of the world are now essentially fixed-land economies.[13] Roughly 1,424 million hectares or about 11 per cent of the total world land area is classified as arable land, fallow, and orchards. Another 3,000 million hectares, or 19 per cent of the total, is used for grazing. The remainder of the world land area, or about 70 per cent, produces little or no food.[12, 17] Potential total arable land has been estimated at about 2,666 million hectares.

The increased world grain yields (Figures 1–4 and 1–6) were achieved by growing more productive varieties and hybrids, greater fertilizer use (Table 1–1), and other improvement practices. Increases in Asia and Latin America (Figure 1–4 and 1–6) are largely a result of new "miracle" wheat and rice varieties and increased fertilizer usage. However, increased agricultural production has not kept pace with population growth in many less-developed countries (Figure 1–2).

Higher acre yields involve technology (applied science) plus capital. Probably 90 per cent of the increase in world food production since 1950 has come from higher yields on present agricultural land. Yield increases may be obtained by improved crops, fertilizers, irrigation, drainage, pesticides, more effective farm implements, multiple cropping, fallow, improved cultural practices, or some combination of these items. One of the preconditions for a yield-per-hectare takeoff

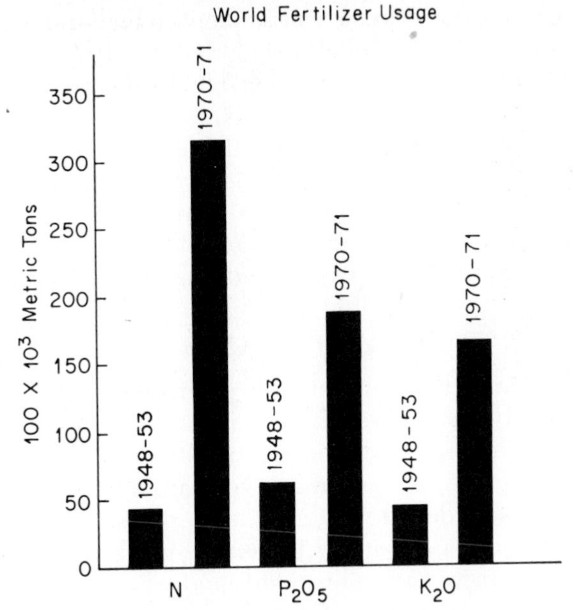

Figure 1-7. N, P_2O_5, and K_2O. World consumption comparisons 1948–53 vs. 1970. Data compiled from 1971 *FAO Production Yearbook.*

usually is literacy of the rural population.[5, 8, 13, 16] World cereal-grain yields per hectare increased 57 per cent between 1948 and 1971 (Figure 1–6). Feed-grain yields in the United States increased 65 per cent from 1959 to 1972 (Figure 1–5). The 22 per cent increase in production was achieved by 39 per cent fewer farm workers. The area of land irrigated on farms of the United States doubled to 40 million acres from 1944 to 1972, with resultant increases in crop yields.

World application of major fertilizer nutrients (N, P_2O_5, and K_2O) increased by percentages of 634, 240, and 267, respectively, between the 1948–53 period and 1970–71 (Figure 1–7).

ORIGIN OF CULTIVATED PLANTS

All basic cultivated plants are believed to have been derived from wild species. However, the exact time and place of origin and the true ancestry of many crops is still as highly speculative as the origin of man. Most crop species were adapted to the needs of man before the dawn of recorded history. Evidence from archeology and from the literature of the ancients bears out the early origin of most present-day crop plants.[20] The centers of origin of both agriculture and culture were in populated areas favored by a more or less equable climate.[31] Vavilov[40] predicted the center of origin of a crop by finding the region where the greatest diversity of type occurred in each crop plant. However, some plants appear to have originated in two or more centers or over wide zones.[24] De Candolle[18] concluded that 199 crops originated in the old world, and 45 in the Americas. The crop plants peculiar to the western hemisphere include maize, potato, sweetpotato, field bean, peanuts, sunflower, Jerusalem artichoke, and tobacco.[18] Eurasia yielded wheat, barley, rye, rice, peas, certain millets, soybeans, sugarbeets, sugarcane and most of the cultivated forage grasses and legumes. Sorghum, cowpeas, yams, pearl millet, finger millet, and teff were domesticated in Africa. Cottons originated in both hemispheres.

VARIATION IN CULTIVATED PLANTS

Cultivated plants have undergone extensive modifications from their wild prototypes as a result of the continuous efforts of man to improve them. The differences between cultivated and wild forms are largely in their increased usefulness to man, due to such factors as yield, quality, and reduced shattering of seed. Through the centuries man selected from among many thousands of plant species the few that were most satisfactory to his needs and which, at the same time, were amenable to culture. Primitive man was a master in making these selections, for modern man has added little of basic importance.

All cultivated plants were divided by Vavilov[40] into two groups: (1) those such as rye, oats, and vetch that originated from weeds, and (2) fundamental crops known only in cultivation. Cultivated rye is believed to have originated from wild rye which even today is a troublesome weed in wheat and winter-barley fields in certain parts of Asia. Oats are said to have come into culture as a weed found among ancient crops such as emmer and barley. Maize is known only in cultivation.

SPREAD OF CULTIVATED PLANTS

In their migrations, people invariably have taken their basic cultivated plants with them to insure a permanent food supply and support their culture. This happened in prehistoric as well as in recorded times. Man also transported weeds and disease and insect pests along with the crops. Pre-Columbian American agriculture was based on strictly native American plants and animals. None of the many plants involved were known in Europe or Asia prior to 1492, nor were cultivated plants native to Eurasia known in America before that time.[31]

CLASSIFICATION OF CROP PLANTS

Crop plants may be classified on the basis of a morphological similarity of plant parts.[7, 24, 25, 33] (See Chapter 3.) From the agronomic standpoint they may be partly classified on the basis of use, but some crops have several different uses.

Agronomic Classification

1. *Cereal or grain crops.* Cereals are grasses grown for their edible seeds, the term cereal being applied either to the grain or to the plant itself. They include wheat, oats, barley, rye, rice, maize, grain sorghum, millets, teff, and Job's tears. *Grain* is a collective term applied to cereals. Buckwheat is used like a grain but it is not a cereal. Quinoa (*Chenopodium quinoa*) is grown for its edible seeds in the Andes highlands of South America, for consumption like a food grain of more than 50,000 metric tons annually.

2. *Legumes for seed (pulses).* These include peanuts, field beans, field peas, cowpeas, soybeans, lima beans, mung beans, chickpeas, pigeonpeas, broadbeans, and lentils.

3. *Forage crops.* Forage refers to vegetable matter, fresh or preserved, utilized as feed for animals. Forage crops include grasses, legumes, crucifers, and other crops cultivated and used for hay, pasture, fodder, silage, or soilage.

4. *Root crops.* Crops designated in this manner are grown for their enlarged roots. The root crops include sugarbeets, mangels, carrots, turnips, rutabagas, sweetpotatoes, cassava, and yams.

5. *Fiber crops.* The fiber crops include cotton, flax, hemp, ramie, phormium, kenaf, and sunn hemp. Broomcorn is grown for its brush fiber.

6. *Tuber crops.* Tuber crops include the potato and the Jerusalem artichoke. A tuber is not a root; it is a short, thickened, underground stem.

7. *Sugar crops.* The sugarbeet and sugarcane are grown for their sweet juice from which sucrose is extracted and crystallized. Sorghum as well as sugarcane is grown for sirup production. Dextrose (corn sugar) is made from corn and sorghum grain.

8. *Drug crops.* The drug crops include tobacco, mint, wormseed, and pyrethrum.

9. *Oil crops.* The oil crops include flax, soybeans, peanuts, sunflower, safflower, sesame, castorbean, mustard, rape and perilla, the seeds of which contain useful oils. Cottonseed is an important source of oil, and corn and grain sorghum furnish edible oils.

10. *Rubber crops.* The only field crop grown that has been used for rubber in the United States is guayule, but other plants such as koksagyz (Russian dandelion) have been tested.

11. *Vegetable crops.* Potatoes, sweetpotatoes, carrots, turnips, rutabagas, cassava, Jerusalem artichokes, field pumpkins and many of the pulses are utilized chiefly as vegetable crops.

Special-Purpose Classification

1. *Cover crops.* Cover crops are those seeded to provide a cover for the soil. Such a crop turned under while still green would be a *green-manure* crop. Important green-manure crops are the clovers, alfalfa, the vetches, soybeans, cowpeas, rye, and buckwheat.

2. *Catch crops.* Catch crops are substitute crops planted too late for regular crops or after the regular crop has failed. Short-season crops such as millet and buckwheat are often used for this purpose.

3. *Soiling crops.* Crops cut and fed green, such as legumes, grasses, kale, and maize, are soiling crops.

4. *Silage crops.* Silage crops are those preserved in a succulent condition by partial fermentation in a tight receptacle. They include corn, sorghum, forage grasses, and legumes.

5. *Companion crops.* Sometimes called nurse crops, companion crops are grown with a crop such as alfalfa or red clover in order to secure a return from the land in the first year of a new seeding. Grain crops and flax are often used for this purpose.

6. *Trap crops.* Planted to attract certain insect or phanerogamous parasites, trap crops are plowed under or destroyed once they have served their purpose.

BOTANICAL CLASSIFICATION OF CROP PLANTS

Method of Botanical Classification

Botanical classification is based upon similarity of plant parts. Field crops belong to the Spermatophyte division of the plant kingdom, in which reproduction is carried on by seeds. Within this division the common crop plants belong to the subdivision of angiosperms, which are characterized by having their ovules enclosed in an ovary wall. The angiosperms are divided into two classes, the monocotyledons and the dicotyledons. All the grasses, which include the cereals and sugarcane, are monocotyledonous plants. The legumes and other crop plants except the grasses are classified as dicotyledonous plants because the seeds have two cotyledons. These classes are subdivided into orders, families, genera, species, subspecies, and varieties.

Crop-Plant Families

Most field crops belong to two botanical families, the grasses (Poaceae, recently changed from Gramineae), and the legumes (Fabaceae, recently changed from Leguminoseae).

THE GRASS FAMILY: The grass family includes about three fourths of the cultivated forage crops and all the cereal crops. They are either annuals, winter annuals, or perennials. Grasses are almost all herbaceous (small nonwoody) plants, usually with hollow cylindrical stems closed at the nodes.[34, 38] The stems are made up of nodes and internodes. The leaves are two-ranked and parallel-veined. They consist of two parts, the sheath, which envelops the stem, and the blade. The roots are fibrous. The small greenish flowers are collected in a compact or open inflorescence, which is terminal on the stem. The flowers are usually perfect, small, and with no distinct perianth. The grain or caryopsis may be free, as in wheat, or permanently enclosed in the floral bracts (lemma and palea), as in oats.

THE LEGUME FAMILY: Legumes may be annuals, biennials, or perennials. The leaves are alternate on the stems, stipulate, with netted veins, and mostly compound. The flowers are almost always arranged in racemes as in the pea, in heads as in the clovers, or in a spike-like raceme as in alfalfa. The flowers of field-crop species of legumes are papilionaceous or butterfly-like. The irregular flowers consist of five petals, a standard, two wings, and a keel that consists of two petals more or less united. The calyx is normally four- or five-toothed. The fruit is a pod that contains one to several seeds. The seeds are usually without an endosperm, the two cotyledons being thick and full of stored food. Legumes have taproots. Often the roots have abnormal growths called nodules caused by the activities of a bacterium, *Rhizobium*, which has the ability to fix atmospheric nitrogen in their bodies and eventually in the plant residues.

The principal genera of legume field crops, all of which belong to the suborder Papilionaceae, are *Trifolium* (clovers), *Medicago* (alfalfa, burclovers, and black medic), *Glycine* (soybean), *Lespedeza*, *Phaseolus* (field bean), *Pisum* (field pea), *Melilotus* (sweetclovers), *Vigna* (cowpea), *Vicia* (vetches), *Stizolobium* (velvetbean), *Lupinus* (lupines), *Crotalaria*, *Lotus* (trefoils), and *Pueraria* (kudzu).

OTHER CROP FAMILIES: Among the other botanical families that contain crop plants are (1) Cannabaceae (hops and hemp), (2) Polygonaceae (buckwheat), (3) Chenopodiaceae (sugarbeets, mangels, and wormseed), (4) Cruciferae (mustard, rape, and kale), (5) Linaceae (flax), (6) Malvaceae (cotton), (7) Solanaceae (potato and tobacco), and (8) Compositae (sunflower, Jerusalem artichoke, safflower, and pyrethrum).

Binomial System of Nomenclature

In a botanical classification, each plant species is given a binomial name. This provides two names for a plant: the genus and species. The binomial system of nomenclature is founded upon the publication by the Swedish botanist, Carl Linneaus, of his *Species Plantarum* in 1753. The name of the man who first gave the accepted name is affixed by a letter or abbreviation. For example, the letter L following the botanical name of corn or maize—*Zae mays* L.—means that Linneaus named it (see Appendix Table A–1). The binomial system provides a practically universal international designation for a plant species, which avoids much confusion. Some crops, e.g., proso and roughpea, are known by several different common names in the United States, but are immediately identifiable by their botanic names.

A species is a group of plants that bear a close resemblance to each other and usually produce fertile progeny when intercrossed within the group. Nearly every crop plant comprises a distinct species or, in some cases, several closely related species of the same genus. Within a species, the plants usually are closely enough related to be interfertile. Interspecies crosses are infrequent in nature, but many of them have been made artificially.

Varietal names are sometimes added to the species name to make a trinomial, but ordinarily crop varieties are given a common name or serial number to designate them. Classifications of agricultural varieties have done much to standardize variety names. The American Society of Agronomy has adopted a rule to use a single short word for a variety (cultivar) name, and the variety is not to be named after a living person. That society also registers properly named improved varieties of several field crops.

A comprehensive list of native and cultivated plants in American use or commerce, giving both the scientific and common names and the names of many varieties of cultivated plants, is found in the book *Standardized Plant Names*, published in 1942 by J. Horace McFarland Company, Harrisburg, Pennsylvania.

Life Cycles

The life cycles of plants provide us with a simple yet universal means of plant classification. All higher plants can be classified as annuals, winter annuals, biennials, or perennials. An annual is a short-lived plant which completes its entire life cycle from seed to seed in a single growing season and then dies. A winter annual utilizes parts of two growing seasons in completing its life cycle. Winter annuals are planted in the fall, vernalized during the fall and winter after which they produce seed and die the following summer. A biennial on the other hand normally utilizes two complete growing seasons to complete its life cycle. Vegetative growth occurs during the first season resulting in a rosette form. This is then followed by flowering and fruiting, a process known as bolting, which occur in a second growing season. A stalk emerges from the center of the rosette and a flowering head forms on the terminal end of the stalk. Perennials have an indefinite life period. They do not die after reproduction but continue to grow indefinitely from year to year.

THE LEADING FIELD CROPS

The world production of the leading field crops (exclusive of forage crops) is shown in Table 1–2 and Figure 1–8. Wheat, rice, corn, barley, sorghum, soybeans, cotton, oats, beans, and potatoes occupied the largest areas. The acreage, yield, and production of the different field crops in the United States and Canada are shown in Tables 1–3, 1–4, and 1–5. Corn, wheat, soybeans, alfalfa, and sorghum occupy the largest acreages of specific crops. Soybean, grain sorghum, corn, wheat, and alfalfa production has increased markedly since about 1940. The production of buckwheat, sirup sorghum, cowpeas, and broomcorn continues to decline. Former domestic crops that apparently were not grown in 1969 include fiber flax, teasel, hemp, and chicory.

The total land area in the 50 states of the United States is about 2,264 million acres. More than half of this area, 1,176 million acres, was in farms in 1969. Of this, 475 million acres were classed as crop land (Table 1–6). The total area for grazing was 889 million acres, not counting the harvested crop land in grains and forages that were grazed for part of the season or after harvest.

The labor required in 1971 to produce the major field crops, except

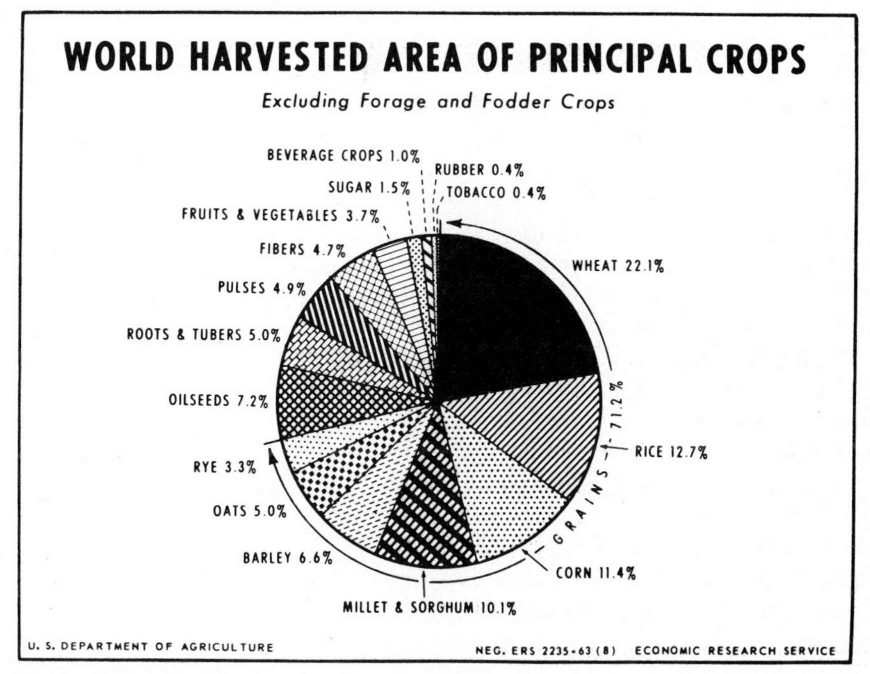

Figure 1–8. Field crops (exclusive of forages) occupy nearly 96 per cent of the crop acreage of the world.

TABLE 1–2. Area, Production and Yield of the Principle Crops, Except Forages, in the World and the United States (1969 to 1971 Average)†

CROP	AREA HECTARES (MILLIONS)		PRODUCTION (MILLION METRIC TONS)		YIELD (METRIC TONS PER HECTARE)	
	World	*United States*	*World*	*United States*	*World*	*United States*
Wheat:	214.4	19.0	281.3	41.5	1.31	2.19
Rice[*]	134.8	0.7	309.4	3.8	2.22	5.25
Maize[*]	109.5	25.1	288.9	135.0	2.69	5.40
Barley[*]	79.5	4.0	146.1	9.5	1.84	2.38
Sorghum grain[*]	58.0	5.9	74.8	20.4	1.29	3.45
Millets[*]	52.0	0.23	32.0	0.2	0.60	0.81
Oats[*]	32.0	6.5	56.1	12.1	1.76	1.87
Rye	19.2	0.6	29.3	1.0	1.53	1.70
Buckwheat	2.2	0.007	1.5	0.008	0.7	1.14
Canaryseed	0.15		0.1		0.9	–
Soybeans	35.4	16.9	45.5	31.0	1.32	1.83
Peanuts (in shell)	18.7	0.6	17.6	1.3	0.94	2.17
Dry beans	22.8	0.6	11.4	0.8	0.50	1.38
Dry peas	9.0	0.1	10.8	0.2	1.20	1.90
Broad beans	4.6		5.1	T	1.10	
Cowpeas	3.1	0.3	1.1	0.2	0.4	0.61
Pigeonpeas	2.9		2.0	T	0.68	
Lentils	1.7	0.03	1.1	1.3	0.63	0.035

Vetch	1.6	0.003	2.3	0.008	1.4	0.28
Lupins	0.9	0.001	0.8	0.0005	0.8	0.5
Cottonseed*	33.2	4.6	22.7	3.8	0.68	0.84
Rapeseed	9.8		7.2	T	0.74	
Sunflower seed*	8.6	0.9	10.0	10.5	1.2	0.9
Linseed*	6.3	0.75	3.1	0.54	0.5	0.72
Sesame seed	5.9	0.001	1.9	0.005	0.33	0.53
Castorbean	1.4	0.015	0.8	0.014	0.58	0.94
Hemp seed			0.4			
Hops	0.6	0.11	0.1	0.02	0.16	1.8
Tobacco	4.1	0.36	4.6	0.83	1.14	2.3
Potatoes	22.5	0.6	302.3	14.5	13.4	25.4
Sweetpotatoes & yams	16.2	0.06	139.2	0.6	8.6	11.1
Cassava	9.7		91.2	T	9.4	
Sugarcane	10.9	0.23	576.5	21.4	53.2	93.4
Sugarbeets	7.7	0.6	233.4	24.0	30.2	40.8
Cotton (lint)*	32.4	4.8	11.5	2.5	0.35	0.51
Fiber flax	1.6		0.65		0.42	0
Fiber hemp	0.6		0.28		0.47	0
Jute	2.8		3.60		1.25	0
Agave	0.8		0.62		0.8	0

† Computed from reports in the Monthly Bulletin of Agricultural Economics and Statistics. Food and Agricultural Organization of the United Nations. Rome and other sources. T = trace (small amount).

* Average for 1970–72.

tobacco, was only 1/6 to 1/19 of that required in the 1910–1914 period (Table 1–7). This resulted from almost complete mechanization and higher crop yields. The 2,585,051 farms were equipped with 4,469,000 field tractors, 725,000 combines, 666,000 pick-up hay balers, 593,000

TABLE 1–3. Acreage, Production, Yield, and Principle States of Farm Crops in the United States. Average 1971–73

CROP	HARVESTED ACREAGE (1000)	ACRE YIELD	PRODUCTION (1000)	LEADING STATES
Corn				
grain (bu.)	61,076	92.2	5,619,229	IA, IL, IN, NB, MN
silage (tons)	8,604	12.7	109,012	WI, IA, MN, NY, NB
forage	604			SD, ND, FL, GA, AL
Sorghum				
grain (bu.)	15,203	57.7	873,868	TX, KS, NB, OK, MO
silage (tons)	900	11.4	10,193	KS, NB, TX, OK, MS
forage	2,455			TX, KS, OK, CO, SD
Oats (bu.)	14,369	51.4	745,703	MN, SD, IA, ND, WS
Barley (bu.)	10,128	43.2	437,148	ND, MT, CA, ID, MN
Wheat (bu.)				
winter	35,203	32.8	1,199,014	KS, WA, NB, OK, CO
spring	14,409		425,027	ND, MN, MT, SD, ID
Hard Red Winter			822,598	KS, OK, TX, NB, CO, MT
Soft Red Winter			198,401	OH, IN, IL, MO, AR
Hard Red Spring			324,631	ND, SD, MT, MN
Durum (spring)	2,796	29.7	83,192	ND, MT, SD, MN, CA
White (w & s)			196,159	WA, OR, ID, MI, NY, CA
Rice	1,935	4542 lbs.	88,010 cwt.	AR, TX, LA, CA
Rye (bu.)	1,292	26.8	34,956	SD, ND, MN, NB, GA
Soybeans (bu.)	48,272	27.7	1,337,712	IL, IA, IN, MO, AR
Flax (bu.)	1,474	11.1	16,181	ND, SD, MN, MT, TX
Peanuts (lbs.)	1,480.2	2189	3,242,590	GA, TX, AL, NC, VA
Popcorn (lb.)	159.0	3267	519,576	IN, IA, NB, OH, IL
Cotton	12,147.9	488 lbs.	12,380	TX, MS, CA, AR, LA
Hay (all) (tons)	61,139	2.14	130,780	WS, MN, CA, NB, SD
alfalfa +	27,391	2.84	77,851	WS, CA, MN, IA, NB
other hay	33,748	1.57	52,929	MO, TX, NB, NY, ND
Dry Beans	1,369	1237 lbs.	16,946 cwt.	MI, CA, ID, NB, CO
Dry Peas	159.1	1239 lbs.	2,566 cwt.	WA, ID, MN, OR
Peppermint (lb.)	59.9	55	3,318	OR, WA, ID, IN, WS
Spearmint (lb.)	26.5	60	1,598	WA, IN, ID, MI, WS
Sugarbeets (tons)	1,297.6	20.6	26,682	CA, ID, CO, WA, MN
Sugarcane (total)	699.4	38.2	26,682	HI, FL, LA, TX
Potatoes (cwt.)	1,316.1	231	304,220	ID, ME, WA, CA, ND, MN
Sweetpotatoes (cwt.)	113.7	107	12,182	NC, LA, VA, MS, TX
Tobacco (lbs.)	857.2	2,031	1,764,076	NC, KY, SC, VA, GA, TN
Broomcorn	47.1	314 lbs.	7.2	CO, OK, NM
Hops	30.0	1,730	51,914	WA, OR, ID, CA
Wheat (all) (bu.)	49,412	32.8	1,624,708	KS, ND, WA, MT, NB
Cottonseed (tons)	12,147.9	0.41	4,965	

Table 1–4. Miscellaneous Crops and Seeds Grown in the United States in 1969 (U.S. Census)

CROPS	FARMS (*No.*)	ACRES (*1000*)	PRODUCTION (*1000*)	ACRE YIELD	LEADING STATES
Grass					
silage	41,665	1,364	9,118 tons	6.7	
soilage		695	4,263 tons		WS, NY, MI
Wild hay		7,549	7,730 tons	1.02	NB, SD, ND
Small grain hay		2,249	3,443 tons	1.53	CA, TX, MT
Coastal bermuda hay		903	2,182 tons	2.42	
Lespedeza hay		768	1,164 tons	1.52	KY, TN, MO
Peanut vine hay		312	237 tons	0.76	TX, OK, GA
Mixed grains	8,308	260	12,193 bu.	47	MN, SD, ND
Emmer & spelt	4,378	54	2,601 bu.	48.0	OH, MI, ND
Buckwheat	1,080	38	680 bu.	17.7	MN, NY, PA
Proso millet	4,556	317	6,691 bu.	21.1	CO, ND, NB
Cowpeas	1,117	31	333 bu.	10.7	TX, CA, SC
Guar	767	65			TX, OK
Lentils	414	66			WA, ID
Mung beans	396	23	9,597 lbs.		OK, TX
Velvetbean seed	60	1.2			TX, GA, SC
Sunflower seed	1,481	144			ND, MN
Safflower seed	857	223	451,608 lbs.	2,030	CA, AZ, MT
Castorbeans	400	39			NM, TX
Dill (for oil)	50	3			OR
Mustard seed	105	7.8	5,277 lbs.	680	CA, MT, ND
Sugarcane sirup	201	3.8			LA, GA, FL
Sorghum sirup	656	3.7			IA, TN, TX
Sugarbeet seed	225	8.3	21,799 lbs.	2640	OR, AZ, CA
Alfalfa seed	8,596	404,702	89,905,712 lbs.	222	CA, WA, ID, OK, OR, KS, UT, SD, TX, ND
Red clover seed	20,652	324,703	34,490,819 lbs.	106	OR, IL, MI, MN, OK, MO, WS
Aust. winter peas	270	19,252	32,120,097 lbs.	1,668	ID, OR, TX, OK
Bahiagrass seed	709	48,186	5,310,371 lbs.	110	FL, GA, MS, AL
Bentgrass	709	48,186	5,310,371 lbs.	110	OR (10 farms other states)
Birdsfoot trefoil	287	7,421	692,048 lbs.	93	NY, CA, VT, WS
Merion bluegrass	180	17,832	5,290,840 lbs.	297	OR, WA, ID, MN, SD
Kentucky bluegrass	709	77,450	29,547,131 lbs.	381	OR, WA, MN, ID, SD, KY
Bromegrass	1,764	36,021	7,119,731 lbs.	198	KS, NB, OR, ID, WA
Alsike clover	66	2,046	416,786 lbs.	203	MN, ID, WI, other
Crimson clover	684	23,392	7,002,413 lbs.	300	OR, AL, GA, MS
Ladino clover	136	17,166	4,934,522 lbs.	287	CA, OR, OK, AR
Sweetclover	1,366	35,449	7,511,137 lbs.	212	MN, SD, ND, TX, KS, OH, OR
White clover	185	10,273	2,414,831 lbs.	235	ID, LA, OR, AL, TX
Chewings fescue	588	25,798	9,519,802 lbs.	370	OR, MO, TN, AR
Red fescue	170	11,630	4,834,824 lbs.	415	OR, MO, other
Tall fescue	6,176	214,316	50,625,579 lbs.	236	NC, OR, LA, TN, AL, GA, AR, SC
Other fescue	417	13,170	3,258,589 lbs.	247	OR, MO, TN, WA

TABLE 1–4 Continued

CROPS	FARMS	ACRES	PRODUCTION	ACRE YIELD	LEADING STATES
	(No.)	*(1000)*	*(1000)*		
Lespedeza	5,908	137,622	36,083,200 lbs.	262	KY, MO, TN, KS, AR
Orchardgrass	1,534	43,567	15,092,562 lbs.	346	OR, VA, MO, KY
Lupine	34	1,284	1,194,473 lbs.	930	SC, FL
Redtop	304	11,164	1,409,293 lbs.	126	MO, IL, KY
Ryegrass	1,196	178,337	192,066,957 lbs.	1,078	OR, LA, TX, CA, MS
Sudangrass seed	143	10,288	12,029,301 lbs.	1,169	CA, TX, CO, OR, KS
Timothy	5,772	134,410	22,866,052 lbs.	170	MN, MO, OH, WS, IA, ID
Hairy vetch	510	21,088	4,976,106 lbs.	236	OR, TX, OK, NB
Other vetch	202	8,850	2,177,058 lbs.	246	CA, OR, NB, PA, TX
Wheatgrass	290	14,696	1,716,398 lbs.	117	SD, MT, WA, OR, CO, ID

corn pickers and shellers, and 298,000 field forage harvesters. Yields of the different crops increased 50 to 300 per cent since 1900. The higher yields resulted from increased fertilizer and pesticide use, more irrigation, more productive cultivars, improved cultural practices, more timely field operations, and shifts of many of the crops to more productive areas or to areas where they could be produced more economically with less labor.

The use of chemical pesticides is not the only method for controlling diseases and insects. Breeding crops for resistance to diseases

TABLE 1–5. Acreage, Production, and Yield of Crops in Canada (Average 1969–70)

CROP	ACRES (1000)	PRODUCTION (1000)	ACRE YIELD
Wheat (bu.)	18,726	507,898	27.1
Barley (bu.)	10,089	397,644	39.4
Oats (bu.)	7,402	369,818	50.0
Fall rye (bu.)	849	17,168	20.2
Spring rye (bu.)	123	2,293	18.7
All rye (bu.)	972	19,461	20.0
Mixed grains (bu.)	1,840	92,956	50.5
Shelled corn (bu.)	1,088	87,176	81.3
Buckwheat (bu.)	126	2,264	17.9
Flaxseed (bu.)	2,855	38,240	13.4
Rapeseed (bu.)	3,031	52,800	17.4
Mustardseed (lbs.)	234	223	953
Sunflower seed (lbs.)	59	45	755
Tame hay (tons)	13,113	26,992	2.1
Fodder corn (tons), green	692	8,978	13.0
Sugarbeets (tons)	74	998	13.5
Field roots (tons)	10.5	127	12.1
Potatoes (cwt.)	313	53,449	171

TABLE 1–6. Land Utilization on the 2,585,051 Farms in the United States in 1969

	MILLION ACRES		MILLION ACRES
Cropland*		Grazed Area	
Harvested	300	Grassland pasture	452
Fallow or crop failure	36	Woodland pasture	62
Cover crops or idle	51	Grazed (not on farms)	287
Pastured	88	(Cropland)	88
Total	475	Total grazed	889
Woodland—not grazed	50	Aftermath and part	
Farmsteads, roads, waste	25	season crop	?
Total land in farms	1,176**		

* 1971—Cropland harvested 306 million acres, 34 million fallow, 7 million crop failure.

** 1969—Land not in farms 1,200 million acres.

World—Arable area, 1,424 million hectares (3,517 million acres); prairie and pasture 3,001 million hectares (7,424 million acres).

began after 1890 and is a much expanded and necessarily continuous enterprise today. Selection and breeding for resistance to insects has been in progress for about 50 years. Sex attractants are being developed to permit sterilization of male insects by radiation. Other attractants are used as baits for trapping insects. Natural enemies of insect pests, and disease organisms and viruses may help to eliminate some of the pests. It will be many years, if ever, before pesticides can be replaced for protecting all farm crops. Chemicals are not the only hazard to mankind. Sunshine, in excess, can cause sunburn, skin cancer, and sunstroke, but we cannot live without it.

Destructive insects were a problem before the beginning of agriculture in an area. The following item, dated July 19, 1806, appears in the diary of Captain Wm. Clark while in Montana enroute east from the Pacific Coast:—". . . emence sworms of grass hoppers have de-

TABLE 1–7. Labor Hours to Produce and Harvest Crops

CROP	LABOR HOURS		
	Unit	*1910–14*	*1970–71*
Corn	100 bu.	135	7
Sorghum grain	"	104*	7
Wheat	"	106	7
Soybeans	"	126*	17
Hay	ton	10.3	1.7
Potatoes	"	25	4
Sugarbeets	"	12.1	1.6
Cotton	bale	276	26
Tobacco	100 lbs.	44	24

* 1925–29.

stroyed every sprig of grass for many miles on this side of the [Gallatin] river."

Even as late as 1951–1960, losses in various crops ranged from 3 to 28 per cent from plant diseases, 3 to 20 per cent from insects, and 3 to 17 per cent from weed competition. This justified greater pesticide use. Current fertilizer use has not restored the fertility levels of most of the virgin soils. Water pollution has been in vogue since the earth was created, as is evident from the presence of considerable amounts of mercury deep in the Greenland icecap. Many other countries are modernizing their agriculture along lines similar to those in North America.[9, 10]

FUTURE INCREASES IN PRODUCTIVITY

"Give a man a fish and he can live for a day; teach him to fish and he can live forever!"

This philosophy is reflected in the growing realization that less developed countries of the world must have the desire to succeed in modernizing agriculture. Leaders of these nations must make an effort to provide the multiple factors involved in modernizing agriculture with help, where available and necessary, from outside.

Several promising research developments are pointing toward continued productivity increases in the future:

(1) Collection of plant species and broadening of germ plasm bases.
(2) Improved breeding techniques.
(3) Crop physiology advances in understanding plant efficiency in utilization of light and nutrients.
(4) Interactions between plant breeders, geneticists, soil chemists, engineers, etc., is increasingly bringing together the best of many disciplines.
(5) Extension of areas of adaptation of previously restricted varieties.
(6) Improved disease and insect control with advances in chemical and biological methods.
(7) Improved protein quality factors such as hi-lysine corn.
(8) Improved medication and breeding principles in livestock.
(9) Reduced production costs of nitrogen fertilizer.
(10) Improved winter hardiness of varieties to extend crop production.

These are but a few of the areas in which progress is being made. One combination of enhanced agricultural production and adjustment of his numbers by man to meet available subsistence must be a primary goal of all peoples of the world.

REFERENCES

1. Aldrich, D. G. *Research for the World Food Crisis.* AAAS, Washington, D.C., pp. 1–320. 1971.
2. *Agricultural Research: Impact on Environment. Iowa State University Spec. Rpt.* 69, pp. 1–84. 1972.
3. Anonymous. "Six billions to feed," *FAO World Food Problems,* 4:1–41. 1962.
4. Anonymous. "Third world food survey," *Freedom from Hunger Campaign, FAO Basic Study,* 11:1–102. 1963.
5. Army, Thomas J., Greer, Frances A., and San Pietro, Anthony. *Harvesting the Sun; Photosynthesis in Plant Life,* Academic Press, New York, 1967.
6. Asimov. Isaac. *Life and Energy,* Doubleday, Garden City, N.Y., 1962.
7. Bailey, L. H. *Manual of Cultivated Plants,* Rev. ed., Macmillan, Inc., New York, pp. 1–1116. 1949.
8. Beiser, Arthur, and Krauskopf, Konrad. *Introduction to Physics and Chemistry,* McGraw-Hill, New York, 1964.
9. Black, C. A., and others, Eds. *Agronomy in a Changing World and Research Needs for the Seventies,* Am. Soc. Agron. Spec. Pub., 19:1–65. 1971.
10. Borlaug, N. E. *Mankind and Civilization at Another Crossroad,* Wisc. Agri-Business Council, Inc., Madison, Wis., 1–48. 1971.
11. Brown, Harrison. *The Challenge of Man's Future,* Viking, New York, 1954.
12. Brown, L. R. "Man, land, and food," *USDA Foreign Agricultural Economic Rpt.,* 11:1–153. 1963.
13. Brown, L. R. "Increasing world food output," *USDA Foreign Agricultural Economic Rpt.,* 25:1–140. 1965.
14. Carrier, L. J. *Beginnings of Agriculture in America,* McGraw-Hill, New York, pp. 1–323. 1923.
15. Clayton, Roderick K. *Light and Living Matter: A Guide to the Study of Photobiology,* Vol. 2, *The Biological Part,* McGraw-Hill, New York, 1971.
16. Columella, L. Junius Moderatus. *Of Husbandry* (translated into English with several illustrations from Pliny, Cato, Varro, Palladius, and other ancient or modern authors), London, pp. 1–600. 1797.
17. Cook, R. C. "Population and food supply," *Freedom from Hunger Campaign, FAO Basic Study,* 7:1–49. 1962.
18. De Candolle, A. *Origin of Cultivated Plants,* 2nd ed. (1886), Hafner, New York (reprint), pp. 1–468. 1959.
19. Fisher, R. A. "The over-production of food," *The Realist* (Pt. 4): 45–60. 1929.
20. Flannery, K. V. The Origins of Agriculture. *An. Rev. of Anthropology.* 2:271–310. An. Reviews Inc., Palo Alto, Cal. 1973.
21. Galston, Arthur W. *The Life of the Green Plant,* Prentice-Hall, Englewood Cliffs, N.J. 1964.
22. Gates, David M. *Energy Exchange in the Biosphere,* Harper & Row, New York. 1962.

23. Geiger, Rudolf. *The Climate near the Ground,* Harvard University Press, Cambridge, Mass. 1965.
24. Harlan, J. R. "Agricultural origins: Centers and non-centers," *Science,* 174(4008):468–74. 1971.
25. Hitchcock, A. S. *A Text Book of Grasses,* Macmillan, Inc., New York, pp. 1–276. 1914.
26. Hitchcock, A. S., and Chase, A. "Manual of the grasses of the United States," *USDA Misc. Pub.* 200, pp. 1–1051. 1950.
27. Janick, Jules., Scherz, Robert W., Woods, Frank W., Ruttan, Vernon W. *Plant Science: An Introduction to World Crops,* Freeman, San Francisco, 1969.
28. Krauss, Harold P. *Discovering Physics,* Addison-Wesley Press, Cambridge, Mass. 1951.
29. Leonard, W. H. "World population in relation to potential food supply," *Sci. Mo.,* 85:113–125. 1957.
30. Malthus, T. R. *An Essay on the Principle of Population,* 1798–1803, Macmillan, Inc., New York, pp. 1–134. (reprint). 1929.
31. Merrill, E. D. "Domesticated plants in relation to the diffusion of culture," *Bot. Rev.,* 4:1–20. 1938.
32. Nourse, Alan E. *Universe, Earth, and Atom: The Story of Physics,* Harper & Row, New York, 1969.
33. Pawley, W. H. "Possibilities of increasing world food production," *Freedom from Hunger Campaign, FAO Basic Study 10,* pp. 1–231. 1963.
34. Robbins, W. W. *Botany of Crop Plants,* 3rd ed., Blakiston, Philadelphia, pp. 1–639. 1931.
35. Salisbury, Frank B., and Cleon, Ross. *Plant Physiology,* Wadsworth, Belmont, Cal. 1969.
36. Schuster, Richard P. *The Next Ninety Years,* California Institute of Technology, Pasadena, Cal. 1967.
37. Stiles, Walter, and Cocking, E. D. *An Introduction to the Principles of Plant Physiology,* Methuen, London. 1969.
38. Theophrastus. *Enquiry into Plants* (with an English translation by Sir Arthur Hort), Putnam, New York. 1916.
39. *U.S.D.A. Yearbook, Contours of Change,* pp. 1–366. 1970.
40. Vavilov, N. I. *Studies on the Origin of Cultivated Plants,* Institute of Applied Botany and Plant Breeding, Leningrad, pp. 1–248. 1926.
41. Williams, Virginia R., and Williams, Hulen B. *Basic Physical Chemistry for the Life Sciences,* Freeman, San Francisco. 1967.
42. Winter, Stephen S. *The Physical Sciences: An Introduction,* Harper & Row, New York, 1967.

Chapter 2

Crop Plants in Relation to Environment

FACTORS IN CROP DISTRIBUTION

Staple agricultural crops show a marked tendency to geographic segregation despite the fact that they may grow well over wide areas. Thus corn and oats, although concentrated in the Corn Belt, are both grown successfully in forty-eight states in the union. The principal factors that influence localization are climate, topography, character of the soil, insect pests, plant diseases, and economic conditions.[25]

Crops are generally profitable only when grown in regions where they are well adapted. The best evidence of adaptation of a crop is a normal growth and uniformly high yields.[23] Adapted crops usually produce satisfactory yields even on the poorer soils in a region.[43] The farther a crop is removed from its area of good adaptation, the more care is necessary for satisfactory production.

CLIMATE

Climate is the dominant factor in determining the suitability of a crop for a given area. Knowledge of the crops and crop varieties grown in a given region is a better measure of climate, as applied to crop production in that region, than are complete climatic records. Thus, it is known that the climate in the Puget Sound region of Washington is more mild than that of eastern Maryland, and that the latter region is more mild than northern Texas, because winter oats survive the winter more regularly. It is also known that conditions in the winter-wheat region of Sweden are not so severe as in many winter-wheat regions in the United States, because the Swedish varieties are less resistant to cold. Climatic counterparts of American agricultural regions are found in various parts of the world. The Great Plains region is similar to the Ukraine because the same crop varieties thrive in both places. Many crop varieties are equally adapted to Australia and California. The irrigated regions of southern California, southern

Arizona, and southern Texas are comparable to irrigated sections of the Mediterranean region. The wheats of Great Britain and Holland fail miserably in the United States except in the mild cool humid Pacific Northwest, where they thrive. The Corn Belt of the United States has a counterpart in the Danube Valley of Europe.

General Types of Climate

Climatic differences are due chiefly to differences in latitude, altitude, distances from large bodies of water, ocean currents, and the direction and intensity of winds.

Continental climates, which occur in interior regions, are characterized by great extremes of temperature between day and night and between winter and summer. The temperature ranges increase, in general, with the distance from the ocean. Some of these regions, such as the steppes of Russia and the Great Plains of North America, are further characterized by an irregular approach of seasons, deficient rainfall, low humidity, and generally unobstructed winds.[21] The limited rainfall that occurs is usually sporadic and often torrential. The great wheat areas of the world are found in such climates. Other hardy crops also are grown there.

Oceanic or marine climates are more equable. Moderate temperature changes occur between day and night, and between winter and summer. The temperature changes are influenced predominantly by the sea or other large bodies of water. This influence results from differences in the specific heat of water and land. Water takes up and gives off heat only one fourth as rapidly as does land.

Three distinct climatic regions are recognized in the United States.[43] The first is a narrow strip of territory from the Pacific Coast to the Cascade and Sierra Nevada mountains, a purely oceanic climate in which the rainfall ranges from less than 10 inches in southern California to over 100 inches per year in the Northwest. In this region, the winters are mild, while the summers in the northern part and along the southern coastline are cool. The second region is the upland plateau from these mountains eastward to the 100th meridian. The climate is continental over most of this area. The third region is from the 100th meridian, where a continental climate prevails, to the Atlantic, where conditions again are modified by the ocean. The change from one type to the other is gradual in this area.

Precipitation

Rainfall has a dominant influence in crop production. In semiarid regions, such as the Great Plains and Great Basin, the conservation and utilization of the scanty rainfall is so important that it relegates all other factors, including soil fertility, to minor positions.

In regions of low rainfall, the deficiency may be partly overcome by moisture-conserving tillage and rotation practices, or by irrigation. Under dry farming conditions red clover and sugarbeets are failures, and alfalfa is unprofitable except on bottomlands. When such lands are irrigated, these crops may be highly successful in the same region.

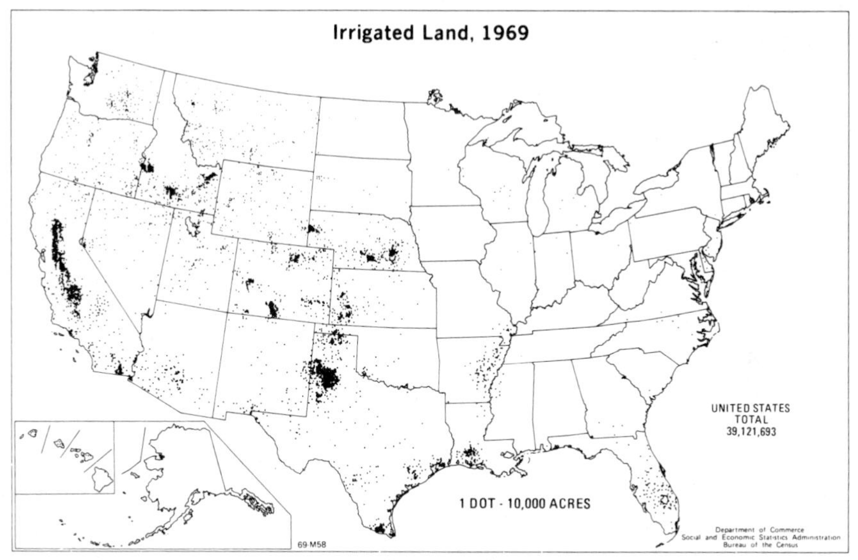

Figure 2–1. Irrigated land in farms. About 40 million acres of land in the United States and one-eighth of the world's crop land is irrigated.

Crop areas based upon rainfall: Crop regions are frequently classified on the basis of average annual rainfall. It is obvious that these regions are arbitrary, since the actual boundaries may fluctuate from year to year. (1) The arid region is that in which the average annual rainfall is 10 inches or less. Irrigation is necessary for successful crop production in most of such areas (Figure 2–1). (2) The semiarid region is arbitrarily considered to be that in which the rainfall varies from 10 to 20 inches. Tillage methods that conserve moisture, or crop varieties adapted to dry farming regions, or irrigation are necessary for successful crop production. (3) Annual rainfall in most subhumid areas varies from 20 to 30 inches (Figure 2–2). This amount of rainfall often is inadequate for satisfactory crop yields unless methods which utilize the rainfall to best advantage are followed in regions such as the southern Great Plains where seasonal evaporation is high. (4) The humid region is regarded as that in which the annual precipitation is more than 30 inches. There conservation of moisture is not the dominant factor in crop production.

Effectiveness of rainfall: The effectiveness in crop production of a given quantity of rainfall depends upon the time of year that it

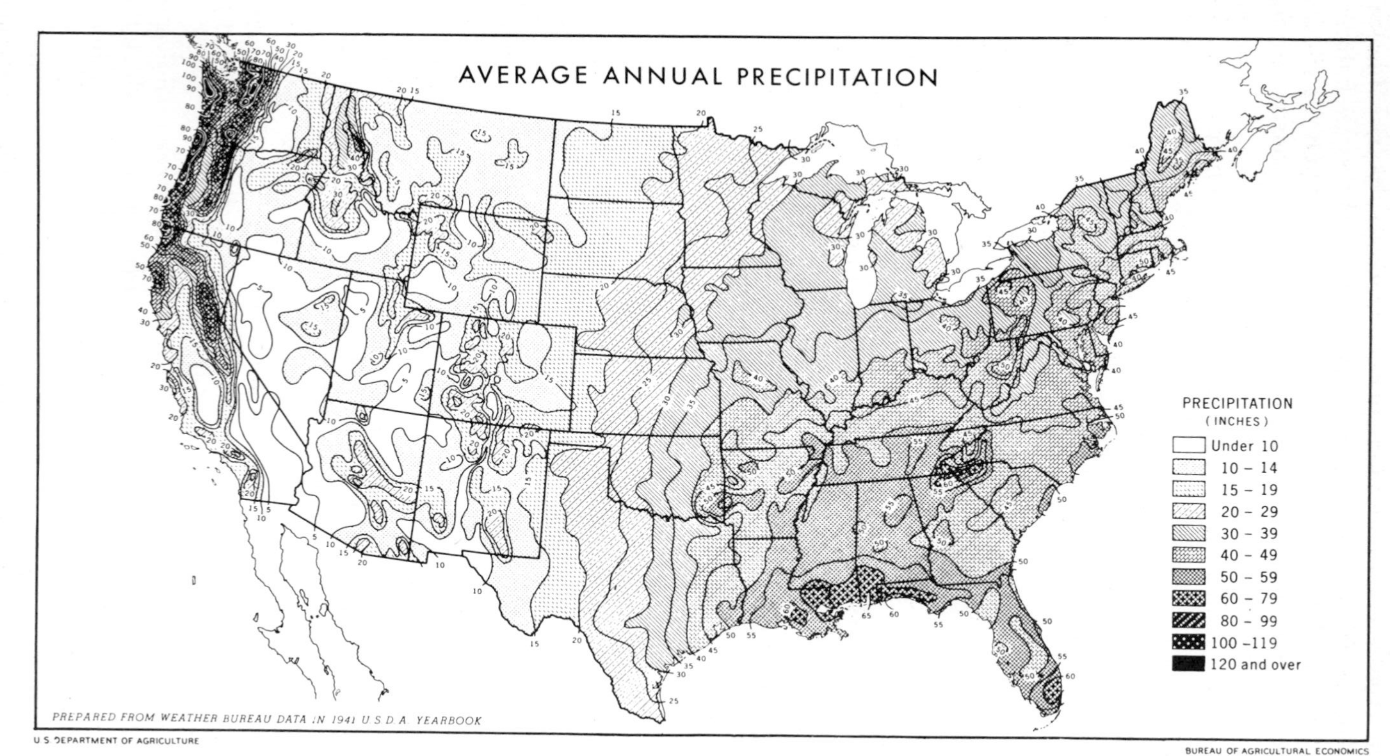

Figure 2–2. Precipitation in the United States. (*a*) Average annual precipitation.

falls, and the rapidity and intensity of individual rains. Of still more importance is the seasonal evaporation.

The total rainfall fluctuates widely from year to year. Over a 10-year period, the annual precipitation at North Platte, Nebraska, in the Great Plains area, varied from 10 to more than 40 inches.[7] In general, the lines of equal rainfall from the Rocky Mountains eastward lie north and south, but shift eastward as they extend to the north. In the

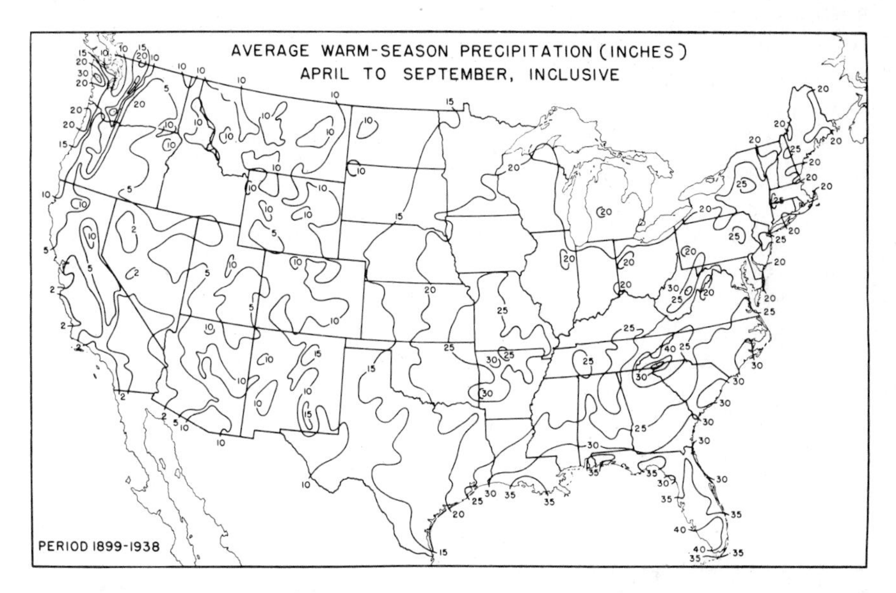

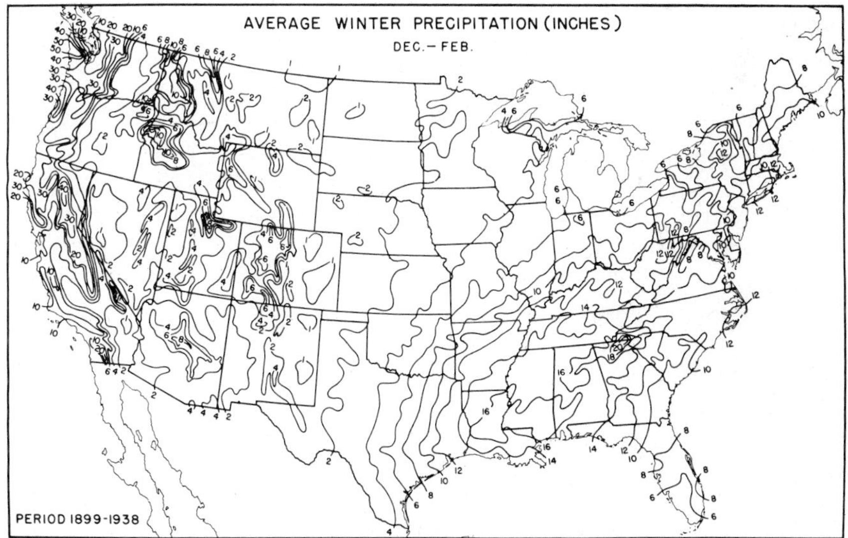

Figure 2–2. [CONCLUDED] Precipitation in the United States. (*b*) Average warm-season precipitation. (*c*) Average winter precipitation.

southern states east of Texas, equal rainfall lines follow the outline of the Gulf coast.

Rainfall has its greatest value when it falls during the growing season, ordinarily between April 1 and September 30, for summer crops. The critical period for moisture for most crops occurs just before or after flowering.[19] In corn, the ten days after tasseling and silking have an almost dominant effect on yield. In regions of winter rainfall, a greater portion of the water can be conserved by fallowing than in regions of summer rain because it falls during a period of low evaporation. Winter rainfall regions are especially suited to the growing of winter grains, although these crops thrive in summer rainfall regions also. Corn and sorghum are poorly adapted to winter rainfall regions because their greatest demands for moisture occur during the dry hot season.

East of the Mississippi River, the monthly precipitation distribution is rather uniform throughout the year, especially north of the Cotton Belt.[20] However, in Florida the six months from November to May are comparatively dry, the rainfall being heavy in the summer months. On the Great Plains, approximately 70 to 80 per cent of the annual precipitation falls from April to September, inclusive. Most of the precipitation in the Pacific Coast and intermountain regions comes in the winter months. In Arizona and adjacent sections, rainy periods frequently occur in both summer and winter.

Ordinarily, summer showers of less than ½ inch are of very little value in storing soil moisture for the use of crops unless they immediately precede or follow other rains. Light showers are largely lost by evaporation from plant foliage and soil surface. However, such showers may benefit crops by reducing evaporation and cooling the atmosphere. Frequently, rain comes in torrential showers in the Great Plains and states to the east, with the result that much of it is lost through runoff, especially on steep lands with heavy soils.

Water is the chief component of living plant tissues. It is involved in the transfer of nutrient elements from the soil into the roots and upward throughout the plants. It supplies hydrogen as a component of organic materials formed in photosynthesis. The transpiration of water cools the plants under high temperature conditions.

Temperature

Temperature also is an important factor in limiting the growing of certain crops (Figure 2–3). The 50° F. isotherm of monthly mean temperature during each summer month marks the growth limits for most plants. The great crop areas have 4 to 12 months in which the mean monthly temperature is between 50° and 68° F. This is in the temperate belt. Temperature is influenced by latitude and altitude as well as by exposure. A difference of 400 feet in altitude or 1 degree in latitude causes a difference of approximately 1° F. in mean annual and mean July temperatures and 1.5° F. in mean January temperatures in the United States.[16] Slopes that face south and west receive the most sunshine, and are regularly warmer than slopes facing north and east. Crops can be grown at much higher mountain elevations on the warm slopes.[32]

EFFECT OF TEMPERATURE ON PLANTS: Each crop plant has its own approximate temperature range, i.e., its minimum, optimum, and maximum for growth. Although they are subjected to a rather wide range of temperature, most crop plants make their best development between 60° and 90° F. They either cease growth or die when the temperature becomes either too low or too high. Temperatures of 110° to 130° F. will kill many plants. Most cool-weather crop plants cease growth at temperatures of 90° to 100° F., while annual crops are killed by low temperatures that range from about 32° F. down to −40° F. The minimum temperature at which any plant can maintain life during the growing season is the approximate freezing point of water. The threshold value, or point at which appreciable growth can be detected, is usually taken at 40° or 43° F. This value varies for different crops. Winter rye in the sun may show signs of growth even when the temperature recorded in a shaded shelter is below freezing. Sorghum practically ceases growth at a temperature of 60° F,[26] sugarcane at 70° F, and tomatoes at 65° F.

Crop areas have more or less definite boundaries that extend in an

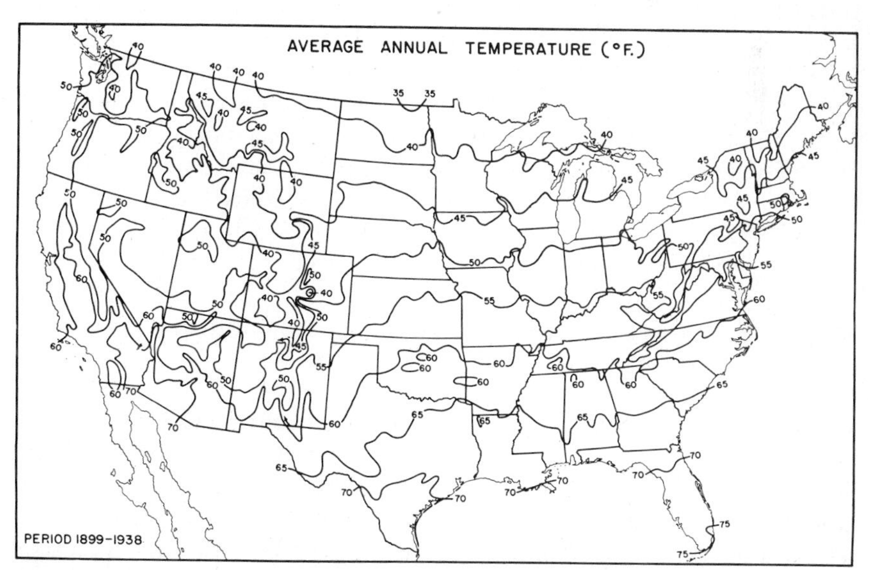

Figure 2–3. Temperatures in the United States. (*a*) Average annual temperature.

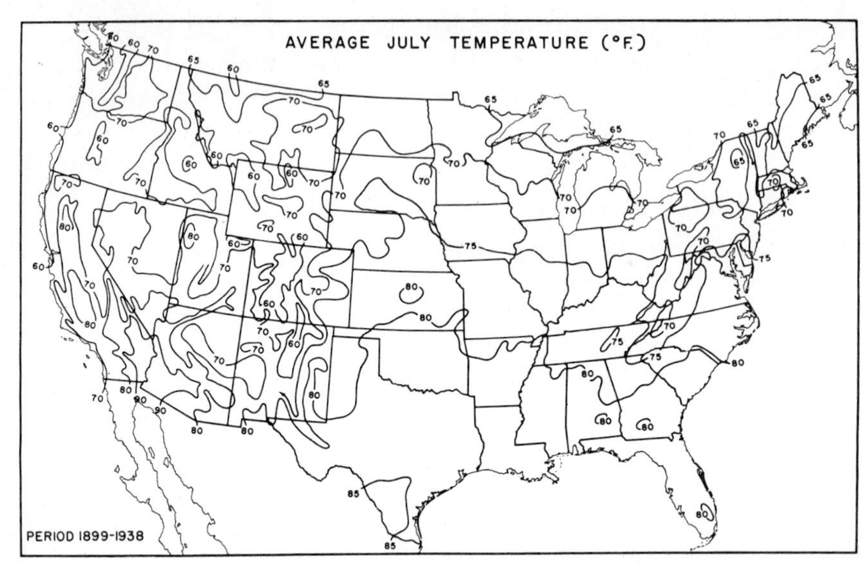

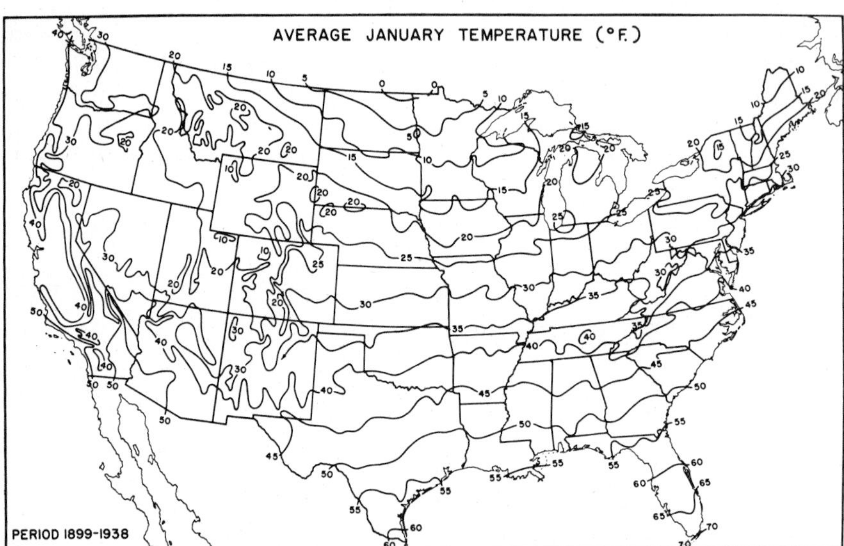

Figure 2–3. [CONCLUDED] Temperatures in the United States. (*b*) Average July temperature. (*c*) Average January temperature.

east-west direction in conformity with the isothermal trend.[19] The time to plant crops is defined by the temperature conditions of the locality, but the planting time is generally later as one proceeds northward. According to the Hopkins bioclimatic law,[16] such events as seeding time vary 4 days for 1 degree of latitude, 5 degrees of longitude, and 400 feet of altitude, being later northward, eastward,

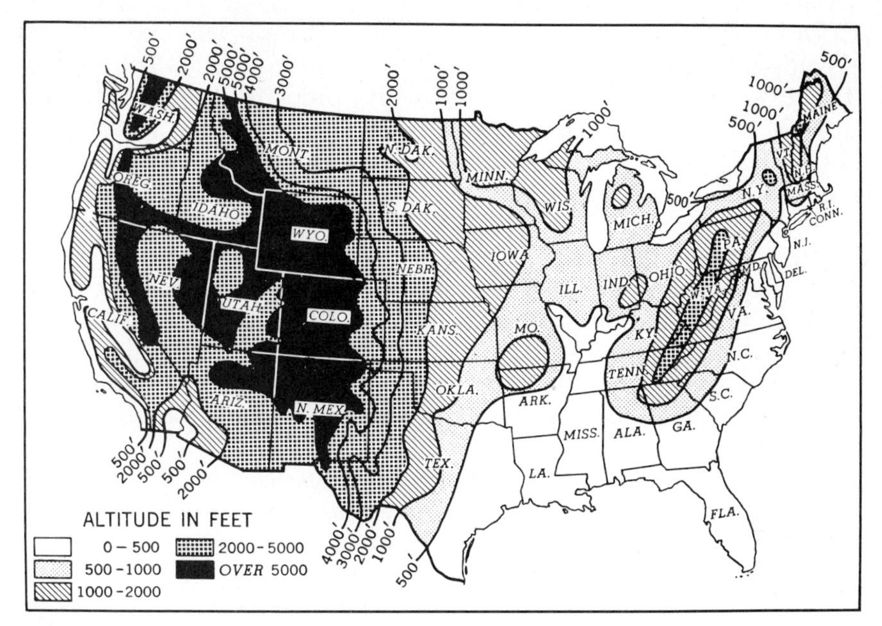

Figure 2–4. Approximate general altitudes in the United States.

and upward in the United States (Figure 2–4). The normal daily temperature for spring planting of a given crop is rather uniform throughout the country. The seeding of spring wheat begins when the normal daily temperature rises to about 37° F., of spring oats at 43° F., of corn at 55° F., and of cotton at 62° F.

Within suitable temperature ranges, growth increases with each rise in temperature, often in approximately proportional relationships corresponding to the Van't Hoff-Arrhenius law for monomolecular chemical reactions, i.e., a doubling in activity (or growth) for each rise in temperature of 10° C. (18° F.).

Crops differ in the number of heat units required to mature them. Heat units are usually computed by a summation of degree days for a growing season. A degree day is the number of degrees F. above an established minimum growing temperature, as for example, 50° F. for corn. Thus, where the sum of maximum daily temperatures above that minimum is 3,000 during a 100-day growing season, there are 3,000 total heat units. The number of heat units necessary to mature a given crop differs with the region. Increased growth is not directly proportional to increased temperatures. Also, fewer units are required in a cool region where temperatures do not become so high as to retard plant growth. This relationship is recognized in the Linsser law, which states that in different regions the ratio of total heat units to the heat units necessary to mature a given crop is nearly constant.[40]

COOL- VERSUS WARM-WEATHER CROPS: Crops may be classed broadly as warm-weather or cool-weather crops (Appendix Table A–1). The crops that make their best growth under relatively cool conditions, but are damaged by hot weather, include wheat, oats, barley, rye, potatoes, flax, sugarbeets, red clover, vetches, field peas, and many grasses. Field peas, the small grains just mentioned, flax, and northern grasses may withstand freezing temperatures at any time up to about the flowering stage. The 10° F. isotherm for daily mean temperatures for January and February coincides remarkably well with the northern boundary of the Winter Wheat Belt, beyond which spring wheat is more important.[34] The isotherms for 20° and 30° F. coincided closely with the northern limits of winter barley and winter oats, respectively, up to about 1950 when more hardy varieties were available.

Among the crop plants that prefer a warm season are corn, cotton, sorghum, rice, sugarcane, peanuts, cowpeas, soybeans, velvetbeans, dallisgrass, and bermudagrass. Warm-weather crop plants, after they reach appreciable size, are killed by temperatures slightly below freezing, and sometimes at several degrees above freezing when exposed for prolonged periods.

Length of Growing Season

The number of days between the average date of last spring frost and first fall frost usually is considered as the length of the growing season (Figure 2–5). A frost-free season of less than 125 days is an effective limitation to the production of most crops. Wheat, oats, and barley mature under a shorter frost-free period than corn or sorghum because wheat, oats, and barley can be sown safely two months before the average last spring frost. Cotton requires 200 days of frost-free weather, a fact which restricts its growth to the South. The length of the period for growth of a crop may be limited also by drought.

Humidity

Humidity is water vapor in the air. Relative humidity is the vapor pressure in the air in terms of the percentage necessary to saturate the atmosphere at the particular temperature. A saturated atmosphere that causes fog, dew, or rain has a theoretical relative humidity of 100 per cent. The lower the relative humidity at a given temperature the more rapidly will the air take up water transpired by the leaf or evaporated from a moist soil surface. At the same relative humidity, the vapor pressure and drying capacity at 95° F, is 9.2 times that at 32° F.

Evaporation and transpiration increase with increases in temperature and decreases in relative humidity. A high seasonal evaporation from a free-water surface is generally reflected in a high water requirement of crops. Evaporation determines the efficiency of rainfall,

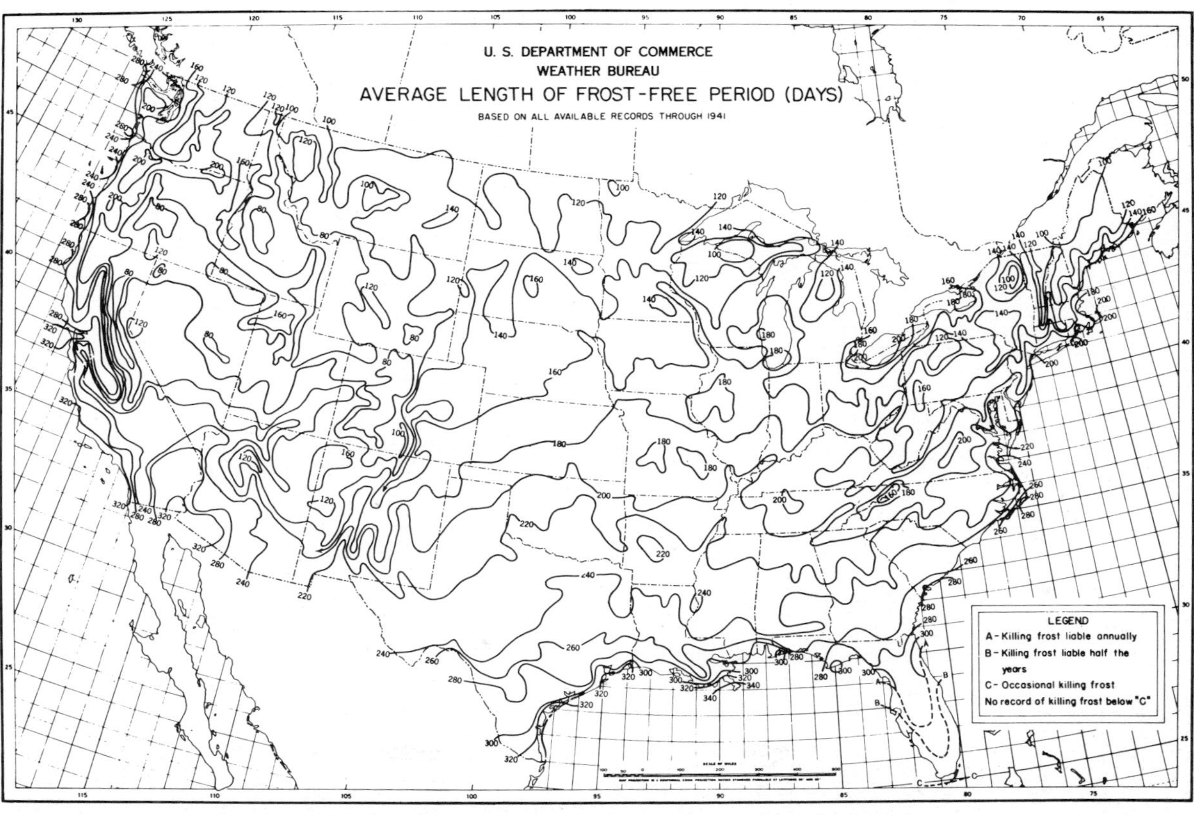

Figure 2–5. Frost-free periods with frost dates. (*a*) Average length of frost-free period.

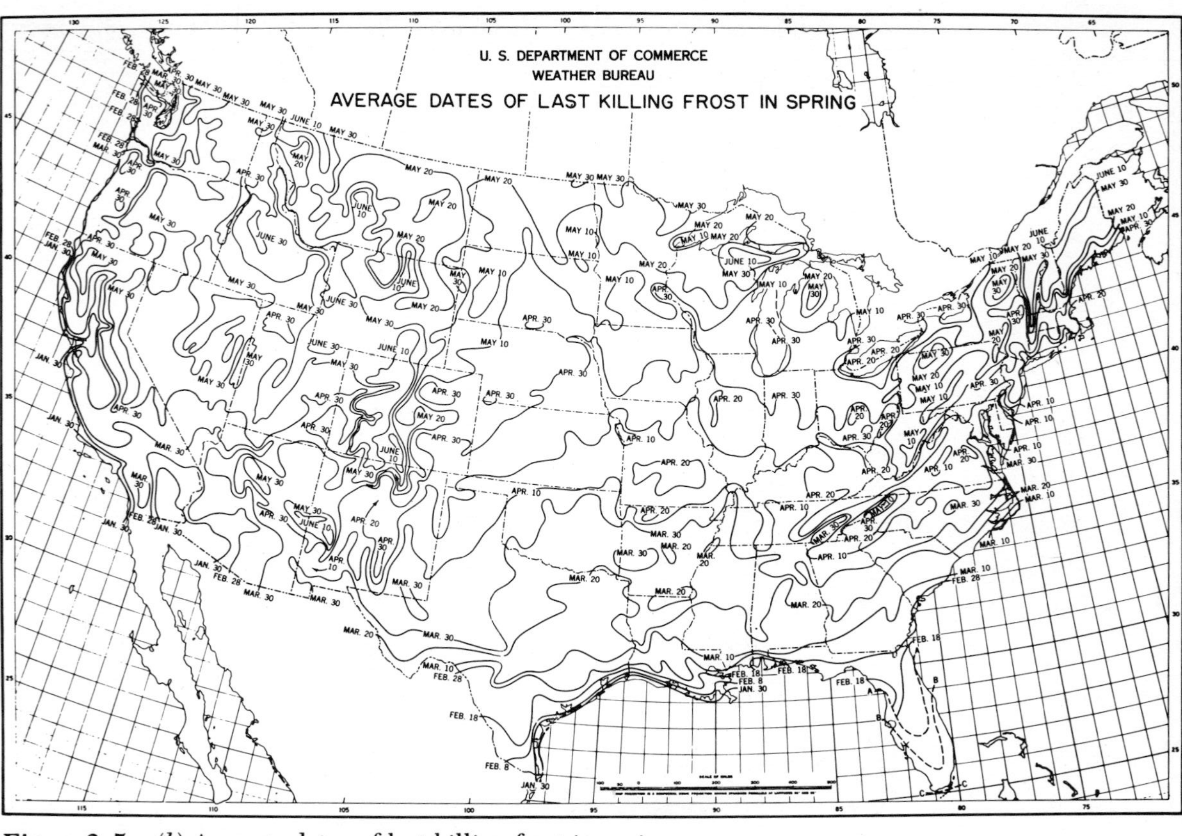

Figure 2–5. (*b*) Average dates of last killing frost in spring.

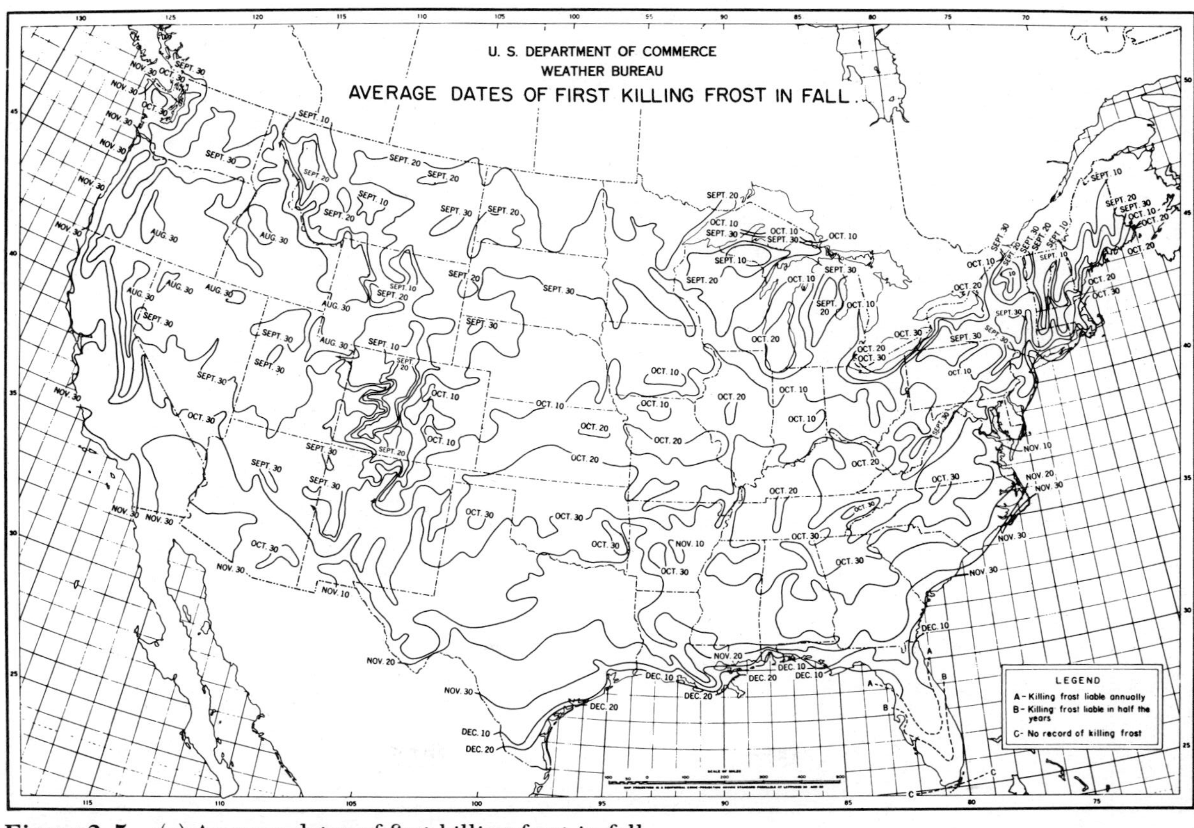

Figure 2–5. (*c*) Average dates of first killing frost in fall.

particularly under a rainfall of less than 30 inches per year. There is an increase in the water requirement of crops and in the seasonal evaporation from north to south in the Great Plains. In North Dakota a ton of alfalfa can be produced with 500 tons of water, while in the warmer Texas Panhandle 1,000 tons would be required.[36]

Each additional inch (2.54 cm.) in seasonal evaporation increases the water requirement to produce a short ton of dry matter about 14 to 18 tons.[37] The maximum transpiration for young sorghum plants occurred in soils well above field capacity (a soil moisture tension of 0.04 to 0.2 atmospheres) with a rapid decline to the permanent wilting point at a soil moisture tension of about 14 atmospheres.[29] The consumptive use of water (transpiration plus evaporation from the soil) by different irrigated crops varied from 13 to 74 inches in Arizona.[10] This varied with transpiration efficiency, total plant yield, and the length and time of year of the growing season for the particular crop. Short-season crops and those grown during the cool winter period required less water.

The rainfall-evaporation ratio is sometimes used as an index of the external moisture relations of plants, a ratio of 100 per cent meaning that rainfall and evaporation are equal.[44] This ratio ranges from 110 to 130 per cent along the Atlantic Coast, but westward it drops gradually to 20 per cent at the base of the Rocky Mountains. Only xerophytic shrubs and grasses can thrive where rainfall is so far below evaporation. True forests occur mostly where rainfall exceeds evaporation.

LIGHT

Light is necessary for the formation of chlorophyll in green plants and for photosynthesis, the process of manufacturing food essential to their growth. Plant growth may increase with greater light intensities of up to 1800 foot-candles (19,375 luxes or meter-candles) or more, an intensity of about one-fourth to one-half of normal summer sunshine. Maximum solar radiation may exceed 10,000 foot-candles. The hours of sunlight each day vary from a nearly uniform 12-hour day at the equator to continuous light or darkness throughout the 24 hours for a part of the year at the poles (Table 2–1). The day length is approximately 12 hours everywhere at the time of the vernal and the autumnal equinoxes. At the Canadian border of the United States (latitude 49° N.), the day length (sunrise to sunset during the year) ranges from about 8 hours 12 minutes to about 16 hours 6 minutes. At the southern tip of Florida (latitude 25° N.) the seasonal range is only from 10 hours 35 minutes to 13 hours 42 minutes. The long summer days at high latitudes enable certain plants to develop and mature in a relatively short time. When timothy was grown from Georgia to Alaska, the

TABLE 2–1. Approximate Length of Day on Various Dates at Different Latitudes North of the Equator

LATITUDE (NORTH)	DECEMBER 21 (WINTER SOLTICE)		MARCH 21 (SPRING EQUINOX)		APRIL 21		MAY 21		JUNE 21 (SUMMER SOLSTICE)		JULY 21		AUGUST 21		SEPTEMBER 21 (AUTUMN EQUINOX)	
	(hr.)	*(min.)*	*(hr.)*	*(min.)*	*(hr.)*	*(min.)*	*(hr.)*	*(min.)*	*(hr.)*	*(min.)*	*(hr.)*	*(min.)*	*(hr.)*	*(min.)*	*(hr.)*	*(min.)*
0°	12	7	12	7	12	7	12	7	12	7	12	7	12	7	12	7
10°	11	33	12	7	12	24	12	37	12	43	12	38	12	24	12	8
20°	10	56	12	9	12	42	13	9	13	19	13	11	12	43	12	10
25°	10	35	12	10	12	53	13	28	13	42	13	29	12	54	12	10
30°	10	11	12	10	13	4	13	47	14	4	13	48	13	6	12	12
35°	9	48	12	12	13	17	14	9	14	32	14	12	13	18	12	12
40°	9	20	12	12	13	31	14	35	15	2	14	36	13	32	12	13
45°	8	46	12	14	13	48	15	4	15	38	15	6	13	49	12	15
50°	8	4	12	15	14	9	15	41	16	24	15	44	14	9	12	17
55°	7	10	12	17	14	34	16	29	17	24	16	33	14	35	12	19
60°	5	52	12	18	15	8	17	38	18	54	17	41	15	9	12	21

Source: Data from Allard and Zaumeyer, in *USDA Tech. Bul.*, 867, 1944.

vegetative period was progressively shortened from south to north.[11] Flowering was hastened by the longer days.

The life processes of many plants were found by Garner and Allard[13] to be influenced by the relative lengths of day and night (which they called *photoperiodism*). This led to the distinction between long-day and short-day plants. Long-day plants require a relatively long day for the formation of inflorescences, but they increase in vegetative growth when the days are short. Among the long-day plants are red clover and the small grains (except rice), which ordinarily flower in the long days of early summer (Appendix Table A–1). Long days hasten flowering and maturity of these crops, but reduce vegetative growth. Short-day plants are stimulated to vegetative growth, with delayed flowering and maturity, when the days are long, and produce flowers and fruits when the days are relatively short. Corn, sorghum, rice, millet, and soybeans are short-day plants. In one experiment, Biloxi soybean plants grew 9 inches tall and flowered in 23 days under a 10½-hour photoperiod, but were 30 inches tall and flowered in 60 days under a 14½-hour photoperiod. The length of the night (dark period), rather than the length of day, is the critical determinant,[6] although light also is essential for the complete photoperiodic reaction.

Short-day plants require daily prolonged darkness to induce flowering, but a short interval of light during the night may prevent flowering. Long-day plants may flower under continuous light. Flowering in long-day plants is hastened by even short light exposures during the night.

Other crop plants and certain varieties of some of the above crops tend to be rather indifferent in their responses to photoperiodic influences. Cotton, cultivated sunflowers, and buckwheat are in the day-neutral, indeterminate, or intermediate group.

The distribution and time of maturity of different crops and crop varieties are influenced by photoperiodic responses. Corn or sorghum varieties from the tropics seldom mature when planted in the field in regions where the days are long. Corn hybrids and soybean cultivars are adapted to particular zones of latitude, i.e., day length. When planted outside of these zones, they usually are too early or too late for best performance. Native grass species that extend from the northern to the southern portions of the United States are differentiated into numerous local strains (ecotypes), each adapted to its particular photoperiodic range. The northern strains are small and early when grown in the South, whereas southern strains are large and late when moved to the North. Wild rice and gramagrasses exhibit this variation to a marked degree.

Day length and temperature compensate in their effects to some extent.[17] Photoperiodic responses of a number of plants may be modified by temperature.[33] Quick-maturing sorghum varieties may grow larger in the North where days are long than they do in the South where temperatures are more favorable. The time a crop is planted affects the periods required to reach maturity. This is due to temperature and photoperiodic influences. The growing period for spring-sown crops decreases as seeding becomes later, except in the case of very late seeding, which forces the crop to complete its growth at a slower rate induced by cool autumn conditions. Most cool-weather crops have a long-day response, whereas most warm-weather crops have a short-day response.

Supplementary light to shorten or to interrupt the dark period usually is effective in the control of flowering at intensities of 30 foot-candles or even less. Red light of wave lengths of about 660 millimicrons is most effective in inducing flowering in long-day plants, or in preventing flowering of short-day plants. Far red light (wave length about 730 millimicrons) nullifies the effect when applied immediately after a short exposure to red light. Red or far red rays are absorbed by *phytochrome*, a photo-reversible blue pigment of protein nature. Phytochrome is present in minute amounts in all flowering plants, even in seedlings without chlorophyll that were grown in the dark. It is present in some seeds. The pigment behaves like an enzyme when it triggers flowering reactions.[6]

AIR

Air not only supplies carbon dioxide for plant growth, but also furnishes nitrogen indirectly. It supplies oxygen for respiration of the plant as well as for chemical and biological processes in the soil. Air sometimes contains gaseous substances in concentrations harmful to plant growth; these gases usually come from fuel combustion or industrial fumes. Of these ozone and PAN (peroxyacetyl nitrate) are photochemical oxidants formed by sunlight acting on products of fuel combustion. Ozone excess produces flecks on the leaves of tobacco, maize, and other plants; and PAN causes young leaf tissue to collapse. Sulfur dioxide produces blotches between leaf veins. Nitrogen dioxide suppresses plant growth. Fluorides develop chlorotic patterns or the death of the margins and tips of the leaves. Industrial smokes and auto exhausts contribute an average of 50 pounds of nitrogen per acre over the whole state of Connecticut.[12] A concentration of sulfur dioxide of one part or more in two million parts of air is injurious to plants.

Carbon monoxide levels in the air over city streets are no higher with the present heavy motorized traffic than they were in the earlier years when coal stoves and furnaces, artificial gas fuel, and kerosene lamps were in vogue. The pretty blue flame in a wood fireplace displays the release of carbon monoxide. Artificial (coal) gas is a mixture of carbon monoxide and methane. Lead levels in human tissues, despite the use of leaded motor fuels, are lower than they were two generations previously. Then, coal burning, lead paints, lead plumbing, pewter utensils and thickly soldered tin cans prevailed in the homes, and lead arsenate was a popular pesticide.

All air pollutants, including dust, soot, and ashes, return to the earth eventually in rain, snow, or by gravity or downward air currents. The solid particles become a part of the mineral or organic soil constituents. Ashes and dust contain essential plant nutrients such as potassium, phosphorus, and calcium. Gaseous compounds of nitrogen and sulfur are converted into solid plant nutrients but some atmospheric nitrogen is released by soil organisms that use the oxygen (derived from nitrogen oxides). Carbon dioxide, the chief component of consumed fuels is absorbed directly by plants in photosynthesis. Some of the carbon monoxide is converted to carbon dioxide or carbonates by lightning or solar energy in the air. Phosphorus and its compounds from the air, soils, and sewage, when emptied into still waters promote the growth of algae. When algae die they use the oxygen from the water to promote decay. This process, called eutrophication, reduces the oxygen supply which suffocates the fish. Eutrophication also occurs in ponds remote from farm lands.[41] Eutrophication has been operating since the creation to produce peat. Heavy pressure in later geological ages converted some of the peat into coal and petroleum.[5] Most of the nitrogen and phosphorus that reach the waters are in industrial and sewage wastes rather than from farms.[39]

Wind may cause mechanical lacerations and bruises to the tissues of crop plants. High winds may bruise and tear the leaves or, as often happens with small grains grown on fertile soils, the plants may be blown over or lodged. The rate of transpiration or water loss from plants increases in proportion to the square root of the wind velocity. In high winds crop plants may be cut off by moving sand particles, completely uprooted, or buried under soil drifts.

SOIL REQUIREMENTS*

Although less important than climate, soil texture and soil reaction (acidity) play a major role in determining which crops are grown. Despite popular belief, soil texture and reaction have only a minor effect in determining the variety of a particular crop that is grown. Surveys of crop varieties show that their distribution is determined largely by climatic and soil-moisture factors.

* For more complete information on soil requirements for crops, see *Soil,* USDA Yearbook, 1957. Also see Russell, E. W., *Soil Conditions and Plant Growth,* 9th ed., Wiley, New York, 1961.

Soils provide naturally, or after treatment, certain favorable conditions for plant growth. Soil not only furnishes a stratum for germination of the seed and spread and anchorage of the plant roots, but, in addition, maintains a reasonably satisfactory balance of soil moisture and essential mineral elements for nutrition of the plants.

Soil is not essential for plant growth, however, as is evidenced by the frequent growing of many crop species in sand and water cultures, to which nutrients are supplied artificially. Spearmint was grown in water cultures by John Woodward[45] as early as 1699. Some 240 years later, the process of growing plants in water cultures was named hydroponics.[14, 2] The growing of crops on heavily fertilized sandy lands in the high-rainfall southeastern region of the United States is, in effect, largely a sand culture.

Texture

Soil texture has an important influence upon crop adaptation. Medium or heavy soils are best for fine-rooted grasses, wheat, and oats, whereas the coarser rye, corn, and sorghum plants can thrive, and are commonly grown, on the light sandy soils. Rice demands a reasonably heavy soil with a nearly impervious subsoil that prevents excessive water losses from leaching.

Water percolates more quickly and more deeply into light soils, and this checks runoff, although heavy soils have the greater water storage capacity per cubic foot. Partly for this reason, crops in semiarid regions are less subject to drought on sandy soils than on heavy soils, whereas in regions of high rainfall where much of the water may percolate downward below the root zone, crops on sandy soils are more subject to drought.

Texture of a soil refers to the size of its individual grains or particles, grouped on the basis of diameter as sand (2 to 0.05 millimeter), silt (0.05 to 0.002 millimeter), and clay (0.002 millimeter and less). Soil class is recognized on the basis of the relative percentages of these separates. The principal classes as to texture are sand, loamy sand, sandy loam, silt loam, clay loam, silty clay loam, and clay, in increasing order of their content of fine separates. The farmer commonly calls a soil with a large proportion of clay particles a heavy soil, and one with a large proportion of sand a light soil, the difference being due to the ease of cultivation. The so-called hard and soft lands are a popular classification based on this same condition. In weight per unit of volume light soil is heavier than heavy soil.

The finest particles, having a diameter of 0.001 millimeter or less (usually less than 0.0001), constitute the colloidal material which has been aptly designated the protoplasm of the soil. It retains available plant nutrients and water by adsorption. The degree of adsorption depends upon the amount of surface. Fine particles have an enormous surface area per unit weight of soil. One pound of average soil colloidal particles has about five acres of surface.

In any soil, the fine clay and colloidal fractions contain most of the available nutrients. They play a major role in determining its properties, which include ion exchange. The percentage of particles of clay size and smaller approximates the percentage of moisture at the field capacity of many soils. At field capacity, where the water is retained against the force of gravity, the soil-moisture tension is ½ atmosphere or less. At the permanent wilting point for plants, the tension is 12 to 15 atmospheres. Plant roots take up water readily from loam soils with tensions of up to about 8 atmospheres (50 per cent of field capacity), but water stress, or even wilting, may develop above this tension in hot dry weather. Irrigation is advisable at moisture tensions of about six atmospheres, at the stage at which about one third of the available water capacity of the soil remains.

Structure

Soil structure refers to the manner in which the individual particles are arranged. An aggregated or compound structure, in which the particles are grouped into crumbs or granules from 1 to 5 millimeters in diameter, provides irregular spaces through which water and air can circulate. This also favors water infiltration, good seedbed preparation, ease of cultivation, and protection from wind or water erosion. Soil aggregation normally occurs when ample organic matter is present. It increases sharply with increases in the soil carbon content from zero to 2 per cent or more. A moist soil of good structure crumbles in the hand. This granular condition or tilth may be destroyed by flooding or by packing or working the soil when it is too wet. It is then puddled or deflocculated, i.e., the soil is made up of single unaggregated grains.

Relation of Soil Groups to Crop Production

A line drawn from western Minnesota southward through central Texas roughly divides the country into two parts with regard to soil characteristics. East of this line (Figure 2–6), the soils called Pedalfers are leached so that no accumulation of lime is found in the soil depths occupied by the roots of most crop plants. Thus, eastern soils and those in the humid Pacific Northwest are acid because basic materials (sodium, potassium, calcium, and magnesium) are leached out by the heavy rainfall. Lime applications are beneficial when clover or alfalfa are being seeded on Pedalfers. In most of the West, soils called Pedocals have an accumulation of calcium carbonate in either the subsurface or subsoil. The higher the rainfall, the deeper the carbonate zone is found. Pedocals tend to have an accumulation of soluble mineral salts due to the fact that the low rainfall is not adequate to

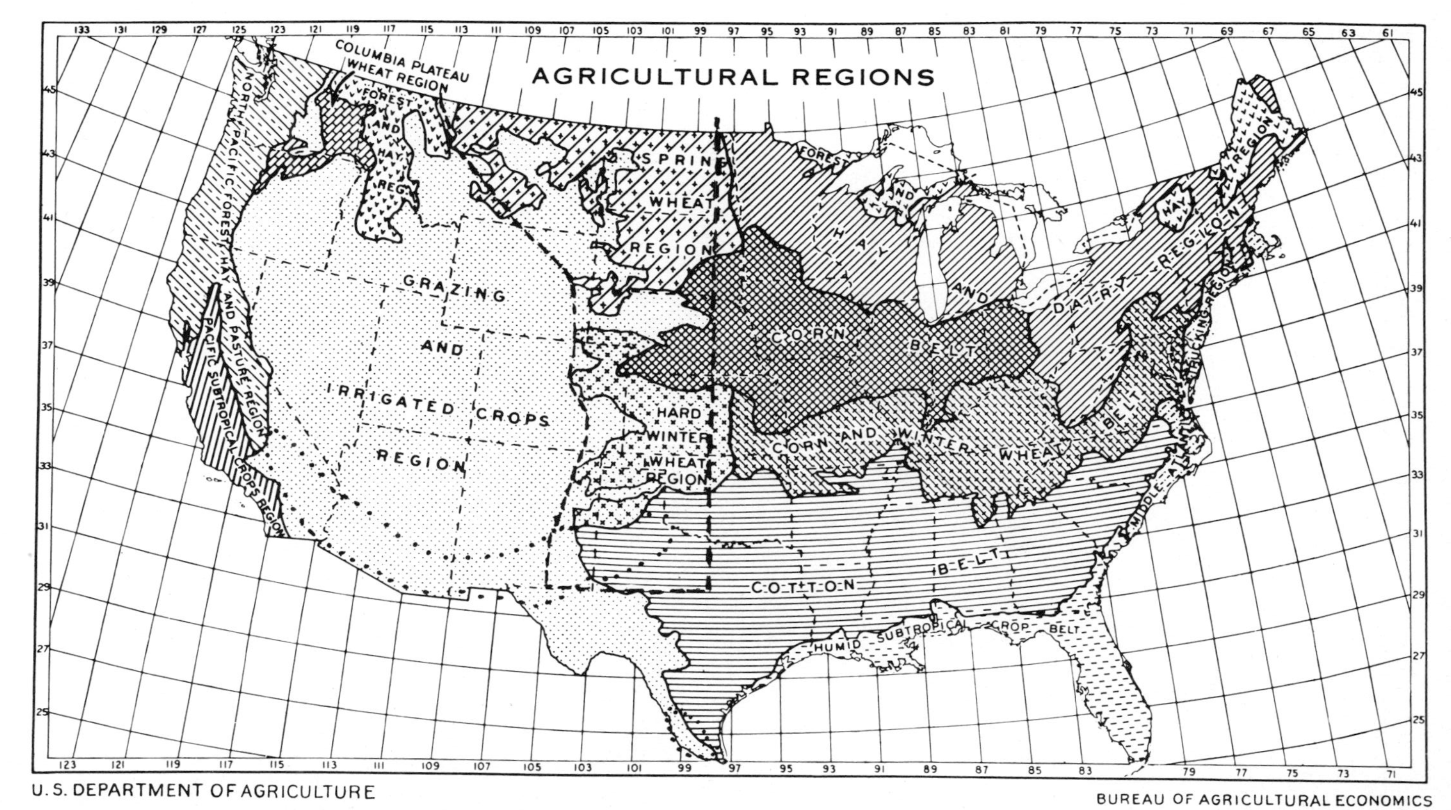

Figure 2–11. Agricultural regions of the United States. The Great Plains is the west-central area outlined by dashes. The western irrigated cotton-growing area lies within the dotted outline.

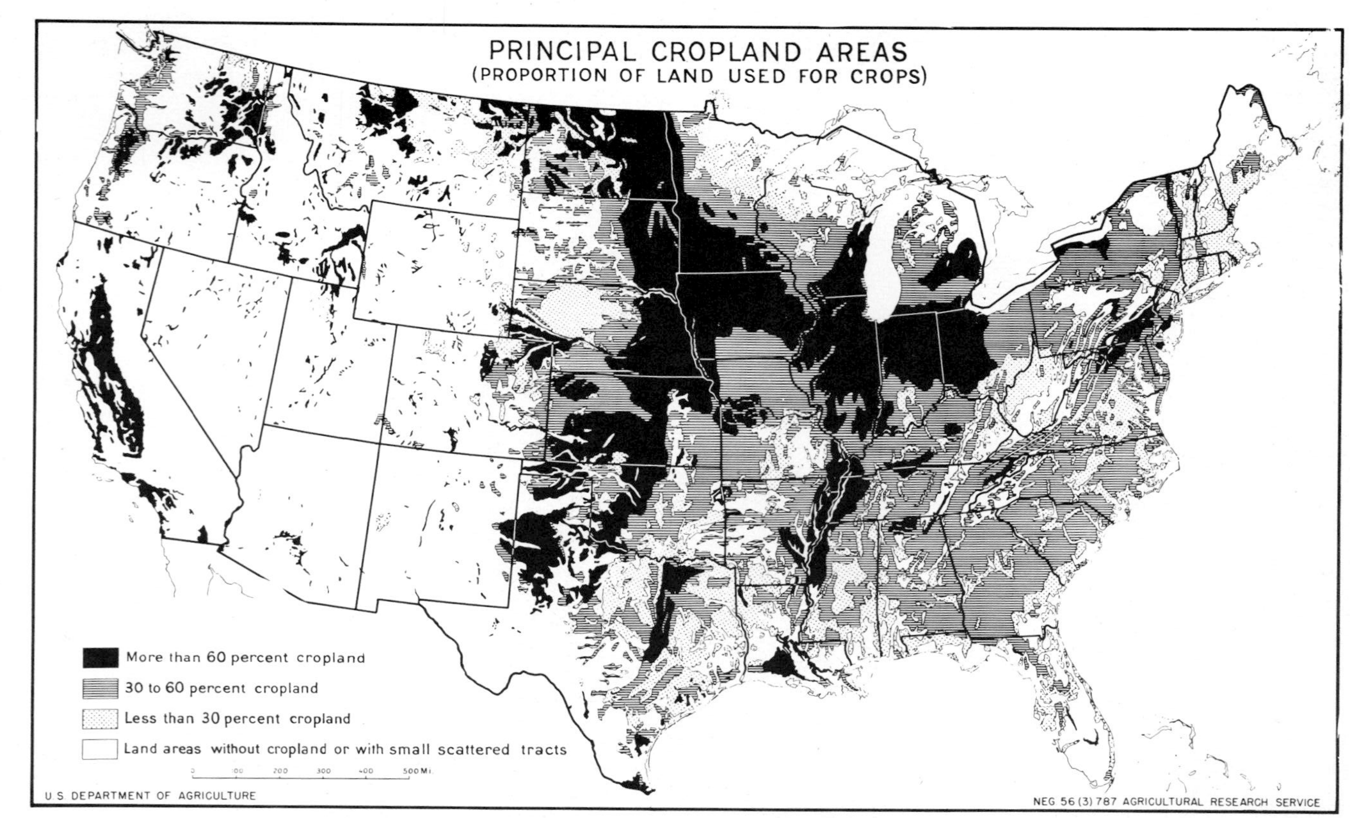

Figure 2–12. Deserts, mountains, and infertile soils limit the areas suitable for crops in parts of the United States.

although wheat is well adapted. In fact, wheat is grown on many farms in rotation with corn throughout the prairie and deciduous forest climaxes. Wheat is higher in protein when grown in a dry climate. Oats are concentrated in the northern portion of the Corn Belt because of the cooler climate.

The region southeast of the 100 per cent rainfall-evaporation ratio line is known ecologically as the southeast evergreen center. Cotton is the principal crop in this area because of the high temperatures and long growing season. In the Cotton Belt, cotton competes with cereals, forage crops, tobacco, soybeans, and peanuts, and, on some soils, with sugarcane and rice.

The northern evergreen forest, north and east of the 110 per cent rainfall-evaporation ratio line, is devoted agriculturally to tame pastures and hay crops. The need for the bulkier hay and fodder crops, which grow well under the cool temperature conditions, makes cereal production less profitable in this region.

Great Plains

In the western half of the country, evaporation consistently exceeds precipitation two or three times. To survive under such droughty conditions, the cultivated plants of the plains must be grown under the best known cultural methods for conservation and utilization of the moisture captured by the soil. Irrigation is practiced where stream flow or underground water supplies can be diverted to supplement the natural rainfall. Wheat is the major crop. Sorghums are grown in some of the more arid portions. Cotton, wheat, and sorghum predominate in the southern areas of the plains. Small grains, corn, flax, sorghum, and hay are grown in the northern parts.

Pacific Coast

The third distinct crop region, the Pacific Coast, is a winter-rainfall area, while the portion near the ocean has mild winters and cool summers. The northern half of this region is devoted primarily to seed, feed, special purpose, and horticultural crops. Wheat is an important crop in the intermountain section east of the Cascade Mountains in Oregon and Washington. The southern half of the Pacific Coast is devoted to fruits, vegetables, and many different field crops, especially under irrigation.

REFERENCES

1. Aldous, A. E., and Shantz, H. L. "Types of vegetation in the semiarid portion of the United States and their economic significance," *J. Agr. Res.*, 28:99–128. 1924.
2. Anon. "Soilless growth or hydroponics," *Can. Dept. Agr. Pub. 1357*, 1–7. 1968.

3. Baldwin, M., Kellogg, C. E., and Thorp, J. "Soil classification," in *Soils and Men*, USDA Yearbook, 1938, pp. 979–1001.
4. Bernstein, L. Salt tolerance of plants. *USDA Inf. Bul. 2838*, 1–23. 1964.
5. Borlaug, N. E. "Ecology fever," *Seventh Memorial Lecture at the 16th FAO Conference*, pp. 21–25. 1971.
6. Borthwick, H. A., and Hendricks, S. B. "Photoperiodism in plants," *Science*, 132(3435):1223–1228. 1960.
7. Burr, W. W. "Contributions of agronomy to the development of the Great Plains," *J. Am. Soc. Agron.*, 23:949–959. 1931.
8. Chapman, H. D. "Diagnostic criteria for plants and soils," *Univ. Calif. Div. Agr. Sci.*, pp. 1–793. 1966.
9. Donahue, R. L., and others. *Soils: An Introduction to Soils and Plant Growth*, Prentice-Hall, Englewood Cliffs, N.J., pp. 1–587. 1971.
10. Erie, L. J., and others. "Consumptive use of water by crops in Arizona," *Ariz. Tech. Bul. 169*, pp. 1–41. 1965.
11. Evans, M. W. "Relation of latitude to time of blooming of timothy," *Ecol.*, 12:182–187. 1931.
12. Frink, C. R. *Frontiers of plant science*, Conn. Agr. Exp. Sta., New Haven. 1969.
13. Garner, W. W., and Allard, H. A. "Effect of the relative length of day and night and other factors of the environment on growth and reproduction in plants," *J. Agr. Res.*, 18:553–606. 1920.
14. Gericke, W. F. *The Complete Guide to Soilless Gardening*, Englewood Cliffs, N.J., Prentice-Hall, pp. 1–285. 1940.
15. Hargrove, T. R. "Agricultural research: impact on environment," *Ia. Agr. Exp. Sta. Special Rpt. 69*, pp. 1–84. 1972.
16. Hopkins, A. D. "Bioclimatics—a science of life and climate relations," *USDA Misc. Pub. 280*, pp. 1–188. 1938.
17. Hurd-Karrer, A. M. "Comparative responses of a spring and a winter wheat to day-length and temperature," *J. Agr. Res.*, 46(10):867–888. 1933.
18. Kellogg, C. E. "Development and significance of the great soil groups of the United States," *USDA Misc. Pub. 229*. 1936.
19. Kincer, J. B. "The relation of climate to the geographic distribution of crops in the United States," *Ecol.*, 3:127–133. 1922.
20. Kincer, J. B. "Precipitation and humidity," in *Atlas of American Agriculture*, USDA, Pt. II, Sec. A, pp. 1–48. 1922.
21. Kincer, J. B. "Temperature, sunshine, and wind," in *Atlas of American Agriculture*, USDA, Pt. II, Sec. B, pp. 1–34. 1928.
22. Klages, K. H. "Crop ecology and ecological crop geography in the agronomic curriculum," *J. Am. Soc. Agron.*, 20:336–353. 1928.
23. Klages, K. H. "Geographical distribution of variability in the yields of field crops in the states of the Mississippi valley," *Ecol.*, 11:293–306. 1930.
24. Klages, K. H. "Geographical distribution of variability in the yields of cereal crops in South Dakota," *Ecol.*, 12:334–345. 1931.
25. Klages, K. H. *Ecological Crop Geography*, Macmillan, Inc., New York, pp. 1–615. 1942.
26. Martin, J. H. "Climate and sorghum," in *Climate and Men*, USDA Yearbook, 1941, pp. 343–347.

27. McMurtrey, J. E., Jr., and Robinson, W. O. "Neglected soil constituents that affect plant and animal development," in *Soils and Men,* USDA Yearbook, 1938, pp. 807–829.
28. Nikiforoff, C. C. "Soil organic matter and soil humus," in *Soils and Men,* USDA Yearbook, 1938, pp. 929–939.
29. Pallas, J. E., Jr., and others. "Research in plant transpiration: 1961," *USDA Prod. Rsh. Rpt. 70,* pp. 1–37. 1962. (279 references.)
30. Piper, C. V. *Forage Plants and Their Culture,* Macmillan, New York, pp. 1–671. 1924.
31. Richards, L. A., editor. "Diagnosis and improvement of saline and alkali soils," *USDA Handbook 60.* 1964.
32. Robbins, W. W. "Native vegetation and climate of Colorado in their relation to agriculture," *Colo. Agr. Exp. Sta. Bul. 224.* 1917.
33. Roberts, R. H., and Struckmeyer, B. E. "The effect of temperature and other environmental factors upon the photoperiodic responses of some of the higher plants," *J. Agr. Res.,* 56:633–678. 1938.
34. Salmon, S. C. "The relation of winter temperature to the distribution of winter and spring grain in the United States," *J. Am. Soc. Agron.,* 9:21–24. 1917.
35. Shantz, H. L. "Native vegetation as an indicator of the capabilities of land for crop production in the Great Plains area, "*USDA Bul. 201.* 1911.
36. Shantz, H. L. "Plants as soil indicators," in *Soils and Men,* USDA Yearbook, 1938, pp. 835–860.
37. Shantz, H. L., and Piemeisel, L. N. "The water requirement of plants at Akron, Colo.," *J. Agr. Rsh.,* 34(12):1093–1190. 1927.
38. Sprague, H. B., ed. *Hunger Signs in Crops,* 3rd ed., McKay, New York, pp. 1–461. 1964.
39. Vavilov, N. I. *World Resources of Cereals, Leguminous Seed Crops, and Flax, and Their Utilization in Plant Breeding,* Offices of Technical Services, U.S. Department of Commerce, Washington, D.C. 1960.
40. Viets, F. G., Jr., and Hageman, R. H. "Factors affecting the accumulation of nitrate in soil, water and plants." *USDA Agr. Handbk. 413.,* pp. 1–63. 1971.
41. Vinall, H. N., and Reed, H. R. "Effect of temperature of other meteorological factors on the growth of sorghums," *J. Agr. Res.,* 13(2):133–147. 1918.
42. Wadleigh, C. H., and Britt, C. S. "Conserving resources and maintaining a quality environment," *J. Soil and Water Cons.,* 23(5):172. 1969.
43. Wallace, T. *The Diagnosis of Mineral Deficiencies in Plants by Visual Symptoms,* His Majesty's Stationery Off., London, pp. 1–116; also Supp., pp. 1–48. 1944.
44. Waller, A. E. "Crop centers of the United States," *J. Am. Soc. Agron.,* 10:49–83. 1918.
45. Weaver, J. E., and Clements, F. E. *Plant Ecology,* 2d ed., McGraw-Hill, New York, pp. 333–417. 1938.
46. Wilsie, C. P. *Crop Adaptation and Distribution,* Freeman, San Francisco, 1962.
47. Woodward, J. "Thoughts and experiments on vegetation." *Phil. Trans.,* pp. 382–392. 1699.

Chapter 3

Botany of Crop Plants

GENERAL NATURE OF CROP PLANTS

Most of the important crop plants are reproduced by seeds, and all are classified as spermatophytes or seed plants. In the life cycle of these plants the seed germinates and produces a seedling. The vegetative phase is characterized by increases in the number and size of roots, stems, and leaves. Finally the reproductive phase is reached in which the plant flowers and produces seeds.

THE PLANT CELL

Plants are composed of units called cells. A living cell (Figure 3–1), containing 85 to 90 per cent water, is composed of two main regions—the cell wall, which is made up of cellulose and is nonliving, and the protoplast, which is the living portion of the cell and surrounded by a differentially permeable membrane (plasma membrane). The cell wall provides rigidity to the plant's structure while there are vacuoles in the protoplasm that provide a means to remove and store materials, thereby effectively removing them from chemical transformations within the cell. The protoplasm is the site of the living activity that a plant conducts. The protoplast contains many organelles and structures which are very specific in their contribution to the living cell and plant. The differentially permeable plasma membrane provides a selectivity of materials entering the cell. The nucleus contains the chromosomes, the carriers of heredity. The chromosomes contain molecules of DNA (deoxyribonucleic acid) in a linear aggregate of four building blocks called nucleotides. The order of these nucleotides determines the genetic message carried by the chromosome. Via this mechanism, plants control the synthesis of enzymes and other proteins.

Mutations occur from chemical changes in the structure of DNA molecules during reproduction of the molecules or as a reaction to radiation or other exposures. The mutations result in changes in the hereditary characters of the plants.

Ribosomes consist of ribonucleic acid (RNA)-rich material on the surface of which amino acids are connected by peptide bonds to form polypeptide units and finally proteins. Ribosomes are essential for protein synthesis; plastids are globular cytoplasmic bodies, the most significant of which is the chloroplast where photosynthesis occurs; mitochondria are very specialized structures which are involved primarily in energy release. They oxidize organic molecules and make energy available for the life processes of the cell.

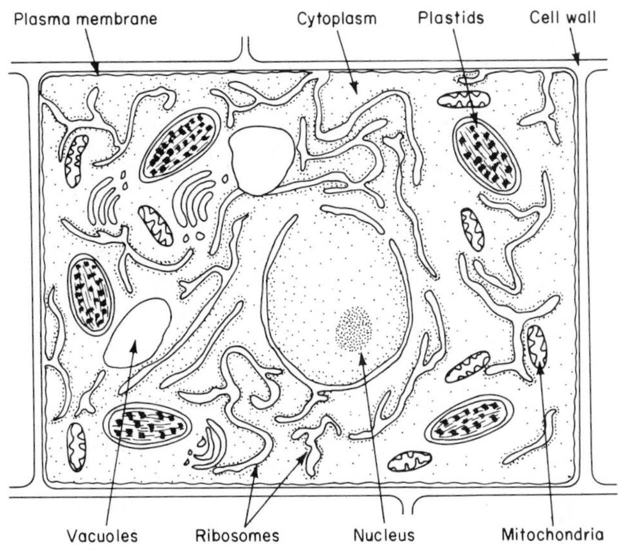

Figure 3–1. The plant cell. [As adapted from Samual N. Postlethwait, Henry D. Telinde, and David D. Husband, *A-T Botany, Matter and Mechanics of Cells*. Minneapolis, Minn.: Burgess Publishing Co., 1963.]

Other components of the living cell include microbodies (peroxisomes and glyoxysomes) which are involved in the decomposition of the toxic hydrogen peroxide and play a major role in fat catabolism. Dictyosomes function in the assembly and secretion of products outside the cell leading to cell-wall synthesis via polysaccharides produced inside the cell. Spherosomes are involved in the storage and transfer of lipids in plant cells. Cells differ in size and shape in accordance with their special functions (Figure 3–2).

All living cells give off carbon dioxide and absorb oxygen. This exchange of gases, known as respiration, is accomplished by the breakdown of sugars as well as by the oxidation of the carbon in the sugar into carbon dioxide. The most important aspect of respiration is the consequent release of energy necessary for growth and maintenance of the plant.

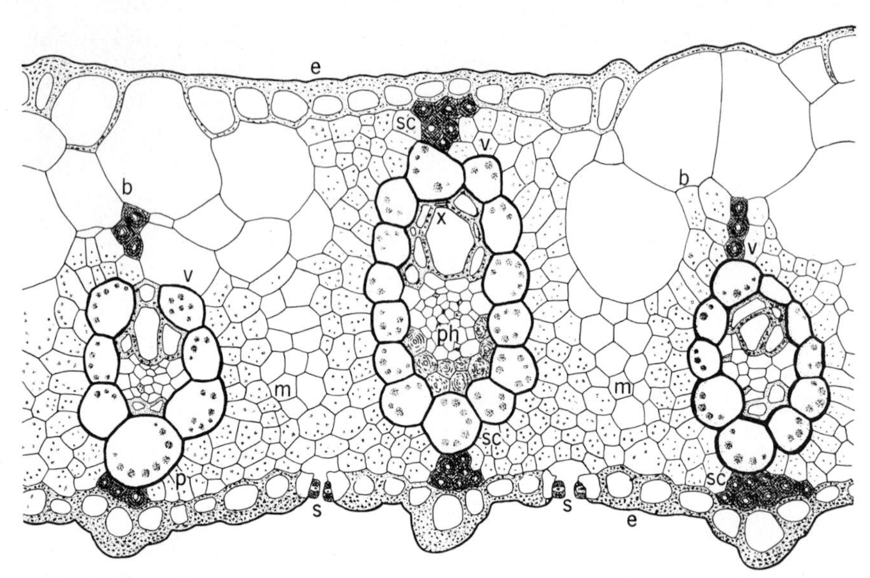

Figure 3–2. Cells observable in a cross-section of a sugarcane leaf, a typical grass leaf: (b) bulliform or motor cells that shrink when the leaf wilts and cause the leaf to roll up, and later take up water, expand, and flatten out the leaf; (e) epidermal cells; (m) mesophyll cells; (p) palisade-shaped cell of mesophyll; (ph) phloem; (s) stoma; (sc) sclerenchyma (stiffening) cells; (v) vascular bundle containing phloem and xylem; (x) xylem. [After Artschwager.]

STRUCTURE AND FUNCTIONS OF CROP PLANTS

Roots

Sometimes as much as one half, but in certain root crops more than one half, of a crop plant is underground. The underground plant parts are fully as important as the tops because nearly all of the water and all of the mineral nutrients except some nitrogen compounds used by the plant are absorbed from the soil by the roots.[52]

KINDS OF ROOTS: Two general types of roots are found in crop plants: fibrous roots and taproots. Certain weeds, e.g., Canada thistle (Figure 3–3), field bindweed, and leafy spurge, also have lateral roots that send up new shoots which in turn establish new root systems.

Plants of cereals and other grasses have fibrous roots, i.e., many slender roots similar in diameter and length, with somewhat smaller branches. The seminal or primary roots develop when the seed germinates, growing from the lower end of the embryo. These roots sometimes have been erroneously called temporary roots in the belief that they function only in the earlier growth stages. However, they often

Figure 3–3. Lateral roots of Canada thistle.

remain active until the plant matures.[32, 34, 53] The seminal roots of corn often penetrate to a depth of 5 or 6 feet. In wheat they are the chief support of the main stem of the plant. Sorghum, rice, and proso produce only a single-branched seminal root. After the young plant unfolds a few leaves, the coronal (crown or nodal) roots that arise from stem nodes underground develop into an elaborate root system. Coronal roots rise 1 inch, more or less, below the soil surface, even though the seed was planted much deeper. The subcrown internode between the seed and crown elongates until stopped by light striking the emerging coleoptile. At that stage the crown node is still below the surface. Frequently, roots grow from nodes above the soil surface as, for example, brace (or aerial) roots of corn. They are unbranched above the ground but are the same as other roots once they enter the soil.

Other crop plants, such as the sugarbeet, other root crops, and the legumes, have a main taproot. The main root develops and pushes straight downward. It may branch throughout its length. The thickened taproot of biennial or perennial plants often serves as an organ for food storage.

The smallest subdivisions of the roots are the root hairs, the single-celled structures found on roots and root branches of both fibrous-rooted and tap-rooted plants.

EXTENT OF ROOT SYSTEMS: Elongation at the rate of ½ inch (1.25 cm.) per day is common in roots of many grasses. Seminal roots of winter wheat[52] may grow at that rate for 70 days. The main vertical roots of corn grow 2 to 2½ inches (5 to 12 cm.) per day for a period of three or four weeks. Plant roots may extend as far or farther below the ground than the plant does above. The root system of a corn plant, given ample space, may occupy 200 cubic feet (5.66 cu. m.) of soil. A corn plant grown five weeks in a loose soil in Nebraska produced 19 main roots with 1,462 branches of the first order. Upon these were 3,221 branches of the second and third orders. The main roots consisted of 4 to 5 per cent of the total root length, the primary branches 47 to 67 per cent, and the finer secondary and tertiary laterals 29 to 48 per cent of the total. A rather high correlation between root and top growth is found in corn.[54] Given ample space, the roots of a single plant may have a total spread of 4 feet in wheat, 8 feet in corn, and 12 feet in sorghum. Roots of small grains, corn, sorghum, and perennial grasses normally penetrate downward 3 to 6 feet, but sometimes they extend 8 or 10 feet. Alfalfa roots may penetrate more than 30 feet (9 meters). The length of the entire root system, which includes all branches of a single large isolated plant of crested wheatgrass, may be as much as 360 miles.[39]

The roots grow in moist soil but most crop plants do not extend roots into water-logged soil. The roots can penetrate short distances into dry soil when moisture is supplied from roots penetrating adjacent moist soil.[9]

ABSORPTION: Plants absorb water and also the substances dissolved in it—namely, nitrogen and other mineral elements, largely through the root hairs. The root hairs absorb water by osmosis, while their walls act as differential membranes in the selection of different ions that the plants absorb. When the cell sap is more concentrated than the soil solution, water passes into the root hairs; this occurs except under very abnormal soil conditions. The rate of absorption of water may vary with the osmotic pressure of the cell.[36]

The more a plant needs water the more vigorously it is absorbed, provided the water supply remains ample. Water taken up in excess of the ability of the plant to transpire water vapor through the stomata, and especially at night when stomata are closed, is forced out through pores in the leaf called hydathodes. Most of what are called dew drops are actually water globules from hydathodes. Plasmolysis (or cell collapse) results when the soil solution is more concentrated than that in the root hair and other cells. Such a condition may be found in highly saline soils where water is actually withdrawn from the plant.

Nutrients are absorbed independently of the rate of water intake, being taken into the plant as ions by diffusion. Thus, it is possible for mineral nutrients to be absorbed when the atmosphere is saturated and little water is being taken up and none is being transpired. Minerals in solution tend to pass through the differential cell membrane and reach an equilibrium. For instance, potassium nitrate (KNO_3) dissolves and becomes dissociated into K^+ and NO_3^- ions.

These ions tend to reach an equilibrium regardless of the total concentration in the two solutions inside and outside the cell membrane.

Stems and Modified Stems

After a grain of wheat germinates, the plumule internodes elongate into the first young stem. Stem elongation continues by growth (cell division and cell elongation) at the growth ring (intercalary meristem) above each successive node while the cells are young and active. Growth in the diameter of the stem of cereals and other grasses is due to cell enlargement, not cell division, after the essential stem structures have been formed.

In dicotyledonous crop plants, i.e., those other than cereals and other grasses, growth is much like that of a tree. New branches arise from buds or adventitious cells, while the branches elongate by cell division and cell enlargement near the tips. Growth in diameter comes from cell division in the cambium layer under the bark or periderm, followed by cell enlargement.

GENERAL NATURE OF STEMS: The stem is divided by joints or nodes, the stem section between the nodes being called internodes (Figure 3–4). In grasses the internodes when young are solid, i.e., filled with

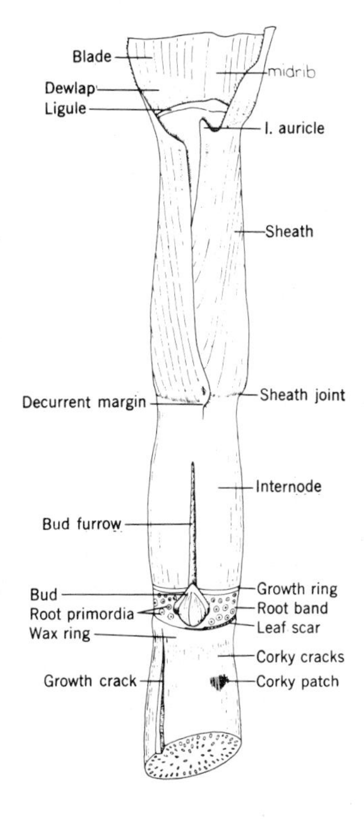

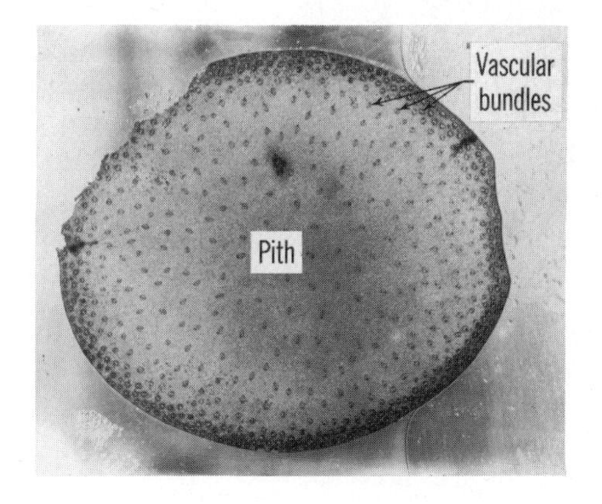

Figure 3–4. (*Left*) A portion of the stem bud at the node. (*Right*) Cross-section of of sugarcane: A branch may arise from the corn stalk internode showing vascular bundles distributed through the pith but more concentrated in the outer rind.

pith.[24] In the mature plant the pith usually disappears, sugarcane, corn, and sorghum being exceptions (Figure 3–4). The stem or culm of a grass is usually cylindrical or nearly so. The culms of most grasses die down to the soil surface each year, or, usually, after the seed has matured. In addition to the main stem, several branches known as tillers arise from the lower nodes (Figure 3–5). Large numbers of tillers may arise under favorable environmental conditions, each tiller developing a root system of its own.

Figure 3–5. Tillers and tiller buds on a sorghum plant. The swollen section of the stalk is a node.

The type of stem may not remain the same during the entire life of the plant.[41] Sometimes the primary axis within the plumule remains so short that the leaves arise crowded together, usually in the form of a rosette. Winter annuals, e.g., winter wheat, exhibit this habit until some time in the spring. Sugarbeet and mangel plants have short contracted internodes during the first year. In the second year, the growing point of the stem, hidden in the center of the rosette, elongates and produces a shoot with long internodes and with the leaves at considerable intervals. Seed is produced on these shoots.

Temperature and day length may affect the growth habit of a plant.

Thus, certain varieties of winter oats may show a spreading habit when sown in the fall but an erect habit when sown in the spring.

A bud is an undeveloped stem, leaf, or flower. Buds may be terminal on the stem, axillary (side), supernumerary, or adventitious (arising out of order). Many buds are dormant.

MODIFIED STEMS: Several types of modified stems are recognized.

1. *Rhizomes.* Rhizomes are horizontally elongated underground stems that bear scales at the nodes. They may arise extravaginally, i.e., the stems may burst through leaf sheaths of older stems and elongate horizontally below the soil surface. Rhizomes produce new stem shoots from their joints, which serve to propagate the plant. Some of the most pernicious weeds, quackgrass and johnsongrass (Figure 3–6), have rhizomes. Many desirable pasture plants, such as bromegrass and Kentucky bluegrass, are extravaginal or sod-forming grasses.

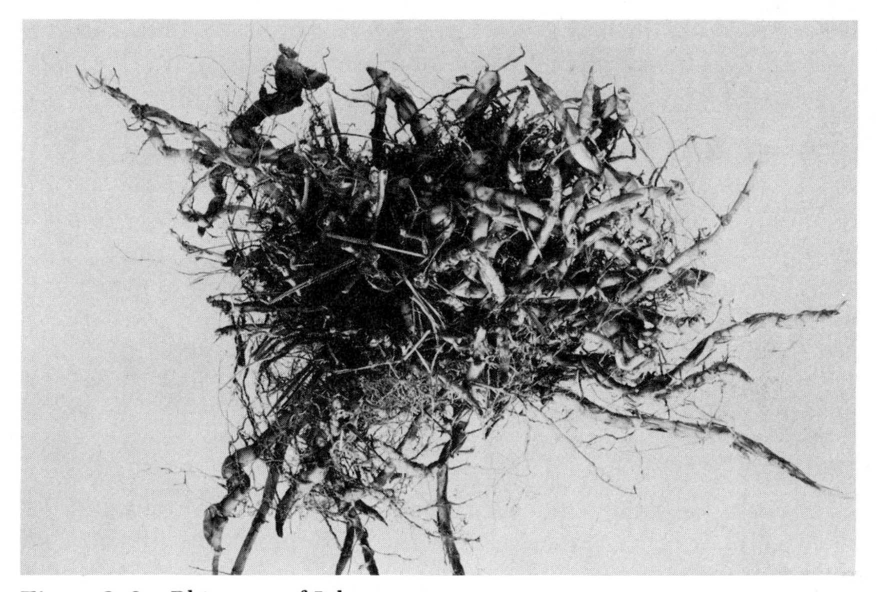

Figure 3–6. Rhizomes of Johnsongrass.

2. *Stolons.* Stolons are modified propagative stems produced above ground, as in buffalograss.

3. *Tubers.* A fleshy underground stem is known as a tuber. Examples are the potato and the Jerusalem artichoke.

LODGING OF STEMS: Resistance to lodging, or the capacity of stems to withstand the adverse effects of rain and wind, is an important quality in cereals. Resistance to lodging often is a varietal character, but its expression is modified by environment. A thick heavy growth keeps much of the sunlight from reaching the base of the stems. Reduced light or shading causes the cell walls of the stems to be thin and weak. No single morphological factor can be used as an index of standing power.[1] Abundant soil nitrogen may produce a lush growth that leads to increased lodging. Lodging may result from a low content of dry matter per unit length of culm,[55] or from a reduced content of lignin or of certain reserve di- and polysaccharides in the stem. Short, thick, heavy stems with thick cell walls are the best insurance against lodging.[3]

When a grass stem lodges before maturity it usually bends upward as a result of cell elongation on the lower side of the nodes. This in turn is the result of a response to an apogeotropic (opposed to gravity) stimulus. When a stem of a dicotyledonous plant is lodged the terminal portion tends to become erect also because of the same response. The thickened nodal tissue, which bends up the stem, occurs in both the sheath and culm of some grass plants, e.g., sorghum and corn, but is found only at the base of the sheath in wheat, oats, barley, and rye.

Lodging also may occur by stem breakage, especially after maturity but some crop plants may fall because of weak root anchorage.[15]

FUNCTIONS OF STEMS: The functions of stems are as follows:[44]

1. Conduction of water and mineral solutes from the soil chiefly through the vessels of the xylem, and of synthesized food materials through the phloem from the leaves to other parts of the plant.
2. Support and protection of plant organs and display of leaves and flowers to the sunlight to aid pollination.
3. Storage of food materials, as in sugarcane and sorgo.
4. Manufacture of carbohydrates in the young stems that contain chlorophyll.

Leaves

Leaves arise from buds and are side or lateral appendages of the plant stem. Leaves are generally involved in photosynthesis and transpiration.[17]

GENERAL CHARACTERISTICS: The principal parts of a leaf are the blade (or lamina) and the petiole or leaf stalk. Some leaves have stipules, which are two small leaves at the base of the petiole. The leaf is said to be sessile when the petiole is absent. Ordinary green foliage leaves may be classified as:

1. Parallel-vein leaves of the grasses, which contain many veins about equal in size that run parallel and are joined by inconspicuous veinlets. Growth of such leaves takes place by elongation near the base.

2. Netted-vein leaves, which have a few prominent veins with a large number of minor veins. Netted-vein leaves are found in all crop plants other than grasses.

Photosynthesis

The 17th-century physician, Van Helmont, reported that "164 pounds of wood, bark and roots arose out of water only," because photosynthesis was then unknown. Joseph Priestley in the 18th century found that plants could "restore" air that had been made "undesirable" for mice by burning a candle in a confined container. The Dutch physician Jan Ingenhousz established the fact that sunlight reaching the green parts of a plant restored the air. In 1804, De Saussure showed that the leaves of plants in the presence of light absorb carbon dioxide and evolve oxygen in equal proportions. He also found that the uptake of carbon dioxide through plant stomates could not account for the total increases in plant weight. Hydrogen from water, and mineral nutrients, contributed the additional weight. Hydrogen comprises about 5 percent of the plant tissue. The overall equation for photosynthesis is as follows:

$$CO_2 + H_2O + \text{light} + \text{green plants} \rightarrow \begin{matrix} \text{plant materials} \\ + \\ \text{restored air } (O_2) \end{matrix}$$

Photosynthesis is the most significant enzyme-mediated plant process. It supplies the energy for maintenance of plant and animal life.

ENERGY CAPTURE, TRANSFORMATION, AND STORAGE: Light absorbed by chlorophyll in the green plant (Figure 3–7) induces the chemical reactions shown in Figure 3–8.

The splitting of water (the Hill reaction) releases hydrogen to produce NADPH. The phosphate compounds participate in the synthesis of sugar. This is followed by the formation of other carbohydrates, as well as proteins and fatty compounds.

The so-called "dark reactions" in photosynthesis (Figure 3–9) occur during both day and night. This involves the addition of CO_2 to a 5-carbon compound. Ribulose diphosphate (*Step 1*a) which is split to two 3-carbon compounds (*Step 2*); these are next changed with the addition of hydrogen (*Step 3*) (NADPH is consumed) to two compounds which are further converted to fructose (*Step 4*) a 6-carbon sugar. Fructose then, through a series of conversions, leads to the production of other sugars. This photosynthetic process is called the C_3 system.

In an additional process (Step 1b—Figure 3–9) CO_2 which will eventually combine with ribulose diphosphate is fixed first by a

3-carbon compound and then transferred to ribulose diphosphate for "normal" sugar synthesis. This 3-carbon acceptor process is referred to as the C_4 system. The C_4 system is a much more efficient user of CO_2. The photosynthetic process which begins with Step 1a is referred to as the Calvin cycle. Whereas the process which first uses

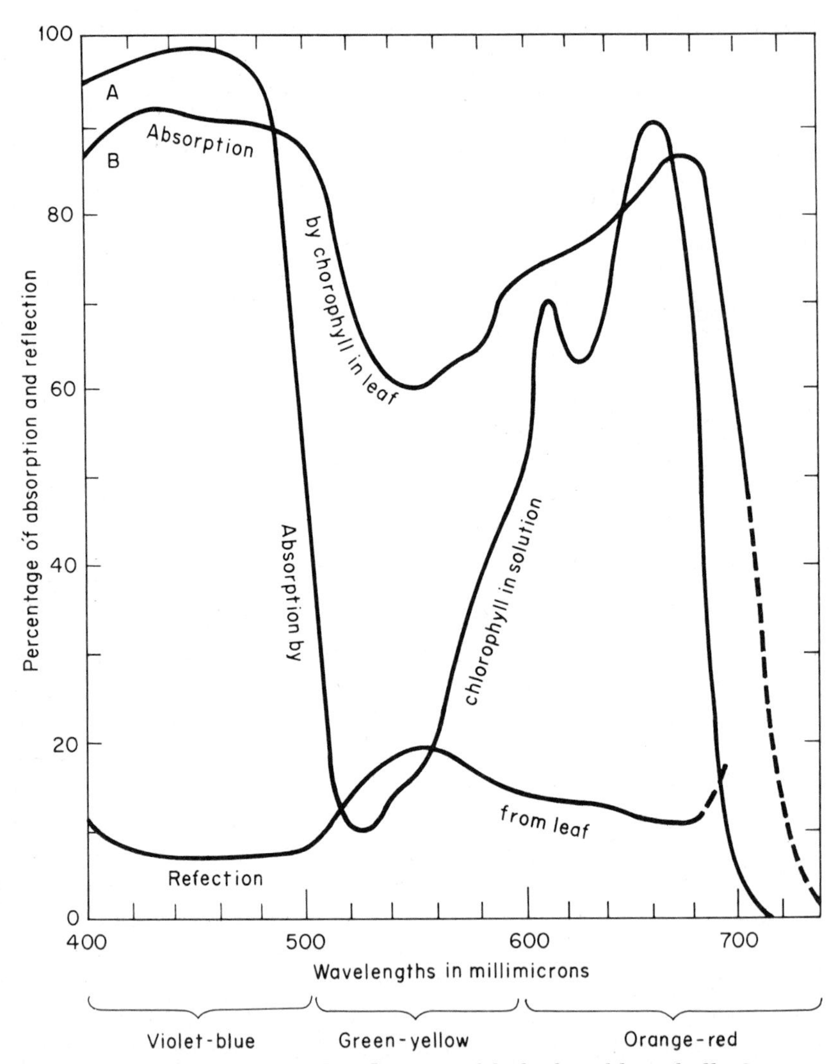

Figure 3–7. Absorption and reflection of light by chlorophyll. Curve A: Absorption of light by chlorophyll in solution. Curve B: Absorption by the same concentration of chlorophyll in a leaf. Chlorophyll in the leaf absorbs several times more green light than the same chlorophyll in solution. Curve C: The green color of a leaf is due to a slightly greater reflection of green than of other light. [Adapted from Carl L. Wilson and Walter E. Loomis, *Botany*, Rev. Ed. New York: Holt, Rhinehart & Winston, 1957.]

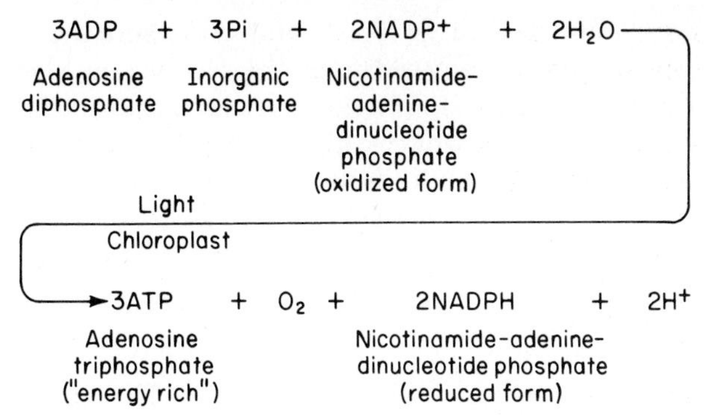

Figure 3–8. Photosynthetic phosphorylation.

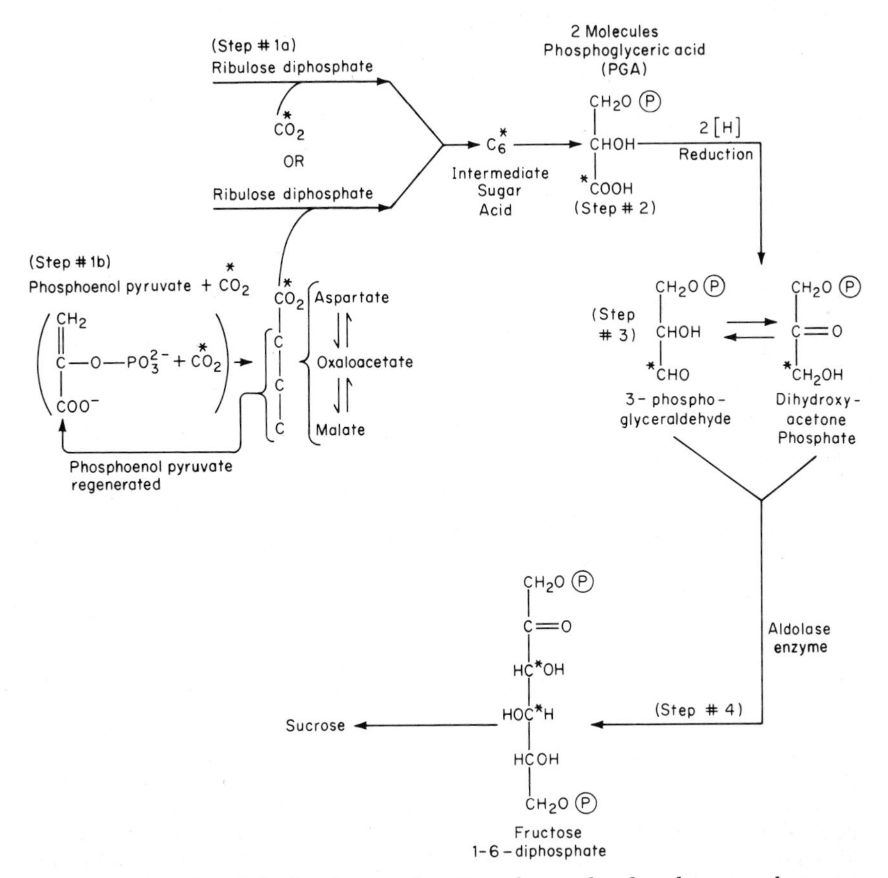

Figure 3–9. Simplified scheme showing the path of carbon in photosynthesis. C* denotes C^{14}-labeled carbon.

phosphoenol-pyruvate as a CO_2 acceptor and then transfers CO_2 to ribulose diphosphate is referred to as the Hatch and Slack pathway. High dry matter yielding crop plants such as corn, sorghum, and sugarcane are C_4 while the lower yielding wheat, soybeans, and alfalfa are C_3 plants. The higher yields from the C_4 cycle result from the trapping of some of the CO_2 that would otherwise escape into the atmosphere. Incorporation of the C_4 system into a C_3 plant type is a goal of many crop scientists.

Light Energy Utilization

The solar energy reaching 1.5 square miles (3.75 square kilometers) of the earth's surface in one day is equivalent to a "Hiroshima model" atomic bomb. About a third of the total solar energy is reflected into outer space or absorbed in our atmosphere.

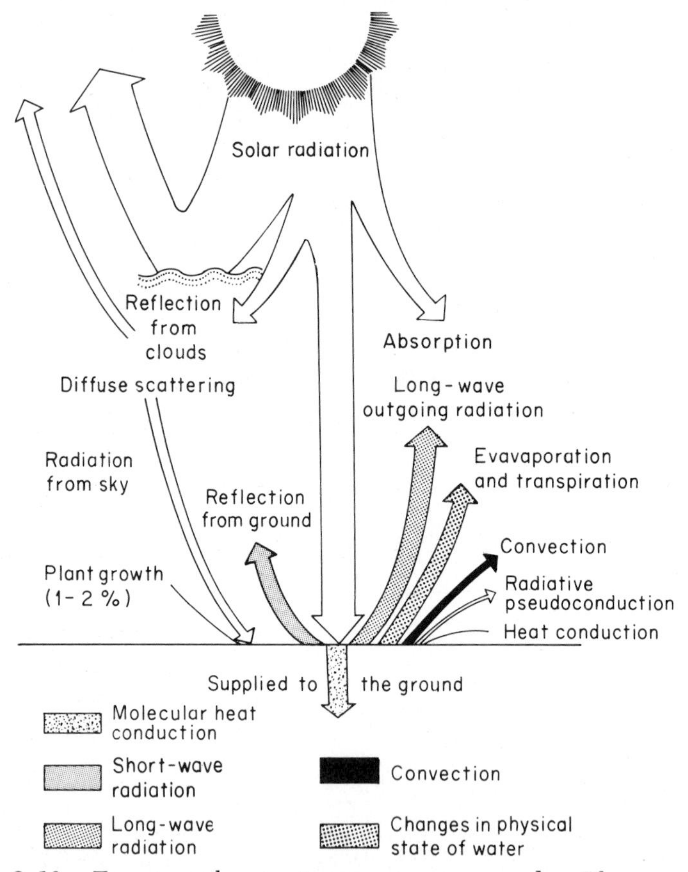

Figure 3–10. Energy exchange at noon on a summer day. The arrow width indicates relative amounts of energy transferred. Note that plant growth accounts for a very small part of the total energy budget. [After Geiger, *The Climate near the Ground.* Cambridge: Harvard Univ. Press, 1950.]

Only about 1–2 per cent of the solar energy received at a cropped field during the growth of the plants is stored by photosynthesis (Figure 3–10).

Of the total energy falling on the earth's upper atmosphere (3.3 cal./cm.^{2}/minute) only .146 cal./cm.^{2}/minute is available for photosynthesis.

The visible spectrum of 390–700 nanometers is that largely used by plants (Figure 3–11). The longer red rays are more effective than the higher-energy blue rays in promoting plant growth. Light travels through space as a stream of compact units referred to as "photons." The energy present in one photon is referred to as a "quantum." The energy in a quantum is directly proportional to the frequency of the radiation and inversely proportional to the wavelength. Heavy X-ray dosages modify or destroy chromosomes or kill plants. Infrared wavelengths cause molecular excitation, and long exposure to intensive ultraviolet radiation is fatal to plants and seeds.

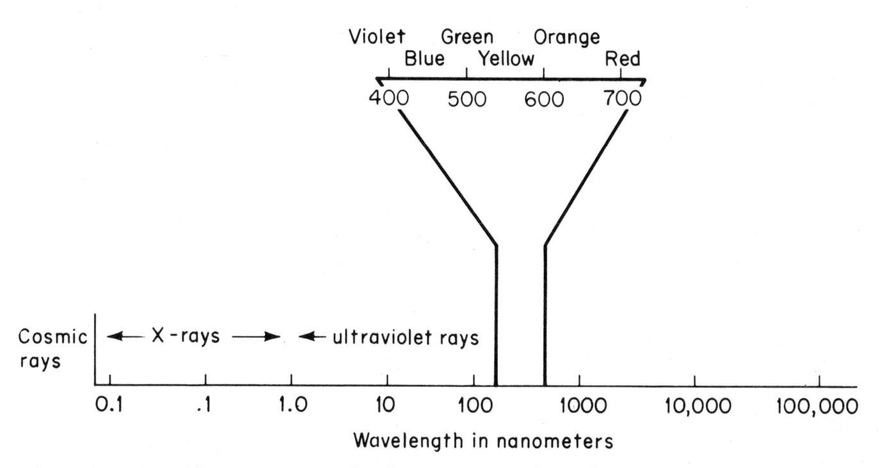

Figure 3–11. The spectrum of radiant energy plotted on a log scale. [Adapted from Galston, *The Life of the Green Plant.* Englewood Cliffs, N.J.: Prentice-Hall, 1964.]

ENVIRONMENTAL DIFFERENCES: The photosynthetic rate varies with environmental and nutritional factors affecting plant growth, and with plant characters such as the leaf area, and the position, density, and location of the leaves.

Other Plant Responses to Light

Light is essential, also, for the formation of chlorophyll and of anthocyanin pigments in plants; depth of coronal nodes in grass plants; hypocotyl unhooking in cotyledons; plant growth responses to light (phototropism); dormancy or germination in certain seeds; and photoperiodism.

Chlorophyll formation occurs through a direct light effect upon protochlorophyll. Anthocyanin production is affected by light in the red and far red ranges of the spectrum and in wavelengths between the two as well as light in the blue region of our spectrum.

RESPIRATION: Respiration consists of the oxidation of carbon and hydrogen accompanied by the release of carbon dioxide and water. This process, the reverse of photosynthesis, is continuous in living plant cells. It provides the energy for plant growth during the night when photosynthesis has ceased. As a result, the dry weight of many growing plant species decreases each night, but this loss is more than replaced by photosynthesis the next day. The heavily shaded lower leaves on a plant often consume more carbohydrates by respiration than they can manufacture.

Transpiration

Transpiration is the loss of water in vapor form from a living plant or plant part. Nearly all the water taken in by plants is transpired, but a plant is unable to grow unless it has sufficient water at its disposal. Transpiration may take place from any part not covered by layer cork or cuticle, but it is usually through the stomata. The harmful effects of transpiration, i.e., wilting of the plant and exhaustion of soil moisture, generally outweigh the possible benefits attributed to the process. The possible benefits are cooling of the leaves, rapid conduction of soil solutes, and more rapid dissolving of carbon dioxide.[11]

Transpiration occurs as follows:[28]

1. The cell wall imbibes water from the cell contents. The water in the cell wall is then at a higher concentration than the surrounding gaseous environment of the substomatal cavity.

2. The water evaporates from the cell wall and raises the concentration of water vapor in the cavity.

3. This water vapor diffuses out through the stomata since the air in the substomatal cavity is more highly saturated than the outside air. Guard cells, which surround the stomatal opening, open and close with changes in the water balance, and also with light and darkness in many species.

The following conditions influence the rate of transpiration:

1. Capacity of the leaves to supply water.

2. Power of the aerial environment to extract water from the leaves, i.e., influence of humidity, light, temperature, and wind.

3. Structure of the transpiration parts, i.e., the number of stomata and the sizes of the openings.

4. Modifications that tend to check excessive evaporation, including reduced plant surfaces, sunken stomata in special epidermal cavities, thickened cuticle, and waxy bloom on the cuticle of leaves

and stems. A waxy bloom is conspicuous on sugarcane, sorghum, wheat, and barley.

Crop plants transpire 200 to 1,000 kilos of water for every kilo of dry matter produced, which means that the transpiration ratio usually ranges from 200 to 1,000 for different crops. Thus each centimeter of rainfall (100 metric tons per hectare) stored in the soil and protected from evaporation would produce from 1,000 kilos down to 200 kilos of dry matter in the form of crops. For a given crop, the transpiration ratio depends not only upon the previously mentioned factors that influence transpiration rate, but also upon the growth as well as adaptation of the crop to the particular environment. Soil of high fertility, and a large volume of soil in which the roots can feed, contribute to a lower transpiration ratio than occurs with less favorable soil conditions.[29] Also, oats have a comparatively high ratio when the temperatures are too high for optimum growth, under which conditions the warm-weather crops—sorghum and cotton—are relatively more efficient in the use of water.[46] In a cool season, the reverse may be true.

Crops that have the lowest transpiration ratios include millets, sorghum, and corn; the small grains are intermediate, whereas legumes such as alfalfa have relatively high transpiration ratios. The transpiration ratio is not necessarily a measure of the ability of a crop to produce[6] well under drought conditions.[36] It merely measures the water transpired by the plants when the soil is supplied with water for optimum plant growth. For example, field peas and field beans (which have a high ratio) are grown successfully in rows under dry-land conditions where fair yields of seed are obtained from relatively small plant growth. The high value of the product compensates for the low production. Corn, sorghum, and certain millets that have a low transpiration ratio, also have the efficient C_4 type of photosynthesis for dry matter production.

The adaptation of a crop to drought conditions depends also upon:

1. Drought escapement or evasion, i.e., ability to complete the growth cycle before soil moisture is exhausted. Early-maturing varieties of small grain, for example, have made this adaptation.
2. Drought tolerance, i.e., the ability of the plant to withstand drying and recover later when moisture becomes available; e.g., sorghum.
3. The ability of the plant to adjust its growth and development to the available water supply. Thus, cotton not only regulates its growth with the moisture supply but also sheds squares (flower buds), flowers, some of its bolls and, finally, many of its leaves when moisture continues to be deficient. A so-called drought-resistant crop or crop variety shows one or more of these three characteristics.

A summary of the transpiration ratios of several crops and pigweed (*Amaranthus retroflexus*) at Akron, Colorado,[46] is shown in the following table. The lowest transpiration occurred in 1915, a cool year of high rainfall.

Transpiration Ratios of Crops and Pigweed at Akron, Colorado

CROP	1913	1914	1915	1916	1917	AVERAGE
Alfalfa	834	890	695	1047	822	858
Oats	617	599	448	876	635	635
Cowpeas	571	659	413	767	481	578
Cotton	657	574	443	612	522	562
Barley	513	501	404	664	522	521
Wheat	496	518	405	636	471	505
Corn	399	368	253	495	346	372
Blue gramagrass	389	389	312	336	290	343
Pigweed	320	306	229	340	307	300
Millet	286	295	202	367	284	287
Sorghum	298	284	203	296	272	271

Weighted mean transpiration ratios of other crops and weeds were as follows: proso 267, sudangrass 380, sugarbeet 377, buckwheat 540, potato 575, spring rye 634, red clover 698, bean 700, hairy vetch 587, soybean 646, sweetclover 731, field pea 747, flax 783, crested wheatgrass 678, bromegrass 828, Russian thistle 314, purslane 281, and lambsquarter 658. All these figures are based upon total dry matter in the plant above the ground, except for the sugarbeet and potato, in which case the roots or tubers also were harvested.

The water balance is the relation between absorption and loss of water by the plant.[36] There is a deficit when absorption is exceeded by water loss. A deficit often occurs at midday when transpiration exceeds absorption, i.e., plants show signs of wilting and curling leaves. Increased and permanent wilting may cause injury and lead to complete ruin. *Permanent wilting* is defined as the point at which a plant cannot recover when placed in a saturated atmosphere. Plant species that endure great water loss without injury (e.g., guayule) are called xerophytes; those that require medium supplies of water (most crop plants) are called mesophytes; while those growing best with abundant water, including aquatics (e.g., wild rice), are called hydrophytes. Important characteristics of xerophytes are high concentrations of cell sap or high content of colloidal material or coverings that retard evaporation and ability to recover after drying. The *consumptive* use of water (transpiration plus evaporation from the soil) for growing a crop usually ranges from 8 to 40 inches in total depth, but may be up to 8 feet for long-season irrigated crops on very permeable soils.

REPRODUCTIVE PROCESSES IN CROP PLANTS

Crop plants reproduce either asexually or sexually, i.e., by the fertilization of flowers to produce seeds.

Asexual Reproduction

Crop plants normally produced asexually, together with the plant part used in propagation, include:

1. Roots—sweetpotato, cassava, kudzu.
2. Tubers—potato, Jerusalem artichoke.
3. Stolons—buffalograss, creeping bent, bermudagrass.
4. Rhizomes—bermudagrass.
5. Stem cuttings—sugarcane, napiergrass.

Asexual reproduction perpetuates uniform progeny except for an occasional bud mutation such as the sudden appearance of a red tuber in a white variety of potato. Many varieties that are uniform when propagated asexually are found to be extremely heterozygous (or variable) in their hereditary make-up when propagated from seeds.

Sexual Reproduction

The types of floral arrangement involved in sexual reproduction are as follows:

1. Perfect or bisexual flowers containing both stamens and pistils. Most crop plants have perfect flowers although some of them are more or less self-sterile. Some unisexual flowers are found on crop plants that are bisexual, e.g., sorghum and barley.

2. Imperfect or unisexual flowers containing either stamens or pistils, but not both.

When the separate staminate and pistillate flowers are borne on the same plant, the plants are called monoecious. Examples of these are corn, castorbean, and wild rice.

When the two kinds of flowers are borne on separate plants, the plants are dioecious. These include hemp, hops, and buffalograss.

An exception to ordinary sexual reproduction is a process called apomixis, a form of parthenogenesis or the production of seed by a stimulation of the ovary without fertilization. This is a common method of seed formation in Kentucky bluegrass, dallisgrass, and species of *Pennisetum* and *Paspalum* and numerous other plant species.[6] Occasional apomixis occurs in sorghum, and many other species. In one type of apomixis, called apospory, somatic cells in the nucleus enlarge and the nuclei divide to form an embryo sac containing an egg and polar nuclei.[18] When Kentucky bluegrass is cross-pollinated, apospory may also occur and increase the number of chromosomes in the cross.

MODE OF POLLINATION: Sexually propagated crop plants may be grouped into three general classes, on the basis of their normal habit of pollination. These are (1) naturally self-pollinated, (2) often cross-pollinated, and (3) naturally cross-pollinated.

1. *Naturally self-pollinated.* Crop plants naturally self-pollinated, including wheat, oats, barley, tobacco, potatoes, flax, rice, field peas, cowpeas, and soybeans, usually show less than 4 per cent cross-pollination. The percentage of cross-pollination in these crops varies with variety and season or environment.[7, 12, 16, 45, 48] Any environmental or hereditary factor that interferes with normal pollination or fertilization may result in a high proportion of cross-pollination. In naturally self-pollinated plants, the pistil usually is pollinated by pollen from the same flower.

2. *Often cross-pollinated.* These include cotton, sorghum, foxtail and proso millets, and several cultivated grasses. The pistil may be pollinated by pollen from the same flower or from another flower on the same plant or from another plant. Crossing among cotton flowers usually is due to insects, while that in the grass crops mentioned above is largely from air-borne pollen. Self-pollination usually occurs for 94 to 95 per cent in sorghum.[47]

3. *Naturally cross-pollinated.* Cross-pollination is the normal mode of reproduction in many crop plants, including maize, rye, clovers, alfalfa, buckwheat, sunflowers, some annual grasses, and most

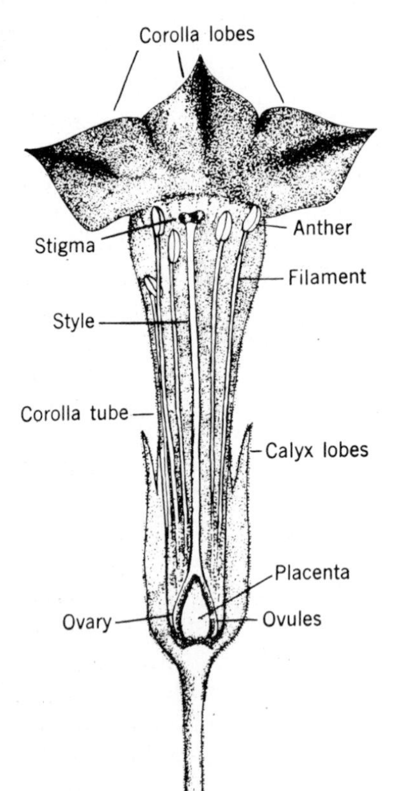

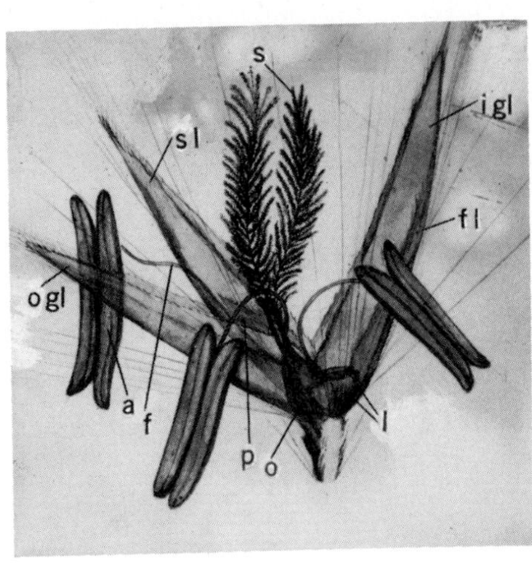

Figure 3–12. (*Left*) Tobacco flower showing typical floral parts. (*Right*) parts of an opened grass (sugarcane) flower: (a) anther, (f) filament, (fl) fertile lemma, (igl) inner glume, (l) lodicules, (o) ovary, (ogl) outer lemma, (p) palea, (s) stigmas and (sl) sterile lemma. [After Artschwager.]

perennial cultivated grasses. The extent of self-pollination in corn is less than 5 per cent,[19] and sometimes as low as 0.7 per cent.[29] Most of the grasses are wind-pollinated, but buckwheat and legumes such as clovers and alfalfa are adapted to pollination by insects. Cross-pollination is essential to seed production in many plants of red clover, rye, and common buckwheat because of self-sterility.

DESCRIPTION OF FLOWERS: The flower arises from a bud either at the apex of the stem or in the leaf axil. Flowers may be grouped into an inflorescence on a more or less compact special shoot or axis. The leaves of the inflorescence are known as bracts, while the axis is usually a rachis or a peduncle. An individual flower stalk is a pedicel.

A flower is composed of the stamens (male organs), pistils (female organs), calyx, and corolla (Figure 3–12). The latter two structures are replaced by glumes, lemma, and palea in the grasses.[8]

The stamens and pistils are the organs of fertilization. The stamen is composed of a filament or stalk at the upper or outer end of which is the anther. The interior lobes of the anthers are hollow spaces or pollen sacs in which the pollen is produced in the form of loose round pollen grains. Finally, each sac splits open (dehisces) to allow the pollen to escape. The pistil is composed of the stigma, style, and ovary. The ovary is the swollen hollow base of the pistil (Figures 3–13 and 3–14). The elongated portion is the style, at the apex of which is the stigma. The ovary contains ovules, generally egg-shaped, attached to the ovary walls.

In the grasses the flowers at the time of fertilization are forced open by the swelling of two small organs called lodicules, which lie between the ovary and the surrounding lemma and palea.

The perianth, composed of the calyx and corolla, is a nonessential part of the flower, although it may aid in the attraction of insects. The calyx is made up of sepals which enclose the other floral parts in the bud and thus protect the young flower. The corolla is generally the showy part of the flower, being composed of petals (Figure 3–15).

PROCESS OF FERTILIZATION: The ovule is composed of a nucellus surrounded by an inner and outer integument. The embryo sac is within the nucellus (Figure 3–14). It contains eight nuclei: two synergids and one egg cell at the micropylar end, three antipodal cells at the opposite end, and two endosperm or polar nuclei near the center. Pollen grains that fall on the stigma of the pistil germinate in the sticky stigmatic excretion. The germinated pollen grain contains a tube nucleus and two sperm nuclei. The tube nucleus develops a pollen tube that elongates down through the style to the embryo sac. The pollen tube usually grows in the intercellular spaces of the style. One sperm nucleus unites with the egg cell while the other fuses with

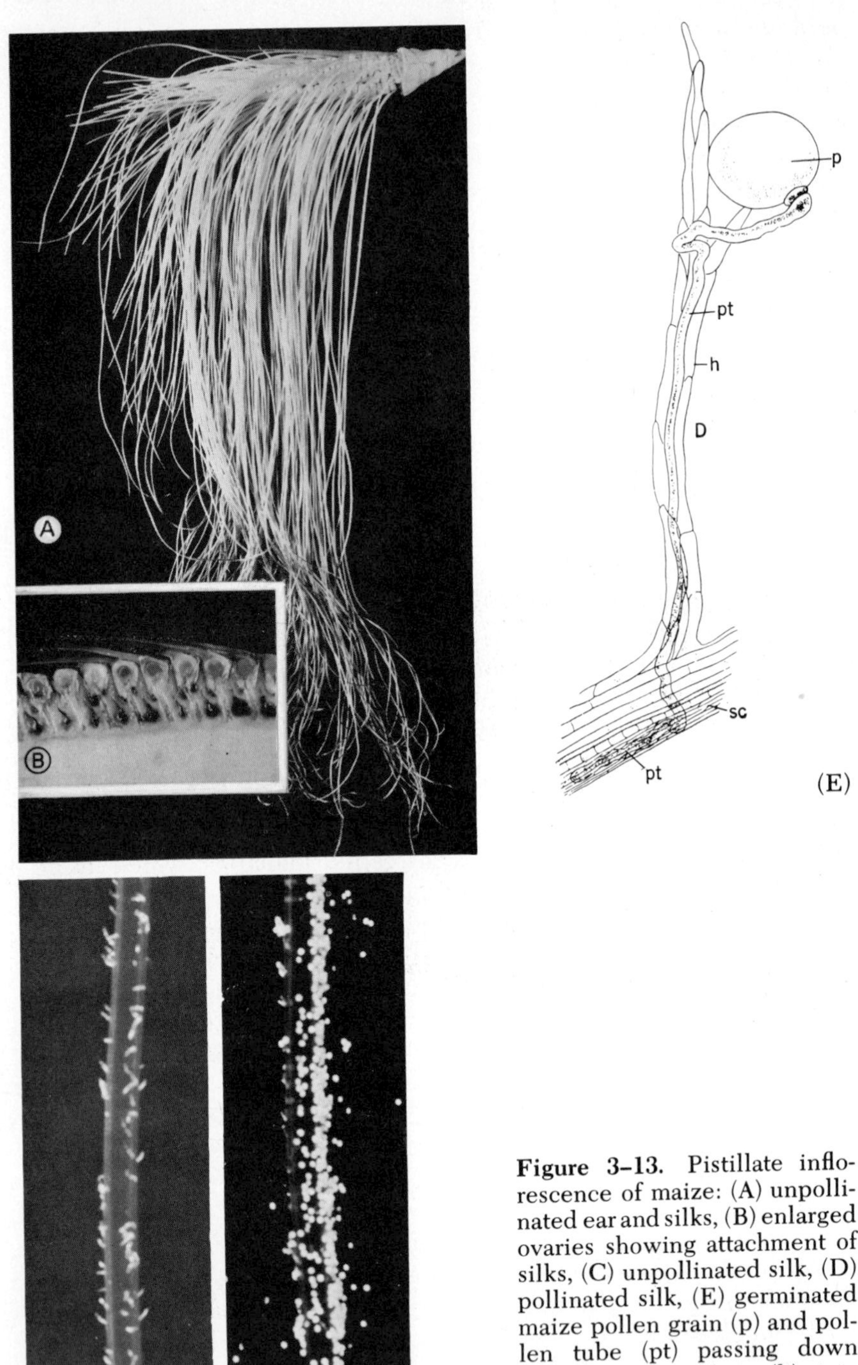

Figure 3–13. Pistillate inflorescence of maize: (A) unpollinated ear and silks, (B) enlarged ovaries showing attachment of silks, (C) unpollinated silk, (D) pollinated silk, (E) germinated maize pollen grain (p) and pollen tube (pt) passing down through stigma hair (h) into silk (sc).

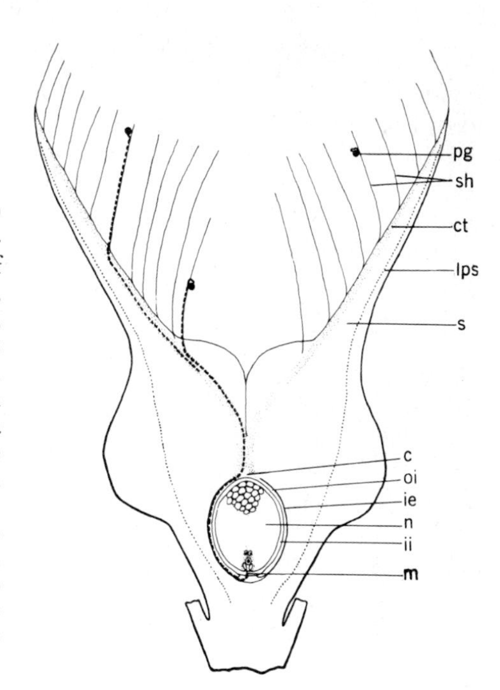

Figure 3–14. Barley pistil: (c) cone-shaped tip of outer integument, (ie) inner epidermis of ovary wall, (ii) inner integument, (lps) lateral procambial strands, (m) micropyle, (n) nucellus which contains the embryo sac, (oi) outer integument, (pg) pollen grain, (s) style, and (sh) stigma hair. The broken line shows the course of the pollen tube from pollen grain to micropyle (m). Above the mycropyle are two synergids and the egg nucleus, and above the latter are the two endosperm nuclei. The three antipodal cells higher up have already divided several times.

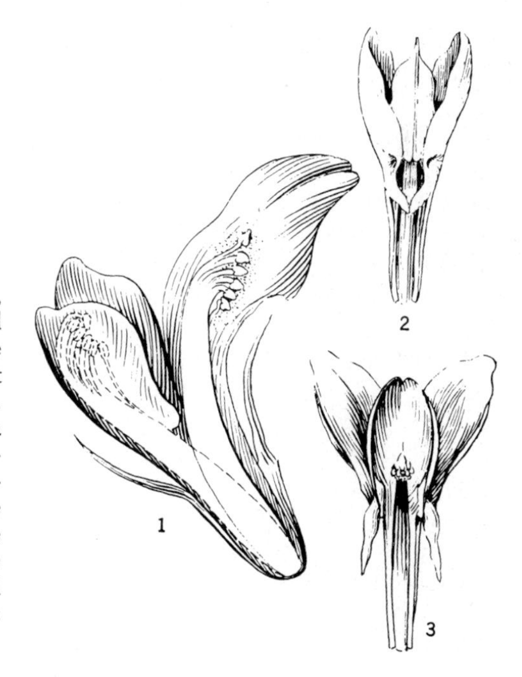

Figure 3–15. Alfalfa flower, typical of legumes. (1) Opened flower showing position of style and anthers before tripping at left and after tripping at right. The large segment of the corolla at right is the "banner." (2) Unopened flower with banner removed. (3) Open flower with the banner removed to show the two wings (left and right), the two "petals" of the keel fused at rear, and the style and anthers at front.

two polar nuclei. The fusion of the second male nucleus with the polar nuclei has been termed double fertilization, which is believed to occur in all flowering plants.

The mechanism of double fertilization accounts for the phenomenon of xenia, or the immediate observable effect of foreign pollen on the endosperm. The classical example is the various endosperm colors in maize produced by foreign pollen, which are evident as soon as the out-crossed grains ripen. Xenia also has been observed in rye, rice, barley, sorghum, peas, beans, and flax. The process is called metaxenia when the pericarp or other maternal tissue also is modified by double fertilization as a result of adjustment to a change in size or shape of the affected endosperm.

Fertilization in corn takes place within 26 to 28 hours after pollination.[38] In barley, the pollen germinated within 5 minutes after reaching the stigma, the male nuclei had entered the egg sac within 45 minutes, and the fertilized egg had begun division within 15 hours.[43] Flax pollen germinates and the pollen tube penetrates to the base of the style in about 4 hours after pollination.

Crop seeds develop rapidly after pollination, being fully mature within four to six weeks in many crops and up to nine weeks in others.

SEEDS AND FRUITS

A representative mature seed consists of the seed coat (testa), nucellus, endosperm, and embryo. The embryo or germ is composed of the plumule (leaves), hypocotyl (stem), radicle (root), and one or two cotyledons. The seed may be defined as a matured ovule. The bean is a good example of a seed. It is attached to the ovary wall (pod) by a short stalk (the funicle). The hilum is an elongated scar on the bean where the funicle was attached. Close to one end of the hilum is a small opening called the micropyle. A seed coat or testa covers the bean, beneath which are two fleshy halves called cotyledons. A typical endosperm is lacking in the bean, the cotyledons occupying most of the space within the seed coat.

A fruit is a matured ovary. It contains the seeds or matured ovules. The mature ovary wall is known as the pericarp. The entire bean pod is a fruit (Figure 3–16), but the *beans* are seeds. Such fruits may contain several seeds.

In the indehiscent dry fruits, the pericarp is dry, woody, or leathery in texture, and does not split or open along any definite lines. The achene is a one-seeded fruit in which the pericarp can be readily separated from the seed as, for example, in buckwheat. The caryopis of cereals and grasses is a one-seeded fruit. The seed within it is united with the ovary wall or pericarp. Percival[41] describes the wheat kernel as a kind of nut with a single seed. The wheat kernel (caryopsis)

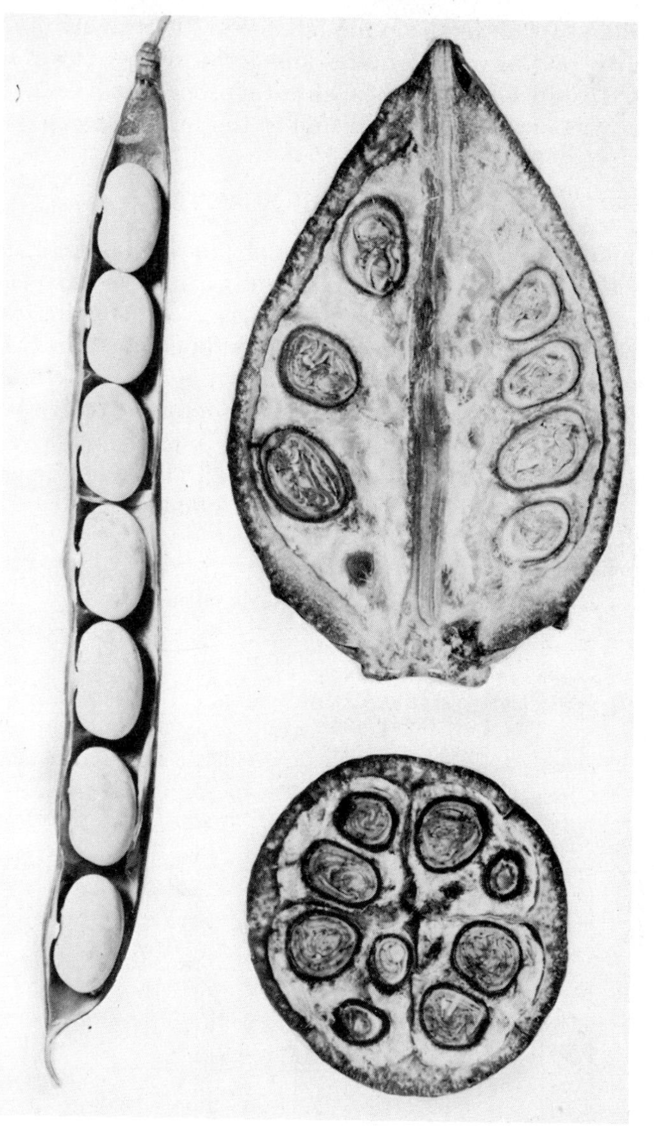

Figure 3–16. (*Left*) Pod (fruit) and seeds (ovules) of the Pinto bean. (*Right*) Half of fruit (boll or capsule), and seeds (ovules) of cotton; the cross-section (*below*) shows four locks.

consists of the pericarp, endosperm, and embryo. The outer rim of the embryo is the scutellum, which is a flattened, somewhat fleshy, shield-shaped structure that lies back of the plumule and close to the endosperm. The scutellum is regarded as the single cotyledon of a grass "seed."

In the dehiscent dry fruits, the pericarp splits in various ways or opens by pores. The seeds on the interior of the fruit are thus set free. The legume pod or fruit dehisces along two sutures, as in field peas or beans. In red clover, the capsule dehisces transversally, that is, the upper part of the carpels falls off in the form of a cap or lid.

GROWTH PROCESSES IN CROP PLANTS

Growing plants pass through a gradual and regular vegetative cycle followed by maturity and reproduction.[27] The vegetative growth of most plants is characterized by three phases, namely, an initial slow start, a period in which the growth rate becomes gradually faster, and finally, a period when it slows down again. Growth curves of sorghum[5] are shown in Figure 3–17. The processes involved in growth are extremely complex. Growth is usually measured by the amount of solid material or dry matter produced. This is often expressed agriculturally as yield of all or part of the plant.

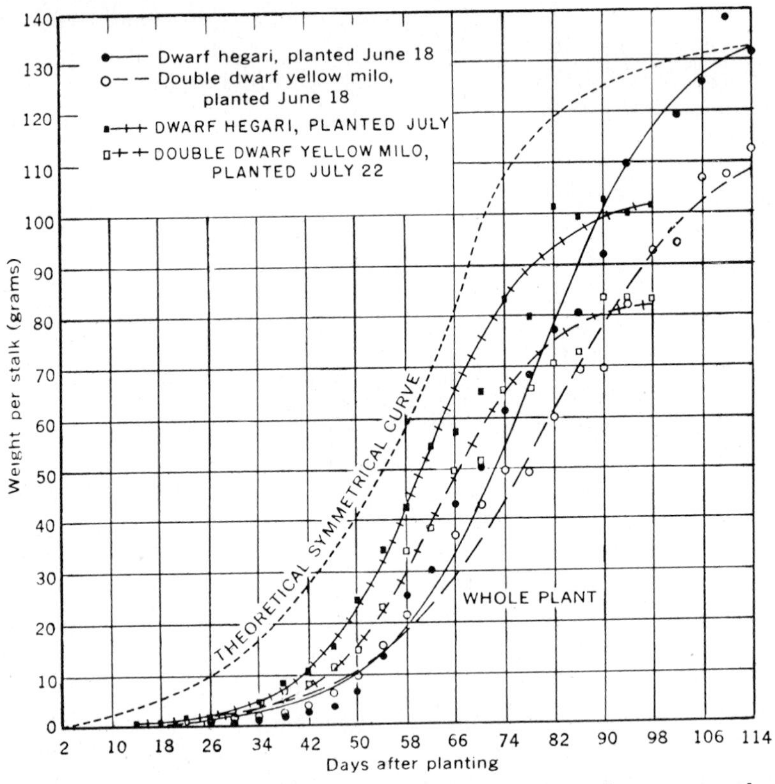

Figure 3–17. Most crop plants, e.g., corn and barley, show a growth rate similar to the theoretical symmetrical curve. The sorghums, Dwarf hegari and Double Dwarf Yellow milo, have smaller seeds in proportion to the ultimate plant size and therefore show an apparent slower growth rate at the beginning.

Vegetative development in plants is interrupted by the reproductive phase in which the flowers and seeds are formed. Different species and crop varieties have a characteristic though highly variable vegetative period. However, the onset of the reproductive period may be altered greatly by changes in environmental conditions. Thus, a soil rich in nitrogen will tend to keep down the carbon-nitrogen ratio, which delays flowering in *nitronegative* crops such as wheat, barley, oats, most pasture grasses, mustard, alfalfa, and clover. The excess of carbohydrates over nitrogen is usually very great, i.e., the carbon-nitrogen ratio is high after the plant commences to flower. However, changes in the carbon-nitrogen ratio in the cotton plant may be merely incidental rather than fundamental to flowering.[13] *Nitropositive* crops, such as corn, sorghum, millet, cotton, tobacco, lupines, sunflower, perilla, and pepper are reported to flower earlier with abundant nitrogen. The flowering time of *nitroneutral* crops, such as buckwheat, hemp, soybeans, peas, and beans is not affected appreciably by supplies of available nitrogen. Moisture shortage may cause plants to flower abnormally early, as is often observed in a dry season. The length of the daily illumination period affects the time that plants flower. Temperature alone, or in combination with the other factors, also affects the onset of flowering.

PLANT COMPETITION

Regulation of the kind and amount of plant competition is an important ecological phase of crop production. Weeds are destroyed to keep them from competing with crop plants. Crops are spaced to obtain maximum yields of suitable quality. A moderate amount of competition between plants is not detrimental on an acre basis. A "struggle for existence" results when plants are grouped or occur in communities in such a way that their demands for an essential factor are in excess of the supply. Competition is a powerful natural force that tends to eliminate or wipe out the weak competitors. It is chiefly a physical process, i.e., it is essentially a decrease in the amount of water, nutrients, and light that otherwise would be utilized by each individual plant. Competition increases with the density of the plant population. Excessive populations often reduce the yield of seeds, while stimulating vegetative growth.

Numerous chemicals, called growth regulators, modify the growth rate, transpiration, height, branching, rooting, flowering, or maturity of plants or plant parts.[4] A number of these are useful in ornamental greenhouses, home gardens, and lawns. In certain field crops chemicals are used as desiccants, or defoliants, or to suppress further branching or flowering.

Weed Competition

In farm fields, crops often are confronted with serious competition with weeds. Success of the crop depends on the readiness and uniformity of germination under adverse conditions, ability to develop a large assimilation surface in the early seedling stage, the possession of a large number of stomata, and a root system with a large mass of fibrous roots close to the surface but with deeply penetrating main roots.[40] Crop plants have been classified in order of decreasing competitive ability against weeds as follows: barley > rye > wheat > oats > flax. Common wild mustard (*Brassica arvensis*) and wild oats (*Avena fatua*) are vigorous competitors among the annual weeds. Wild oats plants may adapt their growth to the height of the grain with which they are competing.[19] However, in a flax field, the wild oat plants quickly outstrip the shorter flax plants.

An important application of plant competition in agriculture is the use of what are called smother crops for weed control.

Competition Among Cultivated Crops

Competition is greatest between individuals of the same or closely related species because they make their demands for moisture, nutrients, and light at about the same time and at about the same level (Figure 3–18). Competition for moisture is especially important in dryland regions. In the grass crops competition for light reduces the number of heads, causes great irregularity in the number of tillers produced, diminishes the amount of dry matter formed, encourages shoot growth at the expense of root growth, and reduces the tillering, length of culm, and number of kernels per head.

The reduction in the yield of corn is not proportional to the reduction in stand because of the adjustment of nearby plants that benefit by feeding in the space occupied by blank hills.[30] Considerable fluctuation in stand may occur without material effect on the final yield.

Sugarbeets may respond to additional space so that they can increase in weight sufficiently to compensate for 96 per cent of the loss due to a blank hill.[10] Very close spacing in flax reduces the growth of plants materially.[31] As the space between plants increases there is an increase in number of bolls, yield per plant, weight per plant, and number of stems per plant.

In sorghum, within certain limits, medium-thick planting makes the plants taller because of greater competition for light. Very thick stands make the plants shorter because of inadequate moisture or nutrients to support a good growth in so many plants.

Competition between different crop varieties causes marked changes in population after a few years, once the original mixture

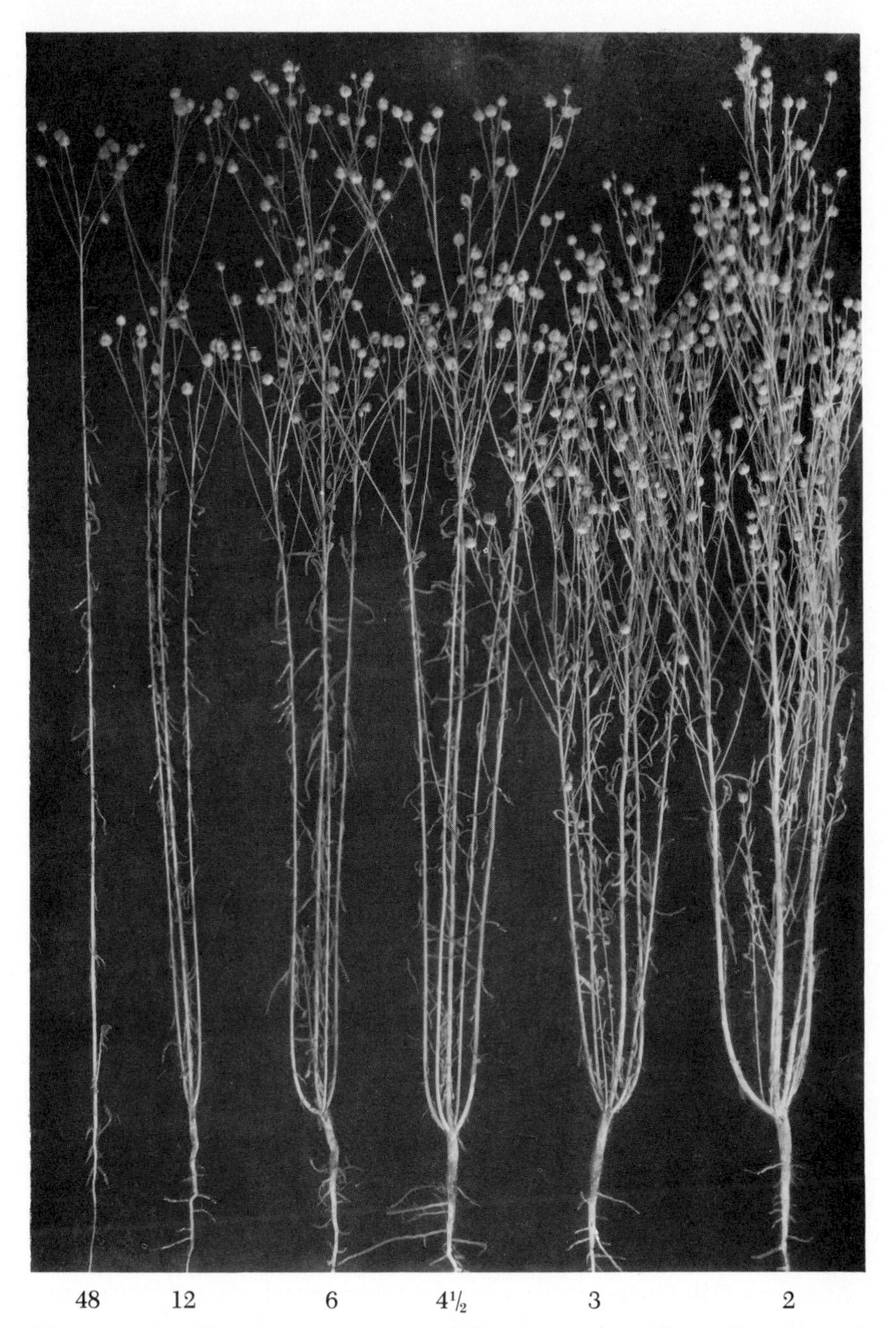

Figure 3–18. Flax plants grown at six different spacings. The single-stemmed plant at the left suffered from competition with other flax plants. The figures indicate the number of plants that occupy a square foot of soil.

comprises equal numbers of seeds of each variety. The ascendant varieties are not necessarily those that yield best when seeded alone.[20]

Crop Association

Competition between different crop species may not be too severe because one plant has an opportunity to fit in among some others. For example, in a mixture of timothy and clover, the two crops make different demands on the habitat at different times and at different soil depths. The more unlike plants are, the greater the differences in their respective needs.

In irrigated pastures in northern Colorado, it was found that competition played an important role in pasture mixtures. The most vigorous seedlings, especially of orchardgrass, soon gained dominance even when seeded in amounts as small as 4 pounds in a total of 30 pounds of seed per acre. Kentucky bluegrass, white clover, and timothy suffered greatly in this competition.

Companion crops are annuals planted with new seedlings of alfalfa, red clover, and similar perennial crops in order to secure a return from the land the first year. These companion crops, which compete with the young seedlings, usually are planted at about half the usual rate to reduce competition.

Mixtures of a small grain with a trailing vine legume such as vetch and field peas are very popular because the grain stems help support the legume vines. Korean lespedeza often is grown with small grains. The planting of corn or sorghum with cowpeas, soybeans, or velvetbeans has been practiced in the Southeast with the object of increasing total yields over the yields produced by the same crops grown alone. Some increase in yield is possible because of the different growing habits of the crops. In crop mixtures in which the cultural requirements for the individual crops are different, there may be sacrifice in quality or yield of one or both crops as well as an increase in production costs.

REFERENCES

1. Anonymous. *Lodging of Cereals, Bibliography,* Imperial Bureau of Plant Genetics, Cambridge. 1930.
2. Arnon, D. I. "Changing concepts of photosynthesis," *Bul. Torr. Bot. Club,* 84(4):215–259. 1961.
3. Atkins, I. M. "Relation of certain plant characters to strength of straw and lodging in winter wheat," *J. Agr. Res.,* 56(2):99–120. 1930.
4. Audus, L. J. *Plant Growth Substances,* Leonard Hill, London, pp. 1–553. 1959.
5. Bartel, A. T., and Martin, J. H. "The growth curve of sorghum," *J. Agr. Res.,* 57(11):843–849. 1938.

6. Bashaw, A. W., and others. "Apomixis, its evolutionary significance and utilization in plant breeding," *XI International Grassland Congress Proceeding*. 1970.
7. Beachell, H. M., and others. "Extent of natural crossing in rice," *J. Am. Soc. Agron.*, 27:971–973. 1935.
8. Bonnett, O. D. "Inflorescences of maize, wheat, rye, barley and oats—their initiation and development," *Ill. Agr. Exp. Sta. Bul. 721*, pp. 1–105. 1966.
9. Breazeale, J. F. "Maintenance of moisture equilibrium and nutrition of plants at or below the wilting percentage," *Ariz. Agr. Exp. Sta. Tech. Bul. 29*. 1930.
10. Brewbaker, H. E., and Deming, G. W. "Effect of variations in stand on yield and quality of sugarbeets grown under irrigation," *J. Agr. Res.*, 50:195–210. 1935.
11. Clements, H. F. "Significance of transpiration," *Plant Physiol.*, 9:165–172. 1934.
12. Dillman, A. C. "Natural crossing in flax," *J. Am. Soc. Agron.*, 30:279–286. 1938.
13. Eaton, F. M., and Rigler, N. E. "Effect of light intensity, nitrogen supply and fruiting on carbohydrate utilization by the cotton plant," *Plant Physiol.*, 20(3):380–411. 1945.
14. Evans, L. T. *Environmental Control of Plant Growth*, Academic Press, New York, pp. 1–499. 1963.
15. Fellows, H. "Falling of wheat culms due to lodging, buckling, and breaking," *USDA Cir. 769*. 1948.
16. Garber, R. J., and Odland, T. E. "Natural crossing in soybeans," *J. Am. Soc. Agron.*, 18:967. 1926.
17. Gausman, H. W., and others. "The leaf mesophylls of twenty crops, their light spectra and optical and geometrical parameters," *USDA Tech. Bul. 1405*, pp. 1–59. 1973.
18. Hanna, W. W., and others. "Apospory in *Sorghum bicolor* (L.) Moench," *Science*, 17:338–339. 1970.
19. Harlan, H. V. "The weedishness of wild oats," *J. Hered.*, 20(11):515–518. 1929.
20. Harlan, H. V., and Martini, M. L. "The effect of natural selection in a mixture of barley varieties," *J. Agr. Res.*, 57:189–200. 1938.
21. Hayes, H. K. "Normal self-fertilization in corn," *J. Am. Soc. Agron.*, 10:123–126. 1918.
22. Hector, J. M. *Introduction to the Botany of Field Crops:* Vol. I Cereals; Vol. II Non-Cereals, Central News Agency, Johannesburg, South Africa. 1936.
23. Hill, R., and Whittingham, C. P. *Photosynthesis*, Wiley, New York, 1957.
24. Hitchcock, A. S. *A Textbook of Grasses*, Macmillan, Inc., New York, pp. 1–276. 1914.
25. Hitchcock, A. S., and Chase, A. "Manual of the grasses of the United States," 2nd ed., *USDA Misc. Pub. 200*, pp. 1–1051. 1950.
26. Idso, S. R. "A holocoenotic analysis of environment-plant relationships," *Minn. Agr. Exp. Sta. Tech. Bul. 264*, pp. 1–147. 1968.

27. James, W. O. *An Introduction to Plant Physiology*, 2nd ed., Oxford U. Press, New York, pp. 1–263. 1933.
28. Kiesselbach, T. A. "Transpiration as a factor in crop production," *Nebr. Agr. Exp. Sta. Res. Bul. 6*. 1916.
29. Kiesselbach, T. A. "Corn investigations," *Nebr. Agr. Exp. Sta. Bul. 20*. 1922.
30. Kiesselbach, T. A. "Competition as a source of error in comparative corn yields," *J. Am. Soc. Agron.*, 15:199–215. 1923.
31. Klages, K. H. "Spacing in relation to the development of the flax plant," *J. Am. Soc. Agron.*, 24:1–17. 1932.
32. Krassovsky, I. "Physiological activity of the seminal and nodal roots of crop plants," *Soil Sci.*, 21:307–325. 1926.
33. Larson, K. L., and Eastin, J. D., Eds. *Drought Injury and Resistance in Crops, CSSA Spec. Pub. 2*, pp. 1–95. 1971.
34. Locke, L. F., and Clark, J. A. "Normal development of wheat plants from seminal roots," *J. Am. Soc. Agron.*, 16:261–268. 1924.
35. Maryland, H. F., and Cary, J. W. "Frost and chilling injury to growing plants," *Adv. Agron.*, 22:203–234. 1970.
36. Maximov, N. A. *The Plant in Relation to Water; a Study of the Physiological Basis of Drought Resistance*, Allen and Unwin, London. 1929.
37. Maximov, N. A. *Textbook of Plant Physiology*, 2nd ed., McGraw-Hill, New York, pp. 1–473. 1938.
38. Miller, E. C. "Development of the pistillate spikelet and fertilization in *Zea mays*," *J. Agr. Res.*, 18:255–265. 1919.
39. Pavlychenko, T. "Root systems of certain forage crops in relation to the management of agricultural soils," *Natl. Res. Council Canada and Dominion Dept. Agr. N. C. R. 1088*, Ottawa, pp. 1–46. 1942.
40. Pavlychenko, T. K., and Harrington, J. B. "Competitive efficiency of weeds and cereal crops," *Can. J. Res.*, 10:77–94. 1934.
41. Percival, J. *Agricultural Botany*, 7th ed., Duckworth, London, pp. 7–24, 34–36, 40–42, 52–60, 113–121, 136–140, 622–646. 1926.
42. Poehlman, J. M. *Breeding Field Crops*, Holt, New York, pp. 1–427. 1959.
43. Pope, M. N. "The time factor in pollen-tube growth and fertilization in barley," *J. Agr. Res.*, 54(7):525–529. 1937.
44. Robbins, W. W. *Botany of Crop Plants*, Blakiston, Philadelphia, pp. 22–39. 1931.
45. Robertson, D. W., and Deming, G. W. "Natural crossing in barley at Fort Collins, Colo.," *J. Am. Soc. Agron.*, 23:402–406. 1931.
46. Shantz, H. L., and Piemeisel, L. N. "The water requirement of plants at Akron, Colo.," *J. Agr. Res.*, 34(12):1093–1190. 1927.
47. Sieglinger, J. B. "Cross-pollination in milo in adjoining rows," *J. Am. Soc. Agron.*, 13:280–282. 1921.
48. Stanton, T. R., and Coffman, F. A. "Natural crossing in oats at Akron, Colo.," *J. Am. Soc. Agron.*, 16:646–659. 1924.
49. Wareing, P. F., and Phillips, I. D. J. *The Control of Growth and Differentiation in Plants*, Pergamon Press, Elmsford, N.Y., pp. 1–303. 1970.
50. Watson, J. D. *The Double Helix*, Penguin Books, Harmondsworth, England, pp. 1–175, 1970.

51. Weaver, J. E. *Root Development of Field Crops,* McGraw-Hill, New York, pp. 1–291. 1926.
52. Weaver, J. E. "Investigations of the root habits of plants," *Am. J. Bot.,* 12:502–509. 1925.
53. Weaver, J. E., and Zink, E. "Extent and longevity of the seminal roots of certain grasses," *Plant Physiol.,* 20(3):359–379. 1945.
54. Weihing, R. M. "The comparative root development of regional types of corn," *J. Am. Soc. Agron.,* 27:526–537. 1935.
55. Welton, F. A., and Morris, V. H. "Lodging in oats and wheat," *Ohio Agr. Expt. Sta. Bul. 471.* 1931.

Chapter 4

Crop Improvement

POSSIBILITIES IN CROP IMPROVEMENT

The creation of each new superior hybrid or variety (cultivar) is a definite advance in human welfare. The breeding of more productive types is the only stable method of increasing crop yields. Although crop yield and quality can be improved by disease and pest control,[29] fertilizer application, and better cultural methods, such practices must be repeated each season. Improved cultivars of many crops are rapidly replacing inferior local varieties in most countries of the world. The maintenance, production, and appropriate distribution of seed stocks of improved types has resulted in striking increases in crop yields in many countries.[7]

Crop improvement has been in progress since primitive man first exercised a choice in selecting seed from wild plants for growing under cultivation. The greatest advances were made before the dawn of civilization, but material progress also was made thereafter. However, not until the present century, when some knowledge of genetics was acquired, did crop breeding become a science with the outcome of breeding methods reasonably predictable.[14a] Much further progress is still possible by utilizing material from world collections of seeds and plants that are being widely distributed to breeders. Remarkable increases in yield followed the breeding and wide distribution of semidwarf, photoperiod-insensitive cultivars of rice and wheat that resist lodging under heavy nitrogen fertilization.[3]

Striking benefits from crop breeding have been obtained since about 1940.[32] The increased farm income from hybrid corn and hybrid sorghum exceeds the cost of all agricultural research in the United States.

Crop breeders are concerned chiefly with more abundant, stable, and economical crop production, but they also are deeply motivated by social and humane wants.[24] Farmers now have awnless and smooth-awned cereals that do not irritate the eyes, mouth, or skins of the farm workers or livestock. Consumers now have some food products that are higher in protein or certain essential amino acids.

They also have sweeter and more tender sweet corn as well as lighter bread made from stronger wheats. Crops that are easier for man to harvest, and feed crops that are more palatable or less toxic to livestock, also have been developed. Improved cultivars have helped to alleviate crop failures and their accompanying human and livestock starvation.

OBJECTIVES IN CROP BREEDING

A perfect crop variety or hybrid has not yet been created. Every one has several weaknesses or defects that curtail maximum yields and quality except under rare ideal conditions. The objective of the crop breeder is to correct these defects while developing cultivars with higher yield capabilities. New high-yielding cultivars are being bred in many countries.[30] First-generation hybrids with increased vigor, growth, and reproduction have markedly increased the yields of several crops.

Some of the chief objectives in crop breeding are the following improvements in characteristics:

1. Resistance to diseases, insects, drought, cold, heat, lodging, and alkali.
2. Adaptation to variable photoperiods, shorter seasons, longer seasons, heavy grazing, or frequent cutting.
3. Feeding quality—e.g., palatability, leafiness, hull percentage, nutritive value, and texture.
4. Market quality—e.g., higher content of textile fiber or of protein, sugar, starch, or other extractives; better processing quality for textiles, foods, beverages, and drugs; and better color.
5. Seed quality—e.g., higher or lower seed-setting tendency, greater longevity, higher viability, and larger size.[2]
6. Growth habit—e.g., more erect or prostrate stems, more or less tillering or branching, more uniform flowering and maturity, more uniform height, longer life, and better ratio of tops to roots.
7. Harvesting quality—e.g., stronger, shorter, or taller stalks; erect stalks and heads; nonshattering qualities; easier processing; and freedom from irritating awns and fuzz.
8. Productive capacity—e.g., greater vigor, higher fertility, and faster recovery after cutting.

METHODS OF CROP BREEDING

Three general methods of crop improvement are commonly listed, namely, (1) introduction, (2) selection, and (3) hybridization. These methods are not wholly distinct because hybridization almost always must be preceded, or followed, or both preceded and followed, by some scheme of selection. Irradiation of plant material occasionally produces a new character that is useful in crop improvement.

Introduction

Crop introductions from other countries may be of superior productivity, and they often provide better foundation stocks for breeding. Foreign varieties, that seemingly are worthless, may possess resistance to some disease or insect, or may have some other useful characteristic than can be transferred to adapted varieties by hybridization.[13] Numerous foreign and domestic wild plant species are being tested for possible cultivation. Collections of exotic and domestic wild and cultivated species and strains of various crop plants are being screened for disease and insect resistance, and other characteristics that might be useful in breeding.

When some variety, strain, or inbred line is found to be outstanding for some desired character such as disease resistance, heterosis capacity, or cytoplasmic sterility, it is likely to be used by most plant breeders. This results in a narrow genetic base, which is disastrous when a new or minor race of a parasite multiplies and almost destroys previously resistant varieties or hybrids that are widely grown. Severe losses have occurred from epidemics of a previously unimportant race of a fungus parasite attacking a large number of varieties or hybrids of wheat, oats, or corn that had a common resistant ancestry.[2a, 27a] This necessitates a search for plants with genes for resistance to the new race.

Resistance to a specific race is referred to as vertical resistance which provides good protection until other races arise and then multiply rapidly with little competition from any existent race.[33] The alternative is to incorporate into the crops genes for horizontal resistance (race nonspecific or generalized resistance). Horizontal resistance provides partial protection against, or tolerance to, a number of races of a fungus but the plants are not wholly free from the disease.[34] Horizontal resistance often is polygenic but it sometimes is controlled by only a few genes and occasionally is monogenic.

Selection

MASS SELECTION: Mass selection is a quick method of purifying or improving mixed or unadapted crop varieties. It is done by selecting a large number of plants of the desired type and then increasing the progeny. It serves to eliminate undesirable types. Most of the open-pollinated varieties of corn were products of mass selection.[17] Natural selection through survival of the more vigorous strains accomplishes a similar objective. Winter barley and winter rye were improved in winter hardiness by natural selection while being grown under cold conditions in the United States. Mass selection accomplishes nothing when selection is confined to vigorous plants that are merely favored by a good environment such as thin stands or more favorable soil or moisture locations in the field.[21]

PURE-LINE OR PEDIGREE SELECTION: Pure-line, pedigree, or individual plant selection consists of growing individual progenies of each selected plant so that their performances can be observed, compared, and recorded. Only a few superior strains among the numerous original selections are saved for advanced testing.[15] In crops that are largely self-fertilized, reselection is not necessary except for progenies resulting from occasional natural hybrids or mutations that are still segregating. In crops that are partly or largely cross-fertilized, selection must be repeated until the strains appear to be uniform. Also, cross-fertilized crop plants selected for propagation must be self-pollinated under controlled conditions. This may be done by covering the floral parts to exclude foreign pollen or by hand pollinating, or both, or in some cases by isolating the plants. Pure-line selection offers a quick means of segregating desired types from mixed varieties.[28] Numerous varieties were originated by this method.[8, 9, 31] Simple selection fails to introduce new desired characters.

In cross-pollinated crops, or when the original variety and the selections differ by several genetic factors, several (five or more) generations of selection may be required to purify the strains so that each will continue to breed true thereafter while being tested for yield.

Ear-to-row continued selection and progency testing was a popular breeding method for the cross-pollinated corn crop from about 1900 to 1920, but the method was discarded. Ear-to-row testing is effective in isolating some of the better strains of an unadapted variety during the early years of selection, but progressive increases in yield have never been obtained.[17] The chief limitations to the method are the partial inbreeding that reduces yields, as well as the lack of complete control of the pollen parentage of the seed ears. Continued selection in cotton has met with some success because complete purity is never obtained when the selections are grown without protection from insects that can effect intercrossing.

The breeding of cross-pollinated grasses and legumes such as bromegrass, crested wheatgrass, and alfalfa[11, 19] usually has involved continued selection and self-pollination to isolate and purify improved strains. Some of these strains later may be combined into a synthetic variety and allowed to intercross.[19]

PURE-LINE CONCEPT: The modern methods of handling, testing, and increasing individual selections of self-fertilized crops had as their basis the pure-line concept proposed in 1903 by W. L. Johannsen who worked with beans. According to this concept, variations in the progeny of a single plant of a self-fertilized species are not heritable but are due to environmental effects. Failure to recognize this principle before, or in some countries as late as 50 years after, its discovery resulted in some futile work in continuously selecting self-fertilized crops such as wheat and barley.

When homozygous self-fertilized plants are selected, the progeny is pure for all characters until outcrossing or mutation occurs.[14a] It should be recognized, of course, that varieties that appear to be pure may be pure only for the characters that are observed under a particular environment. A pure-line was originally defined as the descendants of a single, homozygous, self-fertilized individual. The definition commonly used today is that a pure-line comprises the descendants of one or more individuals of like germinal constitution that have undergone no germinal change.

INBREEDING OF CROSS-FERTILIZED CROPS: Before making hybrids in cross-fertilized crops, it may be desirable to select and inbreed the varieties for several generations or until they are reasonably pure for the characters desired. Without such inbreeding, the hybrids that are obtained from cross-fertilized crops cannot be reproduced. Artificial self-pollination in a normally cross-pollinated crop leads to segregation into pure uniform (homozygous) lines. There is often a rapid reduction in vigor when self-pollination is practiced. In corn the reduction in vigor in the first generation of selfing is approximately one half the difference in vigor between the original corn and that of the homozygous inbred progeny. When appropriate lines of self-pollinated plants are intercrossed there usually is restoration of vigor, a fact that is applied in hybrid corn production.[18] In addition, certain abnormalities, such as sterility, poor chlorophyll development, dwarf habit, lethal seedlings, and susceptibility to diseases may appear as a result of inbreeding. These types must be discarded. Such abnormalities are mostly recessive characters which under open-pollinated conditions are largely suppressed as a result of crossing, usually with normal plants.

Hybridization

Hybridization, the only effective means of combining the desirable characters of two or more crop varieties, offers far greater possibilities. It has been the standard crop improvement method in advanced countries for 50 years. Most cereal crop varieties grown in the United States by 1970 were selected from crosses.

The first step in breeding by hybridization is to choose parents that can supply the important character or characters that a good standard variety lacks. It is important to have definite characteristics in mind, but also to test the material for these characteristics repeatedly. Success occasionally has followed haphazard crossing between productive varieties, but failure to accomplish any improvement by such a procedure is traditional.

HYBRIDIZATION TECHNIQUES: In crossing self-fertilized crops the flowers are first emasculated, i.e., the anthers are removed or killed before they have shed pollen. The flowers are then covered with a small paper, glassine, or cellophane bag to exclude insects and foreign pollen (Figures 4–1, 4–2, 4–3, 4–4, 4–5). The anthers usually are removed with tweezers, a needle, or a jet of air or water, but rice, sorghum, and certain grass pollens can be killed by immersing the panicles in hot water at about 47° C for 10 minutes. Pollen from the male parent is applied by brushing or dusting it on the pistil when the stigmatic surface of the latter is receptive. This receptiveness may be immediately after emasculation or 1 or 2 days later. It is indicated by a sticky exudate covering the stigma.

The seeds that develop from the cross-pollinated flowers, when planted, produce plants of the first filial (or F_1) generation. These plants should be all alike, only the dominant, and mutual recessive, characters being expressed. In the second (F_2) generation, the plants break up or segregate into all possible combinations of the dominant and recessive characters of the two parents. Plants of the types desired are selected; these and several subsequent generations are handled as previously described under selection methods. Reselection continues thereafter until the desired strains are uniform, usually for three to six generations.

The size of the F_2 population may vary from 200 to 10,000 plants, the proper number depending upon the number of genetic factor differences involved in the cross. For example, when closely related

Figure 4–1. Cross-pollinating wheat (*right*) after emasculation by removal of the three anthers from flower as shown at left. The white bags protect emasculated flowers from fertilization by undesired foreign pollen.

Figure 4–2. Oat floret showing the anthers and stigma within the lemma and palea, with a detached ovary at the right.

varieties are crossed to combine two single-factor character differences, one from each variety, an F_2 population of 200 to 500 plants should be ample because approximately one sixteenth of this number, or about 13 to 31 plants, should breed true for the desired recombination. On the other hand, with a six-factor difference between the two parents, e.g., two factors for smut resistance, two for seed color, and two for plant height, one would expect only $^1/_{4096}$, or not more than 3 in 10,000 F_2 plants, to breed true for the desired recombination. However, additional true-breeding plants of the desired recombination would be obtained in later generations from segregating F_3 lines.

Figure 4–3. Pollinating a tobacco flower by brushing the stigma with ripe anthers from another species.

The expected proportion of true-breeding strains of a particular recombination in the F_2 generation is determined by the formula $^1/_{4n}$ in which n is the number of genetic factor differences involved in the cross. Thus, with one pair of factors (i.e., $n = 1$) $^1/_{4^n} = {}^1/_4$, and six factors ($n = 6$) $^1/_{4^6} = {}^1/_{4096}$. It is obvious that in wide crosses, i.e., crosses between widely unrelated varieties, the number of factor differences

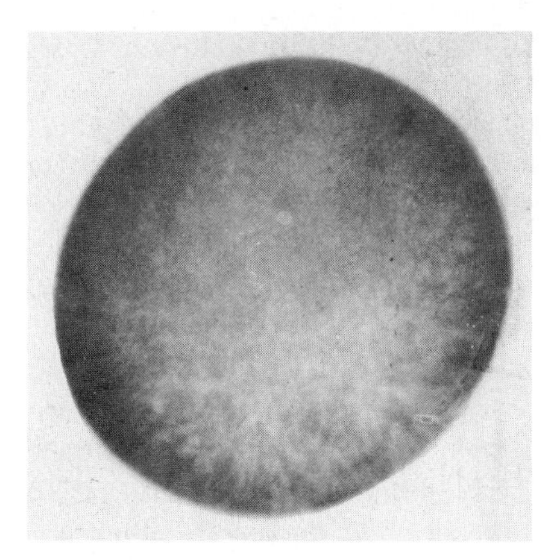

Figure 4–4. Barley pollen grain.

Figure 4–5. Sorghum florets in bloom. The exposed stigma often permit stray foreign pollen to produce outcrosses unless the panicles are covered with bags.

is so large that the chances of finding the desired recombination in the first cross are extremely remote. In such cases the strain obtained that most closely approaches the desired recombination may be crossed with one of the parents, or with some other variety, and selection then resumed.

The success of the above procedures in hybridization and selection is illustrated by the results from crossing two oat varieties, one resistant to smut, and the other resistant to rust. Selections from such a cross are resistant to both smut and rust.[9]

BACKCROSSING: Backcrossing is a useful method of breeding when it is desired to add only one or two new characters to an otherwise desirable variety.[4, 12] In this method the good variety is crossed with one having the other character desired. First-generation plants are backcrossed with the good variety. Thereafter in each segregating generation a number of plants approaching the desired new recombination are selected and backcrossed with the good commercial parent. Usually, with repeated rigid selection, the improved type is recovered in approximately five backcrosses. Furthermore, little testing is necessary because the recovered variety is practically identical with the original adapted variety, except that it has a new character such as disease resistance. Thus, by omitting several years of testing, the recovered variety may be quickly increased and distributed to farmers. Unfortunately, the procedure fails to introduce new factors for basic higher yields. A specific type of selection that accompanies backcrossing has been termed convergent improvement.[17]

BULK PROPAGATION OF HYBRID MATERIAL: In the bulk method of handling hybrid material the entire population from a cross is grown and harvested in mass each year. Successive crops of seed are sown for five to ten years to allow natural selection for characters, such as cold resistance or insect resistance, to take place. By this time the poorly adapted strains are largely eliminated, while the remaining strains are mostly true breeding (or homozygous). Plants are then selected, and the progenies tested as previously described. Testing of large numbers of unadapted and heterozygous strains is thus avoided. However, it is recognized that desirable types may be lost during the interim because of competition with aggressive but otherwise undesirable plants.

PRODUCTION OF HYBRID SEED

Nearly all of the corn (Figure 4–6), sorghum, sugarbeets, sunflowers, and castorbeans grown in the United States are first-generation hybrids, either single crosses, double crosses, or three-way crosses.

Hybrids of wheat, barley, and other crops are under development for possible commercial production. Some hybrid wheat was grown in the United States in 1974, mostly in the southern states for pasture.

Commercial hybrid seed of corn can be produced by detasseling the seed-parent rows in a cross-pollination field, or by use of cytoplasmic male-sterile lines. Pearl millet hybrids have been produced by planting a mixture of four selected lines. They also can be produced by clipping (topping) the upper part of the heads on the seed-parent plants before they shed pollen, but after the stigmas are exserted, or by pollinating cytoplasmic male-sterile lines. Hybrid seed of crops that naturally are wholly or largely self-pollinated can be produced economically only by the use of male-sterile lines. Plants with cytoplasmic male-sterility have been discovered or developed in several crops. Genetic factors in different varieties or strains of these crops determine whether their progeny is male-sterile or male-fertile when they are used as pollinators on cytoplasmic male-sterile

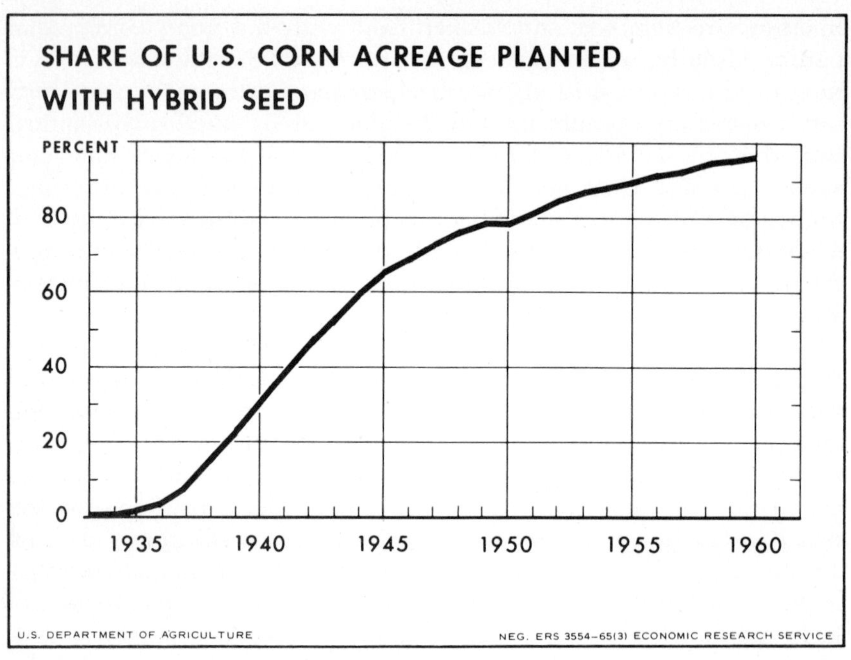

Figure 4–6. About 98 per cent of the corn acreage of the United States was planted with hybrid seed by 1969.

plants. A related (fertile counterpart) strain that carries the recessive sterile-producer gene is used as a pollinator to maintain or to increase the male-sterile line. An unrelated variety or strain that carries the dominant fertility-restorer gene is used as a pollen parent to produce commercial hybrid seed. The pollen parent chosen is one that produces a suitable productive hybrid.[2]

BREEDING FOR DISEASE RESISTANCE

Some of the most outstanding accomplishments in crop breeding have been in the development of disease-resistant varieties of wheat, oats, sugarcane, sugarbeets, flax, potatoes, sorghum, and alfalfa. The plant-breeding methods are the same as for other objectives except that the progeny must be exposed to the particular diseases either by artificial inoculation or by choosing conditions favorable for natural infection. Plants that resist the disease and are otherwise desirable are selected for further progeny tests.

The breeding of varieties resistant to diseases sometimes is highly complex because of the existence of different physiologic races of the disease organism. Rust-resistant varieties of wheat and oats suffer heavy damage when other races or subraces of the rust parasite become abundant and are able to attack them. Such races may be new

estimate of the uniformity or reliability of the experiment, particularly the reliance that can be placed upon differences found between varieties or treatments. When one variety outyields another by a sufficiently consistent margin so that the computed odds for a real difference are 19 to 1 or better it is commonly assumed to be superior. With lower odds, the difference observed between the two varieties is regarded as insignificant, i.e., largely due to chance. Since this arbitrary arrangement may accidentally favor certain varieties, the varieties and treatments almost always are arranged in a random or designed order within definite experimental blocks of plots. Special experimental designs, especially the so-called lattice designs,[15] are used as a means of partly adjusting the plot yields to smooth out differences due to soil variation. When the plots of the different varieties or treatments are repeated in the same systematic order with similar or most promising varieties or treatment grouped together, a more accurate observation and comparison is obtained but the entire experiment shows a higher error measurement.

The size of plot is determined largely by convenience and by the availability of uniform land for the experiment. Nursery rows give reliable comparisons between varieties when they are repeated several times and protected from competition with adjacent varieties by the planting of border rows. Field plots give a better view of the behavior of the variety under field culture. They also furnish more material for further experimentation or more seed for increase. Field plots occupy more land than do nursery rows, but they are only slightly more reliable in yield tests. Field plots of different varieties of small grain or forage crops are usually 20 to 40 meters long and one drill width wide. Corn usually is tested in plots two hills wide and five to ten hills in length, and sorghum in two-row plots 9 to 15 meters in length. Small-grain nurseries usually are planted in rows 1 foot apart with the rows approximately 1 rod in length.

REFERENCES

1. Aamodt, O. S. "Breeding wheat for resistance to physiologic forms for stem rust," *J. Am. Soc. Agron.*, 19:206–218. 1927.
2. Airy, J. M., Tatum, L. A., and Sorenson, J. W., Jr. "Producing seed of hybrid corn and grain sorghum," in *Seeds*, USDA Yearbook, 1961, pp. 145–153.

2a. Anonymous. *Genetic Vulnerability of Major Crops.* Natl. Acad. Sci., Washington, D.C. pp. 1–307. 1972.

3. Borlaug, N. E. Mankind and civilization at another crossroad. Seventh Biennial Conference of FAO. Rome, Italy, pp. 1–48. 1971.
4. Briggs, F. N. "The backcross method in plant breeding," *J. Am. Soc. Agron.*, 27:971–973. 1935.

5. Briggs, F. N., and Allard, R. W. "The current status of the backcross method of plant breeding," *Agron. J.*, 45:131–138. 1953.
6. Briggs, F. N., and P. F. Knowles. *Introduction to Plant Breeding.* Reinhold, New York, pp. 1–426. 1967.
7. Burgess, S., Ed. *The National Program for Conservation of Crop Germ Plasm.* Univ. Ga., Athens. pp. 1–73. 1971.
8. Clark, J. A. "Improvement in wheat," in *USDA Yearbook*, 1936, pp. 207–302.
9. Coffman, F. A., Murphy, H. C., and Chapman, W. H. "Oat breeding," in *Oats and Oat Improvement*, Academic Press, New York, Agronomy Monograph 8:263–329. 1961.
10. Frey, K. J., Ed. *Plant Breeding.* Iowa State Univ. Press. Ames, Iowa. pp. 1–430. 1966.
11. Hanson, A. A., and Carnahan, H. L. "Breeding perennial forage grasses," *USDA Tech. Bul. 1145.* 1956.
12. Harlan, H. V., and Pope, M. N. "The use and value of the backcross in small grain breeding," *J. Heredity*, 13:319–322. 1922.
13. Harlan, J. R. "Genetics of disaster." *J. Environ. Quality.* 1(3):212–15. 1972.
14a. Hayes, H. K. "Green pastures for the plant breeder," *J. Am. Soc. Agron.*, 27:957–962. 1935.
14b. Hayes, H. K., and others. "Thatcher wheat," *Minn. Agr. Exp. Sta. Bul. 325*, 1936.
15. Hayes, H. K., Immer, F. R., and Smith, D. C. *Methods of Plant Breeding*, 2nd ed., McGraw-Hill, New York, pp. 1–551. 1955.
16. Kappert. H., and W. Rudorf, Eds. Handbuch der Pflanzenzuchtung, (Manual of plant breeding). Parey, Berlin and Hamburg. 1959.
17. Jenkins, M. T. "Corn improvement," in *USDA Yearbook*, 1936, pp. 455–522.
18. Jones, D. F. "Crossed corn," *Conn. Agr. Exp. Sta. Bul. 273.* 1926.
19. Kirk, L. E. "Breeding improved varieties of forage crops," *J. Am. Soc. Agron.*, 19:225–238. 1927.
20. LeClerg, E. L., Leonard, W. H., and Clark, A. G. *Field Plot Technique*, 2nd ed., Burgess, Minneapolis, pp. 1–373. 1962.
21. Leighty, C. E. "Theoretical aspects of small grain breeding," *J. Am. Soc. Agron.*, 19:690–704. 1927.
22. Love, H. H. "A program for selecting and testing small grains in successive generations following hybridization," *J. Am. Soc. Agron.*, 19:702–712. 1927.
23. Martin, J. H. "Sorghum improvement," in *USDA Yearbook*, 1936, pp. 523–560.
24. Martin, J. H. "Breeding sorghum for social objectives," *J. Heredity*, 36(4):99–106. 1945.
25. Martin, J. H., and Salmon, S. C. "The rusts of wheat, oats, barley, rye," in *Plant Diseases*, USDA Yearbook, 1953, pp. 329–343.
26. Martin, J. H., and Yarnell, S. H. "Problems and rewards in improving seeds," in *Seeds*, USDA Yearbook, 1961, pp. 113–118.
27a. Melchers, L. E., and Parker, J. H. "Rust resistance in winter wheat varieties," *USDA Bul. 1046*, pp. 1–32. 1922.

27b. Nelson, R. R., Ed. Breeding Plants for *Disease Resistance: Concepts and Applications.* Penn. State Univ. Press, University Park, Pa., pp. 1–401. 1973.
28. Noll, C. F. "Mechanical operations of small grain breeding," *J. Am. Soc. Agron.*, 19:713–721. 1927.
29. Painter, R. G. Crops that resist insects provide a way to increase world food supply. *Kans. Agr. Exp. Sta. Bul. 520.* pp. 1–22. 1968.
30. Pal, B. P. Recent cereal research and future possibilities. Science and culture. 34:136–149. 1968.
31. Quinby, J. R. and Martin, J. H. "Sorghum improvement," in *Advances in Agronomy*, Vol. 6, Academic Press, New York, pp. 305–359. 1954.
32. Salmon, S. C., Mathews, O. R., and Leukel, R. W. "A half-century of wheat improvement in the United States," in *Advances in Agronomy*, Vol. 5, Academic Press, New York, pp. 1–151. 1953.
33. Robinson, R. A. Vertical resistance. *Rev. Plant Path.* 50(5):233–39. 1971.
34. Simons, M. D. Polygenic resistance to plant disease and its use in breeding resistant cultivars. J. Environ. Quality. 1(3):232–38. 1972.
35. Sprague, G. F. "Corn breeding," in *Corn and Corn Improvement*, Academic Press, New York, Agronomy Monograph 5:221–292. 1955.
36. Tysdal, H. M., Kiesselbach, T. A., and Westover, H. L. "Alfalfa breeding," *Nebr. Agr. Exp. Sta. Res. Bul. 124.* 1942.
37. White, G. A., and others. Agronomic evaluation of prospective new crop species. *Econ. Bot.* 25(1):22–43. 1971.

Chapter 5

Tillage Practices

DEVELOPMENT OF TILLAGE OPERATIONS

Tillage began before the earliest written records of mankind. The first implements were hand tools to chop or dig the soil, usually made of wood, bone, or stone. They were used to subdue or destroy the native vegetation, make openings in the soil to receive seeds or plants, and reduce competition from native plants and weeds growing among the crops. The next stage of tillage, application of the power of domestic animals,[23] occurred in parts of the world before the dawn of history. This made possible development of implements with a steady forward movement. Among these were the crooked-stick plow to stir the soil, and the brush drag to pulverize the surface. Little further progress was made for many centuries except that eventually some plows were fitted with iron shares despite a common misapprehension that iron poisoned the soil. The development of steel in the 19th century resulted, in 1833, in a plow with sharp edges that cut the soil layer and in a curved polished surface that permitted the plow to scour. That straight-line movement of the plow has since been supplemented with the rotary movement in such implements as disk plows, harrows, the rotary hoe, and various pulverizing and stirring tools.

Very little was known about the effects of cultural operations in the Middle Ages.[70] In 1731, Jethro Tull published his *New Horse-Houghing Husbandry* in England. He believed that plants took up the minute soil particles, i.e., the more finely the soil was divided the more particles would be absorbed by the roots.

During the 19th century it became evident that nutrition of plants depended on certain chemical elements from the soil minerals, organic matter, water, and air. The foundation for this concept was laid by Justus von Liebig and others. The idea became widespread that tillage, by increasing the aeration in the soil, increased the oxidation of chemical compounds in the soil and made them more soluble.

Early American writers believed that tillage allowed the roots to penetrate more deeply, or refined the soil to make a greater surface to hold moisture, although certain others recognized the importance of tillage in weed control. That harmful effects might result from excessive tillage, particularly from greater oxidation of organic matter or from increased erosion, was also pointed out by early writers in this country. Not until after 1890 did experimental evidence begin to show the basic reasons for tillage.

PURPOSES OF TILLAGE

The fundamental purposes of tillage[23] are (1) to prepare a suitable seedbed, (2) to eliminate competition from weed growth, and (3) to improve the physical condition of the soil. This may involve destruction of native vegetation, weeds, or the sod of another crop. It may further involve removal, burial, or incorporation in the soil of manures or crop residues. In other cases, the tillage operation may be solely to loosen, compact, or pulverize the soil. The best system of tillage is, then, the one that accomplishes these objectives with the least expenditure of labor and power.

The effect on soil erosion also must be considered in planning all tillage operations. The kind of tillage for seedbed preparation on dry lands is governed almost entirely by the effects of tillage upon the conservation of soil moisture and the prevention of surface runoff and wind erosion.

In seedbed preparation for grain sorghum under dryland conditions in Kansas, land cultivated with a disk harrow or duckfoot cultivator to control spring weed growth gave an average yield of 18.5 bushels.[81] Land left uncultivated, with the weeds allowed to grow until it was listed at planting time, produced only 10.8 bushels per acre.

Under most conditions, a desirable seedbed is one that is mellow yet compact enough so that the soil particles are in close contact with the seed, but the seedbed must be free from trash and vegetation that would interfere with seeding, unless special planting equipment that cuts through the residues is used. The seedbed should contain sufficient moisture to germinate the seed when planted, and support subsequent growth. In irrigated regions, it is occasionally necessary to plant the crop and then irrigate the field in order to supply sufficient moisture for germination, but this practice is avoided whenever possible because of frequent irregular stands. Irrigation before planting is preferable.

IMPLEMENTS FOR SEEDBED PREPARATION

An acre of land originally was the area tilled by one ox-drawn plow in one day. Tractor-drawn implements cover 0.3 to 0.6 acres per hour per foot of width, or 0.4 to 0.8 hectares per hour per meter of width.

The moldboard plow breaks loose or shears off the furrow slice by forcing a triple wedge through the soil. This action inverts the soil and breaks it into lumps. Some pulverization takes place (Figure 5–1). Several different shapes of moldboard plow bottoms are used. At one extreme is the breaker type with a long moldboard adapted to virgin or tough sod for completely inverting without pulverization. At the

Figure 5–1. Pulling a five-bottom moldboard plow, a disc-harrow, and a grain drill in one operation reduces soil compaction. [Courtesy John Deere & Co.]

other extreme is the stubble bottom with a short, abrupt moldboard which pulverizes the furrow slice while turning old ground, such as small-grain stubble, so as to mix the stubble with the soil. The general-purpose or mellow-soil plow is intermediate between these two types in length, slope, and action. The two-way plow is adapted to steep hillsides because it throws the soil downhill when drawn across the slope in either direction. It is often used on irrigated land to avoid dead furrows. Better coverage of trash is facilitated on all plows by use of attachments such as coulters, jointers, rods, and chains. The majority of moldboard plows range in size from 7 to 18 inches in width of furrow cut.[37] The 14-inch width is very common.

Disk plows (Figure 5–2) are important for use in loose soils but also in those soils too dry and hard for easy penetration of moldboard plows. In some sticky soils neither type of plow will scour when the soil is wet. In this case, scrapers help keep the disks clean. The disks vary in size from 20 to 30 inches, while the depth of plowing may be varied from 4 to 10 inches. The one-way disk plow (Figure 5–3) has been widely used in the stubble fields of the wheat belt for seedbed preparation. When there is sufficient stubble, straw, weeds, or aftermath available, the one-way disk plow leaves the residues mixed with the surface soil with enough exposed to check wind erosion, except on light or other soils that tend to blow badly. The advantages[85] of the one-way disk plow are (1), with the same tractor it will cover about three times as much ground as the moldboard plow, (2) the draft is

Figure 5–2. Disk plow. [Courtesy Allis-Chalmers Manufacturing Co.]

Figure 5–3. The one-way disk plow leaves about one-third of the crop residue on the soil surface. [Courtesy John Deere & Co.]

less for the work done because the soil is moved a shorter distance, (3) the speed of operation makes early plowing possible, and (4) because of the rough surface left by this implement, snow is caught and soil blowing is prevented. Wheat yields may be less when the one-way is used than when the land is plowed.[65] The one-way should be avoided for bare summer-fallowed land, or dry fields sparsely covered with stubble or weeds where it tends to pulverize the soil so that the soil blows easily.

Blade or subtillage implements, which leave crop residues on the surface of the soil, are being used in semiarid areas.[87] Wide-sweep blades with staggered mountings undercut the stubble and weeds, but do not pulverize the soil surface (Figure 5–4). This so-called stubble mulch reduces soil erosion as well as runoff, but it may not increase soil moisture storage. Yields on subtilled land often average no greater than those on plowed land. Special planting equipment is needed to cut open furrows on land covered with heavy crop residues unless a disc cutter is used to chop up the straw.

The lister (middlebuster) resembles a double plow with a right-hand and a left-hand bottom mounted back to back. In some of the southern humid sections, it is used to throw up beds on which to plant cotton, potatoes or other row crops. In the semiarid regions, the lister formerly was widely used to furrow the land. It replaced the moldboard plow in seedbed preparation there because it covered the field in half the time. The ridges also checked wind erosion, snow

Figure 5–4. (*Top*) Noble blade that cuts 13 feet. (*Bottom*) Gang of Noble blades cutting a 56-foot strip, with the grain stubble left on the surface. [Courtesy Noble Cultivators, Ltd.]

drifting, and runoff when the furrows ran on the contour or across the slope. Row crops are planted in the moist soil at the bottom of the lister furrows. For drilled crops such as small grains, the lister ridges or middles are broken down or leveled with ridge busters. The lister has been largely replaced by sweep-blade implements or the one-way plow for the initial seedbed preparation in the semiarid areas.

SURFACE IMPLEMENTS IN FINAL SEEDBED PREPARATION

Tillage to prepare a seedbed after the land has been plowed is accomplished with harrows, field cultivators, or other machines equipped with disks, shovels, teeth, spikes, sweeps, knives, corrugated rolls, or packer wheels. Under most conditions, a smooth, finely pulverized seedbed should not be prepared until just before a crop is to be planted.

Harrows, which smooth and pulverize plowed soil, compact it, and destroy weeds, require less power than do plows or listers. The disk harrow cuts, moves, and pulverizes the soil and destroys weeds. The spike-tooth harrow breaks clods, levels the land, and kills small weeds. Both disk and spike-tooth harrows are a menace on soil subject to blowing. The spring-tooth harrow consists of flexible spring-steel teeth about 2 inches wide which will penetrate sufficiently to tear up deep clods or bring them to the surface while destroying weeds. The corrugated roller is used to mash clods as well as to compact the seedbed,[65] particularly for small grain and grass seeds to improve the chances for germination.

Figure 5–5. Rotary rod weeder cultivating summer fallow. [Courtesy John Deere & Co.]

Several types of cultivators, chiefly the duckfoot or field cultivator and the rod weeder, are used in summer fallowing and seedbed preparation. The field cultivator has sweeps or shovels on the ends of stiff or spring bars. The bars are staggered in two rows, with the result that the weeds are cut off by the overlapping sweeps. This implement leaves the soil surface rough and cloddy, a condition necessary to check wind erosion. The principal feature of the rod weeder is a rotating horizontal square rod at right angles to the direction of travel at a depth of 3 to 6 inches below the soil surface. This rod pulls out the weeds. The rod is driven by gears or sprocket chains from one of the support wheels, so that it revolves in a direction opposite the direction of travel, which keeps the rod in the ground (Figure 5–5). The rotation of the rod keeps it from clogging, while weeds, straw, and clods are lifted toward the surface, where they check soil blowing.

PLOWING IN SEEDBED PREPARATION

Plowing buries green or dried material, loosens the soil so other implements can operate, removes or delays competition with weeds, and roughens the soil surface so as to check runoff of rain water.[16] Other effects of plowing are control of certain insects and diseases and promotion of nitrification. After a clean-cultivated row crop, the seedbed can often be double-disked and harrowed to give adequate preparation without plowing.

Time to Plow

Spring wheat, oats, and barley in the northern Great Plains yield slightly more after spring plowing than after fall plowing, when the time of seeding is the same.[18] This difference is assumed to result from the increased moisture saved, because stubble and weeds retard the drifting of snow from the field. Despite this advantage, the farmer finds it impractical to delay much of the plowing until spring because of the extreme importance of early seeding of spring small grains. Thus the distribution of labor may become a controlling factor in tillage practices.

For corn or other crops planted later, plowing need not be so early, except that fall-plowed land is warmer in the spring because bare soil, which usually is darker than the plant vegetation, absorbs more solar heat and thus hastens corn emergence and growth.[2]

In the western half of the United States, where moisture and nitrates are limiting factors in crop production, and winter wheat follows any small grain, plowing as soon as possible after harvest has been found desirable.[13, 22, 47, 81] This permits a period of about three months in which the soil may accumulate moisture and nitrates. Plowing, together with subsequent tillage controls weeds to conserve moisture, hastens crop residue decay and puts the soil in condition for rapid nitrification.

Land to be summer-fallowed usually is not disturbed until the spring following the harvest except when it is very weedy, the weeds and stubble being left to hold snow and check wind and water erosion. Early spring tillage for summer fallow is highly important[65] in order to stop weed and volunteer growth and thus conserve moisture and permit accumulation of nitrates.

In the more humid regions, early seedbed preparation may be inadvisable because of water erosion.[29, 31] More than half the annual soil erosion may take place between July 15 and September 15 while the land is bare.

Land plowed in spring is regularly higher in nitrates than that left unplowed, but the increase is not always evident immediately after plowing.[1] The difference becomes pronounced during the summer season but almost disappears during the winter when nitrates are leached downward. Significant increases in nitrates follow early fall plowing, with the greatest increases in soil plowed in the middle of July.

Depth of Plowing

The theories that favor very deep plowing are not based upon experimental evidence.[70] At the Pennsylvania Experiment Station deep 12-inch plowing was compared with ordinary 7½-inch plowing for corn, wheat, oats, barley, alfalfa, clover, and timothy.[64] The average yields failed to show any advantage for the deeper plowing. Shepperd and Jeffrey[76] in 1897 found that the yields of wheat in North Dakota from shallow and deep plowing were nearly the same, but fall plowing gave higher yields than spring plowing.

In Kansas, soil plowed to a depth of 7 inches in July gave better yields than soil plowed to a depth of 3 inches.[71] The deeper plowing did not influence soil moisture conservation, but showed greater nitrate accumulation. Differences due to depth of plowing are less than from plowing on different dates[47] (Table 5–1). A gradual increase in yield may be expected as the depth of plowing increases from 4 to 7 inches. Plowing as deep as 10 inches is not considered to be advantageous,[65] and in general plowing deeper than 7 to 8 inches does not increase crop yields.[70]

An experiment under semiarid conditions conducted at Nephi, Utah, for 20 years[7] gave wheat yields shown in Table 5–2.

The effect of depth of tillage on soil erosion was studied in Missouri.[31,59] On land where no crop was grown the annual erosion from tilled soil was greater than from soil that was not stirred. Deep tillage (8 inches) reduced the soil loss only 13.4 per cent as compared to shallow tillage (4 inches), indicating that deep plowing is less effective in control of erosion than is usually supposed.

TABLE 5–1. Effects of Depth of Plowing upon Grain Yield of Winter Wheat in Nebraska (1922–26)

DATE OF PLOWING	DEPTH OF PLOWING			
	4 Inches (*bu.*)	*5.5 Inches* (*bu.*)	*7 Inches* (*bu.*)	*10 Inches* (*bu.*)
September 15	20.5	–	20.1	21.4
August 15	23.0	24.8	25.5	27.2
July 15	25.7	27.5	28.4	27.4
Average July and August	24.4	26.2	27.0	27.3
Average three months	23.1	–	24.7	25.3

TABLE 5–2. Effect of Depth of Tillage on Wheat Yields in Utah

TREATMENT	BUSHELS PER ACRE
Plowed 5 inches deep	22.7
Plowed 8 inches deep	25.2
Plowed 10 inches deep	25.0
Subsoiled 15 inches deep	24.1
Subsoiled 18 inches deep	23.9

For fully 100 years, advocates of deeper plowing recommended that the depth of plowing be increased gradually so as not to turn up too much raw subsoil in any one year. The reasoning behind such a practice seems logical, but the writers have not been able to find any report of an experiment that would either prove or disprove the soundness of the recommendation. Varying the plowing depth from year to year may avoid establishing a compact "plow-sole" in heavy soils, but heavy soils commonly have compact subsurface layers. The more friable and the muck soils do not form plow soles.

Plows with very large moldboards that penetrate 1 to 4 feet deep are used for land reclamation operations such as turning under brush or turning up topsoil that has been buried under water-borne sand.

SUBSOILING

Tillage below the depth reached by the ordinary plow formerly was advocated widely. There was a popular belief that plants utilize fully only the soil moved by man, i.e., unusually deep tillage provides a greater opportunity for root development and moisture storage. A less extreme belief is that merely to loosen, stir, pulverize, or invert the deeper soil layers permits more effective plant growth. No general increase in yields resulted from subsoiling or other methods of deep tillage in the Great Plains in experiments that covered a wide range of crops, soils, and environments in ten states.[19]

The same general results were obtained under other conditions.[7, 61, 64] Increased water infiltration and better crop yields sometimes have followed the subsoiling of land that had a dense hardpan subsurface layer or a plowsole that tended to be rather impervious to water. Such benefits may be evident for one or two years after subsoiling, but the cost of the operation may offset any gain in yield.[42] Subsoiling is ineffective unless the hardpan is dry enough to be shattered.[66]

The reasons why subsoiling and deep tillage often are ineffective seem rather obvious. Heavy soils shrink and leave wide cracks when they dry out, thus providing natural openings for periodical water absorption and aeration. When such soils are wetted, they swell

Figure 5–6. The chisel, a tool for deep tillage. [Courtesy Caterpillar Tractor Co.]

tightly and the cracks close up so that any effect of subsoiling could be only temporary. Light soils do not shrink and swell much, but they are always rather open. Since crop plant roots ordinarily penetrate several feet below any tilled layer, deep tillage usually is not essential to deep rooting. The chisel, an implement with a series of points that break up the soil to a depth of 10 or 18 inches without turning it, in an operation similar to subsoiling, is sometimes used on heavy soils (Figure 5–6). As a substitute for plowing, the chisel opens and roughens the soil so that heavy rains can be absorbed quickly without danger of wind or water erosion. Such deep tillage may favor the growth of sugarbeets. More than 7 million acres were chiseled or deep tilled in the United States in 1972.

SUMMER FALLOW

Land that is uncropped and kept cultivated throughout a growing season is known as summer fallow or summer tillage. Under humid conditions, however, land that merely lies idle for a year or two often is referred to as fallow. The most important function of summer fallow is storage of moisture in the soil, but fallowing also promotes nitrifica-

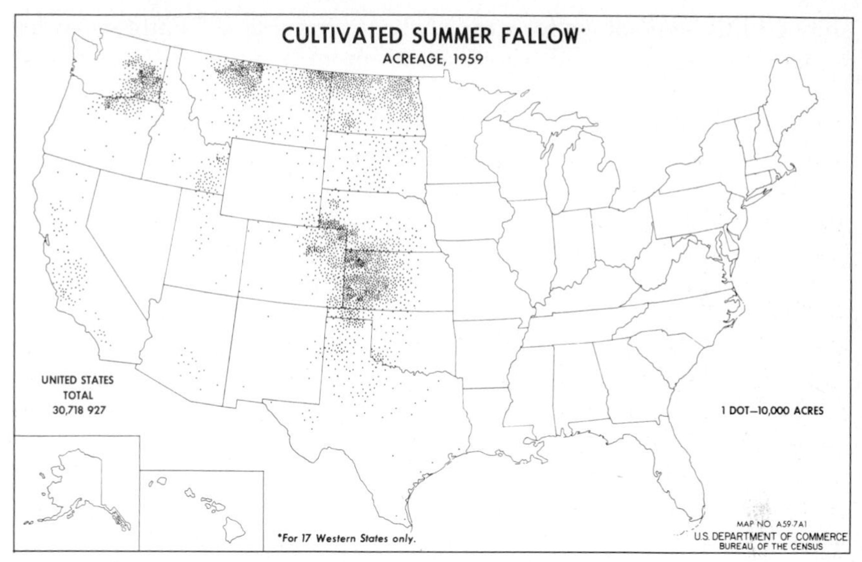

Figure 5–7. Acreage of cultivated summer fallow in 1959.

tion and is a means of controlling weeds. Fallowing is a common practice in semiarid sections where the rainfall is insufficient to produce a satisfactory crop every year, especially where the annual precipitation is less than 15 inches (Figure 5–7). Averages of 15 to 31 per cent of the precipitation that fell during the fallow period were stored in the soil at different locations in the dry farming regions.[33, 79] Moisture conservation by fallow is most effective in winter rainfall areas and where summer evaporation is low. Fallowing conserves little moisture in coarse sandy soils or in high rainfall areas.

In the semiarid regions, fallow has been most widely used in alternation with small grains. In southwestern Kansas, consistently higher yields of winter wheat were obtained on fallow than on land continuously cropped.[81] Fallow also was the best preparation for winter wheat at the Central Great Plains Field Station near Akron, Colorado, where the rainfall is about 18 inches per year.[5, 7] In South Dakota, where the mean annual rainfall averaged 16 inches, spring wheat produced 74 per cent more on fallowed land and 46 per cent more on disked corn land than on land continuously cropped to wheat.[55] The respective increases were 109 per cent and 48 per cent for winter wheat. The relative yields of barley and oats on fallow were much the same as those of spring wheat. Similar results were obtained at Havre, Montana, where the annual precipitation averaged 11 inches.[60] In comparison with small grains grown continuously, fallow gives a better distribution of yields between years, with much less frequent failures. At Mandan, North Dakota, fallowing a light soil showed little advantage for spring-wheat production.[69] Fallow may be replaced entirely by cultivated crops under such conditions, wheat being alternated with corn.

In the Great Basin region, where most of the precipitation occurs in the winter months, fallowing has been standard practice for wheat production since the 19th century. Yields of winter wheat after fallow are nearly double those from continuous cropping. Continuous cropping is impractical in the driest sections. Fallowing operations seldom conflict with the seeding or harvesting of wheat, which fact enables a farmer to grow as large an acreage of wheat each year as he could by continuous cropping. Fallowing often involves only one to three additional rapid tillage operations. The higher yields from fallowing as compared with continuous cropping much more than compensate for the investment in double the acreage of tillable land.[39]

Trashy fallow, or stubble mulch, is excellent for erosion control but excessive amounts of straw are detrimental to wheat yields. Nitrogen fertilizers applied to the soil avoid yield decreases from medium amounts of straw. A continuous stubble mulch practice favors increased infestation of the weedy cheatgrass (*Bromus tectorum*). A skew treader is used to chop through the mulch and cut out the cheatgrass. Special seeding equipment must also be used to cut through the heavy residues. Trashy fallow is formed by tillage with a field cultivator or sweep-blade implement that leaves the crop residues on the surface.[29, 30, 43] Subsequent tillage is with a rod weeder or a blade implement.[67]

The alternation of row crops with fallow is seldom advised, even under conditions where fallow has been found to be profitable for small-grain production. However, grain sorghum responds well to fallow on the heavier soils of the southern Great Plains, the yields being almost double those from continuously cropped land.[81] In South Dakota corn produced only 35 per cent more on fallow than on continuously cropped land.[55] Sorgo showed little response to fallow in comparison with continuous cropping.

For successful summer fallowing certain principles must be observed: (1) the surface of the soil must be kept sufficiently rough to absorb rains and prevent wind erosion, (2) weed growth must be suppressed to conserve soil moisture, and (3) the operations must be accomplished at a low cost. Fallow should be plowed or listed early in the spring,[65] worked as little as necessary to control weeds, and—when not seeded to winter grains—also ridged in the fall at right angles to the prevailing winds where soil blowing is likely to occur. Field cultivators or rod weeders are most satisfactory for summer fallow tillage because they cut off the weeds and leave the surface rough and cloddy. The field cultivator also has been used successfully as the sole tillage implement on binder-harvested stubble in Wyoming.[63] Yields of wheat after such a plowless fallow were as high as from land plowed for fallow.

SOIL MULCHES FOR MOISTURE CONSERVATION

Soil mulches are created by stirring the surface soil until it is loose and open. The effectiveness of the soil mulch in conserving soil moisture has been investigated for many years.[49] A soil mulch reduces evaporation from the soil surface when free water is present only a few feet below the surface of the soil.[74] Water moves upward by capillarity 15 to 120 inches during a season.[75] Capillary movement to appreciable heights is very slow[56] and movement decreases as the soil moisture decreases. Water in sufficient quantities to support crop plants can be raised only a few inches from a moist subsoil.[50] Most of this movement takes place at moisture contents well above the wilting point. Capillarity and gravity combined may move moisture downward for greater distances and in about double the quantity occurring in the upward movement against gravity.

In eastern Kansas, plots with soil mulches actually lost more water than did bare undisturbed plots where the weeds were kept down with a hoe.[16] In experiments in California where the water table was from 18 to 40 feet below the surface, mulching by thorough cultivation failed to save moisture.[80] On upland soils in Pennsylvania where there was no water table in the soil mantle, frequent cultivation did not decrease the evaporation loss materially.[57]

In Nebraska,[12] in soils partly dry and away from a source of free water, capillary movement was so slow that it could rarely be detected. In the northern Great Plains no water moved up appreciable distances to replace that removed by roots. Water was supplied to wheat roots only by the soil they occupied, and only that part of the soil suffered exhaustion or reduction of its water content.[54] Water once in soil deep enough to escape rapid drying at the surface largely remains until it is reached and removed by plant roots.[12]

In semiarid regions, the rainfall is sufficient to wet the soil to a depth of only a few feet.[54] This usually is exhausted each year by growth of native vegetation or crops. The lower layers are dry except in the wettest of seasons. The water table may be 20 to 2,000 feet below the surface. Formerly a dust mulch was advocated to prevent moisture loss by capillary rise. After some diastrous experiences with soil blowing, the dust mulch theory was abandoned and a clod mulch was advocated. Soon thereafter the effectiveness of any soil mulch to check moisture loss under dryland field conditions was refuted.

Thus a soil mulch can reduce surface evaporation where the water table is so shallow that drainage rather than moisture conservation may be desirable. In field practice the mulch formed by tillage is merely incidental to weed control.

CULTIVATION IN RELATION TO SOIL NITRATES

Cultivation, by drying, aerating, and thereby promoting nitrification, seems to be a positive benefit to a heavy water-logged soil. Controlling weeds by cultivation permits nitrates to accumulate, since nitrates are used up when a crop or weeds occupy the land. Weed control by cultivation also saves soil moisture and thus favors nitrification, which does not operate in dry soil. Cultivation otherwise has produced no regular increase in the accumulation of nitrates, especially in light soils. Greater nitrification occurred in a compacted than in an uncompacted Kansas soil up to the point where the moisture content reached two-thirds of saturation. In Arkansas varied depths of cultivation had little effect on accumulation of nitrate nitrogen in a soil of rather open structure.[68] In Pennsylvania experiments, scraped soil and soil cultivated three to eight times contained almost equal quantities of nitrate nitrogen.[57] Natural agencies appeared to promote sufficient aeration to admit needed oxygen in the soil. Excessive cultivation reduced the nitrates in Missouri soil where the cultivation kept so much surface soil continually dry that nitrate production apparently was retarded in the upper 7 inches.

Other experiments have shown either slight or distinct increases in nitrate accumulation from surface cultivation as compared with soil that was merely scraped.[1, 14, 48, 51, 72] The higher nitrate content in these cultivated soils may have been due to increased aeration.[54] In Kansas experiments ample accumulation of nitrates took place in the uncultivated soil to insure an excellent growth of wheat, provided weed growth was prevented, despite the lower nitrate content.[73] Nitrates leach downward so rapidly after rains that their measurement in the surface foot is of little significance.[48]

OTHER EFFECTS OF CULTIVATION

Cultivation may conserve soil moisture by prevention of runoff. Under semiarid conditions, 88 per cent of the water in a dashing rain sometimes has been lost by runoff. A cultivated surface retains more water from a rain than does an uncultivated surface.[12] The faster the rain falls, the greater is the difference between a cultivated and uncultivated surface in the amount of water held. On the other hand, excessive pulverization of the surface soil is likely to result in a quickly puddled condition with great runoff losses. Under subhumid conditions there is little or no relation between the type of tillage treatment for a given crop and the amount of soil moisture. Fallow areas merely scraped to control weeds have proved to be slightly less effective in moisture conservation than those that received normal cultivations, probably as a result of increased runoff on the scraped plots.[48]

A loosened soil surface acts as an insulator, i.e., a cultivated soil is slightly cooler than uncultivated soil.[57] Many soils naturally have sufficient aeration for optimum bacterial and chemical activity without cultivation.[70]

INTERTILLAGE OR CULTIVATION

The primitive husbandman hoed or pulled out the weeds that grew among his crops planted at random in his small clearings. In ancient and medieval field husbandry, field crops were planted at random or in close rows with the seed dropped in plow furrows. These crops were later weeded by hand or with crude hoes or knives. Jethro Tull introduced intertillage into English agriculture in 1731 when he applied it to crops like turnips planted in rows.

Purposes of Intertillage

The primary purpose of intertillage is weed control. Intertillage also breaks a crust which otherwise might retard seedling development, and in some cases roughens the soil sufficiently to increase water infiltration. Some have claimed, without substantial proof, that intertillage brings about aeration of the soil, with the result that plant foods are more readily available because of increased bacterial and chemical action in the soil.

The crops that generally require planting in rows with sufficient space between them to permit cultivation during their growth include corn, cotton, grain sorghum, sugarbeets, sugarcane, tobacco, potatoes, field beans, and broomcorn. Intertillage controls weeds that grow in the open spaces before the crop can shade the ground.

Crops with relatively slender stems, such as small grains, hay crops, and flax, that thrive under close plant spacing, cover the land rather uniformly. Without cultivation after they have been sown, they tend to suppress weeds by root competition and shading. Yields of these crops usually are low when they are planted in cultivated rows, because they do not utilize the land fully.[76] Where conditions are so severe that small-grain and hay crops succeed only in cultivated rows they are generally unprofitable.

Intertillage Implements

Many different types of implements are used for intertillage. They vary in size from those that cultivate only one side of one row to those that cultivate several rows at once. They come equipped with shovels, disks, teeth, sweeps, or knives.

ORDINARY CULTIVATORS: Shovel or sweep cultivators are the types most generally used for intertillage, being suited to practically all soil conditions. Cultivators for more than one row should correspond with the planter in width so that they operate over a group of rows that were planted simultaneously. Sweeps of various widths are used for shallow cultivation (Figure 5–8). Shovels or sweeps are sometimes replaced or supplemented by disks or disk hillers when considerable soil is to be moved or where a considerable amount of vines, trash, or roots is to be cut. Shovel cultivators are likely to clog under such conditions. In weedy corn, a cultivator with six pairs of disk hillers and one pair of sweeps has given the best results.[37]

ROTARY HOE: The rotary hoe (Figure 5–9) consists of a series of 18-inch hoe wheels each of which is fitted with teeth shaped like fingers. As the wheels rotate these teeth penetrate and stir the soil. This hoe is useful for uprooting small weeds by *blind* cultivation before the crop is up, as well as for early cultivations of young corn

Figure 5–8. Cultivating six rows of soybeans with sweeps that cut off weeds about 1 inch below the soil surface. [Courtesy International Harvester Co.]

Figure 5–9. Rotary hoe that breaks a soil crust and kills weed seedlings; for use before or soon after crop seedlings emerge. [Courtesy John Deere & Co.]

and other row crops. Its success depends on its use before weeds have made as much growth as the crop; otherwise the crop plants would be uprooted also.[37] It also is effective in breaking crusts that result from hot sunshine after a torrential rain.

OTHER CULTIVATION IMPLEMENTS: The ordinary spike-tooth harrow is used with little injury to the corn to kill small weeds in corn when the corn plants are small and when the field is comparatively free of trash. Special cultivators for listed crops planted in furrows are necessary for following the furrows and turning the soil along their sides. The lister cultivator is equipped with disks or knives, or both, and sometimes with disks and shovels. For the first cultivation, the disks are set to cut the weeds on the sides of the furrow, the soil being thrown away from the crop row. The shovels may be set to stir the soil near the plants. Hooded shields prevent the soil from rolling in on the young plants. For the second and usually final cultivation, the disks are set to roll the soil into the trench around the plants. This buries the weeds in the row while the rows are being leveled. Knife cultivators (go-devils or knife sleds) cut off the weeds below the surface while slicing the lister ridges.

Cultivation for Weed Control

Corn is practically a failure unless weeds are kept under control, as was shown by experiments in Nebraska[48] (Table 5–3).

In the Corn Belt about four cultivations are required for surface-planted corn. Experiments at the Illinois Agricultural Experiment

TABLE 5-3. Effect of Cultivation on Yield of Corn in Nebraska (1922–27)

CULTIVATION	YIELD OF SHELLED CORN PER ACRE	
	(bu.)	(%)
No cultivation (weeds allowed to grow)	7.1	19
1 normal cultivation	21.6	58
2 normal cultivations	33.6	90
3 normal cultivations	35.9	97
4 normal cultivations	37.2	100
4 normal cultivation and 2 continued late	35.2	95
Scraped to control weeds	35.1	94

Station[84] as early as 1888–93 showed that corn yielded as well when the weeds were merely cut off with a hoe as when it received four or five deep cultivations.

These results were not taken seriously because of the widespread firm belief that cultivation was necessary for reasons other than weed control.[83] Later, results were summarized[17] for 125 experiments with corn carried on for six years in 28 states, in which regular cultivation was compared with scraping with a hoe to destroy weeds. The average of all the tests showed that the scraped plots produced 95.1 per cent as much fodder and 99.1 per cent as much grain as did the cultivated plots. Cultivation was not beneficial to the corn except in destruction of weeds. These results were confirmed by additional experiments.[48, 57, 61] Other crops responded in a similar manner.[57] Competition for soil moisture is not the only reason for the low yield of corn in weedy fields, because irrigation of such fields increases the corn yield only slightly.[61]

Cultivation of Drilled Crops

Small-grain fields sometimes have been harrowed in the spring, but in general this has been of no benefit to the grain, nor has it eliminated all the weeds.[65] Cultivation of alfalfa sod with the toothed renovator or disk harrow to destroy grassy weeds, which formerly was widely advocated, was discontinued after it was found that the resulting injury to the alfalfa crowns permitted entrance of the bacterial wilt organism.

STRAW AND OTHER ARTIFICIAL MULCHES

Moisture loss from the surface foot of both irrigated and dryland soils in Utah[34] was retarded by a straw mulch. A depressive effect on nitrate accumulation in soils may occur where straw mulches are heavy enough to suppress weed growth.[1]

Implements with large sweeps or blades leave the stubble and straw on the soil surface to form a partial mulch. This process, called mulch tillage or subsurface tillage,[29, 30, 43] must be repeated during the season to control weeds, because the straw and stubble from a single crop are insufficient to smother all the weeds. However, this partial mulch checks soil blowing, stops runoff, and increases the surface infiltration of rainfall. Soils are cooler under a straw mulch in summer. Mulches, as well as sod and cover crops, reduce the frequency of alternate freezing and thawing in winter. This retards the disintegration of soil aggregates.[77] However, mulch tillage may keep the soil too cool in the spring for favorable germination of early planted corn, sorghum or soybeans. It also fails to control certain weeds and grass especially cheatgrass (*Bromus tectorum*) where herbicides are not fully applicable, unless a machine such as the skew treader or rotary tiller is used to chop through the mulch and cut out the weeds. Damage to corn from mice, slugs, and birds may occur. Special equipment such as coulters, chisels, or scalpers to push residue aside are used on planters for mulch tillage fields. Heavy nitrogen applications are necessary to maintain crop yields and hasten the ultimate decay of the residues.[34, 35, 41, 53]

Zero tillage of sod land involves killing the grass plants with chemicals and using a modified planter that cuts through the sod while planting corn or some other crop. The dead sod mulch conserves some soil moisture, and the yield of corn may equal that following 8-inch plowing and discing.[6] Zero tillage also reduces water runoff and soil erosion. Minimum and zero tillage is practiced on 24 million acres.

Plastic mulches as well as asphalt-coated paper mulches increase the mean temperature but also decrease the temperature range of the soil. This may hasten germination and reduce the deflocculation of the soil due to freezing and thawing. These mulches reduce soil moisture loss and check weed growth but may fail to increase yields.[52, 74] The cost of these mulches limits their use for most field crops.

TILLAGE IN RELATION TO WATER EROSION

Probably 75 per cent of the cultivated land in the United States has a slope greater than 2 per cent on which the wasteful processes of accelerated soil removal and runoff can occur when the land is not managed properly.[4, 5, 30] There are three types of water erosion; sheet, rill, and gully erosion. Gully erosion causes deep channels in the field which interfere with cultivation; rill erosion produces small channels; sheet erosion is the more nearly uniform removal of the topsoil from the entire slope. Severe sheet erosion on cultivated fields leads to gully formation.

Erosion probably is greatest in the southeastern states where the heavy rainfall, rolling topography, unfrozen soil, and predominance of intertilled crops combine to bring about excessive soil washing. The widely spaced crops, such as corn, cotton, tobacco, and soybeans, are most generally associated with erosion.[21, 27, 28, 58] The numerous fibrous roots of sorghum and its more complete drying of the soil may account for the lower erosion on sorghum land than is observed on the land devoted to the taprooted cotton plant.[27] The greatest losses are on bare land,[27] being about twice those on corn land.[56] However, on continuously fallowed land the soil is nearly saturated with moisture.[78] Consequently, excess water is lost by runoff. Land in small-grain crops erodes very little except during periods when the soil is bare or nearly so.[30] Pasture and grass crops in general are very efficient in soil conservation[70] unless they are overgrazed and trampled. The erosion from continuous ungrazed untrampled sod or meadow is almost negligible. Heavily grazed pasture may have more than three times the runoff of a moderately grazed pasture.[78] A three-year corn, wheat, clover-timothy rotation showed only twice as much runoff as did grass land comparable to moderately grazed pasture.

Natural vegetative cover usually prevents or greatly retards erosion.[44] Davis[26] stated: "A field abandoned because of erosion soon shows these efforts of nature to prevent devastation."

The chief methods of retarding soil erosion[11] are (1) reversion of 30 million acres of steep slopes and abandoned cropland to pastures and woodlands, (2) contour tillage and planting on 50 million acres, (3) strip cropping on 23 million acres, (4) terracing one million miles of terraces, (5) cover crop on many million acres, (6) damming of gullies, and (7) minimum tillage as previously described.*

Strip cropping has been practiced in Pennsylvania for generations.[58] The alternate strips of thick-growing crops catch the soil watershed from the areas occupied by cultivated row crops. The land is tilled and crops on sloping land are planted on the contour. A regular rotation may be followed when the strips are of approximately equal width.

The bench terrace long used in Europe, Asia, and the Philippine Islands is not feasible in the United States because of its excessive costs. Terracing became a common practice in the Southeast after the Mangum terrace was devised by P. H. Mangum of Wake Forest, N.C., in 1885. Later many of the terraces were abandoned because of frequent failures of the terraces, high costs of maintenance, and questionable benefits from terracing. Many of these early terraces were built poorly or the land was managed improperly.[65a] When a terrace breaks, erosion losses exceed those on unterraced land. The broad-base terraces permit farm machines to pass over them. The Nichols

* See Summary of activities of the Soil Conservation Service for the fiscal year 1972.

terrace, a broad-channel type, is constructed with less labor. This broad-base terrace[65a] is 15 to 25 feet wide at the base and 15 to 24 inches high. The runoff water is carried away in a gradually sloping broad shallow channel at a low velocity. Level terraces are advised in parts of Texas where the rainfall is less than 30 inches per year, so as to hold all the water on the field until it is absorbed.[27] The conservation bench is even more effective because the water above the levee spreads back over a wide bed. The terracing of steep slopes (more than 8 per cent) no longer is recommended universally because of the expense, because all the topsoil may be scraped up to build the terraces.[21] and because steep terrace banks erode badly. Two million acres of grassed waterways carry surplus water from the terraces and sloping fields.

The nitrogen, phosphorus, calcium, and sulfur in the soil eroded from corn or wheat land may equal or exceed the amounts taken off in the crops.[31] Much of the soil carried by streams comes from stream or gully banks, deserts, and similar areas that have been eroding since time immemorial. Most important of all, it comes from bad lands and breaks. This eroded soil is not chiefly topsoil. A combination of computed estimates of the quantity of suspended soil carried into the sea by rivers of the United States,[5, 26] indicates that the total is about 1,450 pounds per acre per year. Should half of this be topsoil, the equivalent of the total topsoil is washed to the sea every 2,759 years. Thus, only a small part of the erosion losses reported[4, 21] represent soil washed from the continent. Estimates of soil losses sometimes greatly exceed losses actually measured experimentally under the same conditions.[43] Estimates for 1971 suggest 12 tons of soil per acre lost by water erosion on farms.[38]

TILLAGE IN RELATION TO WIND EROSION

Soil blowing has been going on at times in various parts of the Great Plains for at least 25,000 years. It also occurs elsewhere in the United States when a high wind (20 miles per hour or more) strikes a bare, smooth, loose, deflocculated soil.[15, 46] The organic matter and fine soil particles in the topsoil contain much of the nitrogen.[4, 25] They may be carried great distances and then deposited in some humid area where the additional fertility is needed. When accumulated drifts are stirred and allowed to blow back onto eroded cultivated fields where the soil is caught by vegetation, the aggraded soil that results may be very productive.[82] Uniform sand moves readily with wind where the surface is unprotected. Peat land is subject to serious wind damage. Heavy soils also are subject to movement under certain conditions. The ideal structural condition to prevent erosion is coarse aggregations (clods and crumbs) too large to blow, and yet not so large as to interfere with cultivation and plant growth. There is no advantage in further pulverization.

Summer-fallowed land, except when protected by alternate strips of crops, is particularly vulnerable to high winds unless it is cultivated to a minimum necessary to control weed growth, cultivated when the soil is moist, and tilled with tools that do not pulverize the soil but bring clods and plant residues to the surface. Land can be ranked as follows for decreasing vulnerability to blowing: (1) bean fields, (2) corn and sorghum stubble, (3) cornstalks, (4) fallow seeded to winter rye or winter wheat, and (5) small-grain or hay stubble.[9] Small-grain stubble land rarely blows unless the organic debris is very scanty or has been turned under.

Tillage that furrows the ground or protects the surface soil with small clods and trash is especially desirable in wind-erosion control. Furrows should be at right angles to the direction of the wind.[15,20,24] Among the useful implements are the rod weeder, lister, field cultivator, shovel cultivator, and spring-tooth harrow. The one-way disk plow is satisfactory only on land with a heavy stubble, because it leaves the residues mixed with and protruding from the surface soil. Cultivation checks wind erosion only temporarily,[15] but may delay destructive action until rains start a vegetative cover. Then blowing ceases to be a serious problem.

A permanent vegetative cover is the best protection against soil blowing.[20,24] The soil is too dry for grass seedings to become established when soil blowing conditions prevail, and when rainfall is ample, weeds cover the land and protect it from erosion without seeding. During severe droughts Russian thistles are the predominating weed cover in the Great Plains. These break off and blow away during the winter. In moist seasons, weeds such as wild sunflower and pigweed take root and protect the soil while perennial weeds and grasses become established.

Strips of sorghum planted perpendicular to the prevailing wind direction check soil blowing.[20,45] Tree belts check soil blowing and accumulate drifted soil. They are, however, of little economic benefit to field crops under semiarid conditions because they preclude crop growth for 4 rods on each side and seldom grow more than 20 feet high. Tree belts protect farmsteads, gardens, and livestock because they check the wind and catch drifting snow.

REFERENCES

1. Albrecht, W. A. "Nitrate accumulation in soil as influenced by tillage and straw mulch," *J. Am. Soc. Agron.*, 18:841–853. 1926.
2. Allmaras, R. R., and others. "Fall versus spring plowing and related heat

balance in the western Corn Belt," *Minn. Agr. Exp. Sta. Tech. Bul. 283*, pp. 1–22. 1972.
3. Barrows, W. C., and others. *International Conference on Mechanized Dry Land Farming*, John Deere & Co., Moline, Ill., pp. 1–343. 1970.
4. Bennett, H. H. "Cultural changes in soils from the standpoint of erosion," *J. Am. Soc. Agron.*, 23:434–454. 1931.
5. Bennett, H. H., and Lowdermilk, W. C. "General aspects of the soil-erosion problem," in *Soils and Men*, USDA Yearbook, 1938, pp. 581–608.
6. Blevins, R. L., and Cook, D. "No-tillage, its influence on soil moisture and soil temperature," *Ky. Agr. Exp. Sta. Prog. Rpt.* 187, pp. 1–15. 1970.
7. Bracken, A. F., and Stewart, George. "A quarter century of dry farming experiments at Nephi, Utah," *J. Am. Soc. Agron.*, 23:271–279. 1931.
8. Brandon, J. F. "Crop rotation and cultural practices at the Akron Field Station," *USDA Bul. 1304*. 1925.
9. Brandon, J. F., and Kezer, A. "Soil blowing and its control in Colorado," *Colo. Agr. Exp. Sta. Bul. 419*. 1936.
10. Brandon, J. F., and Mathews, O. R. "Dry-land rotation and tillage experiments at the Akron (Colorado) Field Station," *USDA Circ. 700*, pp. 1–53. 1944.
11. Brill, G. D., Slater, C. S., and Broach, V. D. "Conservation methods for soils of the northern Coastal Plain," *USDA Inf. Bul. 271*. 1963.
12. Burr, W. W. "Storage and use of soil moisture," *Nebr. Agr. Exp. Sta. Res. Bul.* 5. 1914.
13. Call, L. E. "The effect of different methods of preparing a seedbed for winter wheat upon yield, soil, moisture, and nitrates," *J. Am. Soc. Agron.*, 6:249–259. 1914.
14. Call, L. E. "The relation of weed growth in nitric nitrogen accumulation in the soil," *J. Am. Soc. Agron.*, 10:35–44. 1918.
15. Call, L. E. "Cultural methods of controlling wind erosion," *J. Am. Soc. Agron.*, 28:193–201. 1936.
16. Call, L. E., and Sewell, M. C. "The soil mulch," *J. Am. Soc. Agron.*, 9:49–61. 1917.
17. Cates, J. S., and Cox, H. R. "The weed factor in the cultivation of corn," *USDA B.P.I. Bul.* 257. 1912.
18. Chilcott, E. C. "The relations between crop yields and precipitation in the Great Plains area," *USDA Misc. Circ. 81*, Supp. I, pp. 1–163. 1931.
19. Chilcott, E. C., and Cole, J. S. "Subsoiling, deep tilling, and soil dynamiting in the Great Plains," *J. Agr. Res.*, 14:481–521. 1918.
20. Chilcott, E. F. "Preventing soil blowing on the southern Great Plains," *USDA Farmers Bul. 1771*. 1937.
21. Clark, M., and Wooley, J. C. "Terracing, an important step in erosion control," *Mo. Agr. Exp. Sta. Bul. 400*. 1938.
22. Cole, J. S., and Hallsted, A. L. "Methods of winter-wheat production at the Fort Hays Branch Station," *USDA Bul. 1094*, pp. 1–31. 1922.
23. Cole, J. S., and Mathews, O. R. "Tillage," in *Soils and Men*, USDA Yearbook, 1938, pp. 321–328.
24. Cole, J. S., and Morgan, G. W. "Implements and methods of tillage to

control soil blowing on the northern Great Plains," *USDA Farmers Bul. 1797.* 1938.

25. Daniel, H. A., and Langham, W. H. "The effect of wind erosion and cultivation on the total nitrogen and organic matter content of soils in the southern High Plains," *J. Am. Soc. Agron.*, 28:587–596. 1936.
26. Davis, R. O. E. "Economic waste from soil erosion." *USDA Yearbook*, 1913, pp. 207–220.
27. Dickson, R. E. "Results and significance of the Spur (Texas) run-off and erosion experiments," *J. Am. Soc. Agron.*, 21:415–422. 1929.
28. Duley, F. L. "Soil erosion of soybean land," *J. Am. Soc. Agron.*, 17:800–803. 1925.
29. Duley, F. L. "Yields in different cropping systems and fertilizer tests under stubble mulching and plowing in eastern Nebraska," *Nebr. Agr. Exp. Sta. Res. Bul. 190.* 1960.
30. Duley, F. L., and Fenster, C. R. "Stubble-mulch farming methods for fallow areas," *Nebr. Agr. Exten. Circ. 54–100* (rev.), pp. 1–19. 1961.
31. Duley, F. L., and Hays, O. E. "The effect of the degree of slope on run-off and soil erosion, *J. Agr. Res.*, 45:349–360. 1932.
32. Duley, F. L., and Miller, M. F. "Erosion and surface run-off under different soil conditions," *Mo. Agr. Exp. Sta. Res. Bul.* 63. 1923.
33. Evans, C. E., and Lemon, E. R. "Conserving soil moisture," in *Soil*, USDA Yearbook, 1957, pp. 340–359.
34. Fenster, C. R., and McCalla, T. M. "Tillage practices in western Nebraska with a wheat-fallow rotation," *Nebr. Agr. Exp. Sta. SB 507*, pp. 1–20. 1970.
35. Fenster, C. R., and McCalla, T. M. "Tillage practices in western Nebraska with a wheat-sorghum-fallow rotation," *Nebr. Agr. Exp. Sta. SB 515*, pp. 1–23. 1971.
36. Gainey, P. L., and Metzler, L. F. "Some factors affecting nitrate nitrogen accumulation in the soil," *J. Agr. Res.*, 11:43–64. 1917.
37. Gray, R. B. "Tillage machinery," in *Soils and Men*, USDA Yearbook, 1938, pp. 329–346.
38. Hargrove, T. R. "Agricultural research impact on environment," *Iowa Agr. Exp. Sta. Special Rpt. 69*, pp. 1–64. 1972.
39. Harris, F. S., Bracken, A. F., and Jensen, I. J. "Sixteen years' dry farm experiments in Utah," *Utah Agr. Expt. Sta. Bul. 175*, p. 43. 1920.
40. Harris, F. S., and Turpin, H. W. "Movement and distribution of moisture in the soil," *J. Agr. Res.*, 10:113–155. 1917.
41. Hayes, W. A. "Mulch tillage in modern farming." *USDA Leaflet 554*, pp. 1–7. 1971.
42. Hobbs, J. A., and others. "Deep tillage effects on soils and crops," *Agron. J.*, 53(5)313–16. 1961.
43. Horning, T. R., and Oveson, M. M. "Stubble Mulching in the Northwest," *USDA Inf. Bul.* 253. 1962.
44. Jacks, G. V., and Whyte, R. O. "Erosion and soil conservation," in *Imperial Bur. Pastures and Forage Crops Bul.* 25, Aberwystwyth, England, pp. 1–206. 1938.
45. Jardine, W. M. "Management of soils to prevent blowing," *J. Am. Soc. Agron.*, 5:213–217. 1913.

46. Kellogg, C. E. "Soil blowing and dust storms," *USDA Misc. Pub. 221.* 1935.
47. Kiesselbach, T. A., and others. "The seedbed factor in winter wheat production," *Nebr. Agr. Exp. Sta. Bul. 223.* 1927.
48. Kiesselbach, T. A., Anderson, A., and Lyness, W. E. "Tillage practices in relation to corn production," *Nebr. Agr. Exp. Sta. Bul. 232.* 1928.
49. King, F. H. *Physics of Agriculture,* 5th ed., F. H. King, Madison, Wis., pp. 158–203. 1910.
50. Lewis, M. R. "Rate of flow of capillary moisture," *USDA Tech. Bul. 579.* 1937.
51. Lyon, T. L. "Inter-tillage of crops and formation of nitrates in the soil," *J. Am. Soc. Agron.,* 14:97–109. 1922.
52. Magruder, Roy. "Paper mulch for the vegetable garden," *Ohio Agr. Exp. Sta. Bul. 447.* 1930.
53. Mannering, J. V., and Burwell, R. E. "Tillage methods to reduce runoff and erosion in the corn belt," *USDA Inf. Bul. 330,* pp. 1–14. 1968.
54. Mathews, O. R., and Chilcott, E. C. "Storage of water by spring wheat," *USDA Bul. 1139.* 1923.
55. Mathews, O. R., and Clark, V. I. "Summer fallow at Ardmore (South Dakota)," *USDA Circ. 213.* 1932.
56. McLaughlin, W. W. "Capillary movement of soil moisture," *USDA Bul. 835.* 1920.
57. Merkle, F. G., and Irwin, C. J. "Some effects of inter-tillage on crops and soils," *Penna. Agr. Exp. Sta. Bul. 272.* 1931.
58. Miller, M. F., "Cropping systems in relation to erosion control," *Mo. Agr. Exp. Sta. Bul. 366.* 1936.
59. Miller, M. F., and Krusekopf, H. H. "The influence of systems of cropping and methods of culture on surface run-off and soil erosion," *Mo. Agr. Exp. Sta. Res. Bul. 177.* 1932.
60. Morgan, G. W. "Experiments with fallow in north-central Montana," *USDA Bul. 1310.* 1925.
61. Mosier, J. G., and Gustafson, A. F. "Soil moisture and tillage for corn," *Ill. Agr. Exp. Sta. Bul. 181,* pp. 563–586. 1915.
62. Musgrave, G. W., and Dunlavy, H. "Some characteristics of an eroded soil," *J. Am. Soc. Agron.,* 23:245–252. 1931.
63. Nelson, A. L. "Method of winter wheat tillage," *Wyo. Agr. Exp. Sta. Bul. 161.* 1929.
64. Noll, C. F. "Deep versus ordinary plowing," *Penna. Agr. Exp. Sta. Ann. Rpt.,* pp. 39–47. 1912–1913.
65. Oveson, M. M., and Hall, W. E. "Longtime tillage experiments on Eastern Oregon wheat land," *Oreg. Agr. Exp. Sta. Tech. Bul. 39.* 1957.
65a. Ramser, C. E. "Farm terracing," *USDA Farmers Bul. 1669.* 1931.
66. Raney, W. A., and Zingg, A. W. "Principles of tillage," in *Soil,* USDA Yearbook, 1957, pp. 277–281.
67. Robins, J. S., and Blakely, B. D. "Management of fallow," in *Power to Produce,* USDA Yearbook, 1960, pp. 142–147.
68. Sachs, W. H. "Effect of cultivation on moisture and nitrate of field soil," *Ark. Agr. Exp. Sta. Bul. 205.* 1926.
69. Sarvis, J. T., and Thysell, J. C. "Crop rotation and tillage experiments

of the northern Great Plains Field Station," *USDA Tech. Bul. 536,* pp. 1–75. 1936.
70. Sewell, M. C. "Tillage: A review of the literature," *J. Am. Soc. Agron.,* 11:269–290. 1919.
71. Sewell, M. C., and Call, L. E. "Tillage investigations relating to wheat production," *Kans. Agr. Exp. Sta. Bul. 18.* 1925.
72. Sewell, M. C., and Gainey, P. L. "Nitrate accumulation under various cultural treatments," *J. Am. Soc. Agron.,* 24:283–289. 1932.
73. Shaw, C. F. "The effect of a paper mulch on soil temperature," *Hilgardia,* 1:341–364. 1926.
74. Shaw, C. F. "When the soil mulch conserves soil moisture," *J. Am. Soc. Agron.,* 21:1165–1171. 1929.
75. Shaw, C. F., and Smith, A. "Maximum height of capillary rise starting with soil at capillary saturation," *Hilgardia,* 2:399–409. 1927.
76. Sheppard, J. H., and Jeffrey, J. A. "A study of methods of cultivation," *N. Dak. Agr. Exp. Sta. Bul. 29.* 1897.
77. Slater, C. S. "Winter aspects of soil structure," *J. Soil and Water Conservation,* 6(1):38–41. 1951.
78. Smith, D. D., and others. "Investigations in erosion control and reclamation of eroded Shelby and related soils at the Conservation Experiment Station, Bethany, Mo., 1930–42," *USDA Tech. Bul. 883,* pp. 1–175. 1945.
79. Thysell, J. C. "Conservation and use of soil moisture at Mandan, N. Dak.," *USDA Tech. Bul. 617,* pp. 1–40. 1938.
80. Veihmeyer, F. J. "Some factors affecting the irrigation requirements of deciduous orchards," *Hilgardia,* 2:125–284. 1927.
81. Von Trebra, R. L., and Wagner, F. A. "Tillage practices for southwestern Kansas," *Kans. Agr. Exp. Sta. Bul. 262,* pp. 1–17. 1932.
82. Whitfield, C. J., and Perrin, J. A. "Sand-dune reclamation in the southern Great Plains," *USDA Farmers Bul. 1825,* pp. 1–13. 1939.
83. Williams, C. G. "The corn crop," *Ohio Agr. Exp. Sta. Bul. 140.* 1903.
84. Wimer, D. C., and Harland, M. B. "The cultivation of corn," *Ill. Agr. Exp. Sta. Bul. 259.* 1925.
85. Wirt, F. A. "Some recent developments in tillage machinery," *Agr. Engr.,* 9:121–123. 1928.
86. Zingg, A. W., and Hauser, V. L. "Terrace benching to save potential runoff for semiarid land," *Agron. J.,* 51:289–292. 1959.
87. Zingg, A. W., and Whitefield, C. J. "A summary of research experience with stubble-mulch farming in the Western States," *USDA Tech. Bul. 1166,* pp. 1–56. 1957.

Chapter 6

Fertilizer, Green Manuring, and Rotation Practices

OBJECT OF FERTILIZATION

Fertilizers are applied to the soil to promote greater plant growth and better crop quality. Farmers endeavor to apply the kind and amount of each fertilizer constituent that gives the most profitable response.[47] A residue not utilized by the crop usually remains in the soil for the following crops. Fertilizer needs are established by field experiments and farm trials for each soil type, environment, crop and crop variety. Fertilizer needs can be estimated by laboratory tests of soils and plant tissues, by pot cultures, and by observation of deficiency symptoms in plants.[70] Fertilizer recommendations based on suitable tests usually are dependable when previous experience with the soil type and crop behavior in the locality is available. Public or private soil testing facilities are available to all farmers. More than a century ago, Justus von Liebig proposed that we should return to the soil that which is removed in crops. This policy is not valid for soils that are especially high in their content of certain of the elements. It also does not apply to semiarid conditions where only light applications of fertilizer give a yield response, and these may even reduce grain yields in a year of subnormal precipitation.[10, 31]

Soluble compounds of nitrogen, potassium, calcium, and magnesium are leached downward into the soil, but they contact water supplies only if they reach a drainage outlet or the underlying water table. Phosphorus compounds become fixed in the upper soil levels and they enter streams or ponds only from fields subjected to water erosion. Very few fertilized fields are as high in fertility as they were in their virgin state. The chemical elements comprising the earth have been entering the seas, streams, and ponds ever since the earth was created.

Excess phosphates and nitrates in standing or slow-moving waters stimulate the growth of algae. This may form a scum on the water surface which limits the oxygen supply and thus restricts the fish population. However, many ponds remote from any agricultural area exhibit this scum formation. More than one million artificial ponds are distributed over the United States. Many of these have been fertilized to stimulate algae growth in order to produce fish. The algae provide food for minute members of the animal kingdom. These are devoured by a series of larger species, which in turn provide food for the fish.

Soil is the cheapest source of most fertilizer elements. High yields of crops remove more fertilizer elements from the soil than do low crop yields, but this fact does not justify crop neglect. The soil depletion that occurs under good management need not be viewed with alarm. Columella wrote in the 1st century A.D.: "The earth neither grows old nor wears out, if it be dunged." This philosophy is confirmed on fields of Europe and Asia that have been farmed for centuries. Fertilizers can restore depleted soils.

Any practice other than fertilizer application that improves total crop yields—even crop rotation, and green manuring—speeds up depletion of the mineral elements of the soil.

The use of improved varieties and other practices that bring profitable crop yields may facilitate soil improvement when they encourage greater use of fertilizers. This has resulted in striking increases in crop yields in the United States since 1940.

The total depth of soil in which annual crop-plant roots customarily feed is 3 to 6 feet. Phosphorus usually is distributed rather uniformly through the surface, subsurface, and subsoil layers. Potash also is distributed rather uniformly throughout the different soil horizons except in sandy soils and under high rainfall conditions, where the potash content even increases with depth. Although nitrates leach downward several feet into the soil, the total nitrogen is mostly in the organic residues of the upper horizons.

Squanto, the first American extension agronomist, advised the Massachusetts colonists to put two dead fish in each hill of corn planted on new land just cleared from the forest. His advice was based upon the known fact that this treatment increased corn yields. Squanto did not know that the fish were rich in nitrogen, phosphorus, and calcium. George Washington found that his land in the coastal plain of northern Virginia would produce two crops before it needed to be fertilized or rested for further cropping to succeed. Forest soils in high rainfall regions on a rolling topography may be expected to benefit from fertilizers immediately or soon after they are cleared and farmed, and forever thereafter. On the other hand, the deep black soil of the level, drier prairie of the Red River Valley showed no decrease in productivity in 60 years of continuous cropping to wheat without fertilizer or green manure.[72] Despite this fact, wheat grown there in rotations or with fertilization outyielded continuous wheat without fertilizer. Crops grown on newly irrigated desert land usually respond to heavy applications of nitrogen.

The heaviest use of fertilizers has been in western Europe, Japan, and Taiwan. The consumption of fertilizers has increased materially in North and South America, Asia, Africa, and the Soviet Union since 1950. The world use of primary nutrients, N, P_2O_5 and K_2O was about 60 million metric tons in 1972, one-fourth of which was applied in the United States. An estimated 15 per cent of the domestic use is not on farms. The consumption of the three primary nutrients increased 15-fold in the United States between 1920 and 1972. The greatest increases were in the Great Plains, the Corn Belt, and the Lake states. The average nutrient content of the 41 million short tons of fertilizer exceeded 40 per cent. About 30,000 short tons of secondary and micronutrients and 25 million tons of liming materials were applied. Fertilizer price increases were much less than those of farm equipment.

Nearly all of the intertilled crops in the irrigated, humid and subhumid areas receive mineral fertilizers except sometimes when they follow legumes or green manuring.

EFFECT OF NUTRIENT ELEMENTS ON PLANT GROWTH

Fertile soils supply 16 chemical elements essential to plant growth. These are nitrogen, phosphorus, potassium, calcium, sulfur, magnesium, boron, manganese, zinc, iron, molybdenum, sodium, copper, chlorine, aluminum, and silicon.[12] At least 6 others—cobalt, arsenic, seleniun, lead, lithium, and vanadium—stimulate certain plants under some conditions. Carbon and additional oxygen are drawn from the air, and oxygen and hydrogen are supplied in water. Most of these elements, in addition to iodine, fluorine, chromium, and tin, are essential to human or animal nutrition.[*, 1, 9] The use of major elements, nitrogen, phosphorus, potassium, and calcium are described later. The minor (secondary or micro) elements—copper, iron, manganese, zinc, cobalt, chlorine, sodium, and molybdenum—are included in mixed fertilizers for certain crops in soils in which these elements are deficient. Magnesium is supplied in dolomitic limestone applications. Cobalt, iodine, selenium, or fluorine may be applied for crops to correct their nutritional balance when used as a livestock feed.[9, 17] Iron and manganese become unavailable in calcareous or alkaline soils, but calcium applications are used to correct manganese toxicity on highly acid soils. Fertilizers containing sodium are beneficial for sugarbeets, rape, and mustard and for several other crops when potassium is deficient. Chlorine containing fertilizers are required for flue-cured tobacco. Silicon supplies are usually ample. Figure 6–1 shows some effects of certain mineral deficiencies in tobacco.

[*] *Science News*, 102(14):216. 1972.

Figure 6–1. Effect of mineral deficiencies on tobacco plant growth: plants grown in complete nutrient solution (*A*), or without nitrogen (*B*), phosphorus (*C*), potassium (*D*), boron (*E*), calcium (*F*), or magnesium (*G*).

Soil-Nitrogen Relations

Except for peat and muck, which are high in organic matter, nitrogen fertilizers cause a response in crops grown on most soils of the humid, subhumid and irrigated areas (Figure 6–2). Nitrogen is absolutely essential to plant growth. Plants grown on soils with sufficient amounts of available nitrogen in the soil make a thrifty, rapid growth with a healthy deep green color. Ample nitrogen has a tendency to encourage stem and leaf development. Deficiency of this element results in plants of poor color, poor quality, and low production. An oversupply of nitrogen in the soil tends to cause lodging, late maturity, poor seed development in some crops, and greater susceptibility to certain diseases. On semiarid lands, surplus available nitrates cause excessive plant growth that exhausts the soil moisture in dry seasons before grain is produced. In humid areas, abundant nitrogen may prevent firing of the leaves during dry periods.

The nitrogen in the soil is derived originally from the air, which contains about 34,000 tons over each acre of land. Aside from being fixed artificially, the gaseous nitrogen is fixed with other elements through the activities of soil bacteria.[23] It is introduced into the soil as organic nitrogen of plants or plant residues. Upon decomposition of the organic matter, some of the nitrogen passes into the air as elemental nitrogen or ammonia, while the part that remains in the soil is converted into ammonia and nitrites, and finally into nitrates.

Ammonia and nitrogen oxides in the air are returned to the soil in

Figure 6–2a. Effect of fertilizers on the growth of oats on a fertile Iowa prairie in the early 1900's. (*1*) no fertilizer, (*2*) 400 pounds per acre of superphosphate, (*3*) 140 pounds of ammonium sulfate, and (*4*) superphosphate and ammonium sulfate.

Figure 6–2b. Effect of fertilizer on the growth of corn in the Willamette Valley of Oregon. (*Left*) fertilized; (*right*) unfertilized.

rain or snow. Ammonia and nitrates are used by plants, but some are lost in the drainage waters. Some of the nitrogen remains in the soil a long time in the organic matter that is not fully broken down. The nitrogen supply of the soil may be maintained by growth of legumes, use of manures, and by addition of nitrogen fertilizers.

ROLE OF BACTERIA IN NITROGEN FIXATION: Bacteria that multiply in nodules on the roots of legumes fix nitrogen from the air into forms that the plant can utilize (Figure 26–3). This is called symbiotic fixation of nitrogen. An average of 40 to 200 pounds of nitrogen per acre is added to the soil by legume bacteria annually if the crop is plowed under.[16, 42] Alfalfa and most clovers fix considerably more nitrogen than do the large-seeded legumes. These Rhizobium bacteria, which require molybdenum, cobalt, and iron in order to function, are most effective in soils low in nitrogen.

The species and specific strain of groups of *Rhizobia* are as follows (also referred to as cross-inoculation groups):[23]

Rhizobia meliloti—alfalfa, other *Medicago* species, sweetclover, and fenugreek

R. trifolii—clovers of the genus *Trifolium,* except *T. ambiguum* (kura clover)

R. leguminosarum—peas of genera *Pisum* and *Lathyrus,* and vetches of the genus, *Vicia*

R. lupini—lupines (genus *Lupinus*), and serradella (*Arnithopus*)

R. phaseoli—beans of the species *Phaseolus vulgaris* and *P. coccineus*

R. japonicum—soybeans

Cowpea group—cowpeas, lespedezas, crotalarias, other species of *Phaseolus,* peanuts, velvetbeans, kudzu, pigeonpea, guar, alyceclover, black gram, hairy indigo, and others.

Specific strains for each species—birdsfoot trefoil, big trefoil, kura clover (*Trifolium ambiguum*) sainfoin, crownvetch, garbanzo, tropical kudzu, sesbania, and others.

A number of leguminous trees also fix nitrogen by specific *Rhizobia* strains. Some legume varieties of one species require different strains or cultures for effective nitrogen fixation. Also strains of bacteria isolated from the same legume differ in nitrogen-fixing effectiveness. Some strains form nodules but fix no nitrogen. The interior of an active nitrogen-fixing nodule is pink or red.[23]

Effective legume inoculation is provided by using the correct, fresh, refrigerated culture applied to the seed along with a little water, milk, or sirup or other adhesive substance, by thorough mixing. The treated seed should be sown in moist soil that is pressed around the seed, within 24 hours if possible. Acid soils should be limed. The culture may not survive in dry soil unless the seed is pelleted. The culture should not be exposed to direct sunlight either. The *Rhizobium* bacteria enter through the root hairs of the germinated seedling and form the nodules in the cortex of roots. Legume plants obtain their nitrogen from the soil when the bacteria are absent.

Some 15 other genera of bacteria, including *Azotobacter* and *Clostridium* are reported to convert nitrogen into organic combinations by a process called nonsymbiotic fixation.[65] A number of genera of blue green algae growing in ponds or flooded rice fields also fix nitrogen. Root nodulation has been observed on several leguminous and nonleguminous trees and plants.

NITROGEN DISTRIBUTION IN SOILS UNDER NATURAL CONDITIONS

Peat soils may contain as much as 200,000 pounds of nitrogen per acre. The nitrogen content is highest in soils of the Prairie and Chernozem types of the central states, where it is estimated to average 16,000 pounds per acre to a depth of 40 inches (Figure 6–3). From this region the amount of nitrogen decreases down to 4,000 pounds per acre eastward, westward, and southward. The distribution of nitrogen in soils is closely related to climatic conditions.[34] In the East the high rainfall leaches out nitrates. In the drier West sparse plant growth retards accumulation of nitrogen-containing humus. The total nitrogen content of the soil in the United States decreases from north to south as temperature increases. For every 10° C rise in mean annual temperature, the average nitrogen content of the soil decreases one-half to two-thirds. The nitrogen in soils ranges from 0.01 to 1.0 percent or more in the United States, while the annual temperature ranges from 32° to 72° F. Because of favorable conditions for its preservation, the nitrogen content of soils in the North may be built up. In the South, high temperatures, which favor decomposition, make it practically impossible to increase, except temporarily, the nitrogen content of soils by green-manuring practices while the land is being cropped.

Under natural soil conditions there exists an equilibrium between the formation of organic matter by vegetation and its decomposition by micro-organisms, the balance being determined primarily by climatic factors. Cultivation depletes both organic matter and nitrogen. In the wheat production areas, from 20 to 40 per cent of the original soil nitrogen was lost in from 20 to 40 years.[35] The end result is a new equilibrium point at a decidedly lower level than the original natural nitrogen content.

The upper 7 inches of semiarid Kansas soil lost an average of 23 per cent of its nitrogen and 36 per cent of the organic carbon in 44 years of various cropping methods. Losses from row-cropping and fallowing greatly exceed those from continuous wheat culture.[30]

Soil-Phosphorus Relations

Adequate amounts of phosphorus in soils favor rapid plant growth and early fruiting or maturity, and often improve the quality of vegetation. Liberal supplies of phosphorus enable late crops to mature in the North before they are injured by early freezes. Soils in much of the United States usually give a response to phosphorus applications (Figure 6–2). An exception is parts of the bluegrass region of Kentucky and Tennessee where the soils were formed from rocks high in

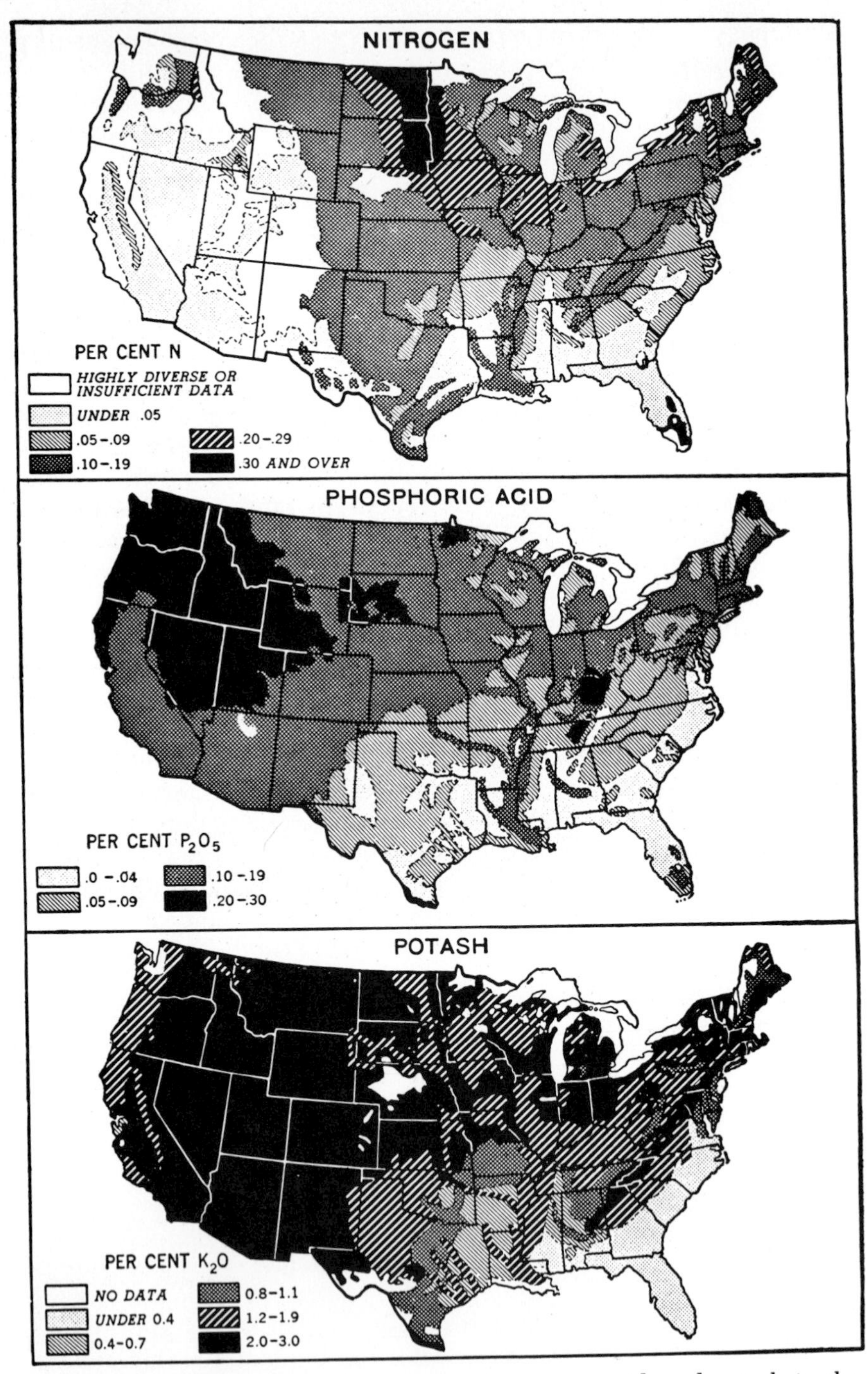

Figure 6–3. Distribution of nitrogen, phosphoric acid, and potash in the surface foot of United States soils.

phosphorus (Figure 6–3). Acid clay soils low in organic matter usually give a greater response to phosphate fertilizers than do other soils of the same region.[51] Crops respond to phosphorus on irrigated lands in the West, particularly after 3 to 30 years of cropping. Calcareous irrigated soils are most likely to respond to phosphorus.[41] Certain western soils are high in both phosphorus and calcium, but the *availability* of phosphorus is low, probably because of the high calcium content.[61]

Soil-Potassium Relations

An adequate supply of potassium in the soil improves the quality of the plant, insures greater efficiency in photosynthesis, increases resistance to certain diseases, helps to balance an oversupply of nitrogen, and aids plants to utilize soil moisture more advantageously. It also insures the development of well-filled kernels and stiff straw in the cereals, encourages growth in leguminous crops, assists in chlorophyll formation, and is particularly helpful in the production of starch- or sugar-forming crops.[15, 17] Potassium is particularly beneficial to tobacco, potatoes, cotton, sugarbeets, and certain cereals. Potatoes grown with too little potassium may be watery, low in starch, and generally poor in quality. With insufficient potassium the leaves of the tobacco plant may have poor color and flavor, and lack burning qualities. Crops usually respond to potash fertilizers in the Atlantic coastal plain, in regions of high rainfall in the eastern third of the United States (except in certain portions of the piedmont), and on sandy, muck, and peat soils of the Great Lakes region. Potash usually is abundant in soils of volcanic origin. The ordinary range of potassium expressed as potash (K_2O) in the plow surface of mineral soils is 0.15 per cent in sands to 4.0 per cent or more in clay soils.[17]

COMMERCIAL FERTILIZERS*

Nitrogenous Fertilizers

1. Nitrate fertilizers are materials with their nitrogen combined in the nitrate form, as in sodium nitrate. Nitrate nitrogen, readily soluble in water, is rapidly utilized by most crop plants. Sodium nitrate tends to make the soil alkaline.

2. Ammonium sulfate, ammonium phosphate, anhydrous ammonia, and aqua ammonia carry nitrogen only in the ammonium form. Ammoniacal nitrogen, although soluble in water, is less readily leached from the soil than nitrate nitrogen. It can be used directly by crops, although it is often converted by bacteria to the nitrate form before

* See Nelson, L. B., "Advances in fertilizers," in *Advances in Agronomy*, Vol. 17, Academic Press, New York, pp. 1–84. 1965.

being utilized by plants. Ammonium sulfate tends to make soils acid because of the presence of the sulfate ion.

The ammoniacal form is better utilized in wet water-logged soils than the nitrate form. Dentrification of the ammoniacal form does not occur under anaerobic conditions, whereas the nitrate form is lost as gaseous nitrogen under the same conditions.

3. Organic materials, such as tankage or castorbean pomace, contain nitrogen in a complex organic form largely insoluble in water. The nitrogen becomes soluble and available to plants after decomposition of the materials.

4. Fertilizers that contain nitrogen in amide form include urea and calcium cyanamide. These simple nonprotein compounds dissolve in water, while the nitrogen usually is converted rapidly to the ammoniacal or nitrate forms by soil bacteria. Dissolved nitrogen is considered to be available to plants.

The principal nitrogen fertilizer materials in the United States, together with their average nitrogen contents in percentages are anhydrous ammonia (82), nitrogen solutions (20–40), ammonium nitrate (33.5), ammonium sulfate (20.5), ammonium phosphates (11–27), aqua ammonia (20.5), urea (45), and sodium nitrate (16).

Phosphate Fertilizers

Phosphorus must be dissolved in the soil solution before it can be taken up by plants. The phosphorus in fertilizers usually is in the form of phosphates, mostly of calcium. Phosphoric acid in phosphates of organic origin is conceded to be more readily available to plants than that in the raw mineral phosphates. The superphosphate industry began in Great Britain in 1843 when it was found that phosphoric acid in rock phosphate could be made available to plants by treatment with sulfuric acid. The principal phosphatic fertilizer materials used in the United States are as follows:

1. Concentrated or triple superphosphate with 23 to 48 per cent phosphoric acid.

2. Ammonium phosphate, chiefly monoammonium phosphate with 11 per cent nitrogen and about 48 per cent phosphoric acid. Since this has a tendency to increase soil acidity, it is sometimes necessary to add lime to the soil.

3. Superphosphate with 16 to 20 per cent phosphoric acid (P_2O_5).

4. Other materials: bone meal, basic slag, and finely ground raw rock phosphate.

When phosphate fertilizer is applied with the soil, it soon reacts to form compounds that are only slightly available. This is known as fixation. Under ordinary conditions, 10 to 20 per cent of the phosphorus in broadcast applications is recovered in the first crop. Some residual phosphorus is recovered in subsequent years. Phosphorus seldom leaches from soils. Accumulations may take place from heavy applications year after year, as in certain tobacco areas. More efficient use of phosphates has been obtained for row crops when the material is distributed in narrow bands at the side of the seed instead of being broadcast. Broadcast phosphates are fixed rapidly because the particles come in contact with a larger amount of soil.

POTASSIUM FERTILIZER MATERIALS

The principal potash fertilizer materials used commercially are potassium chloride with 47 to 61 per cent potash (K_2O), potassium sulfate with 47 to 52 per cent, and the manure salts with 19 to 32 per cent. Much of the potash is used for cotton, potatoes, tobacco, and hay crops. The greatest consumption is in the southern Atlantic Coast and Great Lakes states.

Fertilizer Usage

Commercial fertilizer usage in the United States was about 32,000 tons in 1860; 1,400,000 tons in 1890; 2,730,000 tons in 1900; 8,425,000 tons in 1930; 12,468,000 tons in 1944 and 41 million tons in 1972. The average plant-food content was 14.5 per cent in 1900, 20.6 per cent in 1944, and 41 per cent in 1972. The NPK (Nitrogen-Phosphorus-Potassium) ratio for all mixed or unmixed fertilizers used in the United States is roughly 18.5–12.4–10.3 and 18.5–5.3–8.4 on the elemental basis.

Fertilizer is generally used in humid, subhumid, and irrigated areas (Figure 6–4) but is less frequently used in semi-arid regions. Eighty per cent of the cornfields and practically all of the tobacco, rice, potato, and sugarbeet fields received fertilizers by 1964.[32]

Nitrogen is the chief nutrient applied to the irrigated cornfields of the West.[9] Tobacco land receives over 300 pounds of nutrients per acre. Less nitrogen is used by the flue-cured and Maryland bright tobaccos, but 200 to 250 pounds of nitrogen per acre is applied to the fields of shade-grown cigar tobacco in Connecticut and Massachusetts. Cotton receives an average of about 150 pounds of nutrients per acre. The Mississippi Delta area and the southwestern irrigated cotton areas use mostly nitrogen. The average nutrient application for potatoes is about 275 pounds per acre, nearly 120 pounds of which is potash. Very little potash is applied to potato fields in the intermountain or Pacific states, but the New England potato crop uses more than 200 pounds per acre. Nearly 200 pounds of phosphorus, but also more than 100 pounds of nitrogen per acre, are applied to the potato fields of the Northeast. Dairy farms use less commercial nitro-

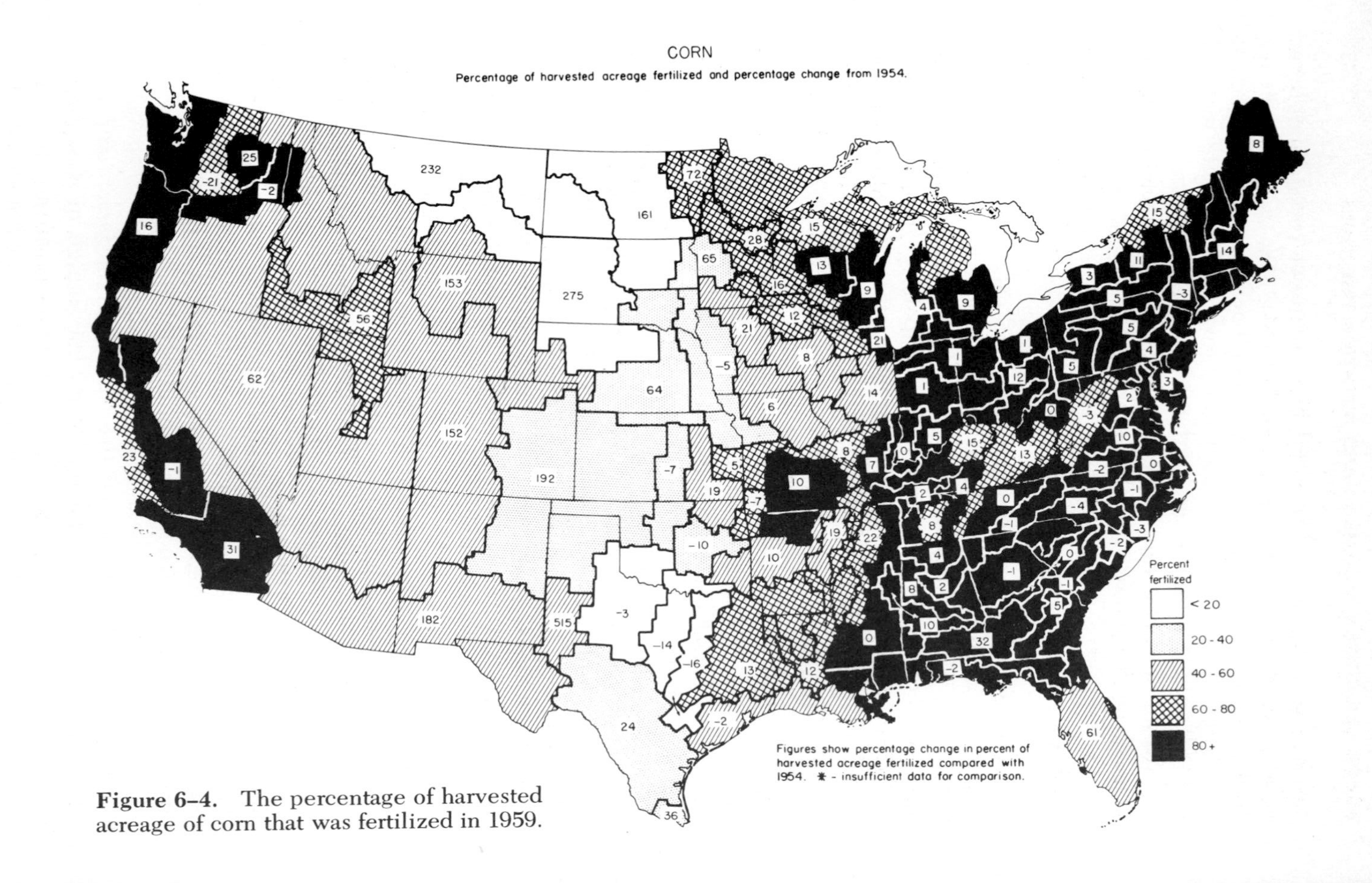

Figure 6–4. The percentage of harvested acreage of corn that was fertilized in 1959.

gen than do other farms because of the abundance of animal manure. Less potash and phosphorus are needed by alluvial soils than by upland soils in the same locality.

Mixed Fertilizers

A mixed fertilizer contains two or more fertilizer elements. About 55 per cent of the fertilizer distributed through commercial channels is a mixture. The cost of application for mixed fertilizers is less than when the various components are applied separately. The fertilizer is sold on the basis of its composition as ascertained by chemical analysis. The fertilizer formula indicates the composition of the mixture in terms of the three important elements often designated as NPK. For example, a 10–10–10 fertilizer contains 10 per cent each of nitrogen (N), phosphoric acid (P_2O_5), and potash (K_2O). The remainder consists of other elements, usually calcium, sulfates, chlorides, inert materials, and often some micro-nutrients. It usually is more economical to purchase high-grade mixtures that contain 30 points (30 per cent) or more of available plant food constituents.

Mixtures with low nitrogen percentages, usually are applied to the soil before or during planting. A top dressing or side dressing with a liquid or granulated nitrogen fertilizer follows later. Winter cereals usually receive a top dressing in the spring. Row crops, such as corn, would receive a side dressing after the plants had made some growth. The 0–20–20 and low nitrogen mixtures are especially useful for legume crops or legume-grass mixtures which require little or no supplementary nitrogen after the plants are established.

The percentages of phosphorus (P) and potassium (K) in the above formulas are actually percentages of phosphorus pentoxide (P_2O_5) and potassium oxide (K_2O). This procedure complies with general fertilizer recommendations as well as with the labeling requirements of state fertilizer laws in effect up to 1964 or later. However, some countries label the phosphorus and potassium content of fertilizers on the elemental basis. This procedure was recommended by the American Society of Agronomy in 1963.

Some companies report the analysis of a fertilizer in both forms. A conversion chart for adjusting the percentages of the two elements is shown in Figure 6–5. Thus, a 10–10–10 fertilizer under the new system would be a 10–4.4–8.3 fertilizer for NPK. About 500,000 tons of organic fertilizers are marketed each year.

Application of Fertilizers

Commercial fertilizer is applied by such methods as (1) injecting liquid materials into soil with machines, (2) broadcasting or drilling dry materials before or after plowing the soil, (3) placing in bands or hills during planting, (4) side dressing between or beside the crop

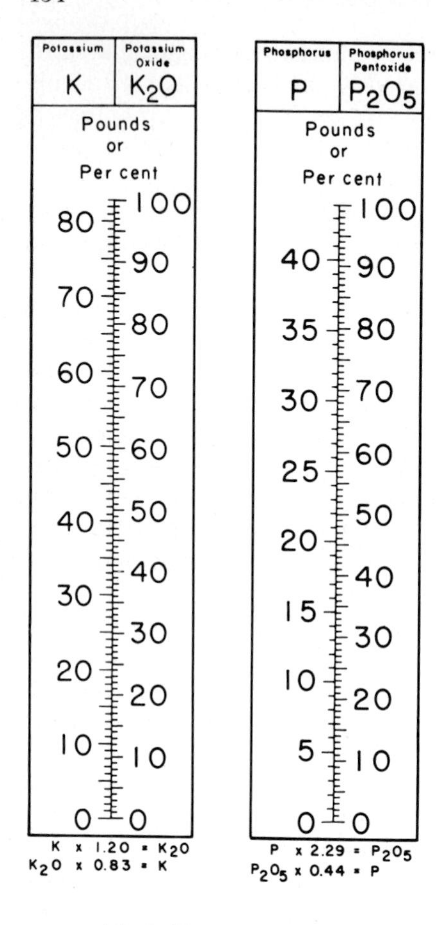

Figure 6–5. Fertilizer conversion scales for phosphorus and potassium.

rows, (5) drilling into slits being opened by deep tillers, or (6) metering liquids into irrigation water being applied with sprinklers or by gravity flow.

Some crops will utilize nitrogen from dilute solutions of urea sprayed on the leaves, but this method of fertilization for field crops is rarely followed. Zinc, copper, iron, and manganese also can be applied by foliar spraying.

Localized applications of fertilizers near the plant favor early, rapid growth. Localized applications also cause a greater response to small amounts of fertilizer. Advantages of localized placement are less marked in soils of high natural fertility or when heavy applications are made. Application of fertilizer at the sides of and below the seed or plant has been most efficient. This superiority has been explained by the tendency of fertilizer salts to move up and down in the soil with only a slight horizontal movement. The fertilizer bands vary from ¾ inch to 2 or more inches in width, the narrower bands being used for relatively light applications of 100 to 200 pounds. The bands are placed 1 to $2\frac{1}{2}$ inches to the side and 2 inches below the seed. The distance between the bands of seed and fertilizer ordinarily should be greater for heavy rates of application on coarse soils and for crops more sensitive to fertilizer injury. Plow-sole applications, made by fertilizing before plowing, place the fertilizer at a depth more accessible to the roots of certain crop plants than is obtained from surface applications after the land is plowed. Placement of fertilizers at still greater depths is seldom advantageous.[36]

FERTILIZER-DISTRIBUTING MACHINERY: Fertilizer-distributing machinery is classified as (1) machines used only for fertilizer application, (2) combination machines for applying fertilizer as well as planting the crop and sometimes applying pesticides in one operation (Figures 7–6 and 36–4), (3) combination machines for applying fertilizer and tilling the soil in one operation, and (4) machines for applying fertilizer in solution with irrigation water. Combination planter-fertilizer distributors for side-band placement of fertilizers are now available.

Some transplanters for setting tobacco carry attachments for applying fertilizer at the side of the plant.

High-pressure tanks and equipment are needed for injection of anhydrous ammonia into the soil.[1] The liquid exerts a gauge pressure of 211 pounds per square inch at a temperature of 104° F., and 126 pounds at 75° F. The tank must be able to withstand a pressure of 250 pounds. The liquid is injected with a field applicator into the soil at a depth of 4 to 6 inches in furrows cut by chisels or knives to which the outlet pipes are attached (Figure 36–14). A press wheel or slide covers the furrow with soil to check the escape of ammonia gas. The soil should be friable as well as moist (but not saturated) to retain the ammonia. The ammonium ion soon becomes attached to the clay minerals in the soil.

Anhydrous ammonia should be applied one or two weeks before a crop is to be planted, in order to avoid injury to the seedlings. Side dressings should be applied between the crop rows. Fall applications to land for a spring crop should be limited to retentive soils in areas of low or medium rainfall with low winter temperatures. Heavy losses of nitrogen are thus avoided.

Injection of aqua ammonia, nitrogen solutions, or mixed solutions requires only low-pressure tanks and equipment. The pressures are 25 pounds or less at 104° F. The low-pressure liquids should be injected at a depth of 2 inches or more, but liquid solutions may be spread on the soil surface. Liquid solutions may contain ammonium nitrate, urea, phosphoric acid, or ammonium phosphates, alone or in mixtures.

Liquid fertilizers are applied with less labor than are dry fertilizers.

Anhydrous ammonia usually is the cheapest source of fertilizer nitrogen. Much of the liquid fertilizer has been applied direct to the field by local dealers on a custom basis. This is more economical than for a small farmer to buy a tank and applicator.

The use of granular or coated granule fertilizers eliminates the dust that accompanies the application of dry powdered fertilizers. Coated granules are less subject to caking than are powdered fertilizer materials, while coating also reduces the fire hazard associated with some of the nitrogen compounds.

LIME AS A SOIL AMENDMENT

Marl, which is high in calcium, was applied to soils by ancient Greeks. About 25 million tons of limestone or equivalent lime materials are applied annually in the United States. Lime usually is regarded as a soil amendment rather than a fertilizer, although calcium, the essential ingredient in lime, may be deficient in some cases. The principal effect of liming in some cowpea-wheat experiments[45] was to increase the amount of soil nitrogen available for crop use. Soil acidity can be corrected by applications of lime. Soils low in lime are frequently acid, a condition that occurs in humid regions where calcium is leached from the soil or removed by plants.[68] Porous sandy soils leach readily and are more likely to be acid than are heavy soils under the same conditions. In South Carolina, approximately 40 per cent of the soils are strongly to extremely acid (*p*H 5.5 or less).[14] This acidity is too high for satisfactory growth of even the somewhat acid-tolerant crops, tobacco and cotton, without liming. In Missouri only small yield increases were obtained by liming soils to raise the *p*H values above 5, 5.5, 5.5 to 6 and 6.0 for cotton, corn, soybeans and alfalfa, respectively.[25]

Forms of Agricultural Lime

The chief forms of agricultural lime are crushed limestone, burned lime, and hydrated lime. Most lime originally comes from limestone rock where it occurs as calcium carbonate ($CaCO_3$), the active element being calcium. Dolomitic limestone contains considerable quantities of magnesium carbonate. The value of crushed limestone depends upon the purity as well as the degree of fineness of the crushed particles. Limestone becomes more effective as the rock is ground finer. A common recommendation is for the material to be sufficiently fine for all of it to pass through a 10-mesh sieve and for 40 per cent of it to pass through a 100-mesh sieve.

Burned lime or quicklime (calcium oxide, CaO) is prepared by burning limestone. Hydrated lime [$Ca(OH)_2$] is formed when water is added to burned lime. Since limestone is only slightly soluble, it

TABLE 6–1. Amount of Finely Ground Limestone Needed to Raise the *p*H from 4.5 to 5.5 and from 5.5 to 6.5 in a 7-inch Layer of Soil

SOIL TYPE	LIMESTONE REQUIREMENTS PER ACRE			
	Cool and Temperate Regions[a]		*Warm and Tropical Regions*[b]	
	pH 4.5–5.5 (tons)	*pH 5.5–6.5 (tons)*	*pH 4.5–5.5 (tons)*	*pH 5.5–6.5 (tons)*
Sand, loamy sand	0.5	0.6	0.3	0.4
Sandy loam	0.8	1.3	0.5	0.7
Loam	1.2	1.7	0.8	1.0
Silt loam	1.5	2.0	1.2	1.4
Clay loam	1.9	2.3	1.5	2.0
Muck	3.8	4.3	3.3	3.8

[a] Gray or brown Podsol soils
[b] Red or yellow soils

reacts slowly over a long period. Because of its lower cost, however, it is the form used for field application in acid soil regions, except where marl or oyster shells are available locally. The relative neutralization values of liming materials, assuming pure calcium carbonate to be 100 per cent, are high-calcium limestone (75 to 95), dolomite limestone (90 to 110), marl (90 to 95), oyster shells (95), hydrated lime (125 to 145), burned lime (150 to 185), and magnesium carbonate (119).

Application of Lime

The limestone requirements to raise the *p*H of soils are shown in Table 6–1.[4] The annual losses of limestone from leaching or crop removal are 100 to 500 pounds per acre in the North Central States, but as much as 50 per cent more in the higher rainfall areas of the Southeast. An application of two tons per acre on average soils should last from 5 to 20 years.

Certain chemical tests may be used as an indication of the lime requirement of a soil. Limestone can be applied at any time of the year and to a crop at any stage without injury. Burned lime should not come in contact with the foliage of plants. Broadcast applications are generally made in amounts as large as 1,000 pounds per acre. The material can be applied with a lime spreader. It should be thoroughly incorporated in the surface soil. A common practice is to plow the soil early, disk, and then apply the lime.

BARNYARD MANURE

Value of Manure

Approximately a billion tons of manure are produced annually by livestock on American farms, but only about one fourth to one third of its potential value is taken up by cropland and pastureland.[59] The remainder is lost by misplacement, drainage, leaching, or fermentation.

Farm manure is a mixture of animal excrements and stable litter. Its value for maintenance of soil productivity has been recognized since very early times. Manure brings about soil improvement because it contains fertilizer materials and possibly certain growth-promoting organic constituents, and because it adds humus to the soil.

Composition of Manure

The composition of manures[58] is given in Table 6–2. Manure exposed to rain may lose a large proportion of the nitrogen, phosphorus, and potash by leaching.[56]

Comparisons of equal weights of fresh and rotted manure as fertilizer invariably favor the rotted product. Manure increases in concentration of nutrients with rotting.[62] This increase is obtained by the loss of more than half the organic matter and a considerable loss of nitrogen.[59] The claim that rotted manure does not *burn* the crop merely reflects the fact that the highly available ammonia nitrogen has been sacrificed in the rotting process.

Application of Manure

Maximum returns from manuring are obtained when certain facts are recognized:[58]

1. Manure is relatively deficient in phosphoric acid, a ton being equivalent to 100 pounds of 3–5–10 or 4–5–10 commercial fertilizer. A phosphorus supplement often is profitable.
2. Returns from a given quantity of manure usually are greater with lighter application to a larger area. The common rate of eight tons per acre is equivalent in value to 1,000 pounds of a 20-unit mixed fertilizer. Residual effects beyond the first year are greater with heavier applications.
3. Because of serious storage losses during the summer months, manure should be applied to spring crops rather than held in storage for fall application.
4. Top dressing may result in improved stands of grass and legume seedlings despite the loss of available nitrogen from the manure.
5. Often a supplementary application of fertilizer in the hill or row is much more economical than the excessive amount of manure that would be required to meet the early demands of the crops equally well.
6. The greatest profit may be expected from applications to crops of high acre value, and to rotations that include such crops as tobacco or potatoes. Corn responds well to manure.

TABLE 6–2. Average Daily Amount and Composition of Solid and Liquid Excrement of Mature Animals

ANIMAL	DAILY PRODUCTION PER ANIMAL		COMPOSITION OF THE FRESH EXCREMENT									
			Dry Matter		*Nitrogen*		*Phosphoric Acid*		*Potash*		*Lime*	
	Solid (*lb.*)	*Liquid* (*lb.*)	*Solid* (%)	*Liquid* (%)	*Solid* (%)	*Liquid* (%)	*Solid* (%)	*Liquid* (%)	*Solid* (%)	*Liquid* (%)	*Solid* (%)	*Liquid* (%)
Horses	35.5	8.0	24.3	9.9	0.50	1.20	0.30	Trace	0.24	1.50	0.15	0.45
Cattle	52.0	20.0	16.2	6.2	0.32	0.95	0.21	0.03	0.16	0.95	0.34	0.01
Sheep	2.5	1.5	34.5	12.8	0.65	1.68	0.46	0.03	0.23	2.10	0.46	0.16
Hogs	6.0	3.5	18.0	3.3	0.60	0.30	0.46	0.12	0.44	1.00	0.09	0.00
Hens	0.1	–	35.0	–	1.00	–	0.80	–	0.40	–	–	–

7. The returns from a ton of manure applied to land that is low in fertility may be expected to be greater than from an equal quantity applied to land already highly productive.

8. It sometimes is more economical to purchase and apply commercial fertilizers than to hire labor to load, haul, and spread barnyard manure and then later apply the necessary supplemental mineral fertilizers.

9. Manuring of semiarid land usually results in an increase in average total crop yield. However, in grain yield, manure usually produces increases only in wet years, with decreases in dry years and little or no benefit in average seasons.[13] Thus in dryland regions, manure should be applied to fields to be planted to forage crops such as sorghum.

GREEN MANURE

Purpose of Green Manuring

A green-manure crop is grown to be turned under for soil improvement while in a succulent condition. A cover crop is one used to cover and protect the soil surface, especially during the winter. Cover crops are commonly turned under for green manure. Green manures are used primarily to increase the yield of subsequent crops as well as to improve the friability of the soil. This is brought about by (1) an increase in the content of organic matter in the soil to counterbalance losses of organic matter through cultivation, (2) prevention of leaching of plant nutrients from the soil during periods between regular crops, (3) an increase in the supply of combined nitrogen in the soil when leguminous plants are turned under, and (4) mobilizing mineral elements.

Crops such as buckwheat, rye, lespedeza, and sweetclover are able to grow fairly well on poor soils. When plowed under to decay, they may release nutrients that were unavailable to other crops.

The use of rye for green manure on very sandy soils being brought under irrigation adds organic matter that helps hold moisture and retard sand drifting while alfalfa is being established. Heavy green manuring helps cotton as well as sweetpotatoes to overcome injury from the *Phymatotrichum* root-rot disease.

Turning under a leguminous green manure may be expected to add from 50 to 200 pounds per acre of nitrogen, but neither legumes nor nonlegumes add other mineral elements to the soil. When either type of green manure increases crop yields it also hastens depletion of phosphorus, potash, calcium, magnesium, and sulfur in the soil. A nonlegume speeds up nitrogen depletion also.

Green manuring was practiced by the Greeks and Chinese before the Christian Era. Lupines, peas, vetch, lentils, and weeds were being turned under as green manures 2,000 years ago.[52] Leguminous green-manure crops, as well as buckwheat, oats, and rye, were used by the American colonists.

The use of cover crops is most widespread along the Atlantic seaboard as well as in the southern states where cover crops are planted especially to check soil erosion, runoff, and leaching. They are finally turned under for green manures. Their value for soil improvement from the standpoint of yields of subsequent crops appears to be greatest in this general region.[52, 55] Some 3 million acres of cover crops are planted annually in the United States.

Crops for Green Manures

The most important green-manure crops in this country are hairy vetch, rye, crimson clover, lupines, sweetclover, and alfalfa.[52] Buckwheat and other crops are used in limited areas.

Along the Atlantic seaboard, crimson clover, vetch, and burclover are being used, especially from Virginia southward to South Carolina. Potato growers in New Jersey use rye as a green-manure crop and depend upon commercial fertilizers for their nitrogen supply.

In the Cotton Belt, vetch has been popular because of its winter hardiness. Other crops used in this region include crotalaria, lespedeza, cowpeas, soybeans, lupines, kudzu, and velvetbeans. Lespedeza as a cover crop has been effective in reducing erosion in North Carolina.

In the Corn Belt, clover, sweetclover, and alfalfa have been the chief green-manure crops. Clover is superior to soybeans for addition of nitrogen.[54]

Rye, field peas, sweetclover, and cowpeas have been tried as green-manure crops in the semiarid regions of the western states. Their use is seldom beneficial in such regions.[11]

Sesbania is a popular green-manure crop on irrigated lands of the Southwest (Figure 6–6).

In the humid or irrigated regions of the Pacific Coast states, vetches, Ladino clover, burclover, sourclover, field peas, rye, and other crops are used for green-manure and winter-cover crops.

On a dry-matter basis, Austrian winter peas, hairy vetch, and alfalfa green manures contain 3 to 4 per cent nitrogen, other legumes 2 to 3.5 per cent, and cereal plants and ryegrass 1.2 to 1.4 per cent.[37] The average availability of nitrogen in green-manure material harvested, applied to other lands, and turned under, is about 50 to 80 per cent.

Decomposition of Green Manures

The green manure must be decomposed before its nutrients become available for plant growth or the organic residues become a

Figure 6–6. Plowing under sesbania for green manure on irrigated land of the lower Colorado River basin.

part of the soil humus.[71] Young plants and substances high in nitrogen decompose most rapidly. As decomposition progresses, it becomes slower because of the comparatively greater resistance of the residual organic matter to decay. Green manures furnish energy to the micro-organisms that bring about decomposition. The water-soluble constituents, largely sugars, organic acids, alcohols, glucosides, starches, and amino acids, are decomposed most rapidly and completely. The free-living, nitrogen-fixing bacteria in the soil use most of these substances as a source of energy. Green manures may contain 20 to 40 per cent of the total dry matter in water-soluble form. The micro-organisms that decompose hemicelluloses and celluloses take up nitrogen from the soil in the process. The proteins decompose quickly, but the nitrogen liberated is immediately assimilated by micro-organisms that attack the celluloses and hemicelluloses. The lignins are very resistant to decomposition and usually add to the soil humus.

Postponement of turning under fall cover crops until spring, rather than in late fall, delays decomposition, conserves moisture, and retards leaching, runoff, and erosion.[55] Decomposition of organic matter in soil continues until the carbon-nitrogen ratio drops from perhaps 50:1 or 20:1 down to about 11:1, the humus stage. The decomposition of organic matter by micro-organisms soon brings about the release of gums, slimes, and other products when the organisms die. These, or other materials from decaying organic matter, leach down in the soil where they increase soil aggregation in clays and loams, possibly by cementing soil particles together. Thus, the soil structure is improved. Proteinaceous organic materials react with clay minerals, especially montmorillonite and illite, and are retained in the soil. Montmorillonite and illite clay crystals are composed of silica and alumina sheets bonded together in a 2:1 ratio. In montmorillonite the entire surface of a crystal unit is accessible for surface reactions. In illite parts of the space between units is accessible. The nonswelling kaolinite clays that predominate in the southeastern states have a low exchange capacity and do not retain much carbon.

Organic matter increases the water- as well as the mineral-holding capacity of sandy soils.

Effects on Subsequent Crop Yields

In the Great Plains, because of moisture shortages, as good or better crop yields are obtained after fallow as after green manure, in experiments conducted for many years.[13, 26] There, the less growth the green-manure crop makes before it is turned under, the less moisture it uses up, and the greater are the yields of the subsequent crop. Where moisture is the chief limiting factor in crop yields, green manuring may be expected to remain unprofitable as long as the organic matter in the soil continues at the prevalent general level. However, green manures are beneficial under irrigation, e.g., sweetclover turned under has increased yields of sugarbeets.

In Georgia and in Louisiana, increases in cotton yields after legumes were turned under ranged from approximately 22 to 100 per cent. Winter legumes plowed under generally increase corn yields from 24 to 78 per cent in Georgia, Mississippi, South Carolina, and Virginia. Similar results have been obtained in the Corn Belt when sweetclover for green manure precedes corn. Some data from Georgia[39] are given in Table 6–3.

Turning under summer legumes also has increased yields. In

TABLE 6–3. Influence of Winter Green-Manure Crops on Yield of Cotton and Corn in Georgia, 1926–37[a]

GREEN MANURE	AVERAGE YIELD PER ACRE	
	Corn (*bu.*)	*Cotton* (*lb.*)
Monantha vetch	45.8	1,286
Hairy vetch	45.7	1,266
Austrian peas	50.3	1,351
Abruzzi rye	30.6	1,079
None	31.2	772

[a] No supplemental nitrogen fertilizers added.

Alabama, corn after velvetbeans plowed under, without addition of phosphate, yielded 58 per cent more than without velvetbeans but with phosphate. The increased yields of corn in Arkansas after turning under the following legumes were cowpeas, 62 per cent; soybeans, 7 per cent; and velvetbeans, 27 per cent. When these legumes were cut for hay, with only the stubble turned under, the increased corn yields were 30, 14, and 24 per cent, respectively. In Tennessee,[45] wheat yielded 8 bushels more per acre as a 20-year average where cowpeas were turned under than where the land was similarly fertilized but the cowpeas removed. Also in Tennessee, the yield of corn was increased 52 to 151 per cent by sweetclover, and 27 to 132 per cent by lespedeza green manure. Corn after lespedeza in North Carolina gave increased yields of 74 to 310 per cent in various trials.

The turning under of rye has often resulted in decreased yield of the unfertilized subsequent cotton or corn crop because of the temporary exhaustion of the available soil nitrogen. The depressing effect is increased as the rye approaches maturity before it is turned under.

In a Georgia experiment, rye turned under March 1 increased the yield of the subsequent cotton crop by 39 per cent, while the increase after Austrian winter peas was 76 per cent. In the same test, rye turned under March 15 depressed the yield of corn grown with nitrogen fertilization by 1 per cent, and without the nitrogen by 6 per cent. The yields of cotton were depressed by 12 per cent in Louisiana when oats were turned under March 20, and by 8 per cent by rye similarly treated. Rye 10 to 14 inches high contained 2.5 per cent total nitrogen, while mature plants contained only 0.24 per cent under the same conditions.

Immediate benefits from a green manure usually are proportional to its nitrogen content. Turning under a good legume growth has increased crop yields by as much as have applications of 30 to 80 pounds of nitrogen per acre. Crops after rye or other nonleguminous green manures need nearly as much nitrogen fertilizers as without the green manure.

Utilization of Green Manures

A crop is seldom economically feasible for green manure when it requires the entire crop season for its growth, although velvetbeans grown on sandy soils in the South may be an exception. The most desirable green-manure crop occupies the land for a part of the season without interference with the regular crops in the rotation. Winter cover crops serve effectively as green manures in the South, while catch crops are often used for this purpose in the North. Legumes grown with grain crops can be turned under in the fall.

In the South, when a legume is plowed under in the fall, it should be followed by a winter cover crop to hold the nutrients released by the decomposition of the summer green manure until the crop planted next season can utilize them. On the Norfolk coarse sands of South Carolina, the subsequent yields of both cotton and corn were improved when summer leguminous green manures were followed by winter rye.[42]

When large quantities of green materials are turned under, some time should elapse before a subsequent crop is planted, in order to avoid seedling injury from the decomposition products. In the South, a green manure should be plowed under about two weeks before corn is planted, and three weeks before cotton.[52]

Heavy nitrogen applications (150 pounds per acre or more) to corn land, where all residues are turned under, produce high yields of corn grown continuously without green manures.

ORGANIC FARMING

Organic farming was devised by our aboriginal ancestors. With a few exceptions, such as applying ashes (potassium nitrate) or marl to the soil, it continued in vogue until the 19th century. Fertilizers and pesticides came into use following remarkable scientific discoveries. The heaviest use of these materials is in advanced countries with high rates of literacy and longevity, and the absence of famines. It is estimated that a complete world reversion to organic farming would solve the population explosion problem because hundreds of millions of people would die from starvation each year.[43, 76]

Plant root hairs absorb ions of mineral nutrient elements from the soil. The root hairs absorb the ions nonpreferentially regardless of whether they come from the soil minerals or mineral fertilizers, or from putrid decayed organic matter. Soil productivity might be maintained by heavy applications of organic matter, but a much larger area of land would be required to produce the organic material needed by the cropped fields. The fire-culture system practiced in primitive localities leaves a field idle for 3 to 7 years between croppings.

All pesticides are purposely toxic to some forms of plant or animal life. Most of them are dangerous to humans when absorbed to excess as a result of misuse or accidents. Several elements essential to human life or health, such as iodine, chlorine, selenium, fluorine and copper, may be fatal if taken to excess or in a wrong chemical compound. Vehicles, machines, tools, electricity, and water are fatal when used improperly or when accidents occur. DDT is thought to be destructive to bird life but it has saved millions of humans from malaria and typhus diseases. The repression and reoccurrence of malaria in Ceylon is seen as associated with the level of DDT usage. The question agronomists must face is how to increase food production without fertilizers and pesticides. Until we can accomplish this challenge a starving world dictates continued intelligent use of agricultural chemicals. Mercurial pesticides are banned because some ignorant people disregarded labeled directions for proper utilization and used mercury-treated seed grain for food or animal feed. Mercury levels in sea foods are the same as they were 70 years ago. These data are substantiated by analysis of museum specimens collected 70 to 100 years ago.

In the United States the use of agricultural chemicals expanded rapidly in the past three decades. During this period the death rate decreased, and the younger generations grew larger and matured earlier than did their parents.

CROP ROTATION

Principles of Crop Rotation

Crop rotation may be defined as a system of growing different kinds of crops in recurrent succession on the same land.[22, 36, 38, 66, 73] A rotation may be good or bad as measured by its effects on soil productivity or on its economic returns. A good rotation that provides for maintenance or improvement of soil productivity usually includes a legume crop to promote fixation of nitrogen, a grass or legume sod crop for maintenance of humus, a cultivated or intertilled crop for weed control, and fertilizers. Perennial legumes and grasses may leave 2 to 3 tons of dry weight per acre of roots residues in the soil when plowed down.

Modern crop rotation was established about the year 1730 in England. The famous Norfolk four-year rotation consisted of turnips, barley, clover, and wheat. On a particular field turnips would be grown the first year, barley the second, clover the third, and wheat the fourth. The wheat was followed by turnips in the fifth year to repeat the rotation. In some rotations a crop may occupy the land two or more years.

The Rothamsted (England) experiments included rotations that were continued for more than 100 years. In the United States, investigations were conducted in Pennsylvania, Ohio, Illinois, and Missouri for many years.

Factors That Affect Crop Rotations

The choice of a rotation for a particular farm depends upon the crops adapted to the particular soil, climate, and economic conditions. In addition, weeds, plant diseases, and insect pests may limit the kinds of crops to be grown in a locality.

Rotations, except in the case of an alternate grain-fallow system, provide some diversification of crops. Diversification may assure more economical use of irrigation water as well as other facilities. The risk of complete failure due to weather, pests, and low prices is less with several crops than with one. Crops may be selected so as to spread labor throughout the year. Seasonal labor requirements conflict with certain crops such as alfalfa, corn, and winter wheat in Kansas and Nebraska.

Maintenance of Soil Productivity

Loss of organic matter as a result of continuous growth of the same crop has a bad effect on tilth. Growth of grass, pasture, and deep-rooted legume crops in rotation tends to correct this condition through maintenance of organic matter. Well-arranged systems of crop rotation make practicable the application of manure and fertilizers to the most responsive crops or to those with high cash value. The alternation of deep- and shallow-rooted crops prevents continuous absorption of plant nutrients from the same root zone year after year. Deep-rooted plants like alfalfa improve the physical condition of the subsoil when the underground parts decay.[38]

The nitrogen requirements of nonleguminous crops may be provided by legumes in the rotation, but rotations cannot supply other plant nutrients in which the soil may be deficient. The production of larger crops, made possible by rotation, depletes the soil more rapidly than does continuous cropping. These larger yields cannot continue indefinitely without application of manures and fertilizers. In a corn-wheat-clover rotation in Indiana, the yields of unfertilized plots dropped one third in ten years.[74]

Legumes in Rotations

For satisfactory growth of legumes, deficient soils may require application of minerals such as lime and phosphorus. Some experiments in Pennsylvania indicate that crop rotation is unable to maintain yields at a high level unless the soil is fertile enough to maintain production of clover also. When legumes such as soybeans and lespedeza are harvested from fertile land, more nitrogen may be taken from the soil than is added to it by the legume.

On poor lands in the south, rotations without a legume still produce low crop yields because of nitrogen deficiency. In an unfertilized cotton-oats-corn rotation, the acre yields were[26] cotton, 168 pounds; oats, 6.9 bushels; and corn, 11.3 bushels. When legumes were included as cover and green-manure crops, the yields were cotton, 706 pounds; oats, 9.0 bushels; and corn, 22.7 bushels.

Legumes are more efficient in fixation of nitrogen on soils with low rather than high nitrogen content because they obtain nitrogen from the air only to the extent that the supply in the soil is insufficient. For this reason, a legume will be more effective as a nitrogen gatherer when two or more crops come between applications of barnyard manure. It is usually customary to grow legume crops previous to crops that require large amounts of nitrogen. For example, the dark tobaccos in Ohio and Wisconsin are grown on heavily manured clover sods.

Increased yields of crops that follow alfalfa may be due chiefly to the addition of nitrogen to the soil contributed by the alfalfa crop.[27]

Crop Sequences

The preceding crop has an important influence on crop yields. A Rhode Island experiment was designed to study crop sequence in which 16 different crops were grown for two seasons, followed the third year with one of the crops grown over the entire area.[28] Alsike clover gave the lowest yields after clover and carrots, and the highest yields after rye and redtop. The yields probably were influenced by soil acidity.

Growth of crops in descending order of their lime requirements sometimes is advisable.[8] For example, for a soil heavy limed for alfalfa, clover, or soybeans the sequence might be (1) alfalfa, sugarbeets, barley; (2) red clover, tobacco, wheat; (3) alsike clover, corn, oats; or (4) soybeans, potatoes, rye.

The amount of nitrogen left in the soil by a crop may influence the yield of the crop that follows it. In a West Virginia experiment[20] the yields of wheat, oats, and corn were higher after soybeans harvested for hay than the yields after oats harvested for grain.

Corn that follows deep-rooted legumes, such as kudzu or sweetclover, may yield more as a result of better root penetration as well as from the nitrogen residues. Grasses in a rotation improve the soil tilth after it has been impaired by excessive tillage and compaction incident to the growing of an intertilled crop like corn. Heavy fertilization often may restore most of the productivity lost by soil depletion.

Crop sequences are very important under dryland conditions because of the differences in residual soil moisture left by different crops, as well as the length of the fallow period for moisture storage between crops. Thus small grains yield more after corn than after small grains or sorghum in the Great Plains region, because corn leaves more moisture in the soil.[14] Most dryland crops yield poorly after alfalfa because the excessive nitrogen and depleted soil moisture cause the crops to burn except in wet seasons.

Relation of Rotation to Pest Control

Crop rotation aids in control of many plant diseases. Certain parasites that live over in the soil tend to accumulate when a susceptible crop is grown year after year. Finally the disease may become too severe for profitable crop production. Rotation is particularly effective in the control of this group of parasites.

Many insects are destructive to only one kind of crop. The life cycle is broken when crops are grown that are unfavorable to the development of the insect pest. The sugarbeet nematode is partly controlled by a rotation in which sugarbeets are grown on the land only once in four or five years. Cotton root-knot can be reduced by the growth of immune crops in the crop sequence.

Crop Rotation and Weed Control

Crop rotation is the most effective practical method for control of many farm weeds. Some weed species are particularly adapted to cultivated crops, others to small grains, while another group thrives in meadows. The continuous growth of small grains on the same land encourages weeds. In Utah, plots continuously cropped to wheat for seven years became so infested with wild oats as to reduce the yield seriously.[64] The chief practical difficulty affecting continuous growth of wheat on Broadbalk field at Rothamsted, England, has been the weed problem.[57]

Except for some 30 noxious species, weeds, of which there are more than 1,200 species, are not able to thrive indefinitely on crop-rotated land.[38] Annual weeds are restricted by a rotation that includes a small-grain crop, a cultivated crop, and a meadow crop. Rotations that include smother crops such as alfalfa, rye, buckwheat, sorghum hay, and sudangrass offer another means of controlling certain weeds.

Cropping Systems and Erosion Control

In Missouri a corn-small grain-clover rotation was effective in reducing erosion in comparison with continuous corn.[44] A one-year rotation of lespedeza and winter grain also was effective. Grass-legume mixtures in rotations have been very effective in the reduction of erosion.[22]

Crop Rotation Compared with Continuous Culture

Crop rotation alone may be 75 per cent as effective as fertilizers for increasing crop yields as a whole, or 90 per cent when only the results for corn, wheat, and oats are averaged.[73] The long-time experiments at the Rothamsted Experimental Station in England indicate the trend in yields when a crop is grown in rotation or in continuous culture. Some results with wheat are shown in Table 6–4.

It is obvious that the highest yield occurred where both crop rotation and fertilization were practiced. Similar results were obtained with corn in the Morrow plots in Illinois, operated since 1888.[4, 18]

TABLE 6–4. Yields of Wheat at Rothamsted from 1851–1919

TREATMENT	AVERAGE YIELD PER ACRE (*bu.*)
Continuous wheat unfertilized	12.33
Continuous wheat fertilized	23.58
Wheat in a four-year rotation without fertilizer (Turnips, barley, clover, wheat)	24.05
Wheat in a four-year rotation fertilized	32.49

Rotations in Practice

CORN BELT: Crop rotations that supply organic matter and nitrogen are built around corn as the major crop.[38] Continuous corn has resulted in reduced yields where it has been practiced for any length of time. The corn-oats-clover rotation is used in the northern part of the region. A four-year rotation of corn, oats, wheat, and alfalfa or clover or a grass-legume mixture is followed in the northeastern part of the Corn Belt where corn is a less important crop. A rotation of corn, wheat, and hay is widely used in central Indiana. The grass hay crops in the Corn Belt are mostly smooth bromegrass and orchard grass.

Where soybeans are grown, a five-year rotation of corn, corn, soybeans, wheat, and hay is popular.[74] Yields can be maintained when cover crops can be seeded in both corn crops and when manure and phosphate can be applied to the corn.

In more recent years a corn-soybean rotation with corn for 2 years followed by 1 to 2 years of soybeans has become increasingly popular on better lands of less than 3–5 per cent slope.

COTTON BELT: Definite crop sequences are less common in the Cotton Belt than in many other regions.[26] On relatively productive soils, a typical cropping system appears to be one year of corn or an annual hay crop followed by two to four years of cotton. Interplanted legumes are used in this sequence in some instances. Two suggested rotations are as follows:

1. Three-year rotation for land equally adapted to corn and cotton: cotton; summer legumes (cowpeas or soybeans) for hay or seed, followed by winter legumes; and corn interplanted with cowpeas, soybeans, or velvetbeans.
2. Four-year rotation for areas where half of the cropland is planted to cotton: cotton, followed in part by a winter legume; cotton; summer legumes for hay or seed followed by winter legumes; corn interplanted with summer legumes.

HAY-PASTURE REGION: Hay and pastures predominate in the northeastern states. A three-year rotation of corn for silage, oats, and hay is popular on dairy farms of the Great Lakes states where considerable land is available for pasture. Potatoes are sometimes substituted for corn as an intertilled crop. Grass-legume mixtures may be sown with spring small grains and kept two years for hay.[75] A large percentage of the cropland is devoted to hay on many New England farms.

WHEAT REGIONS: In the eastern central area from Missouri to Virginia, various modifications of the corn, wheat, clover, or grass rotation are used. In the northeastern part of the winter-wheat area, a rotation that can be used is corn for two years; oats or barley, one year; and wheat, two years. Some hay crops may be included as a modification of this sequence. Popular hay crops are clover, alfalfa, lespedeza, tall fescue, and orchardgrass. In the spring-wheat area of the Red River Valley, a rotation of small grains for two years, alfalfa for two years, and corn or potatoes or sugarbeets for one year is often approximated. Definite crop rotations are not prevalent in the major wheat-production areas.

In the Pacific Northwest, dryland wheat has been produced under a wheat-summer fallow rotation. Winter or spring wheat is the most important cash crop of the Great Plains and intermountain dryland areas. The rotations in this area are characterized by summer fallow periods and absence of sod crops. In the southern Great Plains a summer fallow-wheat-grain-sorghum rotation may be adapted. Alternate wheat and fallow is a common cropping system in the drier areas (12–18 inches of precipitation per year), while alternate corn and wheat is a practical sequence with slightly more precipitation. Continuous cropping has given satisfactory results in many parts of the semiarid region.

IRRIGATED REGIONS: In Utah, a crop rotation that included alfalfa and cultivated crops with rather heavy applications of manure effectively maintained high yields of wheat, potatoes, sugarbeets, field peas, oats, and alfalfa.[64] A satisfactory irrigated rotation for western Nebraska[29] is alfalfa (three years), potatoes, sugarbeets (manured), sugarbeets, and oats.

MULTIPLE CROPPING

Multiple cropping means growing and harvesting two or sometimes three or four crops from the same land in one year. This is possible in the warm climate areas of all continents. Two crops of rice a year are frequently harvested from fields in Southeast Asia. In Taiwan two rice crops with two intervening green vegetable crops have been grown where the rice is transplanted. In a trial in the Philippines four transplanted crops of early rice varieties were harvested from one field in a year. In southeastern, southwestern, and southern Corn Belt regions of the United States other crops such as sorghum or soybeans can be planted after a small-grain crop is harvested in May or early June and reach maturity in the late autumn.

REFERENCES

1. Adams, J. R., Anderson, M. S., and Hulburt, W. C. "Liquid nitrogen fertilizers," *USDA Handbook 198*. 1961.
2. Adams, J. R., and others. "Liquid nitrogen fertilizer for direct application," *USDA Handbook 198,* pp. 1–47. 1965.
3. Allaway, W. H. "*p*H, soil acidity, and plant growth," in *Soil,* USDA Yearbook, 1957, pp. 67–71.
4. Allaway, W. H. "Cropping systems and soil," in *Soil,* USDA Yearbook, 1957, pp. 386–395.
5. Anonymous. "Growing summer cover crops," *USDA Farmers Bul. 2182,* pp. 1–16. 1967 Rev.
6. Anonymous. "Superphosphate: its history, chemistry and manufacture," USDA and Tennessee Valley Authority, Gov't Printing Office. 1964.
7. Anonymous. "Fluorine and fertility," *Science Newsletter,* 102(14), p. 216. 1972.
8. Bear, F. E. "Some principles involved in crop sequence," *J. Am. Soc. Agron.,* 19:527–534. 1927.
9. Beeson, K. C. "Soil management and crop quality," in *Soil,* USDA Yearbook, 1957, pp. 258–268.
10. Brengle, K. G., and Greb, B. W. "The use of commercial fertilizers with dryland crops in Colorado," *Colo. Agr. Exp. Sta. Bul. 516–S.* 1963.
11. Brown, P. L. "Legumes and grasses in dryland cropping systems in the northern and central Great Plains," *USDA Misc. Pub. 952,* pp. 1–64. 1964.
12. Chapman, H. D., Ed. "Diagnostic criteria for plants and soils," Univ. Calif. Div. Agr. Sciences, pp. 1–793. 1966.
13. Chilcott, E. C. "The relations between crop yields and precipitation in the Great Plains area: Supplement 1—Crop rotations and tillage methods," *USDA Misc. Cir. 81,* pp. 1–164. 1931.
14. Cooper, H. P. "Fertilizer and liming practices recommended for South Carolina," *S. C. Agr. Exp. Sta. Cir. 60,* pp. 1–23. 1939.
15. Cooper, H. P., Schreiner, O., and Brown, B. E. "Soil potassium in relation to soil fertility," in *Soils and Men,* USDA Yearbook, 1938, pp. 397–405.
16. Dalrymple, D. G. "Survey of multiple cropping in less developed nations," USDA and USAID FEDR-12, pp. 1–108. 1969.
17. Dean, L. A. "Plant nutrition and soil fertility," in *Soil,* USDA Yearbook, 1957, pp. 80–94.
18. De Turk, E. E., Bauer, F. C., and Smith, L. H. "Lessons from the Morrow plots," *Ill. Agr. Exp. Sta. Bul. 300,* pp. 105–140. 1927.
19. Dinauer, R. C., Ed. *Micronutrients in Agriculture,* Soil Sci. Soc. Amer., Madison, Wis., pp. 1–666. 1972.
20. Dodd, D. R., and Pohlman, G. G. "Some factors affecting the influence

of soybeans, oats, and other crops on the succeeding crops," *W. Va. Agr. Exp. Sta. Bul. 265.* 1935.

21. Englestad, O. P., Ed. "Nutrient mobility in soils: Accumulation and losses," Soil Sci. Soc. Amer., Madison, Wis., pp. 1–81. 1970.
22. Enlow, C. R. "Review and discussion of literature pertinent to crop rotations for erodible soils," *USDA Cir. 559,* pp. 1–51. 1939.
23. Erdman, L. W. "Legume inoculation: What it is; what it does," *USDA Farmers Bul. 2003,* pp. 1–20 (revised). 1953.
24. Erdman, L. W. "Legume seed inoculation," in *Grassland Seeds,* Van Nostrand, Princeton, N.J., pp. 129–130. 1957.
25. Fisher, T. R. "Crop yields in relation to soil pH as modified by liming soils," *Mo. Agr. Exp. Sta. Bul. 947,* pp. 1–25. 1969.
26. Funchess, M. J. "Crop rotation in relation to southern agriculture," *J. Am. Soc. Agron.,* 19:555–556. 1927.
27. Gardner, R., and Robertson, D. W. "The beneficial effects from alfalfa in a crop rotation," *Colo. Agr. Exp. Sta. Tech. Bul. 51.* 1954.
28. Hartwell, B. L., and Damon, S. C. "The influence of crop plants on those that follow," *R. I. Agr. Exp. Sta. Bul. 175,* pp. 1–32. 1917.
29. Hastings, S. H. "Irrigated crop rotations in western Nebraska, 1912–32," *USDA Tech. Bul. 512,* pp. 1–36. 1936.
30. Hobbs, J. A., and Brown, P. L. "Effects of cropping and management on nitrogen and organic carbon contents of a western Kansas soil." *Kans. Agr. Exp. Sta. Tech. Bul. 144,* pp. 1–37. 1965.
31. Horner, G. M., and others. "Effect of cropping practices on yield, soil organic matter, and erosion in the Pacific Northwest wheat region," *Ida., Wash., Oreg., and USDA Bul. 1.* 1960.
32. Ibach, D. B., and Adams, J. R., "Fertilizer use in the United States by crops and areas," *USDA Stat. Bul. 408,* pp. 1–384. 1967.
33. Ibach, D. B., Adams, J. R., and Fox, E. L. "Commercial fertilizer used on crops and pasture in the United States," *USDA Stat. Bul. 348.* 1964.
34. Jenny, H. "Relation of climatic factors to the amount of nitrogen in soils," *J. Am. Soc. Agron.,* 20:900–912. 1928.
35. Jenny, H. "Soil fertility losses under Missouri conditions," *Mo. Agr. Exp. Sta. Bul. 324,* pp. 1–10. 1933.
36. Johnson, T. C. "Crop rotation in relation to soil productivity," *J. Am. Soc. Agron.,* 19:518–527. 1927.
37. Larson, W. E., and others. "Effect of subsoiling and deep fertilizer placement on yields of corn in Iowa and Illinois," *Agron. J.,* 52:185–189. 1960.
38. Leighty, C. E. "Crop rotation," in *Soils and Men,* USDA Yearbook, 1938, pp. 406–430.
39. Lewis, R. D., and Hunter, J. H. "The nitrogen, organic carbon, and *p*H of some southeastern coastal plain soils as influenced by green manure crops," *J. Am. Soc. Agron.,* 32:586–601. 1940.
40. Lie, T. A., and Mulder, E. G., Eds. *Biological Nitrogen Fixation in Natural and Agricultural Habitats,* Martinus Nijhoff, The Hague, pp. 1–590. 1971.
41. Lyons, E. S., Russel, J. C., and Rhoades, H. F. "Commercial fertilizers for the irrigated sections of western Nebraska," *Nebr. Agr. Exp. Sta. Bul. 365,* pp. 1–29. 1944.

42. McKaig, N., Carns, W. A., and Bowen, A. B. "Soil organic matter and nitrogen as influenced by green manure crop management on Norfolk coarse sand," *J. Am. Soc. Agron.,* 32:842–852. 1940.
43. McVickar, M. H. *Using Commercial Fertilizers* (Ed. 3). Interstate, Danville, Ill., pp. 1–353. 1970.
44. Miller, M. F. "Cropping systems in relation to erosion control," *Mo. Agr. Exp. Sta. Bul. 366,* pp. 1–36. 1936.
45. Mooers, C. A. "Effects of liming and green manuring on crop yields and on soil supplies of nitrogen and humus," *Tenn. Agr. Exp. Sta. Bul. 135,* pp. 1–64. 1926.
46. Nelson, L. B. "Advances in fertilizers," in *Advances in Agronomy,* Vol. 17, Academic Press, New York, pp. 1–84. 1965.
47. Nelson, L. B., and Ibach, D. B. "The economics of fertilizers," in *Soil,* USDA Yearbook, 1957, pp. 267–276.
48. Okuda, A., and Takahashi, E. *The Role of Silicon in the Mineral Nutrition of the Rice Plant,* Johns Hopkins Press, Baltimore, 1965.
49. Olson, R. A., and others, Eds. *Fertilizer Technology and Use,* Soil Sci. Amer., Madison, Wis., pp. 1–611. 1971.
50. Pearson, R. W., and Adams, F. Eds. *Soil Acidity and Liming,* Am. Soc. Agron. Pub., Madison, Wis. 1967.
51. Pierre, W. H. "Phosphorus deficiency and soil fertility," in *Soils and Men,* USDA Yearbook, 1938, pp. 377–396.
52. Pierre, W. H., and McKee, R. "The use of cover and green-manure crops," in *Soils and Men,* USDA Yearbook, 1938, pp. 431–444.
53. Ripley, P. O. "Crop rotation and productivity," *Can. Dept. Agr. Pub. 1376,* pp. 1–78. 1969.
54. Roberts, G. "Legumes in cropping systems," *Ky. Agr. Exp. Sta. Bul. 374,* pp. 119–153. 1937.
55. Rogers, T. H., and Giddens, J. E. "Green manure and cover crops," in *Soil,* USDA Yearbook, 1957, pp. 252–257.
56. Russell, E. J., and Richards, E. H. "The changes taking place during the storage of farmyard manure," *J. Agr. Sci.,* 8:495–563. 1917.
57. Russell, E. J., and Watson, D. J. "The Rothamsted field experiments on the growth of wheat," *Imperial Bur. Soil Sci. Tech. Comm. 40,* Harpenden, England, pp. 1–163. 1940.
58. Salter, R. M., and Schollenberger, C. J. "Farm manure," in *Soils and Men,* USDA Yearbook, 1938, pp. 445–461.
59. Salter, R. M., and Schollenberger, C. J. "Farm manure," *Ohio Agr. Exp. Sta. Bul. 605.* 1939.
60. Schreiner, Q., and Brown, P. E. "Soil nitrogen," in *Soils and Men,* USDA Yearbook, 1938, pp. 361–376.
61. Scott, S. G. "Phosphorus deficiency in forage feeds of range cattle," *J. Agr. Res.,* 38(2):113–120. 1929.
62. Shutt, F. T. "The preservation of barnyard manure," *Rept. Min. Agr. Canadian Exp. Farm,* pp. 126–137. 1898.
63. Stanford, G., and others. "Fertilizer use and water quality," *USDA ARS.* 41–168. 1–19. 1970.
64. Stewart, G., and Pittman, D. W. "Twenty years of rotation and manuring experiments," *Utah Agr. Exp. Sta. Bul. 228,*pp. 1–32. 1931.

65. Stewart, W. D. P. "The nitrogen-fixing plants," *Science,* 158:1426–1432. 1967.
66. Throckmorton, R. I., and Duley, F. L. "Soil fertility," *Kans. Agr. Exp. Sta. Bul. 260,* pp. 1–60. 1932.
67. Tisdale, S. L., and W. L. Nelson. *Soil Fertility and Fertilizers,* 2nd ed., Macmillan, Inc., New York, pp. 1–694. 1966.
68. Truog, E. "Liming—The first step in improving acid soils," *Crops and Soils,* 7(7):12–13. 1955.
69. Utz, E. J., and others. "The problem: The nation as a whole: The nature and extent of soil losses," in *Soils and Men,* USDA Yearbook, 1938, pp. 84–110.
70. Viets, F. G., and Hanway, J. J. "How to determine nutrient needs," in *Soil,* USDA Yearbook, 1957, pp. 172–184.
71. Waksman, S. A. "Chemical and microbiological principles underlying the decomposition of green manures in the soil," *J. Am. Soc. Agron.,* 21:1–18. 1929.
72. Walster, H. L., and Stoa, T. E. "Continuous wheat culture versus rotation wheat culture," *N. Dak. Agr. Exp. Sta. Bimonthly Bul. 5(1),* pp. 2–8. 1942.
73. Weir, W. W. "A study of the value of crop rotation in relation to soil productivity," *USDA Bul. 1377,* pp. 1–68. 1926.
74. Wiancko, A. T. "Crop rotation in relation to the agriculture of the corn belt," *J. Am. Soc. Agron.,* 19:545–555. 1927.
75. Wiggins, R. G. "Experiments in crop rotation and fertilization," *Cornell Univ. Agr. Exp. Sta. Bul. 434,* pp. 1–56. 1924.
76. Wiggans, S. C., and Williams, B. "Exploding the myths of organic farming," *Crops and Soils,* 24(4):8–11. 1972.
77. Youngberg, H. "Inoculating alfalfa and clover seed," *OR Ext. Fact Sheet FS 184.* 1972.

Chapter 7

Seeds and Seeding*

IMPORTANCE OF GOOD SEEDS

Reasonably good seed is a prime essential to successful crop production, whereas poor seed is a serious farm hazard.[65] The variety and the approximate germination and purity of seed should be known before it is planted.[85] Introduction of weeds in the seed often increases the labor for production of the crop, reduces crop yields,[78] and contaminates the current product as well as the seed and soil in future seasons.

SEED GERMINATION

External Conditions for Germination

The most important external conditions necessary for germination of matured seeds are ample supplies of moisture and oxygen, a suitable temperature, and, for some seeds, certain light conditions. A deficiency in any factor may prevent germination.

Good seed shows a germination of 90 to 100 per cent in the laboratory. Some sound crop seeds, particularly small grains, show a seedling emergence of as high as 90 per cent of the seed when sown under good field conditions. Even corn, which is a rather sensitive seed, often produces stands of 90 per cent or more in the field. Sorghum and cotton give a lower percentage of emergence because they are more susceptible to attack from seed-rotting fungi. Treated sorghum seed may give a field emergence of 75 per cent, but 50 per cent emergence is all that is normally expected from untreated seed with a 95 per cent laboratory germination, even in a good seedbed. However, when the seed germinates only 60 to 70 per cent in the laboratory, many of the

* For more complete information, see (1) *Seeds*, USDA Yearbook, 1961, pp. 1–591; (2) *Agricultural and Horticultural Seeds*, FAO Agricultural Studies, 55, FAO, Rome, 1961, pp. 1–531; (3) *Testing Agricultural and Vegetable Seeds USDA Handbook 30*, 1952, pp. 1–440 and (4) Wheeler, W. A., and Hill, D. D., *Grassland Seeds*, Van Nostrand, Princeton, N.J., 1957, pp. 1–734.

sprouts will be so weakened that a field emergence of 20 to 25 per cent is all that can reasonably be expected. In a poor seedbed, the emergence may be much less. Seeds that germinate slowly may produce weak seedlings. However, the strong seeds in a low-germinating sample may give good yields, provided enough seed is planted.[48]

Small seeded legumes and grasses are sown at heavy rates to compensate for poor germination and low seedling survival resulting from the necessary shallow seeding, and for hard seeds, many of which fail to grow immediately when sown.

Seeds of legumes are sometimes pelleted for aerial sowing to repel pests, but especially to maintain the viability of the *Rhizobia* inoculant on the seed. The most important ingredients of the pellet material for this purpose are lime or dolomite when sowing in acid soils. Phosphates are helpful for promoting seedling growth. The *Rhizobia* to be added to the seed are most frequently carried in peat. These materials combined with gum arabic or methyl cellulose are mixed with the seed in a revolving drum to make the pellets.[81]

In seeded grass-pasture mixtures, the species with the most viable seeds often predominates in the immediate stand.[24] Small-seeded grasses usually are sown at rates greatly in excess of the rates that would be required if all of the seeds were to produce seedlings, because the mortality of the seeds and seedlings is likely to be high. Thus, in a bluegrass pasture a seeding rate of 25 pounds per acre provides more than 1,000 seeds per square foot, whereas 100 plants per square foot would soon provide dense turf.

Commercial seed of Kentucky bluegrass and that of certain other grasses often does not germinate more than 70 per cent. This low germination is due to harvesting when many of the panicles are immature and to inadequate drying. Any dicotyledonous plant with an indeterminate flowering habit, or a grass that sends up new tillers and panicles over a considerable period, will not mature its seed uniformly. With such crops, immature seed is gathered even though harvesting is delayed until the ripest seeds have already been lost by shattering.

MOISTURE: Abundant water is necessary for rapid germination. This is readily supplied by damp blotters or paper towels in a germinator or by soil that contains about 50 to 70 per cent of its water-holding capacity. Field crop seeds start to germinate when their moisture content (on a dry basis) reaches 26 to 75 per cent, e.g., 26 per cent in sorghum, millet, and sudangrass; 45 to 50 per cent in small grains;[69] and as high as 75 per cent for soybeans. The minimum moisture for germination of corn is approximately 35 per cent in the whole grain and 60 per cent in the embryo.[97]

The water usually enters the seed through the micropyle or hilum, or it may penetrate the seed coat directly. Water enters certain seeds, such as castorbean and sweetclover, through the strophiole or caruncle, an appendage of the hilum.[66] Water inside the seed coat is imbibed by the embryo, scutellum, and endosperm. The imbibed water causes the colloidal proteins and starch of the seed to swell. The enormous imbibitional power of certain seeds enables them to draw water from soil that is even below the wilting point, but not in sufficient amounts to complete germination, because the adjacent soil particles become dehydrated. Seeds sown in dry soil therefore may fail to germinate, or they may absorb sufficient moisture to swell and partly germinate. Wheat, barley, oats, corn, and peas have been sprouted, allowed to dry, and resprouted three to seven times before germination was fully destroyed.[115] However, germination was lower with each repeated sprouting. Wheat seeds can absorb water from a saturated atmosphere until they reach a moisture content exceeding 30 per cent on a wet basis, but this is not high enough to start germination.[19]

OXYGEN: Many dry seeds, particularly peas and beans, are practically impervious to gases, including oxygen. Absorption of moisture may at the same time render the seed permeable to oxygen. Seeds planted too deeply or in a saturated soil may be prevented from germinating through an oxygen deficiency. Rice apparently is less restricted in its oxygen needs than most seeds since it will germinate on the soil surface under 6 inches of water. However, an atmosphere of pure oxygen is as harmful to seeds as it is to man.

TEMPERATURE: The extreme temperature range for the germination of field crop seeds is from 32° to 120° F. In general, cool-season crops germinate at lower temperatures than warm-season crops.

Wheat, oats, barley, and rye may germinate somewhat at the temperature of melting ice.[13] Buckwheat, flax, red clover, alfalfa, field peas, soybeans, and perennial ryegrass germinate at 41° F. or less. The minimum temperature for germination of sorghum and corn is about 48° F. Tobacco seeds[41] germinate slowly below 57° F. Of the commonly grown crops, seeds of alfalfa and the clovers will germinate more readily at low temperatures than any others. Since starchy seeds appear to be more easily destroyed by rots, they are less likely to produce sprouts at low temperatures than are oily or corneous seeds of the same species.[21, 56] Smooth hard hybrid seed corns give better stands than do rough softer types.

A temperature of 59° F. is about optimum for wheat, with progressively decreasing germination at higher temperatures.[103] Mold attack increases directly with increased temperatures. Soybeans of good quality may germinate almost equally well at all temperatures from 50° to 86° F., but seeds of low vitality germinate best at 77° F.

The most favorable temperature for germination of tobacco seed is about 88° F.[41] The optimum laboratory germination for seeds of most cool-weather crops is about 68° F. (20° C.), but certain fescues and other grasses require a somewhat lower temperature. Warm-weather crops, particularly the southern legumes and grasses such as crotalaria and bermudagrass, germinate best at 86° to 97° F. Most crop seeds are germinated at alternating night and day temperatures of 68° and 86° F. in laboratory tests. The alternation of temperatures, which simulates field conditions, favors better germination.

Maximum temperatures at which seeds will germinate are approximately 104° F. or less for the small grains, flax,[20] and tobacco,[41] 111° F. or less for buckwheat, beans, alfalfa, red clover, crimson clover, and sunflower, and 115° to 122° F. for corn, sorghum, and millets. At temperatures too high for germination, the seeds may be killed or be merely forced into secondary dormancy. The killing has been ascribed to destruction of enzymes and coagulation of cell proteins. These reactions as a rule are not observed at temperatures as low as 122° F., but might occur over a time as long as the 24 to 28 hours or more necessary to start germination. The secondary dormancy induced by heat may be an oxygen relationship.

LIGHT: Most field-crop seeds germinate in either light or darkness. Many of the grasses germinate more promptly in the presence of, or after exposure to light, especially when the seeds are fresh. Among these are bentgrass, bermudagrass, Kentucky bluegrass, Canada bluegrass, and slender wheatgrass. Light is necessary for germination of some types of tobacco, except at low temperatures of about 57° F. Most standard American varieties will germinate in its absence, although the rate and percentage of germination may be considerably retarded. The light requirement in all cases is small.[41] Even a flash of light may induce germination in seeds that are wet and swollen. Most weed seeds require light for germination, whereas the absence of light enables such seeds to remain dormant when buried in the soil. Red light initiates the germination, but far red light of 730 millimicron wave length inhibits germination.[99]

Process of Germination

When placed under the proper conditions, seeds capable of immediate germination gradually absorb water, until, after approximately 3 days, their moisture content may be 60 to 100 per cent of the dry weight. Meanwhile the seed coats have become softened and the seeds swollen. Soluble nutrients, particularly sugars, go into solution.

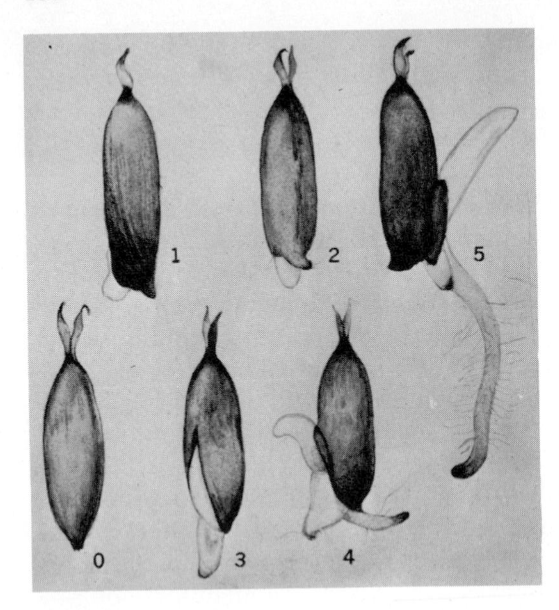

Figure 7–1. Six successive stages in the germination of the sugarcane seed.

The soluble glucose is transported to the growing sprout chiefly by diffusion from cell to cell, and there is synthesized into cellulose, nonreducing sugars, and starch. Proteins broken down by proteolytic enzymes into amides—e.g., asparagin—and into amino acids, then build proteins in the seedling. Fats, which occur chiefly in the cotyledons of certain oil-bearing seeds and in the embryos of cereal seeds, are split by enzymes called lipases into fatty acids and glycerol. These in turn undergo chemical changes to form sugars, which are used to build up the carbohydrates and fats in the seedlings. Energy for the chemical and biological processes of germination and growth is supplied by respiration, or biological oxidation of carbon and hydrogen into carbon dioxide and water.[99] During germination respiration proceeds rapidly at a rate hundreds of times that in dry seeds.

The energy consumed during germination may amount to one-half the dry weight of the seed.[77] The germination of a bushel of wheat utilizes the equivalent of all the oxygen in 900 cubic feet of air and

Figure 7–2. [OPPOSITE] Stages in corn germination: (*1*) Before germination; (*2*) germinated 36 hours, (*3*) 48 hours, (*4*) 4 days, and (*5*) 8 days. In the two upper views the seedcoat has been removed to expose the embryo. In germinating the radicle or first seminal root (*r*) pushes out quickly; the nodal region (*n*) swells; the coleoptile, which encloses the first leaves and has a vent at the tip (*c′*), grows upward; additional seminal roots (*se*) arise, usually in pairs above the radicle, after 3 days. Finally the coronal or crown roots (*cr*) develop and the food substance in the seed (*s*) is practically exhausted. [Courtesy T. A. Kiesselbach.] At (*6*) a wheat germ enlarged about 25 times, shows the scutellum (*sc*), vent in coleoptile (*v*) epiblast (*e*), seminal root swellings (*se*), and radicle (*r*) which is enclosed in the coleorhiza.

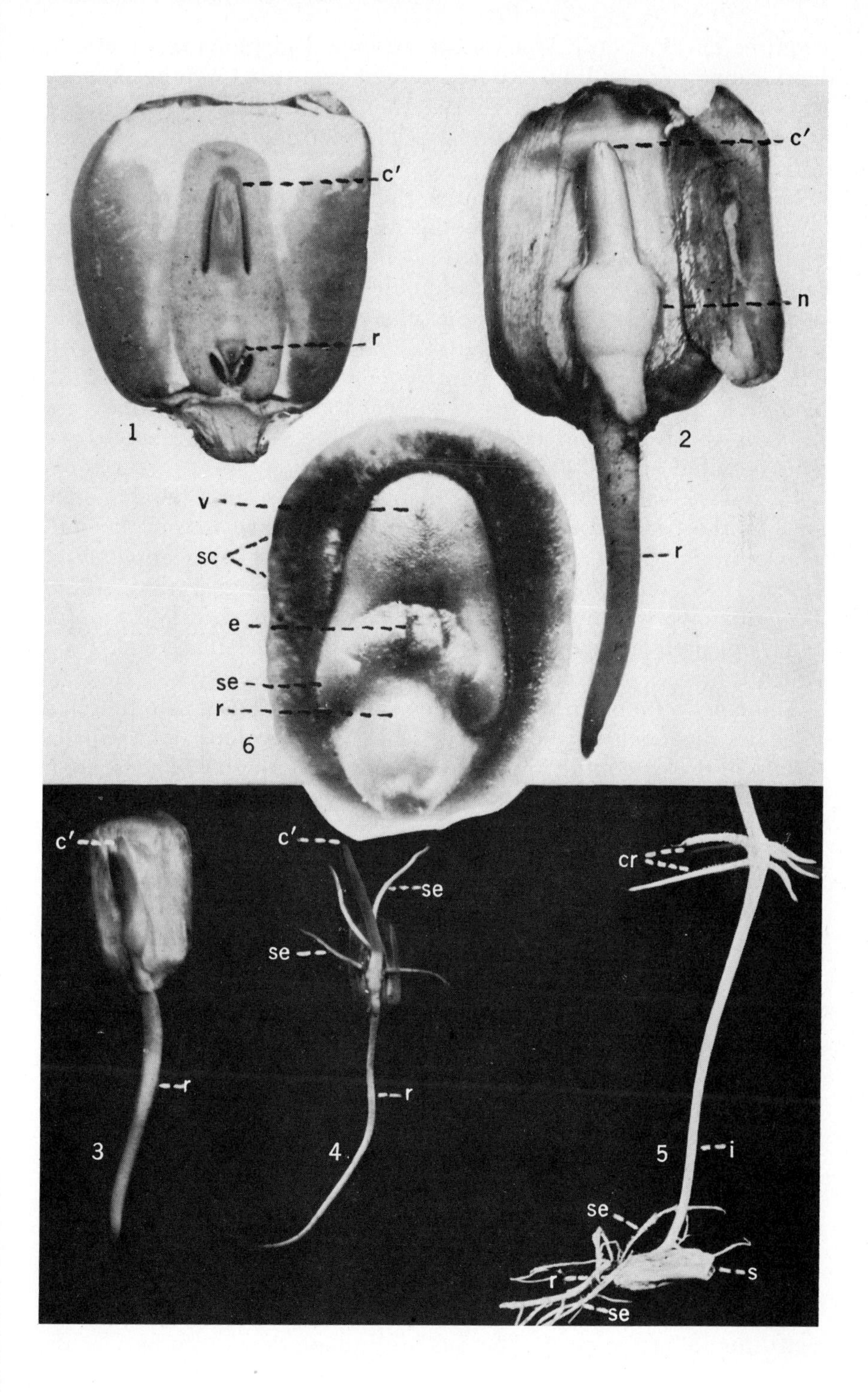
c′
r
1
c′
n
2
r
v
sc
e
se
r
6
c′
r
3
c′
se
se
r
4
cr
i
5
se
r
s
se

requires energy equivalent to that expended in plowing an acre of land. Emerging seedlings exposed to light begin photosynthesis early, but even then their dry weight may not equal the dry weight of the seed until 7 to 14 days or more after the seedling appears above the soil surface.

In seeds that germinate promptly, the growing embryo ruptures the seed coat within 1 or 2 days after the seeds are wetted. The radicle, or embryonic root, is the first organ to emerge in nearly all cases. The radicle is soon followed by the plumule or young shoot (Figures 7–1 and 7–2). In many dicotyledonous plants such as the bean and flax, the cotyledons emerge from the soil and function as the first leaves. The plumule emerges in a bent or curved position (Figure 7–3). The arch thus formed serves to protect the cotyledons as they are brought above the surface of the soil by the elongating hypocotyl. This is called epigeal germination.

In grasses (monocotyledonous plants), and also in a few legumes such as the pea and the vetches, the cotyledons remain in the soil. The plumule grows or is pushed upward by the elongation of an epicotyl or a subcrown internode. This is called hypogeal germination. The subcrown internode of different grasses has been called a mesocotyl, epicotyl, or hypocotyl, depending upon the seedling node from which it arises.

The coleoptile of grasses emerges from the soil as a pale tubelike structure that encloses the first true leaf. A slit develops at the vent on the tip of the coleoptile and the leaf emerges through it. Then photosynthesis begins and the seedling gradually establishes independent metabolism as the stored food of the seed nears exhaustion. The roots are well developed by that time.

Qualities in Seeds for Germination

WHOLE VERSUS BROKEN SEEDS: A marked decrease in germination of mutilated wheat, corn, and alfalfa seeds occurs when the germ is injured.[113] Broken seeds that contain the embryo germinate less, have a higher seedling mortality, and produce smaller plants than whole seeds.[11, 95] Breaks in the seed coat of cereals are deleterious to germination, injury at the embryo end being most serious.[60] Broken or cracked seeds mold more than do whole seeds.[56] Mechanical injury due to broken seed coats and splitting frequently occurs in field peas. Cracked peas that consist of the embryo and a single cotyledon, or a part of one, may fail to germinate.[35]

The viability of seeds may be destroyed quickly by molding or heating as a result of the growth of fungi and bacteria on damp seeds stored in a warm place. These organisms break down and absorb the constituents of the seed. The fats are broken down into fatty acids,

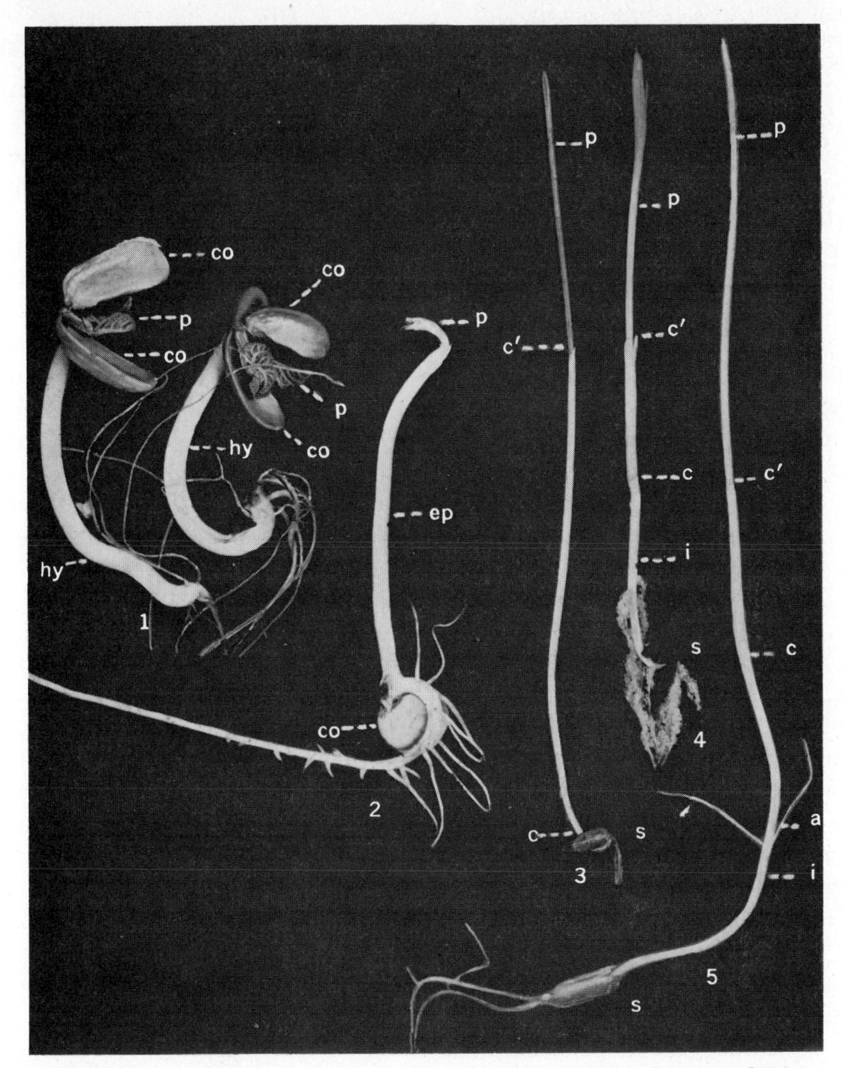

Figure 7–3. Seedlings of (*1*) bean, (*2*) pea, (*3*) rye, (*4*) sorghum, and (*5*) oats. During germination, the cotyledons (*co*) of the bean are pushed up by the elongating hypocotyl (*hy*); then the cotyledons separate and the plumule or true leaves (*p*) emerges. In the germinating pea, the epicotyl (*ep*) grows upward from the cotyledons (*co*) and the plumule (*p*) grows out from the tip of the epicotyl. In the cereals, the coleoptile grows or is pushed to the soil surface and then the plumule (*p*) grows out through a slit (*c'*) at the tip of the coleoptile. In the rye, as in wheat and barley, the coleoptile base (*c*) arises at the seed (*s*) and the crown node lies somewhere within the coleoptile. In sorghum, as in corn, the coleoptile base (*c*) is at the crown node which is carried upward from the seed (*s*) by the elongating subcrown internode (*i*). In oats, as in rice, the node at the coleoptile base (*c*) stands just below the crown node and also is carried upward from the seed (*s*) by the elongating subcrown internode. Occasionally adventitious roots (*a*) arise from the subcrown internode (*i*) in oats and other cereals. (The irregular direction of roots resulted from germination between blotters instead of in the soil.)

and germination drops as fat acidity goes up. After planting, the seeds are exposed to organisms in the soil as well as to those on the seed. The organisms utilize the food materials in the seed, thus starving the young sprout, and certain organisms even invade and kill the young sprouts. Seed-borne and soil-inhabiting organisms often prevent seedling emergence. Sowing sound seeds at the optimum temperature for germination helps to retard seed rots and seedling blights. Thus, small grains and field peas should be sown when the soil is cool, and planting of corn, sorghum, cotton, peanuts, soybeans, and millet should be delayed until the soil is warm. The best protection against seed rots and seedling blights is treatment of the seed with approved disinfectants containing a chemical that is toxic to fungi and bacteria.[26] The fungi commonly associated with the molding and rotting of seeds and the blighting of seedlings are mostly prevalent species of several genera including *Pythium, Fusarium, Rhizopus, Penicillium, Aspergillus, Gibberella, Diplodia, Helminthosporium, Cladysporium, Basisporium,* and *Collectotrichum*.[3, 56]

SEED MATURITY: Mature seed is preferable to immature seed, but occasionally growers are obliged to plant seeds that have failed to reach full maturity. Prematurely harvested barley kernels have germinated and produced small seedlings when the seeds had attained only one-seventh of their normal weight.[28, 29] Corn seeds grew when gathered as early as 20 days after fertilization of the silks, provided they were carefully dried.[43] Such poorly developed seeds obviously are unsatisfactory for field planting. Table 7–1 shows that corn gathered as early as the denting stage is suitable for seed.[46]

Mature corn produces heavier sprouts than that harvested at immature stages.[104] Immature corn shows more disease infection and yields slightly less than mature corn.[50]

Immature seed,[5] because of its small size, has a low reserve food supply, and usually produces poor plants when conditions are ad-

TABLE 7–1. Effect of Maturity of Seed upon the Grain Yield of Dent Corn (5-Year Average)

WEEKS BEFORE RIPE	DATE SEED HARVESTED	DAYS SINCE FERTILIZATION	CONDITION OF GRAIN	FIELD GERMINATION (%)	YIELD OF SHELLED CORN PER ACRE (bu.)
Ripe	September 28	51	Mature	94	55.8
1	September 21	44	Glazing	94	54.4
2	September 14	37	Denting	93	54.9

verse at planting time. Immature seeds, high in moisture, are vulnerable to frost injury.

SEED SIZE: Small seeds invariably produce small seedlings. The logarithms of seedling and seed weights are directly proportional[6] (Figure 7–4).

In Nebraska experiments[44] small seeds of winter wheat, spring wheat, and oats yielded 18 per cent less than the large when equal *numbers* of seeds were sown per acre at an optimum rate for the large seed, but only 5 per cent less when equal *weights* of seed were sown, also at an optimum rate for the large seed. Grain drills sow about equal volumes per acre of large or small seeds of any particular grain, so the latter comparison is of the most practical significance. When unselected seed was used, it yielded 4 per cent less than the large when equal numbers were sown per acre, but only 1 per cent less when equal weights of seed were sown.

In comparisons of fanning-mill grades of winter wheat over a 17-year period, the heaviest quarter yielded 0.3 per cent more, and the lightest quarter yielded 2.0 per cent less, than the unselected seed. Similar results were obtained with oats.

There is no material or practical gain in grain yield from grading normally developed small-grain seed that is reasonably free from trash. Large seeds produce more vigorous seedlings, which survive adverse conditions better, but this advantage within certain limits is largely offset by the greater number of plants obtained from an equal weight of smaller seeds.

The germination and seedling size of shrunken and plump spring

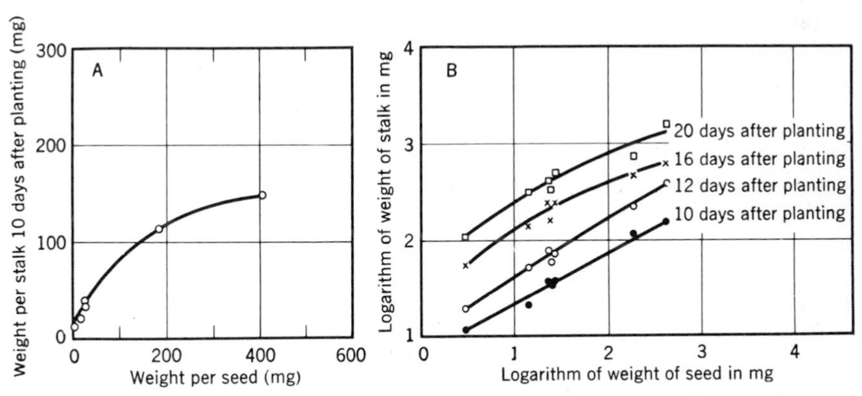

Figure 7–4. The seedling weight (10 to 12 days after planting) shows a direct logarithmic relation to seed weight: (*A*) relation between weight per seed and weight per stalk 10 days after planting of corn, sorghum, and proso; (*B*) relation between logarithms of the seed weights and of the stalk weights taken at intervals up to 20 days after planting.

wheat, with test weights that ranged from 39.5 to 60.8 pounds per bushel, was determined.[55] Test weight shows little relation to viability, but shrunken seeds produce such small weak seedlings that sowing of wheat testing less than 50 pounds per bushel is not recommended. Kernels of wheat testing 60 pounds per bushel are about twice as heavy as those testing 50 pounds. A reduction in test weight of one-third, i.e., from 60 down to 40 pounds per bushel, reduces the weight of an individual kernel nearly two thirds,[103] and seedling weights are reduced nearly as much. In general, matured seeds less than one-half normal size are unsuitable for sowing.

Formerly, the usual recommendation was to throw away the butt and tip kernels of the corn ear, and plant only seed from the middle portion. Extensive experiments showed that the average yield of butt seed was 103 per cent of that from the middle of the ear, while the yield from tip kernels was 105 per cent.[54] In a similar comparison, seeds from the tips, butts, and middles of corn ears averaged 55.3, 55.4, and 56.0 bushels per acre, respectively.[47] Thus, butt and tip kernels are not inferior for seed purposes but may give irregular stands with ordinary corn-planter plates. Shelled butt and tip kernels of hybrid corn separated with a seed-corn grader are planted with special plates. However, they sell at a lower price than the more popular seed from the middle of the ear called large flat seed.

Dormancy in Seeds

Seeds of some crop species show dormancy,[16] i.e., they fail to grow immediately after maturity even though external conditions favor germination, until they have passed through a rest or after-ripening period. This is more common in wild plants, but varieties of cultivated plants differ considerably in dormant tendencies. Dormancy in cereal seeds is indicated by inability to germinate at higher temperatures when it germinates well at 36° to 50° F.

CAUSES OF DORMANCY: Dormancy may result from seed characteristics or environmental conditions as follows:[8, 5]

1. Thick or hard seed coats prevent intake of water and probably also of oxygen. The *hard seeds* in many legumes are an example.
2. Seed coats interfere with the absorption of oxygen, e.g., cocklebur, oats, and barley.
3. In some species the embryo is still immature and has not yet reached its full development at harvest.
4. The embryos in still other seeds appear to be mature but must undergo certain changes before they will germinate. Wheat and barley seeds harvested at certain stages before maturity, i.e., 12 to 24 days after flowering, and dried quickly, may retain their green color, but will germinate poorly and produce weak seedlings. Such green seeds

also are found in wheat that has been frosted before maturity. Dormant varieties of winter barley apparently acquire dormancy during ripening or drying because seeds of such varieties sprouted in the head before maturity when the seed coats were kept wet by artificial watering.[83]

5. Germination inhibitors which must undergo natural or applied chemical changes to permit germination.
6. High temperatures during seed maturity may induce dormancy.

HARD SEEDS IN LEGUMES: Hard or impermeable seeds prevent penetration of water and cause an apparent enforced dormancy period. Such seeds are common in alfalfa, but also are found in most small-seeded legumes.[100] Hard seeds in alfalfa are due to inability of the palisade cells to absorb water.[61] Apparently the cuticle does not restrict the intake of water. The percentage of hard seeds varies among different branches of the same plant. Probably as a result of some scarification, machine-threshed seed contains fewer hard seeds than that which is hand-threshed. Plump seed is more likely to be dormant than is shriveled or immature seed.

AFTER-RIPENING: Seeds of peanuts, alfalfa, clover, and lupines planted soon after maturity under conditions nearly optimum for germination frequently show dormancy ranging up to two years. The rest period appears to be one of after-ripening in peanuts.[36]

Seeds of many small-grain varieties require a short period of dry storage after harvest in order to after-ripen and give good germination at temperatures as high as 20° C. Seeds usually are stored several months before germination tests are made, but in winter cereals it may be necessary to test the seeds immediately after threshing to determine their viability for fall planting. Storage of oats at 40° C. for three months largely eliminated dormancy,[9] and also destroyed most of the molds on the seed.

The embryos of cereals are never essentially dormant, the dormancy being imposed by the seed coats. Artificial dry heating, opening of the coat structures over the embryo, and cutting off the brush ends admit oxygen which induces germination of non-after-ripened or partially after-ripened seeds of wheat, oats, or barley. A temperature considerably below 20° C. is most satisfactory for germination of freshly harvested seed of these crops.[32]

Immature, poorly cured wheat has a higher percentage of dormancy.[113] Dormancy may decrease to a minimum after 4 to 12 weeks of storage, and can be broken immediately by placing the seeds in an ice chest at 4° to 6° C. for five days, and transferring them to an alternation of temperature from 20° to 30° C. for three days. Some wheats grown under high altitude conditions may be dormant as long as 60 days after harvest.[59] Seedsmen have repeatedly encountered difficulty in getting satisfactory germination in laboratory tests of sound plump durum wheat, especially in the fall and early winter. After-ripening is completed during warm spring weather. Good stands are obtained in field planting in cool soil even though durum wheat germinates slowly in laboratory tests until spring.[110] After-ripening of mature corn is coincident with loss of moisture.[97] It may be necessary to reduce the moisture content of immature seeds to approximately 25 per cent before normal germination occurs. The mechanism that inhibits normal germination of such seeds is believed to occur in the scutellum rather than in the endosperm or pericarp.

Slow germination of freshly harvested seeds is extremely desirable in a wet harvest season, when heavy losses occur from grain sprouting in the field.

Dormant varieties do not sprout appreciably.[18, 33] All degrees of prompt, slow, and delayed germination occur in freshly harvested common oats, but cultivated red oats regularly show slow or delayed germination.[13] Dormancy disappears in most oat varieties after 30 days. Grain sorghums often sprout in the head in the field during rainy periods before harvest.

The usual dormancy in buffalograss seed can be broken by soaking the seed in a 0.5 per cent solution of potassium nitrate for 24 hours, then chilling it at 41° F. for six weeks in a cold-storage room, and drying it.[111] However, hulling of the seed is fully as effective, and the treatment is more simple and economical.

SECONDARY DORMANCY: High temperatures in storage or in the germinator or seedbed may throw seeds of cereals or grasses into a secondary dormancy. Such seeds usually germinate later at normal temperatures after they have been subjected to cold treatment.

Very dry cotton seeds may fail to germinate as promptly or as vigorously as those with a moisture content of approximately 12 per cent when planted. A marked increase of hard seeds occurs when they are dried down to 5 or 6 per cent moisture.[109] Excessively dry seed gives satisfactory germination when moistened at planting time with about 2 gallons of water per 100 pounds of seed. Certain Texas samples entered into a secondary dormancy and failed to produce a crop when the proper seedbed conditions were not present.

Scarification of Hard Seeds

Germination of impermeable seeds in legumes such as alfalfa, sweetclover, and the true clovers may be brought about in several ways.[58]

In the mechanical scarifier, the seed is thrown against a roughened surface to scratch the seed coats. The scarified seeds imbibe water and germinate in a normal manner. However, mechanical scarification is practiced less now than formerly. In tests in Utah[101] scarification of alfalfa seed increased germination about 30 per cent, but there were more weak and moldy seedlings from the scarified seed. Scarification injured about as much good seed as it made hard seeds germinable. Mechanically scarified seed has been found to deteriorate rapidly in storage. When scarified at all, seed should be planted immediately.

In New York, fall-sown sweetclover seeds softened and grew the next spring, with 50 to 75 per cent of the seeds producing plants. This is about as high a percentage as is ordinarily obtained in field seeding. In fact, 90 to 100 per cent hard sweetclover seeds fall-sown will yield better results under such conditions than scarified seeds sown at any time of the year.

Other Hard-Seed Treatments

Aging brings about slow natural deterioration of the seed coat in dry storage. In certain experiments one third to two thirds of the hard seeds in red clover were still impermeable after four years, but a majority of the impermeable seeds of alfalfa and hairy vetch became permeable before they were two years old.[31] In another experiment one half of the impermeable seeds in alfalfa germinated after 1½ years, while all germinated after 11 years in storage.[61] The percentage of hard seeds in Korean lespedeza is high when the seeds are tested for germination immediately after harvest, but most of them become permeable during the winter.[70] The average percentages of hard seeds in tests made in November, January, and March were 47.25, 12.25, and 11.05, respectively.

Alternate freezing and thawing sometimes stimulates germination of hard seeds of alfalfa and sweetclover, but may also destroy some seeds that germinate normally.[91] The breaking of dormancy while in the soil may be due to a period of cold.

The germination of hard seeds of alfalfa or clover also can be achieved by exposure for 1 to 1.5 seconds or less to infrared rays of 1180 millimicrons wave length,[117] or by exposure for a few seconds to high frequency electric energy.

Vernalization of Seeds

Temperatures affect the flowering time of many plants. Winter wheat sown in the spring fails to produce heads unless the sprouting seeds or growing plants are subjected to cold or cool conditions. Winter varieties of cereal crops head normally from spring sowing and behave like spring varieties when the seeds are germinated at temperatures slightly above freezing before they are sown. The degree of sprouting during the cold treatment can be restricted by wetting the seeds only enough to start germination. This process is called vernalization.

Winter varieties of cereals are treated by soaking the seed for 12 to 24 hours and then storing them for four to nine weeks at a temperature of about 2° C. (35.6° F.), while maintaining the moisture content of the grain at approximately 50 per cent (on a dry-weight basis). Winter-annual legumes and grasses may respond to similar treatments. Vernalization is so laborious and complicated that it is useful only in certain experiments. The sprouted seeds are difficult to store and sow, germination often is damaged by the treatment,[64] and drying and storing the seed at warm temperatures often causes the seeds to lose much of the effect of the cold treatment. Vernalized spring-sown winter grain yields much less than when sown in the autumn in the field, and about the same or somewhat less than adapted spring varieties sown at the same time as the vernalized grain. Certain spring varieties of cereals that have a partial or intermediate winter growth habit usually respond to vernalization treatment when sown late in the spring,[105] but true spring varieties are not affected. Russian workers reported that special vernalization of corn, sorghum, millet, and other warm-temperature crops is effective in hastening flowering. This treatment consists of germinating the seeds in the dark at normal temperature while restricting the sprouting by adding only limited quantities of water, largely in the form of dilute salt solutions. Extensive experiments in other countries failed to substantiate claims for this type of vernalization.[3] The seeds often mold and lose their viability during treatment.

Longevity of Seeds

Most farm crop seeds are probably dead after 25 years, even under favorable storage conditions. The alleged germination of seeds after prolonged storage in ancient tombs is known to be a myth, although fresh seeds that have been placed in tombs shortly before the visits of gullible travelers often germinate very well. Authentic seeds from the ancient tombs are highly carbonized and have lost much of their original substance.

SEEDS IN DRY STORAGE: The optimum conditions for storing seeds that will endure drying are a 5 to 7 per cent moisture content, sealed storage in the absence of oxygen, and a temperature of 23° to 41° F.[6, 76] In moist, hot climates seeds often can be kept viable between seasons only by storing them, well dried, in airtight containers.[22] It has been suggested that seed life is doubled for each drop of 1 per cent in moisture content and for each drop of 10° F. in temperature.[30] Under cool semiarid conditions in Colorado, wheat, oats, and barley germinated about 10 per cent lower when 10 years old than when 1 year old.[89] The germination of soybeans decreased about 10 per cent in 5 years, but Black Amber sorgo germinated about 97 per cent after 17 years. Yellow Dent corn germinated well for 5 years, but declined to 32 per cent after 20 years. After 15 years of storage[90] other approximate germination percentages were rye, 8; corn, 36; naked barley, 74; wheat, 80; and unhulled barley, 96. After 20 years the germination percentages of wheat, barley, and oats were 15, 46, and 50 respectively.[91]

In a semiarid area of eastern Washington, certain varieties of barley, oats, and wheat germinated from 84 to 96 per cent after 32 years of storage. Corn germinated 70 per cent, but rye had lost its viability.[24]

Under Nebraska conditions, corn 4 years old was satisfactory for seed.[46] Kafir seed retained its germination well for 10 years in western Texas, but deteriorated almost completely during the next 7 years.[42]

Flaxseed of good quality stored under favorable conditions may be expected to maintain its viability for 6 to 8 years. Seeds 9, 12, 15, and 18 years old germinated 99, 89, 56, and 58 per cent, respectively.[20]

Tobacco seed usually retains ability to germinate over many years. One lot of seed germinated 25 per cent after 20 years.[41] However, there was a marked retardation in rate of germination in seed more than 10 years old.

While many farm seeds in ordinary storage in drier climates may have a life span of 15 years, they might live 50 or even 100 years in sealed storage at low temperatures in the absence of oxygen and after proper desiccation.[16] Seeds of *Albizzia* germinated after storage for 149 years in a British herbarium.

BURIED SEEDS: Seeds of some species of wild plants retain their vitality in moist soil for 50 years or more. In the United States Department of Agriculture buried-seed experiment a total of 107 species of seed were mixed with sterilized soil and buried in pots in 1902. None of the cereals or legumes whose seeds are used for food germinated on being dug up after 20 years. The seeds of wild plants grew better than those of cultivated plants. Several persistent weeds showed high germination after 20 years in the soil, some of them being the docks, lambsquarters, plantains, purslane, jimsonweed, and ragweed. Seeds still alive when finally dug up 39 years after they were buried included Kentucky bluegrass, red clover, tobacco, ramie, and some 25 species of weeds.[84] Wild morning glory germinated within 2 days after it was dug up after being buried 30 years.

The classical experiment with buried seeds was started in 1879 by Dr. W. J. Beal of Michigan State College. Seeds of 20 species were placed in sand in bottles and buried under 18 inches of soil. The chess and white-clover seeds were dead after 5 years, while common mallow survived until the 25-year test. Curled dock and mullein germinated 70 years after burial.[84]

Dry arctic lupine seeds found buried in lemming burrows under 10 to 20 feet of frozen soil in Yukon Territory, Canada, were able to germinate. Their assumed age was about 14,000 years.*

GERMINATION AND PURITY TESTS

The real value of a seed lot depends upon its purity and the proportion of pure seed in it that will grow. Seeds after threshing usually contain foreign materials such as chaff, dirt, weed seeds, and seeds of other crop plants. These can be removed to a large extent but not entirely by cleaning machinery.[23, 30]

The object of laboratory seed testing is to determine the percentages of germination, pure seed, other crop seed, inert matter, and weed seed present, as well as the kinds of weed seeds and, in so far as possible, the kind and variety of the seed sample. This serves as an aid in selecting suitable seed, in adjusting seeding rates to germination percentages, in giving warning of impending weed problems, and in reducing dissemination of serious weeds. Seed testing should be done by well-trained seed analysts. Then the label is an accurate guide to the value of the seed.

Germination tests should include 400 seeds counted from the sample indiscriminately and divided into four or more separate tests.[15] Purity tests should be made by hand separation of a sample that contains approximately 3,000 seeds. The size of sample specified for purity tests is indicated in Table 7–2

The final germination count for legumes with hard seeds in the sample is extended for 5 days, as indicated, when partly sprouted seeds are present. Fresh seeds may require prechilling, or treatment with 0.1 per cent or 0.2 per cent potassium nitrate, or even scarifying. Different seeds are germinated between blotters, on top of blotters, in paper toweling, in rolled towels, in sand, or in soil.

Special Tests

In many cases tests other than germination and purity and tests for vigor may be needed. Tests for the absence or presence of fungicides, diploidy or tetraploidy, X rays to detect "hollow seeds," and any other factors which may be of help to a seed buyer or seller could be classified under the area of special tests.

The viability and vigor of the embryos of many seeds can be evaluated quickly by a staining (tetrazolium) test.[15] A fluorescence test is useful to identify or separate seeds of certain species or varieties that are indistinguishable otherwise.[74]

* *Crops and Soils* 23(8):30–31. 1971.

TABLE 7–2. **Methods for Testing Typical Seeds**

CROP SEED	SEEDS PER GRAM	MINIMUM WEIGHT FOR NOXIOUS WEED EXAMINATION (*grams*)	MINIMUM WEIGHT FOR PURITY ANALYSIS (*grams*)	GERMINATION TEST *Temperature* (°*C.*)	*First Count* (*days*)	*Final Count* (*days*)	SPECIAL TREATMENTS
Alfalfa	500	50	5	20	3	7–12	–
Alsike clover	1500	50	2	20	3	7–12	–
Bahiagrass	366	50	10	30–35	3	21	Light; hulling
Barley	30	500	100	20	3	7	–
Bean (field)	4	500	500	20–30	5	8–13	–
Buckwheat	60	300	50	20–30	3	6	–
Crimson clover	330	50	10	20	3	7–12	–
Kentucky bluegrass	4800	25	1	20–30	7	28	Light
Meadow fescue	500	50	5	20–30	5	14	–

SEED LAWS AND REGULATIONS

Federal Seed Act

For many years, low-grade, foul, and adulterated seeds from abroad were dumped in this country. The Federal Seed Act was passed in 1912 to protect the American farmer from such seed.[76] The act was amended in 1916 to include a minimum requirement of live pure seed. Another amendment was passed in 1926 to provide for staining imported red clover and alfalfa seed so as to indicate its origin. Investigations have shown that many foreign importations are unadapted to certain regions in this country. Red clover seed from Italy is strikingly unadaptable to general use in this country, as is alfalfa seed from Turkestan or South Africa. The Federal Seed Act of 1939 was much more drastic and inclusive than the original act. The 1939 act required correct labeling as to variety as well as to purity and germination, established heavier penalties, and permitted penalties for mislabeling without the government's having to prove fraudulent intent. False advertising is prohibited. Punishments for violation of the act include fines up to $2,000, seizure of the seed, and orders to "cease and desist" violations of the act.

The act prohibits importation of seed that is adulterated or unfit for seeding purposes. Adulterated seed is considered to be that with more than 5 per cent mixture of other kinds of seed, except for mixtures that are not detrimental. Seed specified as unfit for seeding purposes is that with more than one noxious seed in 10 grams of small-seeded grasses and legumes; one in 25 grams of medium-sized seeds such as sudangrass, sorghum, and buckwheat; and one in 100 grams of grains and other large seeds. The act also prohibits importation of seed that contains more than 2 per cent weed seeds or (with several exceptions)* less than 75 per cent pure live seed. Noxious weeds specified are whitetop (*Lepidium draba, L. repens*, and *Hymenophysa pubescens*), Canada thistle (*Cirsium arvense*), dodder (*Cuscuta* species), quackgrass (*Agropyron repens*), johnsongrass (*Sorghum halepense*), bindweed (*Convolvulus arvensis*), Russian knapweed (*Centaurea picris*), perennial sowthistle (*Sonchus arvensis*), leafy spurge (*Euphorbia esula*), and other seeds or bulblets of any other kinds which, after investigation, the Secretary of Agriculture finds should be included in the list.

The act also prohibits shipment in interstate commerce of agricultural seeds that contain noxious weeds in excess of quantities allowed by the laws of the state or territory to which the seed is shipped, or as established by the Secretary of Agriculture. Such seeds must be labeled as to kind and variety or type, lot number, origin (of certain kinds), percentage of weed seeds including noxious weeds, kinds and rate of occurrence of noxious weeds, percentages of mixtures of other seeds, germination, hard-seed percentage, month and year of germination test, and the name and address of the shipper and consignee.

In enforcement regulations of the Federal Seed Act, tolerances are specified to allow variations in determining germination and purity.

It is now possible for an individual or organization to obtain a federal plant patent on a new distinct variety or hybrid of any crop which they have originated.

State Seed Laws

All states have seed laws designed to regulate the quality of agricultural seeds sold within their borders. These laws have done much to reduce the spread of weed seeds, especially of noxious weeds. Most of the laws were patterned after the Uniform State Seed Law drawn up by the Association of Official Seed Analysts in 1917, and later were revised to conform more closely with the present Federal Seed Act.[92] There is considerable variation in the details of the laws adopted by the various states, but all require seeds in commerce to be labeled. The laws embody many of these specifications as reflected by information required on the labels: (1) the commonly accepted name of

* Admittable germination percentages for these are bahiagrass, 50; bluegrass, 65; carrots, 55; chicory, 70; dallisgrass, 35; guineagrass, 10; molassesgrass, 25; and rhodesgrass, 35 per cent.

the agricultural seeds, (2) the approximate total percentage by weight of purity, i.e., the freedom of the seeds from inert matter and from other seeds, (3) the approximate total percentage of weight of weed seeds, (4) the name and approximate number per pound of each of the kinds of noxious weed seeds and bulblets, (5) the approximate percentage of germination of such agricultural seed, together with the month and year that the seed was tested.

The weed seeds regarded as noxious by ten or more states include dodders, Canada thistle, quackgrass, wild mustards, buckhorn plantain, corn cockle, wild oats, wild onion, narrow-leaved plantain, wild carrot, ox-eye daisy, leafy spurge, Russian knapweed, bindweed, perennial sowthistle, and curled dock.

The seed laws usually are enforced by designated state officials. Field inspectors draw samples from seeds offered for sale. These are tested in official laboratories. Penalties are inflicted on dealers who sell seeds in violation of the state law.

SEED ASSOCIATIONS

Most states have seed or crop improvement associations of growers who produce quality agricultural seeds under strict regulations. These associations usually cooperate closely with their state agricultural colleges to bring superior crop varieties into widespread use at a reasonable cost.[77]

REGISTERED OR CERTIFIED SEED

The seed associations, whose rules and regulations differ somewhat among the states, supervise the growing of seeds by their members for certification and registration. The International Crop Improvement Association defines the classes of seed as follows:

1. *Breeder seed* shall be seed or vegetatively propagated material directly controlled by the originator, or, in certain cases, by the sponsoring plant breeder or institution, which provides the source for the initial and recurring increase of foundation seed.

2. *Foundation seed,* which includes elite seed in Canada, shall be seed stocks that are so handled as to most nearly maintain specific genetic identity and purity and that may be designated or distributed by an agricultural experiment station. Production must be carefully supervised or approved by representatives of an agricultural experiment station. Foundation seed shall be the source of all other certified seed classes, either directly or through registered seed.

3. *Registered seed* shall be the progeny of foundation or registered seed that is so handled as to maintain satisfactory genetic identity and purity, and that has been approved and certified by the certifying

agency. This class of seed should be of a quality suitable for production of certified seed.

4. *Certified seed* shall be the progeny of foundation, registered, or certified seed that is so handled as to maintain satisfactory genetic identity and purity, and that has been approved and certified by the certifying agency.

Requirements for Registration or Certification

The majority of seed associations require a grower to start with registered seed or foundation seed from an experiment station or equally reliable source.[78] The seed field is inspected before harvest for varietal purity, freedom from disease, and freedom from noxious weeds. An inspector either takes a bin sample after the crop is threshed, or the grower is instructed to send in a representative sample for purity and germination tests. The seed that comes up to the standard set by the association for the particular crop is registered or certified. This seed is sound, plump, and of good color, has high germination, and is free from noxious weeds. It is usually well cleaned and graded. Registered or certified seed is sold under specific tag labels which carry the necessary pedigree information as well as that required under the state seed law.

SOURCES OF FARM SEEDS

A farmer who desires a new variety or a fresh seed supply may secure high-quality pure seed of the standard varieties of most field crops from members of crop improvement associations at reasonable prices. These certified seeds often are available in the community. Seed lists issued each year give the name and address of the grower, the crop and variety grown, the amount of seed available, and the price. In general, certified seed is the highest quality of seed available for field-crop production. However, the yields obtained from certified seed are not measurably higher than those from uncertified good-quality seed of the same variety containing only a small admixture of other varieties. For commercial crop production, an appreciably higher price for certified seed may not be justified when good uncertified seed is available, unless the grower wishes to obtain a new improved variety or to be assured of seed relatively free from disease.

Good seed of most adapted crops can be grown on the home farm with care to prevent admixtures and weed contamination. The seed can be cleaned with an ordinary fanning mill unless it contains weed seeds that can be removed only with the special equipment described in Chapter 9. Consequently, a farmer's home-grown seed often contains many weed seeds.

The farmer usually finds it necessary to purchase hybrid seeds and small seeds such as clover, alfalfa, and forage grasses. Most commercial seed houses endeavor to sell correctly labeled seeds, but use a disclaimer clause for their protection because the crop produced by the farmer is entirely beyond the control of the seed seller. The disclaimer usually is stated as follows:

> The ____ Company gives no warranty, express or implied, as to the productiveness of any seeds or bulbs it sells and will not in any way be responsible for the crop.

This statement on letterheads and seed tags may not exempt dealers from legal redress when the seed is misbranded or fraudulently represented.

SEEDING CROPS

Implements used in Seeding

Seeders and planters used for field crops vary in size from one-row horse-drawn planters to multiple-row tractor-drawn implements.

GRAIN DRILLS: Combination grain and fertilizer drills are often used. The single-disk furrow opener is best for penetrating a hard seedbed or cutting through trash (Figure 7–5a). Double-disk and shoe openers are best for mellow seedbeds that are firm below. The hoe-type opener is best in loose soil or where soil blowing is likely to occur. The hoe drill turns up clods and trash and does not pulverize the soil to any extent. The surface drill is used under humid conditions, while the furrow and semi-furrow drill, are widely used for winter wheat, particularly in the Great Plains under dryland conditions.[34] The furrow drill places the seed deep in moist soil. The ridges and furrows hold snow, reduce winter killing, and tend to protect the young plants from wind erosion.

In comparative tests with winter wheat, the furrow drill has shown no advantage over the surface drill except where winterkilling is frequent.[47, 67, 93]

ROW PLANTERS: Multiple-row tractor planters are widely used for planting intertilled-crops. The rows are spaced 20 to 48 inches apart. By proper selection of drill plates, a planter can be used to plant corn, sorghum, beans, and other row crops. The lister planter is a combination tillage machine and planter which has been used under semiarid conditions to plant sorghum, corn, cotton, and other row crops in the bottom of furrows.[34] Lister planters may plant two to ten rows. No-till planting (Figure 7–5b) is growing in popularity. This results in a reduction of soil compaction and reduced labor.

Figure 7–5a. Disk drill equipped with packer wheels. [Courtesy John Deere & Co.]

The cotton grower requires a planter that will plant cotton, corn, and other row crops (Figure 7–6). A variable-depth planter which plants cotton in depths of 0 to $1\frac{1}{2}$ inches in cycles of about 18 inches usually does away with the necessity of replanting, because suitable emergence are likely to be present at some depths.

Method of Planting

Surface or level seedbed planting prevails where moisture conditions are favorable. On irrigated fields and in high rainfall sections of the South, row crops often are planted on elevated beds made with a lister. In the semiarid Great Plains, row crops may be planted in the bottom of lister furrows. This places the seed in contact with moist (and often too cool) soil, but the chief reason for lister planting is the saving of labor in controlling weeds within the row.

Listed corn and sorghum start growth slowly, and flower and mature later than from the level and furrow planting.[40, 93] Crop stands frequently are destroyed by heavy rains that wash soil into the bottoms of the lister furrows, where it buries the seedling or covers the ungerminated seeds too deep for emergence. Most of the advantages of lister planting without its disadvantages can be achieved with a semi-lister or a surface planter equipped with disk furrow openers.

Figure 7–5b. No-till planter. [Courtesy John Deere & Co.]

Time of Seeding

The time to seed or plant various field crops is governed not only by the environmental requirements for the crop, but also by the necessity of evading the ravages of diseases and insect pests.

SMALL GRAINS: Spring small grains, like other cool-season crops, generally are seeded early in the season to permit maximum growth

Figure 7–6. Planting row crops and applying fertilizer and herbicides in bands is accomplished in one operation. [Courtesy J. I. Case Co.]

and development toward maturity before the advent of hot weather, drought, and diseases. Nineteen hundred years ago Columella wrote:

> If the conditions of the lands and of the weather will allow it, the sooner we sow, the better it will grow, and the more increase we shall have.

For winter wheat in the western states,[63] the optimum date of seeding occurs when the mean daily temperature lies between 50° and 62° F. the higher temperatures prevailing in the South and the lower ranges in the North. In the cold semiarid regions, the sowing of winter wheat early enough in the fall to allow the seedlings to become established before the soil freezes gives maximum protection against cold.

In humid regions, it is essential that the plants be well rooted in order to avoid winter injury from heaving. Columella recommended seeding between October 24 and December 7 in temperate regions, while for colder regions he advised seeding October 1, "so the roots of the corn [grains] grow strong before they be infested with winter showers, frost or hoar frosts." Plants heaved up have many of their roots broken and then[38] are killed by desiccation. Losses from heaving are a common occurrence in late-sown small grains in the eastern half of the United States. Seeding late enough to escape severe injury from Hessian fly is important for winter wheat where that pest is prevalent. In the case of cereal crops, practices that lead to maximum average yields are also satisfactory from the standpoint of crop quality.[45]

Early-sown crops mature earlier than those sown later but they require a longer growing period. Consequently the difference in harvest date is less than the difference in planting date. This reaction in broomcorn, which is typical of both long-day and short-day spring-planted crops, is shown in Figure 7–7. The growing period becomes longer from very late planting of summer crops because of cool weather in the fall.

OTHER CROPS: Corn is a warm-weather crop that generally utilizes the full season. In Nebraska, early planting of corn in a normal season insured earlier maturity, lower grain moisture content, and higher grain viability when the crop was exposed to low temperatures.[45] Under Colorado conditions[87] corn was better in both yield and quality when planted early, light frosts doing less damage than delayed planting.

Three planting dates for cotton, i.e., early, medium, and late, were studied under Georgia conditions.[25] Late-planted cotton (late May) yielded only about half as much lint per acre as that planted in March and April. Stands were poorest on the cotton planted earlier and thickest in the May planting. In Texas experiments, early-planted cotton had a longer period of development before being attacked by

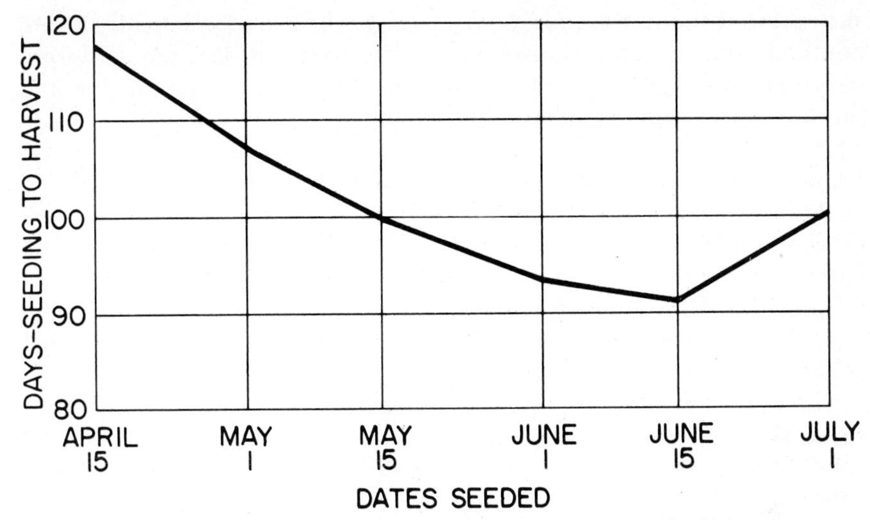

Figure 7–7. Effect of planting date on the growing period of broomcorn.

root rot (*Phymatotrichum omnivorum*), but development of the disease was more rapid in early than in late plantings. The greatest losses were sustained by early plantings.[17]

Perennial legumes may be seeded either in spring or fall. When sown in fall, they should be seeded early enough to permit satisfactory root development before the ground freezes, or else so late that the seeds do not germinate until spring.

With many crops, it is a safe practice to seed at heavier than normal rates when seeding has been delayed materially beyond the optimum time determined for the region.

Rate of Seeding

The objective in spacing crop plants is to obtain the maximum yield on a unit area without sacrifice of quality. The rate of seeding is governed by the ultimate stand desired. Most crops are seeded at lighter rates under dryland than under humid or irrigated conditions. Small short-season varieties of corn require thicker planting than long-season varieties.[71] Corn is generally planted in rows 18 to 48 inches apart, because it has been determined in many experiments that rows more widely spaced resulted in lower yields, even under most dryland conditions.[8, 71, 118] Under semiarid conditions in western Nebraska the highest yields were obtained when single plants were spaced 24 inches apart in the row.[118]

In general, there is a rather definite limit for various regions of the country beyond which heavier seeding rates of small grains fail to produce increased crop yields. However, heavy seeding seldom reduces the yields. Extensive investigations under western conditions showed that the optimum rate of seeding wheat was practically independent of soil type, moisture, locality, date of seeding, cultural treatment, and variety,[63] where rates of 4 to 6 pecks per acre in general produced the highest net yields of both winter and spring wheat. Thin seeding of hard red winter wheat (20 to 30 pounds per acre) is a common practice in parts of the Great Plains. This is feasible for wheat sown early in the season, because of heavy tillering.[39] Irregularity of distribution of barley plants in the drill row may cause only slight variations in yield.[98] Thus American grain drills, when functioning properly, distribute seed of small grains satisfactorily so far as total crop yield is concerned, even though some variation in stand occurs in the drill rows.

The optimum rate of seeding barley may vary over a range of 1 to 3 bushels per acre, depending on variety.[107] An increase in rate of seeding beyond this optimum range can cause a great reduction in tillering, length of head, and number of kernels per head. There also can be a tendency for grain seeded at the heavier rates to lodge. Most cereals have remarkable ability to adjust themselves to the environment.[37]

Sudangrass, which, in the subhumid regions, is normally drilled at the rate of 22 pounds per acre, was planted at one-fourth normal, one-half normal, normal, twice normal, and three times the normal rate.[80] The yields of hay in tons per acre were: 3.25, 3.96, 4.50, 4.44, and 4.22, respectively.[58]

In millet,[58] as in other grass crops, the closer the plants are in the row within practical limits, the higher will be the yield per row. With a free-tillering variety, thin stands are often compensated for by an increase in number of tillers. With a nontillering variety, thin stands often cause a decrease in yield, even when there is a material increase in plant size.

In cotton, closely spaced plants complete their fruiting earlier than do widely spaced ones. The general optimum spacing appears to be from 12 and 16 inches between single plants,[71] although wider spacing may be warranted under certain conditions,[108] and closer spacing and narrow rows under favorable conditions.

INFLUENCE ON CROP QUALITY: Grain quality is affected only to a slight extent by usual variations in seeding rate. In flax, differences in plant spacing show no consistent influence on the oil content of the seed.[49] In forage crops, in which fineness of stems adds to palatability with a consequent reduction in waste when fed to animals, it is desirable to seed more thickly than is necessary to get maximum yields. Ordinarily, the forage will be finer and more leafy without reduction in yield. Sorgo (forage sorghum) can be seeded at rates as heavy as 120 to 150 pounds per acre without reduction in yield. The yield is not materially affected by increases in the seeding rate above 50 pounds per acre, but the quality of hay is improved because of the finer stems. For silage purposes, corn may be spaced closer than for grain in order to obtain maximum tonnage of feed. An unduly heavy rate results in no gain in yield, and produces silage containing little grain. In cotton, thick spacing may contribute to decreased size of boll.[108]

Depth of Seeding

Seeds will emerge from greater depths in sandy soil than in clay soil, and in warm soil than in cold soil. It is customary to plant deep in dry soil in order to place the seeds in contact with moisture. Peas will emerge from a greater depth than will beans when the seeds are the same size, because the bean seedling must push the cotyldons up above the soil surface, whereas the pea cotyledons remain where planted.

In general, the larger the seed the deeper it can be planted and still emerge from an arable soil (Figure 7–8). Approximately one-quarter inch in heavy soil or one-half inch in sandy soil is the most satisfactory depth for seeding small-seeded legumes and grasses under optimum conditions.[2, 73] These include alfalfa, sweetclover, red clover, alsike clover, white clover, timothy, bromegrass, crested wheatgrass, reed canarygrass, and Kentucky bluegrass. Satisfactory emergence of reed canarygrass was obtained from a 1-inch depth and bromegrass from a 2-inch depth on all soil types. A reduction in the stand of soybeans followed seeding deeper than 2 inches in fine sandy loam and 1 inch in a clay soil.[102] However, satisfactory stands were secured at depths up to 4 inches in loam and 2 inches in clay soil. Depth of planting may be an important factor determining the seedling emergence of many grasses and small-seeded legumes.[73]

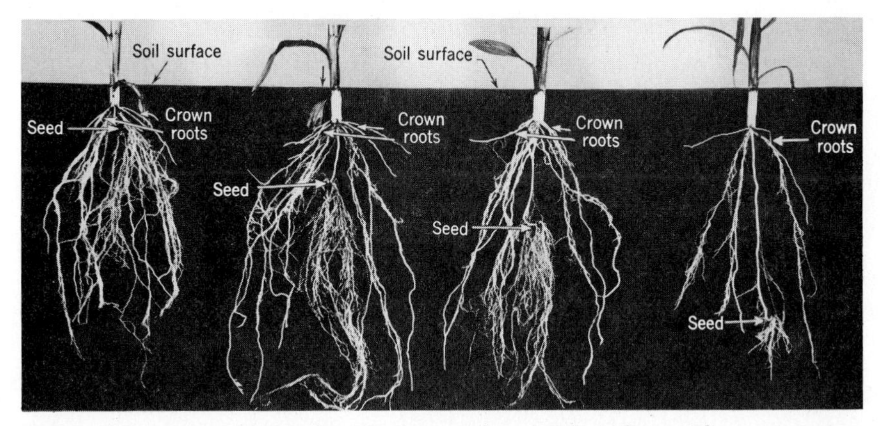

Figure 7–8. Corn planted at 2, 4, 6, and 10 inches deep. The crown was formed at nearly the same depth regardless of planting depth.

TABLE 7–3. Seeding Depths for Seeds of Different Sizes

NORMAL DEPTH OF SEEDING (*in.*)	USUAL MAXIMUM DEPTH FOR EMERGENCE (*in.*)	SEED SIZE (*no./lb.*)	REPRESENTATIVE CROPS
0.25 to 0.50	1 to 2	300,000 to 5,000,000	Redtop, carpetgrass, timothy, bluegrass, fescues, white clover, alsike clover, and tobacco.
0.50 to 0.75	2 to 3	150,000 to 300,000	Alfalfa, red clover, sweetclover, lespedeza, crimson clover, ryegrass, foxtail millet, and turnip.
0.75 to 1.50	3 to 4	50,000 to 150,000	Flax, sudangrass, crotalaria proso, beet (ball of several seeds), broomcorn, and bromegrass.
1.50 to 2	3 to 5	10,000 to 50,000	Wheat, oats, barley, rye, rice sorghum, buckwheat, hemp, vetch, mung bean.
2 to 3	4 to 8	400 to 10,000	Corn, pea, and cotton.
4 to 5		4 to 20 (tubers or pieces)	Potato and Jerusalem artichoke.

The seeding depths for seeds of different sizes under field conditions are shown in Table 7–3. The logarithms of seed size and typical planting depth of most seeds are directly proportional.

REFERENCES

1. "Agricultural and horticultural seeds," FAO Agricultural Studies 55, FAO, Rome, pp. 1–531. 1961.
2. Ahlgren, H. L. "The establishment and early management of sown pastures," *Imperial Bur. Pastures and Forage Crops Bul. 34,* Aberystwyth, England, pp. 139–160. 1945.
3. Anderson, A. M. *Handbook on Seed-borne Diseases,* Assn. Off. Seed corn," *J. Am. Soc. Agron.,* 196–197. 1920.
4. Anonymous. "Vernalization and phasic development of plants," *Imperial Bur. Plant Genet. Bul. 17,* Cambridge. 1935.
5. Bartel, A. T. "Green seeds in immature small grains and their relation to germination," *J. Am. Soc. Agron.,* 33(8):732–738. 1941.
6. Bartel, A. T., and Martin, J. H. "The growth curve of sorghum," *J. Agr. Res.,* 57(11):843–847. 1938.

7. Barton, L. V. *Seed Preservation and Longevity,* Interscience, New York, pp. 1–216. 1961.
8. Barton, L. "Seed dormancy," *Encyclopedia Plant Phys.,* 15:699–727. 1965.
9. Brandon, J. F., "The spacing of corn in the West Central Great Plains," *J. Am. Soc. Agron.,* 29:584–599. 1937.
10. Brown, E., and others. "Dormancy and the effect of storage on oats, barley, and sorghum," *USDA Tech. Bul. 953,* pp. 1–30. 1948.
11. Brown, E. B. "Relative yields from broken and entire kernels of seed corn," *J. Am. Soc. Agron.,* 196–197. 1920.
12. Ching, T. M., Parker, M. C., and Hill, D. D. "Interaction of moisture and temperature on viability of forage seeds stored in hermetically sealed cans," *Agron. J.,* 51:680–684. 1959.
13. Coffman, F. A. "The minimum temperature of germination of seeds," *J. Am. Soc. Agron.,* 15:257–270. 1923.
14. Coffman, F. A., and Stanton, T. R. "Variability in germination of freshly harvested Avena," *J. Agr. Res.,* 57:57–72. 1938.
15. Colbry, V. L., and others. "Tests for germination in the laboratory," in *Seeds,* USDA Yearbook, 1961, pp. 433–441.
16. Crocker, W., and Barton, L. V. *Physiology of Seeds,* Chronica Botanica Co., Waltham, Mass., pp. 1–267. 1953.
17. Dana, B. F., Rea, H. E., and Dunlavy, H. "The influence of date of planting cotton on the development of root rot," *J. Am. Soc. Agron.,* 24:367–377. 1932.
18. Deming, G. W., and Robertson, D. W. "Dormancy in seeds," *Colo. Agr. Exp. Sta. Tech. Bul. 5,* pp. 1–12. 1933.
19. Dillman, A. C. "Hygroscopic moisture of flax seed and wheat and its relation to combine harvesting," *J. Am. Soc. Agron.,* 22:51–74. 1930.
20. Dillman, A. C., and Toole, E. H. "Effect of age, condition, and temperature on the germination of flax seed," *J. Am. Soc. Agron.,* 29:23–29. 1937.
21. Dungan, G. H. "Some factors affecting the water absorption and germination of seed corn," *J. Am. Soc. Agron.,* 16:473–781. 1924.
22. Duvel, J. W. T. "The vitality and germination of seeds," *USDA Bur. Plant Industry Bul. 58.* 1904.
23. Gregg, B. R., and others. *Seed Processing.* Miss. State Univ. Nat'l Seeds Corp. and USAID, Avion Printers, New Delhi, India, pp. 1–396. 1970.
24. Haferkamp, M. E., Smith, E. L., and Nilan, R. A. "Studies on aged seeds. I. Relation of age of seed to germination and longevity," *Agron. J.,* 45:434–437. 1953.
25. Hale, G. A. "The effect of variety, planting date, spacing, and seed treatment on cotton yields and stands," *J. Am. Soc. Agron.,* 28:364–369. 1936.
26. Hanson, E. W., Hansing, E. D., and Schroeder, W. T. "Seed treatments for control of diseases," in *Seeds,* USDA Yearbook, 1961, pp. 272–280.
27. Hanson, H. C. "Analysis of seeding mixtures and resulting stands in irrigated pastures of northern Colorado," *J. Am. Soc. Agron.,* 21:650–659. 1929.
28. Harlan, H. V., and Pope, M. N. "The germination of barley seeds harvested at different stages of growth," *J. Hered.,* 8(2):72–75. 1922.

29. Harlan, H. V., and Pope, M. N. "Development in immature barley kernels removed from the plant," *J. Agr. Res.*, 32(7):669–678. 1926.
30. Harmond, J. E., and others. "Mechanical seed cleaning and handling," USDA Handbook. 354, pp. 1–56. 1968.
31. Harrington, G. T. "Agricultural value of impermeable seeds," *J. Agr. Res.*, 6:761–796. 1916.
32. Harrington, G. T. "Forcing the germination of freshly harvested wheat and other cereals," *J. Agr. Res.*, 23:79–100. 1923.
33. Harrington, J. B. "The comparative resistance of wheat varieties to sprouting," *Sci. Agr.*, 12:635–645. 1932.
34. Hudspeth, E. B., Jr., Dudley, R. F., and Retzer, H. J. "Planting and fertilizing," in *Power to Produce*, USDA Yearbook, 1960, pp. 147–153.
35. Hulbert, H. W., and Whitney, G. M. "Effect of seed injury upon the germination of *Pisum sativum*," *J. Am. Soc. Agron.*, 26:876–884. 1934.
36. Hull, F. H. "Inheritance of rest periods of seeds and certain other characters in the peanuts," *Fla. Agr. Exp. Sta. Bul. 314*. 1937.
37. Hutchinson, R. E. "Rates of seeding wheat and other cereals with irrigation," *J. Am. Soc. Agron.*, 28:699–703. 1936.
38. Janssen, G. "Effect of date of seeding of winter wheat on plant development and its relationship to winterhardiness," *J. Am. Soc. Agron.*, 21:444–466. 1929.
39. Jardine, W. M. "Effect of rate and date of sowing on yield of winter wheat," *J. Am. Soc. Agron.*, 8:163–166. 1916.
40. Jenkins, M. T. "A comparison of the surface, furrow, and listed methods of planting corn," *J. Am. Soc. Agron.*, 26:734–737. 1934.
41. Johnson, J., Murwin, H. F., and Ogden, W. B. "The germination or tobacco seed," *Wisc. Agr. Exp. Sta. Res. Bul. 104*. 1930.
42. Karper, R. E., and Jones, D. L. "Longevity and viability of sorghum seed," *J. Am. Soc. Agron.*, 28(4):330–331. 1936.
43. Kiesselbach, T. A. "Corn investigations," *Nebr. Agr. Exp. Sta. Res. Bul. 20*. 1922.
44. Kiesselbach, T. A. "Relation of seed size to the yield of small grains," *J. Am. Soc. Agron.*, 16:670–682. 1924.
45. Kiesselbach, T. A. "The relation of seeding practices to crop quality," *J. Am. Soc. Agron.*, 18:661–684. 1926.
46. Kiesselbach, T. A. "Effects of age, size, and source of seed on the corn crop," *Nebr. Agr. Exp. Sta. Bul. 305*. 1937.
47. Kiesselbach, T. A., and Lyness, W. E. "Furrow vs. surface planting winter wheat," *J. Am. Soc. Agron.*, 26:489–493. 1934.
48. Kiesselbach, T. A., and Weihing, R. M. "Effect of stand irregularities upon the acre yield and plant variability of corn," *J. Agr. Res.*, 47:399–416. 1933.
49. Klages, K. H. "Spacing in relation to the development of the flax plant," *J. Am. Soc. Agron.*, 24:1–17. 1932.
50. Koehler, B. Dungan, G. H., and Burlison, W. L. "Maturity of seed corn in relation to yielding ability and disease infection," *J. Am. Soc. Agron.*, 26:262–274. 1934.
51. Kozlowski, T. T., Ed. Seed Biology, Vol. 1 Importance, Development and germination, 416 pp. Academic Press, New York and London. 1972.

52. Kozlowski, T. T., Ed., Vol. II, *Germination, Control, Metabolism and Pathology,* 447 pp. Academic Press, New York and London. 1972.
53. Kozlowski, T. T., Ed., Vol. III *Insects, and Seed Collection, Storage, Testing and Certification.* 442 pp. Academic Press, New York and London. 1972.
54. Lacy, M. G. "Seed value of maize kernels: butts, tips, and middles," *J. Am. Soc. Agron.,* 7:159–171. 1915.
55. Leukel, R. W. "Germination and emergence in spring wheats of the 1935 crop," *USDA B.P.I. Div. Cereal Crops and Diseases,* Feb. 1936 (processed).
56. Leukel, R. W., and Martin, J. H. "Seed rot and seedling blight of sorghum," *USDA Tech. Bul. 839,* pp. 1–36. 1943.
57. Li, H. W., and Meng, C. J. "Experiments on the planting distance in varietal trials with millet (*Setaria italica*)," *J. Am. Soc. Agron.,* 29:577–583. 1937.
58. Love, H. H., and Leighty, C. E. "Germination of seed as affected by sulfuric acid treatments," *Cornell Agr. Exp. Sta. Bul. 312,* pp. 294–336. 1912.
59. Lute, A. M. "A special form of delayed germination," *Proc. Assn. Off. Seed Analysts,* 16:23–29. 1924.
60. Lute, A. M. "Some notes on the behavior of broken seeds of cereals and sorghums," *Proc. Assn. Off. Seed Analysts,* 17:33–35. 1925.
61. Lute, A. M. "Impermeable seed in alfalfa," *Colo. Agr. Exp. Sta. Bul. 326.* 1928.
62. Mann, A., and Harlan, H. V. "Morphology of the barley grain with reference to its enzyme-secreting areas," *USDA Bul. 183.* 1915.
63. Martin, J. H. "Factors influencing results from rate and date of seeding experiments with winter wheat in the western United States," *J. Am. Soc. Agron.,* 18:193–225. 1926.
64. Martin, J. H. "The practical application of iarovization," *J. Am. Soc. Agron.* (note), 26(3):251. 1934.
65. Martin, J. H., and Yarnell, S. H. "Problems and rewards in improving seeds," in *Seeds,* USDA Yearbook, 1961, pp. 113–118.
66. Martin, J. N., and Watt, J. R. "The strophiole and other seed structures associated with hardness in *Melilotus alba L.* and *M. officinalis* Willd.," *Iowa State College Jour. Sci.,* 18(4):457–469. 1944.
67. May, R. W., and McKee, C. "Furrow drill for seeding winter wheat in central Montana," *Mont. Agr. Exp. Sta. Bul. 177,* pp. 1–24. 1925.
68. Mayer, A. M., and Polyakoff-Mayber, A. The Germinations of Seeds, Pergamon, New York, pp. 1–226. 1963.
69. McKinney, H. H., and Sando, W. J. "Russian methods for accelerating sexual reproduction in wheat: Further information regarding iarovization," *J. Hered.,* 24:165–166. 1933.
70. Middleton, G. K. "Hard seeds in Korean lespedeza," *J. Am. Soc. Agron.,* 25:119–122. 1933.
71. Mooers, C. A. "Planting rates and spacing for corn under southern conditions," *J. Am. Soc. Agron.,* 12:1–22. 1920.
72. Mooers, C. A. "The effect of spacing on the yield of cotton," *J. Am. Soc. Agron.,* 20:211–230. 1928.

73. Murphy, R. P., and Arny, A. C. "The emergence of grass and legume seedlings planted at different depths in five soil types," *J. Am. Soc. Agron.*, 31:17–28. 1939.
74. Musil, A. F. "Testing seeds for purity and origin," in *Seeds*, USDA Yearbook, 1961, pp. 417–432.
75. Nikolaeva, M. G. *Physiology of Deep Dormancy in Seeds*, Nat'l. Sci. Foundation, pp. 1–220. 1969.
76. Owen, E. B. "The storage of seeds for maintenance of viability," *Commonwealth Bureau Pastures and Field Crops Bul. 43*, Hurley, Berks., England, pp. 1–81. 1956.
77. Palladin, W. *Plant Physiology*, Blakiston, Philadelphia, pp. 1–320. 1918.
78. Parsons, F. G., Garrison, C. S. and Beeson, K. E., "Seed certification in the United States," in *Seeds*, USDA Yearbook, 1961, pp. 394–401.
79. Pavlychenko, T. K., and Harrington, J. B. "Competitive efficiency of weeds and cereal crops," *Can. J. Res.*, 10:77–94. 1934.
80. Peralta, Ferdinand de. "Some principles of competition as illustrated by Sudan grass," *Ecolog. Monos.*, 5:355–404. 1935.
81. Plucknett, D. L. "Use of pelleted seed in crop and pasture establishment," Hawaii Univ. Ext. Circ. 446, pp. 1–15. 1971.
82. Pollock, B. M., and Toole, V. K. "Afterripening, rest period, and dormancy," in *Seeds*, USDA Yearbook, 1961, pp. 106–112.
83. Pope, M. N., and Brown, E. "Induced vivipary in three varieties of barley possessing extreme dormancy," *J. Am. Soc. Agron.*, 35(2):161–163. 1943.
84. Quick, C. R. "How long can seed remain alive?" in *Seeds*, USDA Yearbook, 1961, pp. 94–99.
85. Robbins, W. W., and others. "Cleaned, tested, and treated seed for Colorado," *Colo. Exp. Sta. Bul. 238*. 1918.
86. Roberts, E. H., Ed. *Viability of Seeds*, pp. 1–448. Syracuse Univ. Press, Syracuse, N.Y. 1972.
87. Robertson, D. W., and Deming, G. W. "Date to plant corn in Colorado," *Colo. Agr. Exp. Sta. Bul. 238*. 1930.
88. Robertson, D. W., and Lute, A. M. "Germination of the seed of farm crops in Colorado after storage for various periods of years," *J. Agr. Res.*, 46:455–462. 1933.
89. Robertson, D. W., and Lute, A. M. "Germination of seed of farm crops in Colorado after storage for various periods of years," *J. Am. Soc. Agron.*, 29:822–834. 1937.
90. Robertson, D. W., Lute, A. M., and Kroeger, H. "Germination of 20-year-old wheat, oats, barley, corn, rye, sorghum, and soybeans," *J. Am. Soc. Agron.*, 35:786–795. 1943.
91. Rodriquez, G. "Study of influence of heat and cold on the germination of hard seeds in alfalfa and sweetclover," *Proc. Assn. Off. Seed Analysts*, 16:75–76. 1924.
92. Rollin, S. F., and Johnston, F. A. "Our laws that pertain to seeds," in *Seeds*, USDA Yearbook, 1961, pp. 482–492.
93. Salmon, S. C. "Seeding small grain in furrows," *Kans. Agr. Exp. Sta. Tech. Bul. 13*, pp. 1–55. 1924.

94. Salmon, S. C. "Corn production in Kansas," *Kans. Agr. Exp. Sta. Bul.* 238. 1926.
95. Sando, W. J. "Effect of mutilation of wheat seeds on growth and productivity," *J. Am. Soc. Agron.*, 31(6):558–565. 1939.
96. Seeds, USDA Yearbook. 1961, pp. 1–591.
97. Sprague, G. F. "The relation of moisture content and time of harvest to germination of immature corn," *J. Am. Soc. Agron.*, 28:472–478. 1936.
98. Sprague, H. B., and Farris, N. F. "The effect of uniformity of spacing seed on the development and yield of barley," *J. Am. Soc. Agron.*, 23:516–533. 1931.
99. Stanley, R. G., and Butler, W. L. "Life processes of the living seed," in *Seeds*, USDA Yearbook, 1961, pp. 88–94.
100. Stevenson, T. M. "Sweet clover studies on habit of growth, seed pigmentation and permeability of the seed coat," *Sci. Agr.*, 17:627–654. 1937.
101. Stewart, G. "Effect of color of seed, of scarification, and of dry heat on the germination of alfalfa seed and some of its impurities," *J. Am. Soc. Agron.*, 18:743–760. 1926.
102. Stitt, R. E. "The effect of depth of planting on the germination of soybean varieties," *J. Am. Soc. Agron.*, 26:1001–1004. 1934.
103. Stoa, T. E., Brentzel, W. E., and Higgins, E. C. "Shriveled lightweight wheat: Is it suitable for seed?" *N. Dak. Agr. Exp. Sta. Cir. 59*, pp. 1–11. 1936.
104. Tascher, W. R., and Dungan, G. H. "Seedling vigor and diastatic activity of dent corn as related to composition of endosperm and stage of maturity," *J. Am. Soc. Agron.*, 20:133–141. 1928.
105. Taylor, J. W., and Coffman, F. A. "Effects of vernalization on certain varieties of oats," *J. Am. Soc. Agron.*, 30:1010–1018. 1938.
106. "Testing agricultural and vegetable seeds," *USDA Handbook. 30*, pp. 1–440. 1952.
107. Thayer, J. W., and Rather, H. C. "The influence of rate of seeding upon certain plant characters in barley," *J. Am. Soc. Agron.*, 29:754–760. 1937.
108. Tisdale, H. B. "Effect of spacing on yield and size of cotton bolls," *J. Am. Soc. Agron.*, 20:298–301. 1928.
109. Toole, E. H., and Drummond, P. L. "The germination of cottonseed," *J. Agr. Res.*, 28:285–292. 1924.
109a. USDA. "Rules and regulations under the Federal Seed Act," *AMS Service and Regulatory Announcements 156* (amended), pp. 1–79. 1963.
110. Waldron, L. R. "Delayed germination in durum wheat," *J. Am. Soc. Agron.*, 1:135–144. 1908.
111. Wenger, L. E. "Buffalo grass," *Kans. Agr. Exp. Sta. Bul. 321*, pp. 1–78. 1943.
112. Wheeler, W. A., and Hill, D. D. *Grassland Seeds*, Van Nostrand, Princeton, N.J., pp. 1–734. 1957.
113. Whitcomb, W. O. "Dormancy of newly threshed grain," *Proc. Assn. Off. Seed Analysts*, 16:28–33. 1924.
114. Whitcomb, W. O., and Hay, W. D. "Notes on the germination of broken seeds," *Proc. Assn. Off. Seed Analysts*, 17:38–39. 1925.

115. Widtsoe, J. A. *Dry Farming*, Macmillan, Inc., New York, pp. 1–445. 1911.
116. Wilson, H. K. "Wheat, soybean, and oat germination studies with particular reference to temperature relationships," *J. Am. Soc. Agron.*, 20:599–619. 1928.
117. Works, D. W., and Erickson, L. C. "Infrared radiation, an effective treatment of hard seeds in small seeded legumes," *Ida. Agr. Exp. Sta. Rsh. Bul. 57*, pp. 1–22. 1963.
118. Zook, L. L., and Burr, W. W. "Sixteen years' grain production at the North Platte Substation," *Nebr. Agr. Exp. Sta. Bul. 193*. 1923.

Chapter 8

Harvest of Field Crops

SMALL GRAINS AND OTHER SEED CROPS

Stage of Harvest for Small Grains

Grains cease growing and gaining in dry weight when they reach, approximately the hard-dough stage, or when the moisture content of the grain drops below about 40 per cent.[28, 51, 70] Further ripening consists of desiccation unaccompanied by transport of nutrients into the kernel. Ripening is not entirely uniform among different heads or different grains within a head. Consequently, growth may continue until the average moisture content is appreciably below 40 per cent. Small grains at the hard-dough stage have a moisture content of 25 to 35 per cent, the heads are usually light yellow, and the kernels are too firm to be cut easily with the thumbnail.

EFFECTS OF PREMATURE HARVEST: Premature harvest reduces both yield and quality even when rust is damaging the grain.[56] Underdeveloped grains are low in test weight, starch content, and market value.[4]

Grain seed crops may be cut with a swather 7 days or more before they are ripe and allowed to dry under cool humid conditions, or 3 to 4 days early under warm dry conditions, without appreciable loss in yield or quality.[3] Wheat grain sometimes draws material from the straw after it is cut when nearly ripe. Considerably more growth has occurred when immature barley grains were left to dry in the head than when they were threshed immediately.[29] However, others[4] have found no significant transfer of material from the straw to the grain during the curing process.

Most forage grasses and legumes ripen irregularly, and the early ripened seeds often shatter. The maximum recovery of marketable seed is obtained by mowing or windrowing when the average seed moisture content is somewhere within the range of 25 to 45 per cent, depending upon the species.[39] The effect of moisture content at the time of harvest upon field losses in perennial ryegrass is seen in Figure 8–1.

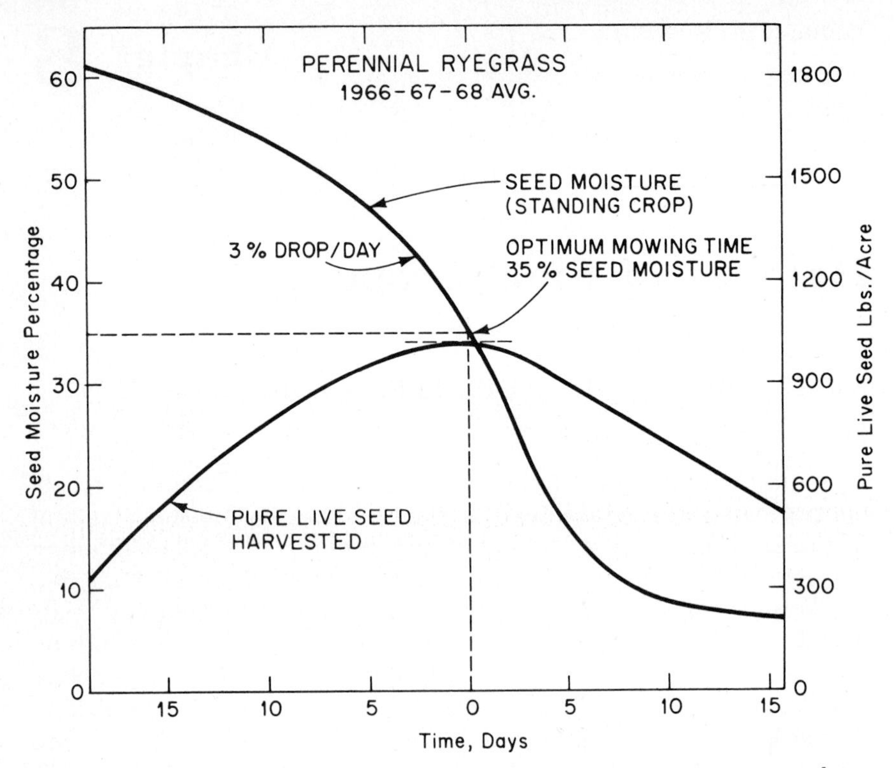

Figure 8–1. The influence of seed moisture at mowing time on combine yields of perennial ryegrass.

EFFECTS OF DELAYED HARVEST: When harvested with a combine, the small grains must stand in the field for 5 to 10 days past the period of windrow harvest or until the moisture content of the grain has dropped to 14 per cent or less, a necessity for safe storage without artificial drying. When wheat grain has a moisture content of 13 per cent or less, the rachis of the wheat spike breaks readily, and the straw will burn freely. Losses from the delayed harvest are caused by shattering, crinkling, lodging, and leaching.

Cereal stems are likely to crinkle down or break over soon after maturity, especially in damp weather.[43, 54, 57] Oats are more susceptible than barley to crinkling, while barley is more susceptible than wheat or rye. Most tall-stalked grain sorghums go down soon after maturity or after a frost, whereas flax stands erect for long periods.

Weathered or sun-bleached grain is unattractive and often brings a lower grade and price on the market. Ripe grains exposed to wet and dry weathering for long periods in the field are lowered in test weight because the grains swell when damp and do not shrink to their original volume after drying. The kernels then are less dense but the baking quality of the flour from such grains may not be impaired.[7, 22]

Methods of Harvesting

In the harvest of grain, the hand sickle, scythe, and cradle were replaced by the reaper, which was widely used until the binder was put on the market between 1880 and 1890.[47] The header soon introduced still greater economies in the harvest of small grains. Now com-

Figure 8–2. (*Top*) Harvesting wheat with combines. Reels of the slat type are adequate for cutting upstanding crops. (*Bottom*) Harvesting soybeans. The pick-up type reel is advisable for low-growing or lodged crops. [Courtesy J. I. Case Co.]

bines are generally used in the important grain-growing regions of the world.

Horse-drawn combines (combined harvester-threshers) were used in the Pacific Coast States to harvest wheat and barley before 1880. Their use, usually tractor-drawn, spread across the intermountain states 25 to 35 years later. Combine harvesting began in the Great Plains states during World War I and in other grain-growing states by 1930. By 1971 fully 96 per cent of all combines sold were self-propelled.

Combines (Figure 8–2) are used for the harvest of nearly all small grain, grain sorghum, flax, soybeans, cowpeas, and safflower in the United States. It was used also in harvesting about 95 per cent of the dry peas, beans, and seed crops of the forage legumes and grasses by 1960. About 20 per cent of the combine harvesting was done by custom operators. With special attachments, called corn-heads, the combine also harvests and shells a majority of the corn.[19, 37, 23] The grain is collected in a tank from which it is conveyed or spouted into a truck or trailer. The straw usually is spread evenly over the ground. Where it is desired to save the straw for feed or bedding, it may be deposited in piles or windrows. A self-propelled combine is illustrated in Figure 8–3.

WINDROW PICK-UP MODIFICATION: The windrower or swather cuts the crop and places it on the stubble in a windrow (Figure 8–4). After drying, the windrow is gathered with a combine, which has a pick-up device attached to the cutter bar (Figure 8–5). The most satisfactory stubble height for windrowed grain is about one-third the total length of the straw.[57] The short-stemmed heads will drop to the ground when cutting is too high. A fairly heavy windrow may stay in good condition for 30 days, while a light windrow works down into the stubble where the grain may start to sprout.[62] The windrowed grain is usually dry enough to thresh with a combine with pick-up attachment in about 5 days.

The windrower is used particularly for flax, peas, beans, buckwheat, rapeseed, crambe, and small-seeded legumes and grasses.[21] Swathing small grains is advisable in fields that contain many green weeds, or that are ripening irregularly, or need to be cut early to avoid shattering or crinkling. Green weeds dry in the windrow and thus do not increase the moisture content of the grain or interfere with threshing.

Usually approximately half the excess moisture in weeds is transferred to the wheat in the first 24 hours of storage and the remainder within 7 days.[57] Damp grain often is dried with forced heated air (Figure 8–6).

The fuel shortage may decrease the amount of artificially dried

Figure 8–3. A sectional view of a self-propelled combine. The numbered parts referred to are: (1) feeder, (2) cylinder and concave, (3) straw deflector, (4) straw walker, (5) straw throw or lifter, (6) conveyor to move threshed grain to the cleaner, (7) cleaner, (8) fan, (9) cutter bar, and (10) reel. [Courtesy John Deere & Co.]

Figure 8–4. Self-propelled swather cutting alfalfa. [Courtesy International Harvester Co.]

grain. High moisture corn storage with addition of organic acids or other materials as preservatives is being investigated.

Binders and separators usually caused greater losses of grain than did combines because of the extra handling.[54] However, where cereal stems broke soon after maturity, the losses from combining exceeded those from binder harvesting.

CORN BINDER AND OTHER HARVESTING MACHINES: The corn binder, or row binder, is used for harvesting corn or sorghum for fodder and some other crops growing in cultivated rows. The binder cuts one row at a time and ties the stalks into bundles. The bundles are shocked and allowed to dry in the field. A majority of the corn crop is field shelled by combines equipped with corn heads, and most of the remainder is gathered with picker-shellers. Hand husking, or snapping of unhusked ears, and even machine husking without shelling, is becoming obsolete, except for the harvest of hybrid seed corn, inbred lines, and breeding stocks. Also, some corn is snapped for farm storage before later feeding because the husks help to protect the ears from insect damage.

Losses from late mechanical harvesting of corn after some stalk lodging and ear dropping occur are reduced by growing a hybrid adapted to the machine as well as by early completion of harvesting, i.e., early in November in the Corn Belt.

Corn combines and picker-shellers which harvest two or eight rows at a time reduce harvest labor and expense, but a considerable percentage of the grain must be dried artificially.[33, 47]

Figure 8–5. Combine with pick-up attachment threshing from windrow. [Courtesy John Deere & Co.]

Figure 8–6. Portable grain drier forcing heated air through perforated flues distributed through a bin of grain.

Bean harvesters cut off the roots of beans and similar legumes just below the soil surface and turn the crop from a pair of rows into a single windrow. After curing, the beans can be threshed with a combine equipped with a pick-up and other attachments.

Field silage cutters and forage harvesters (Figure 8–14) are in wide use. They cut the crop, then cut it into short lengths in one operation.

Seed of some forage legumes and grasses can be harvested directly with a combine.[30] Alfalfa, clover, and certain other seed crops may be harvested with a swather or a mower with a windrowing attachment. A suction seed reclaimer, which sometimes is attached to the combine, can gather much of the seed that has shattered on the ground (Figure 8–7). Considerable seed of alfalfa, clovers, and certain grasses is combined about 3 to 5 days after the crop has been sprayed with a chemical desiccant. This treatment stops growth and dries the leaves and stems. Endothal or DNEP is applied by an airplane at a rate of 1 to 3 pints in 10 to 15 gallons of weed oil per acre. Chemical desiccation before harvest is effective in warm dry climates.

The seed of several grasses is gathered with a forage harvester, which chops up the heads and stems and blows them into a trailer. The chopped material is dried and then is ready for planting with a picker-wheel type of cotton planter, or is passed through a hammer mill to permit sowing with conventional seeders.

The seed of Kentucky bluegrass and sometimes of other grasses formerly was harvested with strippers. These machines have revolv-

Figure 8–7. Suction seed reclaimer that gathers seed that fell on the ground before or during harvest.

ing spiked cylinders which knock the seed from standing plants into hoppers. The pneumatic type of stripper blows the heads of the plants into the cylinder. The stripped material is then dried. Some forage seeds are cut with a grain binder, and the bundles shocked and later hauled to a stationary thresher or standing combine.

About 750 million pounds of seed of small-seeded legumes and grasses are harvested annually in the United States. A part of this seed is needed for lawns and other turf areas, or for export. The cleaning of seeds is discussed in Chapter 9.

Straw and Stover

Straw comprises the dried stalks or stems and other parts of various crops from which the seed has been threshed in the ripe or nearly ripe stage. Some of the small-grain straw produced is gathered or saved, usually with a pick-up baler. Considerable additional quantities are recovered by livestock that graze the stubble fields. Most of the harvested straw is used for bedding livestock or for mulching.

Stover is the corn or sorghum plant that remains after the ear or head has been removed. Typical mature corn is 50 to 65 per cent stover, while grain sorghum of different heights is 45 to 75 per cent stover.

Straws and stovers are the least nutritious of all substances commonly used as feed for livestock. Animals maintained on straw alone are scarcely more than kept alive unless considerable waste grain has been left in the straw. However, straw is an important supplement in livestock maintenance. Oat and barley straw are usually considered more valuable for feed than either wheat or rye straw. Corn stover is more digestible than the small grain straws, but the waste is large unless it is shredded. Fully ⅓ of the total digestible nutrients produced in a cornfield remain in the stover after the grain is harvested.

Straw is bulky, and that of small grain is of low manure value. It supplies organic matter when returned to the land. One ton of wheat straw contains on the average about 10 pounds of nitrogen, 2.6 pounds of phosphoric acid, and 14.8 pounds of potash.[42] The straw of oats and barley contain slightly more nitrogen and phosphoric acid and about twice as much potash as does wheat straw. About one million tons of straw were fed in the United States in 1963.[19]

In 1924 it was estimated that approximately 15 per cent of the straw of small grains was burned, chiefly in areas of the United States where the livestock population was small, or where it could not be added to the land advantageously.[42] Western farmers often burned their straw and stubble before plowing because of the difficulty in disposing of a large volume of straw. At the Nephi Substation in Utah, where the annual rainfall is about 13.33 inches, the wheat yield was not reduced by burning the straw and stubble, nor increased by plowing the straw

Figure 8–8. Experimental mechanical field burner (sanitizer) showing supplemental propane fuel tanks. [Courtesy Oregon State University.]

under. Burning of straw is objectionable because of air pollution, and not recommended because repeated burning increases a tendency to soil erosion or blowing. Most of the straw is now scattered over the land by straw spreaders attached to combines. It is then plowed under or left on the surface by sweep-tillage implements. An application of nitrogen fertilizer usually precludes the depression in yields that often follows in the first crop after a heavy straw residue.

Field burning of residues from the harvest of seed of perennial ryegrass and other grasses has been practiced widely in Oregon and Washington as a means of controlling weeds, blind seed, ergot, nematodes, and rust as well as other leaf diseases. In California, burning is used to destroy rice straw residues. Burning after the harvest also aids in the control of the sod webworm, silvertop, meadow plant bugs, thrips, mites, and other insects. Urban residents and passing motorists object to the smoke. Other methods of straw disposal under investigation include its use in feeds, paper and fiber board manufacture, and as fuel. Mechanical field burners (Figure 8–8) consume the straw with a greatly lessened release of smoke. Burning of grass fields after seed harvest removes the residues to provide more light to the surviving plants. This induces more tillering and higher subsequent yields.

The straws of cowpeas, soybeans, and other leguminous crops are superior to cereal straws in feed and fertilizer value. They are comparatively high in nitrogen and calcium, but similar to wheat straw in other elements. One ton of soybean straw contains 18 pounds of nitrogen, 2.4 pounds of phosphoric acid, and 17.8 pounds of potash.

HAY AND HAY MAKING

Principal Hay Crops

The distribution of forage crops for hay in the United States is shown in Figure 8–9. The principal hay crops are (1) alfalfa, including

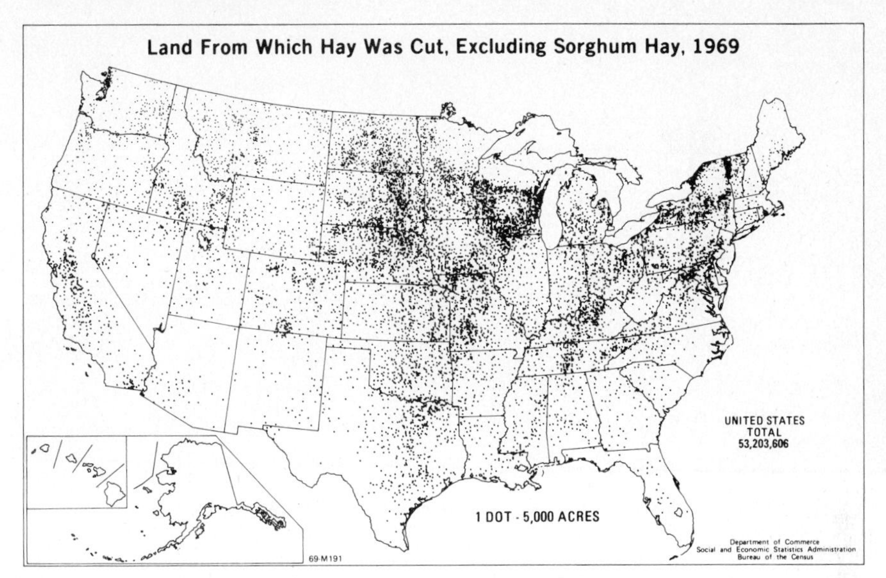

Figure 8–9. Land from which hay was cut, excluding sorghum hay, in 1969.

alfalfa-grass mixtures, (2) clover and timothy or other grass mixtures, (3) wild hay (Figure 8–10), (4) grain hay, (5) coastal bermudagrass and (6) lespedeza. Additional important hay crops formerly were soybeans, cowpeas and peanut vines. Other hay crops include tall fescue, bromegrass, orchardgrass, sudangrass, johnsongrass, crested wheatgrass, sweetclover, vetch, and peas. The area harvested for hay

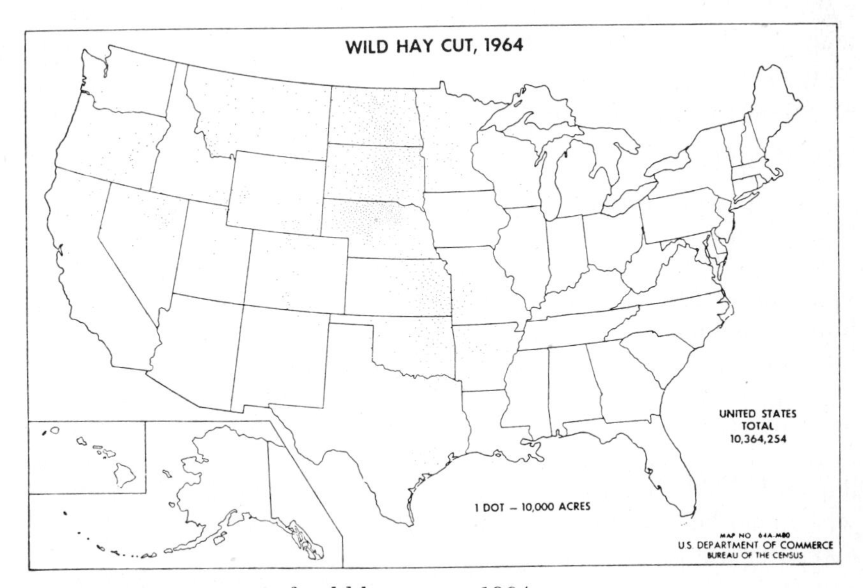

Figure 8–10. Acreage of wild hay cut in 1964.

Figure 8–11. Hay conditioner that hastens drying in the field. The crop is cut, crushed, and windrowed in one operation. [Courtesy International Harvester Co.]

in the United States in 1971, more than 62 million acres, was 4 million acres less than 60 years earlier, but the production was 55 per cent higher. The acreage of clover exceeded that of alfalfa before 1951, but the 1970 acreage of alfalfa was more than double that of clover.

Hay is by far the most valuable harvested forage because it does not deteriorate appreciably in storage, it can be shipped and marketed, and it provides a large intake of dry-matter nutrients for livestock.

Quality in hay really means nutritive value. The important physical factors[71] that can be gauged in a practical way are color, leafiness, maturity of plants when cut, amount of foreign material, condition, and texture. These factors are generally correlated with palatability as well as chemical composition.

Hay Making

Hay is cut with tractor mowers, mower-conditioners, swathers, or self-propelled mower-crusher-windrowers (Figure 8–11). Hay is cured (nearly 40 per cent of it conditioned by crushing) in the mowed swath or windrow, or in windrows after raking, and occasionally in bunches gathered from the windrow with a stacker, hay rake, or buck-rake. Mobile stacking machines load the hay on mechanized trailers that can unload or reload stacks containing 1 to 6 tons of hay. The windrowing of mowed swaths is done mostly with side-delivery or finger wheel rake (Figure 8–12), or with a buncher attached to the mower.[1, 47] When the hay is wetted by rain while in the swath, or windrow, it may be turned over to dry, preferably with a side-delivery rake after partial drying.

Figure 8–12. Side delivery rake. [Courtesy Allis-Chalmers Manufacturing Co.]

Windrowing relatively green alfalfa hay materially extends the curing period as compared with swath curing.[37, 38] Prolonged swath curing may result in an undue loss of leaves with a reduction of almost 10 per cent in weight, whereas in hay cured entirely in the windrow, or after 3 to 6 hours' initial swath curing, only 1 to 2 per cent loss in leaves may occur. A combination of partial swath curing, followed by windrowing, seems to be the better practice.[37, 38] Alfalfa can be best cured by windrowing after it has dried down to 50 per cent moisture. The crop then should be raked before it becomes dry enough to shatter the leaves, which sometimes occurs before the hay is dry enough to store.

Only about 2 per cent of the 1967 hay crop was dried by artificial aeration.[19]

Stage of Maturity for Harvesting

The stage of maturity at which hay crops are cut affects color, leafiness, and other factors of quality. It is impossible to produce high-quality hay from late-cut grasses and legumes, because lignin (tough fiber) increases as the plants mature.

PLANT STAGE FOR HARVEST: The more immature a forage crop is when cut, the smaller the yield and the more palatable and nutritious the product. Because of smaller yields along with increased labor costs from frequent cutting, some sacrifice of quality must be made in the interest of greater yields. Consequently, plants are harvested for hay at an intermediate stage when neither yield nor quality is at its maximum. The best time to harvest the grasses, clover, and alfalfa for hay is ordinarily some time between early bloom and full bloom.[71] Sweetclover should be cut in the bud stage to avoid subsequent coarse growth.[64] Crops that are heavy producers of seed, such as cereals, should be cut for hay when the grain is in the soft- to medium-dough stage,[58] soybeans when the beans are about half grown, and cowpeas when the first pods are mature. When a hay of very high quality is required, grasses and alfalfa are sometimes harvested before they bloom. Care should be taken so as not to cause injury from too frequent cutting or consistently cutting alfalfa before the bud to $^1/_{10}$th bloom stage. Excessive depletion of food reserves in the roots may cause a reduction in the vigor of the stand.[18, 45]

CHEMICAL COMPOSITION: The composition of hay changes with advance toward maturity. The amounts of protein and minerals are higher, and the amounts of the less valuable crude fiber is lower, in young than in old plants. Changes in alfalfa[36] and timothy[62] are given in Table 8–1.

The crude fiber (chiefly cellulose) of young plants, as in immature pasture herbage, appears to be largely digestible. As the plant matures, a progressively greater proportion of the crude fiber is made up of less digestible lignin, which lowers the net nutritive value of the plant. Increased growth of small grains beyond the medium-dough stage is more than offset by shattering of kernels, loss of leaves, leach-

TABLE 8–1. Effect of Stage of Maturity at Cutting on Chemical Composition of Alfalfa and Timothy Hay

STAGE OF MATURITY	ASH (%)	CRUDE PROTEIN (%)	CRUDE FIBER (%)	NITROGEN-FREE EXTRACT[a] (%)	ETHER EXTRACT (FAT) (%)
Alfalfa					
Prebloom	11.24	21.98	25.13	38.72	2.93
Initial bloom	10.52	20.03	25.75	40.67	3.03
One-tenth bloom	10.27	19.24	27.09	40.38	3.02
One-half bloom	10.69	18.84	28.12	39.45	2.90
Full bloom	9.36	18.13	30.82	38.70	2.99
Seed stage	7.33	14.06	36.61	39.61	2.39
Timothy					
No heads showing	8.41	10.18	26.31	50.49	4.61
Beginning to head	7.61	8.02	31.15	49.14	4.07
Full bloom	6.10	5.90	33.74	51.89	2.38
Seed formed	5.54	5.27	31.95	54.12	3.13
Seed all in dough	5.38	5.06	30.21	56.48	2.87
Seed fully ripe	5.23	5.12	31.07	55.87	2.72

[a] Mostly carbohydrates; chiefly starches, hemicelluloses, and sugars.

ing, and general deterioration of plant structure. Close associations are evident between the contents of fiber and stems; of total ash and calcium, and leaves; and of phosphorus, crude protein, and nitrogen-free extract, and heads.[58]

The carotene content, a measure of vitamin A potency, is higher in early stages of leafy growth because the leaves contain more carotene than the stems, and old leaves lose carotene. In alfalfa, clover, and timothy hays, vitamins B and G decrease as the plant matures.[34] In general these vitamins are correlated with the leafiness, greenness, and protein content of the plant.

Cutting hay in late afternoon may avoid the depletion of food materials from the leaves and stems by translocation to the roots and by respiration during the night when photosynthesis is suspended. Practical conditions of curing hay favor morning cutting despite these assumed small losses.

Field Curing of Hay

The aim of the hay maker is to dry the crop to 25 per cent moisture or less with as little loss of leaves, green color, and nutrients as possible. Hay crimpers, crushers or conditioners crush the stems between rolls, which shortens the drying time by one third to two thirds. This reduces curing losses. The crushers are towed by a tractor during or after mowing, or a combination mower-crusher is used.[47] Loss of leaves is more pronounced with legumes than with grasses, especially when the moisture content is 30 per cent or less. The loss of alfalfa leaves begins when the moisture content of the hay drops below 40 per cent.

In the 1st century A.D., Columella described the curing of hay in a manner which indicates that hay making has changed very little.

> It is best to cut down hay before it begins to wither; for you gather a larger quantity of it, and it affords a more agreeable food to cattle. But there is a measure to be observed in drying it, that it be put together neither over-dry, nor yet too green; for, in the first case, it is not a whit better than straw, if it has lost its juice; and, in the other, it rots in the loft, if it retains too much of it; and often after it is grown hot, it breeds fire and sets all in a flame. Sometimes also, when we have cut down our hay, a shower surprises us. But, if it be thoroughly wet, it is to no purpose to move it while it is wet; and it will be better if we suffer the uppermost part of it to dry with the sun. Then we will afterwards turn it, and, when it is dried on both sides, we will bring it close together into cocks, and so bind it up in bundles; nor will we, upon any account, delay to bring it under a roof.

The leaves of alfalfa apparently do not function in the withdrawal of moisture from the cut stems. The moisture loss in alfalfa stems with the leaves attached is about the same as when the leaves are removed.[37, 65] In legumes with large stems, such as sweetclover and soybeans, the leaves may aid in the withdrawal of water from the stems.[28]

Losses in Field Curing of Hay

Profound changes in composition occur during the curing of hay in the field.[41, 71] There is an inevitable loss of nutrients when crops are made into hay. Curing losses are small when drying is rapid. Under adverse conditions the loss may be up to 40 per cent or spoilage may be complete. Under normal weather conditions, the dry-matter loss in field curing of hay may be 10 to 25 per cent in different parts of the country. (Table 8–2). A loss of 17.6 per cent of the dry matter and 21.5 per cent of the protein has occurred during the ordinary field curing of alfalfa[41] due to shattering of leaves, pods, seeds, and stems, to leaching and fermentation, and to the respiration that continues in green plants for some time after cutting. Included in the above losses was 30 per cent of the leaves, with a correspondingly large loss of nutrients. The leaves are the portion of the plant richest in protein, vitamins, phosphorus, and calcium. The leaf loss in alfalfa ranges from 6 to 9 per cent of the weight of the total crop. In cereal hays,[58] the leaves constitute one-fifth the weight of the plant. After the milk stage, cereal leaves contain nearly 50 per cent of the mineral and 40 per cent of the fat of the whole plant.

During curing the carotene starts to decompose at once, due chiefly to oxidation. Alfalfa hay exposed in the swath and windrow for 30 hours in good hay-making weather has lost 60 to 65 per cent of its carotene, 25 per cent of the protein, 15 per cent of the leaves, and 10 per cent of the total dry matter. The losses in artificially dried hay are less than in sun-cured hay of alfalfa, lespedeza, sorgo, and soybeans. Baled alfalfa, timothy, and clover hays stored in a dark, unheated barn may lose 3 per cent of their carotene per month in the winter, but the losses are much higher with higher temperatures.[35] The percentage rate of loss of carotene was much more rapid than that of the natural

TABLE 8–2. Losses in Harvesting and Preserving Forages

	LOSSES (%)		
	Field Loss	*Storage Loss*	*Total*
Direct cut grass silages	2–3	18–22	20–25
Haylage – 65% moisture	11–13	8–12	19–25
Haylage – 50% moisture	11–13	3–8	14–21
Baled alfalfa – raked at baling	30–35	2–4	32–39
Baled alfalfa – direct windrowed or raked at 50–60% moisture	12–15	2–4	14–19
Legume-grass – raked at baling	18–23	2–4	20–27
Legume-grass – windrowed or raked at 50–60% moisture	12–18	2–4	14–22

Figure 8–13. Pick-up baler. [Courtesy International Harvester Co.]

green color. Hay exposed to rain loses a considerable proportion of its vitamin C. Synthesis of vitamin D occurs only when alfalfa is cured in the sun, a condition that results in a loss of vitamin A.

The protein content of alfalfa hay exposed to 1.76 inches of rain distributed over a period of 15 days was only 11 per cent compared with prime hay that contained 18.7 per cent protein.[31] Leached bur-clover hay, oat hay, and naturally cured range forage showed the greatest percentage loss in minerals.[27] The loss of crude protein ranged from 1 to 18 per cent of the total, and the loss of nitrogen-free extract, from 6 to 35 per cent of the total according to the nature of the forage. Further losses in nutritive value are probably reflected in impairment of palatability and in loss of color.

Hay Processing

Less than 15 per cent of the 1939 hay crop in the United States was baled. By 1959, 82 per cent was baled, 11 per cent was chopped and only 7 per cent was loose hay stored in outdoor stacks or in barns.[63] Most of the loose hay is wild hay on western ranches where livestock feed on the stacks during the winter.

Baling is done with an automatic field baler, which gathers the hay from the windrow or swath (Figure 8–13). It compresses the hay to a density of 200 to 250 cubic feet per ton, or to about 40 to 50 per cent of the density of settled, stacked loose hay. The rectangular or round bales are mostly tied with twine, and usually weigh 25 to 85 pounds. The throw-type baler, which requires only one man to operate the machine, ejects the bales into a trailer towed behind the baler.[47] Other types force the bales up a shute into a trailer or truck, or drop the bales on the ground. Many of the dropped bales are later loaded mechanically. Recent equipment produces giant round bales. These are picked up and loaded onto trucks with special mechanical devices.

Field baling exposes the hay for so long that the quality of the hay is impaired in the event of wet weather, unless facilities for artificial drying are available so that the hay can be baled while still damp.

Chopping is done with some types of field-forage harvesters which gather the hay from the swath or windrow, chop it, and blow the material into a trailer. When it is still damp, chopped hay can be dried artificially.

Considerable baling and chopping is done by custom operators because this system is more economical than owning the machines unless at least 100 to 200 tons of hay are harvested each season.

Machines for cubing, pelleting or wafering hay in the field have been devised for processing hay that contains 12 to 16 per cent moisture. The mobile field machines have a lower capacity than field balers or forage harvesters.[47]

Larger stationary machines are also operating. The cubing or pelleting of alfalfa has become popular because cattle eat more with less waste than when fed alfalfa hay, feeding can be done mechanically, and the cubes can be stored in less than half of the space required for baled hay.[17, 20] The cubes and pellets also are used in mixed ground feeds. Alfalfa cubes measure $1\frac{1}{4} \times 1\frac{1}{4} \times 2\frac{1}{2}$ inches in size. Pellets are $\frac{1}{4} \times \frac{3}{8}$ inches. Cubing is done in either mobile field machines, or in stationary units which also may cube baled hay after the harvest season. In cubing hay of about 10 per cent moisture content, it is sprayed with water to a moisture content of 14 to 16 per cent as it enters the cuber. Then the hay is rolled, chopped, and compressed through dies. The emerged cubes are then dried down by heated air to a moisture content of 12 to 14 per cent.

Spontaneous Heating of Hay

Some heating is likely to occur when slightly damp hay is stored. Micro-organisms, which are able to multiply when the hay is not too dry, consume the material for growth and energy, and release heat and moisture. When this moisture is sufficient to be detectable the hay is said to be going through a *sweat*. Very dry hay and hay in small piles that allow the heat and moisture to be dissipated are not known to sweat. The moisture released by organisms stimulates further microbiological activity and fermentation, and contributes to increased heating and sweating. Loose hay that contains less than 25 per cent moisture is sufficiently dry at the time of storage to heat or sweat only moderately. Such hay retains its green color and nutritive value. The storage of alfalfa hay of high moisture content causes heavy losses of organic substances, chiefly fats, sugars, and hemicelluloses. The loss in total weight may be as high as 22 per cent.[32]

Stored wet hay is subject to excessive spontaneous heating. A moisture content somewhat greater than 25 per cent may cause the hay to become brown or black. The green color of hay is destroyed when heating temperatures exceed 122° F. Clean brown hay suitable for feeding is formed at temperatures above 131° F., but below 158° F. The dry-matter losses may be very little in brown hay but they may be heavy when the hay becomes black. In Kansas experiments[61] alfalfa hay stacked with 53 per cent moisture sustained a dry-matter loss of 39 per cent, and a large proportion became black.

Damp hay stored in large masses may develop temperatures sufficiently high to produce ignition.[10, 41] The initial production of heat is due mainly to the action of micro-organisms which are capable of raising the temperature of the hay to as high as 158° F., or slightly higher. Temperatures above this (the death point of micro-organisms) probably are the result of fermentation which produces volatile substances. These gaseous substances may ignite when the temperature rises to 374° F. or higher.

Dehydrated Hay

Some alfalfa and, occasionally, other hays are dehydrated to produce a high-quality feed product. In dehydration, hay is put through a drier immediately after it is cut and chopped (Figure 8–14). Forced air in driers conducts heat from the burner to the hay.[3] The rotary drum or high-temperature drier with an inlet temperature, usually

Figure 8–14. Forage harvester for gathering green material for silage, soilage, or dehydration. [Courtesy Allis-Chalmers Manufacturing Co.]

between 1400° and 1500° F. dehydrates the hay in a few minutes. The drum assembly is rotated slowly while the hay is agitated.

The dry meal may be pelleted or mixed with other feeds. In some cases urea is added to make a high concentrate protein supplement. In addition separation of the leaves and stems is a frequent operation, with the leaves being used as a protein and xanthophyll source by poultry feed manufacturers.

Baled hay with less than 20 per cent moisture usually keeps without spoilage. It dries down until it reaches an equilibrium with the moisture in the air. Hay with more moisture may require artificial drying. Molding and heating occur when the moisture content is 30 per cent or more.

Artificial Drying of Hay

A high quality of hay usually is achieved when the crop is allowed to dry for a day or less down to a 35 to 40 per cent moisture content in the windrow. Then it is baled, chopped, or left loose, and hauled in for drying down to a moisture content of about 15 per cent. Some hay is dried in batches with heated air, either on a drying platform or in specially equipped drying wagons. Most of the drying is done in a barn by use of forced unheated air supplied by a large blower.[33] A system of ducts or slotted panels placed on a tight barn floor is covered with 8 feet of loose or chopped hay, or 10 to 14 layers of baled hay. The air that enters the system is forced upward through the hay. The volume of air should be no less than 2 cubic feet per minute per cubic foot of loose hay and 2½ cubic feet per minute for chopped or baled hay. This requires air-gauge pressures of ¾ inch to 2 inches of water. Less pressure is required for shallower layers of hay or for hay with 30 per cent or less moisture. Drying with heated air is faster than with unheated air, but it involves a fire hazard in barn drying. Prolonged heating of hay at temperatures above 140° F. lowers the digestibility of the proteins.

Freshly cut alfalfa usually contains 75 per cent moisture. This requires the removal of 4800 pounds of water to make one ton of hay that contains 15 per cent moisture. When the hay is allowed to wilt in the field down to 40 to 50 per cent moisture, then only 800 to 1400 pounds of water are removed by the drier. The temperatures permissable range from 250° to 275° F. where the gases contact the driest hay.

The nutrient materials in dehydrated hay are about equal to those in the fresh green material, except for the unavoidable loss of some carotene and the more volatile nitrogen compounds in the drying process. The dehydrated crop is 2 to 2.2 per cent higher in protein and also usually higher in carbohydrates and ether extract than the same crop when sun-cured.[3] The natural green color, an indicator of carotene content, is retained to a greater extent in artificially dried hay.

The cost of hay dehydration has been high enough to preclude its use by the average farmer.

Hay dehydration has wider uses where losses from sun curing exceed 25 per cent, especially when hay prices are relatively high.[71]

SILAGE

Economy of Silage

Silage is a moist feed that has been preserved by fermentation in the absence of air. The forage, usually green, is commonly chopped into small sections and stored in a silo. The principal use of the silo is to preserve succulent roughage for winter feedings as well as to save forages that otherwise would be largely wasted, damaged, or lost. While it is impossible to preserve forage crops as silage as cheaply as in the form of hay, properly prepared silage will preserve a greater proportion of the nutritive value of the green plant.[71]

The modern practice of ensiling green forages traces directly to the process of making sour hay in Germany in the 19th century. The green grasses, clover, and vetches were stored in pits, salted at the rate of 1 pound per 100 pounds, thoroughly trampled, and covered. The first attempt to ensile green maize was made in Germany in 1861.[14] The first American silo was built in Maryland in 1876. The silo became popular and spread to all parts of the country.

Crops Used for Silage

Corn (maize) is the principal silage crop in the United States, more than 8 million acres, 100 million tons, being put up annually. Nearly 1 million acres, producing more than 9 million tons of sorghum silage, are harvested in average years. More than 1¼ million acres or 9 million tons of grass, legume, and small-grain silage is produced annually. The best silage is made from carbohydrate-rich crops, i.e., those that contain more than two parts of carbohydrates to one part of protein.

CORN: Corn is nearly the ideal crop for silage. The acre yields of silage range from 4 to 20 tons. A corn crop that produces 50 bushels of mature corn per acre yields 8 to 12 tons of silage.[71] The maximum nutrient value and the largest yield of carotene are generally obtained in corn harvested when the grains are in the glazed or dent stages[69] (Figure 8–15).[46] The grain types usually are superior to late-maturing "silage" corn, since the nutritive value of corn silage is closely associated with the proportion of grain it contains. Late-maturing varieties yield a greater weight of silage per acre but less dry matter than do the grain varieties.[49] The best silage variety is one that

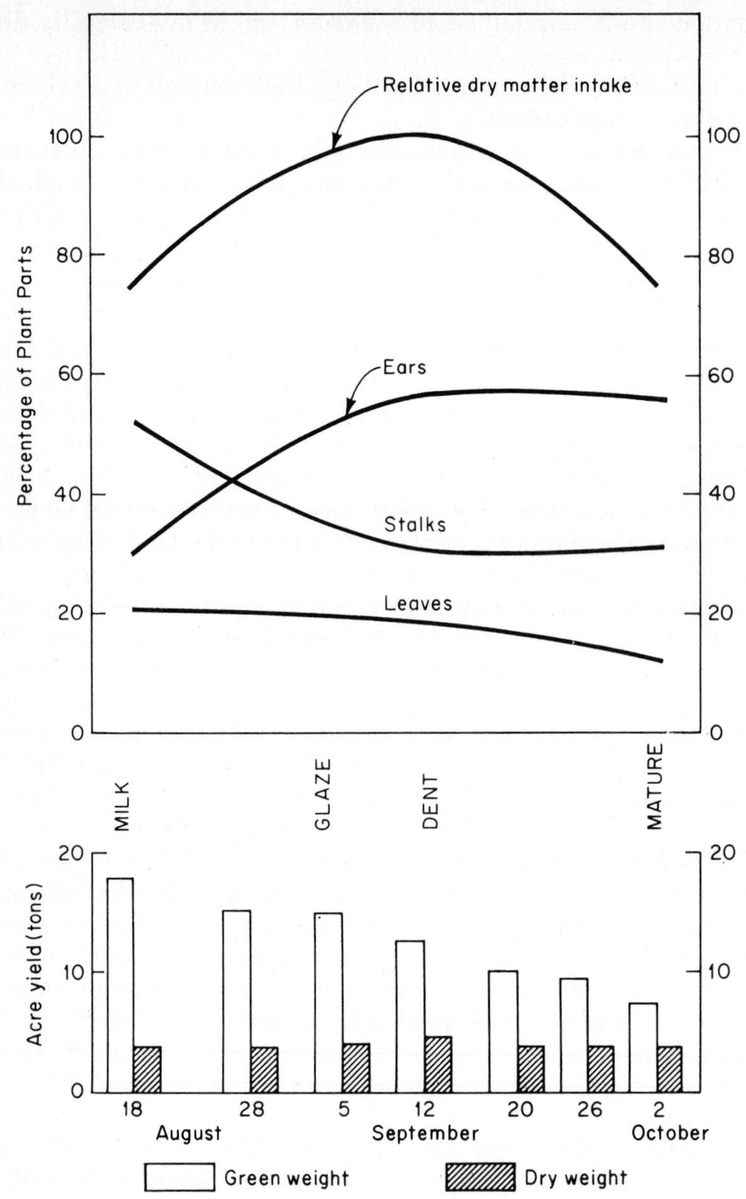

Figure 8–15. Corn nutrient yields as related to kernel development.

utilizes the growing season to the best advantage in production of dry matter but at the same time reaches, at least three years in five, a stage of maturity that may be loosely described as the dough stage. There is less loss of dry matter in storage when corn is ensiled at the more mature stages.

SORGHUM: Sorghum often replaces corn as a silage crop under dry or hot conditions, particularly in the southern Great Plains. The sorgos are more satisfactory from the tonnage standpoint, but grain sorghums make a high-quality silage. Sorghum silage appears to be slightly less palatable than corn silage. Cured sorghum and even damp threshed sorghum grain frequently have been ensiled successfully. Water is added to the grain to raise the moisture content to about 30 to 50 per cent, the material is covered, and the usual silage fermentation process begins. Sorghums for silage should be cut when the seed is in the stiff-dough stage. The addition of 10 pounds of urea to each ton of sorghum silage increases its nutritive value.[8]

OTHER CROP PLANTS: Forage grasses and legumes are widely used as silage crops, especially when weather conditions make it difficult to cure good hay. Grasses can be siloed alone or with legumes almost as readily as can corn, provided the crop is dried for several hours before it is put in the silo. The material should be finely chopped, packed to exclude air, and possibly weighted. Timothy and other grasses ensiled this way make highly palatable silage with an agreeable odor without excessive losses of dry matter. They are cut for silage when heading begins.

Legumes (alfalfa, soybeans, clover, field peas, and vetch) have been ensiled successfully. Excellent silage can be made from alfalfa, and palatability tests show it to rank next to corn.[66] Alfalfa is cut at a stage of one-tenth to one-fourth bloom, clover at one-half bloom, and the large-seeded legumes when seeds are forming. Sweetclover, while less palatable as silage than alfalfa, is satisfactory when harvested at a more mature stage than for hay. In one experiment sweetclover silage contained 16 per cent digestible protein and 57 per cent total digestible nutrients on a dry basis, as compared with 5 per cent digestible protein and 70 per cent total digestible nutrients in corn silage.[15]

Sugarbeet tops often are made into silage. The chief characteristic of this silage is the enormous crude-ash content (ash plus dirt) (Chapter 36) ranging from 20 to 58 per cent on a water-free basis. The composition of dirt-free tops on the basis of percentages in water-free material is ash, 15; protein, 15; nitrogen-free extract, 49; fat, 3; and fiber, 18.

Sunflowers formerly were grown as a silage crop where the season is too short or too cool for corn or sorghums.[50] Addition of salt at the time of ensiling apparently reduces or eliminates the resinous odor and flavor, and improves the palatability of sunflower silage.[56] The crude protein content of the silage is highest when the seeds are three-fourths mature. Silage from sunflowers harvested at the bud or full-bloom stage is less palatable than that cut at later stages of maturity. The nutritive value increases as the plants approach maturity.

WEEDS AND OTHER PLANTS: Silages made from wild sunflowers and certain other weeds with a rather strong characteristic odor, such as cocklebur and ragweed, are eaten by livestock with reluctance. Among the silages entirely refused are those made from woody plants with a strong undesirable characteristic odor or taste such as gumweed and wild sage.[66]

Russian thistles have been used for making silage. They should be harvested when the spines start to feel prickly but while still fairly soft. Analyses have shown Russian thistles to be equivalent to alfalfa in protein and fat content and superior in their carbohydrate-crude-fiber ratio.[18] Russian thistles have a high mineral content with over 8 per cent of potash (K_2O). They are not recommended for silage when reasonably good yields of corn, sorghums, or sunflowers can be obtained. In a majority of cases, weeds prove to be an expensive source of silage.

Silage Formation

In a tight silo, respiration by micro-organisms in the fresh silage uses up the oxygen in a few hours.[51] Then anaerobic bacteria start fermentation. Carbon dioxide increases rapidly for 2 to 3 days until it comprises 60 to 70 per cent of the gaseous atmosphere, the remainder being mostly nitrogen. Acid-forming lactic acid bacteria multiply rapidly from 2 to 4 thousand up to a billion in a gram of good silage.[40]

The temperature of the silage rises for about 15 days after which it subsides and then fluctuates somewhat with the air temperature. Top silage exposed to the air soon spoils. It may reach a temperature of 140° F. The temperature of the interior mass of silage usually does not exceed about 100° F.

The sugars are the principal food for the bacteria that cause fermentation, but the marked destruction of the pentosans and starches indicates that these substances are also used in part. These carbohydrates are converted through the process of fermentation to alcohols and then to acids, principally lactic acid with some acetic and succinic acids and a trace of formic acid.[48] The lactic fermentation is the one desired.[71] Crops such as corn and sorghum have a high content of readily fermentable carbohydrates. Because of their content of bases, they soon develop an acidity of *p*H 4 or below. Forage grasses develop less acidity because of a lower amount of soluble carbohydrates and a higher content of basic elements. The legumes develop still less acid because of a small amount of soluble carbohydrates and a very high calcium content. Good-quality silage has a *p*H of 4.6 or lower. Poor-quality silage contains less lactic acid, but considerable butyric, acetic, propionic, and succinic acids, and ammonia. Butyric acid and ammonia impart bad odors and low palatability to the silage.

The average changes in composition induced in four silage crops by fermentation are shown in Table 8–3. The chief difference after fermentation is a reduction in carbohydrates (nitrogen-free extract). This results in a proportionately higher content of the other components. Biological actions release moisture, but the surplus liquid often seeps out of the silo, carrying away dissolved substances. The *p*H drops from about 5.4 to 6.6 in freshly chopped materials down to 3.6 to 4.6 in good silage. Poor silage fermentation breaks down much of the protein and amino nitrogen into less digestible ammonia forms.[22, 23]

During the ensiling process, some of the green chlorophyll fades, but a fair proportion of the carotene is retained. Considerable carotene is lost in the swath during wilting before the silage is chopped.[71] Good silage has a pleasant aromatic, alcoholic odor and a sour taste. Silage development usually is completed within 12 days.[40]

Losses in dry matter during the fermentation of good silage range from 5 to 20 per cent, with an average of 10 per cent or more. From 5 to 10 per cent of the dry matter is lost as gasses, while seepage losses of similar amounts occur in silages with moisture contents of 72 to 82 per cent. Little or no seepage loss occurs in wilted forage. Losses from spoilage depend largely upon the access of air to the silage. This may range from less than 1 per cent to 2 per cent in a good silo and up to 40 per cent in uncovered silos.[22, 23]

Spoilage is greatly reduced when the silage, except in airtight silos, is covered with a plastic sheet to exclude air. The sheet is then weighted down with sawdust or other waste material. Any hole or crack in the silo or in the covering causes spoilage for some distance from the opening.[2]

TABLE 8–3. Typical Composition of Four Crops and Their Silages

MATERIAL	MOISTURE	COMPOSITION OF THE DRY MATTER				
	(%)	*Crude Protein* (%)	*Ether Extract* (%)	*Crude Fiber* (%)	*Ash* (%)	*N-free Extract* (%)
Corn	62.9	8.1	2.9	16.8	5.1	67.0
Corn silage	69.5	8.3	2.8	22.5	7.1	59.4
Alfalfa	75.9	19.6	2.1	26.9	7.9	45.1
Alfalfa silage	76.0	20.1	3.4	30.4	8.4	38.0
Orchardgrass	78.1	16.8	3.4	27.0	8.0	44.8
Orchardgrass silage	78.8	17.3	5.0	30.4	8.4	38.6
Grass-clover	81.7	18.1	2.7	22.0	10.3	47.8
Grass-clover silage	76.9	19.7	4.9	27.7	9.3	40.6

Types of Silage

Direct-cut grass and legume silages usually contain 72 to 82 per cent moisture. This involves the hauling and filling of a large volume of water with each ton of dry matter. Such a high moisture content causes large dry matter losses and often, but not always, results in poor silage of high *p*H and considerable butyric acid. Good high-moisture silage is obtained by using additives such as ground shelled corn or dried beet pulp (100 to 150 pounds per ton), ground ear corn (175 to 200 pounds per ton), liquid blackstrap molasses at 60 to 100 pounds per ton of silage, sodium metabisulfite at 8 to 10 pounds per ton, or 5 to 10 pounds of a calcium formate-sodium nitrate mixture. Acid additives, such as phosphoric, hydrochloric, and sulfuric acids, lower the *p*H. They prevent undesirable fermentations but also lower the palatability of the silage. These materials are spread over each load of silage before dumping. When tower silos are being filled, the molasses usually is diluted with water to facilitate its movement through the blower, but this increases the moisture content and seepage losses of the silage.[2] The addition of dry feeds, such as corn meal or dried beet pulp lowers the moisture content and provides carbohydrates that ensure good silage.[22] Fermentation may eliminate up to 20 per cent of the nutrients in the added materials. An addition of 0.25 to 0.50 per cent of an enzyme, cellulase, to silage increases the digestibility of the cellulose.

Grasses and legumes for silage are cut when the grasses are just starting to head and alfalfa and clover are in early bloom. Their moisture content does not drop to 70 per cent until the seed is ripe and the leaves begin to dry. The nutritive value and palatability are then greatly reduced. Corn can stand in the field until its moisture content is low enough to make good silage by direct cutting. Sorghum silage often is more moist than typical corn silage but its high sugar content permits it to ferment satisfactorily from direct cutting.

Wilting appears to be the preferred method for preparing grass silage. Wilted forage with 62 to 70 per cent moisture does not require additives to make good silage. Silage with this moisture content favors a larger intake of dry matter by cattle, with a resultant higher milk production. However, the intake and milk production may still be less than that obtained from hay. This system of ensiling usually involves additional labor as well as some loss of nutrients in the field during wilting.[2] Compaction of the silage and the exclusion of air is more difficult than with silages of high moisture content.

Haylage is the name applied to fermented forages with 40 to 60 per cent moisture. It often is difficult to preserve haylage that contains less than 50 per cent moisture.[24] Haylage is best preserved in an air-tight silo because of difficulty in compaction and exclusion of air in other types of silos. Decreasing the moisture content of chopped forage from 80 to 50 per cent produces silage with increasing lactic-acid content and decreasing percentages of total acid and other undesirable acids, and of ammonia nitrogen. Barn-dried hay often has a higher feeding value for dairy cows than silages of haylage made from the same crop.

The moisture content of chopped green feed can be estimated by a squeeze test.[2] Squeeze a handful into a ball, hold it for 20 to 30 seconds, and then release the grip quickly. The condition of the ball upon release indicates the approximate moisture content as follows:

MOISTURE CONTENT (%)	CONDITION OF BALL
75+	Holds shape; considerable free juice
70–75	Holds shape; little free juice
60–70	Falls apart; no free juice
60–	Falls apart readily; no free juice

Making Silage

Corn and sorghum row crops for silage are harvested with a field silage cutter or a forage harvester. Grasses, legumes, and small grains are gathered with a forage harvester. These machines cut and chop the material and blow it into a trailer or truck. Wilted silage and haylage from grasses and legumes are cut with a mower or windrower. Later, the material is gathered and chopped with a forage harvester equipped with a pick-up attachement. The chopped material is hauled to a silo. It is dumped directly into trench and bunker silos. A blower is used to elevate the material into tower and stack silos from a hopper into which the loads are dumped. The material in trench and bunker silos is packed with a tractor that operates continuously during the filling operation.[2] Tower silos may be filled without packing, but tramping or compacting the material in the upper half or two thirds of the silo helps to exclude air and thus improves the quality of the silage. A revolving distributor placed in the silo also compacts the silage. Stack silage usually is tramped manually.

Well-packed settled silage in a trench or bunker silo weighs about 45 pounds per cubic foot. The density is influenced by depth of silage, the moisture present, the proportion of grain to stalk, and the diameter of the silo. Settled silage in a tower silo weighs 45 pounds per cubic foot at a depth of 17 to 18 feet, but the weight varies with the depth in a slightly curvilinear relationship as seen in Figure 8–16.

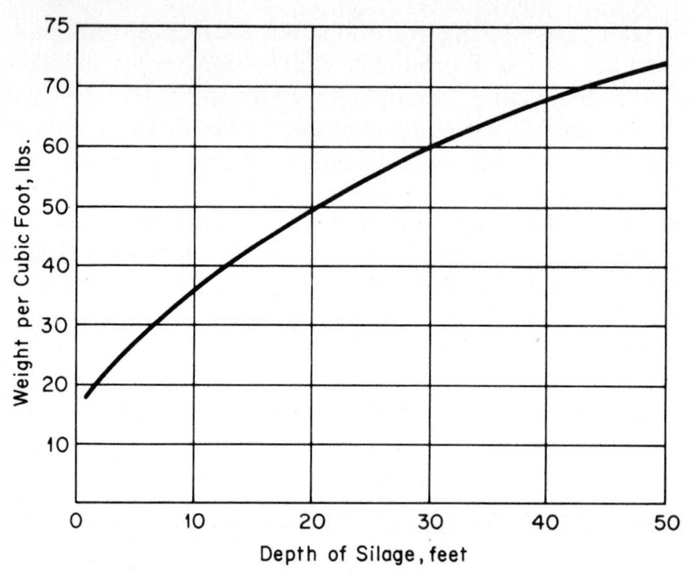

Figure 8–16. Weight per cubic foot of silage at different depths.

Types of Silos

Tower silos are constructed of concrete, brick, vitreous tile, wood staves, or glazed steel.[2] The latter type is airtight with a top seal and has a mechanical unloading device at the bottom. Tower silos are circular, with vertical walls. They range from 12 to 30 feet in diameter and 30 to 60 feet in height with capacities of 50 to 1000 tons. A 2-inch depth from the top of the silage should be used up each day during cold weather, and 3 inches in summer, to avoid spoilage.

Trench silos usually are 10 to 30 feet wide, 8 to 30 feet deep, and any desired length. A few hold as much as 10,000 to 50,000 tons of silage. Many trench silos are dug into the side of an earth bank with one end at ground level. This permits drainage of excess silage juice and also facilitates the removal of silage. Trench silos usually have a concrete floor and the sides also are often lined with concrete.

Bunker silos are built on the ground with concrete floors and either concrete or wooden walls.[2] The walls are braced with pillars. The sides of trench and bunker silos have a slope of 1 or 2 inches per foot. The sloping sides permit the silage to settle and shrink without opening a gap next to the wall to start spoilage. The silage in both bunker and stack silos is covered with a heavy impervious sheet.

Stack silos are temporary circular structures set on the ground. The walls are made of strips of snow fence or of woven wire-fence material spliced at the ends. Two to five overlapping tiers of this fencing give it sufficient height. The walls are lined with sheets of impervious plastic or paper.

About 69 per cent of the silos are of the upright type. Most of the newer silos are of the bunker type because of a lower construction cost. Trench silos are popular in the semiarid and arid regions because of their low cost and because there is less interference from rain during filling or emptying the silos in those areas. The trenches are excavated with a bulldozer or power shovel. The silage usually is covered with straw or other waste material, after which some of the excavated soil is spread over the top. Much of the trench silage is held over for emergency feeding in drought years when feed production and grazing are limited.

Many trench and bunker silos are adapted to self-feeding by means of movable feeding racks at the end. The animals reach through the rack for feed and move it forward as the feed is consumed. Many others are unloaded with mechanical equipment.

SILAGE CONSUMPTION

Cows (900 to 1200 pounds)	30–40 pounds per day
Yearling cattle	15–20 pounds per day
Sheep	4–5 pounds per day
Fattening cattle	25–35 pounds per 1000 pounds live weight

REFERENCES

1. Anonymous. "High quality hay," *USDA Agricultural Research Service Spec. Rept. ARS*, pp. 22–52. 1960.
2. Anonymous. "Making and feeding hay-crop silage," *USDA Farmers Bul. 2186*. 1962.
3. Arny, A. C. "The influence of time of cutting on the quality of crops," *J. Am. Soc. Agron.*, 18:684–703. 1926.
4. Arny, A. C., and Sun, C. P. "Time of cutting wheat and oats in relation to yield and composition," *J. Am. Soc. Agron.*, 19:410–439. 1927.
5. Barnett, A. J. G. *Silage Fermentation*, Academic Press, New York, pp. 1–208. 1954.
6. Bechdel, S. I., and others. "Dehydrated and sun-cured hay," *Penn. Agr. Exp. Sta. Bul. 396*. 1940.
7. Bechdel, S. I., and Bailey, C. H. "Effect of delayed harvesting on quality of wheat," *Cereal Chem.*, 5(2):128–145. 1928.
8. Becker, R. B., and others. "Silage investigations in Florida," *Fla. Agr. Exp. Sta. Bul. 734*, pp. 1–31. 1970.
9. Briggs, L. J. "Harvesting wheat at successive stages of ripening," *Mich. Agr. Exp. Sta. Bul. 125*. 1895.
10. Browne, C. A. "The spontaneous combustion of hay," *USDA Tech. Bul. 141*, pp. 1–28. 1929.
11. Burnett, L. C., and Bakke, A. L. "The effect of delayed harvest upon grain yield," *Iowa Agr. Exp. Sta. Res. Bul. 130*. 1930.
12. Burton, G. W. "The performance of various mixtures of hybrid and parent

inbred pearl millet, *Pennisetum glaucum L. R. Br.*," *J. Am. Soc. Agron.*, 40:908–915, 1948.
13. Camburn, O. M., and others. "Legume and grass silages," *Vt. Agr. Exp. Sta. Bul. 434.* 1938.
14. Carrier, L. "The history of the silo," *J. Am. Soc. Agron.*, 12:175–182. 1920.
15. Christensen, F. W., and Hopper, T. H. "Digestible nutrients and metabolizable energy in certain silages, hays, and mixed rations," *J. Agr. Res.*, 57:477–512. 1938.
16. Clanton, D. C., and Matsushima, J. K. "Sorghum and alfalfa silages for wintering beef cattle," Neb. Agr. Exp. Sta. S.B. 461, pp. 1–15. 1960.
17. Dobie, J. B., and Curley, R. G., "Hay cube storage and feeding," Calif. Agr. Exp. Sta. Circ. 550, pp. 1–16. 1969.
18. Donaldson, F. T., and Goering, J. J. "Russian thistle silage," *J. Am. Soc. Agron.*, 32:190–194. 1940.
19. Ferguson, W. L., and others, "Hay harvesting practices and labor used" Econ. Res. Svc. and Stat. Rpt. Govt. Print. office. 1967.
20. Forester, E. M., and Wilson, E. B., "The feasibility of operating a stationary hay cubing installation in the Columbia Basin," Wash. Ext. E. M. 3423, pp. 1–24. (Rev.) 1971.
21. Friesen, O. H. "Combine operation and adjustment," *Can. Dept. Agr. Publ. 1464,* pp. 1–30. 1972.
22. Gordon, C. H., and others. "Some experiments in preservation of high moisture hay-crop silages," *J. Dairy Sci.*, 11(7):789–799. 1957.
23. Gordon, C. H., and others. "Comparisons of unsealed and plastic sealed silages for preservation efficiency and feeding value," *J. Dairy Sci.*, 14(6):1113–1121. 1961.
24. Gordon, C. H., and others. "Preservation and feeding value of alfalfa stored as hay, haylage, and direct-cut silage," *J. Dairy Sci.*, 14(7):1299–1311. 1961.
25. Grandfield, C. O. "The trend of organic food reserves in alfalfa roots as affected by cutting practices," *J. Agr. Res.*, 50(8):697–709. 1935.
26. Gray, R. B. "Harvesting with combines," *USFA Farmers Bul. 1761.* 1955.
27. Guilbert, H. R., and others. "The effect of leaching on the nutritive value of forage plants," *Hilgardia,* 6:13–26. 1931.
28. Harlan, H. V. "Daily development of kernels of Hannchen barley from flowering to maturity at Aberdeen, Idaho," *J. Agr. Res.*, 19(9):393–429. 1920.
29. Harlan, H. V., and Pope, M. N. "Development of immature barley kernels removed from the plant," *J. Agr. Res.*, 27:669–678. 1926.
30. Harmond, J. E., Smith, J. E., Jr., and Park, J. K. "Harvesting seed of grasses and legumes," in *Seeds,* USDA Yearbook, 1961, pp. 181–188.
31. Headden, W. P. "Alfalfa," *Colo. Exp. Sta. Bul. 35.* 1896.
32. Hoffman, E. J., and Bradshaw, M. A. "Losses of organic substance in the spontaneous heating of alfalfa hay," *J. Agr. Res.*, 54:159–184. 1937.
33. Hukill, W. V., Saul, R. A., and Teare, D. W. "Outrunning time; combatting weather," in *Power to Produce,* USDA Yearbook, 1960, pp. 183–199.
34. Hunt, C. H., and others. "Effect of the stage of maturity and method of curing upon the vitamin B and vitamin G content of alfalfa, clover, and timothy hays," *J. Agr. Res.*, 51:251–258. 1935.

35. Kane, Edward A., and others. "The loss of carotene in hays and alfalfa meal during storage," *J. Agr. Res.*, 55:837–847. 1937.
36. Kiesselbach, T. A., and Anderson, A. "Alfalfa investigations," *Nebr. Agr. Exp. Sta. Res. Bul. 36.* 1926.
37. Kiesselbach, T. A., and Anderson, A. "Curing alfalfa hay," *J. Am. Soc. Agron.*, 19:116–126. 1927.
38. Kiesselbach, T. A., and Anderson, A. "Quality of alfalfa hay in relation to curing practice," *USDA Tech. Bul. 235.* 1931.
39. Klein, L. M., and Harmond, J. E. "Seed moisture – A harvest timing index for maximum yields," *Trans. ASAE* 14(1):124–26. 1971.
40. Langston, C. W., and others. "Microbiology and chemistry of grass silage," *USDA Tech. Bul. 1187.* 1958.
41. LeClerc, J. A. "Losses in making hay and silage," in *Food and Life*, USDA Yearbook, 1939, pp. 992–1016.
42. Leighty, C. E. "The better utilization of straws," *J. Am. Soc. Agron.*, 16:213–224. 1924.
43. Martin, J. H. "The influence of the combine on agronomic practices and research," *J. Am. Soc. Agron.*, 21:766–773. 1929.
44. McCalmont, J. R. "Silo types and construction," *USDA Farmers Bul. 1820*, pp. 1–62. 1939.
45. McCalmont, J. R. "Farm silos," *USDA Misc. Pub. 810*, pp. 1–28. 1967.
46. McCulloch, M. E. "Silage research at the Georgia Station," *Ga. Agr. Exp. Sta. Rsh. Rpt. 75*, pp. 1–46. 1970.
47. Miller, H. F., Jr. "Swift untiring harvest help," in *Power to Produce*, USDA Yearbook, 1960, pp. 164–183.
48. Neidig, R. E. "Chemical changes in silage formation," *Iowa Agr. Exp. Sta. Res. Bul. 16.* 1914.
49. Nevens, W. B. "Types and varieties of corn for silage," *Ill. Agr. Exp. Sta. Bul. 391.* 1933.
50. Odland, T. E., and Henderson, H. O. "Cultural experiments with sunflowers and their relative value as a silage crop," *W. Va. Agr. Exp. Sta Bul. 204.* 1926.
51. Olson, G. A. "A study of factors affecting the nitrogen content of wheat and the changes that occur during the development of wheat," *J. Agr. Res.*, 24:939–953. 1923.
52. Ragsdale, A. C., and Turner, C. W. "Silage investigations – Loss of nutrients during field curing of corn," *Mo. Agr. Exp. Sta. Res. Bul. 65.* 1924.
53. Reynoldson, L. A., and others. "The combined harvester thresher in the Great Plains," *USDA Tech. Bul. 70.* 1928.
54. Reynoldson, L. A., Humphries, W. R., and Martin, J. H. "Harvesting small grains, soybeans, and clover in the corn belt with combines and binders," *USDA Tech. Bul. 244.* 1931.
55. Salmon, S. C., Swanson, C. O., and McCampbell, C. W. "Experiments relating to time of cutting alfalfa," *Kans. Agr. Exp. Sta. Tech. Bul. 15*, pp. 1–50. 1925.
56. Schoth, H. A. "Comparative values of sunflower silages made from plants cut at different stages of maturity, and the effect of salt on the palatability of the silage," *J. Am. Soc. Agron.*, 15:438–442. 1923.

57. Schwantes, A. J., and others. "The combine harvester in Minnesota," *Minn. Agr. Exp. Sta. Bul. 256*. 1929.
58. Sotola, J. "The chemical composition and nutritive value of certain cereal hays as affected by plant maturity," *J. Agr. Res.*, 54:399–419. 1937.
59. Stoa, T. E. "The early harvest of rusted wheat," *J. Am. Soc. Agron.*, 16:41–47. 1924.
60. Strickler, P. E., and Kendall, J. R. "Silos, silage handling practices and minor feed products," *USDA Stat. Bul. 417*, pp. 1–16. 1968.
61. Swanson, C. O., and others. "Losses of organic matter in making brown and black alfalfa hay," *J. Agr. Res.*, 18:299–304. 1919.
62. Trowbridge, P. F., and others. "Studies of the timothy plant, Part II," *Mo. Agr. Exp. Sta. Res. Bul. 20*. 1915.
63. Van Arsdall, R. N. "Forage mechanization," *Crops and Soils*, 15(6):10–12. 1962.
64. Walster, H. L. "Sweet clover as a hay crop," *J. Am. Soc. Agron.*, 16:182–186. 1924.
65. Westover, H. L. "Comparative shrinkage in weight of alfalfa cured with leaves attached and removed," *USDA Bul. 1424*. 1926.
66. Westover, H. L. "Silage palatability tests," *J. Am. Soc. Agron.*, 26:106–116. 1934.
67. Whitcomb, W. O., and Johnson, A. H. "Effect of severe weathering on certain properties of wheat," *Cereal Chem.*, 5(2):117–128. 1928.
68. Wiant, D. E., and Patty, R. L. "Combining grain in weed-free fields," *S. Dak. Agr. Exp. Sta. Bul. 251*. 1930.
69. Wiggans, R. G. "The influence of stage of growth on the composition of silage," *J. Am. Soc. Agron.*, 29:456–467. 1937.
70. Woodman, H. E., and Engledow, F. L. "A study of the development of the wheat grain," *J. Agr. Sci.*, 14:563–586. 1924.
71. Woodward, T. E., and others. "The nutritive value of harvested forages," in *Food and Life*, USDA Yearbook, 1939, pp. 956–991.

Chapter 9

Handling and Marketing Grain, Seeds, and Hay

MARKETING GRAIN

Grain from the farm is trucked mostly to elevators or mills built at railroad sidings. An elevator (Figure 9–1) is equipped with a truck scale combined with a dump that tilts the truck or trailer so that the bulk grain flows out the rear end of the vehicle into a pit. The grain is picked up from the pit by buckets attached to an endless belt which carries it up through a *leg* to the top (cupola) of the elevator, from which it is spouted to bins. It later is spouted into a car from a bin or the cupola.[11, 14] Successful country elevators handle grain equivalent

Figure 9–1. Grain elevator with concrete silo-type bins.

to five or more times their storage capacity during a marketing season. Consequently, they can store only a small proportion of the total crop.

An elevator may be owned by a farmers' cooperative association, an independent operator, or by a corporation operating several elevators. The three types are commonly referred to as co-op, independent, and line elevators, respectively. The same prices are nearly always paid by all elevators for grain of a given grade at a particular shipping point. The local price is based upon a card price that is received daily from the terminal market to which the grain usually is shipped. The card price is determined by the terminal price for a particular grade and quality of grain on that day, less the freight to the terminal market, less the fees for handling the grain at the local elevator and the terminal market.

The farm price of grain in any region in the United States is based on Chicago prices. For example, wheat in the surplus-producing regions brings a lower price as the shipping distance west of Chicago increases. The highest prices are paid in the chief deficiency areas, the southern and the Atlantic seaboard states. The lowest prices are in the intermountain states. In the Pacific Coast states, prices are higher than in the intermountain states because of the sea outlet.

Before a railroad car is loaded with bulk grain, all cracks and holes are coopered to prevent leakage. Grain doors are nailed on the inside of the car door up to within 2 or 3 feet of the top of the door. The car is loaded up to the allowable weight, but a clearance of about $2\frac{1}{2}$ feet is left at the top to permit accurate sampling. The car door is then sealed. A car loaded with two kinds, classes, or grades of bulk grain separated by a board partition is called a bulkhead car.

Most grain is shipped to a commission firm, elevator firm, or cooperative agency at a terminal market. The shipper, who uses the bill of lading as security, draws a draft upon the consignee for a substantial portion of the expected selling price of the grain. The draft may be payable on sight or when the grain arrives at the market. The consignee sells the grain. He remits any balance due after deducting his commission, weighing fees, inspection fees, freight, and other charges.

MARKETING SEEDS

Hybrids and varieties of crop seeds are often produced by seed companies, frequently under contract with growers. Large quantities, however, are purchased directly from farmers, or indirectly from local or wholesale seed buyers. Later, individual seed companies may trade with others by buying or selling to provide the quantities required for their expected market. Considerable quantities of seeds are exported to, or imported from, other countries. The regulations concerning seed marketing are described in Chapter 7.

HANDLING GRAIN AT A TERMINAL MARKET

Upon arrival at the market, the cars of grain are sampled by licensed inspectors, while other samples taken by authorized representatives are furnished to the consignee. The latter samples, with accompanying grade cards, are placed on tables and offered to buyers on a large cash-grain trading floor at the Grain Exchange.[19]

At the terminal elevator or mill, the grain is unloaded with a power dragline scoop or by a dump that tilts up the car and pours the grain into a hopper or pit. The terminal elevators usually are equipped with power windlasses and cables to move cars. Inside the terminal elevator the operations are much the same as in a country elevator except that for the horizontal movement of grain an endless belt about 3 to 5 feet wide (Figure 9–2) is used in a large terminal elevator, whereas an auger conveyor ordinarily is used in a country elevator.

Figure 9–2. The movable automatic "dump" takes grain from the belt conveyor and spouts it into the top of a bin.

Grain and seeds are moved by pneumatic systems[12] in some elevators, and some bulked field seeds are moved by vibrating conveyers.[28] Terminal elevators along a waterfront have marine legs which enclose pneumatic systems or conveyor buckets on an endless belt. These are lowered into the hold of a boat to unload bulk grain. Modern terminal elevators have many large circular bins or silos, each with a capacity of 25,000 to 110,000 bushels. Such an elevator may hold several million bushels of grain. Each bin is equipped with wired thermocouples (usually called the Zeleny System), about 5 feet apart from top to bottom. Thus, temperatures in all parts of each bin are measured with a potentiometer located inside the building. Whenever a hot spot (90° F. or more) is detected, the grain is aerated, or moved to another bin, so that the moist or heating portion is mixed with the larger bulk of cool sound grain.

At a terminal market cash grain is sold on the basis of the sample and the official grade. The agreed price is usually a stated premium above the future price for that particular grade of grain. The future price is based upon grain of contract grade, i.e., grain that barely meets the minimum requirements for the particular grade, except for a small allowance for variations. Grain received from a country elevator may grade No. 2 because of a test weight per bushel below the requirement for No. 1, but it may more than meet the minimum requirements for the No. 1 grade with respect to other grading factors such as foreign material, damage, moisture content, and mixtures of grain of other kinds or classes. Consequently, such grain commands a premium over the price for contract grain of No. 2 grade. These expected premiums are reflected back in the posted card price at a country shipping point. In a typical transaction, a lot of corn of No. 3 grade may be sold at the terminal market for 5 cents a bushel over the July, September, December, or May future price for that grade. The actual sale price is determined soon thereafter when the hedge is closed by the purchase of futures at the prevailing price.

HEDGING

Speculative risks in buying grain, which result from changes in prices, are avoided by the practice of hedging the purchases for each day.[35] Hedging consists of selling futures whenever cash grain is bought and later closing the hedge by buying futures when the grain is sold. The hedging is done for the country dealer by his broker or other representative at the terminal market. For example, a country dealer has just bought 5,000 bushels of wheat from local farmers. He wires his broker to sell 5,000 bushels of wheat futures on the grain exchange. Some days or weeks later the 5,000 bushels of wheat arrive at the terminal market. The broker sells the wheat and buys back the futures for his client. In the meantime, when the price of wheat has advanced 10 cents a bushel, the country dealer makes an extra profit of $500 on the cash grain. But the futures, bought back at the advanced price of wheat, bring a loss of $500. However, when the price of wheat has dropped 10 cents a bushel, the dealer loses $500 on the cash grain. But the futures are bought back for $500 less than the amount for which they are sold, or a profit of $500. Losses or gains due to changes in the market price of the actual (cash) grain are thus balanced by corresponding gains or losses in the transactions with grain futures. A small fee is charged for the futures transactions. Millers and grain dealers at the terminal markets likewise hedge their purchases of grain. This permits a miller to sell flour for delivery several months later without a speculative risk.[33]

FUTURES TRADING

Dealing in grain futures began in Chicago in 1848. The transactions are conducted by men standing in or around a pit located in the room in which cash grain also is sold (Figure 9–3). The smallest unit for futures trading of grain usually is 5,000 bushels. With two motions of the arm accompanied by hand signals, a dealer may offer to buy or sell as much as 25,000 bushels of grain at an indicated price. A signal accompanied by a nod indicates acceptance. The shouting around the pit merely helps attract the attention of a buyer or seller. The transactions are recorded as well as conducted under strict rules established by the particular Grain Exchange or Board of Trade. These exchanges, which operate the marketing facilities, are subject to federal regulation under the Commodity Exchange Act. Only members of the particular exchange may conduct future trading operations, but members may act as brokers for others. Speculators as well as grain dealers may trade in futures through their brokers. Prices of grain futures fluctuate constantly with changes in present and prospective supplies, demands for the different cash grains, changes in the volume, and trends in futures trading. The prices of cash grain roughly follow the futures prices. The volume of futures trading appears to be very large in proportion to the quantity of grain marketed. This is due partly to speculating in grain futures. However, in hedging, the buyer of each lot of cash grain sells futures and then later buys them back as a separate transaction. Some lots of grain change hands several times, with hedging transactions each time.

Although the speculative operations may tend either to disrupt or stabilize the market at any given time, they offer additional outlets for, and lend flexibility to, ordinary hedging operations. Without speculators it might be difficult to hedge large sales of grain for export or processing. A speculator is in a *long* position when his purchases of

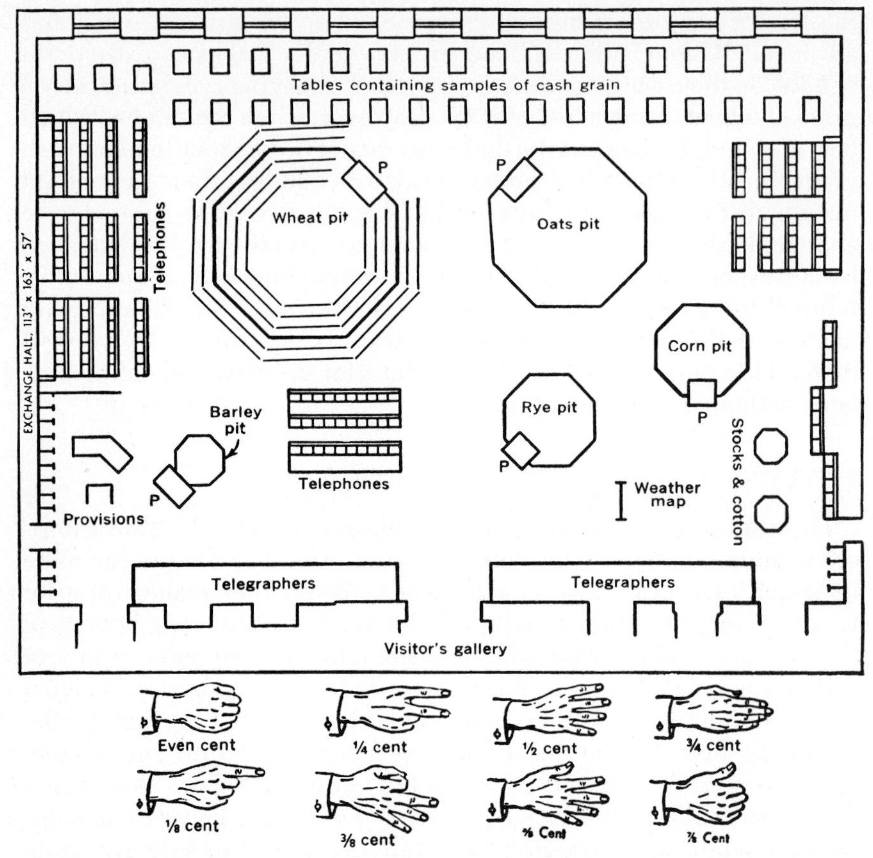

Figure 9–3. Diagram of grain trading floor at Chicago, Illinois, which shows the tables for displaying samples of cash grain offered for sale, as well as the step-bordered pits for futures transactions. Changes in price are recorded at the platforms marked *P*. Below are shown the hand signals used to indicate the fractional prices in futures transactions.

futures exceed those that he has sold. In that case he expects the price to rise. When he expects the price to go down he sells *short*. All sales of grain futures are in fact contracts to deliver the grain on or before the end of the month (May, July, September, or December) indicated in the transaction. Only occasionally is delivery called for, cash settlements being much more simple. When delivered on a future sale, the grain nearly always is of a contract grade stored in a terminal elevator. Contract grain frequently consists of varieties or lots of mediocre quality for processing.

The leading terminal grain markets of the United States are Chicago, Kansas City, Minneapolis, and Omaha.

GRAIN GRADING

Grain-Grading Factors

The purpose of grading grain is to facilitate its marketing, storage, financing, and future trading.[1,5] Grain grading standardizes trading practices and reduces the hazards of merchandising. The factors upon which grades are based determine the quality or market value for the purpose for which the particular grain is generally used. Thus plumpness, which usually is measured by the test weight per bushel, indicates a well developed endosperm which in wheat may be expected to give a high yield of flour. Plump barley gives a higher yield of malt than does thin barley. In all grains a high test weight indicates a low percentage of hull or bran or both, i.e., a low percentage of crude fiber, which reflects an enhanced feeding value.

The test weight of a lot of wheat depends upon its plumpness, shape, density, and moisture content. Plump rounded grain packs together more closely than does wrinkled, shriveled, or angular grain. Dense vitreous kernels are heavier than starchy kernels. In a given lot of wheat, the test weight varies inversely with the moisture content. Wetting causes the starch to swell. Furthermore, the wheat grain has a specific gravity of about 1.41 compared with that of water, which is 1.00. When wetted wheat grains dry out they shrink, but not entirely back to their original volume. Each repeated wetting and drying makes the grain lighter in bushel weight. Fully ripe wheat that tests 62 pounds per bushel may test only 55 pounds per bushel when soaked. After several wettings it may test only 59 pounds after it is thoroughly dried, yet it may have deteriorated only slightly in milling and baking quality.

Other grains likewise swell and decrease in test weight when they are wetted. Flaxseed, on the other hand, increases in test weight as its moisture content increases because it contains no starch to absorb water and swell. Furthermore, the 40 per cent of oil in the seed has a lower specific gravity than water, about 0.93, and thus the added water increases the weight of the seed.

The test weight per bushel of any grain is increased by cleaning out chaffy or shriveled kernels and other light material. Thereafter additional cleanings will further increase the test weight of oats and barley by breaking off awn remnants, hairs, and glume tips, which permits closer packing in the test kettle. The test weight of oats is increased merely by transferring the grain from one concrete bin to another because of the abrasive effects just mentioned. The test weight of clean wheat is not altered appreciably by such handling. With flaxseed, however, each successive cleaning or handling reduces the test weight because the resultant roughening of the originally smooth and glossy seed coat causes the seeds to pack less closely.

In addition to plumpness or test weight per bushel, the chief factors that contribute to high quality are soundness, dryness, cleanliness, purity of type, and general condition. Grain that is sour, musty, heated, or moldy is unpalatable, disagreeable in odor, and often unfit for food or feed. Binburnt, severely sprouted, or severely frosted wheat makes flour of poor baking quality. Dry grain keeps well in storage, whereas wet or damp (tough) grain is likely to spoil. Furthermore, the moisture is valueless because the shrinkage in weight when drying occurs is a complete loss to the purchaser. A lot of grain containing 20 per cent moisture (80 per cent dry matter), when dried down for safe storage to a moisture content of 13 per cent (again on a wet basis), shrinks more than 8 per cent in total weight. Most of the slightly damp grain that reaches the terminal markets is not dried artificially, but is mixed thoroughly with drier grain.

Cleanliness is important except in grain containing easily separated foreign material that has a value in excess of cleaning costs plus cleaning shrinkage. In the grading of wheat, barley, rye, flax, and grain sorghum the foreign material removed by dockage testers or appropriate screening devices is considered dockage. The weight of the dockage is subtracted from the total weight of these grains in marketing transactions except when the dockage amounts to less than 1 per cent of the total weight of the grain. Impurities, i.e., foreign material, that cannot be separated readily may detract seriously from the value of the grain. Certain foreign seeds that often remain in wheat after cleaning, such as garlic, rye, and kinghead affect the quality (color, flavor, or texture) of bread made from the wheat. Stones and cinders the size of grains are particularly objectionable in grain because they are not separated readily by cleaning machines. As a result, they may injure grinding machinery or contaminate the milled products.

Purity of type is important, except when grain is used for feed. A mixture of grains of different sizes is difficult to clean without heavy shrinkage losses. Different classes of wheat require different temper-

Figure 9–4. Sampling grain with metal probe that has partitioned slots in the core.

ing and grinding procedures for milling. Admixtures of inferior-quality varieties lower the quality of the entire lot. In toasted wheat products, which are made from white wheats, any red grains present become so dark in color that they appear to be scorched. Mixed barley types do not sprout uniformly during the malting process. Yellow corn mixed with white corn makes an unattractive hominy or meal. Toasted flakes of yellow corn kernels, which are very dark in color, detract from the appearance of corn flakes made from white corn.

Cracked grain is unsuitable for certain grain products. Bleached, weathered, discolored, or off-colored grain is unattractive.

Grain Sampling and Inspection

Grain sampling is the first essential step in grain inspection and grading[15] (Figure 9–4). A carload of bulk grain is sampled by inserting a trier or probe through the grain to the bottom of the car at five or more places and withdrawing it after it fills with grain (Figure 9–5). The ordinary probe consists of a 63-inch double brass tube with partitioned slots in the core, and corresponding slots on one side in the outer shell. Turning the core opens or closes the openings. Sacked grain and seeds are sampled with a small pointed bag trier. A falling stream of bulk grain from an elevator spout is sampled by cutting

Figure 9–5. Equipment for grading grain to determine: test weight (*left on table*), dockage and foreign material (*middle*), and a sampler or divider (*right*). In the right background are a pearler and moisture meter.

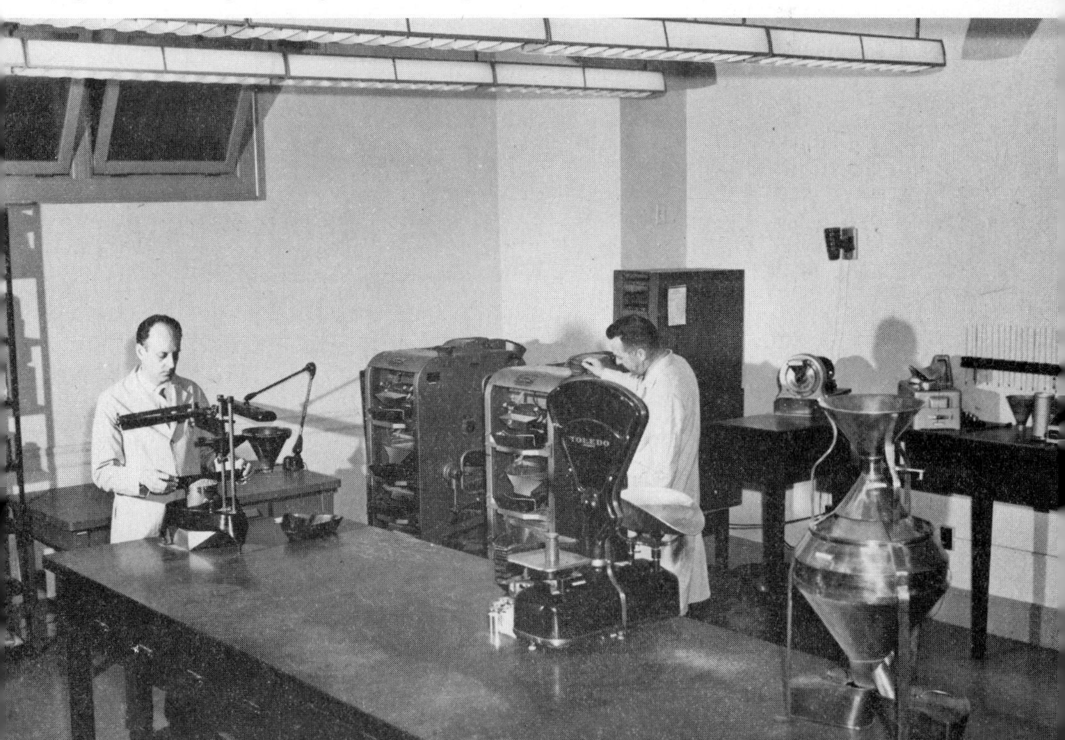

across the grain stream at intervals with a dipperlike device called a *pelican*. After a brief examination for quality and uniformity, the probe or trier samples are bulked together. Before grading, all samples are mixed and cut down to a suitable size in a Boerner divider.

At the inspection laboratory, the market class, test weight, and the dockage (or foreign material in corn and oats) are determined on all samples. (Figure 9–6). Any other factor that seems likely to determine the grade also is analyzed. For example, the moisture content is determined when the grain feels even slightly damp or tough. Likewise, when damaged grains, or foreign material left after the dockage is removed, or grains of other market classes appear to be present in sufficient amount to modify the grade, these factors also are measured. Grain is graded by inspectors licensed and supervised by the United States Department of Agriculture under the authority of the United States Grain Standards Act passed in 1916 (Table 9–1).

The inspection certificate that gives the grade and test weight also designates the car number or elevator bin in which the grain was sampled. Both seller and buyer have access to these records.

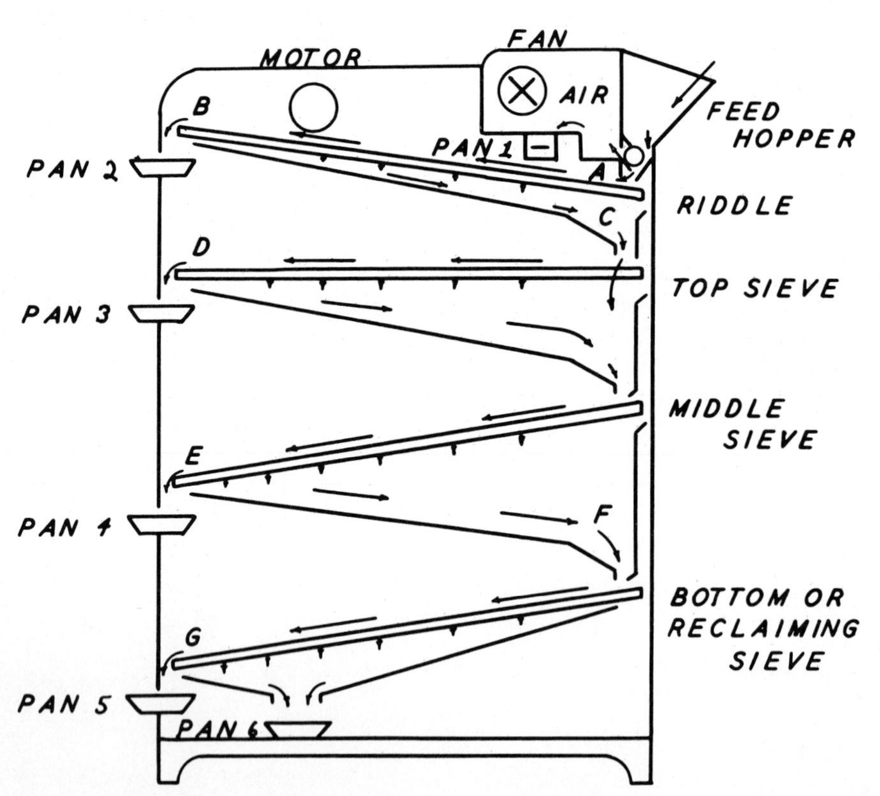

Figure 9–6. Screens and a suction fan separate dockage, broken grains, and foreign materials from a sample.

TABLE 9–1. Wheat: Grade Requirements for All Classes of Wheat Except Mixed Wheat[a]

GRADE	MINIMUM TEST WEIGHT PER BUSHEL		MAXIMUM LIMITS OF: DEFECTS					MAXIMUM LIMITS OF: WHEAT OTHER CLASSES[b]	
	Hard Red Spring Wheat (*lb.*)	*All Other Classes* (*lb.*)	*Heat-Damaged Kernels* (%)	*Damaged Kernels* (*Total*) (%)	*Foreign Material* (%)	*Shrunken and Broken Kernels* (%)	*Defects* (*Total*) (%)	*Contrasting Classes* (%)	*Wheat Other Classes* (*Total*) (%)
1	58	60	0.1	2	0.5	3	3	0.5	3
2	57	58	0.2	4	1.0	5	5	1.0	5
3	55	56	0.5	7	2.0	8	8	2.0	10
4	53	54	1.0	10	3.0	12	12	10.0	10
5	50	51	3.0	15	5.0	20	20	10.0	10

[a] *Sample Grade:* Sample grade is wheat that does not meet the requirements for any grades from No. 1 to No. 5 inclusive; or which contains stones; or which is musty, sour, or heating; or which has any commercially objectionable foreign odor except of smut or garlic; or which contains a quantity of smut so great that any one or more of the grade requirements cannot be applied accurately; or which is otherwise of distinctly low quality.

[b] Red durum wheat of any grade may contain not more than 10 per cent of wheat of other classes.

TABLE 9–2. Corn: Grade Requirements for Classes of Yellow Corn, White Corn, and Mixed Corn[a]

GRADE	MINIMUM TEST WEIGHT PER BUSHEL	MAXIMUM LIMITS OF			
	(*lb.*)	*Moisture* (%)	*Cracked Corn and Foreign Material* (%)	*Damaged (Total)* (%)	*Heat-Damaged Kernels* (%)
1	56	14.0	2	3	0.1
2	54	15.5	3	5	0.2
3	52	17.5	4	7	0.5
4	49	20.0	5	10	1.0
5	46	23.0	7	15	3.0

[a] *Sample Grade:* Sample grade is corn that does not meet the requirements for any of the grades from No. 1 to No. 5 inclusive; or which contains stones; or which is musty, sour, or heating; or which has any commercially objectionable foreign odor; or which is otherwise of distinctly low quality.

TABLE 9–3. Oats: Grade Requirements for the Classes White Oats, Red Oats, Gray Oats, Black Oats, and Mixed Oats[a]

GRADE	MINIMUM LIMITS OF		MAXIMUM LIMITS OF		
	Test Weight per Bushel (*lb.*)	*Sound Cultivated Oats* (%)	*Heat-Damaged Kernels* (%)	*Foreign Material* (%)	*Wild Oats* (%)
1[b]	34	97	0.1	2	2
2[c]	32	94	0.3	3	3
3[d]	30	90	1.0	4	5
4[e]	27	80	3.0	5	10

[a] *Sample Grade:* Sample grade is oats that do not meet the requirements for any of the grades No. 1 to No. 4 inclusive; or which contain more than 16.0 per cent of moisture; or which contain stones; or which are musty, or sour, or heating; or which have any commercially objectionable foreign odor except of smut or garlic; or which are otherwise of distinctly low quality.

[b] The oats in grades No. 1 White oats may contain not more than 5.0 per cent of red oats, gray oats, and black oats, singly or in combination, of which not more than 2.0 per cent may be black oats.

[c] The oats in grades No. 2 White oats may contain not more than 3.0 per cent of black oats.

[d] Oats that are slightly weathered shall be graded not higher than No. 3.

[e] Oats that are badly stained or materially weathered shall be graded not higher than No. 4.

The moisture content of the grain is measured almost instantaneously in a moisture meter which depends for its action upon a lower electric resistance in grain high in moisture.

Grades of Grain

For grading, wheat is divided into seven classes: Hard Red Spring, Durum, Red Durum, Hard Red Winter, Soft Red Winter, White, and Mixed. Mixed Wheat contains less than 90 per cent wheat of one class.[7] Contrasting classes of wheat are those of different color, or durum-common wheat mixtures.

Summarized tabulations of the grades of wheat, corn, and oats are shown in Table 9–1, 9–2, and 9–3.

STORAGE AND DRYING OF GRAIN AND SEEDS

Threshed or shelled grain and some seeds are stored in comparatively tight bins that allow little moisture to escape. Also, adjustment of temperature to that of the outside takes place slowly. The temperature of the air in the center of a full 1,000 bushel bin 14 feet in diameter lags one to two months behind the temperature of the outside air. Consequently, the keeping of the grain or seed depends largely upon their condition when placed in storage. Binned grain that contains more than 13 to 14 per cent moisture is likely to go out of condition in warm weather. Grain that deteriorates in storage may heat, become musty, sour or "sick." Seeds containing excessive moisture soon lose their viability. Dry seed stored in woven bags fluctuates in temperature and moisture with the temperature and humidity of the surrounding air.

Moldy corn, peanuts, and several other grains and seeds sometimes contain dangerous quantities of poisonous aflatoxin. This is formed mostly by the growth of *Aspergillus flavus*, a mold fungus.

During heating, grain usually reaches a temperature of 90° to 160° F. At the higher temperatures it acquires a browned or mahogany bin-burnt color and a burned taste. At still higher temperatures that approach spontaneous combustion grain acquires a charred appearance and taste. Formerly it was believed that the initial heating was caused entirely by the respiration of the embryo of the live grain. Later it was determined that moist grain heats at about the same rate when dead as it does when viable (alive). Thus most of the heating in grain is caused by the respiration of living and multiplying fungi or mold organisms always present on and in the grain.[18] Damp dead grain does not heat when it is sterile.

The presence of live weevils or other stored-crop insects also causes grain to heat. Both molds and insects respire. They release heat, moisture, and carbon dioxide during the respiration process. This released heat and moisture continues to accelerate the heating of the grain. When the temperature increases to about 130° F., the insects are killed and bacteria and molds are largely inactivated, but spore-forming organisms are not all killed. Enzymes are inactivated at about 140° to 150° F. Heating beyond that point must result from molecular oxidation of organic materials in the grain.

Grain may become musty or moldy when it contains enough moisture for the growth of molds but not enough to cause heating, or when the heat is dissipated rapidly by exposure to air. For example, grain can become musty in the shock, or in the stack before threshing, or in sacks, shallow piles, or bins from which heat escapes readily. Grain of very high moisture content may become sour from fermentation similar to that which occurs in a silo where alcohol and then organic acids are formed. A barrel of grain soaked for hog feed sours in a few days. Wet shelled corn that contains as much as 24 to 30 per cent moisture is likely to sour.

Wheat becomes sick when the moisture content is about 14 to 16 per cent, or slightly above that necessary for good keeping quality. It occurs in deep bins, usually in terminal elevators, where the oxygen supply becomes depleted by respiration of the grain and of the fungi and bacteria on the grain. The oxygen content of the air in such bins may drop to 5 or 10 per cent while the carbon dioxide content rises from an initial 0.04 per cent up to 10 or 12 per cent.

The growth of anaerobic fungi or bacteria may be the cause of sick wheat in bins.[16] Anaerobic organisms thrive in the absence of air because they are able to obtain oxygen by breaking down the carbohydrates in the grain. Sick wheat has a dull lifeless appearance with the germs dead and discolored. Such grain has a high acidity caused by the breaking down of the fats of the embryo into fatty acids. Sick wheat is unsuitable for making bread. Damaged corn that shows similar germ injury is not designated as sick.

Some spoilage may occur in a bin of grain as a result of convection currents which arise with the onset of cold weather. Air around the warm grain at the bottom of the interior portion of the bin moves upward through the grain. It is replaced by cold air that passes downward next to the cold outer walls of the bin. The rising warm air contains moisture, which often condenses on the cool grain near the top of the bin. This results in some molding or caking of the top portion. Such condensation can be avoided by aeration of the grain. A small exhaust fan installed at the top of the bin draws air up through a metal pipe in the center of the bin, which has perforations near the bottom. The fan draws warm air from the bottom. Then cold outside air passes down through the grain from above. Thus the whole mass of grain is cooled.[32]

DRYING GRAIN AND SEEDS

Natural drying before threshing or shelling, when possible, is most economical for the farmer. Damp grain often is sold to buyers who can mix it with dry grain or dry it artificially. Some grain is dried by spreading it out in shallow piles on a barn or bin floor or on the ground outside. About 73 per cent of the rice crop and 27 per cent of the corn was dried with heated or unheated air in 1966.[37]

The material used in the construction of the bin has little effect on the keeping quality of grain, i.e., ability to maintain its grade while in storage. Metal bins both absorb and radiate heat from the sun more rapidly than do wooden or masonry bins. Insulated and underground bins retard the absorption of summer heat, but they also retain heat that should be dissipated from warm grain in cool weather. Ventilated steel bins with a wind-pressure cowl at the top with either perforated sides or bottoms, or containing perforated pipes or screen-covered flues, will dry grain that is slightly damp or tough. The moisture content of grain may drop only 1 to 3 per cent during several months of storage but the grain will remain cool and sound. Grain with a moisture content of 2 per cent or more above that essential for good keeping quality is likely to go out of condition in a ventilated bin. Any type of ventilated bin permits seeds to absorb moisture during rains or in damp weather. Many farm driers using either unheated or heated forced air are now in operation[3, 4, 25, 32] (Figures 8–6, 12–6). Corn should be cleaned before drying to remove trash that obstructs air movement through the grain.[9, 30]

DRYING WITH UNHEATED AIR: Forcing unheated air through a bin of grain or seeds is satisfactory when the grain contains no more than 3 to 4 per cent of excess moisture. The air fans are operated only on days, or parts of days, when the relative humidity is less than about 70 per cent. The humidity should be below 60 per cent for final drying of the grain down to a safe moisture content. Grain to be dried with unheated air is stored in a bin with a false-bottom, perforated floor, or with a connected system of perforated ducts or tunnels set on the bin floor. The forced air enters the grain at the bottom of the bin, passes upward through a 4-foot to 8-foot depth of grain, and escapes through ventilators at the top.[4] The air flow ranges from 1 to 6 (usually 2 to 4) cubic feet per minute per bushel of grain. Drying may be completed in one to four weeks. The use of a portable fan permits the drying of several storage bins of grain during and after harvest.

DRYING WITH HEATED AIR: Drying with heated air is more effective, but also more expensive, than with unheated air. About 2000

BTU's are required to evaporate 1 pound of water.[30] One gallon of liquid fuel dries 5 bushels of corn.[37] Heated air generally is used for drying grain that contains 18 per cent moisture or more. Much of the corn that is harvested with a corn combine or picker-sheller requires heated air. Drying with heated air usually is done in a special bin to which the burner and fan are attached.[3] Drying is completed in a few hours, after which the grain can be moved to a storage bin. Then another batch of grain is dried. The burner uses either gas or petroleum fuels. The temperature of the drying grain should not exceed 110° F. when the grain is to be used for seed, or 130° F. when it is to be processed into starch, or 180° F. when used for feed. A reduction in the digestibility of proteins in the grain sometimes occurs from prolonged heating even at temperatures of 140° to 170° F., but quick drying at temperatures above 180° F., may not be harmful. For most damp field seeds the temperature usually should not exceed 90° to 100° F.

Direct heaters utilize the combustion gases mixed with air to dry the grain. Indirect heaters have a heat exchanger so that only hot air enters the bin. This reduces fire hazard, but requires about 50 per cent more fuel to dry a bin of grain.

Portable driers that use either heated or unheated air can be used to dry ear corn in a crib. The two ends of the crib are covered with impervious panels or sheets; the air is blown into the crib through a canvas funnel that covers one side of the crib.

Farm grain driers, which are common in the rice-producing regions, usually are large enough to handle the crop from several farms (Figure 19–9). These are essentially like the commercial driers installed by grain elevators, where drying is done by blowing heated air into chambers between thin layers of grain that are passing down over baffles or through screen-covered flues. Rotary driers similar to hay dehydrators also are used.

Dry seeds can be preserved for long periods in moisture-resistant containers.

Aeration

In elevators it formerly was customary to turn grain during cold weather by conveying it from one bin to another. This exposure to cold air may lower the temperature of the grain 20 degrees or more. The moisture content occasionally may be lowered as much as 0.25 to 0.50 per cent by turning when the atmosphere is dry. Often little or no change in moisture content results from turning. Cooling of the grain retards respiration, heating, and insect damage. The conditioning of grain is now being done more economically by aeration. A large exhaust fan at the bottom of the bin draws cold outside air down through the mass of grain until all portions are about the same temperature. A large silo-type bin of grain can be aerated in about a week when the fan draws 0.1 cubic foot of air per bushel of grain per minute.[31]

Wet grain, especially grain sorghum, for animal feed can be ensiled at a 50 to 60 per cent moisture content. Sorghum grain containing 20 to 35 per cent moisture can be preserved for feed by adding 0.75 to 1.5 per cent (by weight) of propionic acid. Both treatments destroy the germination of the grain and the acid treatment prevents molding.

CLEANING GRAIN AND SEEDS

Most grain ultimately requires cleaning before use for any purpose except feed. Market grain rarely is cleaned on the farm because the loss from shrinkage and the cost of cleaning usually exceed the value of the screenings or the dockage deduction. Very little market grain is cleaned at a country elevator except where it is desired to reduce the dockage content to just below 1 per cent so the grain can be sold on a dockage-free basis. Such partial cleaning is done to a considerable extent in Canada. Screenings have little sales value in a surplus grain area. Many screenings cannot be shipped without processing, because of weed laws. Screenings have a substantial value at a terminal grain market where the edible ingredients can be added to ground mixed feeds sold on a guaranteed chemical-analysis basis without specifying the individual ingredients. The value of the dockage is taken into consideration when an offer is made for cash grain at the terminal market.

Seed warehouses have elaborate equipment for cleaning various kinds of seeds.[24, 28]

Terminal elevators usually have suitable equipment for separating some different seeds. Wild oats screened from wheat or barley formerly found a good market for mule feed in the southeastern states where wild oats are not a weed pest. They were sold in carload lots under the designation of mill oats. Seed of mustard and Frenchweed have a market value for oil extraction when separated from screenings (Chapter 39).

Shrinkage losses of grain in cleaning usually range from 2 to 5 per cent, in addition to the foreign material removed, even when the best equipment is used. The invisible loss, that is, dust, moisture evaporation, spilled grain, and grain lodged in the equipment during cleaning in an elevator is likely to be about 0.25 per cent. Appreciable cleaning losses are avoided only when the operation is a mere scalping to remove excess dockage. The moving of oats or barley through elevator machinery knocks off portions of the glumes or awns that then become screenings, foreign material, or dockage. Wheat grains lose hairs and bran, and suffer cracking during handling.

Types of Cleaners

Separations of different grains, seeds, and foreign material are made mechanically on the basis of differences in length, thickness, shape, or specific gravity.[29] The most common cleaning device is the fanning mill found on many farms. It consists of two or more screens to sift out particles or seeds both larger and smaller than the particular seed being cleaned, together with an air blast that blows out the material lighter than the desired clean seed.

Larger machines similar to the fanning mill (Figure 9–7), often

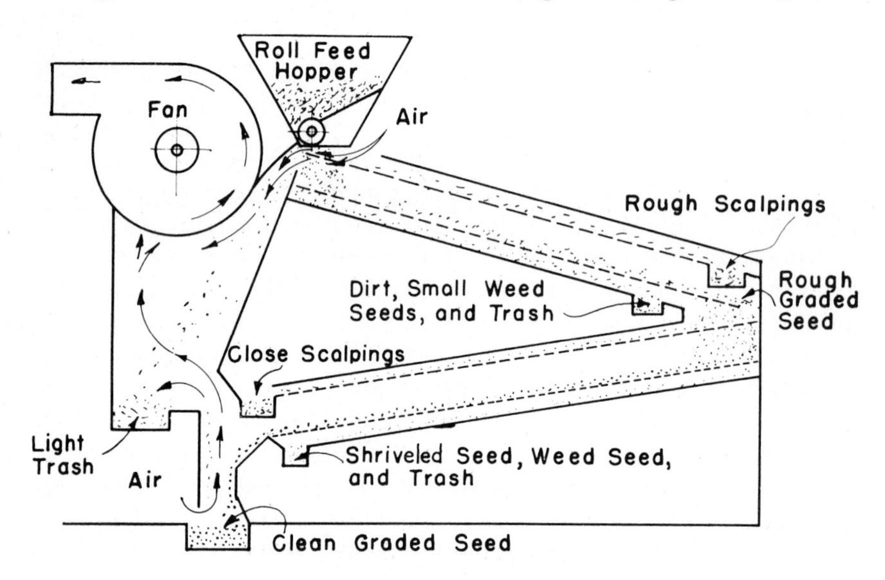

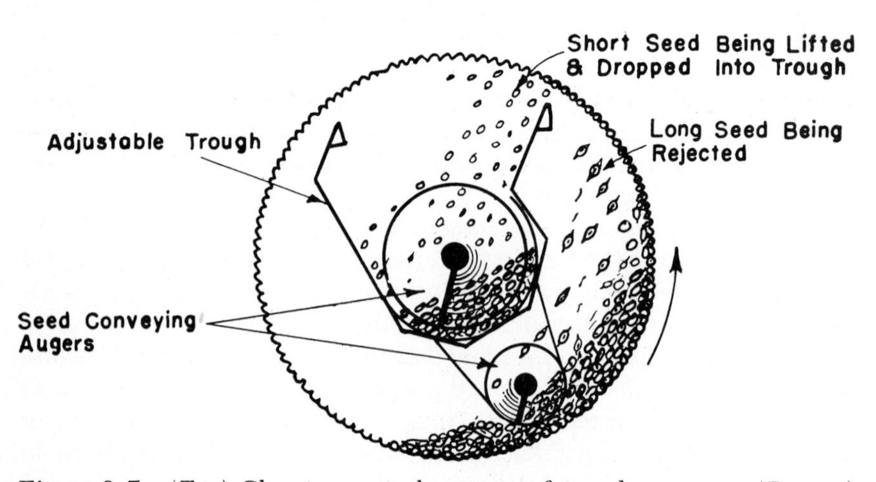

Figure 9–7. (*Top*) Cleaning grain by suction fan and screening. (*Bottom*) Indent cylinder separates seeds or grains of different lengths.

called receiving separators, are found in many elevators and in all mills and seed warehouses. These machines, which use suction rather than an air blast to remove light or chaffy material, make more accurate separations than are accomplished with a fanning mill. Such machines are excellent for the removal of trash, chaff, certain fine seeds, and cracked or shrunken grain, but they are unsatisfactory for the separation of grains or seeds of similar diameter or specific gravity. For example, oats and wild oats can be separated from wheat, rye, or barley only in part with a screen cleaner because their diameter is much the same, while the differences in specific gravity are too small. Since vetch seed has almost the same diameter and specific gravity as wheat, it cannot be separated with a fanning mill. Mustard seeds will pass through fine screens or sieves that grain cannot penetrate, but many of them roll rapidly down the sloping screen of the cleaner where they are discharged with the cleaned grain. On the other hand, the triangular wild buckwheat seeds, which are appreciably larger than mustard seeds, are screened out readily with special sieves because they do not roll.

Indented or cylinder cleaners (Figure 9–7) separate seed of different length. The pockets on revolving disks, or indentations inside a slowly revolving cylinder, pick up short particles, seeds or grains that can be held in shallow pockets or indentations while they reject the longer ones. The separated fractions are discharged into different spouts. Such machines have a series of disks or cylinders with different size pockets or indentations which permit several excellent separations. Thus, cockle seed (Figure 9–8) can be separated from wheat because of its shorter length. Oats and barley are separated from wheat because of their greater length. A fair separation of mixtures of two varieties of even the same kind of grain can be made when the grains differ appreciably in length. Special pockets or indentations separate seeds of different shape.[28] Large cleaners of the screen, disk, or cylinder types, may have a capacity of 300 to 600 bushels per hour.

Ring graders are equipped with rotating spaced spring-steel rings to separate grains differing in diameter. With these machines, oats can be separated from barley, and thin barley kernels can be separated from plump grains. Accurate separations are accomplished by different spacings between rings. The needles machine makes similar separations by allowing the grain to flow over a series of baffles bearing the needles (fine spring rods), for slender grains to fall between. Spiral, horizontal disk and inclined draper separators remove globular seeds from seeds of other shapes that roll, merely slide, or rest.[28]

Two other interesting types of machines, called gravity and magnetic separators, are used especially for cleaning small-seeded legumes such as alfalfa and clover. The gravity separator removes the lighter grains infested by Angoumois moths from sound seed corn.

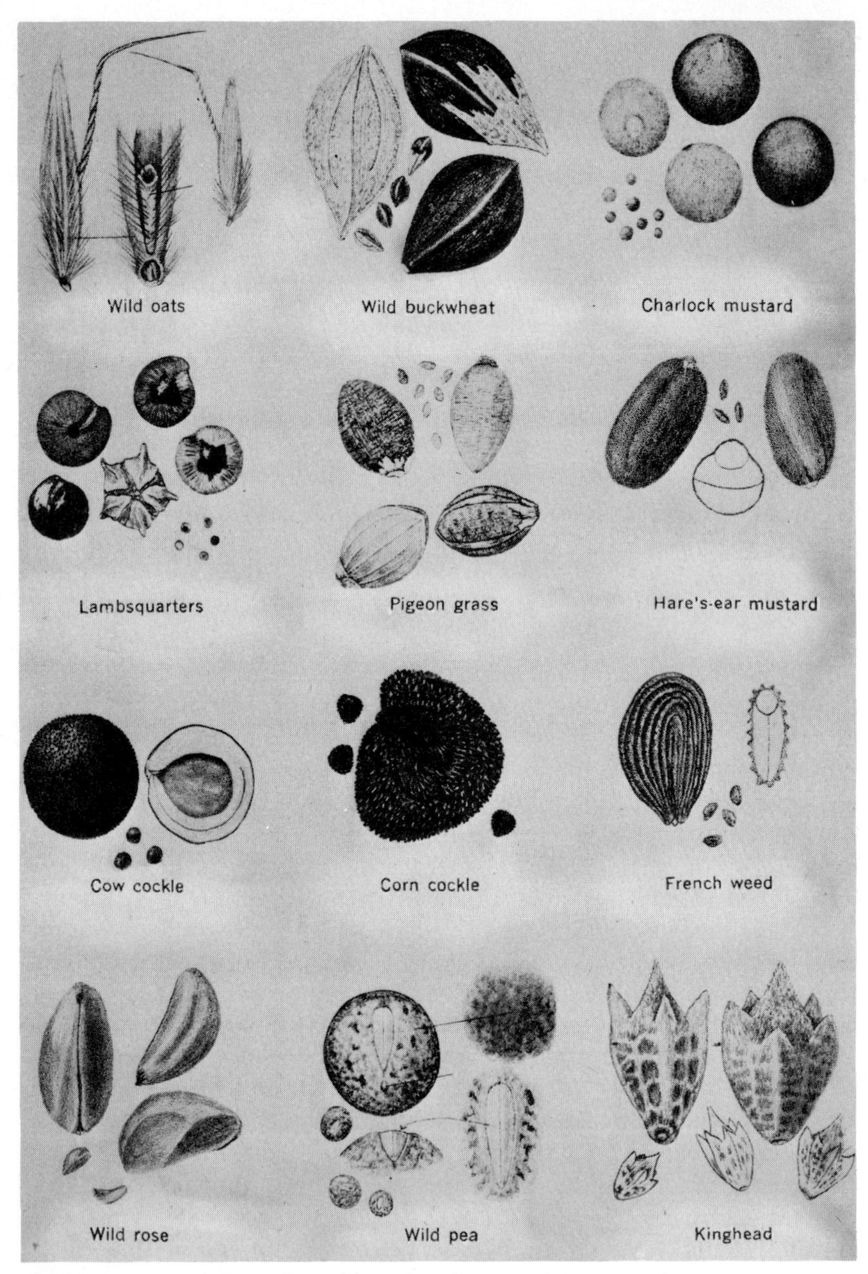

Figure 9–8. Weed seeds (enlarged and comparative sizes).

The gravity machine has a shaking table with air blowing up through numerous small openings in the table top which *floats* the seeds and material differing in specific gravity over into different compartments or spouts. Moist sawdust added to clover seed sticks to the mucilaginous seed coats of buckhorn seeds. A gravity separator then removes the sawdust particles from the clover seed, taking with them the otherwise inseparable weed seeds. Dodder seeds have a very rough surface to which small steel or iron filings will cling (Figure 9–9). Filings are mixed with clover seed that contains dodder, after which a magnetic separator is used to draw out the metal-coated dodder seeds. Another type of magnetic separator equipped with powerful electromagnets is placed in grain and seed spouts to catch any nails, pieces of iron, or steel scrap in a moving stream of grain. This not only prevents damage to grinding or other machinery, but also reduces the hazard of fires that might start from accidental sparks arising from striking metals.

Other cleaning machines include the velvet-roll separator which removes rough seeds from smooth ones and electrostatic separators which operate on the differences in electrical properties of seeds. A machine with a photoelectric cell can separate seeds of different colors. In some admixtures, other varieties or discolored seeds can be removed with this equipment. Awns are removed from some grass seeds with beaters or hammer mills. Other seeds require hulling in abrasive machines.

Insects in Stored Grain

Grain in storage is subject to damage from the attack of several species of insects.[6] The chief insect pests of stored grain are the rice weevil (*Sitophilus oryzae*), the granary weevil (*Sitophilus granarius*), and the Angoumois grain moth (*Sitotroga cerealella*). The lesser grain borer or Australian wheat weevil (*Rhizopertha dominica*) is destructive to grain in elevators, but is not often present in farm bins. The Indian meal moth (*Plodia interpunctella*) damages corn and sometimes other grains also. The cadelle (*Tenebroides mauritanicus*) not only consumes stored grain but also burrows into the walls of wooden bins. The khapra beetle (*Trogoderma granarium*) is very destructive to stored grain in warm climates. It appeared in the United States about 1955, but has since been largely eradicated. Other insects, including beetles, moths, and mealworms, that are found in grain but that subsist largely on broken grains and milled products, are popularly referred to as bran bugs. Among the more common bran bugs are the saw-toothed grain beetle (*Oryzaephilus surinamensis*), the flat grain beetle (*Laemophloeus minutus*), the confused flour beetle (*Tribolium confusum*), and the red flour beetle (T. *castaneum*).

Species of weevils infest the seeds of beans, peas, and several other legumes.

MEANS OF INSECT INFECTION: The rice weevil, broad-nosed grain weevil, Angoumois grain moth, flour beetles, and certain other insects often fly to the field and attack grain still standing in the field. When

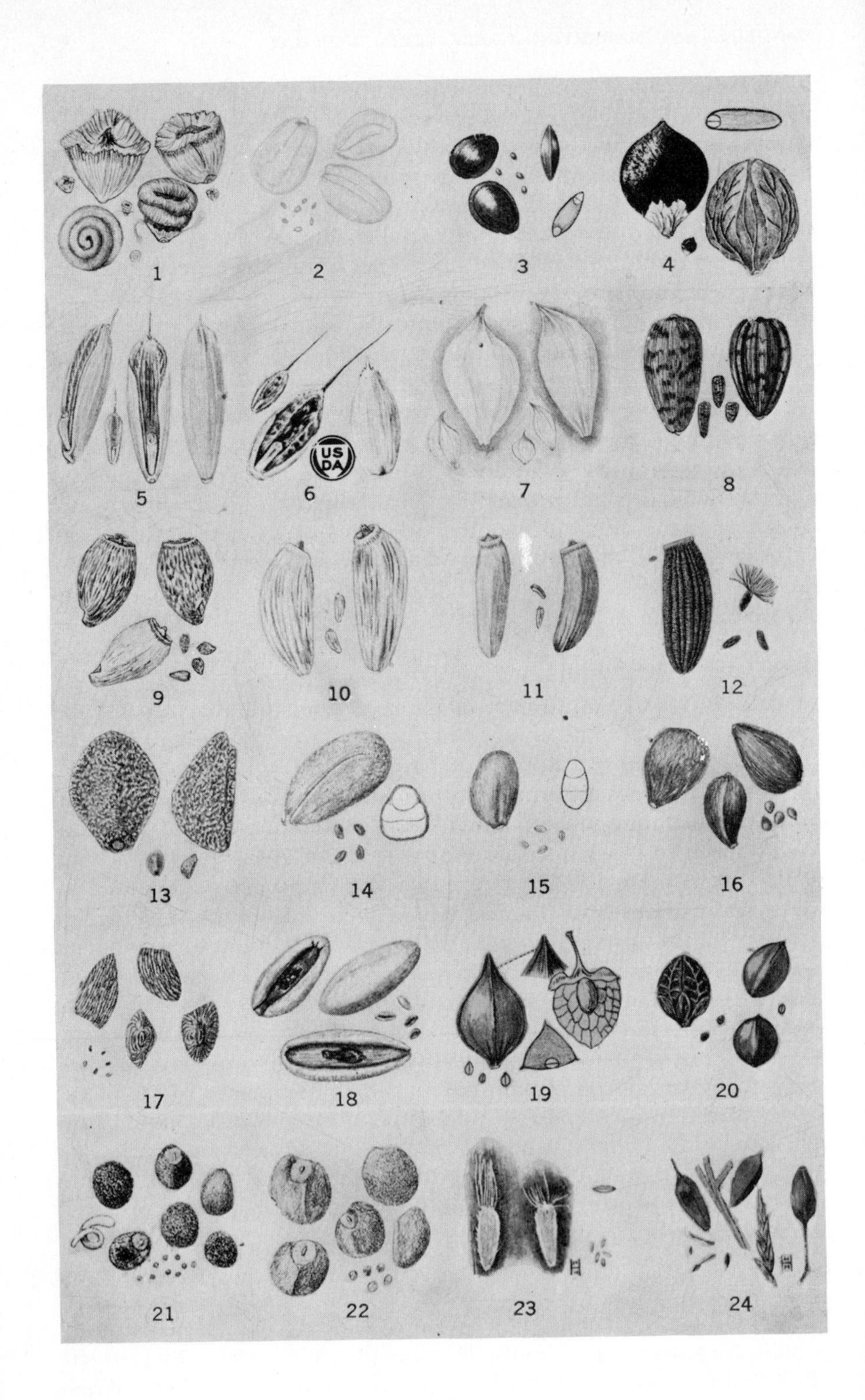
1
2
3
4
5
6
US
DA
7
8
9
10
11
12
13
14
15
16
17
18
19
20
21
22
23
24

harvest or threshing is delayed, the grain may be heavily infested by the time it is placed in storage. The granary weevil has no wings under its wing covers and therefore is unable to fly.

The rice weevil and granary weevil are snout beetles that deposit eggs in small borings in the kernels. The Angoumois grain moth deposits eggs on the surface of the kernels. When the larvae of these insects hatch, they bore inside the kernels, consume the contents, pupate, and then emerge as adults. The grain borers have similar feeding habits. Other insects such as the cadelle, square-necked grain beetle (*Cathartus quadricollis*), and the Indian meal moth eat the germ out of the grain.

Temperature is the chief factor determining the abundance of stored-grain insects in a region, and the greatest precautions to protect grain from stored insects are necessary in the South, where temperatures are high. The rice weevil is dormant at 45° F. or below, and the granary weevil at 35° F. and below. These insects die at prolonged temperatures below 35° F. Therefore, grain is moved (or run) to another bin during cold weather to break up hot spots due to insect infestation.

Damp grain is very subject to insect injury. Grain-infesting insects either die or fail to multiply when the moisture content of the grain is below 9 per cent.

CONTROL OF INSECTS IN STORED-GRAIN: Grain or seeds for storage should be cleaned to remove cracked material and live insects and then placed in clean bins free from insects. The walls and floors of the cleaned bin should be sprayed at the rate of 2 gallons per 1,000 square feet with a 2.5 per cent concentration of methoxychlor, or 0.5 per cent concentration of pyrethrins, in combination with the synergist piperonyl butoxide. The pyrethrum mixture also is useful as a protectant for spraying on the grain as it is placed in the bin.

Figure 9–9. [OPPOSITE] **Seeds of common weeds:**

1. Russian thistle
2. Tumbling mustard
3. Rough pigweed or redroot
4. Pennsylvania smartweed
5. Chess or cheat
6. Darnel
7. Wild garlic or onion
8. Wild sunflower
9. Star thistle
10. Bull thistle
11. Canada thistle
12. Perennial sowthistle
13. Field bindweed
14. Peppergrass
15. Whiteweed
16. Poverty weed
17. Broad-leaved plantain
18. Buckhorn or narrow-leaved plantain
19. Curled dock
20. Sheep sorrel
21. Clover dodder
22. Field dodder
23. Horseweed
24. Nutgrass

Stored grain or seeds should be fumigated when they become infested with insects. Some effective fumigants for farm-stored grain are as follows:

FUMIGANT	GALLONS PER 1,000 BUSHELS	
	Wooden Bin	*Metal or Concrete Bin*
Carbon tetrachloride (100%)	6	3
Carbon tetrachloride-carbon disulfide (80–20%)	4	2
Ethylene dichloride-carbon tetrachloride (75–25%)	6	3
Carbon tetrachloride-ethylene dichloride-ethylene dibromide (60–35–5%)	4	2

The above mixtures are sold in the prepared form. The fumigants are applied as a coarse spray on the surface of the grain while the sprayer operator stands outside the bin to protect himself against the toxic vapors. Since the vapors of these sprays are heavier than air, they settle to the bottom of the bin.

Hydrocyanic acid gas, methyl bromide, or chloropicrin (tear gas) are used for fumigation of flour mills and warehouses. A gas mask is advisable when handling any of these toxic materials.

Heating grain in a dryer at 130° to 140° F. for 30 minutes will kill all insect life.

A concentration of 35 to 60 per cent of carbon dioxide in a tight bin atmosphere kills the insects in 7 to 4 days, respectively. Diatomaceous earths or silica gels applied to grains prevent insect attack.*

MARKETING HAY

The chief users of market hay are producers of market milk, owners and breeders of horses. The dry sunny climates in the irrigated valleys of the West are favorable to the quick curing of bright green hay. The Black Belt of Alabama has produced large quantities of johnsongrass hay for the market. Only about 14 per cent of the hay crop of the United States moves off the farm on which it is produced. About two thirds of the market hay is produced in the western half of the country. Fully half of the hay that is produced in California, Arizona and New Mexico is sold.

Baled hay is sold at the farm, trucked to markets, or shipped to commission dealers. Much of the hay marketed in the western range areas is sold in stacks or bales to livestock raisers or feeders.

The market price of hay depends upon palatability, nutritive value,

* USDA, ARS 51:44, 1972.

and appearance. Hay of a bright green color has been cut before it was too ripe, was cured promptly, and is high in carotene. Leafy legume hay that contains 40 to 50 per cent or more of its weight in leaves is high in protein. Hay leached by rain loses water-soluble carbohydrates and amino acids, which lowers its nutrient value. Foreign material may be offensive, inedible, of low nutrient value, or even injurious to livestock. Hay cut when too ripe contains seeds that may be rejected or lost by shattering. Seed development withdraws nutrients from the leaves and stems, making the latter less valuable. The stems of overripe hay usually are coarse, woody, and high in crude fiber. Coarse stems usually are rejected, especially by cattle. Horses may eat the coarser stems but refuse shattered seeds and leaves.

GRADING HAY

The grading of hay for shipment in interstate commerce is optional, i.e., it is graded only when one or more parties interested in the hay request an official inspector to grade it and issue a grade certificate. Contracts for the purchase of hay of a given grade require official grading. The grading is done by licensed inspectors who operate on a fee basis at primary and terminal markets.

The chief factors in the grading of hay are leafiness (for legume hays), color, foreign material, maturity, and fineness. Hay that contains more than a trace of injurious foreign material, or has an objectionable odor, or is undercured, heating, hot, wet, musty, moldy, caked, badly broken, badly weathered, badly overripe, or very dusty, is placed in Sample grade in the Official Standards for Hay. Hay containing more than 35 per cent foreign material or 35 per cent moisture is not even classed as Hay. Peanut straw, i.e., peanut vines from which the peanuts have been removed, is classed as Hay.

Standards have been established for 11 groups or kinds of hay. Each group includes one or more classes based upon the kind of hay or mixture of various kinds. Three numerical grades (1, 2, and 3), in addition to Sample Grade, are provided for each group or class. Special grade designations in addition to the numerical grades are provided for certain groups and classes. These include Extra Leafy, Leafy, Extra Green, Green, Fine, and Coarse.[2]

Grades of Hay

The standards for Group I, Alfalfa and Alfalfa-Mixed Hay; Group II, Timothy and Clover Hay; and Group III, Prairie Hay, are shown in the following table.

Prairie hay consists largely of native grasses and sloughgrasses. Group X, Grass Hay, includes cultivated, introduced, and some native grasses.

Injurious foreign material consists of species with awns, seeds, or leaves that cause punctures, lacerations, or abrasions on the membranes of livestock. It includes squirreltail grass (*Hordeum jubatum*), mature ripgut or bronchograss (*Bromus rigidis*), needlegrasses (*Stipa* spp.), and three-awn grass (*Aristida oligantha*).

The United States Department of Agriculture grade requirements for Alfalfa, and Alfalfa-Mixed, Timothy, and Clover, and Prairie hays are as follows:

CLASS AND USDA GRADE	MINIMUM LEAVES	MINIMUM GREEN COLOR	MAXIMUM FOREIGN MATTER
	(%)	(%)	(%)
Alfalfa and Alfalfa-Mixed hays:			
No. 1	40	60	5
No. 2	25	35	10
No. 3	10	10	15
Timothy and Clover hays:			
No. 1	40	40	10
No. 2	25	30	15
No. 3	10	10	20
Prairie hay:			
No. 1	–	50	10
No. 2	–	35	15
No. 3	–	–	20

The relative nutrient values of different grades of alfalfa hay, compared with leaves and stems, are indicated by the analyses of typical samples shown below:

HAY GRADE, LEAVES, OR STEMS	CRUDE PROTEIN	NITROGEN-FREE EXTRACT	CRUDE FIBER	CAROTENE PER POUND
	(%)	(%)	(%)	(mg.)
No. 1	22.5	45.2	22.8	–
No. 2	16.9	43.2	30.8	–
No. 3	16.8	36.8	38.3	–
Alfalfa leaves	25.3	–	17.1	35.5
Alfalfa stems	11.3	–	44.6	5.5

The No. 1 grade is high in protein and nitrogen-free extract, but low in crude-fiber content. The high nutritive value of the leaves also is evident. Dairy cows eat more, and yield more milk, when fed high-grade hay.

CROP JUDGING

Exhibition of crop products has considerable educational, cultural, and recreational value. In premium awards for the best products, consideration is given chiefly to appearance, uniformity, size, and trueness to type, based upon weighted point scores adopted by exhibit officials or judges. Such scores usually reflect market values for the particular crop commodity. However, they have little relation to productivity or to the inherent value of the variety or type being shown. Prize-winning crops usually indicate two things: (1) favorable environment for plant development, and (2) meticulous selection and preparation of the exhibit material. The best plants often are found where thin stands have permitted optimum individual plant development at the expense of yield per unit area of land. However, good cultural conditions along with freedom from disease and insect injury are essential to the production of acceptable crop exhibits.

Standards of excellence in crop products are largely arbitrary. In some cases, they are based upon characters that actually are deleterious to crop yield or quality. For example, deep rough kernels of corn indicate late maturity, slow drying, and susceptibility to kernel rots. Compact heads of sorghum favor worm damage and moldy grain. Tall corn stalks make harvesting difficult. Certain varieties of wheat known to be of inferior quality for making bread regularly take prizes at local, state, and international exhibits. These prizes are won because the grain has a high test weight as well as plump, vitreous, and smooth kernels. The substitution of utilitarian or quality standards does not fully remedy this incongruous situation. The merits of a crop or crop variety are determined by comparative experimental tests and by established performance on farms and in markets and processing plants. Experimental facts outweigh the opinions of the best crop judges, who examine only a few selected exhibit samples. However, it is incumbent upon crop students to learn to judge and exhibit crops because the present type of agricultural fair will continue in vogue for many years. People adore a contest. Furthermore, the judging of exhibits develops skill in the appraisal, identification, and grading of crop products.

REFERENCES

1. Anonymous. "The service of Federal grain standards," *USDA Misc. Pub. 328*, pp. 1–18. 1938.
2. Anonymous. "Handbook of official hay and straw standards," USDA Agr. Marketing Service, Grain Div., pp. 1–55. 1958.
3. Anonymous. "Drying shelled corn and small grain with heated air," *USDA Leaflet 331*. 1952.

4. Anonymous. "Drying shelled corn and small grain with unheated air," *USDA Leaflet 332*. 1952.
5. Anonymous. "Grain grading primer," *USDA Misc. Pub. 740*. 1957.
6. Anonymous. "Stored grain pests," *USDA Farmers Bul. 1260* (rev.). 1958.
7. Anonymous. "Official grain standard of the United States," *USDA SRA-AMS-177* (rev.). May 1964.
8. Anonymous. "Controlling insects of stored rice," *USDA Handbook 129*, pp. 1–19. 1971 (Rev.).
9. Anonymous. "Drying shelled corn and small grains," *USDA Farmers Bul. 2214*, pp. 1–12. 1971 (Rev.).
10. Atkins, M. D., and Smith, J. E. "Grass seed production and harvest in the Great Plains," *USDA Farmers Bul. 2226*, pp. 1–30. 1967.
11. Bouland, H. D., and Smith, L. L. "A small country elevator for merchandizing grain," *USDA Marketing Res. Rpt. 387*. 1960.
12. Brandenburg, N. R., and Harmond, J. E. "Fluidized conveying of seed," *USDA Tech. Bul. 1315*, pp. 1–41. 1964.
13. Brandenburg, N. R., and Harmond, J. E. "Separating seeds by length with special indent cylinders," *Ore. Agr. Exp. Sta. Tech. Bul. 88*, pp. 1–20. 1966.
14. Bruce, W. M., and others. "Planning grain storage elevators," *U. Georgia Bul. 51(7d)*. 1951.
15. Carlson, E. *Biennial Rpt. Kans. State Grain Insp. Dept.*, pp. 1–29, Kansas City, Mo. 1944.
16. Carter, E. P., and Young, G. Y. "Effect of moisture content, temperature, length of storage on the development of sick wheat in sealed containers," *Cereal Chem.*, 22(5):418–428. 1945.
17. Christensen, C. M. "Deterioration of stored grain by fungi," *Bot. Rev.* 23(3):108–134. 1957.
18. Christensen, C. M., and Kaufman, H. L. *Grain Storage*. Univ. Minn. Press, Minneapolis. 1969.
19. Clough, M., and Browning, J. W. "Feed grains," in *Marketing*, USDA Yearbook, 1954, pp. 403–413.
20. Collier, G. A. "Marketing hay by modern methods," *USDA Farmers Bul. 1700*, pp. 1–25. 1933.
21. Couch, H. B. *Diseases of Turf Grasses*, Reinhold, New York, pp. 1–289. 1962.
22. Goldblatt, L. A., Ed. *Aflatoxin: Scientific Background, Control and Implications*, Academic Press, New York, pp. 1–472. 1969.
23. Gould, F. W. *Grass Systematics*, McGraw-Hill, New York, pp. 1–382. 1968.
24. Gregg, B. R., and others. *Seed Processing*, Miss. State Univ., Nat'l Seeds Corp & USAID, Avion Printers, New Delhi, India, pp. 1–396. 1970.
25. Hall, C. W. *Drying Farm Crops*, Agricultural Consulting Associates, Inc., Reynoldsburg, Ohio, pp. 1–336. 1957.
26. Hanson, A. A. "Grass varieties in the United States," *USDA Handbook. 170*, pp. 1–24. 1972 (Rev.).
27. Hanson, A. A., and Juska, F. V. "Turfgrass science," *Am. Soc. Agron. Mon. 14*, Madison, Wis., pp. 1–715. 1969.

28. Harmond, J. E., and others. "Mechanical seed cleaning and handling," *USDA Handbook, 354,* pp. 1–56. 1968.
29. Harmond, J. E., Klein, L. M., and Brandenburg, N. R. "Seed cleaning and handling," *USDA Handbook 179.* 1961.
30. Heard, R. F., and others. "Shelled corn handling systems," *Ont. Dept. Agr. & Food Pub. 551,* pp. 1–33. 1969.
31. Holman, L. E. "Aeration of grain in commercial storages," *USDA Mktg. Rpt. 178.* 1960.
32. Hukill, W. V., Saul, R. A., and Teare, D. W. "Outrunning time; combating weather," in *Power to Produce,* USDA Yearbook, 1960, pp. 183–199.
33. Mehl, J. M. "The futures markets," in *Marketing,* USDA Yearbook, 1954, pp. 323–331.
34. Milner, A. R. *Grain Marketing Pricing, Transportation,* West-Camp Press, Inc., Westerville, Ohio, pp. 1–287. 1970.
35. Rowe, H. B. "The danger of loss," in *Marketing,* USDA Yearbook, 1954, pp. 309–323.
36. Semple, A. T. *Grassland Improvement,* Leonard Hill, London, pp. 1–400. 1970.
37. Strickler, P. E., and Hinson, W. C. Jr. "Crop drying in the United States, 1966, quantity, equipment and fuel used," *USDA Stat. Bul. 439,* pp. 1–22. 1969.
38. Weintraub, F. C. "Grasses introduced into the United States," *USDA Forest Serv. Agr. Handbk. 58.* 1953.

Chapter 10

Pastures and Pasturage

IMPORTANCE OF PASTURES

About 3000 million hectares (7.5 billion acres) of the earth's uncultivated land area is used for pasture. The grazing area in the United States in 1969 was about 889 million acres (360 million hectares). Of this, 452 million acres on farms are grassland pasture, 62 million pastured woodland and 88 million cropland pastured, 287 million acres are grazed areas not on farms. The Bureau of Land Management and the Forest Service, provide 262 million acres for grazing. The remainder is mostly on Indian reservations and state-owned lands. In addition, millions of acres of harvested crop lands are grazed for part of the season to salvage feed from aftermath, stubble, winter grain and hay, and forage-seed fields. Pastures provided 35 per cent of the feed units for all livestock and 82 per cent for sheep and goats in 1965.

Development of improved methods of pasture management, together with a better realization of the value of pasturage, have been responsible for an intense interest in the establishment and maintenance of productive pastures since about 1925. In addition, national agricultural policies have favored grassland farming for erosion control.[58] This situation has encouraged renovation of old pastures by tillage, herbicide application, mowing to destroy less desirable plants, fertilization[44a] and reseeding with productive palatable species. On unplowable pastures the land may be worked with a disk or springtooth harrow or other implement. In some sections a cutaway disk implement called a *bush-and-bog* is used for this purpose. Improvement of pastures may involve replacement of Kentucky bluegrass and white clover, which are persistent under heavy grazing, with more productive taller-growing but less permanent crops such as alfalfa, Ladino clover, bromegrass, orchardgrass, tall fescue, and timothy. Such replacement was not advantageous in northern Virginia.[10]

More productive forages, with heavy fertilization, can easily double or treble the carrying capacity of pastures in the humid areas. Heavy grazing eventually suppresses the productive species. The pastures then revert to shorter, persistent plants such as Kentucky bluegrass,

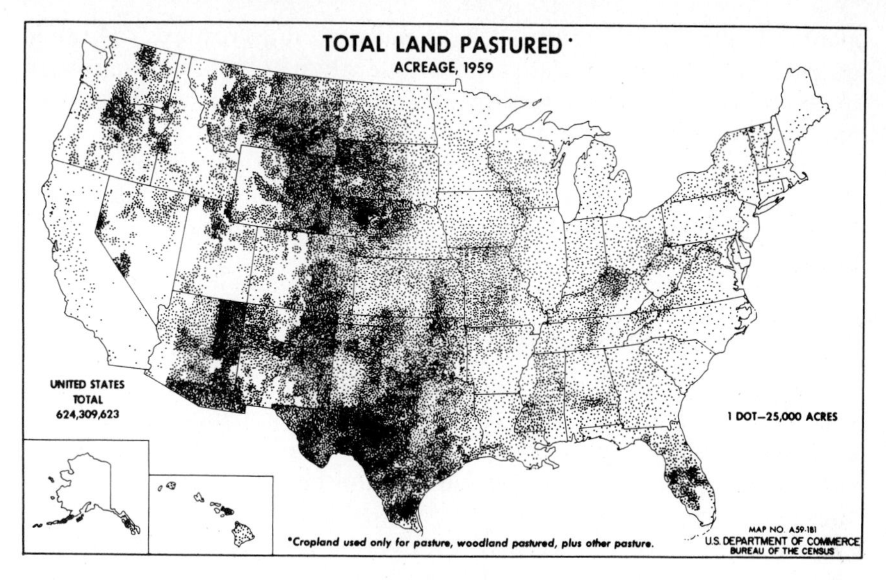

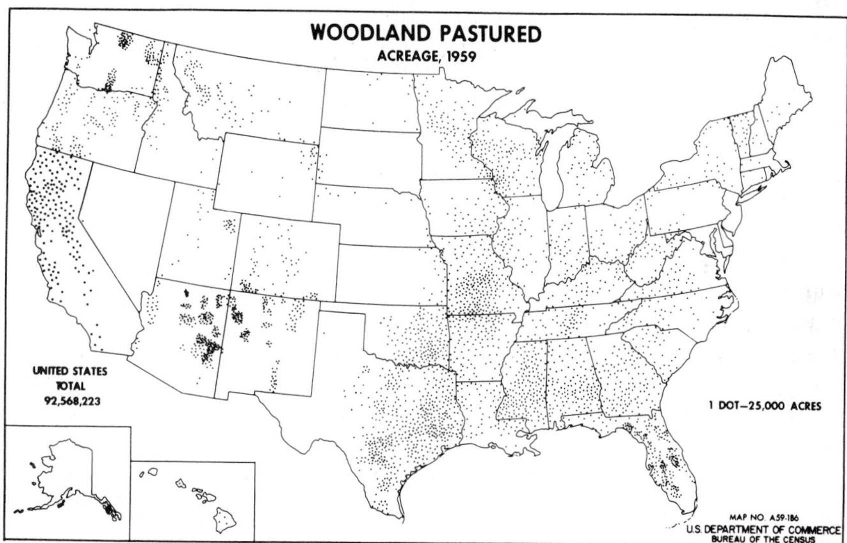

Figure 10–1. (*Top*) Total land pastured in the United States in 1959. (*Bottom*) Woodland pastured in 1959.

white clover, common bermudagrass, or carpetgrass. Reseeding of improved pastures every two to five years often is necessary.

Millions of acres were plowed for crop production in the semiarid Great Plains and Great Basin regions between 1910 and 1930. The more unfavorable sites were soon abandoned; eventually they reverted to grassland. The natural process of range restoration on plowed land may require 25 to 40 years,[56] although good grazing may be available after 8 or 10 years.[64] Artificial reseeding restores the range quickly, but it is more costly and frequently results in failure except in favorable wet seasons. Improved methods of establishing perennial grasses facilitate range improvement in the drier areas.[18]

The seeding of better grasses, such as crested wheatgrass or wild ryegrasses, at suitable sites on western rangeland, may double the forage yield, extend the grazing season and supplement the native pastures.[38, 44, 45, 50]

ADVANTAGES OF PASTURES

The chief object of good pasture maintenance is to provide green succulent feed for livestock during the entire grazing season. Livestock kept on pastures are more comfortable and sanitary than those kept in dry lots. The animals harvest the crop at a minimum cost. Good pastures produce about two thirds as much dry matter as the same area in cultivated crops.[59] In Virginia experiments the yield of pasturage was 40 to 65 per cent that of the same crops allowed to mature and then cut for hay.[68] However, the dry matter of immature grasses is higher in protein and vitamins and is more digestible than the dry matter from mature grasses harvested for hay. Consequently, pasturing produces about three fourths as much in digestible nutrients as do the same crops cut for hay[59] (Table 10–1).

Dairy cows produced more milk per unit of dry matter intake from strip grazing than from green feeding or stored feed, due to selective grazing. They rejected 33 per cent of the pasture growth in selecting the more tender, palatable and nutritious herbage. They rejected

TABLE 10–1. Digestible Nutrients in Harvested Crops and Pasturage

CROP	ROUGHAGE YIELD PER ACRE	TOTAL DIGESTIBLE NUTRIENTS	
	(*tons*)	*Harvested* (*lb.*)	*Grazed* (*lb.*)
Alfalfa	2.61	2793	2067
Clover: red, alsike, and crimson	1.46	1504	1113
Sweetclover	1.82	1867	1382
Lespedeza	1.82	1170	866
Clover and timothy mixed	1.40	1393	1030
Timothy	1.26	1210	895
Grains cut green	1.31	1292	956
Annual legumes	1.02	1102	815
Millet, johnsongrass, and sudangrass	1.23	1225	906

only 2 per cent of the green feed, and 8.5 per cent of the coarser portion of the stored feed.[40]

KINDS OF PASTURES

Tame Pastures

Tame pastures are lands once cultivated that have been seeded to domesticated pasture plants and are used chiefly or entirely for grazing by livestock. The principal kinds of tame pastures are as follows:

1. *Permanent pastures* are grazing lands occupied by perennial pasture plants or by self-seeding annuals which remain unplowed for long periods (five years or more).

2. *Rotation pastures* are fields used for grazing which are seeded to perennials or self-seeding annuals, but which form a unit in the crop rotation plan and are plowed within a five-year or shorter interval.

3. *Supplemental pastures* are fields used for grazing when the permanent or rotation pastures do not supply enough feed for the livestock on the farm. Such pastures may be provided by the aftermath of meadows, small-grain stubble, seeded small grains, annuals such as sudangrass, lespedeza, and crimson clover, or biennials such as sweetclover.

4. *Annual pastures* are pastures that are seeded each year to take the place of permanent pasture wholly or in part. Such pastures may include a series of crops such as rye, oats, barley, sudangrass (Figure 10–2), Italian ryegrass, vetch, soybeans, and rape.

5. *Renovated pastures* are those restored to former production by tillage, mowing, reseeding, or fertilization.

Natural or Native Pastures

Natural or native pastures are uncultivated lands occupied wholly or mainly by native or naturally introduced plants useful for grazing. The chief types are as follows:

1. *Ranges* are very extensive natural pastures.*

2. *Brush pastures* are areas covered largely with brush and shrubs, where a considerable portion of the feed obtained by livestock comes from woody plants.

* The plants occurring on the western ranges are well described and illustrated in the book *Range Plant Handbook*, prepared by the Forest Service, USDA. The carrying capacity of different range types is shown in *Food and Life*, 1939, pp. 925–955. A complete report entitled *The Western Range*, prepared by the United States Forest Service in 1936, was issued as *Senate Document 199*. Descriptions of range forage plants also are contained in "99 Range Forage Plants of the Canadian Prairies," in *Canada Department of Agriculture Publication 964*, 1956, pp. 1–99.

3. *Woodland pastures* are wooded areas with grass and other edible herbage growing in open spaces and among trees.

4. *Cut-over* or *stump pastures* are lands from which the salable trees have been cut but on which there are stumps and usually also some new tree growth.

PERMANENT PASTURES

Pasture Areas

Most of the area in the eastern half of the United States, as well as the north Pacific coastal region, originally was in forest. As the trees were removed, the pastured land was occupied by introduced plants. The western area between these general regions includes the prairie lands or tall-grass region, the native short-grass region of the Great Plains, and the native desert grasses and shrubs of the intermountain region.

Environmental Conditions

The 60° F. annual isotherm marks approximately the northern limit of usefulness of southern pasture plants. The exceptions to this are mostly annuals such as lespedeza and sudangrass. North of this line, southern grasses are subject to winter injury, while to the south bluegrass, orchardgrass, timothy, redtop, and most clovers are unable to thrive during the long period of high temperatures. This temperature effect is particularly important in the humid regions.

Rainfall differences largely determine the flora in the Great Plains and intermountain area. In the intermountain region, the annual precipitation is so low that desert or semidesert conditions prevail except in the extreme northern part and at higher altitudes in the mountains. The Pacific Northwest has a fairly abundant winter rainfall and a mild climate due to the Japanese Current.

Although mountain valleys, parks, and clearings furnish considerable pasturage (Figure 10–1), the higher mountain slopes are unimportant from the national pasture standpoint, because in most sections they are rather completely forested or are composed of rock masses with very little productive soil. The most valuable grazing lands are level or rolling areas, particularly those in the Corn Belt. The level grasslands of the Great Plains are also good pasture but are less productive than those in the Corn Belt, due to limited rainfall (Figure 10–3).

Next to climate and topography, soil characteristics have the most

Figure 10–2. [OPPOSITE] (*Top*) Sheep range in Colorado. (*Middle*) Improved pasture in Mississippi. (*Bottom*) Temporary sudangrass pasture in Nebraska.

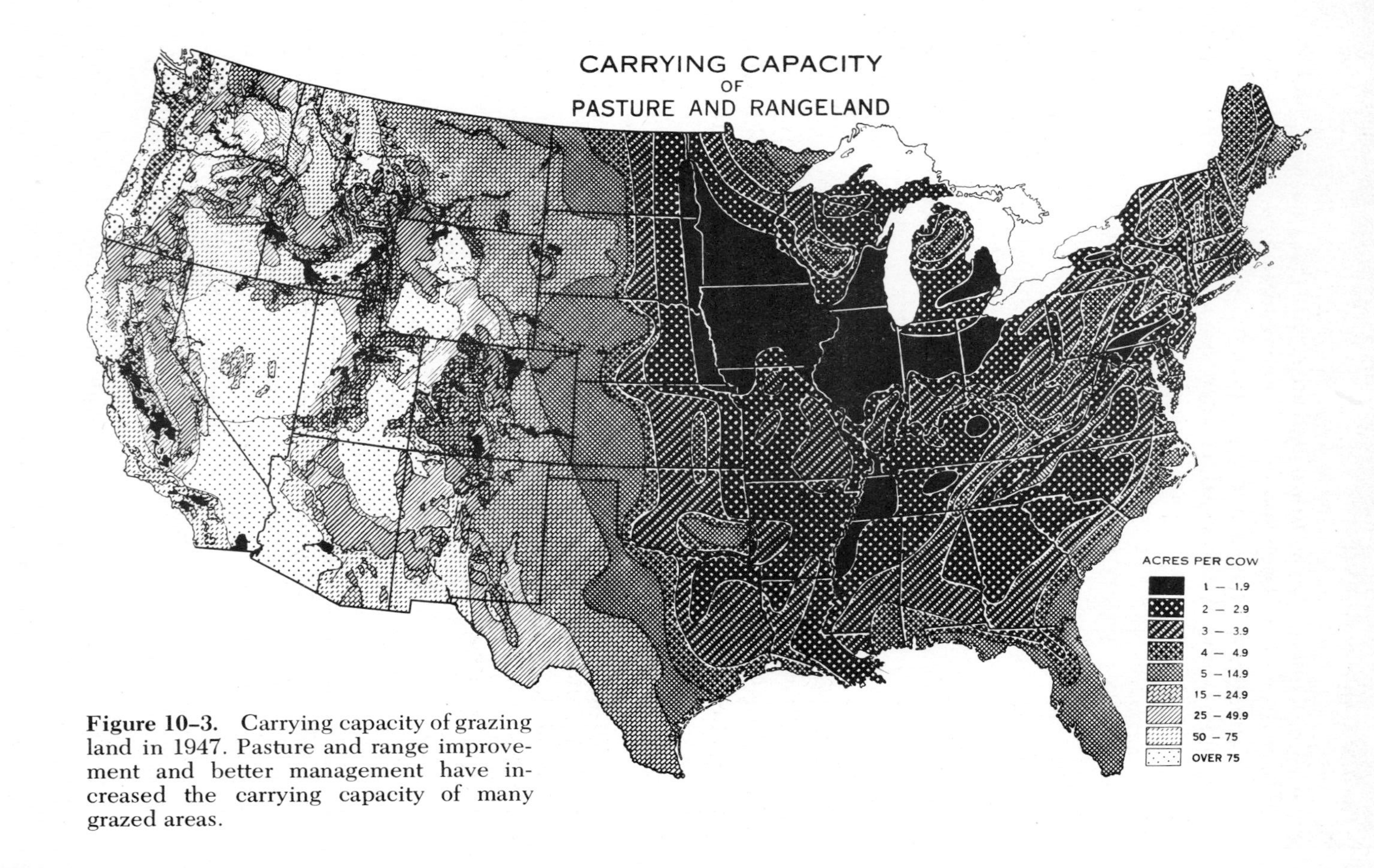

Figure 10–3. Carrying capacity of grazing land in 1947. Pasture and range improvement and better management have increased the carrying capacity of many grazed areas.

influence upon the type of pasture plants that occupy the land. Lespedezas will grow on acid and infertile soils, alsike clover on acid and poorly drained soils, and reed canarygrass on land that is wet. Native wheatgrasses and buffalograss prevail on the heavier soils in the Great Plains, while the gramagrasses are more abundant on the sandy loams.

ADAPTED PASTURE SPECIES

In the northeastern region, Kentucky bluegrass (*Poa pratensis*) probably is the most important pasture grass. In mixtures with white clover it is nearly always present on productive soils in permanent pastures. However, taller hay-type grasses are commonly seeded in pasture mixtures. These include orchardgrass (*Dactylis glomerata*), smooth bromegrass (*Bromus inermis*), tall fescue (*Festuca arundinacea*), timothy (*Phleum pratense*), and reed canarygrass (*Phalaris arundinacea*). Other grasses are redtop (*Agrostis alba*) and wild ryegrass (*Elymus* species). The principal pasture legumes are alfalfa, clovers (Ladino, red, and alsike), and birdsfoot trefoil (*Lotus corniculatus*). Sweetclover is used in the western part of the region, while annual lespedezas are widely grown in the southern portion.

Warm-weather grasses and legumes prevail south of the 60° F. isotherm, except for some tall fescue and alfalfa in the cooler portions. White clovers, vetch, and Austrian peas also are grown there as winter annuals. Tropical grasses are grown on the South Atlantic and Gulf Coastal plains. Native southern grasses also are utilized.[43]

Much of the pasturage on the Great Plains is supplied by native short grasses (Figure 10–4). The principal grass association in the northern Great Plains consists of blue grama (*Bouteloua gracilis*), green needlegrass (*Stipa viridula*), and western needlegrass (*Stipa comata*) with a small admixture of junegrass (*Koeleria cristata*). Seeded crested wheatgrass may have twice the grazing value of native grasses.[44, 67] Wild ryegrasses (*Elymus* species) also are productive. The central Great Plains has a very typical association of buffalograss (*Buchloë dactyloides*) and blue grama. Farther south below the Texas Panhandle, the black grama (*Bouteloua eripoda*) and curly mesquite (*Hilaria belangerii*) are the most important pasture plants. Just east of this region is the tall-grass region with the bluestems, *Bothrichloa* (*Andropogon gerardi*) and *B. scoparius*, switchgrass (*Panicum virgatum*), dropseeds (*Sporobolus* species), sideoat grama (*Bouteloua curtipendula*) being among the predominant species. Switchgrass gave much greater steer gains per acre than the seeded native pasture mixture in the Central Great Plains.[42] In both the Great Plains and intermountain areas, the clovers thrive only in the cooler irrigated

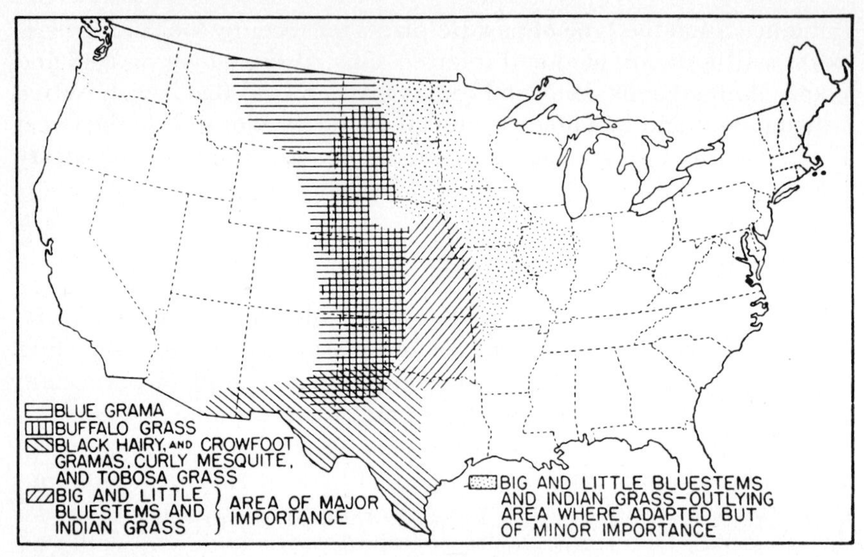

Figure 10–4. Leading native pasture grasses of the Great Plains and of prairie regions (dotted area at right). The seeding also of introduced grasses—such as crested wheatgrass, Russian wildrye, and smooth bromegrass—has improved some of the grazing areas.

sections. Most of the alfalfa grown there also is irrigated. Sainfoin is being sown for pasture.

The intermountain area has the lowest pasture production per acre because of the low rainfall. In the northern part, thin stands of wheatgrass (*Agropyron*) *spicatum* little bunchgrass (*Festuca idahoensis*), and Sandberg bluegrass (*Poa sandbergii*) are found.[67] In the central portion of the region (Utah and Nevada) there exists a true desert shrub vegetation characterized by large sagebrush (*Artemisia tridentata*) and shadscale (*Atriplex confertifolia*). During rainy periods, the introduced annual bromegrass (*Bromus tectorum*) and similar plants spring up to furnish some grazing. Associated with shadscale is the winterfat (*Eurotia lanata*), one of the most valuable grazing plants in the Great Basin. In the southern part of the region, the best grazing lands are found in western New Mexico and eastern Arizona. Drouth-resistant grasses that grow among the shrubs include galletagrass (*Hilaria jamesii*), tobosagrass (*Hilaria mutica*), black grama (*Bouteloua eripoda*), and wild ryegrass (*Elymus* species). Disk plowing of the more fertile sagebrush lands, followed by sowing of crested wheatgrass, increases range production several times.[18, 50]

Abundant rainfall and a mild cool climate along the north Pacific coast permit the establishment of permanent pastures of introduced plants. Grasses that thrive there include tall fescue, ryegrasses, Kentucky bluegrass, orchardgrass, tall oatgrass, timothy, meadow foxtail, reed canarygrass, and the bentgrasses. Clovers and also alfalfa do well in this area.

The rainfall in central and southern California is so low that most of the pastures in the valleys are irrigated. Important pasture plants are alfalfa, Ladino clover, hardinggrass, orchardgrass, subclover, medics, and strawberry clover. Alfileria (*Erodium cicutarium*) and various native grasses provide grazing in the uplands.

CHARACTERISTICS OF PASTURE PLANTS

Short young herbage low in fiber content is eaten in preference to old, tall, stemmy, highly fibrous herbage. All grasses are palatable when closely clipped.[62] They become unpalatable when growth stops. A dense sward with a height of about 4 inches approaches the ideal.[36] Animals exercise a choice between species when the grass is 4 to 6 inches high, but show little discrimination when it is only 2 to 4 inches high.[8] Perennial ryegrass leaves grow fast enough and long enough to have a long period of palatability.[62] Orchardgrass leafage grows very rapidly and is very palatable while in active growth, but is unpalatable when growth ceases. Hairy and scabrous plants tend to be unpalatable. The more mature a leaf, the more these conditions are accentuated. Timothy, bromegrass, and Italian ryegrass are regarded as most palatable, followed by white clover and orchardgrass. Timothy may be almost suppressed in a pasture because of its extreme palatability. In one test in Massachusetts, cattle selected herbage in order as follows: white clover, timothy, redtop, and Kentucky bluegrass.[8] In another test, the preference was, in descending order, timothy, redtop, Italian ryegrass, English ryegrass, tall oatgrass, meadow fescue, red fescue, and reed canarygrass.

The high palatability of big bluestem and little bluestem causes these tall grasses to be replaced by the less palatable but more persistent short buffalograss under heavy grazing in the Great Plains region. Where the land is protected from grazing these palatable species are restored and the grass again grows "as high as a horse's belly," as pioneers reported when they observed it for the first time.

BASIC PRINCIPLES OF PLANT BEHAVIOR

Certain important principles of pasture plant behavior are as follows:[62]

1. A plant excessively grazed will have its leaf development and root system reduced proportionally. Moreover, the new leaves are

reduced in size, with the result that their functional efficiency is lowered. Sod blocks of seven prairie grasses in Nebraska clipped every two weeks produced only 13 to 47 per cent as much dry matter as did the unclipped sods.[9] Plants weakened by clipping renewed growth slowly.

2. Plants can withstand defoliation in proportion to their ability to develop side shoots and tillers in spite of defoliation. Clipped plants may fail to produce new rhizomes.[9]

3. Plants suffer punishment in direct proportion to their yielding capacity.

4. Other things being equal, plants are grazed in proportion to their erectness or accessibility.

5. Plants with an erect growth habit usually are less efficient than prostrate plants in the production of side shoots and tillers.

6. Severe defoliation when the plants commence active growth is more harmful than at other times of the year. Consequently, plants that start active growth exceptionally early in the spring are most subject to damage. The clipping of seedling grass plants[52] has resulted in a decrease of 80 to 96 per cent in dry weight of forage.

PLANTS IN PASTURE MIXTURES

It is seldom desirable to seed land intended for permanent pasture to a single species for the following reasons:[58]

1. Legumes in pastures tend to maintain the nitrogen content of the soil.[14]

2. Mixtures result in a more uniform stand and higher production because several soil conditions are often represented in a pasture.[51]

3. Mixtures provide a more uniform seasonal production because the growth and dormancy periods vary among different plants.[19]

4. Mixtures of grasses and legumes provide a better balanced ration since legumes are richer in both proteins and minerals.[63] However, animals often are subject to bloat when the proportion of alfalfa or clover exceeds 50 per cent.

Addition of wild white clover to a seeding of Kentucky bluegrass increased the yield of herbage more than 500 per cent in New York experiments.[36] The protein content of Kentucky bluegrass grown alone averaged 18 per cent while the same grass grown in association with wild white clover averaged 25 per cent. A similar gain may be achieved by a heavy application of nitrogen fertilizer.

The number of pounds of the different seeds in a mixture determines only in part the percentages of the different species that emerge

TABLE 10–2. Rates of Seeding Grasses and Legumes in Pasture Mixtures

GRASS	RATE (lbs./acre)	GRASS	RATE (lbs./acre)
Cool-Weather Grasses		Boer lovegrass	0.5
Tall fescue	8–10	Weeping lovegrass	0.5
Timothy	3–6	Galletagrass	5
Orchardgrass	5–8	Sand dropseed	0.5–1
Bromegrass (awnless)	4–10	Hardinggrass	2–4
Mounting bromegrass	2–6	Carpetgrass (seeds naturally)	
Reed canarygrass	3–10	Bermudagrass (stolons)	
Tall oatgrass	3–4	Big bluestem	5–6
Meadow fescue	8–10	*Cool-Weather Legumes*	
Kentucky bluegrass	1–10		
Canada bluegrass	1–3	Alfalfa	2–10
Bulbous bluegrass	1–2	Ladino clover	1–4
Redtop	2–10	White clover	1–3
Perennial ryegrass	6–10	Red clover	3–10
Crested wheatgrass	4-11	Alsike clover	2–6
Slender wheatgrass	2–5	Strawberry clover	2–6
Russian wild rye	5	Sweet clover	2–10
		Birdsfoot trefoil	4–6
Warm-Weather Grasses		*Warm-Weather Legumes*	
Dallisgrass	4–10		
Blue grama	4–8	Lespedeza (Korean and common)	5–15
Sideoat grama	3–4	Low hop clover	1–3
Rothrock grama	3–4	Persian clover	1–3
Buffalograss	1–2	California burclover	2–8
Lehman lovegrass	0.5	Black medic	3

or the relative stands that result later, due to differences in the size of seeds, the percentage of live seeds, and the plant competition.[29, 30]

Rates of Seeding

The rates of seeding grasses and legumes in pastures are shown in Table 10–2.

Seeding rates per acre of grasses and legumes are affected by the other species included in the mixture. Seed quality is also a factor. Rates are affected by the number of seeds per pound for a species, pure live seed ratio, amount of foreign material, and other conditions.

PASTURE MIXTURES FOR DIFFERENT REGIONS

The rates of seeding of pasture mixtures, the adapted pasture plants, and some of the recommended mixtures for different regions and conditions in the United States follow:*

* From *Grass*, USDA Yearbook, 1948, pp. 1–892.

REGION	ADAPTED PASTURE PLANTS
Northeast (20–25 pounds per acre)	Orchardgrass, bromegrass, timothy, reed canarygrass, tall fescue, Kentucky bluegrass, redtop, perennial ryegrass, Ladino clover, alfalfa, red clover, alsike clover, birdsfoot trefoil, white clover, Korean lespedeza.

Recommended Mixtures:

(1) General use: Orchardgrass; tall fescue, or timothy; and alfalfa, or Ladina clover, or red clover.

(2) Poorly drained land: Reed canarygrass or redtop; and alsike clover or Ladino clover.

(3) Poorly controlled grazing areas: Kentucky bluegrass and white clover included with No. 1.

North Central (11–20 pounds per acre)	Bromegrass, orchardgrass, timothy, tall fescue, reed canarygrass, redtop, Kentucky bluegrass, alfalfa, Ladino clover, red clover, alsike clover, birdsfoot trefoil, Korean lespedeza, sweetclover.

Recommended Mixtures:

(1) Bromegrass; orchardgrass or tall fescue or timothy; and alfalfa or Ladino clover.

(2) Timothy, Ladino clover, red clover, alfalfa.

(3) For poorly drained soils: Timothy, redtop, reed canarygrass, alsike clover, birdsfoot trefoil.

Northern Great Plains Eastern portion (12 pounds per acre)	Crested wheatgrass, intermediate wheatgrass, bromegrass, alfalfa, wild rye, feather bunchgrass.
Western portion (10 pounds per acre)	Crested wheatgrass, sweetclover, blue grama, sideoat grama, buffalograss, western wheatgrass, wild rye, feather bunchgrass.
Northern intermountain irrigated (16–25 pounds per acre)	Bromegrass, orchardgrass, alsike clover, Ladino clover, alfalfa, strawberry clover, alsike clover, and reed canarygrass for wet lands.
Northwestern irrigated (16–25 pounds per acre)	Bromegrass; tall fescue; orchardgrass; tall oatgrass; alfalfa; and red, alsike, Ladino, New Zealand white and strawberry clover; birdsfoot trefoil; big trefoil.
Southwestern irrigated (13–30 pounds per acre)	Orchardgrass, perennial ryegrass, tall fescue, Ladino clover, alfalfa, burclover, dallisgrass, rhodesgrass, bromegrass, hardinggrass, yellow sweetclover.
Western range lands (5–13 pounds per acre)	Crested wheatgrass, bluestem wheatgrass, slender wheatgrass, bromegrass, tall oatgrass, sand dropseed, intermediate wheatgrass, wild rye.
Southwestern range lands (8–10 pounds per acre)	Crested wheatgrass, western wheatgrass, bromegrass, galletagrass, yellow sweetclover, Indian ricegrass, domestic ryegrass, weeping lovegrass, subterranean clover, hardinggrass, alfilaria, burnet.
Pacific Northwest Coastal wetlands (12–20 pounds per acre)	Tall fescue, reed canarygrass, meadow foxtail, birdsfoot trefoil, big trefoil.
Humid uplands (14–20 pounds per acre)	Perennial ryegrass, orchardgrass, tall oatgrass, tall fescue, Chewings fescue, New Zealand clover, subterranean clover, big trefoil, birdsfoot trefoil.

REGION	ADAPTED PASTURE PLANTS
Great Basin semiarid (6–12 pounds per acre)	Crested wheatgrass, Siberian wheatgrass, pubescent wheatgrass, slender wheatgrass, bromegrass, big bluegrass.
Southern Great Plains (9–19 pounds of unprocessed seed per acre)	Blue grama, sideoat grama, sand lovegrass, sand bluestem, switchgrass, indiangrass, little bluestem, weeping lovegrass.
Southeast (9–32 pounds per acre)	Coastal bermudagrass, tall fescue, hardinggrass, rescuegrass, dallisgrass, bahiagrass, pangolagrass, St. Augustine grass, lespedeza, white clover, hop clover, crimson clover, burclover, kudzu.

Recommended Mixtures:

(1) Upper South:	Orchardgrass or tall fescue with Ladino clover or alfalfa.
	Kentucky bluegrass and Ladino clover.
(2) Lower South:	Coastal bermudagrass and crimson clover.
	White clover and dallisgrass.
	Carpetgrass establishes itself naturally in Florida and vicinity.

ESTABLISHMENT OF PERMANENT PASTURES

Since most of the plant species used for permanent pastures have small seeds, it is necessary to prepare a firm seedbed. The use of good seed of small-seeded grasses and legumes is important in order to provide good pasture coverage relatively free from weeds. The seed should be drilled rather than broadcast on a new seedbed. It should be seeded not over 3/4 inch to 1 1/4 inches deep. Special drills for seeding grasses and legumes have attachments which open furrows, drop fertilizer in bands, and cover the fertilizer with a layer of soil compacted with a press wheel. In the same operation, the seed is drilled into the compacted soil band about 1 1/4 inches above the fertilizer.[4, 14, 47] Forage-crop seeders with two sets of corrugated rollers are very effective on prepared seedbeds.

Ladino clover can be established in closely grazed or clipped grass sod either by broadcasting or drilling.

In general, it is advisable to sow cool-weather grasses in early fall, but early spring seeding is preferable on extremely heavy soils in the northern states where winter heaving would eliminate many seedlings. Early spring sowing also is advisable in the southern states where the perennial grasses are warm-weather species. Small-seeded legumes may be broadcast in late winter or early spring on the fall-sown grass and companion crop fields. Alternate freezing and thawing and rains help to cover the seeds with soil. When a farmer is seeding with a companion crop, the grasses and legumes are seeded with the small grain. The pasture plants are more productive the first year when seeded alone.[58] When grazed properly, they may provide a larger net return than the grain crop, or even offset the grazing value of the companion crop. However, a high-yielding small-grain crop is profitable. It also suppresses weed growth.

When irrigation is used, it is necessary to keep the soil surface moist until the seedings are well started. They may require irrigation every 7 to 15 days. Dryland grass seedings succeed only during favorable moist periods.

Several grasses adapted to Florida are established by vegetative methods, such as root or stem cuttings, or sod pieces. These grasses include napier, bermuda, para, St. Augustine, pangola, and centipede.

A new pasture should be grazed lightly the first season. The young plants must have time to develop a good root system so as to withstand drought and the strain of grazing.

Pasture maintenance or improvement may be brought about by reseeding, fertilization, and good grazing management.

FERTILIZATION OF PASTURES

A ton of dry matter in legume hay contains about 50 to 75 pounds of nitrogen, 15 to 20 pounds of P_2O_5, 20 to 25 pounds of K_2O, and 20 to 25 pounds of CaO. Grass hays contain about 30 to 40, 10 to 20, 40 to 50, and 10 to 15 pounds per ton of the respective mineral nutrients.

Most pasturelands in the humid regions of the country are deficient in calcium, phosphorus, nitrogen, and sometimes potassium.[44a] The minerals calcium, phosphorus, and potassium must be applied before much response can be expected from application of commercial nitrogen.[58] Pastures on soils of fair natural fertility, particularly those that have been neglected several years, can be improved by fertilization.[47, 65, 66] An application of 400 to 600 pounds of a 6–12–6 fertilizer gave the best increases in southern Georgia.[63]

The effect of heavy fertilizer applications is pronounced.[15] In Wisconsin, the completely fertilized areas of a bluegrass pasture produced about three times as much herbage as did the unfertilized areas during the first year.[22] The fertilizers used were 100 pounds per acre each of phosphates, potash, and nitrogen. The turf was greatly thickened, the weeds had largely disappeared, and white-grub injury was practically eliminated. In Ohio, fertilizer treatment doubled the carrying capacity and also lengthened the grazing season by about 15 per cent.[6]

Applications of superphosphate alone generally give the greatest response because they encourage the legumes that supply nitrogen to the grasses. However, an excess of clover or alfalfa increases the risk of bloating of cattle.

Applications of nitrogen usually increase the protein content of herbage. Use of nitrogen has increased the crude protein of bluegrass-white clover herbage as much as 12 per cent, and of orchardgrass from 50 to 100 per cent.[26, 60] Nitrogen may promote growth enough so that pastures will be ready for grazing as much as three weeks earlier than those without such applications. Nitrogen applications often discourage the growth of legumes in grass-legume mixtures.

Nitrogen is relatively ineffective on Ohio sods with a high clover content.[20] Heavy applications of nitrogen (60 to 200 pounds per acre) to grass pastures, supplied with ample amounts of other elements, may increase herbage yields from 20 to 100 per cent.[16, 65]

When added to a pasture that had been unfertilized for 40 years, all fertilizers increased the nitrogen content of the herbage.[13] The averages for each year were about the same for the phosphorus-lime as for the phosphorus-nitrogen treatments. In Ohio, run-out bluegrass and white-clover pastures have been treated with superphosphate, lime being added where the *p*H was below 5.5.[6] In the case of lime deficiency, it may be desirable to replace certain pasture plants with others not sensitive to acid soils. For this purpose, lespedezas or birdsfoot trefoil may be used in place of clovers in areas where they are adapted.

Mineral fertilizers, limestone, and barnyard manure can be applied in the fall, winter, or early spring. Commercial nitrogen should be applied at least two weeks before increased growth is desired. Applications of nitrogen are rarely effective except in the presence of adequate soil moisture,[32] but moderate applications are beneficial in the more favorable parts of the semiarid Great Plains.[53]

RENOVATION AND RESEEDING

Cultivation of pastures to secure improved grazing is of little value unless accompanied by reseeding or application of fertilizers, or both.[16]

Practices in Humid Regions

Cultivation in connection with reseeding and fertilization resulted in improved pastures in Vermont and Iowa by elimination of weeds, covering the seed, and mixing the fertilizer in the soil. Grasses and clovers that make a quick growth may be seeded on old pasture sod that has been well disked and fertilized.[21] In Wisconsin,[23, 27] run-out bluegrass pastures can be improved by the scarification of the sod with a disk or spring-tooth harrow, after which legumes such as alfalfa, sweetclover, and red clover are seeded in the thinned pasture sod. In 27 different pasture renovations the average total weed population (mainly ragweeds and horseweeds) was reduced 85 per cent after two or three years. Reseeding alone may be desirable in some instances in connection with the improvement of old pastures, but it is seldom a complete remedy. Pastures on tillable land may be plowed, tilled, fertilized, and reseeded, either immediately or after an intervening crop (Figure 10–5). In pastures that are too rough or stony for plowing, the broadleaf weeds can be killed with herbicides and then followed by heavy summer grazing. Dalapon at 4 to 8 pounds per acre is applied in late summer to kill the grasses. The dead sod is disked 3 to 5 weeks later and the pasture is then reseeded.

Revegetation of Rangelands

The return of abandoned cultivated land to grass is a difficult problem on the Great Plains as well as in other areas of moisture shortage.[55] From 20 to 50 years are required for buffalograss to become reestablished naturally on abandoned farmland.

Artificial range reseeding in the native sod often has been unsuccessful in the semiarid region unless the seeding was favored by unusual rainfall. However, the seeding of blue grama, buffalograss,[63] and other species sometimes has been fairly successful. Seedings of cultivated grasses (crested wheatgrass, smooth brome, and slender wheatgrass) thrived on areas where the original native vegetation consisted of grama and fescue grasses.[28] The land was disked, the seed sown broadcast early in the spring, and the field protected from grazing the first season. Range grasses have been restored by reseeding on depleted mountain meadows, alluvial bottoms, and the better sites of mountain slopes where soil moisture conditions were above average. For cultivated grasses, an annual precipitation of 15 inches or more appears to be essential. In Colorado, range seedings have been unsuccessful on areas that receive less than 10 inches annually.[34]

An overgrazed oak-brush range in Utah was seeded successfully with crested wheatgrass, smooth brome, and mountain brome.[49] The best stands were obtained from seed broadcast on plowed furrows spaced approximately 3 feet apart, the seed being covered with a brush drag. During a period of seven years this method resulted in an increase of 360 to 900 per cent in grazing capacity compared with open grazed unseeded areas. Some degree of soil preparation was necessary to assure successful reseeding.

Sagebrush lands, as well as other southwestern range areas that have a rainfall of 15 inches or more, were seeded successfully after the land was tilled with a disk plow. Grass production was increased from two- to ten-fold. Adapted grasses were crested wheatgrass, other wheatgrasses, big bluegrass, and smooth bromegrass.[18, 50]

Dropping pelleted grass seed from an airplane on burned or plowed mountain rangeland has been unsuccessful, whereas drilling has resulted in good stands.[35]

Unless moisture and other conditions are unusually favorable, two or three years are required to establish a new range. When livestock are kept off for such a period, scattered seedlings have an opportunity

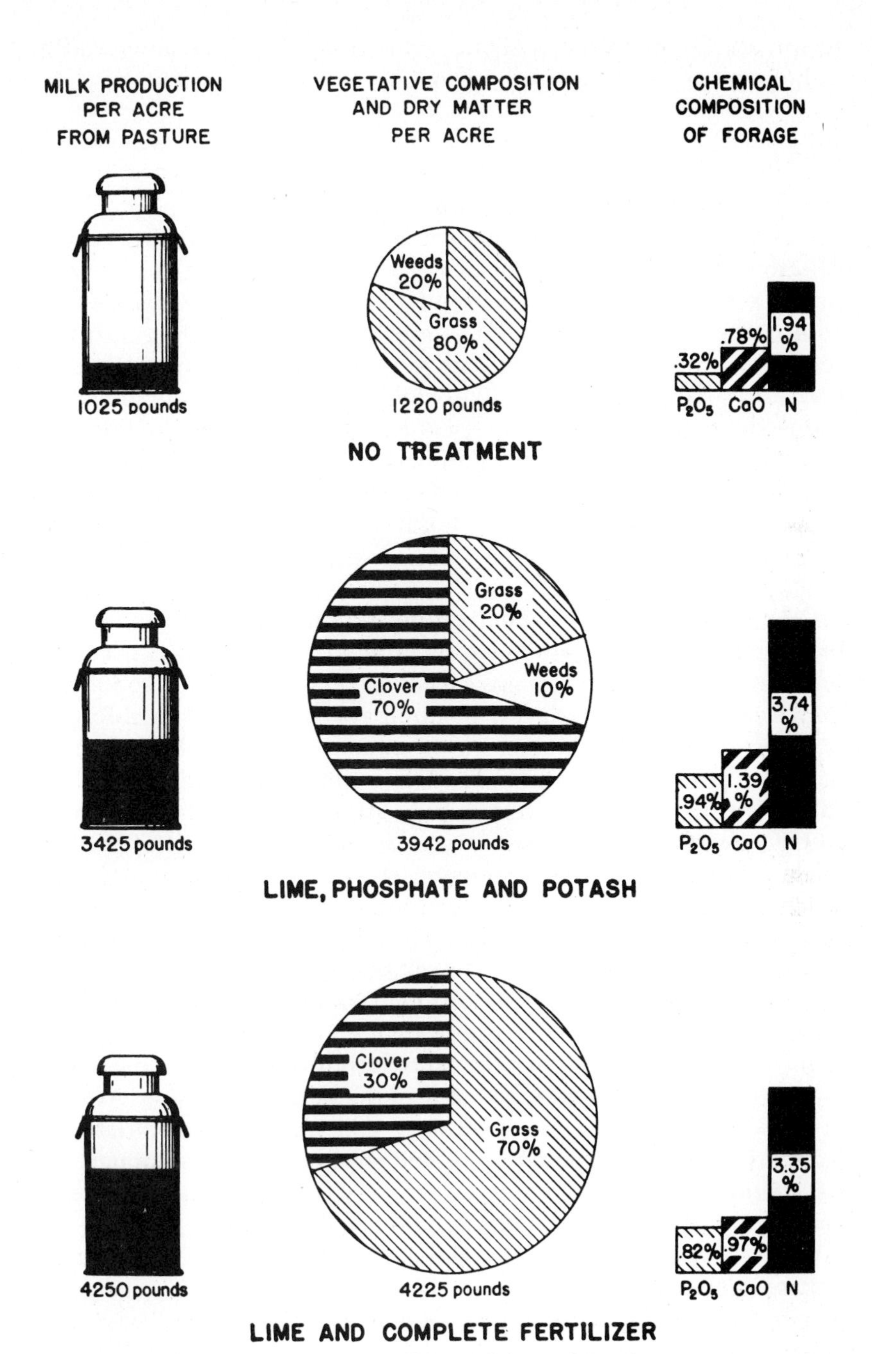

Figure 10–5. Pasture improvement by seeding and fertilization. Nitrogen in the complete fertilizer stimulated grass growth at the expense of clover.

to spread. Dormant seeds also germinate and produce new seedlings to help increase the ground cover.

GRAZING SYSTEMS

The capacity of native pastures has been increased as much as 50 per cent by good grazing management. Controlled grazing is necessary to give the palatable species an opportunity to recuperate and to produce seed. Persistence of vegetation through a dry summer and a cold winter is directly related to root development. Overgrazing has resulted in poor root growth with very little food accumulation, with the result that the plants are likely to die either from drought or cold.[61] Large decreases in weight of roots have followed overgrazing.[62] The total decrease from the early to the late stage of grazing was from 2.17 to 0.95 tons of roots per acre in the 0 to 4-inch depth, and from 0.86 to 0.34 ton in the 4- to 12-inch depth on upland Nebraska soil.

Influence of Grazing on Species

One of the first signs of an overgrazed range is that the most palatable grasses such as bluegrass, needlegrass, junegrass, wheatgrass, bromegrass, and the fescues become less vigorous, produce less forage, and become reduced in numbers. The less palatable plants and poisonous plants increase in numbers. As the condition becomes more severe, annual weeds tend to replace perennial weeds and shrubs. In the last stage, bare spots appear and gradually increase, which in turn causes an increase in soil erosion. After the perennial grasses have perished and erosion started, a long time may be required to restore the range to its original productivity.

Blue bunchgrass and slender wheatgrass are two valuable range species in Montana. In order to maintain their vigor, these grasses should not be utilized beyond 60 to 70 per cent of their foliage production by early summer, or more than 80 or 85 per cent at the close of the summer grazing period.[17] The higher the successional stage of the vegetation, the greater the value of the range for grazing. As shown in Table 10–3, the rangeland tends to take on the characteristics of a more arid type as it becomes depleted. The higher successional stages were characterized by greater density of stand, a higher percentage of grass in the stand, and greater grazing capacity.

Grazing management influenced the composition of cultivated grasses grown in mixtures in the British Isles[37] in a pasture composed of perennial ryegrass, rough-stalked meadowgrass, and wild white clover. The plot grazed heavily in March, April, and May had white clover as the most important constituent of the pasture by the middle of the third season. Another plot was made grass-dominant when grazing was deferred until April 15. The perennial ryegrass became more

TABLE 10–3. Productivity of Successional Types of Range Vegetation

ITEM	BLUE BUNCHGRASS	SLENDER WHEATGRASS	PORCUPINE GRASS	RABBIT BRUSH
Density of vegetation (%)	60–80	40–60	30–50	20–40
Grasses (%)	75	65	85	25
Weeds (%)	20	25	10	50
Browse (%)	5	10	5	25
Palatable (%)	62	54	49	25
Surface acres to feed 1 cow 1 month	2	3	4	11

vigorous while the white clover was considerably checked. A third plot not grazed before April 15, but completely pastured down at each subsequent grazing and then rested for a month, maintained a good balance between perennial ryegrass and white clover.

Deferred Grazing

On the western range, the problem is to maintain the important palatable native range plants. Ranges may be injured more by too early grazing than by any other faulty practice. Grasses grazed too early in the spring may be pulled up by the roots or damaged by trampling in a wet soil. In the spring, when the new grasses commence their growth, the water content of the herbage may be as high as 85 per cent, with a low feeding value. Excessive early grazing each year may delay satisfactory development of the palatable plants by as much as six weeks. It is desirable to delay grazing until the important forage plants have reached a height of 6 inches or, in the case of the shorter grasses, until the flower heads are in the boot.[17] Normal stand and vigor of bluestem pastures in eastern Kansas were maintained when grazing was deferred until June 15.[3] The deferred system gave an increase of approximately 25 per cent in carrying capacity.

Rotation Grazing

Rotation grazing consists in grazing two or more pastures in regular order with intervening rest periods for each pasture.[54] An experimental area on the Jornada Reserve, moderately grazed under the deferred system, was more than four times as valuable as the outside range heavily grazed all year long.[17] As applied to western ranges, the system of deferred-rotation grazing provides for reservation until after seed maturity of about one fourth or one fifth of the entire area used by the herd. Then that area is grazed. A different area is so reserved each year. The result is an increase in carrying capacity of the range, a chance for improvement when the range is depleted, and better growth of animals without losses through nonuse of feed. Continuous grazing is superior to the rotation of range pastures at approximately monthly intervals.

Rotation grazing of Kentucky bluegrass in western Missouri increased beef gains per acre only 5 per cent.[15] Larger benefits from rotation grazing occur with taller, heavier yielding forages. At Beltsville, Maryland, rotation grazing increased the yield of total digestible nutrients by 10.4 per cent, heavy fertilization increased the yield 16.4 per cent, and both combined increased the yield 28.6 per cent.[69]

An intensive plan of grassland management, known as the Hohenheim system,[69] was developed in Germany in 1916. This system involves (1) division of the pasture into from four to eight paddocks, about equal in size, (2) heavy applications of fertilizers, especially nitrogen, (3) separation of the herd into two groups: producers and nonproducers, and (4) frequent rotation of these groups. The cattle are moved progressively to other paddocks at intervals of about one week, or whenever the grass reaches a height of 4 or 5 inches. Young stock and dry cows, kept separate from the milk cows, follow into each paddock as the milkers are advanced.

Strip grazing is a modification of rotation grazing. Temporary fences confine the cattle to small areas which they can graze down to the desired height in 1 or 2 days. The fences are moved across the field as the herbage is consumed. The animals are returned to the original strips after sufficient new growth has occurred. The strips being grazed usually are bordered with electric fences. Strip grazing largely eliminates selective grazing by the animals, but requires more attention than does rotation grazing.

Both rotation and strip systems permit mowing, fertilization, and irrigation of the areas not being grazed. Any herbage in excess of current grazing needs can be cut for hay or silage. The closeness of grazing is under control, the plants not being grazed again until they have made sufficient growth recovery. Milk production from rotation and strip grazing is about equal.[25]

Soiling or Green Feeding

Soiling or green feeding was practically discontinued in the United States by 1918 because of the hand labor required to cut, load, and feed the heavy green forage. The development of forage harvesters and mechanical unloading trucks revived some interest in the practice. Soiling also is called green chop, zero grazing, and mechanical grazing. Soiling eliminates problems of trampling, fencing, and water facilities for pastures. It also eliminates the rejection of the coarser portions of the herbage. The risk from bloating is reduced because the cattle are forced to eat grass and stems along with the tender legume herbage. The disadvantages of soiling are the cost, operation and possible breakdown of machines, as well as the difficulty of getting into the field in extremely wet weather. Animal production per acre from green feeding is superior to that from well managed rotation or strip grazing, but labor and machinery costs are higher.[25, 66]

BURNING GRASSLANDS AND BRUSH

Burning of grasslands has been practiced for many years, particularly in the Southeast, but authorities differ as to its effectiveness in grassland improvement. Those who advocate burning claim that it (1) brings about an earlier growth of vegetation, (2) results in more palatable vegetation than that from unburned areas, (3) increases the productivity of the soil through liberation of the lime, phosphoric acid, and potash contained in the ash, (4) improves the character of the herbage by control of weeds and brush, and (5) controls chinch bugs.

Realization of such benefits is more likely to occur in humid regions. Burning is practically necessary in the Gulf Coast region as long as the land is used for both grazing and lumbering, particularly when only the native grasses and legumes are grazed. Burning increases the number of legumes and grasses by destruction of the heavy ground cover of pine needles and leaves from other trees.

Burning often is used in western Colorado to destroy relatively dense stands of big sagebrush. Controlled burning is usually done in late summer or early fall when a clean burn can be obtained.[34]

Burning frequently has detrimental effects, particularly in the west. Some of these effects are (1) removal of the extra vegetative material that would add humus and nitrogen to the soil, (2) destruction of old vegetation in the soil which functions to increase the water-holding capacity, and (3) injury to the living vegetation. Burning is detrimental to the vigor of short grasses, the total yield for the season usually being less on burned areas. Shallow-rooted grasses, like bluestem and the fescues, may be killed by a single burning. The dry forage left protects the young growth from too close grazing.

In Kansas, continued annual burning of native bluestem pastures decreased the total production of grass.[2] Yields from burned and unburned areas are given in Table 10–4. Burning should be carried out in early spring after hard freezing weather is over, but before the native grasses start growth, and only in occasional years when an excess amount of dry material is on the pasture from the previous season. Burning has had very little effect on control of weeds unless done late in the spring. Burning has stimulated early spring growth because of the warmed soil.

In Kansas, buckbrush (*Symphoricarpos* species) was greatly retarded when burned late in the spring.[1] Sagebrush has been eradicated from rangelands when burned in the late fall when it was dry.[28]

TABLE 10–4. Effects of Burning on Bluestem Pastures in Kansas

DATE OF ANNUAL BURNING	SIX-YEAR AVERAGE YIELDS PER ACRE	
	Grasses and Weeds (tons)	*Weeds (tons)*
Late fall (December 1)	1.02	0.20
Early spring (March 20)	1.12	0.15
Medium spring (April 10)	1.18	0.18
Late spring (May 5)	1.48	0.03
Unburned	1.68	0.14

Removal of the sagebrush permitted grasses such as the fescues, wheatgrasses, and arid bluegrasses to become more productive. Some pastures increased their foliage two to five times from this practice. It took several years for the sagebrush to become re-established.[48] Burning was equally damaging to grasses and sand sagebrush in western Oklahoma.[56] Burning as a management practice will most likely be decreased if not eliminated in the future, on the basis of adverse aesthetic effects upon the environment.

ERADICATION OF WEEDS OR BRUSH

Mowing western Oklahoma pastures in June, in two consecutive years, eliminated much of the sand sagebrush (*Artemisia filifolia*). Resting the pastures for 2½ months after each mowing permitted the grasses to recover. As a result, beef production per acre was more than doubled.[56]

Buckbrush and sumac were eradicated when cut in the flower stage, i.e., about May 10 and June 8, respectively, in Kansas.[1] In Connecticut experiments,[12] July was the best time to mow brush consisting of soft maples, alders, white birches, and blackberries. In the northern regions the critical period for destroying the brush appears to be when the roots contain the smallest amount of starch, generally when the plants are in blossom. In the southern states woody shrubs must be grubbed out or killed with herbicides.

Spraying with herbicides is preferable to either mowing or burning for destroying buckbrush, big sagebrush, sand sagebrush, skunkbush, rabbitbrush, and certain other pasture shrubs.[57] This treatment also kills many broad-leaved weeds as well as poisonous plants such as loco, halogeton, deathcamas, orange sneezeweed, woody aster, silvery lupine, water hemlock, princes' plume, and two-grooved milk vetch. Mesquite, oaks, and other woody plants, as well as tall larkspur, are controlled with other herbicides. Reseeding the range after weed control often is helpful.

The sprays are most effective when applied in the spring after growth is well established. The herbicides usually are applied with aerial spraying equipment. A repeated spraying after two or more years usually is necessary.

STOCK-POISONING PLANTS

Poisonous plants cause estimated losses of 4 per cent pėr year to range livestock. Animals usually eat poisonous plants in harmful quantities only when the more nutritious and palatable plants are inadequate to meet their requirements.

Poisonous plants vary greatly with respect to (1) the condition under which animals are poisoned by them, (2) the portion of the plant that is poisonous, (3) changes in the toxicity of the parts of the plant during growth and drying, (4) the susceptibility of different species of animals to being poisoned by them, and (5) the effects on the poisoned animals.

Some of the important poisonous plants are arrowgrass (*Triglochin maritima*), deathcamas (species of *Zygadenus*), horsetail (species of *Equisetum*), larkspur (species of *Delphinium*), loco (species of *Oxytropis* and *Astragalus*), lupine (species of *Lupinus*), whorled milkweed (*Asclepias galioides* and *A. mexicana*), poison vetch (some species of *Astragalus*), water hemlock (species of *Cicuta*), white snakeroot (*Eupatorium urticaefolium*), crazyweed (species of *Oxytropis*), and sneezeweed (species of *Helenium*). Halogeton (*Halogeton glomeratus*) is now present in western ranges that cover an area of 11 million acres.

Poisonous principles found in these plants include cyanogenetic glucoside in arrowgrass; an alkaloid, zygadenine, in deathcamas; an alkaloid (equisetin), and aconitic acid, and fungi, in horsetail; alkaloids (delphinine and others) in larkspurs; a toxic base, locoine, in loco weeds; alkaloids (lupinine and others) in lupines; an alcohol-soluble resin in whorled milkweed; selenium in *Astragalus* species; a resin-like substance, cicutoxin, in water hemlock; oxalic acid in halogeton; and a higher alcohol, tremetol, in white snakeroot.[24, 44]

Grasses with rough awns that injure the mouth, eyes and noses of grazing animals include foxtail (*Hordeum jubatum*), cheatgrass (*Bromus tectorum*) and the unpalatable medusahead (*Elymus caput-medusae*). Control of medusa head involves burning, herbicides or close grazing followed by discing and seeding of a suitable grass such as crested wheatgrass.[68]

CROPS FOR TEMPORARY OR ANNUAL PASTURES

The growth of perennial grasses and legumes stops or is greatly retarded, and the carrying capacity is reduced, each season when the temperature becomes too hot or too cold for the particular species. It is often desirable to grow annual or biennial crops to supplement permanent pastures during unproductive periods. A suitable succession of annual crops may provide pasturage for a long season in nearly all parts of the country. Some of the advantages of annual pastures are greater production per acre, a longer gowing season, less trouble from internal parasites, and less danger from noxious weeds. Disadvantages are the greater labor requirement, cost of seed, greater danger of erosion, impracticability of grazing such crops on clay soils in wet weather, and frequent inability to produce good stands.

In many of the humid sections of the North, clover and alfalfa may be pastured temporarily by sacrifice of one cutting of hay. Alfalfa hayfields in the irrigated Southwest are pastured during the hot summer months when the growth is too meager for hay production.

Other temporary pasturing schemes include Italian ryegrass in early spring; winter or spring small grains from April to July; rape, field peas, or vetch in early summer; and soybeans or sudangrass from midsummer to fall. Sweetclover may be pastured in the fall in its first year and in early spring of its second year. Some of these crops have two to four times the carrying capacity of comparable permanent pastures in the early spring, midsummer, or late fall.

An example of an effective arrangement of annuals for southeastern North Carolina, to provide pasturage for dairy herds, is as follows:[58] (1) Abruzzi rye, sown in September and grazed from November 15 to March 15, (2) crimson clover and hairy vetch sown August 15 to September 1 and grazed from March 1 to May 15, (3) sudangrass sown April 1 and grazed from May 15 to November 15, and (4) Biloxi soybeans sown March 15 and grazed from June 1 to November 15.

Also in the South, legumes such as lespedeza can be maintained in association with dallisgrass but cannot compete satisfactorily with carpetgrass or bermudagrass. Rye and vetch can be drilled into bermudagrass sod in the fall to provide temporary winter pasture.[63] White clover, a cool-weather perennial legume, formerly considered only for the North, provides excellent early pasture in the South where it is maintained largely as a winter annual. Fall-sown rust-resistant small grains, alone or with crimson clover or burclover, provide excellent grazing during the fall, winter, or early spring in the South. A mixture of Italian ryegrass and crimson clover also provides excellent winter pasture. Pearl (cattail) millet furnishes abundant, summer pasture in the South.

In the southern Corn Belt, Korean lespedeza grown with winter small grain provides summer grazing after grain harvest. Then the lespedeza reseeds the field after plowing in preparation for a succeeding small-grain crop. Oat pasturage is more palatable than barley, while barley is more palatable than wheat or rye.

In the Great Plains, grazing is provided for much of the season by (1) native pasture in late spring, early summer, and late fall, and (2) by sudangrass from July to September. In eastern Colorado, winter rye furnishes late fall as well as spring pasture. Native grass pastures or spring-sown small grains provide grazing from late spring to about July 1. Sudangrass can be grazed from early July until frost (Figure 10–2).

WINTER WHEAT FOR PASTURE

Winter wheat is pastured extensively from October to December and again in the spring in the central and southern Great Plains states.[21, 60, 65] From 20 to 65 per cent of the winter-wheat acreage may be pastured to some extent in favorable years. From 60 to 120 days of grazing are available during the period from November to April. Moderate grazing causes little or no reduction in grain yield in fields of winter wheat well established and well supplied with soil moisture. Yields have been reduced from 5 to 40 per cent by heavy grazing or by grazing when the wheat growth was scanty and soil moisture was limited. Wheat-pasture experiments were conducted in western Kansas for five years on both fallowed land and on land previously cropped to wheat.[65] The grain yields are shown in Table 10–5. Grazing of irrigated wheat in west Texas from autumn to March 20 reduced the wheat yield 20 per cent.[60]

Moderate timely grazing of the rank growth of wheat on fallowed land even improved the grain yield. Spring grazing may be started when growth is resumed in the spring, but should be discontinued when the plants start to grow erect just previous to jointing. This usually occurred about April 10 at the Hays Station.[65] The wheat plants may be injured by grazing at any time after their growing points

TABLE 10–5. Effect of Pasturing Winter Wheat on Grain Yield, 1926–30

METHOD PASTURED	TIME PASTURED	AVERAGE GRAIN YIELDS PER ACRE	
		Fallow Land (bu.)	*Wheat Land (bu.)*
Fall pasturing	October 15 to December 15	29.6	25.3
Moderate seasonal pasturing	October 15 to December 15 and March 1 to April 15	28.9	25.0
Check (unpastured)	–	26.6	26.2
Spring pasturing	March 1 to April 15	25.6	23.5
Severe seasonal pasturing	October 15 to December 15 and March 1 to April 15	23.5	21.3
Late spring pasturing	April 15 to May 1	20.4	–

are above the ground line.[39] Permanent injury resulted when the culm tips were grazed, as severed culms never produced heads. The sowing of an additional peck of seed per acre is recommended where grazing is contemplated.[65] Moderate pasturing is especially beneficial where wheat has been sown very early or has made an excessive growth. Pasturing checks rank growth and tends to prevent lodging.

Samples of wheat plants taken late in the fall contained 27 to 28 per cent protein and 12 to 15 per cent ash. The high-protein content accounts for the high nutritive value of wheat pasture. However, cattle or sheep grazing in cool weather on small grains or other lush grasses growing on soils that are high in nitrogen and potassium may suffer from grass tetany (hypomagnesaemia, nitrate poisoning or so-called wheat-pasture poisoning). The plants under these conditions have a low content of magnesium in the foliage which causes the grass tetany attack. It is prevented by adding magnesium to any supplementary feed or to the salt blocks, and treated by injection of a calcium gluconate solution containing magnesium. Most cases develop after 60 to 150 days of grazing on the lush forage by cows in late pregnancy or when nursing calves under 60 days of age.[31] The animals suffer convulsions, with periods of relaxation, and eventually die.

Legumes, but not grasses may take up enough molybdenum from soils high in molybdenum content to be toxic to cattle and sheep.[69]

REFERENCES

1. Aldous, A. E. "The eradication of brush and weeds from pastures," *J. Am. Soc. Agron.*, 21:660–666. 1929.
2. Aldous, A. E. "Effect of burning on Kansas bluestem pastures," *Kans. Agr. Exp. Sta. Tech. Bul.* 38. 1934.
3. Aldous, A. E. "Management of Kansas bluestem pastures," *J. Am. Soc. Agron.*, 30:244–253. 1938.
4. Anonymous. "Band seeding in the establishment of hay and pasture crops," in *USDA Agricultural Research Service Spec. Rept. ARS 22–21.* 1956.
5. Anonymous. "Utilizing forage from improved pastures," in *USDA Agricultural Research Service Spec. Rept. ARS* 22–53. 1960.
6. Barnes, E. E. "Ohio's pasture program," *J. Am. Soc. Agron.*, 23:216–220. 1931.
7. Bateman, G. Q., and Keller, W. "Grass-legume mixtures for irrigated pastures for dairy cattle," *Utah Agr. Exp. Sta. Bul.* 382. 1956.
8. Beaumont, A. B., and others. "Some factors affecting the palatability of pasture plants," *J. Am. Soc. Agron.*, 25:123–128. 1933.
9. Biswell, H. H., and Weaver, J. E. "Effect of frequent clipping on the development of roots and tops of grasses in prairie sod," *Ecol.*, 14:368–390. 1933.

10. Blaser, R. E., and others. "Pastures for Florida," *Fla. Agr. Exp. Sta. Bul. 409*, pp. 1–78. 1945.
11. Blaser, R. E., and others. "Managing forages for animal production, *Va. Poly. Inst., Rsh. Div. Bul. 45*, pp. 1–88. 1969.
12. Brown, B. A. "Effect of time of cutting on the elimination of bushes in pastures," *J. Am. Soc. Agron.*, 22:603–605. 1930.
13. Brown, B. A. "The effects of fertilization on the chemical composition of vegetation in pastures," *J. Am. Soc. Agron.*, 24:129–145. 1932.
14. Brown, E. M. "How to seed new pastures," *Mo. Agr. Exp. Sta. Bul. 739*. 1959.
15. Brown, E. M. "Managing Missouri Pastures," *Mo. Agr. Exp. Sta. Bul. 750*. 1960.
16. Brown, E. M. "Improving Missouri pastures," *Mo. Agr. Exp. Sta. Bul. 768*. 1961.
17. Chapline, W. R. "Range research of the U.S. Forest Service," *J. Am. Soc. Agron.*, 21:644–649. 1929.
18. Cornelius, D. R., and Talbot, M. W. "Rangeland improvement through seeding and weed control on east slope Sierra Nevada and southern Cascade mountains," *USDA Handbook 88*. 1955.
19. Currie, P. O., and Smith, D. R. "Response of seeded ranges to different grazing intensities in the Ponderosa pine zone of Colorado," *U.S. Forest Service Prod. Rsh. Rpt. 112*, pp. 1–41. 1970.
20. Dodd, D. R. "The place of nitrogen fertilizers in a pasture fertilization program," *J. Am. Soc. Agron.*, 27:853–862. 1935.
21. Fink, D. S. "Grassland experiments," *Maine Agr. Exp. Sta. Bul. 415*. 1943.
22. Fink, D. S., Mortimer, G. B., and Truog, E. "Three years' results with an intensively managed pasture," *J. Am. Soc. Agron.*, 25:441–453. 1933.
23. Fuellman, R. F., and Graber, L. F. "Renovation and its effect on the populations of weeds in pastures," *J. Am. Soc. Agron.*, 30:616–623. 1938.
24. Gilkey, H. M. "Livestock-poisoning weeds of Oregon," *Oregon Agr. Exp. Sta. Bul. 564*, pp. 1–74. 1958.
25. Gordon, C. H., and others. "A comparison of the relative efficiency of three pasture utilization systems," *J. Dairy Sci.*, 42(10):1686–1697. 1959.
26. Gordon, C. H., Decker, A. M., and Wiseman, H. G. "Some effects of nitrogen fertilizer, maturity, and light on the composition of orchardgrass," *Agron. J.*, 54:376–378. 1962.
27. Graber, L. F. "Evidence and observations on establishing sweetclover in permanent bluegrass pastures," *J. Am. Soc. Agron.*, 20:1197–1205. 1928.
28. Hanson, H. C. "Improvement of sagebrush range in Colorado," *Colo. Exp. Sta. Bul. 356*. 1929.
29. Hanson, H. C. "Analysis of seeding mixtures and resulting stands on irrigated pastures of northern Colorado," *J. Am. Soc. Agron.*, 21:650. 1929.
30. Hanson, H. C. "Factors influencing the establishment of irrigated pastures in northern Colorado," *Colo. Exp. Sta. Bul. 378*. 1931.
31. Holt, E. C., and others. "Production and management of small grains for forage, *Tex. Agr. Exp. Sta. Bul. B-1082*. 1969.

32. Houston, W. R. "Range improvement methods and environmental influences in the Northern Great Plains," *USDA Prod. Rsh. Rpt. 130,* pp. 1–13, 1971.
33. Hull, A. C., Jr., and others. "Progress report—Seeding depleted mountain rangelands," *USDA CR–5–62.* 1962.
34. Hull, A. C., Jr., and others. "Seeding Colorado rangelands," *Colo. Agr. Exp. Sta. Bul. 498–S.* 1962.
35. Hull, A. C., Jr., and others. "Pellet seeding on western rangeland," *USDA Misc. Pub. 922.* 1963.
36. Johnstone-Wallace, D. B. "The influence of grazing management and plant associations on the chemical composition of pasture plants," *J. Am. Soc. Agron.,* 29:441–455. 1937.
37. Jones, A. G. "Grassland management and its influence on the sward; II, The management of a clovery sward and its effects," *Emp. Jour. Exp. Agr.,* 1:122–127. 1933.
38. Keller, W. "Breeding improved forage plants for western ranges," in *Grasslands,* Am. Assn. Adv. Sci., pp. 335–344. 1959.
39. Kiesselbach, T. A. "Winter wheat investigations," *Nebr. Agr. Exp. Sta. Res. Bul. 31.* 1925.
40. Larsen, H. J., and Johannes, R. F. "Summer forage: Stored feeding, green feeding and strip grazing," *Wisc. Agr. Exp. Sta. Rsh. Bul. 257,* pp. 1–32. 1965.
41. Launchbaugh, J. L. "The effect of stocking rate on cattle gains and on native shortgrass vegetation in west-central Kansas," *Fort Hays Branch Exp. Sta. Bul. 394.* 1957.
42. Launchbaugh, J. L. "Upland seeded pastures compared for grazing steers at Hays, Kansas," *Kan. Agr. Exp. Sta. Bul. 548,* pp. 1–29. 1971.
43. Leithead, H. L., and others. "100 native forage grasses in 11 southern states," USDA Handbk. 389, pp. 1–216. 1971.
44. Lodge, R. W., and others. "Managing crested wheatgrass pastures," *Can. Dept. Agr. Publ 1473,* pp. 1–20. 1972.

44a. Mays, D. A., Ed., Forage Fertilization Am. Soc. Agron., Madison, Wisc., pp. 1–621. 1974.

45. McLean, A., and Bawtree, A. H. "Seeding grassland ranges in the interior of British Columbia," *Can. Dept. Agr. Pub. 1444,* pp. 1–15. 1971.
46. Munscher, W. D. Poisonous plants of the United States, 2nd ed., Macmillan, Inc., New York. 1951.
47. Park, J. K., and others. "Establishing stands of fescue and clovers," *S. Car. Agr. Exp. Sta. Circ. 129.* 1961.
48. Pechanec, J. F. "Sagebrush control on rangelands," *USDA Agr. Handbk.* 277, pp. 1–40. 1965.
49. Price, R. "Artificial reseeding on oak-brush range in central Utah," *USDA Circ. 458.* 1938.
50. Reynolds, H. G., and Springfield, H. W. "Reseeding southwestern range lands with crested wheatgrass," *USDA Farmers Bul. 2056.* 1953.
51. Ritchey, G. E., and Henley, W. W. "Pasture value of different grasses alone and in mixture," *Fla. Agr. Exp. Sta. Bul. 289.* 1936.
52. Robertson, J. H. "Effect of frequent clipping on the development of certain grass seedlings," *Plant Phys.,* 8:425–447. 1933.

53. Rogler, G. A., and Lorenz, R. J. "Pasture productivity of crested wheatgrass as influenced by nitrogen fertilization and alfalfa," *USDA Tech. Bul. 1402,* pp. 1–33. 1969.
54. Sarvis, J. T. "Effects of different systems and intensities of grazing upon the native vegetation at the Northern Great Plains Field Station," *USDA Dept. Bul. 1170.* 1923.
55. Savage, D. A. "Methods of reestablishing buffalo grass on cultivated land in the Great Plains," *USDA Circ. 328.* 1934.
56. Savage, D. A. "Grass culture and range improvement in the central and southern Great Plains," *USDA Circ. 491,* pp. 1–55. 1939.
57. Sawyer, W. A., and others. "Range robbers – Undesirable range plants," *Oreg. Ext. Bul. 780,* pp. 1–13. 1963.
58. Semple, A. T., and others. "A pasture handbook," USDA Misc. Publ. 194, pp. 1–88 (Rev.). 1946.
59. Semple, A. T. "Grassland Improvement," Leonard Hill, London, pp. 1–400. 1970.
60. Shipley, J., and Regier, C. "Optimum forage production and the economic alternatives associated with grazing irrigated wheat: Texas High Plains," *Tex. Agr. Exp. Sta.* MP 1068, pp. 1–11. 1972.
61. Sims, F. H., and Crookshank, H. R. "Wheat pasture poisoning," *Tex. Agr. Exp. Sta. Bul. 842.* 1956.
62. Stapledon, R. G. "Four addresses on the improvement of grassland," *U. Col. Wales,* Aberystwyth, England. 1933.
63. Stephens, J. L. "Pastures for the Coastal Plain of Georgia," *Ga. Coastal Plain Exp. Sta. Bul. 27,* pp. 1–57. 1942.
64. Swain, F. G., Decker, A. M., and Retzer, H. J. "Sod-seeding rye, vetch extends Bermuda pasture," *Crops and Soils,* 13(3):20. 1960.
65. Swanson, A. F., and Anderson, K. "Winter wheat for pasture in Kansas," *Kans. Agr. Exp. Sta. Bul. 345.* 1951.
66. Swanson, E. W., McLaren, J. B., and Chapman, E. J. "A comparison of milk production and forage crop utilization from strip grazing vs. green chop feeding," *Tenn. Agr. Exp. Sta. Bul. 292.* 1959.
67. Tisdale, E. W., and others. "The sagebrush region in Idaho, a problem in range resource management," *Ida. Agr. Exp. Sta. Bul. 512,* pp. 1–15. 1969.
68. Torell, P. J., and Erickson, L. C. "Reseeding medusahead-infested ranges," *Ida. Agr. Exp. Sta. Bul. 489,* pp. 1–17. 1967.
69. Turelle, J. W., and Austin, W. W. "Irrigated pastures for forage production and soil conservation in the West," *USDA Farmers Bul. 2230,* pp. 1–22. 1967.
70. Vinall, H. N., and Wilkins, H. L. "The effect of fertilizer applications on the composition of pasture grasses," *J. Am. Soc. Agron.,* 28:562–569, 1936.
71. Weaver, J. E. "Underground plant development in its relation to grazing," *Ecol.,* 11(3):543–557. 1930.
72. Weaver, J. E., and Harmon, G. W. "Quantity of living plant materials in prairie soils in relation to run-off and soil erosion," *Conserv. and Survey Div., U. Nebr. Bul. 8.* 1935.

73. Wenger, L. E. "Buffalo grass," *Kans. Agr. Exp. Sta. Bul. 321*, pp. 1–78. 1943.
74. Whitman, W. C., Hanson, H. T., and Loder, G. "Natural re-vegetation of abandoned fields in western North Dakota," *N. Dak. Agr. Exp. Sta. Bul. 321*, pp. 1–18. 1943.
75. Willhite, F. M. "How to improve mountain meadows in the west," *Plant Food Rev.*, 9(2):5–8. 1963.
76. Willhite, F. M., Lewis, R. D., and Rouse, H. K. "Improving mountain meadow production in the west," *USDA Info. Bul. 268*. 1963.
77. Williams, R. M., and Post, A. H. "Dry land pasture experiments at the Central Montana Branch Station, Moccasin, Mont., *Mont. Agr. Exp. Sta. Bul. 431*, pp. 1–31. 1945.
78. Wolfe, T. K. "The yield of various pasture plants at different periods when harvested as pasturage and as hay," *J. Am. Soc. Agron.*, 18:381–384. 1926.
79. Woodward, T. E., Shepherd, J. B., and Hein, M. A. "The Hohenheim system in the management of permanent pastures for dairy cattle," *USDA Tech. Bul. 660*. 1938.

Chapter 11

Weeds and Their Control

ECONOMIC IMPORTANCE

More than 3000 species of herbaceous and woody plants of the world are regarded as weeds. Weeds cost the farmers of America $5 billion annually in crop losses and in the expense of keeping them under control. Much of the cost of intertillage of row crops, maintenance of fallow, seedbed preparation, and seed cleaning is chargeable to weed control. Another expensive item is suppression of weeds along highways and railroad right-of-ways, and in irrigation ditches, navigation channels, yards, parks, grounds, and home gardens. Ragweed pollen is a source of annual periodic distress to several million hay-fever sufferers. Poison ivy, poison oak, poison sumac, nettles, thistles, sandburs, and puncturevine bring pain to other millions.[16, 19] The barberry bush, which spreads the black stem rust to grains and grasses, can be regarded as a weed. Weeds also serve as hosts for other crop diseases as well as for insect pests.

In some locations, weeds play a beneficial role in checking soil erosion, supplying organic matter to the soil, and in furnishing food and protection to wild life.

TYPES OF FARM WEEDS

A weed has been defined as a plant that is useless, undesirable, or detrimental, or simply as a "plant out of place."[23, 28] Some plants, e.g., Canada thistle, are always considered weeds, whereas red clover would be considered a nuisance under some circumstances but hardly a weed at any time. Nevertheless, more than 30 American crop plants frequently have been listed as weeds. Among the more troublesome crop plants that get out of place are johnsongrass, sweetclover, bermudagrass, sesbania, vetch, rye, tartary buckwheat, black medic, burclover, hemp, black mustard, chicory, and kudzu.[24]

Annual weeds live for a single year, mature, seed, and die. Most of the common farm weeds belong to this group, among them pigweed, ragweed, fanweed, Russian thistle, wild oats, mustard, lambsquarters, pigeongrass, peppergrass, and dodder. Other annual weeds are wild barley, witchgrass, canarygrass, crabgrass, barnyardgrass, darnel, smartweeds, summer cypress, puncturevine, mallow, false flax, horseweed, wild buckwheat, prickly lettuce, purslane, corn cockle, and sunflower.[21]

Biennial weeds require two seasons to complete their growth. They grow from seeds and devote the first season to food storage, usually in short fleshy roots. During the next spring they draw on the stored food to produce a vigorous vegetative growth and to mature seeds. Among the biennial weeds are burdock, bull thistle, mullen, wild carrots, and wild parsnip.

Perennial weeds live for more than two years. The majority of simple perennials possess root crowns that produce new plants year after year, being supported like the dandelion by a fleshy taproot, or by means of fibrous roots. Except in a few instances, plants of this type depend upon production of seed for their spread. Creeping perennials propagate by means of underground parts, but often also by seeds. Field bindweed, Canada thistle, perennial sowthistle, leafy spurge, and whitetop spread by horizontal roots. Quackgrass and johnsongrass spread by underground stems (rhizomes). Other perennial weeds are nutgrass, poverty weed, horsetail, bracken fern, Russian knapweed, St. Johnswort, horsenettle, milkweeds, curled dock, sheep sorrel, wild rose, blue vervain, chicory, and buckhorn plantain.

Weeds also may be classified as common and noxious. Common weeds are annuals, biennials, or simple perennials that are readily controlled by ordinary good-farming practices. Noxious weeds are those which are difficult to control because of an extensive perennial root system or because of other characteristics that make them persistent.

LOSSES CAUSED BY WEEDS

Weeds may cause losses in several ways.[2, 3, 6, 28]

1. *Decrease in crop yields.* Weeds decrease yields by removal of moisture needed by crop or pasture plants. Most weeds require as much water to produce a pound of dry matter as do the cereal crops. In addition, weeds compete with crop and pasture plants for light and soil nutrients.[29] In a Kansas experiment, yields of close-drilled sorghum on bindweed-infested land and on clean land averaged 2.06 and 3.92 tons per acre, respectively. Barley produced an average acre yield of 7½ bushels on infested land, but 21½ bushels on land free from bindweed.[25]

2. *Impairment of crop quality.* The presence of weed seeds in small grains may materially lower the quality of the grains. Green weed pieces in threshed grain raise the moisture content so that the

grain may not keep in storage. Weeds such as wild garlic, mustard, fanweed, yarrow, chicory, or ragweed consumed in hay or pastures impart undesirable flavors to dairy products.

3. *Harboring of plant pests.* Many weeds act as hosts to organisms that carry plant diseases. Curly top, a serious virus disease of the sugarbeet, is carried from such weeds as the common mallow, chickweed, and lambsquarters to the sugarbeet by the beet leafhopper, which breeds upon these weeds. The sugarbeet webworm prefers to deposit its eggs on the Russian thistle and similar weeds. Weeds of the family *Solanaceae* contribute to the spread of such pests as the Colorado potato beetle.

4. *Increase of irrigation costs.* Weeds on ditch banks or growing in the ditches may seriously impair the efficiency of irrigation channels. Windblown weeds often obstruct headgates and diversion boxes. As a result, ditches must be cleaned every year.

5. *Injury to livestock.* Some poisonous weeds described in Chapter 10 may cause illness or death to livestock. Mature plants of sandbur, three-awned grass, porcupine grass, downy brome (cheatgrass), and squirreltail grass may cause injury to stock that eat them.

6. *Decrease of land values.* Land values may be reduced by the presence of weeds, particularly noxious weeds.

PERSISTENCE OF WEEDS

Weeds usually are able to survive in competition with crop plants because of a wide range of adaptability as well as effective means of propagation, such as by underground parts.

Many weeds produce large amounts of seeds per plant. The possible numbers for several species are as follows:[27] fanweed 7,040; pigweed 117,400; Russian thistle 24,700; lambsquarters, 72,450; green foxtail 34,000; perennial sowthistle 9,750; and tumbling mustard 80,400. Seeds of some weeds escape notice because they are so small.

Many weed seeds remain viable in the soil for many years. Some weed seeds, such as those of bindweed, wild oats, and cocklebur may exhibit dormancy. Their prolonged viability in the soil explains the sudden appearance of certain weeds after years of good cultural practices.

DISSEMINATION OF WEEDS

Natural Agencies

Seeds with barbs or hooks may become attached to animals and carried long distances. Weed seeds eaten by birds and animals may pass through the digestive tract uninjured. Cactus is spread principally by jack rabbits that eat the fruits that contain indigestible seeds. Seeds equipped with tufts of hair, e.g., Canada thistle, bull thistle, or sow-thistle, may be disseminated by wind for distances of 2 to 15 miles. Weed seeds often are carried to other locations by rains or in streams. Tumbleweeds of various species, e.g., Russian thistle, tumbling amaranth, and tumbling mustard, may roll in the wind for long distances, scattering seeds as they go. Weeds with horizontal roots or rootstocks spread from one to several feet each year. Bindweed patches may double their area every five years.

Man-made Agencies

Weeds are widely spread in impure farm seeds. Some of the most serious weed pests have been introduced from foreign countries in wheat, alfalfa, clover, sugarbeet, and other crop seeds. Hay, straw, and other forages also have contributed to the spread of weed seeds.

In irrigated sections, seeds of weeds that grow on the bank of reservoirs, canals, and ditches may fall into the water, which carries them to cultivated fields. Several million weed seeds of 81 species floated down a 12-foot irrigation ditch in Colorado in one day.[11]

Farm machinery even may spread weed seeds. Plows, harrows, and cultivators may drag roots or seed-bearing portions of perennial plants to other parts of a field.

Spreading of fresh barnyard manure on cultivated fields may disseminate weed seed. In one experiment,[13] 6.7 per cent of the weed seeds fed to farm animals were viable when recovered in the fresh manure. After burial for one month in manure, velvet weed, bindweed, and whiteweed seeds were still viable. Practically all seeds were dead after being buried in manure for three months.

CONTROL OF COMMON FARM WEEDS

The popular method of controlling weeds is by the application of herbicides (Figures 7–6, 11–1, and 11–2).[2, 4, 7, 12, 17, 18, 30]

General Control Measures

Crop seeds of high purity, free from noxious weed seeds, should be sown. An important control method is early and frequent cultivation of the land with field cultivators, harrows or rotary hoes. Row crops should be grown on the land periodically to permit intertillage. Weeds are killed most easily when they are seedlings. The greatest benefit of bare fallow in the semiarid region is prevention of weed growth. Annual weeds growing in uncultivated areas of a farm may be sprayed with herbicides, or mowed to prevent them from going to seed.

Crop rotation is a valuable aid in weed control because many weeds are associated with certain crops. Dodder is troublesome in alfalfa.

Wild oats and the mustards become serious pests on land cropped continuously to small grains. Green foxtail and giant foxtail are often troublesome in flax and corn fields. Smother crops make a rapid growth that shades out other plants when grown in thick stands. Common smother crops are alfalfa, foxtail millet, buckwheat, rye, sorghum, and sudangrass.

All types of community farm machinery, particularly separators, combines, and hay balers, should be cleaned free of weed seeds before they are brought to the farm.

Clean Cultivation for Noxious Weed Control

A combination of clean cultivation and smother crops may either weaken or eliminate perennial weeds. Land clean-cultivated until about July 1, followed by a smother crop of sorghum or sudangrass repeated for several successive seasons, has resulted in almost complete control of bindweed. Winter rye and winter wheat have been used as smother crops in some regions, with the land clean-cultivated between harvest and seeding. A smother crop is most effective when it follows a full season of clean cultivation.

Clean cultivation for two years, with 20 to 25 cultivations, is generally required for complete eradication of bindweed. Repeated cultivation reduces the food materials stored in the roots. Allowance of

Figure 11–1. A liquid pre-emergence herbicide from the two large tanks is applied in bands over the planted rows during the planting operation. [Courtesy J. I. Case Co.]

Figure 11–2. Control of weeds in sugarbeets by the postemergence application of three herbicides at rates per acre shown on the label. The weed growth in the untreated area is evident in the background. [Courtesy Great Western Sugar Co.]

some top growth for 6 to 8 days after the shoots emerge, before repeating cultivation, aids in bindweed eradication.[15] Cultivation every two weeks to a depth of 3 inches throughout the growing season is desirable. The best time for plowing is when the weeds are in flower because root reserves are low at that time.[8, 9, 25] Subsequent cultivation is preferably with sweep or blade implements.[27] Herbicide use greatly reduces the necessary cultivations.

Flaming

A flaming implement has been used to kill weeds in a cottonfield,[22] and in corn and sorghum fields (Figure 11–3). The flaming of cotton is started a month after planting when the cotton plants are large enough, i.e., have a stem diameter of 3/16 inch or more, to escape injury from the flame. The weeds are held in check during the first month by cultivation but thereafter may be controlled satisfactorily by flaming without injury to the cotton. The flame is directed toward the base of the cotton plants, and is most effective in killing small weeds.

Other Weed-Control Methods

Klamath weed (St. Johnswort) is being controlled with parasitic beetles, *Chrysolina* species, imported from Australia. Special parasitic insects also have been used to control cactus in Australia, lantana in Hawaii, and yellow toadflax in Canada. Pasturing land with sheep sometimes is an effective method for the control of certain weeds. Sheep are able to suppress field bindweed on land seeded to sudangrass for pasture. They eat the bindweed in preference to the sudangrass, but they make good gains on the latter after eating down the weeds. In certain Texas rangelands, sheep eat many herbaceous weeds that were rejected by grazing cattle, then goats follow to consume woody shrubs. Fish consume algae in flooded rice fields.[30] The development and multiplication of pathogens that attack specific weeds is a rather remote possibility.

Chemical Control

The control of annual, biennial, and perennial weeds through the use of chemical herbicides has expanded remarkably. Selective herbicides were applied to more than 90 million acres of crop, grazing, and other land in the United States by 1965. The stage of growth of the crop at which the herbicide is effective, but with the least injury to the crop, is very important. Some of the chemicals are effective in granular form as well as in the usual spray formulas.

About 140 selective herbicides were available for use in 1973, and many more being tested over the world will be marketed thereafter. There is a need for greater selectivity for specific weed and crop

Figure 11–3. Simultaneous shovel cultivation, flame cultivation, and insecticide application in a cotton field.

plants. The frequent changes in herbicides and in their most appropriate use makes definite recommendations advisable for only limited periods. Crop growers should rely upon the latest local suggestions as to the best herbicide for particular weeds and crops, and for the best rate, time, and method of application.

GROUPS OF HERBICIDES: Inorganic herbicides especially sodium borate compounds, ammonium sulfamate, and sodium chlorates have been used chiefly as soil sterilants. These as well as petroleum oils may prevent the growth of weeds or other plants for several years.

Arsenic compounds: Calcium arsenate; sodium arsenite.

The main groups of organic herbicides and some representatives of each are listed below.

Phenoxyaliphatic acids: 2,4-D; 2,4,5-T; MCPA; 2,4-DB; silvex; MCPB; 4,CPA; 3,4-DA.

Substituted amides: CDAA; CDEA; diphenamid; propanil; dicryl; solon; NPA; cypromid.

Nitroanilines: trifluralin; benefin; nitralin.

Ureas: diuron; monuron; fenuron; linuron; chloroxyron; fluometuron; metobromuron; cycluron; siduron.

Carbamates: IPC; CIPC; barban; terbritol.

Thiocarbamates: EPTC; pebulate; vernolate; molinate; diallate; triallate; CDEC.

Bipyridyliums: diquat; paraquat.

Saturated aliphatics: allyl alcohol; acrolein.

Miscellaneous: endothall; bensulide.

Nitrogen heterocyclics: symmetrical triazines; simazine; atrazine; propazine; prometone; prometryne; ametryne.

Substituted pyridazinones: pyrazon.

Substituted pyridines: picloram; pyriclor.

Substituted uracils: bromacil; terbacil.

Substituted aliphatic acids: TCA; dalapon.

Arylaliphatic acids: 2,3,6-TBA; ofenac; dicamba, amiben; DCPA; TCBC.

Phenol derivatives: pentachlorophenol; DNC; DNBP.

Substituted nitriles: dichlorobenil; diphenatrile; ioxynil; bromoxynil.

New or better herbicides are being discovered every year. Consequently many current herbicidal recommendations soon become obsolete. Each new herbicide must be tested before it is marketed, to determine any possible injury to man, animals, or plants that might result from its use.

The 2,4-D herbicide is used extensively. It kills many of the broad-leaved annual weeds as well as emerging weedy grass seedlings. It kills many plants of noxious perennial weeds, but also retards the growth of others so that crops can be grown on the infested land. Repeated treatments are necessary to destroy the most persistent noxious perennial weeds. The ester forms of 2,4-D are more toxic and more volatile than the amine or salt forms. They also are effective at lower rates of application. The amine form usually requires twice the amount of 2,4-D active ingredient as does the ester form to be effective.

Certain herbicides including diallate, ordram, EPTC, vernam, IPC, and trifluralin may be applied and incorporated in the soil before planting the crop, especially for controlling wild oats and other grasses.

Preemergence herbicide treatments provide effective weed control for corn, but sometimes also for potatoes, peas, beans, and soybeans. These large-seeded crops are planted deep enough to escape injury from herbicides applied to the surface, except when heavy rains on sandy soils wash the chemicals downward. Certain preemergence treatments are suitable for sugarbeets and rice. Preemergence treatments are applied within 2 days after planting, or often during planting immediately after the crop seed is covered. The sprayers either are mounted on the tractor or planter or are trailed behind the planter. A preemergence spray often is applied in bands 8 to 14 inches wide directly over the planted row.[14] This reduces the cost of chemicals for an acre of land. Weeds between the rows are controlled by cultivation. Preemergence applications serve chiefly to kill weed seedlings that have sprouted on or near the soil surface. Their effectiveness is greatly impaired by any subsequent cultivation that not only buries some of the chemical but also raises other weed seeds up to the soil surface where they germinate.

Postemergence applications of chemicals are timed to control weeds with the least injury to the crop. Carbon banding over the seeded rows protects the seeds and later emerging seedlings, while controlling weeds with nonselective herbicides. Most crop plants are killed or injured in the seedling stage. They are damaged before they reach a height of 5 or 9 inches. Applications in the flowering stage, or shortly before, usually induce sterility. Treatments between the above stages are least injurious to the crop. Small weeds are easier to kill than large ones, hence early spraying is desirable. Grain crops that are near to maturity are not harmed by the herbicides. Grains can be sprayed then to kill or desiccate green weeds that would otherwise interfere with harvest operations and add moisture to the threshed grain.

Lower rates of herbicide application are more effective on sandy soils than on heavy soils, particularly on organic soils, because less of the chemical is adsorbed on the coarse soil particles. Some of the chemicals are volatile. They have injured cotton plants when hot weather followed their application. DNBP becomes volatile at 85° F. and CIPC at 90° F. The drifting of sprays with their volatile constituents may injure susceptible crops in near-by fields.

Injuries to crops from herbicides may be evident in stunted or distorted growth or dead leaves. Stem bending occurs in sorghum, corn, and other crops that have received a heavy application of 2,4-D on the growing plants. Such treatment with 2,4-D also suppresses the development of brace roots and induces stalk brittleness in sorghum and corn so that lodging or stalk breakage occurs.

Most herbicides applied at recommended rates do not have any residual effect in the soil.[27] Even 2,4-D decomposes in about six weeks or less, while other chemicals are effective for even a shorter time. Simazine, atrazine, propazine, monuron, and diuron may remain active for nearly a year and cause injury to a succeeding crop. Spring oats, sugarbeets, and soybeans planted on corn ground, where a preemergence spray of simazine or atrazine had been applied the previous year, often show retarded growth.

Most postemergence treatments are applied with field sprayers that may cover a strip from 7 to 30 feet in width. Aerial spraying is common for brushy, woody, or rough pastures and ranges,[1] and for flooded rice fields.[29] It can be used on weedy cornfields that are too wet for cultivation after repeated rains. Aerial spraying of crops that are nearing maturity avoids the damage that would result from operation of a tractor-drawn or tractor-mounted field sprayer over the field.

SOME SERIOUS PERENNIAL NOXIOUS WEEDS

The most serious noxious weeds in the United States probably are quackgrass, Canada thistle, bindweed, leafy spurge, whiteweed (whitetop), perennial sowthistle, johnsongrass, Russian knapweed, and nutgrass. These perennial species, serious weed pests in several states, are found in cultivated fields, pastures, and meadows.[28]

Field or European bindweed (*Convolvulus arvensis*) (Figure 11–4) is distributed throughout the United States. It is a serious pest in Iowa, South Dakota, Minnesota, Nebraska, Kansas, Colorado, Utah, Idaho, Washington, and California. This plant, a member of the morning-glory family, is a perennial that propagates by seeds and by lateral creeping roots from which rhizomes and then new stems and roots arise. The roots may grow 1 inch a day and penetrate to a depth of 12 to 20 feet or even more. The lateral roots, which can spread several feet in a year, are found mostly at a depth of 12 to 30 inches. The stems are smooth, slender, slightly angled vines that spread over the ground or twine around and climb any erect crop plant (Figure 11–4). The vines may be 1 to 6 feet in length.

Whiteweed (*Lepidium draba*) also is known as perennial pepper-

Figure 11–4. Field bindweed climbing on barley and oat plants.

grass, hoary cress, and whitetop. Whiteweed propagates both by seed and creeping roots. The plant is erect, 10 to 18 inches high, and grayish white in color. Two other species, very similar to *L. draba* and known by the same common names, are *L. repens* (or *L. draba* variety *repens*), and *Hymenophysa pubescens*. Whiteweed makes a vigorous growth on the irrigated alkaline soils of the West. It is a serious pest in the Rocky Mountain region and on the Pacific Coast.

Canada thistle (*Cirsium arvense*) (Figure 3–3) propagates both from seeds and creeping roots. The stems are erect, hollow, smooth to slightly hairy, 1 to 4 feet tall, simple, and branched at the top. The leaves are set close on the stem, slightly clasping, typically green on both sides, sometimes white and hairy beneath, deeply and irregularly cut into lobes tipped with sharp spines, on sometimes entire or nearly entire. Canada thistle is widely distributed in the northern half of the United States. A smooth-leaved type of Canada thistle that is more difficult to control is spreading rapidly in various sections.

Russian knapweed (*Centaurea picris*) reproduces both by seeds and black creeping roots. The plant is erect, rather stiff, branched, and 1 to 3 feet high. The young stems are covered with soft gray hairs. Russian knapweed is generally distributed in the western states.

Perennial sowthistle (*Sonchus arvensis*) (Figure 11–5) spreads both by seeds and creeping roots. The stems are erect, smooth, 2 to 5

Figure 11–5. Perennial sowthistle.

feet high, and unbranched except at the top. The leaves are light green, the lower ones being 6 to 12 inches long, deeply cut, with the side lobes pointing backwards; the upper leaves are smaller, clasping, with margins slightly toothed, and prickly. This weed is found throughout northern United States.

Leafy spurge (*Euphorbia virgata*) (Figure 11–6) propagates by seed and creeping roots. The plants are pale green, erect, 1 to 3 feet high, and unbranched except for flower clusters. The leaves are long and narrow with smooth margins. The plant is characterized by a milky sap. It is rather widely scattered throughout the United States.

Quackgrass (*Agropyron repens*) propagates by seed as well as by long, jointed yellow rhizomes or rootstalks. This grass is erect, 1 to 3 feet high, with slender stems. The flat, narrow leaves are rough above but smooth beneath. The seed is borne in spikes 3 to 7 inches long. Quackgrass is found throughout the northern half of the United States. Although it has forage value, quackgrass becomes a very bad weed because of its persistence. It is related to and resembles some of the native wheatgrasses which are valuable forage and pasture plants.

Johnsongrass (*Sorghum halepensis*) (Figure 3–6) propagates by seeds and rhizomes. It is an erect plant, 5 to 6 feet tall. It is closely related to sorghum and sudangrass. Johnsongrass is widespread throughout the southern states. The most economical control method is heavy pasturing or frequent mowing to starve the rootstocks, followed by plowing and the planting of a clean cultivated crop.

Figure 11–6. Leafy spurge.

Nutgrass (*Cyperus rotundus*), also called cocograss, is a sedge rather than a grass. It propagates by deepset rootstocks as well as by small tubers (nuts) borne on the rootstocks. The nuts are $^1/_4$ to $^3/_4$ inch in diameter. Nutgrass is widely distributed, especially from Virginia to Kansas and southward. It is one of the most serious weed pests in the South.

The following have been listed as the 10 worst weeds in the United States: barnyardgrass (*Echinchloa crus-galli*), cocklebur (*Xanthium* species), foxtail (*Setaria* spp.), johnsongrass (*Sorghum halpense*), lambsquarters (*Chenopodium* spp.), Morning-glory (*Ipomea* spp.), nutsedge (*Cyperus* spp.), panicum (*Panicum* spp.), pigweed (*Amaranthus* spp.) and quackgrass (*Agropyron repens*). In addition, wild oats (*Avena fatua*), wild mustard, or charlock (*Brassica arvensis*) are perhaps the most prevalent weed pests in the spring grain regions. Downy brome or cheatgrass (*Bromus tectorum*) is a serious pest on minimum-tillage grain fields and on range lands of the northwestern quarter of the United States. Crabgrass in lawns is very difficult to control.

Some noxious perennial weeds, bindweed, Canada thistle, and sowthistle, are retarded by herbicides such as 2,4-D, MCPA, dicamba, and amitrole. When combined with fallow frequent tillage and smother crops they are largely suppressed. Complete eradication is obtained with sterilants such as chlorates, borates, and fenac. Quackgrass is suppressed with atrazine, amitrole, or MH followed by plowing and planting corn. Johnsongrass is reduced with dalapon, PBC, TCA, or EPTC. Herbicides for nutsedge control include butylate, vernolate, MSMA, and DSMA; for fall panicum, alachlor and butylate; and for leafy spurge, amitrole.[5] Whitewheed is controlled with DNBP. Russian knapweed is controlled with amitrole or dicamba.

Lambsquarters, mustards, Russian thistle, cocklebur, and many other broadleaf annuals are controlled with 2,4-D, MCPA, diuron, DNBP, EPTC, dicamba, and numerous other herbicides. Foxtails are controlled with butylate, propachlor, alachlor and cypromid. Barnyardgrass is controlled with propanil, molinate, and PCP. Cheatgrass is partly controlled with amitrole + 2,4-D, terbutryn, linuron, dicamba, and prometryne. Crabgrass controls include siduron, cypromid, PCP and silvex. Wild oats is controlled with barban and diallate.

Atrazine is the chief herbicide for weed control in corn, and propazine is favored in sorghum fields. Some of the herbicides used for weed control in various other crops are as follows:

Wheat—2,4-D, dicamba, bromoxynil, linuron

Cotton—nitralin, trifluralin, diuron, MSMA

Soybeans—chloramben, alachlor, chloroxuron, dinoseb, treflan, ramrod

Peanuts—alachlor, ametryne, dinaseb, 2,4-DEP
Sugarbeets—TCA, pyrazon, endothall, diallate
Sugarcane—terbacil, MSMA, trifluralin
Beans and peas—EPTC, CDEC, CDAA, MCPB
Flax—MCPA, TCA, dalapon, barban

PHANEROGAMOUS PARASITES

Dodder (Figure 11–7), broomrape, and witchweed are flowering plants that parasitize certain crop plants. A herbicide such as Dacthal (DPA) is taken up by alfalfa plants without injury, but it suppresses the dodder. It must be used only where the alfalfa is to be harvested for seed.

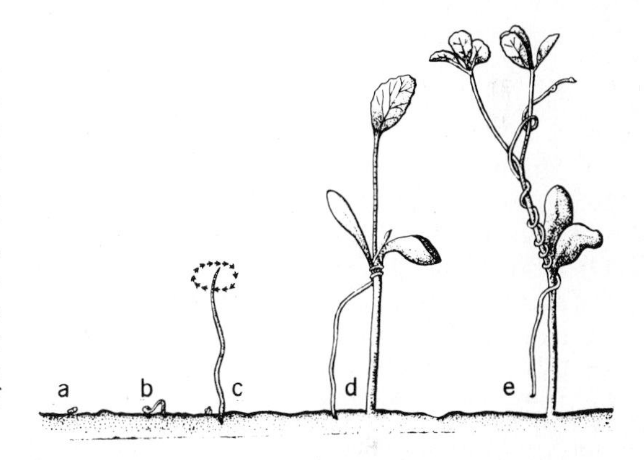

Figure 11–7. A dodder or "lovevine" (*Cuscuta* species): Seedling comes up and sways around until it encounters a host plant, which it entwines. Haustoria or suckers penetrate the stem of the host and take up nutrients. Then the lower stem of the dodder dies, but the plant proceeds to thrive at the expense of the host.

Witchweed (*Striga lutea* or *S. asiatica*) is an annual flowering (phanerogamous) plant parasite of sorghum, corn, and some other grass species. It is prevalent over Africa and India, but also has invaded North Carolina and South Carolina in the United States. Haustoria from germinating seedlings enter the crop roots where they sap the growth of the plants before the witchweed stems emerge above the soil surface. The witchweed is killed by spraying with 2,4-D or other phenoxy compounds, but the damage has already occurred. The planting of trap crops to be plowed under, such as cowpeas, which stimulate the germination of witchweed seeds but are not parasitized, is helpful where the practice is feasible. Complete control is difficult because witchweed attacks grasses in uncultivated areas and produces numerous seeds where it is not easily detected.[10]

REFERENCES

1. Anonymous. "Chemical control of brush and trees," *USDA Farmers Bul. 2158*. 1961.

2. Anonymous. "Suggested guides for weed control," *USDA Handbook,* 1966.
3. Anonymous. "A survey of extent and cost of weed control and specific weed problems," in *USDA Agricultural Research Service Spec. Rept. 34–23.* 1962.
4. Anonymous. *Principles of Plant and Animal Pest Control,* Vol. 2, *Weed Control,* Nat'l Acad. Sci., Washington, D.C., pp. 1–471. 1968.
5. Anonymous. "Suggested guide for weed control," *USDA Handbk. 332,* pp. 1–70. 1969.
6. Anonymous. *FAO International Conference on Weed Control.* Weed Sci. Soc. Amer., Urbana, Ill., pp. 1–668. 1970.
7. Ashton, F. M. and Harvey, W. A. "Selective chemical weed control," *Cal. Agr. Exp. Sta. Circ. 558,* pp. 1–17. 1971.
7a. Ashton, F. M., and Crafts, A. S. Mode of Action of Herbicides. Wiley, New York, pp. 1–504. 1973.
8. Bakke, A. L., and others. "Relation of cultivation to depletion of root reserves in European bindweed at different soil horizons," *J. Agr. Res.,* 69(4):137–147. 1944.
9. Barr, C. G., "Preliminary studies on the carbohydrates in the roots of bindweed," *J. Am. Soc. Agron.,* 28:787–798. 1936.
10. Cardenas, J., and others. *Tropical Weeds,* Inst. Colombiano Agropecuaro, Bogata, Colombia, pp. 1–341. 1972.
11. Egginton, G. E., and Robbins, W. W., "Irrigation water as a factor in the dissemination of weed seeds," *Colo. Agr. Exp. Sta. Bul. 253.* 1920.
12. Gerlow, A. R. "The economic impact of cancelling the use of 2,4,5-T in rice production," *USDA ERS 510,* pp. 1–11. 1973.
13. Harmon, G. W., and Keim, F. D. "The percentage and viability of weed seeds recovered in the feces of farm animals and their longevity when buried in manure," *J. Am. Soc. Agron.,* 26:762–767. 1934.
14. Heddon, O. K., and others. "Equipment for applying soil pesticides," *USDA Handbk. 287,* pp. 1–37. 1967 (Rev.).
15. Kiesselbach, T. A., Petersen, N. F., and Burr, W. W. "Bindweeds and their control," *Nebr. Agr. Exp. Sta. Bul. 287.* 1934.
16. Kingsbury, J. M. *Poisonous Plants of the United States and Canada,* Prentice-Hall, Englewood Cliffs, N.J. 1964.
17. Klingman, G. C. *Weed Control as a Science,* Wiley, New York, pp. 1–421. 1966.
18. Klingman, D. L., and Shaw, W. C. "Using phenoxy herbicides effectively," *USDA Farmer's Bul. 2183,* pp. 1–24. 1971.
19. Klingman, D. L., and Shaw, W. C. "Using phenoxy herbicides effectively," *USDA Farmers Bul. 2183.* 1962.
20. Kommedahl, T., and Johnson, H. G. "Pesky plants," *Minn. Ext. Bul. 287,* pp. 1–32. 1972.
21. McWarter, C. G. "A summary of methods for johnsongrass control in soybeans in Mississippi," *Miss. Agr. Exp. Sta. Bul. 788,* pp. 1–8. 1971.
22. Muenscher, W. C. L. *Weeds,* Macmillan, Inc., New York, pp. 1–560. 1961 (Rev.).
23. Neeley, J. W., and Brain, S. G. "Control of weeds and grasses in cotton by flaming," *Miss. Agr. Exp. Sta. Circ. 118,* pp. 1–6. 1944.

24. Pieters, A. J. "What is a weed?" *J. Am. Soc. Agron.*, 27:781–783. 1935.
25. Reed, C. F., and Hughes, R. O. "Selected weeds of the United States," *USDA Handbk. 333*, pp. 1–463. 1970.
26. Seely, C. I., Klages, K. H., and Schafer, E. G. "Bindweed control," *U. of Idaho, Coll. Agr. War Circ. 41*. 1945.
27. Shaw, W. C., and others. "The fate of herbicides in plants," in *USDA Agricultural Research Service Spec. Rept. 20–9*:119–133. 1960.
28. Stevens, O. A. "The number and weight of seeds produced by weeds," *Am. J. Bot.*, 19:784–794. 1932.
29. Thornton, B. J., and Harrington, H. D. "Weeds in Colorado," *Colo. Agr. Exp. Sta. Bul. 514-S*. 1964.
30. Weldon, L., and others. "Common aquatic weeds," *USDA Handbk. 352*, pp. 1–43. 1969.
31. Wicks, G. A. and Smika, D. E. "Chemical fallow in a winter wheat-fallow rotation," *Weed Sci.* 21(2):97–102. 1973.

PART TWO

CROPS OF THE GRASS FAMILY

Chapter 12

Indian Corn or Maize*

ECONOMIC IMPORTANCE

Corn ranks third, following wheat and rice, in the world production of cereal crops. The annual area devoted to corn from 1970 to 1972 was 270 million acres (109 million hectares). Corn grain production averaged 42 bushels per acre (2650 kg./ha.) for a total of 11.4 billion bushels (290 million metric tons). About 45 per cent of the world corn crop was grown in the United States. Other important countries in corn production are China, the Soviet Union, Brazil, Mexico, Rumania, Argentina, and Yugoslavia. Maize is a highly important food plant in Mexico, Central America, and the tropics of South America (Figure 12–1).

In the United States field corn was harvested on an average of about 70,284,000 acres annually from 1971 to 1973. The leading states in corn production were Iowa, Illinois, Indiana, Nebraska, and Minnesota. About 61 million acres were harvested for grain with a production of 5,619 million bushels or 92.2 bushels per acre. About 8,604,000 acres were harvested for silage with a production of 109 million tons, or 12.7 tons per acre. An additional area of 604,000 acres was harvested for forage, either dry or as soilage, chiefly in North and South Dakota.

Corn yields in the United States averaged about 26 bushels per acre from 1910 to 1914 and increased to an average of 92 bushels per acre from 1970 to 1972. In the meantime commercial corn production was almost completely mechanized. An average of 135 hours of labor was required to produce 100 bushels of corn during the earlier period, compared with the recent average of only 7 hours.

Corn is grown in every state except Alaska, and on half of all cropped farms in the United States (Figure 12–2).

* The word *corn* merely means *grain* to the British, Germans, and many other nationalities, and often is used to designate the predominant grain in a country or section. Corn usually refers to wheat in England; oats in Scotland; barley in North Africa; grain sorghum, locally, in southern California; broomcorn in the vicinity of Lindsay, Oklahoma; but maize in most of the rest of the United States.

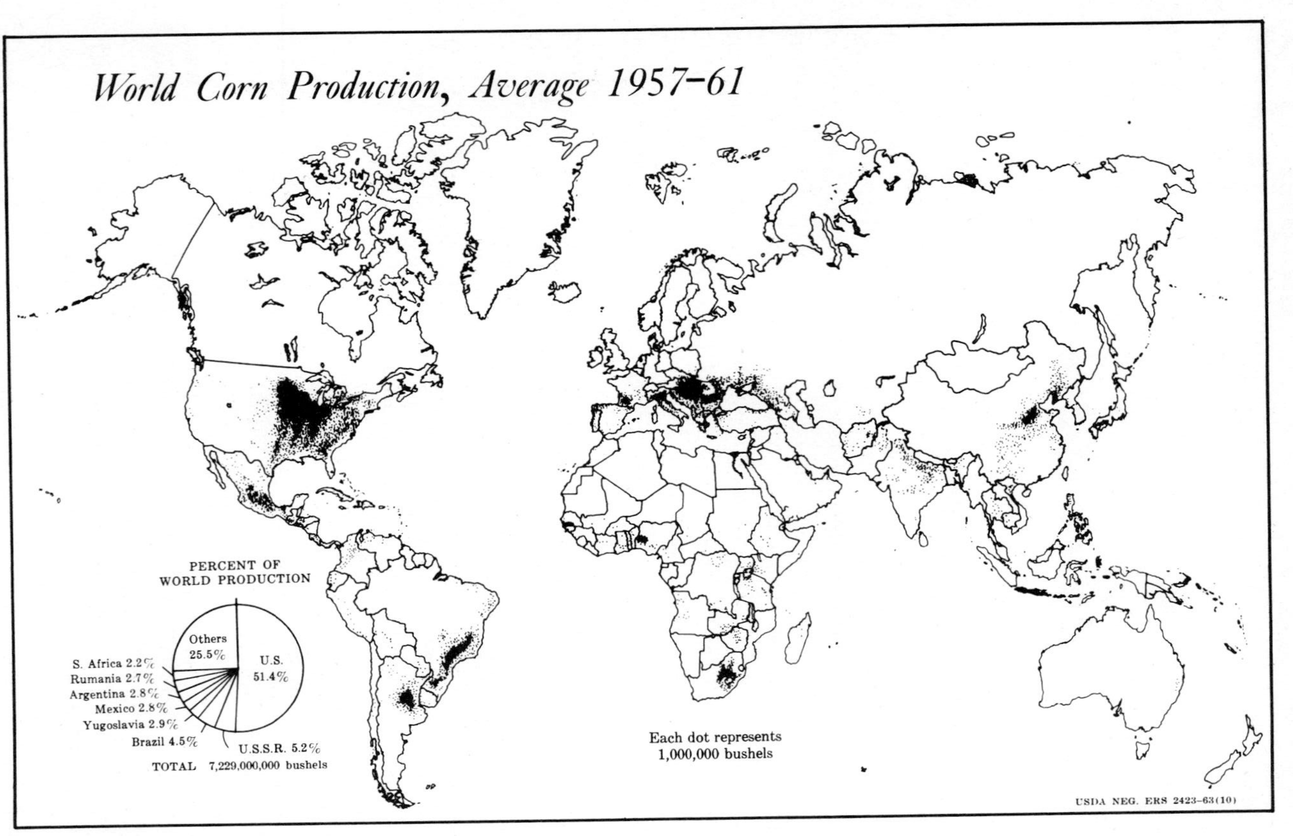

Figure 12–1. World corn production.

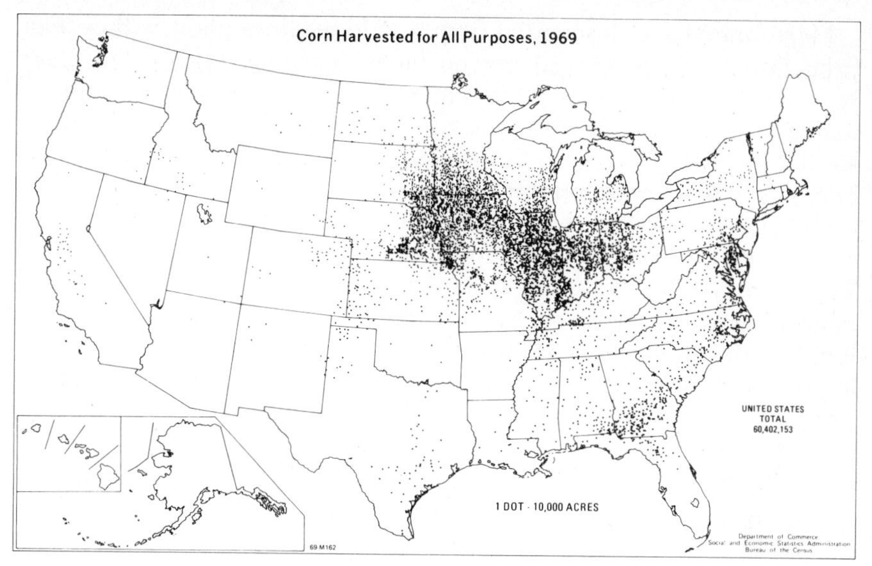

Figure 12–2. Acreage of corn in the United States.

Corn supplies three-fourths of the nutrients derived from feed grain and over 80 per cent of that from silage fed in the United States. The processing of corn furnishes some 3 million tons of human food, 60 million gallons of beverage alcohol, and many industrial products.

The popcorn supplies of 236,000 tons were grown on nearly 160,000 acres from 1970 to 1972. The average yield was nearly 3000 pounds of dry ears per acre. The leading producing states were Indiana, Iowa, and Nebraska.

More than 600,000 acres of sweetcorn were grown annually for canning, freezing, and the fresh market. The average yield was about 4.6 tons of fresh unhusked ears per acre. More than 2 million tons were processed. The leading states in the production of processed sweetcorn were Minnesota, Wisconsin, and Oregon. Florida and California produced nearly half of the 640,000 tons of sweetcorn for the fresh market.

ORIGIN AND HISTORY

Maize (*Zea mays*) is perhaps the most completely domesticated of all field crops. Its perpetuation for centuries has depended wholly upon the care of man. It cannot exist as a wild plant. Corn was seen by Columbus in Cuba on November 5, 1492, on his first voyage to America. The explorations of the 16th and 17th centuries showed that corn extended from Chile to the Great Lakes. The most likely center of origin of corn is Mexico or Central America, with a possible secondary origin in South America.

Maize may have descended from a wild teosinte plant, rather than from either a hypothetical but no longer existent wild maize plant or a *Tripsacum* species.[21a] A suggested origin of teosinte from a maize-*Tripsacum* hybrid[14] seems questionable. Small ears and grains, similar to those of popcorn but enclosed in floral bracts, were found in deposits in a bat cave in New Mexico. Carbon radiation indicates that these specimens are about 5,600 years old. Numerous specimens, which date from 5200 B.C. to 1500 A.D., were found in caves in Mexico. They portray the evolution from the wild to the cultivated types of maize. Marked increases in ear size are evident as are reduced glume lengths and departures from the podded types.[68] Cultivated maize appears to have been first selected by American Indians during the period between 3400 and 2300 B.C. with continued improvement up to 1500 A.D.

ADAPTATION

Maize has a remarkable diversity of vegetative types, with the result that sorts adapted to a wide range of environmental conditions are in cultivation. From latitude 58° N in Canada, the culture of maize, passes without interruption through the tropical regions and on to the frontier of agriculture in the Southern Hemisphere (latitude 35° to 40° S). It is grown from below sea level to altitudes of 13,000 feet. Some small, early sorts only 2 feet tall bear eight to nine leaves and are able to produce mature grain in 50 days. Others with 42 to 44 leaves and growing 20 feet tall require as many as 330 days to come to maturity. This shows how corn has been modified to meet the conditions of many environments. The phenomenon has been well described by Jenkins[43] who states:

> The greatest plant-breeding job in the world was done by the American Indians. Out of a wild plant, not even known today, they developed types of corn adapted to so wide a range of climates that this plant is now more extensively distributed over the earth than any other cereal crop. Modern breeders are carrying on this work and making important discoveries of their own.

In the United States, corn hybrids or varieties grown in the northern states are 3 to 8 feet tall, mature in 90 to 120 days, and may develop several tillers. In the central and southern Corn Belt, the hybrids may grow 8 to 10 feet tall, mature in 130 to 150 days, and usually have few or no tillers. Hybrids in the south Atlantic and Gulf Coast states may grow to a height of 10 to 12 feet on fertile soils and require 170 to 190 days to reach maturity. They often tiller profusely and produce two or more ears per stalk.

Climatic Requirement

The region of greatest production of corn in the United States has a mean summer temperature of 70° to 80° F., (21–27° C.), a mean night temperature exceeding 58° F. (13° C.), and a frost-free season of over 140 days.[43] An average June–July–August temperature of 68° to 72° F. (20–21° C.) seems to be most favorable for maximum yields. Corn flowers and ripens more quickly at 80° F. Very little corn is grown where the mean summer temperature is less than 66° F., or the average night temperature below 55° F. The minimum temperature for the germination and growth of corn is 50° F. or slightly below. Corn plants will withstand a light freeze in the seedling stage up to the time when they are 6 inches tall. Thereafter, a light freeze will kill the plants except for those of certain hardy strains. Prolonged low temperatures below 45° F. but above freezing will kill many strains of corn. Corn is grown extensively in hot climates, but yields are reduced where the mean summer temperatures are above about 80° F.

The misconception has long existed that corn grows best when nights are hot. Actually accelerated respiration in warm nights reduces the carbohydrates that can accumulate and consequently the dry weight of the plants. Cell division and enlargement may continue in darkness when the temperature exceeds 14° C.

The annual precipitation where corn is grown ranges from 10 inches in the semiarid plains of Russia to more than 200 inches in the tropics of India. In the United States, the best corn regions have an annual precipitation of 24 to 40 inches, except where the crop is irrigated. General corn production is limited to sections with an annual precipitation of 15 inches or more, and a summer (June to August) precipitation of 8 inches or more. A heavier precipitation is required in southern latitudes and low altitudes. Some corn is grown at high altitudes where the summer precipitation is less than 6 inches. Indians of New Mexico and Arizona grow corn where the annual precipitation does not exceed 8 inches by planting in widely spaced hills in dry sandy washes that receive extra water whenever a rain is heavy enough to cause runoff.

The corn plant utilizes about half its seasonal intake of water during the five weeks following attainment of its maximum leaf area, which is at about the tasseling stage.[50]

Corn is a short-day plant, i.e., flowering is hastened and vegetative growth retarded by long nights. Long days increase the leaf number, plant size, and length of growing period of corn. Northern varieties moved southward to where the days are shorter mature quickly with a reduced plant growth. Southern varieties moved to the North make a large vegetative growth but do not reach the silking stage until the short days of autumn. A variety becomes progressively 1 day earlier or later for about every ten miles it is moved southward or northward, respectively, where the altitude is the same. The varieties that succeed best in any region are those fitted to the particular length of day. Quick-maturing hybrids are grown in the North. Progressively longer-season hybrids are grown to the southward in the Northern Hemisphere.

Adapted types from 43 states, when grown in Nebraska, showed wide variations in vegetative characters. In maturity, the extreme variation was 155 days for Mexican June corn from New Mexico, to 96 days for Northwestern Dent from North Dakota. There was a marked transition from large late types to small early types from south to north across the country. This was associated with a progressively shortened frost-free season and increased maximum length of day from south to north.[51, 54] The number of leaves on the main stalk is the most reliable measure of the length of season required for maturity.[62] The earlier the variety, the fewer the leaves on the main stem. The number of leaves among some 8,000 world varieties and strains ranges from 8 to 48. Canadian varieties vary from 9 to 18 leaves, American from 10 to 27, and the Peruvian from 14 to 33. American Corn Belt hybrids generally have from 18 to 21 leaves.

Soil Requirements

Corn requires an abundance of readily available plant nutrients and a soil reaction between *p*H 5.5 and *p*H 8.0 for the best production.[69] The prairie soils seem to be among those most favorable for corn production. Corn makes its best growth on fertile, well-drained loam soils. On poor soils, vegetative growth may use most of the available nutrients at the expense of grain production later in the season. Under short-season conditions, light sandy soils are preferable for corn because they warm up more quickly in the spring.

The highest authentic acre yield of corn is 305 bushels of shelled grain (based upon a 15 per cent moisture content), grown in Mississippi.

Adjustment to Environment

In most cases open-pollinated corn grown for a period of years becomes adjusted to the local conditions of moisture supply, temperature, and length of frost-free season.[25] This self-adaptation is brought about by natural selection in the mixed or heterozygous population of any open-pollinated corn variety. Adaptation of corn to a region of moisture shortage consists of a reduction in vegetative development with a consequent reduction in the amount of water used by the individual plant.[59] As climatic conditions become more adverse, the stalks in adapted hybrids or varieties are shorter, the leaf area less, the ears shorter, the kernels more shallow, and the yield lower.

Seed of locally adapted strains will ordinarily produce the highest yields. A variety or hybrid from another region sometimes may give increased yields, particularly when moved from a less favorable to a more favorable climatic region, provided the length of growing season permits satisfactory maturity.[46] Ordinarily, corn hybrids adapted within a latitude range of 100 miles are most likely to produce maximum grain yields.

BOTANICAL CHARACTERISTICS

Maize (*Zea mays*) is a coarse annual grass, classified in the tribe Maydeae which has monoecious (separate staminate and pistillate) inflorescences.[107] Among other members of the Maydeae tribe, teosinte (*Euchaena mexicana*), a native of Mexico, is an annual rank-growing grass grown in the southeastern part of the United States as a forage crop. A perennial species of this genus (*E. perennis*) also is known. Gamagrass (*Tripsacum dactyloides*) is distributed over the eastern and southern United States and is a useful pasture and meadow grass. Other species of *Tripsacum* occur in Mexico and Central America. Teosinte and gamagrass spikes break up into sections upon threshing, revealing a seed enclosed in the thick hulls of each section. Job's tears (*Coix lachryma-jobi*), from southern Asia is grown in the United States for ornamental purposes. The fruits are used as beads. Thin-hulled strains have been grown in the Philippine Islands and South America as a cereal under the name of *adlay.*

Vegetative Characters

CULMS OR STALKS: The culm or stalk of maize ranges in length from 2 to 25 feet (0.3 to 7.6 meters) and in diameter from 1/2 to 2 inches (1.3 to 5 cm.) in different regional types. The internodes are straight and nearly cylindrical in the upper part of the plant, but alternately grooved on the lower part. A bud occurs (usually in the groove) at the base of each internode except the terminal one. When the buds develop, they produce ear shoots. Buds below the soil surface develop the tillers. Modern field corn hybrids seldom develop tillers in optimum spaced plantings. The outside of the culm is protected by an epidermis that renders it fairly impervious to moisture. Inside the epidermis is the cortex and sclerified parenchyma (rind) containing many vascular bundles. The stem inside the rind is filled with ground tissue (pith) in which vascular bundles are interspersed (Figure 3–4). Tillers, when present, contribute to grain formation on the main stalk.[19] Removal of tillers often may injure the plant and reduce the yield.

Commercial strains and inbred lines of corn vary considerably in lodging. Lodged stalks are considered as those that break over or lean 30 degrees or more from the vertical. Most commercial corn hybrids are rather resistant to lodging. Increases in percentages of broken stalks occur when they are infected with diseases such as *Diplodia, Gibberella,* or *Cephalosporium,* or when starchy seed susceptible to scutellum rot is planted.[61] Corn has stood more erect when planted at the rate of two kernels per hill than when three kernels per hill were planted. Much of the lodging in corn is due to weak roots. Erect plants have about twice as large a root system as lodged plants.[27] Broken or bent shanks cause the ears to hang down or drop to the ground which results in greater harvest losses in either case. Growing corn plants increase in height during the night as well as in daytime, although photosynthesis occurs only during daylight.[57]

Corn stalks contain about 8 per cent sugar before the grain is formed, and as much as 10.5 per cent sugar when pollination fails or is prevented.[93]

LEAVES: The corn leaf consists of the blade, sheath, and a collar-like ligule. Liguleless lines of corn have more erect leaves, which permits increased light penetration to the lower leaves, while at the same time resulting in greater photosynthetic efficiency. The leaf blade tapers to a point at the tip and also tapers from the central midrib to the thin edges. At the base of the blade, the two edges extend to form two auricles. The sheath clasps the culm.

A typical leaf of a corn plant grown at Ithaca, N.Y.,[86] was about 80 centimeters long, 9 to 10 centimeters wide, with an average thickness of ¼ millimeter, and contained more than 140 million cells.

The structure of the leaf blade[85] is similar to that of the sugarcane leaf (Figure 3–2). The upper surface bears scattered hairs and large stomata. The lower surface is free from hairs and has smaller but more numerous stomata. A cross section of the leaf shows both the upper and lower epidermis to consist of single layers of cells. Between the upper and lower epidermal layers are five or six layers of mesophyll cells, and within the mesophyll are occasional fibrovascular bundles containing both the strengthening and conducting tissues of the leaf. Chloroplasts in the mesophyll tissues surrounding the vascular bundles are a characteristic of plants such as corn and sugarcane (Figure 3–2), which have the C_4 pathway for high photosynthetic efficiency. At intervals along the upper epidermal layer are groups of large bulliform or hygroscopic cells. Whenever evaporation from the leaf appreciably exceeds the water intake, these bulliform cells shrink and the leaves roll up, thus reducing the surface area exposed to evaporation. When water becomes more abundant, the bulliform cells absorb water, swell, and flatten out the leaf blade again. In the sorghum leaf, in contrast, the bulliform cells are concentrated chiefly in large groups near the midrib, so the drying leaf usually folds up like butterfly wings instead of rolling.

The reduction in yield resulting from the loss of leaves during storms or other natural hazards becomes progressively less with advances in the development of the plant.[18]

ROOTS: The corn plant has three types of roots: (1) seminal, (2) coronal or crown, and (3) brace, buttress, or aerial roots. The seminal roots, usually three or five, grow downward at the time of seed germination. Shortly after the plumule has emerged, the coronal roots form at the nodes of the stem, usually about 1 to 2 inches below the soil surface.

These coronal roots, which are ultimately 15 to 20 times as numerous as the seminal roots, develop at the first seven or eight nodes at the base of the stem. Brace roots arise from the nodes above ground, but their function is similar to that of the coronal roots after they enter the soil.

The usual spread of the main roots is approximately 3½ feet in all directions from the stalk.[33] The lateral spread ceases about one or two weeks before the tasselling stage,[22] the later-season growth being in depth. Although most of the lateral roots are found in the first and second feet of soil, some may extend downward as far as seven to 9 feet. The seminal root system may live to maturity of the plant and penetrate 5 or 6 feet deep. The size of the root system increases with that of the top growth. Compared with a small type, a large type of corn may have a 50 per cent greater spread, 10 per cent deeper penetration, 65 per cent more functional main roots, 92 per cent greater combined length of main roots per plant, 311 per cent greater root weight, and 29 per cent larger diameter of main roots.[109] The rate of root growth as well as top growth is approximately the same for all types until the small varieties tassel. Thereafter, continued vegetative growth causes the medium and large types to surpass small early types. Similarly, as the medium types commence tasseling, their vegetative development is exceeded by that of the large type.

Inflorescence

The corn plant is normally monoecious, i.e., the staminate and pistillate flowers are borne in separate inflorescences on the same plant. The staminate flowers are borne in the *tassel* at the top of the stalk, while the pistillate flowers are located in spikes which terminate lateral branches arising in the axils of lower leaves. The mature pistillate inflorescence is called the *ear* (Figure 12–3). Occasionally, off-type plants produce seed in the tassel, or a tassel on the ear tip.

The central axis of the staminate panicle is the continuation of the

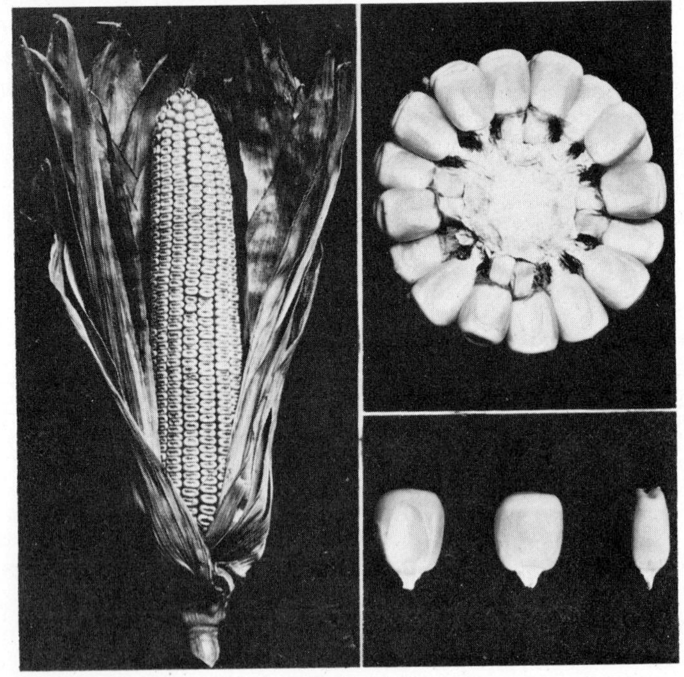

Figure 12–3. [OPPOSITE] *Top:* (*Left*) Pistillate inflorescence—young ear in silking stage. (*Right*) Staminate inflorescence or tassel of the corn plant.
Bottom: (*Left*) Mature ear with husks. (*Upper right*) Cross section of ear showing kernel attachment to cob. (*Lower right*) Kernels—the germ side of the kernel faces the tip of the ear except in those that develop from the lower floret of the spikelet.

main vegetative stem. The branches of the panicle are spirally arranged around the axis. The spikelets are usually arranged in pairs, one sessile and the other pediceled. Groups of three or four may be found occasionally. The spikelet is completely enclosed by two firm, more or less pubescent, ovate glumes. There are two florets per spikelet, the upper being the more advanced in development. Each floret contains three stamens, two lodicules, and a rudimentary pistil. Each anther produces about 2,500 pollen grains or approximately 15,000 for a single spikelet. Thus, a tassel containing 300 spikelets produces 4,500,000 pollen grains to fertilize the 500 to 1,000 silks produced on an ear.

The pistillate inflorescence, or ear, is a spike with a thickened axis. The pistillate spikelets are borne in pairs in several longitudinal rows. This paired arrangement explains the customary even number of rows of grains on the ear. An occasional ear with an odd number of rows has lost one member of the pair. The individual spikelet is two-flowered, only one floret ordinarily being fertile. When the second flowers of the spikelet also are fertile, the crowding of the kernels destroys the appearance of rows. The result is an irregular distribution of kernels as seen in the Country Gentleman variety of sweet-corn (Figure 12–4).

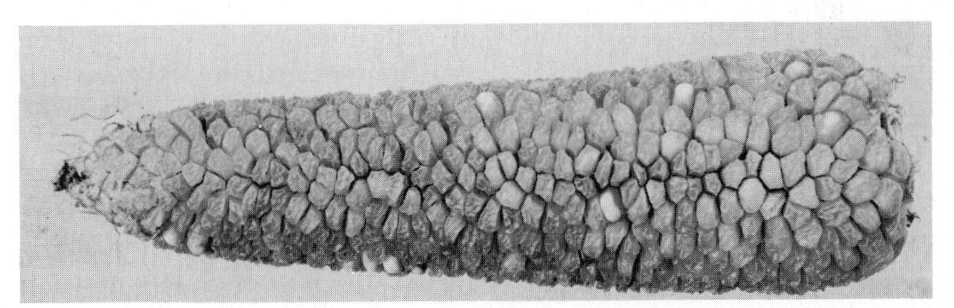

Figure 12–4. Ear of Country Gentleman sweet corn which develops kernels in both florets of the spikelet; the resultant crowding prevents the kernels from forming straight rows. The smooth white-capped kernels show xenia that resulted from outcrossing by pollen from field corn.

The two thick and fleshy glumes are too short to enclose the other parts of the spikelet. The lemma and palea, which together constitute the *chaff*, are thinner and shorter than the glumes. The single ovary in a fertile floret bears a long style or *silk* which is forked at the tip (Figure 3–13). The silk has a sticky stigmatic surface and is receptive to pollen for about two weeks and throughout its length. The remnant of the silk is evident on the crown of each mature kernel. The silks ordinarily are 4 to 12 inches (10 to 30 cm.) in length. Unpollinated ears may produce silks 20 to 30 (50 to 75 cm.) inches long.

Pollination

Corn is normally cross-pollinated,[51] being well adapted to wind pollination by bearing the male flowers at the top of the plant and the female flowers about midway up the stalk.

Cold wet weather retards the shedding of pollen, while hot dry conditions tend to hasten it. The emergence of the silks is delayed by drought and inadequate nitrogen. Protandry is the shedding of pollen before the appearance of the silks and protogyny is maturation of the stigmas first. Corn is somewhat protandrous, but the pollen shedding and the receptivity of the silks overlap sufficiently to permit some self-pollination.

In the fertilization of the maize flower[74, 85] the pollen grain germinates and establishes a pollen tube within 5 to 10 minutes after it falls on the silk. The pollen tube enters the central core of a hair on the silk and passes down through the silk (Figure 3–13). The two sperm nuclei migrate into the pollen tube and pass down to the embryo sac. Fertilization is accomplished within 15 to 36 hours after pollination, the slower rates occurring at low temperatures and when the silks are long. One of the two sperm nuclei unites with the egg cell to produce the embryo, while the other unites with first one and then the other of the two polar nuclei of the embryo sac to give rise to the endosperm. The union of sperm nuclei with both the egg cell and the polar nuclei is called double fertilization. The endosperm often shows certain characteristics of the pollen parent because it develops directly from the endosperm nucleus after union with the male nucleus. The pericarp, on the other hand, does not display characters derived from the pollen parent that season. Any effects are shown when a new plant develops the next season. The immediate effect of the pollen parent on the characteristics of the endosperm, embryo, or scutellum is called xenia (pronounced zee′-nee-ah). Xenia occurs when an ear of sweet corn is pollinated with flint corn and the kernels are smooth and starchy instead of wrinkled and sugary (Figure 12–4). Also, when pollen from a plant with yellow endosperm fertilizes a plant with white endosperm, the kernels have a yellow endosperm because of the Mendelian dominance of yellow over white. When yellow corn is pollinated by white corn, the kernels are yellow but are lighter in color and often capped with white.

Pollen of dent on sweetcorn silks may cause a 25 per cent increase in average kernel weight because of the formation of a starchy instead of a sugary endosperm.[51] The effect on kernel size is negligible when one commercial variety of dent corn is crossed on another.[52] However, hybrid kernels on pure-line ears have increased as much as 11 per cent in weight.

The flowers near the middle of the ear develop the silks early, and usually are pollinated first. The flowers in the lower third of the ear develop at about the same time but because of their remote position take more time to extend the silks out of the husk. Those at the tip of the ear are last to develop. Under unfavorable conditions the earlier-pollinated kernels appear to take precedence over the others in food supplies, and the tips and butts of the ears are filled very poorly and bear smaller grains.

Development of the Corn Kernel

In the development of the corn kernel or caryopsis, the ovary wall grows by cell division and cell enlargement to form the pericarp of the kernel.[44, 86] The endosperm and the entire kernel grow rapidly in length and diameter during the period of 10 to 20 days after pollination. This is followed by a correspondingly rapid growth of the embryo during the period from 20 to 40 days after pollination. The embryo grows and differentiates until the various rudimentary structures of a new plant are apparent. The endosperm cells divide rapidly and nearly fill the space within the growing ovary wall within 18 days after pollination. The endosperm cells at first divide in all directions. Later, cell divisions occur largely around the periphery, and rows of cells are forced inward. Final endosperm growth is due chiefly to cell enlargement. The mature endosperm is 50 times as long and wide as it is in the initial stages of its observable formation.

The endosperm cells at first contain simple sugars but later the sugars are converted to starch.

The testa, often called the nucellar layer, arises just inside of the pericarp where the cells of the integuments and nucellus of the embryo sac have degenerated.

The kernel ceases growth about 45 days after pollination. The shape of the individual kernel is determined in part by the degree of crowding on the ear. Isolated grains and many of those at tip and butt usually are somewhat rounded. Grains in crowded rows on the ear are often angular in outline.

About 1,200 to 1,400 kernels of typical dent corn weigh one pound (2,600 to 3,000 per kilogram). The kernels usually range from 1/4 to 3/8 inch (6 to 9 mm.) wide, 1/8 to 1/4 inch (3 to 6 mm.) thick, and 3/8 to 5/8 inch (9 to 16 mm.) deep. The kernels of Cuzco corn from Peru are about .58 inch (16 mm.) wide.

The parts of the corn kernel, as shown in Figures 7–2 and 12–5, include a pedicel or tip cap over the closing layer of the hilar orifice

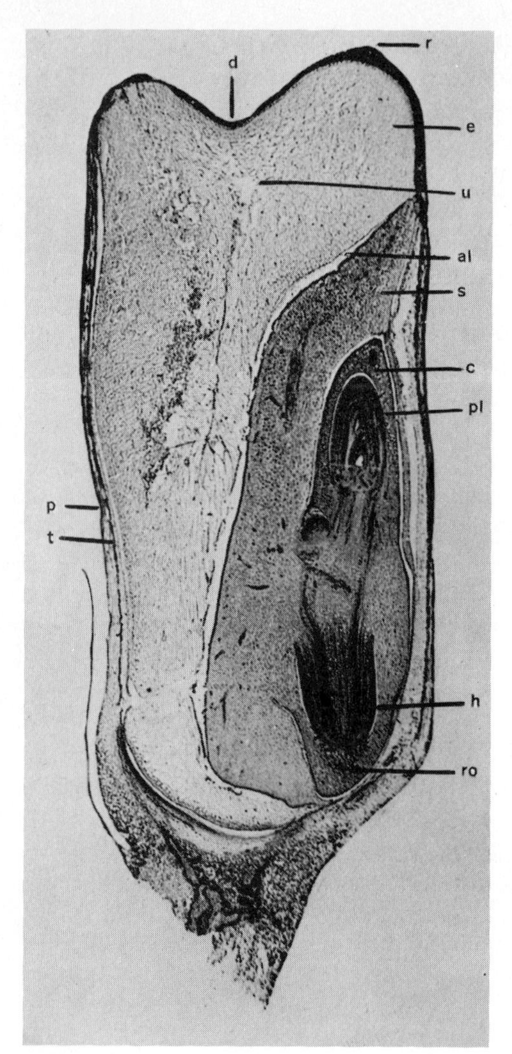

Figure 12–5. Half of nearly mature dent corn kernel that shows dent (*d*), remnant of style (*r*), endosperm (*e*), unfilled portion of immature endosperm (*u*), aleurone (*al*), scutellum (*s*), plumule (*pl*), hypocotyl and radicle (*h*), testa or true seed coat (*t*), and pericarp (*p*). The germ comprises the entire darker area which includes the scutellum (*s*), coleoptile (*c*), and rootcap (*ro*).

at the end of the kernel where it is attached to the cob.[34] The embryo (germ) lies near one face of the kernel. The outer rim of the embryo is the scutellum. The remainder of the kernel inside the testa is occupied by the endosperm. The outer layer of cells of the endosperm is the aleurone which comprises about 8 to 12 per cent of the kernel and is high in nitrogenous substances. The aleurone may be red or purple in certain varieties. In dent corn the endosperm may be partly corneous (horny) and partly soft or starchy white. The corneous portion may be one-tenth to one-fourth higher in protein. The germ, comprising about 11 per cent of the kernel, contains five-sixths of the oil, three-fourths of the ash, and one-fifth of the protein in the kernel.

Outside of the aleurone is the testa or true seed coat, and this in turn is surrounded by the pericarp. The hull (pericarp and testa), comprising about 6 per cent of the kernel, is high in cellulose and hemicelluloses. The hull protects the kernel from mechanical damage and the entrance of disease organisms.

The Ear

The normal ear of corn contains 8 to 28 rows of grains, while each regular row bears 20 to 70 grains. An occasional freak ear may have only four rows when other rows are aborted. Ears with 14, 16, 12, and 18 rows predominate among American dent corns. A large ear of corn may contain about 1,000 kernels, but ordinary good ears contain about 600.

The average well-dried ear of Corn Belt corn weighs about 8 ounces (227 gm.) and contains nearly 500 grains. The shelling percentage of dry well-filled ears averages about 80 to 85 per cent. In average dry corn, 70 pounds of ears yield 56 pounds (1 bushel) or 80 per cent shelled corn. In volume, 70 pounds of ear corn occupy 1½ heaping bushel measures or nearly 2 level bushels.

GROUPS OF CORN

Corn is often classified into seven groups or types, based largely upon endosperm and glume characteristics. These are dent, flint, flour, popsweet, waxy, and pod corns (Figure 12–6). The Latin names listed are presented because the different types sometimes have been regarded as distinct botanical subspecies. The use of botanical subspecies names is no longer valid in the light of our knowledge of corn heredity. The sweet, waxy, and pod types each differ from dent corn by only a single hereditary factor, and the flint, flour, and popcorn types may differ by relatively few hereditary factors.

Dent Corn (*Zea mays indentata*)

Both corneous and soft starch are found in dent corn, the corneous starch being extended on the sides to the top of the kernels (Figure 12–7). When the grain dries, a pronounced wrinkle or dent forms on the top of the kernel because of the shrinkage of the soft starch. Dent corn is the most widely cultivated type, being by far the most important one in the United States and almost the only type in the Corn Belt. The kernels have a wide range of colors but are yellow or white in most commercial varieties.

"High lysine" dent corns have a modified opaque endosperm with a high content of lysine, an essential nutritional amino acid for humans, swine, and poultry.

Figure 12–6. Ears of (*left to right*) pop, sweet, flour, flint, dent, and pod corn.

Flint Corn (*Zea mays indurata*)

The endosperm of flint corn usually is soft and starchy in the center but completely enclosed by a corneous outer layer. The kernels are usually rounded at the tip. Most flint corns formerly grown in the northern part of the United States had eight rows of kernels on the ear and were early in maturity. The ears of the tropical flint corns of Louisiana and Florida usually had 12 or 14 rows of kernels. Flint corn is widely grown in Argentina.

Flour Corn (*Zea mays amylacea*)

Kernels of flour corn are composed almost entirely of soft starch. They are shaped like flint kernels because they shrink uniformly as they ripen. The kernels exist in all colors, but white, blue, and variegated are the most common. Flour corns are grown, chiefly for food, by some of the American Indian tribes and by many growers in Central and South America.

Popcorn (*Zea mays everta*)

This group is characterized by a very hard corneous endosperm and small kernels. There are two subgroups of popcorn: rice, with pointed kernels, and pearl, with rounded kernels. The grains may differ greatly in color and size. Additional ears may be borne on the main stalk or tillers.

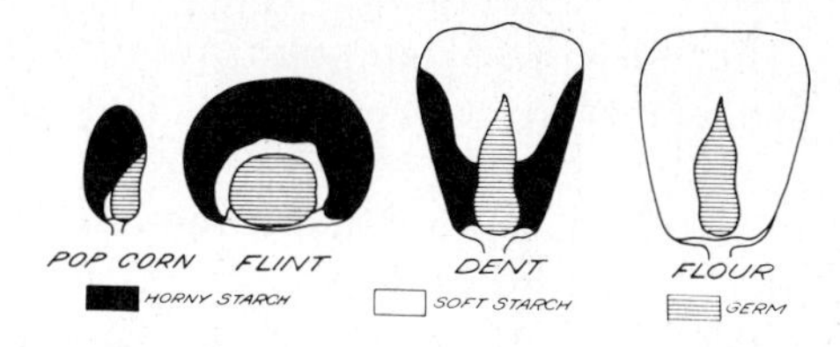

Figure 12–7. Typical proportions of hard and soft starch in kernels of four corn groups.

Sweet Corn (*Zea mays saccharata*)

The kernels of sweet corn are translucent and more or less wrinkled at maturity. Sweet corn, before it is ripe and dry, has a sweeter taste than do other types because the endosperm contains sugar as well as starch. It lacks the ability to produce fully developed starch grains. Sweet corn also may bear two or more ears per plant.[21]

Waxy Corn

The kernels of waxy corn have a uniformly dull, rather soft endosperm, showing neither white-starchy nor corneous-translucent layers. The endosperm breaks with a wax-like fracture. The starch in waxy corn consists of amylopectin, which has a branched-chain molecular structure and a high molecular weight (50,000 to 1,000,000). The starch, which resembles tapioca starch, is used for making special foods or adhesives. The endosperm starch and also the pollen of waxy corn are stained red by an iodine–potassium iodide solution. Starch of ordinary corn, which stains blue in the iodine test, consists of about 78 per cent amylopectin and 22 per cent amylose. The latter substance has a straight-chain molecular structure and a molecular weight of 10,000 to 60,000.[32]

Pod Corn (*Zea mays tunicata*)

In pod corn, each kernel is enclosed in a pod or husk. The ear formed is also enclosed in husks as in other types. Pod corn may be dent, sweet, waxy, pop-, flint, or flour corn in endosperm characteristics. It is merely a curiosity and is not grown commercially.

The typical podded ear as described never breeds true. When planted, half the crop is pod corn, a fourth is normal without pods, and the remaining fourth are long-podded ears that contain no grain, although a few seeds may develop in the tassel. Thus, common pod corn is genetically a heterozygous plant, the progeny of which segregates in a typical 1:2:1 Mendelian ratio with the podded character dominant. True-breeding types of pod corn have been developed recently.

OPEN-POLLINATED VARIETIES

An unknown number in excess of 1,000 varieties of corn were grown in the United States before the spread of corn hybrids (Figure 4–6).

They originated through hybridization, segregation, and selection, either natural or controlled. Few of them are grown at the present time.

Varietal names of corn mean less than those of almost any other crop. All are heterozygous. Selection and local adaptation have produced numerous strains and types within varieties. Frequently there are larger differences among strains within varieties than between the varieties.[51] In general, the prolific varieties and hybrids that produce more than one ear per stalk were grown in the southern states.

The Indian tribes maintained their own strains of corn, to a large degree, for many decades. Many of the Indian varieties were a colorful mixture of red, purple, yellow, and white kernels, mostly of the flint and flour types.[8] Such mixing helps to avoid the deleterious effects of inbreeding. Ears with mixed kernel colors are now grown and marketed for decorative purposes.

A characteristic of hybrids grown in the South is a long husk on the ear. Long husks partly protect the ears from the corn earworm, weevils, moths, and birds.

CORN HYBRIDS

State agricultural experiment stations, the United States Department of Agriculture, and a considerable number of commercial firms are engaged in the development of new corn hybrids. Several hundred different corn hybrids are being grown on a commercial scale. These are being produced from a large number of inbred lines. However, certain outstanding lines enter directly or indirectly into the pedigrees of a large percentage of the hybrids that have gone into production. The line WF 9 is still used as the single-cross seed parent of many hybrids because of its attractive kernels. The original inbred lines were selected largely from some of the most productive open-pollinated varieties. Newer inbred lines are derived mostly from hybrids or synthetic varieties. Many of the varieties that contributed these lines had been carefully selected for many years.

Publicly supported institutions usually release "open-pedigree" hybrids in which the inbred lines and single crosses are revealed. Commercial hybrid seed-corn producers often designate their hybrids by their own firm numbers, without disclosing the pedigrees. These are called "closed-pedigree" hybrids.

Popcorn and Sweetcorn

The popcorns and sweetcorns grown in the United States have shorter, more slender stalks, and smaller ears, and are earlier in maturity than the field-corn hybrids typical of the Corn Belt. Because of this smaller size and also because of the premature harvesting of sweetcorn, planting usually is at a thicker rate than for field corn.

The leading popcorn hybrids were derived from varieties with large yellow kernels that have a mushroom shape after popping; others have smaller yellow kernels of high popping expansion, or pointed white kernels. Some have short thick ears with 30 to 40 irregular rows of slender pointed white kernels.[12] The most popular hybrids which are three-way crosses, outyield all open-pollinated varieties, and may have a popping expansion of 31 volumes.

Most of the sweetcorn consists of yellow hybrids that have in their ancestry a phenomenal inbred line from the Golden Bantam variety. This line is resistant to the bacterial wilt (Stewart's disease) of corn, has an extremely high quality, and is much more productive than the open-pollinated variety from which it was derived.

Hybrid sweetcorn is especially advantageous to growers of corn for processing because of its more uniform growing period. Since the ears usually are gathered from the field at a single picking, this uniformity of maturity permits a larger portion of the ears to be in prime condition at harvest time.

FERTILIZERS

The highest yields of corn are obtained on heavily fertilized land. In 1970 from 95 to 100 per cent of the corn acreage was fertilized in the eastern, central, and southern states. In Texas, only 84 per cent and in South Dakota only 61 per cent of the corn fields received fertilizers.

Barnyard manure is commonly applied to land to be planted to corn when the manure is available and not needed for more intensive crops such as potatoes or sugarbeets. Much of the corn land in the northeastern states, Great Lakes states, and western irrigated regions is manured.

Corn producing at the rate of 150 bushels per acre will remove from the soil the amounts of fertilizer elements shown in Figure 12–8.

Nitrogen – (Figure 12–9)

One bushel of corn contains about 1 pound of nitrogen with another one-half present in the associated stover. Ample nitrogen should be available throughout the growing season. Usually part of the nitrogen is applied in a mixed fertilizer during or before planting, and a second application follows about 20 to 30 days after emergence of the seedlings.

High plant populations require heavy fertilization. In Illinois the most profitable yield responses occur from 160 to 180 pounds per acre of N (for continuous corn); 100 to 150 following soybeans, a small grain, or one year of corn in a rotation that includes a forage legume

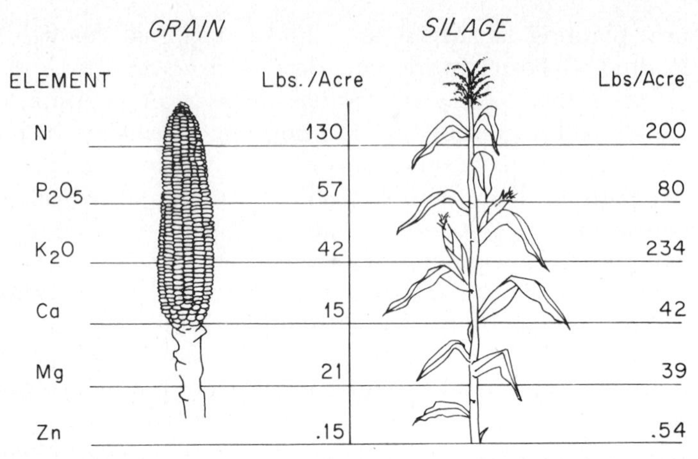

Figure 12–8. Corn plant nutrient removal at yield of 150 bu. acre.

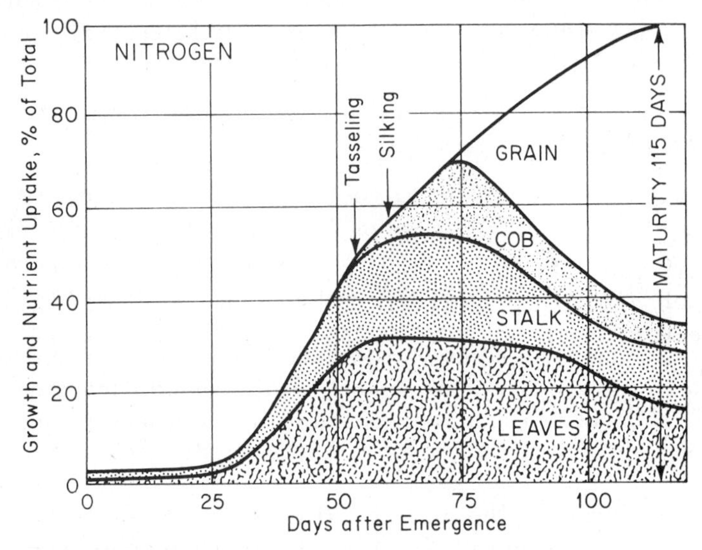

Figure 12–9. Nitrogen uptake and distribution in corn.

once in 5 to 6 years; and 75 to 100 pounds N on corn following a good legume sod or an application of 10 tons of manure per acre.

Phosphorus – (Figure 12–10)

Corn accumulates phosphorus throughout the growing period. Phosphorus deficiency (purpleing of leaves) symptoms usually appear in plants 20 to 40 days old. Phosphorus is best applied in a mixture near the row at planting time. A 150-bushel corn crop removes about 25 pounds of elemental phosphorus.

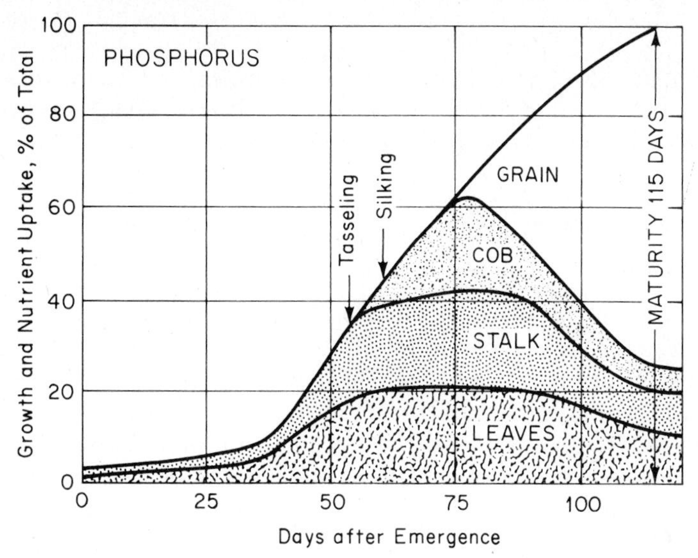

Figure 12–10. Phosphorous uptake and distribution in corn.

Soils receiving manures or other organic residues require less P in the fertilizer.

Potassium

Potassium deficiencies (yellow leaf margins) also appear in young plants. The majority of the potassium is taken up by the plant before the tassel stage (Figure 12–11). The total potassium uptake by a corn plant is distributed $\frac{1}{3}$ in the grain and $\frac{2}{3}$ in other plant parts.

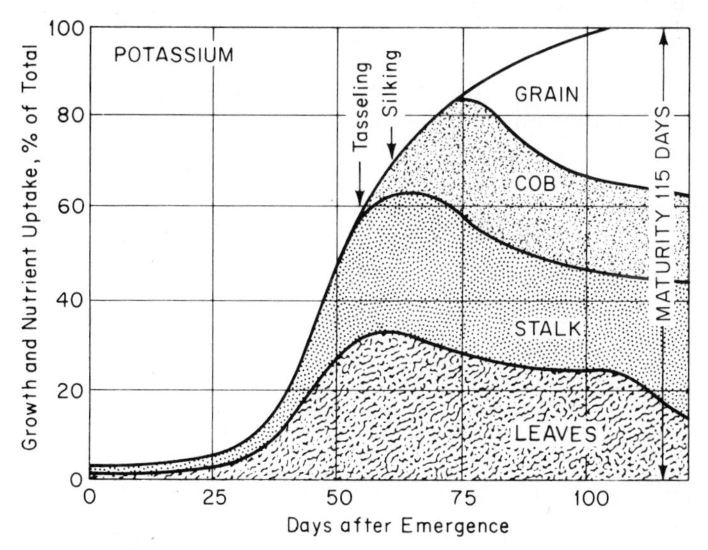

Figure 12–11. Potassium uptake and distribution in corn.

Micronutrients

The addition of secondary or micronutrients may be necessary on soils that are strongly weathered, coarse textured, alkaline, very acid, or peat and muck. Corn grows best on soils with a *p*H of 6.0–7.0.

CULTURAL METHODS

Crop Rotations

Rotations in the Corn Belt include a two-year rotation of corn and small grain; a three-year rotation of corn, small grain, and clover; a four-year rotation of corn, oats, wheat, and clover; and a four-year rotation of corn, corn, small grain, and clover. Sweetclover, alfalfa, or legume-grass mixtures often replace clover in rotations. Soybeans often replace corn or one of the small grains in the rotation. Most rotations in the irrigated regions of the western states include alfalfa and small grain, with corn, sugarbeets, or potatoes as cultivated crops. Under dryland conditions, alternate corn and wheat, or corn with barley or oats occasionally replacing wheat, appear to be most satisfactory. Corn usually shows a relatively small response to the extra moisture provided by fallow. In much of the South, corn ranks second only to cotton in acreage. Since these two crops do not supplement each other well, it is advisable to grow legumes in a cotton-corn rotation followed by a winter cover crop such as rye or a legume, to be turned under for green manure as a preparation for corn and cotton. In a Rhode Island experiment corn was grown continuously over a 40-year period in a comparison of cover crops versus no cover crops.[78] Clovers as legume cover crops were the most effective in maintenance of corn yields. However, winter rye seeded at the last cultivation of corn in the fall increased the average annual yield by 6 bushels over the adjacent no-cover-crop plot for a 34-year period. Small grains usually yield well following corn.

Continuous Corn

Considerable corn is being grown continuously, with success, by heavy fertilization and retaining the crop residues and controlling insects and weeds with pesticides. The chief hazards are a possible build-up of soil-borne disease or insect organisms and a greater risk from soil erosion. Erosion control procedures are necessary on fields with a slope of 3 to 5 per cent or more.

Seedbed Preparation

Land is usually plowed with a moldboard plow as the first step in seedbed preparation for corn. In the North and in the Corn Belt, where the ground is frozen over winter, land can be plowed equally well in the fall or spring where an annual cultivated crop has been grown the previous season. Fields that have been in alfalfa, clover, or pasture should be fall-plowed to permit the sod to decay. Fall plowing for corn is inadvisable in the South because it promotes water erosion.

In the subhumid prairie region, early fall plowing tends to overstimulate the crop, which renders it more susceptible to drought injury.

The final preparation of plowed land usually consists of pulverization of the top 4 inches of surface so as to provide a soil free from large air spaces in which to plant the seed. This should be done immediately before the crop is planted in order to suppress weeds. Frequently the land can be put in condition to plant by double disking and harrowing. The corrugated roller is sometimes used on heavier soils to break clods and firm the seedbed. Some corn is planted on plowed land with no subsequent tillage in the wheel tracks of the tractor that pulls the planter. This "wheel-track" planting saves labor, machine cost, and soil compaction.

In the drier western Corn Belt tools such as the sweepblade, oneway plow, chisel lister, or buffalo till planter are often used in place of the moldboard plow. Rotary tilling can prepare a seedbed in one operation. This, or other minimum tillage methods, saves labor, improves water infiltration, reduces soil compaction, and may enhance or reduce the weed population.

Planting Methods

Corn is usually planted 2 to 3 inches deep in rows from 30 to 40 inches apart. The average spacing in different states ranged from 35.6 to 39.2 inches in 1970. Experiments in seven states: Iowa, Kansas, Michigan, Nebraska, Ohio, and Georgia, indicate that yields average about 5 per cent more in 30-inch than in 40-inch rows under most conditions. For very early planting in cold soils, 1–1½ inches deep may be advisable, but with a dry surface soil condition, it may be necessary to plant 3–4 inches deep to place the seeds in moist soil. Where soil moisture and fertility is not limiting, narrow rows with higher plant populations may lead to yield increases of 10–15 percent. Narrow row planting requires special planting, cultivating, and harvesting equipment.

Formerly, most of the corn in the Corn Belt was surface-planted in check-rows to permit cross cultivation. Now, weeds are controlled by herbicides or by more effective tillage. The corn is now drilled rather than planted in check-rows because the operation is faster, and somewhat higher yields are possible. The weeds in the corn row are controlled by preemergence band applications of herbicides.

Listing is adapted to regions with limited rainfall and light soil types, with the surface-planting method more suitable for areas with abundant precipitation and heavy types of soil.[92] Listed corn is better protected from spring frosts. In a two-year period in Kansas 9 per cent of the plants were killed by frost in listed corn, and 63 per cent in surface-planted corn. The retarded early growth of listed corn is advantageous in a dry season because the soil moisture is less likely to be exhausted before the silking period is reached.

In the South corn usually has been planted in drills rather than in hills. On poorly drained soils, it is customary to list up the plowed land into a bed for each corn row. The corn is planted on the top of the beds, while the furrows between the beds are left to drain the excess water.

Date of Planting

Corn planting begins about February 1 in southern Texas and progresses northward through the two eastern thirds of the country at an average rate of 13 miles per day (Figure 12–12). In the Corn Belt the planting season is at its height about May 15. At a corresponding latitude, planting may be somewhat later in the western third of the country because of cooler weather at the higher elevations. There are two general periods of corn planting in much of the Cotton Belt, one early and one late. The early date corresponds closely to that for planting cotton, while the late period begins in June. In Texas the highest yields are obtained when corn is planted at the average date of the last spring frost.[67] In western Kansas, planting in June permits the

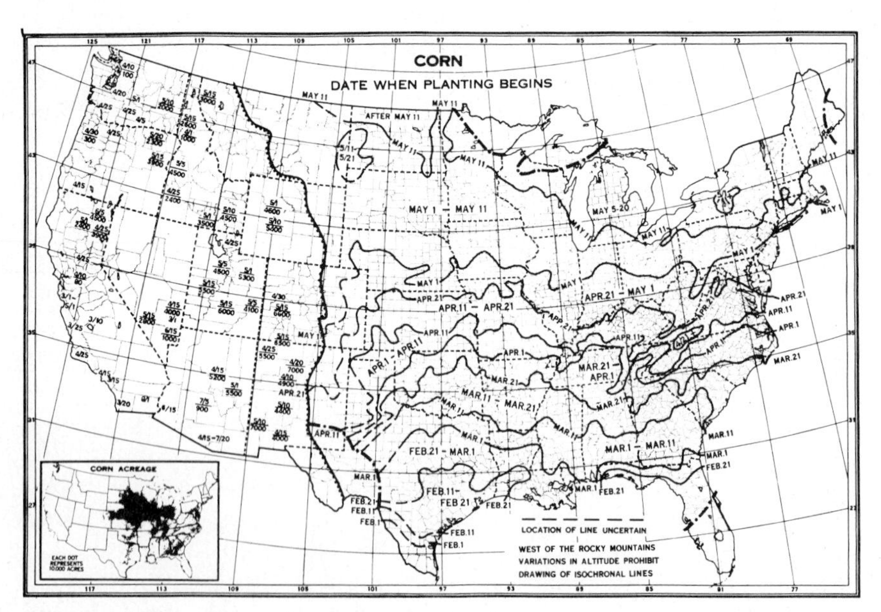

Figure 12–12. Dates when corn planting begins.

corn to tassel and silk after the extreme midsummer drought and heat are past, and this results in higher yields.

The number of days from planting to maturity decreases as the date of planting is delayed.[76] A variety that required 150 days to mature when planted April 1 in Tennessee matured in 113 days when planted on June 1.

Corn can be planted earlier on sandy soil than on the heavier clay soils. It is futile to plant corn early in a cold, poorly prepared seedbed. In Colorado corn planted in cold soil April 20 emerged about the same time as that planted May 1.[91] The popular idea of the Indians not to begin planting corn "until the oak leaves are as big as squirrels' ears" is a fairly safe rule to follow. However, it is unnecessary to delay planting until all danger of spring frosts is past. Permanent frost damage to corn planted early is infrequent.

Rate of Planting

Optimum spacing of corn ranges from about 3,000 to 4,000 plants per acre on poor soils or under extreme semiarid conditions,[10] up to 24,000 or more under optimum growing conditions. About 3,500 seeds per acre should be planted for each 20 to 25 bushels of expected grain yield. Perhaps 85 per cent of the seeds planted will grow into mature plants.

The average plant population per acre in 1970 exceeded 18,000 in the more favorable Corn Belt areas, and 14,000 to 16,000 in other parts. The plant populations in the southeastern states ranged from 9,000 to 12,000. Productive hybrids and greater fertilizer rates justify high plant populations of 12,000 to 24,000 plants per acre in humid regions. About 6,000 to 12,000 plants per acre are ample under subhumid conditions. Early hybrids usually have a smaller size and should be planted thicker than late hybrids.

The average rate of planting corn in the United States in 1970 was 12.9 pounds per acre. Fodder corn is planted more thickly. A stand of 24,000 plants per acre (60,000 per hectare) requires the planting of 18 to 20 pounds of seed per acre (20 to 22 kilos per hectare). Large long-season prolific varieties, with tillering stalks and multiple ears per plant, adapted to the southern states, produce maximum yields at thinner rates than do smaller single-ear hybrids.

Thicker planting of up to 24,000 kernels per acre of productive hybrids on good well-fertilized land is becoming common. Plant populations above 24,000 per acre may require supplementary irrigation.

Under favorable western irrigated conditions, thick rates of planting have resulted in the highest yields (Figure 12–13). The highest grain yields were obtained with stands of 15,000 to 17,000 plants per acre.[63, 84] Very thick rates tend to produce weaker stalks with increased

Figure 12–13. Irrigating corn rows by the use of syphon tubes.

lodging[41, 91] and sometimes reduced yields as well as a larger percentage of barren stalks.

Under semiarid conditions, the rate is adjusted to the limited moisture available. At Akron, Colorado, the highest average yield of ear corn was obtained from plants spaced 24 inches apart in 44-inch rows.[10] The yield from double-spaced, or 88-inch rows, was 28 per cent less. The double-spaced rows had markedly decreased yields in years of good production and did not increase the sureness of production in poor years.

In Arkansas, 3,000 to 3,500 plants per acre in double-width rows produced 66.6 per cent of the yield of the same rate in ordinary rows.[81] In Nebraska the yield from double-width rows was reduced 23 per cent when the stand of plants per row remained the same, and 14 per cent when the number of plants per row was doubled so as to provide the same number of plants per acre. In Ohio, wide-row (7-foot) planting of corn facilitated the establishment of stands of winter wheat or legumes drilled between the corn rows. The yields of corn were about 15 per cent less than from 38-inch rows when the number of kernels planted per acre was the same.

Sweetcorn and popcorn often are planted in rows 30 to 40 inches apart and thick enough in the row so that the number of plants per acre ranges from one-fourth more to twice as many as the usual spacing for field corn. Popcorn usually is planted at a rate of 3 to 6 pounds per acre and sweetcorn at 8 to 16 pounds per acre.

Weed Control

The principal reason for the cultivation of corn is to control weeds (Chapter 5), as shown by many comparisons of corn from cultivated plots and corn from plots merely scraped with a hoe to control weeds. Extensive experiments have shown no important differences in favor of deep cultivation. Root pruning has frequently resulted from cultivation deeper than 3 inches.[78] Early in the season weeds are controlled economically by herbicides or cultivation. Such implements as the harrow or rotary hoe can be used without damage to the corn plants until they are approximately 4 inches high. Subsequent cultivation—frequent enough to control weeds but shallow enough to avoid injury to the corn roots—can be done satisfactorily with ordinary cultivators. Cultivation of corn in listed furrows saves subsequent labor by burying the small weeds in the row.

The use of herbicides may eliminate much or possibly all of the cultivation of corn. A preemergence application of a suitable herbicide controls most of the annual broad-leaved weeds as well as the germinating seedlings of weedy grasses. It may be applied also in bands with a sprayer immediately behind the planter to control only the weeds that come up in the corn row. Late weeds that emerge after the final cultivation can be suppressed by applications of other herbicides along the row.

Harvesting

About 87 per cent of the corn acreage in the United States was harvested for grain from 1970 to 1972; 12 per cent was cut for silage; and 1 per cent was hogged down, grazed, or cut for fodder. The harvest of corn for grain begins in late August in the Gulf Coast region and in early October in the northern states (Figure 12–14). Nearly all the corn is harvested for grain with a corn combine, picker-sheller, or corn picker (Figure 12–15). Early harvest with a corn combine or a picker-sheller followed by artificial drying in a ventilated bin is a popular practice. This method not only saves labor but also avoids greater field losses than result from delayed harvest. Some corn is still husked or snapped by hand from standing stalks, or from the shock by hand, or with a husker-shredder.

In the South, corn for home farm feeding is snapped from the standing stalk with a picker. It is cribbed without husking as a protection against stored-grain insect pests. It is husked later, at the time it is fed or shelled. Cutting and shocking formerly was the prevailing method of harvesting corn in the east central states where winter wheat was to be sown between the rows of shocks. Much of the winter

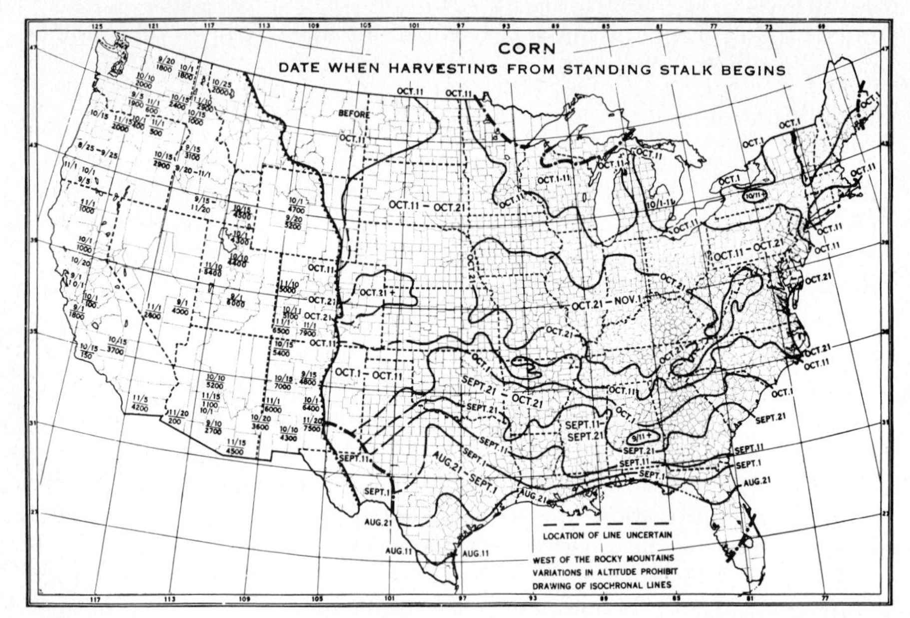

Figure 12–14. Dates for harvesting corn for crib storage without artificial drying.

wheat following corn now is sown after corn has been harvested for silage, or after the grain harvest of an early-maturing hybrid. Corn to be shocked is cut with a binder, or by hand, or with cutting sleds.

While corn is mature at a moisture content of 30–32 per cent, harvesting is seldom done at that time. The best time to harvest for minimum yield loss is at a grain moisture content of 26–28 per cent. Early harvesting results in less lodging, less shelling loss in mechanical picking, and there is less delay due to bad weather. A delay from October to December resulted in average losses of 4.6 to 11.8 per cent in four Corn Belt states.

Safe long-term storage requires that shelled corn moisture be reduced to at least 13 per cent but it can be kept over winter in cold climates at 15 per cent moisture content. Cribbed ear corn on the other hand may be safely stored at 20 per cent moisture in the Corn Belt provided the moisture falls to 18.5 per cent by late February or early March. Corn keeps 10 times as long at 15 per cent moisture than at 20 per cent moisture, 3 times as long at 20 per cent than at 25 per cent moisture by retarding mold growth. Moldy corn may develop aflatoxin which is poisonous.[6]

Corn picked, husked (or shucked) from the standing stalks, at a grain moisture content of 20 per cent or less, usually is dry enough for safe storage in cribs. Mechanical corn pickers do very satisfactory work where the plants stand erect in the field and when the husks are

Figure 12–15. (*Top*) Two-row self-propelled corn combine that conveys the shelled grain from the grain tank to a truck. [Courtesy International Harvester Co.]

(*Bottom*) Two-row tractor-mounted cornpicker harvesting and husking corn. [Courtesy Allis-Chalmers Manufacturing Co.]

slightly damp. The picker may be wasteful where the plants are lodged unless the machine is equipped with lifter guards. Most strains of hybrid corn have strong root systems and stand well for mechanical picking.

Harvesting of corn at a moisture content above that recommended for safe storage requires artificial drying or storage in a high moisture silo. Corn can be dried by one of several methods. They are: (1) dry and store in the same structure; (2) dry in one bin as a batch and transfer to another location leaving the final batch in place; (3) portable batch drying, enabling easy movement and transfer of the corn or drier; (4) continuous flow dryer which dries the corn as a continuous stream of corn moves through the drier. Corn for seed should not be dried at a temperature over 100–110° F., for processing this temperature should not exceed 140° F., while corn for feed can be dried at temperatures of up to 200° F. with no apparent reduction in feed value.

Corn for feed containing 20 to 30 per cent moisture may be treated by dryeration. In this process the corn is dried down to 16 per cent moisture at temperatures of about 200° F., then held for "tempering" (adjustment of moisture within the kernel). Then the grain is allowed to cool, after which it is dried for 10 hours down to 14–15 per cent moisture and again allowed to cool. Grain with 25 per cent moisture or less may be dried with a strong blast of unheated air.[4]

Corn can be shelled when it contains 27 per cent moisture or less, but the shelled grain cannot be stored safely in bins until it is dried down to less than 13 to 14 per cent moisture. Most of the husked ears are stored in cribs until dry enough for safe storage of the shelled grain. Drying often is hastened by ventilating the ears in the crib with forced air, either heated or unheated (Figure 12–16). Most cribs are 4 to 10 feet wide, about 8 to 10 feet high, and of any convenient length.[49] For natural drying, the recommended maximum width for corn cribs for the Corn Belt ranges from 6 feet in the northern part to 9 feet in the southwestern part.[94] Corn cribs have slatted sides and ends to provide free circulation of air. Wet corn that contains more than 30 per cent moisture is likely to spoil when ears in cribs or piles are more than 2 feet from the open air. For a crib wider than 4 feet, natural drying is aided by inserting an A-shaped or rectangular ventilator in the middle of the crib with the ventilator extended the full length of the crib. In the Great Plains region much of the husked ear corn is stored in temporary outdoor cribs constructed of woven wire fencing or picket snow fencing.

Damp ear corn, with a grain moisture content exceeding 20 per cent or more, should be dried artificially for safekeeping. The screening out of shelled grain and the blowing out of husks and silks improves the circulation of air through the crib and hastens drying. Cleaning,

Figure 12–16. Drying ear corn in a crib with portable drier.

combined with forced air ventilation, usually takes care of corn that contains 25 to 30 per cent moisture.

When corn is killed by freezing before the grain is mature, and while it is high in moisture, the damaged damp grain is referred to as *soft* corn. Even after soft corn is dried, so that it will keep in storage, it is low in test weight per bushel. It also is shrunken or chaffy. However, such corn is nearly equal in feeding value (pound for pound) to plump sound corn. The lower starch content of soft corn makes it undesirable for processing. The best utilization of soft corn is to feed it promptly to avoid spoilage or the expense and labor of drying it.

Some corn is sold in the ear for immediate feeding before it is dry enough for shelled-corn bin storage. Such damp corn will not yield the 56 pounds of dry shelled corn normally obtained from 70 pounds of dry ear corn. Consequently the corn is sold on the basis of an agreed-upon number of pounds of ear corn to constitute a bushel. The relation between the moisture content of the grain and the number of pounds of ear corn required to produce 56 pounds of shelled corn of 15 per cent moisture content, as reported by the Missouri Agricultural Extension Service, is shown in Table 12–1.

TABLE 12–1. Weight of Ear Corn Required to Produce 56 Pounds of Shelled Corn at 15 Per Cent Moisture Content

MOISTURE IN GRAIN	EARS FOR 1 BUSHEL OF SHELLED CORN AT 15 PER CENT MOISTURE	MOISTURE IN GRAIN	EARS FOR 1 BUSHEL OF SHELLED CORN AT 15 PER CENT MOISTURE
(%)	(*lb.*)	(%)	(*lb.*)
15	70	45	108
20	75	50	120
25	80	55	132
30	85	60	149
35	92	65	170
40	100		

Sweetcorn is ready to harvest for the market or for canning when the grain is in the late milk stage. It then contains about 70 per cent moisture,[17] 5 to 6 per cent sugar, and 10 to 11 per cent starch. That stage is reached about the time the silks become brown and dry, which usually is about 21 days after silking. The corn remains in prime condition only about 2 days in hot summer weather and 5 days in cooler autumn weather. After that, the sugar content decreases, the starch content increases, the appetizing flavor is lost, and the pericarp becomes tough.[13]

CORN STOVER

The leaves and stalks contain about 30 per cent of the total nutrients in the corn plant.[47] Stover from corn husked or snapped from the standing stalk usually is utilized only for pasture. The harvested stover is used for feed or bedding. The best utilization of corn stover is obtained by putting the stalks through a husker-shredder. The coarse parts of the unshredded stalks are not consumed, but they furnish bedding and can be returned to the soil mixed with manure. Even where all the corn stalks are returned to the land they do not maintain the soil organic matter, except on heavily fertilized fields with high plant populations.[101]

CORN FODDER

About one per cent of the domestic corn acreage is cut for fodder, grazed, or hogged off. Some corn is cut and fed green. The feeding of corn fodder is popular in the semiarid as well as in the northern border sections where corn often fails to mature grain. Such stalks are more palatable and higher in protein than when they have produced mature ears.[57]

Old practices of gathering fodder by hand by stripping the leaves or topping the stalks above the ear before the kernels are mature checks grain development and reduces yields.

CORN SILAGE

The most complete utilization of the corn plant is as silage, cut when the kernels are in the glazed stage and dented, which is before a serious loss of leaves occurs. The yield may be 20 per cent less if the corn is ensiled when only half of the kernels are dented. Up to two-thirds of the digestible nutrients of corn silage is in the grain. Corn that fails to produce grain because of drought should be allowed to mature in the field as completely as possible without undue loss of leaves. Nearly all of the corn for silage is harvested with a field silage cutter or a forage harvester (Figure 12–17).

Figure 12–17. One-row tractor-drawn field silage cutter. [Courtesy International Harvester Co.]

Corn harvested for silage, because of drought stress, from soils high in nitrates and low in magnesium may subject cattle to nitrate poisoning (grass tetany) which sometimes is fatal to livestock.

HARVEST BY LIVESTOCK

A small acreage of corn is hogged or grazed, mostly in the northern Great Plains. Corn may be pastured by cattle, sheep, or hogs, but cattle usually waste a large amount of corn unless they are followed by hogs. A poor crop or a shortage of labor may make it advantageous to pasture the crop. Cornfields often are pastured after harvest.

SEED CORN

Seed corn should be placed where it will start to dry the day it is gathered from the field. A free circulation of air is essential to avoid danger of molds and to prevent heating when the moisture content is high. The crop is dried artificially by a forced draft of heated air by most growers of hybrid seed corn. A reduction in the moisture content of the seed to 12 or 13 per cent at a temperature range of 105° to 110° F. does not reduce the viability.[55]

High-moisture corn exposed to freezing is likely to be reduced in viability. Death from freezing is directly related to the moisture content of the corn kernel and the duration of the exposure to cold.[60] Corn that matures in a natural way becomes progressively tolerant to freezing as the moisture content diminishes. Corn with 10 to 14 per cent moisture will stand the most severe winter temperatures without injury to germination.

Well matured seed corn kept dry retained its viability satisfactorily up to four years under Nebraska conditions.[54] Old seed should be tested for germination before it is planted. In Illinois second-year seed was equal to the first, but the yield from three-year-old seed dropped 7.8 per cent, while the decline in yield from seed more than six years old was extremely sharp. The decreased yields from old seed were due to weak plants as well as to low germination.[20]

Seed Selection

The ideas of the earlier corn growers differed widely as to what constituted the best type of ear to save for seed. This dilemma is eliminated when hybrid seed is used because all ears in a pure hybrid have the same inherent productive capacity, if they are sound, viable, and free from disease. When corn shows came into prominence about 1900, judges formulated the corn scorecard as a guide to excellence. The ear was logically regarded as a thing of beauty but, unfortunately, characteristics associated with beauty were erroneously regarded as important from the production standpoint. Physical differences among good seed ears such as number of kernel rows, weight per kernel, ratio of tip to butt circumference, shelling percentage, and filling of tips and butts are valueless as indications of relative productivity.[64, 80] However, yield sometimes is associated with weight of ear and length of ears.[87, 89] In general, ears that are heavy because they are long are more likely to be productive than are heavy ears of a large circumference. Ears with heavier cobs, fewer rows, and heavier kernels with rounded corners and smoother indentation are more productive than the old standard show type.[12] Extremely rough, starchy ears are more susceptible to the rot diseases than the smoother, more flinty types.

Close selection for a particular set of characters of a uniform type tends to reduce yields because of a decrease in vigor similar to that obtained by inbreeding.[24]

Field selection of sound seed ears from healthy vigorous plants at about the average date of the first fall frost appears to be the most practical method for yield maintenance or improvement of an open-pollinated variety.

Corn improvement by ear-to-row tests of selections was introduced by C. G. Hopkins at the Illinois Station in 1897 in experiments designed to modify the protein and fat content of the corn kernel. Later, extensive experiments failed to show yield increases by ear-to-row selection, particularly where practiced continuously.[51]

Hybrid Seed Corn

About 98 per cent of the corn acreage of the United States was planted to hybrid corn by 1965 and almost all of it was in hybrids by 1973. In 1933 only 0.1 per cent of the corn acreage was of hybrids. This increased to 14.9 per cent by 1938, and the expansion thereafter was rapid. Since 1950, hybrid seed corn has been produced in the Soviet Union, Yugoslavia, Italy, France, Egypt, India, South Africa, Mexico, Argentina, Chile, and many other countries.

First-generation hybrid seed for growing the hybrid crop must be obtained each year. Yields are reduced when seed is saved from the hybrid crop. The yield of the second generation of double crosses is about 84 per cent of that of the first generation, or about the same as open-pollinated corn.[77, 88]

The chief concern of the hybrid-corn grower is to obtain seed of the hybrid best suited to his conditions. The published results of state yield tests are a valuable guide in making the proper selection. According to Jenkins,[41]

> Corn hybrids are as specific in their adaptation to the soil and climate as open-pollinated varieties and will perform in a satisfactory manner only when grown under the conditions to which they are suited.

The first four hybrids released by the Iowa Agricultural Experiment Station yielded 11 to 26 per cent more than the average of open-pollinated varieties in the tests from 1935 to 1940.[95] In Minnesota, hybrids yielded 9 to 12 per cent more than the check variety.[29] The better-recommended hybrids distributed for commercial production have given yield increases that range from 15 to 35 per cent above open-pollinated varieties. Many superior hybrids are also outstanding for their strength of stalk and lodging resistance.

The advantages from the use of hybrid seed are greatest where yields are high. For example, a 20 per cent increase above a 75-bushel yield is five times as valuable as the same percentage increase above varieties grown where they yield only 15 bushels an acre. Hybrids that shed their pollen during a hot dry period may give very low yields, whereas open-pollinated varieties, which are irregular in pollen shedding, may partly escape the adverse weather period and produce a better yield. Corn hybrids were first developed under favorable growing conditions. Hybrids designed especially for adverse conditions later extended the adaptation of hybrid corn.

Production of hybrid seed corn requires the use of well over 200,000 acres to produce 17 million bushels of hybrid seed for planting. About 16 million bushels are required for planting 69 million acres in the United States. Workers are employed for a period of two to three weeks to detassel the seed-producing rows. Large high-framed motor-propelled carriers permit the crews to ride along the rows while they pull out the tassels before the tassels shed pollen.[1] Four to six rows are thus detasseled in one trip across the field. The use of male-sterile lines eliminates much of the detasseling.

The general methods of developing and producing corn hybrids[29] consist essentially of (1) isolation by self-fertilization and selection of lines that breed more or less true for certain characters, (2) determination of the inbred lines with best combining ability, and (3) utilization of these selfed lines in the production of hybrid seed.

Inbred lines are produced by self-pollination of plants selected from among hybrids or varieties. In self-pollinating or selfing, the pollen from the tassel is placed on the silks of the same plant under controlled conditions. This is done by placing paper bags over the tassels. The ear shoots are covered with small bags before the silks extrude from the husks, and usually the silk is trimmed so that it will grow out as a uniform brush. Usually 1 or 2 days later the pollen is jarred into the tassel bag and then poured on the silks, and the ear shoot is then covered with a large tassel bag. The seed from the better selfed ears is planted in progeny rows. Self-pollinations are then made on several plants in each row, and selection is continued. The vigor of these selfed lines is much less than that of the parent field variety, but the inbreeding process reveals many undesired abnormalities that can be eliminated. Such abnormalities include sterility, excessive suckers, tassel ears, silkless ears, susceptibility to diseases, and lethal seedlings. After five to seven years of repeated self-pollination and selection the inbred lines breed relatively true for most plant characters.

Frequently, after one to three years of inbreeding, some of the seed of each line is planted in a special crossing block with every fourth, third, or alternate row planted to an adapted corn variety or hybrid to be used as a pollen parent. The tassels are removed from all the inbred rows before they shed pollen so that all plants are pollinated by the one-parent variety. This top-crossed seed produced on each row is planted separately to determine which inbred lines have the best general combining ability, i.e., are likely to produce the highest yields in subsequent crosses. The better lines as shown by the top-cross test are retained and inbred until uniform. Most of the inferior inbred lines meanwhile have been discarded.

Small groups of 10 to 15 high combining lines are then intercrossed in all possible single-cross combinations. A total of 45 single crosses* is obtainable with 10 lines, and 105 with 15 lines. These single crosses are tested thoroughly for yield, grain quality, and other characteristics. The performance of all possible double crosses may be predicted from the performance of the single crosses. Six different single crosses are obtainable with any four inbred lines, e.g., A, B, C, and D. These are $(A \times B)$, $(A \times C)$, $(A \times D)$, $(B \times C)$, $(B \times D)$, and $(C \times D)$. The predicted yields of the double cross $(A \times B) \times (C \times D)$ is computed as the average of the yields of the four nonrecurrent single crosses, $(A \times C)$, $(A \times D)$, $(B \times C)$, and $(B \times D)$. Seed of only the double crosses predicted to be best is then produced for field testing.† The double-crossed hybrids which prove most productive and otherwise desirable are made commercially available for farm use.

A single cross involves a cross of two inbred lines, e.g., $A \times B$. Commercial single crosses (Figure 12–18) of field corn were not popular before 1963 because the seed yield is low and the seed often is small. The three-way cross is made from a single cross and an inbred line, i.e., $(A \times B) \times C$. The double cross is a cross between two single crosses, for example, $(A \times B) \times (C \times D)$. The commercial field-corn hybrids in production are double crosses, three-way crosses, or single crosses.

For commercial hybrid-seed production the single crosses or inbred lines are grown in alternate rows or groups of rows in isolated fields. In these crossing fields all the plants of the seed (or female) parent strain are detasseled before they shed any pollen except when male-sterile lines are used. The male or pollen parent rows are not detasseled, and the ears are not gathered for seed. The ratio of male to female rows depends upon the region and the cross to be made, and particularly on the abundance of pollen produced by the male parent. Inbred lines are usually low pollen producers. When they are used to

* This number is determined by the formula $n(n-1)/2$, in which n is the number of inbred lines.

† The number of possible double crosses is too large to permit field testing of all of them. The number, which is 630 for 10 inbred lines and 4,095 for 15 inbred lines, is computed by the formula $n(n-1)(n-2)(n-3)/8$, in which n equals the number of inbred lines.

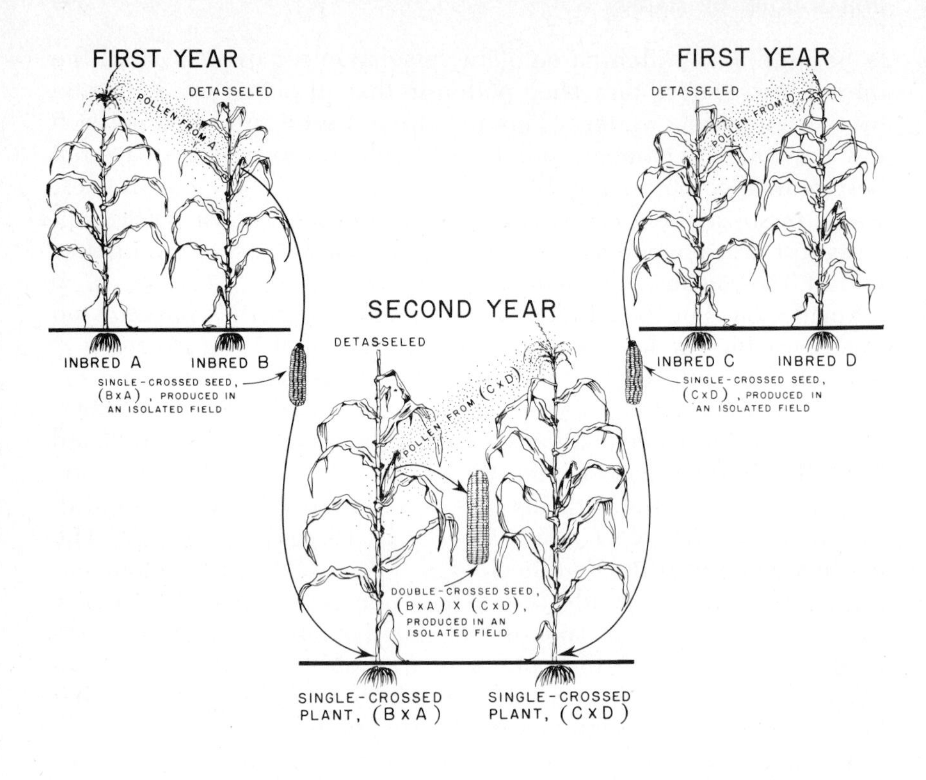

Figure 12–18. (*Top*) Method for producing double-cross hybrid seed from four inbred lines of corn. (*Bottom*) Six seed parent (detasseled) rows alternate with pairs of rows of the pollen parent to produce hybrid seed.

supply the pollen, one row of the male parent is generally planted to every two rows of the female parent. When single-cross plants supply the pollen, either one row of male parent to three or four rows of female parent, or a two- and six-row combination (Figure 12–18) is satisfactory under favorable climatic conditions.

Inbreeding experiments with maize were started by E. M. East and G. H. Shull before 1905. In 1908 Shull suggested use of hybrids between inbred lines as a basis for practical corn improvement. The use of double crosses, or crosses between crosses, was advocated by D. F. Jones in 1918. The first commercial production of double-cross hybrid corn was in 1921.

USES OF CORN

About 45 per cent of the corn grain produced in the United States from 1970 to 1972 was kept on the farms on which it was produced for use as feed and food. Much of that sold, which includes by-products, also is used for feed. The approximate domestic utilization of field corn in 1970 was as follows:

USE	MILLIONS OF BUSHELS
Feed	3526
Wet-process milling	229
Dry milling	119
Alcohol and spirits	24
Breakfast foods	22
Seed	17
Total	3937

Corn used for feed is for hogs, cattle, poultry, horses, and sheep in that order. Waxy corn is superior to common field corn in weight gains of steers and lambs; however, reduced yields are not offset by corresponding increases in weight gains.

Exports of corn from 1958 to 1971 ranged from 3 to 750 million bushels annually.

A minor but interesting use of the corn plant is in the manufacture of corncob pipes. Much of this industry is localized near Washington, Missouri, where special hybrids of Missouri Cob Pipe corn are grown for their large thick cobs. In this case the grain is the by-product.

At times large quantities of corn stalks have been used in manufacturing cellulose products, which include plastic panels. This industry provides a market for the corn stalks but deprives the land of needed organic residues. Pulverized corncobs are used extensively as a mild abrasive for removing the carbon from airplane motors. Corncobs occasionally are used for home fuel or for other purposes.

TABLE 12–2. Human Consumption of Grain and Grain Products in the United States (Pounds per Capita for Civilian Population, 1972)

PRODUCT	WHEAT	CORN	BARLEY	OATS	RYE	RICE
Flour	111	–	–	–	1.2	–
Meal	–	7.0	–	3.2	–	–
Breakfast food	2.9	1.8	–	–	–	–
Sugar & sirup	–	20	–	–	–	–
Starch	–	1.9	–	–	–	–
Other	–	10.0	1.2	–	–	7.4
Total	120	35.7	1.2	3.2	1.2	7.4

The consumption of corn and other grains for food in the United States is shown in Table 12–2.

White corn usually is used for toasted breakfast foods made from corn because the yellow endosperm pigments turn too dark unless heated only slightly. Processors pay a substantial premium for white corn of suitable quality. A shift from white to yellow corn occurred about 1920 after the discovery of the higher vitamin-A value of the latter type. A majority of the consumers in the North long ago showed strong preference for corn meal from yellow corn[2] because it "looks richer." This choice was further enhanced by the discovery of the difference in vitamin values. Consumers in the South, on the other hand, use white corn meal almost exclusively because it "looks purer," although they consider yellow corn very appropriate for feeding hogs.

Corn meal has never been a generally acceptable food to a majority of northern Europeans even in times of food scarcity.

MILLING OF CORN

Before milling, the corn grain is cleaned by aspiration, sieving, and by electromagnetic removal of any foreign metal fragments. Dry milling is of two types called (1) new process, and (2) old process or water ground.[2] In the milling by the new process the grain is tempered to a 20 per cent moisture content; passed through a degerminator to loosen the germ and hull; and passed through a hominy reel after partial drying. The hull or bran, flour, and germ are removed by sifting and aspiration.[96] The degermed grain may be used for hominy or else ground further into grits or corn meal. Large particles of endosperm are called samp or pearl or cracked hominy. When coarsely ground, the particles are called hominy grits. The bran is used for feed. The germ is pressed to extract the corn oil and the residue used for feed. Grits are used mostly for food and in brewing. The fine particles of corn flour, which come mostly from the soft part of the endosperm, are used for food or industrial purposes. A bushel of corn yields about 29 pounds of grits and meal, 4 pounds of flour, 1.6 pounds of oil, and 21 pounds of feed.

Old-process corn meal is ground between rolls or stone burs usually in small mills, many of which formerly were, or still are, run by water power. The coarser bran particles sometimes are sifted out, but most of the germs remain in the corn meal. Many people relish the flavor of this old-fashioned product, but its keeping quality is impaired by the presence of the oil in the germ which becomes rancid after prolonged storage. Two bushels (112 pounds) of corn yield about 100 pounds of old-process corn meal.

Lye hominy is made by a process learned from the Indians. The grain is soaked in hot water that contains lye or a bag of wood ashes until the hulls are loosened. The hulled grains are washed to remove the alkali.[2]

In wet milling the grain is steeped for about 2 days in warm water that has been treated with sulfur dioxide to prevent fermentation.[3, 4, 66, 96, 111] Then the softened grain is passed through degermination mills that crack the kernel and loosen but do not crush the germ. The mass then passes into germ-separation tanks where it is flooded with *starch milk*, and the light, loosened germs floated off. The remainder of kernel is then ground fine and the particles of hull are sifted out. The starch is separated from the protein (*gluten*) by letting the mixture, now suspended in water, flow down long tables. The particles of starch, being heavier, are freed from adhering bits of gluten as they roll over and over, until they finally settle out and come to rest on the bottom of the table while the gluten passes on over the end. The modern plants separate the starch in a centrifuge. The starch from the tables is washed, filtered, and then dried in kilns. The gluten is dried and used mostly for feed. Zein, a purified alcohol-soluble protein of corn, is used in making certain plastics and paints.

The cornstarch largely is used directly for food, textile and paper sizing, laundry starch, dextrines, and adhesives. Some cornstarch is used in making cores for foundry molds. Large quantities of starch are hydrolized into glucose (corn sugar), corn sirup, and dextrines by being heated in a dilute solution of hydrochloric acid. The converted solution is then blown into a neutralizing tank that contains sodium carbonate. The glucose is crystallized out of the solution much as cane and beet sugar are crystallized. A bushel of corn (56 pounds) containing 16 per cent moisture yields in the wet-milling process about 1.6 pounds of oil and 35 pounds of starch (or 27.5 pounds of glucose or 40 pounds of sirup). The starch of waxy corn is used for food or for making adhesives such as the gums applied to stamps, envelopes, gummed tape, corrugated paper boxes, and plywood. It can be substituted for tapioca starch for food purposes. The amylose fraction of the starch can be made into films and fibers. Inbred lines that are high in amylose content have been developed.

Corn-steep water has been used in the culture medium for growing *Penicillium notatum*, the organism that produces the antibiotic drug penicillin.

MAKING CORN WHISKY

In the manufacture of alcohol or whisky from corn the grain is cleaned and then either ground or exploded by steam, and the germ then removed. The remaining mass, mixed with water to form a slurry, is heated with steam to gelatinize the starch. The resulting material is cooled to a mashing temperature of about 145° F. and malt (usually barley malt) is added. The diastase in the malt converts the starch to sugar. The solution (wort) is drawn off, yeast is added, and fermentation allowed to proceed. The resulting alcohol (or whisky) is recovered by successive distillations. Whisky is aged in oaken barrels that often are charred on the inside. During aging objectionable substances are absorbed by the wood or charcoal, whereas the whisky absorbs colors and flavors from the wood. Bourbon whisky is made largely from corn; rye whisky from rye; and Scotch whisky chiefly from barley. Other grains or potatoes may supply part of the starch for any type of whisky. Damp heating grain can be used for making industrial alcohol.

The dried residue known as distillers' grain is used for feed. Distillers' slops or distillers' solubles likewise may be dried and used for feed.

POPPING CORN

When popcorn grain that contains 10 to 15 per cent moisture is heated sufficiently, this moisture, which is confined in the colloidal protein matrix in which the starch grains are imbedded, is converted into steam. The steam pressure finally is released with explosive force and the grain pops as the starch grains expand. Good popcorn expands 24 volumes or more in popping. Poorer lots may show an expansion as low as 15 volumes. Popcorn that contains less than 8 per cent or more than 15 per cent moisture gives only about half as large a volume on popping as that with 12 to 13 per cent moisture.[110] Corneous types of grain sorghum also can be popped. Sweetcorn and field corn can be expanded and slightly popped by parching in a kettle with hot edible oils.[107]

PROCESSING SWEET CORN

More than 2 million tons of unhusked fresh ears of sweet-corn were processed in the United States in 1972. Nearly three-fourths of this was canned and the remainder was frozen. The ears of sweetcorn used for processing are snapped closely so as to leave little or no shank. The ears usually are snapped in the morning and hauled to the cannery as promptly as possible, because sweetcorn deteriorates rapidly after picking.[7, 16] The grains may lose half their sugar as well as much of their flavor within 24 hours. Piles or truckloads of fresh corn will heat unless they are handled soon. At the processing plant the corn is husked mechanically, sorted, trimmed, and cleaned. The whole kernels may be cut from the cobs by mechanical knives or they may be cut across, the remaining contents scraped out, leaving part of the pericarp on the cob. The latter method produces Maine-style or cream-style canned corn. A ton of ears yields between 600 and 900 pounds of cut corn. The residues consist of husks, cobs, shanks, and silks, which usually are chopped and made into silage either at the cannery or back on the farm of the grower. Most of the processed crop is grown under contract, with the company supplying the seed and notifying the grower when his corn is ready to be gathered and processed. At the stage for best quality the dry weight of the kernels is not more than one half that of mature sweetcorn.

Some of the sweetcorn is frozen and marketed as husked ears.

COMPOSITION OF THE CORN KERNEL

The average composition of the field corn kernel is approximately as follows:

SUBSTANCE	PER CENT		
Water	13.5		
Protein	10.0		
Oil	4.0		
Starch	61.0		Carbohydrates
Sugars	1.4		Carbohydrates
Pentosans	6.0		Carbohydrates
Crude fiber	2.3		Carbohydrates
Ash	1.4		
Potassium		0.40	
Phosphorus		.43	
Magnesium		.16	
Sulfur		.14	
Other minerals		.27	
Other substances (organic acids, etc.)	0.4		
Total	100.0		

The germ contains about 35 per cent oil, 20 per cent protein, and 10 per cent ash.

The comparative vitamin content of yellow corn and wheat is about as follows:

VITAMIN	MILLIGRAMS PER POUND	
	Yellow Corn	*Wheat Grain*
Vitamin A	1990.00	86.00
Thiamin	2.06	2.25
Riboflavin	.60	.51
Niacin	6.40	27.34
Pantothenic acid	3.36	5.83
Vitamin E (as alpha tocopherol)	11.21	16.88

White corn has the same general composition as yellow corn except in being practically devoid of vitamin A. Corn is relatively low in niacin (nicotinic acid), the vitamin that checks pellagra. This disease, which causes a dermatitis (inflamed skin), has occurred in the South among people living on diets that included considerable quantities of corn meal and hominy grits. Consumption of large quantities of corn may leave symptoms of niacin deficiency even with an abundance of supplemental niacin in the diet because of the deficiency of an amino acid, tryptophan, in corn. Lines of corn that are high in lysine are being developed. Corn is higher in oil and somewhat lower in protein than is wheat grain.

In general, about 450 to 500 pounds of corn produces 100 pounds of pork when fed to growing pigs. There is a close relationship between the prices of corn and hogs. Under average free-market conditions the price of 100 pounds of hog has been about equal to the cost of 11.4 bushels of corn. When the price of corn rises so that the hog price per 100 pounds will purchase less than 11 bushels of corn, hog feeding soon decreases.

The relative nutritive value of corn and other grains is indicated in Table 12–3.

The true feeding value of corn is not evident from the figures shown, because corn is higher in fat than the other grains. The fats have a calorific value about 2¼ times as high as that of carbohydrates because of a reduced oxygen content. Aside from this difference the caryopses of the grains are similar to corn in composition and feeding

TABLE 12–3. Nutritive Value of Corn and Other Grains

TEST ANIMAL	PER CENT OF TOTAL DIGESTIBLE NUTRIENTS IN								
	Barley	*Buck-wheat*	*Corn*	*Grain Sorghum*	*Proso or Millet*	*Oats*	*Rice (Rough)*	*Rye*	*Wheat*
Cattle	–	–	78	78	–	72	–	–	–
Sheep	77	62	74	79	69	76	71	79	76
Swine	74	69	80	77	74	74	–	–	73
Poultry	68	65	80	81	72	62	65	60	73

value. Barley, buckwheat, proso, oats, and rice grains are enclosed in hulls composed chiefly of celluloses, lignin, silica, and other constituents of limited nutritive value.

The hull constitutes approximately 13, 22, 22, 28, and 20 per cent, respectively, of barley, buckwheat, proso, oats, and rice.

Dependable high-protein corns are not yet available on the market. The protein content of the opaque-2 type corn is no higher than normal corn, but the higher percentage of lysine results in the increased feeding quality.

DISEASES

The diseases of corn have been described by Dickson,[16] Hoppe,[39] Robert,[90] and Ullstrup.[103–06]

Corn Smut

Corn smut, caused by the fungus *Ustilago maydis* is the most widely recognized disease of corn. It causes complete barrenness in many plants[23] and reduced grain development in others. The average yield reduction from smut was 30 per cent in experiments in Minnesota.[45] Galls on the ear are most destructive while those on the stalk or leaves above the ear cause more damage than when they attack the lower portion of the plant (Figure 12–19).

The disease appears as galls of various sizes on the aerial parts of the corn plant. The galls are whitish at first but become dark with the development of the black smut spores inside the white membrane. At maturity the galls break and scatter the powdery spores. Galls may appear at almost any place where meristematic tissue occurs, but commonly they are found near the midribs of the leaf or at the nodal buds on the stem and on the ears. Vigorously growing plants are the most likely to be attacked. The smut is not poisonous to livestock. In fact, young galls cooked and eaten form a palatable substitute for mushrooms. Indians made this discovery.[8]

The smut organism lives over in soil. Consequently, seed disinfection is ineffective except for preventing the spread of the disease to new localities. Future infection may be reduced by systematic destruction of smut, by crop rotation, and by refraining from applying infested manure to land that is to be planted to corn. Smut-resistant inbred lines have been developed, and some hybrids involving these lines partly escape smut injury. Doubtless highly resistant hybrids eventually will be developed.

Head Smut

Head smut is a distinct disease that produces galls on the ear and tassel, entirely destroying these parts. The disease occurs in local

Figure 12–19. Galls of common corn smut.

areas of the Southwest and Pacific Coast states and rather rarely elsewhere. The fungus *Sphacelotheca reiliana* lives over in the soil or on the seed. It enters the young plant and grows into galls in the developing ears or tassels. The galls break and discharge the powdery spores in the wind. The disease can be controlled when infested land is not planted to corn for two years and the seed is treated so that the disease is not carried to new fields.

Root, Stalk, and Ear Rots

Corn rots may bring about reductions in field stand, reductions in vigor of plants that survive, chlorosis, barrenness, general blighting of the plant, and rotting of ears in the fields. Among the organisms that cause seedling blights are the ear-rot fungi *Gibberella zeae, G. fujikuroi* (*Fusarium moniliforme*), and various species of *Aspergillus, Penicillium,* and *Pythium.* Root rots are caused by *Pythium arrhenomanes, Gibberella zeae, G. fujikuroi,* and *Diplodia zeae.* Stalk rots are caused by *Diplodia zeae, Gibberella zeae, Macrophomina phaseoli* (*Sclerotium bataticola*), and occasionally by *Gibberella fujikuroi,* and *Nigrospora sphaerica* (*Basisporium gallarum*), and by the bacterium *Phytomonas dissolvens.* Ear rots are caused chiefly by *Diplodia zeae, D. macrospora, Gibberella fujikuroi, G. zeae, Nigrospora sphaerica, Rhizoctonia zeae,* and various species of *Penicillium* and *Aspergillus.*

The ear-rot fungi may continue to grow and damage corn after it is harvested. Molding of corn in storage may result from the growth of *Penicillium, Aspergillus,* and *Fusarium* species when moisture content of the grain exceeds 14 per cent. A rot and darkening of the germ caused by species of *Penicillium* causes the condition called blue eye. Nearly all these organisms are carried in the soil or on decaying corn residues. Most of them also are carried on the seed. The molds are evident on the ears. The dark fruiting bodies of certain stalk and root rots are often seen either on the outside or inside of old corn stalks.

Diplodia develops abundantly on infested kernels, and weak infected plants die following the rotting of roots at the crown. The subcrown internodes of plants grown from *Diplodia*-infected seed usually appear dry and brown in contrast to the white tissues in the normal plant. There is little evidence that this fungus advances up the stalk from the rotted roots and rotted crown. Ears may be infected through the shank, and may be reduced to a char-like mass. Ears conspicuously rotted with *Fusarium* may be recognized by the characteristic pinkish color of the kernels. *Gibberella* causes both root rot and seedling blight. Charcoal rot (caused by *Macrophomina phaseoli*) destroys the pith in the roots and the lower part of the stalks, leaving the fibrous strands carrying black spores. This disease, which kills the stalks prematurely and causes them to fall down, is most serious under dry-soil conditions.

The seedlings diseases and some of the rots are best controlled by the use of sound, undamaged, disease-free seed and by seed treatment. The corn should be planted after the soil is warm. Some of the stalk rots are checked when the soil nutrient elements are properly balanced by suitable fertilization. The planting of resistant hybrids reduces damage from these diseases.

Chemical seed-corn disinfectants, such as captan and thiuram, often increase the yield, stand, and vigor through control of the rot diseases when damaged or infested seed must be planted. Especially favorable results from seed treatment have been obtained on corn planted early, which often rots in cool, waterlogged soil before it can germinate. West of the Missouri River where conditions are drier, and rotting organisms less prevalent, significantly increased stands or yields as the result of seed treatment usually are not obtained.[53, 72, 73]

Leaf Diseases

The chief leaf diseases of corn are Northern leaf blight (*Helminthosporium turcicum*), Southern corn leaf blight (*Helminthosporium maydis*), Corn rust (*Puccinia sorghi*), Maize dwarf mosaic, and corn stunt virus.

Northern leaf blight is characterized by infection of the lower leaves and development of boat-shaped grayish lesions which spread to other leaves in cool damp weather. Leaf death may occur. Control of the disease is usually economical only in breeding nurseries, where the value of the crop justifies treatment with an acceptable fungicide. While this disease is usually most severe in the East and South, it has caused serious damage as far West as Nebraska. Resistant hybrids help to control the disease.

Southern corn leaf blight formerly was considered less of a problem than Northern leaf blight. However, in 1970, a severe infestation of this disease occurred throughout the major corn-producing areas of the United States.

Especially susceptible to the disease were hybrids in which the female (seed) parent carried the Texas cytoplasmic male sterility. This disease organism needs warm moist weather to flourish which is why it is more frequently a problem in the South. Lesions form on the leaf as well as the ear shanks and husks, gradually infecting the ear in severe cases. The major control measure is to plant corn without the Texas sterile cytoplasm.

Corn rust caused by *Puccinia sorghi* and southern rust caused by *P. polysporia* are characterized by small reddish spots on the leaf and favored by humid weather. Resistant hybrids are available.

Maize dwarf mosaic virus (MDMV) and corn stunt virus (CSV) symptoms are the appearance of faint yellowish stripes on plants 6 to 7 weeks old and shortened internodes.[99] Yellowing leaves may eventually turn reddish purple. Excessive tillering with many grainless ears may also be observed along with extra long ear shanks.

Maize drawf mosaic was first identified in the Corn Belt in 1963. It occurs from Pennsylvania southwest to Texas. It also attacks sorghum, Johnsongrass, and other grasses. Aphids, particularly the corn leaf aphid (*Rhopalosiphum maydis*), serve as vectors in transmitting the disease. Resistant lines are available.

The corn stunt virus disease occurs in the southern states from the southeast to California. At least three species of leafhoppers serve as vectors.

Wheat streak mosaic virus (WSMV) also attacks corn in the Pacific

Northwest and the Central States and in Ontario, Canada. The vector is the wheat curl mite (*Aceria tulipae*).

Sugarcane mosaic attacks corn from Louisiana to California. It is transmitted by several aphid vectors.

Bacterial wilt or Stewarts disease, caused by *Bacterium stewarti*, sometimes incurs heavy damage especially to sweetcorn. Brown spot, caused by the fungus *Physoderma zeamaydis* attacks maize in the southern states.

INSECTS

The insects that attack corn are reviewed by Dicke.[15]

European Corn Borer

The European corn borer (*Ostrinia nubilalis*) is one of the most serious pests of corn in the United States. The yields of hybrid corn are reduced about 3 per cent for each borer per plant.[82] Losses of market sweetcorn are about 8 per cent per borer per stalk. From 1917, when it was first observed in Massachusetts and New York, to 1946, the European corn borer had migrated as far as Kansas, Nebraska, and the Dakotas. It reached the Rocky Mountains by 1950. The larvae bore within the stalks, tassels, cobs, and ears of corn, eating as they go. The mature larva is about 1 inch (25 mm.) long, with a brown head, a grayish to pinkish body with two dark-brown spots on the back of each body segment (Figure 12–20). The insect spends the winter as a full-grown larva in the stalks, stubble, and cobs of corn or in the stalks of other coarse-stemmed crop or weed plants. The larvae begin to pupate from April to July and emerge as moths two or three weeks later. The period of pupation is early in central latitudes and progressively later to the north. However, the multiple-generation strain of the borer pupates earlier than does the single-generation strain. The moths choose the larger plants upon which to deposit the eggs.

Control measures include complete destruction of all waste plant materials around the cornfields and plowing so as to cover all plant residues in the fields. Planting as soon as the soil is warm enough to permit rapid growth improves the vigor of the plants and helps them to survive corn-borer attack. Extremely early or late planting should be avoided. Locally adapted stiff-stalked, nonlodging, disease-resistant hybrids should be planted.

Sweetcorn or hybrid seed corn can be protected by insecticides with repeated applications at five-day intervals. Treatment should begin when the first eggs start to hatch. Control with insecticides in corn for grain is advisable when three-fourths of the plants show first-generation larvae feeding in the whorl (Figure 12–21).

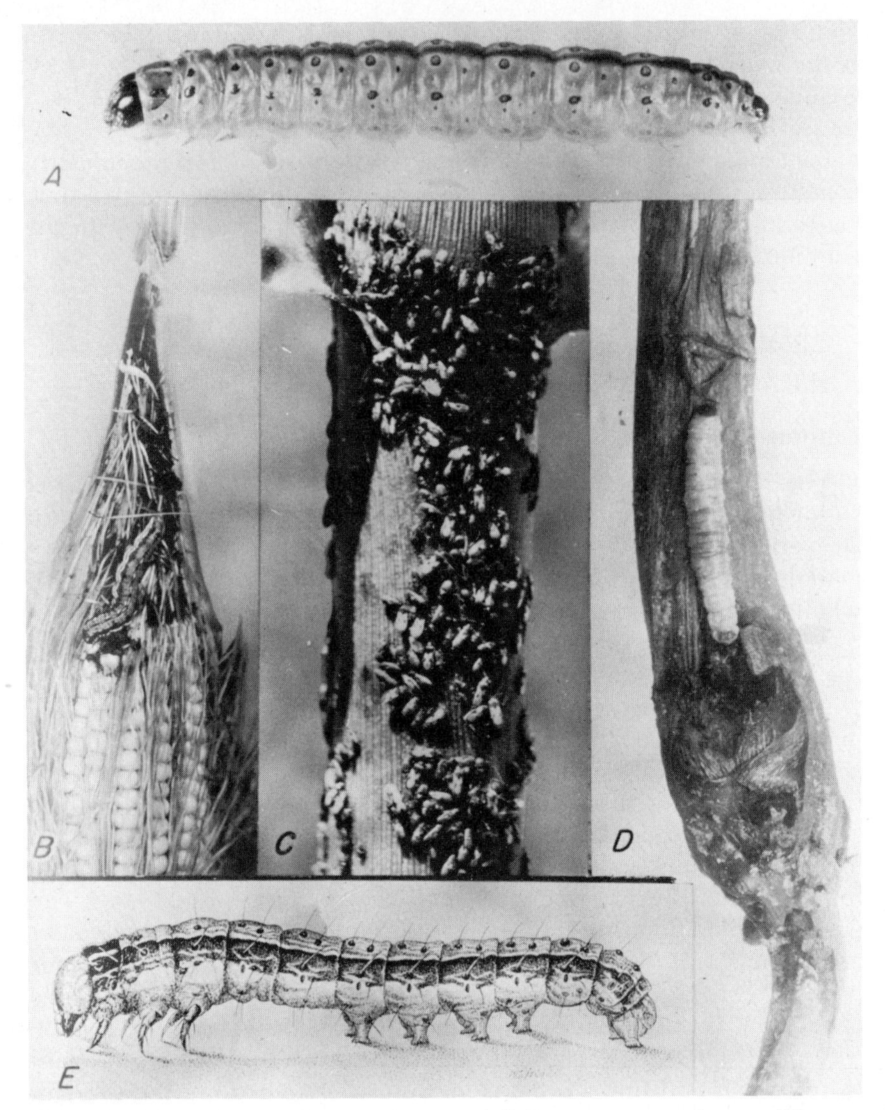

Figure 12–20. Insects that attack corn: (*A*) European corn borer, (*B*) corn earworm, (*C*) chinch bugs, (*D*) corn rootworm, and (*E*) fall armyworm.

Southwestern Corn Borer

The southwestern corn borer (*Diatraea grandiosella*) is a serious pest of corn under some conditions. The damage it does is similar to that produced by the European born borer. Stalks containing several borers often break over and may fail to produce ears. A native of Mexico, the present range of this insect extends from Arizona to the south central and southeastern states.

The larva of the southwestern corn borer is about 1 cm. long and is nearly white, with brown spots scattered over the back. The form that occurs on corn roots in the winter is creamy without the brown spots.

The best method of avoiding loss from this insect is to replace corn with grain sorghum, which suffers much less injury. Midearly planting of corn avoids some damage. The first and second generations of the borer can be controlled by two applications of insecticides 10 days apart.[30]

Chinch Bug

The chinch bug (*Blissus leucopterus*) (Figure 12–20) is so destructive that in some cases it destroys entire fields. The greatest injury occurs when the corn is planted late, adjacent to a field of barley or other small grain. The bugs overwinter in tall grasses, breed in the small grains, and migrate into the corn on foot in June or early July when the small grains ripen and lose their succulence. Chinch-bug losses are greatly reduced by spraying border strips 4 rods wide around the cornfield and also along any adjacent field of small grain. Barriers are ineffective when the bugs reach maturity in the small grains and fly into the corn or sorghum fields. Then the entire field may be sprayed with an insecticide.

The growing of resistant hybrids will retard damage from chinch bugs. Inbred lines from eastern and central Kansas and Oklahoma,

Figure 12–21. Field sprayer for use in controlling the European corn borer.

where chinch bugs usually are abundant, tend to be more resistant than those from the eastern Corn Belt.

Corn Earworm

The corn earworm (*Heliothis zea*) causes more direct damage to the ear of corn than does any other insect. About 2 per cent of the corn crop is destroyed annually by this pest. It occurs throughout the United States wherever corn is grown but is most destructive in the southeastern states. This insect also causes considerable damage to compact-headed grain sorghums. Under the name cotton bollworm it is a serious enemy of cotton. As the tomato fruitworm it eats into the nearly ripe fruits of the tomato plant. It also feeds on many other crops.

The number of generations of the corn earworm per year ranges from one in the extreme North and two to four in the Central States to as many as seven in the extreme South.

The corn earworm lives over winter in the pupal stage in burrows 1 to 9 inches deep in the soil. The moth (adult) emerges from these burrows in the spring and the female deposits about a thousand eggs on the leaves of corn or other plants. The larvae hatch in 2 to 8 days and begin feeding. The larva attains full size in 13 to 28 days during which time they molt (shed their skins) about five times, as they enlarge. The full-grown larva is about 1½ inches (38 mm.) long, of various colors, and often with conspicuous stripes (Figure 12–20). The eggs usually are deposited on the silks of corn. The young larvae eat their own way down to the ear.

The corn earworm devours developing kernels and fouls the ear so that molds develop. Its tunnels permit weevils to gain access to the ear within the husk. The larvae also feed on leaves, tassels, and silks of corn, and often destroy the young growing point or *bud* of the corn stalk.

The best prevention of corn-earworm damage consists of growing hybrids with long heavy husks. Good husk covering retards or prevents the young larvae from reaching the ear. The best protection is given by husks that extend 4 to 6 inches beyond the tip of the ear. Early planting partly permits corn to escape damage before the corn earworms become numerous. Early plowing destroys the burrows in which the pupae are hibernating and prevents emergence of many moths. Market sweetcorn and hybrid-seed fields, because of their high value, can feasibly be protected by the spraying of insecticides into the leaf whorls and later on the silks.

Corn Rootworm

Three species of the corn rootworm (Figure 12–20) attack corn in the United States. These are (1) the southern corn rootworm or 12-

spotted cucumber beetle, *Diabrotica undecimpunctata howardi*, which occurs in the South, the East, and the Central States; (2) corn rootworm, *D. longicornis*, of the Great Plains; and (3) the Western corn rootworm, *D. virgifera*, of the Mountain States. The larvae bore into bases of seedling plants and ruin the bud or growing point. They also feed on the roots of older corn, causing the plants to fall over. The two latter species can be kept under control by crop rotation with at least two years intervening between crops of corn on a particular field. Rootworms are controlled by applications of insecticides to the soil before or during planting. Hybrids with vigorous root formation are less likely to lodge when attacked by rootworms.

Grasshoppers

Several species of grasshopper feed on the foliage of the corn plant. When grasshoppers are very abundant they devour large plants, leaving only the bare stalks or, sometimes, only stubs in the field. Grasshoppers can be controlled with insecticides, preferably applied to the hatching areas when the nymphs are young.

Weevils

Weevils and other pests of stored corn and other grain were discussed in Chapter 9. The rice or black weevil (*Sitophilus oryza*) and the Angoumois grain moth or fly weevil (*Sitotroga cerealella*) attack corn in the field as well as after it is gathered and stored. These two insects breed in stored grain most of the year, but in the South the adults fly to the field and attack the grain when the corn is about in the roasting-ear stage.

The best protection against weevil damage to corn is to grow hybrids with long heavy husks (Figure 12–22). The Angoumois grain moth cannot penetrate the husk to deposit eggs on the grain. The rice weevil usually does not eat through the husk but attacks ears that have a poor husk covering or those with husks damaged by the corn earworm. In the South, corn often is snapped at harvest and stored in the husk until time to feed the grain. Good undamaged husks offer effective protection against the spread of weevils from infested to sound ears. Repeated fumigation of all grain in storage will keep down the number of insects available to attack corn in the field.

Cutworms

Various species of cutworm larvae attack corn. Their chief damage results from the cutting off of the culm of young plants near the soil surface. The common cutworms are nocturnal feeders. They are controlled by band applications of insecticides applied during planting. The pale western cutworm (*Agrotis orthogonia*) is controlled by the use of clean summer fallow, which starves the larvae.

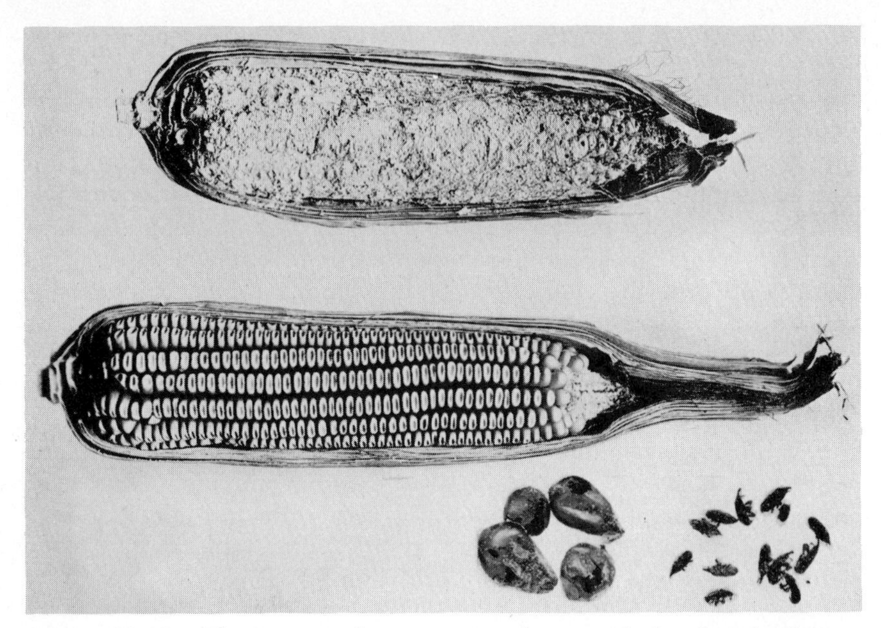

Figure 12–22. The grain in the upper ear of corn with the short husks was destroyed by weevils. The long husks on the lower ear protected the grain from the weevils. (*Below left*) Popcorn kernels showing weevil borings. (*Below right*) Adult rice weevils.

Army Worms

The army worm (*Cirphis unipuncta*) is widely distributed throughout the central latitudes where corn is important. The fall army worm (*Spodoptera frugiperda*) (Figure 12–20) occurs generally in the South, and the moths frequently migrate far to the North where the larvae damage late-planted corn. Both of these pests devour the leaves of corn and the fall army worm often attacks the growing point or bud of the plant. They can be controlled with insecticides.

Corn-Root Aphid

The corn-root aphid (*Aphis maidi radicis*) is a small bluish-green insect found on the roots of corn and usually attended by small brownish ants.[92] The injury is evidenced by weakened plants that turn yellow. The corn-root louse (aphid) can be controlled by late fall plowing, which destroys the nests of the attendant ants, or by rotation. Where these insects are present, corn should not be grown for more than two years in succession on the same field.

Other Insects

Among the numerous other insects attacking corn may be mentioned the corn-leaf aphid (*Aphis maidis*), maize billbug (*Spheno-*

phorus maidis), corn fleabeetle (*Chaetocnema pulicaria*), desert corn fleabeetle (*C. ectypa*), larger corn-stalk borer (*Diatraea zeacolella*), rough-headed corn-stalk beetle (*Euetheola rugiceps*), white grubs (*Phyllophaga* species), corn-root webworm or budworm (*Crambus caliginosellus*), maize billbug (*Calendra maidis*), seed-corn maggot (*Hylemyia platura*), wireworms, webworms, and the Japanese beetle. The Japanese beetle chews off the silks of corn, thus preventing pollination. The corn fleabeetle acts as a carrier (vector) of the bacterial wilt-disease organism. Several corn hybrids are resistant to the cornleaf aphid.[37]

REFERENCES

1. Airy, J. M., Tatum, L. A., and Sorenson, J. W. Jr. "Producing seed of hybrid corn and grain sorghum," in *Seeds*, USDA Yearbook, 1961, pp. 145–153.
2. Anonymous. "Corn and its uses as food," *USDA Farmers Bul. 1236*, pp. 1–24. 1924.
3. Anonymous. "Corn facts and figures," 5th ed., *Corn Industries Research Foundation*, New York, pp. 1–48. 1949.
4. Anonymous. "Corn in Industry," 5th ed., *Corn Industries Research Foundation*, New York, pp. 1–63. 1960.
5. Anonymous. *Bibliography of Corn*, 3 volumes, Scarecrow Press, Metuchen, N.J. 1971.
6. Anonymous. "Guidelines for mold control in high-moisture corn," *USDA Farmers Bul. 2238*, pp. 1–16. 1968.
7. Beattie, J. H. "Growing sweet corn for the cannery," *USDA Farmers Bul. 1634*, pp. 1–18 (rev.). 1945.
8. Biggar, H. H. "The old and new in corn culture," in *USDA Yearbook*, 1918, pp. 123–136.
9. Boswell, F. C., Anderson, O. E., and Stacey, S. V. "Some effects of irrigation, nitrogen, and plant population on corn," *Ga. Agr. Exp. Sta. Bul. 60*. 1959.
10. Brandon, J. F. "The spacing of corn in the West Central Great Plains," *J. Am. Soc. Agron*, 29:584–599. 1937.
11. Brunson, A. M., and Smith, G. M. "Popcorn," *USDA Farmers Bul. 1679*, pp. 1–18. 1948.
12. Brunson, A. M., and Willier, J. G. "Correlations between seed ear and kernel characters and yield of corn," *J. Am. Soc. Agron.*, 21:912–922. 1929.
13. Culpepper, C. W., and Magoon, C. A. "Studies upon the relative merits of sweet corn varieties for canning purposes and the relation of maturity of corn to the quality of the canned product," *J. Agr. Res.*, 28(5):403–443. 1924.
14. De Wet, J. M. J., and Harlan, J. R. "Origin of maize: The tripartite hypothesis," *Euphytica*, 21:271–279. 1972.

15. Dicke, F. F. "The most important corn insects," in *Corn and Corn Improvement,* Academic Press, New York, pp. 537–612. 1955.
16. Dickson, J. G. *Diseases of Field Crops,* 2nd ed., McGraw-Hill, New York, pp. 74–114. 1956.
17. Doty, D. M., and others. "The effect of storage on the chemical composition of some inbred and hybrid strains of sweet corn," *Purdue U. Agr. Esp. Sta. Bul. 503,* pp. 1–31. 1945.
18. Dungan, G. H. "Relation of blade injury to yielding ability of corn plants," *J. Am. Soc. Agron.,* 22:164–170. 1930.
19. Dungan, G. H. "Indications that corn tillers nourish the main stalks," *J. Am. Soc. Agron.,* 23:662–670. 1931.
20. Dungan, G. H., and Koehler, B. "Age of seed corn in relation to seed infection and yielding capacity," *J. Am. Soc. Agron.,* 36(5):436–443. 1944.
21. Evans, D. D., and others. "Growth and yield of sweet corn," *Ore. Agr. Exp. Sta. Tech. Bul. 53,* pp. 1–36. 1960.
21a. Flannery, K. V. "The origins of agriculture," in *Ann. Rev. of Anthropology,* Ann. Rev. Inc. Palo Alto, Calif. vol. 2, pp. 271–310. 1973.
22. Foth, H. D. "Root and top growth of corn," *Agron. J.,* 54:49–52. 1962.
23. Garber, R. J., and Hoover, M. M. "Influence of corn smut and hail damage on yield of certain first generation hybrids between synthetic varieties," *J. Am. Soc. Agron.,* 27:38–45. 1935.
24. Garrison, H. S., and Richey, F. D. "Effects of continuous selection for ear type in corn," *USDA Bul. 1341.* 1925.
25. Goodding, T. H. and Kisselbach, T. A. "The adaptation of corn to upland and bottom land soils," *J. Am. Soc. Agron.,* 23:928–937. 1931.
26. Gurley, W. H. "Growing corn in Georgia," *Ga. Ext. Bul. 547,* pp. 1–45. 1972 (Rev.).
27. Hall, D. M. "The relationship between certain morphological characteristics and lodging in corn," *Minn. Agr. Exp. Sta. Tech. Bul. 103.* 1934.
28. Hayes, H. K. "The commercial use of double crossed corn in Minnesota," *J. Am. Soc. Agron.,* 22:606–613. 1930.
29. Hayes, H. K. *A Professor's Story of Hybrid Corn,* Burgess, Minneapolis, pp. 1–237. 1963.
30. Henderson, C. A., and Davis, F. M. "The southwestern corn borer and its control," *Miss. Agr. Exp. Sta. Bul. 173,* pp. 1–16. 1969.
31. Hinkle, D. A., and Garrett, J. D. "Corn fertilizer and spacing experiments," *Ark. Agr. Exp. Sta. Bul. 635.* 1961.
32. Hixon, R. M., and Sprague, G. F. "Waxy starch of maize and other cereals," *J. Ind. Eng. Chem.,* 34:959–962. 1942.
33. Holbert, J. R., and Koehler, B. "Anchorage and extent of corn root systems," *J. Agr. Res.,* 27(2):71–78. 1924.
34. Hopkins, C. G., Smith, L. H., and East, E. M. "The structure of the corn kernel and the composition of the different parts," *Ill. Agr. Exp. Sta. Bul. 87,* pp. 79–112. 1903.
35. Hoppe, P. E. "Infections of corn seedlings," in *Plant Diseases,* USDA Yearbook, 1953, pp. 377–380.
36. Hortik, H. J., and Arnold, C. Y. "Temperature and the rate of development of sweet corn," *Proc. Amer. Soc. Hort. Sci.,* 69:400–04. 1965.

37. Huber, L. L., and Stringfield, G. H. "Aphid infestation of strains of corn as an index of their susceptibility to corn borer attack," *J. Agr. Res.*, 64(5):283–291. 1942.
38. Hughes, H. D., and Metcalfe, D. S. *Crop Production*, Macmillan, Inc., New York, 627 pp. 1972.
39. Inglett, G. E., ed. *Corn: Culture, Processing, Products*, Avi Publ. Co., Westport, Conn., 369 pp. 1970.
40. Jenkins, M. T. "A comparison of surface, furrow, and listed methods of planting corn," *J. Am. Soc. Agron.*, 26:734–737. 1934.
41. Jenkins, M. T. "Corn improvement," in *USDA Yearbook*, 1936, pp. 455–522.
42. Jenkins, M. T. "Seed corn," *USDA Farmers Bul. 1822*, pp. 1–13. 1939.
43. Jenkins, M. T. "Influence of climate and weather on the growth of corn," in *Climate and Man*, USDA Yearbook, 1941, pp. 308–320.
44. Johann, H. "Histology of the caryopsis of yellow dent corn, with reference to resistance and susceptibility to kernel rots," *J. Agr. Res.*, 51(10):855–883. 1935.
45. Johnson, I. J., and Christensen, J. J. "Relation between number, size, and location of smut infections to reduction in yield of corn," *Phytopath.*, 25:223–233. 1935.
46. Jones, D. F., and Huntington, E. "The adaptation of corn to climate," *J. Am. Soc. Agron.*, 27:261–270. 1935.
47. Jones, W. J., Jr., and Huston, H. A. "Composition of maize at various stages of its growth," *Purdue U. Agr. Exp. Sta. Bul. 175*, pp. 599–630. 1903.
48. Jordan, H. V., Laird, K. D., and Ferguson, D. D. "Growth rates and nutrient uptake by corn in a fertilizer spacing experiment," *Agron, J.*, 42:261–268. 1950.
49. Kelly, M. A. R. "Corn cribs for the corn belt," *USDA Farmers Bul. 1701*, pp. 1–26. 1933.
50. Kiesselbach, T. A. "Transpiration as a factor in crop production," *Nebr. Agr. Exp. Sta. Res. Bul. 6*, pp. 1–214. 1916.
51. Kiesselbach, T. A. "Corn Investigations," *Nebr. Agr. Exp. Sta. Res. Bul. 20*, pp. 1–151. 1922.
52. Kiesselbach, T. A. "The immediate effect of gametic relationship of the parental types upon the kernel weight of corn," *Nebr. Agr. Exp. Sta. Res. Bul. 33*, pp. 1–69. 1928.
53. Kiesselbach, T. A. "Field tests with treated seed corn," *J. Agr. Res.*, 40:169–189. 1930.
54. Kiesselbach, T. A. "Effects of age, size, and source of seed on the corn crop," *Nebr. Agr. Exp. Sta. Bul. 305*. 1937.
55. Kiesselbach, T. A. "Effect of artificial drying upon the germination of seed corn," *J. Am. Soc. Agron.*, 31:489–496. 1939.
56. Kiesselbach, T. A. "The structure and reproduction of corn," *Nebr. Agr. Exp. Sta. Res. Bul. 161*. 1949.
57. Kiesselbach, T. A. "Progressive development and seasonal variations of the corn crop," *Nebr. Agr. Exp. Sta. Res. Bul. 166*. 1950.
58. Kiesselbach, T. A., Anderson, A., and Lyness, W. E. "Tillage practices in relation to corn production," *Nebr. Agr. Exp. Sta. Bul. 232*. 1928.

59. Kiesselbach, T. A., and Keim, F. D. "Regional adaptation of corn in Nebraska," *Nebr. Agr. Exp. Sta. Res. Bul. 19.* 1921.
60. Kiesselbach, T. A., and Ratcliff, J. A. "Freezing injury of seed corn," *Nebr. Agr. Exp. Sta. Res. Bul. 16.* 1918.
61. Koehler, B., Dungan, G. H., and Holbert, J. R. "Factors influencing lodging in corn," *Ill. Agr. Exp. Sta. Bul. 266.* 1925.
62. Kuleshov, N. N. "World's diversity of phenotypes of maize," *J. Am. Soc. Agron.*, 25:688–700. 1933.
63. Leonard, W. H., and Robertson, D. W. "Rate of planting corn under irrigated conditions," *Colo. Agr. Exp. Sta. Bul. 417.* 1935.
64. Love, H. H. "Correlations between ear characters and yield in corn," *J. Am. Soc. Agron.*, 9:315–322. 1917.
65. Macy, L. K., Arnold, L. E., and McKibben, E. G. *Changes in Technology and Labor Requirements in Crop Production,* Works Progress Admin., National Research Project, Report No. A-5, pp. 1–181. 1938.
66. Majors, K. R. "Cereal grains as food and feed," in *Crops in Peace and War,* USDA Yearbook, 1950–51, pp. 331–340.
67. Mangelsdorf, P. C. "Corn varieties in Texas: Their regional and seasonal adaptation," *Tex. Agr. Exp. Sta. Bul. 397.* 1929.
68. Mangelsdorf, P. C., MacNeish, R. S., and Galinat, W. C. "Domestication of corn" *Science,* 143:538–545. 1964.
69. Martin, J. H. "Field crops," in *Soil,* USDA Yearbook, 1957, pp. 663–665.
70. Martin, J. H., and Jenkins, M. T. "New uses for waxy cereal starches," in *Crops in Peace and War,* USDA Yearbook, 1950–51, pp. 142–147.
71. McClelland, C. K. "A comparison between Mexican June and three other varieties for summer planting," *J. Am. Soc. Agron.*, 28:799–806. 1936.
72. McClelland, C. K., and Young, V. H. "Seed corn treatments in Arkansas," *J. Am. Soc. Agron.*, 26:189–195. 1934.
73. Melchers, L. E., and Brunson, A. M. "Effect of chemical treatments of seed corn on stand and yield in Kansas," *J. Am. Soc. Agron.*, 26:909–917. 1934.
74. Miller, E. C. "Development of the pistillate spikelet and fertilization in *Zea mays,*" *J. Agr. Res.*, 18:255–265. 1919.
75. Mooers, C. A. "Planting rates and spacing for corn under Southern conditions," *J. Am. Soc. Agron.*, 12:1–22. 1920.
76. Mooers, C. A. "Varieties of corn and their adaptability to different soils," *Tenn. Agr. Exp. Sta. Bul. 126,* pp. 1–39. 1922.
77. Neal, N. P. "The decrease in yielding capacity in advanced generations of hybrid corn," *J. Am. Soc. Agron.*, 27:666–670. 1935.
78. Nelson, M., and McClelland, C. K. "Cultivation experiments with corn," *Ark. Agr. Exp. Sta. Bul. 219.* 1927.
79. Odland, T. E., and Knoblauch, H. C. "The value of cover crops in continuous corn culture," *J. Am. Soc. Agron.*, 30:22–29. 1938.
80. Olsen, P. J., Bull, C. P., and Hayes, H. K. "Ear type selection and yield in corn," *Minn. Agr. Exp. Sta. Bul. 174.* 1918.
81. Osborn, L. W. "Experiments with varying stands and distributions of corn," *Ark. Agr. Exp. Sta. Bul. 200.* 1925.

82. Parker, J. R., and others. "Comparative injury by the European corn borer to open-pollinated and hybrid field corn," *J. Agr. Res.*, 63(6):355–368. 1941.
83. Poole, C. F. "Improvement of sweet corn," in *USDA Yearbook*, 1937, pp. 379–394.
84. Pumphrey, F. V., and Dreier, A. F. "Grain, silage, and plant population experiments with corn hybrids at the Scottsbluff Experiment Station," *Nebr. Agr. Exp. Sta. Bul. 449*. 1959.
85. Randolph, L. F. "Developmental morphology of the caryopsis in maize," *J. Agr. Res.*, 53(12):881–916. 1936.
86. Randolph, L. F., Abbe, E. C., and Einset, J. "Comparison of the shoot apex and leaf development and structure in diploid and tetraploid maize," *J. Agr. Res.*, 69(2):47–76. 1944.
87. Richey, F. D. "Corn breeding," *USDA Dept. Bul. 1489*, pp. 1–63. 1927.
88. Richey, F. D., Stringfield, G. H., and Sprague, G. F. "The loss in yield that may be expected from planting second generation double crossed seed corn," *J. Am. Soc. Agron.*, 26:196–199. 1934.
89. Richey, F. D., and Willier, J. G. "A statistical study of the relation between seed-ear characters and productiveness in corn," *USDA Bul. 1321*. 1925.
90. Robert, A. L. "Some of the leaf blights of corn," in *Plant Diseases*, USDA Yearbook, 1953, pp. 380–385.
91. Robertson, D. W., Kezer, A., and Deming, G. W. "The date to plant corn in Colorado," *Colo. Agr. Exp. Sta. Bul. 369*. 1930.
92. Salmon, S. C. "Corn production in Kansas," *Kans. Agr. Exp. Sta. Bul. 238*. 1926.
93. Sayre, J. D., Morris, V. H., and Richey, F. D. "The effect of preventing fruiting and of reducing the leaf area on the accumulation of sugars in the corn stem," *J. Am. Soc. Agron.*, 23:751–753. 1931.
94. Shedd, C. K. "Handling and storing soft corn on the farm," *USDA Farmers Bul. 1976*, pp. 1–13. 1945.
95. Sprague, G. F. "Production of hybrid corn," *Iowa Agr. Exp. Sta. and Ext. Serv. Bul. 48*, pp. 556–582. 1942.
96. Sprague, G. F. "Industrial utilization," in *Corn and Corn Improvement*, Academic Press, New York, pp. 613–636. 1955.
97. Sprague, H. B. "The adaptation of corn to climate," *J. Am. Soc. Agron.*, 27:680–681. 1935.
98. Sprague, G. F., and Larson, W. E. "Corn production," *USDA Agr. Handbk. 322*, pp. 1–37. 1966.
99. Stoner, W. N., and others. "Corn (maize) viruses in the continental United States and Canada." USDA ARS 33–118, pp. 1–95. 1968.
100. Stringfield, G. H., and Anderson, M. S. "Corn production," *USDA Farmers Bul. 2073*. 1954.
101. Sutherland, W. N., Shrader, W. D., and Pesek, J. T. "Efficiency of legume residue nitrogen and inorganic nitrogen in corn production," *Agron. J.*, 53:339–342. 1961.
102. Tatum, L. A. "The southern corn leaf blight epidemic," *Science*, 171:1113–15. 1971.

103. Ullstrup, A. J. "Some smuts and rusts of corn," in *Plant Diseases,* USDA Yearbook, 1953, pp. 386–389.
104. Ullstrup, A. J. "Several ear rots of corn," in *Plant Diseases,* USDA Yearbook, 1953, pp. 390–392.
105. Ullstrup, A. J. "Diseases of corn," in *Corn and Corn Improvement,* Academic Press, New York, pp. 465–536, 1955.
106. Ullstrup, A. J. "Corn diseases in the United States and their control," *USDA Handbk. 199.* 1–44. 1966 (Rev.).
107. Weatherwax, P. *Indian Corn in Old America,* Macmillan, Inc., New York, pp. 1–253. 1954.
108. Weaver, J. E. *Root Development of Field Crops,* McGraw-Hill, New York, pp. 180–192. 1926.
109. Weihing, R. M. "The comparative root development of regional types of corn," *J. Am. Soc. Agron.,* 27:526–537. 1935.
110. Willier, J. G., and Brunson, A. M. "Factors affecting the popping quality of popcorn," *J. Agr. Res.,* 35(7):615–624. 1927.
111. Zipf, R. L. "Wet milling of cereal grains," in *Crops in Peace and War,* USDA Yearbook, 1950–51, pp. 142–147.

Chapter 13

Sorghums

ECONOMIC IMPORTANCE

Sorghum grain was harvested on about 58 million hectares annually from 1970 to 1972, with a production of nearly 75 million metric tons or 1.3 metric tons per hectare. The leading producing countries were the United States, India, Nigeria, Argentina, Mexico, and Sudan. It is grown to some extent or occasionally grown in all countries of the world except in the cool northwestern part of Europe. Sorghum ranks fifth in acreage and production among all world crops. The grain production in the United States was about 20.4 million metric tons (795 million bushels), yielding 3.45 metric tons per hectares (55 bushels per acre), harvested from nearly 5.9 million hectares or

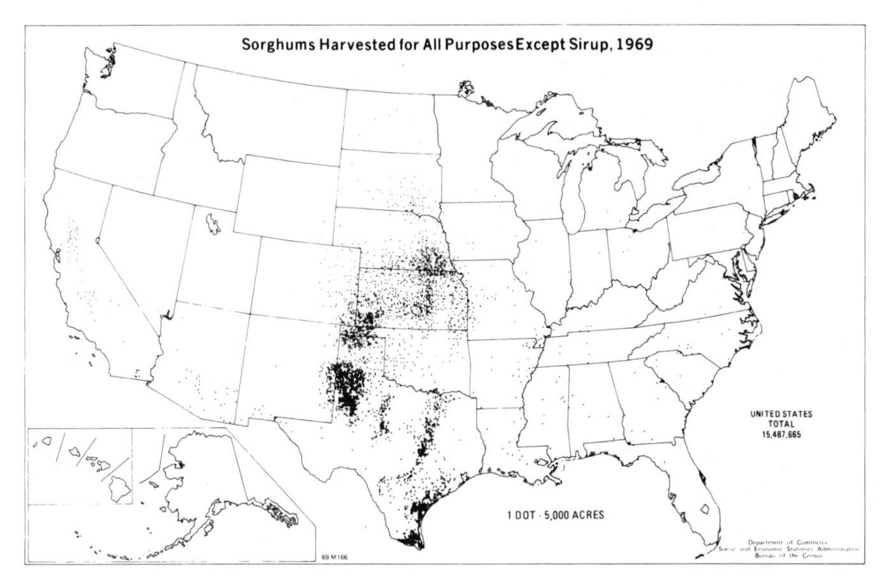

Figure 13–1. Sorghum harvested for all purposes in 1969. Sorghum for sirup is grown in the same areas, and also east and north of the important grain sorghum areas.

14,485,000 acres. The leading states in grain sorghum production were Texas, Kansas, Nebraska, Oklahoma, and Missouri (Figure 13–1). Sorghum ranks fifth in acreage among the crops grown in the United States, being exceeded by corn, wheat, soybeans, and alfalfa. The tonnage of grain sorghum in 1972 and 1973 exceeded slightly the combined tonnages of oats, barley, and rye. Since 1948, the yield of grain sorghum has more than doubled and the world production nearly trebled. Sorghum is also an important forage crop in the United States and some other countries. An average of more than 2,465,000 acres were harvested for fodder in the United States from 1970 to 1972, and an additional 868,000 acres were harvested for the production of 9,400,000 tons of silage. Only 3,709 acres of sorgo were harvested for sirup by 1969. About 800,000 acres were harvested for sirup from 1919 to 1921, with gradual declines thereafter. Broomcorn, also a sorghum, has declined in production from 50,000 tons annually

Figure 13–2. Heads and stalks of a sorgo (sweet sorghum), suitable for silage, fodder, or sirup.

40 years earlier to only 1,000 tons harvested from 7,300 acres in 1974.

Sudangrass, a sorghum subspecies, and sorgrass (sorghum-sudan hybrids) are important silage, pasture, and forage crops.

Sorgo (sweet sorghum) (Figure 13–2) is grown for forage, silage, and sirup, while grain sorghum is grown largely for grain. Texas, Kansas, and Oklahoma lead in the production of sorghum forage. Kansas and Nebraska lead in the production of sorghum silage. Sorgo sirup is produced chiefly in Alabama, Mississippi, Georgia, Tennessee, and Iowa. Broomcorn is produced almost entirely in Oklahoma, Colorado, New Mexico, and Texas.

Sudangrass is used for pasture or hay in nearly all states. It is the leading temporary summer pasture crop in the North Central and South Central states.

A perennial natural hybrid between sorghum and johnsongrass, collected in Argentina, called columbusgrass or *Sorghum almum*, is now grown in the United States, South America, South Africa, and Australia. The seed-shattering habit as well as the perennial nature of this grass threatens to make it a weed pest like johnsongrass. A selected perennial hybrid derivative of columbusgrass, called DeSoto grass, was developed later.[13]

ADAPTATION

Sorghums are grown in warm or hot regions that have summer rainfall as well as in warm irrigated areas.[29] The most favorable mean temperature for the growth of the plant is about 37° C. The minimum temperature for growth is 15° C. Consequently, only a part of the frost-free season may be available to produce the crop. The sorghum plant seems to withstand extreme heat better than other crops, but extremely high temperatures during the fruiting period reduce the seed yield.[64] Sorghum is a short-day plant.

The sorghums are well adapted to summer rainfall regions where the average annual precipitation is only 17 to 25 inches.[29] The plants remain practically dormant during periods of drought but resume growth as soon as there is sufficient rain to wet the soil. This characteristic accounts in large part for the success of sorghum in a dry season and is the reason why it has been called a crop camel. As compared with corn of similar seasonal requirements, sorghum has more secondary roots and a smaller leaf area per plant. Sorghum leaves and stalks wilt and dry more slowly than those of corn,[25] enabling sorghum to withstand drought longer. A waxy cuticle apparently retards drying. However, sorghum is also highly productive on irrigated land and in humid sections. More than 25 per cent of the sorghum acreage in the United States was irrigated in 1972.

Sorghum is grown successfully on all types of soil. In moist seasons the highest yields are obtained on heavy soils, but in dry seasons it does best on sandy soil. Sorghum will tolerate considerable soil salinity.

Its resistance to drought and heat as well as to grasshopper, rootworm, and corn-borer injury accounts for the growing of sorghum instead of corn in many sections of the United States. The chief drawbacks of sorghum as compared with corn are lower yields of succeeding crops, greater uncertainty in getting stands, the necessity of prompt harvesting, greater difficulty in storing the grain, the greater necessity of grinding the grain before feeding, and the usually lower feeding and market value.

ORIGIN AND HISTORY

Sorghum apparently is a native of Africa in the zone south of the Sahara desert, where several closely related wild species are found, and the cultivated types are very diverse. The cultivated type (*Sorghum bicolor* Moench) may have been selected by 3000 B.C. or several centuries later.[11] It appears to have been grown in India by the beginning of the Christian Era or earlier. It was grown in Assyria before 700 B.C. and in southern Europe sometime thereafter.[65] Some sorghum seed was brought to the United States from Africa by imported slaves but noticeable cultivation did not begin until after the introduction of several varieties in 1853 and 1857.

Broomcorn was grown in Europe before 1596 but was not cultivated in the United States until late in the 18th century after its introduction, presumably by Benjamin Franklin.[31] Sudangrass was introduced from Sudan in 1908.[28]

BOTANICAL DESCRIPTION

Sorghum belongs to the family *Gramineae*, tribe *Andropogoneae*. *Sorghum bicolor* includes the annual sorghums with 10 pairs of chromosomes, namely, grain sorghum, sorgo, broomcorn, and sudangrass. Sudangrass is further classified into the subspecies *sudanense*. *Sorghum halepense* (johnsongrass) is perennial in habit, produces rhizomes, and has 20 haploid chromosomes.

Closely related to sudangrass are several species of wild grass sorghums in Africa. One or more of these may be the progenitor of cultivated sorghum. Other related species of sorghum, such as *S. versicolor*, which has only 5 haploid chromosomes, are found in Africa.[33]

The Sorghum Plant

Sorghum is a coarse grass with culms 0.5 to 5.0 meters in height.[4, 65] The culms are similar to those of corn, being grooved and nearly oval.

The peduncle (top internode) is not grooved. The grooves alternate from one side to the other on each successive internode. Young sorghum plants can be distinguished readily from corn plants because of the saw-tooth margins of sorghum leaves. Some varieties have sweet juicy pith in the stalks, others are juicy but not sweet, while still others are deficient in both sweetness and juiciness. A dry-stalked variety has a white midrib in the leaf. A juicy-stalked variety has a dull or cloudy midrib caused by the presence of juice instead of air spaces in the pithy tissues. A leaf arises at each node, the blades being glabrous with a waxy surface. The surface of the culms, sheaths, and leaves is glaucous. Buds at the nodes of the culm often give rise to side (axillary) branches. Crown buds give rise to tillers (Figure 3–5). The total number of leaves on the main stalk, including those formed during the seedling stage, averaged 16 to 27 per stalk in 21 American varieties.[54] The first 10 (more or less) of the small leaves arise from the crown nodes underground. Early maturing varieties have few leaves and consequently are limited in plant yield.

The sorghum inflorescence is a loose to dense panicle, having many primary branches borne on a hairy axis, bearing paired ellipsoidal spikelets (Figure 4–5). The sessile spikelet of each pair is perfect and fertile, while the pedicellate spikelet is either sterile or staminate. Two pedicellate spikelets accompany the sessile spikelet at the end of each panicle branch. The two glumes of the fertile spikelet are usually indurate (leathery). There are two florets in the fertile spikelet, the lower sterile and the upper fertile. The lemma and palea are thin and translucent. The lemma may be awned or awnless. Some sorghums have seeds fully covered by the chaff even after being threshed, while others are more than half exposed and thresh out completely free from the chaff. The position of the panicle is usually erect, but may be recurved. Recurving is the result of heavy thick panicles being forced out the side of a too-narrow sheath while the peduncle is too flexible to support the panicle in an erect position. Later the peduncle stiffens (becomes lignified) in the recurved condition.[26] Erect varieties have slender panicles during the boot stage. A well-developed panicle may contain as many as 2,000 seeds. The number of seeds in a pound is about as follows: grain sorghum, 12,000 to 20,000; sorgo, 20,000 to 30,000; sudangrass, 50,000 to 60,000; and broomcorn, 22,000 to 30,000 (Figure 7–4).

Sorghums are about 95 per cent self-pollinated in the field,[53] but they will cross naturally with other varieties of sorghum, broomcorn, or sudangrass, and occasionally with johnsongrass also.

The pigments of colored sorghum seeds are found in the pericarp, or the subcoat (testa), or both. When the pericarp only is pigmented, the seeds are yellow or red. When the pericarp is white and a testa is present, the seeds are buff-colored, or bluish white in types with a thick starchy mesocarp in the seedcoat. With a colored pericarp and a testa present, the seeds usually are dark brown or reddish brown. Starchy seeds take up water quickly and are subject to invasion by seed-rotting fungi.

Mature seeds of sorghum (like those of maize) have a black spot nearby the base. Immature seeds develop the black spot after drying.

Although considered an annual and usually grown as such, sorghum can survive as a perennial where the temperature is mild and soil moisture is available. It has lived for at least 6 years in the field in southern California, and 13 years in a greenhouse. Sorghum planted early in a mild climate may produce a second (ratoon) crop later in the year.[44]

Each new culm arising from crown buds develops its own root system and a new series of crown buds but remains attached to the old crown.[27] A culm dies after it has flowered and after all active buds at the culm and crown nodes also have elongated into stems. Then its roots also die and decay. In the meantime, new culms have grown alongside the original culm. This process continues as long as conditions are favorable for vegetative reproduction or until the crowns form too high above the soil surface.

PRUSSIC-ACID POISONING

Young plants, including the roots (Figure 7–3) and especially the leaves of older plants of sorghum, sudangrass, and johnsongrass contain a glucoside called dhurrin, which upon breaking down releases a poisonous substance known as prussic acid or hydrocyanic acid (HCN). Some losses of cattle, sheep, and goats occur each year from sorghum poisoning when they graze upon the green plants. Silage and well-cured fodder and hay usually may be fed with safety.[62] Silage may contain toxic quantities of prussic acid, but it escapes in gaseous form while the silage is being moved and fed. The prussic-acid content of sorghum hay and fodder decreases during curing so that it is dangerous only occasionally.

The prussic-acid content decreases as the plant approaches maturity.[35] Small plants, and young branches and tillers, are high in prussic acid. The prussic-acid content of the leaves is 3 to 25 times greater than that of the corresponding portions of stalks of plants in the boot stage. Heads and sheaths are low in prussic acid. The upper leaves contain more prussic acid than the older lower leaves. The amount of prussic acid varies in different sorghums. Sudangrass contains about two-fifths as much as many sorghums grown under the same conditions. Sudangrass rarely kills animals unless contaminated with sorghums or sorghum-sudangrass hybrids, except occasionally in the northern states. Even there it usually is safe after the plants are 18 to 24 inches high.

Freezing does not increase the prussic-acid content of sorghum, but it causes the acid to be released quickly from the glucoside form, thus making frosted sorghum very dangerous until it begins to dry out. An abundance of soil nitrates causes sorghum to be high in prussic acid as well as nitrates. Drought-stricken and second-growth plants are dangerous for ruminants because they are small and consist largely of leaves, which are high in prussic acid.

The toxic level of HCN is above 200 parts per million.[14] A mere half gram of pure prussic acid can kill a cow.

Sorghum is unsafe for pasturing except after the plants are mature and little new growth is present. Individual animals differ in their susceptibility to sorghum poisoning. Sheep seem to be slightly less susceptible to prussic acid than are cattle, while horses and hogs apparently are not injured. However, horses have suffered from a cystitis syndrome when pasturing sudan-sorghum hybrids.[58]

Poisoning is less likely to occur if the animals eat some ground grain before they are turned into the pasture. The growing of varieties of sorgo and sudangrass that are low in dhurrin content reduces losses from poisoning.

The remedy of cyanide poisoning is intravenous injection of a combination of sodium nitrite and sodium thiosulfate. For cattle, 2 to 3 grams of sodium nitrite in water followed by 4 to 6 grams of sodium thiosulfate in water are recommended, and for sheep, up to 1 gram of sodium nitrite and 2 to 3 grams of sodium thiosulfate.

SORGHUM GROUPS

The grain sorghum originally introduced into the United States could once be classified into rather distinct groups. The sorghum hybrids now grown in the United States and many other countries were derived from crosses between different groups. They represent recombinations of group characters.

Grain-sorghum stalks usually are either fairly juicy or comparatively dry at maturity, and the stalk juice is not sweet or is at most only slightly sweet. Grain sorghums, in general, have larger heads and seeds and shorter stalks, and produce more seed in proportion to total crop than do the sorgos. The stalks of grain sorghum hybrids usually range from 3/4 to 1 1/2 meters in height. The seed threshes free from the glumes.

Among the grain sorghum groups, kafirs have thick and juicy stalks, relatively large, flat, dark-green leaves, and awnless cylindrical heads (Figure 13–3). The seeds are white, pink, or red and of medium

Figure 13–3. Panicles of grain sorghum groups: (1) Shallu, (2) milo, (3) kafir, and (4) feterita. (*Below*) Spikelets (left) and kernels (right) of kafir.

size (Figure 13–4) The chaff is either black or straw-colored. Kafirs are grown for both grain and forage. Hegari is similar to kafir in appearance except that the heads are more nearly oval, the seeds are a chalky or starchy white, and the more abundant leaves and sweeter juice make it more prized for forage.

Milos have somewhat curly light-green leaves and slightly smaller leaves and stalks than kafir, and are less juicy. The leaves have a yellow midrib containing carotene. The heads of the true milos are bearded, short, compact, and rather oval in outline, with very dark-brown chaff. The seeds are large and are yellow or white. The plants tiller considerably,[55] and in general are earlier and more drought-resistant than those of kafir.

The cytoplasmic male sterility required for hybrid sorghum production was obtained from milo but many hybrids also have kafir in their parentage.

Feterita has few leaves, slender relatively dry stalks, rather oval, compact heads, and very large chalk-white seeds. It matures early.

Durra has dry stalks, flat seeds, very pubescent glumes, and either

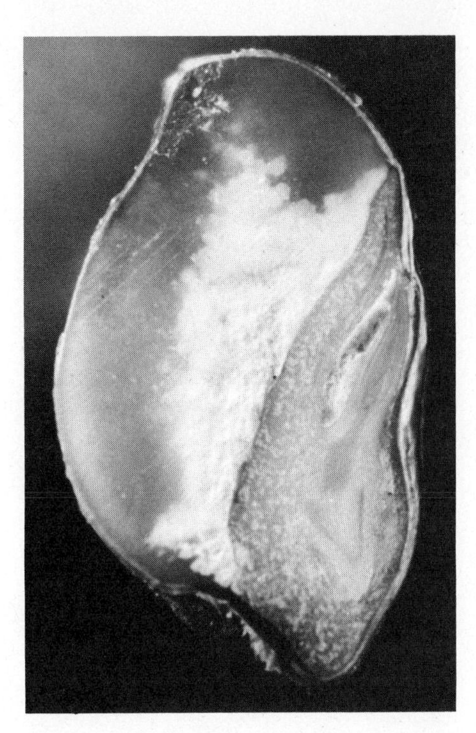

Figure 13–4. Longitudinal section of a sorghum grain. [Courtesy Corn Products Co.]

compact and recurved or loose and erect panicles. Durra is the chief type in North Africa, the Near East, and India.

Shallu has tall, slender, dry stalks, loose heads, and pearly white seeds. It is relatively late in maturity. The loose heads dry out quickly, do not harbor worms, and make it difficult for birds to roost on the slender branches and eat the grain. Shallu is widely grown in India and also in tropical Africa.

Kaoliang has dry, stiff, slender stalks, open bushy panicles with wiry branches, and rather small brown, or white, seeds. Kaoliang is grown almost exclusively in China, Korea, Japan, and southeastern Siberia.

Sorgo or sweet sorghum (often called *cane*) is characterized by abundant sweet juice in the stalks, which usually range in height from 1.5 to 3 meters. The seed of some varieties remains enclosed in the glume after threshing. The heads (panicles) may be loose or dense, and the lemmas awned or awnless, and black, brown, or red, depending upon the variety. The seeds are small or medium-size and are either white or some shade of brown.

Broomcorn produces heads and fibrous seed branches 30 to 90 centimeters long (Figure 13–5) that are used for making brooms and whisk brooms. The stalks range from 1 to 4 meters in height and are dry, not sweet, and of limited value for forage. The lemmas are awned

Figure 13–5. Panicles of broomcorn. [Courtesy Colorado Agricultural Experiment Station.]

with small brown seeds enclosed in very pubescent tan, red, or dark-brown glumes.

Sudangrass has slender leaves and stalks and loose heads with small brown seeds (Figure 13–6). Johnsongrass is similar to sudangrass except in being larger, and it is a perennial with underground stems, and the seeds are different (Figures 13–6 and 13–7).

VARIETIES

Grain sorghum hybrids with a yellow endosperm containing carotene and xanthophyll are now in production in the United States. The chief characteristics of the common varieties of sorghum formerly grown in the United States have been listed.[32, 65] Some of the hybrids are also described.[33]

The time of maturity is the most important factor that determines the adaptation of a sorghum hybrid or common variety to a particular locality.[6, 61] However, choice for any region is based largely upon its intended use and method of harvesting.[37] Differences in yield, when not too large, are only a secondary consideration provided the variety possesses the other characteristics desired. It is essential that a hybrid

Figure 13–6. Sudangrass panicles. [Photo by John Jeter.]

or common variety reach maturity before it is killed by frost. Consequently, as the average frost-free period becomes shorter, progressively quicker-maturing types are grown.

Sorghums that mature later have large leafy stalks and are suited to regions with short days and a long growing season. They fail to head under the long days in northern latitudes. Quick-maturing hybrids are best adapted to long-day conditions. In the South they mature very quickly with limited vegetative growth, but fail to utilize the full growing season.

Varieties of both standard and dwarf broomcorns are grown in the United States. Some newer varieties produce brush free from red stain.

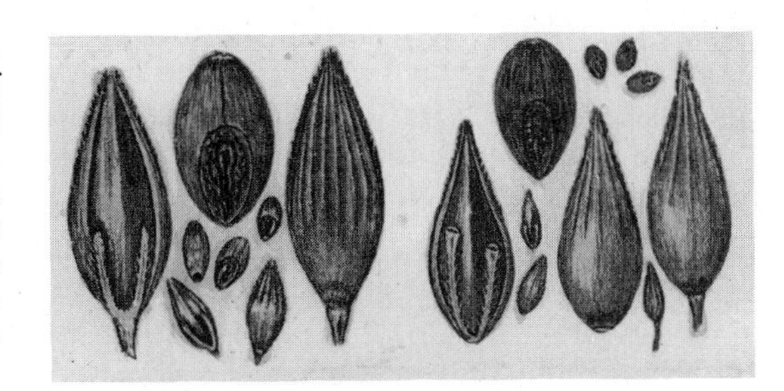

Figure 13–7. Seeds, enlarged and natural size of sudangrass (*left half*) and johnsongrass (*right half*). The swollen tips of the pedicels of the sudangrass seed at extreme left are broken off but they remain on the corresponding johnsongrass seed.

Hybrids that involve lines of sudangrass and sorghum-sudangrass hybrids are replacing sudangrass and sorgo varieties. Varieties were bred that are low in prussic-acid content, resistant to bacterial leaf spots, *Helminthosporium* leaf blight, as well as other fungus leaf spots, and that are juicy and sweet and consequently very palatable pasturage.

The breeding and distribution of new sorghum hybrids and varieties has effected marked changes in production since 1928.[30, 33, 45, 46, 46a, 47, 66] The development of hybrids or varieties with short erect stalks have made it possible to harvest the crop successfully with a combine and at the same time eliminate losses from root rot. Thus grain-sorghum production became adapted to large-scale mechanized methods. The development of quick-maturing common varieties and hybrids extended the grain-sorghum belt into Nebraska, South Dakota, and northeastern Colorado. The breeding of productive hybrids extended grain-sorghum production eastward into the more favorable subhumid and humid areas. Average yields of grain sorghum in the United States were trebled between 1940 and 1970.

ROTATIONS

Sorghum follows other crops readily, but care should be exercised in the choice of a crop to follow sorghum. Grain sorghum often is grown continuously[18] or is alternated with sorgo, sudangrass, broomcorn, or corn. In northwestern Kansas an occasional rotation is sorghum, spring barley, and winter wheat.[51] A three-year rotation of sorghum, fallow, and wheat is popular where fallow is a desirable preparation for winter wheat.[46, 47, 48] In the southern Great Plains grain sorghums usually produce more after fallow, winter small grains, cowpeas, or cotton than in continuous culture. They respond well to the additional moisture conserved by fallow, the yields on fallow in western Kansas being 75 per cent higher than from land continuously in sorghum.[41] In the irrigated areas of southern California and Arizona, the sorghum crop usually follows either wheat or barley in the same season and is followed by some spring-planted crop.

Sorghum is supposed to be "hard on the land," the effect being particularly noticeable when fall-sown grains follow immediately. At Hays, Kansas, average wheat yields were 4.1 bushels per acre lower after sorghum than after corn. Some of the injurious after-effects of sorghum can be explained by the persistence of the sorghum plant—the fact that it keeps growing until killed by frost. Sorghum thus depletes the soil moisture to a greater extent than do other crops.

Sorghum injury to subsequent crops under irrigation, where soil moisture is ample, has been attributed to the high sugar content of sorghum roots and stubble.[8] The sugars in dry roots of different varieties of sorghum at maturity have ranged from 15 to over 55 per cent. Corn varieties ranged from less than 1 to about 4.5 per cent.[10] These sugars furnish the energy for soil microorganisms which multiply and compete with the crop plants for the available nitrogen in the soil and thus retard the crop growth. This condition lasts for only a few months, or until the sorghum residues have decayed. Ground sorghum roots added to the soil have depressed nitrate accumulation and bacterial development more than have corn roots.[68]

The injurious after-effect of sorghum on irrigated land may be overcome by nitrogenous fertilizers, barnyard manure, or by inoculated legume green manures.[9] Both alfalfa and fenugreek made practically normal growth after sorghum in California. The detrimental influence of sorghum on dry land may be avoided by fallowing the next season. Another means of avoiding this effect is to plant spring crops after sorghum, especially in May or June. By that time available nitrates will have accumulated, while much of the soil moisture deficiency caused by the sorghums will have been overcome by normal precipitation.

FERTILIZERS

Sorghum responds to applications of barnyard manure or 22 to 45 kilos per hectare of nitrogen in the semiarid Great Plains and 45 to 65 kilos per hectare in the subhumid areas.[52] Little or no benefit from fertilizers is evident in a very dry season. On irrigated land grain yields usually increase with applications of nitrogen of 100 to 180 kilos per hectare. In humid regions sorghum gives about the same fertilizer response as does corn. There applications of 40 to 120 kilos of nitrogen per hectare, plus 45 to 70 kilos per hectare each of P_2O_5 and K_2O are common.[42] Ample nitrogen induces earlier maturity in sorghum unless the application is excessive. Foliar spraying of an iron chelate prevents leaf yellowing in soils of high *p*H. Zinc chelates may be helpful where top soils have been heavily graded for irrigation or terracing.

SORGHUM CULTURE

Seedbed Preparation

Moldboard plowing is the common practice for seedbed preparation in the humid and irrigated areas. The yields may be increased from 25 to 30 per cent by thorough tillage of medium-heavy soils. A warm mellow seedbed is essential for good seed germination. Fall or winter first tillage is preferable to spring land-breaking. Weed control before planting is desirable. Most of the fields in the semiarid region are tilled by stubble-mulch or one-way disk methods. Mimimum tillage methods make planting for good seed germination more difficult.

Planting Practices

Sorghum for grain, fodder, or silage is usually planted in cultivated rows 0.5 to 1.0 meters apart, with multiple-row equipment. A sorghum planter, or other types of row planters with special sorghum seed plates, are satisfactory. Planting in shallow furrows is desirable (Figure 13–8). Some sorghum is planted with a furrow drill in rows 36 to 54 cm. apart. Such close row spacing is preferable to 1-meter spacing for a suitable rate of planting except under dry conditions (Figure 13–9). Narrow rows provide a more balanced plant distribution, for absorbing light rays and better shading to suppress weed growth and soil moisture evaporation. Narrow rows result in higher yields under humid and irrigated conditions.[56]

Planting in narrow rows requires either matching cultivating equipment or the use of herbicides to control weeds after the plants are a few inches in height. Sorghum may be planted in double rows on beds to facilitate irrigation, drainage, and cultivation. The paired rows are 30 to 36 centimeters apart, on beds spaced at 1-meter intervals.

The amount of seed to plant per acre for a given stand depends upon the condition of the seedbed, seed viability, seed size, and weather conditions at seeding time. A discrepancy between field and laboratory germination of sorghum seed frequently ranges from 30 to 50 per

Figure 13–8. Six-row lister planter suitable for planting sorghum in shallow furrows. [Courtesy J. I. Case Co.]

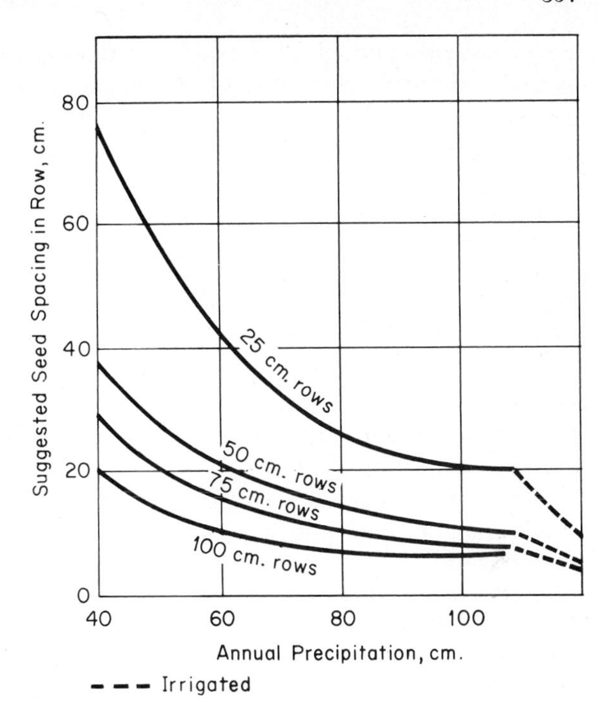

Figure 13–9. Row spacing as related to precipitation. [Adapted from *Kans. Ext. Circ. C447.*]

cent when seed of high viability is used.[59] Marked deficiency in field emergence may be expected when the laboratory germination is 85 per cent or lower. The stands are improved by seed treatment so that a 75 per cent emergence is possible.[22, 52]

Spacing between plants in the row for maximum yields depends upon tillering habits of the hybrid or variety.[55] Grain sorghums that tiller little are planted at higher rates.[5, 20, 24, 36] The dwarf combine types require 2 to 4 pounds of seed per acre under semiarid conditions for a stand of 15,000 or more plants per acre.[24] About 5 to 12 pounds per acre (6 to 14 kilos per hectare) usually are planted on rich land under irrigation, or in humid areas. The average rate of seeding in the United States in 1972 was 7 pounds per acre.

A 2-pound per acre planting rate should provide 13,000 to 22,000 plants per acre. A 4-pound rate should assure an adequate stand of 20–35 thousand plants per acre for subhumid conditions. An 8-pound rate should provide stands of 50,000 to 100,000 plants per acre, which is ample for fertile soils under humid conditions.[56] A 10- to 12-pound planting rate should provide the 100,000 to 120,000 plants per acre essential for maximum yields on heavily fertilized irrigated land.[52]

The highest tonnage of forage sorghums planted in 40-inch rows may be expected when the plants are spaced not more than 4 to 6 inches apart. Sorgo, grown in rows for forage or silage in semiarid areas, should be planted at a rate of 4 to 5 pounds per acre, but at

6 to 8 pounds per acre in humid or irrigated areas. The yields of sorghum planted in close drills for hay are about the same for all rates of seeding between 15 and 75 pounds per acre. Recommended seeding rates are 30 pounds per acre west of the 100th meridian in the Great Plains, 45 pounds between the 98th and 100th meridians, and 60 to 75 pounds east of the 98th meridian.[37]

The seed of sorghum is generally covered with 1 to 2 inches of soil. The percentage and rapidity of germination is reduced by soil temperatures below 77° F. (25° C), and slightly reduced by deep planting (2½ inches).[38]

Sorghum yields usually are highest when the crop reaches maturity shortly before the average date of the first killing frost. To achieve this, quick-maturing varieties are planted late while long-season varieties are planted early.[52]

The date of planting of sorghums should be arranged so that germination and early growth will take place during the period of moderately high soil temperatures, 70° to 80° F. (21°–26° C.), with the blooming and filling at such a time as to avoid the highest temperatures.[64] In the extreme southern parts of the grain-sorghum region, the crop can be planted from late February until August with a good chance to mature. In the northern part of the region, the soil does not become warm until after May 15. In the central as well as in the south central parts of the United States, the best time to plant sorghums is between May 15 and July 1. A safe rule in all localities, except where the sorghum midge is troublesome, is to plant not earlier than about two weeks after the usual corn-planting time. The earliest possible safe planting is advisable where chinch bugs are present.

In the irrigated areas of southern Arizona and California the best yields are obtained from planting in July. Medium-late planting results in better stands, taller stalks, larger heads, and shorter growing periods than does early planting. Planting earlier than necessary for safe maturity usually reduces plant growth. Early planting sometimes is desirable to avoid conflicts with other crops at harvest time, or to clear the land in time for seeding other crops. Sorghum for seed must be ripe before heavy freezes occur, because the germination is damaged by freezing when the seed contains 25 per cent moisture or more.[16]

Weed Control

From two to four cultivations are required to control the weeds in sorghum that is planted in rows. Fewer cultivations are required when the seed is planted in furrows than when level cultivation is practiced. Surface-planted sorghum in wide or narrow rows can be cultivated with a rotary hoe before the seedlings emerge as well as again shortly thereafter. Weed control with herbicides is less satisfactory in sorghum than in corn because the plants are more sensitive to the chemicals. Preemergence herbicides often are effective.

Irrigation

From 20 to 25 inches of water consisting of stored soil moisture, rainfall, and irrigation are required for maximum yields of sorghum.[19, 43] Of this amount, some 8 to 20 inches of water must be supplied by irrigation in the subhumid, semiarid, and arid regions. Preplanting irrigation that wets the soil to its field-carrying capacity to a depth of 6 to 7 feet will reduce the number of summer irrigations.[52] The maximum water usage is during the boot stage of plant growth when 1 inch (2.5 cm.) of water is consumed during 3 or 4 days. Water consumption increases in direct proportion to yield, as seen in Figure 13–10.

Harvesting

Grain sorghum usually is harvested with a combine (Figure 13–11).[20] Flexo-guards on the combine saves some heads from loss in

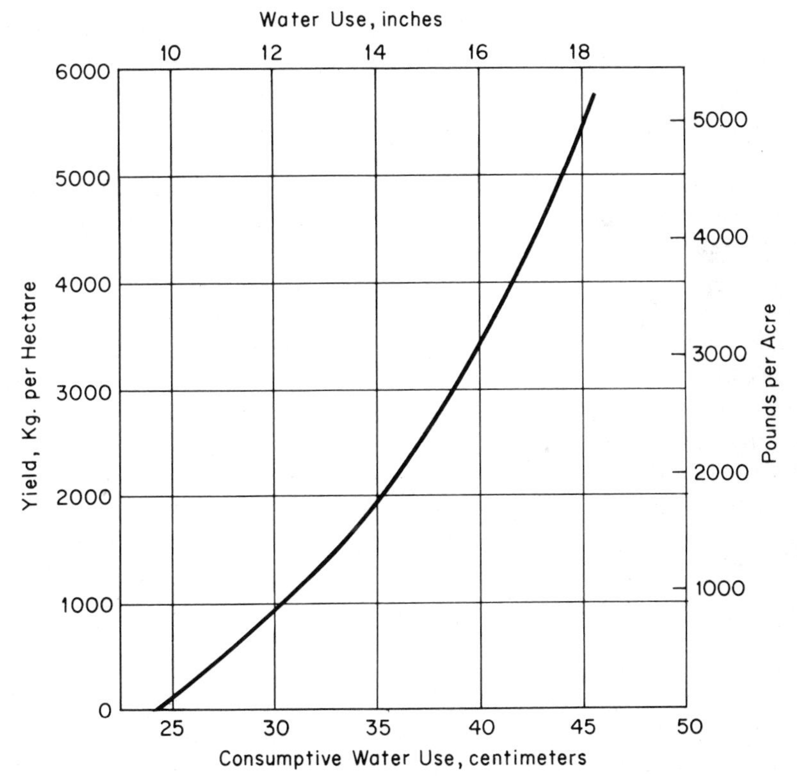

Figure 13–10. Water consumption as related to yield. Data obtained at North Platte, Nebraska.

Figure 13–11. Harvesting a dwarf grain sorghum with a self-propelled combine.

harvesting. Gathering lugs attached to the cutter bar of row harvester combines lift lodged stalks. The grain is mature when the seeds are fully colored, have begun to harden and contain as much as 18 to 20 per cent moisture. For combining, the crop should be allowed to dry until the grain contains 13 per cent or less moisture unless provision is made for drying the grain. The cheapest method of drying is to dump the threshed grain in long low tiers or piles on a clean, dry, sodded area. Artificial drying is practiced frequently (Figure 8–6). Desiccant sprays are used sometimes to dry the crop for combining.

A few days of dry weather after a severe freeze usually will reduce the moisture content of the grain to the point where it can be combined at a moisture content low enough for safe storage. Cured sorghum heads contain about 70 per cent grain.

The seed of sorghum should be fairly mature before the crop is cut for forage[37] for several reasons: (1) the largest tonnage of dry matter is obtained from mature plants, (2) the feed is more palatable, (3) the plants contain less prussic acid, (4) the fodder does not sour in the shock so easily, and (5) the silage made from mature sorghum is drier and thus less acid. The total dry weight of the plants may increase about 40 per cent between first heading and maturity.

For silage, sorghum is best harvested with a field silage cutter. For seed, bundle feed, or fodder, sorgo in rows is usually harvested with the row (or corn) binder, but the wheat binder is more efficient for light crops of short-stalked varieties. Usually two to three months of curing in the shock are necessary before the fodder is dry enough for stacking.

Drilled sorgo is cut for hay with a mower. Because of the thick juicy stems, it takes a long time to cure sorgo sufficiently for stacking unless it is crimped. The crop is often allowed to lie in the swath for a day or two, then raked into windrows, where it is left for 4 or more days for curing, and then usually is bunched with a buck rake or hay-rake. Then it is left in bunches for two weeks or longer. Drilled sorgo should be allowed to form seed before it is harvested, except when it has been severely injured by drought and will not head.

SUDANGRASS CULTURE

Sudangrass will endure considerable drought but, like all sorghums, is sensitive to low temperatures. However, it succeeds in areas where the season is too short to mature sorgos except of the very early types.

The crop is grown much the same as other sorghums, being planted after the soil has become warm, usually about two weeks after corn is planted. The date of planting varies from April 1 to July 1 from the extreme South to the North.[3] Sudangrass is often planted in close drills for hay or pasture, but in the southwestern semiarid region it is planted in cultivated rows for all purposes. Since the plant tillers profusely, the final weight of stems per unit area usually is about the same for rates of seeding, ranging from 15 to 40 pounds per acre. The usual rate for close drills under humid conditions is 20 to 25 pounds per acre, while the rate under semiarid conditions is generally 15 to 20 pounds. When planted in rows 36 to 44 inches apart, 2 to 4 pounds of seed per acre will produce a good stand.

Sudangrass is commonly cut for hay when the first heads appear, being more palatable at this stage than when cut later. Usually two crops are harvested for hay, although three or even four may be cut under favorable conditions. For seed production, the second crop usually is saved for seed, the first crop being cut rather early for hay. Sudangrass grown in rows for seed usually is cut with a windrower. It should be cut after the greatest amount of seed appears to be ripe, and before it shatters. The straw is about as valuable for feed as is prairie hay.

Sudangrass hybrids and sorgrass (sorghum-sudangrass hybrids) are now frequently grown. The sorgrass is used chiefly for silage. It outyields sudangrass but the stalks are much thicker.[3]

BROOMCORN CULTURE

Broomcorn can be grown wherever the temperatures are high enough for dependable corn production.

Seedbed preparation for broomcorn is the same as for other sorghums. The crop is planted in rows about 1 meter apart. In the humid regions, the plants should be spaced 3 inches apart in the row. A thinner spacing of plants, 6 to 9 inches apart in the row, is desirable in the western broomcorn areas because the moisture supply is likely to be limited. The quantity of seed required usually ranges from 2 to 4 pounds per acre. Broomcorn is planted between April 1 and July 15, generally from May 1 to June 15, or after the soil has become warm. June planting is generally advocated in the western areas (Figure 7–7). Two or three periods of planting are helpful in spreading the harvest period.

Broomcorn brush turns from pale yellow to green before maturity. It should be harvested when the entire brush is green from the tip down to the knuckle. The fibers will be weak at the bottom if they are harvested while the lower ends are still yellow. The seeds are about in the milk stage when the brush is ready to harvest. About four or five days after the proper stage is reached, the brush often begins to redden, and is then less flexible. Improved varieties free from red stain are now being grown.

Tall standard broomcorn is tabled before it is cut from the stalk.[31] The tabler walks backward between two rows and breaks the stalks diagonally across each other to form what is called a table of the two rows, which is from 2½ to 3 feet high. In the next operation the brush is cut, pulled out of the boot, and placed on the table to dry for a short time. The brush is usually taken to a curing shed within 24 hours. The heads are cut from the standing stalks of the shorter standard plants in semiarid areas. When dwarf varieties are grown, the heads are jerked or pulled from the stalks. The heads are placed in bunches on the ground or between the stalks to dry somewhat before they are hauled from the field.

Broomcorn may be threshed either before or after curing. It is of better quality when threshed before curing because fewer of the fine branches are knocked off when the brush is moist and flexible. The best quality of broomcorn is cured in 4- to 6-inch layers on slats in sheds. Curing requires from 10 to 20 days, after which the broomcorn is baled. The bales weigh about 330 pounds each. In the western broomcorn area, where the crop is now grown, the brush often remains in the field for several days before it is hauled. The curing usually is completed in outdoor ricks. After the crop is entirely dry, it is threshed, and then baled, in one operation. Heavy rains may stain the brush in the rick.

In hauling, curing, threshing, and baling, the brush is handled in small bunches or armfuls to keep the fibers straight and untangled. These operations plus harvesting may require 120 to 130 man-hours for each ton of shed-cured brush cut from tabled stalks and handled so as to obtain a high-quality product.

This vast labor requirement caused a decline in broomcorn production in the United States. Its production in Illinois ceased in 1967 after a century of successful culture of a high-quality product. Fully half of the domestic needs are now being imported from Mexico.

MARKET FACTORS IN BROOMCORN

High-quality broomcorn brush is pea-green in color, with straight fibers, and is free from discoloration. The fibers should be approximately 20 inches long, pliable, and free from coarseness. Brush that is overripe, reddened, or bleached is of poor quality. Defective brush consists chiefly of spikes, crooks, and twisted, coarse, flat, stemmy fibers.

Broom factories are located mostly in cities or near large centers of population because it is cheaper to ship broomcorn than brooms.

At a factory, the brush is sorted for fiber length and quality and then dipped in a solution of usually green dye. It is then bleached with sulfur dioxide fumes to remove stains and excess dye. It is then worked into brooms while the fibers are still moist and pliable. Brooms are made by placing successive handfuls of the fiber around a handle and wrapping them on with fine wire while the handle revolves in a winder (Figure 13–12). The shoulder of the broom usually is made by winding a bunch of line fibers on each of two opposite sides with the fibers pointed toward the upper end of the handle and then bending the fibers downward. Then the outer or hurl brush is attached to give the broom a covering of smooth fibers. The fibers

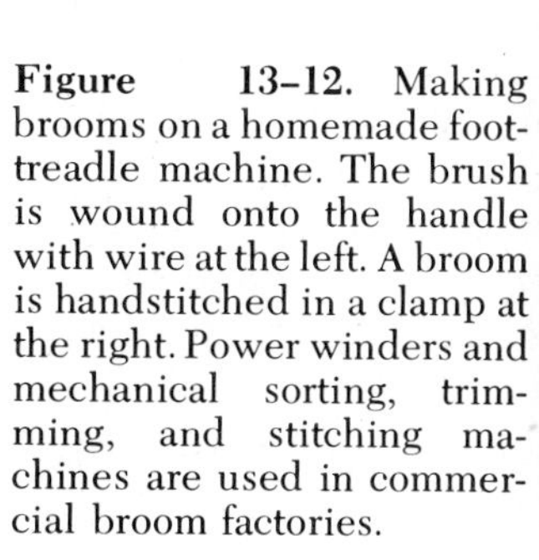

Figure 13–12. Making brooms on a homemade foot-treadle machine. The brush is wound onto the handle with wire at the left. A broom is handstitched in a clamp at the right. Power winders and mechanical sorting, trimming, and stitching machines are used in commercial broom factories.

are then sewed and trimmed. A ton of broomcorn brush makes 80 to 100 dozen brooms.[31]

JOHNSONGRASS AS A CROP

Johnsongrass is such a noxious weed pest that its importance as a pasture and forage crop is often overlooked. It is a perennial hay crop of the South, where it also furnishes an appreciable portion of the pasture. Johnsongrass is widely distributed throughout the South and as far north as the 38th parallel from the Atlantic Coast to the Colorado border. It is found even farther northward in the Potomac and Ohio Valleys and in California, Oregon, and Washington. Johnsongrass seldom is sown. Much of its spread occurred from crop seeds that contained johnsongrass or by grazing livestock.

Johnsongrass was collected as early as 1696 apparently at Aleppo, a town in Syria. It is still known in Europe and North Africa as Aleppo grass or Sorgho d'Alep. It is regarded as a native of the Mediterranean coastal countries of Europe, Africa, and Asia where it extends eastward into India. It was introduced into South Carolina from Turkey in 1830. Ten years later it was popularized by a Colonel William A. Johnson of Selma, Alabama, which accounts for its present name. Despite its weedy habits, johnsongrass was recommended as a crop by certain agricultural authorities as late as about 1890, and again for controlling wind erosion about 1935.

Johnsongrass is most abundant and vigorous on the richer bottomlands. Where such land is reserved for johnsongrass hay meadow it is often plowed every two or three years to break up and cover the old stubble and clear the land temporarily, which stimulates new growth from the rhizomes. Sometimes winter oats are sown in the fall and harvested in the late spring, and the johnsongrass then allowed to grow for hay. When the land is needed for other crops the johnsongrass is pastured heavily or mowed frequently to deplete the food reserves in the roots and rhizomes and retard development of new stalks or rhizomes. The rhizomes that develop under these conditions are mostly rather shallow. The land is then plowed in the spring and planted to a cultivated crop, or is summer-fallowed. Under these circumstances the johnsongrass often is eradicated by six cultivations at two-week intervals on semiarid land and by 10 to 15 cultivations in humid areas. The rhizomes develop mostly after the plant blooms. Rhizomes and roots live only about a year. Control is aided by repeated applications of suitable herbicides.

Johnsongrass usually is cut for hay before it blooms in order to avoid the development and dispersal of seeds.[63]

USES OF SORGHUM

Feed

Most of the sorghum grain grown in Asia and the African tropics is used for human food. Elsewhere it is generally fed to livestock or poultry.

The grain should be ground, steam-rolled or flaked, popped or dry-heated before feeding to increase its digestibility.[49] About 28,000 metric tons are processed by steaming or heating methods each year in the United States. In general, 1 pound of sorghum grain is approximately equal to 1 pound of corn in nutritive value for feeding diary cows or fattening lambs. It is equal to about 95 per cent of the feed value of corn on a pound-for-pound basis for fattening beef cattle or hogs.

Industrial Uses

Considerable grain sorghum replaces corn grits in the brewing and distilling industries, and in the manufacture of alcohol. During World War II, the grain of waxy varieties was used for extraction of starch to manufacture a satisfactory substitute for *minute tapioca.* The important use is in the manufacture of starch, glucose, sirup, oil, gluten feeds, and other products similar to those produced in the wet-milling of corn. Grain-sorghum flour is used for adhesives.[39, 40] The entire plants of sorgo sometimes are chopped, dehydrated, ground, and pelleted for feed.[59]

Composition

The composition of grain sorghum (Appendix Table 2) is similar to that of corn except that it is slightly higher in protein and lower in fat.[21]

Grain sorghum hybrids and varieties differ widely in digestibility when fed to livestock and they generally are lower in digestibility than is corn. The brown-seeded (bird-resistant) types are high in tannin. Those containing more than 1.6 per cent tannin depressed the growth of chicks when the constituted as much as 50 per cent of their diet. Sorghum gluten feed is especially high in tannin because it includes the seedcoats which bear the tannin.[12] In other digestion experiments each additional 1 per cent in tannin content lowered digestible dry matter by about 4 points.

Types with a high lysine and low tannin contents are now being developed.[55]

Sorgo bundles fed as roughage require some supplemental grain, legume hays, or meals to balance the ration. Sorghum silage has about the same composition as corn silage. The comparative feeding value of the two silages varies with the relative moisture content and the proportion of grain they contain. When these two factors are equal, corn silage still has a higher feeding value because the grain is softer and more completely absorbed in the intestinal tract. The silage of sorghum-sudangrass hybrids has about 90 per cent of the nutritive value of corn silage.[3]

Sorgo and sudangrass hay are coarser but similar in chemical composition to other grass hays except in being somewhat higher in protein and ash and lower in fat and crude fiber due to the fact that they usually are cut at an earlier stage of growth.

Sirup

Sorgo for sirup is cut when the seed is nearly ripe, or at least in the stiff-dough stage. Before cutting, the leaves usually are stripped from the stalks by hand by a worker using a special two-pronged tool or a lath, paddle, or pitchfork.[57] After cutting, the stalks are topped with a knife to remove the heads, peduncles, and, often, two or three of the upper internodes. The juice of the upper internodes and peduncle is much higher in starch and mineral matter than that from the remainder of the stalk. The leaves constitute about 10 per cent of the total green weight of the stalk,[67] and the topped portion, 20 to 30 per cent. For large sirup-factory operations, the whole stalks can be topped, after which the stalks are chopped into short sections mechanically. Later, the leaves are blown out by fans. A fair quality of sirup can be made from whole stalks that have been stored as long as about 6 days before crushing.[7] The juice is expressed from the stripped cane by the crushing of the stalks between revolving fluted iron rolls in a cane mill equipped with three (or occasionally two) rolls (Figure 13–13). The strained juice that sometimes is allowed to settle for a time is piped to an evaporating pan where it is boiled down to a sirup containing about 70 per cent sugar. During the boiling, the juice is skimmed constantly to remove floating impurities such as chlorophyll, soil, plant fragments, proteins, gums, fats, and waxes. Often, the juice or the partly evaporated semisirup is treated with lime, clay products, or phosphoric acid to neutralize the acid juice and to precipitate impurities. Treatment with malt diastase, an enzyme product, hydrolyzes the starch into glucose and prevents thickening or jellying due to the 0.5 to 3 per cent of starch usually present in the juice. With such treatment a concentrated sirup containing as much as 70 per cent sugar will pour readily. Sometimes a yeast extract containing another enzyme, invertase, also is added to change part of the sucrose (cane sugar) into dextrose (glucose) and levulose (grape sugar), and thus prevent crystallization (sugaring). The juice of a good variety of sorgo grown under suitable conditions contains 13 to 17 per cent sugar, of which 10 to 14 per cent is sucrose.

A good field should yield 15 tons of fresh sorghum or 10 tons of stripped stalks per acre. A ton of stripped stalks should yield 700 to 1200 pounds of juice or 8 to 20 gallons of sirup. However, the average yield of sirup is less than 80 gallons per acre. Sorgo sirup acquires its flavor or tang chiefly from organic acids that are present in the juice. The sirup is rich in iron, particularly when it has been evaporated in an iron pan or kettle. Sorghum sirup often is referred to as molasses, which is a misnomer, since molasses is a by-product of sugar manufacture. The production of sorgo sirup has declined about 99 per cent since 1920 when the industry reached its peak.

Sugar

Attempts to develop a sorghum sugar industry have not been successful thus far because of certain difficulties and limitations that make the sugar more expensive than that derived from sugarcane or sugarbeets. Sorghum sugar was manufactured on a commercial scale under public subsidy in Kansas and New Jersey about 1890. Extensive experiments on sorghum sugar manufacture were made between 1878 and 1893 and again between 1935 and 1941. The crystallization of the sugar is satisfactory only after the starch and the aconitic acid crystals have been removed by centrifuging. The stalks must be harvested promptly when they reach the proper stage. They may lose sucrose (cane sugar) by inversion to dextrose and levulose when not processed soon. Sorghum often is less uniform in yield and com-

Figure 13–13. Crushing sorgo stalks between rolls to extract the juice for boiling down into sirup.

position than are sugarcane and sugarbeets. The yields of sorghum that is grown during a period of 4 or 5 months obviously are less than are obtained from sugarcane that has a growing period of 8 to 14 months. Further research, and development of varieties which are adapted to the extreme south and which also yield a high percentage of sucrose, may ultimately create an economical source of sorghum sugar.

DISEASES

The principal diseases attacking sorghum in the United States[22] are smuts, leaf spots, and root and stalk rots. Other serious diseases attack sorghum in the Eastern Hemisphere.[66]

Kernel Smuts

The kernel smuts may reduce seed production materially, but probably have less effect on forage yields. All sorgos, kafirs, and broomcorns are susceptible. Each infected ovule becomes a mass of dark-colored spores instead of a sorghum seed (Figure 13–14). Covered-kernel smut (*Sphacelotheca sorghi*) and loose-kernel smut (*S. cruenta*) can be controlled by seed treatment with fungicides. Some hybrids and varieties of grain sorghum are resistant to certain races of these smut organisms.

Head Smut

The head-smut disease is caused by *Sphacelotheca reiliana*. The entire head is replaced by a gall or smut mass (Figure 13–14). The spores are carried in the soil and occasionally on the seed. Seed from fields where the disease is prevalent should be avoided. Certain varieties and hybrids are inclined to have head smut. Seed treatment offers no protection against infection carried in the soil. Removal and burning of diseased plants will eliminate reinfestation of the soil.

Root and Stalk Rots

Root rot, which attacks the roots and crown of a plant, formerly caused extensive damage to some varieties, but resistant strains and hybrids replaced the susceptible varieties. This disease is caused by a widespread soil-borne fungus, *Periconia circinati*. Charcoal rot attacks the base of the stalk and also the larger roots, rotting away the pith and causing the stalk to fall over. The disease is caused by a soil-borne fungus, *Macrophomina phaseoli* (*Sclerotium bataticola*) which also attacks several other crop plants. Damage from this disease is limited when the soil is well supplied with moisture and the sorghum planting is delayed so that the crop reaches maturity under mild, favorable fall conditions.

Figure 13–14. Sorghum smuts. (*Left to right*) Sound head, covered kernel smut, loose kernel smut, and head smut.

Leaf Spots

Bacterial leaf-spot diseases include stripe, caused by *Pseudomonas andropogoni;* spot, caused by *Pseudomonas syringae;* and streak, caused by *Xanthomonas holicola.* Fungus leaf-spot diseases include sooty stripe, caused by *Titeospora andropogonis;* rough spot, caused by *Ascochyta sorghina;* anthracnose, caused by *Colletotrichum graminicolum;* leaf blight, caused by *Helminthosporium turcicum;* zonate leaf spot, caused by *Gleocercospora sorghi;* gray leaf spot, caused by *Cercospora sorghi;* and rust, caused by *Puccinia purpurea.*

Sorghums containing a yellow leaf pigment are resistant to bacterial leaf spots. Others are resistant to several of the fungus leaf-spot diseases. Maize dwarf mosaic virus overwinters in johnsongrass and is carried to sorghum chiefly by aphids.

Other Diseases

Other diseases attacking sorghum include *Rhizoctonia* stalk rot, caused by *Rhizoctonia solani*, anthracnose stalk rot and red rot, caused by species of *Colletotrichum*, downy mildews caused by *Sclerospora* species, and crazy top caused by *Sclerospora sorghi*.

Weakneck is not a true disease, but a decay at the base of the peduncle that occurs after the upper part of the stalk has ceased growth and begun to dry. This decay causes mature heads to fall over and be lost.

INSECTS

The most destructive insect pests on the sorghums in the United States are the chinch bug, sorghum midge, corn-leaf aphids, corn earworm (Figure 12–20), and sorghum webworm. Grasshoppers cause little injury to the foliage of sorghum even where corn is wholly destroyed. However, they seem to prefer the leaves of certain varieties. Grasshoppers eat developing sorghum grain, but they may be controlled with insecticide sprays. Other pests occur in Africa and Asia.[66]

Chinch Bugs

The chinch bug, *Blissus leucopterus*, is a sucking insect that feeds heavily on the leaves and sheaths of sorghum (Figure 12–20). The chinch bug moves to sorghum from nearby small-grain fields after the harvest of the latter. In Kansas and northward, they usually migrate on foot from small grains to sorghum, but later in the season they acquire wings and fly to the sorghum field. There they mate and deposit eggs. These eggs hatch a second generation which continues to feed on the sorghum plants. A third generation develops in the South in the fall. The most serious injury by chinch bugs is caused in seasons of low rainfall. Many chinch bugs die in cool wet weather. The migration of chinch bugs on foot from small-grain fields to sorghum may be prevented by spraying border strips with insecticides along the two fields. South of central Kansas chinch bugs develop wings by the time the small grains are harvested. When they migrate by flight, barriers are ineffective, but injury can be reduced by early planting, resistant varieties, or by spraying with insecticides. The adult chinch bugs hibernate at the base of bunches of grass or in trash.

Common varieties and hybrids of sorghum vary in susceptibility to chinch-bug injury.

Sorghum Midge

The sorghum midge is abundant in the southern states. It formerly made grain-sorghum production unprofitable in that region, but later the damage was less severe, probably because sorghum is more abundant and because more natural enemies of the midge are present. The midge is a small fly with a red body.[37] It lays its eggs inside the glumes at the blossom stage. The egg produces a small white larva that absorbs juices from the sorghum ovary and prevents seed development. The adult midge emerges to start a new generation every 14 to 20 days under favorable conditions. The larva causes the injury to the developing seed.

The midge breeds on johnsongrass in the spring, and infests sorghums as soon as they head. All sorghums are subject to attack. The midge overwinters in the larval stage on sorghum heads, in johnsongrass, and on trash. Planting sorghums very early in the season, so that they come into bloom before the midge is plentiful, is one means of combating this insect.

Other Insects

In some seasons, the corn-leaf aphid (*Aphis maidis*) is a serious pest of sorghum, when large numbers occur in the leaf curl. This hinders the exsertion of the head from the boot. The honey-dew secretion of the aphids forms a sticky mass about the plant heads, which encourages the growth of molds, which in turn prevent the production of high-quality seed. Aphids winter in the southern states and migrate north.

The sorghum webworm (*Celama sorghiella*) is rather destructive periodically when it devours developing grains, especially on late-planted fields. It, like the midge and aphid, can be controlled by spraying with insecticides.

The Banks grass mite (*Oligonychus pratensis*) has acquired tolerance to pesticides and is threatening sorghum in Texas.

REFERENCES

1. Amador, J., and others. "Sorghum diseases," *Tex. Agr. Exp. Sta. Bul. 1085*, pp. 1–21. 1969.
2. Anonymous. *Sorghum: A Bibliography of the World Literature Covering the Years 1930–1963*, Scarecrow Press, Metuchen, N.J., pp. 1–301. 1967.
3. Anonymous. "Sudangrass and sorghum-sudangrass hybrids for forage," *USDA Farmers Bul. 2241*, pp. 1–12. 1969.
4. Artschwager, E. "Anatomy and morphology of the vegetative organ of *Sorghum vulgare*," *USDA Tech. Bul.* 957. 1948.

5. Bond, J. J., Army, T. J., and Lehman, O. R. "Row spacing, plant populations, and moisture supply as factors in dryland grain sorghum production," *Agron. J.*, 56(1):3–6. 1964.
6. Brandon, J. F., Curtis, J. J., and Robertson, D. W. "Sorghums in Colorado," *Colo. Exp. Sta. Bul. 449*. 1938.
7. Coleman, O. H., and Stokes, I. E. "Storage studies of sorgo," *USDA Tech. Bul. 1307*. 1964.
8. Conrad, J. P. "Some causes of the injurious after-effects of sorghums and suggested remedies," *J. Am. Soc. Agron.*, 19:1091–1111. 1927.
9. Conrad, J. P. "Fertilizer and legume experiments following sorghums," *J. Am. Soc. Agron.*, 20:1211–1234. 1928.
10. Conrad, J. P. "The carbohydrate composition of corn and sorghum roots," *J. Am. Soc. Agron.*, 29:1014–1021. 1937.
11. DeWet, J. M. J., and Harlan, J. R. "The origin and domestication of *Sorghum bicolor*," *Econ. Bot.*, 25(2):128–125. 1971.
12. Fuller, H. L., and others. "The feeding value of grain sorghums in relation to their tannin content," *Ga. Agr. Exp. Sta. Bul. N. S. 176*, pp. 1–14. 1966.
13. Gangstad, E. O. "DeSoto grass," *Tex. Res. Foundation Hoblitzelle Ag. Lab. Bul. 22*. 1965.
14. Gillingham, J. T., and others. "Relative occurrence of toxic concentrations of cyanide and nitrate in varieties of sudangrass and sorghum-sudangrass hybrids," *Agron. J.*, 61(5):727–730. 1969.
15. Gray, R. B. "Harvesting with combines," *USDA Farmers Bul. 1761*. 1955.
16. Gritton, E. T., and Atkins, R. E. "Germination of sorghum grain as affected by freezing temperatures," *Agron. J.*, 55(2):139–142. 1963.
17. Harris, H. B. "Grain sorghum production in Georgia," *Ga. Agr. Exp. Sta. Rsh. Rpt. 98*, pp. 1–34. 1971.
18. Hobbs, J. A. "Yield and protein contents of crops in various rotations," *Agron. J.*, 63(6):832–836. 1971.
19. Jensen, M. E., and Musick, J. T. "Irrigating grain sorghums," *USDA Leaflet 511*. 1962.
20. Karper, R. E., and others. "Grain sorghum date of planting and spacing experiments," *Tex. Agr. Exp. Sta. Bul. 424*. 1931.
21. Le Clerc, J. A., and Bailey, L. H. "The composition of grain sorghum kernels," *J. Am. Soc. Agron.*, 9:1–16. 1917.
22. Leukel, R. W., and Martin, J. H. "Four enemies of sorghum crops," in *Plant Diseases*, USDA Yearbook, 1953, pp. 368–377. 1953.
23. Malm, N. R., and Finkner, M. D. "Fertilizer rates for irrigated grain sorghum on the High Plains," *N. Mex. Agr. Exp. Sta. Bul. 523*. 1968.
24. Mann, H. O. "Effects of rates of seeding and row widths on grain sorghum grown under dryland conditions," *Agron. J.*, 57:173–176. 1965.
25. Martin, J. H. "Comparative drought resistance of sorghums and corn," *J. Am. Soc. Agron.*, 22:993–1003. 1930.
26. Martin, J. H. "Recurving in sorghums," *J. Am. Soc. Agron.*, 24:500–503. 1932.
27. Martin, J. H. "The use of the greenhouse in sorghum breeding," *J. Hered.*, 25(6):251–254. 1934.

28. Martin, J. H. "Sorghum improvement," in *USDA Yearbook*, 1936, pp. 523–560.
29. Martin, J. H. "Climate and sorghum," in *Climate and Man*, USDA Yearbook, 1941, pp. 343–347.
30. Martin, J. H. "Breeding sorghum for social objectives," *J. Hered.*, 36(4):99–106. 1945.
31. Martin, J. H. "Broomcorn—the frontiersman's cash crop," *Econ. Bot.*, 7:163–181. 1953.
32. Martin, J. H. "Sorghums, broomcorn, sudangrass, johnsongrass, and millets," in *Grassland Seeds*, Van Nostrand, New York, pp. 605–628. 1957.
33. Martin, J. H. "Sorghum and pearl millet," in *Handbuch der Pflanzenzuchtung*, Paul Parey, Berlin, vol. II, pp. 565–587. 1959.
34. Martin, J. H. "Field Crops," in *Seeds*, USDA Yearbook, 1961, pp. 663–665.
35. Martin, J. H., Couch, J. F., and Briese, R. R. "Hydrocyanic acid content of different parts of the sorghum plant," *J. Am. Soc. Agron.*, 30:725–734. 1938.
36. Martin, J. H., and others. "Spacing and date-of-seeding experiments with grain sorghums," *USDA Tech. Bul. 131.* 1929.
37. Martin, J. H., and Stephens, J. C. "The culture and the use of sorghums for forage," *USDA Farmers Bul. 1844.* 1940.
38. Martin, J. H., Taylor, J. W., and Leukel, R. W. "Effect of soil temperature and depth of planting on the emergence and development of sorghum seedlings in the greenhouse," *J. Am. Soc. Agron.*, 27:660–665. 1935.
39. Martin, J. H., and Jenkins, M. T. "New uses for waxy-cereal starches," in *Crops in Peace and War*, USDA Yearbook, 1950–51, pp. 159–162.
40. Martin, J. H., and MacMasters, M. M. "Industrial uses for grain sorghum," in *Crops in Peace and War*, USDA Yearbook, 1950–51, pp. 349–352.
41. Mathews, O. R., and Brown, L. A. "Winter wheat and sorghum production in the Southern Great Plains under limited rainfall," *USDA Circ. 477*, pp. 1–60. 1938.
42. Musick, J. T., and Dusek, D. A. "Grain sorghum row spacing and planting rates under limited irrigation in the Texas High Plains," *Tex. Agr. Exp. Sta. MP 932*, pp. 1–10. 1969.
43. Musick, J. T., Grimes, J. W., and Harron, G. M. "Irrigation water management and nitrogen fertilization of grain sorghum," *Agron. J.*, 55(3):295–298. 1963.
44. Plucknett, D. S., and others. "Sorghum production in Hawaii," *Haw. Agr. Exp. Sta. Bul. 143*, pp. 1–33. 1971.
45. Poehlman, J. M. *Breeding Field Crops*, Holt, New York, pp. 1–427. 1959.
46. Quinby, J. R. *A Triumph of Research—Sorghum Production in Texas*, Tex. A & M Univ. Press, College Station, Texas, pp. 1–28. 1971.
46a. Quinby, J. R. *Sorghum improvement and the genetics of growth*, Tex. A & M Univ. Press. College Station, Texas, pp. 1–108. 1974.
47. Quinby, J. R., and Martin, J. H. "Sorghum Improvement," in *Advances in Agronomy*, vol. 6, Academic Press, New York, pp. 305–359. 1954.

48. Quinby, J. R., and others. "Grain sorghum production in Texas," *Tex. Agr. Exp. Sta. Bul. 912.* 1958.
49. Riggs, J. K., and others. "Dry heat processing of sorghum grain for beef cattle," *Tex. Agr. Exp. Sta. Bul. 1096,* pp. 1–11. 1970.
50. Riggs, J. K., and others. *Tex. Agr. Exp. Sta. Bul. 1096,* pp. 1–11. 1970.
51. Ross, W. M., and Laude, H. H. "Growing sorghums in Kansas," *Kans. Agr. Exp. Sta. Cir. 319.* 1955.
52. Ross, W. M., and Webster, O. J. "Culture and use of grain sorghum," *USDA Handbk. 385,* pp. 1–30. 1970.
53. Sieglinger, J. B. "Cross pollination of milo in adjoining rows," *J. Am. Soc. Agron.,* 13:280–282. 1921.
54. Sieglinger, J. B. "Leaf number of sorghum stalks," *J. Am. Soc. Agron.,* 28:636–642. 1936.
55. Sieglinger, J. B., and Martin, J. H. "Tillering ability of sorghum varieties," *J. Am. Soc. Agron.,* 31:475–488. 1939.
55a. Singh, R., and Axtell, J. D. "High-lysine mutant gene (hl) that improves protein quality and biological value of grain sorghum," *Crop Sci.,* 13(5):535. 1973.
56. Stickler, F. C., and others. "Row width and plant population studies with grain sorghum at Manhattan, Kansas," *Crop Sci.,* 1(4):297–300. 1961.
57. Stokes, I. E., Coleman, O. H., and Dean, J. S. "Culture of sorgo for sirup production," *USDA Farmers Bul. 2100.* 1957.
58. Sumner, D. C., and others. "Sudangrass, hybrid sudangrass and sorghum × sudangrass crosses," *Ca. Agr. Exp. Sta. Circ. 547,* pp. 1–16. 1968.
59. Swanson, A. F. "Making pellets of forage sorghum," in *Crops in Peace and War,* USDA Yearbook, 1950–51, pp. 353–356.
60. Swanson, A. F., and Hunter, R. "Effect of germination and seed size on sorghum stands," *J. Am. Soc. Agron.,* 28:997–1004. 1936.
61. Swanson, A. F., and Laude, H. H. "Sorghums for Kansas," *Kans. Agr. Exp. Sta. Bul. 304,* pp. 1–63. 1942.
62. Vinall, H. N. "A study of the literature concerning poisoning of cattle by the prussic acid in sorghum, Sudan grass, and Johnson grass," *J. Am. Soc. Agron.,* 13:267–280. 1921.
63. Vinall, H. N., and Crosby, M. A. "The production of Johnson grass for hay and pasturage," *USDA Farmers Bul. 1957,* pp. 1–26. 1929.
64. Vinall, H. N., and Reed, H. R. "Effect of temperature and other meteorological factors on the growth of sorghums," *J. Agr. Res.,* 13:133–147. 1918.
65. Vinall, H. N., Stephens, J. C., and Martin, J. H. "Identification, history, and distribution of common sorghum varieties," *USDA Tech. Bul. 506.* 1936.
66. Wall, J. S., and Ross, W. M., Eds. *Sorghum Production and Utilization,* AVI, Westport, Conn., pp. 1–702. 1970.
67. Walton, C. F., Ventre, E. K., and Byall, S. "Farm production of sorgo sirup," *USDA Farmers Bul. 1791,* pp. 1–40. 1938.
68. Wilson, B. D., and Wilson, J. K. "Relation of sorghum roots to certain biological processes," *J. Am. Soc. Agron.,* 20:747–754. 1928.

Chapter 14

Sugarcane

ECONOMIC IMPORTANCE

Sugarcane (*Saccharum officinarum*) is harvested on more than 11 million hectares each year, with a production exceeding 580 million metric tons of cane or 51 metric tons per hectare. Leading countries in sugarcane production are India, Brazil, Cuba, Mexico, China, and the United States (Figure 14–1). The average production in the United States from 1970 to 1972 exceeded 23 million metric tons of cane harvested from about 650,000 acres (260,000 hectares) or nearly 40 tons per acre (90 metric tons per hectare). The production was in Hawaii, Louisiana, and Florida except for 104,000 tons harvested in Texas in 1972. Between 400,000 and 500,000 additional tons also are harvested in Puerto Rico and the Virgin Islands each year. The lowest yields in the United States are obtained in Louisiana where the fall-planted cane has a growing period of 12 to 15 months and the spring-planted cane perhaps 9 months. The Florida plant cane grows for 15 to 18 months. The ratoon crop makes its growth in one year in both states. In Hawaii where both planted and stubble sugarcane are allowed to grow for 2 to 3 years before harvesting, the average yields exceed 100 metric tons per hectare. The best sugar yields are obtained by harvesting every 20 to 24 months.[24]

Of the average of 25,887,000 short tons of cane harvested from 1970 to 1972, 1,053,000 tons was saved for seed cane. The remaining 24,834,000 tons yielded 3,533,000 tons of raw sugar,* or 2,368,000 tons of refined sugar together with 154,142,000 gallons of molasses. An additional tonnage of sugarcane was processed for sirup in Louisiana, Georgia, Alabama, and Mississippi, and other southern states.

Per capita consumption of sugar in the United States since 1970 exceeds 100 pounds annually. Sugar supplies nearly one-seventh of the energy in the diet. It is an economical source of food energy.

* Raw value is the equivalent in terms of 96° sugar as defined in the Sugar Act of 1948.

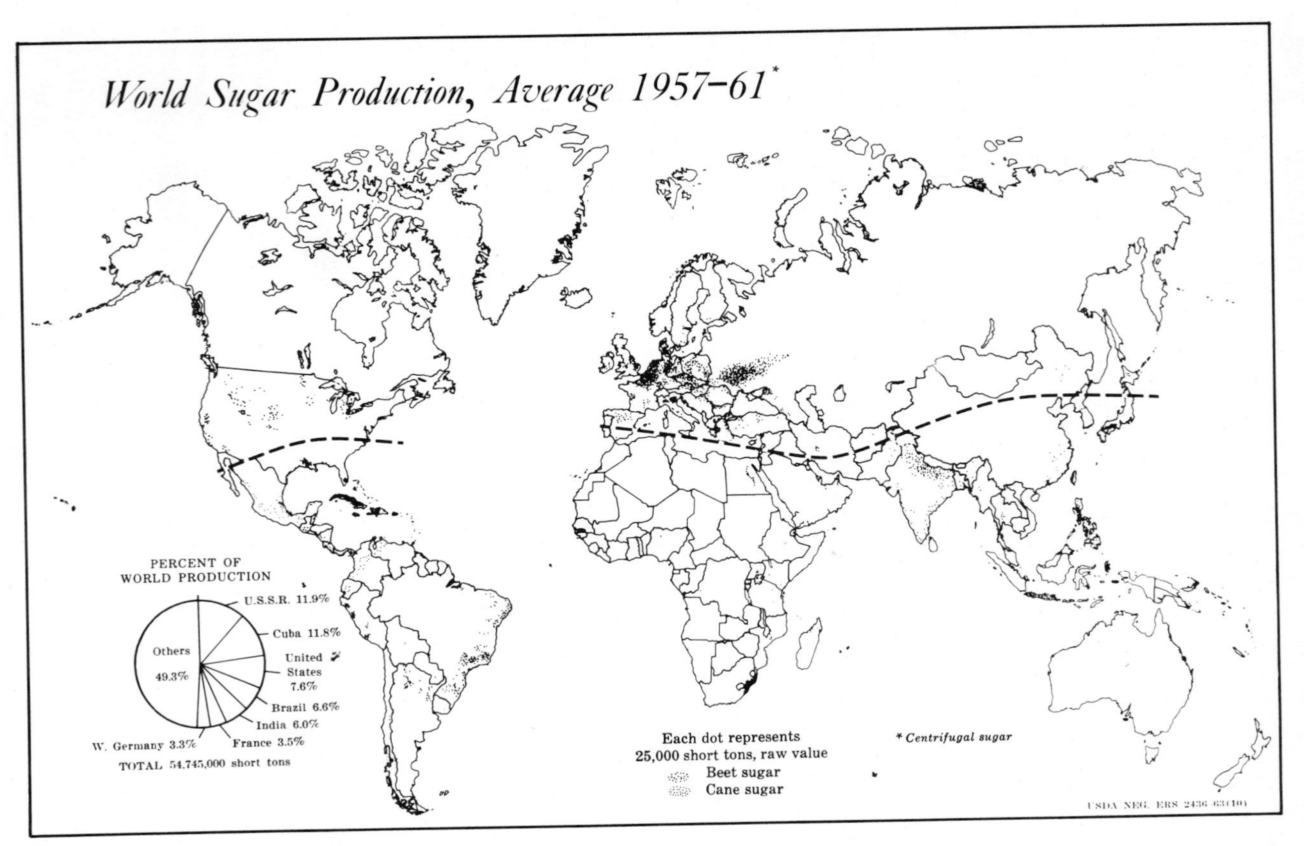

Figure 14–1. World sugar production.

ORIGIN AND HISTORY

Cultivated sugarcanes (*Saccharum* species) probably are derived from indigenous wild species on the islands of Melanesia. It is postulated that they arose by selection from wild canes in New Guinea.[6, 10] They were carried by man as stem cuttings to various centers where they were further modified by natural hybridization with other wild grasses.

Sugarcane of economic usefulness has been transported by man probably for thousands of years. Sugarcane was cultivated in India as early as 1000 B.C., while in China crude sugar was being made from sugarcane by 760 B.C. The first refined white sugar was made in Iran (Persia) about 600 A.D.[32] Most people still prefer refined sugar.

Sugarcane was taken to Santo Domingo by Columbus on his second voyage, in 1493, but the crop was not established permanently until 1506. Introduction of sugarcane to Louisiana dates from 1751, while the first granulated sugar was produced in that state in 1795.[22]

ADAPTATION

Sugarcane is primarily a tropical plant that usually requires 8 to 24 months to reach maturity. The temperatures should be high enough to permit rapid growth for 8 months or more. After a period of rapid growth there is usually, in most cane-producing countries, a period of slow sugarcane growth with increased sugar storage. The

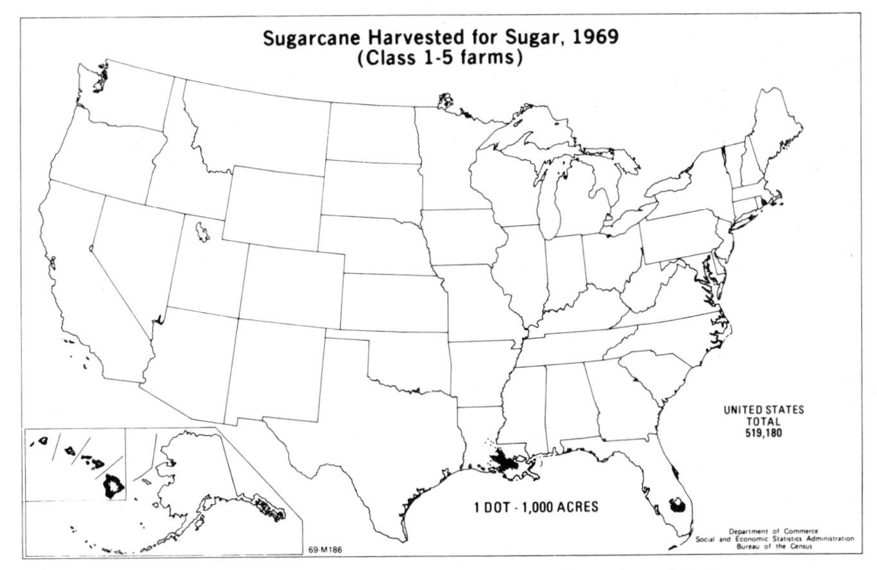

Figure 14–2. Sugarcane harvested for sugar in the United States.

cane is considered to be ripe when the sugar content is at its maximum. Sugarcane culture has gradually migrated from the tropics to the subtropical regions in the United States and other countries (Figure 14–2).

The best soils for sugarcane are well supplied, naturally or artificially, with nutrient minerals and organic matter. In Louisiana, heavy soils are used for sugar production, and lighter soils for cane sirup production. Drainage must be provided for low, wet lands and where the water table is less than one meter below the surface. In heavy and poorly drained soils a low oxygen content of the air at a 70-centimeter soil depth retards root development.[25] In Florida more than 90 per cent of the sugarcane acreage is on muck and peat soils containing 50 to 97 per cent of organic matter. The soils with the lower organic matter contents are preferable. In Hawaii, sugarcane is grown mostly on yellowish-brown or reddish-brown lateritic soils that require heavy fertilization, but are often high in manganese. Many of the fields are irrigated. Irrigation is beneficial and often essential for successful sugarcane production, except where a well-distributed rainfall of 120 cm. or more occurs annually.

BOTANICAL DESCRIPTION

Sugarcane is a tall perennial grass[5, 14, 33] with culms (Figure 3–4) bunched in stools of 5 to 50 culms or evenly scattered. The stalks are 2 to 5 cm. in diameter and may be 5 meters or more in height. The internodes are comparatively short, usually 5 to 7 cm., often being marked by corky cracks; mature canes may be green, yellow, purplish, or reddish. A sugarcane plant has a leafy appearance because of the abundant tillers and the numerous nodes, each of which produces a leaf. The stalks are covered with a layer of wax that forms a band at the top of each internode. In addition to the growth ring, there is a root band and a single bud or *eye* at each node. The leaf blade is usually erect, ascending, and gently curved, but it may be erect with a drooping tip (Figure 3–2). Fully developed blades measure about one meter in length. The leaf blade is green, sometimes with a purplish cast. The leaf sheath folds around the stem and serves as a protection to the bud.

Sugarcane seldom produces seed in the continental United States except in Florida. Consequently it is harvested before the inflorescence appears. Ordinarily the panicles do not appear until at least 12 to 24 months after the cane is planted. Most varieties flower after the shorter summer day lengths when grown in the tropics. A stalk ceases growth when it flowers. The inflorescence is an open, much-branched plume-like panicle known as the arrow, 20 to 60 cm. long (Figure 14–3). The spikelets are arranged in pairs, one spikelet being

Figure 14–3. Sugarcane inflorescence or "arrow."

sessile and the other pedicellate. The spikelets, about 3 millimeters long, are obscured in a tuft of silky hairs two to three times as long as the spikelet. Each flower is subtended by two bracts which form the outer and inner glumes (Figure 3–12). There is also a sterile lemma, but the fertile lemma and palea of the typical grass flower are wanting.[4, 33] The fruit is extremely small, about 1½ millimeters long, with a distinct constriction in the region opposite the embryo. The seed, which is enclosed in the glumes and the sterile lemma, is of low viability (Figure 7–1).

While the flowers of all sugarcane varieties are perfect, many plants are pollen-sterile or at least self-sterile.[13] There are probably no completely sterile varieties, except under extreme conditions.

SUGARCANE SPECIES

The sugarcanes of America are derived from five species and their hybrids. In the good sugar quality *noble* canes, *Saccharum officinarum*, the axis of the inflorescence is without long hairs. The stalk girth is large, the leaves wide, and the fiber content low. The Chinese canes, *S. sinense*, have long hairs on the axis of inflorescence. The northern Indian canes, *S. barberi*, differ from the Chinese canes in having narrower blades and more slender stalks.

The wild canes of Asia, *S. spontaneum*, are used in making hybrids with other species. They are a serious weed pest in India. The stalks are very slender, while the leaves are narrow. The wild cane of New Guinea, *S. robustum*,[11] has stalks of medium thickness (about $2\frac{1}{2}$ centimeters in diameter), very hard, and 25 to 30 feet in height. The nodes are swollen. The leaves are of medium width and rather long. The flowers are very similar to those of *S. officinarum*.

VARIETIES

Disease epidemics resulted in almost total destruction of the commonly cultivated sugarcane varieties in the United States by 1926. Disease-resistant varieties were introduced soon after that, while others were developed in the United States at Canal Point, Florida. These replaced the noble canes susceptible to mosaic, root rot, and red rot.[10] The resistant varieties of *Saccharum officinarum* were developed by crossing the susceptible, high-sugar noble varieties with other species, especially *S. spontaneum* and *S. barberi*, which are resistant to the most serious sugarcane diseases. These hybrids were then back-crossed to the noble varieties. High-yielding disease-resistant plants that were high in sugar percentage were selected from the back-cross progeny and increased by vegetative propagation. Practically all the improved domestic varieties were derived from all three of the above species. Current varieties are resistant to some mosaics and partly resistant to the sugarcane borer, and red rot.

Sugarcane was crossed with sorghum in India, Taiwan, and the United States,[9] but no improved commercial varieties have yet evolved from this wide intergeneric cross.

CULTURE

Sugarcane usually is not grown in a regular rotation with other crops. In Louisiana, the land usually is spring-plowed and then planted to soybeans for green manure or left fallow, after three cane crops have been harvested. This green manure is plowed under in early fall after which the land is again planted to sugar cane. In the

Florida everglades region the intervening green-manure crop is omitted on the muck and peat soils that already contain excessive quantities of organic matter and nitrogen. There is very little Hawaiian sugarcane grown in rotation with other crops.

Fertilizers

Sugarcane responds chiefly to nitrogen in Louisiana and Hawaii,[12, 31] and to some extent on the mineral soils in Florida. Rates of nitrogen application range from 90 to 155 kilos per hectare, with the higher rates for the ratoon crops. About 45 kilos per hectare of P_2O_5 and 90 kilos of K_2O are desirable in Louisiana and Hawaii, and 33 kilos and 160 kilos, respectively, on Florida mineral soils. The Florida peat and muck soils require no nitrogen, but about 20 Kg/hectare of P_2O_5 and 160 Kg/hectare of K_2O are desirable. Excessive phosphorus lowers the sugar content of the cane. An application of 550 Kg/ha of sulfur is helpful on Florida soils of *p*H above 6.5, and lime applications are helpful on Louisiana soils that are below *p*H 6.0.[15]

Several minor elements, particularly copper, zinc, and boron, are required on Florida muck and peat soils, and manganese also is needed on soils of *p*H 5.7 or higher. These elements may be needed elsewhere, and the need can be determined by plant-tissue tests. Dry fertilizers can be placed in the opened furrows during the planting operation. Nitrogen fertilizers can be applied in dry, liquid, or anhydrous forms. In Hawaii liquid nitrogen solutions may be metered into the irrigation water, or urea may be broadcast on the cane fields from an airplane.

Seedbed Preparation

Seedbed preparation for sugarcane is similar to that for corn in the same region. In Louisiana the soil is thrown into beds 1 to 2 feet wide and 5½ to 6 feet apart. A water furrow is left between the beds where the land is inclined to be wet. Level culture is practiced in Florida. The Hawaiian crop usually is grown on low beds. Deep tillage before planting is helpful on some Hawaiian soils.

Planting

Sugarcane is produced vegetatively by planting whole canes or 2- to 3-foot sections of cut canes. The planting cane is stripped and cut by hand or by machines. One to three continuous rows of cuttings or canes are dropped into furrows by men riding a cart or trailer. Even when machine planters with cutters are used the cut pieces are dropped into the newly opened furrows by men riding the planter. Fertilizers may be applied during this operation.[31] The spacing between rows ranges from 4 to 7 feet. The amount of seed cane planted varies from one to four tons per acre, depending upon the stalk di-

Figure 14–4. Planting sugarcane. [Courtesy Hawaiian Sugar Planters' Association.]

ameter as well as upon the number of lines planted in a single furrow (Figure 14–4). The stalks are covered with 2 to 8 inches of soil, only enough to avoid frost injury. A light furrow turned from each side covers the seed canes.

In Louisiana, the greater part of the crop is planted for sugar production from August 1 to November 15.[16] Planting previous to sugar harvest is commonly practiced in order to avoid labor conflicts. In the northern portion of the sugarcane-growing area, the crop may be spring-planted. There the canes are preserved over the winter in mats or windrows, being covered with soil to protect them from the cold. Canes for fall planting are usually cut before the plants are mature enough for sugar harvest. Even when fall-planted, the new crop makes very little growth until spring. In Hawaii and other continuously warm areas the cane may be planted at any time of the year.

Three crops of sugarcane usually are harvested before the field is plowed up and replanted. The first crop is known as plant cane, while subsequent crops are known as stubble or ratoon crops. Yields from ratoon crops usually are lower. A common spring operation in Louisiana and Florida is to bar off, i.e., to throw aside the soil covering from either the stubble or nonemerged fall-planted cane, using a plow, harrow, or other implement. This leaves the cane on a narrow bed that warms up quickly to start spring growth of the sugarcane. After growth is well started, the fertilizer is applied in furrows next to the row and the furrows are levelled. The seasonal cultivation and the bedding move so much soil toward the cane row that it is necessary to bar off each spring.

Cultivation

Cultivation to control weeds is necessary until the plants shade the ground. Weeds within the row are destroyed by hand hoeing. Weeds, chiefly johnsongrass and alligator weed (*Alternanthera philoxeroides*), have been controlled by flaming. Alligator weed and other broad-leaved weeds also can be controlled with herbicides applied before the cane is 2 feet tall.[16]

Harvesting

Sugarcane begins to mature in Louisiana with the advent of cool nights in the fall, usually in October, and is usually harvested between October 15 and January 1. It continues to mature and increase in sugar content during October and November until its growth is checked by frost. Light freezes of −3° C. kill leaf tissue and stop sugar accumulation. A freeze of −4.5° C. damages the upper part of the stalk which must then be topped off and discarded. A freeze below −5° C. kills the whole stalk which then deteriorates.

Figure 14–5. Topping, stripping, and cutting sugarcane by hand.

Hand harvesting consists of stripping off the leaves, removing the tops, and cutting off the stalks at the bottom (Figure 14–5). Specially designed knives and stripping implements are used. About 20 to 30 per cent of the total crop is leaves and tops. In Florida and Hawaii the standing cane is usually fired to burn off the leaves before harvesting begins. Burning the cane reduces harvest and hauling labor, speeds up mill operation, and provides better sugar refinability.[29] In Florida the crop is cut and topped by hand and the stalks laid in windrows because machine harvesting of tall and lodged stalks is not feasible. The crop in Louisiana is cut, topped, and windrowed by machines which pile three rows into one "heap" row. The leaves and other trash are burned after drying 2 to 24 hours. Machines that cut, top, and load the cane (Figure 14–6) are not in general use.[7, 27, 31] The Hawaiian cane is cut off or uprooted and then bunched with a bulldozer-like machine.

Figure 14–6. Cutting and loading sugarcane by machine. [Courtesy Hawaiian Sugar Planters' Association.]

The harvested stalks are hauled to the cane mill where the juice is extracted to make sugar or sirup. Practically all the cane except small lots is loaded mechanically whether it is cut by machine or by hand. The loaders or derricks usually are of the grapple type driven with a gasoline engine and mounted on a wagon or truck (Figure 14–7). The cane is loaded on large trucks, trailers, or rubber-tired carts. It is hauled to the mill on these vehicles or is reloaded for shipment. Some of it is shipped in cars on narrow-gauge private railroads or in standard railroad cars or on barges by water. The cane is processed within a few days to avoid sucrose inversion and deterioration.[18, 21]

Figure 14–7. Mechanical loading of sugarcane. [Courtesy Hawaiian Sugar Planters' Association.]

SUGAR AND SIRUP MANUFACTURE

In sugar manufacture, the sugarcane stalks are cut, sometimes shredded, and then passed between a series of heavy, grooved iron rollers to press out the juice.[16, 22, 30] Water is added between crushings to dissolve more of the sugar. Milk of lime (hydrated lime in water) is added to the strained juice after which the limed juice is heated with steam. The impurities unite with the lime and appear on the surface as scum or at the bottom of the clear juice as sediment. The clear juice, along with the settlings, is conducted into a series of multiple-effect vacuum evaporators where it is concentrated to a semisirup that contains about 50 per cent sucrose. This sirup is then pumped into other vacuum pans where it is crystallized by further evaporation. The mixture of sugar crystals and molasses, called massecuite, is then separated into a centrifuge chamber with perforated walls. The resultant raw sugar, which contains about 96 per cent sucrose, is then shipped to refineries.

A large sugar mill may grind 2,000 to 4,500 tons of cane in 24 hours and process the crop from 5,000 to 10,000 acres in a season. Fresh sugarcane in Louisiana contains an average water content of about 75 per cent. The average extraction of juice from sugarcane under normal conditions is also approximately 75 per cent for the dry-mill process. In the average juice, 15 to 20 per cent is solids, of which some 2 to 3 per cent is impurities (invert sugar, nitrogen compounds, salts, fats, wax, fiber, and so on), which leaves 12 to 17 per cent sucrose.[30] A ton of average cane yields about 170 pounds of raw (96 per cent) sugar in Louisiana, 200 pounds in Florida, 230 pounds in Hawaii, and 240 pounds in Cuba, together with about 5 gallons of blackstrap molasses from each ton of cane.

The juice of sugar plants (sugarcane, sorghum, or sugar beets) usually is tested with a Baume hydrometer, giving an arbitrary reading in degrees, or with a Brix hydrometer having a scale reading that approximates the percentage of total solids (degrees Brix) based upon the weight of sugar.[34] A similar measure, using only a few drops of juice, can be obtained with a refractometer that measures the refractive index of the liquid. It also has a scale giving a reading of the percentage of total solids. The percentage of sucrose in the juice is measured with a saccharimeter, a special type of polariscope with a sugar percentage scale. The purity of the juice in per cent is then determined by the formula, Sucrose/Brix × 100.

The primary manufactured products are sugar, sirup, and edible molasses. By-products consist of blackstrap molasses, bagasse or stalk residue, filter-press cake, and bagasse ashes. Blackstrap molasses is largely used for the production of alcohol and for livestock feed. About 500 pounds of bagasse that contains 50 per cent or more moisture is obtained from a ton of cane. The fiber in bagasse is used for the manufacture of paper, hardboard, and insulating wallboard.[31] Bagasse is burned for fuel in the operation of some sugar mills, the ashes being returned to the sugarcane fields as fertilizer. Filter-press cake has a fertilizer value equal to barnyard manure.

Most of the sugarcane used for sirup is milled on three-roll mills operated by motor power. These have a lower pressure than the 30 tons per square inch used on the rolls of commercial sugar mills, and the juice extraction is less. The juice is then concentrated into sirup in heated evaporating pans.

DISEASES

Mosaic

Mosaic is a virus disease introduced into the United States probably about 1914.[2, 14, 16] Mottling of the leaves is the principal symptom. One type of mosaic appears as longitudinal streaks. The principal damage is caused by stunting, particularly in the noble canes. The disease is transmitted by several species of aphids. The only practical control for mosaic is the use of resistant varieties.

Red Rot

Red rot caused by the organism *Colletotrichum falcatum* is one of the major sugarcane diseases, especially in areas where the crop is produced for sirup.[1] The disease attacks the stalks, stubble, rhizomes, and leaf midribs of the sugarcane plant. The disease is recognized by longitudinal reddening of the normally white internal tissues of the internodes, a discoloration that may extend through many joints of the stalk. Lesions on the leaf midrib are dark, reddish areas. Red rot causes poor stands of both plant and stubble crops, brings about destruction of seed cane in storage beds in sirup-producing states, and induces inversion of sucrose in mature canes. Among the control measures advocated are seed-cane selection, summer planting, and the growing of resistant varieties.[14] Wounds caused by borers and other insects are avenues of entrance for the fungus.

Root Rots

The root rot diseases cause great damage to sugarcane, particularly in Louisiana. The principal cause appears to be *Pythium arrhenomanes*. Root rot is usually manifested by an unthrifty appearance, deficient tillering, closing in of the rows, yellowing of the leaves, severe wilting, and even occasional death of young plants.[6, 14, 28] Poor stands may result in failure of the crop on heavy clay soils due to destruction of the roots. A high water table combined with a fine-textured soil, and low winter and spring temperatures, are probably indirect causes of severe outbreaks. Cultural practices that promote rapid growth are the most beneficial controls. Root-rot damage to susceptible varieties has been markedly reduced by plowing under all cane trash, by moderate applications of filter-press cake or barnyard manure, and by good drainage.

Other Diseases

Chloratic streak, a virus disease, is controlled by planting resistant varieties or by soaking the seed cane in hot water for 20 minutes at 52° C.[2, 11, 31]

The ratoon stunting virus disease is controlled only in part by prolonged heat treatment. The disease develops during drought or under other unfavorable growing conditions.[2, 31] The virus is inactivated by treating the stalks in hot water for 1½ hours at 52° C.[11]

More than 60 additional diseases that attack sugarcane in other parts of the world are not very serious or do not occur in the United States.[23]

INSECT PESTS

The chief insects attacking sugarcane[20] include the sugarcane borer (*Diatraea saccharalis*) and the sugarcane beetle (*Euetheola rugiceps*). Other pests include the mealy bug (*Pseudococcus boninsis*), termite, chinch bug (*Blissus leucopterus*), wireworm, southern cornstalk borer (*Diatraea crambidoides*), southwestern cornborer (*Diatraea grandiosella*), and lesser corn-stalk borer (*Elasmopalpus lignosellus*).

The sugarcane borer may attack the growing point of the plant, causing *dead heart*, or kill the tops of older plants or cause them to break over. Borings in the stalk reduce the sugar yields and contribute to the decay of seed cane. The borers overwinter in sugarcane stalks and stubble and in coarse grasses.[19] The sugarcane borer is controlled by three applications, at three-week intervals, of 1/3 pound per acre of the granular form of endrin.

The fall armyworm is controlled with toxaphene. Wireworms are controlled with chlordane sprayed on the seed pieces in the open furrow at planting time.

REFERENCES

1. Abbott, E. V. "Red rot of sugarcane," *USDA Tech. Bul. 641*, pp. 1–96. 1938.
2. Abbott, E. V. "Sugarcane and its diseases," in *Plant Diseases*, USDA Yearbook, 1953, pp. 526–539.
3. Arceneaux, G., and Stokes, I. E. "Studies of gaps in sugarcane rows and their effect upon yield under Louisiana conditions," *USDA Circ. 521*, pp. 1–20. 1939.
4. Artschwager, E. "Anatomy of the vegetative organs of sugarcane," *Jour. Agr. Res.*, 30:197–242. 1925.
5. Artschwager, E. "Morphology of the vegetative organs of sugarcane," *Jour. Agr. Res.*, 60:503–550. 1940.
6. Artschwager, E., and Brandes, E. W. Sugercane (*Saccharum officinarum*)," *USDA Agriculture Handbook 122*, pp. 1–307. 1958.
7. Atkinson, A. M., and others. "Mechanized sugar cane harvesting in Australia," *World Crops*, 17(1):46–50. 1965.
8. Barnes, A. C. *The Sugar Cane*, Interscience, New York, pp. 1–456. 1964.
9. Bourne, B. A. "A comparative study of certain morphological characters of sugarcane × sorgo hybrids," *Jour. Agr. Res.*, 50:539–552. 1935.
10. Brandes, E. W., and Sartoris, G. B. "Sugercane: Its origin and improvement," *USDA Yearbook*, 1936, pp. 561–623.
11. Brandes, E. W., and Grassl, C. O. "Assembling and evaluating wild forms of sugarcane and closely related plants," *Proc. Int. Soc. Sugar Cane Technologists*, pp. 128–153, 1939.
12. Breaux, R. D., and others. "Culture of sugarcane for sugar in the Mississippi delta," *USDA Handbk. 417*, pp. 1–43. 1972.

13. Cowgill, H. B. "Cross pollination of sugarcane," *J. Am. Soc. Agron.*, 10:302–306. 1918.
14. Edgerton, C. W. *Sugarcane and Its Diseases*, La. State U. Press, Baton Rouge, La., pp. 1–290. 1955.
15. Freeman, C. E., and others. "Sugarcane for sugar production guide," *Fla. Agr. Ext. Circ. 353*, pp. 1–16. 1971.
16. Hebert, L. P. "Culture of sugarcane for sugar production in Louisiana," *USDA Handbook 262*, pp. 1–40. 1964.
17. Hitchcock, A. S. "Manual of grasses of the United States," *USDA Misc. Publ. 200*, pp. 1–1040. 1935.
18. Humbert, R. F. *The Growing of Sugar Cane*, Elsevier, New York. 1968.
19. Ingram, J. W., and Bynum, E. K. "The sugarcane borer," *USDA Farmers Bul. 1884*, pp. 1–17. 1941.
20. Ingram, J. W., Janes, H. A., and Lobdell, R. N. "Sugar cane pests in Florida," *Proc. Int. Soc. Sugar Cane Technologists*, pp. 89–98, 1938.
21. Laurtizen, J. I., and Balch, R. T. "Storage of mill cane," *USDA Tech. Bul. 449*, pp. 1–30. 1934.
22. Martin, L. F. "The production and use of sugarcane," in *Crops in Peace and War*, USDA Yearbook, 1950–51, pp. 293–299.
23. Martin, J. P., and others. *Sugar Cane Diseases of the World*, Elsevier, New York. 1961.
24. May, K. K., and Middleton, F. H. "Age-of-harvest, a historical review," *Hawa. Planters Rec.*, 58(18):241–263. 1972.
25. Patrick, W. G., Jr., and others. "Soil oxygen content and root development in sugarcane," *La. Agr. Exp. Sta. Bul. 641*, pp. 1–20. 1969.
26. Ramp, R. M. "Sugarcane planter development in Louisiana," *The Sugar Bulletin*, 33(13):187–190. 1955.
27. Ramp, R. M. "Development of a cutter-cleaner-loader sugarcane harvester for use in Louisiana," *USDA Agricultural Research Service Spec. Rept. 42–66*. 1962.
28. Rands, R. D., and Dopp, E. "Pythium root rot of sugarcane," *USDA Tech. Bul. 666*, pp. 1–96. 1938.
29. Sloane, G. E., and Rhodes, L. J. "A comparison of the processing of burned and unburned sugarcane," *Hawa. Planters Rec.*, 58(14):173–182. 1972.
30. Spencer, G. L., and Meade, G. P. *Cane Sugar Handbook*, 9th ed. Wiley, New York. 1963.
31. Stokes, I. E. "The world of sugarcane," *Crops and Soils*, 12(8):16–17. 1960.
32. Taggart, W. G., and Simon, E. C. "A brief discussion of the history of sugarcane: its culture, breeding, harvesting, manufacturing, and products," in *La. State Dept. Agr. and Immigration* (4th ed.), pp. 1–20. 1939.
33. Van Dillewijn, C. *Botany of Sugarcane*, Chronica Botanica Co., Waltham, Mass., pp. 1–371. 1952.
34. Walton, C. F., Ventre, E. K., McCalip, M. A., and Fort, C. A. "Farm production of sugarcane sirup," *USDA Farmers Bul. 1874*, pp. 1–40. 1938.

Chapter 15

Wheat

ECONOMIC IMPORTANCE

Wheat is the most important world crop, with a 1970–72 average production of 337 million metric tons on 214 million hectares, or nearly 1.6 metric tons per hectare. The chief producing countries are the U.S.S.R., the United States, China, India, France, and Canada (Figure 15–1). The leading exporting countries are the United States, Canada, Australia, and Argentina. Wheat ranks second to corn (maize) in area in the United States. The production from 1970 to 1972 averaged about 1.5 billion bushels (41.5 million metric tons) on 47 million acres (19 million hectares), or 32.5 bushels per acre (2.2 metric tons per hectare). The leading states in wheat production were Kansas, North Dakota, Washington, Nebraska, and Oklahoma (Figure 15–2). Winter and spring wheat, comprise the totals shown in Figure 15–12. All states except Louisiana, Florida, Alaska, Hawaii, and those in New England are commercial wheat producers. National wheat yields are highest in northwestern European countries having mild winters, cool summers, ample rainfall and heavy rates of fertilization. The average yield of wheat in the United States from 1899 to 1901 and also from 1919 to 1921 was only about 13 bushels per acre. The highest authentic wheat yield is 209 bushels per acre (141 quintals per hectare) on a 2.2-acre, well-fertilized, irrigated field in Kittitas County, Washington, in 1965.

HISTORY

Wheat played an important role in the development of civilization. It is the preferred grain for food in advanced countries. The origin of cultivated wheat is still speculative. A recent proposal is that a tetraploid, emmer, evolved from a wild type (*Triticum dicoccoides*) that carries the A and B genomes. A genome is a group of chromosomes; seven in each wheat genome. Mutations of emmer resulted in other tetraploids—durum, poulard, and polish wheats. Emmer eventually crossed naturally with *Triticum taushii* (*Aegilops squarrosa*), a

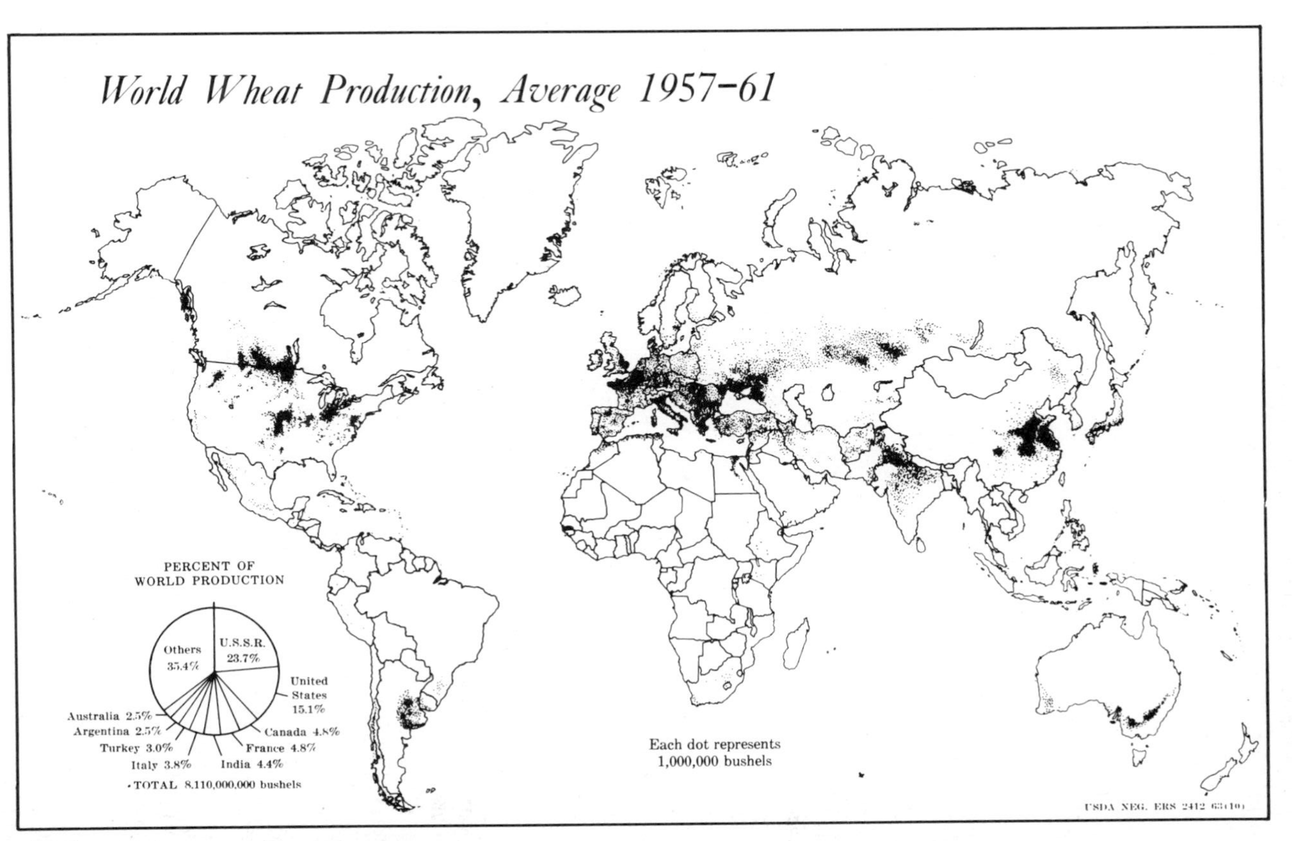

Figure 15–1. World wheat production.

diploid with the D genome, to produce the hexaploid spelt with ABD genomes. Spelt later mutated to common wheat (*Triticum aestivum*), which, in turn, mutated to produce club and shot wheats.[82]

Emmer was cultivated before 7000 B.C. Specimens from Egyptian tombs dated about 3000 B.C. closely resemble the varieties grown in the 20th century. Emmer was grown frequently in the northern Great Plains states before about 1930. A few thousand acres were grown in the United States in 1969. It is still an important crop in the Soviet Union.

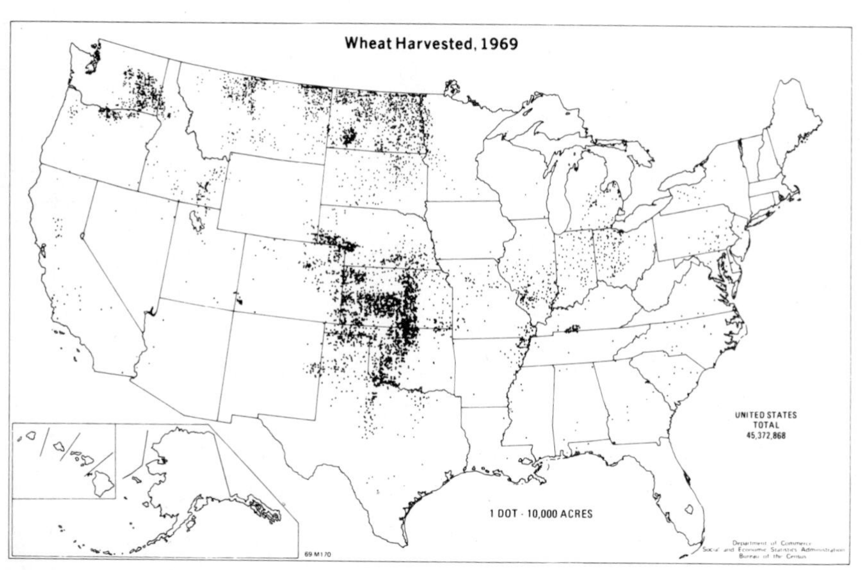

Figure 15–2. Distribution of wheat harvested in the United States in 1969. Each dot represents 10,000 acres.

Wheat was cultivated throughout Europe in prehistoric times, and was one of the most valuable cereals of ancient Persia, Greece, and Egypt. Archaeologists have come upon carbonized grains of wheat in Pakistan and Turkey, in the tombs of Egypt, and in storage vessels found in many other countries. Wheat was already an important cultivated crop at the beginning of recorded history. Wheat culture in the United States began along the Atlantic Coast early in the 17th century and moved westward as the country was settled. Wheat was sown on Elizabeth Island in Buzzards Bay, Massachusetts, by Captain Gosnold in 1602. It was sown in the Jamestown Colony as early as 1611 and in the Plymouth Colony by 1621.[21] Many varieties were introduced from foreign countries during the Colonial period.

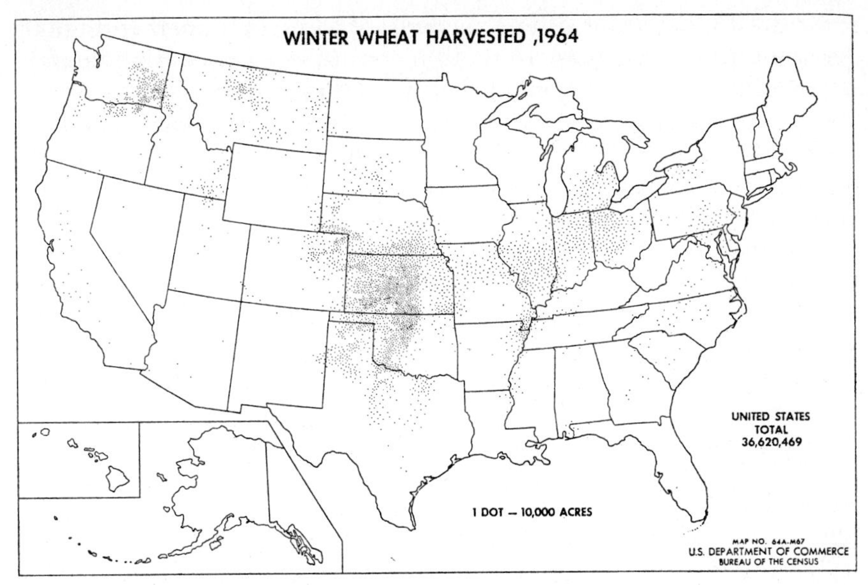

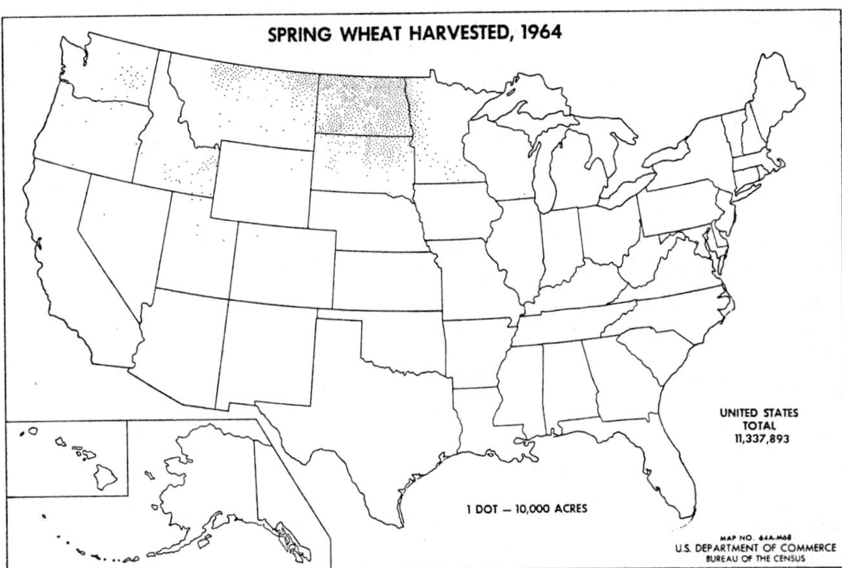

Figure 15–3. *Top:* winter wheat; *bottom* spring wheat harvested in the United States in 1964.

ADAPTATION

Wheat is poorly adapted to warm or moist climates unless there is a comparatively cool dry season which favors plant growth and retards parasitic diseases. The world wheat crop is grown in temperate regions where the annual rainfall averages between 10 and 70 inches, with the exception of drier irrigated areas, being mostly in regions having a precipitation of 15 to 45 inches (38–113 cm). High rainfall, especially when accompanied by high temperatures, is unfavorable for wheat, chiefly because these conditions favor development of wheat diseases. High rainfall also promotes lodging; interferes with harvesting, threshing, storing, seeding, and soil preparation; and leaches fertility elements, particularly nitrates, from the soil.[92] The greater part of the wheat grown in the United States is found where the rainfall averages less than 30 inches per year. The best-quality bread wheats are produced in areas where the winters are cold, the summers comparatively hot, and the precipitation moderate.

In the eastern United States the rainfall in certain seasons is sometimes too high for maximum wheat yields,[109] because it promotes certain parasitic diseases such as *Septoria*. On the other hand, insufficient rainfall is the principal climatic factor limiting wheat yields in the major wheat areas of the world.

Under semiarid Great Plains conditions, good yields are obtained in most years when the soil is wet to a depth of 3 feet (about 1 meter) at seeding time.[4, 22, 34] Failures or low yields usually result when the soil is wet to a depth of 1 foot or less at the time of seeding.

Large acreages of wheat are grown under irrigation in the western parts of the southern and central Great Plains, as well as in the Intermountain and Pacific Coast states. Each additional inch of soil water from rain or irrigation may increase wheat yields from 1.3 to 6 bushels per acre.

Winter wheat in a hardened condition has survived temperatures as low as −40° F. (−40° C.) when protected by snow and probably as low as −25° F. (−32° C.) without snow protection. Spring wheat may survive temperatures as low as 15° F. (9.5° C.) in the early stages of growth. A light frost may cause sterility in wheat that has headed or is about to head. A light frost before wheat is fully mature produces blisters on the seed coat and stops further grain development. In general, spring wheat requires a frost-free period of about 100 days or more for safe production. Quick-maturing varieties may be grown where the frost-free period is 90 days or less in Canada and Alaska, where long days hasten flowering.

Short days or high temperatures stimulate tillering and leaf formation but delay the flowering of wheat plants. Although wheat is a long-day plant, quick-maturing or photo-insensitive varieties can be grown to maturity in any day length ranging from the 12-hour days of the equatorial highlands of Ecuador and Kenya to the 20-hour days of the "land of the midnight sun" at Rampart, Alaska, at latitude 66° N. Wheat plants grown quickly under continuous artificial light are poorly developed and produce small heads and very little grain.

Much of the superior protein content and baking quality of hard wheats is due to the environment in which they are grown. When grown in the soft-wheat regions, the grains of the hard wheats are starchy or "yellow berry"; they are only slightly if at all higher in protein content and baking quality than are the soft wheats.[30] However, differences in protein *quality* among varieties may account for considerable differences in the baking qualities of flours of the same protein *content.* Also, the protein percentage of some varieties may be 1 or 2 points higher than others. In general high-yielding wheat may be expected to be relatively low in protein content except where soil nitrates are abundant.

Most of the plant nutrients derived from the soil are taken up before the plant blooms, being later translocated to the kernel,[1] but nitrogen uptake may continue until the grain is ripe.[37] The ratio of protein to starch in the wheat kernel is largely determined by moisture, temperature conditions, and available nitrates in the soil at the blossom and postfloral periods. The fruiting or postfloral period tends to be prolonged when the weather is cool, and ample soil moisture is available to the plant. A relatively large amount of starch tends to be deposited and a plump starchy kernel of low protein content results.[1, 8] When sufficient nitrogenous material has been absorbed before or at the blossom stage, and when moisture conditions are such as not to prolong the postfloral period, the grain tends to be plump and fairly high in gluten. Starch deposition tends to be upset more than gluten formation in hot, dry weather. This results in small-berried kernels rich in gluten. Dark, hard, and vitreous kernels in the bread wheats are usually an indication of a comparatively high protein content. When the development of wheat on soil containing ample nitrates is stopped prematurely by drought, the shrunken grain is high in protein and low in starch. Grain shrunken as a result of the stem-rust disease is low in both starch and protein because both nitrogenous and carbohydrate compounds are withdrawn from the wheat plant by the growing rust fungus.

In a North Dakota study the protein content of wheat was above 13 per cent when the average July temperature was greater than 70° F. (21° C.).[43] Lower protein content was obtained in years when the July temperature was slightly above 65° F. (18° C.) and the June temperatures relatively low. Temperatures above 90° F. (32° C.) lower the baking quality of high-protein wheats.

The influence of the soil on the protein content of wheat is less marked than is that of climate.[8] In many cases, the supply of available soil nitrogen is insufficient, and part or all of the endosperm is yellow berry (soft and starchy). Wheats of high protein content are found on soils of high nitrogen content with a low moisture content at time of maturity.[8] The addition of irrigation water or prolonged heavy rainfall that leaches out soil nitrates may be expected to result in wheat of lower protein content. Applications of nitrogenous fertilizers, particularly inorganic forms, usually result in wheat with a higher protein content. Nitrogen applications, particularly foliar sprays of urea solutions, as late as the blossom period, generally will increase the protein in the kernel.

Wheat is best adapted to fertile medium to heavy-textured soils that are well drained. The silt and clay loams in general give the highest yields of wheat, but wheat can also be grown successfully on either clay soils or fine sandy loams. In general, wheat is an unsatisfactory crop on very sandy or poorly drained soils. Wheat tends to lodge on rich bottomlands.

BOTANICAL DESCRIPTION

Wheat is classified in the grass tribe *Hordeae* and in the genus *Triticum*. This genus is characterized by two- to five-flowered spikelets placed flat at each rachis point of the spike.

Wheat is an annual or winter-annual grass (Figures 15–4 and 15–5) with a spikelet inflorescence which consists of a sessile spikelet placed at each notch of the zigzag rachis. A spikelet is composed of two broad glumes and one or several florets. A floret consists of a lemma, palea, and caryopsis. The awn arises dorsally on the tip of the lemma in awned sorts. The caryopsis or grain has a deep furrow (or crease) and a hairy tip or brush. The color of the grain may be amber, red, purple, or a creamy white, in different varieties. The purple sorts are not grown in the United States.

The plants normally produce two or three tillers under typical crowded field conditions, but individual plants on fertile soil with ample space may produce as many as 30 to 100 tillers. The average spike (head) of common wheat contains 25 to 30 grains in 14 to 17 spikelets (Figure 15–6). Large spikes may contain 50 to 75 grains. The shape of individual kernels depends in part upon the degree of crowding within a spikelet. The spikelets at the base and tip of the spike usually contain small kernels. Within a spikelet the second grain from the base usually is the largest, and the first grain is next in size but, when present, the third, fourth, and fifth grains are progressively smaller. A pound of wheat may contain 8,000 to 24,000 kernels. A bushel of common wheat averages 700,000 to 1,000,000 grains. The legal weight of a bushel of wheat in business transactions is 60 pounds, but the average test weight is about 58 pounds. The maximum test weight is about 67 pounds and the minimum about 40 pounds per bushel.

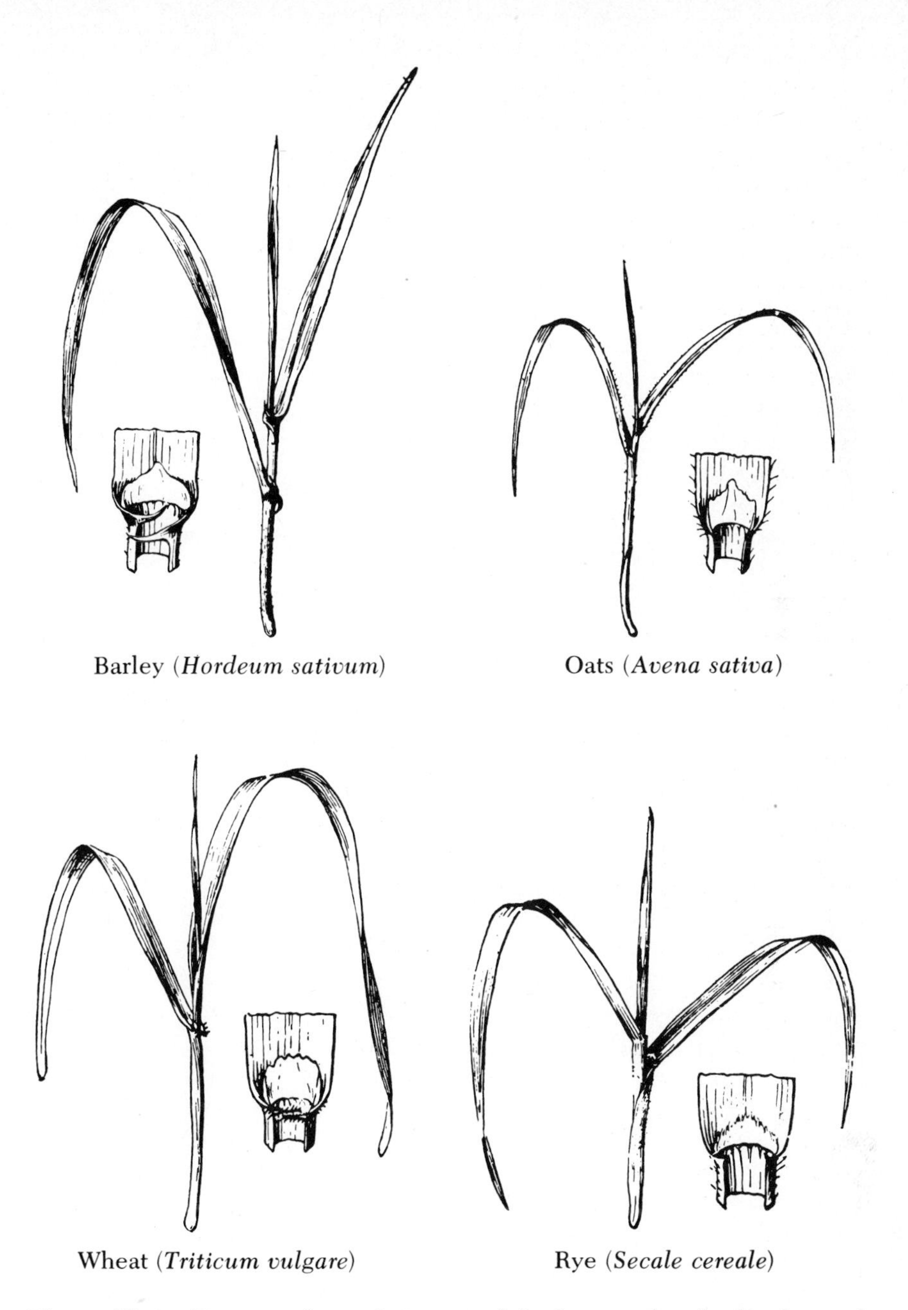

Figure 15–4. Distinguishing characters of the leaves, sheaths, ligules, and auricles of young cereal plants. Barley has long clasping smooth auricles; wheat has shorter hairy auricles; rye has very short auricles; the oat sheath has no auricles. The sheaths of rye and oats and the leaf margins of oats are hairy. The shapes of the shield-like ligules at the base of each leaf blade also are different.

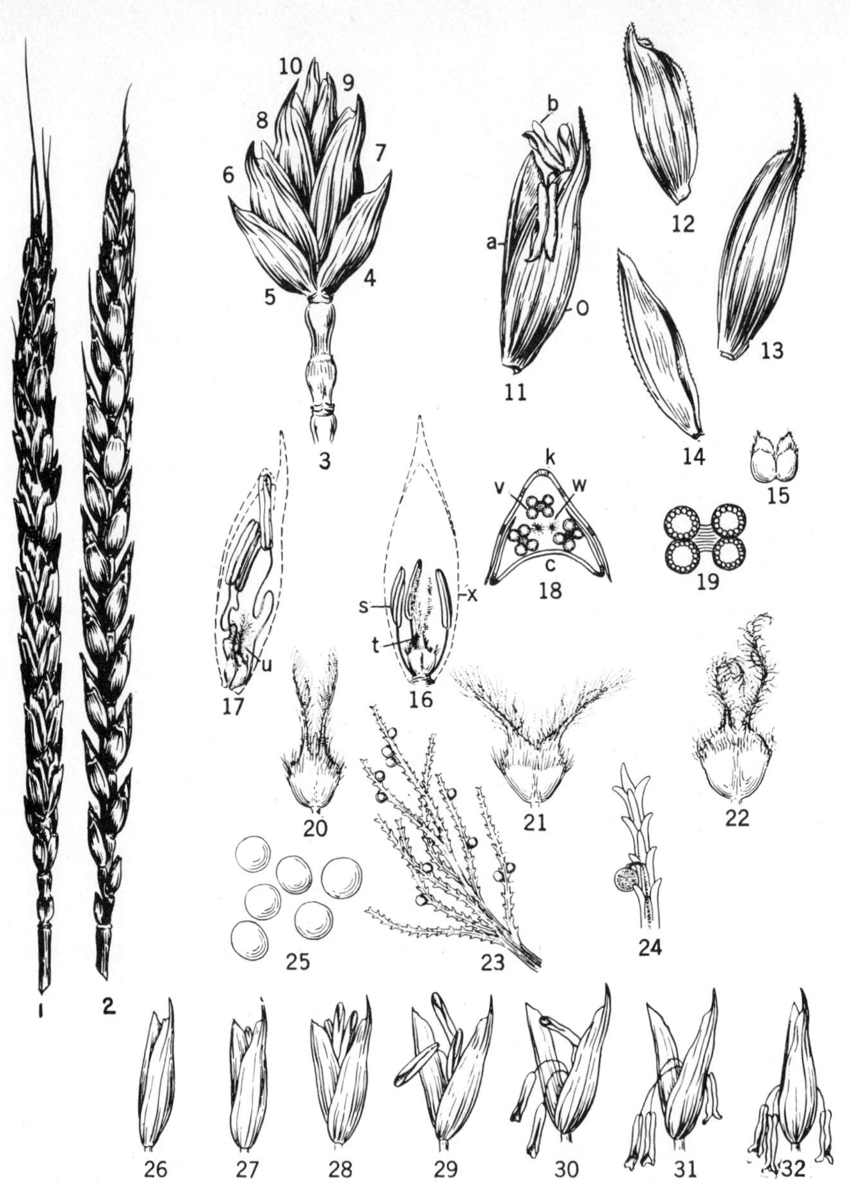

Figure 15–5. The wheat inflorescence. (1) Spike, dorso-ventral view. (2) Spike, lateral view. (3) Spikelet, lateral view and subtending rachis. (4) Upper glume. (5) Lower glume. (6 to 10) Florets: 7 is the largest and 6 second largest; 8, 9, and 10 are progressively smaller. (11) Floret, lateral view, opened for anthesis. (12) Glume, lateral view. (13) Lemma, lateral view. (14) Palea, lateral view. (15) Lodicules, which swell to open the glumes. (16) Floret before anthesis, showing position of stamens (*s*) and pistil (*t*), enclosed in lemma and palea (*x*). (17) Floret at anthesis showing position of pistil (*u*) and elongating filaments of the stamens. (18) Cross section of floret: palea (*c*), lemma (*k*), stamen (*v*), and stigma (*w*). (19) Cross section of anther. (20) Pistil before anthesis. (21) Pistil at anthesis. (22) Pistil after fertilization. (23) Portion of stigma (greatly enlarged) penetrated by germinating pollen

Figure 15–6. Spikes of four varieties of common wheat and one variety of club wheat: (1) fully awnless, (2 and 3) awnless (tip awned), (4) fully awned, and (5) club wheat with tip awns.

POLLINATION

The time required for the wheat flower (Figure 4–1) to open fully and the anthers to assume a pendant position is approximately 3 minutes and 36 seconds.[52] Approximately 86 per cent of the flowers bloom in daylight. Flowers bloom at temperatures ranging from 56° to 78° F. (13° to 25° C.).

Although wheat is normally a self-pollinated plant, natural cross-pollination occurs in 1 to 4 per cent of the flowers.[32] Wheat that is

---

(24) Tip of stigma hair (greatly enlarged) penetrated by germinating pollen grain. (25) Pollen grains (enormously enlarged). (26 to 32) Florets during successive stages of blooming and anthesis. Time required for stages 26 to 31 is about 2 to 5 minutes, for stages 26 to 32 about 15 to 40 minutes.

partly self-sterile from chromosomal irregularities or adverse environment sometimes becomes crossed extensively.[53] In one season, approximately six times as much natural crossing occurred in the secondary heads as in the primary heads.

Blooming begins in the spikelets slightly below the middle of the spike and proceeds both upward and downward. Within a spikelet the upper flowers bloom last. Under ordinary conditions a wheat spike completes its blooming within 2 or 3 days after the first anthers appear.

THE WHEAT KERNEL

The wheat kernel or berry is a caryopsis, 3 to 10 millimeters long and 3 to 5 millimeters in diameter. The structure of the kernel is shown in Figure 15–7. The pericarp of the kernel consists of several layers that have developed from the ovary wall. These layers include the cuticle, epidermis, parenchyma, and cross layer. The latter three layers are sometimes referred to as the epicarp, mesocarp, and endocarp, respectively. The testa (or true seed coat) is often called the perisperm. The thickness of all of these layers may not exceed 3 per cent of the thickness of the grain. The testa of red-kerneled varieties contains a brownish pigment, but that of white or amber varieties is practically unpigmented. The aleurone is the single layer of large cells, rectangular in outline, comprising the outer periphery of the endosperm. The rectangular cells of the aleurone are polyhedral when viewed from the top or outside of the grain. The aleurone cells contain no starch, their chief contents being aleurone grains composed of nitrogenous substances, probably proteins, but not gluten. In milling wheat, the bran fraction consists of all of the pericarp layers, the testa, nucellus, and aleurone, and some adhering endosperm.

A plump well developed kernel consists of about 2.5 per cent germ (plumule, scutellum, radicle, and hypocotyl), 9 to 10 per cent pericarp, 85 to 86 per cent starchy endosperm, and 3 to 4 per cent aleurone. Shrunken wheat of one-half normal kernel weight (testing about 50 pounds per bushel) may contain not more than 65 per cent starchy endosperm. Under the aleurone layer the endosperm consists of cells filled with starch grains cemented together by a network of gluten.[15] Gluten is a cohesive substance usually considered as being comprised of two proteins, glutenin and gliadin, but also reported as being a single but complex protein.[64] Gluten gives wheat flour its strength or ability to hold together, stretch, and retain gas while the fermenting dough expands. Only wheat and rye flour have this ability. Rye contains a protein similar to gluten.

The development of the wheat kernel[23, 29, 65, 99] proceeds by cell division following the fertilization of the egg nucleus to produce

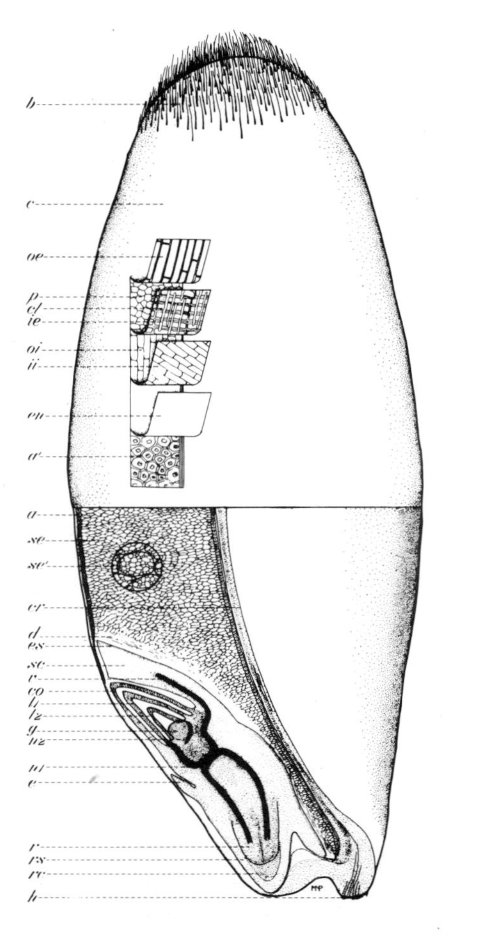

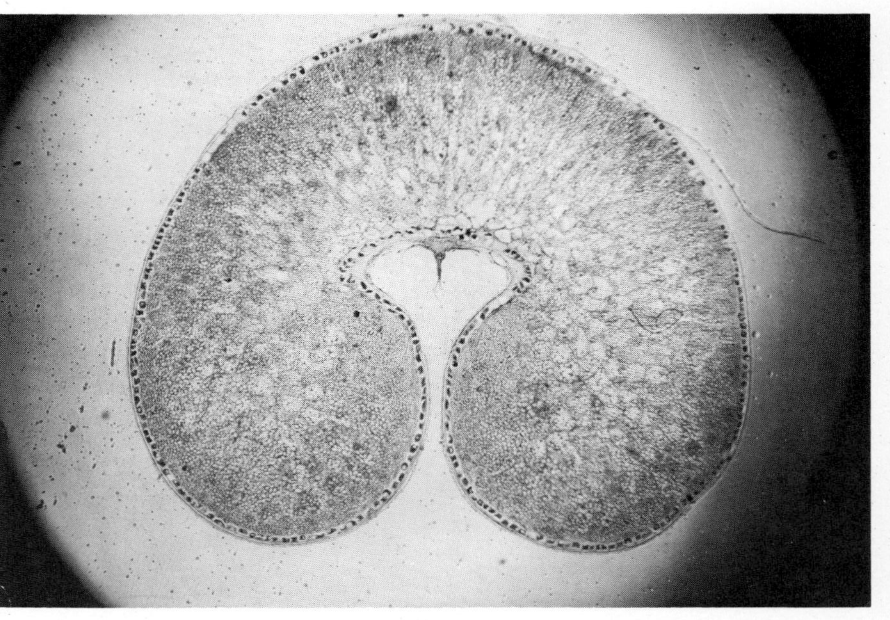

Figure 15–7. (*Left*) Wheat kernel. Brush (*b*). Pericarp consists of cuticle (*c*), outer epidermis (*oe*), parenchyma (*p*), cross-layer (*cl*), and inner epidermis (*ie*). Testa consists of outer integument (*oi*) and inner integument (*ii*). Nucellar layer—epidermis of nucellus (*en*). Endosperm consists of the aleurone (*a* and *a'*), starch, and gluten parenchyma (*se* and *se'*), and crushed empty cells of endosperm (*d*). Germ consists of scutellum (*sc*), epithelium of scutellum (*es*), vascular bundle of scutellum (*v*), coleoptile (*co*), first foliage leaf (l_1), second foliage leaf (l_2) growing point (*g*), second node (n_2), first node (n_1), epiblast (*e*), primary root (*r*), root sheath or coleorhiza (*rs*), and rootcap (*rc*). Hilum (*h*). The bran layer is comprised of the pericarp, testa, nucellar layer, and aleurone. [Drawn by M. N. Pope] (*Above*) Cross section of wheat kernel showing the crease and endosperm cells. The dark cells near the periphery constitute the aleurone layer of the endosperm.

the embryo. The embryo reaches its approximate final size in advance of the full development of the endosperm. The integuments of the ovary become modified and form the layers of the pericarp. Repeated cell division following fertilization of the endosperm nucleus produces endosperm cells. These cells are gradually filled with starch grains and proteins, starting from the outer periphery, including the region of the crease, toward the interior of the endosperm (Figure 15–7). Sugars synthesized in the leaves and culms are transported to the endosperm to be deposited finally as starch. The grain increases in percentage of starch until it is fully mature. Amino acids synthesized in the leaves and transported to the endosperm change to proteins as the grain begins to dry. A part of the nitrogen, phosphorus, potassium, and other mineral constituents absorbed by the plant before the grain is formed are transported to the kernel.

WINTER HARDINESS

Low temperatures cause losses to wheat nearly as great as those from all wheat diseases combined. An average of one eighth of the winter wheat seeded is abandoned, much of it due to winterkilling. Winterkilling may be due to one or more causes which include (1) heaving, (2) smothering, (3) physiological drought, and (4) direct effect of low temperature on plant tissues.[91]

Wheat varieties differ widely in winterhardiness.[81]

FREEZING PROCESS IN PLANTS

Freezing injury occurs when water is withdrawn from the plant cells into the intercellular spaces. This results in concentration of the cell sap, with attendant increase in salt concentration and increase in hydrogen ion concentration. These conditions bring about an irreversible precipitation of the cell proteins as well as other changes in the organization of the cell. Death of the plant or plant part ensues when the protoplasm is disorganized beyond recovery. The plant loses water rapidly after thawing. Quick freezing of unhardened tissues ruptures the cells.

More winter-wheat plants survive low temperatures in moist soils than in dry soils.[47] Plants in moist soils start to kill more quickly, but the killing progresses more slowly. An early erect habit of growth often indicates a lack of hardiness, but a recumbent habit of growth does not always indicate hardiness.[48] Injury in wheat plants progresses from the older to the younger leaves, and the crown and meristematic tissues are the most hardy parts.[57] Late fall-sown wheat with shallow roots heaves more readily than that well rooted before the onset of winter.[41]

A cold-resistant plant must have the ability to harden, an inherited characteristic which comes into expression only after exposure to low temperatures[36] with ample sunlight. Hardening begins at temperatures below about 41° F. (5° C.), i.e., the threshold value above which growth commences. There is a concentration of sugars, of amino nitrogen, and of amide nitrogen in the tissues of hardy wheat varieties which lowers the freezing point to some extent and decreases protein precipitation.[111] A progressive increase in dry matter and sugar content of winter-wheat plants follows low-temperature hardening. The hardier the variety, the lower the moisture content of the leaves, and in hardened-off leaves this moisture is held under a greater force.[57, 67] This condition is believed to be due to the hydrophilic colloid or bound-water content. The more hardy wheat plants lose less water after thawing. Hardened plants become more susceptible to cold injury with advanced age and season.

INFLUENCE OF AWNS ON DEVELOPMENT OF WHEAT KERNELS

In general, the awn of wheat has been considered a useful structure under certain climatic conditions.[6, 50, 66, 72] Awned varieties often predominate in areas of limited rainfall and warm temperatures, but awnless varieties may yield equally well in cool, humid, or irrigated areas. The presence of awns on the florets of wheat has tended toward the production of heavier kernels[89] as well as higher test weights. The awns serve as a depository of ash. Thus, they may remove from the transpiration system of the plant at filling time substances (possibly silicates) that otherwise might interfere with rapid movement of material into the grain.[50] De-awned heads transpire less than do awned heads.[33] Transpiration through the awns helps to dry out the ripe grains. The tips of awnless spikes often are blasted in hot dry weather. This suggests that a cooling effect produced by the awns might be helpful to the developing grains. Awns (beards) contain considerable chlorophyll. They contribute considerable carbohydrates by photosynthesis to the developing grain, particularly during drought stress.[60]

SPECIES OF WHEAT

The species and groups (or subspecies) of cultivated wheat in the world are listed below.[14] Only common, club, and durum wheats, and emmer and spelt are grown in the United States but poulard and Polish wheats have been grown sporadically. The groups shown were long regarded as separate species, but members with the same chromosome number intercross readily. They also differ from each other by only a few hereditary genes, so do not merit separate specific rank. Some of the wild species listed may be ancestors of the cultivated types.

SPECIES	GROUP	COMMON NAME
Triticum monococcum L.	Diploid (2N = 14 chromosomes) AA genomes	einkorn
T. turgidum L.	Tetraploid (2N = 28 chromosomes) AA BB genomes	
	dicoccon (*dicoccum*)	emmer
	durum	durum wheat
	turgidum	poulard (branched) wheat
	polonicum	Polish wheat
	carthlicum	Persian wheat
T. timopheevii Zhuk.	*timopheevii*	none
	Zaukovski	none
T. aestivum	Hexaploid (2N = 42 chromosomes) AABBDD genomes	
	spelta	spelt
	vavilovii	none
	aestivum (*vulgare*)	common wheat
	compactum	club wheat
	sphaerococcum	shot wheat

SOME WILD RELATIVES OF THE WHEATS
Triticum aegilopoides—AA genomes
T. speltoides (*Aegilops speltoides*)—BB genomes
T. taushii (*Aegilops squarrosa*)—DD genomes
T. cylindricum (*Aegilops cylindrica*)
T. diccocum (*T. dicoccoides*)—wild emmer

The spike of common wheat is usually dorsally compressed. The spikelets have from two to five flowers, but usually contain only two or three kernels at maturity. The kernels, which thresh free from the chaff, are either red or white in color and soft or hard in texture. The so-called white kernel is in reality light yellow or amber in color. The common wheats may be winter or spring habit, awned or awnless, and white, brown, or black in glume color.

The plants of durum wheat are of spring habit and tall. The spikes are compact and laterally compressed (Figure 15–8). The glumes are persistent and sharply keeled and the lemmas awned in the commonly grown varieties. The long stiff awns and thick compact spikes and long glumes often cause laymen to confuse durum wheat with barley. The

Figure 15–8. *Left to right:* Spikes of spelt, durum wheat, timopheevi wheat, and winter emmer.

amber or red kernels are usually long and pointed, and the hardest of all known wheats. The kernels of poulard wheat are similar except in being shorter, thicker, and no so hard. Some poulard wheats have branched spikes (Figure 15–9).

Club wheat plants may be either winter or spring habit, and either tall or short. The spikes, which usually are awnless, may be elliptical, oblong, or clavate in shape. They are short, compact, and laterally compressed. The spikelets usually contain five and sometimes as many as eight fertile florets. The kernels of club wheat, which may be either white or red, are small and laterally compressed.

Most of the kernels of spelt, emmer, and einkorn remain enclosed in the glumes after threshing (Figure 15–10). Emmer is distinguished from spelt by the shorter, denser spikes which are laterally compressed. In the United States, emmer and spelt are used for livestock feed, but they have been grown for food in Europe. Einkorn (Figure 15–11) differs from the other species in that many of the spikelets contain only one fertile floret. Einkorn or *one-grain* wheat is grown sparingly in Europe. The wild form (*T. boeticum*) is still abundant in Southwestern Turkey. Emmer, spelt, and einkorn have red kernels.

Because of their unusual appearance, Polish and poulard wheats have been fraudulently exploited in the United States (Figures 15–9

Figure 15–9. Composite or branched head of poulard wheat.

and 15–11). Polish wheat, sometimes called corn wheat, has large, lax, awned heads that sometimes attain a length of 6 or 7 inches. The chaff is extremely long, thin, and papery. The kernels, which are sometimes ½-inch long and very hard, frequently have been confused with rye.

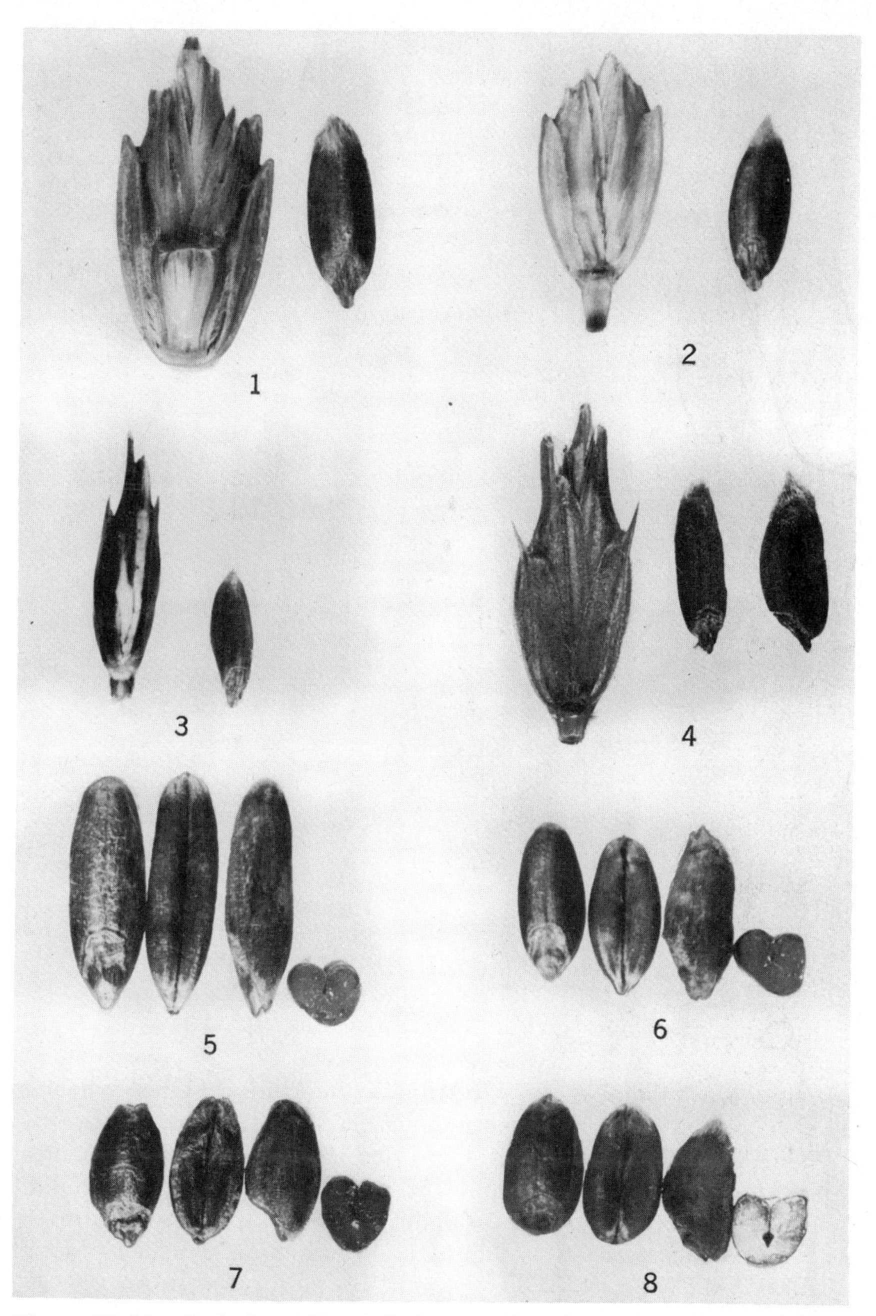

Figure 15–10. Spikelet and kernel of: (1) spelt, (2) spring emmer, (3) einkorn, and (4) timopheevi wheat. Wheat kernels: (5) Polish, (6) durum, (7) poulard, and (8) club.

Figure 15–11. Spikes and glumes of Polish wheat (*left*) and einkorn (*right*).

WHEAT CLASSES

Under the official grain standards of the United States, wheat is now separated into seven commercial classes: (1) Hard Red Spring, (2) Durum, (3) Red Durum, (4) Hard Red Winter, (5) Soft Red Winter, (6) White, and (7) Mixed. Emmer, spelt, einkorn, Polish, and poulard wheat are not considered wheat in the official grades. The distribution of some of these wheats as to principal production areas is shown in Figure 15–12.

The percentage of the production of the different wheat classes in the United States from 1970 to 1972 was approximately: Hard Red Winter, 50; Hard Red Spring, 18.6; Soft Red Winter, 13.8; White, 12.8; and Durum, 4.8. No Red Durum was grown.

Hard Red Spring Wheat

Hard red spring wheat is grown in the North Central States, mostly where the winters are too severe for production of winter wheat.[7] This class of wheat led in North Dakota, Minnesota, Montana, and South Dakota. Hard red spring is the standard wheat for bread flour.

Hard red spring wheats also are grown in Canada, Russia, and Poland.

Durum Wheat

Durum wheat is produced in the same general region where hard red spring is grown, but the chief section lies west of the Red River Valley in North Dakota. Most of the durum wheat is grown in North Dakota, with nearly all of the remainder in Montana, South Dakota, and Minnesota (Figure 15–13). Small acreages of durum were grown in California, Washington, and Oregon in 1969. Durum wheat is used in the making of semolina, from which macaroni, spaghetti, and similar products are manufactured.

Durum wheat is grown extensively in North Africa, southern Europe, and the Soviet Union.

Hard Red Winter Wheat

The hard red winter wheats are adapted especially to the central and southern Great Plains where the annual rainfall is less than 35 inches. Kansas, Oklahoma, Texas, Nebraska, Colorado, and Montana lead in the production of hard red winter wheat (Figure 15–12). As bread wheat, the better varieties of this class are nearly or completely equal to hard red spring wheat.

Hard red winter wheats are grown extensively in the southern part of the Soviet Union, the Danube Valley of Europe, and Argentina.

Soft Red Winter Wheat

Soft red winter wheat is grown principally in the eastern states. The western border of this region coincides roughly with the line of a 30-inch average annual precipitation.[12] Ohio, Missouri, Indiana, Illinois, and Pennsylvania lead in the production of soft red winter wheat (Figure 15–12). The soft red winter wheats are softer in texture and lower in protein than the hard wheats. Wheats of this class are generally manufactured into cake, biscuit, cracker, pastry, and family flours. For bread flour it sometimes is blended with flour of the hard red wheats.

Soft red winter wheats are grown in western Europe and on other continents.

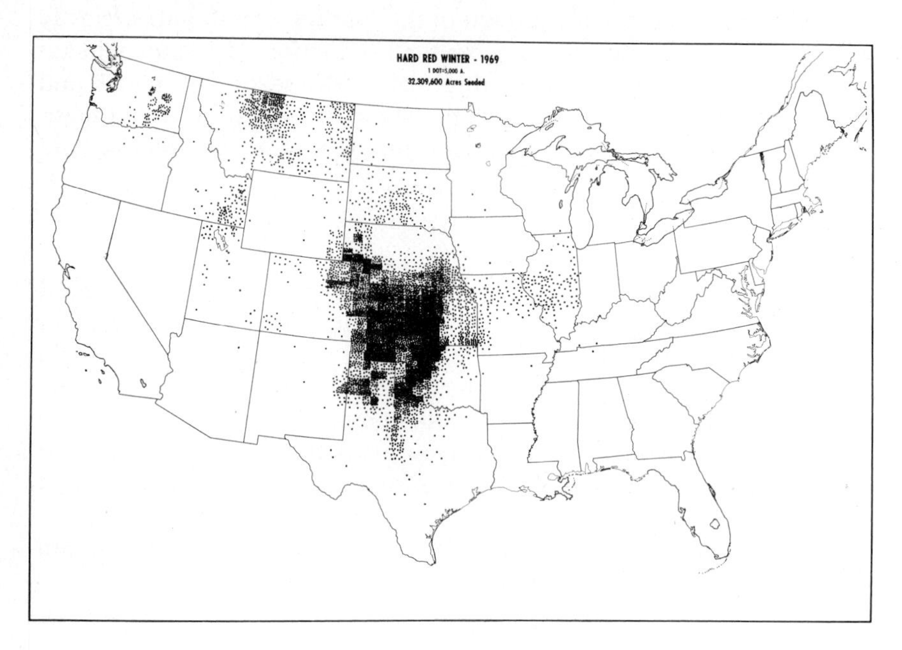

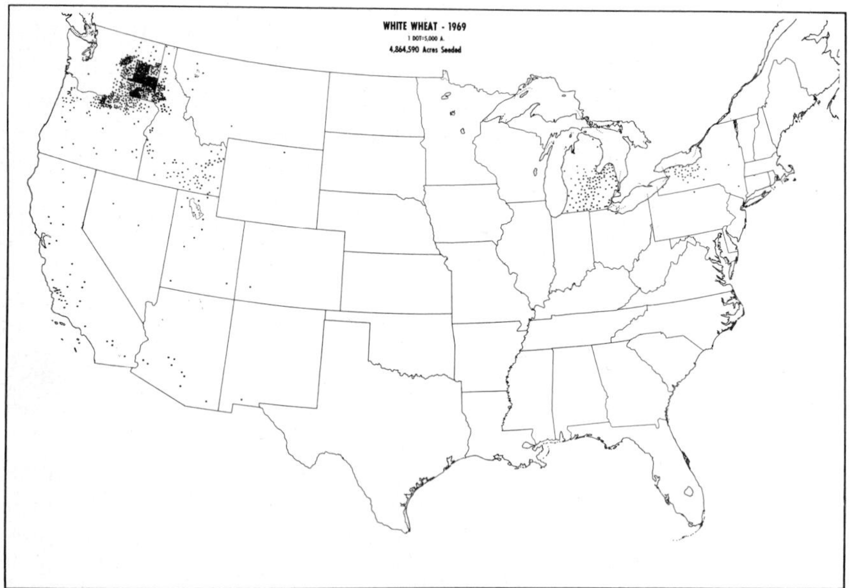

Figure 15–12. Distribution of four classes of wheat in United States in 1969.

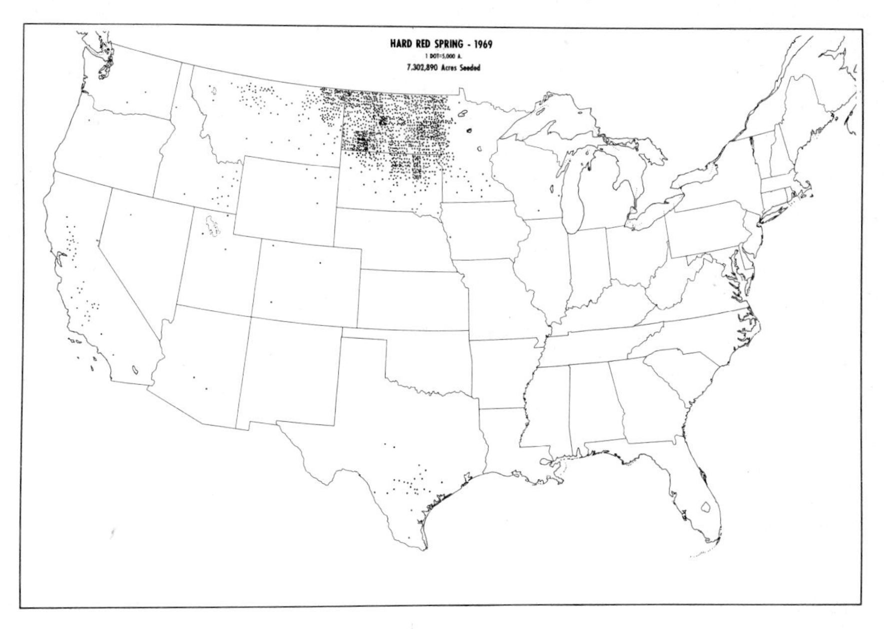

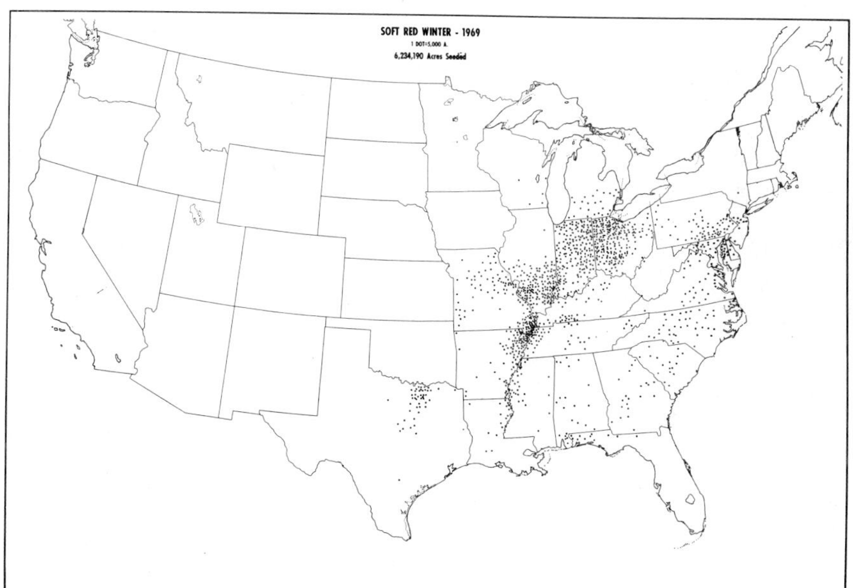

Figure 15–12. *(Continued)*

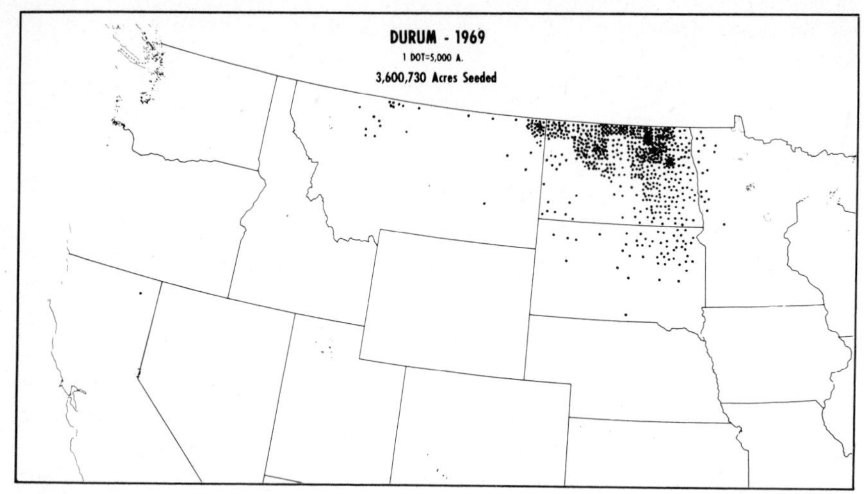

Figure 15–13. Distribution of durum wheat in 1969 (3,600,730 acres).

White Wheat

The white wheats are grown in the far western states and in the northeastern states. The largest acreages are in Washington, Oregon, Idaho, Michigan, New York, and California (Figure 15–12). About 14 per cent of the white wheat consists of varieties of club wheat grown in the northwestern states. The white wheats are used principally for pastry flours and shredded and puffed breakfast foods.

White wheats are grown in northern, eastern, and southern Europe, and in Australia, South Africa, western South America, and Asia.

WHEAT RELATIVES

Wheat (genus *Triticum*) is related botanically to certain other grass genera that can be crossed with wheats. Several species of *Aegilops* crossed with wheat include goatgrass (*Aegilops cylindrica*), a weedy plant. Wheat has been crossed with several species of the genus *Agropyron* (wheatgrass). Crosses of wheat with western wheatgrass (*Agropyron smithii*) and slender wheatgrass (*A. trachycaulum*) have not been successful. Successful crosses with several other species of *Agropyron*, including certain collections of quackgrass (*A. repens*), have been reported. Wheat was crossed with *Agropyron elongatum* in the Soviet Union, Canada, and the United States in an endeavor to produce a perennial wheat. Thus far the long-lived perennial types obtained have been decidedly grass-like, whereas the more wheat-like selections have been annuals or short-lived perennials. The desirability of a perennial wheat is not fully established since a crop that can grow any time in the season when conditions are favorable would leave little opportunity for accumulation of moisture or nitrates in the soil.

HYBRID WHEAT

Extensive effort has been devoted to the development of hybrid seed wheat, but it was not a commercial success before 1974. Only small quantities of hybrid seed have been produced. The chief limitations are the poor sets of seed often obtained in the crossing field, and the larger amounts of seed per acre for optimum yields in comparison with those of hybrid corn and sorghum. Both cytoplasmic and genetic male-sterile lines are now available. Wheat stigmas are less exposed than those of some other crops and probably intercept less air-borne pollen. Hopeful results may yet be obtained.[42, 87]

Wheat has been crossed with rye, and selected progenies, called triticale, have characters and chromosomes of both parents. (See Chapter 16.) Wheat also has been crossed with the grass genus *Haynaldia*. Reputed crosses of wheat and barley have not been authenticated.

Formerly, there was a popular belief that wheat changed into cheat (*Bromus secalinus*) when subjected to winter injury or when the developing spikes were mutilated in the boot stage, although this was refuted by botanists and agricultural authorities more than a century ago. Reports of the reputed change came from sections where cheat (chess) is a common weed in grain fields and waste places, and appears in abundance where wheat is thinned out by winterkilling. Cheat formerly was grown extensively as a hay crop in western Oregon. The botanical relationship between wheat and cheat appears to be too remote to permit intercrossing.

FERTILIZERS

Wheat usually responds to fertilizers except on fertile nonirrigated loam or clay soils in semiarid regions. In humid or in irrigated regions, nitrogen applications of 30 to 120 pounds per acre are applied, except when wheat follows a legume forage crop. The usual response is one bushel of wheat to each 3 to 5 pounds of nitrogen. Applications of 10 to 30 pounds of nitrogen often are beneficial to wheat on sandy soils in semiarid regions, except in the drier seasons. Excessive applications of nitrogen often reduce wheat yields not only by increasing plant lodging, but also by delaying the maturity of the crop so that it is subject to greater damage from rust. Semidwarf wheats with strong culms can utilize more nitrogen without lodging and with consequent high yields.[84] Fall applications of nitrogen are especially beneficial in continuous wheat culture where a heavy growth of straw has been turned under.

Fall applications of 30 to 50 pounds per acre of phosphorus (P_2O_5) are helpful on most soils in the humid region as well as occasionally elsewhere. Potassium applications of 40 to 50 pounds per acre of K_2O are desirable in regions of heavy rainfall.

In humid winter-wheat regions, it is customary to apply a mixed fertilizer that contains some nitrogen in the fall during seedbed preparation or seeding. The remainder of the nitrogen is applied as a top dressing in early spring. Early spring applications give the greatest increases in yield.[27] Later applications may give greater increases in the protein content of the grain.

ROTATIONS

In the eastern states where winter wheat often is seeded on corn stubble, the rotation is likely to contain at least one legume and one or more cultivated crops.

In the Corn Belt an efficient rotation for much of the area is winter wheat; clover and timothy (one or two years); corn; oats. Soybeans may be substituted for oats, while alfalfa, combined with bromegrass or orchardgrass, may replace the clover and timothy. Where lespedeza is an important crop, it may be seeded in the spring on fall-sown wheat. After the wheat harvest, the lespedeza may be permitted to produce a crop of hay or seed. A common rotation is corn; wheat; lespedeza for pasture, hay, or seed; wheat.

In the Cotton Belt, wheat often cannot be sown immediately after cotton because of the late maturity of the latter crop. A rotation satisfactory for this area is wheat, followed by soybeans or cowpeas for hay or seed; well-fertilized cotton, with vetch or Austrian winter peas planted between the rows for green manure; corn, stalks cut and removed from the field.

For Kansas[93] a popular rotation for the northeastern part of the state is clover, corn, oats, and wheat. Red clover or sweetclover is seeded with the wheat to be grown for hay or seed the next year, making a five-year rotation. The most satisfactory rotations in central Kansas are those that contain alfalfa. Because of subsoil moisture depletion by alfalfa, the crop that follows it must be able to endure drought. A satisfactory rotation is alfalfa (two or three years), sorghum, corn, oats, wheat. In western Kansas, rotations have failed to show any marked gain in wheat yields over continuous cropping, except where wheat follows fallow. A suggested rotation is grain sorghum, fallow, wheat (two or three years). Continuous winter wheat with mulch tillage has produced good yields in some areas where weeds and pests are not serious.

Alternate winter wheat and fallow are a practical sequence where the annual precipitation is less than about 22 to 24 inches in the southern Great Plains, 18 to 20 inches in the central Great Plains, and 14 to 16 inches in the northern Great Plains as well as in the intermountain region.[94] A wheat-pea rotation is profitable in parts of the Columbia Basin where the annual precipitation is 15 to 22 inches.

Spring wheat follows an intertilled crop such as corn, potatoes, or sugarbeets most advantageously in the northern spring-wheat region. Much of it is sown after wheat or other small grains, or after fallow, because of the relatively small acreage of intertilled crops. Fallow helps in the control of weeds. Land to be fallowed usually is not tilled before early spring of the year after the previous crop harvest. The undisturbed crop residues help to retain snow and rain as well as to reduce wind and water erosion.[94] Green-manure crops in preparation for wheat have been unprofitable in semiarid regions.[4]

CULTURE

Seedbed Preparation

The time and manner of seedbed preparation for continuous winter wheat influence the elaboration and accumulation of soil nitrates, soil-moisture storage, and the physical condition of the seedbed at seeding time.[93] In experiments in eastern Kansas, soil worked early in July, regardless of method, usually gave comparatively high yields, whereas that worked late in September usually resulted in a low yield.[19] The advantage appeared to be due to the large supply of plant foods, especially nitrates, that were liberated. When the land was kept clean by tillage the time of tillage was the most important factor in the preparation of land for winter wheat.[95] An average difference between July and September plowing of approximately 11.0 bushels per acre was recorded. Early summer tillage increased the amount of soil moisture and nitrates in comparison with late summer tillage. Early tillage also resulted in control of weeds that otherwise would use the nitrates needed by the wheat seedlings. Increased yields resulting from early tillage for wheat grown continuously were obtained regardless of whether the land was plowed, listed, or disked and plowed. Early plowing gave better results than early listing, but early listing proved better than late plowing. Splitting the lister ridges did not increase the yield as compared to single listing.

In experiments in southwestern Kansas, early plowing gave average yields 25 to 50 per cent higher than late plowing.[103] Land plowed late lost considerable moisture due to weed growth, and the seedbed was cloddy, loose, and in poor tilth. In eastern Nebraska plowing on July 15 at a depth of 7 inches, followed by disking on August 15 and disking again prior to seeding, gave an average yield of 33.9 bushels per acre.[46] Omission of the August 15 disking reduced the yield 6.7 bushels per acre. In another experiment, plowing about July 15, followed by disking on August 15 and again prior to seeding, gave a yield of 28.4 bushels per acre. Disking in July, August, and September without plowing returned a yield of 21.6 bushels. Weed control was not the only factor involved because tillage methods equally effective in weed control differed in their effect on the accumulation of nitrates. The type of tillage supplied by the lister or plow was more effective than that by disking.

The lister formerly was widely used in western Kansas for seedbed preparation. Some results from its use are given in Table 15–1.

Most of the land for wheat in the semiarid regions of the United States is prepared with undercutting blade implements, chisels, or the one-way disk plow. Subsequent tillage for weed control before seeding is done mostly with the rotary rod weeder, various blade implements, or repeated one-way tillage.

In the eastern states, the seedbed is often prepared with a disk or spring-tooth harrow when wheat follows corn, soybeans, or cotton. When land is to be plowed, as after a green-manure crop, three or four weeks should intervene between plowing and seeding. Because of potential water erosion in the humid regions, rolling land should not be left bare of vegetation for long periods. Consequently, plowing should not precede seeding by more than four weeks.

When spring wheat is sown after intertilled crops, the land usually is double-disked and harrowed before drilling. The plowing of such land offers little or no advantage. Yields after small grain are appreciably higher when the land is plowed than when it is merely disked, chiefly because of better weed and insect control. Spring plowing usually gives higher yields of spring wheat in the drier western section than does fall plowing. This is because more snow is caught in undisturbed stubble during the winter and more moisture is added to the soil. Fall plowing usually is practiced in the more

Table 15–1. Relation of Method of Preparation to Yields of Wheat in Western Kansas (at Hays), 1907–1927

METHOD OF PREPARATION	AVERAGE YIELD PER ACRE	FAILURES IN 21 YEARS
	(*bu.*)	(*no.*)
Continuous wheat		
Late fall plowing	10.1	11
Early fall plowing	15.7	6
Early listing	19.1	4
Alternate cropping		
Summer fallow	23.9	3

favorable eastern sections to permit earlier seeding, which is of prime importance. Some fall plowing is advisable on farms having a large acreage in spring grain.

Method of Seeding

Nearly all the American wheat crop is sown with a drill. In Nebraska broadcast winter wheat averaged 17 per cent lower in yield than drilled wheat.[46] Drills with disk furrow openers are used most generally in the humid regions. Hoe drills are popular in the semiarid sections of the Pacific Northwest on clean fallowed land. This type of furrow opener turns up clods and does little pulverizing and thus is helpful in preventing soil blowing. The furrow drill has been widely used in the semiarid western Great Plains where winters are severe. It is a partial insurance against winterkilling and soil blowing and it places the seed deep enough to reach moist soil. In Montana average yields of winter wheat from furrow drilling have been higher than from surface drilling.[59] In Kansas the furrow drill has shown no advantage in average seasons. Under eastern Nebraska conditions, furrow drilling in 14-inch rows yielded 81 and 91 per cent as much as surface drilling in rows spaced 7 and 14 inches apart, respectively. The semifurrow drill (Figure 15–14) with rows 10 inches apart later became more popular. The furrow drill has no appreciable merit for seeding spring wheat. The wider row spacing required for making the furrows may be detrimental to yield.[98]

Sowing no-till fields or those covered with a thick stubble mulch requires drills with disk or blade cutters to penetrate the residues.

Depth of Seeding

Wheat usually is sown at a depth of $1\frac{1}{2}$ to 3 inches. Seeding deeper than 1 to 2 inches is advantageous only in permitting germination to occur and roots to develop before the surface soil dries out. For example, at Lind, Washington, where spring rains are rather infrequent; the deepest sowing of spring wheat gave the highest yields.[98] With winter wheat the average yields from shallow (1-inch), medium ($2\frac{1}{2}$-inch), and deep (4-inch) seeding were 10.8, 9.9, and 8.9 bushels per acre, respectively. At Moro, Oregon, the yields of winter wheat from shallow, medium, and deep seeding were 25.7, 26.4, and 26.4 bushels per acre, respectively. These results were obtained on dry land where the seed often did not germinate until some time after seeding. It often happens that when fall rains are insufficient to wet the soil and germinate the shallow-sown wheat, the deeper-sown wheat also is not likely to thrive, except sometimes on fallowed land.

When wheat germinates the plumule grows upward from the sprouting seed enclosed in the elongating coleoptile. It also can be pushed upward by the elongation of the subcrown internode which

Figure 15–14. The semifurrow drill with packer wheels is desirable for seeding winter wheat in the semiarid Great Plains. [Courtesy John Deere & Co.]

arises below the coleoptile node. Sometimes other internodes also elongate. Coleoptile elongation ceases when light strikes its emerging tip. The coleoptile then opens and allows the first leaves to push out. Tillering begins then. Crown roots start to develop at nodes $\frac{3}{4}$ inch to 2 inches below the soil surface. Seeding wheat at a depth of 1 inch or less causes the crown to form just above the seed. Seeding deeper than $1\frac{1}{2}$ inches, except where necessary to insure prompt germination, merely uses up part of the energy in the seed in producing an excessively long sprout, and delays and weakens the seedling accordingly. Deep seeding may cause the crown to form somewhat deeper than when seeding is shallow,[108] but the difference in crown depth is much less than the difference in seeding depth. High soil temperatures during seedling emergence cause the crowns to be formed at shallow depths. Hardy varieties of winter wheat have deeper crowns than do nonhardy varieties, and spring wheats have the shallowest crowns.[108]

Time of Seeding

Factors that determine the best time for seeding winter wheat were discussed in Chapter 7. Medium-season seeding of winter wheat for any locality is usually most favorable. Wheat sown late generally suffers more winter injury, tillers less, and may ripen later the next season. Wheat sown too early may use up soil moisture, joint in the fall, suffer from winter injury and foot rots, and become infested with the Hessian fly where that insect is prevalent. In the semiarid Great Plains the optimum date for seeding winter wheat is about September 1 in Montana, and progressively later in the southward to northwestern Texas, where the best time is about October 15. Planting earlier than the optimum seeding time is a common practice in the Great

Plains either to take advantage of favorable soil moisture conditions when they occur or to provide wheat pasture. The chief disadvantage of too early seeding of winter wheat is the likelihood of injury from foot-rot diseases that develop under warm conditions.[88] Seeding about mid-September is recommended in the Pacific Northwest. November and early December seeding are most favorable in California and southern Arizona. Under the mild climatic conditions in these states nearly all the varieties have a spring growth habit but are grown as winter wheats. Spring wheats also are grown from fall or winter sowing in China, India, southern Europe, Africa, and parts of Latin America. In the central and eastern states the optimum seeding date is about the time of, or a few days earlier than, the fly-free date,[107] as shown in Figure 15–15. The safe date is the earliest date at which wheat can be seeded to escape damage from this insect. Wheat should generally be seeded 7 to 10 days earlier than the safe date where the Hessian fly is not present (Figure 15–16). This gives the plants a better start in the fall.

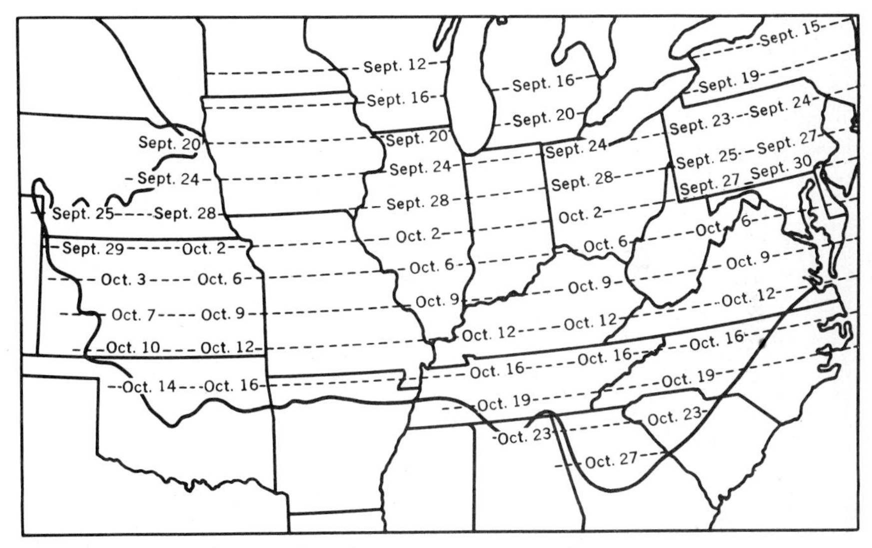

Figure 15–15. Fly-free date for sowing winter wheat.

Winter wheat fails to head when sown in the spring except when sown very early or under cool conditions. The critical seeding date for normal heading of wheats with a winter growth habit depends upon the variety (Figure 15–17). This date was February 15 for the one winter variety at Moro, Oregon, whereas another variety headed when sown fully a month later.[16] The critical date for a spring wheat with a partial winter habit, was about April 30, whereas true spring wheats headed when sown about a month thereafter.

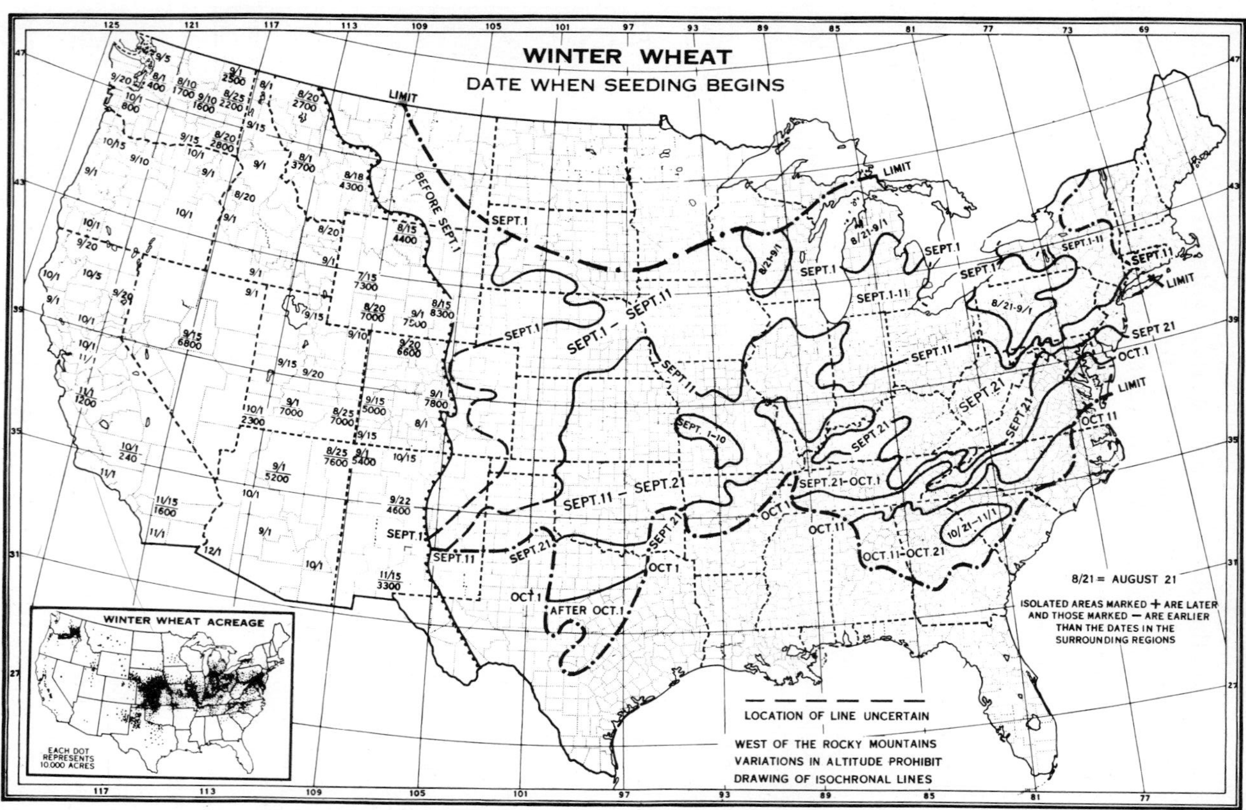

Figure 15–16. Date when winter-wheat seeding begins in the United States.

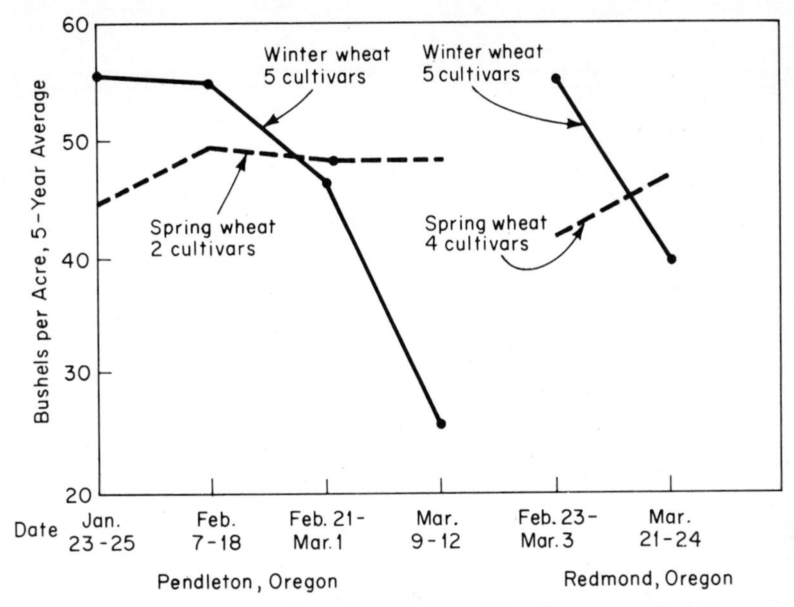

Figure 15–17. Effect of seeding date upon yields of winter and spring wheat.

Early seeding of spring wheat usually results in the highest yields. Early-sown spring wheat is most likely to escape injury from drought, heat, and diseases, which become more prevalent as the season advances.

The usual procedure is to sow spring wheat as early as the soil can be worked into a good seedbed. This time varies widely between seasons, usually arriving in March in Nebraska and Colorado, April in South Dakota and southern Minnesota, and May along the Canadian border (Figure 15–18). Spring seeding usually is accomplished in March in Washington and Oregon, however, because of a climate milder than that east of the Rockies at the same latitude. When wheat is sown during an extremely early mild period that is followed by a cold snap, the wheat remains in the ground and emerges during subsequent warmer weather usually without evidence of having been injured. There is no advantage in extremely early seeding except to insure seeding in ample time. Late winter seeding of spring wheat is not advisable except in warm climates.

Rate of Seeding

The average rate of seeding wheat in the United States in 1972 was 1.13, 1.39, and 1.22 bushels per acre for winter wheat, durum wheat, and other spring wheats, respectively. These rates are equivalent to 76, 93 and 88 kilos per hectare.

The optimum rate of seeding for wheat in the western half of the

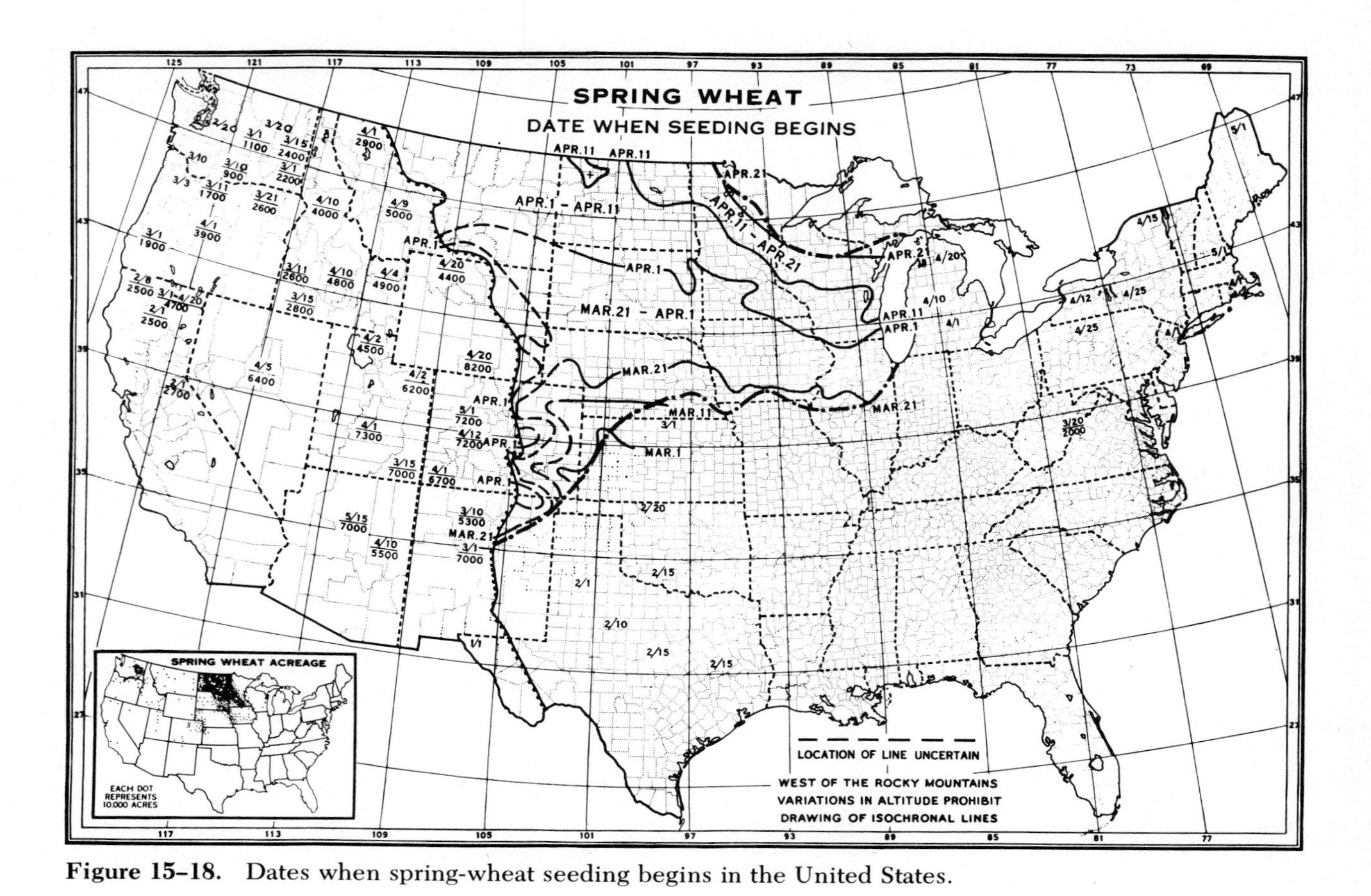

Figure 15–18. Dates when spring-wheat seeding begins in the United States.

TABLE 15–2. Effect of Rates of Seeding on Winter Wheat in Nebraska

RATE OF SEEDING	TILLERS PER PLANT IN		YIELD PER ACRE (8-YEAR AVERAGE)
(pecks)	*Fall*	*Spring*	*(bu.)*
Three	3.8	3.8	24.7
Four	3.4	3.5	26.2
Five	3.2	2.6	26.4
Six	3.0	2.6	27.4
Eight	2.7	1.7	27.4

United States is largely independent of soil type, moisture, locality, date of seeding, cultural treatment, and variety.[56] In general, 4 to 6 pecks per acre* have produced the highest net grain yields of both winter and spring wheat. Yields from 2-peck seedings have been decidedly less than from heavier rates except from relatively early seeding on soil well supplied with moisture and nitrates. Many growers of hard winter wheat in the Great Plains sow only 20 to 30 pounds per acre, but the average rate there is about 40 pounds.[46]

There may be a rather wide range in seeding rate for winter wheat without material effect on the yield (Table 15–2).

Winter wheat in the eastern United States usually is seeded at the rate of 5 to 8 pecks per acre. At Arlington Farm in Virginia, the highest grain yields were obtained from the 6-peck rate.[53] About 6 pecks or more will generally produce a larger yield than a lesser quantity of seed. Heavier seeding than usual is advised where seeding is delayed beyond the normal date because of less opportunity for the plants to tiller.

Extensive experiments in Ohio[110] have shown the highest net yield (yield minus seed sown) to be obtained from a seeding rate of 6 to 8 pecks per acre.

A former belief that wheat must be sown thinly under semiarid conditions to avoid crop failure has been refuted by numerous rate-of-seeding experiments in the western part of the United States.[56] The 12-year average yields, shown in Table 15–3, were obtained on corn land at Akron, Colorado. The low average yields shown are largely the result of moisture deficiency. The heavier seeding rates gave higher average gross yields than did the 1- and 2-peck rates, and the net yields were slightly higher, although the difference probably is not significant. The higher seeding rates give some assurance against loss of stands under adverse conditions at very little cost.

* 1 peck per acre = 17 kilos per hectare.

TABLE 15–3. Rate and Date of Seeding Test of Kanred Winter Wheat at Akron, Colorado (12-Year Average)

RATE SEEDED	ACRE YIELD IN BUSHELS WHEN SOWN					AVERAGE
(pecks)	*Aug. 15*	*Sept. 1*	*Sept. 15*	*Oct. 1*	*Oct. 15*	
1	7.3	11.2	11.4	6.7	5.9	8.5
2	9.1	11.6	11.3	6.7	6.7	9.1
3	10.4	12.1	10.6	6.9	6.9	9.4
4	11.0	12.3	10.7	6.8	7.1	9.5
5	11.6	12.3	10.2	7.6	7.7	9.9
Average	9.9	11.9	10.8	7.0	6.9	9.3

Weed Control

Weeds in wheat fields are controlled to a considerable extent by rotation with intertilled crops or meadow crops, and by summer fallow. Weedy fields of wheat often are sprayed with herbicides to control annual broad-leaved weeds and wild oats. Postemergence herbicides should be applied after the wheat plants are 4 to 6 inches tall, but before the wheat is in the boot stage, in order to avoid injury to the crop. Wild oats are best controlled by preemergence treatments. The wheat field also may be sprayed a few days before harvest to kill or wilt green weeds that otherwise would add moisture to the wheat grain during combine harvest. Mature wheat plants are not injured by the herbicides.

Downy brome (cheatgrass, *Bromus tectorum*) is a serious weed in the wheat fields of the intermountain region. Stubble-mulch culture promotes the population of downy brome plants. Herbicides that kill the weed are also fatal to the wheat plants. It can be controlled to a considerable extent by deep plowing which buries the seeds. Fallowing provides partial control, but adequate control is obtained by applying herbicides to the fallowed field.

HARVESTING

Nearly all of the wheat in the United States has been harvested with a combine since 1960 (Figure 8–2). The binder is used rarely when it is desired to save all of the straw. The header and self-rake reaper are obsolete, and the cradle is feasible only for cutting patches of wheat on mountain farms.

Wheat harvest begins in May in the southern portion of the United States and continues through September in the prairie and Great Plains areas of the northern states and the Canadian provinces (Figure 15–19). Wheat is being harvested in some part of the world throughout the year (Figure 15–20).

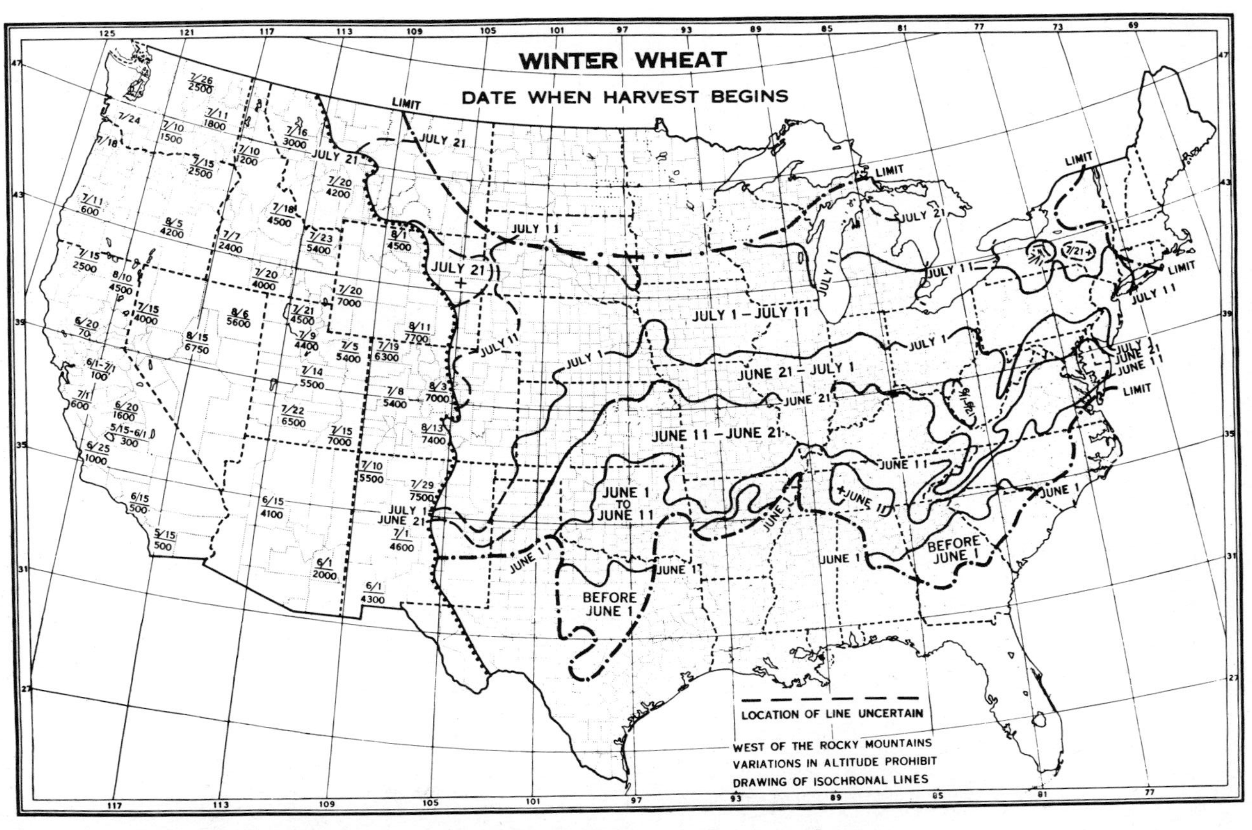

Figure 15–19. Dates when winter-wheat harvest begins in the United States.

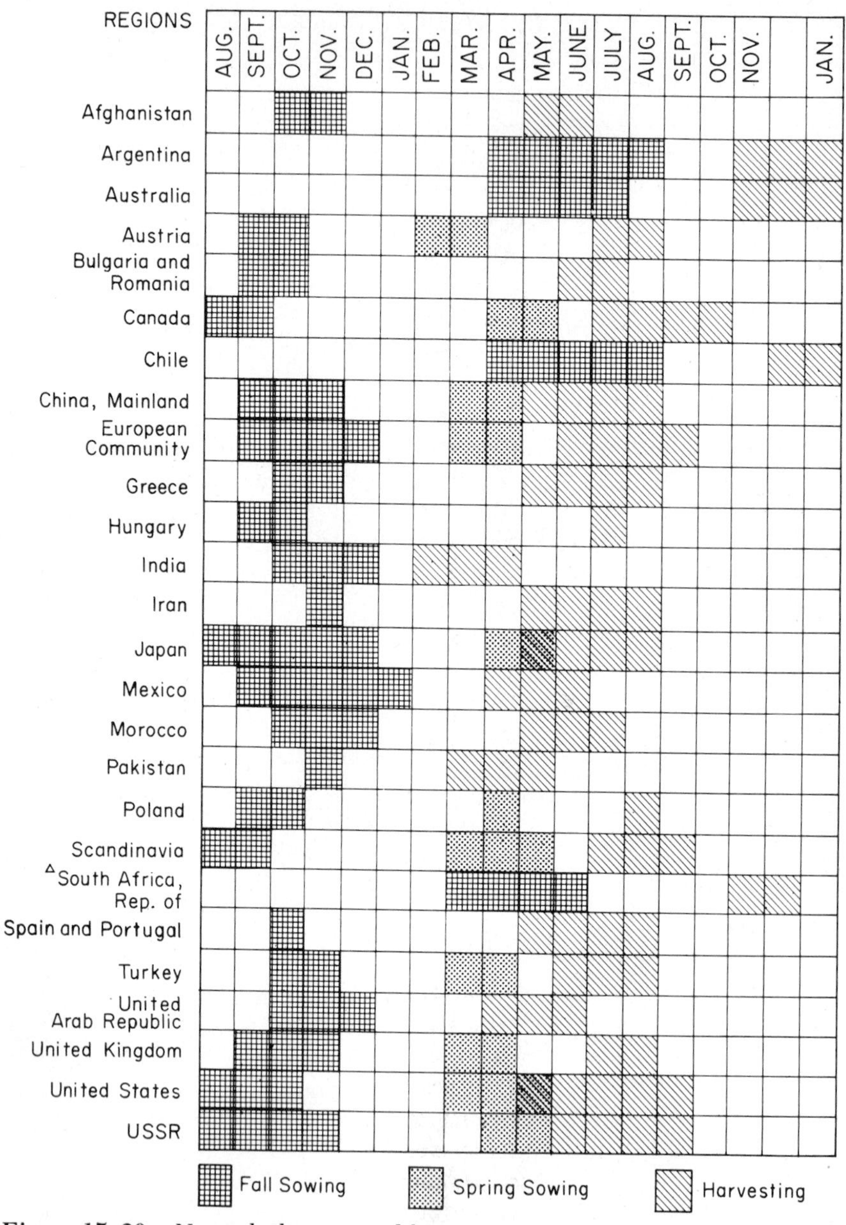

Figure 15–20. Normal planting and harvesting season for wheat at selected locations throughout the world. [Sources:[10, 12, 14, 15] and selected reports from U.S. agricultural attaches.]

USES

An estimated 512 million bushels of wheat was milled for flour in the United States in 1972. More than 14 million bushels was processed into semolina, bulgur, and cereal breakfast foods. More than 63 million bushels were used for seed in 1972, and an equal amount was fed to livestock on the farms where the wheat was grown. Feed wheat is cracked or rolled except for poultry. Wheat is a more efficient feed than other grains in weight gains per feed unit, but, when fed as the only grain, the daily intake and daily gain are less. Wheat should thus constitute no more than half of the grain ration and be mixed with corn or grain sorghum.[16] By-products of wheat milling, which constitute about 30 per cent of the uncleaned wheat, are used largely as livestock feeds. The per capita use of wheat products for food in the United States was about 115 pounds in 1970 compared to 330 pounds in 1910. Wheat exported from the United States consists chiefly of hard red winter wheat and Pacific Coast white wheat.

The chief food use of wheat is in the form of flour for baked products. Other food uses include prepared breakfast foods and the edible pastes made mostly from durum-wheat semolina. Varieties of common or durum wheat with large white kernels are often used for puffing. Shredded wheat products are made from soft white wheat varieties. Hard wheats are used chiefly for bread flours. Soft wheats, both red and white, are used mostly for cake, cracker, pastry, and family flours. Bulgur, chiefly used as a substitute for rice, is prepared from wheat by parboiling, drying, cracking, and removing some of the bran. Since the high bran content of bulgur may induce ulcers or rickets in people who consume large quantities, it is being replaced with a similar product, called wurld wheat, in which all of the bran is removed.[39]

Milling Wheat

PREPARATION FOR GRINDING: Wheat is first cleaned thoroughly at the mill to remove all weed seeds and foreign material. Then it is scoured to rub off hairs and dirt from the kernel. The scourer contains revolving beaters that throw the grains against the roughened walls of a drum. If the wheat is appreciably smutty or dirty, it also is washed. Smutty wheat sometimes is limed before it is scoured. Washed wheat often is dried before milling. The objects of tempering wheat just before grinding are to make grinding easier and to toughen the bran coat so it can be milled off in large flakes and thus be separated from the flour more completely. Wheat is tempered by the addition to sufficient water to raise the moisture content from the usual 12 to 13½ per cent up to about 15 per cent. After the water has been added to a moving mass of wheat, the grain is allowed to stand for

2 to 24 hours, but usually 4 to 6 hours, to permit the added moisture to penetrate the kernel to the proper depth for toughening the bran. Very dry wheat is first moistened and allowed to stand for 2 days or more for the moisture to penetrate the entire kernel. It is then tempered just before milling.

Heat generated during grinding and the aeration of the milled products during their sifting, purifying, and movement through the mill combine to reduce the moisture content in the milled products to about 13 to 13½ per cent. There is an invisible loss in weight in milling due to evaporation when tempered wheat contains a higher moisture than 13½ per cent before tempering. Wheat containing less than 13 per cent moisture before tempering usually shows a gain in milled weight.

THE MILLING PROCESS: In a large mill the milled products are divided into numerous streams, each requiring different treatment or disposition.[26] The grain is first cracked or crushed gradually through a series of four to six pairs of chilled iron break rolls. The surface of the break rolls is roughened by sharp lengthwise corrugations. One member of each pair of break rolls runs about two and a half times as fast as the other, which produces a shearing action on the grain. The grist from each break is sifted and the coarsest particles are conducted to successive break rolls where the process is repeated.

Each successive pair of break rolls has finer corrugations and closer distances between the rolls in order to grind the grain into progressively smaller particles. The finest particles from the breaks are sifted off as flour.

The aim in milling on the break rolls is to get out as large a proportion of middlings as possible. These middlings are granular fractions of the endosperm from which the finer particles (flour), the bran, and shorts, have been separated. The middlings are rebolted and also aspirated in a middlings purifier to remove small light bran particles. The purified middlings pass through a series of paired reduction rolls which revolve at speed ratios of about 1:1½. These rolls, which are smooth, are spaced so that each successive reduction produces finer particles. Flour is sifted out after each reduction. In the milling of straight flour, all the flour streams are combined except for the one called red dog, which is dark in color, due to a high content of fine bran particles and much of the aleurone. In usual milling practice, the finer and whiter fractions comprising 70 to 80 per cent of the total flour are combined into patent flour. The remaining darker grades, usually called clears or baker's flour, are marketed for special baking uses or are mixed with bran, graham, or rye flours. Low-grade flours (clears and red dog) sometimes are later washed to remove the starch, leaving what is known as gluten flour, which is used mostly by diabetics or by people with digestive ailments.

BLEACHING FLOUR: Most flour is bleached to make it whiter and more attractive. The bleaching is done with small quantities of such chemicals as chlorine, benzoyl peroxide, nitrosyl chloride, chlorine dioxide, or nitrogen peroxide. Bleaching destroys the yellow pigments, chiefly xanthophyll, that are naturally present in the wheat flour. Chlorine and its compounds improve the quality of the flour to a degree comparable to that attained from natural aging by storage. Extensive experiments failed to show any deleterious effect on the wholesomeness of the flour as a result of these bleaching processes.

BY-PRODUCTS OF MILLING: The by-products of flour milling, often referred to as *offal,* are usually marketed as bran and shorts, and are valuable dairy feeds and poultry mashes. Some mills make three feed products, namely, bran, shorts (brown shorts), and white middlings. Bran consists mostly of the large particles of pericarp and aleurone with small quantities of starchy endosperm attached.

Shorts (or standard middlings) are the final middlings, consisting mostly of coarse particles of endosperm mixed with, or attached to, considerable quantities of fine bran particles. White middlings (not the intermediate milling product called middlings) comprise red-dog and other low-grade flours mixed with the finer and whiter particles that otherwise would be included in the shorts. Shorts and middlings are often used as hog feed, with enough water added to make a slop. Mill feed from different mills may vary from equal parts to 3:2 proportions of bran and shorts. Bran is a bulky feed, a given volume weighing less than one third as much as wheat. Most of it is fed to dairy cows.

The germ of wheat usually comes from the break rolls in a stream called sizings in which the germ is attached to a large particle of bran. The germ is flattened out and much of the bran detached in the reduction rolls. The germ usually is added to the shorts and thus used for feed. Some of the germ stream or the extracted oil from germs is saved for food purposes because of its high content of vitamins, particularly thiamin (vitamin B_1) and alpha-tocopherol (vitamin E). This latter vitamin is essential for reproduction in rats and for strengthening the walls of red blood cells in premature new-born human infants. Numerous other reported benefits from increased vitamin E consumption are still debatable. Wheat germ bread is now being baked.[78]

Purified middlings of hard spring wheat are sometimes marketed as *Cream of Wheat,* and the corresponding product from hard winter wheat called farina also is used for a cooked breakfast cereal. Purified middlings from durum wheat, called semolina, are used in the manufacture of macaroni, spaghetti, and other edible paste products.

FLOUR YIELDS FROM MILLING: The milling yield of about 72 per cent flour from wheat that consists of nearly 85 per cent endosperm is an indication of the difficulty of making a perfect separation of the kernel parts in the milling process. Particles of endosperm, particularly the aleurone, cling to the bran, and these are carried into the feed or by-products streams.

The average straight flour yield from wheat is about 70 to 74 per cent. Higher extractions result in darker flour such as the 85 per cent dark "victory" flour used in European countries during World War II.

For making about 150 to 160 pounds of freshly baked bread, 100 pounds of flour are sufficient.

FACTORS AFFECTING FLOUR YIELDS: Yields of total flour (including low-grade) range from less than 62 per cent for 49-pound wheat to more than 79 per cent for wheat of a test weight of 64 pounds. With light-weight (shrunken) wheat, which has a low percentage of endosperm, the separation of bran and flour is difficult.

Wetting and drying in the field changes the texture of the wheat kernel, making it softer and more starchy in appearance as well as lighter in test weight. This change is the result of additional air space between the starch grains. The low test weight resulting from weathering does not affect the flour yield in milling, and the softening of the endosperm does not impair the baking quality of the flour. In fact, the baking quality may even have been improved by the wetting, as the flour has the characteristics of having been aged. Weathered wheat thus is of better quality for milling and baking than its test weight and appearance indicate.

Graham or whole-wheat flour is the entire wheat grain ground into flour, and it includes all the bran. Often graham flour is ground in a bur mill or even between old-fashioned stone burs. Apparently the people of the Stone Age ground their wheat into graham flour. However, the ancient Egyptians learned to bolt flour through papyrus sifters, and the tendency through the centuries has been to make the whitest flour possible.[9] The ancient Romans made a fairly white flour. George Washington sold bolted flour at his mill. In 1837, Dr. Sylvester Graham published a book extolling the virtues of flour made from the entire wheat kernel. Such flour has been called graham flour since that time. When the germ is retained in the flour as in graham and certain special flours, the fat in the germ tends to become rancid, which makes the flour unpalatable. Graham flour becomes infested with insects more quickly than does white flour. Graham flour usually sells at a higher price because of these storage risks and because of the small quantity marketed.

White flour, until the 1970s, constituted about 97 per cent of the wheat flour manufactured in the United States. The remainder was graham and other types of dark flour. White flour often is mixed with dark flours in making bread. Consequently, about 93 per cent of the wheat bread baked was made from white flour only, 1 per cent graham bread, and the remainder mixed white and dark flours. These proportions represented the tastes and desires of the American public until health faddists began promoting whole wheat foods.

CHEMICAL COMPOSITION OF WHEAT AND FLOUR

The approximate chemical composition of the wheat kernel[44, 96] in percentages is starch, 63 to 71; proteins, 10 to 15; water, 8 to 17; cellulose, 2 to 3; fat, 1.5 to 2; sugars, 2 to 3; and mineral matter, 1.5 to 2. The protein content of wheat varies for that grown in different regions. The approximate composition of white and graham flours, and of bran and wheat germ, are shown in Table 15–4. It can be seen that white flour is lower in protein, fat, and ash, and higher in starch than is the original wheat or the graham flour.

A patent flour contains roughly 10 to 12 per cent of the total thiamin and niacin, 20 per cent of the riboflavin and iron, 25 per cent of the phosphorus, and 50 per cent of the calcium found in the wheat kernel.[2] Enrichment, as practiced for a majority of the flour used in the United States, restores about 60 per cent of the niacin and 80 per cent of the

TABLE 15–4. Average Composition of Wheat, Flour, Bran, and Germ Containing about 13 Per Cent Moisture

COMPOSITION	WHEAT OR GRAHAM FLOUR (%)	WHITE FLOUR (%)	BRAN (%)	GERM (%)
Carbohydrates				
(nitrogen-free extract)	68	74	50	18
Starch	55	70	10	–
Pentosans	6	3.5	25	6
Dextrins	–	–	4	–
Sugars	2	1.5	1.5	15
Crude fiber	2.3	0.4	9	2
Fat	2	1	4	11
Crude protein	13	11	17	30
Ash (mineral matter)	2	0.45	7	5

thiamin and iron, and about doubles the concentration of riboflavin in the wheat.

The germ fraction is high in thiamin, riboflavin, phosphorus, and iron. Bran is high in niacin, phosphorus, and iron. In milling, most of the thiamin is recovered in the red-dog flour and the shorts, and the niacin is found mostly in the bran. The thiamin of the germ is chiefly in the scutellum rather than in the plumule and the embryonic stem and root tissues.

About 75 to 80 per cent of the thiamin in wheat is found in the bran and shorts (feed) after milling. When these two products are fed to dairy cows and hogs, respectively, not more than about one sixth of the thiamin is returned to human food in the milk and edible pork produced. A distinct rise in the incidence of digestive disorders and of rickets in young children has followed mass consumption of high-extraction flours, i.e., 82 to 100 per cent of the grain. Bran contains considerable percentages of indigestible but partly fermentable material that irritates the intestinal tract. Bran also contains phytic acid (inositol hexaphosphoric acid),[28] which inhibits the absorption of calcium from the intestinal contents and induces the development of rickets. Addition of calcium to the branny diet merely reduces phosphorus absorption. If bread were the only food available, graham flour would doubtless be preferable to white flour from the nutritional standpoint. No one is forced to live by bread alone.

ASH CONTENT AND FLOUR QUALITY: Highly refined patent flour has a low ash content of 0.4 per cent or less. The more bran particles a flour contains, the higher the ash content. Therefore clear and straight flours have a higher ash content than does patent flour. Bakers and flour buyers often purchase flour on a guaranteed maximum allowable ash basis. High ash content is not necessarily an indication of high extraction, poor milling, or dirty wheat because wheat varieties differ in ash content. Hard wheat flours usually have a higher ash content than do soft wheat flours. Environment, chiefly weather conditions, also affects the ash content of wheat and the resultant flour. The chief constituents of wheat ash are oxygen, phosphorus, potassium, magnesium, sulfur, and calcium.

PROTEIN CONTENT AND FLOUR QUALITY: The hard red spring and hard red winter wheats contain an average of about 11 to 15 per cent protein when grown in the Great Plains and northern prairie states.[23] Soft wheats contain 8 to 11 per cent protein when grown in humid areas. In years when the crop is generally low in protein, large premiums may be paid for desirable lots of high-protein hard wheat. Graduated premiums for fractional percentages of protein above 12.5 to 13 per cent usually are offered. High-protein wheat is blended with other wheats in order to bring up the average protein content of the resultant flour to established standards for the particular flour brand. Often wheat and flour are sold on the basis of a specified protein content. Flours of high and low protein content can be obtained from any finely ground flour by air separation.

The hard wheat flours as a rule are high in bread-making quality, i.e., the ability to make a large, light, well-piled loaf of bread of good uniform texture and color. Strong wheats, i.e., those having ample gluten (protein) of good quality, are high in water absorption. They usually produce more baked 1-pound loaves of bread from a barrel of flour because of their greater ability to retain moisture as compared with the weaker flours from soft wheats. A light loaf of bread can be made from a weak flour, but the pores of such bread are large and the loaf tends to dry out quickly. Hard wheats tend to produce flour of a granular texture regardless of their texture or protein content. The granular flour of hard wheats is not well suited for pastry, but soft wheats produce a fine, soft flour well suited to making cakes, crackers, cookies, and hot breads.

The preferred protein percentages in wheats for various domestic purposes are about as follows: macaroni and noodles, 12.5–15; white bread, 11–14; all purpose flour, 9.5–12; crackers, 9.5–11; pie crust and donuts, 8–10.5; cookies, 8–9.5; cake, 8–9.5.[100]

The gluten of the wheat kernel contains about 17.6 per cent nitrogen. The percentage of nitrogen, determined by analysis, is multiplied by 5.7 to determine the protein content ($100 \div 17.6 = 5.7$). The protein of most feed crops contains about 16 per cent nitrogen, and the factor 6.25 is used.

MACARONI

In the manufacture of macaroni and other alimentary paste products the semolina or farina is mixed with hot water and kneaded heavily under corrugated rolls until a stiff dough is formed. The dough is then placed in a cylindrical press where the heavy pressure of a plunger forces it out through a perforated plate or die at the end. The macaroni comes out in a series of continuous strands and is cut into desired lengths and hung up to dry. The pressure drives out the air and compresses the dough, thus leaving a dense product. Various shapes of cut or molded macaroni and noodles, including flat noodles, are made in considerable quantities. The tubular types having a diameter of 0.11 to 0.27 inch (2.8–6.9 mm.) are called macaroni. Small tubes or solid strands with a diameter larger than 0.06 inch (1.5 cm.) and not more than 0.11 inch (2.8 cm.) are called spaghetti. Very small solid strands, not more than 0.06 inch (1.5 cm.) in diameter, are called vermicelli. The making of the holes in macaroni is relatively simple.

Inside each round hole in the die plate is a small round rod attached only at the inside end of the hole. The dough surrounds this rod and forms a tube as it passes through the die plate. When the dies are shaped in the form of letters and the emerging dough strands are cut into very short sections the basic ingredient of the familiar alphabet soup is produced.

Durum semolina of high protein content is preferred for making macaroni because it produces a hard translucent product that is firm after cooking. Also, semolina having a high content of yellow carotenoid pigments, i.e., a gasoline color value of 1.50 or more, is desired because it imparts a desirable rich yellow color to the macaroni.

DISEASES

Rusts

The three rusts attacking wheat are stem rust or black stem rust (*Puccinia graminis tritici*), leaf rust (*P. recondita*), and stripe rust (*P. striiformis* [glumarum]).[58] These are all fungus diseases, known for many centuries. The Romans attributed the crop damage to the rust god Robigus, who resorted to this means of wreaking his vengeance on a wicked people. Others, noting that rust followed damp weather, believed it to be similar to the rusting of their own iron tools. In fact, some laymen once believed that their wheat did not rust before the advent of barbed-wire fences around their fields.

STEM RUST: Stem rust (*Puccinia graminis tritici*) has been the most destructive disease of wheat. It causes such severe shriveling of the grain of susceptible varieties that often the crop is not worth harvesting. The rust produces on the leaves and stems masses of pustules that contain brick-red spores (Figure 15–21). These spores spread the disease to other plants. Rusted plants transpire water at a greatly accelerated rate. Under the mild climate of the southern United States and in Mexico, the red or urediospore stage of the stem-rust organism lives throughout the year on seeded and volunteer wheat and on certain grasses. In the spring the spores multiply and spread to other plants or other spots on the same plant. A new generation of spores is produced every one to three weeks. The spores are caught by wind and air currents and may be carried upward 16,000 feet, and outward for hundreds of miles. By this means, wheat is infected as the season advances northward. A heavy epidemic of stem rust in Texas in April may be expected to produce a similar epidemic in North Dakota in July when the season favors rust development throughout the central wheat-growing region. A close relationship between rust epidemics in Kansas and North Dakota was recognized as early as

Figure 15–21. Rusted wheat culms (*left*) and resistant rust-free culms (*right*).

1910. Moist warm weather favors rust development. A lush growth of wheat produced on soils high in nitrogen and moisture is most subject to rust attack.

In the North the rust organism may pass through additional stages. The red (uredio) spores on the plants are replaced by black (telio) spores as the wheat approaches maturity. These spores remain on the straw and stubble over winter, germinate in the spring, and produce basidiospores, which infect the leaves of the common barberry bush (*Berberis vulgaris*) and wild species such as *B. canadensis* and *B. fendleri*. A type of spore (spermagonium or pycnospore) which develops on the barberry constitutes the sexual reproductive cells of the fungus. The union of these cells produces tissues which give rise to cluster cups (aecia) in which aeciospores are borne. The aeciospores then are blown from the barberry bushes to wheat or grass plants, which they infect. Urediospores are produced on the wheat and the life cycle is completed. The teliospores usually fail to survive the long summer weather south of Nebraska, and consequently barberry bushes seldom are infected in the South. On the other hand, the urediospores of stem rust seldom live over winter north of Texas. Where barberry bushes are present in the northern states, they may be expected to spread the rust to nearby fields of wheat, and the rust then spreads to other fields of wheat. Wheat near barberry bushes becomes infected with rust fully two weeks earlier than wheat infected from spores from the South.

Eradication of the barberry bushes reduces rust infection in wheat as well as in other grain crops. Since the barberry bush also is one medium on which new races of the rust fungus may arise, its eradication is essential. The eradication of the wild species of barberry bushes in isolated valleys of Colorado, Virginia, and Pennsylvania has greatly reduced local damage from rust. However, in the plains and prairies of the Midwest where rust clouds sweep up from the

South, the eradication of barberry bushes can reduce total rust damage only partially.

Dusting or spraying fields of wheat with repeated applications of certain fumigants is possibly but is not economically feasible.

Seed treatment has no effect on rust because the disease is not seed-borne. Early-sown wheat and early-maturing varieties partly evade rust injury.

The breeding and distribution of wheats resistent to stem rust has been a spectacular achievement.[94] Unfortunately, rust resistant wheats suffer heavy losses later after new races appear. Additional sources of disease resistance must then be obtained from wheat varieties grown in other countries to use in breeding for resistance. The breeding and testing of wheats for resistance to new and old races of rust is of necessity a continuous process in the wheat-growing countries of the world.

At least 275 distinct physiological races or biotypes of the stem-rust organism have been discovered, but only a few of these are of economic importance at any one time in a region.

LEAF RUST: Leaf rust (*Puccinia recondita*) occurs chiefly in the eastern half of the United States, but has been less destructive than stem rust. Numerous races of the leaf-rust organism are known. An early infection of leaf rust can reduce the yield of a susceptible wheat variety 42 to 94 per cent.[43] Reduction in yield is due primarily to a reduced number of kernels in the spike. The kernels also are reduced in weight but are not shriveled noticeably. Pustules containing the orange-red spores cover the leaves and part of the stems. The rust fungus lives over winter in wheat plants either in these pustules or as mycelium (vegetative strands) within the leaf tissue. Leaf rust can be controlled by several dustings with sulfur, but this is too expensive for field-scale operations. Practical control is effected by the use of resistant varieties.

STRIPE RUST: Stripe rust (*Puccinia striiformis*) in the United States is confined largely to the western half of the country but is a serious disease in most warm countries of the world. Linear orange-yellow lesions develop on the leaves and, later, on the floral bracts. Damage by stripe rust is similar to that caused by leaf rust, except that stripe rust shrivels the kernels. The urediospores overwinter where the winters are mild or where snow cover protects the foliage. The use of resistant varieties is the most practical control measure in regions where the disease is serious. Many adapted varieties are resistant.

Smuts

BUNT OR STINKING SMUT: Bunt (caused by the fungi *Tilletia caries* and *T. foetida*) is found throughout the country, but is most severe in the Pacific Northwest where the spores are carried over in the soil. The disease usually is carried over from one crop to the next as black spores on the seed or as smut balls mixed with the seed. The smut spores germinate when bunt-infested seed is sown in moist cool soils. The fungus infects the young seedlings, grows within them, and produces smut balls completely filled with black spores instead of kernels in the wheat head (Figure 15–22). The spores have the odor of stale fish. Bunt can be controlled by seed treatment with suitable fungicides where the soil is not infested with the fungus spores, or with hexachlorbenzene where soil infestation occurs.[35]

Figure 15–22. (*Left to right*) Sound wheat, nematode galls, cockle seeds, and bunt balls.

Dwarf bunt (*T. controversa*) causes severe dwarfing of smutted plants,[40] and it cannot be controlled by seed treatment because of heavy soil infestation where it occurs. Soil treatment with hexachlorbenzene in the drill row at seeding time reduces losses from the disease. Dwarf bunt attacks winter wheat only, chiefly in the northern intermountain states. Resistant varieties are grown where the disease is most prevalent.

LOOSE SMUT: Loose smut (caused by the fungus *Ustilago tritici*) may occur wherever wheat is grown, but is most abundant under humid or irrigated conditions. The floral bracts of the spike are almost completely replaced by black smut masses. The spores are soon disseminated by wind, rain, and other agencies, leaving a naked rachis. The spores infect flowers of sound heads while in blossom. When the spores germinate, an internal infection is produced within the new kernel. Infected kernels are indistinguishable from normal kernels at maturity. When the infected seed is sown the fungus grows within the new plant as it develops. Surface disinfectants are ineffective against the internal infection. The modified hot-water treatment will control the disease but it is injurious to the seed and rather expensive. However, it is a practical means of eliminating the disease from a foundation seed supply. The seed wheat is soaked for 4 or 5 hours in cold water, then placed in a warming tank at 49° C. for 1 minute, put in a hot-water tank at 54° C. for 10 minutes, and finally dipped in cold water for 1 minute. Certain water-soak treatments are less injurious to the seed and about as effective in disease control. During such a treatment, the seed is soaked for 6 hours, after which it is held in an air-tight container for 72 hours. The treated seed is spread out to dry and is ready for planting as soon as it is dry enough to feed through a drill. Some varieties are resistant to loose smut.

FLAG SMUT: Flag smut (caused by the fungus *Urocystis tritici*) has been reported in Illinois, Missouri, Kansas, and Washington. Losses in yield of 10 to 20 per cent may occur in susceptible varieties. Flag smut produces dark stripes on the leaves, sheaths, and culms. Infected plants are more or less dwarfed and often fail to form normal heads. The disease is carried to the next crop both on the seed and in the soil of infested fields. Flag smut can be controlled by the same seed treatments used for bunt if the seed is sown on noninfested soil. Several winter-wheat varieties are resistant to the disease.

Scab

Wheat scab, caused by *Gibberella saubinetii*, is a serious disease in the northern and eastern states, particularly in the Corn Belt. The fungus attacks the heads shortly after they emerge. One or more spikelets may be killed, or the development of the kernel prevented. The disease can be identified by the appearance of a pinkish-white fungus growth on or around the dead tissue. The grain in diseased portions is shriveled, almost white, and scabby in appearance (Figure 15–23). The disease is carried from season to season both on the seed and on old crop refuse, such as corn stalks. It is most severe when wheat follows corn in the rotation. Infected seed should be treated with a chemical protectant before it is sown, to prevent seedling blight.

Take-All

Take-all (caused by *Ophiobolus graminis*) is a widely spread disease, but appears to be most serious in Kansas and New York. In some years, it has destroyed 10 to 50 per cent of the crop in infested fields in Kansas. Take-all may kill wheat plants in the rosette stage or later after the heads begin to fill. In the latter case, the plants turn almost white. Nearly all plants in certain spots may be killed. Other plants are damaged by rotting of the roots. The bases of infected plants usually are black to a height of 1 or 2 inches above the soil. The take-

Figure 15–23. Healthy wheat plants and grain (*left*); scab-diseased plants and scabbed grain (*right*).

all fungus lives over in the soil. The only feasible control measure is to keep infested land free from wheat, barley, and rye for at least two to four years.[20]

Mosaics

Wheat mosaics are virus diseases. Soil-borne mosaic produces a severe dwarfing or rosetting of the plants, or a mottling of the leaves, and often kills most of the plants in areas scattered over the fields. The virus persists in the soil for several years. The disease occurs in Maryland, North Carolina, and South Carolina, as well as in several states westward to Nebraska, Kansas, and Oklahoma.[63] It may be controlled by sowing resistant varieties.

Streak mosaic occurs in Kansas, several other central and western states, and in Canada. The virus is spread by the wheat-curl mite (*Aceria tulipae*). It is carried over in volunteer wheat plants as well as in several grasses. The symptoms of the disease are yellow stripes and blotches on the leaves and stunted plant growth. Partial control methods are late seeding of winter wheat and destruction of volunteer wheat.

Barley stripe mosaic is a seed-borne disease, also of wheat. It occurs in several states. Control consists of sowing seed that is free from the virus.

Nematodes

The nematode, or eelworm, disease caused by a small, almost microscopic round worm (*Anguinea tritici*) occurs in some of the eastern and southeastern states, particularly Virginia, Maryland, North Carolina, and South Carolina. The nematodes living in the soil enter the wheat plants through the roots and move up into the young developing heads before the latter emerge from the sheath. The infested ovary does not develop into a grain but instead forms a hard black gall filled with the nematodes. The galls resemble those formed by the bunt disease except that they are not easily crushed. The wheat nematode is controlled by rotation with crops other than rye for one or two years and by the planting of clean seed.

Other Diseases

Ergot (*Claviceps purpurea*) sometimes attacks durum wheats rather severely but causes little damage to other wheats. It is controlled by use of clean seed combined with crop rotation.

Powdery mildew (*Erysiphe graminis tritici*), which occurs under moist conditions, can be controlled by use of resistant varieties.

The foot-rot diseases are caused by *Helminthosporium sativum*, *H. tritici*, *Pythium arrhenomanes*, other species of *Pythium*, species of *Fusarium* and other fungi.[97] The *Helminthosporium sativum* fungus also causes the *black-point* disease of wheat, a discoloration and infection of the germ end, particularly of durum wheat. The western dryland foot rot is associated with a mosaic disease.

Cercosporella footrot, caused by the fungus *Cercosporella herpotrichoides*, damages wheat mostly in southeastern Washington and adjacent areas. The disease is also called straw breaker and eyespot footrot. It spreads to basal leaf sheaths of young plants by rain-borne spores from infected stubble in a wheat-fallow crop system. Damage from the disease is reduced by delayed seeding and by spraying with suitable fungicides.

Leaf spot (*Septoria*) attacks wheat over much of the eastern and northwestern parts of the United States when weather conditions favor the fungus. Black chaff and basal glume blotch occur in the Midwest. There is no special control method for these diseases.

Snow mold is caused by the fungi, *Calonectria graminicola*, *Typhus incarnata*, and *T. idahoensis*. It frequently damages wheat in the northwestern United States, Europe, and Asia. The disease develops under a deep snow that covers unfrozen ground.[62] Remedies are crop rotation, early seeding, and spring application of a nitrogen fertilizer.

Several other diseases are damaging to wheat.[82]

INSECT PESTS

More than 30 insect pests are injurious to growing wheat, among them being the Hessian fly, wheat jointworm, wheat strawworm, chinch bug, army worm, and the false wireworm.[69]

The Hessian Fly

The Hessian fly (*Mayetiola destructor*) is one of the most serious insect enemies of wheat, particularly east of the 100th meridian. Damage occurs both in the fall and the spring. The injury is caused by the maggots (larvae) located between the leaf sheath and stem where they extract the juices of the young stems. Many of the small tillers are killed. The older stems are weakened and break over readily shortly before harvest. The fall brood may kill many plants outright in cases of serious infestation. The larvae (Figure 15–24) are transformed to a resting stage called a puparium or *flaxseed*. In the spring the flaxseed changes into a pupa and then into an adult fly. The female fly deposits eggs on the young wheat plants. Then another generation of larvae hatches out.

The most practical control measure is to sow winter wheat late

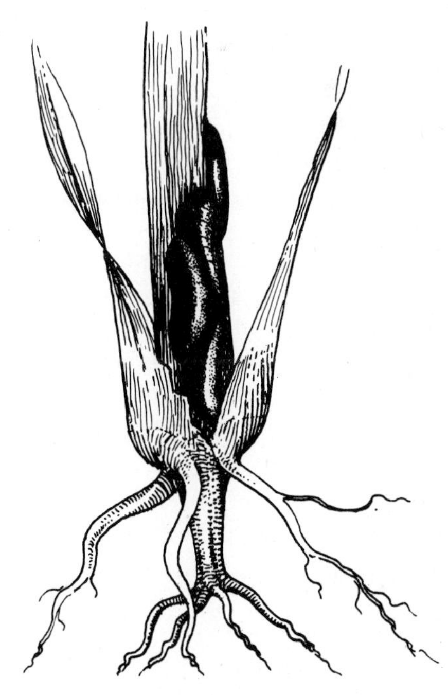

Figure 15–24. Larvae or maggots of the Hessian fly.

enough so that the main brood of flies will have emerged and died before the young wheat plants appear above the ground (Figure 15–17). Other methods are crop rotation (except in strip-cropped fields), and the plowing under of infested stubble soon after harvest. The sowing of resistant varieties also is helpful.[70]

Wheat Jointworm

As a wheat pest, the wheat jointworm (*Harmolita tritici*) ranks second only to the Hessian fly in a majority of the wheat states east of the Mississippi River and in parts of Missouri, Utah, and Oregon. It is a small grub that lives in the stem and feeds on the juices. As a result, wartlike swellings on the stem usually occur above the joint or node. The egg from which the grub hatches is laid in the stem by the adult, which resembles a small black ant with wings. The damage to wheat, caused by the lodging of the infested straws, varies from slight injury to total destruction.

This jointworm can be controlled by plowing wheat stubble under deeply after harvest in order to bury it so that the jointworm adults cannot emerge. This may be impractical where grass, clover, or some other legume is seeded with the wheat, except where heavy infestations occur. Rye may be substituted for wheat in the more northerly states where the jointworm is severe.[77]

Wheat Strawworm

Losses caused by the wheat strawworm (*Harmolita grandis*) may be severe throughout the wheat states east of the Mississippi River. It is also an important wheat pest in other wheat-growing states particularly in western Kansas.[73] Propagation of the worm is favored by infested stubble left on the ground following use of subsurface tillage implements and the one-way plow.

Two complete generations of the insect occur each year. The first generation, or spring form, kills outright each wheat tiller that it infests. As the larva completes its development, the tiller usually becomes bulblike at the point of infestation. The spring form is most injurious to winter wheat. The injury caused by the second generation, or summer form, is less severe except in spring wheat.

The wheat strawworm attacks only wheat. Since the spring form is wingless, it can be controlled where wheat is sown 75 yards or more from wheat stubble or straw of the previous season. Also, where spring wheat is grown, all volunteer wheat should be destroyed when this pest is abundant, to prevent reinfestation.[76]

Grasshoppers

Wheat is attacked by several species of grasshopper that belong mostly to the genus *Melanoplus*.[71] The grasshopper deposits eggs enclosed in sacs about 1 or 2 inches below the surface of cropped, idle, or sod land. Good plowing with a moldboard plow at a depth of 5 inches or more prevents the young grasshoppers from emerging. The young grasshoppers hatch in the spring and soon begin feeding. The grasshoppers eat the leaves and often the stems and heads of the wheat plants.

The chief method for the control of grasshoppers is with insecticides. These materials should be applied to the hatching areas when the nymphs are young.

Green Bug

The green bug or spring grain aphis (*Schizaphis graminum*) occurs in most of the wheat-growing states, and in some years causes severe losses, especially in Texas, Oklahoma, Kansas, and Missouri. It is a sucking insect, adults as well as young bugs feeding on the wheat plants throughout their lives. They reproduce very rapidly during the summer by vivipary (giving birth to the young) and later by eggs, with or without fertilization. One control measure is destruction of all volunteer wheat, oats, and barley in the summer and fall in each community.[105] Recommended insecticides are helpful.

Several wheat varieties are somewhat resistant to light or medium green-bug attacks.[6] Oats and barley are damaged more severely than is wheat.

Chinch Bug

The chinch bug (*Blissus leucopterus*), a sucking insect, often is present in damaging numbers from Illinois and Missouri westward to the central Great Plains. It is more destructive to barley, corn, and sorghum than to wheat.[69] No satisfactory methods of controlling infestation in small grains are yet known. Spraying field borders with insecticides prevents the survival and migration of the chinch bug to fields of corn or sorghum.

Mormon Cricket

The Mormon cricket (*Anabrus simplex*) ranges from the Cascade and Sierra Nevada mountains to the central Dakotas, devouring range grasses, grains, and other crops. This insect is wingless but it migrates on foot in dense hordes for considerable distances. Control is effected by broadcasting a poison bait in advance of the horde. Seagulls devoured the crickets that invaded the first crop in Utah.

Wheat-Stem Sawfly

The wheat stem sawfly (*Cephus cinctus*) frequently causes severe damage in western Canada and in Montana and other states. The adult female fly splits the wheat stem with a pair of sawlike ovipositor appendages and then deposits an egg inside the stem.[69] The cannibalistic larvae feed downward inside the stem, and the survivor finally chews a groove or ring just above the soil surface, which causes the stems to fall over. The chief loss is from the broken stems, which are difficult to save in harvesting. Control measures consist in rotation with other crops, and the turning under of the stubble with a breaking plow that turns the furrow completely. The adults have difficulty in emerging through 6 inches of soil. The use of a pick-up reel on the combine would salvage most of the broken stems. Resistant varieties with solid stems are being grown. Two other species of stem sawflies were introduced from Europe.[104] Similar pests, the black grain-stem sawfly (*Cephus tabidus*), and the European wheat-stem sawfly (*C. pygmaeus*), occur in the East Central States.[102]

Wireworms

The Great Basin wireworm (*Otenicera pruinina* var. *noxia*) destroys considerable wheat in sections of Washington, Oregon, and Idaho having an annual precipitation of less than 15 inches. These wireworms, which live in the soil for three to ten years, eat seeded grain and the underground parts of young wheat plants, thus thinning the stands. The adults (*click beetles*) live only a few weeks. A control method is to treat the seed with a suitable insecticide. Partial control measures are thick seeding and clean fallowing. Keeping the land free from any growing weed or wheat plant during the fallow season will starve the young larvae.

False wireworms (*Eleodes* species) have damaged wheat severely in Kansas and Idaho by devouring the seeded grain before it germinates. Control measures are crop rotation, summer fallow, seeding only when the soil is moist enough to insure rapid germination, and seed treatment.

Other Insects

The fall armyworm (*Laphygma frugiperda*) attacks wheat and numerous other crops in the southern and central states. Control in wheat fields consists in applying insecticides. Delayed seeding of winter wheat, as recommended for Hessian fly control,[106] is helpful.

The armyworm (*Cirphis unipuncta*) has been destructive in many localities in the eastern half of the United States. It is controlled with the insecticides used for the fall armyworm.

Billbugs (*Calendra* species) attack wheat as well as other crops. The chief control measure is crop rotation.

The wheat-stem maggot (*Meromyza americana*) feeds on the stem just above the top node. It cuts off the peduncle, causing the head to turn white. The percentage of plants injured is usually too small to justify control practices.

The pale western cutworm (*Agrostis orthogonia*) destroys considerable wheat in the dryland sections of Montana, North Dakota, Kansas, and other states. The light-gray larvae hatch out in the spring and feed on the underground protions of wheat and grass plants. The moths deposit eggs in the soil in the fall. The best control method is a soil application of an insecticide worked into the soil 6 inches deep before seeding. Summer fallow should be clean to destroy all weeds, grass, and volunteer grain, and to starve the larvae.[70]

The insects attacking wheat in storage are described in Chapter 9.

REFERENCES

1. Alsberg, C. L., and Griffing, E. P. "Environment, heredity, and wheat quality," *Wheat Studies,* Food Res. Inst. Stanford U., 10:229–249. 1934.
2. Andrews, J. S., Boyd, H. M., and Gortner, W. A. "Nicotinic acid content of cereals and cereal products," *J. Ind. Eng. Chem. Anal. Ed.*, 14(8):663–666. 1942.
3. Anonymous. *Bibliography of Wheat,* Vols. I, II, and III, Scarecrow Press, Metuchen, N.J. 1971.
4. Army, T. J., and Hide, J. C. "Effects of green manure crops on dryland wheat production in the Great Plains area of Montana," *Agron. J.*, 51:196–198. 1959.
5. Army, T. J., Bond, J. J., and Van Doren, C. E. "Precipitation-yield relationships in dryland wheat production on medium to fine textured soils of the southern high plains," *Agron. J.*, 51:721–724. 1959.
6. Atkins, I. M., and Dahms, R. C. "Reaction of small-grain varieties to green bug attack," *USDA Tech. Bul. 901,* pp. 1–30. 1945.
7. Ausemus, E. R., and Heerman, R. M. "Hard red spring and durum wheats: Culture and varieties," *USDA Farmers Bul. 2139.* 1959.
8. Bailey, C. H. *The Chemistry of Wheat Flour,* Chemical Catalogue Co., New York, pp. 1–324. 1925.
9. Bailey, C. H. *Constituents of Wheat and Wheat Products,* Reinhold, New York, pp. 1–332. 1944.
10. Bauer, A. "Effect of water supply and seasonal distribution on spring wheat yields," *N. Dak. Agr. Exp. Sta. Bul. 490,* pp. 1–21. 1972.
11. Bayles, B. B., and Martin, J. F. "Growth habit and yield of wheat as influenced by time of seeding," *J. Agr. Res.*, 42:493–500. 1931.
12. Bayles, B. B., and Taylor, J. W. "Wheat production in the eastern United States," *USDA Farmers Bul. 2006.* 1951.
13. Bonnett, O. T. "Inflorescences of maize, wheat, rye, barley and oats—their initiation and development," *Ill. Agr. Exp. Sta. Bul. 721,* pp. 1–105. 1966.
14. Bowden, W. M. "The taxonomy and nomenclature of the wheats, barleys and ryes and their wild relatives," *Can. J. Bot.*, 37:657–684. 1959.
15. Bradbury, D., and others. "Structure of the mature kernel of wheat," Pts. I–IV. *Cereal Chem.*, 33(6):329–391. 1956.
16. Brethour, J. R. "Feeding wheat to beef cattle," *Kans. Agr. Exp. Sta. Bul. 487,* pp. 1–43. 1966.

17. Briggle, L. W., and Reitz, L. P. "Classification of *Triticum* species and of wheat varieties grown in the United States," *USDA Tech. Bul. 1278*. 1963.
17a. Bruehl, G. W., and others. Experiments with Cercosporella footrot (straw breaker disease a winter wheat). *Wash. Agr. Exp. Sta. Bul. 694*, 1–14. 1968.
18. Brumfield, K. *This Was Wheat Farming*, Superior Publ. Co., Seattle, Washington, pp. 1–191. 1968.
19. Call, L. E. "The effect of different methods of preparing a seedbed for winter wheat upon yield, soil moisture, and nitrates," *J. Am. Soc. Agron.*, 6:249–259. 1914.
20. Christensen, J. J. "Root rots of wheat, oats, rye, and barley," in *Plant Diseases*, USDA Yearbook, 1953, pp. 321–328.
21. Clark, J. A. "Improvement in wheat," in *USDA Yearbook, 1936*. pp. 207–302.
22. Cole, J. S., and Mathews, O. R. "Relation of the depth to which the soil is wet at seeding time to the yield of spring wheat on the Great Plains," *USDA Circ. 563*, pp. 1–20. 1940.
23. Coleman, D. A., and others. "Milling and baking qualities of world wheats," *USDA Tech. Bul. 197*, pp. 1–223. 1930.
24. Cowan, F. T., Shipman, H. J., and Wakeland, C. "Mormon crickets and their control," *USDA Farmers Bul. 1928*, pp. 1–17. 1943.
25. Dalrymple, D. G. "Imports and plantings of high-yielding varieties of wheat and rice in the less developed nations," *USDA FEDR-14*. 1972.
26. Dedrick, B. W. *Practical Milling*, National Miller, Chicago. 1924.
27. Doll, E. C. "Effects of fall-applied nitrogen fertilizer and winter rainfall on yield of wheat," *Agron. J.*, 54:471–473. 1962.
28. Dunlap, F. L. *White versus Brown Flour*, Wallace and Tiernan, New York, pp. 1–272. 1945.
29. Eckerson, S. H. "Microchemical studies in the progressive development of the wheat plant," *Wash. Agr. Exp. Sta. Bul. 139*. 1917.
30. Fifield, C. C., and others. "Quality characteristics of wheat varieties grown in the western United States," *USDA Tech. Bul. 887*, pp. 1–35. 1945.
31. Gallun, R. L. "The Hessian fly, how to control it," *USDA Leaflet* 533, pp. 1–7. 1965.
32. Garber, R. J., and Quisenberry, K. S. "Natural crossing in winter wheat," *J. Am. Soc. Agron.*, 15:508–512. 1923.
33. Gauch, H. G., and Miller, E. C. "The influence of the awns upon the rate of transpiration from the heads of wheat," *J. Agr. Res.*, 61:445–458. 1940.
34. Hallsted, A. L., and Mathews, O. R. "Soil moisture and winter wheat with suggestions on abandonment," *Kans. Agr. Exp. Sta. Bul. 273*, pp. 1–46. 1936.
35. Hanson, E. W., Hansing, E. D., and Schroeder, W. T. "Seed treatments for control of disease," in *Seeds*, USDA Yearbook, 1961, pp. 272–280.
36. Harvey, R. B. "Physiology of the adaptation of plants to low temperatures," in *World Grain Exhibition and Conference*, Vol. 2, Canadian Society of Technical Agriculturists, Ottawa, pp. 145–151. 1933.

37. Haunold, A., Johnson, V. A., and Schmidt, J. W. "Variation in protein content of the grain in four varieties of *Triticum aestivum* L.," *Agron. J.*, 54:121–125. 1962.
38. Hayes, H. K., and others. "Thatcher wheat," *Minn. Agr. Exp. Sta. Bul. 325.* 1936.
39. Heid, M. "World wheat in foods of many lands," *ARS 74–61.* 1972.
40. Holton, C. S., and Tapke, V. F. "The smuts of wheat, oats, and barley," in *Plant Diseases*, USDA Yearbook, 1953, pp. 360–368.
41. Janssen, G. "Effect of date of seeding of winter wheat on plant development and its relationship to winterhardiness," *J. Am. Soc. Agron.*, 21:444–466. 1929.
42. Johnson, V. A. and Schmidt, J. W. "Hybrid wheat," *Adv. Agron.*, 20:199–233. 1968.
43. Johnston, C. O., and Miller, E. C. "Relation of leaf-rust infection to yield, growth, and water economy of two varieties of wheat," *J. Agr. Res.*, 49:955–981. 1934.
44. Kent-Jones, D. W., and Amos, A. J. *Modern Cereal Chemistry*, 5th ed., Northern Publ. Co., Liverpool, pp. 1–817. 1957.
45. Kent, N. L. *Technology of Cereals with Special Reference to Wheat*, Pergamon Press, London, pp. 1–266. 1966.
46. Kiesselbach, T. A., Anderson, A., and Lyness, W. E. "Cultural practices in winter wheat production," *Nebr. Agr. Exp. Sta. Bul. 286.* 1934.
47. Klages, K. H. "Relation of soil moisture content to resistance of wheat seedlings to low temperatures," *J. Am. Soc. Agron.*, 18:184–193.
48. Klages, K. H. "Metrical attributes and the physiology of hardy varieties of winter wheat," *J. Am. Soc. Agron.*, 18:529–566. 1926.
49. Kneen, E., and Blish, M. J. "Carbohydrate metabolism and winter hardiness of wheat," *J. Agr. Res.*, 62:1–26. 1941.
50. Lamb, C. A. "The relation of awns to the productivity of Ohio wheats," *J. Am. Soc. Agron.*, 29:339–348. 1937.
51. Leighty, C. E., and Sando, W. J. "The blooming of wheat flowers," *J. Agr. Res.*, 27:231–244. 1924.
52. Leighty, C. E., and Taylor, J. W. "Studies in natural hybridization of wheat," *J. Agr. Res.*, 35:865–887. 1927.
53. Leighty, C. E., and Taylor, J. W. "Rate and date of seeding and seed-bed preparation for winter wheat at Arlington Farm," *USDA Tech. Bul. 38.* 1927.
54. Leggett, G. E. "Relationship between wheat yield, available moisture and available nitrogen in Eastern Washington dry land areas," *Wash. Agr. Exp. Sta. Bul. 609.* 1959.
55. Mangels, C. E. "Pre-harvest factors that affect wheat quality," *Cereal Chem.*, 4:376–388. 1927.
56. Martin, J. H. "Factors influencing results from rate- and date-of-seeding experiments in the western United States," *J. Am. Soc. Agron.*, 18:193–225. 1926.
57. Martin, J. H. "Comparative studies of winter-hardiness in wheat," *J. Agr. Res.*, 35:493–535. 1927.
58. Martin, J. H., and Salmon, S. C. "The rusts of wheat, oats, barley, rye," in *Plant Diseases*, USDA Yearbook, 1953, pp. 329–343.

59. May, R. W., and McKee, C. "Furrow drill for sowing winter wheat in central Montana," *Mont. Agr. Exp. Sta. Bul. 177*, pp. 1–24. 1925.
60. McDonough, W. T. and Gauch, H. G. "The contribution of the awns to the development of the kernels of the bearded wheat," *Md. Agr. Exp. Sta. Bul. A 103*, pp. 1–16. 1959.
61. McFadden, E. S., and Sears, E. R. "The origin of *Triticum spelta* and its free-threshing hexaploid relatives," *J. Hered.*, 37(3):81–89. 1946.
62. McKay, H. C., and Raeder, J. M. "Snow mold damage in Idaho Winter Wheat," *Ida. Agr. Exp. Sta. Bul. 200.* 1953.
63. McKinney, H. H. "Virus diseases of cereal crops," in *Plant Diseases*, USDA Yearbook, 1953, pp. 350–360.
64. Mecham, D. K., and Brother, G. H. "Wheat proteins, known and unknown," in *Crops in Peace and War*, USDA Yearbook, 1950–51, pp. 621–627.
65. Miller, E. C. "A physiological study of the winter plant at different stages of its development," *Kans. Agr. Exp. Sta. Tech. Bul. 47*, pp. 1–167. 1939.
66. Miller, E. C., Gauch, H. G., and Gries, G. A. "A study of the morphological nature and physiological function of the awn of winter wheat," *Kans. Agr. Exp. Sta. Tech. Bul. 57*, pp. 1–82. 1944.
67. Newton, R. A. "Comparative study of winter wheat varieties with especial reference to winter-killing," *J. Agr. Sci.*, 12:1–19. 1922.
68. Osborn, W. M. "Rotation and tillage experiments at the Lawton (Okla.) Field Station, 1917–1930," *USDA Tech. Bul. 330*, pp. 1–35. 1932.
69. Packard, C. M. "Cereal and forage insects," in *Insects*, USDA Yearbook, 1952, pp. 581–595.
70. Packard, C. M., and Martin, J. H. "Resistant crops the ideal way," in *Insects*, USDA Yearbook, 1952, pp. 429–436.
71. Parker, J. R. "Grasshoppers," in *Insects*, USDA Yearbook, 1952, pp. 595–605.
72. Paterson, F. L., and others. "Effects of awns on yield, test weight, and kernel weight of soft red winter wheats," *Crop Sci.*, 2:199–200. 1962.
73. Percival, J. *The Wheat Plant: A Monograph.* Duckworth, London, pp. 1–463. 1921.
74. Peterson, R. F. *Wheat, World Crops*, Hill Books, London, pp. 1–400. 1964.
75. Peterson, R. F. *Wheat: Botany, Cultivation and Utilization*, Hill Books, London, pp. 1–422. 1965.
76. Phillips, W. J., and Poos, F. W. "The wheat strawworm and its control," *USDA Farmers Bul. 1323.* 1937.
77. Phillips, W. J., and Poos, F. W. "The wheat jointworm and its control," *USDA Farmers Bul. 1006.* 1940.
78. Pomeranz, K. "Germ bread," *Bakers Digest*, 44:30–33. 1970.
79. Pomeranz, Y., Ed. *Wheat Chemistry and Technology*, Am. Assn. Cereal Chem., St. Paul, Minn., pp. 1–821. 1971 (Rev.).
80. Pumphrey, F. V. "Winter wheat fertilization in Northeast Oregon," *Ore. Agr. Exp. Sta. Circ. Inf. 610*, pp. 1–8. 1961.
81. Quisenberry, K. S. "Survival of wheat varieties in the Great Plains winter-hardiness nursery," *J. Am. Soc. Agron.*, 30:399–405. 1938.

82. Quisenberry, K. S., and Reitz, L. P., Eds. "Wheat and wheat improvement," *Am. Soc. Agron. Mono.*, 13:1–566. 1967.
83. Reitz, L. P., and Briggle, L. W. "Distribution of the varieties and classes of wheat in the United States in 1964," *USDA ARS Statistical Bul. 369.* 1966.
84. Reitz, L. P. "Short wheats stand tall," *USDA Yearbook*, 1968, pp. 236–39.
85. Reitz, L. P., and others. "Wheat in livestock and poultry feeds," *Proc. Int. Symposium*, pp. 1–200. 1970.
86. Roberts, S., and others. "Fertilizer experiments with winter wheat in Western Oregon," *Ore. Agr. Exp. Sta. Tech. Bul. 121*, pp. 1–203. 1972.
87. Roberts, T. H. "The price of hybrid wheat," *Crops and Soils*, 22(1):5–6. 1969.
88. Robertson, D. W., and others. "Rate and date of seeding Kanred winter wheat and the relation of seeding date to dry-land foot rot at Akron, Colo.," *J. Agr. Res.*, 64(6):339–356. 1942.
89. Rosenquist, C. E. "The influence of awns upon the development of the kernel of wheat," *J. Am. Soc. Agron.*, 28:284–288. 1936.
90. Kydrych, D. J. and Muzik, T. J. "Downy brome competition and control in dryland wheat," *Agron. J.*, 60:279–280. 1968.
91. Salmon, S. C. "Why cereals winterkill," *J. Am. Soc. Agron.*, 9:353–380. 1917.
92. Salmon, S. C. "Climate and small grains," in *Climate and Man*, USDA Yearbook, 1941, pp. 321–342.
93. Salmon, S. C., and Throckmorton, R. I. "Wheat production in Kansas," *Kans. Agr. Exp. Sta. Bul. 248.* 1929.
94. Salmon, S. C., Mathews, O. R., and Leukel, R. W. "A half century of wheat improvement in the United States," in *Advances in Agronomy*, vol. 5, Academic Press, New York, pp. 1–151. 1953.
95. Sewell, M. C., and Call, L. E. "Tillage investigations relating to wheat production," *Kans. Agr. Exp. Sta. Bul. 18.* 1925.
96. Shollenberger, J. H., and others. "The chemical composition of various wheats and factors influencing their composition," *USDA Tech. Bul.* 995. 1949.
97. Sprague, R. "Rootrots of cereals and grasses in North Dakota," *N. Dak. Agr. Exp. Sta. (Tech.) Bul. 332*, pp. 1–35. 1944.
98. Stephens, D. E., Wanser, H. M., and Bracken, A. F. "Experiments in wheat production on the dry lands of Oregon, Washington, and Utah," *USDA Tech. Bul. 329*, pp. 1–68. 1932.
99. Thatcher, R. W. "The chemical composition of wheat," *Wash. Agr. Exp. Sta. Bul. 111*, pp. 1–79. 1913.
100. Thompson, T. W., and others. "Role of hard red winter wheat in the Pacific Northwest," *Ore. State Univ. Ext. Circ. 812*, pp. 1–27. 1972.
101. Tsu, S. K. "High-yielding varieties of wheat in developing countries," *USDA ERS-Foreign 322*, pp. 1–39. 1971.
102. Udine, E. J. "The black grain-stem sawfly and the European wheat-stem sawfly in the United States," *USDA Circ. 607*, pp. 1–9. 1941.
103. Von Trebra, R. L., and Wagner, F. A. "Tillage practices for southwestern Kansas," *Kans. Agr. Exp. Sta. Bul. 262*, pp. 1–17. 1932.
104. Wallace, L. E., and McNeal, F. H. "Stem sawflies of economic import-

ance in grain crops in the United States," *USDA Tech. Bul. 1350,* pp. 1–50. 1966.

105. Walton, W. R. "The green-bug or spring grain aphis," *USDA Farmers Bul. 1217,* pp. 1–6. 1921.
106. Walton, W. R., and Luginbill, P. "The fall armyworm, or grassworm, and its control," *USDA Farmers Bul. 752,* pp. 1–14 (revised). 1936.
107. Walton, W. R., and Packard, C. M. "The Hessian fly and how losses from it can be avoided," *USDA Farmers Bul. 1627.* 1936.
108. Webb, R. B., and Stephens, D. E. "Crown and root development in wheat varieties," *J. Agr. Res.,* 52(8):569–583. 1936.
109. Welton, F. A., and Morris, V. H. "Wheat yield and rainfall in Ohio," *J. Am. Soc. Agron.,* 16:731–749. 1924.
110. Williams, C. G. "Wheat: variety and cultural work," *Monthly Bul. Ohio Agr. Exp. Sta.* 5(7), pp. 195–198. 1920.
111. Zech, A. C., and Pauli, A. W. "Changes in total free amino nitrogen, free amino acids, and amides of winter wheat, crowns during cold hardening and dehardening," *Crop Sci.,* 2:421–423. 1962.

Chapter 16

Rye and Triticale

Rye was grown on about 18 million hectares over the world in 1971–72 with an annual production of over 28 million metric tons or 15.6 quintals per hectare. The leading countries in rye production are the Soviet Union, Poland, West Germany, East Germany, the United States, Turkey, and Canada. The highest rye yields are obtained in Switzerland and northwestern European countries. Rye production is declining because of the food preference for and higher yields of wheat. The average production in the United States from 1970 to 1972 was nearly one million metric tons (38.5 million bushels) on some 600,000 hectares (1.4 million acres) or 1.7 metric tons per hectare (27 bushels per acre). The leading states in rye production were South Dakota, North Dakota, Nebraska, and Minnesota (Figure 16–1).

In the above states rye is grown primarily for grain but occasionally

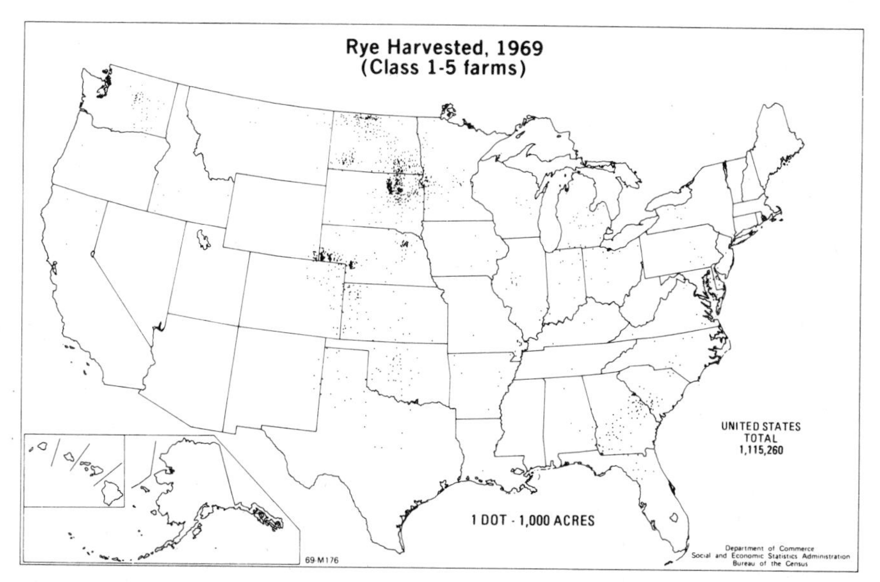

Figure 16–1. Acreage of rye harvested for grain in 1969.

for hay or pasture. In other regions, particularly in the East and Southeast, it frequently is grown for pasture or as a cover and green-manure crop, often in mixtures with vetch or clover.

Rye also serves as an excellent smother crop for weed control. In Virginia, where less than 8 per cent of the seeded acreage is harvested for grain, it may be seeded heavily at 110 to 140 kilos per hectare as a winter cover crop in preparation for continuous sod-planting of corn. The rye and surviving weeds are killed with herbicides in the spring when the rye plants are 30 to 45 centimeters tall. The corn is later planted through the dead sod. Winter rye also is an important smother crop in South Dakota used to assist in suppressing Canada thistle and field bindweed.

HISTORY

Rye evidently has been under cultivation for a shorter time than has wheat. It was unknown to the ancient Egyptians and Greeks, but is supposed to have come into cultivation in Asia Minor more than 4,000 years ago.[7] Rye is found as a weed widely distributed in wheat and barley fields in southwestern Asia, where it has never been grown as a cultivated crop. Rye appears to be indigenous to that part of Asia. Rye, at first, was not intentionally sown with wheat and barley, but when its value became recognized, rye was sown as a separate crop. Cultivated rye (*Secale cereale*) apparently originated from *S. montanum,* a wild species found in southern Europe and the adjoining part of Asia and possibly grown as a cultivated plant in the Bronze Age. Another wild form of rye, *S. anatolicum,* is found in Syria, Armenia, Persia, Afghanistan, Turkestan, and the Kirghiz Steppe.[8]

ADAPTATION

Rye can be grown in all states of the union, but the chief acreage is in the northern states. An apparent cross between the *cereale* and *montanum* species called Michels grass, is an escaped weed in Northeastern California.[27]

Improved winter-hardy rye varieties are the hardiest of all cereals.[4, 12]

The highest yields of rye usually are obtained on rich, well-drained loam soils. Rye is more productive than other grains on infertile, sandy, or acid soils. It is the only small-grain crop that succeeds on coarse sandy soils. It is an especially good crop for drained marshlands and cutover areas of the southeastern states when brought under cultivation. Rye usually yields less grain than winter wheat under conditions favorable for the latter crop because of its shorter growing period, heavier straw growth, and lower spikelet fertility. However, rye usually is sown on poorer soils and with poorer seedbed preparation than is wheat. Much higher yields are obtained on fertile or fertilized soils under good cultural conditions.[19, 21] It is the only successful hay crop for the high desert soils of eastern Oregon.[25]

Rye volunteers freely because the grain shatters readily and the seeds and plants thrive under adverse cultural conditions. For this reason the growing of rye in winter-wheat regions is likely to result in rye admixtures in the wheat, with a consequent depreciation in the market value of the wheat.

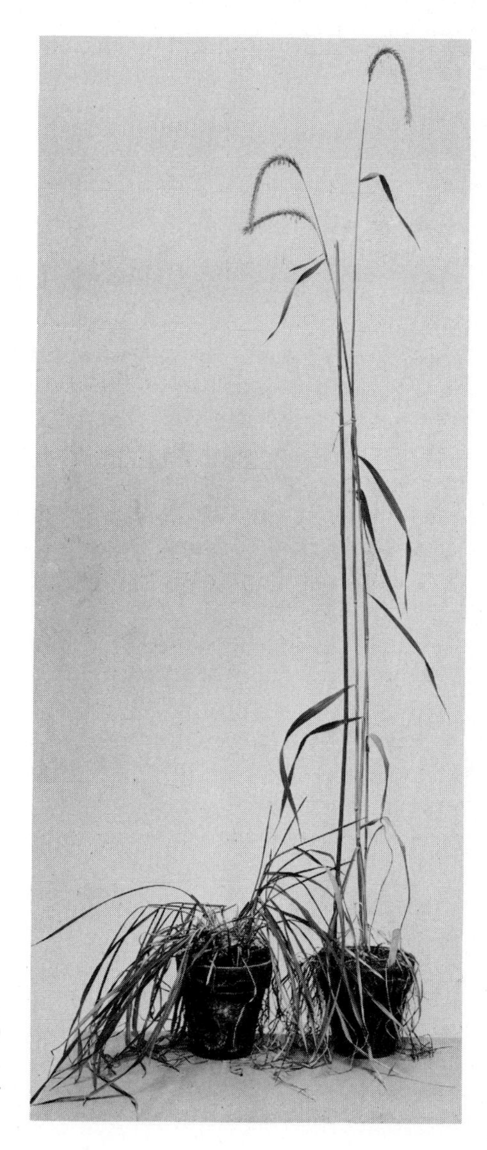

Figure 16–2. (*Right*) Spring rye plant with spikes in flower. (*Left*) A winter variety that failed to head when grown in a warm greenhouse.

BOTANICAL DESCRIPTION

Rye is an annual or winter annual grass classified in the tribe *Hordeae* to which wheat and barley also belong.[5] The stems of rye are larger and longer than those of wheat (Figure 16–2). The leaves of the two plants are similar except that those of rye are coarser and more bluish in color. The roots of rye branch profusely, especially near the soil surface, but some roots may extend to a depth of 5 or 6 feet.[29] This extensive root system has been suggested as one reason why rye grows better than wheat in certain dry climates and on poor soils.

Figure 16–3. Spikes of rye: (*left*) late blooming stage, and (*right*) mature.

The inflorescence of rye is a spike with a single spikelet at each rachis joint (Figure 16–3). The spikelet consists of three florets, two fertile and one abortive. The spikelet is subtended by two narrow glumes. The lemma is broad, keeled, terminally awned, and bears barbs on the keel. The palea is thin and two-keeled. The caryopsis (Figure 16–4) is narrower than the wheat kernel. It usually is brownish-olive, greenish-brown, bluish-green, or yellow. The green or blue pigment, when present, is located in the aleurone, but the brownish pigment is in the pericarp. The grain color is determined by combinations of pigmentation in the two tissues. Varieties with kernels characterized by a white wax-like exterior over a blue aleurone are grayish-blue in appearance. Rye is a long-day plant but continuous lighting may prevent it from jointing.[30]

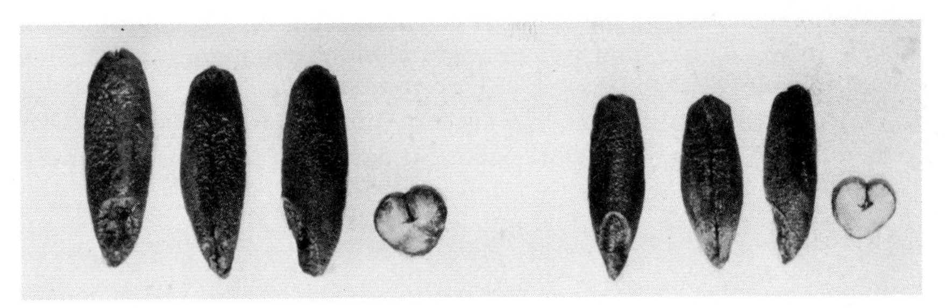

Figure 16–4. Kernels (caryopses) of rye.

POLLINATION

Rye is a naturally cross-pollinated plant. The flower remains open for some time, which facilitates cross-pollination. About 50 per cent cross-pollination between adjacent rows of rye was observed in New Jersey.[26] Sterility is frequent in rye. Approximately one-third of the flowers in Wisconsin rye fields failed to set seed.[13] Considerably more self-sterility is encountered.[3] Selfed lines of rye usually show marked reduction in vigor, shriveling of the grain, and some reduction in plant height, as homozygosity is approached.[3] Some inbred lines are fairly productive.

VARIETIES

Winter varieties of rye are by far the more important.[4] Few varieties of rye maintain distinct characteristics because they comprise a mixture of types produced by cross-pollination. A variety grown in a different environment may quickly adapt itself to the new conditions. Changes in characteristics such as cold resistance, earliness, and color are the result of segregation as well as outcrossing with local varieties. Winter varieties adapted to the northern states are wholly unsuited to conditions in the South where mild temperatures fail to provide sufficient cold to force such true winter varieties into early heading. The nonhardy varieties grown in the South and Southwest actually have a partial or total spring habit of growth. These varieties flower and mature before the advent of unfavorable hot summer weather.

Spring rye is rarely grown from spring sowing in North America, except for grain in cold northern areas where late-spring freezes may kill the flowers of the earlier winter rye in the heading stage. Improved winter varieties with resistance to cold, rusts or mildew are now being grown in many countries. The Balbo variety is popular among dairymen for winter and spring pasture because the milk from cows grazing on this variety is free from the undesirable rye flavor.

Productive, disease-resistant varieties grown in the southeastern States, provide abundant winter pasture and green manure, and good yields of grain.[19]

Tetra-Petkus, a tetraploid variety of winter rye from Germany, was grown for a time in the United States. It has very large bluish kernels.

CULTURE

Rye may replace wheat, oats, or barley in crop rotations that include a small grain. Rye for pasture or green manure often is grown in mixtures with winter legumes such as vetch, Austrian winter peas, or button clover. Cultural requirements are similar to those for wheat. Much of the rye grown in the Soviet Union, or in the western half of the United States is drilled into small-grain stubble without previous soil preparation. This practice, which is economical of labor, also leaves the stubble to hold snow and protect the rye plants from winterkilling. It is satisfactory on land that is reasonably free from weeds. Rye can be grown on disked corn land, fall-plowed land, or on summer fallow in the western states, but usually such land is reserved for wheat or other crops.

Rye responds to fertilizers, especially to nitrogen up to 120 pounds per acre, when grown for pasture. Mixed fertilizers are applied at seeding time. A spring top dressing with nitrogen is desirable where the rye is pastured.[4] Heavy nitrogen applications promote lodging in rye grown for grain.[28]

Winter rye can be seeded at almost any time during the late summer or early fall, but early seeding produces the most fall pasture. Winter rye ordinarily should be sown at about the same time that winter wheat is sown, but the time is less important with rye because it grows better at low temperatures and thus can be sown later. In the eastern states, rye for grain production should be sown from September 1 in the northern part of the country to November 30 in the southern part. It should be seeded two to four weeks earlier when grown for pasture, as a cover crop, or for green manure. Mid-November seeding has resulted in the highest grain yields in the central Cotton Belt. Winter rye is sown in mid-August in Alaska, and about August 15 to September 1 in the northern part of North Dakota and Minnesota, but at later dates farther south. The optimum date for seeding winter rye in central South Dakota is about September 15.[6]

Spring rye, like other spring small grains, should be sown as early as is feasible.

Rye is generally seeded in the western states at the rate of 4 to 5 pecks* per acre.[23] The usual rate in the northeastern states is 7 to 8 pecks, while in the Cotton Belt it varies from 5 to 6 pecks. Heavier rates of seeding prevail where the rye is to be pastured, or for suppressing weeds, but 2 to 4 pecks per acre may be ample for rye grown for grain in the South. The average rate in the United States is 1.4 bushels (36 kilos) per acre.

USES

About 5 million bushels each of rye are used annually for making whisky and alcohol and for food, 7 million bushels for seed, and the balance, 50 to 60 per cent of the grain, is fed to livestock or exported.

The protein content of rye averages somewhat less than that of wheat. Rye flour does not develop true gluten but it has proteins which give it the capacity for making a nutritious leavened bread. The bread is somewhat heavier and darker than that made from wheat flour. Rye usually is mixed with 25 to 50 per cent wheat flour for bread making. Distillers prefer a plump light-colored grain when using rye for whisky.

Rye has a feeding value of about 85 to 90 per cent that of corn. It is not highly palatable and is sticky when masticated, so it usually is ground and fed in mixture with other grains. The proportion of rye in mixed feeds usually is less than a third.

Rye straw formerly was used for packing nursery stock, crockery, and other materials, and for stuffing horse collars.

DISEASES

Ergot

Ergot (*Claviceps purpurea*) causes serious losses in rye. The ergot bodies are poisonous to livestock as well as to human beings, and often cause abortion in pregnant animals. Rye that contains 0.50 per

* 1 peck = 6.36 kilos.

cent or more of ergot is considered unfit for food or feed. Continued consumption exceeding 0.1 per cent may be toxic, inducing lameness and gangrene. Acute ergotism results in convulsions and death.[24] The ergot disease is caused by a fungus that infects the grains at the time of blossoming. About 7 to 14 days after infection, the honey-dew stage becomes evident. The honey-dew contains spores that can infect other grains or grasses in blossom. Soon the fungus threads in the infected grains form ergot bodies (sclerotia), which resemble in shape the grain of the plant on which they develop (Figure 16–5). The purplish-black ergot bodies overwinter in the field or with the seed in storage. They are able, under favorable conditions, to germinate in the spring. Abundant moisture for the germination of the ergot sclerotia, and warm dry weather for blossom infection, are favorable for development of the ergot disease in epidemic form.[2] However, rye ergot is of rare occurrence in the southeastern United States.[19]

Figure 16–5. (*Left*) Two spikes of rye infected with ergot. (*Right below*) ergot sclerotia. (*Right above*) rye grains.

Ergot-infested grain or grasses should be mowed or grazed before the plants flower. Ergot occurrence in grain is as follows: rye > triticale > barley > durum wheat > common wheat > oats. The spores of the sclerotia are disseminated to a great extent by winds.

Ergot can be partly controlled by sowing ergot-free seed on land where rye has not been grown for 1 or 2 years previously. The mowing of ergot-infested grasses adjacent to rye fields also is helpful. Infested rye should be immersed in a 20 per cent solution of common salt to separate the ergot bodies. The grain is stirred and the ergot bodies float to the surface where they can be skimmed off. After the salt treatment, the grain should be washed and dried before seeding or feeding. Much of the ergot can be removed by aspiration cleaning equipment available in mills.

Ergot is used extensively in drug preparations. When produced in sections of low summer rainfall, it can be marketed for that purpose. Under peacetime conditions the supply is largely imported from the Mediterranean countries.

Stem Smut

Stem smut (*Urocystis occulta*) attacks rye wherever it is grown, but particularly in Minnesota and nearby states. Serious losses are caused when infection is high.[15] The smut appears on the culms, leaf sheaths, and blades as long narrow stripes. These stripes first appear lead gray in color but later turn black. Infected plants are more or less dwarfed and the heads fail to emerge from the sheath. The spores are carried both on the seed and in the soil. Seed treatment, as for stinking smut of wheat, will generally control the disease in the northern states. Crop rotation must be used where the spores are soil-borne.

Other Diseases

Anthracnose (*Colletotrichum graminicolum*) is often found on rye in the humid to subhumid eastern states. Infection of the culm results in premature ripening and death of the tillers. Infected tissues are stained brown on the leaf sheath that surrounds the diseased culm. Head infections may occur later to cause shriveled, light-brown kernels. The lower portion of the culm has a blackened appearance where attacked by the disease. Control measures involve proper crop sequences and the plowing under of crop residues.

The early maturity of rye usually enables it to escape serious damage from stem rust caused by *Puccinia graminis secalis*. Leaf rust (*Puccinia recondita*) likewise causes little injury. Severe infections, largely confined to the southern range of rye culture, cause a reduction in tillering and decreased yields. The disease overwinters in the leaves of winter rye as dormant mycelium.[17] Destruction of volunteer rye in stubble fields will aid in control of the disease.

Insects

Rye is attacked by the grasshopper, chinch bug, Hessian fly, jointworm, sawfly, and other common insects of small grains. The total losses are not large. However, winter rye sown early furnishes a favorable environment for the depositing of grasshopper eggs and may thus promote grasshopper injury to other crops.

TRITICALE

Many plant breeders, since 1876, have crossed wheat with rye, sometimes with the hope of transferring the winter hardiness character of rye into wheat. Wheat was the female parent in the successful crosses. Many wheatlike wheat-rye hybrid derivatives merely carried one or more rye chromosomes.[16] These extra rye chromosomes often are lost during meiosis (sexual cell division) in later generations, and the plant reverts to typical wheat when the rye characters are lost. More recently, occasional segments of rye chromosomes have been translocated to a wheat chromosome.

Triticale, which was developed from crosses between wheat and rye, usually carries all of the chromosomes of both parents. The name was coined by combining parts of the parental generic names *Triticum* and *Secale*. Varieties of triticale bred in Mexico, United States, Canada, and other countries[11, 27a, 31, 33] have been grown on a small acreage since 1969,[22] with possibly 200,000 acres in the United States in 1971. Crosses with tetraploid durum wheats have been more successful than those with hexaploid common wheats. The former combination can then be crossed with common wheats and then backcrossed with triticale to retain the rye chromosomes. Triticale kernels are large, and they have a higher content of lysine and of sulfur-containing amino acids than wheat for better nutrition.[10] Most of the triticale strains have shrunken kernels, which accounts for their high protein content. Plump kernels have a total protein content similar to that of wheat but a higher lysine content. Breeders are attempting to develop plump triticales with short, stiff straw and better disease resistance.

REFERENCES

1. Barnes, B. S., and Wells, D. G. "Explorer—a new rye for Mississippi," *Miss. Agr. Exp. Sta. Bul. 600*. 1960.
2. Brentzel, W. E. "Studies on ergot of grains and grasses," *N. Dak. Agr. Exp. Sta. Tech. Bul. 348*. 1947.
3. Brewbaker, H. E. "Studies of self-fertilization in rye," *Minn. Agr. Exp. Sta. Tech. Bul. 40*. 1926.
4. Briggle, L. W. "Growing rye," *USDA Farmers Bul. 2145*. 1959.

5. Deodikar, G. B. "Rye—*Secale cereale* L.," *Indian Council Agr. Res.*, New Delhi, pp. 1–152. 1963.
6. Hume, A. N., Hardies, E. W., and Franzke, C. "The date of seeding winter rye," *S. Dak. Agr. Exp. Sta. Bul. 220.* 1926.
7. Khush, G. S. "Cytogenetics and evolutionary studies in *Secale:* III. Cytogenetics of weedy ryes and origin of cultivated rye," *Econ. Bot.*, 17:60–71. 1963.
8. Khush, G. S., and Stebbins, G. L. "Cytogenetic and evolutionary studies in *Secale:* I. Some new data on the ancestry of *S. cereale,*" *Am. J. Bot.*, 48:723–730. 1961.
9. Klebesdel, L. J. "Winter survival and spring forage yield of winter rye varieties in subarctic Alaska as influenced by date of planting," *Agron. J.*, 61(5):708–712. 1969.
10. Knipfel, J. E. "Comparative protein quality of triticale, wheat and rye," *Cereal Chem.*, 46(3):313–317. 1961.
11. Larter, E. N., and others. "'Rosner,' a hexaploid triticale cultivar," *Can. J. Plant Sci.*, 50:122–124. 1970.
12. Laude, H. H. "Cold resistance of winter wheat, rye, barley, and oats in transition from dormancy to active growth," *J. Agr. Res.*, 54:899–917. 1937.
13. Leith, B. D. "Sterility of rye," *J. Am. Soc. Agron.*, 17:129–132. 1925.
14. Leith, B. D., and Shands, H. L. "Fertility as a factor in rye improvement," *J. Am. Soc. Agron.*, 30:406–418. 1938.
15. Leukel, R. W., and Tapke, V. F. "Cereal smuts and their control," *USDA Farmers Bul. 2069.* 1954.
16. Longley, A. E., and Sando, W. J. "Nuclear divisions in the pollen mother cells of *Triticum, Aegilops,* and *Secale* and their hybrids," *J. Agr. Res.*, 40:683–719. 1930.
17. Mains, E. B., and Jackson, H. S. "Aecial stages of the leaf rusts of rye, *Puccinia dispersa,* and of barley, *P. anomala,* in the United States," *J. Agr. Res.*, 28:1119–1126. 1924.
18. Mooers, C. A. "Balbo rye," *Tenn. Agr. Exp. Sta. Circ. 45.* 1933.
19. Morey, D. D. "Rye, southern style," *Crops and Soils,* 24(7):12–14. 1972.
20. Prato, J. D., and others. "Triticale, what is the latest on this man-made grain," *Crops and Soils,* 23(9):18–19. 1971.
21. Reeves, D. L. "Do you have a rye future ahead," *Crops and Soils,* 23(7):15–17. 1971.
22. Reitz, L. P., and others. "Distribution of varieties and classes of wheat in the United States in 1969," *USDA Stat. Bul. 475,* pp. 1–70. 1972.
23. Robinson, R. R., and others. "Winter rye rate of sowing, row spacing, varietal mixtures and crosses," *Minn. Agr. Exp. Sta. Misc. Rpt. 100,* pp. 1–8. 1970.
24. Seaman, W. L. "Ergot of grains and grasses," *Can. Dept. Agr. Publ. 1438.* 1971.
25. Sneva, F. A., and Hyder, D. N. "Raising dry land rye hay," *Oreg. Agr. Exp. Sta. Bul. 592.* 1963.
26. Sprague, H. B. "Breeding rye by continuous selection," *J. Am. Soc. Agron.*, 30:287–293. 1938.

27. Suneson, C. A., and others. "A dynamic population of weedy rye," *Crop Sci.*, 9(2):121–124. 1969.
27a. Tsen, C. C., Ed. "*Triticale*—First man-made cereal." Am. Assn. Cereal Chem., St. Paul, Minn., pp. 1–300. 1974.
28. Walker, M. E., and Morey, D. D. "Influence of rates of N, P, and K on forage and grain production of Gator rye in South Georgia," *Ga. Agr. Expt. Sta. Cir. (New Series) 27.* 1962.
29. Weaver, J. E. "Root development in the grassland formation," *Carnegie Inst. Washington Publ. 292*, pp. 1–151. 1920.
30. Wells, D. G., and others. "Jointing and survival of rye plants and clones through successive seasons," *Crop Sci.*, 7(5):473–474. 1967.
31. White, G. A. "New crops on the horizon," *Seed World*, 110(4):22–32. 1972.
32. Youngken, H. W., Jr. "Ergot—a blessing and a scourge," *Econ. Bot.*, 1:372–380. 1947.
33. Zillinsky, F. J., and Borlaug, N. E. "Progress in developing triticale as an economic crop," *Int. Maize and Wheat Improvement Center (Mexico) Rsh. Bul. 17*, pp. 1–27. 1971.

Chapter 17

Barley*

ECONOMIC IMPORTANCE

Barley (*Hordeum vulgare*) ranks fourth in area among world crops harvested. About 80 million hectares produced 146 million metric tons (1.84 per hectare) from 1970 to 1972. Both the yield and the area sown to barley have increased sharply since about 1948. The leading producing countries are the USSR, Canada, the United States, China, France, and West Germany. Production in the United States is nearly 9.5 million metric tons (434 million bushels) on some 4 million hectares (nearly 10 million acres), yielding 2.38 metric ton per hectare, or 44 bushels per acre. The leading states in barley production were North

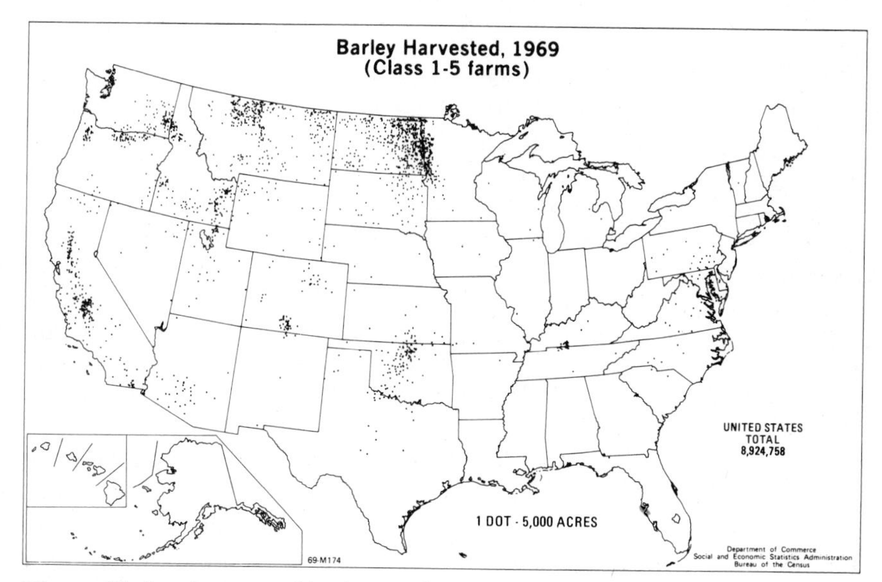

Figure 17–1. Acreage of barley in the United States 1969.

* For complete information and references see Wiebe, G. A., and others. "Barley: origin, botany, culture, winterhardiness, genetics, utilization, pests," *USDA ARS Agr. Handbook 338*, pp. 1–127. 1968.

Dakota, Montana, California, Minnesota, and Idaho (Figure 17–1). Most of the production in Canada is in the southern portion of the three Prairie Provinces, with smaller acreages near the southern border of the northeastern Provinces.

HISTORY OF BARLEY CULTURE

Barley apparently originated by domestication of a wild two-rowed form, *H. vulgare spontaneum*, in or near the area bordering Syria and Iraq with Iran and Turkey. Six-row barley was being cultivated by about 6000 B.C. Barley culture spread to India, Europe, and North Africa during the Stone Age. It was introduced into the western hemisphere by Columbus in 1493.[37] It was first grown in the United States in 1602 and additional introductions by English and Dutch settlers occurred during the next three decades.

The English settlers brought mostly two-row types to the United States but the continental six-row types were introduced by the Dutch. Spanish pioneers introduced the North African six-row type of barley from Mexico into Arizona by 1701 and into California by 1771. Winter barleys arrived later, possibly from the Balkan-Caucasus region or Korea. Barley production shifted westward, mainly from New York, to the North Central States after 1849.

ADAPTATION

Barley is grown throughout the more temperate regions of the world. It thrives in a cool climate. It will stand more heat under semiarid than under humid conditions.[20] In the warmer climates barley is sown in the fall or winter. The best barley soils appear to be well drained loams. It produces a poor crop, especially in grain quality, on heavy poorly drained soils in regions of frequent rains. Light sandy soils are poor for barley because growth often is erratic and, also, the crop is more likely to be ripened prematurely by drought. Barley is the most dependable cereal under extreme conditions of salinity, summer frost, or drought. Barley is unsuited to acid soils below *p*H 6 because of resultant aluminum toxicity which also retards root growth. Calcium applications correct the toxicity and promote root growth.

The effect of adaptation on variety survival in mixtures was determined by growing a mixture of 11 varieties at ten stations for several years.[13] Population counts made each year showed a rapid elimination of the sorts less adapted to the environment and to competitive conditions. The variety that eventually predominated was soon evident at most stations, but a variety that led at one station might be eliminated at another. The varieties that survive best in mixtures are usually, but not necessarily, those that produce the highest yields when grown alone.[32]

BOTANICAL CHARACTERISTICS

Barley belongs to the grass tribe Hordeae, in which the spikes have a zigzag rachis. It belongs to the genus *Hordeum*, section Cerealia. In this section is the cultivated species (*Hordeum vulgare* L.) with a tough rachis, and two uncultivated species with a brittle rachis, *H. agriocrithon* E. Aberg and *H. spontaneum* C. Koch, all with 14 diploid chromosomes. Other wild species of *Hordeum* include *H. murinum*, *H. bulborsum* and *H. jubatum* with 28 chromosomes and *H. nodosum* with 42 chromosomes.

The vegetative portion of the barley plant is similar to that of the other cereal grasses except that the auricles on the leaf are conspicuous (Figure 15–4). The lateral spread of the barley roots usually varies from 15 to 30 centimeters, while the depth of penetration varies from 1 to 2 meters.

The inflorescence is a spike with three spikelets borne at each rachis node (Figures 17–2 and 17–3). Each spikelet contains a single

Figure 17–2. Spikes of six-row barley (*A*) and two-row barley (*B*).

Figure 17–3. Spike of hooded barley.

floret. A spike usually contains 10 to 30 nodes. In six-row forms, all three florets at a node are fertile, while in two-row barley only the central floret is fertile. Each spikelet is subtended by a pair of glumes, which are normally narrow, lanceolate bracts with short bristle-like awns (Figure 17–4). The floret is composed of a lemma and a palea, and a caryopsis when fertile. Except in naked (hull-less) varieties, the lemma may terminate in an awn or hood, or it may be merely rounded or pointed. The awns of barley may be either rough (barbed) or smooth (Figure 17–5), the latter type usually being smooth at the base and slightly roughened at the tip. The hood is a trifurcate (three-forked) appendage that replaces the awn (Figure 17–3). Awnless varieties are comparatively rare. The rachilla is a small, long- or short-haired structure lying within the crease of the kernel (Figure 17–6).

The grains are about 8 to 12 millimeters long, 3 to 4 millimeters wide, and 2 to 3 millimeters thick. A pound of seed contains about 13,000 (8,000 to 16,000) grains. About 10 to 15 per cent (usually 12 to 13 per cent) of the kernel consists of hull, except in the naked varieties which are free from the hull after threshing. Typical bushel

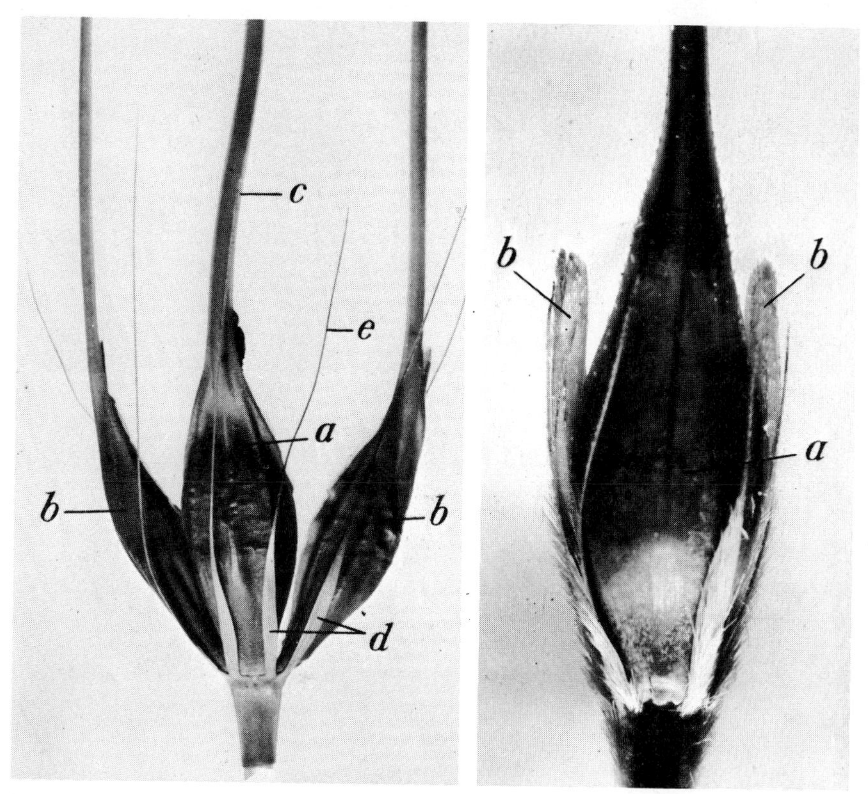

Figure 17–4. Barley spikelets—(*left*) six-row, (*right*) two-row: (*a*) central kernel; (*b*) lateral kernels on six-row barley and empty (sterile) spikelets on two-row barley; (*c*) awn; (*d*) glumes; and (*e*) glume awn.

weights are for the hulled-type kernels, 48 pounds; and for the naked, 60 pounds.

Five color conditions are recognized in the barley grain:[12] white, black, red, purple, and blue. The latter three colors are due to anthocyanin pigments. When pigments occur in the barley hulls, they are red or purple, but when they occur in the aleurone layer, the grains are blue. The red color usually fades. Naked barleys (Figure 17–7) with red in the pericarp and a blue aleurone appear to be purple. Black comes from melanin-like pigments in the hulls or pericarp. The white and blue barleys are the only ones grown extensively.

POLLINATION

Barley is generally self-fertilized because pollination occurs while the head is partly in the boot in many varieties. In Minnesota, occasional natural crosses occurred in some varieties when white and black varieties were grown side by side.[30] Similar results were ob-

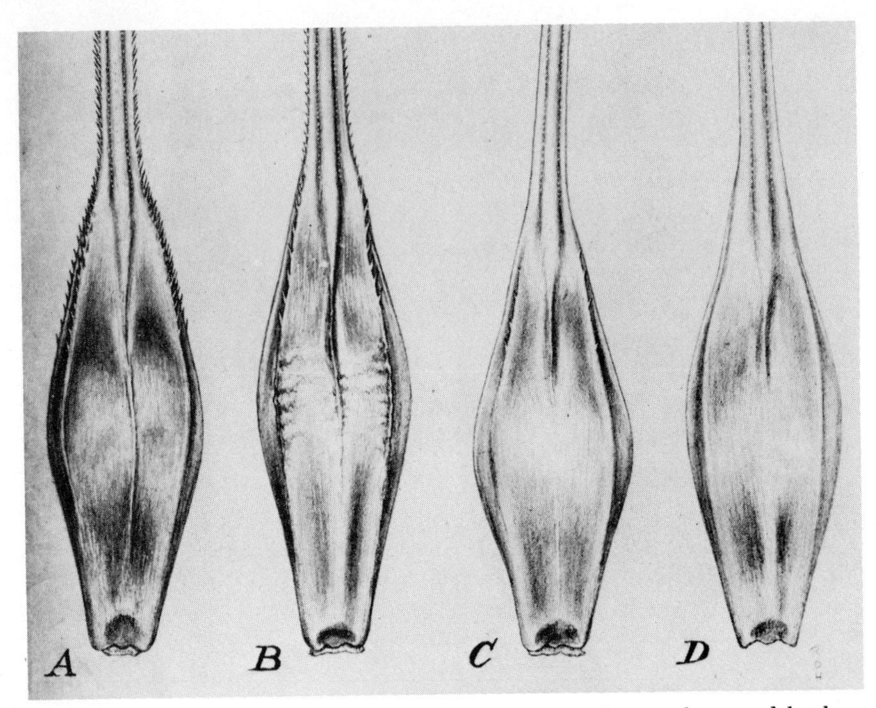

Figure 17–5. Rough-awned barleys (*A* and *B*) and smooth-awned barleys (*C* and *D*). The teeth extend down the veins of the glumes of rough-awned kernels. A few teeth occur on (*C*).

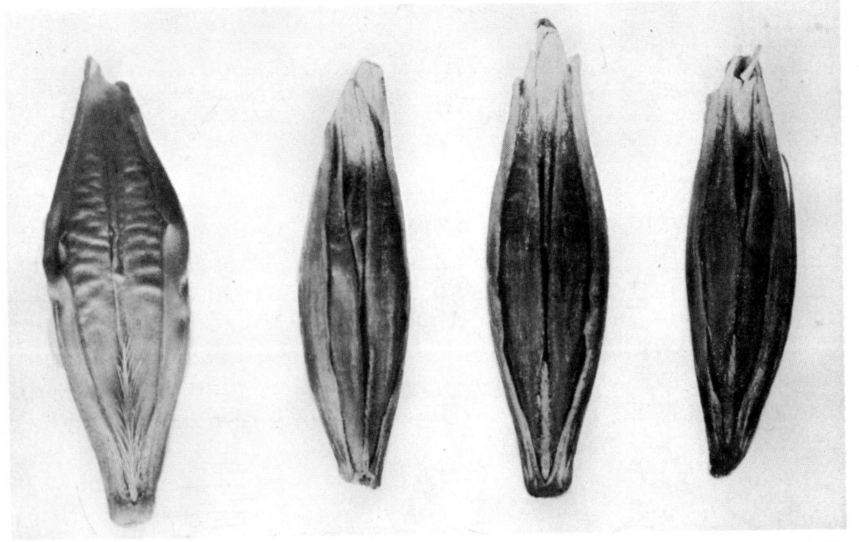

Figure 17–6. (*Left*) Two-row Hannchen barley kernel showing long rachilla hairs and wrinkled palea. (*Right*) The two lateral kernels and the one central kernel of a six-row barley with short rachilla hairs. The lateral kernels are curved and asymmetrical.

Figure 17–7. Kernels (caryopses) of naked barley.

tained in Colorado, where the commercial varieties showed less than 0.15 per cent of natural crosses.[25]

During fertilization pollen may germinate within 5 minutes after it falls on the stigma. Within 6 hours after pollination, fertilization of both the egg cell and endosperm nucleus is completed and cell division has begun[23] (Figures 3–14 and 4–4).

CULTIVATED BARLEY SPECIES

The cultivated barley (*Hordeum vulgare* L, emend. Bowden) includes three subspecies or classes based on the fertility of the lateral spikelets: (1) six-row barley, (2) two-row barley, usually listed as *distichum*, and (3) *irregulare*, irregular barley.[37]

The first class, with all florets fertile, includes (1) the ordinary six-row barleys with lateral kernels slightly smaller in size than the central one (Figure 17–6); and (2) the intermedium group, in which the lateral kernels are markedly smaller in size than the central one.

The second class with only the central florets fertile is divided into two groups: (1) the common two-row type with lateral florets consisting of lemma, palea, rachilla, and reduced sexual parts, and (2) a deficiens group with lateral florets reduced and consisting of lemma, rachilla, and, rarely, palea, but with no sexual parts.

The *H. irregulare* class, of Ethiopian origin, has sometimes been called Abyssinian intermediate. The central florets are fertile and the lateral florets are reduced to rachillae in some cases, and these are distributed irregularly on the spike. The remainder of the lateral florets may be fertile, sterile, or sexless.

The barley varieties grown on farms are predominately the ordinary six-row and two-row forms. So-called four-row barleys are six-row forms in which two-rows of lateral florets overlap to form single rows on each side of the spike. The four-row effect usually occurs only in the upper two-thirds of the spike.

REGIONAL TYPES AND VARIETIES

Probably 4,000 or more varieties are grown over the world.

In the United States those grown in the humid spring barley region of the upper Mississippi Valley include mostly six-row types developed from crosses. Some varieties have smooth awns. Two-row varieties are grown there occasionally.

Both six-row and two-row varieties are grown in the semiarid Great Plains as well as in the intermountain and western regions. The six-row varieties in the Far West are mostly hulled, rough-awned types, largely of North African origin. Some smooth-awned varieties are also grown.

True winter-barley varieties are grown in the area from New York to Colorado and southward, but also occasionally in Utah, Idaho, and the Pacific Coast states. Hooded winter sorts are rarely grown except in the southern region where smooth-awned varieties are also grown.

Hooded varieties in spring or winter types are grown occasionally, generally for hay. Naked barleys, grown rarely, are usually fed to poultry.

BARLEY IMPROVEMENT

An interesting development in barley improvement was the breeding of smooth-awned varieties (Figure 17–5).[12] This eliminated the severe irritation resulting from handling rough-awned varieties previously grown. Awned varieties are valuable because the awns function in transpiration and photosynthesis and as a depository for mineral matter.[10] When the awns are removed, the ash content of the rachis increases, which may account for the tendency of such spikes to break easily. Smooth-awned barleys have no physiological limitations when compared with standard rough-awned varieties.[19]

Breeding for resistance to a majority of the barley diseases has been at least partly successful.

Development of winter-hardy varieties extended winter barley northward from Oklahoma into Nebraska and Colorado, and from Maryland and Kentucky northward to New York and the eastern Great Lakes states. A hybrid barley was released in Arizona in 1968.

A study of barley crosses in Idaho showed that high-yielding selections from hybrids were obtained more frequently from intercrossing six-row varieties than from crossing six-row with two-row varieties. Hooded segregates were definitely inferior to awned ones, and naked segregates were slightly less productive than covered (hulled) strains. Smooth-awned forms averaged greater floret sterility and were slightly lower in yield than the rough-awned forms, but it seemed likely that some of the smooth strains might prove to be equal to the best rough ones.[15]

CULTURAL METHODS

The cultural methods for barley are similar in most respects to those for wheat and oats. Weed-control measures for barley are about the same as those for wheat.

Rotations

In general, barley makes its best growth after a cultivated crop such as corn, sugarbeets, or potatoes. In the humid region, the most practical rotation is one that includes corn, barley, and a leguminous hay or pasture crop, with barley as a companion crop in seeding the legume or grass.[9] Where hay is grown, the grass-seed mixture may be timothy and red clover, which is usually left for two years. This rotation may be modified to include two years of corn or barley when more feed grain is necessary. On the irrigated lands of the West, barley is often used as a companion crop for alfalfa. The latter generally is retained for two years or more before being plowed up; it is followed by one or two years of cultivated crops. In the semiarid Great Plains, a common practice is to grow barley after fallow or on disked corn or sorghum land. In the Corn Belt, barley often follows oats, wheat, or soybeans instead of corn in order to avoid losses from the scab disease, which is carried on corn. The winter-barley–Korean lespedeza combination has been popular in Missouri. The lespedeza is pastured after the barley is harvested, while the land is disked each fall in preparation for barely seeding. The lespedeza reseeds naturally. In the semiarid portions of eastern Washington and Oregon, barley may alternate with fallow or peas. In the South, winter barley occupies the same place in the rotation as wheat. A satisfactory procedure is to sow barley after a summer legume, plowed under about September 1. In the piedmont region, corn (with crimson clover seeded later among the stalks), followed by cowpeas, and then by barley, makes a satisfactory three-year rotation. This may be modified to include a second year of barley or a pasture crop.

Barley usually responds to applications of up to 50 pounds of nitrogen per acre when ample soil moisture is available. Heavy applications may induce lodging and lower the malting quality. Barley yields are increased also by phosphorus and potassium fertilizers on some soils,[8, 22] and phosphorus improves malting quality. Much of the barley is sown on summer-fallowed land or following intertilled crops that had been fertilized so that little fertilization for barley is needed.

Seeding

Maximum yields from spring barley are obtained when the crop is seeded early in the spring. On the northern Great Plains, the most favorable period is from April 1 to 25. In Montana, North Dakota, and South Dakota the loss in yield from seeding after April 25 is more than 1 per cent per day. In southern Minnesota, Iowa, and Wisconsin where the season is slightly earlier, late seeding is more disastrous than in the northern plains. In New England, the cool summer permits a slightly later seeding. In California and southern Arizona maximum yields are obtained from late fall or early winter seeding, i.e., before December 20. The varieties grown there have a spring growth habit but they survive the winter under the mild temperature conditions. Cool temperatures and short photoperiods following late fall seeding prolong the vegetative period for about two or three months beyond that required for the same varieties when sown in the spring. The longer growing period favors greater yields. Good results have been obtained over most of the central and southern winter-barley area with September seeding.

In practice, 8 pecks* per acre of spring barley are usually seeded in the humid regions. On the northern Great Plains, 4 to 6 pecks is the usual rate, the lower rate being followed in the drier localities. In the very dry localities of the Great Basin, 3 pecks are sometimes seeded to advantage. In California, 7 pecks are commonly seeded, while in the winter-barley regions the usual rate is about 6 pecks. In Colorado, about 8 pecks are recommended for irrigated areas, and 4 pecks for the dry lands.[25]

Barley seed should be sown at a depth at which both moisture and air are available, or about $1\frac{1}{2}$ inches in the humid regions, 2 inches on the northern Great Plains, and $2\frac{1}{2}$ to 3 inches in the Great Basin.[9] The crop is generally sown with a grain drill.

Harvesting

Nearly all the domestic barley crop is threshed with a combine. Considerable loss from shattering, particularly in the case of the malting types, occurs when the crop stands until the grain is dry. The windrow or swather–pick-up combine method of cutting the barley when the heads have turned a golden yellow, and while the straw is slightly green, avoids some shattering loss. After the grain has dried 3 or 4 days, the pick-up combine is used. About one third of the acreage is harvested by this method.

Barley is ready for harvest for grain with a windrower when it is physiologically mature, i.e., after material ceases to be added to the kernel (Figure 17–8). Maturity is indicated when a thumb-nail dent

* One peck of covered barley = 12 pounds or 5.45 kilos.

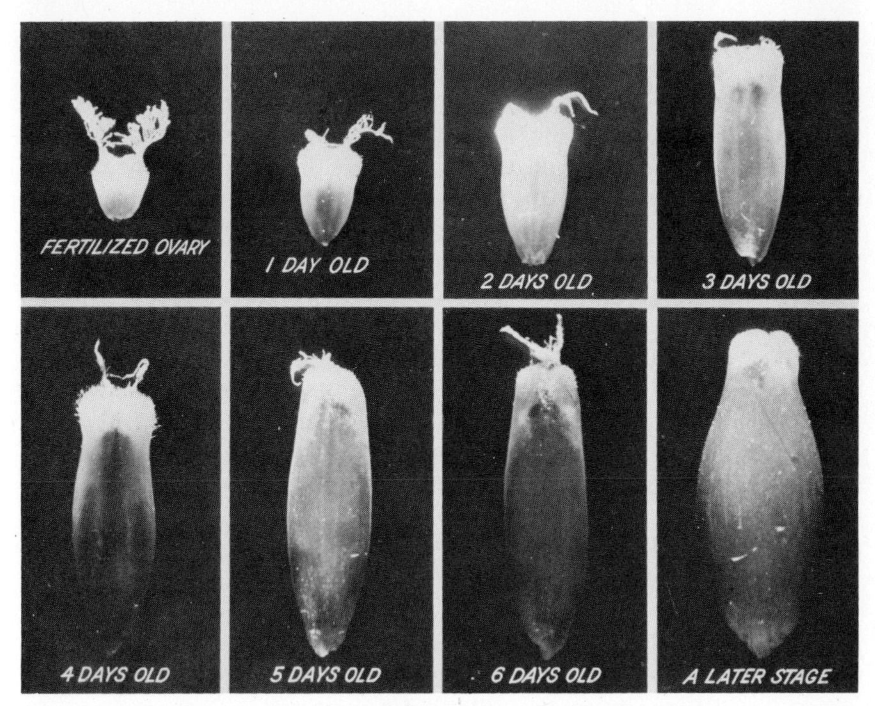

Figure 17–8. Barley kernels at different ages of development.

in the kernel remains visible for some time. At this stage the milky juices will have disappeared from the kernel. The ripening process after this time is principally moisture loss.

USES

More than 100 million bushels of barley are used annually in the United States for producing malt. About 85 per cent of the malt is used in making beer, 10 per cent for making industrial alcohol and whiskey, and the remainder for malt sirups. An average of 25 to 30 million bushels are required for seed. A similar quantity is used for food purposes. The remainder is used for feed or exported.

Barley has a feeding value of about 95 per cent of that of corn. The hulls, comprising about 13 per cent of the kernel, detract from the nutritive value. Barley requires grinding or rolling for satisfactory feeding to animals other than sheep. Barley often is steamed to soften the grain before it passes to the rolls for crushing into flakes.

Barley is a better companion crop for legumes and grasses than are most varieties of oats or wheat because it shades the ground less and matures earlier. Barley has been cut for hay extensively in California. Winter barley is a good winter-cover crop in the southeastern states. Some of it is pastured.

Malting and Brewing

The maltsters in the United States desire a plump, mellow, small-kerneled barley with tight hulls.[16] Such barley should be all of the same type, starchy, mellow, of high germinative capacity, sound, and free from either weather or disease damage. Certain proteins of high solubility are detrimental. Flinty or *glassy* kernels, a condition brought about by interrupted growth or other factors, are objectionable because they contain intermediate products undesirable for malting. All of the malting grades specify 70 per cent or more of mellow kernels which are not *semisteely*. Barley with a blue aleurone appears somewhat *steely*, but the color itself has no known effect upon malting quality. Broken kernels are objectionable because they will not malt. Skinned or peeled grains fail to germinate or convert properly during the malting process. Musty barley is unacceptable. Damaged barley, of which a maximum of 4 per cent is allowed in the malting grades, includes blighted, moldy, heat-damaged, and weathered kernels. The varieties acceptable to most maltsters in the humid spring region are six-row varieties with a white aleurone layer. Certain varieties with a blue aleurone are acceptable but these are graded in a special subclass. Several two-row varieties are acceptable to certain maltsters. Six-row varieties with a partly blue aleurone grown in California also are suitable for malting.

Cultural practices may determine whether the barley crop will meet the requirements for the malting grades. Barley harvested before it is fully ripe in the field may have an objectionable green tinge. The crop should be harvested so as to reduce exposure to unfavorable weather to a minimum. Skinned or broken kernels usually result from threshing at too high a cylinder speed (600 revolutions per minute), from concave teeth set up too close to the cylinder, or from end play in the cylinder. The use of square teeth in the concaves instead of rounded ones is said to result in fewer broken kernels.

MALTING PROCESS: Barley is malted to produce or activate enzymes, chiefly alpha-amylase and beta-amylase, which hydrolize starch into sugars. Enzymes also initiate germination and convert proteins into soluble compounds, and develop the aroma and flavor of malt. In the manufacture of malt, the cleaned and sized barley is soaked (steeped), with occasional draining, until the grain contains 44 to 46 per cent moisture. It then germinates at a temperature of about 20° C. (68° F.) in drums or tanks on aprons with frequent but slow stirring with water added to maintain the moisture content. When the sprouts (acrospires) are about 75 to 100 per cent of the length of the kernel, germination is stopped by kiln-drying the grain down to 4 to 5 per cent moisture. The rootlets (malt sprouts), which in the meantime have been broken off in handling, are screened out. The produce remaining is dried malt. One bushel of barley (48 pounds or 22 kilos) produces about 38 pounds of dried malt. This quantity of malt with some 17 pounds of adjuncts (corn or rice products, sugar, sirup, and so on) and about 0.7 pound of hops and hop extracts is sufficient to make one barrel (31 gallons or 117 liters) of beer.

BREWING PROCESS: The brewing industry usually is separate from the malting industry. In brewing,[28] the malt is first ground and a portion of the ground malt mixed with unmalted cereals (corn grits, broken rice, and so on). These starchy adjuncts are used to furnish additional fermentable material and to reduce the protein content of the mash. This mixture is then cooked in water to gelatinize and liquify the starch, after which it is mixed with water and the remaining malt in a mash kettle or tub. The latter has previously been held for half an hour at approximately 48° C. (118° F.) to favor the breakdown of the proteins. The temperature is raised to 65° C. (149° F.) and held for a few minutes until the conversion of the starch is complete. Then the temperature is raised to 75° C. (167° F.), and the mash is mashed off, i.e., allowed to settle. The liquid (first wort plus spargings) is drawn off and boiled with hops, after which the wort is then drawn off and cooled. The wort is fermented with yeast for 7 to 10 days at temperatures ranging from 6° C. (42.8° F.) to 15° C. (59° F.). Proteins, yeast cells, hop resins, and other insoluble materials settle out during the cooling and the subsequent cool storage, which lasts for three to eight weeks at temperatures that usually range from 0° to 5° C. (32° to 41° F.). The finished beer is then carbonated and filtered, after which it is ready for the market.

USES OF MALT: The spent malt, after brewing is sold for feed, as dried or wet brewers grains. Malt sirup is used in baking, in candies, and in the textile industries. It is also added to medicines for its laxative effect and as an aid in the control of fevers. Many breakfast-cereal manufacturers use malt sirup. Malt is also used for the production of malted milk, alcohol, vinegar, and yeast.

Other Products of Barley

Pearled barley is made by grinding off the outer portions of the kernel (hull, bran, aleurone, and germ) so as to leave the kernel as a round pellet. These outer parts are removed by the abrasive surface of a whirling rough *pearling stone*. The larger, more spherical types of kernel result in a better pearled product with less loss of outer layers in processing.[9] Plump, large-grained, two-row white-grained varieties with a shallow or closed crease are best for pearling. Pearled barley is cooked for human food, and is used in baby foods and other products. Partly pearled grains are marketed as pop barley.

Barley flour contains no gluten and it darkens the bread loaf. Large-grained, two-row varieties with white aleurone layers are best suited to flour milling. Naked barley is grown in parts of Asia for home grinding and cooking. A high lysine barley was found in Sweden in 1968.

DISEASES

Covered Smut

Covered smut (*Ustilago hordei*) attacks the barley heads and causes formation of hard, dark lumps of smut in place of the kernels. The diseased heads often are borne on shorter stems and appear later than the normal heads. Soon after the smutted heads appear, the membranes that enclose the smut galls begin to split, which releases the spores to be carried to sound heads. Infection may occur before the barley is ripe or any time thereafter.[33] Spores that reach the barley kernels frequently germinate and send infection threads beneath the hulls. Covered smut is often found in threshed grain as black, irregular, hard masses.

Treatment with seed disinfectants is effective in control of covered smut. Some varieties are moderately resistant to covered smut.

Brown and Black Loose Smuts

The brown and the black loose smuts attack the barley heads, causing the formation of loose, powdery masses of smut in place of the normal spikelets. The dusty spores are blown over the field at the time the normal heads are in blossom. Some spores fall into the open blossoms, resulting eventually in the infection of the next crop. The spore masses of the nuda or brown loose smut (*Ustilago nuda*) are olive brown, while those of the nigra or black loose smut (*U. nigra*) are dark brown, almost black. The brown loose smut infests the interior of the seed, while the black loose-smut infestation is superficial.

The modified hot-water treatment controls brown loose smut in infected seed. The seed is placed in burlap bags and soaked at about 70° F. for 6 hours, after which it is immersed in hot water at a temperature of 126° F. for 13 minutes. Water-soak treatments described in Chapter 15 also are effective as well as less injurious to germination. These treatments are rather difficult to apply, but are feasible for treating foundation-seed lots. Some varieties are resistant, except when environmental conditions are especially favorable for infection.

Control measures for the shallow-borne black loose smut are the same as for covered smut.

Barley Stripe

Barley stripe (*Helminthosporium gramineum*) is an important disease in the entire United States. It is not unusual for 5 to 15 per cent of the plants in a field to be affected. Each 1 per cent of infection reduces the yield about 0.75 per cent. The disease causes long white or yellow stripes on the leaves. These stripes later enlarge and turn brown. Many of the stripes may run together so as to discolor the entire plant. The affected plants are stunted, the heads fail to emerge properly or at all, and the grain is discolored and shrunken. Because the spores are seed-borne, the disease can be controlled by treatment with seed protectants. Several varieties have been reported as being resistant.[5, 7, 27]

Spot Blotch

Spot blotch (*Helminthosporium sorokinianum*) is widespread in the north, central, and southern portions of the barley area. As a seedling blight, it may cause considerable reduction in stand, while as a leaf spot it may bring about reductions in yield. It also causes *black point* or blight on the kernels, which reduces the market quality of barley. Any portion of the plant may be infected.[4] Dark-brown spots coalesce to form blotches on the leaves, stems, and floral bracts. Root rots often follow, as do shrivelled, discolored kernels. Numerous secondary infections occur up to the time the plant ripens. The disease overwinters on seed and on plant remains in the field.

Both crop rotation and sanitation are important control measures because the organism develops on the crop residues of cereal plants. Seed treatments control only seed infection. Several varieties are resistant to the disease.

Scab or Fusarium Blight

Scab (*Gibberella saubinetti*) is particularly prevalent in the southern part of the spring-barley region where the temperatures are favorable for this disease and where barley frequently follows corn in the rotation. It is a wet-season disease. Scab causes reduction in grain yields as well as causing blight-damaged kernels. The scab fungus in blighted kernels causes nausea when eaten by hogs or men. The scab organism infects barley heads at the blossom stage. The diseased kernels ripen prematurely and turn pinkish to dark brown in color. The kernels are often shrunken. Infected seed produces seedling blight when it is sown. The scab organism overwinters principally on corn stalks and small-grain stubble.

A crop rotation in which barley does not follow either corn or small grains tends to reduce the amount of head blight somewhat, but it is not a complete control. Another partial control measure is to plow under all cereal-crop residues completely. Seed treatment will reduce the amount of seedling blight. Certain varieties have some resistance. Lax-headed varieties are least subject to scab.

Rhyncosporium Scald

Rhyncosporium scald (*Rhyncosporium secalis*) sometimes damages barley seriously in California and in the South. It produces water-soaked lesions and, later, *scalded*-appearing lesions on the leaves, sheaths, and spikes, causing a shrivelling of the grain. The use of resistant varieties is suggested as a control measure.

Powdery Mildew

Powdery mildew of barley caused by the fungus *Erysiphe graminis hordei* attacks many varieties of barley. It often damages barley except in the intermountain area and the Great Plains. The disease can be recognized by growths of mildew on the leaves and stems. Several varieties are resistant to several races of the fungus.

Barley Rusts

Stem rust caused by the fungus *Puccinia graminis* results only occasionally in heavy losses in barley. Some productive varieties are resistant.

Leaf rust of barley caused by the fungus *Puccinia hordei* sometimes is damaging in the South. A few varieties are somewhat resistant. Stripe rust caused by *Puccinia striiformis* is controlled with resistant varieties.

Virus Diseases

Stripe mosaic is caused by a seed-borne virus. This serious disease may occur in any barley region of North America. The symptoms are yellow or light-green stripes, or complete yellowing of the leaves, with eventual stunting of the plants. The control measure is sowing disease-free seed.

Yellow dwarf is a virus disease that is transmitted by several species of aphids. The symptoms are a brilliant golden yellowing of the leaves as well as moderate to severe stunting of the plants. Control measures include early seeding, aphid control, and resistant varieties.[21, 37]

Other Diseases

Among other barley diseases of some importance are ergot (*Claviceps purpurea*) (see ergot of rye, Chapter 16), net blotch (*Helminthosporium teres*), leaf spot (*Septoria passerinii*), anthracnose (*Colletotrichum graminicolum*), and bacterial blight (*Phytomonas translucens*). A root and stem disease caused by *Pythium graminicola* damages barley seriously in the Corn Belt states in some years.

INSECT ENEMIES

Among the chief insect enemies of barley are the chinch bug, green bug (spring-grain aphis), and grasshoppers. Barley is very attractive to chinch bugs. Several barley varieties of Asiatic origin, as well as some American varieties, are resistant to green bugs. Grasshoppers are controlled by insecticides.

REFERENCES

1. Aberg, Ewart. "The taxonomy and phylogeny of *Hordeum L.* sect. *Cerealia* Ands.," *Symb. Bot. Upsalienes*, 4:1–156. 1940.
2. Atkins, I. M., and Dahms, R. G. "Reaction of small-grain varieties to green bug attack," *USDA Tech. Bul. 901*, pp. 1–30. 1945.
3. Atkins, I. M., and others. "Barley production in Texas," *Tex. Agr. Exp. Sta. Bul. B-1087*, pp. 1–23. 1969.
4. Christensen, J. J. "Studies on the parasitism of *Helminthosporium sativum*," *Minn. Agr. Exp. Sta. Tech. Bul. 11*. 1922.
5. Christensen, J. J., and Graham, T. W. "Physiologic specialization and variation in *Helminthosporium gramineum*," *Minn. Agr. Exp. Sta. Tech. Bul. 95*. 1934.
6. Cook, A. H., Ed. *Barley and Malt Biology, Biochemistry, Technology*, Academic Press, New York, pp. 1–740. 1962.
7. Dickson, J. G. *Diseases of Field Crops*, 2nd ed., McGraw-Hill, New York, pp. 1–517. 1956.
8. Foote, W. H., and Batchelder, F. C. "Effects of different rates and times of application of nitrogen fertilizers on the yield of Hannchen barley," *Agron. J.*, 45:532–535. 1953.
9. Harlan, H. V. "Barley: Cultivation, uses and varieties," *USDA Farmers Bul. 1464*. 1925.
10. Harlan, H. V., and Anthony, S. "Development of barley kernels in normal and clipped spikes and the limitations of awnless and hooded varieties," *J. Agr. Res.*, 19:431–472. 1920.
11. Harlan, H. V., Cowan, P. R., and Reinbach, L. "Yields of barley in the United States and Canada, 1927–1931," *USDA Tech. Bul. 446*. 1935.
12. Harlan, H. V., and Martini, M. L. "Problems and results in barley breeding," in *USDA Yearbook, 1936*, pp. 303–346.
13. Harlan, H. V., and Martini, M. L. "The effect of natural selection in a mixture of barley varieties," *J. Agr. Res.*, 57:189–200. 1938.
14. Harlan, H. V., Martini, M. L., and Pope, M. N. "Tests of barley varieties in America," *USDA Bul. 1334*. 1925.
15. Harlan, H. V., Martini, M. L., and Stevens, H. "A study of methods in barley breeding," *USDA Tech. Bul. 720*. 1940.

16. Harlan, H. V., and Wiebe, G. A. "Growing barley for malt and feed," *USDA Farmers Bul. 1732*. 1943.
17. Harlan, J. R. "On the origin of barley: A second look," in *Barley Genetics II*, Wash. State Univ. Press, pp. 45–50. 1971.
18. Hayes, H. K. "Breeding improved varieties of smooth awned barleys," *J. Hered.*, 17:371–381. 1926.
19. Hayes, H. K., and Wilcox, A. N. "The physiological value of smooth-awned barleys," *J. Am. Soc. Agron.*, 14:113–118. 1922.
20. Leukel, R. W. "Seed treatment for controlling covered smut of barley," *USDA Tech. Bul. 207*. 1930.
21. Leukel, R. W., and Tapke, V. F. "Barley diseases and their control," *USDA Farmers Bul. 2089*. 1955.
22. Pendleton, J. W., Lang, A. I., and Dungan, G. H. "Responses of spring barley to different fertilizer treatments and seasonal growing conditions," *Agron. J.*, 45:529–532. 1953.
23. Pope, M. N. "The time factor in pollen-tube growth and fertilization in barley," *J. Agr. Res.*, 54:525–529. 1937.
24. Robertson, D. W., and others. "Barley production in Colorado," *Colo. Exp. Sta. Bul. 431*. 1936.
25. Robertson, D. W., and Deming, G. W. "Natural crossing in barley at Fort Collins, Colo.," *J. Am. Soc. Agron.*, 23:402–406. 1931.
26. Schwarz, R. "Brewing processes," *J. Ind. Eng. Chem.*, 27(9):1031–1037. 1935.
27. Shands, H. L., and Arny, D. C. "Stripe reaction of spring barley varieties," *Phytopath.*, 34(6):572–585. 1944.
28. Shands, H. L., and Dickson, J. G. "Barley—botany, production, harvesting, processing, utilization, and economics," *Econ. Bot.*, 7:3–26. 1953.
29. Smith, F. H. "Control loose smut by the anaerobic (without air) method," *Clemson Agr. Coll. Ext. Cir. 461*. 1959.
30. Stevenson, F. J. "Natural crossing in barley," *J. Am. Soc. Agron.*, 20:1193–1196. 1928.
31. Stoa, T. E. "Barley production in North Dakota," *N. Dak. Agr. Exp. Sta. Bul. 264*. 1933.
32. Suneson, C. A., and Wiebe, G. A. "Survival of barley and wheat varieties and mixtures," *J. Am. Soc. Agron.*, 34(11):1052–1056. 1942.
33. Tapke, V. F. "Studies on the natural inoculation of seed barley with covered smut (*Ustilago hordei*)," *J. Agr. Res.*, 60(12):787–810. 1940.
34. Taylor, J. W., and Zehner, M. G. "Effect of depth of seeding on the occurrence of covered and loose smuts in winter barley," *J. Am. Soc. Agron.*, 23:132–141. 1931.
35. Weaver, J. C. *American Barley Production*, Burgess, Minneapolis, pp. 1–115. 1950.
36. Wiebe, G. A., and Reid, D. A. "Classification of the barley varieties grown in the United States and Canada in 1958," *USDA Tech. Bul. 1224*. 1961.
37. Wiebe, G. A., and others. "Barley: origin, botany, culture, winter hardiness, genetics, utilization, pests," *USDA Handbk. 338*, pp. 1–127. 1968.

Chapter 18

Oats

ECONOMIC IMPORTANCE

Oats were grown on more than 30 million hectares from 1970 to 1972, with a production exceeding 50 million metric tons or 1.65 tons per hectare. The leading countries in oat production are the United States, the U.S.S.R., Canada, and Poland (Figure 18–1). Oats were harvested for grain on an average of about 6.5 million hectares (16 million acres) in the United States from 1970 to 1972. The production was about 12 million metric tons (831 million bushels) or 1.9 metric tons per hectare (52 bushels per acre). The leading five states in oat production, Minnesota, North Dakota, South Dakota, Wisconsin, and Iowa produce two-thirds of the national crop. The production of oats is declining in the United States and other countries due to the replacement of work animals by motorized equipment and because other feed grains—maize, sorghum, and barley—outyield oats.

Oats fit well into many crop rotations. Fall-sown oats are valuable in the South for pasture as well as for retarding soil erosion. Oats are used as a companion crop with red clover or alfalfa in the Corn Belt, and with lespedeza in Missouri and elsewhere. Compared with other grain crops, oats are the easiest and most pleasant to sow, harvest, thresh, handle, and feed. They can be fed to horses, sheep, and poultry without grinding. They can be sown early in the spring on corn ground before plowing for other crops is possible. They also furnish excellent cereal hay as well as the best straw for feed or bedding. These facts account for the popularity of oats among farmers.

ORIGIN OF THE OAT PLANT

Cultivated oats (*Avena sativa*) formerly were believed to have been derived chiefly from two species, the common wild oat (*A. fatua*) and the wild red oat (*A. sterilis*),[32] but later evidence[11] indicates that both *A. sativa* and *A. fatua* probably were derived from *A. byzantina* or its progenitor, *A. sterilis*. Apparently cultivated oats were unknown to the ancient Egyptians, Chinese, Hebrews, and Hindus. Among the

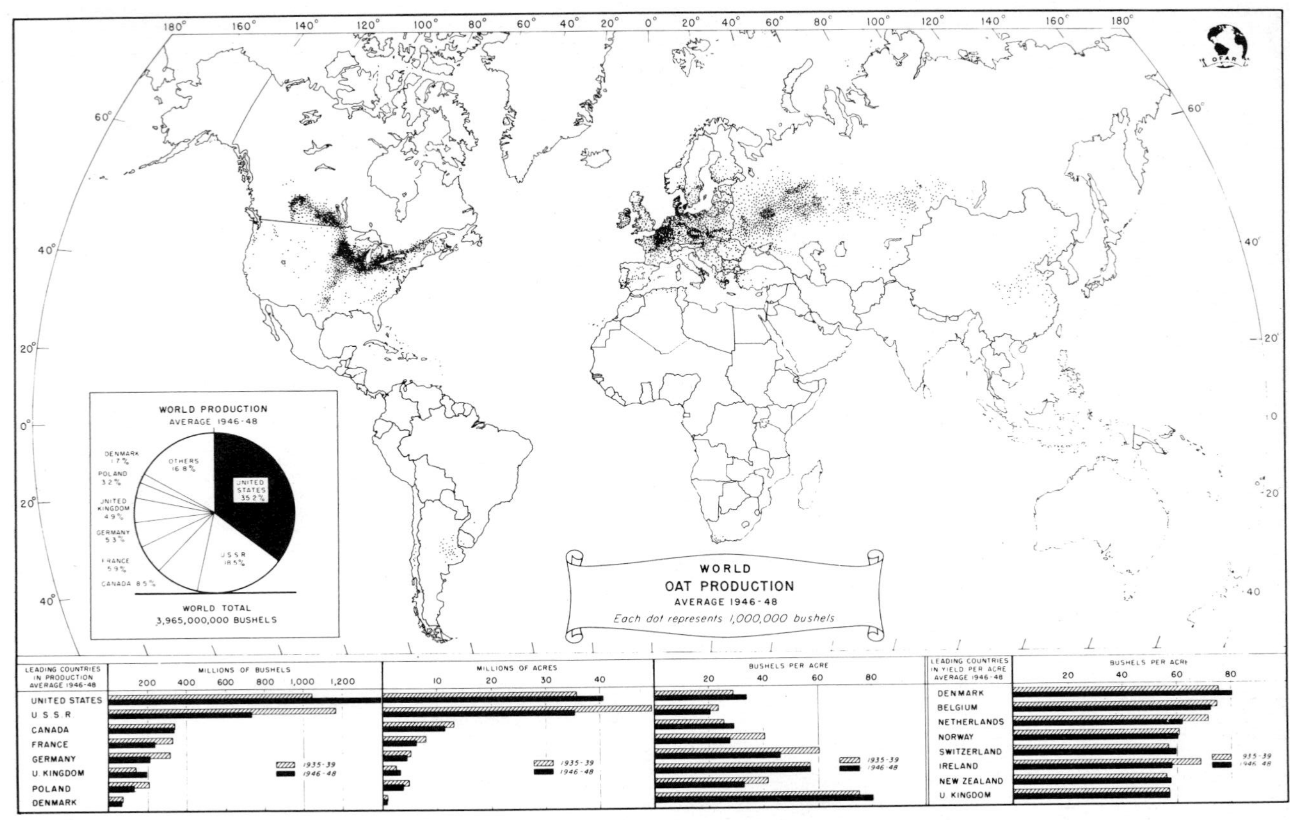

Figure 18–1. World oats production, average from 1946 to 1948. Each dot represents 1 million bushels.

earliest indications of the existence of oats are specimens of *A. strigosa* found in the remains of the Swiss lake dwellers of the Bronze Age. The cultivated species of oats were evident from 900 to 500 B.C., but classical writers in Roman times mention oats only as a weed which was sometimes used for medicinal purposes.[14] Probably it was first distributed as a weed mixture in barley and domesticated later. Authentic historical information on the cultivated oat appears in the early Christian Era. The common oat was reported by writers of this period to be grown by Europeans for grain, while the red oat was grown for fodder, particularly in Asia Minor. The common oat, first found growing in western Europe, spread to other parts of the world. It was believed to have been first cultivated by the ancient Slavonic peoples who inhabited this region during the Iron and Bronze Ages.

Adaptation

Common oats are best adapted to the cooler, more temperate regions where the annual precipitation is 30 inches or more or where the land is irrigated (Figure 18–2). Such areas are found in the northeastern quarter of the United States, in the Pacific Northwest, in valleys of the Rocky Mountain region, in northern Europe, and in Canada.[18] High yields of large, plump grains are possible in these areas. The oat crop often fails in the Great Plains because of drought and heat. Hot, dry weather just before heading causes the oats to blast and, during the heading and ripening period, causes oats to ripen prematurely with poorly filled grain of light bushel weight. Such damage has been

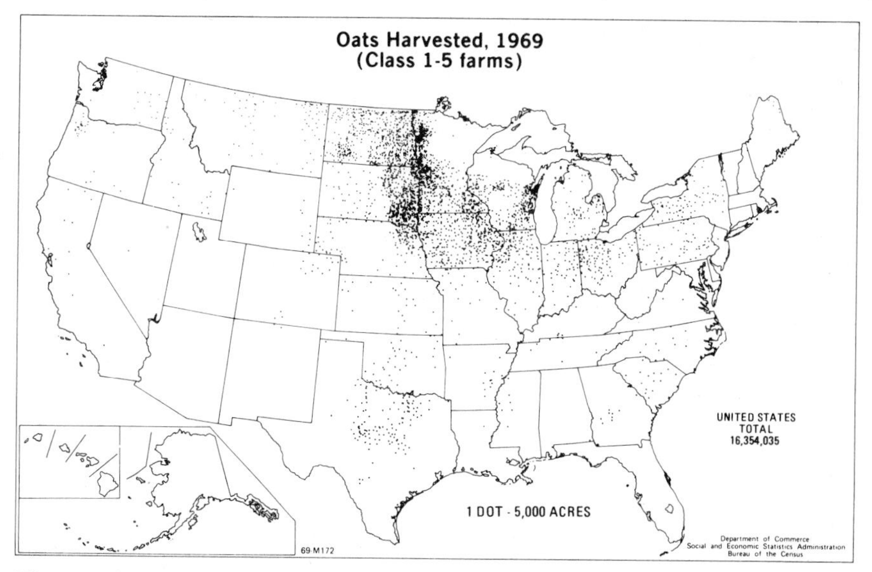

Figure 18–2. Acreage of oats harvested in the United States in 1969. Each dot represents 5,000 acres.

checked to some extent in the warmer regions by growing early varieties of common or red oats. However, red oats as a group are not any more resistant to heat than are certain common winter varieties.[10] The greatest heat resistance was shown by varieties adapted to the South that also were resistant to cold. Red oats are grown primarily in warm climates such as in North Africa and Argentina. Heat-tolerant varieties of red oats are grown in the southern states as well as in the interior valleys of California. Some of these varieties are extremely early. Fall-sown oats are grown successfully where the winters are sufficiently mild, chiefly in the South and on the Pacific Coast.[37] The general oat areas are shown in Figure 18–3.

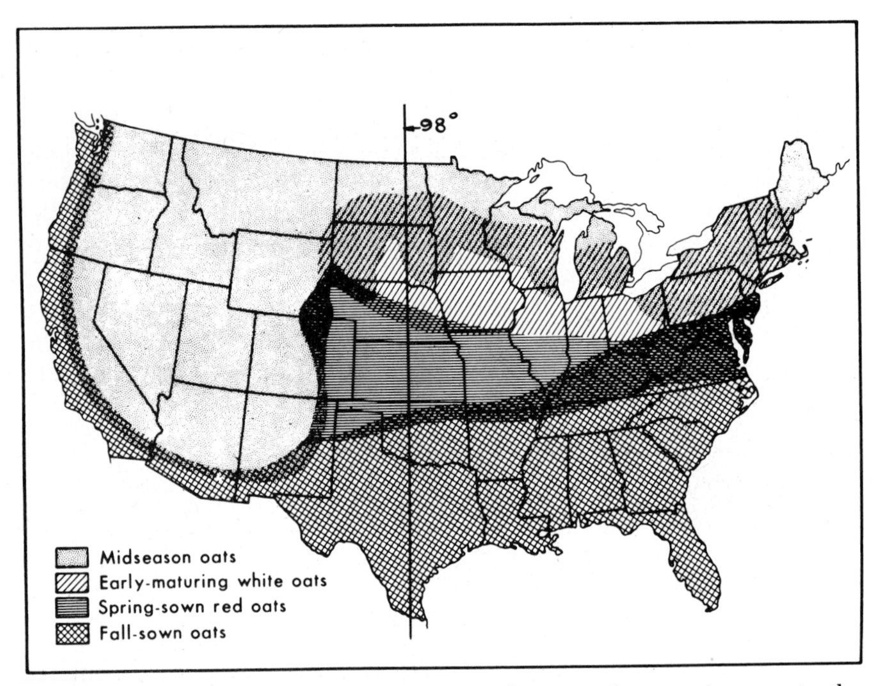

Figure 18–3. Areas in which the principal types of oats are grown in the United States. At the northern border of the winter-oat area the average minimum winter temperature is about −5° F. Midseason oats are grown in the cool climates of the northern states and the intermountain region. Early white and red oats are grown between the winter-oat and midseason-oat areas.

Oats produce a satisfactory crop on a wide range of soil types, provided the soil is well drained and reasonably fertile. In general, loam soils, especially silt and clay loams, are best suited to oats. Heavy, poorly drained clays are likely to cause the crop to lodge or be injured by plant diseases such as rusts and mildew. Excessive available nitrogen and moisture are the chief causes of lodging in oats.

BOTANICAL DESCRIPTION

The oat plant is an annual grass classified in the genus *Avena*. Under average conditions, the plant[32] produces three to five hollow culms from $\frac{1}{8}$ to $\frac{1}{4}$ inch in diameter and from 2 to 5 feet in height (Figure 15–4). Some varieties have hairs on the nodes. The roots are small, numerous, and fibrous, and penetrate the soil to a depth of several feet. The average leaves are about 10 inches long and $\frac{5}{8}$ inch wide.

The inflorescence is either an equilateral (spreading) or a unilateral (one-sided) panicle (Figure 18–4). The panicle is a many-branched determinate inflorescence consisting of a main axis from which arise

Figure 18–4. Oat panicles. (*Left*) Unilateral, "side," or "horse-mane" oat with seven whorls of branches. (*Right*) Equilateral, spreading, or "tree" panicle with five whorls of branches.

lateral axillary branches that are grouped on alternate sides of the main axis at the nodes.[5] The main axis and each of the lateral branches terminate in a single apical spikelet. The axis bears four to six whorls of branches.

The oat spikelet is borne on the thickened end of a slender drooping pedicel that terminates the panicle branch. Each spikelet usually contains two or more florets.[5, 35] The lower two florets usually are perfect, while the third, when present, often is staminate or imperfect.

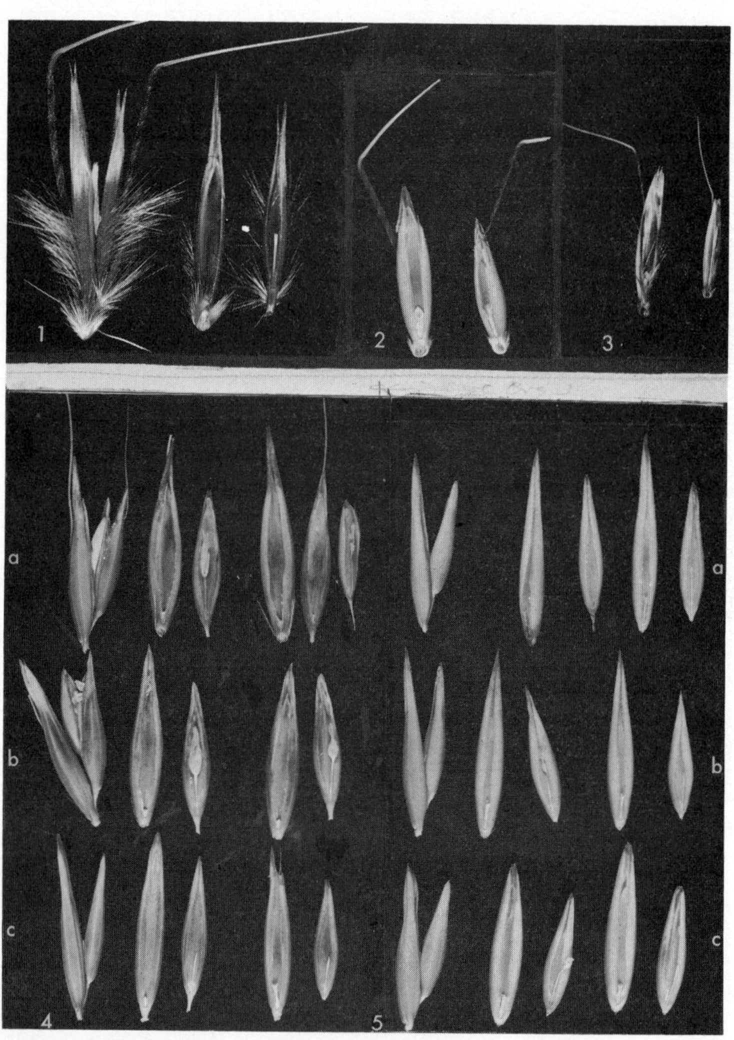

Figure 18–5. Spikelets and kernels of wild and cultivated oats. (1) Wild red oat (*Avena sterilis macrocarpa*); (2) fatuoid from *A. sativa;* (3) wild oats (*A. fatua*); (4) cultivated red oats (*A. byzantina*); (5) cultivated common oats (*A. sativa*).

The two glumes are somewhat unequal, lanceolate, acute, boat-shaped, spreading, glabrous, membranous, and usually persistent. Both usually exceed the lemma in length, except in naked (hull-less) oats. The floret is composed of the lemma, palea, and the reproductive organs, namely, the ovary and three stamens (Figure 4–2). The lemma is the lower of the two bracts that enclose the kernel. It ranges in color from white, yellow, gray, and red to black. The base of the lemma may be extended into a callous, which often bears basal hairs. Most wild-oat species are characterized by hairiness of the callous, lemma, and rachilla (Figure 18–5). The awn of oats is an extension of the midrib of the lemma that usually arises slightly above the middle of the dorsal surface. The awn may vary from almost straight to twisted at the base and bent into a knee (geniculate). Some varieties are awnless. The palea is the inner bract of the floret, thin and parchment-like. In hulled varieties the lemma and palea firmly enclose the caryopsis. The oat caryopsis is narrowly oblong or spindle-shaped, deeply furrowed, and usually covered with fine hairs, especially at the upper end (Figures 18–6 and 18–7).

The natural separation of the lower kernel from the axis of the spikelet is termed disarticulation.[15, 33] In some wild-oat species, as well as in their cultivated derivatives, disarticulation leaves a well defined deep oval cavity commonly called sucker mouth. In most cultivated

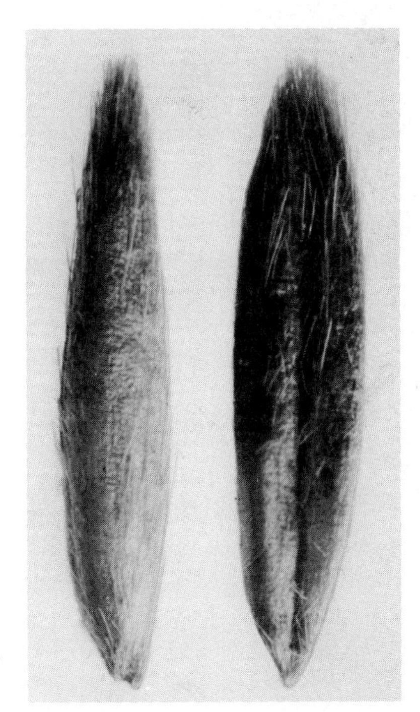

Figure 18–6. Caryopses or "groats" of oats, showing crease in grain at right.

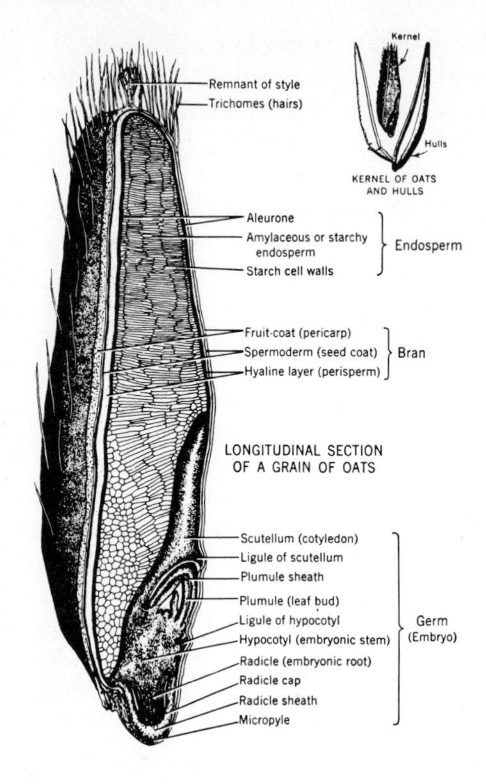

Figure 18–7. The oat caryopsis. Other names for its structures are as follows: starch cell walls = walls of endosperm cells that contain starch; spermoderm = testa; perisperm = nucellus; plumule sheath = coleoptile; ligule of hypocotyl = epiblast; hypocotyl = subcrown internode; radical cap = rootcap; and radical sheath = coleorhiza. [Courtesy The Quaker Oats Company.]

oat varieties, the separation is by a fracture or breaking that results in a roughened tissue with no observable cavity at the base of the lemma. Most varieties of common white and yellow oats lack the scar, commonly found in red oats.

The legal weight per bushel of oats is 32 pounds, but the actual test weight may range from 27 to 45 pounds. Clipped oats have a high test weight because the tip of the hull has been removed.

SPECIES OF OATS

Oats may be classified into three groups on the basis of chromosome numbers.[36]

GROUP I: 7 HAPLOID CHROMOSOMES

A. *brevis*, short oat
A. *wiestii*, desert oat
A. *strigosa*, sand oat
A. *nudibrevis*, small-seeded naked oat

GROUP II: 14 HAPLOID CHROMOSOMES

A. *barbata*, slender oat
A. *magna*[28]
A. *abyssinica*, Abyssinian oat

GROUP III: 21 HAPLOID CHROMOSOMES

A. fatua, common wild oat
A. sativa, common white or northern oat
A. nuda, large-seeded naked or hull-less oat
A. sterilis, wild red or animated oat
A. byzantina, cultivated red oat

Many of the oat varieties grown in the United States were described and also classified by Stanton.[35]

Wild Oats

The common wild oat (*A. fatua*) (Figure 18–5) is a noxious weed in many parts of the country, being particularly troublesome in the hard red spring-wheat region of Minnesota, and Dakotas, and Montana. The wild oat is difficult to eradicate because the seeds shatter readily and because many of the seeds are plowed into the soil, where they lie dormant for one to many years, and then germinate and grow when they are turned up near the surface. However, some collections of wild oats show little or no dormancy.[39] The common wild oat differs from cultivated oats in having taller, more vigorous plants and strongly twisted geniculate awns. The grain has a pronounced sucker mouth at the base, and usually a hairy lemma. In California, wild oats grow so abundantly on wastelands and uncropped fields that they are often harvested for hay.

The wild red oat, *A. sterilis*, occurs infrequently as a weed in the United States. Another wild oat, *A. barbata*, is an important range grass in California. In other parts of the world, the sand oat, *A. strigosa*, is an important forage grass.

Fatuoids or false wild oats (Figure 18–5) appear suddenly in cultivated oat varieties and somewhat resemble the common wild oat (*A. fatua*), although the fatuoids also are similar to the variety in which they occur, particularly as to lemma color and germination habits.[16] The persistence of fatuoids is due to natural crossing and mutation.[17] Many fatuoids originate through natural hybridization between common cultivated oats and wild oats (*A. fatua*).[1]

NATURAL CROSS-POLLINATION

Self-pollination is normal in oats but some natural crossing occurs in different varieties and environments ranging from 0 to nearly 10 per cent, but it usually is less than 1 per cent.[26] Greater numbers of natural hybrids may occur in plants produced from secondary seeds.[25] Open-pollinated fatuoids have shown a maximum of 47 per cent of cross-fertilization in a single season, the five-year average being 11.6 per cent. Under similar conditions a cultivated variety contained less than 0.5 per cent crosses in any season.[17]

DORMANCY IN OATS

Freshly harvested seeds of oats often fail to germinate satisfactorily, while others germinate immediately.[18] All degrees of prompt, slow, and delayed germination occur among oat varieties, but all cultivated red-oat varieties usually show slow or delayed germination[15, 21] and may cause poor field stands from the use of freshly harvested seed. Winter-oat varieties are more likely to show dormancy than are spring varieties.[6] However, there appears to be no association between morphology of the oat kernel and dormancy. Characters associated with wild oats and fatuoids, such as the sucker-mouth base and the presence of basal hairs on the callous are unrelated to dormancy.[16, 39]

OAT TYPES

Seed with white or yellow lemma colors are commonly grown in the North, while red or gray oats are grown chiefly in the South. Black oats are grown rather rarely. Oats are sometimes grouped as early, midseason, and late. A midseason variety may mature 10 to 14 days later than an early variety.

Midseason varieties are well suited to the cooler regions along the northern border and in the irrigated intermountain region. To the South, where hot weather may injure the plants before ripening, the early white and yellow varieties are grown. Further south in the central regions, still earlier spring-sown red varieties are grown,[12] while in the extreme south, fall- or winter-sown red or gray oats are grown, in each case in order that the crop will reach maturity before the advent of severe summer heat or drought.

In the North Central region early maturing common yellow and white varieties are of chief importance, but midseason varieties are grown in the northern portion, and spring-sown red oats lead in the southern portions of the region.

Hull-less (naked) oats can be fed to hogs directly, but they often do not remain sound in storage. Few hull-less varieties yield as well as the hulled varieties and are rarely grown.

The spring-sown red-oat region lies south of latitude 41°.[22] The southern border of this area is transitional between the winter- and spring-oat regions.

Medium-early as well as midseason varieties are grown in the northeastern states.

Fall-sown oats are grown mostly in Delaware, Maryland, West Virginia, Kentucky, Arkansas, Oklahoma, and southward. The winter-oat belt has recently moved northward through the development of winter oats more resistant to cold than those previously available. Winter oats are now grown in states north of those just mentioned.[19]

Midseason varieties prevail in the Rocky Mountain region where oats are grown mostly under irrigation.

Winter gray or white varieties are grown from fall seeding in the western portion of the Pacific Northwest. Varieties of the red-oat group are grown in California, mostly from fall seeding. Midseason varieties are the most extensively grown spring oats in Washington and Oregon.[13]

ROTATIONS

A common rotation in the northeastern states is corn (one year), oats seeded to clover or grass (one year), and clover or grass (two years).

In the northern portion of the spring-sown red-oat region, a widely followed rotation is corn two years, oats and wheat each for one year, and a grass-legume mixture for two years. A one-year rotation of oats and lespedeza has been suggested for Missouri.[22] Oats are sown in the spring on Korean lespedeza sod each year. The lespedeza volunteers each year from the seed produced and shattered to the ground.

In the far west, alternate fallow and small grain is a common rotation in the drier areas. Where a crop can be produced each year, and in the Great Plains, oats usually follow corn, sorghum, cotton or potatoes. Under irrigation, oats are usually a companion crop for alfalfa or clover. A typical rotation is alfalfa for three years, sugarbeets or potatoes or corn for one or two years, and oats seeded to alfalfa. In the humid Pacific area, oats frequently occupy the season between row crops and clover.

In the southern fall-sown oat region, where crop sequences are rather flexible, oats usually follow corn or sometimes tobacco. Corn usually follows a legume winter-cover and green-manure crop. The oats often are harvested in May or early June in time to permit double cropping, i.e., the planting of a summer crop such as soybeans, cowpeas, sorghum, or pearl millet. Winter-oat fields that have been grazed may be plowed for later planting to cotton or tobacco.[27]

FERTILIZERS

Nitrogen is the most essential element for oats, but heavy applications that exceed 30 to 60 pounds per acre are likely to induce lodging.[27] A spring dressing of 20 to 40 pounds of nitrogen per acre is essential to high yields of winter oats in most of the humid areas.

Carry-over nitrogen remaining after intensive corn production may also be at a high enough level to induce extensive lodging.

Oats usually receive, at seeding time, an application of 200 to 400 pounds per acre of a complete or mixed fertilizer in the northeastern and middle Atlantic states and about 300 to 600 pounds in the Southeast. The fertilizer requirements for oats are similar to those for wheat. Oats respond to phosphorus (Figure 6–2), and often to potassium, in most sections of the humid area.

CULTURAL PRACTICES

Seedbed Preparation

Where oats follow corn or some other row crop, the seedbed often is prepared by disking and harrowing without plowing.[29] Where abundant weed growth and heavy soils occur, the corn-stubble land is usually plowed, disked, and harrowed for seedbed preparation. In the Corn Belt, oats usually are sown with a drill (Figure 7–5) but occasionally broadcast on corn land that is disked and harrowed but not plowed. The broadcast fields are harrowed or disked, or both, after seeding. Where the oat crop is irrigated, the land often is fall-plowed and left rough over winter, the land being harrowed in the spring.

Time of Seeding

Spring-sown oats should be seeded as early in the spring as a seedbed can be prepared, but after the danger of prolonged cold weather is past. This is especially important in sections subject to heat or drought. Dates of seeding spring oats are shown in Figure 18–8. Oats should be sown before the average temperature reaches 50° F.[38]

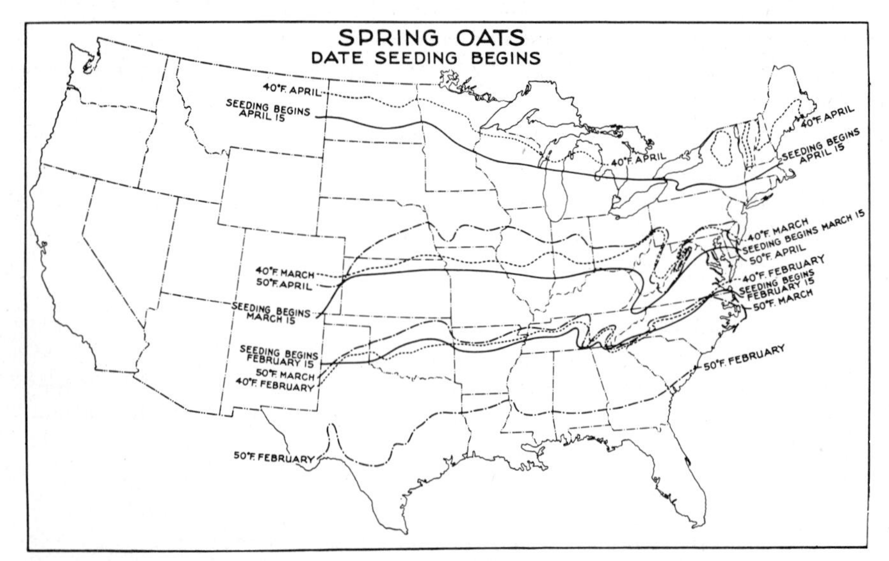

Figure 18–8. Seeding of spring oats in the United States begins when the mean daily temperature is between 40 and 50° F.

In the spring-sown red-oat area, the crop is generally seeded by March 15. Spring seeding is usually later in the more northern regions, but seldom later than May 15, even in New England.

Fall-sown oats may be seeded in the South at any time from October 1 to December 31, but October often is the best month. Earlier seeding is desirable when the oats are to be pastured.[3] Experiments with winter oats sown in Oklahoma and Tennessee showed that seeding about October 1 gave the best yields. North of these states, seeding in September would be desirable. Most oats in California are seeded in November.

Rate of Seeding

The average rate of seeding oats in the United States is 2½ bushels (10 pecks) per acre. Oats sown broadcast on corn stubble in the Corn Belt usually are sown at the rate of 12 to 16 pecks. In the northeastern states, the average rate of seeding is about 10 pecks. In the southeastern states, winter oats often are sown at rates of 12 pecks or more when the crop is intended for pasture.

Under irrigated conditions in the West, 8 to 12 pecks is a common rate. Under conditions extremely favorable for oats in western Oregon and Washington, 14 to 16 pecks are sometimes sown. In the spring-sown red-oat region, 8 to 12 pecks are usually seeded. In the colder part of the fall-sown oat area, 10 to 12 pecks are necessary because of the danger of winterkilling. On the Great Plains, 4 to 8 pecks may be seeded, 5 or 6 pecks being the most common rate. The average yields for nine years under dryland conditions in Colorado (at Akron) for 4, 5, and 6 pecks of seed per acre, were 38.7, 39.6, and 40.1 bushels per acre, respectively.[8] Oats as a companion crop for legumes and grasses should be seeded one-fourth less than the rate where sown alone. The summarized yields in rate-of-seeding experiments with oats in Iowa[7] showed 12 pecks to be preferable to lower rates.

In these experiments, drilled oats yielded 62.8 bushels per acre compared with 58.7 bushels broadcast on disked corn ground.

Weed Control

Many of the broad-leaved weeds in oat fields can be controlled with herbicides. Oats are more sensitive than wheat or barley to 2,4-D. They also are more likely to show injury from a residual herbicide when sown on corn land that had been treated with atrazine the previous season.

Harvesting

About 67 per cent of the 1972 oat acreage sown in the United States was harvested for grain. About 90 per cent of the oats in Minnesota, Wisconsin, and the Dakotas, but less than 40 per cent in the southern states, was harvested for grain. The remainder was pastured, cut for hay, silage, or soilage, or plowed under or abandoned. About 95 per cent or more of the oats for grain are threshed with a combine in the United States, but some of the crop is harvested with a binder in order to save more of the straw. Nearly 40 per cent of the oats are windrowed before combining in order to avoid losses from crinkling and lodging of the stems, or from shattering of the grain. Premature harvesting of oats with a binder before the hard-dough stage reduces both grain and straw yields,[2] even during rust years.[41] In the South, some oats have been fed in the bundle without threshing.

The development of short, stiff-straw varieties facilitated combine harvesting. Some oats are cut for hay with a mower. Oats are best harvested in the dough stage for hay, but in the milk stage for silage. Fall-sown oats often can be pastured moderately in the autumn and winter until February 1, or sometimes later, and still produce a normal crop of grain.

The growing oat plants, or even oat hay, sometimes contain sufficient nitrates to be poisonous to cattle, sheep, and horses.[22a] Oat poisoning occurs occasionally in Colorado, Wyoming, and South Dakota, apparently under soil or climatic conditions in which the uptake of nitrates greatly exceeds the rate of protein synthesis. It possibly results from a relatively slower growth, photosynthesis, and phosphate absorption.[42] Ample phosphate applications may prevent the nitrogen excess. The symptoms of nitrate poisoning include rapid breathing along with a blueing of the mucous membranes. Death occurs from asphyxiation. The remedy is an early intravenous injection of 4 milligrams of methylene blue per pound of body weight in a 4 per cent solution with distilled water.[4]

USES

About 57 million bushels of the oat crop are used for seed. About 119 million bushels are used for food, largely in the form of rolled oats for domestic use or export. More than 80 per cent of the crop is used for feed or export. As horse feed, oats are about 90 per cent as efficient as corn. Oats often are fed whole to horses and sheep. Ground or chopped oats are fed extensively to dairy cattle, breeding stock, and young stock. After the hulls are removed, oats are equivalent to corn for hog feed. Farm-size hullers are used largely. Oats are fed extensively to poultry, usually without hulling, but sometimes they are sprouted. The hulls supply some accessory food factor that checks a malnutrition in chickens inducing feather picking. Oat flour, in part a by-product of the manufacture of rolled oats, is used in breakfast foods and other foods. The flour possesses a property that retards the development of rancidity in fat products. Paper containers coated with oat flour are used for packaging food products having a high fat content. Another oat product is used as an antioxidant and stabilizer in ice cream and other dairy products.[34]

Oats are one of the best crops for mowing while green to supply dried grass products and chlorophyll and carotene extracts for use in food, feed, and medicine. The young leaves are rich in protein and vitamins.

The manufacture of rolled oats consists first of cleaning and sizing the grain.[40] Then the grains are heated in steam-jacketed pans where they dry down to about 6 per cent moisture and acquire a slight roasted flavor. After this drying the hulls are brittle enough to be removed readily by large hulling stones similar to the old-fashioned bur milling stones. The groats are then separated from the hulls and broken pieces. The whole groats are steamed to be made tougher and are then passed between steel rolls to produce the familiar flakes of rolled oats. So-called quick oats are small thin flakes rolled from steel-cut pieces of oat groats. The small size and thinness speed up the cooking process. The hulls, which constitute 27 to 30 per cent of the grain, are used in the manufacture of furfural, a solvent widely used in chemical, fat, and petroleum industries.

DISEASES

Smuts

Loose smut, *Ustilago avenae,* replaces the oat floret by a loose spore mass.[30] In covered smut, *U. kolleri,* the dark-brown spore mass is more or less enclosed within a grayish membrane, frequently within the lemma and palea of the flower. The two smuts cannot be clearly differentiated in the field. Both can be controlled by seed treatment with suitable fungicides.[24] The percentage of oat plants infected with covered smut may be highest from seed sown when the mean temperatures are relatively high.

Losses from oat smuts can be largely eliminated in many sections by growing adapted resistant varieties.

Rusts

Stem rust of oats (*Puccinia graminis avenae*) produces symptoms similar to stem rust of wheat. The alternate host is the common barberry, but in the South the disease overwinters and develops independently of the alternate host and then spreads northward. Control measures consist chiefly of growing resistant varieties.

Crown rust (*Puccinia coronata*) is identified by the bright yellow pustules on the leaves. The alternate host is the buckthorn, but in the South the fungus spreads independently of the alternate host. For each unit increase in coefficient of crown-rust infection (percentage of infection × numerical infection type) the yield is decreased from 0.2 to 0.3 bushel per acre.[27] As the eradication of the buckthorn is incomplete the principal control is resistant varieties.

The breeding of valuable new stem-rust-resistant and crown-rust-resistant varieties is being continued in an effort to avoid losses from still newer rust races.

Victoria Blight

A blight disease, Victoria blight, caused by *Helminthosporium victoriae* caused heavy losses to oat varieties derived from crosses with the Victoria variety. Susceptible varieties are now rarely grown.

Other Diseases

Oats also are attacked by more than 6 virus diseases, 3 bacterial diseases, and more than 20 fungus diseases in addition to those just described.[31] These include septoria blight and barley yellow dwarf (BYDV). The oat cyst nematode (*Heterodera avenae*), a prevalent pest elsewhere, was first observed in the United States in 1974.

INSECTS

Oats are less subject to insect attack than are barley and wheat. The Hessian fly does not attack oats. Chinch bugs, although damaging to oats, greatly prefer barley or wheat.

The bluegrass billbug (*Calendra parvulus*), certain leaf hoppers, the armyworm, the grain bug (*Chlorocroa sayi*), the grasshoppers, and the Mormon cricket attack oats readily. The spring grain aphis (or green bug, *Toxoptera graminum*) sometimes damages oats severely. Methods of controlling most of these insects are discussed in the chapter on wheat. Other oat pests include wireworms, white grubs, cutworms, leafhoppers, fruit flies, thrips, and winter grain mites.[19]

REFERENCES

1. Aamodt, O. S., Johnson, L. P. V., and Manson, J. M. "Natural and artificial hybridization of *Avena sativa* with *A. fatua* and its relation to the origin of fatuoids," *Can. J. Res.*, 11:701–727. 1934.
2. Arny, A. C., and Sun, C. P. "Time of cutting wheat and oats in relation to yield and composition," *J. Am. Sco. Agron.*, 19:410–439. 1927.
3. Atkins, I. M., and others. "Growing oats in Texas," *Texas Agr. Exp. Sta. Bul. B-1091*, pp. 1–28. 1969.
4. Binns, W. "Chemical poisoning," in *Animal Diseases*, USDA Yearbook, 1956, pp. 113–117.
5. Bonnett, O. T. "Morphology and development," in *Oats and Oat Improvement*, Am. Soc. Agron. Monographs, Vol. 8, pp. 41–74. 1961.
6. Brown, E., and others. "Dormancy and the effect of storage on oats, barley, and sorghums," *USDA Tech. Bul.* 953. 1948.

7. Burnett, L. C. "Iogold oats," *Iowa Agr. Exp. Sta. Bul. 247.* 1928.
8. Coffman, F. A. "Experiments with cereals at the Akron (Colorado) Field Station in the 15-year period, 1908 to 1922, inclusive," *USDA Bul. 1287.* 1925.
9. Coffman, F. A. "Species hybridization, a probable method for producing hardier winter oats," *J. Am. Soc. Agron.*, 29:79–81. 1937.
10. Coffman, F. A. "Heat resistance in oat varieties," *J. Am. Soc. Agron.*, 31:811–817. 1939.
11. Coffman, F. A. "Origin of cultivated oats," *J. Am. Soc. Agron.*, 38(11): 983–1002. 1946.
12. Coffman, F. A. "Culture and varieties of spring-sown red oats," *USDA Farmers Bul. 2115.* 1958.
13. Coffman, F. A. "Oat varieties in the western States," *USDA Agricultural Handbook 180.* 1960.
14. Coffman, F. A. "Origin and history," in *Oats and Oat Improvement,* Am. Soc. Agron. Monographs, Vol. 8, pp. 15–40. 1961.
15. Coffman, F. A., and Stanton, T. R. "Variability in germination of freshly harvested Avena," *J. Agr. Res.*, 57:57–72. 1938.
16. Coffman, F. A., and Stanton, T. R. "Dormancy in fatuoid and normal kernels," *J. Am. Soc. Agron.*, 32:459–466. 1940.
17. Coffman, F. A., and Taylor, J. W. "Widespread occurrence and origin of fatuoids in Fulghum oats," *J. Agr. Res.*, 52:123–131. 1936.
18. Coffman, F. A., and Frey, K. J. "Influence of climate and physiological factors on growth in oats," in *Oats and Oat Improvement,* Am. Soc. Agron. Monographs, Vol. 8, pp. 420–464. 1961.
19. Coffman, F. A. "Factors affecting survival of winter oats," *USDA Tech. Bul. 1346,* pp. 1–28. 1965.
20. Dahms, R. G. "Insects and mites that attack oats," in *Oats and Oat Improvement,* Am. Soc. Agron. Monographs, Vol. 8, pp. 391–419. 1961.
21. Deming, G. W., and Robertson, D. W. "Dormancy in small grain seeds," *Colo. Agr. Exp. Sta. Tech. Bul. 5.* 1933.
22. Etheridge, W. C., and Helm, C. A. "Growing good oat crops in Missouri," *Mo. Agr. Exp. Sta. Bul. 359.* 1936.
22a. Gul, A., and Kolp, B. J. "Accumulation of nitrates in several oat varieties at various stages of growth," *Agron. J.*, 52:504–506. 1960.
23. Hancock, N. I., and Long, O. H. "Oat variety studies in Tennessee," *Tenn. Agr. Exp. Sta. Bul. 199,* pp. 1–30. 1946.
24. Hanson, E. W., Hansing, E. D., and schroeder, W. T. "Seed treatments for control of disease," in *Seeds,* USDA Yearbook, 1961, 272–280.
25. Hoover, M. M., and Snyder, M. H. "Natural crossing in oats at Morgantown, W. Va.," *J. Am. Soc. Agron.*, 24:784–786. 1932.
26. Jensen, N. F. "Genetics and inheritance in oats," in *Oats and Oat Improvement,* Am. Soc. Agron. Monographs, Vol. 8, pp. 125–206. 1961.
26a. Kadans, J. M. "*Modern Encyclopedia of Herbs,*" Parker Pub. Co., West Nyack, N.Y. pp. 1–256. 1970.
27. Murphy, H. C., and others. "Relation of crown-rust infection to yield, test weight, and lodging of oats," *Phytopath.*, 30:808–819. 1940.
28. Murphy, H. C., and others. "*Avena magna:* An important new tetraploid species of oats," *Science,* 159:103–104. 1968.

29. Shands, H. L., and Chapman, W. H. "Culture and production of oats in North America," in *Oats and Oat Improvement,* Am. Soc. Agron. Monographs, Vol. 8, pp. 465–529. 1961.
30. Simons, M. D., and Murphy, H. C. "Oat diseases," in *Oats and Oat Improvement,* Am. Soc. Agron. Monographs, Vol. 8, pp. 330–390. 1961.
31. Simons, M. D., and Murphy, H. C. "Oat diseases and their control," *USDA Handbk. 343,* pp. 1–15. 1968.
32. Stanton, T. R. "Superior germ plasm in oats," *USDA Yearbook, 1936,* pp. 347–413.
33. Stanton, T. R. "Maintaining identity and pure seed of Southern oat varieties," *USDA Circ. 562.* 1940.
34. Stanton, T. R. "New products from an old crop," in *Crops in Peace and War,* USDA Yearbook, 1950–51, pp. 341–344.
35. Stanton, T. R. "Oat identification and classification," *USDA Tech. Bul. 1100.* 1955.
36. Stanton, T. R. "Classification of Avena," in *Oats and Oat Improvement,* Am. Soc. Agron. Monographs, Vol. 8, pp. 75–111. 1961.
37. Stanton, T. R., and Coffman, F. A. "Winter oats for the South," *USDA Farmers Bul. 2037.* 1951.
38. Taylor, J. W., and Coffman, F. A. "Effects of vernalization on certain varieties of oats," *J. Am. Soc. Agron.,* 30:1010–1019. 1938.
39. Toole, E. H., and Coffman, F. A. "Variations in the dormancy of seeds of the wild oat, *Avena fatua,*" *J. Am. Soc. Agron.,* 32:631–638. 1940.
40. Western, D. E., and Graham, W. R., Jr. "Marketing, processing, and uses of oats," in *Oats and Oat Improvement,* Am. Soc. Agron. Monographs, Vol. 8, pp. 552–578. 1961.
41. Wilson, H. K., and Raleigh, S. M. "Effect of harvesting wheat and oats at different stages of maturity," *J. Am. Soc. Agron.,* 21:1057–1078. 1929.
42. Wright, M. J., and Davison, K. L. "Nitrate accumulation in crops and nitrate poisoning in animals," *Advances in Agronomy,* Vol. 16, Academic Press, New York, pp. 197–247. 1964.

Chapter 19

Rice

ECONOMIC IMPORTANCE

Rice provides the principal food for about half of the world population. The average production of rough rice from 1970 to 1972 exceeded 303 million metric tons (Figure 19–1.) This tonnage is less than that of wheat, but slightly more than that of maize. However, rough rice consists of about 20 per cent hull so rice ranks third in net kernel production. The rice crop was produced on 133 million hectares with an average yield of 2.28 metric tons per hectare. Leading countries in rice production were China, India, Indonesia, Japan, Thailand, and Burma. Production in the United States from 1970 to 1972 averaged 84,909,000 hundredweight (3.86 million metric tons) with a yield of 4,673 pounds per acre (5.25 metric tons per hectare) on 1,817,000 acres (735,000 hectares).

The domestic rice crop is produced in the four states of Texas, Arkansas, Louisiana, and California (Figure 19–2), with small quantities in Mississippi and Missouri, and occasionally in other states. The United States, despite its small rice production is the leading exporter, followed by Thailand and Burma. Nearly 70 per cent of the domestic production sold by farmers in 1972 was exported. Rice is produced in tropical and subtropical portions of all continents.

Rice, *Oryza sativa,* probably originated in India or southeastern Asia where several wild species are found, possibly evolving from a wild form which no longer exists. Another rice species (*O. glaberrima*), grown in western and central Africa, may have originated there as a derivative of the wild *O. breviligulata.*[60] The so-called wild rice of North America, *Zizania aquatica* (or *Zizania palustris*) bears little resemblence to the cultivated rice. It belongs to a different grass tribe, *Zizanieae.*

Rice culture is believed to have spread into China by 3000 B.C. and westward into Europe by 700 B.C.

The first commercial rice planting in the United States was at Charleston, S.C., about 1685. The crop spread to North Carolina and Georgia. Most of the rice was produced largely by hand methods on

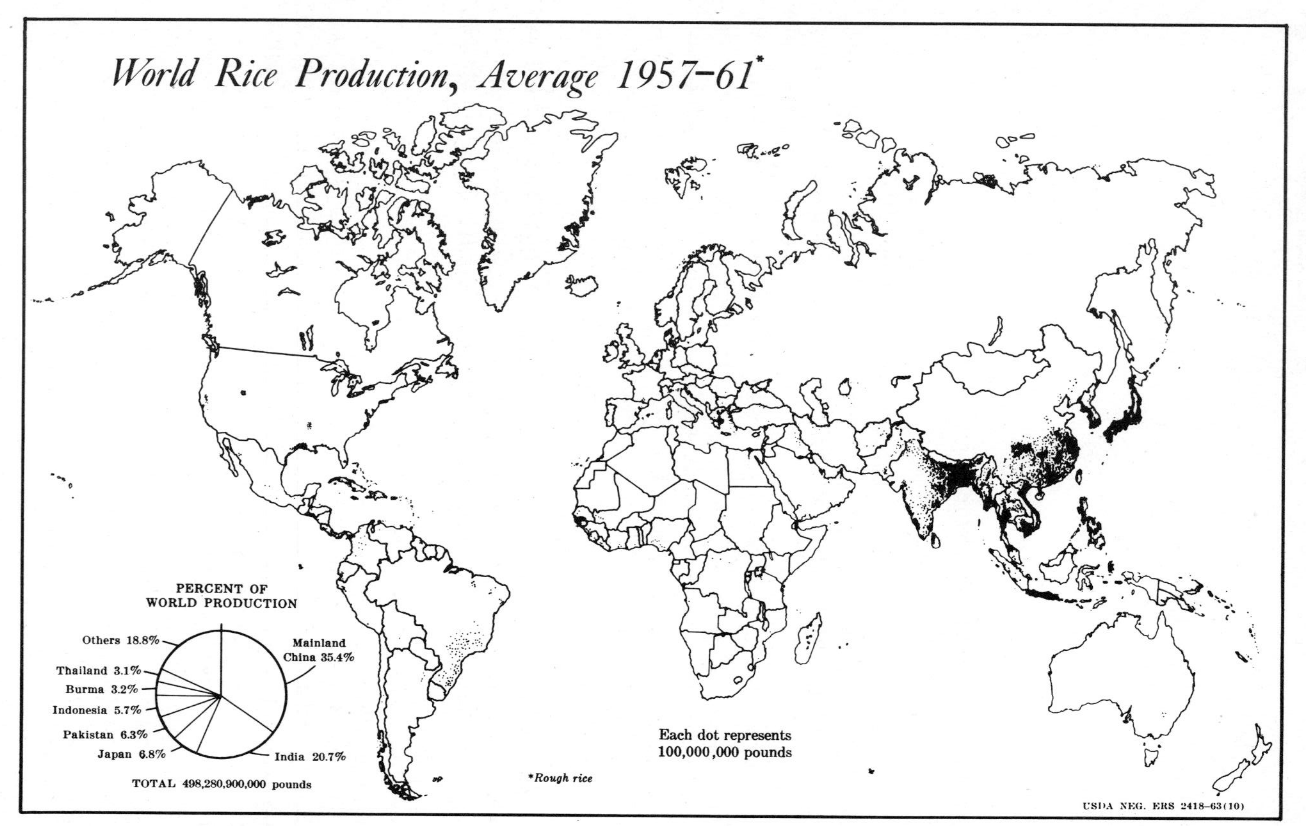

Figure 19–1. World rice production.

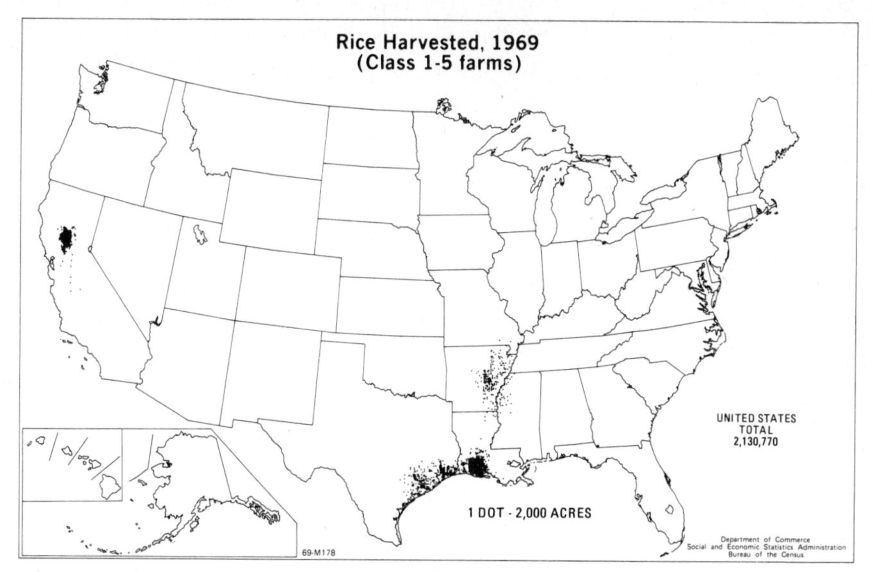

Figure 19–2. Acreage of rice in the United States.

the delta lands of the South Atlantic States until 1888. Rice growing in this region was adversely affected by the Civil War, after which the acreage increased along the Mississippi River in Louisiana. In 1889, Louisiana became the leading state in rice production. In the Gulf Coast prairie section there, it was first grown by animal-powered machine methods. Established rice culture spread from southwestern Louisiana to southeastern Texas about 1900, to eastern Arkansas in 1905, to the Sacramento Valley in California by 1912, and to Mississippi about 1949. Rice has been grown intermittently in Missouri since about 1925.

The yields of rice in the United States by 1970 were four times those obtained at the beginning of the 20th century. Rice yields in many Asiatic countries were greatly improved by growing improved varieties developed by the International Rice Research Institute at Los Blanos in the Philippines. These IRRI varieties have short, stiff culms which enable them to remain erect on soils that are heavily fertilized with nitrogen. Their yield response to nitrogen fertilization of up to 133 kilos per hectare is much superior to that of the tall, weak-stalked varieties formerly grown. The latter types lodge and yield less under even moderate nitrogen fertilization of 34 kilos per hectare.[14]

ADAPTATION

Rice is unique among the cereals in being able to germinate and thrive in water. Other cereal crops are killed if submerged for 2 or 3 days in warm weather, because of lack of oxygen for the roots. The rice plant is able to transport oxygen to the submerged roots from the leaves, where oxygen is released during photosynthesis. The water also contains some oxygen. Thus rice can live in an aquatic environment.[13]

The important factors for rice production are favorable temperatures, a constant supply of fresh water for irrigation, and suitable soils. Rice can be grown most successfully in regions that have a mean temperature of about 70° F. (22° C.) or above during the entire growing season of four to six months.[30] Rice yields are higher in warm temperate regions that have a low summer rainfall with a high light intensity[56] than in the humid tropics where rice diseases and soils of low fertility are more prevalent. The countries with the highest average yields are Australia, Spain, and Italy.

The best soils for rice are slightly acid (*p*H 5.5 to 6.5). Flooding of a rice field increases the *p*H 0.5 to 1.5 units, and thus releases available phosphorus. Rice is grown on soils that range in *p*H from 4.5 to 8.5. It is grown most economically on soils of rather heavy texture with an impervious underlying subsoil at from 1½ to 5 feet from the surface. The loss of water by seepage is small through such soils, which is the chief reason for selecting heavy soils for rice culture.

Upland rice consists of certain varieties that can be grown without irrigation or submergence in regions of high rainfall where the soil is wet much of the time. Yields are much lower than when rice is grown under field submergence. Upland rice is grown mostly on a small scale by hand methods for home use. It is grown rarely in the United States, but is important in parts of some Asian, South American and African countries.

BOTANICAL DESCRIPTION

Rice belongs to the grass tribe *Oryzeae* characterized by one-flowered spikelets, laterally compressed, and two short glumes.

Rice is an annual grass with erect culms 60 to 180 cm tall (Figure 19–3). The plant tillers freely, four to five culms per plant being common. Despite its annual habit rice can be propagated vegetatively for several years by transplanting rooted tillers. Node branching may occur with early maturity and with ample space for plant development.[25]

The rice inflorescence is a loose terminal panicle of perfect flowers (Figures 19–4 and 19–5). At maturity the panicle in different varieties may be enclosed in the sheath, partly exserted, or well exserted. Each panicle branch bears a number of spikelets, each with a single floret (Figure 19–6). The flower has six stamens and two long plumose sessile styles, being surrounded by a lemma and a palea at the base of which are two small glumes. The lemma and palea may be straw yellow, red, brown, or black. The lemmas of various varieties may be fully awned, partly awned, tip-awned, or awnless. The rice grain is enclosed by the lemma and palea, these structures being called the hull. The hulled kernels vary from about 3.5 to 8 millimeters in length, 1.7 to 3 millimeters in breadth, and about 1.3 to 2.3 millimeters in thickness.[47] They may be hard, semihard, or soft in texture. The color of the unmilled kernel may be white, brown, amber, red, or purple. Commercial American varieties are white, light brown, or amber. An average rice panicle contains 100 to 150 seeds.

POLLINATION

In blooming, rice flowers open rapidly and the anthers dehisce when the flower opens or, rarely, before opening. The flowers may

Figure 19–3. A mature rice plant.

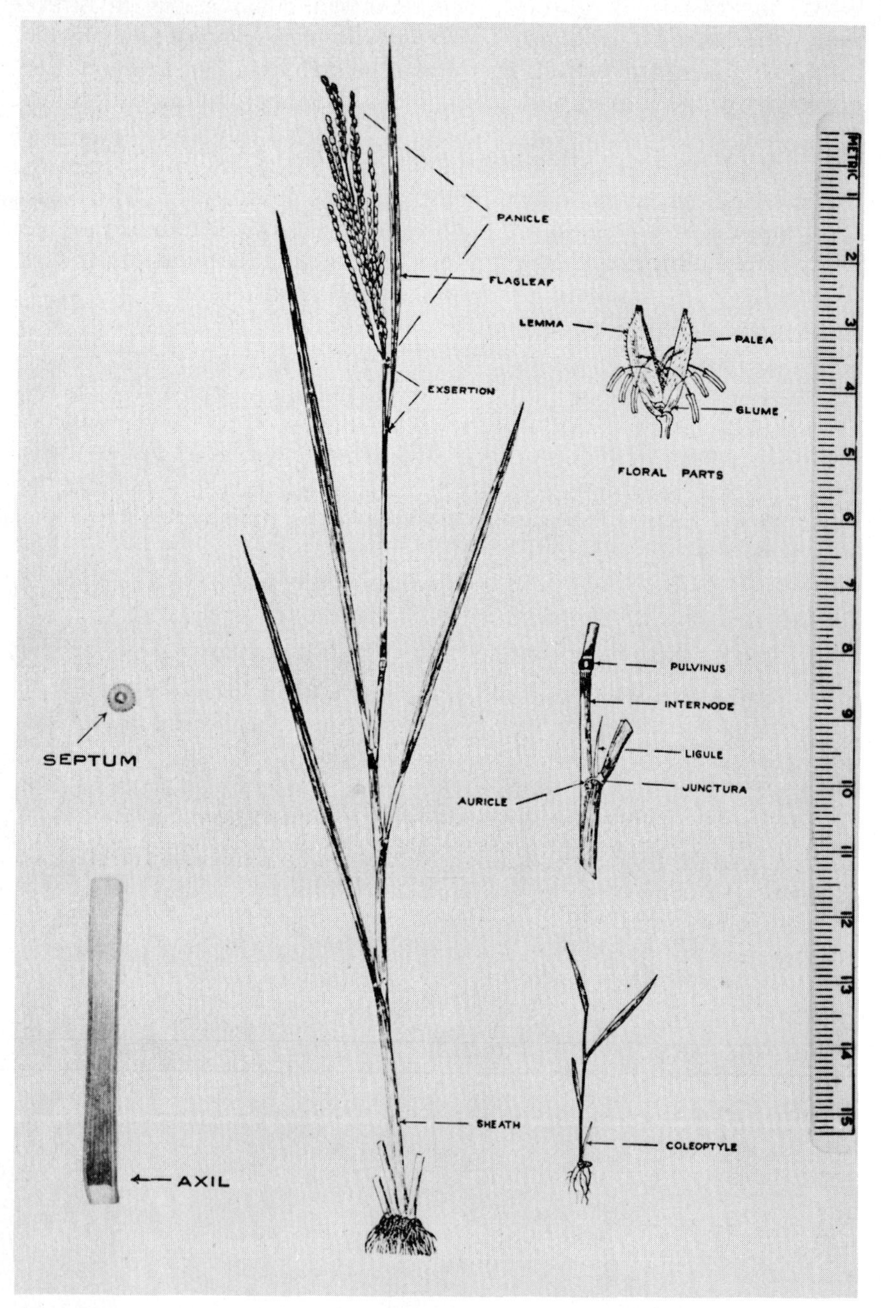

Figure 19–4. Parts of the rice plant.

Figure 19–5. Panicle of short-grain rice (*right*). Section of culm, leaf, and sheath (*left*). Note the long ligule characteristic of the rice plant.

remain open 20 minutes to 3 hours.[24, 36] Blooming starts early in the day when temperatures are high, but slows down materially when the sky becomes cloudy causing a drop in temperature.[1]

Rice is normally self-pollinated, but up to 3 or 4 per cent natural crossing may occur, although less than 0.5 per cent is the average. Varieties differ materially in the extent of natural crossing.[10, 22] Much more occurs in the South than under the higher temperature and lower humidity conditions prevailing in California.

RICE TYPES

Thousands of varieties of rice are known in the Orient but comparatively few are grown in the United States.

The varieties grown include short-, medium-, and long-grain types. The average length of the hulled kernel of the three grain types is: short-grain, 5.5 millimeters; medium-grain, 6.6 millimeters; and long-

Figure 19–6. Spikelets (*top*), seeds (*middle*), and kernels (*bottom*) of four types of rice: (1) slender grain, (2) long grain, (3) medium grain, and (4) short grain.

grain, 7 to 8 millimeters. Early maturing varieties in the southern states require about 120 to 129 days from seeding to maturity, the mid-season varieties about 130 to 139 days, and the late varieties 140 days or more. Several varieties of different maturities may be grown advantageously on one farm to distribute harvest labor over a longer period.

The short-grain rices, classed as the *Japonica* type, have short stiff stalks that resist lodging. They also respond to heavy applications of nitrogen. The long-grain rices of the *Indica* type usually have taller and weaker stalks. The crop often lodges under heavy nitrogen fertilization. First-generation crosses between the *Japonica* and *Indica* types usually are partially sterile. The Ponlai (temperature- and light-insensitive) rices grown in Taiwan were derived from such crosses. The Bulu rices of Indonesia have large kernels and long, but stiff culms.

The short-grain rices have prevailed in Japan, and improved short-grain IR varieties released since 1965 are now widely grown in Asia and other parts of the world. Only about 10 per cent of the United States rice crop is short-grain and nearly all of that is grown in California. Long-grain rices grown in the southern states account for 50 per cent of the production. The remaining 40 per cent is medium-grain grown in the South and also in California.

The endosperm starch may be nonglutenous (common), or glutenous. Glutenous rices have the amylopectin type of starch suitable for preparing special delicacies. Aromatic (scented) rices exude a distinct mouse-like odor during cooking which is prized by some

Asian consumers. The glutenous and aromatic types are grown in the United States only on small acreages for special consumer groups.

All rices are short-day plants but varieties differ greatly in their sensitivity to photoperiods. The more sensitive varieties vary considerably in their length of vegetative growth before flowering when seeded in different months or at different latitudes.

ROTATIONS

Rice yields become stabilized at low-yield levels when grown continuously on the same land. Weeds also become abundant after two or more rice crops. Heavy poorly drained rice soils often are unsuited to other crops which otherwise would be grown in rotation with rice. Consequently, in the Gulf Coast region as well as in California, rice usually is grown on the same land for two or three years. The rice land is then seeded to pasture crops and grazed for two to three years. Grasses, clovers, and lespedeza are often sown in the Gulf Coast area, while Ladino clover, burclover, or strawberry clover may be planted in California. This practice results in higher subsequent rice yields, improved soil structure, higher soil fertility, a lower population of aquatic weeds, and better livestock gains than when the land is pastured without seeding or fertilization.[12] The aquatic weeds on some California rice lands are partly controlled by an occasional year of clean fallow. A *water fallow* sometimes is practiced in Arkansas by maintaining a depth of water of 2 to 4 feet in which fish may be grown for two years. This practice may induce excessive accumulation of ammonium nitrogen, in which case corn or sorghum should precede a rice crop. Crops frequently grown in rotation with rice on the better-drained soils in California include safflower, grain sorghum, wheat, and barley. Purple vetch seed for a green-manure crop may be broadcast on the rice field just before draining the land for harvest. Most of the rice lands in Arkansas, Mississippi, and Missouri are well suited to other crops such as cotton, soybeans, corn, oats, lespedeza, and winter legumes. These crops often are grown in two-year to five-year rotations, in which rice often follows lespedeza or a winter legume for green manure. Two successive crops of rice may be grown in the longer rotations.[55] Rice responds well to green manures.

FERTILIZERS

Dwarf types of rice may respond to applications up to 130 kilos or more per hectare of nitrogen (N), preferably in an ammonium form. Tall, weak-stalked varieties may lodge with consequent reduced yields from applications exceeding 35 to 90 kilos of nitrogen per hectare. Ammonium sulfate is the most popular fertilizer, but ammonium phosphate, anhydrous or liquid ammonia, urea, and cyanamid also are applied to rice fields. Nitrate nitrogen is available to rice before the field is flooded, but it is of limited benefit in submerged soils. It may then be reduced to the unavailable nitrite form as a result of oxygen deficiency. Nitrates also are subject to leaching. Nitrates often are only two-thirds to three-fourths as effective as ammonium forms. Except in excessive amounts, nitrogen fertilizer causes rice to flower earlier. Much of the nitrogen may be applied as a top dressing 30 to 70 days after the rice is seeded. On many soils each 2 to 4 kilos of nitrogen increase the yield of rice by 45 kilos. Heavy applications of nitrogen often are profitable on California soils unless the rice crop follows a legume green manure. Phosphorus and potassium fertilizers usually are not required there. Algae growing in the water on rice fields fix some atmospheric nitrogen.

On some Gulf Coast soils rice responds to 23 to 45 kilos per hectare of P_2O_5 and 23 to 35 kilos of K_2O with increased yields, especially where large amounts of ammonium nitrogen are applied.[11]

In Louisiana, rice yields were increased with phosphorus applications on most soils, but also with potassium on some soils. Response to these elements as well as to zinc or iron[50] tended to be greater where soil tests showed them to be deficient.[46]

CULTURAL METHODS

The land for rice production should be plowed 10 to 15 centimeters (4 to 6 inches) deep in the fall, winter, or spring. Plowing as deep as 9 inches sometimes is helpful. Seedbed preparation for rice should include destruction of weeds as well as sufficient tillage to produce a mellow, firm surface layer. Clods may slacken after flooding, which buries rice seeds too deep in fields that are drilled or broadcast before flooding. A seedbed with small clods is satisfactory where rice is broadcast in the water.

Levees are constructed on contours by use of large diking machines. The land between the levees is leveled with land planes to a gradient of seldom more than 6 to 7 centimeters ($2\frac{1}{2}$ inches). Levelling can be accomplished effectively when the soil is flooded to a depth of 10 to 15 centimeters. The levees may be broad but still low enough for machines to pass over them, but some levees are so high and steep that combines or other implements must operate within individual checks between the levees. The use of plastic levees or asphalt-coated levees permits the field to be harvested as a unit.[37]

Seeding Practices

Seed that contains red rice should be avoided. Red rice, which belongs to the same species as cultivated rice, is sometimes grown as a crop in other countries for local use.[20] Wild red rice is a serious pest in rice fields in the United States because the seeds shatter out early and contaminate the fields. The seed may live over in the soil for several years. Red-rice admixture spoils the appearance of milled white rice.

Rice is sown with airplane seeders, ground broadcast seeders, or with grain drills. The crop usually is sown between April 1 and May 15 in the southern states, but between April 15 and June 1 in California. Seed sown too early may rot because of low temperatures. Rice can be sown over a comparatively long period without serious reduction in yields, but early seeding is desirable.[23] Under very favorable conditions, 80 pounds of seed per acre (90 kilos per hectare) usually is sufficient to give good stands. Under ordinary conditions, 90 to 100 pounds of seed sown with a drill, or 125 to 160 pounds sown broadcast by airplane, will prove adequate. Rice should be drilled 1 to 2 inches deep on good seedbeds, with the shallower depth on heavy soils.

Most of the rice seed in California and also much of that in the southern states, is now broadcast into water by use of an airplane seeder (Figure 19–7). The field is flooded to a depth of 4 to 6 inches before seeding. Such seeding in water helps to control watergrass as well as other weeds on old rice land, but drilling is preferable on land that has not been cropped to rice for many years. The seed for airplane seeding is soaked in burlap bags for 18 to 24 hours and drained for 24 to 48 hours before sowing. A half-pound of sodium hypochlorite ($NaOCl$) per 100 gallons of steep water checks seeding diseases, and deactivates germination inhibitors present in the rice hull. The soaked seeds sink down to the soil, whereas many dry seeds would float on the water and drift with the wind before sinking.

Rice is able to germinate with a lower supply of oxygen than is the case with other cereal seeds. Oxygen is released from the seed by fermentation.[57] This permits rice seedlings to emerge through cool water 6 inches or even more in depth. When the submerged rice seed is covered also with 1 inch or more of soil, the supply of oxygen is then insufficient for germination, so drilled rice fields are irrigated and then drained until after the crop is up.[28, 29]

Irrigation

The water requirement of the rice crop usually varies from 24 to 60 acre inches, 6 to 20 inches of which is supplied by rainfall during the growing season in the southern states. More water, or up to 108 inches, is required for fields with permeable subsoils.[4]

In the southern states, the soil usually contains sufficient moisture for seed germination as well as for seedling growth on drilled fields. In California, drilled rice is usually irrigated after seeding, then drained, again irrigated, and drained at intervals until about 30 days

Figure 19–7. Soaked rice seed broadcast from the airplane is falling into the water that floods the field.

after the rice seedlings emerge. The land is submerged, or flooded, to a depth of 1 to 2 inches when the young rice plants reach a height of 6 to 8 inches. As the plants grow taller, this depth is gradually increased until it reaches 4 to 6 inches. The water level is maintained at this depth for 60 to 90 days, or until the land is drained prior to harvesting. Levees with overflow outlets are built to maintain the water at the proper level.

After final submergence, especially where the seed was broadcast in the water, the fields may be kept flooded until they are drained shortly before harvest. However, occasionally they are drained temporarily for pest control, or to facilitate top dressing with nitrogen fertilizers. The final drainage to permit harvesting occurs about 10 to 15 days before the rice is fully mature. At that stage the plants are fully headed, the panicles turned down, and ripening started in the upper portion.

Some of the water from wells, or from mountain streams, has a temperature of 65° F. (18° C.) or lower. Germination of the seed of most rice varieties is retarded when the water temperature is below 70° F. (22° C.). The most favorable water temperatures for rice growth are 77° to 84° F. (25° C to 29° C.). Often the growth of the rice plant is delayed in the checks at the upper side where the water first enters the field. Shallow basins sometimes are constructed to allow the water to become warmer before it reaches the rice field. Water temperatures that exceed 85° F. (29° C.). cause poor root development in rice.[55]

Water that contains more than 600 parts per million (ppm.) of soluble salts is injurious to the rice plant, especially in the germination, early tillering, and flowering stages.[34, 45] Water from coastal bayous used to irrigate rice may become excessively salty by the invasion of sea water during periods of low rainfall.

Weed Control

Weeds in rice fields include red rice (*Oryza sativa*), barnyardgrass or watergrass (*Echinochloa crusgalli*), Mexican weed (*Caperonia palustris*), tall indigo or coffee weed (*Sesbania exaltata*), redweed (*Melochia corchorifolia*), and spike rush (*Eleocharia* species). Most of the weeds can be suppressed by suitable recommended herbicides. Seeding rice in the water, as described previously, is helpful in the control of barnyardgrass as well as some other weeds. Barnyardgrass (watergrass) is controlled in California by continuous submergence. Barnyardgrass plants are unable to emerge through 8 inches of water, and only a few plants emerge through a 6-inch submergence through which rice seedlings *stretch* to the surface.[27, 29] Crop rotation also is effective.

Excess scum due to blue-green algae is controlled by adding copper sulfate, or 8 to 10 pounds per acre of Delrad 50 to the irrigation water.[42]

Harvesting

Most of the rice crop in the United States is harvested with self-propelled combines equipped with large rubber tires with mud lugs, or with crawler tracks, or halftracks (Figure 19–8). It is then dried promptly with artificial heat.

In order to produce maximum yields of high milling quality, rice should be harvested when the moisture content of the grain of standing rice has dropped to about 18 to 27 per cent.[41, 53] At that moisture, the kernels in the lower portion of the heads are in the hard-dough stage, while those in the upper portions of the head are ripe. Few chalky kernels (a result of immaturity) are found in rice harvested at this stage. Increased shattering and checking of the grains will occur in some varieties when harvested after this stage. During the drying, the air temperature should be kept below 110° F. when the drying is to be completed in one operation. It usually is dried in several stages, being held to allow the grain to reach equilibrium between drying operations. After the grain is partly dried, temperatures as high as 130° F. are not injurious[41] (Figure 19–9).

Other driers include mixing and batch types, and farm bins with portable driers, and also pothole bag driers for seed rice. Aeration between batch dryings is advisable. Drying with unheated air sometimes is successful but heating units should be available for high moisture grain or high humidity conditions.[5]

Rice harvested with a binder is shocked for 10 to 14 days and threshed when the moisture content of the grain is reduced to about 14 per cent.

About 50 to 66 per cent of the 1971 Texas rice crop was harvested a second time (a ratoon or stubble crop) in the same season. This is possible when an early maturing variety is sown early. An application of nitrogen immediately after the first crop is harvested increases the yield appreciably. The yield of the ratoon crop which matures in a shorter season is considerably below that of the first crop.[15]

CULTURAL METHODS IN THE ORIENT

A large part of the Oriental rice that enters commercial trade is grown by transplanting.[17] The reasons for transplanting include better utilization of land by growing two different crops in a year, the saving of irrigation water, and better weed control. In hand transplanting, the rows are spaced for convenient weeding, harvesting, and fertilizer application, all of which also are done by hand. The seed is

Figure 19–8. The self-propelled combine with a pick-up reel is harvesting the rice on the side of a broad levee. [Courtesy J. I. Case Co.]

Figure 19–9. A column-type rice drier: (1) cleaning machinery, (2) columns, (3) dump pit, and (4) hot-air fans.

sown in beds while the fields are still occupied by other crops. When the seedlings are about 30 to 50 days old, preferably 30 days in Peru,[49] they are transplanted in fields (paddies) that have been stirred into a thin soupy mud into which clumps of three or four seedlings can be pushed easily. Experiments show no advantage in yields from transplanting as compared with ordinary seeding methods when rice is sown in fields or beds at the same time, with the weeds controlled to the same degree.[3] Much of the rice in the Orient is harvested by hand and threshed by hand or with small threshers. Considerable quantities of rice for home use are pounded out in a wooden mortar. Rice straw is used for weaving mats, bags, hats, and baskets. It is also used for drinking straws and for making paper and cardboard.

MILLING

The rice kernel enclosed by the hulls as it leaves the thresher is known as rough rice or paddy. It weighs about 45 pounds per bushel. Rough rice is used for seed, but is milled for food. Damaged or low-quality rough rice is used for livestock feed. When milled, rough rice yields about 64 per cent whole and broken kernels, 13 per cent bran, 3 to 4 per cent polish, and 20 per cent hulls. When somewhat undermilled the yields of head rice range from 65 to 70 per cent.

Rough rice is first fanned and screened to prepare it for milling. It is then conveyed to the hulling stones (or sheller) where most of the hulls are detached. The mixture of hulled rice, rough rice, and hulls is then aspirated to remove the detached hulls. The remaining portion of the mixture is then passed to paddy machines (separators) where rough rice is separated from the hulled kernels. The rough rice from the paddy machines is returned to the hulling stones for removal of the hulls, which are used for fuel or packing. The hulled kernels (brown rice) (Figure 19–10), are conveyed to the hullers where the germ, along with a part of the bran layer, is removed by friction. The rice is then passed to a second set of hullers, and in some mills to a pearling cone also. From these machines it is passed to the bran reel for separation of the detached bran from the rice. The rice kernels are then conveyed to the brush for polishing. In this process, more of the bran as well as some of the starch cells are rubbed off and screened out. The light-brown powder from the screens, known as rice polish, is used for feed. The polished rice is next conveyed to a revolving cylinder or trumble where it is steamed. Glucose and talc are also applied when the rice is to be coated. The milled rice is then separated into grades and bagged. A bag containing 100 pounds is called a *pocket*.

Parboiling consists of soaking and steaming the rough rice by various procedures until the endosperm starch is partly gelatinized, and then drying and milling the grain. The advantages of parboiled (or converted) rice are that it is broken less in milling, giving a higher

Figure 19–10. (*Left*) Paddy or rough rice. (*Center*) Brown rice. (*Right*) Milled rice.

yield of head rice and helping the kernels remain whole during long cooking; the rice keeps better (probably because of partial sterilization); and more of the water-soluble vitamins are preserved in the kernel. The B vitamins in the bran and hulls are absorbed by the endosperm from the water used in soaking the rice.[31, 48]

Whole grains as well as nearly whole grains of milled rice are called head rice.[20] Rice that contains mostly half and three-quarter kernels is called second head. That composed mostly of halves and thirds of kernels is called screenings. Finely broken rice, called brewers rice, is used as a starchy adjunct in the brewing and distilling industries or for feed. Rice flour is a by-product sifted or ground from the coarser milled particles and used in foods and face powders.

In the milling of long-grained rice, yields of head and second head of 50 per cent of the weight of rough rice are common. In short-grained and medium-grained types, the yields of head rice are close to 60 per cent.

The greater breakage of long-grained rice was reported by Thomas Jefferson in 1787. Head rice consists of 11 to 12 per cent moisture, about 77 per cent starch, 2 per cent pentosans, 7 to 8 per cent protein, and fractional percentages of fat, minerals, and crude fiber. Like other refined cereal products it is incomplete from the dietary standpoint. This fact is of little consequence in the United States where the average yearly per capita consumption is only 7 pounds, but it is important in parts of the Orient where rice comprises 40 to 80 per cent of the food calories, and annual consumption ranges from 125 to more than 400 pounds per capita. Those who subsist largely on well-milled rice sometimes develop beriberi, a nerve-damaging nutritional disease, because of a deficiency of thiamin (vitamin B_1) in the diet when the rice is not properly supplemented with such foods as meat, fish, soybeans, and vegetables. Brown rice is richer than well milled (polished) rice in thiamin but it tends to become rancid in storage, while steady consumption of brown rice by heavy rice eaters often leads to digestive disturbances. However, undermilled rice as eaten by many Asiatics contains thiamin. Much of the milled rice sold in the United States is enriched by the addition of vitamins and minerals before packaging.

Rice of high milling quality is free from checks, i.e., cleavage lines from stresses resulting from rapid drying; free from pecky grains resulting from insect and disease injury; and free from starchy (opaque) spots caused by immaturity or poor environment, all of which increase the breakage and thus reduce the yield of head rice in milling. Rice of high cooking quality, leaves the kernels flaky, whole, and separated after boiling. Other varieties when cooked are whole but somewhat sticky. However, such rice is highly acceptable to numerous consumers, including those in Europe, the West Indies, Japan, and some other parts of the Orient. Rice grains of low cooking quality break down during cooking. Rice, when cooked, absorbs 3 to 5 times its weight of water. Long-grained rice has a high water absorption.[48]

For use in canned soups, only parboiled grains of high quality varieties are acceptable, because other types are disintegrated by the steam-pressure sterilization incident to canning. A variety of this type has been grown in the United States only on small acreage.

DISEASES

The principal diseases of rice are seedling blights, brown leaf spot, narrow brown leaf spot, blast, and stem rot.[5, 9, 43]

Seedling Blight

Seedling blights, caused by *Sclerotium rolfsii* and other fungi, attack the young seedlings in warm weather. Affected seedlings are slightly discolored. Later, fungus bodies are found on the lower portions of the seedlings. Severely affected seedlings are killed. Rice sown early seems to be more subject to injury by Sclerotium blight than that sown late. Immediate submergence will check the disease. Some fungicides provide considerable protection against seedling blights.

Brown Leaf Spot

Brown leaf spot, caused by *Helminthosporium oryzae* is one of the most serious diseases in Louisiana, Texas, and Arkansas. This fungus attacks the rice seedlings, leaves, hulls, and kernels. It may cause seedling blight until the plants attain a height of 4 inches. Brownish discolorations first appear on the sheaths between the germinated seed and the soil surface, or on the roots. Small circular or elongated reddish-brown spots appear on the leaves, causing them to dry up on severely affected plants. Spots also appear on the hulls and on the kernels. Resistant varieties may be the only satisfactory control. Seed treatments may protect against the seedling blight.

Narrow Brown Leaf Spot

Narrow brown leaf spot, caused by *Cercospora oryzae* is one of the most widespread leaf-spot diseases in the southern states. The spots on the leaves are long and narrow. The disease usually appears late in August and in September. The injury to plants is due principally to the reduction of the leaf area. The most satisfactory control is by the use of resistant or early varieties.[2]

Blast

Rice blast or rotten neck, caused by *Piricularia oryzae*, is a disease long known in many countries; it blights the panicles and rots the stems so that they break over. The fungus lives over on crabgrass and rice straw. The only known remedies are the use of strong-stemmed tolerant varieties and avoidance of heavy nitrogen fertilization, which encourages the disease. It is most severe on new rice land and under hot humid conditions. Certain varieties are resistant to some races of the fungus.

Stem Rot

Stem rot, caused by the fungus *Magnaparthe salvinii*, is an important and widespread disease of rice that causes the plants to lodge. Early seeding of quick-maturing tolerant varieties reduces the damage.

Straighthead

Straighthead is a physiological disorder. Heads of affected plants remain erect and fail to set seed. Plants affected with straighthead have dark-green leaves that remain green after the normal plants have matured. Straighthead is most prevalent on long-submerged heavy soils of new rice land or on land uncropped to rice for several years, on which a heavy growth of weeds has been plowed under. The disease often can be controlled by draining the land before the rice is in the boot. After the soil has dried on the surface, the land should be submerged again. Several varieties are resistant.

Other Diseases

White tip, which is caused by a seed-borne nematode, *Aphelchoides besseyi*, produces white leaf tips under certain soil conditions. It is controlled by planting nematode-free seed of resistant varieties and by water seeding.

The hoja blanca ("white leaf") disease causes a whitening or striping of the leaves and stunted, unproductive plants that tiller profusely. The disease is caused by a virus that is transmitted by a plant hopper, *Sogata orisicola.* Resistant varieties are available.

Rice is attacked also by sheath spot, sheath rots, kernel spots, kernel smut, leaf smut, and other diseases.[5]

INSECTS

Insects cause little injury to rice in California.[17]

Rice Stink Bug

The rice stink bug (*Oebalus pugnaz*) sucks the contents from rice kernels during the milk stage, leaving only an empty seed coat when all the milk is withdrawn and causing pecky rice when the milk is only partly consumed. Pecky rice is discolored and is likely to be broken in milling. Control measures include early mowing of and winter burning of the coarse grasses in which the insect hibernates.

Stalk Borers

Two species of borers, the sugarcane borer (*Diatraea saccharalis*) and the rice-stalk borer (*Chilo plejadellus*), damage rice in Louisiana and Texas. They tunnel inside the rice culms, eating the inner parts, which often causes the panicles to turn white and produce no grain, a condition called whitehead. In other culms, grain formation is not completely inhibited but some of the culms or panicles break off when the culm is badly girdled. The borers prefer large stems. The borers live over winter in the rice stubble. They are killed in their winter quarters by heavy grazing of the stubble, by mashing down of the stubble by dragging, and by flooding the stubble fields by closing up drainage outlets to hold water from winter rains.[5]

Other Insects

The rice water weevil (*Lissorhoptrus oryzaphilus*) can be controlled with insecticides. Before the fields are submerged, the rice plants sometimes are damaged by the fall armyworm (*Laphygna frugiperda*), southern corn rootworm (*Diabrotica duodecimpunctata*), sugarcane beetle (*Euetheola rugiceps*) and the chinch bug (*Blissus leucopterus*). Flooding destroys these insects. Threshed rice is attacked by the rice weevil and other insects that attack stored grains.[7]

Other Pests

In California the tadpole shrimp (*Triops longicaudatus*) digs up and cuts off the leaves of rice seedlings which prevents their emergence through the water. The shrimp is controlled by applying chemicals such as 10 pounds of granular copper sulfate per acre, by airplane about two weeks after the field is submerged.

Additional pests are waterfowl and other birds. Muskrats, nutria, and crayfish burrow through dikes and ditch banks.

WILD RICE

Wild rice (*Zizania aquatica*), also called water oats formerly was seldom sown except in marshes where it was desired to attract waterfowl. By 1972, however, 20,000 acres of wild rice paddies had been established in Minnesota. It is a native aquatic grass found from Lake Winnipeg southward to the Gulf of Mexico and eastward to the Atlantic Ocean.[54] It grows in lakes as well as in tidal rivers and bays where the water does not contain enough salt to be tasted. A wide range of growth types adapted to local day lengths are found at different latitudes. Wild rice grows mostly in water 2 to 4 feet deep. On tidal lands the water may fluctuate through depths from 0 to 3 feet.

Figure 19–11. (1) Wild rice (*Zizania aquatica*) plants growing in water; (2) panicle showing appressed branches of pistillate florets above and spreading branches of staminate florets below; (3) natural-sized unhulled seeds; and (4) hulled seeds.

The seeds shatter quickly when ripe and fall to the muddy bottom where they remain until spring, when germination occurs. The plants can grow up through the water at depths up to 6 feet. The seed fails to germinate when it is allowed to dry. Consequently, it is kept covered with cool water when stored over winter for planting. The plants grow to a height of 3 to 11 feet and are topped by large open panicles that bear pistillate flowers in the upper half and staminate flowers below. This arrangement favors cross-pollination. The seed (Figure 19–11) is harvested mostly by northern Indian tribes and others who tap the tops with sticks and knock the seed into a boat or canoe. Fully satisfactory machines for efficient harvesting are not yet available. It is then partly dried on the ground, parched in an open vessel, and often hulled by pounding or trampling. Because of the labor required to gather and process the grain, wild rice is sold as a high-priced delicacy.

REFERENCES

1. Adair, C. R. "Studies on blooming in rice," *J. Am. Soc. Agron.*, 26:965–973. 1934.
2. Adair, C. R. "Inheritance in rice of reaction to *Helminthosporium oryzeae* and *Cercospora oryzeae*," *USDA Tech. Bul. 772.* 1941.
3. Adair, C. R., and others. "Comparative yields of transplanted and direct sown rice," *J. Am. Soc. Agron.*, 34(2):129–237. 1942.
4. Adair, C. R., and Engler, K. "The irrigation and culture of rice," in *Water*, USDA Yearbook, 1955, pp. 389–394.
5. Adair, C. R., and others. "Rice in the United States: Varieties and production," *USDA ARS Handbk. 289*, pp. 1–154. 1973 (Rev.).
6. Anonymous. "The mineral nutrition of the rice plant," in *Proc. Symposium at the International Rice Research Inst., Feb., 1964*, Johns Hopkins Press, Baltimore, Md., pp. 1–499. 1965.
7. Anonymous. "Controlling insect pests of stored rice," *USDA ARS Handbk. 129.* 1971.
8. Ashrafuzzaman, M. H., and Fredrickson, R. A. "Chemical control of rice blast," *Tex. Agr. Exp. Sta. MP 923*, pp. 1–12. 1969.
9. Atkins, J. G. "Rice diseases," *USDA Farmers Bul. 2120.* 1958.
10. Beachell, H. M., and others. "Extent of natural crossing in rice," *J. Am. Soc. Agron.*, 30:743–753. 1938.
11. Beacher, R. L. "Rice fertilization: results of tests from 1946 through 1951," *Ark. Agr. Exp. Sta. Bul. 552.* 1952.
12. Black, D. E., and Walker, R. K. "The value of pastures in rotation with rice," *La. Agr. Exp. Sta. Bul. 498.* 1955.
13. Chapman, A. L., and Mikkelsen, D. S. "Effect of dissolved oxygen supply on seedling establishment of water-sown rice," *Crop Sci.*, 2:391–395. 1963.
14. Chandler, R. F., Jr. "Dwarf rice—a giant in tropical Asia," *USDA Yearbook*, 1968, pp. 252–255.

14a. Chandler, R. F., Jr. *Rice Breeding*, International Rice Rsh. Inst. Manila. pp. 1–738. 1972.
15. Craigmiles, J. P., and others. "Rice research in Texas, 1971," *Consolidated Progress Rept.*, pp. 3092–3105. 1972.
16. Davis, L. L., and Jones, J. W. "Fertilizer experiments with rice in California," *USDA Tech. Bul. 718*, pp. 1–21. 1940.
17. Douglas, W. A., and Ingram, J. W. "Rice-field insects," *USDA Circ. 632*, pp. 1–32. 1942.
18. Efferson, J. N. *The Production and Marketing of Rice*, Rice Journal, New Orleans, La., pp. 1–534. 1952.
19. Grist, D. H. *Rice*, Longmans, Green, London, pp. 1–331. 1953.
20. Houston, D. F., Ed. *Rice, Chemistry and Technology*. Am. Assn. Cereal Chem., St. Paul., Minn., pp. 1–300. 1974.
21. Jenkins, J. M. "Effect of seeding date on the length of the growing period of rice," *La. Agr. Exp. Sta. Bul. 277*, pp. 1–7. 1936.
22. Jodon, N. E. "Occurrence and importance of natural crossing in rice," *Rice J.*, 62(8):8 and 10. 1959.
23. Jodon, N. E. and McIbrath, W. A. "Response of rice to time of seeding in Louisiana," *La. Agr. Exp. Sta. Bul. 649*, pp. 1–27. 1971.
24. Jones, J. W. "Observations on the time of blooming of rice flowers," *J. Am. Soc. Agron.*, 16:665–670. 1924.
25. Jones, J. W. "Branching of rice plants," *J. Am. Soc. Agron.*, 17:619–623. 1925.
26. Jones, J. W. "Experiments in rice culture at the Biggs Rice Field Station in California," *USDA Bul. 1387*. 1926.
27. Jones, J. W. "Germination of rice seed as affected by temperature, fungicides, and age," *J. Am. Soc. Agron.*, 18:576–592. 1926.
28. Jones, J. W. "Effect of reduced oxygen pressure on rice germination," *J. Am. Soc. Agron.*, 25(1):69–81. 1933.
29. Jones, J. W. "Effect of depth of submergence on control of barnyard grass and yield of rice grown in pots," *J. Am. Soc. Agron.*, 25:578–583. 1933.
30. Jones, J. W. "Improvement in rice," in *USDA Yearbook, 1936*, pp. 415–454.
31. Jones, J. W., Zeleny, L., and Taylor, J. W. "Effect of parboiling and related treatments on the milling, nutritional and cooking quality of rice," *USDA Circ. 752*. 1946.
32. Jones, J. W., Davis, L. L., and Williams, A. H. "Rice culture in California," *USDA Farmers Bul. 2022*. 1950.
33. Jones, J. W., and others. "Rice production in the Southern States," *USDA Farmers Bul. 2043*. 1952.
34. Kapp, L. C. "The effect of common salt on rice production," *Ark. Agr. Exp. Sta. Bul. 465*. 1947.
35. Lange, W. H. and Miller, M. D., Eds. "Insect and other animal pests of rice," *Calif. Agr. Exp. Sta. Circ. 555*, pp. 1–32. 1970.
36. Laude, H. H., and Stansel, R. H. "Time and rate of blooming in rice," *J. Am. Soc. Agron.*, 19:781–787. 1927.
37. Lewis, D. C., and others. "New rice levee concept," *Rice J.*, 65:6–15. 1962.

38. Matsubayashi, M., Ed. *Theory and Practice of Growing Rice,* Fuji Pub. Co., Tokyo, Japan, pp. 1–527. 1967.
39. Matsushima, S. *Crop Science in Rice—Theory of Yield Determination and Its Application,* Fuji Pub. Co., Tokyo, Japan, pp. 1–365. 1967.
40. McNeal, X. "Artificial drying of combined rice," *Ark. Agr. Exp. Sta. Bul. 487.* 1949.
41. McNeal, X. "When to harvest rice for best milling quality and germination," *Ark. Agr. Exp. Sta. Bul. 504.* 1950.
42. Olsen, K. L. "Scum control in rice fields and irrigation canals," *Ark. Agr. Exp. Sta. Rpt. Ser. 69.* 1957.
43. Padwick, G. W. *Manual of Rice Diseases,* Commonwealth Mycol. Inst., Kew, Surrey, England, pp. 1–198. 1950.
44. Padwick, G. W. *Manual of Rice Diseases,* Commonwealth Mycol. Inst., Kew, Surrey, England, pp. 1–198. 1950.
45. Pearson, G. A., and Bernstein, L. "Salinity effect at several growth stages in rice," *Agron. J.,* 51:654–657. 1959.
46. Peterson, F. J., Sturgis, M. B., and Miears, R. J. "Response of rice to fertilizer phosphorus and potassium," *La. Agr. Exp. Sta. Bul. 569.* 1963.
47. Ramiah, K. "Rice breeding and genetics," *Indian Council Agr. Res. Science Monograph 19,* New Delhi. 1953.
48. Rao, R. M. R., and others. "Rice processing effects on milling yields, protein content and cooking qualities," *La. Agr. Exp. Sta. Bul. 663,* pp. 1–51. 1972.
49. Sanchez, P. A., and Larrea, N. "Influence of seedling age at transplanting on rice performance," *Agron. J.,* 64(6):828–832. 1972.
50. Sedberry, J. E., Jr., and others. "Effects of zinc and other elements on the yield of rice, and nutrient content of rice plants," *La. Agr. Exp. Sta. Bul. 653,* pp. 1–8. 1971.
51. Smith, R. J., Jr., and Shaw, W. C. "Weeds and their control in rice production," *USDA ARS Handbk. 292,* pp. 1–66. 1966.
52. Smith, W. D. "Handling rough rice to produce high grades," *USDA Farmers Bul. 1420,* pp. 1–21. 1940.
53. Smith, W. D., and others. "Effect of date of harvest on yield and milling quality of rice," *USDA Circ. 484,* pp. 1–20. 1938.
54. Steeves, T. A. "Wild rice—Indian food and modern delicacy," *Econ. Bot.,* 6:107–143. 1952.
55. Sturgis, M. B. "Managing rice soils," in *Soil,* USDA Yearbook, 1957, pp. 658–662.
56. Tanaka A., and others. "Growth habit of the rice plant in the tropics and its effect on nitrogen response," *IRRI Tech. Bul. 3,* pp. 1–80. 1964.
57. Taylor, D. L. "Effects of oxygen on respiration, fermentation and growth in wheat and rice," *Science,* 95:116–117. 1942.
58. Wickizer, V. D., and Bennett, M. K. *The Rice Economy of Monsoon Asia,* Food Res. Inst., Stanford U., Stanford, Cal., pp. 1–358. 1941.
59. Yeh, B., and Henderson, M. T. "Cytogenetic relationship between cultivated rice, *O. sativa* L., and five wild diploid forms of *Oryza,*" *Crop Sci.,* 1:445–450. 1961.
60. Yeh, B., and Henderson, M. T. "Cytogenetic relationships between African annual diploid species of *Oryza* and cultivated rice, *O. sativa* L.," *Crop Sci.,* 2:463–467. 1962.

Chapter 20

Millets

ECONOMIC IMPORTANCE

The millets are minor crops except in parts of Asia, Africa, and the Soviet Union.[21] The total world seed production in 1970 was approximately 31.7 million metric tons on some 53.9 million hectares, or 0.58 metric tons per hectare. The leading producing countries are China, India, and the U.S.S.R. About 40 per cent of the world production consists of pearl millet (*Pennisetum glaucum*), 24 per cent of foxtail millet (*Setaria italica*), 15 per cent of proso (*Panicum miliaceum*), 11 per cent of finger millet (*Eleusine coracana*), and the remainder of other species. Perhaps 480,000 hectares of millets are grown in the United States. About half of the acreage in the United States is pearl millet, one-fourth is proso, one-sixth is foxtail millet, and the remainder is browntop millet (*Panicum ramosum*) and Japanese barnyard millet (*Echinochloa frumentaceae*). Other millet species include Koda millet (*Paspalum scrobiculatum*), and *Echinochloa colonum, Panicum miliare, Digitaria exilis, Digitaria iburua* and teff, (*Eragrostis tef*). The genus *Eleusine* belongs to the tribe Chlorideae; the *Eragrostis* genus to the tribe Festuceae; and the other millets belong to the Paniceae tribe. The term millet thus is applied to various grass crops whose seeds are harvested for food or feed. Sorghum is called millet in many parts of Asia and Africa. Broomcorn is called broom millet in Australia.

FOXTAIL MILLET

History

Foxtail millet is one of the oldest of cultivated crops. It was grown in China as early as 2700 B.C. The crop probably had its origin in southern Asia, but its culture spread from there westward to Europe at an early date. It is grown mostly in eastern Asia.[21] Foxtail millet was rarely grown in the United States during Colonial times, but it became a rather important crop in the Central States after 1849. Its acreage decreased materially when it was largely replaced by sudangrass as a late-sown hay crop after 1915. Foxtail millets are now grown on about 200,000 acres annually, chiefly in Colorado, North Dakota, and Nebraska.

Adaptation

For its best growth, foxtail millet requires warm weather during the growing season. It matures quickly when grown in the hot summer months of the most northern states. The crop is most productive where the rainfall is fairly abundant, but a large part of the acreage is found in the semiarid region. It has a low water requirement.

Foxtail millet lacks the ability to recover after being injured by drought.[30] Because of its shallow root system it is one of the first crops to show the effects of drought. The millets succeed in localities subject to drought almost entirely through their ability to escape periods of acute drought because of their short growing season. Some early varieties will produce a hay crop in 65 to 70 days. When grown in the dry sections of New Mexico and western Texas, foxtail millet is planted in cultivated rows to permit the crop to succeed. Other crops produce more forage than millet does under irrigation and at high altitudes in Colorado.[9] A fertile loam soil is best for millet, but good drainage is essential.

Botanical Description

Foxtail millet or Italian millet (*Setaria italica*) is an annual grass with slender, erect, leafy stems. The plants vary in height from 1 to 5 feet under cultivation. The seeds are borne in a spike-like or compressed panicle (Figure 20–1). There is an involucre of one to three bristles at the base of each spikelet. The small convex seeds are enclosed in the hulls (lemma and palea). The color of the hulls differs with the variety, some of the colors being creamy white, pale yellow, orange, reddish orange, green, dark purple, or mixtures of various colors.[23] The common annual weeds, yellow foxtail (*Setaria lutescens* or *S. glauca*), green foxtail (*S. viridis*), and giant foxtail (*S. faberi*) resemble foxtail millet.

The foxtail millets are largely self-pollinated, with 0 to 10 per cent natural crossing.[19, 21, 26]

Ten or more distinct varieties of foxtail millet have been grown in the United States.[9, 15, 18] In Nebraska the hay yields of the better varieties were 75 per cent of that of sorghum.

Rotations

Foxtail millet usually is grown as a catch crop since it matures in 75 to 90 days. It can be sown when the date is too late for seeding most other crops. Some crops do not yield well after millet under dryland conditions. Winter wheat yielded 12.4 bushels per acre after corn, but

Figure 20–1. Spikes of foxtail millet varieties. (*Left to right*) Kursk, Hungarian, Common, Siberian, Turkestan, and Golden.

only 9 bushels after foxtail millet in eastern Colorado.[9] Late spring-planted crops (corn or sorghum) follow millet in rotation better than do fall or early spring-sown crops (wheat, oats, or barley). It is usually advisable to grow foxtail millet after small grain or corn.

Culture

Seedbed preparation for foxtail millet is similar to that for spring-sown small grains. A firm seedbed is essential because of the small size of millet seeds. Weeds should be controlled up to the time the crop is planted because the millet seedlings are small and compete poorly with weeds until they have attained some size.

Millet seed (Figure 20–2) should be planted when the soil is warm, or two to three weeks after corn-planting time. For the late dates, a 60- to 70-day season before the normal date of fall frost should be allowed. In Colorado, the crop is generally planted between May 15 and July 1 whenever there is sufficient soil moisture to germinate the seed.

Millet usually is seeded with an ordinary grain drill, since close spacing helps the crop to suppress weeds.[9] The plants should be spaced not more than 2 inches apart in the row.[19] The seed should be placed in moist soil even though it is necessary to plant it 1 inch deep or slightly deeper.

Figure 20–2. Millet seeds, enlarged and natural size: (*a*) proso, (*b*) foxtail millet, (*c*) Japanese millet, and (*d*) barnyardgrass.

Foxtail millet should be sown at the rate of 25 to 30 pounds per acre (28–33 kg./ha.). On clean land in semiarid regions 10 to 15 pounds (11–17 kg./ha.) is sufficient. For cultivated rows, 4 or 5 pounds (4.4–5.6 kg./ha.) of seed are ample.[1] The best quality of hay is obtained when it is cut just as the first heads appear, since it is more palatable then than when fully mature. For seed, the crop may be harvested with a binder and allowed to stand in the field until the seed can be rubbed from the head. The crop often is windrowed and the seed threshed with a combine with a pick-up attachment. Direct combining allows seeds to shatter before the later seeds are ripe.

Utilization

Millet hay is usually considered inferior to that of timothy. To some extent this is due to its being less palatable, but the hay contains a glucoside, called setarian, which acts as a diuretic on horses that consume it continuously as the sole roughage.[13] The hay can be fed to cattle or sheep without danger but the awns may induce sore eyes or lumpy jaws.

Foxtail millet is seldom utilized as a grain crop, except as bird seed, in the United States because the seed is slightly less palatable, with only about 83 per cent of the feeding value of proso.[9] The seed should be ground finely before being fed to livestock. It is used for food or feed in Mainland China, Manchuria, Japan, and India. Ripe heads of golden millet are sold in the United States for feeding caged birds.

Diseases

Foxtail millet is subject to several diseases which include mildew, bacterial blight, and leaf spots. Kernel smut (*Ustilago crameri*) may be controlled by seed treatment.[21]

PROSO

History

Proso (*Panicum miliaceum*) has been grown since prehistoric times as a grain crop for human food. Records of its culture in China extend back for 20 centuries or more. It is grown in the Soviet Union, Mainland China, India, the Balkan countries, and western Europe. Proso was introduced into the United States from Europe during the 18th century. It was grown sparingly along the Atlantic seaboard, but began to assume some importance in the North Central States after about 1875 when it was introduced into the Dakotas. It is now grown on about 280,000 acres annually, chiefly in North Dakota, Colorado, Nebraska, and South Dakota.[28a] Proso also is called broomcorn millet, hog millet, and Hershey millet. A related species (*Panicum milare*) is grown in Southeast Asia.

Adaptation

Proso is adapted only to regions where spring-sown small grains are fairly successful. It is a short-season plant that often requires only 60 to 65 days from seeding to maturity. Proso is readily injured by frost and is not adapted to regions where summer frosts occur. Moderately warm weather is necessary for plant growth.

Proso exhibits the efficient C_4-photosynthetic pathway like sorghum except that it exudes CO_2 during darkness.

While proso has the lowest water requirement of any grain crop, it is less resistant to drought than well-adapted varieties of other grains, largely because of its shallow root system.[22] Sometimes it is a complete failure when wheat or barley produces a fair crop. Proso yields less than corn under irrigated conditions.[12, 28]

Proso grows well on all except coarse sandy soils.[8, 25]

Botanical Description

The proso inflorescence is a large open panicle. Proso has coarse, woody, hollow stems from 12 to 48 inches (30.5–122 cm.) high but usually they are about 24 inches (61 cm.) high. The stems are round or flattened and generally about as thick at the base as a lead pencil.

The stems and leaves are covered with hairs. The stem and outer chaff are green, or sometimes yellowish or reddish green, when the seed is ripe. When threshed, most of the seed remains enclosed in the inner chaff or hull. The seed of proso is larger and not so tightly held in the hull as are those of the millets of the foxtail group. The hulls are of various shades and colors, including white, cream, yellow, red, brown, gray, striped, and black. The bran, or seed coat, of all varieties is a creamy white (Figure 20–2).

A large percentage of proso flowers are self-fertilized, but considerable cross-fertilization occurs.

Varieties

The varieties of proso are divided into three main groups on the basis of the shape of the panicle (Figure 20–3). These shapes are (1) spreading, (2) loose and one-sided, and (3) compact and erect.

Plump, bright, reddish seeded types are the most attractive for use in chick feed and bird seed mixtures.

Rotations

Proso may follow any other crop. It is a late-seeded, short-season summer catch crop. Its response to additional soil moisture is insufficient to justify its production on fallow land.[3]

Fertilizers are of limited value under semiarid conditions but 20–40 pounds (22–44 kg./ha.) of N per acre may be helpful when proso follows wheat and soil moisture is ample.[12]

Culture

Proso requires a firm seedbed that has been kept free from weeds up to seeding time. It will not germinate in a cold soil. Since the plants are easily killed by spring frosts, seeding should be delayed until the danger of frost is practically over. Most varieties require from 60 to 80 days from seeding to maturity. Generally the crop can be seeded safely from two to four weeks after corn is planted. In eastern Colorado good yields of proso were obtained when it was seeded between June 15 and July 1.[8] Earlier seedings were very weedy, while later seedings often resulted in crop failure.

Proso is sown with an ordinary grain drill. A 30- to 35-pound rate of seeding has been most satisfactory.[8, 28] Lighter seeding rates are noticeably weedier, but 16 to 20 pounds, or even less, is ample on clean semiarid soils.[14]

Proso is ready to be harvested when the seeds in the upper half of the heads are ripe. At this stage the plant is still green.

Windrowing is the best method for harvesting the crop. Proso is not adapted to direct combine harvesting because (1) it shatters soon

Figure 20–3. Panicles of one-sided proso (*center*) and compact (*right*). Proso seeds at left.

after it is ripe, (2) it lodges when left standing, and (3) the straw contains too much moisture at harvest time.

Diseases

Proso is relatively free from diseases. A bacterial stripe disease (*Phytomonas panici*) has been found on proso in Wisconsin and South Dakota.[10] Affected plants have brown water-soaked streaks on the leaves, sheaths, and culms. The long narrow lesions show numerous thin white scales of the exudate. The disease is probably seedborne. Head smut caused by the fungus *Sphacelotheca destruens* attacks proso.

Utilization

Proso is a common article of food in the Orient. The hulled seed is used in soups and the ground meal is eaten as a cooked cereal. Digestion experiments with bread made from proso flour indicate that the carbohydrates are as well utilized as in other cereals, but that only about 40 per cent of the protein is digestible.[17]

Proso is eaten readily by all kinds of livestock. It should be ground before being fed. It has about 95 per cent of the feeding value of corn or barley for hogs or poultry but only 90 per cent or less of the value of corn for fattening cattle or sheep. Proso, although occasionally cut green and cured, is a poor hay crop because of the coarse stems and hairy leaves. The chief commercial use of proso is as an ingredient of chick feeds and bird-seed mixtures. Much of the grain grown is fed locally to cattle, sheep, and hogs.

PEARL MILLET

Pearl millet *(Pennisetum glaucum)* or (*P. typhoideum* or *P. americanum*), also called cattail millet, bullrush millet, candle millet, and penicillaria, is grown in the southern United States.[24] It has become the leading temporary summer pasture crop on the southern coastal plain. In India it is grown extensively as a food grain, being usually called bajri, bajra, or cumbo. Pearl millet also is grown widely for food in countries of the Near East as well as in the savannah zones of Africa.

History

The origin of pearl millet probably was in the African savannah zone, having been grown in Africa and Asia since prehistoric times. Pearl millet was introduced in the United States at an early date but it was seldom grown until after 1875. About 575,000 acres were grown in Georgia and other southern states in 1969.

Botanical Description

Pearl millet is a tall, erect, annual grass that grows from 6 to 15 feet in height. The stems are pithy, while the leaves are long-pointed, with finely serrated margins. The plant tillers freely. The inflorescence is a dense spike-like panicle 6 to 14 inches long, and 1 inch or less-in diameter (Figure 20–4). The mature panicle is brownish in color. The spikelets are borne in fascicles of two, being surrounded by a cluster of bristles. Each spikelet, subtended by two unequal glumes, has two florets. The lower floret is staminate, being represented in many cases only by a sterile lemma. The upper floret is fertile, the caryopsis being enclosed in the lemma and palea from which it threshes free at ma-

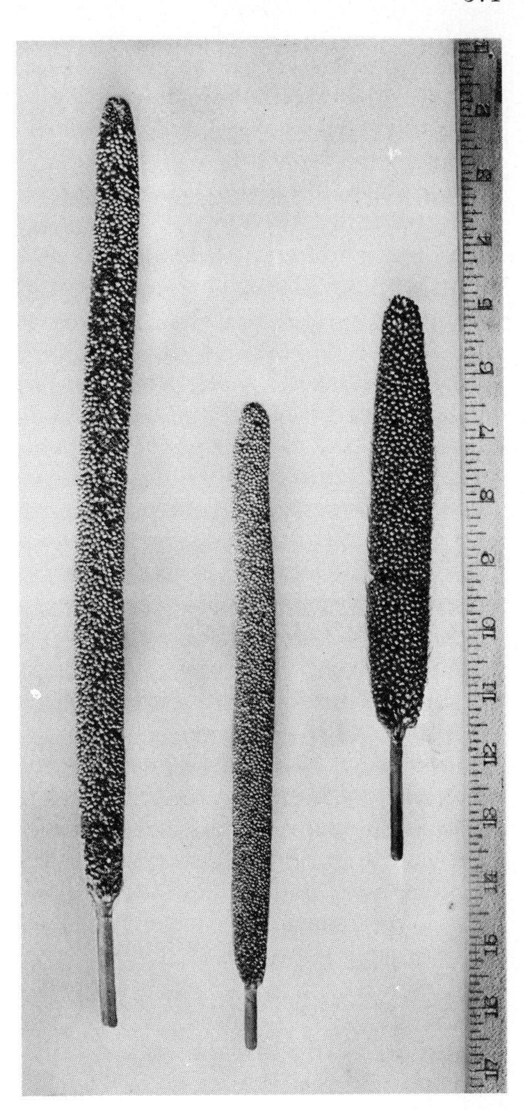

Figure 20–4. Panicles of pearl millet.

turity. In some cases, spikelets contain two fertile florets. The plant is protogynous, i.e., the stigmas appear several days before the anthers are protruded. Consequently, pearl millet is largely cross-pollinated. The spike flowers from the tip downward in four or five days. The late stigmas appear at the base of the spike shortly after the first shedding of pollen starts at the tip of the spike.[24] It is partly self-sterile. Varieties and hybrids have been developed in Georgia.[6] Many varieties are known in the Eastern Hemisphere where hybrids are being developed.

Culture

Sandy loam soils are best suited to the crop. Pearl millet is the most productive grain in extremely dry or infertile soils of India and Africa. However, it responds well to heavy fertilization.[4,5] It should be seeded after all danger of frost is past in warm soils. The date of planting varies from April in the Gulf states to May in the northern states. The seed is generally planted in drills 20 to 42 inches (51–107 cm.) apart, with the plants 4 to 6 inches (10–15 cm.) apart in the row. Close planting at a rate of 10 to 20 pounds per acre (11–22 kg./ha.) is practiced when the crop is intended for grazing or hay. From 4 to 6 pounds of seed per acre (4.4–6.6 kg./ha.) are planted for seed production. The seed should be planted at about ½ inch (1.27 cm.) deep. For the best quality of hay, pearl millet is harvested when the heads begin to appear. Hybrid pearl millet seed is produced by using a cytoplasmic male-sterile seed parent or by planting seed of a mixture of four inbred lines. Most of the hybrid seed is produced under irrigation in Arizona, New Mexico, and Texas to supply seed for growing forage in the southeastern states.[11]

Utilization

In the United States pearl millet is utilized primarily for pasture and occasionally for silage. The plant becomes woody as it approaches the flowering stage. The stover is of little value after the seed has matured. Yields as high as 16 tons of dry forage per acre have been recorded under warm favorable conditions. In Nebraska, pearl millet produced only 85 per cent as much forage as Black Amber sorgo.[16] In India, more than 2½ million tons of the seed are used annually for food. Large quantities also are consumed in Africa.

JAPANESE BARNYARD MILLET

Japanese barnyard millet, or billion-dollar grass (*Echinochloa frumentacea*) is grown on about 25,000 acres as a forage grass, mostly in Pennsylvania, New York, Iowa, Maine, and sometimes Oregon. It is grown for feed or for its edible seeds in Australia, Japan, Egypt, and other countries. It has been called poor man's millet in Korea, sava millet in India, and also Japanese millet. This millet probably originated from the common weed, barnyardgrass (*E. crusgalli*), which it closely resembles. A related species (*E. colona*), called shama millet or jungle rice, is grown in India and other countries.

Japanese barnyard millet is an annual that grows 2 to 4 feet (61 to 122 cm.) tall. The inflorescence is a panicle made up of from 5 to 15 sessile erect branches (Figure 20–5). The spikelets are brownish to purple in color, being crowded on one side of the rachis. The spikelet is subtended by two glumes within which are two florets. The lower

Figure 20–5. Plant and detached raceme of Japanese millet.

floret is staminate while the upper one is perfect. Within the glumes is the lemma of the staminate floret followed by the lemma, caryopsis, and palea of the fertile floret. The caryopsis remains enclosed in the lemma and palea. This millet differs from barnyardgrass in its more erect habit, more turgid seeds, and in being awnless.

The cultural methods are similar to those for the foxtail millets. The chief value of Japanese barnyard millet in the United States is as a pasture or soiling crop, or as food for game birds.[29] The plant is difficult to cure for hay because of its thick stems. Japanese barnyard-millet hay is palatable when cut before the plant heads, but much less so as it approaches maturity. High yields are obtained when moisture is ample.

BROWNTOP MILLET

Browntop millet (*Panicum ramosum*) is grown in the southeastern part of the United States for hay or pasture as well as to provide feed for quail, doves, and other wild game birds. Seed of this plant, a native of India, was introduced into the United States in 1915.

Browntop millet is a quick-growing annual, 2 to 4 feet (61–122 cm.) tall, with a 2- to 6-inch open panicle that ranges in color from yellow to brown. Shattering of seed occurs in sufficient amounts to reseed the crop. Consequently, it can become a weed pest on cropped land.

Browntop millet is sown from May to July. For hay or pasture, it is drilled at a rate of 10 to 20 pounds per acre (11–22 kg./ha.). When grown for seed, it is sown in cultivated rows at rates of 4 to 10 pounds per acre (4.4–11 kg./ha.). It is harvested with a combine.

Browntop millet can be sown in a disked crimson-clover sod after that crop is harvested. This millet will reseed itself thereafter, while a hard-seeded type of crimson clover also will reseed. Natural reseeding of browntop millet also occurs in game preserves.[7]

OTHER MILLETS

Teff (*Eragrostus tef*) (*abyssinica*) is grown for food on about 2 million hectares in East Africa, chiefly in Ethiopia where its production exceeds that of all other grains combined. The iron content of teff is nearly double and the calcium content is nearly 20 times that of other grains, which favors its nutritional value.[27]

Finger millet (*Eleusine coracana*) is grown extensively in India, east Africa, and elsewhere, with a total production exceeding 3 million metric tons. It also is called ragi and birdsfoot millet. The heads are forked like the feet of a bird.

Paspalum scroficulatum is grown in India as koda millet and in New Zealand as ditch millet. Little millet or kuthki samai (*Panicum miliare*) is grown in India and Pakistan. Fonio (*Digitaria* species) are grown as millets in Africa and Asia.

REFERENCES

1. Anderson, E., and Martin, J. H. "World production and consumption of millet and sorghum," *Econ. Bot.*, 3:265–288. 1949.
2. Anonymous. *The Millets. A Bibliography of the World Literature Covering the Years 1930–1963*, Scarecrow Press, Metuchen, N.J., pp. 1–154. 1967.
3. Brandon, J. F., and others. "Proso or hog millet in Colorado," *Colo. Agr. Exp. Sta. Bul. 383*. 1932.
4. Broyles, K. R., and Fribourg, H. A. "Nitrogen fertilization and cutting management of Sudangrasses and millets," *Agron. J.*, 51:277–279. 1959.
5. Burton, G. W. "The adaptability and breeding of suitable grasses for the southeastern states," in *Advances in Agronomy*, Vol. 3, Academic Press, New York, pp. 197–241. 1951.
6. Burton, G. W. "Cytoplasmic male-sterility in pearl millet (*Pennisetum glaucum*) (L.) R.Br.," *Agron. J.*, 50:230–231. 1958.
7. Craigmiles, J. P., and Elrod, J. M. "Browntop millet in Georgia," *Ga. Agr. Exp. Sta. Leaflet 14*. 1957.
8. Curtis, J. J., and others. "Proso or hog millet," *Colo. Agr. Exp. Sta. Bul. 438*. 1937.
9. Curtis, J. J., and others. "Foxtail millet in Colorado," *Colo. Agr. Exp. Sta. Bul. 461*. 1940.
10. Elliott, C. "Bacterial stripe disease of proso millet," *J. Agr. Res.*, 26:151–160. 1923.
11. Garrison, C. S. "How we get stock seed of field crops," in *Seeds*, USDA Yearbook, 1961, pp. 369–378.
12. Grabowski, P. H. "Growing proso in Nebraska," *Neb. Agr. Exp. Sta. SC 110*. 1968.
13. Hinebauch, T. D. "Feeding of millet to horses," *N. Dak. Agr. Exp. Sta. Bul. 26*. 1896.
14. Hinze, G. "Millets in Colorado," *Colo. Agr. Exp. Sta. Bul. 553 S*, pp. 1–12. 1972.
15. Hume, A. N., and Champlin, M. "Trials with millets and sorghums for grain and hay in South Dakota," *S. Dak. Agr. Exp. Sta. Bul. 135*. 1912.
16. Kiesselbach, T. A., and Anderson, A. A. "Annual forage crops," *Nebr. Agr. Exp. Sta. Bul. 206*. 1925.
17. Langworthy, C. F., and Holmes, A. D. "Experiments in the determination of the digestibility of millets," *USDA Bul. 525*. 1917.
18. Leonard, W. H., and Martin, J. H. *Cereal Crops*, Macmillan, Inc., New York, pp. 740–769. 1963.
19. Li, H. W., and Meng, C. J. "Experiments on the planting distance in varietal trials with millet, *Setaria italica*," *J. Am. Soc. Agron.*, 29:577–583. 1937.
20. Li, H. W., Meng, C. J., and Liu, T. N. "Problems in the breeding of millet (*Setaria italica*)," *J. Am. Soc. Agron.*, 27:963–970. 1935.

21. Malm, N. R. and Rachie, K. O. "The *Setaria* millets—a review of the world literature," *Neb. Agr. Exp. Sta. Bul. SB 513*, pp. 1–133. 1971.
22. Martin, J. H. "Proso or hog millet," *USDA Farmers Bul. 1162*. 1937.
23. Martin, J. H. "Sorghums, broomcorn, Sudangrass, Johnsongrass, and millets," in *Grassland Seeds*, Van Nostrand, New York, pp. 1–734. 1957.
24. Martin, J. H. "Sorghum and pearl millet," in *Handbuch der Pflanzenzüchtung*, II Bd., 2 Aufl., Paul Parey, Berlin, pp. 565–589. 1959.
25. McGee, C. R. "Proso—a grain millet," *Mich. Agr. Ext. Bul. 231*. 1941.
26. McVicar, R. M., and Parnell, H. R. "The inheritance of plant color and the extent of natural crossing in foxtail millet," *Sci. Agr.*, 22:80–84. 1941.
27. Mengesha, M. H. "Chemical composition of teff (*Eragrostis tef*) compared with that of wheat, barley and grain sorghum," *Econ. Bot.*, 20(3):268–73. 1966.
28. Nelson, C. E., and Roberts, S. "Proso grain millet as a catch crop or second crop under irrigation," *Wash. Agr. Exp. Sta. Cir. 376*. 1960.
28a. Nelson, L. A. "Producing proso in western Nebraska," *Neb. Agr. Exp. Sta. Bul. 526*, pp. 1–12. 1973.
29. Schoth, H. A., and Rampton, J. H. "Sudan grass, millets, and sorghums," *Oreg. Agr. Exp. Sta. Bul. 425*. 1945.
30. Vinall, H. N. "Foxtail millet: its culture and utilization in the United States," *USDA Farmers Bul. 793*. 1924.

Chapter 21

Perennial Forage Grasses*

ECONOMIC IMPORTANCE

About 90 species of perennial pasture and hay grasses have been seeded in the United States. Some 40 of these are of considerable economic importance.[23, 24] Most of these are introduced species, but some 15 native grasses, including little bluestem (Figure 21–1), have been seeded on appreciable acreages.[48] The native species, except for reed canarygrass, have been sown mostly on western ranges or abandoned cropland. Grasses that are fertilized heavily with nitrogen have a protein content similar to that of legume forages.[9, 31]

More than 700,000 acres of forage and turf grasses are harvested for seed each year in the United States (Table 1–3). Much of this seed production is in Oregon. The leading grasses are tall fescue, ryegrass, timothy, bluegrass, orchardgrass and bromegrass.[3a, 10a, 20a, 32a, 32b]

GRASS TYPES

Perennial grasses, with a few exceptions, are largely cross-pollinated, and thus highly variable. Buffalograss plants are dioecious. Kentucky bluegrass and dallisgrass produce much of their seed by *apomixis,* without pollination. Coastal bermudagrass, gramagrass, guineagrass, paragrass, and napiergrass produce little or no seed, being propagaged only vegetatively. Most bermudagrass pastures are established vegetatively.[2]

Perennial grasses can be roughly divided into two classes: bunchgrasses and sod-forming grasses. The bunchgrasses send up tillers from the crown of the plant and form dense tufts or clumps. Sod-forming grasses spread laterally to form a more or less solid sod either by underground stems known as root stalks (rhizomes) or, in a few species, by above-ground stems (stolons).[23] The nodes of rhizomes and stolons give rise to roots as well as to stems. Tillers, stolons, and rhizomes all arise from buds in the crown of the grass plants below

* For additional information see *Grasses in Agriculture,* FAO, Rome, 1959, pp. 1–417.

Figure 21–1. Little bluestem plant (*left*), panicle (*center*), enlarged spikelet (*right*).

the surface of the soil. When the shoot from the growing bud is on a plant of a species that forms tillers (i.e., intravaginal), the growth is erect and a stem (culm) is formed. When in another species the shoot is slender and pliant and is ageotropic (not responsive to geotropic stimuli), a stolon is formed and both roots and culms arise from the nodes. When, as in still other species, the shoot bears scales with a hard, sharp, pointed tip it is a rhizome that grows along under the surface of the soil and likewise sends up both roots and culms. These distinctions are not absolute, however, because the shoots of certain tillering grasses, e.g., timothy, may grow somewhat like rhizomes and form roots[13, 17a] when buried deeply in soil. Also some grasses, e.g., bermudagrass, often may bear rhizomes below the soil surface in loose soil and stolons above the surface in compact soil. Nearly any of the sod-forming grasses can be established by transplanting pieces of turf.

Numerous improved varieties of forage grasses are now available as a result of breeding, selection, and introductions. A total of 29, 25, and 24 varieties of orchardgrass, timothy, and bromegrass were listed in 1972.[19]

ADAPTATION

Some grass species such as reed canarygrass can be grown where water stands on the land part of the time, while others are drought-resistant but may not withstand prolonged flooding. Many grasses adapted to the South are limited in their northern range because of lack of winter hardiness. Many typical sod-forming northern grasses assume a bunch type of growth and lack aggressiveness when grown in Oklahoma.[26] This probably represents an adaptation to photoperiod and temperature.

Many species require cool weather for their establishment while others such as the gramagrasses and bermudagrass make their growth in warm weather. Nearly all perennial grasses have small seeds (Figure 21–2). They are difficult to establish from seeding unless the seedbed is firm enough to avoid covering the seed too deeply and to form a close contact between the seed and soil. In order to obtain the desired seedbed, plowed land should be harrowed several times or compacted with a corrugated roller before or during seeding (Figure 21–3). When the seed is sown in the spring in a field of grain to be grown as a companion crop, the soil usually is sufficiently compact.

In dryland regions, the sowing of grass is attended with several hazards, chiefly drought and blowing soil. Plowed land that is harrowed or rolled is subject to blowing. Stubble land should be disked and harrowed to destroy weeds before grass is sown in the spring. Often a seedbed suitable for spring seeding is secured on fallow

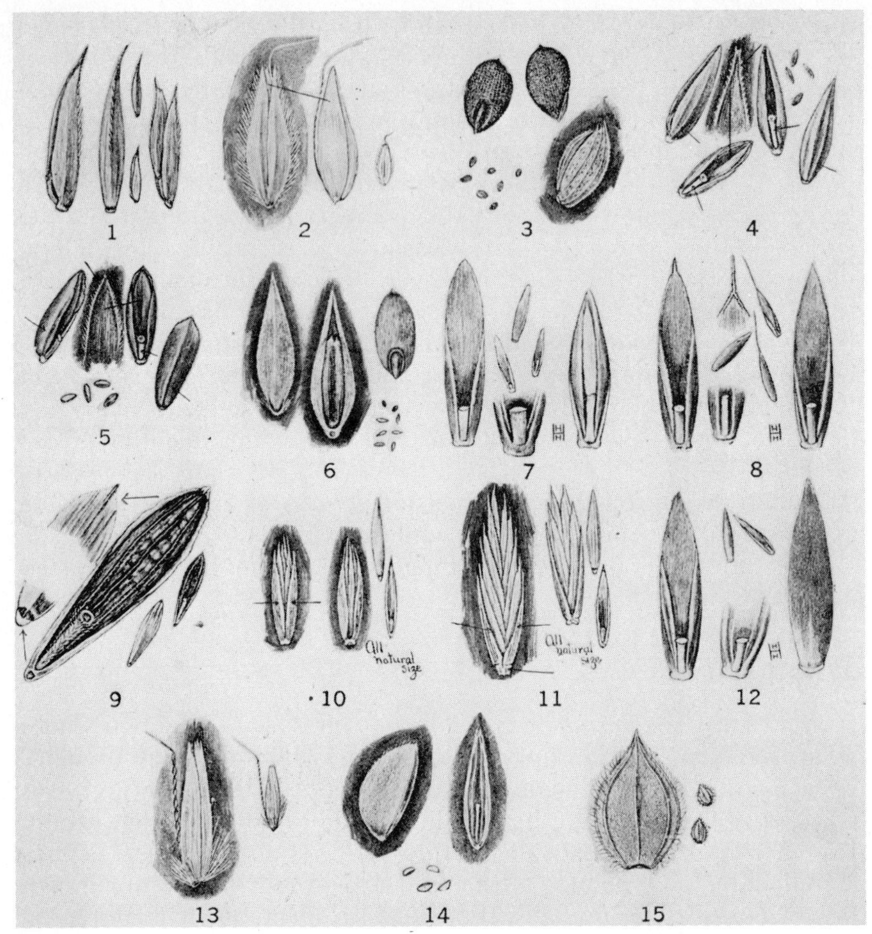

Figure 21–2. Seeds of perennial grasses (natural size and enlarged): (1) orchardgrass, (2) meadow foxtail, (3) timothy, (4) Kentucky bluegrass, (5) Canada bluegrass, (6) redtop, (7) perennial ryegrass, (8) Italian ryegrass, (9) smooth bromegrass, (10) slender wheatgrass, (11) western wheatgrass, (12) meadow fescue, (13) tall meadow oatgrass, (14) Bermudagrass, (15) dallisgrass.

land that has settled during the winter or on corn or potato land that is disked and harrowed. In the northern Great Plains, considerable success has followed the drilling of certain cool-weather grasses such as crested wheatgrass in weedy fields or unworked grain stubble in fall or early winter, preferably seeding in snow before the ground is frozen. The seed settles into the soil during thawing and germinates in the spring. In the southern Great Plains the most certain method of establishing grasses has been to grow a crop of drilled sudangrass or sorghum, mowing often enough to keep it from going to seed, and

Figure 21–3. Drill for sowing grass seeds. Sets of corrugated rolls in front of and behind the seed spouts compact the soil before and after the seed is dropped.

leaving the swaths on the ground over the winter. The grass seed is drilled into this protective cover the next spring[36] (Figure 21–4).

Many broadleaved weeds can be controlled by applications of herbicides when the young grass seedlings have three to five leaves.

TIMOTHY

Timothy (*Phleum pratense*) is an important cultivated grass in the United States. It is grown on about 12 million acres for hay, mostly in mixtures with clover. This mixture yields more than either crop grown alone. It is also higher in feeding value than timothy alone. Perhaps 4 million acres of timothy are seeded each year. Timothy is harvested or grazed for two years or longer before it is plowed up for planting other crops in the rotation.[16]

The timothy and clover, which usually are sown with a small-grain companion crop, furnish some pasturage the first year. In the second year, the mixed hay is mostly clover, whereas the third-year crop is largely timothy. When left longer the crop is nearly all timothy. Some 125,000 acres of timothy are cut for seed each year. The states leading in timothy production are New York, Ohio, Illinois, Wisconsin, and Pennsylvania. Most of the timothy seed is produced in Minnesota, Missouri, and Ohio.

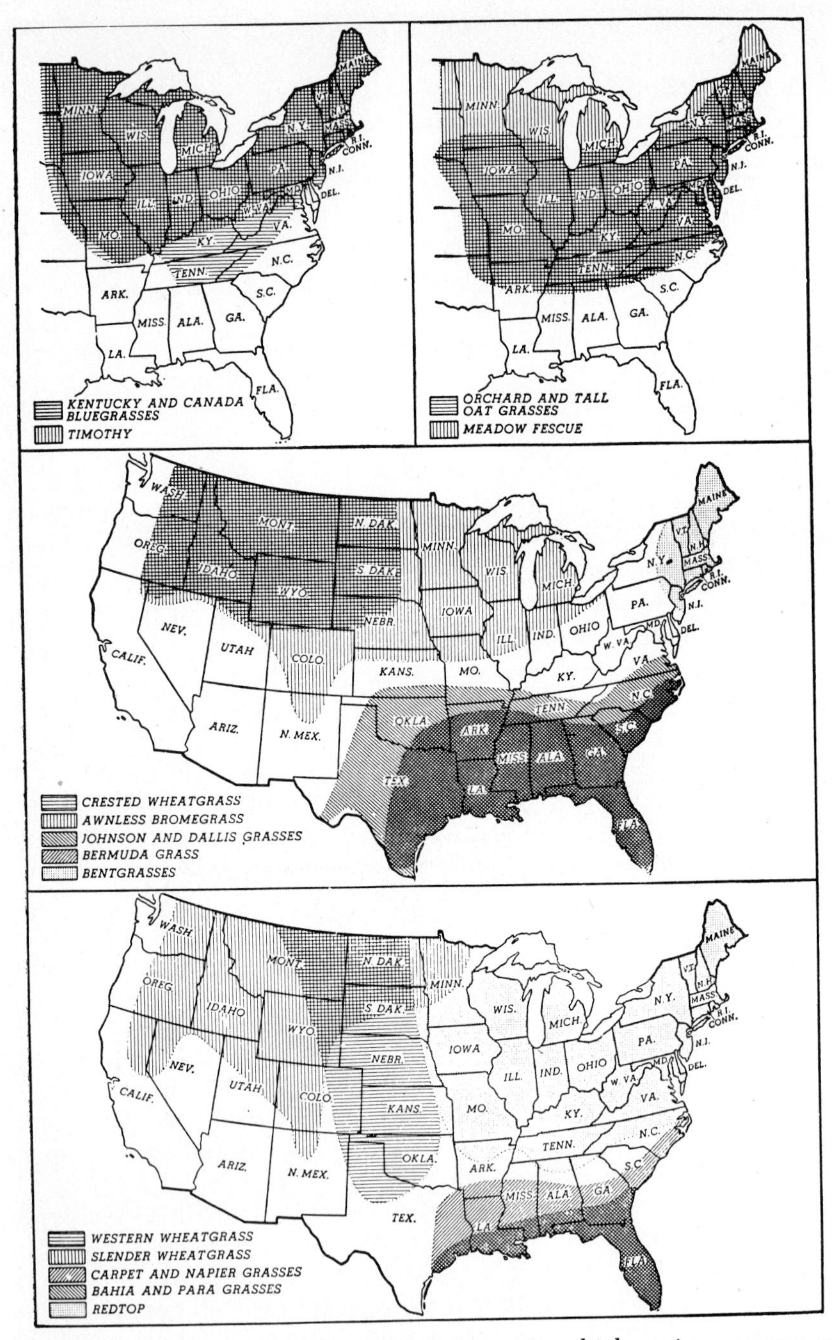

Figure 21–4. Sections of the United States in which various grasses are well adapted and of primary importance.

Timothy, a native of Europe, was first cultivated in the United States by John Herd, who found it growing wild in New Hampshire in 1711. First called Herd's grass, timothy acquired its present name from a Timothy Hanson, who grew it in Maryland in 1720.

Adaptation

Timothy is best adapted to the cool humid climate of the northeastern quarter of the United States, but it is grown also to some extent in the cool mountain valleys in the Rocky Mountain region, as well as in the coastal area of the Pacific Northwest. It is unadapted to the drier regions, nor does it thrive where the average July temperature exceeds 77° F. Timothy grows better on clay soils than on the lighter-textured sandy soils.

Description

Timothy is easily recognized by its erect culms and dense cylindric spike-like inflorescence (Figure 21–5). The spikelets are one-flowered. The plant is largely cross-pollinated under natural conditions. Timothy, a bunchgrass, differs from most grasses in that

Figure 21–5. Timothy plants.

one of the lower internodes is swollen into a bulblike base, a haplocorm,[13] often called a corm. Organic food reserves, which accumulate in the haplocorm up to the stage of seed maturity, furnish nutrients for new shoots. Timothy survives for three years or more, but it gradually becomes contaminated with weeds and other grasses. Consequently it usually is plowed up after the second year.

The leaves of certain selected strains of timothy remain green longer than those of the ordinary timothy, and also yield as well or better.[15]

Culture

Timothy usually is sown with a grain drill equipped with a grass-seeding attachment. It is sown either in the fall with winter grain or in early spring with spring oats or barley. Usually 10 pounds per acre of clover seed are sown in the spring in the same field. When the clover is broadcast in winter-wheat or winter-rye fields in the spring the timothy may be seeded with it. Seedlings that start in the fall are less likely to be injured by dry weather during late spring and early summer than are those from spring seeding.[24] The usual seeding rates for timothy are 3 to 5 pounds per acre in the fall, or 10 pounds in the spring.[16]

The yields of timothy are improved by application of commercial fertilizers, particularly those that contain phosphorus and nitrogen. Complete fertilizer is essential in the Northeast. The protein content of timothy is increased by very heavy application (33 pounds or more) of nitrogen fertilizers. The fertilizer is best applied in the spring.[16]

The largest yield of timothy hay consistent with high quality is obtained by cutting during early blooming or even earlier.[15, 16] The digestibility of timothy declines steadily as the plant develops, beginning as early as when the plants are in full head. Moreover, the more mature plants are usually less palatable because of the increase in crude-fiber content. Timothy cut prior to full bloom contains 70 per cent or more of the total protein of the hay in the leaf blades, leaf sheaths, and stems of the plant, while that left uncut until 10 per cent of the heads are straw-colored contains at least 50 per cent of the plant protein in the heads.[25] Much of this protein may be lost in the ripe seeds that shatter during handling and feeding. The total yield increases only slightly after the full-bloom stage.[16] Cutting before blooming reduces the vigor and growth of subsequent cuttings because lower food-reserve storage is present in the haplocorms.

Decreased yields that often occur in crops following timothy are due largely to a temporary deficiency of available nitrogen in the soil until some of the crop residue has decomposed.

Timothy usually is harvested for seed with a combine after the seed is fully ripe and dry. In threshing, care is required to avoid hulling the seeds. Hulled seeds germinate satisfactorily at first but they gradually lose their viability after several months.[16]

Utilization

Timothy is a popular hay for feeding horses. As compared with legumes, timothy hay is relatively low in protein and minerals, especially calcium. In mixture with clover the hay is excellent for cattle and sheep.

Timothy also is one of the most palatable pasture grasses, usually grazed in preference to redtop, orchardgrass, or even Kentucky bluegrass. It is often included in mixtures of grasses and legumes sown in permanent pastures. As the pasture becomes older, the timothy is gradually replaced by other grasses. Its coarse, bunchy, erect, and semipermanent growth makes it unsuited for lawns.

SMOOTH BROMEGRASS

Smooth bromegrass or awnless bromegrass (*Bromus inermis*), introduced into the United States about 1884, soon became a valuable forage plant in the West. The total acreage was small for many years because native grasses, cereals, and legumes furnished most of the forage. Later, bromegrass became a highly important crop in the northern states from the Dakotas and Kansas eastward to Ohio. In 1969 it was grown on nearly 9 million acres, chiefly in mixtures with alfalfa.

Adaptation

Smooth bromegrass is especially adapted to regions of moderate rainfall and cool to moderate summer temperatures, particularly where the native vegetation consists of tall or medium-tall grasses, as in the North Central and Great Lakes states. It is one of the best introduced grasses in the eastern parts of the Dakotas, Nebraska, and Kansas. Bromegrass is moderately drought-resistant but the more drought-resistant crested wheatgrass outyields and outlasts it in the drier areas.[46]

Bromegrass makes its best growth on moist, well-drained clay to silt loam soils, but it produces satisfactorily on sandy soils when there is sufficient moisture. It is particularly vigorous on fertile soils high in nitrogen. Bromegrass has only a small degree of alkali or salt tolerance.

Description

Smooth bromegrass is an erect, sod-forming perennial that spreads by rhizomes. The stems vary in height from 2 to 4 feet. The plant produces numerous basal and stem leaves that vary in length from

4 to 10 inches. Frequently, the rough (scabrous) leaves are marked by a transverse wrinkling a short distance below the tip. The inflorescence is a panicle 4 to 8 inches long that spreads when in flower. The flower head develops a characteristic rich purplish-brown color when mature. There are several florets in a spikelet (Figure 21–6). The seeds are long, flat, and awnless (Figure 21–2).

Solid stands of pure bromegrass are likely to develop a sod-bound condition after two or three years, unless they are heavily fertilized with nitrogen. The plant starts growth early in the spring and continues growth late into the fall.

Northern types of smooth bromegrass are adapted to the northern border states as well as to higher elevations in the intermountain region. Southern types are adapted to the Corn Belt and the eastern Great Plains.[28, 19] A noncreeping strain grown in Canada is somewhat finely stemmed and more leafy than common brome.

Culture

Late-summer seeding after a small-grain crop gives the best results in eastern Nebraska.[17] Smooth bromegrass is usually seeded there between August 20 and September 15 at the rate of 15 to 20 pounds per acre. It is often included in a mixture that uses 15 pounds of bromegrass and 3 pounds of alfalfa. The bromegrass and alfalfa are ready to cut at the same time. Spring seeding, when practiced, should be early. Inclusion of a legume increases both the growth and the protein content of the grass.[33]

Figure 21–6. Smooth bromegrass: (1) plant, (2) panicle, and (3) spikelet.

Maximum yields of bromegrass are usually obtained the second and third year after the stand is established. The yield decreases when the plants become sod-bound. The sod-bound condition may be avoided or delayed by applying nitrogen fertilizers, or by seeding legumes with the bromegrass.[7]

The best quality of hay is obtained when bromegrass is cut in the bloom stage.

Utilization

Bromegrass makes hay or silage of excellent quality, the protein content being high and the crude-fiber content relatively low. It is one of the most palatable pasture plants, especially for spring and fall grazing.

Bromegrass has been used to retard soil erosion because of its heavy mass of roots. It is suitable for steep slopes, buffer strips in strip cropping, small sod dams, drainage outlets, terraces, and waterways through cultivated fields.

Other Bromegrasses

Mountain brome (*Bromus marginatus*) is a native grass of the Rocky Mountain region, being adapted to the cooler sections of the West. An improved variety that is taller, later, more leafy, and more disease-resistant than common mountain brome is grown in Washington and Oregon.[18, 20]

Rescuegrass (*B. catharticus*) is a biennial or winter annual introduced from Argentina to the Southeastern states where it is grown as a winter pasture.

Two introduced annual weedy grasses with some forage value are cheat (*B. secalinus*) and downy brome, or cheatgrass (*B. tectorum*). The latter provides grazing when the plants are small, but later becomes useless because of its irritant awns.

ORCHARDGRASS

Orchardgrass or cocksfoot (*Dactylis glomerata*), a native of Europe, is grown to some extent in nearly every state. It grows naturally in many localities. It was grown for pasture, hay, or silage on 6.7 million acres in 1969, mostly in mixtures with clover or alfalfa. States that lead in orchardgrass acreage were Kentucky, Ohio, California, Tennessee, Virginia, and Illinois. This illustrates its wide adaptation. Most of the seed was produced in Virginia, Missouri, and Kentucky.

Adaptation

Orchardgrass thrives under cool, humid, moist, or irrigated conditions, but it is more tolerant to heat and drought than is timothy. In the northern states, orchardgrass provides forage during the spring and summer. In the South it furnishes grazing during late fall, winter, and early spring. Since it tolerates some shade, orchardgrass often is seeded in woodland or logged-off pastures. It is less resistant to cold than timothy or bromegrass. Consequently, its acreage in the colder northern areas is limited.

Description

This grass is distinguished by its large circular bunches, folded leaf blades, compressed sheaths, and, particularly, by the spikelets grouped in dense one-sided fascicles borne at the ends of the panicle branches (Figure 21–7).

Perhaps a majority of the orchardgrass acreage in the United States is of the common type but improved varieties for different areas are now available.[19, 20a]

Culture

Orchardgrass usually is sown at a rate of 3 to 10 pounds per acre in mixtures with alfalfa or Ladino clover in the northern and western states, but at 10 to 15 pounds per acre with lespedeza or clovers in the South. Other grasses or legumes may be included in the mixtures.

Figure 21–7. Orchardgrass during anthesis.

Orchardgrass may be sown with a small-grain companion crop in the fall in the South, or in early spring in the North. The field may be fertilized with 40 to 60 pounds of phosphate (P_2O_5) at seeding time.[45] When orchardgrass is grown without a legume, the field should receive nitrogen also, i.e., up to 100 pounds per acre each year.

Potash and lime may be required by the legumes in the mixture in some areas.

Utilization

Orchardgrass is primarily a pasture crop because much of the growth is in the lower leaves, but it is widely harvested for silage. As a hay crop, it should be cut at first bloom because older stems become coarse and unpalatable. It often reaches this stage in the spring before the arrival of good curing weather. Consequently, it is put into the silo to avoid loss. Orchardgrass loses its palatability and digestibility much more rapidly than smooth bromegrass.

TALL FESCUE

Tall fescue (*Festuca arundinacea*) is grown on nearly 13 million acres in the United States, mostly south of the Ohio river, in the southeastern states, in Missouri and Kansas, in the Pacific coastal area and under irrigation in several western states. Most of the seed is produced in Missouri, Kentucky, Oregon, and Tennessee.

Adaptation

Tall fescue, a native of Europe, is a perennial tufted grass suited to various soil and climatic conditions. It is adapted to semiwet conditions, but tolerates temporary drought as well as both acid and alkaline soils.[32] Tall fescue is not suited to semiarid conditions.

Description

Tall fescue plants are 2 to 5 feet tall with dark-green leaves. Since they have a medium-bunch habit, the plants form a fair turf when grazed or mowed regularly. The inflorescence is a long narrow panicle with several florets per spikelet. The seeds tend to shatter upon ripening.[22a]

Tall fescue consists mostly of the Kentucky 31 variety. Other varieties are adapted to the southwestern irrigated areas or, for reclaiming saltgrass meadows in Oregon.[32]

Culture

Tall fescue is sown in late summer or early fall at the rate of 2 to 4 pounds of seed per acre in mixtures, but up to 15 pounds or more per acre when sown alone. It usually is sown in mixtures with legumes such as (1) Ladino clover, alfalfa, or alsike clover in the cooler humid regions, (2) lespedeza or white clover in the Southeast, and (3) subclover in the western coastal areas. Since it grows best under cool conditions, it is pastured in the South in the winter. Tall fescue has a long growing season being tolerant of both temporary drouth and excessive moisture conditions.

Utilization

Tall fescue is grown chiefly for pasture. It is less palatable than most of the cultivated grasses, but it usually is grazed adequately in mixtures with legumes. High-producing dairy cows may not consume enough for maximum milk yield.[43] Close grazing along with nitrogen fertilization, maintains the plants in a succulent, more palatable condition. Grazing animals sometimes develop disorders when confined to tall-fescue pasturage for long periods. These disorders are fescue foot, fat necrosis, and the accumulation of excessive alkaloids especially perloline.[27a]

Other Fescues

Meadow fescue (*Festuca elatior*) was introduced from Europe many years ago. It was a common ingredient of pasture and meadow mixtures, especially in the southern part of the Corn Belt.[24] It has been replaced largely by tall fescue since 1945, partly because of its susceptibility to crown rust.[27] The plants differ from tall fescue in being about 12 inches shorter, but they also lack fine hairs on the leaf auricle. Moreover, meadow fescue has 7 pairs of chromosomes, while tall fescue has 14 pairs.

Sheep fescue (*F. ovina*), red fescue (*F. rubra*), and chewings fescue (*F. rubra* ssp. *commutata*) are turf grasses. The latter two species, the seed of which is produced mostly in Oregon, are used in shaded lawn areas. Chewings fescue is tufted but does not spread to form a dense turf.

REED CANARYGRASS

Reed canarygrass (*Phalaris arundinacea*) is native to the northern part of both hemispheres. It was cultivated first in Oregon about 1885, being among the later grasses to assume importance under cultivation.[38]

Adaptation

Reed canarygrass is grown in the Pacific Coast areas of Oregon, Washington, and northern California, as well as in the North Central states. It makes its best growth in a moist cool climate, being sensitive to neither heat nor cold. It is not very successful where the average mean winter temperature is above 45° F., or the summer temperature above 80° F.[88] This grass is productive on fertile moist soils, being especially suited to swamp or overflow lands. It does well on peat soils, and on land that is too wet for other crops.[3, 22] In Iowa, reed canarygrass produced higher hay yields than did timothy, brome, meadow fescue, tall oatgrass, redtop, or orchardgrass.[51]

Description

Reed canarygrass is a long-lived perennial that spreads by rhizomes. The plants are 2 to 8 feet tall with leafy stems. They tend to grow in dense tussocks 2 to 3 feet in diameter. The leaves are broad, smooth, and light green in color. The inflorescence is a semidense spike-like panicle, 2 to 8 inches long (Figure 21–8). The stems become coarse after the panicles begin to appear. The seeds are enclosed in blackish-brown or gray lemmas, while the paleas are sparsely covered with long hairs. The seeds mature from the top of the panicle downward and shatter readily.

A distinct strain, selected in Oregon, has a more upright growth, greater leafiness, stiffer stems, and better seeding habits than ordinary reed canarygrass.[38] It is able to grow on uplands that often become dry in spring, summer, or fall. It thrives as well as the ordinary strain in the lowlands.

Culture

Reed canarygrass is grown from fall or spring seeding on the Pacific Coast, but early spring seeding is the most common practice in the North Central states. It is usually seeded at the rate of 5 to 8 pounds

Figure 21–8. Reed canarygrass: (1) plant, (2) panicle, (3) spikelet, and (4) floret.

per acre in close drills. It is seldom sown with other grasses, but it may be seeded with small grains on moist, fertile soils. The crop is sometimes seeded in cultivated rows 18 to 20 inches apart on rather dry uplands in some areas. The first crop should be cut for hay as soon as the panicles begin to appear. The later crops usually do not produce panicles.

Utilization

Reed canarygrass furnishes abundant pasturage or silage where soil moisture is adequate because it starts growth early in the spring and continues until late in the fall. It is more palatable than other wetland grasses when grazed closely; otherwise it becomes coarse. Likewise, the hay is palatable when cut before the stems are too coarse.

Related Species

Hardinggrass (*Phalaris tuberosa*, var. *stenoptera*) is a productive hay crop in California.[2] Canarygrass (*P. canariensis*), a native of the Mediterranean region, is established in many parts of the United States. Canarygrass furnishes the canary seed used so generally for feeding canaries. Most of the seed for this purpose is grown in Europe and Argentina.

KENTUCKY BLUEGRASS

Kentucky bluegrass (*Poa pratensis*) is the best-known grass in the United States. It is a native of the old World, but has spread naturally or by direct seeding over the humid and subhumid sections in the northern half of the United States to such an extent that it is the dominant species in most of the older pastures. Most of the seed is harvested in the Pacific Northwest, Minnesota and Kentucky.

Adaptation

Kentucky bluegrass is adapted throughout the northern half of the United States, except where the climate is too dry.[14] It does best under cool, humid conditions on highly fertile limestone soils, but it also thrives on noncalcareous or slightly acid soils with a reaction as low as *p*H 6. It seems to prefer the heavier soils. An ample supply of nitrogen and phosphorus is essential for high production.

Description

Kentucky bluegrass is a dark-green sod-forming grass, and a long-lived perennial. The stems are 1 to 2 feet in height when allowed to grow uncut. They usually are numerous in a tuft. The plants have narrow leaves 2 to 7 inches in length.

Figure 21–9. Kentucky bluegrass plant and panicle.

The inflorescence is a pyramid-shaped panicle about 2 to 8 inches long (Figure 21–9). The spikelets are composed of three to five florets. Kentucky bluegrass reproduces by rhizomes as well as by seed. New tufts with their roots arise from the nodes along the rhizomes, thus continually occupying the spaces left by the death of the older tufts that probably fail to survive more than two years. It is classed as a day-neutral plant. Flower heads may be initiated in short, cool, autumn days, but blooming occurs in the longer days of late spring and summer.

Merion bluegrass, a selected strain, is a popular lawn grass.[19, 20] The seed is produced most in the Far West.

Culture

In pasture mixtures, bluegrass is generally seeded with other grasses, and either clovers or alfalfa, at a rate of 1 to 5 pounds per acre. About 20 to 40 pounds per acre are sown for pure stands. Usually two to three years are required to produce a good sod from seeding. Annual applications of nitrogenous fertilizers increased the average yield of dry matter 72.1 per cent in Wisconsin.[1]

Bluegrass seed is usually harvested from the ripening plants with swathers or with mechanical strippers equipped with revolving spiked beaters.[44] It is cured in ricks or windrows, being turned frequently to prevent heating. Temperatures above 140° F. during curing destroy the germination. Partial heating, together with the presence of immature seed and empty chaff, are responsible for much of the commercial bluegrass seed germinating no more than 70 per cent. Sound plump seed germinates over 90 per cent.

Utilization

Kentucky bluegrass is not a hay crop. It is a leading grass in permanent pastures because of its persistence as well as its ability to withstand close and continuous grazing. It starts growth early in the spring, and thus furnishes succulent forage for early grazing. It becomes nearly dormant during the heat of midsummer when daily maximum temperatures approach 90° F. However, it resumes growth with the advent of cool weather in the fall, when it furnishes additional grazing. It supplies almost continuous summer grazing in the cooler sections of the northeastern states. For rotation pastures, as well as for many seeded pastures in the northeastern and North Central States, Kentucky bluegrass has been partly replaced with orchardgrass, tall oatgrass, and smooth bromegrass. These taller grasses grow more in hot weather. Because of its dense turf, bluegrass is the most popular lawn grass in America. Lawns are established either by seeding or by transplanting sod. It is sown either in very early fall or very early spring.

Other Bluegrasses

Canada bluegrass (*Poa compressa*), a native of Europe, is grown in Canada. It is distinguished from Kentucky bluegrass by its shorter, compressed stems, which long remain green; its single shoot at the end of each rhizome; and its narrower panicles. Canada bluegrass is found throughout the Kentucky bluegrass region, often on the less fertile soils, but it is generally recognized as being decidedly inferior.

Rough-stalked meadowgrass or birdgrass (*Poa trivialis*), also a native of Europe, has bright green leaves and thrives in the shade. It is sown in shady locations in lawns.

Among the range grasses native to the United States are big bluegrass (*Poa ampla*), bulbous bluegrass (*P. bulbosa*), Texas bluegrass (*P. arachnifera*), Sandberg bluegrass (*P. secunda*), and mutton or Fendler bluegrass (*P. fendleriana*). Big bluegrass is being seeded on semiarid rangelands of the Pacific Northwest. The Sherman variety, adapted to varied conditions, is a bunchgrass.

Bulbous bluegrass (*Poa bulbosa*) has become well established in certain areas of northern California and southern Oregon. It normally produces bulbs at the base of the stem and bulbils in the inflorescence. It is adapted to a wide range of soil types.[40] The chemical composition of the hay is similar to that of timothy.

Annual bluegrass (*P. annua*) is a serious pest in lawns and seed fields.

BERMUDAGRASS

Bermudagrass (*Cynodon dactylon*) is a native of Asia but is now distributed throughout the tropical and subtropical parts of the world.

Adaptation

Bermudagrass is distributed from Maryland to Kansas, southward to the Gulf of Mexico, and westward from Texas to California. In general, it is adapted to the same area as cotton. The best use of bermudagrass is in pasture mixtures with crimson, white, hop, and Persian clover, or with lespedeza. Its growth in the northern states is limited by its lack of winter hardiness, although it is sufficiently hardy to endure the winter conditions in north central Oklahoma.[26] Bermudagrass makes its best growth on fertile lands that are well drained. It shows a marked preference for clay soils, but it grows more or less abundantly on sandy soils. Bermudagrass grows luxuriantly during hot midsummer weather, but does not start growth until late spring. It stops growth and becomes bleached with the onset of cold weather in the fall.

The spittlebug (*Prosapia bicincta*) damages bermudagrasses in the southeastern states, especially in wet years.

Description

Bermudagrass is a long-lived perennial with numerous branched leafy stems that vary from 4 to 18 inches in height. Although the stems of bermudagrass, like those of other grasses, have only one leaf at a node, it may appear to have two to four, due to several contiguous short internodes (Figure 21–10). In ordinary bermudagrass, there are numerous stout rhizomes which in very hard soil grow along above the surface as stolons for 1 to 3 feet. The flowers are borne in slender, spreading spikes arranged in umbels of four to six. Bermudagrass produces very little seed in humid regions.

Common bermudagrass has white rootstocks as large as goose quills, besides the leafy stems or stolons that creep on the soil surface. St. Lucie grass is identical in appearance but lacks rootstocks and rarely survives the winter north of Florida. Coastal bermudagrass is an improved hybrid strain that produces very little seed. Consequently, it must be propagated entirely by planting stolons.[8] It spreads rapidly, withstands light frosts, and produces well in the southern three-fourths of the Cotton Belt. Its tall growth makes it a productive hay and silage crop as well as a pasture crop. Coastal bermuda is an important hay crop in the southern United States. Other varieties are hardier, more drought-resistant, or are adapted to soils of low fertility.[21]

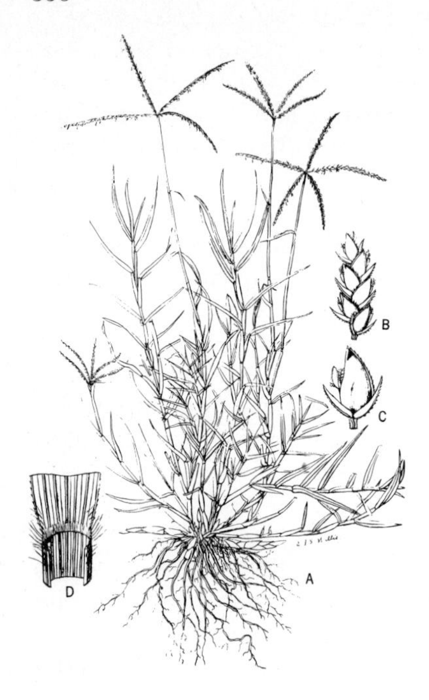

Figure 21–10. Bermudagrass: (*A*) plant, (*B*), panicle branch, (*C*) spikelet, and (*D*) ligule. Stolons are shown at the crown region.

Still other varieties are well-suited for lawns or for grassed waterways. A heavy bermudagrass turf prevents soil erosion.

Culture

Bermudagrass is established by planting sod, crowns, or pieces of runners (stolons). The most common practice is to break up the sod into small pieces. The pieces are dropped, 2 to 3 feet apart each way, into furrows on plowed land. These sod pieces may be planted any time from spring until midsummer—whenever wet weather conditions are likely to prevail. Common and Greenfield bermudagrass can be established by seeding about ½ inch deep, at a rate of about 5 pounds per acre, in the spring after the daily mean temperature reaches 65° F.[30]

Bermudagrass stands tend to become sod-bound. When this occurs, the yields are improved by disking, plowing, or harrowing. Bermudagrass often spreads onto cropped land, where frequent cultivation is necessary to keep it under control. Field burning in early April improves the yields of coastal bermudagrass by reducing disease, insect, and weed infestations.

Utilization

Bermudagrass is an excellent summer-pasture plant, either alone or in mixtures, particularly when closely grazed. It becomes wiry and tough as it approaches maturity. Sometimes it is planted in mixtures with lespedeza. Bermudagrass sown along the highway borders of Florida furnishes much of the summer pasture for cattle that graze the roadsides of that state.

REDTOP

Redtop (*Agrostis alba*) is a cultivated perennial grass introduced during the Colonial period.

Adaptation

Redtop will grow under a great variety of conditions as it is one of the best wetland grasses among the cultivated species. It is especially adapted for growth on acid soils low in lime where most other grasses fail. It has only slight tolerance to soil salinity or drought. It is best adapted to cool or moderate temperatures.

Description

Redtop spreads by rhizomes, making a coarse, loose turf. The leaves are narrow and the stems slender. The panicle is loose, pyramidal, and reddish in color (Figure 21–11). The spikelets are small and contain one flower. Redtop matures at about the same time as timothy.

Utilization

Redtop is a wetland hay crop used as a part of pasture mixtures under humid conditions, as a soil binder, and as an ingredient of hay mixtures.[24] It may be seeded either early in the spring or late in the summer at the rate of 4 to 5 pounds per acre when used with other grasses for hay. In lawn-grass mixtures, redtop does not make a fine,

Figure 21–11. Redtop panicle and plant.

smooth turf, but serves as a companion grass while bluegrass is becoming established.

Related Species

Colonial or Rhode Island bent (*Agrostis tenuis*) is found in meadows and pastures in New England, New York, and the Pacific Northwest. Improved varieties are important turf grasses, some of which are injured rather than improved by applying lime to the lawn.

Creeping bent (*A. palustris*) is distinguished by its dense panicle and by its creeping stolons, which may grow as much as 4 feet in a single season. It is common in pastures in many places where the soil is moist and is found in seaside meadows on both the Atlantic and Pacific coasts. Improved varieties and velvet bent (*A. canina*) are sown on golf-course putting greens to provide a fine turf. Other selected strains of creeping bent are established by planting stolons.

CRESTED WHEATGRASS

Crested wheatgrass is a hardy, drought-resistant perennial bunchgrass native to the cold dry plains of Russia. It was first introduced to the United States in 1898, but failed to become established from that introduction. After 1915, when experiments with new lots of seed from Russia at several points had demonstrated its promise in the northern Great Plains, it was increased for distribution.[24, 34]

Adaptation

Crested wheatgrass is especially well adapted to the northern Great Plains and the intermountain and Great Basin regions where winter temperatures are severe and the moisture supply limited. This grass makes its best growth in cool climates, forage production being reduced by the higher temperatures in the southern Great Plains. There is no known instance in which the plants were killed either by drought or cold under field conditions.[50] Crested wheatgrass is productive on practically all types of soil, but it is less tolerant of salinity than is western wheatgrass.

Description

Crested wheatgrass is an extremely long-lived perennial. The fine stems, which vary in height from $1\frac{1}{2}$ to 3 feet, occur in dense tufts. The leaves are flat, somewhat lax, narrow, and sparsely pubescent on their upper surface. The dense spikes are 2 to 4 inches long, being considerably broader than those of most other species of wheatgrass. The spikelets are closely crowded on the tapered head and tend to stand out from the axis of the spike.

Standard, the ordinary commercial strain of crested wheatgrass, introduced as *Agropyron desertorum*, consists of a mixture of many different types or strains that vary in leafiness, stiffness of stems, and size of spike. The short awned types are considered the most desirable (Figure 21–12). The Nordan variety produces more and better forage and has better seedling establishment than does the ordinary type.[19, 20]

Figure 21–12. Crested wheatgrass spikes. The two at left are Fairway and the original *A. cristatum*. The three at right are awned and awnless types of the Standard type, *A. desertorum*.

A selected strain of crested wheatgrass, Fairway, introduced as *Agropyron cristatum*, is popular in Canada, but the taller Standard strain is more generally grown in the United States because of its higher yields and better palatability. The two strains do not intercross. The Standard type has 14 pairs of chromosomes whereas the Fairway and the original *A. cristatum* type have 7 pairs. Fairway plants are finer-stemmed, more leafy, and tiller more than do the Standard plants. The leaves of the Fairway plants are covered with fine hairs and are bright green in color, while the leaves of the Standard vary from dark green to grayish green. The seeds of Fairway are smaller, bear more awns, and are lighter than those of Standard.

Culture

Crested wheatgrass may be sown either in the fall or spring, in close drills for hay or pasture, or in cultivated rows for seed.[10] Row plantings may be in single or double rows, the rows being 36 to 42 inches apart in either case. Recommended seeding rates are 5 to 8 pounds per acre for close drills, 1 to 2 pounds for single cultivated rows, and 2 to 3 pounds for double cultivated rows.[34] The seed has been drilled on sagebrush rangelands that have been tilled with a disk plow. It may be drilled directly in small-grain stubble where soil blowing is probable. For hay, it should be cut at least by the time it starts to blossom.

Utilization

Crested wheatgrass is highly palatable either for hay or pasture. The hay compares favorably with that of western wheatgrass in quality and palatability. The grass becomes harsh as it matures. Crested wheatgrass furnishes pasture earlier in the spring and later in the fall than do other cultivated grasses. In the northern Great Plains it has two to three times the carrying capacity of the native range, especially in early spring.[46] It should be supplemented with other grasses because it tends to become more or less dormant during hot dry weather. This grass has been utilized effectively for wind or water erosion control in the northern intermountain region as well as in the northern Great Plains because of its persistence and its tough, fibrous root system.

WESTERN WHEATGRASS

Western wheatgrass (*Agropyron smithii*), often called bluejoint, is a native perennial introduced into cultivation. It is distributed generally throughout the United States, except in the more humid southeastern states, but is most prevalent in the northern and central Great Plains.

Adaptation

Western wheatgrass is a cool-season grass adapted to a wide range of soil types. It is extremely drought resistant as well as alkali tolerant. Nearly pure stands of this grass occur in South Dakota on heavy gumbo soils, in swales or flats where additional moisture from runoff had been received.

Description

Western wheatgrass has strong creeping rhizomes (Figure 21–13), and the stems usually are 1 to 2 feet in height. The leaves usually are 4 to 12 inches long and less than 1/4 inch wide. The upper surfaces of the leaves are scabrous (rough) and prominently ridged lengthwise,

Figure 21–13. Western wheatgrass. Note rhizomes at lower left.

while the underside of the leaf is relatively smooth. The leaves are rather stiff and erect and when dry roll up tightly to give the plant the appearance of having scanty foliage. The entire plant is usually glaucous, which gives it a distinctive bluish-green coloration. The wheat-like but more slender spikes are 2 to 6 inches long.

Culture

The threshed seed is relatively free from awns or hairs so that it can be sown with a grain drill without previous treatment. Western wheatgrass is best seeded in the fall or early spring at the rate of 10 to 12 pounds per acre. It is usually included in a mixture with crested wheatgrass. Germination of the seed is frequently delayed. Consequently, stands of this grass are often slow in becoming established. For this reason, it is most valuable when seeded in a mixture.

Utilization

Pioneer ranchers early recognized western wheatgrass as one of the most valuable native forage grasses in the low rainfall areas of the northern Great Plains. It is high in palatability and nutritive value. It should be cut in the very early bloom stage for highest-quality hay. Western wheatgrass provides high-quality pasturage during the early part of the growing season. It has the ability to cure while standing and still retain much of its palatability and nutritive value for winter pasture under semiarid conditions.

Because of its excellent sod-forming qualities, western wheatgrass is valuable for protecting terrace outlets, dam faces, and waterways through cultivated land.[24]

SLENDER WHEATGRASS

Slender wheatgrass (*Agropyron trachycaulum* or *A. pauciflorum*) is another native American grass to become established as a cultivated crop. It is adapted to the northern Great Plains and intermountain regions. It is very short-lived and produces well for only two or three years. The inflorescence is a slender, greenish spike on which the closely appressed spikelets are some distance apart. Recommended seeding rates range from 10 pounds to 35 pounds of seed per acre. For hay, it should be cut just before blooming. Slender wheatgrass should be used in pasture or hay mixtures, rather than sown alone. About 2 pounds per acre should be included in mixtures sown on dry land, but about 4 pounds per acre when sown on irrigated land.

OTHER WHEATGRASSES

Intermediate wheatgrass (*Agropyron intermedium*) is grown for pasture or hay on more than 350,000 acres in the northern Great

Plains, the intermountain region, and the Pacific Northwest. It was introduced from the Soviet Union in 1932.[48] The plants are sod-forming, while the large seeds facilitate the establishment of stands. It is adapted to parts of the Great Plains and intermountain regions where the rainfall exceeds 15 inches, as well as to the Pacific Northwest where it often is grown in mixtures with alfalfa.

Pubescent wheatgrass (*A. trichophorum*), a native of Europe, is grown to some extent in much the same area as intermediate wheatgrass. The glumes are awnless and pubescent.

Tall wheatgrass (*A. elongatum*) was introduced from the Soviet Union.[47] It is a tall coarse bunchgrass that tolerates soil salinity. It is adapted to sagebrush as well as mountain lands in the central intermountain states. Its long growing period enables it to provide green herbage in midsummer. It has been crossed with wheat in attempts to produce a perennial wheat.

Siberian wheatgrass (*A. sibiricum*), introduced from the Soviet Union, is grown on about 19,000 acres in the Pacific Northwest. It is similar to crested wheatgrass, but inferior to it except on infertile or arid soils.

Streambank wheatgrass (*A. riparium*) is a native sod-forming species, a selected variety of which is grown on a small acreage in the Pacific Northwest.

Three native bunchgrasses which have been seeded on western rangelands are beardless wheatgrass (*A. inerme*), bluebunch wheatgrass (*A. spicatum*), and thickspike wheatgrass (*A. dasystachyum*). The Whitmar variety of beardless wheatgrass is grown in Washington, Oregon, and Wyoming.

PERENNIAL RYEGRASS

Perennial ryegrass, or English ryegrass (*Lolium perenne*), is one of the first perennial grasses to be cultivated for forage.

Adaptation

Perennial ryegrass is less hardy than many other grasses.[41] It is grown on about 1.1 million acres chiefly in California, Indiana, Illinois, and Washington. While it is considered a wetland grass in some regions, production usually declines as drainage becomes poor. Hot dry weather affects plant growth adversely.

Description

Perennial ryegrass is a tufted, short-lived perennial that persists for three to four years. The plants grow 1 to 2 feet in height. There are numerous long, narrow leaves near the base of the plant, the seed stems being nearly naked. The inflorescence is a spike, the spikelets

Figure 21–14. Perennial ryegrass. Plant (*left*), spike (*right*).

being set edgewise to the rachis (Figure 21–14). The lemmas of the florets are entirely or nearly awnless.

Culture

Perennial ryegrass may be seeded either in the fall or early spring, but usually in the spring where the winters are severe. The crop is sown at the rate of 20 to 25 pounds per acre where used alone for forage or seed production. Common ryegrass is sometimes seeded with small grain at the rate of 8 to 10 pounds per acre for annual pastures. For the best quality of hay, it should be cut in the soft-dough stage.

Utilization

Perennial ryegrass is used primarily in permanent pasture mixtures to furnish early grazing while long-lived grasses are becoming established. It is used occasionally for hay or winter cover. It is often used for seeding lawns, particularly in mixtures, and is easily established, but the turf is coarse and not permanent. It usually survives for only one year in the eastern half of the United States.

ITALIAN RYEGRASS

Italian ryegrass (*Lolium multiflorum*), together with the so-called domestic or common rye grass, is grown on about 2.8 million acres in the United States, chiefly in Arkansas, California, and the southeastern states. Domestic ryegrass is a mechanical and hybrid mixture of Italian ryegrass with some perennial ryegrass. Ninety per cent of the nation's supply of ryegrass seed is produced in the Willamette Valley in Western Oregon.

Italian ryegrass is a hardy, short-lived annual or winter annual but some plants live into the second season. It makes a rapid growth when seeded in the spring, late summer, or fall.[24]

Awns are present on the seeds of Italian ryegrass and usually absent in perennial ryegrass (Figure 21–2). The culm of Italian ryegrass is cylindrical, whereas perennial ryegrass culms are slightly flattened. The leaves of Italian ryegrass are rolled in the bud, while those of perennial ryegrass are folded. The plants of Italian ryegrass are yellowish at the base, but those of perennial ryegrass are commonly reddish. Also, the roots of annual ryegrass seedlings exude a substance that emits a bright fluorescent glow when subjected to near-ultraviolet light (300–400 mu.). Perennial ryegrass roots lack this characteristic.

Italian ryegrass is a very palatable and productive pasture plant. It grows so rapidly that it can be grazed in a short time after seeding. The quick germination and large seeds (and seedlings) account for its prompt establishment.

Italian ryegrass is used as a companion crop for spring-seeded permanent pastures. Sown in combination with winter grains or crimson clover for temporary pasture, it makes a desirable bottom grass and increases the length of the grazing season. It is grown to a considerable extent for hay in western Oregon and Washington. It also is an important winter pasture in the Gulf Coast states.[47] Italian ryegrass often is seeded in lawns in the South in the fall in order to have a green turf after the warm-weather grasses have ceased growth and turned brown.

WILDRYE GRASSES

A number of wildrye grasses of species of *Elymus* are native to North America or Europe. Canada wildrye (*Elymus canadensis*), a bunchgrass, is widely distributed over the United States and Canada.[24] It is adapted to the northern Great Plains where it supplies green feed during the summer, except during drought. An improved leafy variety is established readily from seeding.

Russian wildrye (*Elymus junceus*) is adapted to the drier areas of the northern Great Plains, but also to somewhat saline irrigated lands.[35] It provides palatable summer grazing. It responds well to nitrogen fertilization on irrigated land. This grass was introduced from the Soviet Union in 1927. The improved Vinall variety has better seed quality than does the original type.

Two species, native to the western United States, are sown occasionally. They are giant wildrye (*E. condensatus*) and blue wildrye (*E. glaucus*).

MEADOW FOXTAIL

Meadow foxtail (*Alopecurus pratensis*), a native of Eurasia, is a bunchgrass. It has become established in the humid and subhumid areas of the northern half of the United States. It is best adapted to cool moist climates where it thrives on wetlands.[39] For many years the seed has been included in pasture mixtures, but sometimes it has been cut for hay.

TALL OATGRASS

Tall oatgrass (*Arrhenatherum elatius*) is grown generally over the United States, except in the drier areas, but is not important in any locality. It is adapted to well-drained soils, especially those that are sandy. It makes very poor growth in shade. This grass is a hardy, short-lived perennial bunchgrass, growing to a height of 2 to 5 feet. The inflorescence is an open panicle similar to that of cultivated oats, but the seed is much smaller.

Tall oatgrass is often sown in mixtures with red clover, alsike clover, orchardgrass, sweetclover, or occasionally with alfalfa. A stand is difficult to obtain because the seed is of low viability.[24] It is necessary to use 30 to 50 pounds of seed per acre when this grass is sown alone.

Tall oatgrass is used for pasture and hay. Although succulent, the grass has a peculiar taste to which grazing animals must become accustomed. It is considered palatable and highly nutritious. This grass will furnish an abundance of grazing from early spring to late in the fall. For hay, the crop should be cut at about the bloom stage.

BLUE GRAMA

Blue grama (*Bouteloua gracilis*) probably is the most important range grass in the Great Plains.

Adaptation

Blue grama occurs generally throughout the dry portions of the Great Plains as a component of the short-grass prairie. It predominates in drier sections and on sandier soils than those that favor buffalograss. Southern strains of blue grama tend to make a greater vegetative growth and a lower seed yield when moved northward, while northern strains grow too sparingly for southern conditions. Also, southern strains often lack resistance to cold.

Description

Blue grama is a low, warm-season, sod-forming perennial with fine, curling basal leaves of a grayish-green color. The leaves are 2 to 5 inches long and less than 1/8 inch wide, with the ligules sparsely (or occasionally distinctly) hairy (Figure 21–15).

Figure 21–15. Blue gramagrass plant and (1) floret, (2) spikelet, and (3) glumes.

The spikelet consists of an awned fertile floret as well as an awned, densely bearded, rudimentary floret. These appendages, which make the seed light and fluffy, interfere with drill seeding unless they are removed by processing. Blue grama frequently is confused with the staminate plants of buffalograss.

Culture

Stands of blue grama are established by seeding in the spring. When disked lightly and packed before seeding, blue grama has had considerable success when seeded in drilled sudangrass stubble that has produced some aftermath. Seeding may be done with a packer-seeder. Unprocessed seed is sown with a grain drill equipped with cotton planter boxes, or is broadcast. The land is packed again after seeding. Processed seed 60 to 70 per cent pure may be sown at the rate of 10 to 12 pounds per acre. It is desirable to include 2 to 4 pounds per acre of buffalograss burs and some side-oats grama with the blue grama seed. The seed should not be covered with more than $\frac{1}{2}$ inch of soil. Blue grama seed is harvested with a bluegrass stripper or a combine.[10]

Utilization

Blue grama is highly palatable and provides choice forage during the summer grazing period. The mature grass cures on the range and retains some of its nutritive value, thus providing excellent fall and winter grazing. The protein content of blue gramagrass decreases toward maturity. At the early-bloom stage the protein content is about two-thirds that of alfalfa at the same stage.

Related Species

A related species, sideoats grama (*Bouteloua curtipendula*), also has been introduced into cultivation. This is a bunchgrass suitable for pasture or hay.[29] Like blue grama, it is adapted to the western Great Plains and parts of the intermountain region. Other gramagrasses are black (*B. eriopada*), hairy (*B. hirsuta*), and Rothrock (*B. rothrockii*).

BUFFALOGRASS

Buffalograss (*Buchloë dactyloides*) is a native, long-lived, perennial, sod-forming grass.[4] It is one of the most important range grasses in the Great Plains, probably being second only to blue grama. It has been cultivated only since about 1930.

Adaptation

Buffalograss is found in the Great Plains region from Texas to North Dakota, but is most abundant on the heavier soils of the central Great Plains. It is not well adapted to sandy soils.[46] It is highly resistant to drought and heat. Although resistant to cold, it is a warm-weather grass that starts growth late in the spring and ceases growth when cold fall weather arrives. Buffalograss forms a dense tough sod under suitable growing conditions and withstands close grazing and tramping better than almost any other grass. Buffalograss furnished abundant forage to buffaloes, antelopes, and Indian ponies in the Great Plains. White settlers used the turf to build their sod houses. With the opening of the region to cattle grazing, buffalograss increased in importance because it soon replaced some of the taller grasses such as the bluestems (*Andropogon* species) that could not withstand heavy grazing, which resulted from their higher palatability.

Description

Buffalograss produces fine grayish-green leaves usually 2 to 4 inches long. The plant is largely dioecious. The staminate plants send up spikes 4 to 8 inches in height, while the pistillate plants produce seeds in burs borne down in the turf just above the surface of the soil. It produces rapid-growing long stolons that enable the plants to sod over bare spots quickly. Small clumps may cover an area of 4 square feet or more in a single season.

Culture

Buffalograss seed is difficult to harvest because it is borne among the leaves close to the ground, but combines with beater and suction attachments save much of the seed. Buffalograss seed is low in germination due both to unsound seed and a high percentage of dormancy.[49] The dormancy can be broken by hulling the seed or by special treatments. Seeding of buffalograss is now more feasible than is sodding. About 4 or 5 pounds of treated seed per acre are sufficient because of the rapid spreading of the plants.

Utilization

Buffalograss is too short to cut for hay, but the acre yields in the central and southern Great Plains are about as large as those of many of the taller grasses. It furnishes palatable nutritious pasturage. In seasons of drought, the stand or cover may become very thin, but when moisture conditions again become favorable the few plants that survive soon spread to form a thick turf. Furthermore, dormant seeds in pastures that appear to be ruined germinate and reestablish an adequate ground cover. Buffalograss lawns composed of pistillate plants are popular because they require little mowing or watering. The grass withstands drought better than any other grass that produces a dense turf. Its color, however, is less attractive than that of other turf grasses. It is more cold-resistant than bermudagrass and fully as hardy as Kentucky bluegrass.

OTHER RANGE GRASSES

Bluestems

Three native grass species of bluestem [*Bothriochloa* (*Andropogon*)] which are distributed over the Great Plains and western prairie states, along with two introduced species, have been seeded for pasture, hay, or erosion control.[2]

Big bluestem (*B. gerardi*) is a coarse bunchgrass commonly found in the eastern half of the six states that extend from North Dakota to Texas. It provides late spring and summer pasture, good ground cover, and also good hay when cut before the plants head.

Little bluestem (*B. scoparius*) occurs more generally in the semiarid Great Plains area. It is like big bluestem except that the plants are smaller.

Sand bluestem (*B. halli*) is distributed over the western Great Plains, particularly on the lighter soils.

Turkestan or yellow bluestem (*A. ischaemum*) was introduced from Asia. It has finer stems than the native bluestems, being especially adapted to the southwestern region from Texas to Arizona. The King Ranch variety is more productive but less hardy and drought-resistant than is the common type. It is adapted to Texas, particularly the southern part.

Caucasian bluestem (*A. caucasicus*) is adapted to the same conditions as little bluestem.

Lovegrasses

Three species of lovegrass (*Eragrostis*) have been sown on appreciable acreages.[2]

Weeping lovegrass (*E. curvula*) was introduced from Africa. It is adapted to the area from Oklahoma to Arizona.[2, 24] The grass is palatable in spring and fall, but not in summer unless mowed or closely grazed.

Lehman lovegrass (*E. lehmanniana*), also introduced from Africa, is adapted to the semidesert areas of New Mexico, Arizona, and Texas.

Sand lovegrass (*E. trichoides*) is a palatable native bunchgrass adapted to sandy soils in the central and southern Great Plains.

Other Western Grasses

Indian ricegrass (*Oryzopsis hymenoides*) is distributed over the western half of the United States, particularly on dry sandy soils. Its high palatability subjects it to overgrazing. The edible seeds resemble rice grains.

Switchgrass (*Panicum virgatum*) is a sod-forming species grown principally in the central and southern Great Plains. It provides spring grazing as well as summer grazing, or hay when soil moisture is ample. Switchgrass often is sown for erosion control.

Vine mesquite (*Panicum obtusum*) is a vigorous native perennial in the southwestern states. It spreads by means of numerous long stolons. The herbage is palatable when green and succulent.

Blue panicgrass (*Panicum antidotale*) is a tall, coarse, sod-forming grass with thick bulbous stolons. It is adapted to fertile well-drained soils in Texas, New Mexico, and Arizona. This grass provides palatable pasturage in early spring, especially in mixtures with alfalfa or sweetclover. It is also used to hold soil in washes, while the stalks can serve as a windbreak since they may reach a height of 9 feet.

Green needlegrass, or feather bunchgrass (*Stipa viridula*), occurs commonly over the northern and central Great Plains states. It provides leafy grazing or hay.

Indiangrass (*Sorghastrum nutans*) is a tall coarse grass that occurs in pastures or open woodlands over the eastern three-quarters of the United States. It provides a quick ground cover after seeding. It is a productive hay crop.

Dropseeds are grasses of the genus *Sporobolus* which shatter their seeds as soon as they are ripe. They occur over the hot dry sections of the southwestern United States. Two species, sand dropseed (*S. cryptandrus*) and sacaton (*S. airoides*), have been used for reseeding rangelands and eroded lands.

CARPETGRASS

Carpetgrass (*Axonopus affinis*) is a native of the West Indies, but is now widespread in the tropics of both hemispheres.

It is grown on the coastal-plain soils from southern Virginia to Texas. It is especially adapted to sandy or sandy loam soils, particularly where the moisture is near the soil surface. The plants require abundant heat and moisture. Carpetgrass grows throughout the year except when damaged by severe drought or heavy frosts. It tends to become established naturally on pastured land in the South, much as Kentucky bluegrass does in the North, and often replaces more palatable grasses.

Carpetgrass is a perennial creeping grass that forms a dense turf. It is readily distinguished by the compressed two-edged creeping stems that root at each joint, as well as by the blunt leaf tips (Figure 21–16). The flower stems grow to a height of about 1 foot. It is less productive than some of the taller grasses. It is now being planted only to a limited extent.

Carpetgrass is one of the most common perennial grasses for permanent pastures over much of the area where it is adapted. It seeds abundantly and forms a dense turf. It can stand heavy continuous grazing.[5] It may be pastured in the South from May to November. Lespedeza and white clover mixed with carpetgrass improve the pasturage but are difficult to maintain because of the dense carpetgrass turf. Carpetgrass lawns are common in the Southeast.

NAPIERGRASS

Napiergrass (*Pennisetum purpureum*), a native of tropical Africa, is grown in the warmer regions of the United States. It is adapted to the southeastern states and southern California.

This grass is a robust, cane-like, leafy perennial which grows 5 to 7 feet or more in height. It grows in clumps of 20 to 200 stalks. The inflorescence is a long, narrow, erect, golden spike about 7 inches

Figure 21–16. Carpetgrass: (1) plant, (2) panicle branch, (3) glume, (4) ligule, and (5) floret.

long. It is propagated by planting root or stem cuttings. It also is grown occasionally from seed.

Napiergrass is utilized primarily as a soiling crop in tropical countries. The mature plants are rather woody, but silage made from the mature crop is eaten readily. Napiergrass is palatable and nutritious when grazed rotationally in Florida.[6] The pasture should be stocked so that most of the grass blades are consumed in 5 to 8 days, after which 20 or more days are allowed between grazings for the grass to recover. It does not survive continuous close grazing.

DALLISGRASS

Dallisgrass (*Paspalum dilatatum*), introduced from South America,[24] now occurs abundantly from North Carolina to Florida and west to Texas. It also is grown under irrigation from west Texas to California. Farther north it is too tender for survival. Dallisgrass favors heavy soils that are too wet for bermudagrass.

Figure 21–17. Dallisgrass plant and culm.

Dallisgrass is a perennial with a deep root system. It grows in clumps or bunches 2 to 4 feet in height. The leaves are numerous near the ground, but the stems are practically leafless (Figure 21–17). The slender stems usually droop from the weight of the flower clusters.

Since the seed is very light, the seedbed must be carefully prepared for dallisgrass. Usually 5 to 10 pounds of hand-picked seeds per acre are sown. In the southern states, the crop is generally sown in October or November. The production of dallisgrass seed is difficult because of heavy ergot attack and a high percentage of sterile florets. Frequent mowing will prevent ergoty seeds from maturing. The ergot is poisonous to cattle.

Dallisgrass is primarily a pasture crop because of its tendency to lodge when left for hay. It is a good summer pasture grass for heavy, moist, fertile soils when grown with legumes such as white, hop, or Persian clover. White clover should be sown after the dallisgrass is established. Dallisgrass is not injured by persistent grazing, the leaves being quickly renewed. Permanent pastures of carpetgrass and bermudagrass are made more valuable when this grass is included.

OTHER SOUTHERN GRASSES

Bahiagrass (*Paspalum notatum*) is a low-growing perennial pasture grass introduced from South America. It differs from dallisgrass not only in being less hardy, but also in producing heavy runners that form a dense sod on sandy soils. Bahiagrass is a tender, warm-weather species adapted to the South Atlantic and Gulf Coast areas. Related species are vaseygrass (*P. urvillei*), and ribbed paspalum (*P. malacophyllum*).[24]

Rhodesgrass (*Chloris gayana*), introduced from South Africa in 1902, is a leafy nonhardy grass adapted to the Gulf Coast as well as to southern Arizona and California. It produces abundant seed, but also spreads by long stolons. It is a pasture and hay crop that is very tolerant to soil salinity.

Pangolagrass (*Digitaria decumbens*) was introduced from South Africa. It resembles a giant crabgrass. It is a nonhardy, rapid growing, leafy grass grown on about 500,000 acres in central and southern Florida for pasture, silage, and hay. Pangolagrass is very palatable, grows well on sandy soils, and is drought-resistant. It must be propagated vegetatively by planting sprigs or stems.

St. Augustinegrass (*Stenotaphrum secundatum*) is a native of the West Indies, Mexico, Africa, and Australia. It is a pasture and lawn grass grown in the states of the Gulf Coast. This grass forms a dense turf which withstands trampling but requires ample moisture as well as heavy fertilization. It is established by planting rooted stolons.

Paragrass (*Panicum purpurascens*), a native of Africa, is a tropical perennial grass grown to some extent on wetlands in southern Florida. It is propagated by planting pieces of stolons or stems.

Buffelgrass (*Pennisetum ciliare*) is a warm-season bunchgrass introduced from Africa. It is adapted to the Gulf Coast states and south central Texas. This grass is drought-resistant, but lacks cold resistance. It provides summer pasture and hay. A variety known as Blue buffelgrass is larger, more erect, more tolerant to cold, and especially adapted to heavy soils.

Smilograss (*Oryzopsis miliacea*), a warm-season bunchgrass from the Mediterranean region, is grown in southern California. Kikuyiigrass (*Pennisetum clandestinum*) is a leading pasture grass in Hawaii. It forms a dense sod from spreading stolons and rhizomes.[27b]

REFERENCES

1. Ahlgren, H. L. "Effect of fertilization, cutting treatments, and irrigation on yield of forage and chemical composition of the rhizomes of Kentucky bluegrass (*Poa pratensis*)," *J. Am. Soc. Agron.*, 30:683–691. 1938.

2. Anonymous. "Grasses and legumes for forage and conservation," *USDA Agricultural Research Service Spec. Rept., ARS* 22-43. 1957.
3. Arny, A. C., and others. "Reed canary grass," *Minn. Agr. Exp. Sta. Bul. 252.* 1929.
3a. Atkins, M. D., and Smith, J. E. "Grass Seed production and harvest in the Great Plains." *USDA Farmer's Bul. 2226,* pp. 1-30. 1967.
4. Beetle, A. A. "Buffalograss—a native of the shortgrass," *Wyo. Agr. Exp. Sta. Bul. 293.* 1950.
5. Blaser, R. E. "Carpetgrass and legume pastures in Florida," *Fla. Agr. Exp. Sta. Bul. 453.* 1948.
6. Blaser, R. E., Krik, W. G., and Stokes, W. E. "Chemical composition and grazing value of Napier grass, *Pennisetum purpureum,* grown under a grazing management practice," *J. Am. Soc. Agron.,* 34:167-174. 1942.
7. Bourg, C. W., and others. "Nitrogen fertilizer for bromegrass in eastern Nebraska," *Nebr. Agron. Cir. 97.* 1949.
8. Burton, G. W. "Coastal Bermudagrass," *Ga. Agr. Exp. Sta. Bul. N52.* 1954.
9. Burton, G. W., and Jackson, J. E. "Effect of rate and frequency of applying six nitrogen sources on Coastal Bermudagrass," *Agron. J.,* 54:40-43. 1962.
9a. Bush, L. A., and Buckner, R. C. "Tall Fescue Toxicity." In: *Antiquality Components of Forages,* CSSA Special Pub. No. 4. Madison, Wisc., pp. 1-155. 1973.
10. Cooper, H. W., and others. "Producing and harvesting grass seed in the Great Plains," *USDA Farmers Bul. 2112.* 1957.
10a. Couch, H. B. *Diseases of Turfgrasses.* Reinhold, New York, pp. 1-289. 1962.
11. Cowan, J. R. "Tall fescue," in *Advances in Agronomy,* Vol. 8, Academic Press, New York, pp. 283-319. 1956.
12. Denman, C. E., and others. "Performance of weeping lovegrass under different management practices," *Okla. Agr. Exp. Sta. Tech. Bul. T-48.* 1953.
13. Evans, M. W. "The life history of timothy," *USDA Bul. 1450,* pp. 1-52. 1927.
14. Evans, M. W. "Kentucky bluegrass," *Ohio Agr. Exp. Sta. Res. Bul. 681.* 1949.
15. Evans, M. W., and Thatcher, L. E. "A comparative study of an early, a medium, and a late strain of timothy harvested to various stages of development," *J. Agr. Res.,* 56:347-364. 1938.
16. Evans, M. W., Welton, F. A., and Salter, R. M. "Timothy culture," *Ohio Agr. Exp. Sta. Bul. 603,* pp. 1-54. 1939.
17. Frolik, A. L., and Newell, L. C. "Bromegrass production in Nebraska," *Nebr. Agr. Exp. Sta. Circ. 68.* 1941.
17a. Gould, F. W. *Grass Systematics.* McGraw-Hill, New York, pp. 1-382. 1968.
18. Hafenrichter, A. L. "New grasses and legumes for soil and water conservation," in *Advances in Agronomy,* Vol. 10, Academic Press, New York, pp. 349-406. 1958.

19. Hanson, A. A. "Grass varieties in the United States," *USDA Handbook 170,* pp. 1–124, Rev. 1972.
20. Hanson, A. A., and Carnahan, H. L. "Breeding perennial forage grasses," *USDA Tech. Bul. 1145.* 1956.
20a. Hanson, A. A., and Juska, F. V. *Turfgrass Science.* Am. Soc. Agron. Mon. 14, Madison, Wisc., pp. 1–715. 1969.
21. Harlan, J. R. "Midland Bermudagrass," *Okla. Agr. Exp. Sta. Bul. B-416.* 1954.
22. Harrison, C. M. "Reed canarygrass," *Mich. Agr. Exp. Sta. Bul. 220.* 1940.
22a. Heath, M. E., Metcalfe and Barnes, R. F., Eds. "Forages". IA. State Univ. Press, Ames, IA. 1973.
23. Hitchcock, A. S. *Manual of the Grasses of the United States,* 2nd ed. (rev. by Agnes Chase), USDA Misc. Pub. 200, pp. 1–1051. 1950.
24. Hoover, M. M., and others. "The main grasses for farm and home," in *Grass,* USDA Yearbook, 1948, pp. 639–700.
25. Hosterman, W. H., and Hall, W. L. "Time of cutting timothy: Effect on proportion of leaf blades, leaf sheaths, stems, and heads and on their crude protein, other extract, and crude fiber contents," *J. Am. Soc. Agron.,* 30:564–568. 1938.
26. Klages, K. H. "Comparative ranges of adaptation of species of cultivated grasses and legumes in Oklahoma," *J. Am. Soc. Agron.,* 21:201–223. 1929.
27. Kreitlow, K. W. "Diseases of forage grasses and legumes in the northeastern states," *Pa. Agr. Exp. Sta. Bul. 573.* 1953.
27a. Mays, D. A., Ed. *Forage Fertilization.* Am. Soc. Agro., Madison, Wisc., pp. 1–621. 1973.
27b. Mears, P. T. "Kikuyii (*Pennisetum clandestinum*) as a pasture grass. A review." *Trop. Grassl.* 4:134–152. 1970.
28. Newell, L. C., and Keim, F. D. "Field performance of bromegrass strains from different regional seed sources," *J. Am. Soc. Agron.,* 35:420–434. 1943.
29. Newell, L. C., and others. "Side-oat grama in the central Great Plains," *Nebr. Agr. Exp. Sta. Res. Bul. 207.* 1962.
30. Nielsen, E. L. "Establishment of Bermuda grass from seed in nurseries," *Ark. Agr. Exp. Sta. Bul. 409.* 1941.
31. Ramage, C. H. "Yield and chemical composition of grasses fertilized heavily with nitrogen," *Agron. J.,* 50:59–62. 1958.
32. Rampton, H. H. "Alta fescue production in Oregon," *Oreg. Agr. Exp. Sta. Bul. 427.* 1949.
32a. Rampton, H. H., and Jackson, T. L. "Orchardgrass seed production in Western Oregon." *Ore. Agri. Exp. Sta. Tech. Bul.* 108:1–39. 1969.
32b. Rampton, H. H., and others. "Kentucky bluegrass seed production in Western Oregon." *Ore. Agri. Exp. Sta. Tech. Bul.* 114:1–27. 1971.
33. Rather, H. C., and Harrison, C. M. "Alfalfa and smooth bromegrass for pasture and hay," *Mich. Agr. Exp. Sta. Cir. 189.* 1944.
34. Rogler, G. A. "Growing crested wheatgrass in the western states," *USDA Leaflet 469.* 1960.

35. Rogler, G. A., and Schaaf, H. M. "Growing Russian wildrye in the western states," *USDA Leaflet 524*. 1963.
36. Savage, D. A. "Grass culture and range improvement in the central and southern Great Plains," *USDA Circ. 491*, pp. 1–56. 1939.
37. Scholl, J. M., and others. "Response of orchardgrass to nitrogen and to time of cutting," *Agron. J.*, 52:587–589. 1960.
38. Schoth, H. A. "Reed canary grass," *USDA Farmers Bul. 1602*. 1938.
39. Schoth, H. A. "Meadow Foxtail," *Oreg. Agr. Exp. Sta. Bul. 433*. 1947.
40. Schoth, H. A., and Halperin, M. "The distribution and adaptation of *Poa bulbosa* in the United States and in foreign countries," *J. Am. Soc. Agron.*, 24:786–793. 1932.
41. Schoth, H. A., and Hein, M. A. "The ryegrasses," *USDA Leaflet 196*. 1940.
42. Seamands, W., and Lang, R. D. "Nitrogen fertilization of crested wheatgrass in southern Wyoming," *Wyo. Agr. Exp. Sta. Bul. 364*. 1960.
43. Seath, D. M., and others. "Comparative value of Kentucky bluegrass, Kentucky 31 fescue, orchardgrass, and bromegrass as pasture for milk cows. I. How kind of grass affected persistence of milk production, TDN yield, and body weight," *J. Dairy Sci.*, 39:574–580. 1956.
43a. Semple, A. T. *Grassland Improvement*, Leonard Hill, London, pp. 1–400. 1970.
44. Spencer, J. T., and others. "Seed production of Kentucky bluegrass as influenced by insects, fertilizers, and sod management," *Ky. Agr. Exp. Sta. Bul. 535*. 1949.
45. Wagner, R. E. "Influence of legume and fertilizer nitrogen on forage production and botanical composition," *Agron. J.*, 46:167–171. 1954.
46. Walster, H. L., and others. "Grass," *N. Dak. Agr. Exp. Sta. Bul. 300*. 1941.
47. Weihing, R. M., and Evatt, N. S. "Seed and forage yields of Gulf ryegrass as influenced by nitrogen fertilization and simulated winter grazing," *Tex. Agr. Exp. Sta. Prog. Rpt. 2139*. 1960.
48. Weintraub, F. C. "Grasses introduced into the United States," *USDA Handbook 58*. 1953.
49. Wenger, L. E. "Buffalo grass," *Kans. Agr. Exp. Sta. Bul. 321*, pp. 1–78. 1943.
50. Westover, H. L., and others. "Crested wheatgrass as compared with bromegrass, slender wheatgrass, and other hay and pasture crops for northern Great Plains," *USDA Tech. Bul. 307*. 1932.
51. Wilkins, F. S., and Hughes, H. D. "Agronomic trials with reed canary grass," *J. Am. Soc. Agron.*, 24:18–28. 1932.

PART THREE

LEGUMES

Chapter 22

Alfalfa*

ECONOMIC IMPORTANCE

Alfalfa is probably the world's leading hay crop, with the largest production in Argentina and the United States. It is grown on some 20 million hectares on all continents, but seldom in humid tropical and subtropical areas or on unlimed acid soils or marshland. In the United States from 1970 to 1972, alfalfa, alone or in mixtures, was harvested for hay on 11 million hectares (27 million acres). It ranked fourth in domestic crop acreages and is also the leading forage legume in Canada. The average production in the United States was nearly 77 million tons (70 million metric tons) with a yield exceeding 2.8 short tons per acre. The leading states in alfalfa hay production were Wis-

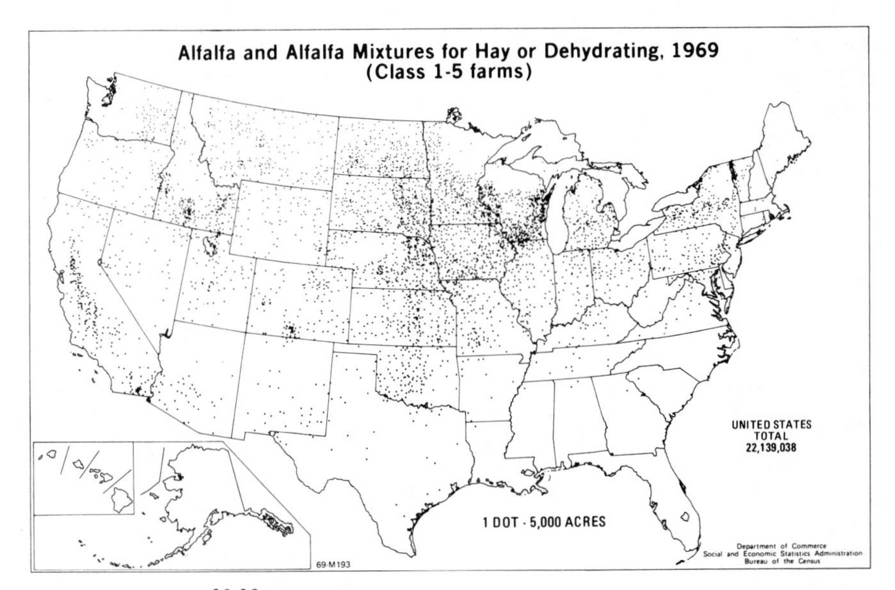

Figure 22–1. Alfalfa acreage.

* For more complete information see Bolton, J. L., *Alfalfa: Botany, Cultivation, and Utilization*, Interscience, New York, 1962, pp. 1–474.

consin, California, Minnesota, Iowa, Nebraska, and South Dakota (Figure 22–1). Alfalfa usually is grown alone in the western irrigated areas, but often in grass mixtures in the humid areas.

Alfalfa was harvested for seed on an average of more than 500,000 acres in the United States in 1968 and 1969, with a production exceeding 100 million pounds or more than 200 pounds per acre (224 kilograms per hectare). The leading states in seed production were California, Idaho, Washington, Oklahoma, and Oregon. About 90 per cent of the production in 1970 consisted of named improved varieties.

HISTORY OF ALFALFA CULTURE

The name alfalfa, which comes from the Arabic language, means *best fodder*. In Europe it is usually called lucerne. Most authorities believe that alfalfa originated in southwestern Asia, although forms from which it could have arisen are found in China and Siberia.[76] Alfalfa was first cultivated in Iran before 700 B.C., and from there it was carried to Arabia, the Mediterranean countries, and finally to the New World. Evidence of the ancient introduction of alfalfa into Arabia is found in the strongly marked characteristics of Arabian varieties,[32] which apparently represented centuries of acclimatization in an arid region.

The first recorded attempt to grow alfalfa in the United States was in Georgia in 1736. Although alfalfa later was tried in the eastern states from time to time, it was not always successful. Introductions into California from Peru in 1841 and from Chile about 1850 started a tremendous expansion.[77] The introduction of a winter-hardy alfalfa into Minnesota by Wendelin Grimm during the mid-1800s further increased the range of utilization of this crop.

ADAPTATION

Alfalfa has a remarkable adaptability to various climatic and soil conditions, as is shown by its wide distribution.

The alfalfa plant makes its best growth in relatively dry climates where water is available for irrigation. Irrigation requirements per year may exceed 2 meters in hot desert areas. It will withstand long periods of drought due to a very deep root system but is unproductive under such conditions. Alfalfa is now grown successfully not only in the central and eastern United States, but also in the South.

Alfalfa tolerates extremes of heat and cold. Yellow-flowered alfalfa has survived temperatures as low as −64° C. Common alfalfa has been grown in Arabia, as well as in Death Valley, California, where maximum summer temperatures are as high as 120° F. but the humidity is low. However, alfalfa is relatively dormant during the summer in very hot regions.

Alfalfa is best adapted to deep loam soils with porous subsoils. Good drainage is essential. The plant requires a large amount of calcium for satisfactory growth. It survives on almost all soils in the semiarid region except those high in alkaline salts, or those that have a shallow water table.

Alfalfa has spread to the more humid eastern states in recent years with the increased knowledge of the requirements for lime, for inoculation with nitrogen-fixing bacteria, and for certain plant nutrients, and also the breeding of hardier, disease-resistant varieties. Except on a few limestone soils, applications of lime are essential for satisfactory growth of alfalfa east of the Mississippi River.

BOTANICAL DESCRIPTION

Alfalfa is an herbaceous perennial legume that may live 15 to 20 years or even more in dry climates unless insects or diseases destroy it. The most commonly cultivated species is *Medicago sativa*. Yellow-flowered alfalfa (*M. falcata*) is sometimes regarded as a subspecies (*M. sativa* ssp. *falcata*) of common alfalfa. Yellow-flowered alfalfa is distinguished by its yellow flowers, sickle-shaped seed pods, decumbent growth habit, low-set branching crowns, and a preponderance of branched roots. It is not satisfactory for hay in America because it is somewhat prostrate and usually yields only one cutting a season in the northern states. It is grown on only about 3,000 acres, but is of interest primarily for hybridization with common alfalfa to produce hardier varieties.[10]

The Alfalfa Plant

The alfalfa seedling emerges with the two cotyledons, produces one unifoliolate leaf, followed by alternate pinnately trifoliolate leaves thereafter. The plant varies in height from 2 to 3 feet. It has 5 to 20 or more erect stems that continue to arise from the fleshy crown as the older branches mature and are harvested. Several short branches may grow from each stem. The oblong leaflets are sharply toothed on the upper third of the margin, the tip being terminated by the projected midrib (Figure 22–2). About 48 per cent of the weight of the plant may consist of leaves.[45]

The root system consists of an almost straight taproot which, under favorable conditions, penetrates the soil to a depth of 25 to 30 feet or more. There are a few side branches that extend short distances from the main taproot. The main root normally persists during the entire life of the plant. All varieties of alfalfa develop branch roots in compact soil, while taproots predominate in porous soil.[8]

Figure 22–2. Branches and flowers of alfalfa. Leaves at lower right.

The flowers of common alfalfa, borne in axillary racemes, are purple except in the variegated types. The fruit is a spirally twisted pod that usually contains from one to eight small kidney-shaped seeds (Figure 22–3). The seeds are normally olive-green in color (Figure 22–4).

Alfalfa can be propagated vegetatively from stem or crown cuttings.

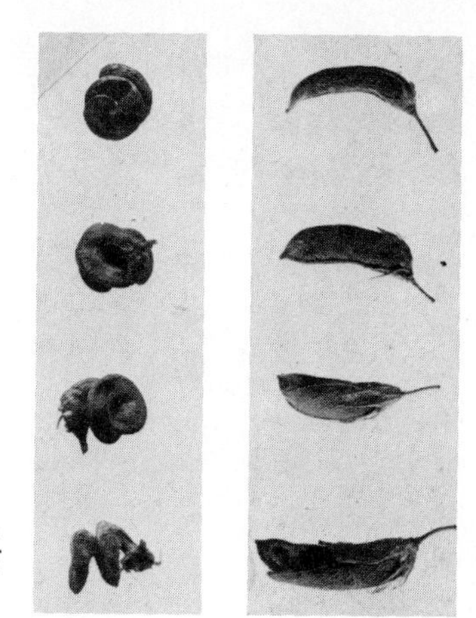

Figure 22–3. Coiled seed pods of common alfalfa (*left*), and sickle-shaped pods of yellow-flowered alfalfa (*right*).

Pollination

The plants of commercial varieties of alfalfa are slightly to almost completely self-sterile. Under ordinary climatic conditions, they are nearly incapable of automatic self-pollination.[7] From 7 to as high as 80 per cent crossing between closely associated plants of purple-flowered and yellow-flowered alfalfa has been observed.[67, 74] A decrease in both forage and seed yields usually occurs when flowers are self-fertilized, the decrease being marked in most cases.

The external flower structures, the keel, wings, and standard enclose the reproductive tissues and serve to attract insects. Tripping is necessary for seed set except in a small percentage of the flowers. Tripping is the release of the sexual column from the keel to the flower.[67, 68] The sexual column includes the style, stigma, and part of the ovary enclosed and surrounded by ten stamens and diadelphous filaments. Tripping, which takes place when the flower is in a turgid condition, is accompanied by an explosive action as though a spring under tension has been released. Tripping naturally, or by mechanical jarring, wind, rain, or sun induces mostly self-pollination and little seed production. Partial self-incompatibility and ovule abortion result in a low percentage of fertilization when alfalfa plants are self-pollinated.[11] The net fertility 144 hours after pollination may be about six times as high in crossed plants as in selfed plants. Insects carry pollen from different plants when tripping the flowers, causing cross-pollination and seed development.

Wild bees, chiefly leaf-cutter bees (*Megachile* spp.) and alkali or ground bees (*Paranomia* spp.) collect pollen and are the most effective pollinators, but bumble bees (*Bombus* spp.) are fairly effective trippers.[51]

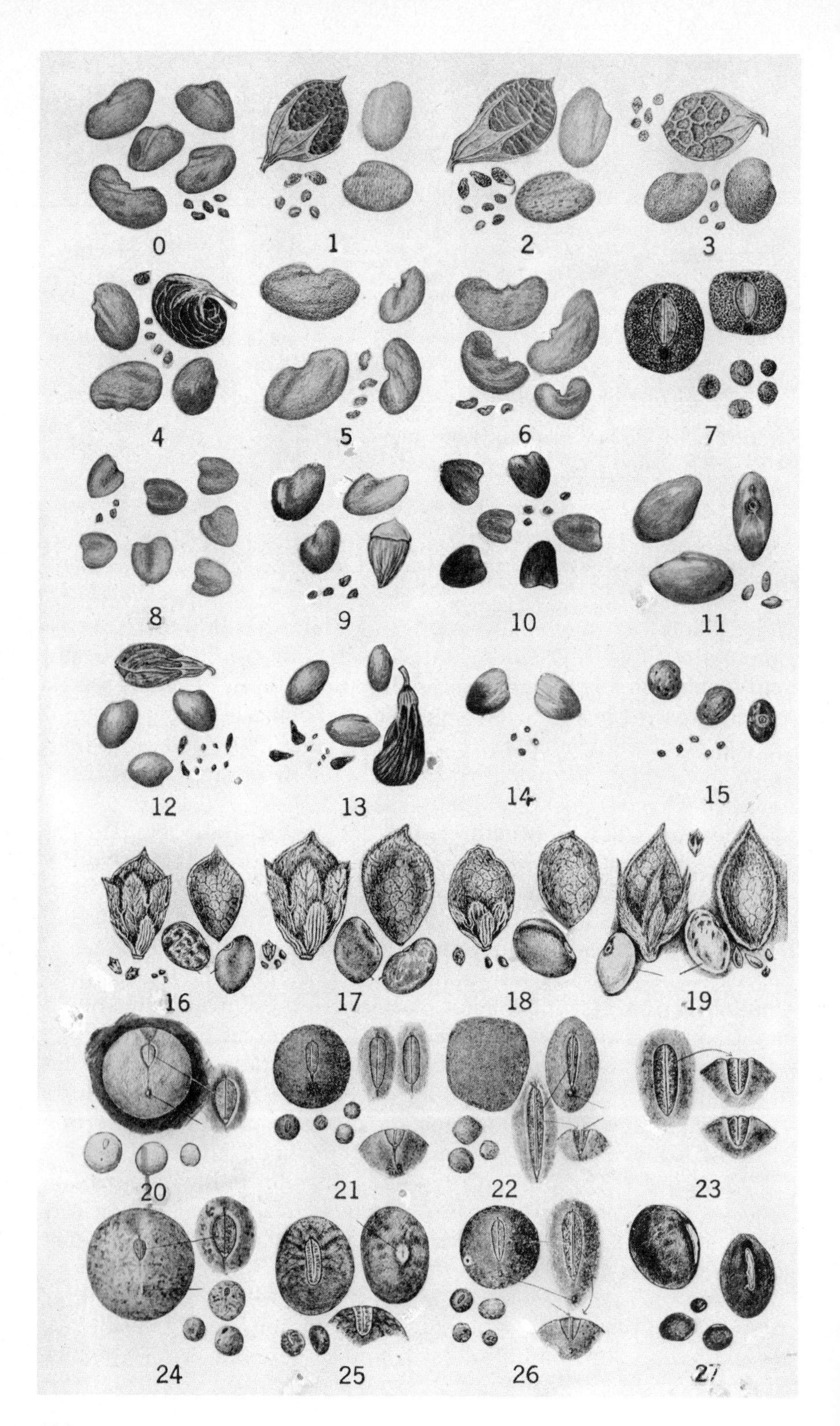
0
1
2
3
4
5
6
7
8
9
10
11
12
13
14
15
16
17
18
19
20
21
22
23
24
25
26
27

About 38 million flowers must be tripped and cross-pollinated to produce seeds for a 500-pound-per-acre crop. Two to three good honey bee colonies per acre provide the 2 to 3 bees per square yard essential for such high seed yields. Only 5 to 20 per cent of the bees are pollen collectors. Nectar collectors, however, trip some of the flowers and thus spread pollen. The drones refuse to participate in the pollination.[50] The alfalfa seed growers usually temporarily rent the colonies from bee keepers, placing them in or adjoining the fields, within 800 to 1400 feet from the most distant plants. The wild bee population may be ample for pollination in localities that are only partly under cultivation. From a distance bees select alfalfa flowers by color, with a preference for white, followed by dark reddish purple and least by brilliant yellow. When nearby they are attracted by aroma, and finally by the amount and sweetness of the nectar.[40]

The small alfalfa leaf-cutting bee (*Megachile rotundata* F.) is being propagated in Canada and the United States for use in pollination. Some 2,500 to 5,000 female bees are needed for each acre of alfalfa. Half of the cocoons produce males, which are ineffective pollinators.[31] Alkali bees are propagated in the United States. Specially treated artificial soil beds are maintained by water and salt control for bee propagation.[64]

Other flowers in bloom may attract bees away from alfalfa and thus reduce the seed set. Lack of tripping by beneficial insects probably is the chief cause of poor seed yields. Insects rarely collect alfalfa pollen in cool or humid weather. Thin stands of alfalfa, as well as restricted irrigation at flowering time, encourage honey-bee visitations. Thickly planted and lodged plants give poor seed sets. Ample food reserves in alfalfa roots also contribute to good seed sets.

Rupture of the stigmatic surface by tripping is essential to the penetration of the pollen tubes.[5] After tripping occurs, the proper moisture relationship for pollen germination and pollen-tube growth must be maintained to effect fertilization.

Lygus bugs cause bud damage and flower dropping, and often cause a poor seed set in alfalfa.[9, 65]

Figure 22–4. [OPPOSITE] Seeds of forage legumes (enlarged and natural size): (0) alfalfa, (1) white sweetclover, (2) yellow sweetclover, (3) sourclover, (4) black medic, (5) California burclover, (6) spotted burclover, (7) roughpea, (8) white clover, (9) red clover, (10) alsike clover, (11) crimson clover, (12) small hop clover, (13) large hop clover, (14) hop clover, (15) birdsfoot trefoil, (16) common lespedeza, (17) Kobe lespedeza, (18) Korean lespedeza, (19) sericea lespedeza, (20) smooth green pea, (21) hairy vetch, (22) common vetch, (23) woolypod vetch, (24) mottled field pea, (25) Hungarian vetch, (26) narrowleaf vetch, and (27) purple vetch.

GROUPS

Five somewhat distinct groups of commercial alfalfas grown in the United States are recognized.[76] These are Common, Turkistan, Variegated, and Nonhardy.

Common Alfalfa Group

The common group includes the ordinary purple-flowered, smooth alfalfa. Hardy types adapted to the northern states tend to recover more slowly after cutting than those produced farther south. The strains produced in the Southwest recover rapidly after cutting, but they are very susceptible to winter injury except when grown in the southern states. Adaptation, especially as to cold resistance, is extremely important in common alfalfa. Seed should be procured from a source where the winters approximate in severity the conditions where it is to be planted.

Several varieties[4] are resistant to bacterial wilt. These varieties of common alfalfa are adapted to the eastern states, the deep South and Southwest, or California.

Flemish Group

Flemish type varieties originated in western Europe or were developed elsewhere from imported European varieties. The plants are similar to those of the common group, but they start spring growth earlier and are less hardy than the standard varieties. They also recover more rapidly after cutting but are less persistant, being more susceptible to bacterial wilt (*Corynebacterium insidiosum*). Flemish types are adapted to the mild humid climates of the eastern and Pacific coastal states.[31] Quick recovery is not advantageous in semiarid areas where only 1 or 2 cuttings are obtained.

Turkistan Group

The Turkistan group is derived from alfalfas that originated in Turkistan. They are characterized by slow recovery after being cut, early fall dormancy, susceptibility to leaf diseases, winter hardiness, resistance to bacterial wilt, and low seed yields. Turkistan alfalfa is practically indistinguishable from Common. The growth is somewhat shorter and more spreading, while the leaves are smaller with slightly more pubescence. Rabbits and hogs graze it in preference to common alfalfa.

Few of its varieties are important in United States, but several improved varieties have Turkistan in their parentage.

Variegated Group

Introduction, breeding, testing, and distribution of cold-resistant variegated varieties are largely responsible for the successful culture of alfalfa in several northern states.

The variegated alfalfas appear to have resulted from natural crossing between the purple-flowered and yellow-flowered species. The predominant flower color is purple, but some brown, green, greenish-yellow, yellow, and smoky to nearly white flowers occur. The seed pods vary in shape from sickle-shaped to coiled. Such variegation is characteristic of many selected plant progenies owing to repeated cross-pollination. The alfalfas of the variegated group are more cold-resistant than common alfalfas. Their hardiness is due to the presence of yellow-flowered alfalfa in their ancestry as well as to natural selection under severe climatic conditions. The variegated group sometimes has been considered a separate species, called *Medicago media.*

Variegated alfalfa varieties predominate in the North Central states.

Synthetic varieties, developed by composite crossing of several strains to obtain resistance to bacterial wilt and winter hardiness, have a variegated flower color with a limited number of yellow flowers. They are adapted to the northern United States.[26, 73]

Nonhardy Group

Nonhardy varieties introduced from Peru, Africa, or India are adapted to the deep Southwest.[3] The Hairy Peruvian variety has pubescent leaves and stems.

Creeping Alfalfas

Varieties of creeping alfalfa,[23] which can multiply from rhizomes, were grown on 50,000 acres of pastureland in the United States in 1972. They often yield less than upright varieties but spread where stands are thin. They also are grown in western Canada and persist longer under pasturing.[30]

Hybrid Alfalfa

Hybrid alfalfa is produced by selection of clonal lines that combine well, after which they are increased from stem cuttings. A group of plants propagated from a single plant is called a clone. Lines to be crossed are grown in alternate strips from transplanted cuttings. Inbreeding of these lines is unnecessary because they can be reproduced at will by maintaining them as perennials and by propagating them vegetatively. Probably 90 per cent of the seed produced is hybrid because the lines used are largely self-sterile.[70]

HARD OR IMPERMEABLE SEED

Impermeable seeds, or those known as hard seeds occur commonly in alfalfa, the average in Colorado-grown seed being 22 per cent,[18] and in all areas ranging from 0 to 72 per cent. The percentage of hard seed increases with maturity. There appears to be no relation between climate as influenced by altitude and production of impermeable seed. Hardy varieties exhibit a higher percentage of impermeable seeds than do the less hardy ones. Mature fresh, hard alfalfa seeds are alive and there is no important difference between permeable and hard seeds in the rate of loss of viability when stored the same length of time.[34] Most of the hard seeds of alfalfa that fail to germinate in the laboratory, will germinate sooner or later when sown in the field.[75] Alfalfa seed lots with many hard seeds have almost the same agricultural value as those with only a few.

WINTER HARDINESS

Resistance of alfalfa to low winter temperatures is important for maintenance of stands in the northern states and Canada. Alfalfa plants have no autonomous rest period, but become dormant because of environmental conditions unfavorable for growth.[63] Winter hardiness is largely a varietal characteristic.[71] In general, only cold-resistant variegated varieties, or northern strains of common alfalfa, are suitable north of the 40° latitude. The nonhardy varieties are suitable where winter temperatures seldom are lower than −8° C.

Hardy alfalfas become dormant earlier in the fall and harden more rapidly than do nonhardy alfalfas.[56] A hardened plant would be one which, due to certain environmental changes such as cooler temperatures and shorter days, goes through a hardening process and becomes more capable of withstanding cold and severe weather conditions frequently experienced during winter months. The hardening process in alfalfa is cumulative over the fall and early winter, and then hardening decreases toward spring. Hardy varieties retain their hardening longer than do nonhardy ones. The plants harden best at 5° C., but much greater hardiness develops under alternating temperatures of 0° C. for 16 hours and 20° C. for 8 hours.[68] Hardening progresses with decreasing day length and alternating temperatures in the fall.

Premature or too frequent cutting of alfalfa depletes the organic root reserves, with a subsequent reduction in stand either by winter injury or disease.[2, 47, 78] These reserves are the carbohydrate and nitrogen compounds elaborated in the leaves, stored mostly in the roots, and later utilized by the plant for maintenance and for future growth. New foliage growth is made partly at the expense of these root reserves. The root reserves are increased during the blossoming period.[19]

Cool weather in the fall is especially conducive to food storage. The protection afforded by some fall top growth aids in maintenance of stands.[2, 21] In Michigan the roots of alfalfa plants cut in September were lower in percentage of dry matter than when cut in October.[62]

In areas where winter hardiness is a problem, alfalfa management is usually best practiced when the plant is not cut during a 2- or 3-week period before the final killing frost. Cutting immediately before this period insures adequate buildup of root reserves to insure maximum winter survival. From the standpoint of survival alfalfa plants should not be cut before the blossom stage, nor permitted to go into the winter in the northern states without some top growth.[79]

ROTATIONS

Alfalfa is an important crop in irrigated rotations, where it usually is sown with a small-grain companion crop. It is cut for hay or seed for two or three years, and often longer when good stands are maintained. The alfalfa field is then plowed in preparation for an intertilled crop such as corn or potatoes. The roots and aftermath furnish nitrogen for the crops following alfalfa. A good alfalfa crop may provide as much as 60–100 pounds of carryover nitrogen for a successive crop. Usually after two years devoted to the above intertilled crops or field beans, the land is again sown to alfalfa and grain. When alfalfa sod is turned under, the land is often plowed shallow in the fall to cut off the roots below the crown of the plant, a procedure called crowning. The land is again plowed deeper the following spring. Ordinary spring plowing without crowning fails to kill many of the alfalfa plants and these are difficult to destroy by cultivation.

When alfalfa is seeded in the humid regions, it is desirable to precede it in the rotation with some cultivated crop for one or two years to eliminate weeds. The alfalfa or alfalfa-grass mixture usually is plowed up after two or three years, after which the field often is planted to corn.

Alfalfa leaves the soil high in available nitrogen, which is used to advantage by the next crop where sufficient soil moisture is available. In eastern Nebraska the increased nitrates after alfalfa bring favorable results in seasons of relatively high rainfall, but may prove detrimental when the rainfall is deficient.[71] The injury results from overstimulation and excessive vegetative growth early in the season with the result that the crop need for moisture is increased beyond the supply, and the plants fire or burn. Intervals between alfalfa in rotations should be lengthened where burning is a problem. Sorghum is a good crop to follow alfalfa. After alfalfa the soil may be but little drier within the zone of root depths of annual crops than it is after small grains.

In nonirrigated areas where the annual precipitation is less than 30 inches, alfalfa exhausts the subsoil moisture. Yields on land sown to alfalfa for the first time have declined abruptly after four or five years due to subsoil moisture depletion. Subsequent growth is dependent upon current rainfall which is not sufficient for maximum production under subhumid conditions. In Nebraska and Kansas alfalfa roots have penetrated 30 to 40 feet in the soil and used the available subsoil moisture at these great depths.[15, 44, 45, 71] In Nebraska experiments, alfalfa hay yields averaged 5.62 tons per acre during the first three years before the deep subsoil moisture was exhausted. During the five years thereafter, the yields averaged only 2.43 tons per acre, even though good stands were maintained.

When alfalfa is sown again on moisture-depleted land the yields are far from satisfactory. In Nebraska, only 30 to 53 per cent of the moisture removed by the alfalfa was restored during the 13 to 15 years of cropping to annual crops. In eastern Kansas two years of fallow were necessary to restore subsoil moisture on old alfalfa ground to the point where the roots of a newly seeded crop could penetrate through moist soil to a depth of 25 feet or more.[22] Soil-moisture needs explain the success of alfalfa on bottomlands subject to overflow or subirrigation, in the semiarid regions.

CULTURE

A firm, moist, well-prepared seedbed is favorable for obtaining a stand of alfalfa.[53, 76] The land may be fall-plowed for spring seeding, or, in the northeastern states, merely disked where alfalfa is to be summer-seeded after small grains.

Fertilizers

On most soils east of the 95th meridian alfalfa responds to lime, to commercial fertilizers, particularly superphosphate, often to potash, and to light applications of boron.

Seed inoculation with effective strains of *Rhizobium meliloti* is required for ample nitrogen fixation in any field that has not produced good alfalfa growth during previous years. Inoculation is especially important in the humid areas. Alfalfa responds to phosphorus applications on many irrigated calcareous soils that have been cropped for some time. A low level of available soil phosphorus is one of the causes of poor yields of alfalfa. An application of 300 pounds of P_2O_5 per acre was required on deficient soils to supply the phosphorus needs for a companion crop and three or four years of alfalfa.[59] Ample potash is essential for high yields. In addition zinc, magnesium, and molybdenum are sometimes needed. Liming and fertilization are required in the humid areas of the Pacific Northwest. Alfalfa often will take up and store in its tissue more potash than is actually needed. This process, known as luxury consumption, may be occurring if at harvest time the leaf content of K is greater than 2 per cent by weight. K contents as high as 3.5 per cent have been recorded, but the excess did not contribute significantly to yield.

Seeding

Alfalfa is grown alone for seed, market hay, or alfalfa meal in the western states. It is often seeded in mixtures with grasses for pasture, silage or home-grown hay, especially in the humid areas.

It is a common practice to seed alfalfa 1/4 to 1 1/2 inch deep on heavy soils and slightly deeper on light soils, in close drills.[53] Alfalfa is seldom grown in cultivated rows, except strictly for seed production. It is not an economically successful hay crop where the climate is so dry that production in cultivated rows is essential for satisfactory growth.

In the northern half of the country, alfalfa generally is seeded in the early spring with a companion crop where rainfall is abundant or irrigation water is available. Sometimes it is sown in August or early September. In the Southeast the most favorable time ranges from August 15 in the latitude of Washington, D.C., to October or November along the Gulf Coast. In the southern Great Plains, as well as in the southern parts in the Corn Belt, late summer seeding or early fall seeding is often practiced. October is the best month for seeding in the irrigated areas of the Southwest, although good stands are obtained at other times from August to April 15.[76]

The general rate of seeding east of the Appalachian Mountains has been 15 to 20 pounds per acre because of frequent difficulties in establishing stands. Most Corn-Belt experiment stations advise 8 to 12 pounds, which should be ample to produce satisfactory stands on well-prepared seedbeds, even in the east. Seeding rates of 10 to 15 pounds per acre with an occasional 20 pound rate,[14] are sown under irrigation in the West. Under dryland conditions, 6 to 8 pounds is the recommended rate for hay production. For seed production under irrigation, the usual rate is about 1 to 3 pounds per acre in cultivated rows 2 to 3 1/2 feet apart.[54] Alfalfa is seeded in mixtures with bromegrass, orchardgrass, or tall fescue for forage at 4 to 10 pounds per acre.

Cutting

Alfalfa produces one or two cuttings a season under semiarid conditions, two or three cuttings in the northern states, three to five in the Central and South Central states, and seven or eight cuttings of hay on irrigated land in southern California and Arizona. In the latter region, the fields usually are pastured during July and August when growth is retarded by hot weather. Too frequent cutting reduces root reserves.

In tests in Wisconsin, alfalfa cut only twice in the season and then left over winter with a protective vegetative cover and abundant stored-food reserves yielded 3.3 tons per acre of dry hay as an average for the three subsequent seasons.[20] After a third or fall cutting, the yields were 2.4 tons per acre, or a reduction of 26 per cent. The stands were badly thinned by the fall cutting. On soils of optimum fertility in Michigan the forage yield of the first crop in the spring was significantly less from alfalfa cut the previous September than from that not cut or cut in October after the food reserves had moved into the roots.[62]

As a crop of alfalfa matures, the yield increases to approximately the full-bloom stage, the percentages of protein and of carotene decrease rather regularly, and the protein and carotene per acre reach a maximum at approximately the early-bloom stage.[79] The lignin and crude fiber content increases and the digestible nutrients decrease as the plants approach maturity.[36]

At the Nebraska station, the average yields of common alfalfa were 3.00, 3.04, 3.35, 3.43, 3.19, and 2.82 tons per acre for cutting in the prebloom, initial bloom, one-tenth bloom, one-half bloom, full-bloom, and seed stages of growth, respectively.[41]

> From the combined standpoints of acre yields of hay and feed constituents, quality of hay, and permanency of stand, harvesting alfalfa during the period from one-tenth to one-half bloom or at approximately the new growth stage should prove the most desirable practice. A modification of this practice to satisfy local conditions may often prove desirable but frequent cutting in more immature stages should be avoided.[71] [Figure 22–5]

In practice, most alfalfa-grass mixtures for hay may be cut as though they were pure alfalfa.[79]

Quality in alfalfa hay depends upon color, leafiness, fineness of stems, and freedom from foreign material. Protein content is closely correlated with leafiness, while vitamin-A content is closely associated with green color. These factors are definitely involved in the curing process. The leaves, which contain about 70 per cent of the total protein in the plant, are easily lost by shattering when the hay is handled improperly. Extensive experiments[41, 42, 43, 61] indicate that crimping and partial swath curing to hasten the rate of drying, followed by windrowing before the leaves dry sufficiently to cause such shattering and by prompt baling or storage, appears to be the best farm practice.[14]

Annual broadleaved weeds in alfalfa or alfalfa-grass mixtures can be partly controlled by applying suitable herbicides immediately after gathering the first crop. A herbicide applied in winter when the alfalfa is dormant controls many annual weeds, including grasses.

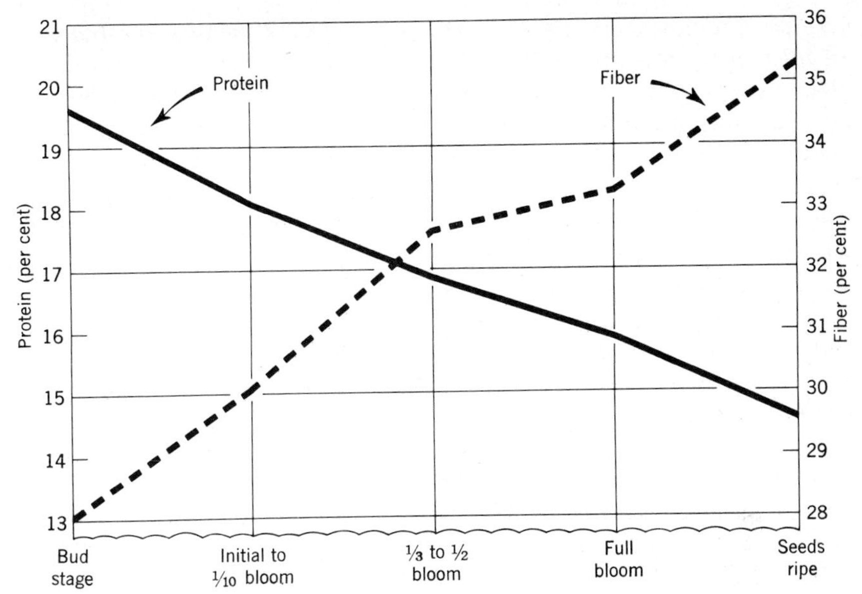

Figure 22–5. Alfalfa protein and fiber content at different growth stages.

Dehydrated Alfalfa

Alfalfa dehydration began in the early 1930's. Some 40 years later, 1,700,000 tons of dehydrated alfalfa meal were produced and another 300,000 tons of alfalfa meal were made by the earlier process of grinding sun-cured hay. By modern methods, the alfalfa is cut, chopped, and loaded into a dump truck or trailer with a self-propelled or tractor-drawn harvester. The chopped forage is hauled immediately to a near-by central plant where it is dried quickly in a rotary-drum dehydrator that usually is heated with gas or fuel oil. The dried product then is ground to a meal in a hammer mill. Much of the meal is pelleted immediately or cubed while the remainder is bagged. The pellets may be stored, shipped, and handled in bulk. Some of the pellets later are ground for mixing with other processed feeds.

Dehydrated alfalfa pellets contain about 17 per cent protein and 100,000 International Units (I.U.) of carotene per pound. The xanthophyll content is reduced by dehydration. The alfalfa is cut in the early-bloom stage when the protein and vitamin contents are still high. The carotene is often preserved by storage with inert gasses in sealed tanks, or by the addition of vegetable oils or antioxidants.[60]

ALFALFA AS A PASTURE CROP

Alfalfa is an excellent pasture for hogs when supplemented with grain. Despite the hazard from bloating, alfalfa in mixture with grasses has become increasingly important as a cattle pasture, particularly in the northern central and northeastern states. It is more productive than the usual permanent grass pastures, especially during hot dry periods. Summer pasturing of alfalfa is a common practice in the hot irrigated area of southern Arizona. In Michigan[57] alfalfa pastured continuously has been as productive as when pastured after a first cutting taken for hay each year. Pasturing the first growth and harvesting the second growth for hay increased the number of annual weeds. Heavy grazing in September proved to be injurious to stands.

Alfalfa is less likely to cause bloat when sown in mixtures with adapted grasses.[57] Consequently, alfalfa usually comprises not more than 50 per cent of the pastured forage. Bloating in ruminants grazing on alfalfa, clover, and most other forage legumes except birdsfoot trefoil, sainfoin, and crownvetch probably is caused by the occurrence of a saponin or chloroplastic protein in the plants, particularly young growth. Bloating is less frequent if the animals graze the legume only briefly at first, and are prefed before pasturing, or are provided with supplementary hay or straw in the pasture. Salt-molasses blocks containing 30 grams of Poloxalene (a surfactant) per pound of block placed in the pasture to provide salt usually prevents severe bloat.[17] Fatal bloat is caused by excessive intrarumenal physical gas pressure which prevents diaphragmatic contractions, and therefore no respiration, which is followed by heart failure.

Alfalfa, clover, and other legumes also contain coumestrol, an estrogen, particularly in diseased, old, spoiled, or ensiled plants, which may cause infertility in sheep and other animals.[6] Saponins also inhibit growth in chicks that are fed large quantities of alfalfa meal. Certain alfalfas contain less saponin than other varieties.

SEED PRODUCTION

Alfalfa-seed production is most successful where the climate is relatively dry, as in the semiarid and irrigated regions of the West.[53, 54] Acre yields of alfalfa seed west of the Rocky Mountains, where about 80 per cent of the American seed crop is produced, average several times those in midwestern states. California leads in alfalfa-seed production, mostly for shipment to other states. The most consistent seed production occurs when the crop makes a steady, continued growth. Abundant moisture results in rank vegetative growth unfavorable to seed production.[66] The insects important in seed production slow down their activity when the weather is cloudy, but stop entirely when it is cold and rainy.[71] As a rule, when seed production is incidental to hay production, growers leave the cutting for seed that matures during the hottest and driest part of the summer.

When alfalfa is grown specifically for seed, the highest average seed yields sometimes have been obtained where the crop is planted in rows, or spaced in hills, under either irrigated or dryland conditions. Maximum seed yields in Utah were obtained from alfalfa sown at the rate of 1 pound per acre in 24-inch rows.[53] Thin stands increased nectar secretion, bee visitation per flower, and also the percentage of cross-pollinated flowers that set seed. Sowing 1 pound per acre in 40-inch rows gave the highest seed yields in California. A growth regulator, sprayed on alfalfa plants at a rate of 1 or 2 ounces per acre, increased the number of plant branches and consequent seed yield.[24] Cultivation between the rows suppresses weeds and maintains thin stands. Herbicides kill dodder without injury to alfalfa.

Alfalfa is generally cut for seed when about two-thirds of the pods are brown or black. The crop may be harvested with a windrower. Later it is threshed with a combine equipped with a pick-up attachment. Alfalfa seed can be combined directly where the crop has matured uniformly. Often the field is sprayed with a desiccant, such as DNBP, to dry the crop just before combining.

DISEASES

Among the most destructive alfalfa diseases in the United States, and the causal organisms, are bacterial wilt (*Corynebacterium insidiosum*), leaf spot (*Pseudopeziza medicaginis*), yellow leaf blotch (*Pseudopeziza jonesii*), blackstem disease (*Ascochyta imperfecta*), crown wart (*Urophylctis alfalfae*), mildew, rust, sclerotinia stem rot, and fusarium root rots.[18, 39, 46] Other diseases include witches' broom, caused by a virus which is transmitted by leafhoppers, and alfalfa dwarf, a virus disease in California transmitted by leafhoppers and spittlebugs,[69] and numerous other virus diseases.[12]

Nematodes also attack alfalfa.[35] Alfalfa varieties now grown are resistant to one or more of the following: bacterial wilt, leaf spot, fusarium root rots, mosaic, downy mildew, alfalfa dwarf, stem rot, stem nematodes, spotted alfalfa aphid, and pea aphids.[4, 39]

Bacterial Wilt

Bacterial wilt is the most serious alfalfa disease in the United States, being found throughout the country.[39] In Nebraska it occurs wherever moisture conditions are favorable for a rapid, vigorous growth of the crop.[55] It is less prevalent in dry and hot areas.

The most conspicuous symptom of bacterial wilt, in addition to reduced stands, is a dwarfing of the infected plants. A bunch-growth effect is produced by an excessive number of shortened stems. The leaves are small and yellowish, and appear curled. Actual wilting occurs only in the advanced stages of the disease, especially during hot weather. A yellow-to-brown ring is observed under the bark when the roots are cut crosswise.

The disease is spread from infected to healthy plants by mowers, drainage water, and irrigation water. Diseased plants are more susceptible to winter injury than are healthy plants of the same variety.[38] Infection enters the plant through wounds. Longevity of infected plants varies, but the greater number that show conspicuous symptoms of the disease die in the second year after infection. The chief control measure is the use of resistant varieties of alfalfa. Breeding for resistance to wilt has been very successful.[4, 16, 71]

Leaf Diseases

Leaf spot is one of the most widespread of alfalfa fungus diseases, being most prevalent in the humid regions in cool wet weather. It causes small circular dark-brown spots to appear on the leaves and may cause them to drop. To avoid loss of leaves, the crop should be cut when leaf-spot infection becomes severe. Resistant varieties are available.

Leaf blotch, common in all large alfalfa-growing regions, also attacks the leaves. The disease is characterized by long yellow blotches sprinkled with minute brown dots. Where blotch is serious, the crop should be cut promptly.

Other Diseases

Crown wart occurs in greatest abundance in California, being characterized by the appearance of galls on the crown at the base of the stems.[76] Affected plants are seldom killed outright but produce low yields. Blackstem is most severe in the Southeast and West. It is most serious after open winters and during long wet springs. The disease is characterized by large irregular brownish or blackish lesions that occur on the leaves and petioles as well as on the stems. Bacterial stem blight causes considerable damage in some sections of the United States.

Stem Nematode

The stem nematode (*Ditylenchus dipsaci*) occurs on alfalfa in North America, South America, and Europe. It has been a serious pest in Nevada and other western states. This colorless eelworm is about $^1/_{20}$ inch long. It infests seedlings at the cotyledonary node, but it also infests older plants in the crown, young buds, stem bases, and sometimes the stems. Infestation results in decay.

INSECT PESTS

The spotted alfalfa aphid (*Therioaphis maculata*), alfalfa weevil (*Hypera postica*), potato leafhopper (*Empoasca fabae*), Lygus bugs

(*Lygus* species), grasshoppers, alfalfa caterpillar (*Colias philodice eurytheme*), three-cornered alfalfa hopper (*Stictocephala festina*), and the clover-seed chalcid (*Bruchophagus gibbus*) are among the most destructive insect pests of alfalfa.

The spotted alfalfa aphid was first observed in the United States in New Mexico in 1954. It had spread over much of the central and southern United States by 1962. It caused an estimated loss of $42 million in 1956. Parthenogenetic adults overwinter in the South. In the fall, sexual forms deposit eggs which survive the winter in the Central States. Several productive varieties are resistant to the spotted alfalfa aphid.[29] Insecticides also are recommended for control.[34]

The alfalfa weevil, an immigrant from Europe and Africa, is found throughout the United States. The alfalfa weevil (*Hypera pastica* [Gyllenhall]) was the first of its kind identified as a pest in the United States. It was discovered in 1904 near Salt Lake City, Utah. By 1960 this pest had spread to New Mexico, Kansas, and all states to the north. In 1951 an eastern strain of the weevil was detected near Baltimore, Maryland, and by 1970 had spread throughout the northeastern United States and was found in Minnesota, Iowa, Kansas, and south to Texas. The Egyptian alfalfa weevil (*Hypera brunneipennis* [Bohemal]), discovered near Yuma, Arizona, in 1939 now threatens 1,000,000 acres in California. The weevil alone causes an estimated $57,000,000 damage to alfalfa. The alfalfa weevil is now reported in all 48 of the continental United States. Tolerant alfalfa varieties are available.

Early first harvest, which is conducive to high yield and quality, causes larvae to feed on crown buds and small regrowth. While a light infestation at this stage can be successfully controlled, the danger lies when the first growth is harvested early and the larval feeding is shifted to the regrowth of the second crop.

Most damage is done by the larvae which feed on plant tips, new leaves as they open, and flower foliage. All but the main veins may be consumed causing damaged fields to appear whitish. Insecticides are available for weevil control in addition to tolerant varieties.

The potato leafhopper may cause injury to alfalfa in the eastern states. It punctures the leaves and causes a yellowing of the plants. When the first crop is cut near the full-bloom stage, the potato-leafhopper eggs attached to the plants of the first crop fail to develop on the cured hay, while the young insects also are deprived of succulent food. The alfalfa may be sprayed with insecticides.

Most of the damage caused by the three-cornered alfalfa hopper occurs south of latitude 36° in the United States. Cleaning up hibernation quarters, i.e., weeds, brush, grass bunches, and rubbish, keeps down the insect population.

The *Lygus* bug causes serious damage to alfalfa-seed production.[9, 65] The bugs attack young alfalfa buds, floral parts, and immature seeds.

Individual flowers are shed soon after injury by the *Lygus* bug. The injured seeds turn brown and in many cases shrivel and become papery. The application of insecticides during the early-bud stage protects the seed crop to a considerable extent.

Grasshoppers cause extensive damage to alfalfa. They are controlled with insecticides.

The alfalfa strain of the clover-seed chalcid causes extensive damage to the alfalfa-seed crop. The wasplike adult chalcid deposits a single egg in each partly developed alfalfa seed. The resulting hatched larva devours the interior of the seed. Several generations may develop in one season. Late-developing larvae hibernate over the winter inside dry seeds that are harvested or in those produced by plants along the ditches and borders of the field. Control measures include destruction of seed-bearing alfalfa and burclover plants along the irrigation-ditch banks and field borders during the winter, and harvesting the crop for hay when heavy infestation threatens. Damage is reduced if the same cutting is saved for the seed crop throughout a locality so that the breeding period for the insect does not extend over the entire season. A heavy cleaning of the seed, which removes the lighter infested seeds for prompt destruction, prevents the direct spread of the pest to new fields. Certain strains of alfalfa are resistant to the chalcid.[25]

The beautiful yellow butterflies that flit about the alfalfa flowers in nearly all of the alfalfa fields in the western four-fifths of the United States are the adults of the destructive alfalfa caterpillar. They cause the most damage in the southwestern states. The larvae devour the leaves and buds of the alfalfa plants. This pest overwinters in the pupal stage on alfalfa stems, chiefly those of alfalfa standing along the ditch banks and field borders. The chief control measure is frequent irrigation to maintain a high humidity in the field, which promotes the development of a disease of the caterpillars. Insecticides can be applied as sprays to control the alfalfa caterpillar, spittlebugs, and other insects.

Dodder (*Cuscuta* species) is a parasitic twining weed that attacks alfalfa, lespedeza and other crops. Dodder seed contamination makes commercial alfalfa seed undesirable or unmarketable. Control methods include flaming, cutting and burning, and tillage of spots of infested alfalfa, and also the use of suitable herbicides.[13]

stop

REFERENCES

1. Aamodt, O. S., "Climate and forage crops," in *Climate and Man*, USDA Yearbook, 1941, pp. 439–458.
2. Albert, W. B. "Studies on the growth of alfalfa and some perennial grasses," *J. Am. Soc. Agron.*, 19:624–654. 1927.

3. Anonymous. "Alfalfa varieties and areas of adaptation," *USDA Leaflet 507*. 1962.
4. Anonymous. "Varieties of alfalfa," *USDA Farmer's Bull. 2231*, pp. 1–12. 1968.
5. Armstrong, J. M., and White, W. J. "Factors influencing seed-setting in alfalfa," *J. Agr. Sci.*, 25:161–179. 1935.
6. Bickoff, E. M., and others. "Studies of the chemical and biological properties of coumestrol and related compounds," *USDA Tech. Bul. 1408*, pp. 1–95. 1969.
7. Bohart, G. E. "Insect pollination of forage legumes," *Bee World*, 41:57–64, 85–97. 1960.
8. Carlson, F. A. "The effect of soil structure on the character of alfalfa root systems," *J. Am. Soc. Agron.*, 17:336–345. 1925.
9. Carlson, J. W. "Lygus bug damage to alfalfa in relation to seed production," *J. Agr. Res.*, 61:791–815. 1940.
10. Clement, W. M., Jr. "Chromosome numbers and taxonomic relationships in *Medicago*," *Crop Sci.*, 2:25–28. 1962.
11. Cooper, D. C., and Brink, R. A. "Partial self-incompatibility and the collapse of fertile ovules as factors affecting seed formation in alfalfa," *J. Agr. Res.*, 60:453–472. 1940.
12. Crill, P., and others. "Alfalfa mosaic: The disease and its virus incitant; A literature review," *Wisc. Agr. Exp. Sta. Rsh. Bul. 280*, pp. 1–40. 1970.
13. Dawson, J. H., and others. "Controlling dodder in alfalfa," *USDA Farmer's Bull. 2211*, pp. 1–16. 1969.
14. Dennis, R. E., and others. "Alfalfa for forage production in Arizona," *Ariz. Agr. Exp. Sta. Bul. A-16*, pp. 1–40. 1966 (Rev.).
15. Duley, F. L. "The effect of alfalfa on soil moisture," *J. Am. Soc. Agron.*, 21:224–231. 1929.
16. Elling, L. J., and Frosheiser, F. I. "Reaction of twenty-two alfalfa varieties to bacterial wilt," *Agron. J.*, 52:241–242. 1960.
17. Foote, L. E., and others. "Controlling bloat in cattle grazing clover," *La. Agr. Exp. Sta. Bul. 629*, pp. 1–28. 1968.
18. Froshheiser, F. I. "The worst diseases of crops: Alfalfa," *Crops and Soils*, 25(1):18–21. 1972.
19. Graber, L. F., and others. "Organic food reserves in relation to the growth of alfalfa and other perennial herbaceous plants," *Wisc. Agr. Exp. Sta. Res. Bul. 80*. 1927.
20. Graber, L. F., and Sprague, V. G. "The productivity of alfalfa as related to management," *J. Am. Soc. Agron.*, 30:38–54. 1938.
21. Grandfield, C. O. "The effect of the time of cutting and of winter protection on the reduction of stands in Kansas Common, Grimm, and Turkistan alfalfas," *J. Am. Soc. Agron.*, 26:179–188. 1934.
22. Grandfield, C. O., and Metzger, W. H. "Relation of fallow to restoration of subsoil moisture in an old alfalfa field and subsequent depletion after reseeding," *J. Am. Soc. Agron.*, 28:115–123. 1936.
23. Graumann, H. O. "Creeping alfalfa," *Crops and Soils*, 10(4). 1958.
24. Hale, V. Q. "TIBA on alfalfa can help plants produce more seed," *Crops and Soils*, 23(8):7. 1971.

25. Hanson, C. H. "Report of the eighteenth alfalfa improvement conference," *USDA CR-71-62*. 1962.
26. Hanson, C. H., and others. "Saponin content of alfalfa as related to location, cutting, variety, and other variables," *USDA Agricultural Research Service Spec. Rept. ARS 34-44*. 1963.
27. Hanson, C. H., Garrison, C. S., and Graumann, H. O. "Alfalfa varieties in the United States," *USDA Handbook 177*. 1960.
28. Hanson, C. H., Ed. *Alfalfa Science and Technology*, American Society of Agronomy, Inc., Agronomy No. 15, Madison, Wisconsin, pp. 1-812. 1972.
29. Harvey, T. L., and others. "The development and performance of Cody alfalfa, a spotted-alfalfa-aphid resistant variety," *Kans. Agr. Exp. Sta. Tech. Bul. 114*. 1960.
30. Heinrichs, D. H. "Creeping alfalfa," *Adv. Agron.*, 15:317-37. 1963.
31. Heinrichs, D. H. "Alfalfa in Canada," *Can. Dept. Agr. Pub. 1377*, pp. 1-28. 1968.
32. Hendry, G. W. "Alfalfa in history," *J. Am. Soc. Agron.*, 15:171-176. 1923.
33. Hobbs, G. A. "Domestication of alfalfa leaf-cutter bees," *Can. Dept. Agr. Pub. 1313*. 1967.
34. Howe, W. L., and others. "Studies of the mechanisms and sources of spotted alfalfa aphid resistance in Ranger alfalfa," *Nebr. Agr. Exp. Sta. Res. Bul. 210*. 1963.
35. Hunt, O. J., and Peaden, R. N. "Alfalfa nematodes," *Crops and Soils*, 24(6):6-7. 1972.
36. Jensen, E. H., and others. "Environmental effects on growth and quality of alfalfa," *Nev. Agr. Exp. Sta. Western Regional Rsh. Pub. T9*, pp. 1-36. 1967.
37. Jewett, R. H. "The relation of cutting to leafhopper injury to alfalfa," *Ky. Agr. Exp. Sta. Bul. 348*. 1934.
38. Jones, F. R. "Development of the bacteria causing wilt in the alfalfa plant as influenced by growth and winter injury," *J. Agr. Res.*, 37:545-569. 1928.
39. Jones, F. R., and Smith, O. F. "Sources of healthier alfalfa," in *Plant Diseases*, USDA Yearbook, 1953, pp. 228-237.
40. Kaufeld, N. M. and Sorenson, E. L. "Interrelations of honeybee preference of alfalfa clones and flower color," *Kansas Agr. Exp. Sta. Rsh. Publ. 163*, pp. 1-14. 1971.
41. Kiesselbach, T. A., and Anderson, A. "Alfalfa investigations," *Nebr. Agr. Exp. Sta. Res. Bul. 36*. 1926.
42. Kiesselbach, T. A., and Anderson, A. "Curing alfalfa hay," *J. Am. Soc. Agron.*, 19:116-126. 1927.
43. Kiesselbach, T. A., and Anderson, A. "Quality of alfalfa hay in relation to curing practice," *USDA Tech. Bul. 235*. 1931.
44. Kiesselbach, T. A., Russell, J. C., and Anderson, A. "The significance of subsoil moisture in alfalfa production," *J. Am. Soc. Agron.*, 21:241-268. 1929.
45. Kiesselbach, T. A., Anderson, A., and Russell, J. C. "Subsoil moisture and crop sequence in relation to alfalfa production," *J. Am. Soc. Agron.*, 26:422-442. 1934.

46. Leath, K. T., and others. "The fusarium root rot complex of selected forage legumes in the northeast," *Pa. Agr. Exp. Sta. Bul. 777*, pp. 1–66. 1971.
47. Leukel, W. A. "Deposition and utilization of reserve foods in alfalfa plants," *J. Am. Soc. Agron.*, 19:596–623. 1927.
48. Lute, A. M. "Impermeable seed of alfalfa," *Colo. Agr. Exp. Sta. Bul. 326*. 1928.
49. Lute, A. M. "Laboratory germination of hard alfalfa seed as a result of clipping," *J. Am. Soc. Agron.*, 34:90–99. 1942.
50. McGregor, S. E., and others. "Beekeeping in the United States," *USDA Agr. Handbk. 335*, pp. 1–147. 1971 (rev.).
51. Medler, J. T. and Lussenhop, J. F. "Leafcutter bees of Wisconsin," *Wisc. Agr. Exp. Sta. Rsh. Bul. 274*, pp. 1–40. 1968.
52. Nelson, S. O. "A hot foot for hard seed," *Crops and Soils*, 22(1):18–19. 1969.
53. Pedersen, M. W., and others. "Growing alfalfa for seed," *Utah Agr. Exp. Sta. Cir. 135*. 1955.
54. Pedersen, M. W., and others. "Cultural practices for alfalfa seed production," *Utah Agr. Exp. Sta. Bul. 408*. 1959.
55. Peltier, G. L., and Jensen, J. H. "Alfalfa wilt in Nebraska," *Nebr. Agr. Exp. Sta. Bul. 240*. 1930.
56. Peltier, G. L., and Tysdal, H. M. "Hardiness studies with 2-year-old alfalfa plants," *J. Agr. Res.*, 43:931–955. 1931.
57. Rather, H. C., and Dorrance, A. B. "Pasturing alfalfa in Michigan," *J. Am. Soc. Agron.*, 27:57–65. 1935.
58. Rather, H. C., and others. "A mixture of alfalfa and smooth bromegrass for pasture," *Mich. Agr. Exp. Sta. Cir. 159*. 1937.
59. Schmehl, W. R., and Romsdal, S. D. "Materials and method of application of phosphate for alfalfa in Colorado," *Colo. Agr. Exp. Sta. Tech. Bul. 74*. 1963.
60. Schrenk, W. G., and others, "Dehydrated alfalfa," *Kans. Agr. Exp. Sta. Bul. 409*. 1959.
61. Sheperd, J. B., and others. "Experiments in harvesting and preserving alfalfa for dairy cattle feed," *USDA Tech. Bul. 1079*. 1954.
62. Silkett, V. W., Megee, C. R., and Rather, H. C. "The effect of late summer and early fall cutting on crown bud formation and winter-hardiness of alfalfa," *J. Am. Soc. Agron.*, 29:53–62. 1937.
63. Steinmetz, F. H. "Winterhardiness in alfalfa varieties," *Minn. Agr. Exp. Sta. Tech. Bul. 38*. 1926.
64. Stephen, W. P. "Artificial beds for alkali bee propagation," *Ore. Agr. Exp. Sta. Bul. 598*, pp. 1–20. 1965.
65. Stitt, L. L. "Three species of the genus *Lygus* and their relation to alfalfa seed production in southern Arizona and California," *USDA Tech. Bul. 741*. 1940.
66. Tysdal, H. M. "Influence of light, temperature, and soil moisture on the hardening process in alfalfa," *J. Agr. Res.*, 46:483–515. 1933.
67. Tysdal, H. M. "Is tripping necessary for seedsetting in alfalfa?" *J. Am. Soc. Agron.*, 32:570–585. 1940.
68. Tysdal, H. M. "Influence of tripping, soil moisture, plant spacing, and

lodging on alfalfa seed production," *J. Am. Soc. Agron.*, 38:515–535. 1946.
69. Tysdal, H. M., and Westover, H. L. "Alfalfa improvement," in *USDA Yearbook, 1937*, pp. 1122–1153.
70. Tysdal, H. M., and Kiesselbach, T. A. "The differential response of alfalfa varieties to time of cutting," *J. Am. Soc. Agron.*, 31:513–519. 1939.
71. Tysdal, H. M., and Kiesselbach, T. A. "Alfalfa in Nebraska," *Nebr. Agr. Exp. Sta. Bul. 331*. 1941.
72. Tysdal, H. M., and Kiesselbach, T. A. "Hybrid alfalfa," *J. Am. Soc. Agron.*, 36:649–667. 1944.
73. Tysdal, H. M., Kiesselbach, T. A., and Westover, H. L. "Alfalfa breeding," *Nebr. Agr. Exp. Sta. Res. Bul. 124*, pp. 1–46. 1942.
74. Waldron, L. R. "Cross-pollination in alfalfa," *J. Am. Soc. Agron.*, 11:259–266. 1919.
75. Weihing, R. M. "Field germination of alfalfa seed submitted for registration in Colorado and varying in hard seed content," *J. Am. Soc. Agron.*, 32:944–949. 1940.
76. Westover, H. L. "Growing alfalfa," *USDA Farmers Bul. 1722* (rev.), 1941.
77. Wheeler, W. A. "Beginnings of hardy alfalfa in North America." Published in *Seed World* in 7 installments in 1950. Reprinted in pamphlet form by Northrup, King, Minneapolis, Minn., pp. 1–31. 1951.
78. Willard, C. J. "Root reserves of alfalfa with special reference to time of cutting and yield," *J. Am. Soc. Agron.*, 22:595–602. 1930.
79. Willard, C. J. "Management of alfalfa meadows after seeding," *Advances in Agronomy*, Vol. 3, Academic Press, New York, pp. 93–112. 1951.

Chapter 23

Sweetclover

ECONOMIC IMPORTANCE

Sweetclover was used as a green-manure crop as well as a honey plant in the Mediterranean region 2,000 years ago. Asia Minor appears to be its native habitat. It was first observed in the United States as a wild plant in Virginia about 1739 but its real value was not recognized for many years. As late as 1910 sweetclover was legislated against as a weed in some states. Formerly it often was called Bokhara clover.

Sweetclover is grown mostly in the North Central states of the United States for green manure, pasture, seed, silage, hay, and as food for honey bees. Production has declined since 1940 as it is being replaced by alfalfa.

ADAPTATION

Sweetclover may be grown where the annual precipitation, properly distributed, is 17 inches or more. It also is grown under irrigation in drier sections of the northwestern states. The biennial species tolerate drought once they have been established, but a good supply of moisture and cool temperatures are essential for germination and early seedling growth.[1] Although the plant may persist under dry conditions, little vegetative growth is made. It is more drought-resistant than alfalfa or any of the true clovers in the Great Plains region, but it is not a dependable crop under dryland conditions in the central or the southern Great Plains unless moisture and temperature conditions are favorable for some time after seeding. Its consistent survival in Canada, Montana, Michigan, and other northern regions proves that some varieties are very winter hardy.[9, 10]

Sweetclover will grow on a wide range of soils provided they are not acid. The soil reaction should be *p*H 6.5 or higher. It will grow on infertile eroded areas, poorly drained soil, heavy muck and peat soils high in organic matter, or on clay soils with almost no organic matter. Most sweetclover failures in humid sections are due to lack of lime in the soil. Sweetclover grows on soils high in salts, even on alkali seepage lands[17] where cereal crops cannot be grown. Its seed-shattering tendency and hard seeds help sweetclover to maintain itself as a weed in waste places and some cropped fields. Sweetclover in alfalfa fields harvested for seed contaminates the crop because sweetclover seeds cannot be screened out of alfalfa.

BOTANICAL DESCRIPTION

The sweetclovers are legumes classified in the genus *Melilotus.* They are tall, erect, annual or biennial herbs.

The plants grow from 2 to 6 feet or more in height. The stems are generally well branched, coarse, and succulent. During the first year only one central stem per plant develops in the biennial sweetclovers, except in certain new strains, which have several. In the second year, numerous branches develop from crown buds formed the previous fall. New growth after cutting arises from buds in leaf axils rather than from crown buds. The leaves are pinnately trifoliolate and possess pointed stipules. The middle leaflet is stalked or petiolated. The leaflets are finely toothed along the margins. Young sweetclover plants can be distinguished from those of alfalfa by the coumarin taste and usually by the presence of serrations (teeth) along the entire margin of the leaflet rather than on the tip only, as in alfalfa.

Sweetclover has a fleshy taproot that penetrates the soil to a depth of 5 to 8 feet in the biennial forms.[38] A few to many side roots develop. The taproot of the annual types is relatively short and slender. In the biennial forms the taproot increases in weight slowly until the end of the growing season. The roots may double in weight after September 25, due to food storage.[28, 42] During the second year the roots are depleted rapidly of their stored food for the production of top growth (Table 23–1).

White or yellow sweetclover flowers are borne in short-pedicelled racemes that arise in the axils of the leaves (Figure 23–1). As many as 100 flowers, each with ten stamens, may occur on a single large raceme.[30] The inflorescence is indeterminate. Flowering occurs during the first year in the annual forms, but usually not until the second year in biennial types. Some plants of the biennial types may flower the first year under favorable growing conditions. Long days hasten flowering and restrict root growth.[20] Biennial types flowered the first year at Palmer, Alaska, where the maximum day length was 19½ hours.

The seeds of sweetclover mature in small, ovoid pods that are indehiscent or finally two-valved. There usually is only one seed per pod, but as many as four may occur in some strains. The pods are dark brown when mature. Sweetclover sets seed readily, but it shatters easily when mature. The mature seeds of white sweetclover are golden yellow in color, but many of those of the yellow varieties are mottled.

TABLE 23–1. Yield and Composition of Tops and Roots of Sweetclover[a] at Successive Dates in the Seeding Year and the Year after Seeding

DATE OF SAMPLING	AIR-DRY YIELD PER ACRE		NITROGEN IN		NITROGEN PER ACRE			PLANTS PER SQUARE YARD
	Tops (lb.)	*Roots (lb.)*	*Tops (%)*	*Roots (%)*	*Tops (lb.)*	*Roots (lb.)*	*Total (lb.)*	
	Development in the Seeding Year							
July 1	515	–	3.25	–	–	–	–	–
July 15	588	75	2.85	2.30	16	2	18	–
August 1	659	182	2.54	2.53	17	4	21	–
August 15	963	310	2.89	2.90	28	9	37	–
Sept. 1	1431	577	2.88	3.02	41	18	59	–
Sept. 15	1544	884	3.09	3.05	48	27	75	–
Oct. 1	1881	1273	2.87	3.12	54	40	94	–
Oct. 15	1714	1721	2.59	3.25	44	56	100	–
Nov. 1	1616	2115	2.38	3.50	39	74	113	–
Nov. 8–10	1397	2324	2.32	3.45	32	77	109	–
	Development in the Year After Seeding							
April 1	420	2130	4.35	4.33	19	92	111	125
April 15	690	1750	4.21	4.15	29	73	102	125
May 1	1930	1360	3.90	3.57	75	49	114	117
May 15	3360	1280	3.27	2.90	110	37	147	102
June 1	4940	1200	2.70	2.38	133	29	162	92
June 15	6030	1110	2.35	2.14	142	24	166	82
July 1	7380	880	1.95	1.94	144	17	161	66
July 15	7990	790	1.77	1.72	141	14	155	50
August 1	7290	760	1.54	1.64	112	13	115	40

[a] Sweetclover sown in early oats which were harvested for grain; sweetclover not clipped or pastured after oat harvest.

They are ovoid in shape but not laterally compressed at the ends like alfalfa seeds (Figure 22–4). A large percentage of the seeds may be hard or impermeable. Selected strains have a high percentage of permeable seeds.[27]

SWEETCLOVER SPECIES

Common biennial white sweetclover (*Melilotus alba*) is an aggregate of various plant types, mostly tall-growing, high-branching, and stemmy with low leaf percentages. The pods are slightly netted. The commonly grown types of biennial yellow sweetclover (*M. officinalis*) are somewhat less upright in growth habit than the white-flowered types, blossom earlier, have finer stems, and tend to be more persistent under pasture conditions. They usually are less productive and mature earlier than do the white-blossomed varieties.[27] Yellow

Figure 23–1. Stems and flowers of white sweetclover.

varieties are somewhat less winter hardy than white ones in Canada.[22]

A yellow-blossom sweetclover (*M. suaveolens*) has growth habits similar to Common White. It usually is a biennial, but an annual form also is known. A nonbitter sweetclover (*M. dentata*) includes both annual and biennial types which are practically coumarin-free.[12] Sourclover (*M. indica*), a winter annual yellow *Melilotus* occurs mostly as a volunteer crop in grain fields, where the seed is obtained, in the South and Southwest. It is prevalent in California and Arizona, where it was first observed as an escaped introduction about 1830. The seeds are small, rough, and dark green (Figure 22–4). It is a green-manure crop on irrigated lands of the Southwest, and along the western Gulf Coast.

POLLINATION

Under natural conditions, pollination of sweetclover is effected principally by honeybees, except for the species, varieties, and individual plants that are spontaneously self-fertilized.[23]

Common white sweetclover appears to be highly self-fertile. Plants may set seed when insects are wholly excluded.[22] Self-pollination takes place readily when the pistils and stamens are of the same length, but very little occurs when the pistil is longer than the stamens. Some natural crossing takes place under normal field conditions.[7]

Yellow sweetclover (*M. officinalis*) is largely, but not completely, self-sterile.[27, 30] It may be unable to set seed unless the flowers are visited by insects.[20]

White sweetclover has been crossed with *M. suaveolens* successfully, but the cross, *M. alba* × *M. officinalis*, failed to produce any viable seeds.[33] White sweetclover has been crossed with *M. dentata*. The plants from the hybrid seed contain very little chlorophyll. Consequently, they must be propagated by embryo culture, or else grafted on white sweetclover plants to enable them to produce flowers.[11, 30]

VARIETIES

The majority of the biennial white-blossom sweetclover is the common unnamed type, but several improved varieties are produced. Some Canadian varieties bloom too early to produce high yields under the shorter summer days of most of the United States,[15] but are extremely winter hardy. Improved varieties of biennial yellow sweetclover are grown in the United States and Canada.

Annual white sweetclovers are grown as winter annuals in the southern states or as summer annuals in the intermountain states.[15] Annual white sweetclover can be sown in the fall and turned under in the spring in time for planting cotton in central and southern Texas. It also can be sown in the late winter or early spring for use as pasture in that region. In the Pacific Northwest it can be sown in the fall and then be turned under in early spring in preparation for potatoes or sugarbeets. It is sown in the spring in the central and northern states for green manure, pasture, and bee pasture.[6] It grows 2 to 5 feet high, and has numerous strong stems with few leaves. The roots are small and without crown buds. It blooms from about July 15 to September 15. The annual types produce larger yields of tops but lower yields of roots than do the biennials during the first season of growth.[2] Biennial sweetclover produces about 25 per cent more total dry matter, and the protein content of the roots is more than double that of the annual variety under Minnesota conditions. The protein content of the different types of biennial sweetclover is very similar.[21]

ROTATIONS

One of the most common crop sequences that includes sweetclover in the northern Corn Belt is corn; small grain—usually oats or barley—

seeded to sweetclover; and sweetclover for pasture, hay, or seed. Many grain farms use a shorter rotation of corn and small grain—barley, oats, or wheat—seeded to sweetclover. The sweetclover is plowed under in the spring for corn. This rotation has resulted in increases in corn yields of 30 to 70 per cent as compared with the same two-year rotation without the sweetclover in experiments in Ohio. Small increases also are obtained in the small-grain crop.[4]

Where rainfall is adequate, sweetclover may be underseeded with winter wheat in the spring. The next year the sweetclover is grazed or harvested for seed, or both; the residue is plowed under in preparation for fall-sown wheat.

A rotation sometimes followed under irrigation is small grain seeded with sweetclover; sweetclover for pasture or silage; sugarbeets; and sugarbeets. Sweetclover also is used as a green-manure crop in potato rotations.

CULTURAL METHODS

The seedbed, inoculation, and lime requirements for sweetclover are similar to those for alfalfa. In experiments in Oklahoma,[13] light applications of finely pulverized limestone mixed with either rock phosphate or superphosphate and applied with a grain drill equipped with a fertilizer distributor resulted in excellent yields of sweetclover on land where a very poor growth occurred on unfertilized soil.

In Indiana[41] an application of two tons per acre of ground limestone was recommended for soils of *p*H 6.0 to 6.2; three tons per acre for medium-acid soils (*p*H 5.5 to 5.9); and 5 tons per acre for strongly acid soils (*p*H 5.0 to 5.5). About 300 pounds per acre of superphosphate for thin pastureland was recommended. About 300 pounds per acre of 0–20–10 fertilizer also was suggested for soils highly responsive to potash. In other states much smaller applications of limestone have been satisfactory on acid soils. Most of the sweetclover in the Corn Belt is seeded with small grain as a companion crop. The seeding of sweetclover alone might be advantageous there, except for the growth of annual weeds which develop freely in the unclipped crop. Spring grains are more satisfactory than winter grains as companion crops under Ohio conditions.[42] Field peas were a better companion crop for sweetclover in the nonirrigated areas of the Pacific Northwest than any of the small grains.[17] Sweetclover should be seeded alone where moisture is a limiting factor, as in many parts of the Great Plains. In the Red River Valley of Minnesota, the maximum total weight of roots, tops, and stubble in October of the first year was obtained when sweetclover was sown alone in April.[6]

Owing to the high percentage of hard seed, scarified seed is used most widely, especially for spring seeding. The sowing of unhulled unscarified seed on winter wheat in January or February has been recommended,[43] but is not generally successful.

Sweetclover is almost always seeded in close drills, except under extremely dry conditions or for seed production. In Oklahoma, sweetclover seeded in 36- to 42-inch rows, and cultivated to control weeds during the first season, produced a yield of roots similar to that from seedings in rows 7 inches apart.[13] The wider rows also were advantageous in the western part of the state where the available moisture is relatively low. However, it is inadvisable to grow sweetclover where such methods are necessary. Deep seeding (more than 2.5 cm.) may be responsible for some sweetclover failures[32] (Table 23–2).

Sweetclover is generally sown early in the spring at about the same time as and usually along with small grains. Sometimes it is seeded broadcast on winter-wheat fields in February or March. In Ohio, summer seeding produced lower yields of nitrogen and organic matter for plowing under early in May of the second year, than did spring seeding.[42] In Oklahoma and Kansas sweetclover frequently is seeded in the fall.

The rate of seeding sweetclover usually ranges from 4 to 25 pounds, 10 to 12 pounds per acre of scarified seed being a popular rate. For Michigan conditions, 15 pounds of scarified seed, 18 to 20 pounds of unscarified seed, or 25 pounds of unhulled seed per acre have been recommended.[24] For the Pacific Northwest, 10 pounds per acre were recommended for nonirrigated lands, and 20 pounds for irrigated lands.[18] Three to six pounds per acre were recommended for 36- to 42-inch rows in Oklahoma.[13] In an Ohio experiment the final stand of sweetclover in the second year was about 30 plants per square yard, regardless of the original stand.[42] Little change occurred in the stand during the first year unless the stand was abnormally thick or the season unusually dry.

TABLE 23–2. Effect of Depth Seeding Sweetclover upon Emergence of Seedlings

DEPTH OF SEEDING	DAYS FROM SEEDING TO EMERGENCE	EMERGENCE
(in.)	*(no.)*	(%)
0.5	4	98
1.0	5	88
1.5	6	46
2.0	7	20
2.5	9	2
3.0	None emerged	0

UTILIZATION

Pasture

Sweetclover is a valuable pasture crop, especially in the Great Plains where it is reported to carry more livestock per acre than other plants common to the area. In the Corn Belt the average period of grazing for the second-year growth was reported as 111 days. The best grazing is from about May 1 to July 20.

When conditions are favorable, sweetclover may be pastured lightly for 60 to 75 days during the fall of the seeding year. Some top growth should be left for winter protection in the northern states. The second-year plants are generally grazed early enough in the spring to avoid coarse woody growth. The danger of bloat is less than on alfalfa, red clover, or alsike clover.[15] When not overgrazed, yellow sweetclover often reseeds itself and thus is more desirable for sowing in permanent grass pastures. Cattle prefer to graze the grasses, which permits some yellow sweetclover plants to produce seed.[25] This seed production, together with the hard seeds that germinate after the first year, often maintains stands for several years.

Hay

Sweetclover hay production declined because of damage from the sweetclover weevil, and also because of its recognized poor hay quality. A good first-year growth makes excellent hay when it can be cured in bright fall weather. In the South, the first-year crop is cut for hay, but in the northern Great Plains the first-year crop seldom attains sufficient height. The coarse juicy stems of the second-year crop are difficult to cure without a heavy loss of leaves except in dry regions. Crushed stems cure more quickly. Improperly cured sweetclover hay (hay that is allowed to mold) is dangerous to feed because it may cause bleeding in animals. Sweetclover is unable to compete with alfalfa as a forage crop under irrigation.

Sweetclover has about the same feeding value as alfalfa when cut at the proper stage and cured into bright-green hay. When cut too late, the coarse woody stems are largely rejected by livestock and the feeding value is low.

In West Virginia, cutting biennial sweetclover the first year resulted in a reduced hay yield in the second-year crop, although the total yield was greater when cut both years (Table 23–3).[8] Clipping reduces root development.[36]

Sweetclover should be cut for hay in the bud stage, or at least by the time the first blossoms open. Plants cut the second season at the pre-bud stage of growth make a more vigorous recovery than when cut later.[8] Delayed cutting until after the plants are in full bloom results in a coarse woody hay of relatively poor quality.

TABLE 23–3. **Effect on Yield of Cutting Sweetclover the First Year of Growth**

DATE CUT IN 1931	AVERAGE HEIGHT		AVERAGE YIELD DRY HAY		
	May 25 (*in.*)	*June 30* (*in.*)	*1931* (*gm.*)	*1932* (*gm.*)	*Total* (*gm.*)
Not cut	27.4	49.2	–	1912	1912
August 1	23.2	44.6	919	1571	2490
August 20	20.8	45.2	1121	1542	2663
Sept. 10	20.4	44.6	1190	1385	2575
Sept. 30	19.4	43.0	1095	1370	2465

Sweetclover should be cut at a height of 8 to 12 inches above the ground where a second crop is desired. Since the new growth arises from buds in the axils of the branches and leaves on the lower stem, the plants are destroyed when cut below these buds. In Utah no second growth occurred on a 4-inch stubble, a 10 per cent crop appeared on an 8-inch stubble, and a full crop on a 12-inch stubble.[35]

Silage

Sweetclover makes a good silage crop when it is first wilted in the swath and then ensiled with molasses or other additives in a tight silo. As in the case of hay, spoiled sweetclover silage is a risky feed. The crop can be made into silage when the weather is too damp for curing hay.

Green Manure

Sweetclover is excellent for green manure, especially in the humid regions.[3, 36] As is the case with other green-manure crops, it has not resulted in any increase in average crop yields in dryland areas. The annual sweetclover may be grown where the crop is to be plowed under in the fall of the seeding year because of the difficulty in eradication of the biennial types at that time. The annual type can be grown for soil improvement in areas of Oklahoma where summer drought interferes with the normal development of the biennial varieties.[13] The annual type is used also in mild sections of Texas and the Pacific Northwest irrigated sections where it survives the winter after fall seeding. It can be plowed under in preparation for crops planted in late spring. However, the biennial type produces more total dry matter in the roots and tops as well as more nitrogen under the same conditions.[36]

The best time to plow under biennial sweetclover for soil improvement is in the spring. Under Ohio conditions, plowing between April 20 and May 10 secured 80 per cent of the maximum amount of nitrogen accumulated during the season.[42] Rates of 100–140 pounds of carryover nitrogen per acre are possible. This permits the planting of corn on the land.

Seed Production

Seed yields vary from 120 to 600 pounds per acre and occasionally higher. The seed crop harvested the second season should be cut when three-fourths of the pods have turned brown or black. The seed pods shatter badly when mature, but this can be reduced by cutting the plants when they are damp from rain or dew. The seed crop may be combined directly and then dried artificially, or it may be windrowed and then threshed with a combine with a pick-up attachment. The seed does not ripen uniformly. Consequently, the ripest seeds have shattered while others are still green and damp when the crop is harvested.

Coumarin in Sweetclover

The leaves and flowers, as well as the seeds, stems, and roots of sweetclover contain considerable quantities of *cis*-O-hydroxycinnamic acid. This converts into a lactone, called coumarin which constitutes the bitter taste and low palatability of sweetclover. The maximum coumarin content occurs at the late-bud or early-flower stage, after which it decreases. The content is increased by higher temperatures and light intensity and longer photoperiods.[40] Coumarin is converted to dicumarol in moldy or decayed sweetclover hay and silage.

Dicoumarol is responsible for the sweetclover disease of cattle.[24, 32] In this disease, blood fails to clot normally when the animals are wounded. Affected animals may bleed to death from minor wounds. The dicoumarol reduces the enzyme, thrombin, in the blood of the animals that eat the spoiled hay. Thrombin is essential to the clotting of the blood, by converting a soluble plasma protein, fibrinogen, into fibrin. Fibrin consists of an insoluble network of fibers which entrap cells and serum. The use of second-year sweetclover for hay is hazardous because of the possibility of poisoning. Apparently, bright hay free from mold is a safe feed.

The hay of nonbitter sweetclover (*M. dentata*) does not become toxic on being spoiled in a similar manner.[29] The low coumarin character of *M. dentata* has been bred into biennial white sweetclover (*M. alba*) by crossing, followed by back-crossing during selection for low coumarin content along with good forage production.[9, 10, 11] Low coumarin content is inherited as a simple recessive character.[16]

Synthetic dicoumarol is used as an anticoagulant for treatment of blood clots in humans, as well as in the making of a rat poison called warfarin.

DISEASES

Diseases affecting sweetclover are more serious in the humid regions than west of the Iowa-Nebraska boundary.

The blackstem disease caused by *Mycosphaerella lethalis* (or *Ascochyta lethalis*), is characterized by blackened stems that become evident late in the spring on second-year stems.[19] The leaves, as well as the young stems, may be killed when injury is severe. The disease appears to be most serious when wet weather occurs while the first 6 to 8 inches of growth are being made.

Other diseases of sweetclover include root rot caused by *Phytophthora cactorum* and gooseneck stem blight caused by *Ascochyta caulicola.*

INSECTS

The most serious pest of the sweetclover crop is the sweetclover weevil, *Sitona cylindricollis.*[15, 26] The insect, a small, dark-gray snout beetle, feeds upon the leaves of the young seedlings. These insects either kill the plants in a short time or weaken them to such an extent that the plants die from unfavorable summer weather conditions. When the weevil feeds upon the second-year growth, the damage is relatively slight because of the rapid rate of growth of the shoots.

Overwintering and newly emerged adult sweetclover weevils migrate from old stands to new seedings. Consequently, sweetclover should not be planted close to an old stand when it can be avoided.

Insecticides are effective against adult sweetclover weevils.

REFERENCES

1. Aamodt, O. S. "Climate and forage crops," in *Climate and Man*, USDA Yearbook, 1941, pp. 439–458.
2. Arny, A. C., and McGinnis, F. W. "The relative value of the annual white, the biennial white, and the biennial yellow sweetclovers," *J. Am. Soc. Agron.*, 16:384–396. 1924.
3. Bowren, K. E., and others. "Yield of dry matter and nitrogen from tops and roots of sweetclover, alfalfa and red clover at five stages of growth," *Can. J. Plant Sci.*, 49(1):79–81. 1969.
4. Cook, E. D., and Rector, C. V. "The effect of fertilizer and sweetclover on oat forage and grain yields, Blackland Experiment Station, 1958–62," *Tex. Agr. Exp. Sta. Prog. Rpt. 2203*, 3 pp. 1964.
5. Crosby, M. A. "Sweetclover in Great Plains Farming," *USDA Tech. Bul. 380.* 1933.
6. Dunham, R. S. "Effect of method of sowing on the yield and root and top

development of sweetclover in the Red River Valley," *J. Agr. Res.*, 47: 979–995. 1933.
7. Fowlds, M. "Seed color studies in biennial white sweetclover, *Melilotus alba*," *J. Am. Soc. Agron.*, 31:678–686. 1939.
8. Garber, R. J., Hoover, M. M., and Bennett, L. S. "The effect upon yield of cutting sweetclover (*Melilotus alba*) at different heights," *J. Am. Soc. Agron.*, 26:974–977. 1934.
9. Goplen, B. P. "Yukon sweetclover," *Can. J. Plant Sci.*, 51(2):178–179. 1971.
10. Goplen, B. P. "Polara, a low coumarin cultivar of sweetclover," *Can. J. Plant Sci.*, 51(3):249–251. 1971.
11. Hanson, A. A. "Development and use of improved varieties of forage crops in the temperate humid region," in *Grasslands*, Publ. No. 53, Am. Assn. Adv. Sci., pp. 1–406. 1959.
12. Harmann, W. "Comparison of heights, yields, and leaf percentages of certain sweetclover varieties," *Wash. Agr. Exp. Sta. Bul. 365*. 1938.
13. Harper, H. J. "Sweetclover for soil improvement," *Okla. Agr. Exp. Sta. Circ. C-94*. 1941.
14. Hodgson, H. J., and Bula, R. J. "Hardening behavior of sweetclover (*Melilotus* spp.) varieties in a subarctic environment," *Agron. J.*, 48:157–160. 1956.
15. Hollowell, E. A. "Sweetclover," *USDA Leaflet 23* (rev.). 1960.
16. Horner, W. H., and White, W. J. "Investigations concerning the coumarin content of sweetclover. III. The inheritance of the low coumarin character," *Sci. Agr.*, 22:85–92. 1941.
17. Hulbert, H. W. "Factors affecting the stand and yield of sweetclover," *J. Am. Soc. Agron.*, 15:81–87. 1923.
18. Hulbert, H. W. "Sweetclover: Growing and handling the crop in Idaho," *Idaho Agr. Exp. Sta. Bul. 147*. 1927.
19. Johnson, E. M., and Valleau, W. D. "Black-stem of alfalfa, red clover, and sweetclover," *Ky. Agr. Exp. Sta. Bul. 339*. 1933.
20. Kasperbauer, M. J., Gardner, F. P., and Johnson, I. J. "Taproot growth and crown development in biennial sweetclover as related to photoperiod and temperature," *Crop Sci.*, 3:4–7. 1963.
21. Kirk, L. E. "A comparison of sweetclover types with respect to coumarin content, nutritive value, and leaf percentage," *J. Am. Soc. Agron.*, 18:385–392. 1926.
22. Kirk, L. E., and Davidson, J. G. "Potentialities of sweetclover as plant breeding material," *Sci. Agr.*, 8:446–455. 1928.
23. Kirk, L. E., and Stevenson, T. M. "Factors which influence spontaneous self-fertilization in sweetclover (*Melilotus*)," *Can. J. Res.*, 5:313–326. 1931.
23a. Matches, A. G., Ed., "Anti-quality components of forages." CSSA, Madison, Wisc. 1973.
24. Megee, C. R. "Sweetclover," *Mich. Agr. Exp. Sta. Spec. Bul. 152*. 1926.
25. Miles, A. D. "Sweetclover as a range legume," *J. Range Mgt.*, 23(3):220–222. 1970.
26. Munro, J. A., and others. "Biology and control of the sweetclover weevil," *J. Econ. Ent.*, 42:318–321. 1949.

27. Pieters, A. J., and Hollowell, E. A. "Clover improvement," in *USDA Yearbook*, 1937, pp. 1190–1214.
28. Smith, D., and Graber, L. F. "The influence of top growth removal on the root and vegetative development of biennial sweetclover," *J. Am. Soc. Agron.*, 40:818–831. 1948.
29. Smith, W. K., and Brink, R. A. "Relation of bitterness to the toxic principle in sweetclover," *J. Agr. Res.*, 57:145–154. 1938.
30. Smith, W. K., and Gorz, H. J. "Sweetclover improvement," in *Advances in Agronomy*, Vol. 17, Academic Press, New York, pp. 163–231. 1965.
31. Stevenson, T. M. "Sweetclover studies on habit of growth, seed pigmentation, and permeability of the seed coat," *Sci. Agr.*, 17:627–654. 1937.
32. Stevenson, T. M. "Sweetclover in Saskatchewan," *Sask. Agr. Ext. Bul.* 9 (rev.). 1938.
33. Stevenson, T. M., and Kirk, L. E. "Studies in interspecific crossing with *Melilotus* and in intergeneric crossing with *Melilotus, Medicago* and *Trigonella*," *Sci. Agr.*, 15:580–589. 1935.
34. Stevenson, T. M., and White, W. J. "Investigations concerning the coumarin content of sweetclover. I. The breeding of a low-coumarin line of sweetclover—*Melilotus alba*," *Sci. Agr.*, 21:18–28. 1940.
35. Stewart, G. "Height of cutting sweetclover and influence of sweetclover on succeeding oat yields," *J. Am. Soc. Agron.*, 26:248–249. 1934.
36. Stickler, F. C., and Johnson, I. J. "The comparative value of annual and biennial sweetclover varieties for green manure," *Agron. J.*, 51:184. 1959.
37. Walster, H. L. "Sweetclover as a hay crop," *J. Am. Soc. Agron.*, 16:182–186. 1924.
38. Weaver, J. E. "Root development in the grassland formation," *Carnegie Inst. Wash. Pub.* 292. 1920.
39. White, W. J., and Horner, W. H. "Investigations concerning the coumarin content of sweetclover. II. Sources of variation in tests for coumarin content," *Sci. Agr.*, 21:29–35. 1940.
40. Whited, D. A., and others. "Influence of temperature, light intensity and photoperiod on the *o*-hydroxycinnamic acid in plant parts of sweetclover," *Crop Sci.*, 6(1):73–75. 1966.
41. Wiancko, A. T., and Mulvey, R. R. "Sweetclover: Its culture and uses," *Purdue U. Agr. Exp. Sta. Circ. 261*, pp. 1–14. 1940.
42. Willard, C. J. "An experimental study of sweetclover," *Ohio Agr. Exp. Sta. Bul. 405*. 1927.
43. Wolfe, T. K., and Kipps, M. S. "Comparative value of scarified and of unhulled seed of biennial sweetclover for hay production," *J. Am. Soc. Agron.*, 18:1127–1129. 1927.

Chapter 24

The True Clovers

The farmer can ameliorate 100 acres with clover more certainly than he can 20 from his scanty dung heap. While his clover is sheltering *the ground,* perspiring *its excrementitious effluvium on it, dropping its putrid leaves, and* mellowing *the soil with its tap roots, it gives full* food *to the stock of cattle, keeps them in heart, and increases the dunghill.*

—*J. B. Bradley,* Essays and Notes on Husbandry and Rural Affairs, *Philadelphia, 1801.*

ECONOMIC IMPORTANCE

The introduction of clover into customary crop rotations in the 16th century revolutionized agricultural practices throughout the world. The clovers (*Trifolium* species) are grown for hay, pasture, or soil improvement, usually in mixtures with grasses except when intended for seed production. Clover and clover-grass mixtures were cut for

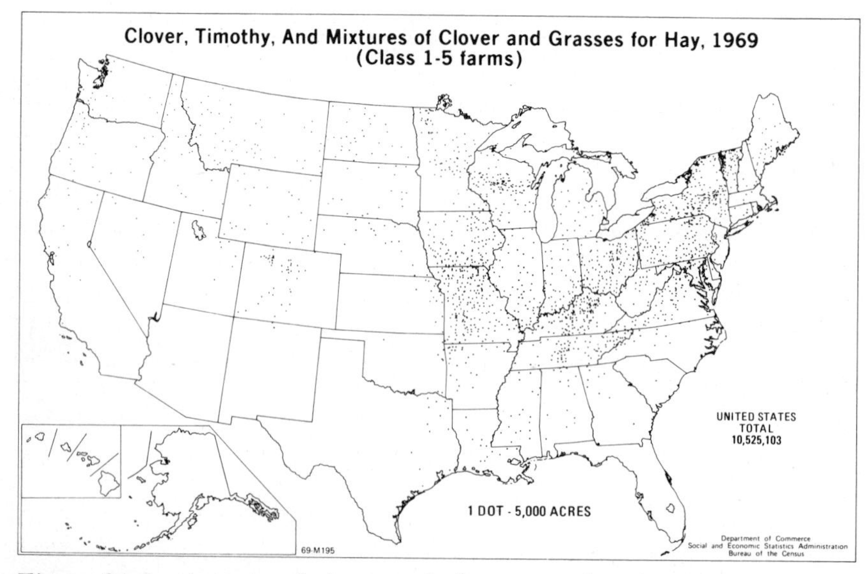

Figure 24–1. Acreage of clover and clover mixed with timothy (or other grasses) and cut for hay.

hay on about 13 million acres annually in the United States from 1968 to 1970. Production was over 23 million tons or 1.8 tons per acre. The leading states were New York, Missouri, Pennsylvania, Wisconsin, and Ohio (Figure 24–1).

IMPORTANT CLOVER SPECIES

The true clovers are members of the genus *Trifolium* of the plant order *Rosales* of the family *Leguminoseae*. About 250 species of *Trifolium* are known over the world. About 80 of these are native to the United States, mostly in the mountain regions of the West. Zigzag clover (*T. medium*), a perennial species grown in England, has strong creeping rootstocks. It was naturalized in the New England States. Kura clover (*T. ambiguum*), a native of Asia Minor, is being tried as a perennial honey plant in the United States. Twenty species of cultivated clover were described by Hermann.[26] Sweetclover, burclover, sourclover, buttonclover, alyceclover, and lespedeza (formerly called Japan clover) are not true clovers.

Red, white, and crimson clovers are of primary importance in the United States. Clovers important in certain localities include the alsike, strawberry, hop, sub, Persian, berseem, ball, lappa, and cluster species. The more important native clovers include seaside (*T. willdenovii*), white-tipped (*T. variegatum*), and long-stalked (*T. longipes*).

RED CLOVER

History

Red clover (*Trifolium pratense*), which includes the type called Mammoth clover, is grown alone or in grass mixtures on about 10 million acres for hay, pasture, seed or green manure. It grows wild throughout most of Europe and ranges far into Siberia. The plant was generally cultivated nearly 400 years ago in the Netherlands, whence it was carried to England in 1645. Red clover was grown by the American colonists as early as 1663. The crop has spread to nearly all regions in the United States where there is sufficient rainfall to grow it but most of it is in the northeastern fourth of the country. Under irrigation in the West, it is inferior to alfalfa for hay, except in some of the cooler intermountain valleys.

Adaptation

In general, red clover is adapted to the humid sections in the northern half of the United States (Figure 24–2). It thrives in a cool, moist climate. In Ohio, yields increased with increases in rainfall for April to June, inclusive.[62] Rainfall probably is the most important climatic

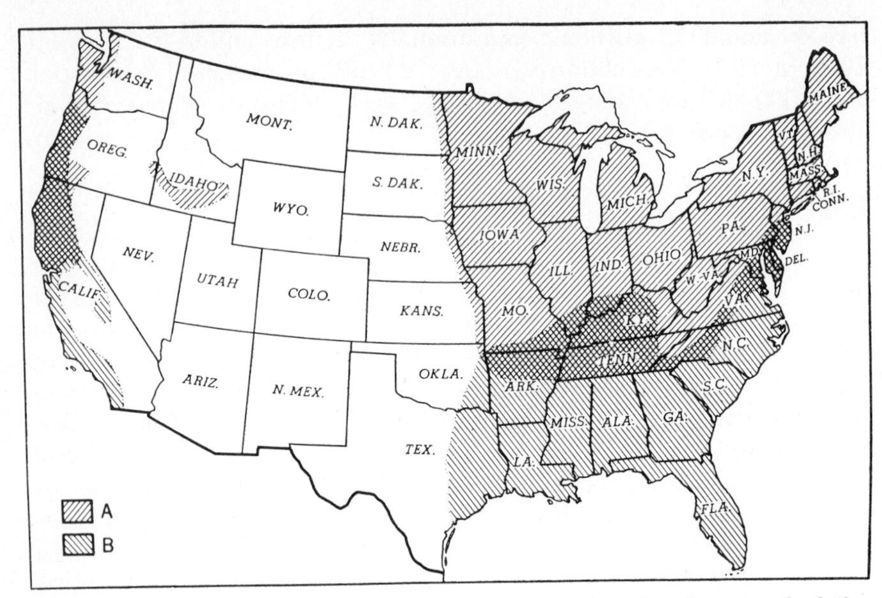

Figure 24–2. General adaptation of clovers: (*A*) red, white, and alsike clovers; (*B*) white, crimson, and other winter annual clovers (except that crimson clover is seldom grown along the Pacific Coast). Sweetclover is grown mostly in the Great Plains states from Montana and North Dakota southward to Oklahoma. Strawberry clover is grown on wet saline soils in the intermountain region.

factor that determines clover adaptation in the northern states, although stands are often lost by winterkilling. Strains adapted to this area must be able to tolerate a long period of dormancy which at times may be accompanied by very low temperatures. High summer temperatures apparently cause little injury, provided moisture is abundant.[1]

Red clover makes its best growth on fertile, well-drained soils of *p*H 6.6–7.6,[8] that contain an abundance of lime. Addition of phosphorus is necessary on most soils, while potash sometimes is beneficial. Red clover is not so well adapted to poorly drained or acid soils as is alsike clover. Clover failure, the partial or complete loss of stands either in the seeding year or second year, may be caused by an unfavorable soil condition. The addition of lime, phosphorus, and potash, alone or in combination, solves the soil difficulty.[19, 48]

Regional Strains

Many clover failures formerly were due to the use of seed of unadapted strains, especially when the seed was imported from countries where the climatic conditions are less severe. Foreign clovers, with few exceptions, are unsatisfactory for use in the Northeast.[49, 62] Strains from southern Europe are particularly unproductive. Foreign clovers that overwinter may produce a fair first crop, but a poor second cutting. They often die immediately after the first cutting. The stands generally go out because of winter injury, diseases (chiefly anthracnose and powdery mildew), or insects (chiefly leafhopper). Tests in the Midwest show that seed lots from Europe or South America yielded only 30 to 80 per cent as much as did the locally adapted strains. Foreign strains generally are badly winterkilled in Minnesota[11, 12] and Wisconsin. The results shown in Table 24–1 typify the behavior of unadapted clovers from foreign and domestic sources grown in the northern part of the clover region.[12]

Locally adapted seed often is preferable to that from other states,[18] unless the seed brought from another area is an adapted variety produced from foundation-seed stocks. Plants in the southern region must be resistant to southern anthracnose (*Colletotrichum trifolii*). Fully half of the acreage is devoted to improved varieties.

Botanical Description

Wild red clover is extremely variable, most of the plants being short-lived perennials. Forms exist that are early, late, smooth, hairy, prostrate, erect, and semierect. One of these forms probably is the ancestor of the clover first used in agriculture.[51]

Red clover is a herbaceous plant composed of three to ten or more leafy, hollow, erect stems that arise from a thick crown. The stems generally attain a height of less than 75 centimeters. Each leaf bears three oblong leaflets, usually with a pale spot in the center of each. The taproot, which has numerous laterals, may penetrate the soil to a depth of 1.5 meters. Red clover usually is a short-lived perennial in the cooler northeastern and high-altitude areas, a biennial in lower-altitude irrigated areas, and a winter annual in the warm southeastern parts of the United States.[8]

TABLE 24–1 Hay Yields of Red-Clover Strains in Minnesota (Two-Year Average)

SOURCE	HAY YIELDS PER ACRE			
	Loss in Stand (%)	*15% Moisture Basis* 1st year (lb.)	2nd year (lb.)	total (lb.)
Domestic				
Northern United States	18.1	2862	1122	3983
Central, southern, and western United States	27.3	2409	1264	3673
Foreign				
Western and central Europe	21.3	2228	1016	3244
Southern Europe, England, Wales, and Chile	85.4	488	256	744

The flowers are borne in compact clusters or heads at the tips of the branches (Figure 24–3). Each head contains 50 to 100 flowers. The flower consists of a green pubescent calyx with five pointed lobes and an irregular rosy purple corolla of five petals.[26] The flowers are pea-like except that they are much more elongated than those of peas and smaller, i.e., about ½ inch long. The claws of the petals are more or less united to the staminal tube. Nectar, secreted at the bases of the stamens, accumulates in the staminal tube around the base of the ovary. The pods are small, short, and break open transversally. Generally they contain a single kidney-shaped seed about 2 mm. long (Figure 22–4). The seeds vary in color from yellow to purple, the color variation being due largely to a complex heredity. Hard seeds are most frequent among those that are purplish.[57]

Pollination

In the pollination of red clover the florets develop and open from the base toward the top of the head.[9] The pistil usually is curved, with the stigma extended beyond the anthers, but florets have been found with greatly shortened styles. The ovary has two ovules, one of which develops normally while the other aborts. Although the anthers shed their pollen in the bud stage, both stamens and pistil remain in the

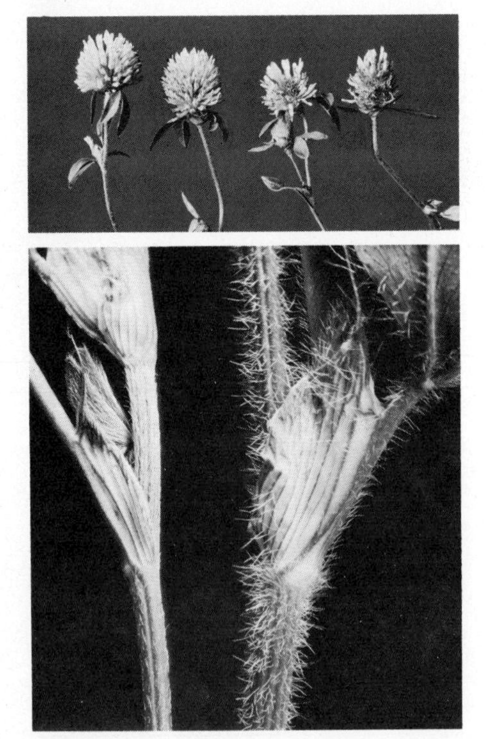

Figure 24–3. (*Top*) Red clover heads, the two at right damaged by clover midge. (*Bottom*) Smooth (or appressed hair) European red clover (*left*). American domestic red clover with hairy stems (*right*).

keel after the floret opens. The pistil is exposed only after the floret has been tripped.

Nearly all red clover plants are self-sterile. Because of the slow growth of the pollen tube in the style, the ovules disintegrate before the generative nucleus from the same plant reaches the egg. Self-fertile progenies have been obtained that continue to set selfed seeds. Other plants produce seeds when selfed, but the progenies may not necessarily be self-fertile. This phenomenon has been termed pseudo-self-fertility. A self-fertile line isolated in Minnesota was much less vigorous than commercial strains.[54] Cross-pollination is necessary for commercial seed production.

Cross-pollination of red clover is effected mainly by bumblebees and honeybees. They visit the floret for pollen or nectar or both. Honeybees principally obtain pollen. The honeybee is the major factor in pollination in western red-clover-seed fields. In Colorado,[53] plants caged with honeybees produced 61.5 seeds per head as compared with 67.3 from the open field. Bees trip the florets and carry pollen from plant to plant. The nectar of red-clover strains with short corolla tubes is more accessible to the honeybee, but less honeybee activity and less seed production has occurred in the Zofka (short-corolla tube) strain than in the American or Canadian strains.[10, 64] About 2 hives of bees per acre located in or within 1/4 mile of the clover field result in maximum seed yields.[46]

Types, Varieties, and Strains

There are two distinct forms of red clover: (1) the early or double-cut that gives two hay crops in a season, and (2) the late or single-cut that produces only one hay crop per season. The double-cut is known generally in this country as medium red clover, while the single-cut is commonly referred to as Mammoth red clover. Mammoth is grown on only 5 per cent of the red-clover acreage. Several types of each form are known. The single-cut clovers are taller, bloom 10 to 14 days later, and yield a heavier crop in the first cutting.[49] Except in unimportant particulars, the two forms do not differ in leaf shape, habit, root growth, or flower head.

Rough pubescent strains typify the red clovers of North America, with the hairs attached at right angles to the stems and leaf petioles. On the European forms the hairs are generally appressed or absent, the plants being commonly classified as smooth. The two forms can be distinguished in the seedling stage.[32] The American form may have developed its characteristic hairiness by the process of natural selection as a defense against leafhoppers which prefer the smooth forms.

Regional strains of red clover, which differ in productivity, winter-hardiness, and disease resistance, have developed as a result of the action of local environment on a highly variable plant. Naturalized varieties originated in Maryland, Louisiana, Pennsylvania, and Mississippi. Improved varieties bred at experiment stations are resistant to anthracnoses and powdery mildew.

Cultural Methods

Red clover is well adapted to three-year rotations. A common practice is to seed red clover with small grains the first year, harvest a hay crop the second year, and plow the clover under in preparation for corn the third year. The aftermath following the hay crop may be pastured before it is plowed under.[37]

The seedbed for red clover should be fine, firm, and moist. Red clover is seeded most frequently either on stands of winter wheat or rye or with small grain in the spring. It is usually broadcast at the rate of 8 to 10 pounds per acre,[19] or is drilled with either a special clover drill or a grain drill with a grass-seeder attachment. The clover usually is sown at three-fourths of the above rates in mixtures with grasses. The maximum safe depth of planting is about 0.6 cm on heavy soils or 1.2 cm on light soils. The crop is generally seeded early in the spring, although late-summer seedings have been successful as far north as central Indiana. Fall seeding prevails in the South and in the Pacific Coast states.[18] Seed inoculation is beneficial in many cases. Weed control methods are similar to those for alfalfa.

Red clover is cut in the early-bloom stage for the best yield and quality of hay. Many of the leaves are lost when clover is allowed to become too dry either in the swath or windrow. The curing practices are similar to those for alfalfa. The use of the hay crusher, together with forced-air drying, greatly facilitates the curing into a good-quality hay. The growth of clover is improved by mowing the stubble and removing the residue of a companion grain crop immediately after combining. The straw, clover tops, and weeds have some feed value. Mowing the first-year growth in late August provides some forage. It also may improve the yields of the second-year crop.[61]

Utilization as Pasture

Red clover is an excellent pasture plant but close, early grazing is injurious to the stand. Some pasture is ordinarily available in the fall of the seeding year, but some fall top growth should be allowed to remain to prevent winterkilling. There is some danger of bloat from plants that are pastured when young or wet.

Seed Production

Most of the red-clover seed is produced in the Corn Belt, but nearly all of the certified seed is produced in the western states on irrigated land.

The production of red-clover seed, especially in the Corn Belt, is generally incidental to hay production. An average set of 25 to 30 seeds per head indicates a fair seed crop, while 60 seeds indicates a good seed set. The general conditions that favor seed production are[19] (1) a vigorous recovery after cutting, (2) clear warm weather when the second crop is in bloom, (3) an abundance of bees for pollination, (4) absence of injurious insects such as the clover-flower midge and the chalcis fly, and (5) good harvesting weather. Wide extremes in soil moisture do not prevent setting of red-clover seed.[27]

High seed yields have been obtained in some of the western states.[15] In order to control injurious insects that affect seed, the first crop should be cut early for hay, or the first growth grazed.[28] Clover for seed should be irrigated sufficiently to produce a vigorous growth, but water should be withheld after the seeds are about half mature. Introduction of honeybees in the field when the plants are in full bloom will increase seed formation.

Red clover is ready to cut for seed when the heads have turned brown and the flower stalks are a deep yellow. The crop may stand until the heads are all black, where cutting is done when the atmosphere is damp.[42]

Much of the seed crop is threshed with a combine and pick-up from windrows, but occasionally, under ideal conditions, by direct harvest of the mature crop.

Diseases

Southern anthracnose (*Colletotrichum trifolii*) is one of the most important factors in loss of stands south of the Ohio River, especially in Tennessee. This disease is characterized by a series of elliptical shrunken areas on the stems and leaf petioles which spread until the death of the plant. The disease is particularly virulent in damp hot weather. Northern anthracnose (*Kabatiella caulivora*) is often severe on the first crop of the second year in the northern Clover Belt, but seldom is it so virulent as the southern form.

Powdery mildew (*Erysiphe polygoni*) occurs on red clover and causes some reduction in quality and yield throughout the Clover Belt. It appears as a powdery dust on the leaves of infected plants.

A number of other diseases that attack the leaves cause losses in the quality of the hay. Varieties resistant to the three major diseases are now in production. Root rots and viruses also attack red clover.[21, 23, 24]

Insect Pests

Several insects cause damage to red clover, among them the clover root-borer (*Hylastinus obscurus*), the clover-seed chalcis fly (*Bruchophagus gibbus*), the clover-flower midge (*Dasyneura leguminicola*), and the root curculio (*Sitonia hispidula*).

Fields frequently are heavily infested with the clover root-borer.

The larvae tunnel in the roots, especially in the summer of the second year. One control measure is to plow under the clover after the first hay crop has been removed, preferably between June 15 and August. A three-year rotation has also been suggested.

The clover-seed chalcis fly is one of the worst seed pests in this country. It is a black, wasplike insect about the size of a red-clover seed. The larva develops within the seed. To reduce the damage from this pest, all debris from threshing should be destroyed and the first spring growth should be grazed lightly.

The seed production of red clover is greatly reduced at times by the clover-flower midge (Figure 24–3). The maggots of this insect attack the florets, with the result that many fail to develop seed. Brown withered petals among the normally pink blossoms are a conspicuous symptom of attack by this pest. Midge may be controlled by cutting the first crop early enough to catch the maggots before they are fully developed. This is usually at the time they begin to change color from creamy white to salmon pink, i.e., early in June in most affected areas.

The root curculio gnaws and sometimes girdles the roots, causing wilting or death of the plants.

Other insects that may cause considerable damage are the clover aphids[29] and the potato leafhopper. The former infest the clover heads, secreting what is called *honey-dew* at the base of the florets. They bring about serious reduction in seed yield in the western states as well as causing the seed to be sticky when threshed. The potato leafhopper (*Emposaca fabae*) frequently attacks seedling or second-year plants in the humid regions, often stunting them. The European smooth clovers are particularly susceptible to leafhopper attacks.

The weevils, aphids, leafhoppers, spittlebugs, and grasshoppers that attack clover can be controlled with certain insecticides.

ALSIKE CLOVER

Alsike clover (*Trifolium hybridum*), a native of northern Europe, has been cultivated in Sweden since the 16th century. It was introduced into England in 1832 and in the United States about 1839. The production of alsike clover in the United States has declined since 1948, but is still an important export seed crop in western Canada.

The geographical distribution of alsike clover is roughly in the red clover-timothy area. It grows well in the intermountain states, especially under irrigation. Alsike clover requires a cool climate with an abundance of moisture. It rarely winterkills, being able to withstand more severe winters than does red clover. Alsike clover makes its best growth on heavy silt or on clay soils that are moist. It is less sensitive to soil acidity or alkalinity and better adapted to low, poorly drained soils than is red clover. Alsike clover sometimes thrives where clover failure is common with red clover.

Alsike clover is a perennial that persists for four to six years in a favorable environment, but usually lasts only two years in the Midwest. The stems may reach a length of 90 cm. The stems and leaves are smooth. The flower heads are partly pink and partly white. Each pod may contain two to four seeds. The main axis continues to grow while single flower-bearing branches, each with one or more flower heads, arise successively from each leaf axil.[26, 47] The seeds of alsike clover are various shades of green mixed with yellow, and are about one third as large as those of red clover (Figure 22–4). Alsike clover plants are generally self-sterile and 3700 bees per hectare provide pollination for seed production.

The cultural methods are similar to those for red clover. The crop is seeded at the rate of 4 to 6 pounds per acre when grown alone. It is more often seeded with timothy or red clover or both. The crop is used for the same purposes as red clover.

Alsike usually yields less hay than does red clover. It produces only one crop in a season. The plant is especially suited for pasture mixtures, particularly in permanent pastures on wet, sour, or occasionally flooded land.

WHITE CLOVER

White clover (*Trifolium repens*), which apparently originated in the eastern or Asia Minor region, is now widely distributed in every continent. It is an important crop in New Zealand. It was brought to the United States from Europe by the early settlers. It is a valuable addition to both humid-region pastures and irrigated pastures in America.[22] It is a leading pasture legume grown chiefly in mixtures on about 13 million acres chiefly from Arkansas and Tennessee northward.

Adaptation

The most favorable habitat for white clover is a moist, cool region under which growth is continuous. It will withstand greater temperature extremes than will either red or alsike clover and even survives in Alaska. In the Southeast[7] it is grown as a winter annual because most plants die there in early summer after they have produced seed.

White clover is adapted to moist soils, especially to clays or loams abundantly supplied with phosphorus and potash. Growth on strongly acid or alkaline soils is not thrifty. Applications of phosphorus, potash, and lime are beneficial to growth on most soil types in humid regions, and often magnesium and sulfur also are helpful.

Botanical Description

White clover is a low, smooth, long-lived perennial plant with solid stems that root at the nodes.[26] The roots are shallow. The flower heads arise from the leaf axils on long flower stalks. The flowers are small and white to pink in color. They generally turn brown when mature. The small seed pods usually contain 3–5 yellow seeds. White clover is extremely variable in leaf size, color, and markings; in size of runners; and in persistence.[3, 51] The spreading habit of the plant is a result of extensive stolon development (frequently referred to as runners). Occasional plants or branches bear four leaflets that have a V-shaped white mark near the middle of the leaflet.

White clover is practically self-sterile. The florets require cross-pollination for seed formation, usually done by bees.

Types or Varieties

White clovers of agricultural value have been grouped into three types, namely (1) large (*T. repens* var. *giganteum*), (2) intermediate, and (3) small.

Large varieties include Ladino and others. Ladino clover appears to have been introduced from the vicinity of Lodi in northern Italy and is about two to four times as large as common white in all parts except the seeds.[26] The growth habits are similar to those of common white. It often establishes itself by natural reseeding. Ladino is adapted to a temperate climate where moist soils favor its growth. It will also grow on poorly drained as well as mildly acid soils.[4] Ladino is grown in the eastern states from New Hampshire to North Carolina; westward to Iowa, Missouri, and Arkansas; and also to the irrigated regions of the West. It is the leading irrigated pasture legume in California, where most of the Ladino seed is produced.[41]

Intermediate white clover includes Common, White Dutch, or other northern types and southern types adapted to warm climates. The cool-weather type adapted to the northern states for pasture mixtures or lawns may be more persistent than Ladino, but produces less herbage. Southern types thrive as winter annuals in the warmer southeastern states, but lack the hardiness to survive in the North.[2, 29]

Small white clover includes the types commonly found in closely grazed pastures. The small type is represented by what is called English wild white clover.[51] The small or low-growing type is more likely to show the presence of a cyanophoric glucoside, but the quantity of hydrocyanic acid present is so small that it is harmless.[55] The development of varieties or strains has resulted from the action of environment on a variable species.

Cultural Methods

White clover is usually seeded with grasses, either on new seedings or on an established turf. In new seedings, the white clover is mixed with grass seed and the entire mixture seeded at one time, usually early in the spring, although in some places it is sown in early fall.[25, 44] The grass should be clipped short when clover is seeded in turf. The rate of seeding is 2 to 3 pounds of clover seed per acre. When seeded in the early fall for winter pasturage in Florida and southern Georgia, scarified seed germinates better, especially at high temperatures.[14] The yield of top growth was significantly higher in New Jersey experiments where a complete fertilizer of lime, phosphorus, and potash was used.

Cultural methods for Ladino clover are very similar to those for common white.[32] Ladino clover may be sown in mixtures with orchardgrass, timothy, bromegrass, tall fescue, or reed canarygrass. Kentucky bluegrass soon crowds out Ladino clover. For Connecticut, a recommended Ladino seed mixture for pasture is orchardgrass, 6 pounds; Ladino clover, 2 pounds.[13] For hay, the mixture is timothy, 6 pounds; red clover, 6 pounds; and Ladino clover, 1 pound. These mixtures may be seeded with small grain in the spring, or without a companion crop between July 15 and August 20. For seed production in California, Ladino clover is sown in the autumn or late winter, without a companion except on blow soils.[41] The best stands are maintained when not mowed until the plants are 6 to 8 inches high, but after mowing, the stubble should be at least 4 inches (10 cm.) above the ground. Close cutting in October is particularly harmful. The crop should be cut for hay when one-tenth of the flower heads have turned brown.[16] Continuous close grazing of Ladino clover is inadvisable. Ladino clover is not suitable for lawns because of its tall growth and inability to withstand close clipping. Ladino clover responds to applications of manure and to phosphate and potash fertilizers in sections where those elements are deficient.

Utilization

The intermediate white clovers are almost always grown in grass mixtures for pasture or lawns. Ladino will produce hay or silage under favorable conditions. Ladino clover pastures carried 30 to 40 per cent more stock than an equal acreage of red clover, alsike clover, or ordinary white clover under Idaho conditions.[58]

Bloating of cattle is less likely to occur if the clover is 30 cm. tall or if the plants are growing slowly.[20]

STRAWBERRY CLOVER

Strawberry clover (*Trifolium fragiferum*) has assumed considerable importance in the United States. It is a native of the eastern Mediterranean region, but its culture has extended to every continent. It is a profitable pasture crop in the intermountain and Pacific Coast states in places where the subsoil water table is so high that other crop plants are largely or wholly eliminated.

Strawberry clover is adapted to the climate of the Pacific Coast, the intermountain region, and western Nebraska.[32] It is unable to compete with white clover in the eastern states. The primary requirement of strawberry clover is an abundance of water. It is of particular value on wet saline soils. Good growth has been observed where the salt content of the soil was more than 1 per cent. Established stands have survived salt concentrations of more than 3 per cent for long periods.[33] It tolerates these high salt concentrations when ample water is present. It has survived on soils flooded for one to two months.

Strawberry clover is a perennial, low-growing plant that spreads vegetatively by creeping stems that root at the nodes. Its leaves, stems, and habit of growth are similar to those of white clover. It is readily identified by the characteristic strawberry-like flower heads. The flower heads are generally round, being pink to white in color (Figure 24–4). As the seed matures, the calyx around each seed becomes inflated. The mature head is gray to light brown in color. When ripe, the capsules break from the head. The seed is reddish brown, or yellow flecked with dark markings. The seeds are much larger than those of white clover but slightly smaller than the seeds of red clover. The flowers are self-fertile. An improved variety is adapted to areas of mild winters such as central and southern California.

Strawberry clover is ordinarily spring-seeded at the rate of 2 to 5 pounds per acre. Unhulled seed is generally advised for late winter planting, and scarified seed for spring planting in places where the soil is too wet to cultivate. Growth is aided by mowing to reduce the

Figure 24–4. Strawberry clover.

competition of rushes and similar plants, particularly on unprepared seedbeds.

Strawberry clover is utilized almost entirely for pasture, but it also has been used for green manure on saline soils. It offers promise in the reclamation of saline, alkaline soils now considered wastelands in the western states. Strawberry-clover herbage is as palatable to livestock as is white clover. It survives close grazing, but it is more productive when grazed moderately.

CRIMSON CLOVER

Crimson clover (*Trifolium incarnatum*) is the most important winter annual among the true clovers. It is a native of Europe, being introduced to the United States as early as 1819. It became agriculturally important about 1880. This clover is grown on about 800,000 acres in the central belt of the eastern and southeastern states and on the Pacific Coast.[51, 9]

Crimson clover is adapted to cool humid weather, where winter temperatures are moderate. It is sown in late summer or early fall in time to become established before winter. Crimson clover will grow on both sandy and clay soils. While the plant is tolerant of ordinary soil acidity, it is difficult to obtain a stand on very poor soils. This clover is being used effectively as a summer annual in Maine.

In general, the leaves and stems of crimson clover resemble those of red clover, but the leaflet tips of crimson clover are more rounded. There also is a greater covering of hair on both leaves and stems. When sown in the fall, the young plants form a rosette that enlarges under favorable conditions. The inflorescence is an elongated spike-like head (Figure 24–5). The flowers are bright crimson in color, while the seeds (Figure 22–4) are yellow. White-flowered types have been selected.

Crimson clover is a long-day, self-fertile plant that is less variable than either red or white clovers. Insects increase the amount of pollination since the florets are not self-tripping.

About half of the crimson clover acreage consists of self-seeding varieties that have hard seeds. The hard seeds do not germinate before autumn when conditions for plant growth are favorable. This provides winter cover and spring pasture where plants are allowed to go to seed. Common crimson clover and some other varieties seldom reseed naturally.

Crimson clover is generally seeded in August or September between the rows of cultivated crops, either broadcast and covered with a cultivator, or with a drill. Crimson clover is sometimes seeded after a small-grain crop, or with rye, vetch, Italian ryegrass, and fall-sown grain crops. The seeding rate is 15 to 20 pounds of hulled seed or 45

Figure 24–5. Flower heads of crimson clover.

to 60 pounds of unhulled seed per acre. Inoculation with suitable cultures often is desirable.[5, 25] Crimson-clover seed is produced in Oregon and in the Southeast.

Crimson clover is used for hay, pasture, winter cover, and green manure. The crop furnishes an abundance of early spring grazing, as well as some fall and winter pasture under favorable growth conditions. It seldom causes bloat. This clover makes excellent hay when cut at the early-bloom stage. As crimson clover reaches maturity, the hairs on the plant become hard and tough. When such hay is fed, hair balls (phytobezoars) may be formed in the stomachs of horses. These occasionally cause death. As a green manure, crimson clover is generally plowed under two or three weeks before time to plant the next crop. As an orchard cover crop, it is generally allowed to mature, after which it is disked into the soil. It is harvested for seed by combining when the hulls are dark brown or by swathing when light brown.

OTHER WINTER-ANNUAL CLOVERS

Many other winter-annual species of *Trifolium* are found throughout the southern, as well as the south central and Pacific Coast states. The principal clovers are large hop (*Trifolium procumbens*), small or low hop (*T. debium*), called suckling clover in England (Figure 24–6), Persian (*T. resupinatum*), cluster (*T. glomeratum*), sub (or subter-

Figure 24–6. Plant of small hop clover.

ranean) (*T. subterraneum*), hop (*T. agrarium*) (Figure 24–7), ball (*T. nigrescens*), rose (*T. hirtum*), striate or knotted (*T. striatum*), rabbitfoot (*T. arvense*), Carolina (*T. carolinianum*), buffalo (*T. reflexum*), and lappa (*T. lappaceum*). Hop, large hop, and small hop clover have yellow flowers. Arrowleaf (*T. vesiculosum*), as well as

Figure 24–7. Branch of hop clover.

bigflower (*T. michelianum*), are established in pastures and meadows in the Southeast. Arrowleaf clover is replacing crimson clover in Alabama.[35]

Winter annuals, seeded in the spring in the Cotton Belt or Corn Belt, make only a small growth before the increasing length of day induces them to flower. Many of these winter annuals are perpetuated from year to year by self-seeding in the early summer, from which they volunteer during the fall months.[25] They become important agriculturally only when sufficient germinable seed is present to establish good stands. The seed coats of many of these species are hard.[60] The seeds of the large hop, cluster, sub, and Persian clovers germinate when scarified. Even after scarification, germination of the seed of all species is inhibited in varying degrees by temperatures as high as 30° to 35° C.

Large hop and small hop clovers are both valuable pasture plants.[30] The association of hop clover with grass appears beneficial to the establishment of the clover, but tall northern grasses may shade it out. The clover is favored when the southern grasses become dormant in the fall. Most nearly complete stands are obtained with September and October seedings. Hop clover (*T. agrarium*) is becoming established naturally in the northern central states. It is a native of Europe, probably introduced as a mixture in other clovers, but George Washington ordered seed of hop clover from England in 1786.

Sub (subterranean) clover is being used in pasture mixtures in the Pacific Coast states, but it also shows promise in the Southeast. It requires cool, mild, moist winters and dry summers together with well-drained soils well supplied with phosphorus. It provides winter and early spring pasture. Subclover has prostrate stems 30 cm. or more in length and white flowers with rose stripes is best adapted as a pasture crop to nonirrigated, foothill, cut-over lands. It requires phosphorus fertilization on these soils and some response is obtained from small molybdenum supplements. Lime was beneficial on the more acid soils when applied separately from phosphorus. Seed inoculation is essential.[36] The plant buries its maturing seed heads in the soil or in plant residues on the surface. The seed can be gathered with a suction type of seed reclaimer. Sub clover is important in Australia[43] as well as in Chile.

Persian clover is a constituent of pasture mixtures in the southeastern states. Sometimes it is also cut for hay or silage, or is turned under for green manure. Applications of phosphorus, potash, and, often, lime to the soil are beneficial. The flowers are light purple.[6]

Berseem or Egyptian clover (*T. alexandrinum*) is grown to a slight extent as a winter annual, chiefly for pasture, in southern portions of Arizona, California, and Texas. It is the leading winter legume in Egypt. The flowers are yellowish. Temperatures of 20° to 25° F. may kill the plants.

Rose clover is grown in California for pasture or for soil conservation. The flowers are purplish red.[38]

Ball clover, introduced from Turkey, reseeds itself when growing in pasture-grass mixtures in the southeastern states. It withstands late heavy grazing. The flowers are white to yellowish.

Bigflower clover, also from Turkey, is adapted to the southeastern states for hay or pasture. It has roseate flowers.

Striate clover is adapted to heavy soils. It furnishes grazing before the stems become harsh at flowering time. The flowers are reddish to rose.[25]

Rabbitfoot clover grows on sandy soils. It has grayish silky heads.

Carolina clover is a small early species with purplish-red flowers. It is found in pastures as well as on roadsides on poor soils.

Buffalo clover is widely distributed in the South. It has large light-brown maturing seed heads, with its flowers reflexed or turned down.

Lappa clover is adapted to the Black Belt of Alabama where it has become naturalized.[59] It is able to reseed itself when used either for pasture or hay. This species has inconspicuous lavender-rose self-fertile flowers.

REFERENCES

1. Aamodt, O. S. "Climate and forage crops," in *Climate and Man*, USDA Yearbook, 1941, pp. 439–458.
2. Aamodt, O. S., Torrie, J. H., and Smith, O. F. "Strain tests of red and white clovers," *J. Am. Soc. Agron.*, 31:1029–1037. 1939.
3. Ahlgren, G. H., and Sprague, H. B. "A survey of variability in white clover (*Trifolium repens*) and its relation to pasture improvement," *N.J. Agr. Exp. Sta. Bul. 676*. 1940.
4. Ahlgren, G. H., and Fuelleman, R. F. "Ladino clover," in *Advances in Agronomy*, Vol. 2, Academic Press, New York, pp. 207–232. 1950.
5. Anonymous. "Growing crimson clover," *USDA Leaflet 482*. 1961.
6. Anonymous. "Persian clover—A legume for the South," *USDA Leaflet 484*. 1960.
7. Anonymous. "White clover for the South," *USDA Leaflet 498*. 1961.
8. Anonymous. "Growing red clover," *USDA Leaflet 571*, pp. 1–8. 1968 (Rev.).
9. Anonymous. "Growing crimson clover," *USDA Leaflet 482* (Rev.), pp. 1–10. 1971.
10. Armstrong, J. M., and Jamieson, C. A. "Cross-pollination of red clover by honeybees," *Sci. Agr.*, 20:574–585. 1940.
11. Arny, A. C. "Winterhardiness of medium red clover strains," *J. Am. Soc. Agron.*, 16:268–278. 1924.
12. Arny, A. C. "The adaptation of medium red clover strains," *J. Am. Soc. Agron.*, 20:557–568. 1928.
13. Brown, B. A., and Munsell, R. I. "Pasture investigations: Ladino clover experiments, 1930 to 1940," *Conn. (Storrs) Agr. Exp. Sta. Bul. 235*. 1941.
14. Burton, G. W. "Factors influencing the germination of seeds of *Trifolium repens*," *J. Am. Soc. Agron.*, 32:731–738. 1940.
15. Dade, E., and Johansen, C. "Red Clover seed production in central Washington," *Wash. Agr. Exp. Sta. Cir. 406*. 1962.
16. Eby, C. "Ladino clover," *N. J. Agr. Exp. Sta. Cir. 408*. 1941.
17. Elliott, C. R., and Pankiw, P. "Alsike clover," *Can. Dept. Agr. Pub. 1264*, pp. 1–15. 1972.
18. Fergus, E. N. "Adaptability of red clovers from different regions of Kentucky," *Ky. Agr. Exp. Sta. Bul. 318*. 1931.
19. Fergus, E. N., and Hollowell, E. A. "Red Clover," in *Advances in Agronomy*, Vol. 12, Academic Press, New York, pp. 365–436. 1960.
20. Foote, L. E., and others. "Controlling bloat in cattle grazing clover," *La. Agr. Exp. Sta. Bul. 629*, pp. 1–28. 1968.
21. Fulton, N. D., and Hanson, E. W. "Studies on root rots of red clover in Wisconsin," *Phytopath.*, 50(7):541–550. 1960.
22. Gibson, P. B., and Hollowell, E. A. "White clover," *USDA Agr. Hdb. 314*, pp. 1–33. 1966.
23. Hanson, E. W., and Hagedorn, D. J. "Viruses of red clover in Wisconsin," *Agron. J.*, 53:63–67. 1961.
24. Hanson, E. W., and Kreitlow, K. W. "The many ailments of clover," in *Plant Diseases*, USDA Yearbook, 1953, pp. 217–228.

25. Henson, P. R., and Hollowell, E. A. "Winter annual legumes for the South," *USDA Farmers Bul. 2146*. 1960.
26. Herman, F. J. "A botanical synopsis of the cultivated clovers (*Trifolium*)," *USDA Monograph 22*. 1953.
27. Hollowell, E. A. "Influence of atmospheric and soil moisture conditions upon seed setting in red clover," *J. Agr. Res.*, 39:229–247. 1929.
28. Hollowell, E. A. "Red clover seed production in the intermountain states," *USDA Leaflet 93*. 1932.
29. Hollowell, E. A. "White clover," *USDA Leaflet 119* (rev.). 1947.
30. Hollowell, E. A. "The establishment of low hop clover, *Trifolium procumbens*, as affected by time of seeding and growth of associated grass," *J. Am. Soc. Agron.*, 30:589–598. 1938.
31. Hollowell, E. A. "The pubescent characteristic of red clover, *Trifolium pratense*, as related to the determination of origin of the seed," *J. Am. Soc. Agron.*, 32:1–11. 1940.
32. Hollowell, E. A. "Ladino white clover for the northeastern states," *USDA Farmers Bul. 1910*, pp. 1–10. 1942.
33. Hollowell, E. A. "Strawberry clover: A legume for the West," *USDA Leaflet 464*. 1960.
34. Hoveland, C. S. "Arrowleaf clover," *Auburn U. Agr. Exp. Sta. Leaflet 67*. 1962.
35. Hoveland, C. S., and others. "Management effects on forage production and digestibility of Yucki arrowleaf clover (*Trifolium vesiculosum* Savi)," *Agron. J.*, 62:115–116. 1970.
36. Jackson, T. L. "Effects of fertilizers and lime on the establishment of subterranean clover," *Ore. Agr. Exp. Sta. Circ. Information 634*, pp. 1–15. 1972.
37. Justin, J. R., and others. "Red clover in Minnesota," *Minn. Ext. Bul. 343*, pp. 1–15. 1967.
38. Knight, W. E., and Hollowell, E. A. "The influence of temperature and photoperiod on growth and flowering of crimson clover (*Trifolium incarnatum* L.)," *Agron. J.*, 50:295–298. 1958.
39. Love, R. M., and Summer, D. C. "Rose clover," *Calif. Agr. Exp. Sta. Cir. 407*. 1952.
40. MacVicar, R. M. and others. "Studies in red clover seed production": Part I, *Sci. Agr.*, 32:67–80, 1952; Part II, *Canad. J. Agr. Sci.*, 33:437–447. 1953.
41. Marble, V. L., and others. "Ladino clover seed production in California," *Calif. Agr. Exp. Sta. Circ. 554*, pp. 1–33. 1970.
42. McClymonds, A. E., and Hulbert, H. W. "Growing clover seed in Idaho," *Idaho Agr. Exp. Sta. Bul. 148*. 1927.
43. Morely, F. H. W. "Subterranean clover," in *Advances in Agronomy*, Vol. 13, Academic Press, New York, pp. 57–123. 1961.
44. Morgan, A. "Some common clovers: their identification," *J. Dept. Agr. Victoria (Australia)*, 30:105–112. 1932.
45. Park, J. K., and others. "Establishing stands of fescue and clovers," *S. C. Agr. Exp. Sta. Cir. 129*. 1961.
46. Peterson, A. G., and others. "Pollination of red clover in Minnesota," *J. Econ. Ent.*, 53(4):546–550. 1960.

47. Pieters, A. J. "Alsike clover," *USDA Farmers Bul. 1151* (rev.). 1947.
48. Pieters, A. J. "Clover problems," *J. Am. Soc. Agron.*, 16:178–182. 1924.
49. Pieters, A. J. "The proper binomial or varietal trinomial for American mammoth red clover," *J. Am. Soc. Agron.*, 20:686–702. 1928.
50. Pieters, A. J., and Morgan, R. L. "Field tests of imported red clover seed," *USDA Cir. 210*. 1932.
51. Pieters, A. J., and Hollowell, E. A. "Clover improvement," in *USDA Yearbook*, 1937, pp. 1190–1214.
52. Rampton, H. H. "Growing subclover in Oregon," *Oreg. Agr. Exp. Sta. Bul. 432*. 1952.
53. Richmond, R. G. "Red clover pollination by honeybees in Colorado," *Colo. Exp. Sta. Bul. 391*. 1932.
54. Rinke, E. H., and Johnson, I. J. "Self-fertility in red clover in Minnesota," *J. Am. Soc. Agron.*, 33:512–521. 1941.
55. Rogers, C. F., and Frykolm, O. C. "Observations on the variations in cyanogenetic power of white clover plants," *J. Agr. Res.*, 55:533–537. 1937.
56. Schoth, H. A., and Rampton, H. H. "Ladino clover for western Oregon," *Oreg. Agr. Exp. Sta. Bul. 519*. 1952.
57. Smith, D. C. "The relations of color to germination and other characters of red, alsike, and white clover seeds," *J. Am. Soc. Agron.*, 32:64–71. 1940.
58. Spangler, R. L. "Ladino clover seed production and the value of Ladino as a pasture crop," *J. Am. Soc. Agron.*, 17:84–86. 1925.
59. Sturkie, D. G. "A new clover for the black lands in the South," *J. Am. Soc. Agron.*, 30:968. 1938.
60. Toole, E. H., and Hollowell, E. A. "Effect of different temperatures on the germination of several winter annual species of *Trifolium*," *J. Am. Soc. Agron.*, 31:604–619. 1939.
61. Torrie, J. H., and Hanson, E. W. "Effects of cutting first year red clover on stand and yield in the second year," *Agron. J.*, 47:224–228. 1955.
62. Welton, F. A., and Morris, V. H. "Climate and the clover crop," *J. Am. Soc. Agron.*, 17:790–800. 1925.
63. Wiggans, R. G. "Local, domestic, and foreign-grown red clover seed," *J. Am. Soc. Agron.*, 23:572–579. 1931.
64. Wilsie, C. P., and Gilbert, N. W. "Preliminary results on seed setting in red clover strains," *J. Am. Soc. Agron.*, 32:231–234. 1940.

Chapter 25

Lespedeza

ECONOMIC IMPORTANCE

Lespedeza is a major crop in the eastern half of the United States, principally for pasture, either alone or in mixtures, but also for hay, seed, silage, or soil improvement. More than 1 million acres of lespedeza were harvested for hay in 1970, yielding nearly 1.5 tons per acre (Figure 25-1).

Lespedeza provides a legume to help maintain soil productivity in the southeastern states where lime-deficient soils are too acid for economical production of either alfalfa or red clover. The temperatures there also are too high for perennial growth of red clover. It is especially valuable for growing on badly eroded soils in that region. Few crops make as much growth as sericea lespedeza on soils of low fertility.

The three cultivated species of lespedeza grown in the United

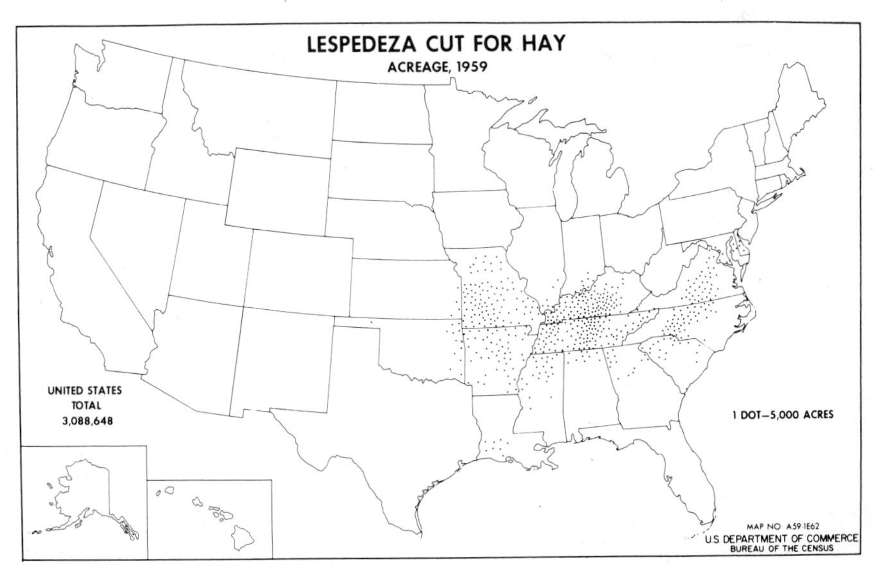

Figure 25-1. Acreage of lespedeza hay in the United States.

States are of Asiatic origin. The earliest record of common lespedeza or Japan clover, as it was called, indicates its presence in the southeastern states in 1846. The introduction of Korean and Kobe lespedezas from Korea in 1919 has been the chief factor in the widespread use of lespedeza. The agricultural use of sericea lespedeza dates from an introduction from Japan in 1924, although the crop was first tried in North Carolina in 1896.[19] Ten species of native perennial lespedezas have been collected in Kansas.[2]

ADAPTATION

High rainfall well distributed throughout the growing season, relatively high humidity, and rather high temperatures favor the growth of lespedeza. The annual lespedezas are more drought-resistant than alfalfa or clovers[20] in the early-growth stages, and the perennial sericea will endure extreme drought after it is well established. However, growth is so limited under dry conditions that lespedeza is not suitable for the semiarid Great Plains.

The lespedezas are warm-season plants that begin growth slowly in the spring. Except for Korean, which stops growth with seed maturity, the lespedezas continue to grow until frost.

All the annual lespedezas are sensitive to cold. Seedlings that start growth during warm periods in late winter may be killed by late spring frosts. The perennial sericea lespedeza has survived winter temperatures as low as 2° C. but it may be killed down to the crowns by heavy spring freezes when in a growing condition. Lespedeza is most cold-resistant in the cotyledon stage and becomes more tender as the plants develop.[25] In alfalfa and red clover, cold resistance increases with age, a reversal in this respect. A temperature of −5° C. for 16 hours has killed all lespedezas.[25]

Lespedeza may fail to set seed under extreme long-day conditions. Photoperiod also affects the adaptation of varieties. Fruiting is prevented in Korean when the day length is more than 14 hours.

The lespedezas will grow on most soils, especially on soils too acid for clover without lime applications. However, plant growth is improved by use of lime or fertilizers, or both, on soils deficient in the mineral elements. Lespedeza makes little growth on the sandy soils of the coastal plain unless supplied with phosphorus.[23] In Florida, it was necessary to add calcium, phosphorus, and potassium to the sandy soils of the flat pinelands to produce a satisfactory growth of lespedeza.[5] From 40 to 60 kilos per hectare of both P_2O_5 and K_2O are recommended for soils of the Gulf Coastal Plain.[13] The distribution of the crop is not limited by soil acidity even though lime applications increase growth.

Sericea lespedeza can be grown on most soils without fertilizers or lime,[4] but it will respond to additions of lime and fertilizers on poor soils.[19] It fails on poorly drained soils.[20] It appears to thrive on the poor, eroded clays, silts, or silt loams of the piedmont region and on similar soils elsewhere.

BOTANICAL DESCRIPTION

The annual species of lespedeza in the United States are common (*Lespedeza striata*) and Korean (*L. stipulacea*), while sericea (*L. cuneata*) is perennial. The prominent veins or furrowed surface of the leaflets serve to distinguish lespedezas from other legume crop plants. All varieties have a higher leaf percentage than does alfalfa.

Annual Lespedezas

The annual lespedezas are erect or spreading, small, branched, and short-day plants. They may attain a height of 60 to 75 cm., but the growth usually ranges from 10 to 30 cm.[9] The leaves are small and trifoliolate. The roots are medium-deep and numerous. The plants produce two kinds of small flowers: petaliferous flowers with purple or bluish petals, and the more numerous but inconspicuous apetalous flowers. The apetalous (cleistogamous) flowers actually have very small petals, while self-pollination occurs before the calyx opens. The petaliferous (chasmogamous) flowers may be cross-pollinated. Chasmogamous flowers predominate when temperatures are high. The plants bloom from midsummer to early fall. The seeds are about the size of, or slightly larger than, those of red clover, being borne in pods that retain the seeds when threshed (Figure 22–4). The hulls are brown.

Common striate lespedeza is a slender plant, prostrate except in dense stands.[19] It usually grows to a height of 10 to 15 cm., but in southern latitudes it reaches 30 to 38 cm. The flowers are very small, purple, and inconspicuous. The seeds are borne in the axils of the leaves along the entire length of the stem. The seeds are dark purple mottled with white. The hairs on the stem are appressed downward.

Korean lespedeza (Figure 25–2) is coarser and earlier than common, and has broader leaflets, larger stipules, and longer petioles. The seed is borne in the leaf axils at the tips of all branches, rather than along the stem as in common. The seeds are a solid dark purple. At maturity the leaves of Korean turn forward so that the tips of the branches resemble small cones. The leaves of common do not turn forward. The hairs on the stems of Korean are appressed upward.

Sericea Lespedeza

Sericea is a short-day perennial that produces coarse, stiff, tough erect stems. A single stem on each plant, 30 to 45 cm. in height, is

Figure 25–2. Branch of Korean lespedeza.

produced the first season. Additional stems, 0.6 to 1.5 meters in height, arise from crown buds in subsequent years. The woody, widely branched roots penetrate the soil to a depth of 1 meter or more.[20] Both chasmogamous and cleistogamous flowers occur in sericea, most of the seed being produced from cleistogamous flowers. Pods from chasmogamous flowers are larger and more acute.[13] The sericea flowers are yellow or purple. The species varies in width of leaflets, height, coarseness and number of stems, and earliness (Figure 25–3). A majority of the chasmogamous flowers are cross-pollinated.

HARD SEEDS IN LESPEDEZA

The annual lespedezas often contain little or no hard seed, so that scarification is unnecessary. In tests made soon after harvest, considerable hard seed was found in Korean lespedeza,[15] but a rapid decrease in hard-seed content occurred from November through January. Most of the hard seeds are small.[16]

Sericea lespedeza has a high percentage of hard seeds. Germination of unscarified seed ranges from 10 to 20 per cent, so that scarification is necessary for prompt germination.[20]

Figure 25–3. Branch of sericea lespedeza.

VARIETIES

Varieties of common lespedeza are best adapted to the general region from northern Tennessee to the Gulf of Mexico because they require high temperatures.

Korean lespedezas mature earlier than those of common lespedeza, and are adapted to more northern latitudes (Figure 25–4). Certain varieties are somewhat resistant to bacterial wilt, or to root-knot nematodes and powdery mildew.[12, 13]

Sericea lespedeza is adapted to the zone between the Ohio river and the northern part of Florida where the annual rainfall exceeds 30 to 35 inches.[10] Some species of shrubby perennial lespedezas are being grown for soil or wildlife conservation or for ornamental purposes.

CHEMICAL COMPOSITION

At the usual hay-cutting stage, the annual lespedezas contain significantly more dry matter than do the other common legumes.[21] Thus,

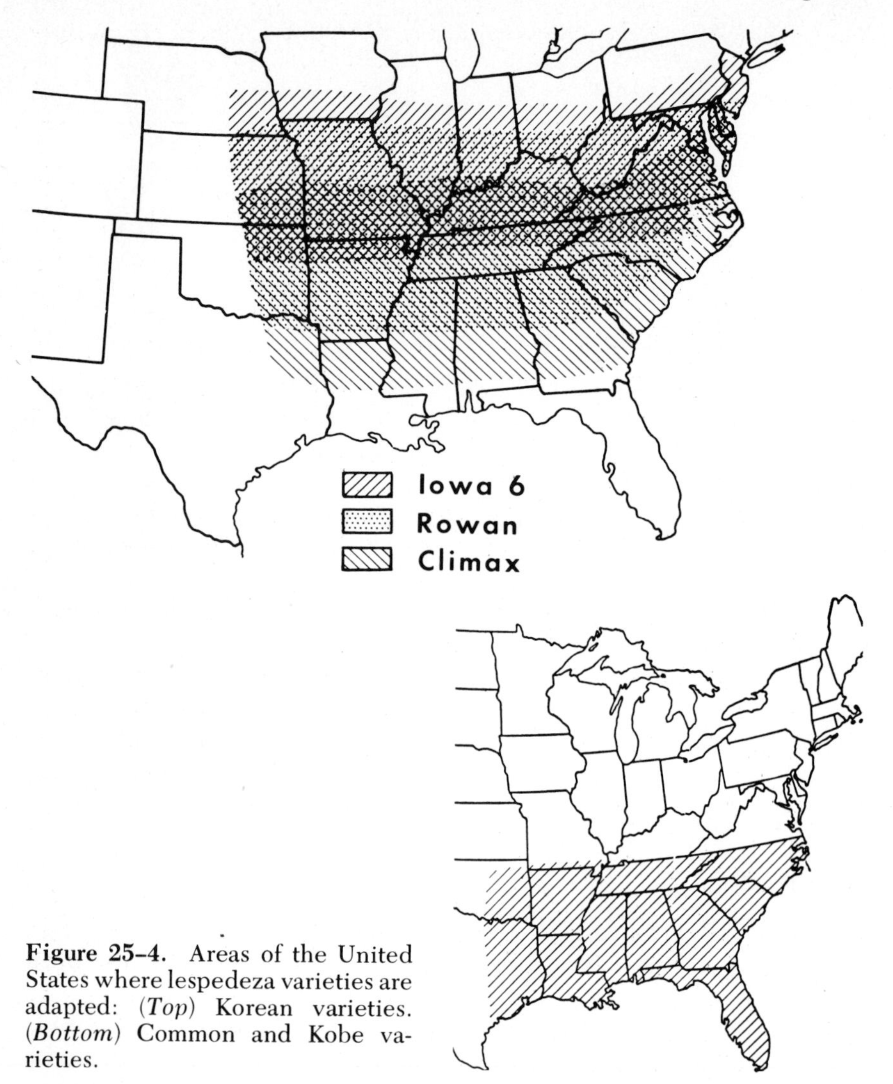

Figure 25–4. Areas of the United States where lespedeza varieties are adapted: (*Top*) Korean varieties. (*Bottom*) Common and Kobe varieties.

they cure more rapidly. The lespedezas contain 36 to 49 per cent dry matter; alfalfa (first three cuttings), 27 to 29 per cent; and Laredo soybeans, 28 to 30 per cent.

The annual lespedezas are nearly equal to other legume hays in feeding value. In general, the digestible protein is slightly lower than in alfalfa while the digestible carbohydrate equivalent is higher. The protein content ranges from 9 to 17 per cent. In sericea it varies with the stage of cutting. Sericea becomes very woody when the plants are 3 feet high, this being reflected in a high crude-fiber content. The chemical composition of lespedeza in Illinois[18] is given in Table 25–1.

TABLE 25–1. Chemical Composition of Lespedeza and Other Legume Hays

CROP	MOISTURE (%)	ETHER EXTRACT (%)	PROTEIN (%)	CRUDE FIBER (%)	ASH (%)
Alfalfa	8.6	2.3	14.9	28.3	8.6
Red clover	12.9	3.1	12.8	25.5	7.1
Soybean	8.6	2.8	16.0	24.9	8.6
Common lespedeza	11.8	2.8	12.1	25.9	5.8
Korean lespedeza	7.2	3.3	16.2	26.0	7.4
Sericea lespedeza	5.9	1.8	12.3	30.1	6.0

All species of lespedeza contain some tannin, an astringent substance found mostly in the leaves.[24] The annual lespedezas have about 3 to 5 per cent tannin while sericea contains much more. Tannin may occur in sufficient quantity in sericea to lower its palatability and digestibility as pasturage. In sericea, tannin is relatively low in the new growth but it may increase from 7.5 up to 18 per cent between May and August. The whole plant may increase from 5 to 8 per cent in tannin content during this period. The tannin content decreases later in the season.[24, 26]

The variations in tannin content appear to offer a plausible explanation of the apparently contradictory experience with the palatability of sericea as a forage plant.[6] However, the coarse woody stems of plants more than 10 to 12 inches tall likewise are unpalatable. Strains with finer stems and less tannin are being developed.

ROTATIONS

In general, lespedeza rotations include a row crop followed by small grain with an annual lespedeza. The lespedezas will volunteer after small grain each year as long as desired, until they are finally plowed under in preparation for a row crop.

In the South, winter grain may be seeded on disked cotton, tobacco, or corn land, with annual lespedeza seeded in the grain.[13] Winter legumes, such as vetch or crimson clover, may follow volunteer lespedeza. These may be pastured or plowed under in the spring in preparation for the planting of corn or cotton.

Farther north, Korean lespedeza may replace red clover in the standard rotations on soils not well suited to red clover. The volunteer stand of lespedeza will take the place of the second-year clover. A one-year small grain-lespedeza rotation has been recommended for Missouri.[8] The harvested grain crop is followed by an abundance of lespedeza pasturage. The lespedeza will produce seed and volunteer each year, the land being disked or cultivated and the small grain seeded each fall.

CULTURAL METHODS FOR ANNUAL LESPEDEZAS

Like all other crops, lespedeza will respond favorably to good cultural practices. A firm seedbed is essential whether the crop is seeded alone or with small grains. On meadows or pastures seeded with lespedeza, a common practice is first to loosen the soil with a spring-tooth harrow or disk. Inoculation is seldom necessary for most varieties in the South, except possibly on badly eroded soils. It is generally advisable to inoculate lespedeza seed when growing it for the first time, in Kentucky and further northward.

The annual lespedezas are seeded either broadcast or ½ inch deep or less with a grain drill. In general, lespedeza is seeded in the spring about two weeks before the last freeze is expected in the locality. In North Carolina, Tennessee, and farther south, it is usually sown between February 15 and March 15. Farther north, seedings are made from March 15 to April 15. Early seedings are subject to late freezes.

Korean or common should be seeded at the rate of 25 to 30 pounds per acre, and Kobe at 30 to 40 pounds.[12, 13] For Missouri conditions, Korean should be sown at the rate of 20 to 25 pounds per acre on good land, or 15 to 20 pounds on poor land.[8] On pastures, or in pasture mixtures, 5 to 8 pounds of seed are generally sufficient.

Utilization for Pasture

The annual lespedezas grow actively during the hot summer months when grasses make little growth. They sometimes can be maintained in grass-legume mixtures for permanent pastures with grasses that do not form a dense sod. In many places, however, the pasture must be cultivated and lespedeza resown every year or two. In a perennial-grass pasture close grazing of the grasses in the spring is often practiced to favor the lespedeza. Common lespedeza appears to survive longer in grass mixtures than do the improved varieties. The lespedezas persist in pastures because they set their seed near the ground. When grown with small grain, lespedeza is ready for grazing soon after the grain is removed. The principal grazing months are July, August, and September.[13] The fall-sown grain may be pastured in winter, spring, and early summer. The lespedeza is grazed later.

Harvesting for Hay

Annual lespedezas are used as hay crops, particularly where soil conditions are unfavorable for perennial legumes. Hay of excellent quality can be made from lespedeza, especially when cut in early bloom or just before the bloom stage. This usually occurs before August 1 from Missouri to Virginia, or August 1 to 15 in the latitude of North Carolina for the Korean varieties. By this time a growth of 8 to 10 inches is attained. Korean may be cut earlier than the other annual varieties because it makes a more rapid early growth. A stubble of 3 to 5 inches is generally left to permit new growth. Since lespedeza cures rapidly because of its low moisture content, it usually is raked into windrows 2 to 4 hours after cutting. A windrower may be used to put the crop in windrows as it is being cut.[8]

Harvesting for Seed

Lespedeza seed is generally harvested when 65–75 per cent of the hulls turn brown because the seeds shatter very rapidly thereafter. The usual harvest method is with a combine. Korean varieties shatter less readily than do those of the common type. A windrowed seed crop is threshed with a pick-up combine. Cutting should be done when the plants are damp, except when direct combining.

CULTURAL METHODS FOR SERICEA LESPEDEZA

Sericea lespedeza usually is sown alone, but it may be sown on a field of winter grain. Seeding should be done in early spring. Self-seeding crimson clover may be sown on established stands of sericea. A satisfactory seeding rate for sericea is 30 to 40 pounds of hulled scarified seed per acre, broadcast, and followed by a cultipacker on noncrusting soils. The soil should be mulched when sericea is seeded on eroded knolls or in gullies. Seedings on established meadows and pastures have failed.[17]

Harvesting

Sericea is cut for hay when the plants are 10 to 15 inches high. At this stage the hay is comparatively high in protein and low in tannin but its feeding value is only about 80 per cent of that of alfalfa hay.[10] The plants develop little woodiness up to a height of 12 inches, but woody tissue increases rapidly in the later stages. As sericea cures quickly, it should be windrowed within 1 hour after cutting, or be cut and windrowed in one operation. It should be baled within 24 hours after cutting to avoid severe leaf shattering. Two or three cuttings may be harvested under favorable conditions. In Tennessee, yields of 4.14, 3.39, and 3.29 tons of hay per acre were obtained when sericea was cut two, three, and four times per season, respectively.[17] Four cuttings may seriously reduce the stand. Only one cutting is advisable on very poor soils. When sericea is being cut, a stubble of 3 to 5 inches should be left because the new growth comes from the stems, as in sweetclover.

The first or second growth may be harvested for seed, usually after most of the pods have turned brown. The seed crop is harvested with a combine, a windrower, or a binder.

Sericea is ready to be grazed when it reaches a height of 6 inches, being palatable at this stage. The pasture should be grazed or mowed down to a height of 3 inches to avoid coarse growth, which is unpalatable. Grasses suitable for mixed sericea pastures are bahiagrass in the Coastal Plains and tall fescue in northward areas.

ROLE IN SOIL CONSERVATION

Lespedeza is effective in the control of soil erosion because of its dense growth, its ability to establish a cover on poor soil, and the high retention of winter rains by old lespedeza sod. During a 12-month period the runoff from Korean lespedeza land was 11.7 per cent, with a soil loss of 1.6 tons per acre.[8] Under comparable conditions continuous corn showed a loss of 60.8 tons of soil per acre and a 30.3 per cent runoff. While the stubble of annual lespedeza tends to prevent erosion during the winter, a winter cover crop after lespedeza will improve the effectiveness.

Sericea, alone or with grasses, is well suited for seeding buffer strips, critical slopes above the flow lines of terraces, small gullies, and depressions for water outlets. The annual lespedezas are used in some of these places, especially in combination with grass.

DISEASES

The chief diseases of annual lespedezas are bacterial wilt, caused by *Xanthomonas lespedezae;* tar spot, caused by *Phyllachora* species; damping off, caused by *Rhizoctonia solani;* southern blight, caused by *Sclerotium rolfsii;* and powdery mildew, caused by *Microsphaera diffusa.* Root-knot nematodes (*Meloidogyne* species) also damage annual lespedeza. Losses from bacterial wilt, powdery mildew, and root-knot nematodes can be reduced by growing resistant or tolerant varieties. Crop rotation reduces damage from tar spot and southern blight.

Sericea lespedeza is relatively free from disease losses, but it is attacked by root-knot nematodes. It is susceptible to cotton root rot, but it is seldom grown where that disease is prevalent.

INSECTS

The chief insects that attack lespedeza are a grasshopper (*Schistocerca americana*), armyworm (*Pseudodaletia unipuncta*), fall armyworm (*Laphygma frugiperda*), three-cornered alfalfa hopper (*Spissistilus festinus*), and a webworm (*Tetralopha scorealis*). Some of these pests can be controlled by suitable insecticides.

REFERENCES

1. Aamodt, O. S. "Climate and forage crops," in *Climate and Man,* USDA Yearbook, 1941, pp. 439–458.
2. Anderson, K. L. "Lespedeza in Kansas," *Kansas Agr. Exp. Sta. Circ. 251,* pp. 1–15. 1956 (Rev.).
3. Ayers, T. T., Lefebvre, C. L., and Johnson, H. W. "Bacterial wilt of lespedeza," *USDA Tech. Bul. 704.* 1939.
4. Bailey, R. Y. "Sericea in conservation farming," *USDA Farmers Bul. 2033.* 1951.
5. Blaser, R. E., Volk, G. M., and Stokes, W. E. "Deficiency symptoms and chemical composition of lespedeza as related to fertilization," *J. Am. Soc. Agron.,* 34:222–228. 1942.
6. Clarke, I. D., Frey, R. W., and Hyland, H. L. "Seasonal variation in tannin content of *Lespedeza sericea,*" *J. Agr. Res.,* 58:131–139. 1939.
7. Duggar, J. F. "Differences between Korean and other annual lespedezas in root nodule formation," *J. Am. Soc. Agron.,* 26:917–919. 1934.
8. Etheridge, W. C., and Helm, C. A. "Korean lespedeza in rotations of crops and pastures," *Mo. Agr. Exp. Sta. Bul. 360.* 1936.
9. Grizzard, A. L., and Hutcheson, T. B. "Experiments with lespedeza," *Va. Agr. Exp. Sta. Bul. 328.* 1940.
10. Guernsey, W. J. "Sericea lespedeza, its use and management," *USDA Farmer's Bul. 2245,* pp. 1–30. 1970.
11. Hendrickson, B. H., and Crowley, R. B. "Preliminary results with mulches applied to eroded wasteland sown to lespedeza," *J. Am. Soc. Agron.* 33:690–694. 1941.
12. Henson, P. R. "The lespedezas," in *Advances in Agronomy,* Vol. 9, Academic Press, New York, pp. 113–157. 1957.
13. Henson, P. R., and Cope, W. A. "Annual lespedezas—culture and use," *USDA Farmers Bul. 2113* (rev.). 1964.
14. McKee, R., and Hyland, H. L. "Apetalous and petaliferous flowers in lespedeza," *J. Am. Soc. Agron.,* 33:811–815. 1941.
15. Middleton, G. K. "Hard seed in Korean lespedeza," *J. Am. Soc. Agron.,* 25:119–122. 1933.
16. Middleton, G. K. "Size of Korean lespedeza seed in relation to germination and hard seed," *J. Am. Soc. Agron.,* 25:173–177. 1933.
17. Mooers, C. A., and Ogden, H. P. "*Lespedeza sericea,*" *Tenn. Agr. Exp. Sta. Bul. 154.* 1935.
18. Pieper, J. J., Sears, O. H., and Bauer, F. C. "Lespedeza in Illinois," *Ill. Agr. Exp. Sta. Bul. 416.* 1935.
19. Pieters, A. J. *The Little Book of Lespedeza,* Colonial Press, Washington, D.C. 1934.
20. Pieters, A. J. "*Lespedeza sericea* and other perennial lespedezas for forage and soil conservation," *USDA Cir. 853.* 1950.
21. Smith, G. E. "The effect of photo-period on the growth of lespedeza," *J. Am. Soc. Agron.,* 33:231–236. 1941.
22. Stitt, R. E. "A comparison of the dry matter content of annual lespedezas, alfalfa, and soybeans," *J. Am. Soc. Agron.,* 26:533–535. 1934.

23. Stitt, R. E. "The response of lespedeza to lime and fertilizer," *J. Am. Soc. Agron.*, 31:520–527. 1939.
24. Stitt, R. E. and Clarke, I. D. "The relation of tannin content of sericea lespedeza to season," *J. Am. Soc. Agron.*, 33:739–742. 1941.
25. Tysdal, H. M., and Pieters, A. J. "Cold resistance of three species of lespedeza compared to that of alfalfa, red clover, and crown vetch," *J. Am. Soc. Agron.*, 26:923–928. 1934.
26. Wilkins, H. L., and others. "Tannin and palatability in sericea lespedeza," *Agron. J.*, 45:335–336. 1953.

Chapter 26

Soybeans*

ECONOMIC IMPORTANCE

Soybeans are an important world crop grown on nearly 38 million hectares (Figure 26–1). About two-thirds of the world production is in the United States with mainland China the second largest producer. Soybeans ranked third in area among United States crops with an average acreage of 43 million from 1970 to 1972. The production was nearly 1,200 million bushels with an average yield of 27.3 bushels per acre and a farm value of over 3,000 million dollars. Production rose from less than 5 million bushels in 1924 to 1,547 million bushels in 1973.

HISTORY OF SOYBEAN CULTURE

The soybean is one of the oldest of cultivated crops. Its early history is lost in antiquity. The first record of the plant in China dates back to 2838 B.C.[41] It was one of the five sacred grains upon which Chinese civilization depended. The cultivated soybean (*Glycine max*) probably was derived in China from a wild type, *Glycine ussuriensis*, with small seeds that grows in eastern Asia.[7a, 27, 41, 55]

The soybean was known in Europe in the 17th century, and in the United States in 1804. Little attention was given to the soybean as a crop until 1889 when several experiment stations became interested in it. A large number of varieties were imported by the United States Department of Agriculture in 1898. Since that time there has been a rapid expansion in soybean production, particularly since about 1920.[48] Most of the soybeans were grown in the South prior to 1924, when they began to assume importance in the Corn Belt (Figure 26–2).

* For more complete information on soybeans, see Markley, K. S., *Soybeans and Soybean Products*, Interscience, New York, Vol. I and II, 1950–51, pp. 1–1145. Norman, A. G. (ed.), *The Soybean*, Academic Press, New York, 1963, pp. 1–239.

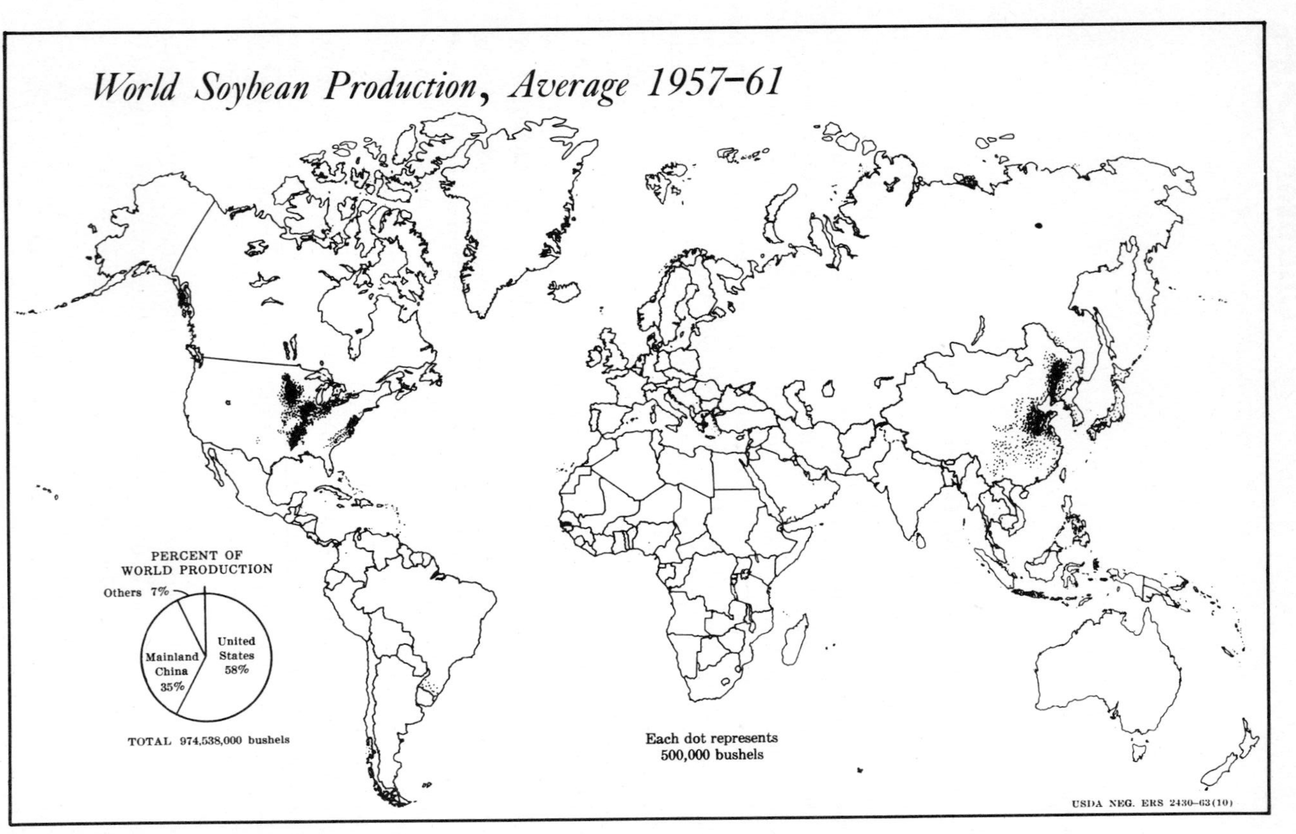

Figure 26–1. World soybean production.

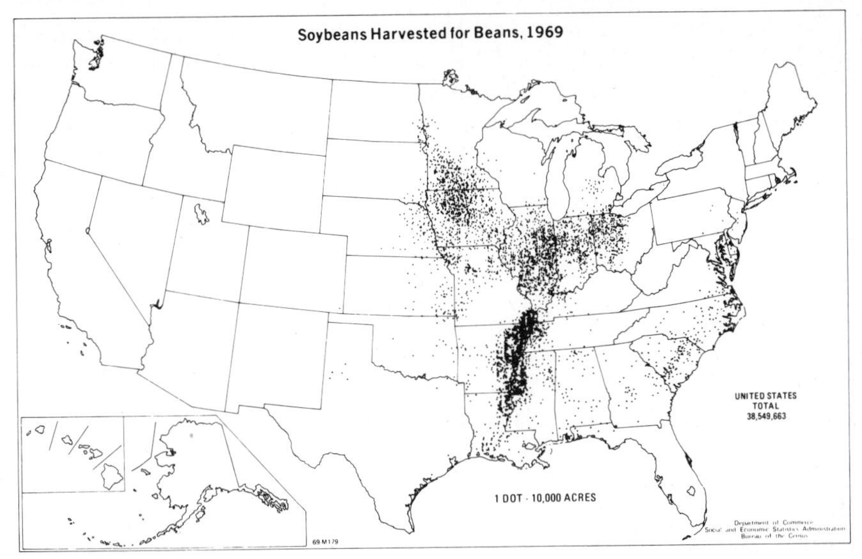

Figure 26–2. Acreage of soybeans in the United States in 1969.

ADAPTATION

The climatic requirements for the soybean are about the same as those for corn. Soybeans will withstand short periods of drought after the plants are well established. Indeed, a well-charged soil profile (6–8 feet) may contain enough moisture to allow the plant to reach bloom stage before additional water is needed after planting. In general, combinations of high temperature and low precipitation are unfavorable. Soybean seed produced under high-temperature conditions tends to be low in oil and oil quality.[9, 24] A wet season does not seriously retard plant growth, but soybeans are sensitive to overirrigation.[50] The period of germination is the most critical stage in soybean plant growth, excess or deficiency of soil moisture at this time being particularly injurious.[23] Soybeans are less susceptible to frost injury than is corn. Light frosts have little effect on the plants when either young or nearly mature. The minimum temperature for growth is about 50° F.

The soybean, a short-day plant, is sensitive to photoperiods.[28] Many varieties require 10 hours of daily darkness in order to flower. Northern varieties mature quickly with little vegetative growth when grown in the South. Within a variety, variations in time of flowering from year to year with the same day length appear to be closely associated with temperature conditions.

A mean midsummer temperature of 75° to 77° F. appears to be optimum for all varieties. Lower temperatures tend to delay flowering. About 4,300 heat units in five months are required to mature soybeans.

Soybeans grow on nearly all types of soil, but they are especially productive on fertile loams. They are better adapted to low fertility soils than is corn, provided the proper nitrogen-fixing bacteria are present. Soybeans will grow on soils that are too acid for alfalfa and red clover.

BOTANICAL DESCRIPTION

Seed and Seedling

A soybean seed is composed of two cotyledons, radicle, testa, hypocotyl and epicotyl. All except the testa are considered to be part of the embryo (Figure 7–3). The cotyledons are the major storage tissue for proteins and oil ranging from 38 to 46 per cent protein and 18 to 20 per cent oil. The hilum varies from buff to yellow, black, brown, clear and imperfect black (slate colored with a brown margin). The seeds are prone to mechanical injury during harvest and handling which reduces germination and/or results in damaged seedlings. The testa is either dull or shiny.

The roots may penetrate 5–6 feet (nearly 2 meters) into the soil, but most of the roots form in the top 8 inches of soil. Nodules (Figure 26–3) begin to form on the roots in 10–14 days after emergence and continue forming throughout the life cycle of the plant. They serve as the main source of nitrogen for the plant. Active nodules are pink inside.

Upon emergence from the soil the cotyledons develop chlorophyll followed by opening to expose the first two true plant leaves. After 5 to 8 days the plant no longer depends upon its cotyledons as a major food supply. At the base of the cotyledons are axillary buds which can

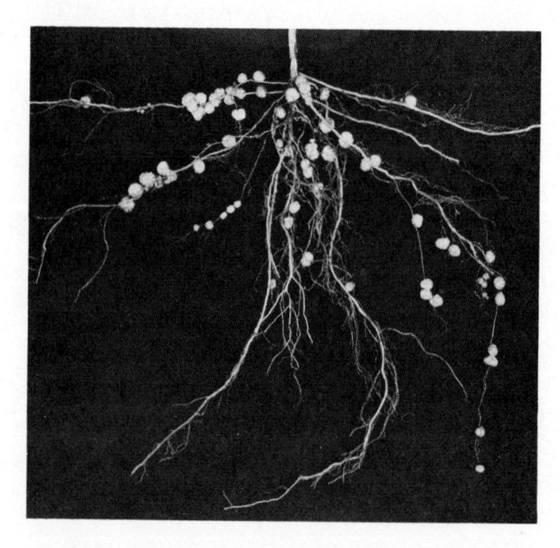

Figure 26–3. Characteristic large nodules on soybean roots which show that proper nitrogen-fixing bacteria are present.

develop into main stems in the event the epicotyl is destroyed by hail or other hazards such as insects or rabbits.

Prereproductive Development

After the two unifoliolate leaves have emerged the additional leaves are trifoliolate, and number from one to six or more until floral initiation occurs about 6 to 8 seeds after seedling emergence.

Reproductive Development

The small white and purple flowers are borne in axillary racemes on peduncles arising at the nodes. The flowers appear first toward the base of the main stem, then progress upward toward the tip. Frequently 65 to 75 per cent of the flowers abort and fail to produce pods. The pods are small, being either straight or slightly curved and covered with long hair (Figure 26–4). They range from very light straw color through numerous shades of gray and brown to nearly black. The pods contain one to four, and, occasionally, five seeds. The seeds are round to elliptical. Most varieties have unicolored seeds that are straw yellow, greenish yellow, green, brown, or black. Bicolored seeds occur in some varieties, the common pattern being green or yellow with a saddle of black or brown.

The seedcoats of many yellow- and green-seeded varieties some-

Figure 26–4. Soybean seeds and pods.

times become splashed or blotched with irregular brown or black markings superimposed on the basic color. This mottling, which is sporadic and never occurs in some areas, is due both to heredity and environment.[47] Rich soils, liberal spacing between plants, and shading are conducive to mottling. Selected strains have a high degree of resistance to mottling,[14] and most of the widely used varieties are not subject to mottling.

The soybean is normally self-fertilized because pollination occurs about as soon as the flower opens or a little before. Although natural crossing is much less than 1 per cent,[19, 61] it may account for many varietal mixtures. Natural cross-pollination appears to be the work of insects such as thrips and bees.

General Development

The soybean will develop into a rather erect, bushy, leafy plant attaining a height of 2 to 4 feet (60 to 120 cm.). Most varieties have a well-defined main stem that branches from the lower nodes when the plants have sufficient space. Determinate varieties cease vegetative growth when the main stem terminates in a cluster of mature pods. Such soybeans are grown mostly in the southern United States where summer nights are long. Indeterminate varieties develop leaves and flowers simultaneously throughout a portion of their reproductive period, with one to three pods at the terminal apex. These are commonly grown in the northern latitudes of the United States.

Varieties having an indeterminate growth habit begin to flower relatively early and continue to increase in height for several weeks after beginning to flower, while determinate type varieties make most of their growth (height) before beginning to flower. At present all commercially grown varieties of Group IV (Table 26–1) or earlier have an indeterminate growth type, while all varieties of Group V maturity or later display determinate growth habits.

As the time of maturity approaches the leaves begin to turn yellow, dropping off before the pods mature, when the seeds still contain about 20 per cent moisture (Fig. 26–5). The leaves and stems of nearly all varieties are covered with a fine lining of either grey or tawny colored pubescence.

VARIETIES

Because of different maturities, soybean varieties are limited in their range of best adaptation to latitude zones from 100 to 150 miles wide. When moved north of their zone in the northern hemisphere, the variety may be too late to mature before frost kills the plants. A southward shift makes the plant mature too early for optimum yields.

Temperature affects the time of flowering as well as time of maturity

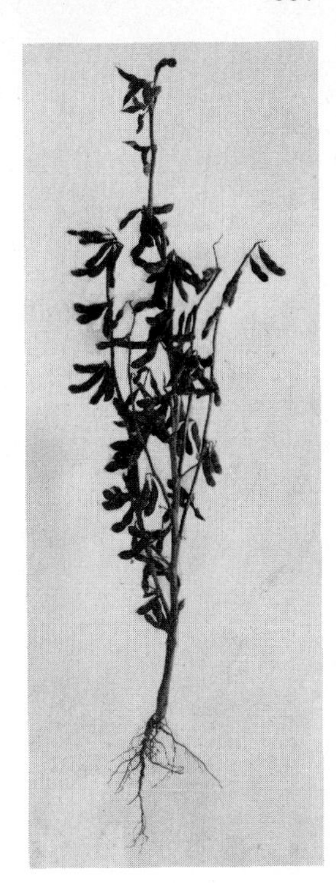

Figure 26–5. Mature soybean plant after the leaves have dropped.

of soybean varieties.[23] The average July temperature decreases to the northward about 1.3° F. for each 100 miles of latitude. This is sufficient to delay the flowering and ripening of soybeans by about 3 days in each 100 miles. Day length and temperature differences together can shift the maturity of a soybean variety from 5 to 6 days with each 100-mile difference in latitude. Altitude as well as proximity to large bodies of water also affect the temperature. Consequently, even when zones of variety adaptation, dates of planting, and times of harvest are similar, the results are not exactly parallel.

More than 50 recommended varieties of soybeans are grown in the United States.

Representative soybean varieties grown in the United States are categorized in 10 maturity classes from 00 to VIII (Table 26–1). Groups 00 and 0 are the earliest maturing varieties adapted to the northern latitudes of the United States and southern Canada (Figure 26–6). The range in maturity within any one given group may vary from 10 to 21 days due to the acceleration of flowering as a result of decreased day length.

The maturity groups of soybeans grown in each state are listed in Table 26–2. All states grow two or more varietal groups. The earlier varieties usually are grown in the northern part of a particular state.[4, 5] Often they are used for late planting or are planted on fields to be

TABLE 26–1. Maturity Date of the Ten Soybean Variety Groups When Grown in Their Areas of Adaptation

GROUP	AVERAGE MATURITY DATE FOR GROUP	DAYS FROM PLANTING	REPRESENTATIVE VARIETIES*
00	September 10	120	Altona, Morsoy, Portage, Flambeau, Sioux, Acme
0	September 28	126	Clay, Mandarin, Merit, Grant, Norchief, Traverse
I	September 30	126	Chippewa 64, Disoy, Dunn, Hark, Rampage, Wirth
II	October 3	130	Amsoy, Beeson, Harosoy 63, Hawkeye 63, Magna, Prize
III	October 3	131	Calland, Ford, Adams, Shelby, Wayne, Wolverine
IV	October 7	136	Carlin, Clark 63, Custer, Polysoy, Wabash, Scott, Kent
V	October 7	139	Hill, Dare, Dorman, S-100 Dortchsoy 67
VI	October 22	148	Lee 68, Davis, Hood, Ogden
VII	October 30	156	Bragg, Ransom, Roanoke, Jackson
VIII	November 9	158	Hampton 266A, Hardee, Bienville, JW 45, Improved Pelican

* The listed varieties are only a random sampling from several varieties available for use by the farmer in his location.

harvested early before the later varieties are mature. This practice not only extends the harvest season, but also permits the planting of fall-sown crops on the early soybean field. Late varieties should be planted early in order to utilize a long growing season and still reach maturity. Varieties that mature early yield more when planted somewhat later. The longer days of late spring and early summer extend the vegetative period and increase the plant size of the early varieties.

CHEMICAL COMPOSITION

The chemical composition of soybeans and soybean products is given in Table 26–3.

The percentage of nitrogen in the leaves of soybean hay is nearly twice as high as that of the stems.[6] The percentage of nitrogen in the total tops decreases during the period of rapid growth and increases as the seed matures. More than half of the nitrogen in the total tops is stored in the seed at maturity.

The oil content of the seeds ranges from 14 to 24 per cent or more,[3]

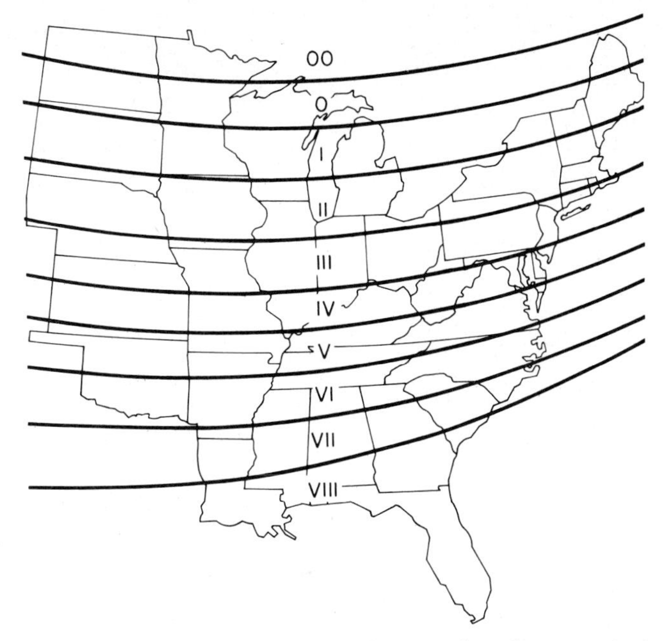

Figure 26–6. There are 10 maturity classes of soybean varieties. Those varieties adapted for use in southern Canada and the northern-most area of the United States are designated 00, and are the earliest maturing. The higher the number, the later the maturity and the further south the variety is adapted for full-season use. The lines across the map are hypothetical. There are no clearly cut areas where a variety is or is not adapted. [Reprinted with permission of SA Publications. Champaign, Ill.]

while the protein may range from 30 to 50 per cent. The breeding of varieties with a high oil content has since raised the average oil content to more than 20 per cent. In general, soybeans with a low fat content are high in protein and vice versa. Soybean protein contains all the essential amino acids for animal feeds and human foods. Soybeans contain two to three times as much ash as wheat, and are a valuable source of calcium and phosphorus. Like other edible legume seeds, they are high in thiamin (vitamin B_1).

ROTATIONS

The soybean is often grown in short rotations with corn, cotton, and small grains. As a full-season crop, it may occupy any place in a rotation where corn is used.[29] A common rotation in the Corn Belt is corn (one or more years), soybeans, small grain, and legumes.[55] In the South, soybeans are usually grown in rotations with cotton, corn, or rice. Sometimes they are planted after the harvest of early potatoes or winter grain. Winter grain may follow soybeans that are harvested

TABLE 26–2. Planting Dates for Soybeans in Major Production States

STATES	MATURITY CLASSIFICATION OF VARIETIES GROWN	FOR BEST RESULTS, PLANT ON:	LATEST SAFE DATE
Alabama	VI, VII, VIII	May 10 to June 15	July 10
Arkansas	V, VI, VII	May 1 to 20	June 30
Delaware	III, IV, V, VI	May 10 to 30	June 30
Florida	VI, VII, VIII	June 1 to 30	July 15
Georgia	VI, VII, VIII	May 1 to June 15	July 10
Illinois	I, II, III, IV	May 5 to 25	June 30
Indiana	I, II, III, IV	" " " "	" "
Iowa	I, II, III	May 1 to 30	June 30
Kansas	III, IV, V	May 10 to June 15	" "
Kentucky	IV, V, VI	May 1 to 30	" "
Louisiana	VI, VII, VIII	May 10 to June 15	July 10
Maryland	IV, V, VI	May 15 to 30	June 30
Michigan	00, 0, I, II	May 5 to 30	" "
Minnesota	00, 0, I, II	May 10 to 30	" "
Mississippi	V, VI, VII	May 1 to June 5	July 5
Missouri	II, III, IV, V, VI	May 5 to 30	June 30
Nebraska	I, II, III	May 15 to June 5	" "
North Carolina	VI, VII	May 1 to 30	" "
North Dakota	00, 0	May 15 to 30	June 20
Ohio	I, II, III	May 5 to 25	June 30
Oklahoma	V, VI	May 10 to 30	" "
South Carolina	VI, VII, VIII	May 1 to 20	" "
South Dakota	00, 0, I, II	May 10 to 30	" "
Tennessee	IV, V, VI	May 1 to 25	" "
Virginia	IV, V, VI	May 15 to June 15	" "
Wisconsin	0, I, II	May 5 to 30	" "

TABLE 26–3. Analyses of Soybean Forage, Seeds, and Meal

ITEM	MOISTURE	ASH	CRUDE PROTEIN	CARBOHYDRATES		OIL OR FAT
				Crude Fiber	*N-Free Extract*	
	(%)	(%)	(%)	(%)	(%)	(%)
Green forage	75.1	2.6	4.0	6.7	10.6	1.0
Hay	8.4	8.9	15.8	24.3	38.8	3.8
Seeds	6.4	4.8	39.1	5.2	25.8	18.7
Oil meal (hydraulic or expelled process)	8.3	5.7	44.3	5.6	30.3	5.7
Oil meal (solvent process)	8.4	6.0	46.4	5.9	31.7	1.6

before time to seed the grain. Soybeans may be substituted for oats in the corn-oats-wheat-clover rotation. They may also be used as a catch crop where new seedings of grass and clover have failed.

On hilly land, soybeans cannot take the place of sod legumes such as the clovers because they do not give sufficient protection against soil erosion, particularly when they are planted in cultivated rows. Erosion from soybean land is nearly as high as from land in corn,[16] unless the soybeans are planted in close drills.[4]

CULTURAL METHODS

In general, the preparation of the land for soybeans should be the same as for corn.

Inoculation

Nitrogen-fixing bacteria (*Rhizobium japonicum*) develop nodules on soybean roots (Figure 26–3) which are capable of providing the entire nitrogen needs of the soybean. Inoculation may not be required in fields where soybeans have been grown and inoculated for many years, but when in doubt seed should be inoculated.

The inoculum is normally mixed with the planting seed. One pint of water to a bushel of dry seed plus the recommended quantity of inoculum applied no sooner than one day before planting is adequate. Treated seed should not be left in direct sunlight since this can destroy the *Rhizobia*.

A 40-bushel soybean crop contains approximately 240–250 pounds of nitrogen; nearly 150 pounds is in the seed. Well-inoculated soybeans usually fail to respond to nitrogen fertilizers although 10–20 pounds per acre of nitrogen in the starter fertilizer may stimulate early planted soybeans before the nodules develop.

Fertilizers

Soybeans do best on soils of high fertility. Calcium and magnesium are applied to acid soils to raise the *p*H to 6.0 or 6.5.

About 20–40 kg. per hectare of P_2O_5 are advisable on soils very low in available phosphorus with smaller rates for soils that tests show to be only low to medium in phosphorus.

Soybeans are heavy users of potassium. Potassium application is recommended where soil tests indicate less than 150 pounds of K_2O available per acre in the southeastern United States or less than 200–250 pounds of K_2O available per acre in the Midwest.

Sulfur or certain micronutrients usually are not applied to soybean fields except on strongly weathered, coarse-textured alkaline or organic soils. In Arkansas and Mississippi molybdenum is frequently applied to soils with a *p*H of 6.0 or less. Iron deficiency in soybeans may occur on soils with a *p*H above 7.5.

Seedbed Preparation

In the northern states, fall plowing of heavy soils is desirable. Excessive tillage of plowed land is not essential for a soybean seedbed.

Planting

Soybeans are best planted in May or early June in most states, as shown in Table 26–2. Earlier planting hastens flowering and reduces yields, and in the north, the soil usually is too cold to plant soybeans before late May or early June. The soil temperature should be 50°–55° F, (10°–12° C.) before planting.

Soybeans are normally planted at a depth of 1 to 2 inches. Usually 8–12 soybean seeds per foot of row in 36–40-inch rows will give an adequate stand. In areas where narrow rows give an increased yield, rates of 6 to 8, 4 to 6, and 3 to 4 plants per foot of row are adequate for 30, 20, and 10-inch rows, respectively, to realize maximum yield potential.

Soybeans usually are planted in cultivated rows. In the Corn Belt soybeans planted in narrow rows normally yield 10 to 15 per cent more than when grown in traditional 40-inch (1-meter) rows. In the south, narrower rows are not beneficial.

The grain drill is occasionally used for planting in rows by closing some of the hopper openings. The corn planter, cotton planter, sugarbeet drill, and pea planter can be equipped with soybean plates for planting soybeans.

Soybeans are usually planted in close drills for hay, soiling, or green manure. A finer quality of hay is obtained by this method. Drilled soybeans are more weedy than those in cultivated rows unless they are cultivated two to three times with a rotary hoe. Seed treatment with a fungicide is generally not recommended because of possible damage to the *Rhizobium* inoculant.

Weed Control

Soybeans' relatively slow early growth necessitates early weed control by cultivation and herbicide applications. Planting in narrow rows restricts weed growth by shading.

Preplant tillage followed by rotary hoeing when the weeds are in the 2 to 3-leaf stage after emergence of soybeans is effective. Shallow sweep cultivation controls later weeds. Chemical weed control can be done on a preplant, preemergence, or postemergence basis.

HARVESTING

The combine is used almost universally to harvest soybeans. Soybeans are combined most efficiently when the moisture content of the seeds drops to 12 per cent. Later harvesting increases shattering loss as well as splitting of the over-dry beans in threshing. Partly shatter resistant varieties are available. To minimize split beans, the cylinder speed of threshing machines should be operated at 300 to 450 revolutions per minute. The seed coat, cotyledons and radicle are very sensitive to mechanical damage during harvest. The soybean at 13 per cent moisture can be combined directly without windrowing and stored without drying. The height of the cutter bar and speed of the combine is very important in limiting harvesting losses. At faster combine speeds the cutter bar rides higher and results in increased shattering of lower pods.

Harvesting for Hay

Soybeans may be cut for hay any time from pod formation until the leaves begin to fall. The best quality of hay is obtained when the seeds are about half developed. The weight of leaves increases until the beans are well formed, remains constant for about three weeks, and then decreases rapidly.[59] The hay contains about 60 per cent leaves when the beans are well formed, and about 50 per cent when the beans appear half grown. In addition to the loss of leaves, the stems become woody when cutting is delayed. Soybeans are more difficult to cure than alfalfa or clover because the thicker stems dry out more slowly. When soybeans are cut with a mower, a common practice is to allow the crop to remain in the swath for a day or two until thoroughly wilted. Use of a hay crusher will facilitate curing.

The average yield of soybean hay in the United States is about 1.4 tons per acre. Yields of up to 3 tons per acre sometimes are obtained. The acreage of soybeans cut for hay declined from about 4 million acres in 1940 down to about 485,000 acres in 1964. In 1924, less than 29 per cent of the soybean acreage was harvested for beans, but in 1964 about 97 per cent of the acreage was grown for the beans.

Soybean Mixtures

Soybeans may be grown in combination with other crops, especially corn. Soybean-corn mixtures are generally used for pasturage or silage. The most desirable rate of planting is one corn plant every 12 inches and one soybean plant every 6 to 8 inches in drilled rows. The soybeans sometimes are planted in alternate rows with corn.

The yield of corn as grain in mixed planting is reduced, but this is partly compensated for by the yield of soybeans as well as by the higher protein content of the mixed crop. As an average for all conditions, the yield of soybean seed was 52 per cent of the reduction in yield of corn grain. When corn and sorghum are drilled together for forage, the yield of dry matter may exceed that of the two crops grown alone by 7 to 15 per cent.[26, 57]

Soybeans grown in corn under more or less reduced light are inclined to lodge. Shaded soybean plants contain less dry matter and total carbohydrates than do those grown in the open.[56]

Soybeans and cowpeas are sometimes grown in mixture for hay, pasture, or green manure. The yield of the mixture is generally greater than for either crop grown alone. Varieties of the two crops should be selected that mature at about the same time. Occasionally, soybeans are grown in combination with sudangrass, small grains, millets, or sorghum.

SOYBEAN-OIL EXTRACTION

The solvent-extraction process has replaced the expeller (Figure 26–7) and the older hydraulic-press methods of soybean-oil extraction.

In the solvent-extraction process, the oil is extracted from the beans by a chemical solvent such as commercial hexane. The solvent, recovered from the oil by distillation, is used again. Oil obtained by this process has superior bleaching properties. The meal, since it contains less oil, is less likely to become rancid. Oil extraction by the solvent method is 95 per cent complete, only 0.5 to 1.5 per cent oil being left in the meal. Before extraction, the beans are cleaned, dried to 10 to 11 per cent moisture, crushed into grits, tempered for 15 to

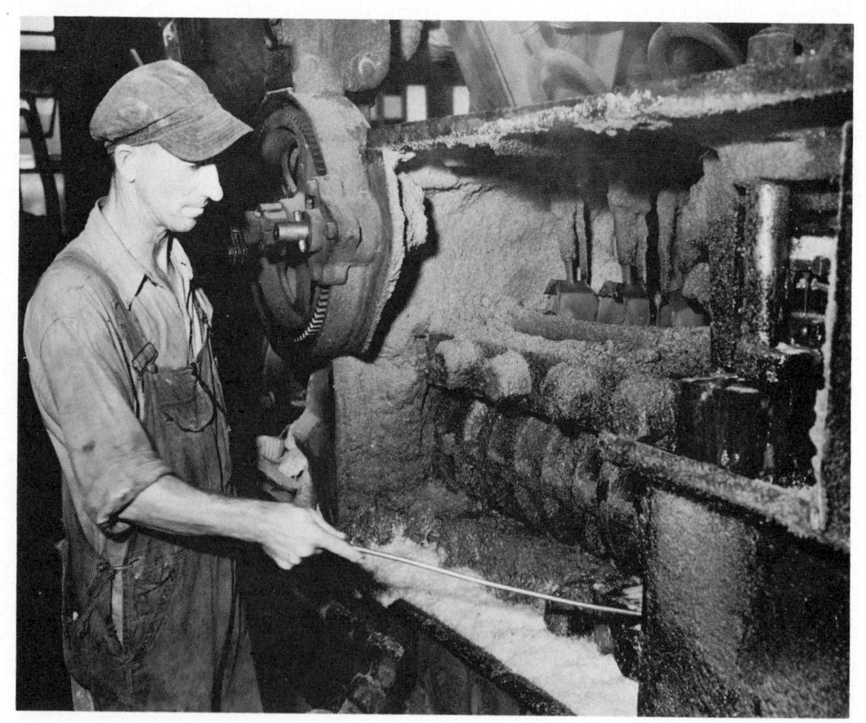

Figure 26–7. Extracting soybean oil in an expeller.

25 minutes at 130° to 170° F., and rolled into flakes about 0.007 inch thick.[13]

QUALITY OF SOYBEAN OIL

One of the important qualities in soybean oil is the drying property as measured by the iodine number. A high iodine number indicates good drying quality for paint purposes. The iodine number of soybean oil ranges from 118 to 141 in different varieties, while that of linseed oil is about 180.[35, 41] However, soybean oil can be fractionated to produce oils of both high and low iodine number. In the hydrogenation process to produce solid fats for shortening or margarine, an oil with a low iodine number is desirable.

SOYBEAN UTILIZATION

World production of soybean oil, 7 million metric tons in 1973, now exceeds that of any other edible vegetable oil, some of the others being peanut, cottonseed, coconut, sunflower, rapeseed, sesame, and olive oils. The United States produced about 8,261 million pounds of soybean oil in 1971. Of this amount, more than 3,432 million pounds each are used in making shortening and margarine. Considerable quantities are used as salad oil as well as in other food products. About 148 million pounds were used in 1972 in the drying-oil industries. Large quantities of soybean oil and soybeans are exported to other countries. Japan, West Germany, the Netherlands, Canada, France, and Spain are the chief importers of United States soybean oil and soybean cake.

Soybean oil is now used primarily for shortening, margarine, and salad oil. The lecithin from the oil is used in baked goods, candies, chocolate, cocoa, and margarine.

Soybean-cake and soybean-meal production in the United States exceeds 18 million tons annually. Soybean cake and meal are the chief high-protein supplements in mixed feed rations for livestock. The meal is also used in plastics, core binders, glue, and water paints. The oil is utilized in the making of candles, celluloid, core coil, disinfectants, electric insulation, enamels, glycerin, insecticides, linoleum, oilcloth, paints, printing ink, rubber substitutes, varnish, and soaps. The casein from vegetable milk produced from the dried bean is used in paints, glue, paper sizing, textile dressing, and for waterproofing.[3, 40, 41]

As a forage crop, the soybean is preserved as hay or silage, or cut and fed green as soilage. Soybeans are a valuable supplement to corn for silage because of the high protein content. They may be used for late summer pasturage when perennial pasture is short. The dried beans are utilized to some extent as high-protein feed for livestock.

Soybean flour is a valuable source of vegetable protein. It may be used in mixture with wheat flour in various baked products such as bread, cake, cookies, and crackers. Soybean flour has been used in ice cream, artificial cream, ice-cream cones, candies, puddings, and salad dressing. Because of its low starch content, it has found a place as a diabetic food. Soybean flour will make good bread when mixed with wheat flour in any combination containing up to 20 per cent soybean flour.[3]

In oriental countries soybeans are used in the production of soybean milk and curd, various soy sauces, fermented products, bean sprouts, and numerous other foods. Soybeans are also used in confections, soups, potted meats, food drinks, breakfast foods (puffed beans, flakes, and prepared meal), and as salted roasted beans, soybean butter, and as a substitute for coffee.[60] Most of the products, except the fermented ones, are now readily available in the United States. Soybeans are used as green shelled beans and dry beans.[15] Vegetable varieties of soybeans cook up more readily than do the field varieties. These vegetable varieties, which are used both as shelled green beans or as dry beans, have large seeds.

DISEASES

Soybeans, while once considered a disease-free crop in the United States, has been grown long enough that several diseases have become a potential problem. A few of the 50 diseases which attack soybeans are described.

Bacterial Blight

Bacterial blight, caused by *Pseudomonas glycinea* and reported from various regions, occurs on the leaves of the soybean plant as small angular spots. These spots at first are yellow to light brown, but later they become dark brown to almost black. Diseased tissues of the leaves may finally become dry and drop out. The spots sometimes spread to the stems and pods. Under favorable conditions, bacterial blight spreads rapidly in the field. The bacteria are seed-borne, but they may overwinter on dead leaves.[30, 31] Crop rotation and the turning under of soybean residues aid in checking the disease. Some varieties are partly resistant.

Bacterial Pustule

The bacterial-pustule disease, caused by *Xanthomonas phaseoli* var. *sojense*, has been reported in several states. The disease is confined largely to the foliage.[30, 31] In the early stages the spots are light green in color. Later the disease is characterized by angular reddish-brown spots on the leaves that may become large irregular brown areas. Portions of the larger spots frequently drop out. The organism overwinters in diseased leaf material as well as on the seed of diseased plants. Some control of the disease has resulted from crop rotation and from disposal of dead diseased leaf material by plowing after harvest. Resistant varieties are available. Bacterial pustule is a warm-weather disease which will occur later than bacterial blight.

Wildfire

Wildfire caused by bacteria, *Pseudomonas tabaci*, is a serious disease in the South. Black or brown spots on the leaves are surrounded by yellow halos and occurs only where bacterial pustules are found; invading areas already infected by bacterial pustule. Some resistant varieties are available.

Brown Stem Rot

Brown stem rot caused by a soil-inhabiting fungus (*Cephalosporium gregatum*) decays the interior of the stem. It attacks the plant early in the season but symptoms are usually not seen till later in the year. The center of the stem in diseased plants is reddish brown. Crop rotations where a given field is out of soybeans for 2 to 3 years is the best cultural control.

Stem Canker

Stem canker is caused by a seed-borne fungus (*Diaporthe phaseolorum* var. *batatatis*) that girdles the stem and kills the plant. A brown lesion which girdles the stem is usually found at the base of a branch or leaf petiole. Resistant varieties are available.

Pod and Stem Blight

Pod and stem blight, caused by the fungus *Diaporthe phaseolorum* var. *sojae*, is found wherever soybeans are grown. The stems and pods are heavily dotted with black spore-filled sacs (*pycnidia*). The fungus may girdle the stem, kill the plant, and prevent seed development. Sometimes the fungus penetrates the seed and destroys subsequent germination. The fungus is carried over winter on diseased stems and infected seeds. Recommended control measures for the disease are crop rotation and the planting of disease-free seed.

Frog-Eye Leaf-Spot Disease

Frog-eye, a leaf-spot disease caused by the fungus *Cercospora sojina*, attacks the leaves, pods, and stems of the soybean plant. It may go through the pod and enter the seed. The lesions are reddish in color when young, but change to brown and then to smoky gray with age. The fungus overwinters on the diseased plant refuse left in the field after harvest. Frog-eye leaf spot is introduced into new fields by diseased seed. Seed disinfectants have failed to give satisfactory control of the disease, but several varieties are resistant.

Brown Spot

Brown spot, caused by a fungus (*Septoria glycines*), attacks leaves, stems, and pods, and causes early defoliation. The small reddish-brown spots on the leaves are angular. The fungus overwinters on the crop residues, which should be plowed under. Crop rotation reduces the disease losses.

Target Spot

Target spot is caused by the fungus *Cornespora casucola*, which produces reddish brown spots that are encircled. Resistant varieties are available.

Downy Mildew

Downy mildew is caused by fungus *Peronospora manchurica*. The disease is seed borne and when infected seed are planted the first leaves to unfold are frequently covered with the characteristic mildew growth. The lack of resistant varieties requires crop rotation, plowing under of residue and planting of disease free seed to control the disease.

Mosaic

Mosaic is one of the most common diseases in Illinois—the number one soybean-producing state of the United States. Infected leaves may show characteristic puckering or wrinkles. A reduction in seed numbers in the pods of affected plants is also noticeable. At temperatures above 80° F. soybeans outgrow the symptoms of the disease. The virus which causes the disease is transmitted by aphids. There is no practical control for the disease.

Purple Stain

This disease is caused by several species of the *Cercospora* genus. The symptom most frequently observed is a discoloration of the seed varying from pink to dark purple. *Cercospora kikuchii* is the most common causal organism. Excessive "purple stain" may lead to reduction of seed value in the market place. Seedlings that develop from diseased seed may die. Control measures include treating the seed and using high-quality seed to avoid the infestation.

Other Diseases

Sclerotial blight, caused by the fungus *Sclerotium rolfsii*, is a rotting of the base of the stem. It occurs chiefly in sandy soil areas in the

South. Losses sometimes are as high as 25 to 30 per cent. The disease attacks many other crop plants.

The charcoal rot disease, caused by the fungus *Macrophomina phaseoli* (*Sclerotium bataticola*), also rots the base of the stems. After the plant is dead, the stem and roots bear numerous black sclerotia (spore bodies). The fungus lives in the soil. No remedy for the disease is known.

Other diseases of soybeans include (1) root rots caused by the soil-inhabiting fungi *Pythium debaryanum, Rhizoctonia solani, Phymatotrichum omnivorium;* (2) stem rot caused by *Sclerotinia sclerotiorum;* (3) anthracnose, caused by *Glomerella glycens;* and (4) fusarium blight caused by *Fusarium oxysporum f. tracheiphilum.* The latter disease occurs on sandy soils in the South; several varieties are resistant.

NEMATODES

Nematodes are a major problem in soybean production primarily in the southern United States. Root knot nematode causes knot-like swellings or galls on the roots. The resulting stunted plants wilt during dry weather and show marginal firing of the leaves. Rotation with nonsusceptible crops such as small grains and sorghum may be the best control.

Cyst Nematode

Plants severely attacked by this nematode are usually stunted and the foliage may be prematurely yellowed. The female, lemon-shaped nematode may be found on the roots of infected plants. Using resistant varieties is the major control recommended. Rotation and prevention of infected soil movement are also important control measures.

INSECT PESTS

One insecticide application is usually sufficient to control insect attacks. Leaf feeding insects including cabbage loopers, green clover worms, velvet bean caterpillars, fall armyworms, beetles and grasshoppers are the major problems in soybeans. As high as 35 per cent foliage loss during blooming and no greater than 20 per cent during pod filling may be tolerated without associated yield reductions before spraying is normally recommended.

Pod and Flower Feeding

Pod and flower feeding insects cause a direct yield reduction, especially if infestation occurs late in the blooming period. Bullworms and stink bugs, mexican bean beetle, grasshoppers, green clover worm and velvetbean caterpillars attack pods or flowers in soybeans.

Stem Feeding

Major stem feeding insects are the three cornered alfalfa hopper and weed borer. Control measures are rarely needed to prevent yield reductions from these pests.

Seed and Seedling Feeding

Seed corn maggot, seed corn beetle, wireworm, grape calapsis, white grubs, thrips, Southern corn rootworm, bean leaf beetle, and garden symphylans all feed on the seed, seedling, plant roots, or leaves. All of these pests except garden symphylans can be controlled by seed treatment or use of a soil insecticide.

RABBITS

Rabbits, especially jack rabbits, have a great fondness for the soybean plant. Where rabbits are numerous they will completely destroy small isolated fields of soybeans unless they are excluded by a tight fence.

REFERENCES

1. Anonymous. "Soybean farming," booklet published by Natl. Soybean Processors Assn., Chicago, pp. 1–43. 1961.
2. Anonymous. "Soybean cyst nematode," *USDA Agricultural Research Service Spec. Rept. ARS 22–72.* 1961.
3. Bailey, L. H., Capen, R. G., and LeClerc, J. A. "The composition of soybeans, soybean flour, and soybean bread," *Cereal Chem.*, 12:441–472. 1935.

3a. Beard, B. H., and Knowles, P. F. Soybean Res. in California, Univ of Cal Press, Davis, Cal., pp. 1–70. 1973.

4. Beeson, K. E., and Probst, A. H. "Soybeans in Indiana," *Purdue U. Agr. Ext. Bul. 231* (rev.). 1961.
5. Bernard, R. L., Mumaw, C. R., and Browning, D. R. "Soybean varieties for Illinois," *Ill. Agr. Exp. Sta. Circ. 794.* 1958.
6. Borst, H. L., and Thatcher, L. E. "Life history and composition of the soybean plant," *Ohio Agr. Exp. Sta. Bul. 494.* 1931.
7. Burris, J. S. "Effect of seed maturation and plant population on soybean seed quality," *Agronomy J.*, 65:440–441. 1973.

7a. Caldwell, B. E., and others. Eds. Soybeans: "Improvement, production and uses." ASA Monograph 16. Am. Soc. Agron. Madison, Wisc. 1975.

8. Cartter, J. L., and Hartwig, E. E. "The management of soybeans," in *Advances in Agronomy*, Vol. 14, Academic Press, New York, pp. 360–412. 1962.
9. Cartter, J. L., and Hopper, T. H. "Influence of variety, environment, and fertility level on the chemical composition of soybean seed," *USDA Tech. Bul. 787*, pp. 1–66. 1942.

10. Clapp, J. G., Jr., and Small, H. G., Jr. "Influence of "pop-up" fertilizers on soybean stands and yield," *Agronomy J.*, 62:802–803. 1970.
11. Cooper, R. L., and Gray, L. E. "Premature plant kill may be limiting soybean yields," *Soybean News*, National Soybean Crop. Imp. Council, 23(2):4. 1972.
12. Criswell, J. G., and Hume, D. J. "Variation in sensitivity to photoperiod among early maturing soybean strains," *Crop Sci.*, 12:664–667. 1972.
13. D'Aquin, E. L., Gastrock, E. A., and Brekke, O. L. "Recovering oil and meal," in *Crops in Peace and War*, USDA Yearbook, 1950–51, pp. 504–512.
14. Dimmock, F. "Seed mottling in soybeans," *Sci. Agr.*, 17:42–49. 1936.
15. Drown, M. J. "Soybeans and soybean products," *USDA Misc. Pub. 534*, pp. 1–14. 1943.
16. Duley, F. L. "Soil erosion on soybean land," *J. Am. Soc. Agron.*, 17:800–803. 1925.
17. Dunleavy, J. M., Chamberlain, D. W., and Ross, J. P. "Soybean diseases," *USDA Handbook No. 302*, pp. 1–38. 1966.
18. Erdman, L. W., Johnson, H. W., and Clark, F. "Varietal responses of soybeans to a bacterial-induced chlorosis," *Agron. J.*, 49:267–271. 1957.
19. Etheridge, W. C., Helm, C. A., and King, B. M. "A classification of soybeans," *Mo. Agr. Exp. Sta. Res. Bul. 131.* 1929.
20. Garber, R. J., and Odland, T. E. "Natural crossing in soybeans," *J. Am. Soc. Agron.*, 18:967–970. 1926.
21. Ham, G. E., Nelson, W. W., Evans, S. D., and Frazier, R. D. "Influence of fertilizer placement on yield response of soybeans," *Agronomy J.*, 65:81–84. 1973.
22. Howell, R. W. "Phosphorus nutrition of soybeans," *Pl. Physiol.*, 29(5): 477–483. 1954.
23. Howell, R. W. "Physiology of the soybean," in *Advances in Agronomy*, Vol. 12, Academic Press, New York, pp. 265–310. 1960.
24. Howell, R. W., and Collins, F. L., "Factors affecting linoleic acid content of soybean oil," *Agron. J.*, 49:593–597. 1957.
25. Howell, R. W., and Bernard, R. L. "Phosphorus response of soybean varieties," *Crop Sci.*, 1:311–313. 1961.
26. Hughes, H. D. "Symposium on soybeans: Soybean-corn mixtures," *J. Am. Soc. Agron.*, 23:1064–1067. 1931.
27. Johnson, H. W., and Bernard, R. L. "Soybean genetics and breeding," in *Advances in Agronomy*, Vol. 14, Academic Press, New York, pp. 149–221. 1962.
28. Johnson, H. W., Borthwick, H. A., and Leffel, R. C. "Effects of photoperiod and time of planting on rates of development of the life cycle," *Bot. Gaz.*, 122(2):77–95. 1960.
29. Johnson, H. W., Cartter, J. L., and Harturg, E. E. "Growing soybeans," *USDA Farmers Bul. 2129.* 1959.
30. Johnson, H. W., and Chamberlain, D. W. "Bacteria, fungi, and viruses on soybeans," in *Plant Diseases*, USDA Yearbook, 1953, pp. 238–247.
31. Johnson, H. W., Chamberlain, D. W., and Lehman, S. G. "Soybean diseases," *USDA Farmers Bul. 2077.* 1955.

32. Kiesselbach, T. A., and Lyness, W. E. "Soybeans in Nebraska," *Nebr. Agr. Exp. Sta. Bul. 322,* 1939.
33. Leffel, R. C. "Planting date and varietal effects on agronomic and seed compositional characters in soybeans," *Md. Agr. Exp. Sta. Bul. A-117.* 1961.
34. Leffel, R. C., and Barber, G. W., Jr. "Row widths and seeding rates in soybeans," *Md. Agr. Exp. Sta. Bul. 470.* 1961.
35. Lloyd, J. W., and Burlison, W. L. "Eighteen varieties of edible soybeans," *Ill. Agr. Exp. Sta. Bul. 453,* pp. 385–438. 1939.
36. Lovely, W. G., Weber, C. R., and Staniforth, D. W. "Effectiveness of rotary hoe for weed control in soybeans," *Agron. J.*, 50:621–625. 1958.
37. Markley, K. S. *Soybeans and Soybean Products,* Vols. I and II, Interscience, New York, pp. 1–1145. 1950–51.
38. McClelland, C. K. "Soybean varieties for hay, seed, and oil production," *Ark. Agr. Exp. Sta. Bul. 334.* 1936.
39. McClelland, C. K. "Methods and rates of planting soybeans," *Ark. Agr. Exp. Sta. Bul. 390.* 1940.
40. McClelland, C. K. "Soybean utilization," *USDA Farmers Bul. 1617.* 1930.
41. McClelland, C. K., and Cartter, J. L. "Improvement in soybeans," *USDA Yearbook, 1937,* pp. 1154–1189.
41a. National Fert. Development Center, Tennessee Valley Authority. Bulletin Y-69, Muscle Shoals, Ala., 1974.
42. Nelson, C. E., and Roberts, S. "Effect of plant spacing and planting date on six varieties of soybeans," *Wash. Agr. Exp. Sta. Bul. 639.* 1962.
43. Norman, A. G. (ed.). *The Soybean,* Academic Press, New York, pp. 1–239. 1963.
44. Odland, T. E. "Soybeans for silage and for hay," *W. Va. Agr. Exp. Sta. Bul. 227.* 1930.
45. Ohlroggs, A. J. "Mineral nutrition of soybeans," in *Advances in Agronomy,* Vol. 12, Academic Press, New York, pp. 229–263. 1960.
46. Osler, R. D., and Cartter, J. L. "Effect of planting date on chemical composition and growth characteristics of soybeans," *Agron. J.*, 46:267–270. 1954.
47. Owen, F. V. "Hereditary and environmental factors that produce mottling in soybeans," *J. Agr. Res.*, 34:559–587. 1927.
48. Pendleton, J. W., Bernard, R. L., and Hadley, H. H. "For best yields grow soybeans in narrow rows," *Illinois Research* (Ill. Agr. Exp. Sta.). Winter, 1960.
49. Piper, C. V., and Morse, W. J. *The Soybean,* McGraw-Hill, New York, pp. 1–331. 1923.
50. Robertson, D. W., Kezer, A., and Deming, G. W. "Soybeans under irrigation in Colorado," *Colo. Agr. Exp. Sta. Bul. 392.* 1932.
51. Scott, W. O., and Aldrich, S. R. *Modern Soybean Production,* SA Publications, Champaign, Ill. n.d.
52. Sears, O. H. "Soybeans: Their effect on soil productivity," *Ill. Agr. Exp. Sta. Bul. 456,* pp. 547–571. 1939.
53. Sherman, W. C., and Albrecht, H. R. "Edible soybeans," *Ala. Agr. Exp. Sta. Bul. 255,* pp. 1–16. 1942.

54. Stitt, R. E. "The effect of depth of planting on the germination of soybean varieties," *J. Am. Soc. Agron.*, 26:1001–1004. 1934.
55. Weiss, M. G. "Soybeans," in *Advances in Agronomy*, Vol. 1, Academic Press, New York, pp. 77–157. 1949.
56. Welton, F. A., and Morris, V. H. "The lodging of soybeans," *J. Am. Soc. Agron.*, 22:897–902. 1930.
57. Wiggans, R. G. "Soybeans in the northeast," *J. Am. Soc. Agron.*, 29:227–235. 1937.
58. Wiggans, R. G. "The influence of space and arrangement on the production of soybean plants," *J. Am. Soc. Agron.*, 31:314–321. 1939.
59. Willard, C. J. "The time of harvesting soybeans for hay and seed," *J. Am. Soc. Agron.*, 17:157–168. 1925.
60. Woodruff, S., and Klaas, H. "A study of soybean varieties with reference to their use as food," *Ill. Agr. Exp. Sta. Bul. 443*, pp. 425–467. 1938.
61. Woodworth, C. M. "The extent of natural cross-pollination in soybeans," *J. Am. Soc. Agron.*, 14:278–283. 1922.

Chapter 27

Cowpeas

ECONOMIC IMPORTANCE

The world production of cowpeas is nearly 1.2 million metric tons grown on more than 3 million hectares yielding about 0.38 metric tons per hectare. Countries leading in cowpea production are Nigeria, Niger, and Upper Volta in central Africa. Other countries in Africa, southern Europe, and southern Asia also produce cowpeas. In the United States cowpea seed was harvested on about 30 thousand hectares in 1970. Average yields were about 9 bushels per acre or 0.6 metric tons per hectare. In addition, about 29 thousand metric tons of blackeye types of cowpeas were produced in California.

The cowpea was for years the leading legume in the southeastern United States but since 1941 it has been gradually replaced by soybeans, clovers, and other legumes. Low yields of hay and seed, together with high production costs, make cowpeas less desirable than other crops. About 50,000 tons of cowpea hay was harvested in the United States in 1969, but later statistics on the declining crop are unavailable.

HISTORY

The cowpea apparently is native to central Africa where wild forms are found at the present time. Hybrids of the wild plant and the cultivated cowpea are readily obtained.[7] The cowpea has been cultivated for human food since ancient times in Africa, Asia, and Europe. The cowpea is probably the *Phaseolus* mentioned by the Roman writers.[5] It was introduced into the West Indies by early Spanish settlers. The crop was grown in North Carolina in 1714 and spread throughout the southern states later.

ADAPTATION

Cowpeas are a short-day, warm-weather crop grown primarily under humid conditions. They have been grown to some extent as far north as southern Illinois, Indiana, Ohio, and New Jersey.[1] The cowpea plant is similar to corn in its climatic adaptation except that it has a greater heat requirement. It is sensitive to frost both in the fall and spring. Severe drought generally prevents formation of seed in most varieties.[6] It can be grown in rows for hay or seed in the southern Great Plains where the average precipitation is as low as 17 inches, but the yields obtained scarcely justify the labor and expense involved in that method of production.

The cowpea is adapted to a wide range of soils. It grows as well on sandy soils as on clays, but Fusarium wilt and root knot are more prevalent in the sandy soils of Florida than in soils that contain more clay.[2] The plant thrives better than clover on either infertile or acid soils but saline and alkaline soils are unsuitable.[9] There is a general tendency for it to produce a heavy vine growth with few pods on very fertile soils or when planted early. The primary soil requirements are good drainage and the presence of, or inoculation with, the proper nitrogen-fixation bacteria cultures.

BOTANICAL DESCRIPTION

The cowpea (*Vigna sinensis*) differs from the common bean (*Phaseolus* sp.) in that the keel of the corolla is only slightly curved instead of twisted or slightly coiled.[7] Two related subspecies are the asparagus bean (*Vigna sinensis* var. *sesquipedalis*) and the catjang (*V. sinensis* var. *cylindrica*).

The cowpea is an annual herbaceous legume. The plants are viny or semiviny, and fairly leafy with trifoliolate leaves (Figure 27–1). The leaflets are relatively smooth and shiny. The growth habit of the cowpea is indeterminate. The plant continues to blossom and produce seed until checked by adverse environmental conditions.

The white or purple flowers are borne in pairs in short racemes. The pods are smooth, 8 to 12 inches long, cylindrical, and somewhat curved. They are usually yellow, but brown or purple pods are found in some varieties. The seeds are generally bean-shaped but usually short in proportion to their width. The Crowder varieties produce seeds more or less flattened on the ends as a result of being crowded in the pods. The seeds are either uniformly colored or multicolored. The more common solid colors are buff, clay, white, maroon, purplish, or nearly black, with a second color usually concentrated about the hilum[7] (Figure 27–2). The multicolored seeds may be variously spotted, speckled, or marbled. The seeds weigh about 60 pounds per bushel. A pound (.454 kilograms) contains 1600 to 4400 seeds.

The cowpea is largely self-pollinated. Natural crosses rarely occur in the field in most regions.

Catjang cowpeas are erect semibushy plants with small oblong

Figure 27–1. Plant of Victor cowpeas.

seeds borne is small pods 3 to 5 inches long. The pods are erect or ascending when green, and usually remain so when dry. Varieties of this group are very late and not very prolific in this country, so they are seldom grown here.

The asparagus bean or yard-long-bean plants are very viny with pendant pods 12 to 36 inches long that become more or less inflated, flabby, and pale before ripening. The seeds are elongated and kidney-shaped. None of the varieties produces as much seed or forage as do

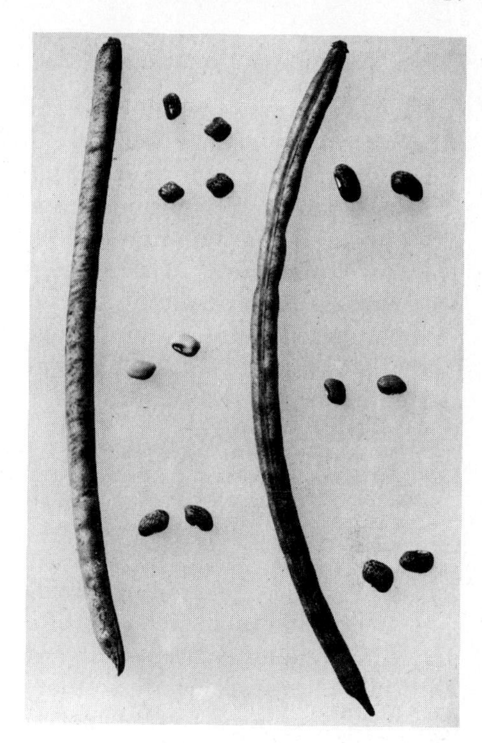

Figure 27–2. Pod and seeds of Brabham cowpea (*left*) and Groit cowpea (*right*).

the better cowpea varieties. Seed salesmen offer the asparagus bean as a novelty vegetable that has pods so large that one suffices for a meal. It is not grown extensively because other beans are more palatable.

VARIETIES

Good forage cowpea varieties are vigorous, erect, prolific, and disease-resistant, bear the pods well above the ground, and retain their leaves late in the season.[6, 11] Early maturity is important in the northern part of the cowpea region. The viny habit is considered desirable when the crop is planted in mixtures with corn or sorghum. White-seeded varieties and the Crowder and Blackeye types are commonly grown for table use, or canning as green beans.[2]

Viny varieties that mature late are preferred in the Mississippi Delta and near-by areas and for forage in Florida.[4] In the northern part of the cowpea region earlier varieties resisting wilt and nematodes are desired. Chinese Red, an early erect dwarf variety with small seeds, is preferred in the semiarid western part of Oklahoma. Improved strains of California Blackeye are grown for food and seed in California.[9]

CROP ROTATIONS

The cowpea is grown in corn and cotton rotations in the South to supply nitrogen and organic matter to the soil. Cowpeas often precede or follow winter oats or winter wheat in rotations.

At the Arkansas Experiment Station, cowpeas were interplanted with corn to determine their effect on the yields of subsequent crops. They were planted (1) in the same row with corn, (2) between 44-inch corn rows at laying-by time, and (3) before laying-by time in wide 58-inch rows. The nine-year average yields of oats following the three methods were 15.5, 9.5 and 5.9 per cent, respectively, higher than for oats following corn alone.[3] Cotton showed the greatest increase when it followed cowpeas planted in the same row and at the same time as corn.

CULTURAL METHODS

Cowpeas succeed on a seedbed prepared as for corn. Sometimes they are seeded as a catch crop after small grains, with disking as the only soil preparation, especially where the land is free from weeds. Fertilizer requirements are similar to those for soybeans. Inoculation is generally unnecessary, except where the crop is being grown for the first time. Seed treatment may reduce damage from seed rots and seedling blight.[2]

Seeding Practices

For best results, cowpeas must be planted in warm soil after all danger of frost is past. For hay or seed production the crop is usually planted at about the same time as corn or up to two weeks later. The crop may be planted in the South for green-manure, pasture, or hay as late as August 1, or up to 90 days before the first frost. In northern Virginia they should be planted in May or early June for seed and in late June for best yields of hay. In California they are best planted between May 1 and June 15.[9]

Cowpeas may be planted with either a grain drill or a corn planter. For seed production, the crop is commonly planted in rows 24 to 40 inches apart with the seeds 2 to 3 inches apart in the row. The rate of seeding varies from 20 to 45 pounds of seed per acre. For forage or green manure, cowpeas are generally seeded in close drills at the rate of 75 to 120 pounds per acre.

Harvesting for Hay

Cowpeas may be harvested for hay when the pods begin to turn yellow. The best quality of hay is obtained when the pods are fully grown and a considerable number of them are mature. The vines are difficult to cure when harvested earlier. Delay beyond this stage results in tough woody stems as well as in excessive loss of leaves. A hay crusher facilitates curing of the coarse stems. Hay yields range from one to two tons per acre under good growing conditions, a single crop being obtained in a season.

Harvesting for Seed

Cowpeas should be harvested for seed when one-half to two-thirds of the pods have matured. Large fields are usually harvested with a combine after most of the seed is ripe. The mower with a bunching or windrowing attachment also is satisfactory. Cowpeas grown for home food in the South are usually picked by hand.

Cowpeas in Mixtures

Cowpeas often are grown in mixtures with corn for silage. They also are grown for hay in combination with such crops as sorghum, sudangrass, and johnsongrass. When grown for silage with corn, the cowpeas and corn are usually planted by the corn planter in one operation. Cowpeas grown in combination with other crops for hay produce a larger yield of more readily curable hay than cowpeas grown alone.

UTILIZATION

A considerable portion of the cowpeas threshed are used for planting. Surplus seed or that damaged or otherwise unfit for planting or human consumption is fed to livestock. Although high in feed value, cowpea seed is generally not an economical livestock feed. The seed of certain varieties is a popular food in the South, being used in the pod, shelled green, or shelled dry, in preference to other types of beans.[1,2,5]

Cowpea hay that has been well cured is considered equal to red-clover hay in nutritive value. For a silage crop, cowpeas grown in mixture with corn or sorghum is cut when the first cowpea pods begin to turn yellow. Cowpeas, although high in protein and low in carbohydrates, will make good silage alone when wilted to the proper moisture content of 60 to 68 per cent before being put into the silo.[10]

Cowpeas are utilized for soil improvement in the southern states. They will make a good growth on soils too poor for soybeans. Often the cowpeas are used for pasture or some seed is picked before being plowed under. On very poor soils it is advisable to plow under the entire crop in the green state. About 85 per cent of the fertilizing value is in the hay, and about 15 per cent in the roots and stubble.[5] Thus the soil benefits even when the crop is cut for hay.

DISEASES

Cowpea wilt, caused by the fungus *Fusarium oxysporum* var. *tracheiphilum*, causes the leaves to yellow and fall prematurely,[6] and finally results in the death of the plant. The stems turn yellow, the plants become stunted, and seed setting generally fails. Diseased stems are brown to black inside. Wilt is generally observed about mid-season, being spread by cultivation implements, drainage water, and other agencies. The most satisfactory control measure is the use of resistant varieties.

Cowpea root knot, caused by a nematode (*Meloidogyne* species), is identified by galls over the entire root system. The roots soon turn brown, decay, and often die. The most practical measure is the use of resistant varieties in combination with other immune crops in rotations. Suitable rotation crops that are immune include winter grains, velvetbeans, corn, sorghum, and some soybeans.

Several varieties are resistant to bacterial blight or canker caused by *Xanthomonas vignicola*.

Charcoal rot caused by *Macrophomina phaseoli* (*Sclerotium bataticola*) damages cowpeas severely under certain soil conditions.

Other fungus diseases of the cowpea include zonate leaf spot, caused by *Aristastoma oeconomicum;* red leaf spot, caused by *Cercospora cruenta;* and mildew caused by *Erysiphe polygoni*. Most American varieties are resistant to the bean rust caused by the fungus *Uromyces phaseoli*, that attacks introduced varieties. Red stem canker, caused by *Phytophthora cactorum*, is of minor importance.[10]

Virus diseases include cucumber mosaic, southern bean mosaic, and curly top.[11]

INSECTS

The chief insect enemies of the cowpea are two species of weevil that damage the seed, namely, cowpea weevil (*Callosobruchus maculatus*) and the southern cowpea weevil or four-spotted bean weevil (*Mylabris quadrimaculatus*). These weevils lay their eggs on the pods or in the seeds in the field and later on in the threshed seeds in storage. The larvae bore into the seeds and complete their life cycle there. New generations of the weevil continue to develop unless the temperature falls too low. Control of the weevils consists of insecticide application, fumigation, or heat treatment of the stored seeds.[6] The cowpea curculio (*Chalcodermus aeneus*) infests developing seeds. It is controlled by the application of insecticides to the field when blossoming begins.

Other pests include lygus bugs, corn earworm, lima bean pod borers, mites, cowpea aphids, bean thrips, yellow-striped and beet armyworms, and rootknot nematodes.

REFERENCES

1. Ligon, L. L. "Characteristics of cowpea varieties," *Okla. Agr. Exp. Sta. Bul. B-518.* 1958.
2. Lorz, A. P. "Production of southern peas (cowpeas) in Florida. I. Cultural practices and varieties." *Fla. Agr. Exp. Sta. Bul. 557.* 1955.
3. McClelland, C. K. "Variety and inter-cultural experiments with cowpeas," *Ark. Agr. Exp. Sta. Bul. 343.* 1937.
4. McKee, R., and Pieters, A. J. "Miscellaneous forage and cover crop legumes," *USDA Yearbook, 1937,* pp. 999–1031.
5. Morse, W. J. "Cowpeas: Utilization," *USDA Farmers Bul. 1153.* 1920.
6. Morse, W. J. "Cowpeas: Culture and varieties," *USDA Farmers Bul. 1148* (rev.). 1947.
7. Piper, C. V. "Agricultural varieties of the cowpea and immediately-related species," *USDA Bur. Plant Industry Bul. 229.* 1912.
8. Piper, C. V. "The wild prototype of the cowpea," *USDA Bur. Plant Industry Circ. 124,* pp. 29–32. 1913.
9. Sallee, W. R., and Smith, F. L. "Commercial blackeye bean production in California," *Calif. Agr. Exp. Sta. Circ. 549,* pp. 1–15. 1969.
10. Weimer, J. L. "Red stem canker of cowpea, caused by *Phytophthora cactorum,*" *J. Agr. Res.,* 78:65–73. 1949.
11. Wright, P. A., and Shaw, R. H. "A study of ensiling a mixture of Sudan grass with a legume," *J. Agr. Res.,* 28:255–259. 1924.

Chapter 28

Field Beans

ECONOMIC IMPORTANCE

The field bean (*Phaseolus* sp.) is an important human food of a high protein content. It is grown on more than 23 million hectares, with a production of about 11.5 million metric tons, or 0.5 tons per hectare. The leading producing countries are Brazil, India, Mexico, and the United States. The average production in the United States from 1970 to 1972 was about 778,000 metric tons on 555,000 hectares (1,370,000 acres) or 1.4 metric tons per hectare (1250 pounds per acre). More than 4 per cent of this reported production consists of California blackeye peas and chick peas which belong to other genera. The states leading in bean production were Michigan, California, Idaho, Colorado, and Nebraska (Figure 28–1).

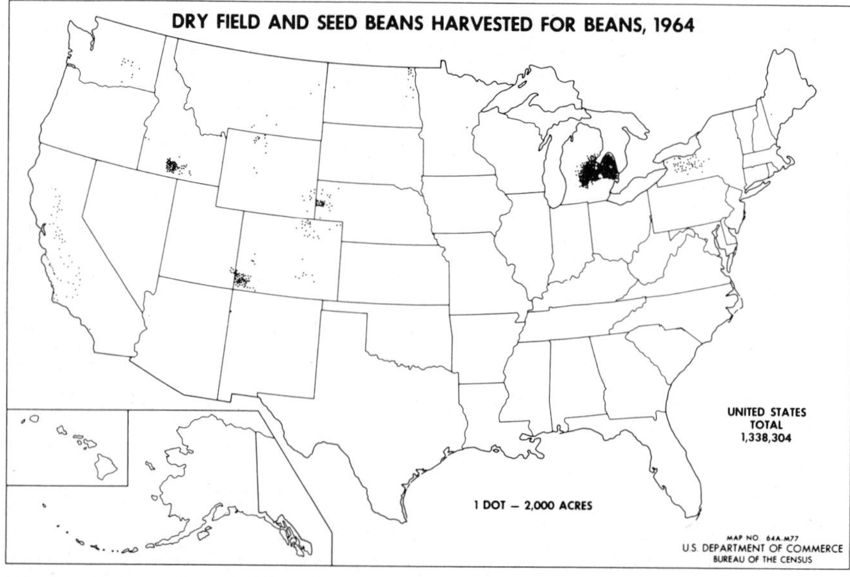

Figure 28–1. Dry-field and seed-bean acreage in the United States in 1964.

HISTORY OF BEAN CULTURE

The common bean (*Phaseolus vulgaris*) probably was domesticated from a wild form having a long slender vine which is found in Mexico and Central America.[10] Kidney beans were being cultivated there some time between 5000 and 3000 B.C. The tepary bean (*P. activolius*) was grown before 3400 B.C. The big and little types of lima bean (*P. lunatus*) are assumed to have arisen in Peru and Mexico, respectively. The common bean was collected by early explorers and was being grown in Europe by 1542. Lima bean seed was brought to the United States from Peru in 1824, and a bush lima bean was found growing along a Virginia roadside in 1875.

ADAPTATION

The bean plant is a warm-season annual adapted to a wide variety of soils. Its optimum mean temperature is 65° to 75° F. High temperatures interfere with seed setting, while low temperatures are unfavorable for growth.[17] Dry beans are produced most successfully in areas where the rainfall is light during the latter part of the season. This reduces weather damage to the mature beans. Under dryland conditions, the bean plant adjusts its growth, flowering, and pod setting to the soil moisture supply to a marked degree. Beans require a minimum frost-free season of about 120 to 130 days in order to mature seed.[22]

In general, white beans are grown in the humid region and in the North, and mottled or colored beans in the dry and warm regions. However, the colored Red Kidney and the mottled Cranberry beans also are grown in the East, small white beans are grown in California and Idaho, and the white-seeded Great Northern variety is grown in the northern irrigated intermountain region. White varieties require a harvest period relatively free from rain to avoid discoloration of the seeds.

Hardshell, a form of dormancy caused when seed coats are impermeable to water, is prevalent in a large number of bean varieties.[11] The condition is accentuated by hot dry winds at harvest time, or by storage in a heated room.

Most types of field beans are short-day plants, but the Boston Yellow Eye and Cranberry types are day-neutral, i.e., they flower in 26 to 39 days after planting at all day lengths from 10 to 18 hours,[1a] and the Red Kidney shows only a partial response to different day lengths. The lima bean (*P. lunatus*) and chickpea (*Cicer arietinum*) are day-neutral, the tepary bean, mung bean, velvetbean, and cowpea are short-day plants, and the scarlet runner bean and yellow lupine are long-day plants.

BOTANICAL DESCRIPTION

Dry edible beans, including the common bean as well as the lima, tepary, runner, and mung bean, belong to the genus *Phaseolus*.[15] All are more or less flat-seeded.

Plants of the common bean (*Phaseolus vulgaris*) may be either bushy or viny (trailing). Bush beans are determinate in growth habit, and stem elongation ceases when the terminal flower racemes have developed.

The intermediate pole beans bear racemes in the leaf axils while the stem continues to elongate. Field beans are either the semipole or bush type. The leaves are pinnately trifoliolate. Both leaves and stems are pubescent. The flowers are white, yellow, or bluish purple. The pods are straight or distinctly curved, 4 to 8 inches long, and end in a distinct spur (Figure 3–16). The seeds may be white, buff, brown, pink, red, blue-black, or speckled in color (Figure 28–2). The immature pods of snap beans are either yellow or green, but those of field varieties are green. The field varieties of the common bean are distinguished by their fibrous pods from the snap beans, which have very little or no fiber in the pods.[29] A satisfactory dry bean bears its pods above the ground, ripens uniformly, and does not shatter appreciably at maturity.

Beans are normally self-pollinated with less than 1 per cent natural crossing.[21]

Dry beans, soybeans and some peanut varieties contain considerable low molecular-weight carbohydrates, especially raffinose and stachyase. Anaerobic bacteria in the human intestine release the gasses—carbon dioxide, methane, ammonia and hydrogen sulfide from the seeds, causing flatulence.

A classification of four species of beans commonly grown in the United States is as follows:[29]

A. Leaves glabrous
- B. Seeds with conspicuous lines radiating from the hilum to the dorsal region; perennial in tropics — *lima bean* [*Phaseolus lunatus* (*limensis*)]
- BB. Seeds without conspicuous lines radiating from the hilum to the dorsal region — *tepary bean* (*P. actifolius* var. *latifolius*)

AA. Leaves pubescent
- B. Roots tuberous, often perennial in warm climates, cotyledons not raised above the ground in the seedlings — *runner bean* (*P. multiflorus*)
- BB. Roots not tuberous, annuals, cotyledons raised above the ground in the seedlings — *common bean* (*P. vulgaris*)

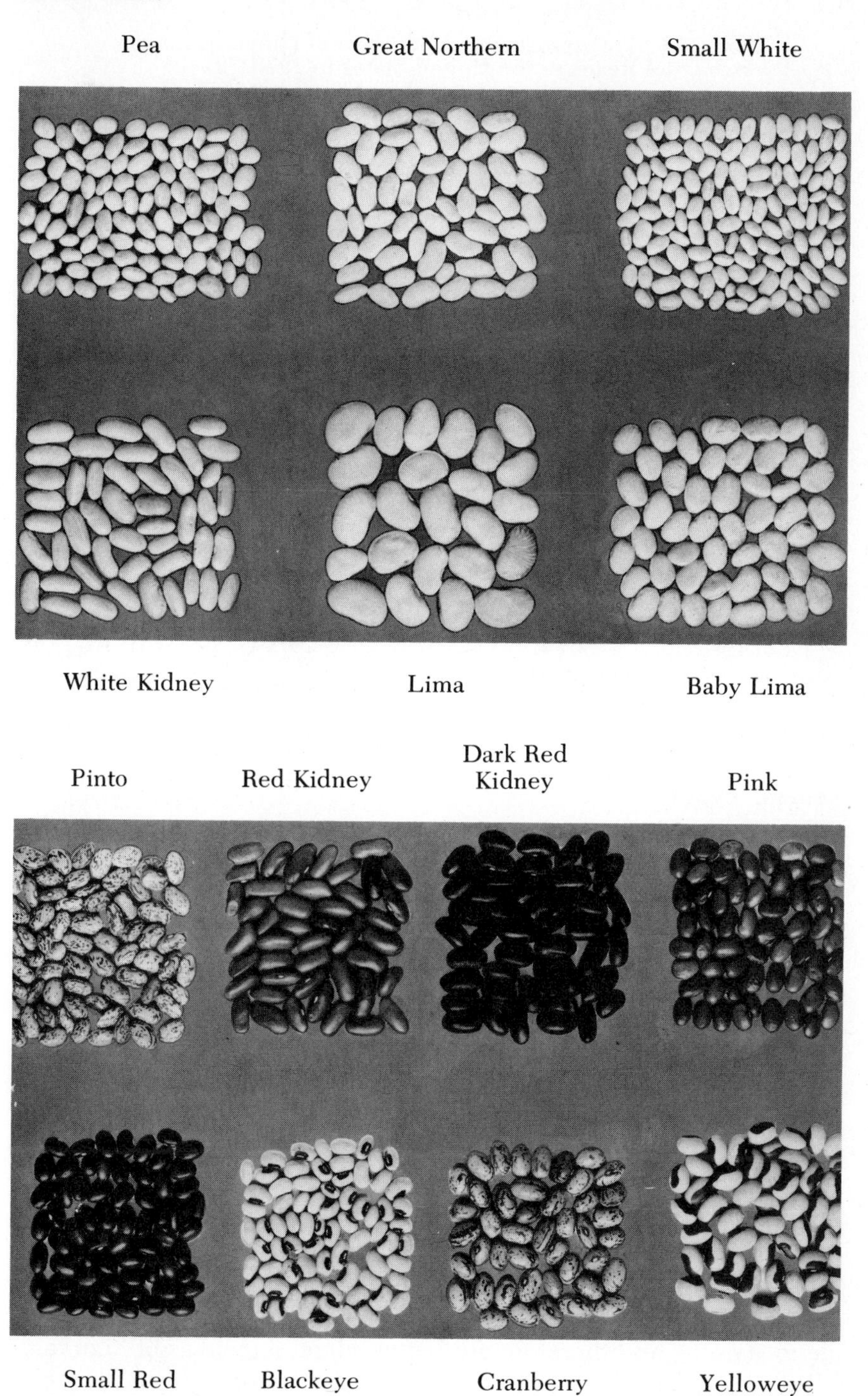

Figure 28–2. Principal types of field beans, showing relative sizes (about two-fifths natural size).

TABLE 28–1. Approximate Production of Different Classes of Beans in the United States in 1970–72

CLASS	PRODUCTION—UNCLEANED (*Millions of lb.*)
Field bean	
Pea	516
Pinto	477
Great Northern	155
Red Kidney	119
Small White	31
Pink	71
Small Red	37
Black Turtle Soup	41
Cranberry	13
Flat Small White	7
Other	14
Mung bean	13*
Lima bean	
Large lima	40
Baby lima	40
California Blackeye (culinary cowpea)	41
Chickpea	
Garbanzo (*Cicer arietinum*)	9

* 1967

General Types

The commonly recognized types of dry beans grown in the United States are listed in Table 28–1.

Pea beans average about 2,300 seeds per pound; and Red Kidney, 1,100.[13]

BEAN TYPES

Field Beans

The White Pea (or Navy) beans are grown mostly in Michigan.[2] Most of the pea-bean acreage in Michigan is planted to improved varieties, which are resistant to common mosaic or anthracnose. Pea beans mature in 110 to 120 days. The White Pea bean is a small semitrailing plant with white flowers and small white seeds (Figure 28–2).

Red Kidney beans require a frost-free season of at least 140 days to mature. Because of the red color of their seeds they are less susceptible to staining and therefore are better adapted to areas of high rainfall than are the white pea beans.[13] The Red Kidney bean plant is a bush type. The flowers are lilac in color. The seeds are large, flattened, pink when newly harvested and mostly dark red when old.

The matured pods are often splashed with purple. They are susceptible to bacterial blights and anthracnose. Consequently, the seed supply for planting in New York and Michigan is produced largely in the dry climates of California and Idaho where these seed-borne diseases are not prevalent.

The Great Northern type is grown chiefly on irrigated land in Nebraska, Idaho, and Wyoming.[5] The plants are short and trailing, the flowers white, and the seeds white, large, and flattened. It is medium-late in maturity. Selected strains of Great Northern beans are grown in southern Idaho for their resistance to common bean mosaic or the curly-top virus.

Pinto varieties are grown in irrigated or dry areas in Colorado, Idaho, and other western states and also in the northcentral states. The Pinto is a semitrailing type with white flowers. The seeds are medium to large, somewhat flattened, buff-colored, and speckled with tan to brown spots and splashes. The Pinto possesses a distinctive flavor not found in other dry beans.

The Red Mexican or small red varieties are adapted to the irrigated and higher rainfall areas of the Columbia and Snake river basins in Washington and Idaho. The Red Mexican bean has bright red seeds. Improved strains of this variety are grown.

The Pink bean is a semitrailing variety with white flowers and medium-size pink seeds. It is very heat resistant. The Cranberry bean, grown in Michigan and California, has deep-red splashes on a buff seed coat. The Cranberry varieties may be of the pole, semipole, or bush type. The Black Turtle Soup, or Venezuelan variety, is a black bean grown in California, New York, and Michigan.

Tepary Bean

The tepary bean (*P. acutifolius* var. *latifolius*) has been grown by the Indians of Arizona and New Mexico since early times but was not tested in field experiments before 1910.[9] It has small leaflets and white seeds somewhat smaller than those of the navy bean. The beans are considered to be harder to cook and less palatable than the common edible beans and therefore have only a limited market. They appear to be very resistant to drought and outyield the Pinto bean in eastern New Mexico and other southwestern areas. For hay, the tepary bean has outyielded cowpeas and other annual legumes by a considerable margin in tests in western Oklahoma.[8]

Mung Bean

The mung bean (green gram or golden gram) (*Phaseolus aureus*) is grown mainly for human food in the southern half of Asia and nearby islands and also in Africa. It is the source of the canned bean sprouts found on the market and used in chop suey and similar foods. The mung bean has been known in the United States since 1835 under such names as Chickasaw pea, Oregon pea, Neuman pea, Jerusalem pea, or chop suey bean. Prior to World War II, nearly all of the mung beans used in the United States were imported. Production was begun in 1945, mostly in Oklahoma. About 34,000 acres were harvested in 1967 and a considerable acreage was plowed under for green manure. Average yields of mung beans are about 400 pounds per acre. The beans are suitable for feed, and the plants make good silage and fair hay. The plants have glabrous leaves and are very similar to those of the cowpea except for being smaller and more bushy (Figure 28–3). The pale yellow flowers are borne in racemes or clusters of 10 to 25. The ripe black or brownish pods are 3 to 5 inches long and contain 10 to 15 small globose or oblong seeds. Most varieties have green, yellow, golden-brown, or marbled seeds.[20]

The urd, black gram, or mungo bean (*Phaseolus mungo* or *Vigna mungo*), which is grown chiefly in India for human food and for hay, is somewhat similar to the mung bean.

Lima Bean

The large lima bean (*Phaseolus limensis*, or *P. lunatus* var. *macrocarpus*) is a perennial grown as an annual. The small (baby) lima (*P. lunatus*) is an annual. These are now consumed mostly as green limas, either frozen, canned, or fresh.

When allowed to mature, lima beans are handled and used in a manner similar to other dry edible field beans. The domestic production of more than 80 million pounds annually is largely in California. Much of this seed is used for planting by growers of green limas in other states. Temperature and moisture extremes may limit the fertilization of the lima bean and induce pod shedding.[17]

Broadbean

The broadbean (horse bean or Windsor bean), *Vicia faba*, is related botanically to the vetches but differs from them in having coarse erect stems, large leaflets, large pods, and large flattened seeds. It is mostly cross-pollinated and was grown by ancient Egyptians. This is a popular crop in Europe, Africa, and Latin America. China is the chief producer, and the world production exceeds 5 million metric tons. In the United States, it is well adapted only to the coastal section of California where it is grown to a limited extent for food or green manuring.[2] It is eaten either green or dry. The broadbean is an important food crop in Egypt and the Sudan, where it is planted at a rate of 70–75 pounds per acre. Small-seeded types are grown for green manuring. It occasionally is fed to horses.

Figure 28–3. Pods and leaflets of mung bean.

Chickpea

The chickpea (*Cicer arietinum*) usually called garbanzo or gram, is grown on a commercial scale in the United States almost entirely in California. Some 7 million pounds are produced annually. The leading countries in chickpea production are India, Pakistan, Spain, Algeria, and Mexico. The world production is about 7 million metric tons. The chickpea is a native of Europe. It was taken to California from Mexico during the Spanish mission period.[2] It is best adapted to warm semiarid conditions, where most of it is grown. The plant is a low bushy annual with hairy stems. Each leaf is comprised of several pairs of small rounded or oblong leaflets. The flowers are white or reddish, small, and borne singly at the tip of axillary branches. The seeds are roughly globular, flattened on the sides and somewhat wrinkled. The chickpea is grown and handled about like the field bean. The threshed seeds are prepared for food in much the same manner as are dried lima beans. Roasted seeds are used as a confection or snack, and sometimes as a coffee substitute. The herbage is low in yield and is toxic to animals.

Other Beans

The culture of cowpeas, including the California Blackeye variety, is discussed in Chapter 27. The adzuki bean, *Phaseolus angularis*, is an important crop in Japan. In the United States it is grown only occasionally for home use.

Lentils

Lentils (*Lentilla lens*) have been an important food item in the Mediterranean region and Asia Minor since the Bronze Age. The world production is about 1 million metric tons. The leading countries in lentil production are India, Pakistan, Turkey and Syria, but they also are grown in Egypt, Ethiopia, the Soviet Union, and adjacent European countries.[34] Lentils were introduced into the United States by 1898,[33] but were grown only to a limited extent, chiefly in home gardens. The commercial culture of lentils in the United States began in the state of Washington in 1937, and from 1968 to 1970 about 57,000 acres were grown annually in Washington, Idaho, Oregon, and Northern California with yields of 950 pounds per acre.

The lentil plant is a branched weakly upright or semiviny annual 18–22 inches (45–55 cm.) tall, with pinnately compound leaves. It has a general resemblance to vetch. The flowers are white, lilac, or pale blue. It is highly self-pollinated.[33] The pods contain 2 or 3 seeds, which are thin and lens-shaped, usually smaller than pea seeds, and of various cotyledon colors including yellow, orange, or both. A new variety, Tekoa, has uniformly colored, larger seeds than the Chilean and Persian types previously grown.[32]

Most of the lentils are grown in rotations preceding fall-sown small grains. It competes poorly with weeds, especially wild oats. The plants require molybdenum, sulfur, and sometimes phosphorus fertilization for good growth.[7] Inoculated seed is best sown in April at rates of 60–80 pounds per acre, depending upon seed size. The crop is swathed when the pods are of golden color, and then threshed with a pick-up combine about 10 days later. Swathing when the pods are dewy and tough reduces seed shattering.

Lentils are susceptible to many viruses that attack peas, clover, or alfalfa, and the fields should be isolated from those of other legumes. The chief insect pests are cowpea and black bean aphids.

CROP ROTATIONS

It is advisable to grow beans in long rotations with other crops in the humid and irrigated regions. An interval of three to four years between bean crops reduces the risk from soil-borne disease infection.[2, 19] Beans succeed well after green-manure crops,[26] legume-grass hay crops, small grains, corn, or potatoes. Typical rotations in New York and Michigan bean areas include such crops as winter wheat, clover and timothy, corn, and potatoes. A rotation of alfalfa, potatoes, beans, sugarbeets, and grain is popular in the irrigated intermountain regions.

In semiarid regions, beans may replace summer fallow in the alternate wheat-fallow cropping system. The yields of wheat after beans are almost as good as those after summer fallow. The beans usually are harvested in time for the fall seeding of wheat. In California beans may be grown continuously or in rotations following alfalfa or an intertilled crop.[22] The continuous culture of beans may result in severe soil erosion. On the drylands of New Mexico and Colorado and Kansas beans may follow small grains, sorghum, or corn.[4] Strip cropping of beans is advisable in order to reduce soil blowing after bean harvest. Thus, strips of 4 to 24 rows of beans may alternate with strips of sorghum or corn of equal width.[12, 25]

CULTURE

A deep, firm seedbed free from clods and coarse debris is desirable for field beans.

Fertilizers

In New York,[13] application of 300 pounds per acre of superphosphate and 500 pounds of 2–12–4 or 4–12–4 have increased bean yields. Fertilizers may be applied before or during planting.

Nitrogen applications of 20 to 60 pounds per acre are helpful in the humid region when beans do not follow a legume crop that has been turned under. In irrigated areas, rates of up to 120 pounds per acre may be profitable on new land or on soils of low residual fertility. Nitrogen may not be needed on fertile irrigated soils when beans follow corn, potatoes, or sugarbeets that have been fertilized or that followed alfalfa. Up to 40 pounds per acre of nitrogen may be helpful after small grains where the straw is left on the land.[19, 23]

Phosphate and potash applications of 40 to 160 pounds per acre each of P_2O_5 and K_2O may increase bean yields on soils that are deficient in available quantities as shown by soil tests. In the irrigated section, ample residual amounts of these elements may be available for the beans from application to the previous crop of alfalfa, potatoes, or other crop. Zinc-deficiency symptoms on bean plants are avoided by applying about 10 pounds per acre of zinc (as zinc sulfate) every four years. Current injury is corrected by spraying the foliage with a 0.3 per cent $ZnSO_4 \cdot 7H_2O$ solution.[23] Iron deficiency may occur on poor soils.[5]

Planting Practices

Beans should be planted in warm soil, preferably above 65° F., after all danger of frost is past. The time of planting in a locality is about the same as that for corn. In the Northwest, the time varies from about May 20 to June 10, in Colorado from May 20 to June 15, in New Mexico from May 15 to July 1, and in California from April 10 to July 10. Cold wet soil is likely to result in low germination when beans are planted in western New York before June 1.[22]

Beans are generally planted in drilled rows 20 to 42 inches apart with a two-row to eight-row bean planter, or with sugarbeet or corn planters equipped with bean plates. The seeds are planted at a depth of 2½ to 4 inches in semiarid areas or where irrigated surface soils may dry out before germination is complete. A planting depth of 1½ to 2 inches is desirable where soil moisture in the seedbed is ample.[22] The desired stand under favorable irrigated or humid conditions is plants spaced 2 to 4 inches apart in rows 20 to 24 inches apart. This requires a planting of 40 to 100 pounds of seed per acre. Large-seeded types such as Red Kidney, Pinto, and Great Northern are planted at a rate of 80 pounds per acre. For the smaller-seeded Red Mexican and Navy types, the desired rates are about 60 and 40 pounds, respectively. Irrigated Pinto beans are planted at a rate of about 60 pounds per acre in Colorado. The planting rate for large lima beans is 80 to 100 pounds per acre, whereas that for baby limas is 40 pounds. As the plant spacing increases, the number of pods per plant increases. Close spacing of plants does not increase the percentage of immature pods.

Dryland beans in southwestern Colorado and northern New

Mexico are planted in 30- to 42-inch rows at a rate of about 30 pounds per acre or less. One plant every 10 inches in the row is ample.

Mung beans are planted from May to July, often on land from which winter grain has been harvested. They may be drilled for hay or planted in cultivated rows for seed.[20]

Lentils are drilled at a rate of 30 pounds per acre or sown in 3-foot rows at a rate of 30 pounds per acre, planted about September 15 to October 15, and harvested in July in California and western Oregon. In eastern Washington and adjacent colder sections they are planted in April at 60 to 80 pounds per acre and harvested about August.

Weed Control

Beans may be cultivated with a rotary hoe before the germinating seedlings emerge, or after the plants are 2 to 4 inches tall. The later two or three cultivations are usually made with sweep implements. Most weeds in the bean fields of some areas can be controlled with herbicides applied as a spray before planting the beans or placed 3 to 5 inches underground with a special subsurface applicator. They may also be applied on the surface and then incorporated with the upper 6 inches of soil with a rototiller or by two double diskings.[6] Preemergence herbicides also are used on bean fields.

Irrigation

At planting time, the soil in an irrigated bean field should be wet nearly up to its field carrying capacity to the depth of the root-feeding zone of 3 feet or more. This often necessitates an irrigation just before final seedbed preparation. The seedbed is then prepared, after which the beans are planted as soon as the soil is dry enough to work. The field should not be irrigated after planting until the seedlings have emerged. Additional irrigations should be applied within 3 to 5 days after the bean foliage turns a dark bluish green. This change in plant color develops as the soil moisture becomes more deficient. Usually three or four irrigations, but sometimes more, are necessary after the beans have emerged. Irrigation for the season should cease when one-fourth of the bean pods have turned yellow.

Beans usually are irrigated by the furrow method or with sprinklers, but basin irrigation is satisfactory on permeable soils where the bean plants are flooded during only short periods.[16, 24, 27]

Harvesting

Field beans are generally harvested when most of the pods have turned yellow but before they are dry enough to shatter from the pods. The vines are cut below the surface with blades attached to two-row or four-row bean harvesters that leave the plants in one or two windrows.[28] It is a common practice to throw two to four windrows together with a side-delivery rake during or immediately after cutting. Some beans are cut above the ground with windrowers. Shattering losses are reduced by cutting and raking when the pods are damp with dew. The beans are threshed with an adjusted pick-up combine after curing in the windrow for 5 to 15 days. Certain special bean combines are equipped with two or three rubber-faced cylinders. Frequently, beans are combined directly (Figure 28–4). The beans should be threshed at low cylinder speeds (700 to 1,000 feet per minute) and with ample concave clearance to avoid cracking and splitting. This is especially important in threshing beans for seed.[30] Rarely, windrowed beans are still bunched with hand forks, and later hauled to a special bean thresher. Choice beans for the retail trade may be handpicked after cleaning. Discolored beans can be removed with an electric-eye sorter.

When threshed beans are being dried with heated air, the temperature should not exceed 100° F. and not more than 3 per cent moisture should be removed in one drying stage.

DISEASES

The most serious losses from bean diseases occur in the humid and irrigated regions. More than 40 parasitic diseases attack the crop.[35]

Bacterial Blights

Blights are among the most serious diseases in the important bean areas. The most striking symptoms are spots that may enlarge rapidly

Figure 28–4. Edible bean combine in operation [Courtesy J. I. Case Co.]

and produce dead areas on the leaf. On the stem a characteristic lesion known as stem girdle appears. This lesion may weaken the stem so that it breaks at the diseased node. Pod lesions, usually reddish, water-soaked, and very irregular in outline, may prevent the proper filling of the pods and cause shriveling of the beans. The causal organisms of common bacterial blight and halo blight are *Xanthomonas phaseoli* and *Pseudomonas phaseolicola,* respectively. The principal control measure is to plant blight-free seed, which can be obtained from certain sections of the intermountain region of the west and from the Pacific Coast states. Some resistant varieties are available. Bacterial wilt and fuscous blight also damage the bean crop.[36]

Anthracnose

Bean anthracnose (*Colletotrichum lindemuthianum*) is a serious disease in the humid East. The disease is usually first noted on the leaves where dark-colored areas appear. The veins may be destroyed by the fungus, while the blade shows numerous cracks or holes with shriveled blackened margins. Large round dark-colored lesions finally appear on the pods. Large oval cankers may also be observed on the stem and may so weaken it that it is easily broken in cultivating or by a strong wind. The disease is most serious in wet seasons. It overwinters in the seed. The most satisfactory control measures are (1) use of disease-free seed grown in the semiarid regions, (2) resistant varieties,[31] (3) crop rotation, (4) keep workers out of the field when the plants are wet, and (5) spraying with suitable fungicides.

Common Bean Mosaic

Common bean mosaic is a seed-borne virus disease that is also spread in the field by several species of aphids. The mottling of the leaves may form various patterns of dark-green and light-green areas.[36] The dark-green areas often occur along the midvein. The leaves of infected plants may be curled downward. Diseased plants are usually paler green and more dwarfed than healthy ones. Early infections may cause a complete failure to form pods. The use of resistant varieties is the only practical control. The yellow mosaic disease spreads to beans from red clover, crimson clover, and sweetclover.

Other Diseases

Bean rust (*Uromyces phaseoli typica*) appears as small brown pustules on the leaves and, to a lesser extent on the stems and pods. The pustules turn black for the winter stage. Control measures include (1) dusting the plants with sulfur, maneb, or zineb, and (2) crop rotation.

Curly top is a virus disease, also of sugarbeets, that causes losses to beans in the area west of the continental divide. It is carried by the beet leafhopper (or white fly) (*Circulifer tenellus*) from plants infected with the virus to other healthy plants. Bean plants are more subject to injury while in the younger stages of growth. The growing point may be killed by the virus and then drop off. The plant will then die. On larger plants the first symptom of curly top is the downward curling of the first trifoliolate leaf. The curly-top virus is not carried in the seed. Several improved varieties are resistant to the disease.

INSECT PESTS

Beans are often damaged seriously in storage and in the field by the bean weevil (*Acanthoscelides obtectus*). It can be controlled by

Figure 28–5. Mexican bean beetle feeding on the underside of a bean leaf. Note the spotted adults, hairy larvae, and the egg mass.

planting weevil-free seed, by fumigation of infected seeds as soon as possible after harvest, and by field sanitation. Fumigation in an air-tight container is done with one part carbon tetrachloride to three parts of ethylene dichloride at the rate of 1 gallon per 100 bushels.

The Mexican bean beetle (*Ephilachna varivestris*) (Figure 28–5) is a serious pest in many bean-growing areas. The larvae feed on the under sides of the bean leaves. They remove the epidermis but seldom cut through to the upper surface. This insect can be controlled by spraying or dusting with insecticides.

Other insects that attack beans include the potato leafhopper, seedcorn maggot, Pacific Coast wireworm, white-fringed beetle, lygus bugs, and cutworms.[3, 18]

REFERENCES

1. Agricultural Statistics, USDA. Government Printing Office, Washington, D.C., pp. 162–184. 1972.

1a. Allard, H. A., and Zaumeyer, W. J. "Responses of beans (*Phaseolus*) and other legumes to length of day," *USDA Tech. Bul. 867*, pp. 1–24. 1944.

2. Allard, R. W., and Smith, F. L. "Dry edible bean production in California," *Calif. Agr. Exp. Sta. Cir. 436*. 1954.

3. Anderson, A. L. "Dry bean production in the Eastern States," *USDA Farmers Bul. 2083*. 1955.

4. Brandon, J. F., and others. "Field bean production without irrigation in Colorado," *Colo. Agr. Exp. Sta. Bul. 482*, pp. 1–22. 1943.

5. Coyne, D. E., and others. "Growing dry edible beans in Nebraska," *Neb. Agr. Exp. Sta. Bul. SB527*, pp. 1–39. 1973.

6. Dawson, J. H., and Bruns, V. F. "Chemical control of annual weeds in field beans," *Wash. Agr. Exp. Sta. Bul. 655*. 1964.

7. Entermann, F. M., and others. "Growing lentils in Washington," *Wash. Ext. Bul. 590*, pp. 1–6. 1968.

8. Finnell, H. H. "The tepary bean for hay production," *Panhandle Agr. Exp. Sta. Bul. 46*, pp. 1–12. 1933.

9. Freeman, G. F. "Southwestern beans and teparies," *Ariz. Agr. Exp. Sta. Bul. 68*. 1912.

10. Gentry, H. S. "Origin of the common bean (*Phaseolus vulgaris*)," *Econ. Bot.*, 23:55–69. 1969.

11. Gloyer, W. O. "Percentage of hardshell in pea and bean varieties," *N.Y. Agr. Exp. Sta. (Geneva) Tech. Bul. 195*. 1932.

12. Greig, J. K., and Gwin, R. E. "Dry bean production in Kansas," *Kans. Agr. Exp. Sta. Bul. 486*, pp. 1–19. 1966.

13. Hardenburg, E. V. "Experiments with field beans," *Cornell U. Agr. Exp. Sta. Bul. 776*. 1942.

14. Harlan, J. R. *Science*, 174:468–474. 1971.

15. Hedrick, V. P., and others. *The Vegetables of New York*, Vol. 1, Part 1, *Beans of New York*, N.Y. Agr. Exp. Sta., Albany, pp. 1–110. 1931.

16. Howe, O. W., and Rhoades, H. F. "Irrigation of Great Northern field beans in western Nebraska," *Nebr. Agr. Exp. Sta. Bul. 459*. 1961.
17. Lambeth, V. N. "Some factors influencing pod set and yield of the lima bean," *Mo. Agr. Exp. Sta. Res. Bul. 466*. 1950.
18. LeBaron, M., and others. "Bean production in Idaho," *Idaho Agr. Exp. Sta. Bul. 282*. 1958.
19. LeBaron, M., Mannering, J. V., and Baker, G. O. "Bean fertilization in southern Idaho," *Idaho Agr. Exp. Sta. Bul. 299*. 1959.
20. Ligon, L. L. "Mungbeans—a legume for seed and forage production," *Okla. Agr. Exp. Sta. Bul. 284*, pp. 1–12. 1945.
21. Mackie, W. W., and Smith, F. L. "Evidence of field hybridization in beans," *J. Am. Soc. Agron.*, 27:903–909. 1935.
22. Mimms, O. L., and Zaumeyer, W. J. "Growing beans in the Western States," *USDA Farmers Bul. 1996*, pp. 1–42. 1947.
23. Morrison, K. J., and Burke, D. W. "Growing field beans in the Columbia Basin," *Wash. Agr. Ext. Bul. 497*. 1962.
24. Myers, V. I., and others. "Irrigation of field beans in Idaho," *Idaho Agr. Exp. Sta. Res. Bul. 37*. 1957.
25. Paur, S. "Growing Pinto beans in New Mexico," *N. Mex. Agr. Exp. Sta. Bul. 378*. 1953.
26. Rather, H. C., and Pettigrove, H. R. "Culture of field beans in Michigan," *Mich. Agr. Exp. Sta. Spec. Bul. 329*, pp. 1–38. 1944.
27. Robins, J. S., and Howe, O. W. "Irrigating dry beans in the West," *USDA Leaflet 499*. 1961.
28. Schrumpf, W. E., and Pullen, W. E. "Growing dry beans in central Maine, 1956," *Maine Agr. Exp. Sta. Bul. 577*. 1958.
29. Steinmetz, F. H., and Arny, A. C. "A classification of the varieties of field beans, *Phaseolus vulgaris*," *J. Agr. Res.*, 45:1–50. 1932.
30. Toole, E. H., and others. "Injury to seed beans during threshing and processing," *USDA Cir. 874*. 1951.
31. Wade, B. L. "Breeding and improvement of peas and beans," in *USDA Yearbook, 1937*, pp. 251–282.
32. Wilson, V. E., and others. "Tekoa lentil and its culture," *Wash. Ext. Circ. 375*, pp. 1–5. 1971.
33. Wilson, V. E., and Law, A. G. "Natural crossing in *Lens esculenta* Moench," *J. Am. Soc. Hort. Sci.*, 97(1):142–143. 1972.
34. Youngman, V. E. "Lentils, a pulse of the Palouse," *Econ. Bot.*, 22(2):135–139. 1968.
35. Zaumeyer, W. J., and Thomas, H. R. "A monographic study of bean diseases and methods of control," *USDA Tech. Bul. 868* (rev.). 1957.
36. Zaumeyer, W. J., and Thomas, H. R. "Bean diseases—how to control them," *USDA Handbook 225*. 1962.

Chapter 29

Peanuts

ECONOMIC IMPORTANCE

The peanut (*Arachis hypogaea*), also called goober, pindar, groundnut, or earthnut, is grown on about 19 million hectares annually, chiefly in India, China, Nigeria, the United States, Senegal, and Brazil (Figure 29–1). The production is 18 million metric tons or 0.95 tons per hectare. About 3,000 million pounds (1.4 million metric tons) were produced annually in the United States from 1970 to 1972 on 1.47 million acres (600,000 hectares) or 2100 pounds per acre (2.36 metric tons per hectare). Also some 230,000 tons of dry vines or straw were salvaged from over 300,000 acres of threshed peanuts to provide peanut "hay." The leading states in peanut production were Georgia, Texas, North Carolina, Alabama, Virginia, and Oklahoma.

The peanut plant is a native of South America where closely related wild species are found. It may have originated in the mountainous parts of northern Argentina,[25] or in the state of Matto Grosso, Brazil,[25, 26] or in Peru where it was being cultivated by 1000 B.C.[39] Peanuts were grown in Mexico by the beginning of the Christian Era. Early slave ships carried peanuts to Africa and years later introduced them to the colonial United States from Africa. Commercial development of a peanut industry began about 1876. A rapid increase in production occurred after 1900 when the boll weevil made serious inroads on the cotton crop.

ADAPTATION

The most favorable climatic conditions for peanuts are moderate rainfall during the growing season, an abundance of sunshine, and relatively high temperatures.[5] The plants need ample soil moisture from the beginning of blooming up to two weeks before harvest. The best crops are obtained where the annual rainfall is between 42 and 54 inches, but peanuts for home use are grown where the precipitation is less than 19 inches. They are grown under irrigation in eastern New Mexico (Figure 29–2). The peanut region has an average frost-

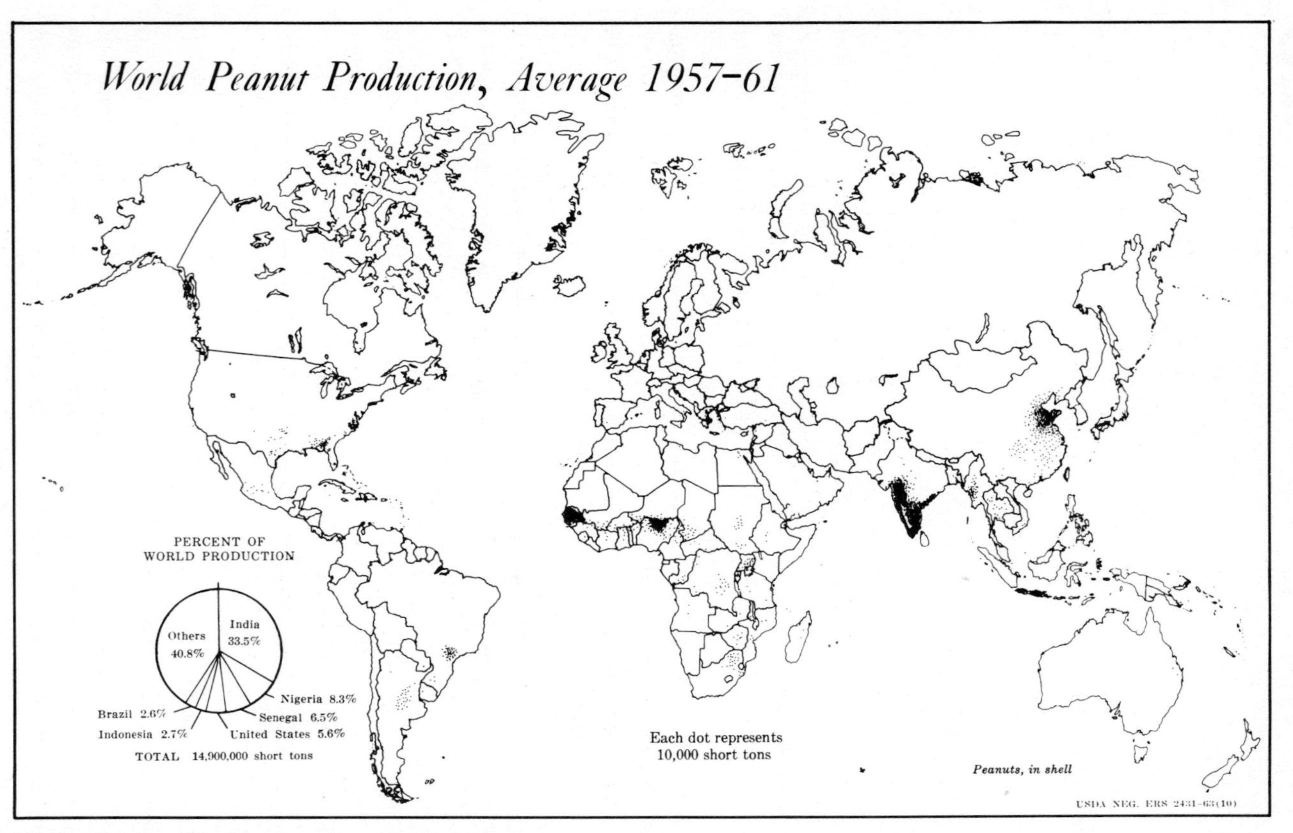

Figure 29–1. World peanut production.

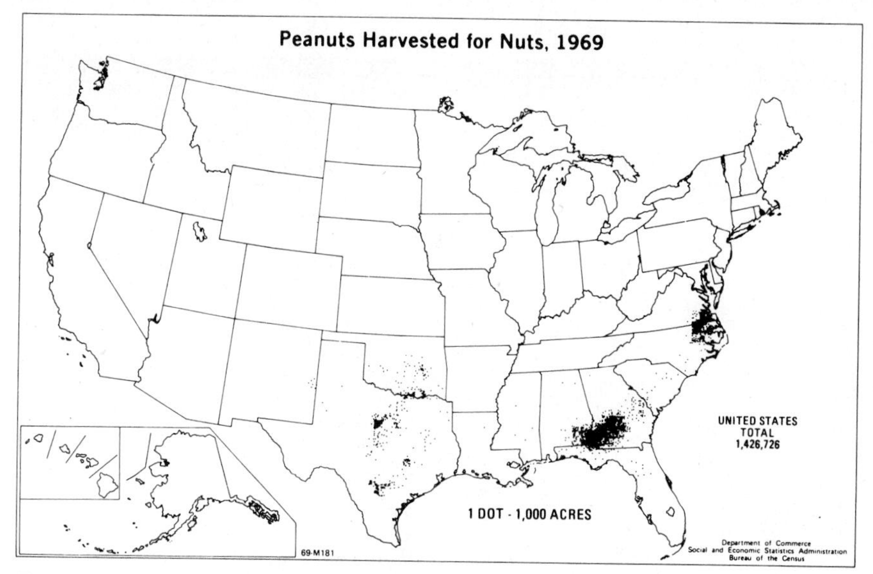

Figure 29–2. Peanut acreage in the United States in 1969.

free period of 200 days or more. The growing periods for different varieties range from 110 to 150 days.

The largest yields of the best quality of market peanuts are secured on well-drained, light, sandy loam soils. Dark-colored soils stain the hulls, which is objectionable for many commercial uses. For forage the crop can be grown on almost any type of soil except heavy clay soils low in organic matter. Peanuts produce the highest yields on soils with a *p*H of 5.8–6.2.[30] Poorly drained soils generally have been unsatisfactory. Light, sandy soils offer less resistance to (1) the penetration of the pegs that must enter the ground in order for the pods to develop, and (2) the recovery of the nuts from the soil when digging. The soil should be free from small stones.

The three major regions of peanut production are (1) the southeastern states, (2) southeastern Virginia–northeastern North Carolina, and (3) Texas and Oklahoma.

BOTANICAL DESCRIPTION

The cultivated peanut is a legume classified botanically as *Arachis hypogaea*. It is a pea rather than a nut. It is unknown in the wild state, but about 15 species that bear little resemblance to the cultivated form are found in South America.

The peanut has been considered as a short-day plant but it is mostly photo-insensitive because floral initials may be present in ungerminated seeds.[46]

Figure 29–3. A prolific peanut plant of the bunch type, showing leaves above and nodule bearing roots below.

The peanut plant is a low annual with a central upright stem. The numerous branches vary from prostrate to nearly erect. The pinnately compound leaf consists of two pairs of leaflets, but occasionally a fifth leaflet is borne on a slender petiole. Peanut varieties are readily separable into bunch and runner types. The nuts are closely clustered about the base of the plant of the erect or bunch type (Figure 29–3). The runner varieties have nuts scattered along their prostrate branches from base to tip. The peanut has a well-developed taproot with numerous lateral roots that extend several inches into the ground. Most roots have nodules but bear very few root hairs (Figure 29–3).

The flowers are borne in the leaf axils, above or below ground, singly or in clusters of about three. Under field conditions it is not uncommon to find the blossoms with their yellow petals 3 inches below the soil surface. The calyx consists of a long slender tube crowned with five calyx tips (Figure 29–4). The corolla is borne at the end of the calyx tube. After pollination takes place, the section immediately behind the ovary (the "gynophore" or peg), elongates and pushes the ovary into the soil, where the pod develops (Figure 29–5). Peanut flowers are perfect and self-pollination is the general rule, but natural crossing between varieties sometimes occurs.[1, 36]

The peanut fruit is an indehiscent pod containing one to six (usually one to three) seeds. The pods form only underground (Figure 29–5). The seed is a straight embryo covered with a thin papery seed coat. The outer layer varies in color, but generally it is brick red, russet, or light tan, and occasionally black, purple, flesh-colored, or white.

Figure 29–4. Portion of peanut plant showing blossom (*b*), calyx tips (*c*), gynophores or "pegs" (*g*), leaflets (*l*), and developing pod (*p*).

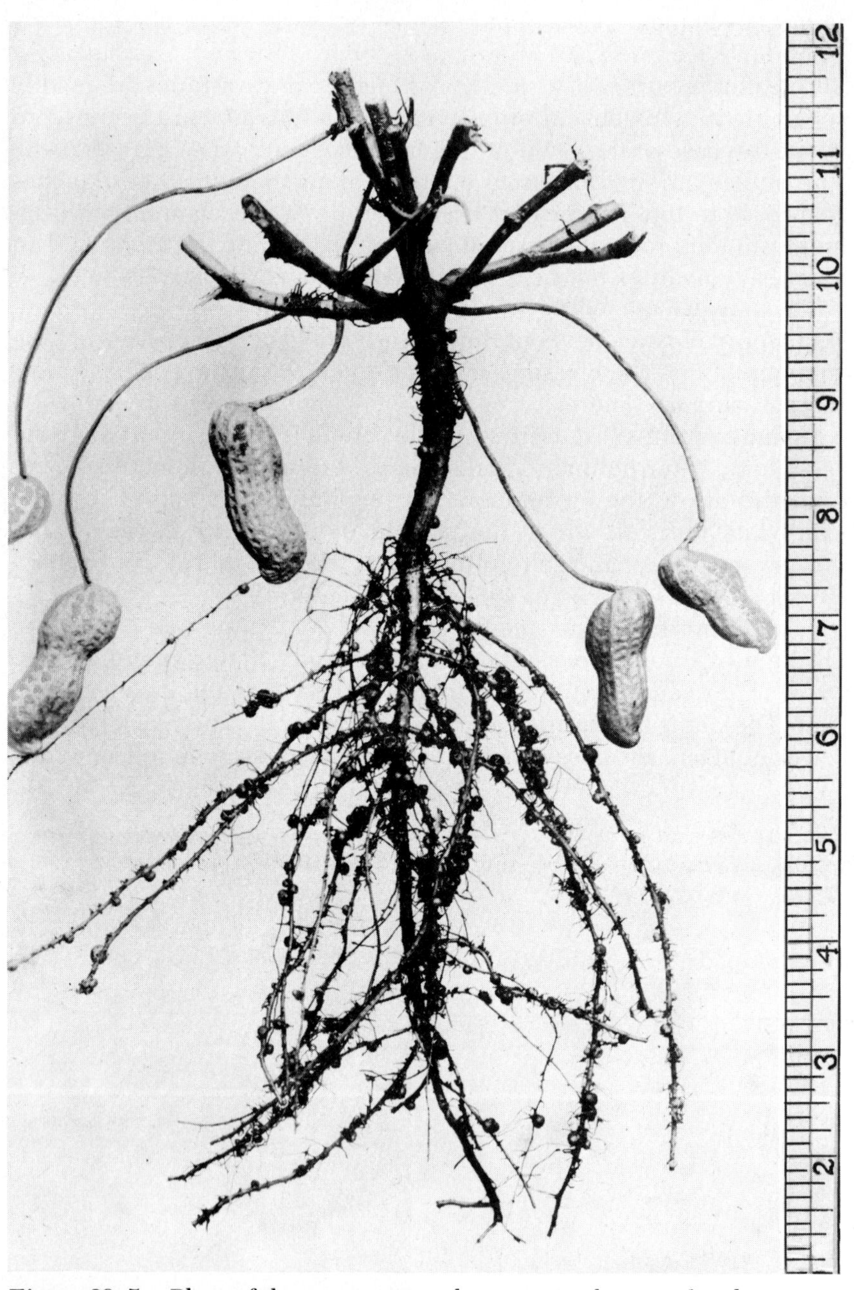

Figure 29–5. Plant of the runner type showing attachment of pods.

The weight per bushel of unshelled peanuts is about 22 pounds for the Virginia type, 28 pounds for the Southeastern Runner types, and 30 pounds for the Spanish type.

Seed dormancy is characteristic of the runner types.[40] Dormant peanuts planted soon after maturity frequently require rest periods ranging up to two years before germination occurs.[22] The average time for emergence of fresh seed of different strains from the Spanish and Valencia groups ranged from 9 to 50 days, while in a more dormant group that included runner peanuts it ranged from 110 to 210 days. The rest period is broken after several weeks or months in dry storage. The Spanish type shows no dormancy.

Peanut seed retains its viability for three to six years under proper storage conditions.

TYPES

American peanuts are classified into four types, namely Virginia (bunch and runner), Runner, Spanish (bunch), and Valencia (bunch). The Spanish type leads in production.

The Virginia varieties have dark-green foliage, large pods, large seeds (approximately 500 per pound), and russet seed coats. Occasional pods have three or four seeds. Virginia varieties are grown in Virginia, in North Carolina, and to some extent in other southeastern states. In Georgia, the Virginia varieties mature in 135 to 140 days after planting.

The Runner varieties[19] mature over a range of 130 to 150 days. A Virginia Bunch variety, which has 700 seeds per pound, is classed commercially as Runner. The Runner peanuts are grown mostly in Georgia, Alabama, and Florida.

The Spanish group consists of erect types with light-green foliage, rarely more than two seeds in a pod, short seeds, and tan seed coats. Both pods and seeds are small. Spanish peanuts are grown in Texas, Oklahoma, and to a large extent in Georgia and Alabama. The varieties have 1,000 to 1,400 seeds per pound. On the Texas coastal prairie, the Spanish types yielded more nuts but less hay than did the Runner types,[40] and more nuts and hay than the Virginia Bunch and Valencia types. The nuts comprised from 19 to 31 per cent of the total crop among the different varieties.

The Valencia type is characterized by many pods with three or four seeds, as well as by very sparse branching habits. The varieties are erect. The foliage is dark green, the seeds long or short, while the seed coats may be purple, red, russet, or tan. Valencia peanuts are now relatively unimportant, except in New Mexico where the acreage is limited.

Well-developed peanuts have a shelling percentage of 70 to 80 per cent, but the seeds usually comprise 60 to 70 per cent of the unshelled product, or average about 67 per cent.

Another groundnut called bambarra (*Boanzeia subterranea* Thouars) is a native African legume grown for food in tropical African countries. The pods and seeds are formed underground as in peanuts.

CROP ROTATIONS

Peanuts harvested for nuts usually deplete the soil and depress the yields of crops that follow them in rotation because (1) fertilizers are applied in limited quantities, (2) no crop residues are returned to the soil when the vines are gathered for hay, (3) clean row cultivation hastens the decay of organic matter, and (4) the soil is left bare over winter. Fertilization of the land for peanuts and the planting of a legume winter cover crop help the yields of the following crop. Peanuts should occupy the land only once in three or four years, while the rotation should include at least two soil-improvement crops, one of them being also a winter cover crop.[5] A Virginia rotation is (1) corn with crimson clover as a winter cover crop, (2) early potatoes followed by cowpeas, and (3) Spanish peanuts followed by rye as a winter cover crop. Peanuts yielded approximately three times the quantity of nuts in this rotation as in continuous culture during the last three years of an experiment that extended over a ten-year period.

In the southern Great Plains peanuts should be planted in strips alternating with strips of crops such as sorghum or corn to check soil blowing, because the removal of the peanut crop leaves the soil loose and bare.

In Georgia the growing of peanuts in rotations reduces attacks by crown rot, Southern blight (white mold), nematodes, and leaf spot. Best yields are obtained following grass pasture. Other alternatives are rye or finally other grains and grasses. Peanuts should not follow soybeans or lupines because of disease hazards. Southern blight is often serious if peanuts follow cotton or if any crop litter is left uncovered in the field.[30]

FERTILIZERS

The peanut plant has a marked ability to use fertilizer residues not utilized by a previous crop, and it sometimes does not respond to additional fertilizer on very fertile soils. The pegs as well as the roots can absorb mineral elements from the soil.[7] About 80 per cent of the peanut acreage was fertilized by 1964. Peanuts usually are grown on leached sandy soils of relatively low fertility. They require abundant potassium[8] and considerable phosphorus but only 12 to 20 pounds per acre of nitrogen except on poor soils. Recommendations in Georgia are 300 to 500 pounds per acre of 4–12–12 or 5–10–15 mixtures preferably applied broadcast before plowing.[29] Also 500 to 1000 pounds per acre of gypsum for Virginia-type peanuts, and 500 pounds for other types are advisable in Georgia chiefly to supply calcium but also magnesium. The exchangeable calcium in the soil should be at least 700 pounds per acre of calcium carbonate equivalent. A calcium deficiency retards the movement of the element through the peg causing a dark plumule[18] or "poppy" peanuts (unfilled shells and reduced yields),[7, 38] especially in dry weather. The calcium can be supplied by dusting the tops of the plants when blooming begins.[3, 14] Molybdenum applications also are sometimes beneficial.

In Florida an application of 3.5 pounds per acre of borax prevents a brownish "hollow heart" in the nuts, as well as stubby shoots, mottled, wilted or abscissed leaves, darkened internodes and cracked stems.[18] Copper treatments also are advisable there to prevent shriveled kernels and abnormal foliage.

CULTURAL METHODS

Seed Peanuts for Planting

Either shelled or unshelled peanuts may be planted, but shelled seed gives quicker and often better stands. Machine-shelled seed often germinates less than hand-shelled seed because of seed-coat injury incident to shelling,[16] but seed treatment may overcome much of this difference. Peanut seed shelled as early as December gives practically the same germination and yield as that shelled just prior to planting.[4] Shelled seed should be stored in a dry place (relative humidity about 60 per cent) that is moderately cool (between 32° and 70° F.). In Mississippi experiments Spanish peanuts germinated better when planted in the shell, while runner peanuts produced better stands when shelled, because the shell is tough and almost waterproof.[43] Breaking the pod into two pieces gives practically the same results as shelling.[31]

Peanuts are graded to screen out the small, immature poorly developed or wrinkled seeds that produce small, weak seedlings when planted.[33] More medium-sized seeds per acre should be planted than are optimum for large seeds.

Treatment of shelled seed with dust fungicides increases the assurance of satisfactory stands.[35] A light coating of a pine tar-kerosine mixture repels crows and squirrels from the planted peanuts.

Seed Inoculation

Inoculation of peanuts with the proper culture of *Rhizobium* bacteria is advisable unless inoculated peanuts, cowpeas, or velvetbeans have been grown on the soil previously. Spanish peanuts may bear a comparatively small number of nodules on certain soils in Alabama.[11]

Artificial inoculation of unshelled Spanish peanut seed has resulted in large increases in numbers of both total and large nodules, as well as a yield of 30 to 40 per cent more nuts per plant, when compared with noninoculated seed.[10] There appeared to be no benefit from inoculation of runner peanuts in the same experiments.

Seedbed Preparation

Thorough seedbed preparation for peanuts is essential. A well-tilled soil aids the penetration of the pegs. Fall or winter plowing is generally practiced except where erosion is serious. Plowing 8 to 9 inches deep, while completely covering the plant residues, greatly reduces losses from the stem- and peg-root disease caused by *Sclerotium rolfsii*.[32] Growers often bed up the peanut fields where there is a tendency for the soil to be wet, but low beds are preferable.

Seeding

Peanuts are planted generally after the danger of late spring frosts is past. In the commercial peanut area the main crop is planted between April 10 and May 10. Planting about March 15 is recommended for southern Georgia. Spanish peanuts will mature seed in the Gulf Coast region when planted as late as July 1, but higher yields result from earlier plantings. In Texas, peanuts are planted about the same time as cotton or a little later. The actual dates of planting range from about March 1 in southern Texas to May 15 or later in the northern part of the state. Peanuts should not be planted in dry soil.

In the commercial areas, shelled peanut planting is done with multiple-row tractor planters, which are either special peanut planters or corn or cotton planters equipped with peanut plates. Slow-moving slanted planter plates with seed cells of adequate size reduce seed injury during planting. Seeding rates for maximum yields range from 35 pounds per acre of machine-shelled small-seeded Spanish types in 36-inch rows, up to 96 pounds per acre of large-seeded types in 24-inch rows. Under favorable growing conditions, better yields are obtained from thick spacing, i.e., 3 to 8 inches apart in 30-inch rows for the larger types, and 3 inches apart in 24-inch rows for the small types. The average seeding rate in the United States is about 67 pounds per acre (unshelled basis) or about 45 pounds of shelled kernels.

Weed Control

The cultivation of peanuts is similar to that of soybeans grown in rows. One or two early cultivations with a rotary hoe or a flexible shank weeder are helpful. Later shallow cultivation is done with row equipment using flat blades that cut few peanut roots or pegs. Level cultivation is essential. Bedding up or moving soil up to the peanut plants during cultivation causes stem and peg rots to develop.[32]

Annual weeds in peanuts may be controlled by preemergence herbicides. They may be applied during planting or any time before crop emergence. These herbicides may cause some temporary injury to peanuts on light sandy soils when later heavy rains leach them into the zone of seed germination.

Harvesting

The proper time for harvesting peanuts is very exacting. The tops are kept green and growing late by the prevalent practice of dusting and spraying, so that the appearance of the tops is not a dependable guide as to when to harvest. Peanuts are ready for harvest when the kernels are fully grown, the skins displaying a distinct texture and the natural color of the variety, and the inside of the shell beginning to color and show darkened veins. Peanuts shrink when harvested too early, and Spanish varieties will sprout in the ground when harvesting is delayed too long. Peanuts are generally harvested before the vines are killed by frost so the vines are often clipped and cut with coulters before digging.

Peanuts in commercial fields are removed from the soil with special diggers equipped with blades or half-sweeps that cut off the roots below the nut zone and lift the plants from the ground. The vines are then windrowed with a shaker-windrower or with a side-delivery rake.

Shaking the peanuts free from dirt is accomplished most readily in light sandy soils free from clods. Small-podded varieties are separated from the soil with less loss than are the large-podded Virginia types. Most growers now use a digger-shaker (Figure 29–6) and a windrower or a combination digger-shaker-windrower. They often reshake the windrow, without digging blades attached to the shaker, in order to remove more of the dirt from the plant roots. Many of the peanuts are threshed with the combine and binned when the kernels are semidry, i.e., with about a 15 to 20 per cent moisture content. At that stage the kernels rattle slightly in the shell. They are immediately artificially dried,[42] down to a moisture content of about 8.5 per cent, using forced air at a temperature below 100° F.[30] Peanuts with 10.49 per cent or more moisture are not salable.[21]

Hand-harvesting methods involve the shaking of the lifted vines with a fork, and later placing the vines in small stacks built around poles set in the field. The bottoms of these stacks around the poles rest on cross bars 12 to 15 inches above the ground. After three or four weeks of curing, the vines may be threshed. Hand methods require eight to ten times as much labor as do the mechanized operations.

Figure 29–6. Peanut harvester.

Rapid curing by direct sunlight reduces seed viability, increases seed breakage as well as skinning during shelling, and impairs the edible quality of the peanuts. Large loose windrows limit the number of pods that are exposed to sunlight, yet permit the pods at the bottom and interior of the windrow to cure without molding.

Modern machines (Figure 29–6) remove the pods from the vines satisfactorily. Market peanuts are generally bagged as they come from the machine. The threshed vines are often baled for hay.

PROCESSING PEANUTS

The operations incident to shelling remove rocks, sticks, and stems as well as the shells. The nuts are passed through various screens and then through a machine sheller. Blanching gives a whiter and more homogeneous appearance to the nuts. This treatment involves removal of the seed coat by a combination of drying, heating, rubbing between soft surfaces, and blowing an air current through the nuts. Shelled nuts may become rancid after storage for two months or more when exposed to the air. Rancidity is retarded when the nuts contain 5 per cent moisture or less, and the relative humidity of the storage room is about 60 per cent. Mechanical sorters equipped with photoelectric cells can remove discolored kernels. More than one-half of the shelled edible peanuts consumed in the United States are used to make peanut butter by grinding dry roasted salted nuts.[44] The remainder are mostly dry roasted until the nuts develop a brown color or are used in making peanut candy. Sometimes the nuts are roasted in oil, especially when they are to be salted.

For oil extraction the peanuts are shelled, cleaned, hulled, and crushed to open the oil cells as much as possible. From the rollers the pulp goes to a cooker where the material is heated at about 235° F. in a humid atmosphere for 90 minutes. The oil is then extracted by the hydraulic press-plate method or by the expeller (cold-press) method. The crude peanut oil is collected in storage tanks, while the residual cake is ground into meal. The crude oil may be filtered to remove the foots (particles of meal), or the residue may be allowed to settle to the bottom of the tank and then removed.

An average ton of cleaned unshelled peanuts yields about 530 pounds of oil, 820 pounds of meal, and 650 pounds of shells.

UTILIZATION

Peanuts that are not harvested are used for livestock feed (hogged off). The equivalent of more than 100 million pounds of unshelled peanuts, or 6 per cent of the threshed crop, is used for planting. An average of more than 450 million pounds of shelled peanuts are crushed, producing more than 180 million pounds of oil. The annual consumption of shelled peanuts in the United States is about 5 pounds per capita. About 2½ pounds per capita are consumed as peanut butter, and more than a pound each as salted peanuts and peanut candy.[28] The world production of peanut oil is about 2.7 million metric tons. It is an important edible vegetable oil in Europe, Asia, and Africa where peanut butter is almost unknown.

Green, leafy peanut hay is as nutritious as good-quality alfalfa hay.[37] Nutrient content of field-cured hay may be reduced 25 per cent or more through loss of leaves, weathering, excessive stems, mustiness, and dirt content.

CHEMICAL COMPOSITION

Peanuts are high in both protein and oil.[31] In general, the vines with the nuts contain 10 to 12 per cent protein, or, without the nuts, about 7 per cent protein. Unshelled peanuts contain about 25 per cent protein and 33 per cent fat. The nuts contain 40 to 48 per cent oil, and from 25 to 30 per cent protein. All of the embryo cells contain oil.[41] The nuts contain the amino acids, including cystine, that are essential for animal growth.[24] They are rich in phosphorus. Like other large legume seeds, peanuts are an excellent source of the vitamins thiamin and riboflavin as well as a good source of niacin. An average thiamin content of 9.6 micrograms per gram has been reported.[20] In thiamin content the skins are highest, followed by the cotyledons, and then the germ. The average niacin content is about 17.2 milligrams per 100 grams.

DISEASES

Peanuts suffer some losses from Cercospora leaf spots.[1] One of these is caused by *Cercospora arachidicola* (*Mycosphaerella arachidicola*) and the other by *C. personata* (*M. berkeleyii*).[5, 23]

The leaf-spot diseases appear as small brown spots on the leaves. They usually become severe in wet weather when the leaves may turn yellow and fall off. Leaf spot often reduces the value of peanut hay and sometimes makes it unfit for feeding. Leaf spot may be controlled with 325-mesh sulfur used as a dust at the rate of 15 to 20 pounds or more per acre for each of three or four applications about 14 days apart. Copper dusts also give satisfactory control.

Stem and peg rots, caused chiefly by the fungus *Sclerotium rolfsii*, are reduced by the cultural practices previously mentioned.[15, 30, 32] Seed rots and seedling blight also are checked by seed treatments. Concealed damage is a mold growth between the cotyledons of peanuts curing in stacks. It seldom develops in windrows. Some varieties are resistant to the disorder. Charcoal rot and *Rhizoctonia* attack peanuts.

INSECTS

Insects that damage peanuts include leafhoppers, velvetbean caterpillars, corn earworms, cutworms, tobacco thrips, and white-fringed beetles. These as well as the fall armyworm and southern corn rootworm are controlled with insecticides.

REFERENCES

1. Anonymous. *The Peanut—The Unpredictable Legume,* National Fertilizer Association, Washington, D.C., pp. 1–333. 1951.
2. Arthur, J. C. Process for the manufacture of Peanut Protein. U.S. Pat 2, 529, 477. 1950.
3. Bailey, W. K. "Virginia type peanuts in Georgia," *Ga. Agr. Exp. Sta. Bul. 267.* 1951.
4. Beattie, J. H., and others. "Effect of planting distances and time of shelling seed on peanut yields," *USDA Bul. 1478.* 1927.
5. Beattie, J. H., and others. "Growing peanuts," *USDA Farmers Bul. 2063.* 1954.

6. Beattie, J. H., and others. "Effect of cold storage and age of seed on germination and yield of peanuts," *USDA Cir. 233*. 1932.
7. Brady, N. C. "The effect of calcium supply and mobility of calcium in the plant on peanut fruit filling," *Soil Sci. Soc. Am. Proc.*, 12:336–341. 1948.
8. Clay, H. J. "Marketing peanuts and peanut products," *USDA Misc. Pub. 416*. 1941.
9. Collins, E. R., and Morris, H. D. "Soil fertility studies with peanuts," *N.C. Agr. Exp. Sta. Bul. 330*. 1941.
10. Doku, E. V., and Karikari, S. K. "Bambarra groundnut," *Econ. Bot.*, 25(3):255–262. 1971.
11. Duggar, J. F. "The effects of inoculation and fertilization of Spanish peanuts on root nodule numbers," *J. Am. Soc. Agron.*, 27:128–133. 1935.
12. Duggar, J. F. "Nodulation of peanut plants as affected by variety, shelling of seed, and disinfection of seed," *J. Am. Soc. Agron.*, 27:286–288. 1935.
13. Duke, G. B. "Progress report on harvesting Virginia type peanuts," *USDA Agricultural Research Service Spec. Rept. ARS 42–11*. 1957.
14. Futral, J. G. "Peanut fertilizers and amendments for Georgia," *Ga. Agr. Exp. Sta. Bul. 275*. 1952.
15. Garren, K. H. "The stem rot of peanuts and its control," *Va. Agr. Exp. Sta. Tech. Bul. 144*. 1959.
16. Gore, U. R. "Culture and fertilizer studies with peanuts," *Ga. Agr. Exp. Sta. Bul. 209*. 1941.
17. Harris, H. C. "Effect of minor elements, particularly copper, on peanuts," *Fla. Agr. Exp. Sta. Bul. 494*. 1952.
18. Harris, H. C. "Calcium and boron effects on Florida peanuts," *Fla. Agr. Exp. Sta. Tech. Bul. 723*, pp. 1–17. 1968.
19. Higgins, B. B., and Bailey, W. K. "New varieties and selected strains of peanuts," *Ga. Agr. Exp. Sta. Bul. New Series 11*. 1955.
20. Higgins, B. B., and others. "I. Peanut breeding and characteristics of some new strains. II. Thiamin chloride and nicotinic acid content of peanuts and peanut products. III. Peptization of peanut proteins," *Ga. Agr. Exp. Sta. Bul. 213*. 1942.
21. Hubbard, E. "Peanut harvesting and marketing costs and practices," *Ga. Agr. Exp. Sta. Rsh. Rpt. 44*, pp. 1–22. 1969.
22. Hull, F. H. "Inheritance of rest period of seeds and certain other characters in the peanut," *Fla. Agr. Exp. Sta. Tech. Bul. 314*. 1937.
23. Jenkins, W. A. "Two fungi causing leaf spot of peanut," *J. Agr. Res.*, 56:317–332. 1938.
24. Jones, D. E., and Horn, M. J. "The properties of arachin and conarachin and the proportionate occurrence of these proteins in the peanut," *J. Agr. Res.*, 40:673–682. 1930.
25. Leppik, E. E. "Assumed gene centers of peanuts and soybeans," *Econ. Bot.*, 25(2):188–194. 1971.
26. Loden, H. D., and Hildebrand, E. M. "Peanuts—especially their diseases," *Econ. Bot.*, 4(4):354–379. 1950.
27. McClelland, C. K. "The peanut crop in Arkansas," *Ark. Agr. Exp. Sta. Bul. 263*. 1931.
28. McGill, J. F. "In a nutshell," *Ga. Agr. Ext. Cir. 517*. 1963.
29. McGill, J. F. "Peanut varieties," *Ga. Agr. Ext. Cir. 518*. 1963.

30. McGill, J. F., and Samples, L. E. "Peanuts in Georgia," *Ga. Ext. Bul. 640*, pp. 1–39. 1969.
31. McNess, G. T. "Peanuts in Texas," *Tex. Agr. Exp. Sta. Bul. 381*. 1928.
32. Mixon, A. C. "Effects of deep turning and non-dirting cultivation on bunch and runner peanuts," *Auburn U. Agr. Exp. Sta. Bul. 344*. 1963.
33. Mixon, A. C. "Effect of seed size and vigor and yield of runner peanuts," *Auburn U. Agr. Exp. Sta. Bul. 346*. 1963.
34. Parham, S. A. "Peanut production in the coastal plain of Georgia," *Ga. Coastal Plain Exp. Sta. Bul. 34*, pp. 1–19. 1942.
35. Perry, A., and others. "Peanut production guide," *N. C. Agr. Ext. Cir. 257*. 1963.
36. Reed, E. L. "Anatomy, embryology, and ecology of *Arachis hypogea*," *Bot. Gaz.*, 78:289–310. 1924.
37. Ronning, M., and others. "Feeding tests with threshed peanut hay for dairy cattle," *Okla. Agr. Exp. Sta. Bul. B-400*. 1953.
38. Scarsbrook, C. E., and Cope, J. T., Jr. "Fertility requirements of Runner peanuts in southeastern Alabama," *Ala. Agr. Exp. Sta. Bul. 302*. 1956.
39. Smith, C. E., Jr. "The new world centers of origin of cultivated plants and the archeological evidence," *Econ. Bot.*, 22(3):253–266. 1968.
40. Stansel, R. H. "Peanut growing in the Gulf Coast Prairie of Texas," *Tex. Agr. Exp. Sta. Bul. 503*. 1935.
41. Stokes, W. E., and Hull, F. H. "Peanut breeding," *J. Am. Soc. Agron.*, 22:1004–1019. 1930.
42. Teeter, N. C., and Givens, R. L. "Technical progress report on curing Virginia type peanuts, 1952–56," *USDA Agricultural Research Service Spec. Rept. ARS 42-12*. 1957.
43. West, H. O. "Peanut production," *Miss. Agr. Exp. Sta. Bul. 366*. 1942.
44. Woodroof, J. G., and Leahy, J. F. "Microscopical studies of peanuts with reference to processing," *Ga. Agr. Exp. Sta. Bul. 205*. 1940.
45. Woodroof, J. G. *Peanuts: Production, Processing, Products*, AVI Pub. Co., Westport, Conn., pp. 1–291. 1966.
46. Wynne, J. C., and others. "Photoperiodic responses of peanuts," *Crop Sci.*, 13(5):511–514. 1973.

Chapter 30

Miscellaneous Legumes

Legumes that are grown for seed, forage, green manure, or as cover crops include field peas, vetches, velvetbeans, burclover, black medic, buttonclover, kudzu, crotalaria, trefoil, sesbania, lupines, guar, Florida beggarweed, roughpea, hairy indigo, alyceclover, crown vetch, pigeon pea, fenugreek, sainfoin, and seradella. They may not occupy a large area in the country as a whole, but are important where they are well adapted. Except for field peas, crown vetches, sainfoin, and trefoil, these crops are grown almost entirely in the southern and Pacific Coast states.

FIELD PEA

Dry edible peas are grown over the world on about 9 million hectares annually. The production is about 10 million metric tons or 1.1 metric tons per hectare. The countries leading in pea production are the Soviet Union, China, India, and the United States. Production in the United States from 1970 to 1972 was 0.14 million metric tons (17 million cwt.) on 0.08 million hectares (198,000 acres) yielding 1.8 metric tons per hectare (1600 pounds per acre). The leading states in dry-pea production are Washington, Idaho, Oregon, Minnesota, and North Dakota. They also are grown in Southern Canada. An appreciable acreage of field peas is also devoted to hay, pasture, silage, and green manure.

Fully 60-per cent of the domestic seed pea production consists of Alaska and other smooth, green, seeded types, used as dry peas or as planting seed by growers of fresh garden, canning, or frozen peas. The smooth yellow types, also produced in Canada are largely consumed as dry split peas, while the white (Canada) types produced mostly in Minnesota and North Dakota serve as feed, or for planting for hay or silage.[61] About 742,000 cwt. of wrinkled green peas for seed are produced for the garden and preserving markets. A considerable portion of the dry pea production is exported, chiefly to Europe. Winter pea seed production in the Pacific Coast states for shipment to the Southeast as a winter cover crop has declined greatly since it was largely replaced by the blue lupine, because of destruction by diseases. The chief disease is black stem caused by the fungi *Ascochyta pinodella* and *Mycosphaerella pinodes.*

Adaptation

A cool growing season is necessary for successful production of field peas. High temperatures are more injurious to the crop than are light frosts, which are injurious only when the plants are in blossom. Climatic requirements limit field-pea production to the northern states as a summer crop, and to the southeastern states and the mild coastal sections of the Pacific Northwest as a winter crop.

Field peas are most productive where rainfall is fairly abundant, but they succeed in cool semiarid regions. Well-drained clay loam soils of limestone origin are best suited to field peas. Inoculation is essential or beneficial except in fields where nodulated peas had been grown in recent years in which case the inoculum content of the soil was adequate.[59]

Botanical Description

The field pea is classified botanically as *Pisum arvense.* Garden and canning peas usually are classified as *P. sativum.* However, several canning and garden varieties also are grown as dry field peas.[25] Garden peas tend to be sweeter and more wrinkled than field peas, but, since this does not hold for all varieties, there can be no sharp distinction between the two species. The pea is an annual herbaceous plant with slender succulent stems 2 to 4 feet long. The foliage is pale green with a whitish bloom on the surface. The leaf consists of one to three pairs of leaflets and terminal branched tendrils. In most varieties, the blossoms are either reddish-purple or white. The pods are about 3 inches long and contain four to nine seeds. These seeds may be round, angular, or wrinkled. The seed cotyledons are mostly yellow or green (Figure 22–4).

Field peas are generally self-fertilized, but some cross-pollination is evident when varieties with white and colored blossoms are grown in alternate rows.[38]

The green types have green seed coats and cotyledons. The yellow and white types have yellow or white seed coats but both have yellow or orange cotyledons. The name Canada or Canadian has been applied to white and yellow-seeded varieties. Varieties with grey, brown or mottled seeds, including the winter types, are grown for soil improvement or feed.[3, 50, 71]

Cultural Practices

Field peas can replace summer fallow to advantage in wheat rotations in the Palouse area of Washington,[63] and to some extent in the more semiarid sections of the Pacific Northwest. However, the harvesting of the two crops conflicts. Peas fit into hay, small-grain, and corn rotations in most northern states.

Fall plowing is generally practiced for seedbed preparation in northern latitudes to facilitate early spring planting. Where grown as a summer crop, field peas are commonly sown with a grain drill as early in the spring as possible, usually in April. In the South and in the coastal section of the Pacific Coast states, the usual practice is to seed winter peas from September 15 to October 15.

Phosphorus fertilization for pea fields is advisable where soil tests show phosphorus availability to be deficient.[2, 59] Sulfur and molybdenum applications sometimes also are beneficial.

Seeding rates with a drill range from 45 to 150 pounds per acre for small-seeded types and 80 to 180 pounds for large-seeded types. The lower rates apply to semiarid conditions while the heaviest rates are suitable for well-fertilized soil in irrigated or humid areas. The desired stand in Idaho and Washington is 6 to 9 plants per square foot. In the Cotton Belt winter peas are seeded at 30 to 50 pounds per acre.[34, 55, 59, 61]

Annual weeds, including wild oats, in pea fields can be controlled with suitable herbicides.

For hay or silage, field peas often are sown in mixtures with oats or barley. The grain stems help to support the pea vines, reduce lodging, and make a better-balanced feed. Field peas are harvested for hay when most of the pods are well formed. Since peas often lodge badly, they are usually cut with a mower equipped with special lifting guards and a windrow attachment. Field peas may be harvested for seed in much the same manner when the pods are mature. The pick-up combine often is used to thresh the crop from the windrow or bunch. Most of the seed crop is combined directly. Defective and weevil infested seeds are separated from sound seed by floating them off in brine.

Utilization

In the Cotton Belt, winter peas have been planted as a fall-sown cover and green-manure crop. Sufficient growth is generally made by March for the crop to be turned under. In northern states field peas may be used for hay, pasturage, and silage. For best results the peas should be allowed to mature before being grazed by hogs or sheep.

Whole dry peas occasionally are soaked and then cooked as a substitute for canned peas. Certain yellow and green varieties listed as dry edible peas are marketed largely as split peas for soups. Split peas are the separated cotyledons of the peas, with the seed coats and most of the embryos removed. The commercial process of splitting peas is a trade secret. Peas can be hulled and split in a burr mill, with the burrs set far enough apart to avoid serious cracking of the cotyledons. The hulls, which have been removed by aspiration, and the embryos and seed fragments, which have been screened out, are useful in mixed feeds. Field-pea seed can be ground, mixed with grains, and fed to livestock. Peas and their by-products are popular pigeon feeds.

Diseases

The Ascochyta blight of leaves and stems is caused by three different fungi: *Ascochyta pisi, Mycosphaerella pinodes,* and *Ascochyta pinodella.* It occurs in all states east of the Mississippi River, and it may cause heavy losses in seasons of abundant rainfall.[77] This disease is characterized in part by the formation of black to purplish streaks on the stem. The lesions are more conspicuous at the nodes. They enlarge into brown or purplish areas scattered from the roots to 10 or more inches up the stem. Spots on the leaves may be small to large, purplish, and irregular or circular. The entire leaf may shrivel and dry up. Similar spots on the pods are shrunken. The fungi that cause ascochyta blight are seedborne, but they also may live over from one season to the next on field refuse.[75] To control the disease, only clean seed should be planted, such as western-grown seed. Crop refuse should be plowed under deep as soon as the pea crop is harvested. A three-year to five-year rotation should be practiced.[77] Resistant varieties are now available.[2]

Bacterial blight, caused by *Pseudomonas pisi,* produces olive-green to olive-brown water-soaked areas on the stems, leaves, and pods. The infected plants may be killed. The bacteria that cause the disease live over winter in the seed. Control measures are rotation and the use of clean seed.

Fusarium wilt of peas is caused by a soil-borne fungus, *Fusarium oxysporum* f. *pisi.* Early plant symptoms are yellowing lower leaves, stunted plant growth, and downward curling of the margins of the leaves. Infected small plants may die without the production of peas, and infected older plants may develop poorly filled pods. The disease can be controlled only by resistant varieties which are now available.[77]

Other diseases that attack field peas and the fungi which cause them include Septoria blight (*Septoria pisi*), powdery mildew (*Erysiphe polygoni*), downy mildew (*Peronspora pisi*), anthracnose (*Colletotrichum pisi*), various root rots, and several virus diseases.

The seed usually is treated with a protectant to reduce damage from seed rots and seedling blights. Graphite may be added to overcome the increased friction caused by the fungicidal dusts, which facilitates the flow of seed through the drill. Insecticides often are applied along with the fungicides.

Insect Pests

The pea weevil (*Bruchus pisorum*) is one of the most serious insect enemies of the field pea. The adult weevil deposits its eggs on the pods while the peas are in the immature stage. The egg hatches and produces a larva that finally bores into the young seed, where it feeds on the embryo. The principal injury is the destruction of the seed, impairing it as food. In regions where the pea weevil is prevalent, the seed should be fumigated with cyanide gas or carbon disulfide as soon as it has been threshed. Crop rotation and the burning of pea residues also is helpful. Application of insecticides will control the weevil when applied to the foliage during the early-bloom period before the eggs are laid.

The pea aphid (*Illinois pisi*) attacks all kinds of peas. It sucks the sap from the leaves, stems, blossoms, and pods of the plants on which it feeds.[15] It can be checked in Austrian winter peas in western Oregon by delayed sowing until after October 16 to 20, and can be controlled with insecticides.

The alfalfa looper and celery looper are controlled with insecticides. Wireworms are partly controlled by broadcasting an insecticide on the field, which is then disked under before the peas are planted.

Other enemies of the pea include the pea moth (*Laspeyresia nigricana*), and the root-knot nematode.

VETCH

Vetch is harvested for seed on about 1.7 million hectares, chiefly in the Soviet Union and Turkey. In the United States vetch was grown on 760,000 hectares (less than 2 million acres) in 1969, primarily for green manure or a winter cover crop, but also for pasture, hay, or silage. Much of the production was in Oklahoma, Texas, Arkansas California, Louisiana, and Alabama. The seed was produced on 55,000 acres mostly in Texas, Oregon, Nebraska, and Oklahoma.

Adaptation

Vetch makes its best development under cool temperature conditions. The vetches vary in winter hardiness, but all are less hardy than alfalfa or red clover. Hairy vetch is the most winter hardy. Hungarian vetch and smooth vetch survive temperatures as low as 0° F. in regions where temperature fluctuations are slight or where snow cover exists. Common vetch will not withstand zero temperatures, and was killed out completely at Stillwater, Oklahoma.[37] Hungarian and woolypod vetches are almost as hardy as smooth vetch. Monantha, Bard, and purple vetches are nonhardy. They are grown as winter crops only in mild climates. Vetches are grown in the winter in the Southeast as well as in the South Central and Pacific Coast states. Hairy and smooth vetches are winter crops in colder northern states. Common vetch is a spring crop in the Northeast and in the Central States. Vetch is unadapted to semiarid conditions.

All vetches grow best on fertile loam soils. Hairy, smooth, and monantha vetches are productive on poor sandy soils, while Hungarian vetch is productive on heavy wet soils. As a group, the vetches are only moderately sensitive to soil acidity.

When grown for seed, some of the seeds shatter and volunteer. The volunteer vetch contaminates wheat grown on the land as long as the vetch continues to reseed itself. Wheat and vetch seed cannot be separated in the thresher or in a fanning mill, but only in special disk, cylinder, and spiral incline machines.

Botanical Description

Some 36 species of native or naturalized vetches are growing in North America.[23]

Most of the commonly grown vetches are annuals. Hairy vetch may be either an annual or a biennial. The common agricultural species are viny.[29] The stems may grow 2 to 5 feet or more in length. All the cultivated species have leaves with many leaflets. The leaves are terminated with tendrils in most of the species. The flowers are generally borne in racemes. The seeds of the vetches are more or less round, while the pods are elongated and compressed. Common vetch (Figure 30–1) is self-pollinated. The flowers of most vetches are violet or purple, the purple vetch having reddish-purple flowers. Hungarian vetch has white flowers.[31]

The most widely grown vetches are smooth, hairy, purple, common, woolypod, monantha, and Bard. Narrowleaf vetch (*V. angustifolia*), often called the wild vetch or wild pea, occurs as an introduced weed in the North Central states, where the seed contaminates wheat. It is sometimes used as an orchard cover crop in the Southeast. The seed usually is salvaged from wheat screenings. Bitter vetch (*V. ervilia*) is grown in southern Europe and in the Near East, but not in the United States. It could be grown along the Pacific coast, but not in the Southeast.[29]

Hairy or sand vetch (*Vicia villosa*) formerly was the most widely grown type in the South, but is has been replaced to a large extent by smooth vetch, which was selected from hairy vetch in Oregon. The stems of hairy vetch are viny and ascend with support. It is characterized by pubescence on the stems and leaves as well as by the tufted growth at the ends of the stems. Smooth vetch is similar to hairy vetch except that it has less pubescence and lacks the tufted growth at the ends of the stems. Although winter hardy in the southern

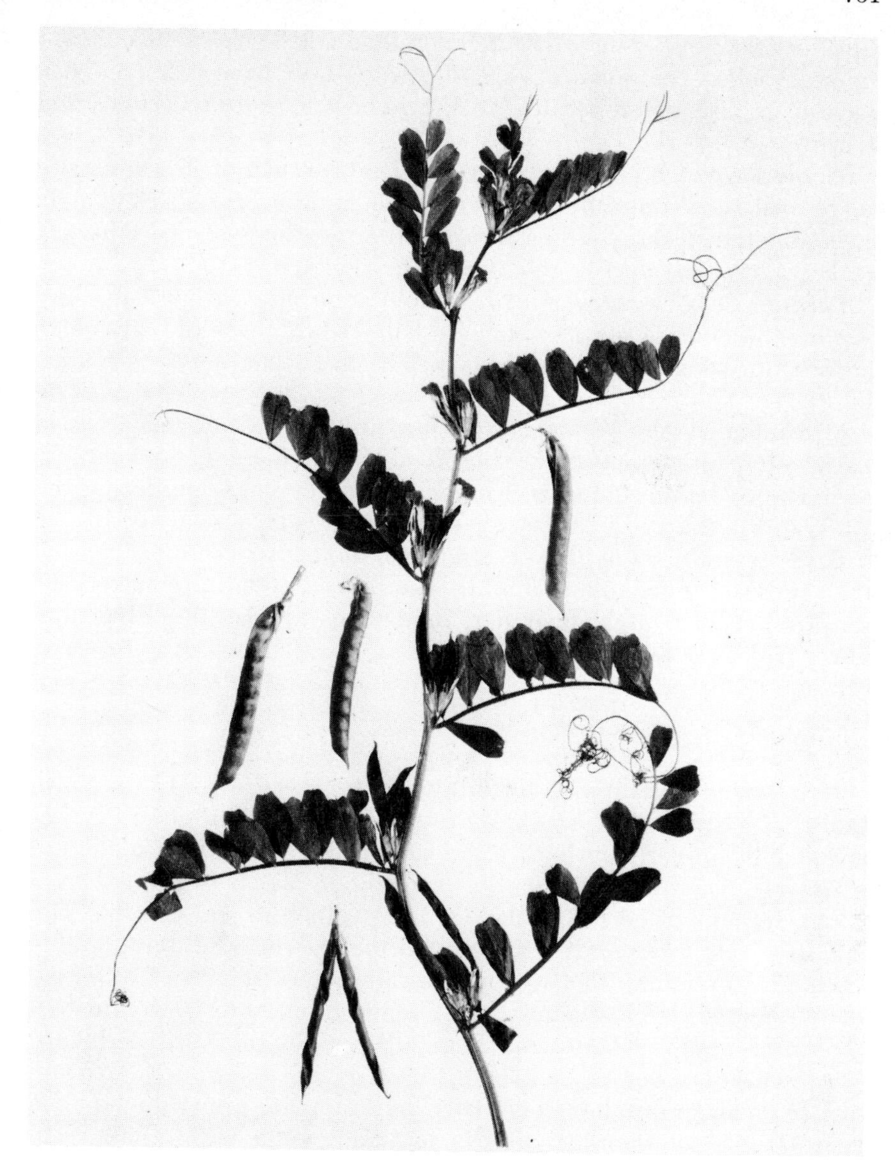

Figure 30–1. Branch of common vetch.

states, smooth vetch will not survive so far north as hairy vetch does.[19, 29]

Common vetch (*V. sativa*) sometimes called spring vetch or tares, is a semiviny plant with slightly larger leaves and stems than those of hairy vetch. One variety has a yellowish seed.

Hungarian vetch (*V. pannonica*), grown commercially in Oregon, is less viny than either hairy vetch or common vetch. The plants tend to be erect when the growth is short or when they have support. They have a grayish color because of the pubescence on the stems and leaves.

Woolypod vetch (*V. dasycarpa*) has slightly smaller flowers and a more oval seed, but otherwise it is very similar to Hungarian vetch. It is adapted to the southeastern and Pacific Coast states, where it makes more early spring growth than hairy vetch.[29]

Purple vetch (*V. bengalensis*) is similar in growth habit to hairy vetch. The pods of purple vetch are hairy while those of hairy vetch are smooth. It is the best adapted vetch for the coastal sections of California.[29]

Monantha vetch (*V. articulata*) has finer stems and leaves than hairy vetch. Single flowers are borne on long stems. It is grown safely as a winter legume only in Florida and other states bordering the Gulf, and in the Pacific Coast states where it is grown for seed.

Bard vetch (*V. monantha*) is similar to monantha vetch except that the seeds are more rounded while the flowers (two in a cluster) are purple instead of light lavender. Bard vetch is adapted only to the lower Colorado River Basin.

Cultural Methods

Hairy vetch is planted in the fall wherever it is grown. In the North (north of latitude 40°), except on the Pacific Coast, all other vetches are planted in early spring. In the Cotton Belt the crop is seeded in early fall, usually in September or October. Vetch is planted from about September 15 to October 15 in Washington and Oregon. The rate of planting depends largely upon the size of seed, but planting 12 seeds per square foot will give a satisfactory stand in California. Rates of seeding of several species range from 20 to 50 pounds per acre in the southern states and 30 to 80 pounds in the North and West. The quantity of vetch seed should be reduced about one-fourth and that of small grain about one-half when they are grown together in mixtures. A common practice is to plant vetch with small grain when it is grown for forage. The grain stems support the vetch vines, which avoids damage from contact with the ground and also makes cutting easier.

Vetch is generally cut for hay when the first pods are well developed. It may be cut with a swather, or a mower equipped with lifter guards and a swather attachment. If grown as a seed crop, common, hairy, and smooth vetches are harvested when the lower pods are ripe to avoid shattering. Purple and Hungarian vetches are usually cut after 75 to 90 per cent of the pods are ripe. The seed crop can be windrowed, dried, and then threshed with a pick-up combine.

Utilization

The vetches are most widely used in the southern states as cover and green manure crops, although occasionally they are cut for hay. In Alabama, vetch or Austrian winter peas grown in a two-year rotation (cotton-winter legume-corn) increased the corn yield 18 bushels per acre.[70] The second-year residue from these legumes increased the cotton yield by 213 pounds per acre. Large increases in the yields of corn and cotton were obtained when vetch was turned under in Alabama as early as March 15. Vetch, or vetch and rye, can be drilled into bermudagrass sod in the fall, and then grazed during the winter.

In the Pacific Northwest the vetches are used for hay either alone or in mixtures with oats. Common and Hungarian are grown most generally for hay. The vetches have been used for soiling. They may also be grazed during the winter, spring, and early summer.

Diseases

Anthracnose, caused by the fungus *Colletotrichum villosum*, attacks the leaves and stems of smooth and Willamette vetch. Other common vetches and Hungarian are resistant. False anthracnose, caused by *Kabatiella nigricans*, girdles the stem and sometimes infects the vetch seeds. Oregon common, Hungarian, and monantha vetches are resistant.

Downy mildew, caused by *Peronospora viciae*, damages vetch in the Pacific Northwest as well as in the South.

Blackstem, caused by *Ascochyta pisi*, discolors the stems. It may occur wherever vetch is grown. The disease is especially serious in thick stands of smooth and hairy vetch.

Root rot, which is caused by several different fungi, damages the plants and often kills vetch seedlings in all areas. Smooth vetch is partly resistant.

Other diseases and the causal organisms include rust (*Uromyces fabae*), leaf and stem spot (*Ovuleria schwarziana*) and *Septoria* scald. Damage from root-knot nematodes, where present, is reduced by late seeding.

Insects

Many insect pests of other forage legumes also attack vetch. Some of these are aphids, corn earworms, grasshoppers, cutworms, armyworms, various weevils, and leafhoppers. These and lygus bugs (*Lygus* species) can be at least partly controlled with suitable insecticides.

A specific pest, the vetch bruchid or vetch weevil (*Bruchus brachialis*) may cause heavy reduction in the seed yield of hairy, smooth, and woolypod vetch. It does not infest the seed of common or Hungarian vetch. It is found throughout the vetch-growing area of the Middle Atlantic States, but it is also considered a serious pest in some of the western states. Dusting the fields with insecticides when the first pods appear is an effective control method.

Heavy infestations of the pea aphid (*Macrosiphum pisi*) destroy the seed crop and seriously damage the quality of the hay. It can be controlled by spraying the field with an insecticide.

CROWNVETCH

Crownvetch (*Coronilla varia*), a native of the eastern hemisphere was introduced into the United States and grown as an ornamental. Its agricultural value was recognized many years later.[26] Crownvetch is a hardy, long-lived perennial that spreads by creeping underground rootstalks. The leafy stems are hollow, and decumbent unless supported by other vegetation. They may reach a length of 60 to 120 cm. The leaves have a superficial resemblance to the true vetch (*Vicia* species). The variegated flowers are white to rose or violet.[26, 53] Insect pollination is essential for good seed yields.[61] Fully mature seeds are mostly hard and require scarification, sulfuric acid soaking, or dipping in liquid nitrogen in order to germinate promptly. Crownvetch is most successful in Pennsylvania but is well adapted to the area from northeastern Oklahoma to northern Iowa and eastward to the Atlantic Ocean. It is grown also in Minnesota where the less hardy varieties may winterkill. It can be grown in the humid areas of the Pacific Northwest. Crownvetch requires good drainage but ample soil moisture. It grows on infertile soils but responds well to applications of phosphorus, potassium, calcium and magnesium. A soil *p*H of 6.5–7.0 is preferable.

Mature, scarified, inoculated, and treated seed should be sown $^1/_4$ to $^1/_2$ inch deep in a well-prepared seedbed.[44] The young plants grow slowly and compete poorly with weeds, or with companion crops unless it is sown in alternate drill rows. From 1 to 4 years are required for a full productive stand to be established.

The chief use of crownvetch has been for erosion control on steep slopes and banks.[53] It provides fair pasturage but recovers slowly after mowing or close grazing. The hay is similar to other forage legumes in protein and fiber content. The plants have a high tannin content and are not highly palatable but livestock become accustomed to it. The development of improved varieties greatly expanded crownvetch as a forage crop.

MILKVETCH

Cicer milkvetch (*Astragalus cicer*), a native legume, is being sown for pasture at high altitudes in Montana. It is more frost tolerant than alfalfa.

SAINFOIN

Sainfoin (*Onobrychis viciaefolia*) is a forage legume native to the Mediterranean region. It has been grown for more than 400 years in Europe and Asia; now chiefly in Russia and Turkey. George Washington introduced sainfoin into the United States in 1786, but it was grown only rarely until improved hardy varieties were developed and released in Montana and western Canada in the 1960's.[24] About 25,000 acres were grown in Montana in 1970. It has also been called Saint Foin, meaning holy grass, because it is reputed to have filled the manger in which Christ was born.

The sainfoin plant is a deep-rooted perennial with numerous coarse hollow stems that arise from the crown and may reach one meter in height. The leaves bear 12–15 leaflets. The rosy-pink flowers are borne on spikelike heads. The pod contains one smooth, kidney-shaped, olive-brown to dark brown seed about 3 millimeters long. Sainfoin forage does not induce bloating.[12] The plants are immune to damage from the alfalfa weevil but they suffer somewhat from seed weevil and sweetclover weevil attacks. Sainfoin is adapted to poor and calcareous soils but cannot tolerate saline, acid or wet soils, or a high water table. It is a suitable dryland hay or pasture crop in western Canada where the annual precipitation is sufficient. Its response to irrigation is less than that of alfalfa. Sainfoin starts spring growth about 2 weeks before alfalfa, and matures 4 to 6 weeks earlier, but yields only 80–90 per cent as much hay. Its slow recovery after mowing permits only one cutting of hay each year.

Podded inoculated seeds are sown at a rate of 30–35 pounds per acre for hay on irrigated land, at 12–17 pounds in 12- to 18-inch rows on dryland for hay or pasture, and at 6–9 pounds in 24- to 36-inch intertilled rows for seed production. It often is sown in mixture with grasses for dryland pastures.

VELVETBEAN

The velvetbean has been grown in the United States as a field crop only since about 1890. Its popularity declined after 1941. Seed yields in 1969 exceeded 800 pounds per acre. The principal producing states are Texas, Georgia, South Carolina, Alabama, Arkansas, and Florida.

Adaptation

The velvetbean is a warm-weather crop that requires a long frost-free season to mature the seed. The crop was confined to Florida and other Gulf Coast states until the introduction of early varieties extended its range to the northern limits of the cotton belt. The velvetbean is well adapted to the sandier, less fertile soils of the southeastern states, especially those of the coastal plain.[4] It makes a poor growth on cold wet soils. It has been used extensively as a green-manure crop on newly cleared woodlands.

Botanical Description

The velvetbean is a vigorous summer annual. With the exception of the bush varieties, the vines attain a length of 10 to 25 feet. The leaves are trifoliate, with large ovate membranous leaflets shorter than the petiole. The flowers are borne singly or in twos and threes in long pendant clusters. The numerous hairs on the pods of velvetbeans sting like nettles and make the crop unpleasant to handle. The pubescence usually sheds soon after maturity. The pods range from 2 to 6 inches in length. The seeds may be grayish, marbled with brown (Figure 30–2), white, brown, or black. The velvetbean has numerous fleshy surface roots that often reach 20 to 30 feet in length.

The Florida velvetbean (*Stizolobium deeringianum*) was the only species grown in this country until 1906. It required 240 to 270 days to mature. Much earlier varieties are now available.

The purple flowers are borne in clusters 3 to 8 inches long.[57] The pods are 2 to 3 inches long, nearly straight, and covered with black velvety pubescence. The seeds are nearly spherical, grayish marbled with brown.

Figure 30–2. Leaves, pod, and seeds of the velvetbean.

The Lyon velvetbean (*S. cochinchinensis*) is a long-season type that matures about 10 days earlier than the Florida. It has white flowers borne in racemes sometimes 2 to 3 feet long. The pods are 5 to 6 inches long, covered with gray hairs, and have a tendency to shatter when mature. The seeds are ash-colored.

Earlier varieties from a cross between the Florida and Lyon species are heavy seed producers.

The Yokohama velvetbean (*S. hassjo*) produces a smaller vine growth than the other species. It matures within 110 to 120 days. The flowers are purple. The pods are 4 to 6 inches long, flat, and covered with gray pubescence. This variety is seldom grown.

Cultural Methods

Since velvetbeans are tender, they should be planted in warm soils after all danger of frost is past. In the northern part of the Cotton Belt the planting time is about the same as for corn. Farther south, late varieties are planted as soon as the soil is in good condition, but planting of early varieties may extend over a period of 40 to 60 days.[50] They should be planted in rows 3 to 4 feet apart at the rate of 30 to 35 pounds per acre of seed when grown for grazing, green manure, or as a smother crop. When grown for seed, they are interplanted with corn at a rate of about 6 pounds per acre or more. Fertilization with phosphorus and, often, potash may be beneficial.

The velvetbean plants produce little seed unless they are grown with an upright crop like corn so that the vines can climb and bear their flowers where there is air circulation. This also avoids pod decay. The pods are picked by hand and threshed later, after drying. No machine is suitable for harvesting the beans. The corn also is harvested by hand because the long, tangled velvetbean vines clog the mechanical harvesters. The heavy labor requirements discourage velvetbean seed production.

Utilization

Velvetbeans are now grown chiefly for fall and winter grazing after the crop has matured. Cattle can chew the beans after they have been softened by rains. As summer pasture, velvetbeans are much inferior to soybeans or pearl millet in palatability and in animal gains.

Velvetbeans are a good green-manure crop. In experiments in Alabama the yield of seed cotton after cotton was 918 pounds per acre; after velvetbeans cut for hay, 1,126 pounds; and after velvetbeans plowed under, 1,578 pounds. In southern Georgia, corn grown after velvetbeans plowed under in the winter has yielded about twice as much as when following corn.

Velvetbeans are an unsatisfactory hay crop because of difficulties in mowing and curing the mass of tangled vines. Moreover, the cured vines and leaves are black and unattractive.

Ground or soaked velvetbeans are sometimes fed to livestock, but they are an uneconomical feed. They are much less nutritious than cottonseed meal. Ground beans soon become rancid.

Crop Pests

The velvetbean is remarkably free from disease or insect enemies.[57] The velvetbean caterpillar (*Anticarsia gemmatilis*) may cause serious injury to the crop south of central Georgia. The caterpillar seldom attacks the crop until the plants begin to bloom. The damage is caused by defoliation, especially on late varieties. The caterpillars are controlled by dusting with insecticides.

ANNUAL MEDICS

Burclovers, black medic, and buttonclover are naturally reseeding winter-annual legumes.[13, 27, 46] They are members of the genus *Medicago* to which alfalfa belongs. All burclovers are native to the Mediterranean region.

Adaptation

The burclovers are primarily adapted to the mild moist winters of the Cotton Belt and the Pacific Coast. Few legumes make more growth in the Gulf Coast area in cool weather than does burclover. The crop will grow on moist soils, but it is most productive on well-drained loam soils. In California, burclover grows vigorously in adobe soils which are often poorly drained. The plant thrives on soils rich in lime but succeeds also on somewhat acid soils.

Most soils on the Pacific Coast are inoculated with the proper legume bacteria, but inoculation is essential in many southern states. In Alabama tests, nodule formation was greater when inoculation bacteria were applied to the unshelled seed than to the shelled seed.[16]

Botanical Description

Burclover plants branch at the crown and have 10 to 20 decumbent branches 6 to 30 inches long. The roots do not extend very deeply into the soil. Burclover has small yellow flowers borne in clusters of five to ten. The coiled pods are covered with spines which form the so-called bur. A vigorous plant may produce as many as 1,000 pods. The seeds resemble those of alfalfa.

Spotted or southern burclover (*Medicago arabica*), widely grown in the Cotton Belt, has a purple spot in the center of each leaflet. The pods contain from two to eight seeds. California or toothed burclover (*M. hispida*) is the most common species on the Pacific Coast. It lacks the spots on the leaflet. The pods have three and sometimes five seeds. The California burclover is less winter hardy than spotted burclover. Tifton burclover (*M. rigidula*), a more recent introduction grown on the coastal plain of Georgia, is more resistant to extremes of temperature, more resistant to disease, and produces more seed than the other common burclovers.[27] Cogwheel burclover (*M. tuberculata*) has spineless pods. This makes it preferable for sheep pasture because the burs do not become entangled in the wool. The burs look like cog-wheels.

Black medic (*M. lupulina*) has slender, finely pubescent procumbent stems, 1 to 2 feet long. The leaves are pinnately trifoliate with long petioles. The leaflets are finely pubescent, obovate, rounded, and slightly toothed at the tips, and are ½ inch or less in length. The small bright-yellow flowers are borne in dense heads about ½ inch long. The seeds (Figure 22–4), which closely resemble those of alfalfa, formerly were imported from Europe as an adulterant of alfalfa seed. Black medic (often called yellow trefoil) is distributed naturally in pastures, waste places, and meadows over much of the humid area of the United States. In certain sections of the Southeast, it furnishes valuable pasturage in late winter or early spring when other pasture plants are largely dormant.[7]

Buttonclover (*M. orbicularis*) has spineless, flat, coiled, button-shaped seed pods. It is adapted to soils relatively high in lime where it is a valuable pasture plant.

Cultural Methods

Burclovers, black medic, and buttonclover are seeded in the fall from September to December, but sufficiently early for the plants to become established before winter.

Seed in the bur is generally broadcast at the rate of 3 to 6 bushels per acre, while the shelled seed may be drilled at the rate of 12 to 20 pounds per acre.[33] Good stands are obtained thereafter without additional seeding, provided the land is plowed after the burs are ripe. The plants reseed indefinitely on established pasture lands. Lighter seeding rates than those mentioned above are advisable when sowing with grasses or in established pastures.

Seed of the annual medics sometimes is harvested with a combine equipped with a tined reel and special lifter guards. In other cases, the vines may be raked and removed after the burs have fallen to the ground. The ripe burs then may be gathered with a combine equipped with brush and suction attachments. Formerly the burs were swept up and gathered by hand.

Utilization

Burclover is used primarily as a pasture crop. It is valuable on California rangelands where animals graze the pods, particularly after the pods are softened by rains, or where the burclover cures and dries after the rainy season is over.[22] Weathering of the cured forage reduces its digestibility. A combination of burclover and bermudagrass is often used as a permanent pasture in the South.

Burclover alone has been widely used in California orchards or ricefields as a green-manure crop. It has proved to be a valuable cover and green-manure crop in the South. It is often seeded in cotton and corn to control soil erosion. Burclover is seldom used for hay.

Black medic sown with locally adapted (Florida or Alabama) seed is recommended for Florida pastures.[7] Imported (European) seed produces slow-growing, prostrate plants that fail to produce seed under high-temperature, short-day conditions. It is adapted to well-drained, open, limed or calcareous soils. It is sown at a rate of 7 to 12 pounds per acre.

Buttonclover is a good hay crop, especially in mixtures with grass or small grains.[13]

Pests

Cercospora leaf spot attacks spotted burclover, while Pseudopeziza leaf spot and rust attacks black medic. Rust also attacks California burclover. Most of the annual medics suffer from virus diseases. Insects that attack these crops are the alfalfa weevil and the spotted alfalfa aphid.

KUDZU

Kudzu is classified as *Pueraria thunbergiana* (or *P. lobata*). It has become an important legume in the southeastern states, where it probably occupied 500,000 acres by 1945.[51]

Adaptation

Kudzu is a warm-weather plant that is somewhat drought-resistant. The leaves and stems are sensitive to frost. Kudzu makes its seasonal growth from the time the soil becomes warm in the spring until the first frost in the fall. The plant will grow on many soil types, being able to thrive on soils too acid for clover. It served as a cover crop for rough, cultivated lands. Kudzu is not adapted to soils with a high water table, nor to the lime soils of the Black Belt in Alabama.

Botanical Description

Kudzu is a perennial leguminous vine native to Japan. The plant has many stems or runners that may grow as much as 70 feet in one season. The stems have nodes that send out roots wherever they come in contact with the soil. The internodes die and the rooted nodes become separate plants called crowns. The leaves resemble those of the velvetbean except that they are hairy. The deep-purple flowers are borne in clusters (Figure 30–3). Seed production usually is very poor. More seed is set on old than on young vines,[68] and more on climbing than on prostrate vines. The seeds are mottled and about one-fifth as large as those of peas.

Tropical kudzu (*P. phaseoloides*) grows in the West Indies and other hot climates.

Cultural Methods

Kudzu can be propagated by seedling plants, crowns, or vine cuttings. Rooted vine cuttings are commonly used. Seedling plants occasionally are reproduced in special seedbeds, the seed (which usually is imported from Japan) being planted from May to July.[68] The plants should be set on well-prepared land in the winter or spring before growth begins, but usually in February or March in the South.[51] They may be set in furrows or in holes dug for the purpose. A stand will be established in from one to five years, this being dependent upon the number of crowns set, soil fertility, and the amount of cultivation. Kudzu covers an area more rapidly where the plants are set 5 to 6 feet apart each way than when set 5 feet apart in 10-foot rows.[66] Where the rows are 10 to 12 feet or more apart, some crop such as soybeans may be interplanted until the kudzu spreads over the field.

Kudzu may be cut for hay after the plants are well established,

Figure 30–3. Leaves and flowers of kudzu.

which is usually in two or three years[51] but may be five years on poor eroded soils. Phosphorus fertilization often hastens the development of the plants. Kudzu may be cut at any time during the season, but should be cut high and not more than once or twice in a season. The root reserves are decreased by excessive cutting.[56] Close, frequent mowing will eradicate kudzu.

Utilization

Kudzu is utilized either for hay or pasture. Its feeding value compares favorably with that of other legumes. Kudzu furnishes grazing during the summer drought period when many other pastures are unproductive. Close grazing usually kills the crop.[52]

Kudzu was also an important soil-improvement crop. In an Alabama experiment plowed-under kudzu increased the average yield of two crops of sorghum hay by 2,536 pounds per acre, four crops of corn by 19.3 bushels, and seven crops of oats by 7.9 bushels.[66]

Kudzu is widely used in the South as an erosion-control plant because of its rank vegetative growth. It does well in gullies, roadside cuts, and in areas destroyed by sheet erosion.[6] It will not thrive on very infertile soil unless phosphate fertilizers are applied. When planted next to a wooded tract, or among scattered trees, it climbs to the tree tops and covers the area. It later became a serious weed pest in wooded and uncultivated areas. It can be controlled by heavy grazing or by spraying with picloram at a rate of 0.34 kg. per hectare.

CROTALARIA

Crotalaria is an annual or perennial that has been grown in the southeastern states as a summer annual cover crop. The five species in the United States are *Crotalaria intermedia*, *C. mucronata* (*striata*), *C. spectabilis*, *C. lanceolata*, and *C. juncea*.[47] *Crotalaria juncea* is called Sunn hemp in India where it is grown frequently as a fiber crop. Crotalaria sowing is now banned or discouraged as even a cover crop because all crotalaria seeds, and also the plants of the *spectabilis* and *juncea* species are poisonous to livestock.[4] It is being replaced by pigeon peas and other legumes.

Adaptation

Crotalaria is native to tropical regions of heavy rainfall. It requires a warm season for vigorous plant growth. The plant is sensitive to frost, being killed or seriously injured at 28° to 29° F.[65] Showy crotalaria (*C. spectabilis*) (Figure 30–4) will mature seed as far north as North Carolina.[47] Crotalaria is adapted to light sandy soils, especially to those of the Coastal Plains area of the southeastern states. Well-drained soils are essential to growth. The soils in this region usually

Figure 30–4. Showy Crotalaria.

are naturally inoculated with the proper legume bacteria. Crotalaria will grow on poor acid soils. It is also resistant to the root-knot nematode.[3]

Botanical Description

The *Crotalaria* species, *C. spectabilis* and *C. striata*, are moderately branched, upright annuals that attain a height of 3 to 6 feet. The leaves are large and numerous. *C. striata* has numerous yellow flowers borne on long terminal racemes.[65] From 40 to 50 seeds are borne in pods similar to those of peas. The seeds are kidney-shaped and vary in color from olive-green to brown. The leaves are trifoliate in this species, but monofoliate in *C. spectabilis*. Both species have a bitter taste in the green state.

Cultural Methods

When used as a cover crop, crotalaria is generally planted from March 15 to April 15 in Florida, or from April 1 to 30 in other Gulf Coast states.[47] As a hay crop, it is often planted in June so as to be ready for harvest by October 1. Crotalaria may be sown broadcast or in close drills at the rate of 15 to 30 pounds of seed per acre. The higher rate is generally followed for unscarified seed, or when seeding conditions are poor. Crotalaria can be seeded in fields of winter oats in the lower South,[47] where it makes a rapid growth after the oats are harvested. Where crotalaria is grown for seed it is commonly planted at a rate of 2 to 6 pounds per acre in cultivated rows.

Crotalaria should be cut for hay at least by the bloom stage and preferably in the bud stage or earlier.[60] To obtain a satisfactory second growth, it is necessary to leave a stubble of at least 8 to 10 inches. Seed may be harvested with a combine, or left to fall on the ground to reseed the crop. Much of the seed requires scarification before it will germinate promptly.

Utilization

The greatest use of crotalaria was as a cover and green-manure crop.[41, 65] Showy (*C. spectabilis*) crotalaria was often planted in pecan and tung groves but striped (*C. striata*) harbors injurious insects in citrus orchards.

Crotalaria is used rarely as a forage crop because it is not very palatable. The coarse fibrous stalks also make a poor quality of hay. A fair quality of hay and silage can be made from *C. intermedia* when the plants are cut early.[60] The green forage, hay, and silage of *C. spectabilis* and *C. juncea* may cause death when fed to livestock. They contain an alkaloid called monocrotaline. The immature plants of *Crotalaria intermedia*, *C. lanceolata*, and *C. striata* are not poisonous.

TREFOIL

Economic Importance

The trefoils are perennial long-day legumes that persist for many years in pastures. They were naturalized, or grown after seeding, on more than 1,700,000 acres of pasture and meadowland in the United States in 1969, chiefly in mixtures with grasses and other legumes. Most of the acreage was in California, Ohio, Iowa, New York, Pennsylvania, and Vermont. It is also of some importance in Illinois, Michigan, Oregon, and Minnesota. Most of the trefoil is the broadleaf birdsfoot trefoil, *Lotus corniculatus*. Narrowleaf trefoil, *L. tenuis*, is grown in New York, but also under irrigation in California, Oregon, Idaho, and Nevada. Big trefoil, *L. uglinosus* (or *L. major*) is grown in Washington and Oregon.[5]

Adaptation

The trefoils are adapted to the cooler climates of United States north of latitude 37°. It is suited to the acid and less-fertile soils and to poorly drained soils.[29] Birdsfoot trefoil has a place on fertile, well-drained soils in permanent pastures, but present varieties are less productive for hay than is alfalfa. It is poorly adapted to the southeastern states, except in the cooler Appalachian Mountains.

Narrowleaf trefoil is well suited to heavy, poorly drained clay soils in the same general region.[29] It is an important species in pastures in New York as well as in California. This species grows well in soils that contain large quantities of soluble salts. It is less hardy than birdsfoot trefoil.

Big trefoil is well adapted to the acid coastal soils of the Pacific Northwest where the winters are mild. It is not tolerant to drought owing to its shallow rooting habit, but grows well on low-lying soils that are frequently flooded during the winter months.[29] Big trefoil is productive on soils of *p*H 4.5 to 5.5. It has promise as a pasture legume in the acid flatwood soils of the southeastern coastal regions when adequately fertilized.[74]

Botanical Description

Birdsfoot trefoil is broad-leaved, with a well-developed branching taproot. It has few to many ascending stems that develop from each crown,[29] and is similar to alfalfa in growth habit. The plants usually are erect or ascending, reaching a height of 30 to 75 cm. The leaves are borne alternately on opposite sides of the stem. They are composed of five leaflets, of which one is terminal, two apical, and two basal, at the base of each petiole. The last two resemble stipules. Two to six flowers are borne in umbels at the extremity of a long peduncle that arises from the leaf axil. The pea-like flowers are yellow with faint red or orange stripes usually present in young flowers. Pods form at right angles at the end of the peduncle in the shape of a bird's foot. The seeds are oval to spherical, and (Figure 22–4) vary in color from light to dark brown. Improved varieties of birdsfoot trefoil are replacing the common type.[55, 58]

Narrowleaf trefoil has narrow, linear lanceolate leaflets on slender weak stems with comparatively long internodes. Flowers are slightly smaller, fewer in number, and usually change from yellow to orange-red at maturity. Otherwise, it is similar to the birdsfoot species.

Big trefoil is similar to birdsfoot trefoil in appearance, but differs in that it has rhizomes.[29] It also has more flowers per umbel than either of the other two species. The seeds of big trefoil are almost spherical. They vary in color from yellowish to olive-green without any speckling. Big trefoil seeds are much smaller than those of the other two species.

Culture

Trefoils may be grown alone or in mixtures with grasses or other legumes. Competition with companion crops, grasses, and weeds often causes poor stands. Shading depressed seedling growth and nodulation more in birdsfoot trefoil than it did in alfalfa or red clover.[45]

Shallow sowing of trefoil on a well-prepared compact seedbed is essential. In California, birdsfoot trefoil sometimes is broadcast from an airplane on flooded irrigated fields. The sprouting seeds settle in the mud when the fields are drained 48 hours after sowing. Early spring seeding is advisable in the northern or central states, while fall or late-winter seeding-occurs in California and Florida. When grown alone, birdsfoot or the narrowleaf trefoils may be sown at 4 to 6 pounds of seed per acre. These rates are based on good-quality seed that is scarified as well as inoculated with a specific trefoil bacterial inoculum. Simple mixtures of a grass, with 4 to 6 pounds of trefoil seed per acre, usually have been successful.[29]

Trefoil seedlings are small and develop slowly the first year being very sensitive to shading. Band seeding above a band of fertilizer helps to establish trefoil stands. Weeds or weedy grasses often restrict their growth unless the weeds are mowed or treated with herbicides.

Trefoil hay yields are comparable to those of red clover under the same conditions. The hay and silage have a composition and quality similar to those of clover and alfalfa. Its value for pasture lies chiefly in its succulent growth in midsummer and late summer when other herbage has nearly ceased growth. Under good management, the stands are maintained when the plants are occasionally allowed to reseed naturally.

Trefoil seed is threshed with a combine after windrowing or after the application of a desiccant to the crop. Clean seed yields of approximately 100 pounds per acre are typical because of uneven flowering and the shattering of seed soon after the pods are ripe.

Pests

Trefoils may be attacked by leaf blight, crown rot, root rots, leaf spot, stem canker, and nematodes. Special control methods are unavailable. Insect pests include spittlebugs, leafhoppers, lygus bugs, stink bugs, grasshoppers, cutworms, and seed chalcids. They are controlled largely by insecticides. They are not damaged by the alfalfa weevil.

SESBANIA

Sesbania (*Sesbania macrocarpa* [*exaltata*]) is a coarse, upright annual legume, native to North America. It is found in various localities in the southern part of the United States from California to Georgia, extending as far north as Arkansas. It is abundant on the overflow lands of the lower Colorado River in Arizona and California where it is referred to as wild hemp or Indian hemp. The Indians there used the fiber for making fish lines and other twines. It also occurs as a weed in the ricefields of southern Texas and Louisiana where it is known as tall indigo.

In hot weather, with ample rainfall or irrigation, and good soil, sesbania grows to a height of 6 to 8 feet in a few weeks (Figure 6–6). The yellow flowers are borne in racemes that arise from the leaf axils. Long slender seed pods are produced later. Sesbania is not relished by livestock. This legume is grown for green manure in the irrigated valleys of southern California, in southern Arizona, and in the lower Rio Grande Valley of Texas. Seeded in June, it provides a heavy growth of green manure to be turned under in late August, in preparation for fall-planted vegetable crops.[4] It should be sown at the rate of about 20 pounds per acre. The seed usually is gathered from wild stands.

Sesbania roots are attacked by nematodes, and the seed by a small weevil. Soil disinfection controls the nematode and seed fumigation controls the weevil. Seed inoculation is unnecessary.

LUPINES

Lupines have been known for 2,000 years. More than 900,000 hectares of seed were harvested in 1970, chiefly, in the U.S.S.R. and Poland. Although cultivated in Europe for 200 years, they did not become well established in the United States before 1943. Extensive acreages have been grown since that time. The culture of lupines for winter cover, green manure, or grazing is limited to the South Atlantic and Gulf Coast states.[30] The lupine seed crop is produced largely in Georgia, South Carolina, and Florida.

Many perennial and annual species of lupines are native to North America. Among them is the bluebonnet, *Lupinus subcarnosus*, the state flower of Texas. The three species now grown commercially in the Gulf Coast area were introduced from Europe, i.e., white lupine (*L. albus*), yellow lupine (*L. lutens*), and blue lupine (*L. angustifolius*). The value of these lupines lies in the fact that the seed can be produced in the South, where they also make a heavy winter growth for turning under. The plants are upright, long-day winter annuals with coarse stems, medium-size digitate (fingered) leaves, and large at-

Figure 30–5. White lupine: plant (*left*), flower raceme (*right*).

tractive flowers of the colors indicated by the names (Figure 30–5).

The bitter blue and white lupines were used only for green-manuring because they contain alkaloids, particularly in the seeds and pods, which are poisonous to livestock. Later, sweet (nonpoisonous) varieties of blue and white lupines were acquired, and are grown for grazing as well as for green manure. Sweet yellow lupines were acquired from Germany.

Inoculated lupine seed is drilled 1 to 2 inches deep into the soil in the fall in the southeastern states, where temperatures do not go below 15° F. The seeding rate is 90 to 120 pounds per acre of white lupine, 60 to 90 pounds per acre of blue lupine, or 50 to 80 pounds per acre of yellow lupine.[27] The seed can be harvested with a combine.

Lupine diseases include brown spot, powdery mildew, Ascochyta stem canker, root rots, seedling blights, southern blight, and virus attacks. Disease-free seed, seed treatment, or crop rotation reduce some of the damage. Insect enemies of lupines include the root weevil, lupine maggot, thrip, and white-fringed beetle.[30]

GUAR

Guar (*Cyamopsis psoralides* or *C. tetrogonolobus*), was introduced into the United States in 1903. About 80 to 90 thousand acres are harvested each year. A similar acreage is used as a leguminous green-manure crop in southern California, Arizona, and Texas,[76] particularly on soils infested with the cotton root-rot organism (*Phymatotrichum omnivorum*). It is the most resistant to this disease of any legume tested. Guar is also partly resistant to root-knot nematode. It

Figure 30–6. Guar plant.

is grown for seed chiefly in northern Texas[10] and southern Oklahoma.[43] Yields average 300 to 700 pounds per acre. The seed contains a gum or mucillage, mannogalactan, which is useful in paper manufacture as a substitute for carob gum. It is also used in textile sizing and as a stabilizer or stiffener in foods and various other products[35] and in drilling muds and ore flotations.[76] The beans have considerable value for livestock feed.

The guar plant is a coarse summer annual. It is an upright drought-resistant herb 2 to 4 feet in height with angular toothed trifoliate leaves (Figure 30–6), small purplish flowers borne in racemes, and long leathery pods. Purple stain is caused by a fungus (*Cercospora kikuchii*) which attacks the stems and seeds of guar.[36]

Guar, sometimes called cluster bean, is a native of India, where it is grown for fodder, for pods used for food or feed, and for beans for export. Guar also is grown in Africa but its use as edible beans in Nigeria is limited because of their high prussic acid content. In the United States, it is sown in late spring at the rate of 40 to 60 pounds per acre drilled, or 10 to 30 pounds per acre in rows. Sheep will graze green guar, but other animals reject it because of the bristle-like hairs on the plants. Cured hay is eaten readily. The seed crop is harvested with a combine.[10, 43, 45]

FLORIDA BEGGARWEED

Florida beggarweed (*Desmodium purpureum*) has been cultivated on a small acreage in the South for more than 60 years. It is a native of tropical America, but is established in the southern United States, chiefly in cultivated fields where it volunteers freely. Florida beggarweed is adapted to the southern coastal plain.[4] When sown in northern states it makes a fair vegetative growth but will not produce seed. It tolerates acid soils and is resistant to the root-knot nematode.

The Florida beggarweed plant is an upright, herbaceous, short-lived perennial legume that attains a height of 4 to 7 feet. It usually lives as an annual in most of the United States. The pubescent main stem is sparsely branched. The leaves are trifoliate with large ovate pubescent leaflets. Racemes of inconspicuous flowers terminate the main stem and lateral branches. The seeds are borne in jointed pods, segments of which adhere to cotton lint and clothing.

The seed is sown in late spring or early summer following a cultivated crop or is planted at the last cultivation of early planted corn. About 10 pounds of hulled or scarified seed or 30 to 40 pounds of unhulled seed are sown on a compact seedbed.

The chief uses of Florida beggarweed are pasturage, hay, green manure, and production of seeds for quail feed. It cannot be maintained in permanent pastures.

ROUGHPEA

Roughpea (*Lathyrus hirsutus*), also known as wild winter pea, caley pea, and singletary pea, is a reseeding winter annual. It is grown in North Carolina, Tennessee, Arkansas, eastern Oklahoma, and states to the south. Roughpea is a native of Europe which escaped to pastures as well as to cultivated land in southeastern United States many years ago.

The roughpea plant resembles the sweetpea (*L. odoratus*), as well as the flatpea or wagnerpea (*L. sylvestris*). Seeds of the roughpea, like those of wagnerpea, are poisonous to livestock. Roughpea has weak stems that are decumbent except in thick stands. Growth is slow until late winter or early spring, when rapid growth begins. The plant matures in late spring or early summer. The crop thrives on lime soils but will grow well on slightly acid soils, heavy clay soils, and poorly drained land.

Roughpea is sown in the fall, alone or in mixtures with grasses or small grains. When roughpea is sown alone, the seeding rate is 20 to 25 pounds of scarified seed or 40 to 60 pounds of unscarified seed. The crop responds to phosphorus fertilizer and sometimes also to potash and lime.

Roughpea may be safely grazed or cut for hay only before pods are formed. The seeds contain a poison that causes lameness in cattle that eat the pods. However, animals recover when roughpea is replaced by other feed.[27] Delayed growth limits the usefulness of roughpea for green manure, except when it precedes a crop that is planted late.

HAIRY INDIGO

Hairy indigo (*Indigofera hirsuta*) is a native of tropical Asia, Australia, and Africa. It was introduced into cultivation in Florida in 1945 after experiments had shown its value as a hay and green-manure crop on sandy soils low in calcium. It is adapted to the coastal-plain area from Florida to Texas.

The hairy indigo plant is a summer-annual legume that attains a height of 4 to 7 feet. The somewhat coarse stems become woody with age. The pinnately compound leaves resemble those of vetch. A late strain of hairy indigo that matures in November is adapted to southern Florida, but a smaller earlier strain is adapted northward to central Georgia. Hairy indigo reseeds naturally when it is allowed to mature. The plant is resistant to root-knot nematode.[4]

The seed of hairy indigo is drilled in the spring at a rate of 3 to 4 pounds per acre. Artificial inoculation is unnecessary. Many of the seeds are hard. In corn fields, the hairy indigo seeds dropped the previous winter will germinate and establish a stand after the last cultivation of corn. The application of 30 to 50 pounds per acre each of phosphorus and potash in a fertilizer mixture is advisable.

Hairy indigo is useful for green manure. It also makes good hay when cut before the plants exceed 3 feet in height.

ALYCECLOVER

Alyceclover (*Alysicarpus vaginalis*) is a summer annual, native to tropical Asia. It is grown in the Gulf region for hay, as an orchard cover crop, or occasionally for pasture. It is not adapted to wet or very infertile soils, or to lands infested with root-knot nematode. The plants are 3 feet or less in height, with coarse leafy stems, broadly oval unifoliate leaves, and a spreading habit of growth except in thick stands. Uninoculated seed is sown in May at the rate of 15 to 20 pounds per acre. The seed shatters quickly after it is ripe. Phosphorus and potash fertilizers improve the yields.[4]

OTHER FORAGE LEGUMES

Pigeon pea, or red gram, [*Cajanus cajan* (*C. indicus*)] is grown on nearly 3 million hectares, chiefly in India where it probably originated, and extensively in Africa and other warm-climate areas, including Hawaii. It is grown chiefly for the seed, which is a popular food. The tall stalks provide abundant fodder for livestock, or an abundance of vegetation for green manure. The pigeon pea is a short-lived perennial, but it must be grown as an annual where frost occurs. Only the earlier varieties can mature seed north of southern Florida. The plants usually are 5 to 7 feet tall, but old perennials may grow to 20 feet in the tropics. It is grown as a forage or green manure crop in southeastern United States to replace the crotalarias that bear poisonous seeds. The coarse stalks make an undesirable hay. Seed yields in Hawaii range from 500 to 1000 pounds per acre. Pigeon peas are planted in cultivated rows at the rate of 10–15 pounds per acre, or in drills at 25 pounds per acre.

Fenugreek (*Trigonella foenumgraecum*) is grown in India, Egypt, and the Near East for forage as well as for the seeds, which are used as a condiment and in horse-conditioning powders. It has been grown occasionally in the milder parts of California as a winter-annual cover crop in orchards.

Partridge pea (*Cassia fasiculata*, Michx.) an annual legume native in the eastern half of United States, was grown as a cover or forage crop from about 1800 to 1870. Strains collected recently in Missouri and Illinois outyielded annual sweetclover, Korean lespedeza, and berseem clover in tests in Illinois but it yielded less than soybeans. It is adapted to eroded low-fertility soils and might be suitable for erosion control, green manuring, wildlife cover or forage in the area where Korean lespedeza is now grown.[17]

Siratro (*Phaseolus atropurpureus* D.C.) is a long-lived perennial legume with large taproots and trailing stems grown in Florida in mixtures with pangolagrass or bahiagrass. It was developed in Australia by crossing two native Mexican strains. This summer-growing crop is suitable also for grazing, green chop, hay, or wildlife.[40, 64a]

Koa Haole (*Leucaena leucocephala*) is a shrubby legume grown on a small acreage in Hawaii for pasture. Many new superior strains tested are capable of high yields.[8]

Townsville "lucerne" (*Stylasanthes humilis*), a self-fertile annual legume indigenous to Mexico and Central America, which was naturalized and grown in Australia after 1900, is now grown in Florida. It is used for pasture, green chop, silage, or hay in mixtures with pangolagrass.[39]

Serradella (*Ornithopus sativus*) is a native of Europe, where it is commonly cultivated. In the United States, it has produced good yields in only a few trials, but it has not been established as a crop. It is an annual legume with a vetch-like appearance with long, often procumbent stems. It grows as a winter annual in warm climates.

Other forage or soil-improvement legumes that have been grown on farms in the United States include Sulla (*Hedysarum cornarum*), and Tangier pea (*Lathyrus tingitanus*).

REFERENCES

1. Aldrich, S. R. "Birdsfoot trefoil," *Farm Quart.*, 4:38–40. 1950.
2. Ali-Khan, S. T., and Zimmer, R. C. "Growing field peas," *Can. Dept. Agr. Pub. 1433*, pp. 1–8. 1972.
3. Anonymous. "Grasses and legumes for forage and conservation," *USDA Agricultural Research Service Spec. Rept. ARS 22–43.* 1957.
4. Anonymous. "Growing summer cover crops," *USDA Farmers Bul. 2182.* 1962.
5. Anonymous. "Trefoil production for pasture and hay," *USDA Farmers Bul. 2191.* 1963.
6. Bailey, R. Y. "Kudzu for erosion control in the southeast," *USDA Farmers Bul. 1840.* 1939.
7. Blaser, R. E., and Stokes, W. E. "Ecological and morphological characteristics of black medic strains," *J. Am. Soc. Agron.*, 38(4):325–331. 1946.
8. Brewbaker, J. L., and others. "Varietal variation and yield trials of *Leucaena leucocephala* (koa haole) in Hawaii." *Hawaii Agric. Exp. Sta. Bull. 166.* 29 pp. 1972.
9. Brindley, T. A., Chamberlain, J. C., and Schopp, R. "The pea weevil and methods for its control," *USDA Farmers Bul. 1971* (rev.). 1952.
10. Brooks, L. E., and Harvey, C. "Experiments with guar in Texas," *Texas Agr. Exp. Sta. Cir. 126.* 1950.

11. Coe, H. S. "Origin of the Georgia and Alabama varieties of velvetbean," *J. Am. Soc. Agron.*, 10:175–179. 1918.
12. Cooper, C. S., and Carleton, A. E., Eds. "Sainfoin symposium," *Mont. Agr. Exp. Sta. Bul. 627*, pp. 1–109. 1968.
13. Davis, J. H., Gangstad, E. O., and Hackerott, H. L. "Button clover," *Hoblitzelle Agr. Lab. Bul. 6.* 1957.
14. Delwiche, E. J. "Field peas for Wisconsin," *Wis. Agr. Exp. Sta. Bul. 329*, pp. 1–24. 1921.
15. Dudley, J. E., Jr., and Bronson, T. E. "The pea aphid on peas and methods for its control," *USDA Farmers Bul. 1945* (rev.). 1952.
16. Duggar, J. F. "Root nodule formation as affected by planting of shelled or unshelled seeds of bur clover, black medic, hubam, and crimson and subterranean clovers," *J. Am. Soc. Agron.*, 26:919–923. 1934.
17. Foote, L. E., and Jacobs, J. A. "Partridge pea management and yield comparisons with other annual forage legumes," *Agron. J.*, 57(6):573–575. 1966.
18. Funchess, M. J. "Time of seeding and turning vetch for cotton and corn," *J. Am. Soc. Agron.*, 20:294–297. 1928.
19. Goodding, T. H. "Hairy vetch for Nebraska," *Nebr. Agr. Exp. Sta. Cir. 89.* 1951.
20. Graham, E. H. "Legumes for erosion control and wildlife," *USDA Misc. Pub. 412*, pp. 1–153. 1941.
21. Greene, S. W., and Semple, A. T. "Fattening steers on velvetbeans," *USDA Bul. 1333.* 1925.
22. Guilbert, H. R., and Mead, S. W. "The digestibility of bur clover as affected by exposure to sunlight and rain," *Hilgardia*, 6:1–12. 1931.
23. Gunn, C. R. "Seeds of native and naturalized vetches of North America," *USDA Handbk. 392*, pp. 1–42. 1971.
24. Hanna, M. R., and others. "Sainfoin for western Canada," *Can. Dept. Agr. Pub. 1470*, pp. 1–18. 1972.
25. Hedrick, U. P., and others. *The Vegetables of New York:* Vol. 1, Part 1, *The Peas of New York*, N.Y. Agr. Exp. Sta., Albany, pp. 1–132. 1928.
26. Henson, P. R. "Crownvetch," *USDA Agricultural Research Service Spec. Rept. ARS 34–53.* 1963.
27. Henson, P. R., and Hollowell, E. A. "Winter annual legumes for the South," *USDA Farmers Bul. 2146.* 1960.
28. Henson, P. R., and Schoth, H. A. "Vetch culture and uses," *USDA Farmers Bul. 1740* (rev.). 1961.
29. Henson, P. R., and Schoth, H. A. "The trefoils—adaptation and culture," *USDA Handbook 223.* 1962.
30. Henson, P. R., and Stephens, J. L. "Lupines," *USDA Farmers Bul. 2114.* 1958.
31. Hermann, F. J. "Vetches in the United States—native, naturalized and cultivated," *USDA Handbook 168.* 1960.
32. Howell, H. B. "Legume for acid soils (*Lotus uliginosus*)," *Oreg. Agr. Exp. Sta. Bul. 456.* 1948.
33. Hughes, H. D., and Scholl, J. M. "Birdsfoot still looks good," *Crops and Soils*, 11:3. 1959.

34. Hulbert, H. W. "Uniform stands essential in field pea variety tests," *J. Am. Soc. Agron.*, 19:461–465. 1927.
35. Hymowitz, T., and Matlock, R. S. "Guar in the United States," *Okla. Agr. Exp. Sta. Bul. B-611.* 1963.
36. Johnson, H. W., and Jones, J. P. "Purple stain of guar," *Phytopath.*, 52(3):269–272. 1962.
37. Klages, K. H. "Comparative winterhardiness of species and varieties of vetches and peas in relation to their yielding ability," *J. Am. Soc. Agron.*, 20:982–987. 1928.
38. Koonce, D. "Field peas in Colorado," *Colo. Agr. Exp. Sta. Bul. 416.* 1935.
39. Kretschmer, A. E., Jr. "*Stylosanthes humilis*, a summer-growing self-generating annual legume for use in Florida pastures," *Fla. Agr. Exp. Sta. Circ. S-184.* 1968.
40. Kretschmer, A. E., Jr. "Sirato (*Phaseolus atropurpureus* D.C.) a summer-growing perennial pasture legume for central and south Florida," *Fla. Agr. Exp. Sta. Circ. S-214.* 1972.
41. Leukel, W. A., and others. "Composition and nitrification studies on *Crotalaria striata*," *Soil Sci.*, 28:347–371. 1929.
42. Matlock, R. S. "Guar variety and cultural studies in Oklahoma, 1950–1959," *Okla. Agr. Exp. Sta. Proc. Ser. P-366.* 1960.
43. Matlock, R. S., Aepli, D. C., and Street, R. B. "Growth and diseases of guar," *Ariz. Agr. Exp. Sta. Bul. 216.* 1948.
44. McKee, G. W., and others. "Seeding crownvetch? Watch seed maturity," *Crops and Soils*, 22(3):8. 1972.
45. McKee, G. W. "Effects of shading and plant competition on seedling growth and nodulation in birdsfoot trefoil," *Pa. Agr. Exp. Sta. Bul. 689*, pp. 1–35. 1962.
46. McKee, R. "Bur clover cultivation and utilization," *USDA Farmers Bul. 1741* (rev.). 1949.
47. McKee, R., and others. "Crotalaria culture and utilization," *USDA Farmers Bul. 1980*, pp. 1–17. 1946.
48. McKee, R., Hyland, H. L., and Ritchey, G. E. "Preliminary information on sweet lupines in the United States," *J. Am. Soc. Agron.*, 38(2):168–176. 1946.
49. McKee, R., and Pieters, A. J. "Miscellaneous forage legumes," in *USDA Yearbook, 1937*, pp. 999–1019.
50. McKee, R., and Pieters, A. J. "Culture and pests of field peas," *USDA Farmers Bul. 1803.* 1938.
51. McKee, R., and Stephens, J. L. "Kudzu as a farm crop," *USDA Farmers Bul. 1923* (rev.). 1948.
52. Miles, I. E., and Gross, E. E. "A compilation of information on kudzu," *Miss. Agr. Exp. Sta. Bul. 326.* 1939.
53. Musser, H. B., Hottenstein, W. L., and Stanford, J. P. "Penngift crownvetch for slope control on Pennsylvania highways," *Pa. Agr. Exp. Sta. Bul. 576*, pp. 1–21. 1954.
54. Oke, O. L. "Nitrogen fixing capacity of guar bean," *Trop. Sci.*, 9(3):144–147. 1967.

55. Pierre, J. J., and Jackobs, J. A. "Growing birdsfoot trefoil in Illinois," *Ill. Exten. Cir. 725.* 1954.
56. Pierre, W. H., and Bertram, F. E. "Kudzu production with special reference to influence of frequency of cutting on yields and formation of root reserves," *J. Am. Soc. Agron.*, 21:1079–1101. 1929.
57. Piper, C. V., and Morse, W. J. "The velvetbean," *USDA Farmers Bul. 1276* (rev.). 1938.
58. Rachie, K. O., and Schmidt, A. R. "Winterhardiness of birdsfoot trefoil strains and varieties," *Agron. J.*, 47(4):155–157. 1955.
59. Reisenauer, H. M., and others. "Dry pea production," *Wash. Agr. Exten. Bul. 582.* 1965.
60. Ritchey, G. E., and others. "Crotalaria for forage," *Fla. Agr. Exp. Sta. Bul. 361,* pp. 1–72. 1941.
61. Robinson, R. J., and Soine, O. C. "Field peas for seed and forage," *Minn. Ext. Bul. 300,* pp. 1–12. 1961.
62. Robinson, R. G. "The shortcomings of crownvetch," *Crops and Soils,* 21(9):18–19. 1972.
63. Schafer, E. G., and Smith, R. T. "Rotation and hogging-off experiments with field peas," *Wash. Agr. Exp. Sta. Bul. 198.* 1926.
64. Seaney, R. R., and Henson, P. R. "Birdsfoot trefoil," *Adv. Agron.*, 22:119–157. 1970.
64a. Siewerdt, L., and Holt, E. C. "Yield components and quality of Sviatro-kleinge association." *Agron. J.* 66(1):65–67. 1974.
65. Stokes, W. E. "Crotalaria as a soil-building crop," *J. Am. Soc. Agron.*, 19:944–948. 1927.
66. Sturkie, D. G., and Grimes, J. C. "Kudzu, its value and use in Alabama," *Ala. Agr. Exp. Sta. Cir. 83.* 1939.
67. Tabor, P. "Seed production by kudzu (*Pueraria thunbergiana*) in the southeastern United States during 1941," *J. Am. Soc. Agron.*, 34:389. 1942.
68. Tabor, P. "Observations of kudzu (*Pueraria thunbergiana*) seedlings," *J. Am. Soc. Agron.*, 34:500–501. 1942.
69. Taylor, T. H., and others. "Management effects on persistence and productivity of birdsfoot trefoil (*Lotus corniculatus* L.)," *Agron. J.*, 65(4):646–648. 1973.
70. Tidmore, J. W., and Sturkie, D. G. "Hairy vetch and Austrian winter peas for soil improvement," *Ala. Agr. Exp. Sta. Cir. 74.* 1936.
71. Varney, K. E. "Birdsfoot trefoil," *Vt. Agr. Exp. Sta. Bul. 608.* 1958.
72. Vinall, H. N., and Davis, W. J. "Winter field peas: Their value as a winter cover and green-manure crop," *USDA Cir. 374.* 1926.
73. Wade, B. L. "Breeding and improvement of peas and beans," in *USDA Yearbook, 1937,* pp. 251–282.
74. Wallace, A. T., and Killinger, G. B. "Big trefoil—a new pasture legume for Florida," *Fla. Agr. Exp. Sta. Cir. S-49.* 1952.
75. Weimer, J. L. "Austrian Winter field pea diseases and their control in the South," *USDA Cir. 565.* 1940.
76. White, G. A. "New crop on the horizon," *Seed World,* 110(4):20–22. 1972.
77. Zaumeyer, W. J. "Pea diseases," *USDA Handbook 228,* pp. 1–30. 1962.

PART FOUR

CROPS OF OTHER PLANT FAMILIES

Chapter 31

Buckwheat

ECONOMIC IMPORTANCE

Buckwheat was harvested on less than 2 million hectares in 1970 with a production of nearly 1.7 million metric tons or 0.84 tons per hectare. About 90 per cent of the production was in the Soviet Union. Other important producing countries were Poland, Canada, France, Japan, Korea, Austria, Germany, and Romania. In the United States buckwheat was harvested on about 7,000 hectares (17,500 acres) yielding some 8,000 metric tons (366,000 bushels) or 1.14 metric tons per hectare (20+ bushels per acre). The leading states in buckwheat production were New York, Pennsylvania, Michigan, and Wisconsin. Production was about 22 million bushels in 1866 and more than a million acres were harvested in the U.S. in 1918, but the acreage has declined steadily since then. Recent exploitation of buckwheat as a "natural"food may increase domestic consumption. Buckwheat yields less than the cereal grains, because of its short growing season, poor response to nitrogen fertilization, and the failure to develop any improved variety until very recently. Also buckwheat has a lower feeding value for livestock than other grains, and the human consumption of buckwheat cakes is now very limited.[5] However, about 137,000 acres were harvested in Canada in 1970; 80,000 acres in Manitoba alone, with smaller acreages in Saskatchewan, Alberta, Ontario, and Quebec. Much of the Canadian production is exported, chiefly to Japan.[1]

ADAPTATION

Buckwheat makes its best growth in a cool moist climate. It is grown in the North as well as in the higher-altitude areas of the East. It is sensitive to cold, being killed quickly when the temperature falls much below freezing. Buckwheat requires only a short growing season of 10 to 12 weeks. The crop is sensitive to high temperatures and dry weather when the plants are in blossom. For this reason, seeding generally is delayed to allow plant growth to take place in warm weather and seed to form in the cooler weather of late summer.

Buckwheat will produce a better crop on infertile, poorly tilled land than other grain crops when the climate is favorable. Its response to fertilizer applications is less than that of other crops.[15] Buckwheat is able to extract more nutrients from raw rock phosphate than are other grain crop plants.[14] It is well suited to light, well-drained soils such as sandy loams or silt loams, and it grows satisfactorily on soils too acid for other grain crops. Buckwheat usually produces a poor crop on heavy wet soils. The crop is likely to lodge badly on rich soils high in nitrogen.[10] Buckwheat is often sown as a catch crop with little regard for the best conditions for its growth. This accounts for the low average yield in the United States.

HISTORY

Common buckwheat appears to have been cultivated in China for at least 1000 years. It probably originated in the mountainous regions of that country. During the Middle Ages, the crop was introduced into Europe and from there it was brought to the United States. It was grown by Dutch colonists along the Hudson River before 1625.

Buckwheat is not a cereal, nor is it really a grain, because it does not belong to the grass family. Because the fruits or seeds are used like grain, buckwheat is commonly regarded and handled as a grain crop. The name buckwheat was coined by the Scotch from two Anglo-Saxon words, *boc* (beech), because of the resemblance of the achene or fruit to the beechnut; and *whoet* (wheat), because of its being used as wheat. In Germany, the name is *Buchweizen*, likewise meaning beech wheat.[15]

BOTANICAL DESCRIPTION

Buckwheat belongs to the *Polygonaceae* or buckwheat family. Common buckwheat, the principal species *Fagopyrum sagittatum (esculentum)* appears to have been derived from *F. cymosum*, a wild species of Asia. A so-called winged buckwheat, designated as *F. marginatum* is obviously a type of *F. sagittatum*. Another species, *F. tataricum*, or tartary buckwheat is also grown.

The buckwheat plant is an annual, 2 to 5 or more feet in height, with a single stem and usually several branches.[8, 9] The stems are strongly grooved, succulent, and smooth, except at the nodes (Figure 31–1). The stems range in color from green to red, but turn brown with age. More reddening is evident when the plants have a poor seed set. The leaf blades, 2 to 4 inches long, are triangular and heart-shaped.

The plant has a taproot with numerous short laterals, which may

Figure 31–1. Plant of Japanese buckwheat.

extend 3 to 4 feet or less in the soil. The root system comprises only about 3 per cent of the weight of the plant, compared with 6 to 14 per cent in the cereal grains.[15]

The inflorescence of common buckwheat consists of axillary or terminal racemes or cymes with more or less densely clustered flowers (Figure 31–2). The inflorescence is partly determinate. The flowers, i.e., the petal-like sepals, are white or tinged with pink. They have no petals. There are eight stamens and a triangular one-celled ovary

Figure 31–2. Branch bearing flowers, immature seeds, and leaves of Japanese buckwheat.

that contains a single ovule, with a 3-parted style with knobbed stigmas.

The buckwheat plant begins to bloom 4 to 6 weeks after seeding. The flowers are dimorphic, i.e., some plants bear "pin" flowers with stamens shorter than the styles, while in others ("thrum") the stamens are longer.[13] The few flowers that have styles and stamens of the same length are usually sterile, but self-fertile lines have been isolated.[6] Sterility occurs following either self-pollination or cross-pollination between flowers of the same type because pollen tube growth is inhibited. Cross-pollinations between the two types usually induces fertility. Common buckwheat thus is largely naturally cross-pollinated and self-sterile.[10] Bees and other insects distribute the pollen, but cross-fertilization can occur where bees are excluded.[6] Tartary buckwheat is self-fertile. Experiments in Europe indicate that boron sprays increase seed setting in buckwheat.

The fruit is an achene that is brown, gray-brown, or black in color. The point of the grain is the stigmatic end, while the persistent calyx lobes remain attached at the base. The hull is the pericarp. The seed (groat) or matured ovule inside the hull has a pale brown testa which is triangular, as is the fruit.

The endosperm is white, opaque, and more starchy than cereal-grain endosperms. The embryo, embedded in the center of the endosperm,[12] possesses two cotyledons which are folded in the form of the letter S.

TYPES

Japanese buckwheat seed is brown and usually large (7 millimeters long and 5 millimeters wide; 15,000 seeds per pound). The seed is nearly triangular in cross section. The plants are tall with large leaves and coarse stems.

Silverhull buckwheat has smaller, glossy, silver-gray seeds – 20,000 per pound. The sides of the seed are rounded between the angles, which makes them appear less triangular. The stems and leaves are smaller than those of Japanese, while the stems are more reddened at maturity. Common Gray is like Silverhull[10] but may have smaller seeds. Lots designated as common buckwheat are mechanical and hybrid mixtures of Japanese and Silverhull.[15] A new tetraploid-seeded variety is now grown in the United States.

Winged or notch-seeded buckwheat occurs only as a mixture. It is merely a type of common Japanese buckwheat in which the angles of the hulls are extended to form wide margins or wings, making the seeds look large.

Tartary or mountain buckwheat (*F. tataricum*) is grown occasionally for feed in the mountains of North Carolina and Maine and in certain other areas. It has escaped from cultivation to become a weed, particularly in the prairie provinces of Canada. It is sometimes called India wheat, duck wheat, rye buckwheat, Mountain Siberian, wild goose, or Calcutta.

Tartary buckwheat is distinguished from the common species by its more indeterminate habit, simple racemes, and smaller seeds (26,000 per pound) that are nearly round in cross section and usually pointed. The seed color ranges from dull gray to black, while the pericarp (hull) varies from smooth to decidedly rough and spiny. The plants are somewhat viny. The flowers are very small and inconspicuous with greenish-white sepals.[10, 15]

A so-called wild buckwheat (*Polygonum convolvulus* L.) is a common weed.

Common buckwheats completed terminal growth 6 to 8 weeks after planting, and tartary in 10 weeks in Pennsylvania.[15] The percentages of hull in the grain ranges from 18 to 22 per cent. Hybrid buckwheat seed is being produced in the Soviet Union by interplanting two strains and setting beehives in the field.

CULTURAL METHODS

Early plowing is advantageous. The soil preparation as well as the seeding depth are about the same as for small grains.[3, 10, 15] Buckwheat land either is not fertilized or 100 to 150 pounds per acre of superphosphate are applied. Recommendations[7] suggest the application of 200 to 300 pounds per acre of superphosphate, or 15 to 30 pounds per acre of phosphoric acid and 10 to 15 pounds each of potash and nitrogen in a mixed fertilizer. A Canadian recommendation is 20 pounds per acre each of N and P_2O_5. Moderate applications of lime have shown some benefit, but heavy liming is detrimental.

Buckwheat is generally seeded in the Northeast between June 24 and July 1, and in mid-June in Manitoba, Canada. It is seldom advisable to seed after July 15. Seeding time in a locality can be calculated fairly accurately by allowing a period of 12 weeks for growth before the average date of first fall frost. Seeding in late May is preferable in northern Europe where the summer climate is cooler.[4] Buckwheat germinates best when the soil temperature is about 80° F., but it will germinate at any temperature between 45° and 105° F. The rate of seeding varies from 3 to 5 pecks per acre for common buckwheat, while 2 pecks is sufficient for the smaller-seeded tartary buckwheat. The crop may be either drilled or broadcast.

Buckwheat often is harvested with a combine, preferably after swathing. It is usually necessary to dry grain harvested by direct combining before it can be stored safely, because the plants do not bloom or ripen uniformly. On small farms or rough lands buckwheat has been harvested with a binder, cradle, scythe, or self-rake reaper.[10]

Injurious effects of buckwheat on subsequent crops on unfertilized land are attributed to excessive removal of mineral plant foods by the rapidly growing, shallow-feeding buckwheat plants.[15] A winter cover crop should follow buckwheat not only to increase organic matter, but also to reduce erosion.[15] Buckwheat stubble land is more subject to erosion than is small-grain land, due to the loose friable condition of the soil.

CHEMICAL COMPOSITION

The percentage chemical composition of the whole grain of buckwheat is about: water 10, ash 1.7, fat 2.35, protein 11, fiber 12, and nitrogen-free extract 60–64. The lysine percentage of the crude protein exceeds that of cereal grains.

Buckwheat has been grown primarily for human food, but since only the grain of better quality is bought by the millers most of it is fed to livestock or poultry. About 6 to 7 per cent is used for seed. Buckwheat flour is consumed largely in the form of buckwheat cakes or in mixed pancake flours. A continued heavy diet of buckwheat cakes results in development of a skin rash in those who are allergic to buckwheat protein. Animals fed a buckwheat ration also develop a rash, but the effect is confined to white-haired animals that are exposed to light.[10]

Buckwheat is one of the best temporary honey crops, since it produces blossoms for 30 days or more. It will supply enough nectar on one acre for 100 to 150 pounds of honey.[3] Buckwheat honey is dark, with a distinctive flavor that some people do not relish. Bees ordinarily do not collect nectar from tartary buckwheat.

The leaves and flowers of buckwheat formerly were the commercial source of rutin, a glucoside used medicinally to check capillary hemorrhages, help reduce high blood pressure, prevent frostbite gangrene, and act as a protection against the after effects of atomic radiation.[7] Synthetic rutin has replaced that from buckwheat.

The whole grains are fed to poultry, but the grain often is hulled for other stock. Buckwheat is a valuable green-manure summer cover crop, smother crop, or catch crop.

Buckwheat shorts or middlings are a valuable constituent in mixed feeds for livestock. Buckwheat hulls have very little feeding value. Most mills burn the hulls but some are sold for packing.

MILLING

Buckwheat may be milled either for flour or for groats (hulled grains).[2] Tartary buckwheat is not milled because the flour has a dark color and a bitter taste.

In flour milling the buckwheat is cleaned and dried to 12 per cent moisture, then scoured and aspirated to remove dust, fuzz, and the calyx that adheres to the fruit. The grain is then passed through break rolls where the hulls are cracked and loosened. Since some moisture may be taken up during this process, the material is again dried to 12 per cent moisture to aid in the separation of the hulls from the kernels. The broken grain, after the hulls have been sifted out, is further ground and sifted. Some buckwheat flour is milled so fine that it is as white as wheat flour. However, mills generally are coarse bolting cloths, through which small particles of hulls pass and remain in the flour. These particles give the flour a characteristic dark color. The dark coarse-particled residue from flour extraction is called middlings or shorts.

About 100 pounds of clean buckwheat may yield 60 to 75 pounds of flour, 4 to 18 pounds of middlings, and 18 to 26 pounds of hulls,[10] but only about 52 pounds of pure white flour are obtained.

In milling for groats, separated uniform medium-sized kernels are passed between two millstones adjusted to crack the hulls without a grinding action. The purified whole groats are used in porridge, soups, and breakfast food, mostly by European immigrants. Broken groats are eaten as roasted broken kernels and farina.

PESTS

Buckwheat suffers relatively little damage from either diseases or insects. The diseases most frequently reported are a leaf spot caused by the fungus, *Ramularia* species, and a root rot caused by *Rhizoctonia.* Wireworms, aphids, birds, deer and rodents attack buckwheat occasionally.

REFERENCES

1. Ali-Khan, S. T. "Growing buckwheat," *Can. Dept. Agr. Publ. 1468,* pp. 1–5. 1972.
2. Coe, M. R. "Buckwheat milling and its by-products," *USDA Cir. 190.* 1931.
3. Cormancy, C. E. "Buckwheat in Michigan," *Mich. Agr. Exp. Sta. Spec. Bul. 151.* 1926.
4. Lewicki, S., Ruszkowski, M., and Kaszlej, Z. "Studies on buckwheat. Pt. 9. Grain and green forage yields as depending on 10 different seeding times," *Off. Tech. Serv. U.S. Dept. Commerce.* 1963.
5. Marshall, H. G. "Description and culture of buckwheat," *Pa. Agr. Exp. Sta. Bul. 754,* pp. 1–26. 1969.
6. Marshall, H. G. "Isolation of self-fertile homomorphic forms in buckwheat, *Fagopyrum sagittatum* Gilib," *Crop Sci.,* 9(5):651–653. 1969.
7. Naghski, J., and others. "Effects of agronomic factors on the rutin content of buckwheat," *USDA Tech. Bul. 1132.* 1955.
8. Percival, J. *Agricultural Botany,* 7th ed., Duckworth, London, pp. 350–355. 1926.
9. Robbins, W. W. *Botany of Crop Plants,* 3rd ed., Blakiston, Philadelphia, pp. 276–286. 1931.
10. Sando, W. J. "Buckwheat culture," *USDA Farmers Bul. 2095.* 1956.
11. Stemple, F. W. "Experiments with buckwheat," *W. Va. Agr. Exp. Sta. Bul. 171.* 1919.
12. Stevens, N. E. "The morphology of the seed of buckwheat," *Bot. Gaz.,* 53:59–66. 1912.
13. Stevens, N. E. "Observations on heterostylous plants," *Bot. Gaz.,* 53:277–308. 1912.
14. Truog, E. "The utilization of phosphates by agricultural crops, including a new theory regarding the feeding power of plants," *Wis. Agr. Exp. Sta. Res. Bul. 41.* 1916.
15. White, J. W., Holben, F. J., and Richer, A. C. "Experiments with buckwheat," *Pa. Agr. Exp. Sta. Bul. 403.* 1941.

Chapter 32

Flax

ECONOMIC IMPORTANCE

Flax (*Linum usitatissum*) was harvested on an average of 6.3 million hectares from 1970 to 1972. The seed production was about 3.2 million metric tons, or roughly 0.51 metric tons per hectare. The leading producing countries were Canada, the United States, the U.S.S.R., India, and Argentina. Perhaps 10 per cent of the flaxseed was a by-product threshed from approximately 1.5 million hectares of flax that was harvested for fiber. The fiber yield was about 92,000 metric tons, or 0.37 metric tons per hectare. The leading countries in fiber flax production are the Soviet Union, Poland, France, Belgium, Czechoslovakia, and the Netherlands. In the United States, flax was harvested for seed on 1,851,000 acres (about 750,000 hectares) yielding 19,885,000 bushels or 11.4 bushels per acre (0.72 metric tons per hectare) (Figure 32–1).

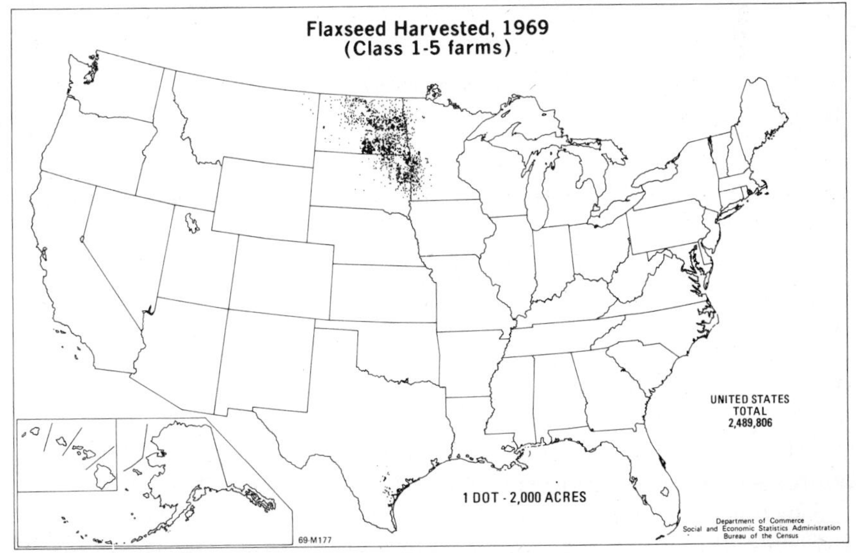

Figure 32–1. Acreage of flax harvested in the United States in 1969.

The production of flax for fiber in the United States was discontinued in 1956, but about 200,000 tons of fiber is decorticated from a million tons of seedflax straw for making cigarette paper. Flax was planted on about 3,368,000 acres in Canada and 2,963,000 acres in the United States in 1970. Only 1,191,000 acres was planted in the United States in 1972. The production is also declining because other products substitute for linseed oil in paints.

HISTORY

Flax was cultivated long before the earliest historical records. Remnants of flax plants were found in the Stone Age dwellings in Switzerland. The art of making fine linen from the flax fibers was practiced by the ancient Egyptians. Primitive man probably used the seeds of wild flax for food.[8]

Flax is an Old World crop that was probably cultivated first in southern Asia as well as in the Mediterranean region. It may have originated from the wild flax (*Linum angustifolium*)[8] native to the Mediterranean region, the only species with which cultivated flax crosses readily.

Cultivation of fiber flax began in America during the colonial period, when it was used to supply linen for hand spinning in the home. The commercial extraction of linseed oil began about 1805. Flax seed production moved westward with the settlement of new lands not only because it was a good first crop on newly broken sod, but also because of the damage from flax wilt on old cultivated fields. The development of wilt-resistant varieties has made flax a dependable crop on old cultivated lands.[3]

ADAPTATION

Flax requires moderate to cool temperatures during the growing season. Seed flax is generally grown where the average annual precipitation ranges from 16 to 30 inches (40 to 75 cm.) but it is also grown under irrigation in dry climates. Drought and high temperatures, about 90° F. (32° C.), during and after the flowering stage reduce the yield, size, and oil content of the seed as well as the quality of the oil.[12] The fiber-flax plant requires adequate moisture as well as a cool temperature during the growing season; but after maturity, dry weather facilitates harvesting, curing, and drying after retting. Cool weather from March to June, followed by warm dry weather in July, affords excellent conditions for fiber-flax production.[31]

Flax may be damaged or killed at temperatures of 18° to 26° F. (−7° to −4° C.) in the seedling stage, while a light freeze of 30° F. (−1° C.) may cause injury in the blossom or green boll stage, but between these stages the plants may survive temperatures of 15° F. (−10° C.) or even lower.[10] Frost injury may occur in the northern states, either when the flax is sown too early in the spring, or when sown so late that it is frosted in an immature stage in the fall. Flax sown in early October in California or southern Texas may be damaged by a sudden freeze in February when it is in the blossom stage. That sown in late November or in December may be damaged in the seedling stage. Varieties released in Texas are a distinctly winter type, resistant to cold.

Flax makes its best growth on well-drained, medium-heavy soils, especially silt loams, clay loams, and silty clays. Light soils are unsuited to seed flax, particularly in regions of deficient rainfall.[24] Since flax has a relatively short root system, it is dependent on moisture largely in the top 2 feet of soil. Fiber flax grown on the heavier soils has consistently outyielded that grown on the lighter soils.[32] Weedy land should be avoided.

BOTANICAL DESCRIPTION

Flax is an annual herbaceous plant that may be grown as a winter annual in warm climates. The plant grows to a height of 12 to 48 inches (30 to 120 cm.). It has a distinct main stem and a short taproot. Two or more basal branches may arise from the main stem just above the soil surface unless the stand is thick. The main stem and basal branches give rise to the primary, secondary, and tertiary branches that bear the leaves, flowers, and bolls (Figure 3–12).

Three principal tissue areas are recognized in the stems: pith, wood, and bark. The bark contains the comparatively long bast or flax-fiber cells that constitute the linen fibers.

The flax flower has five petals and a five-celled boll or capsule that contains ten seeds when each of the cells bears the complete number of two seeds (Figure 32–2).

A typical boll may contain 8 seeds or less at maturity because some of the ovules fail to develop, or cease to develop under stress conditions. Small-seeded varieties produce more seeds per boll than large-seed varieties in order to equalize seed yields.[16]

The flowers open at sunrise on clear warm days, the petals falling before noon. Flowering is indeterminate and continues until growth is stopped. The petals are blue, pale blue, white, or pale pink, depending on the variety. The bolls are semidehiscent in most varieties grown in the North Central states.[6] The bolls of this type rarely dehisce so far as to allow the seeds to fall out. Most varieties of Indian and Argentine origin are indehiscent.

The seeds of flax vary from 3.6 to 5.0 mm. in length. A thousand seeds weigh 3.8 to 7.0 grams. There are 65,000 to 120,000 seeds in a

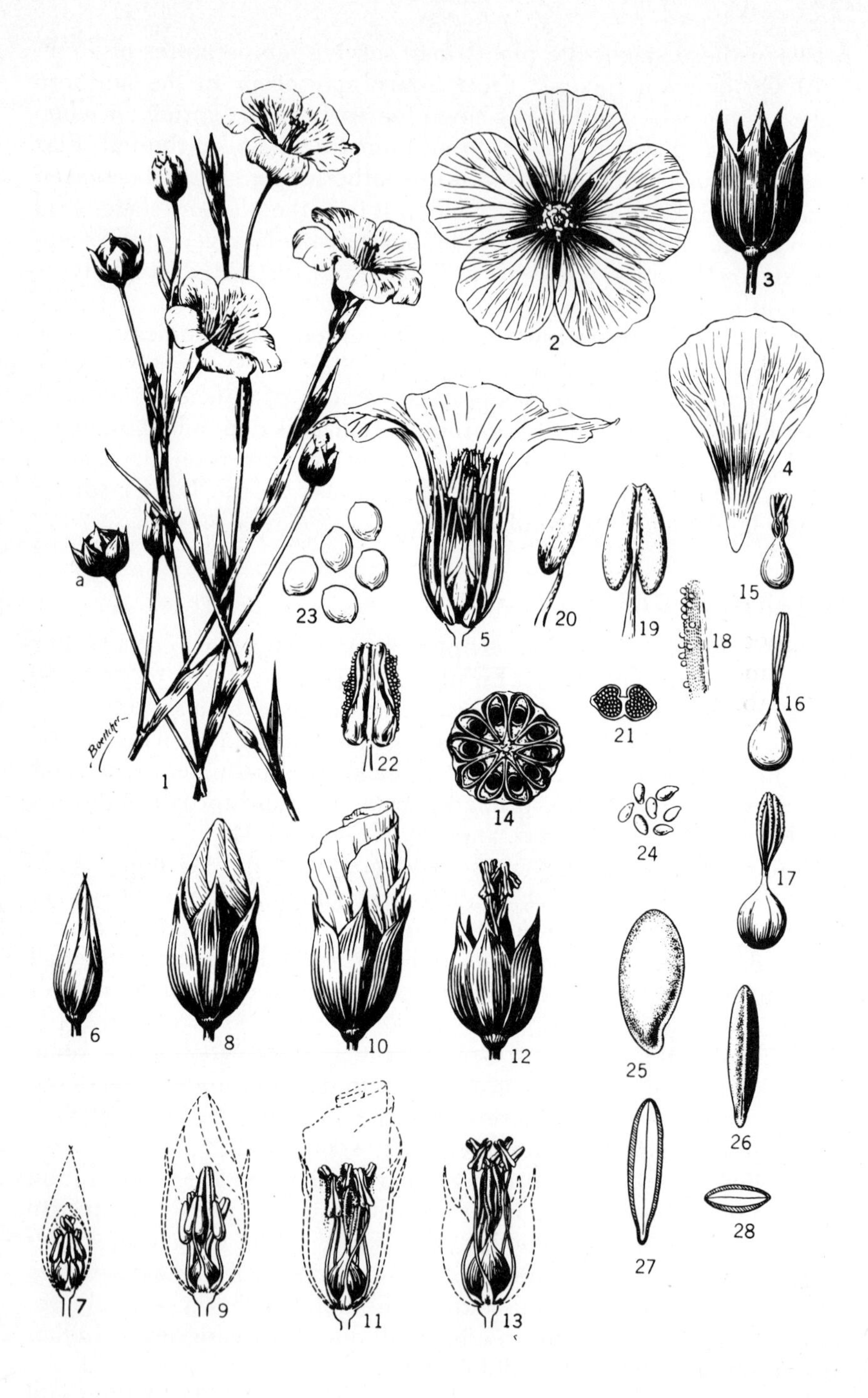
1
2
3
4
5
6
7
8
9
10
11
12
13
14
15
16
17
18
19
20
21
22
23
24
25
26
27
28
a
Boettcher

Figure 32–2. [OPPOSITE] Flax inflorescence: (1) upper stems showing leaves, flowers, and boll (*a*) about natural size; (2) expanded flower; (3) calyx after anthesis and shedding petals; (4) upper surface of a petal; (5) section of flower showing calyx, corolla, the five stamens, and five stigmas; (6 to 13) four stages of flower opening and anthesis—(6 and 7) two days before anthesis, (8 and 9) late afternoon of day before anthesis, (10 and 11) anthesis occurring at sunrise, (12 and 13) three to six hours after anthesis; (14) cross section of boll, showing the ten ovules (seeds) developed in the five carpels; (15 to 17) pistil before and during anthesis; (18) portion of stigma, greatly magnified, showing adhering pollen grains; (19) dorso-ventral view of anther with a portion of the filament; (20) lateral view of anther; (21) cross section of anther; (22) dehisced anther; (23) pollen grains, greatly magnified; (24) seeds, natural size; (25 and 26) seed, magnified; (27) seed dorso-ventral longitudinal (sagittal) section, showing cotyledons and surrounding endosperm; (28) cross section of seed showing cotyledons and surrounding endosperm.

pound. The seeds are usually light brown in color, although in certain varieties they are yellow, mottled, greenish-yellow, or nearly black. The seeds have a smooth shiny surface that results from a mucilaginous covering (Figure 32–3). The embryo is surrounded by a thin layer of endosperm that contains starch in the immature seed. Flaxseed gives satisfactory germination when stored from 5 to 10 years under dry conditions, but that stored from 15 to 18 years shows low viability.[13]

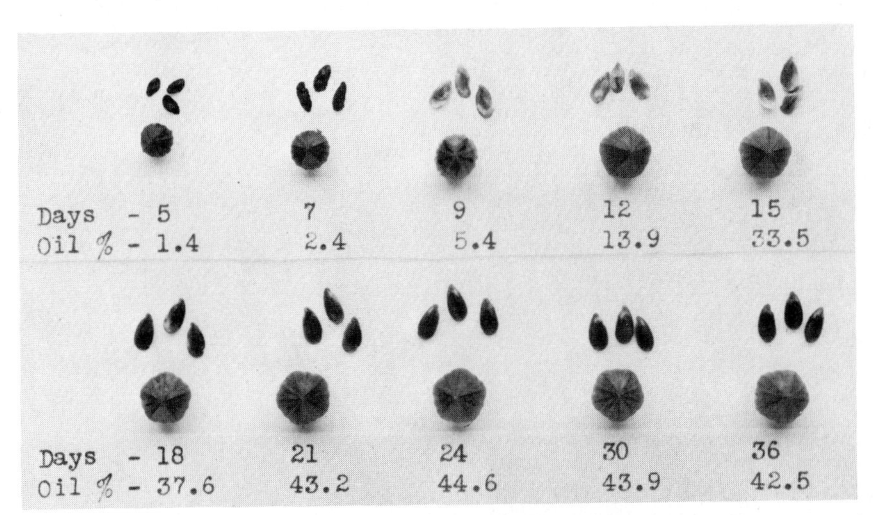

Figure 32–3. Ten stages of seed and boll development, and oil content of flax seeds from five to thirty-six days after pollination.

POLLINATION

Flax is normally self-pollinated, but 0.3 to 2 per cent natural crossing may occur.[9, 20] Insects seem to be important agents of natural crossing. Large-flowered varieties with flat petals show the greatest percentage of crossing.[9] Weather conditions and the distance between plants also influence the amount of out-crossing.[20]

FLAX GROUPS

Several more or less distinct groups of seed flax are recognized.[11] These are (1) wilt-resistant, short-fiber flaxes, (2) common or Russian, (3) Argentine, (4) Indian, (5) Abyssinian, and (6) Golden or yellow-seeded. Most of the varieties grown in the United States belong to the wilt-resistant short-fiber type,[2, 3, 25] and (7) indehiscent short fiber (Indian). Cotyledons are still attached at the bases of the branches in (4) and (7).

The seed flaxes are shorter, more branching, and more productive of seed than the fiber flaxes. They range in height from 15 to 30 inches

Figure 32–4. Flax: (*A*) ripe bundles of (1) fiber type and (2) seed type; (*B*) field in bloom; (*C*) young plants of five varieties – (3) Roman, winter type, (4) Punjab, (5) Cirrus, fiber type, (6) Bison, and (7) Rio. Cotyledons are still attached at the bases of the branches in (4) and (7).

(38 to 76 cm.), while fiber flaxes range from 37 to 48 inches (76 to 112 cm.) (Figure 32–4). Seeds of seed flax may be either large, medium, or small, but those of all fiber varieties are small. The fiber in seed flax is short, and sometimes harsh. Consequently it is not used for production of fine linen yarn. The small stems and the absence of basal branches of fiber flax are merely a result of a thick seeding.[8]

The wilt-resistant, so-called short-fiber flaxes are grown only for seed production. They are much shorter than the fiber flaxes, and have small brown seeds, slender stems, and moderate wilt resistance.[3]

Varieties of the common or Russian group have disappeared from cultivation because of their susceptibility to wilt. These varieties have taller stems and larger seeds than do those of the short-fiber varieties.

Varieties of the Argentine group of seed flax, such as Rio, have large open blue flowers, tightly closed bolls, and large brown seeds.

The Indian flaxes with blue flowers, indehiscent bolls, and medium-large brown seeds of high oil content, are adapted to a long growing season, particularly for fall seeding in mild climates.

The Abyssinian seed flaxes are short, fine-stemmed, much branched, and leafy. They have blue flowers and small brown or yellow seeds.

The typical Golden flaxes have pale-pink flowers and medium to large yellow seeds. Some varieties of this group are very susceptible to the pasmo disease, but have fair wilt resistance and excellent to good resistance to rust.

The typical fiber-flax varieties with blue or white petals are characterized by tall stems, few seed branches, small funnel-shaped flowers, and small brown seeds. Several varieties were formerly grown in the United States.

CROP ROTATIONS

Flax is most productive on clean land because it is a poor competitor with weeds. Flax is a good crop to follow newly broken pastures or meadows. It is best grown after either a clean-cultivated row crop or a legume. In the North Central states, flax grows well when it follows corn. A satisfactory sequence of crops includes a small grain, a legume, corn, and flax. In southeastern Kansas[4] flax after soybeans yielded much better than flax following corn, kafir, or oats. Where it is adapted, flax has proved to be a good companion crop for alfalfa, clover, or grass because it offers less competition for light than do small-grain crops. Flax rarely does well after small grains because of the abundance of weeds, unless the grain stubble is plowed early and then worked to stimulate germination of weed seeds. The weeds then are killed by winter freezes, and the flax is sown in the spring after a shallow disking. Flax sown after potatoes or sugarbeets is likely to be weedy because the digging of these crops brings weed seeds to the surface.

Fiber flax may follow sod or corn.[31] Because of the disease problem, flax should not be planted on the land more often than once in each three to six years.

SEED-FLAX CULTURE

Flax requires a firm, weed-free seedbed. Clean-cultivated land for row crops usually is disked instead of plowed in preparation for flax. Except in the drier areas, fall plowing is generally practiced where plowing is necessary. Commercial fertilizers are seldom applied to flax lands in the North Central states. Manure and nitrogen fertilizers increase flax yields appreciably in southeastern Kansas, while phosphates and lime also are beneficial. Nitrogen fertilizers are often used on irrigated lands in California. Liquid ammonia is added to irrigation water.

Weed Control

Flax does not compete well with weeds. A heavy weed infestation in the field may reduce flax yields by as much as 40 to 70 per cent. Sowing clean seed on clean soil is essential to good flax yields. Wild oats, foxtail, mustard, lambsquarters, canarygrass, and barnyard grass commonly occur in flax fields. These can be partly controlled with herbicides.

Seeding

Flax is generally sown in the spring as early as possible after the seeding of spring small grains is completed. The highest yields are obtained when the crop makes its growth during comparatively cool weather. In the northern Great Plains flax produces well when sown early in May. Flax is sown in April in Iowa, Idaho, and Washington. Fall seeding between November 15 and December 15 is recommended for southern Texas and California. Flax sown in the Imperial Valley of California on September 20, when the soil temperature was above 100° F., gave poor stands.[13]

The general practice is to sow flax with a grain drill at a depth of 1 inch or less. It is sown at the rate of 42 pounds per acre (47 kilos per hectare) or slightly more under humid conditions in Minnesota,[3] Iowa, Kansas, or elsewhere[21] or under irrigation in the western states. Large-seeded varieties are sometimes sown at a rate of 56 pounds.[29] Under drier conditions in Montana and the Dakotas, 28 pounds is a common rate of seeding. Seed treatment is often helpful.

Harvesting Seed Flax

Seed flax is generally harvested after a majority of the bolls are ripe. The bolls of most varieties are semidehiscent when dry enough to thresh readily with a combine. Partial dehiscence occurs when the seeds contain 9 to 11 per cent moisture.[6]

Seed flax may be harvested directly with a combine where the crop is thoroughly dry and free from weeds. Many fields ripen unevenly or contain green weeds. Such fields are harvested with a windrower and pick-up combine, or the field may be sprayed with a desiccant before combining. About 85 per cent of the 1960 crop was windrowed and then combined, while 13 per cent was combined directly. A relative humidity well below 75 per cent appears to be necessary for effective drying.[7] Air-dried flaxseed usually contains about 6 to 10 per cent moisture compared with 10 to 14 per cent in wheat under the same conditions. The lower water absorption of flaxseed, as compared with wheat and other starchy seeds, is characteristic of oleaginous (oil-bearing) seeds. Flaxseed that contains more than 9 to 11 per cent moisture is likely to deteriorate in storage.

LINSEED-OIL EXTRACTION

Flaxseed contain 32 to 44 per cent oil, based on dry weight. Large seeds are highest in oil content. A gallon of linseed oil weighs about 7½ pounds. In commercial crushing about 2.67 gallons, or 20 pounds, of oil are obtained from a bushel (56 pounds) of cleaned flaxseed, chiefly by the solvent process. The oil content of flax continues to increase until 18 to 25 days after flowering, but total oil per seed and total dry weight continue to increase until the seeds are mature[5, 10] (Figure 32–3). The oil content usually is low when drought occurs at the filling stage, i.e., within a period of about 30 days after the flax blossoms. Shriveled seeds are low in oil.

The iodine number is a chemical test for the drying quality of the oil. It is a measure of the quantity of oxygen the unsaturated chemical bonds in the oil will absorb in drying to form the characteristic paint film. The iodine number may be defined as the number of grams of iodine that 100 grams of oil will absorb. The iodine number of linseed oil usually ranges from 160 to 195. Linseed oil must have an iodine number of not less than 177 to meet standard specifications. The unsaturated fatty acids in linseed oil are oleic with one double bond, linoleic with two, and linolenic with three double bonds.

A comparison of the principal vegetable oils is shown in Table 32–1.

The specific gravity of the above oils ranges from 0.91 to 0.93, exceptions being castor oil, 0.96; tung oil, 0.94; and oiticica, 0.97. The plants

TABLE 32–1. **Characteristics of the More Common Oil Seeds and Vegetable Oils**

CROP	BOTANICAL NAME	OIL CONTENT (%)	IODINE NUMBER	WORLD PRODUCTION, OIL 1973 (1000 metric tons)
Drying oil				
Perilla	*Perilla frutescens*	40–58	182–206	–
Flax	*Linum usitatissamum*	35–45	170–195	790
Tung (China wood)	*Aleurites fordii*	40–58	160–170	140
Hemp seed	*Cannabis sativa*	32–35	145–155	–
Safflower	*Carthamus tinctorius*	24–36	140–150	285
Oiticica	*Licania rigida*	60–75	140–148	30
Drying or semi-drying oil				
Soybean	*Glycine max*	17–18	115–140	7170
Semidrying oil				
Sunflower	*Helianthus annuus*	29–35	120–135	3730
Corn (germ)	*Zea mays*	50–57	115–130	305
Cottonseed	*Gossypium hirsutum*	15–25	100–116	2780
Rapeseed	*Brassica napus*	33–45	96–106	2480
Nondrying oil				
Sesame	*Sesamum indicum*	52–57	104–118	605
Peanut	*Arachis hypogeae*	47–50	92–100	3020
Castorbean	*Ricinus communis*	35–55	82–90	365
Coconut	*Cocos nucifera*	67–70	8–12	2675
Olive	*Olea europaea*	–	86–90	1470
Palm	*Elaecis guineensis*	–	49–59	–
Palm kernel	" "	–	204–207	510
Babassu	*Orgignya oleifera*	–		5695
Olive residue	*Olea europaea*	–	10–31	140
Crambe	*Crambe abyssinica*	30–50	95–108	–
Tall oil				35*

* 1970 production.

that produce tung, oiticica, olive, palm, and coconut oils are trees. The other oils are from field crops discussed elsewhere in this book.

The oils from seeds of soybean, cottonseed, peanut, sunflower, sesame, safflower, corn, mustard, and most of the rape are regarded as edible oils. Palm oils include palm kernel, babassu, and coconut, but coconut oil is an edible oil. Industrial oils are linseed, castor, oiticica, hemp seed, tung, perilla, and olive residue.

Tall oil, a by-product of wood pulp contains about 75 per cent of oleic and linoleic acids. It is used in soaps and paints.

Linseed oil is expressed from flaxseed with a hydraulic press, a continuous-screw press, or, more recently, by a combination of pressing and solvent extraction. By the latter method, cleaned linseed is ground to a meal, heated with steam to a temperature of 190° to 200° F., and then forced through a tapered steel bore by continuous screw action. About two-thirds of the oil in the meal is squeezed out through the fine grooves and perforations of the bore. Most of the oil that remains in the press cake is then extracted with a solvent as described for soybean-oil extraction in Chapter 26. The residue left after the oil is expressed is known as linseed cake, or, when ground, as linseed meal. The cake contains from 1 to 6 per cent oil, depending on the method of extraction. The protein content of the oil meal is about 35 per cent.

FIBER-FLAX CULTURE

The seedbed requirements for fiber flax are essentially the same as for seed flax. With some exceptions, phosphorus increases the fiber percentage while nitrogen decreases it.[30] A 4–16–8 fertilizer mixture usually is recommended for fiber flax where little is known regarding the particular soil conditions.

Seeding

Fiber flax is sown in the spring at the rate of 75 to 85 pounds per acre. Such thick seeding produces tall, nonbranching plants. Lodging frequently occurs when the seeding rate is appreciably heavier.

Harvesting

The fiber-flax crop generally is harvested when one-third to one-half of the seed bolls are brown or yellow with fully developed brown seeds. At this stage the stems usually have turned yellow, while the leaves have fallen from the stems two-thirds the distance from the ground.[31] The fibers of flax plants harvested too early tend to be fine and silky but lacking in strength; fibers of flax harvested late are coarse, harsh, and brittle, with poor spinning qualities.

Fiber-flax plants are pulled from the ground with special pulling machines or by hand. The pulled flax is shocked in the field to cure, after which it is stacked or placed in a shed to await processing. Yields of 1¾ tons of cured pulled flax per acre are typical. Such a quantity yields about 400 pounds of scutched fiber (including tow) and about 300 to 500 pounds of seed.

PROCESSING FIBER FLAX

In the mill, fiber flax is first threshed or deseeded. The next steps are retting, breaking, and scutching.[19, 31]

Retting (or partial rotting) dissolves gums that bind the fibers to the wood, and destroys the thin-walled tissues that surround the fibers.

Retting is brought about chiefly by common soil-inhabiting bacteria that are present on the straw when it is harvested.

Two common methods of retting are practiced: dew retting and water retting. Water retting is more satisfactory because of more uniformly favored conditions. In dew retting, the flax is spread on the ground where it is grown. It is retted by molds and bacterial action promoted by frequent rains and dews. Retting usually is completed in 14 to 21 days. The straw must be thinly and evenly spread on the ground for uniform retting. In water retting the straw is placed in tanks of water,[19] or in ponds or sluggish streams. The retting is accomplished in 6 to 8 days when the water is free of impurities and kept at 80° F. The water in the tank should be circulated for uniform retting. Cooler water lengthens the retting period. The bundles of retted straw are taken to a field where they are stood up to dry.

The breaking process consists of breaking the woody portions of straw into fine pieces—shives—by passing the dry retted straw between fluted rollers. At the same time the shives are broken or loosened from the fiber. The shives are then beaten off by a cylinder or wheel in the next process: scutching. The long fibers are strong and flexible enough to resist breaking during these processes. Good fiber averages 20 inches (50 cm.) in length. Single cells average 25 mm. in length and 0.23 mm. in diameter.

USES OF FLAX

About 81 per cent of the linseed oil produced is used in paints and varnishes, 11 per cent in the manufacture of linoleum and oilcloth, 3 per cent in printers' inks, and the remainder in soaps, patent leather, and other products. Linseed cake or meal is used as feed for livestock. Animals that are fed with it have glossy coats.

Seed-flax straw has had a limited market for the manufacture of upholstery tow, insulating material, and rugs. Fiber from flax straw is used in the making of cigarette, Bible pages, currency, and other high-grade papers. Such paper formerly was made entirely from linen rags. To make this paper, straw from fields with a good, dense, uniform growth of flax that is relatively free from weeds and grasses is gathered for processing.[18] Such fields yield about 1000 to 1500 pounds of straw per acre, with a fiber content of about 10 per cent. The straw from combined fields is raked into windrows, sometimes after the stubble has been mowed. It is then baled and hauled to a processing plant. There the straw is crushed and hackled to remove the shives, after which the fiber is baled and shipped to a paper mill.

The feeding value of flax straw compares favorably with that of wheat or oat straw. A flaxseed placed in the eye usually will remove a cinder. Ground flaxseed has long been used in medicine as a conditioner or for making a poultice. Processed edible linseed oil was shipped to Russia during World War II.

Flax fiber is spun into linen yarns, which are used in threads and twines of various kinds. The yarn is also woven into toweling, clothing fabrics, table linen, handkerchiefs, and other textiles. The short tangled fibers, called tow, usually a by-product, are used for upholstering, paper manufacture, and packing. Surplus seed of fiber flax is sold to oil mills for crushing.

DISEASES

Flax Wilt

Wilt is a fungus (*Fusarium lini*) disease that generally causes infected plants to wilt and die. It grows upon the live plant as well as upon the dead plant material in the soil. The fungus may remain in the soil for as long as 28 years. The use of resistant varieties is the only satisfactory control on wilt-infected soil. Varieties grown in the United States are resistant to wilt.

Flax Rust

Flax rust caused by the organism *Melampsora lini* frequently damages flax in wet seasons. Bright orange pustules, the uredineal stage on the leaves, are followed by black shiny areas on the stems (the telial stage) late in the season. The fungus is carried over winter on infected stubble and straw.[15] Crop rotation aids in reducing damage from the disease but resistant varieties are the only effective means of control.

Pasmo

Pasmo, caused by the fungus *Phlyctaena linicola,* appears on the foliage of young plants as yellow-brown circular lesions. As the plant reaches maturity, brown to black blotches are observed on the stems.[15] The disease is seed-borne. Diseased plants produce smaller seeds and sometimes fewer seeds per boll.[17] Some control measures are crop rotation, seed treatment, and use of resistant varieties.

Other Diseases

Anthracnose or canker, caused by the organism *Colletotrichum lini,* has caused some loss as a seedling disease. Seed treatment checks the spread of the disease.

Heat canker is caused by high temperatures at the soil surface when the plants are small. The young stems, girdled at the soil line, finally break over and die. The damage is prevented by early seeding.

Seed treatments improve stands and yields and check damage from seedling blight, particularly when the flaxseed has been damaged during threshing, as is usually the case.

Two virus diseases of flax, aster yellows and curly top, cause losses, but adequate control measures are unavailable. Resistant varieties are being developed. Aster yellows occurs mostly in the northern states where it also attacks several other crops and many weeds. The virus is transmitted by the six-spotted leafhopper, *Macrosteles fascifrons.* Curly top also attacks sugarbeets, beans, and several weeds. The disease is prevalent in the intermountain and Pacific Coast states. The virus is transmitted by the beet leafhopper, *Circulifer tenellus.* Some varieties show tolerance to the disease.

INSECTS

The most common insect injury to flax is from various species of grasshoppers and crickets that chew off the pedicels and allow the bolls to drop to the ground. These pests are controlled with poison baits or sprays. Other insects that sometimes damage flax include cutworms, particularly the pale western cutworm in Montana and the western Dakotas; armyworms, especially the Bertha armyworm in North Dakota and the beet armyworm in California; false chinch bugs in California; stink bugs in Texas; and the flax worm, *Cnephasia longana,* in Oregon. The latter pest is partly controlled by crop rotation. The corn earworm damages flax in Texas. Insecticides control most of these pests.

REFERENCES

1. Anonymous. "What you should know about oilseed crops," *Can. Agr. Exp. Sta. Publ 1448*, pp. 1–10. 1971.
2. Atkins, I. M., Reyes, L., and Merkle, O. G. "Flax production in Texas," *Texas Agr. Exp. Sta. Bul. 957.* 1960.
3. Culbertson, J. O., and others. "Growing seed flax in the North Central states," *USDA Farmers Bul. 2122.* 1958.
4. Davidson, F. E., and Laude, H. H. "Flax production in Kansas," *Kans. Agr. Exp. Sta. Bul. 191*, pp. 1–14. 1938.
5. Dillman, A. C. "Daily growth and oil content of flaxseeds," *J. Agr. Res.*, 37(6):357–377. 1928.
6. Dillman, A. C. "Dehiscence of the flax boll," *J. Am. Soc. Agron.*, 21:832–833. 1929.
7. Dillman, A. C. "Hygroscopic moisture of flax seed and wheat and its relation to combine harvesting," *J. Am. Soc. Agron.*, 22:51–74. 1930.
8. Dillman, A. C. "Improvement in flax," in *USDA Yearbook, 1936*, pp. 745–784.
9. Dillman, A. C. "Natural crossing in flax," *J. Am. Soc. Agron.*, 30:279–286. 1938.
10. Dillman, A. C. "Cold tolerance in flax," *J. Am. Soc. Agron.*, 33:787–799. 1941.

11. Dillman, A. C. "Classification of flax varieties, 1946," *USDA Tech. Bul. 1064.* 1953.
12. Dillman, A. C., and Hopper, T. H. "Effect of climate on the yield and oil content of flaxseed and on the iodine number of linseed oil," *USDA Tech. Bul. 844,* pp. 1–69. 1943.
13. Dillman, A. C., and Toole, E. H. "Effect of age, condition and temperature on the germination of flaxseed," *J. Am. Soc. Agron.,* 29:23–29. 1937.
14. Eastman, W. *The History of the Linseed Oil Industry,* T. S. Denison & Co., Minneapolis, pp. 1–277. 1968.
15. Flor, H. H. "Wilt, rust and pasmo of flax," in *Plant Diseases, USDA Yearbook, 1953,* pp. 869–873.
16. Ford, J. H. "Relation between seed weight and seeds per plant," *Crop Sci.,* 5(5):475–476. 1965.
17. Frederiksen, R. A., and Culbertson, J. O. "Effect of pasmo on the yield of certain flax varieties," *Crop Sci.,* 2:434–437. 1962.
18. Goldsborough, G. H. "Seed flax straw—potential for extending its utilization," *USDA Agricultural Research Service Spec. Rept. ARS 20–11.* 1963.
19. Harmond, J. E. "Processing fiber flax in Oregon," in *Crops in Peace and War,* USDA Yearbook, 1950–51, pp. 484–488.
20. Henry, A. W., and Tu, C. "Natural crossing in flax," *J. Am. Soc. Agron.,* 20:1183–1192. 1928.
26. Hill, D. D. "Seed-flax production in Oregon," *Ore. Agr. Exp. Sta. Cir. 133.* 1939.
22. Jamieson, G. S. *Vegetable Fats and Oils,* Reinhold, New York. 1943.
23. Johnson, I. J. "The relation of agronomic practices to the quantity and quality of the oil in flaxseed," *J. Agr. Res.,* 45:239–255. 1932.
24. Klages, K. H. W. "Flax production in Idaho," *Idaho Agr. Exp. Sta. Bul. 224.* 1938.
25. Knowles, P. F., Isom, W. H., and Worker, G. F. "Flax production in Imperial Valley," *Calif. Agr. Exp. Sta. Cir. 480.* 1959.
26. Kommedahl, T., and others. "A half century of research in Minnesota on flax wilt caused by *Fusarium oxysporum,*" *Minn. Agr. Exp. Sta. Tech. Bul. 273,* pp. 1–35. 1970.
27. Magness, J. R., and others. "Food and feed crops of the United States," *N.J. Agr. Exp. Sta. IR-4. Bul No. 1.,* pp. 1–255. 1971.
28. Ray, Charles, Jr. "Anthracnose resistance in flax," *Phytopath.,* 35(9):688–694. 1945.
29. Reddy, C. S., and Burnett, L. C. "Flax as an Iowa crop," *Iowa Agr. Exp. Sta. Bul. 344.* 1936.
30. Robinson, B. B. "Some physiological factors influencing the production of flax fiber cells," *J. Am. Soc. Agron.,* 25:312–328. 1933.
31. Robinson, B. B. "Flax-fiber production," *USDA Farmers Bul. 1728.* 1940.
32. Robinson, B. B., and Cook, R. L. "The effect of soil types and fertilizers on yield and quality of fiber flax," *J. Am. Soc. Agron.,* 23:497–510. 1931.
33. Zachary, L. G., and others. *Tall Oil and Its Uses,* McGraw-Hill, New York, pp. 1–136. 1965.

Chapter 33

Cotton

ECONOMIC IMPORTANCE

Cotton is the major textile fiber used by man despite the expanding use of synthetic fibers. It is grown on 33 million hectares, chiefly in the United States, the U.S.S.R., China, India, Brazil, Mexico, Pakistan, the UAR, and Turkey (Figure 33–1). Cotton lint production is nearly 12 million metric tons or 0.35 metric tons per hectare. Cottonseed production exceeds 24 million metric tons or 0.7 metric tons per hectare. From 1970 to 1972 cotton was grown in the United States on 11.9 million acres (4.8 million hectares) (Figure 33–2) with a lint production of 11.4 million bales of 480 pounds each (2.49 million metric tons) or 457 pounds per acre (0.51 metric tons per hectare). Cottonseed production was about 4.5 million metric tons. The average ratio of lint to seed production in United States is about 36 to 64. The states leading in cotton production were Texas, Mississippi, California, and Arkansas. Cotton production declined in the South Atlantic states until less than 20 per cent of the crop is produced east of the Mississippi Valley. There, the soil is less fertile and the cotton boll weevil is more prevalent. Also the smaller fields and more rolling terrain make it less suited to mechanized field operations. Synthetic fiber use in the United States is about 40 per cent greater than that of cotton.

HISTORY OF COTTON CULTURE

Cotton (*Gossypium hirsutum*) probably was cultivated in Mexico about 3500 B.C. and *G. barbadense* was grown in Peru about 2500 B.C.[63]

Cotton has been grown in India for making clothing for more than 2000 years, and in certain other countries for several hundred years. Early European travelers returned from southern Asia with weird tales of seeing wool growing on trees. Early herbalists sometimes illustrated the cotton plant by drawings of sheep hanging from the branches of a tree. Apparently tree types of cotton were grown to a

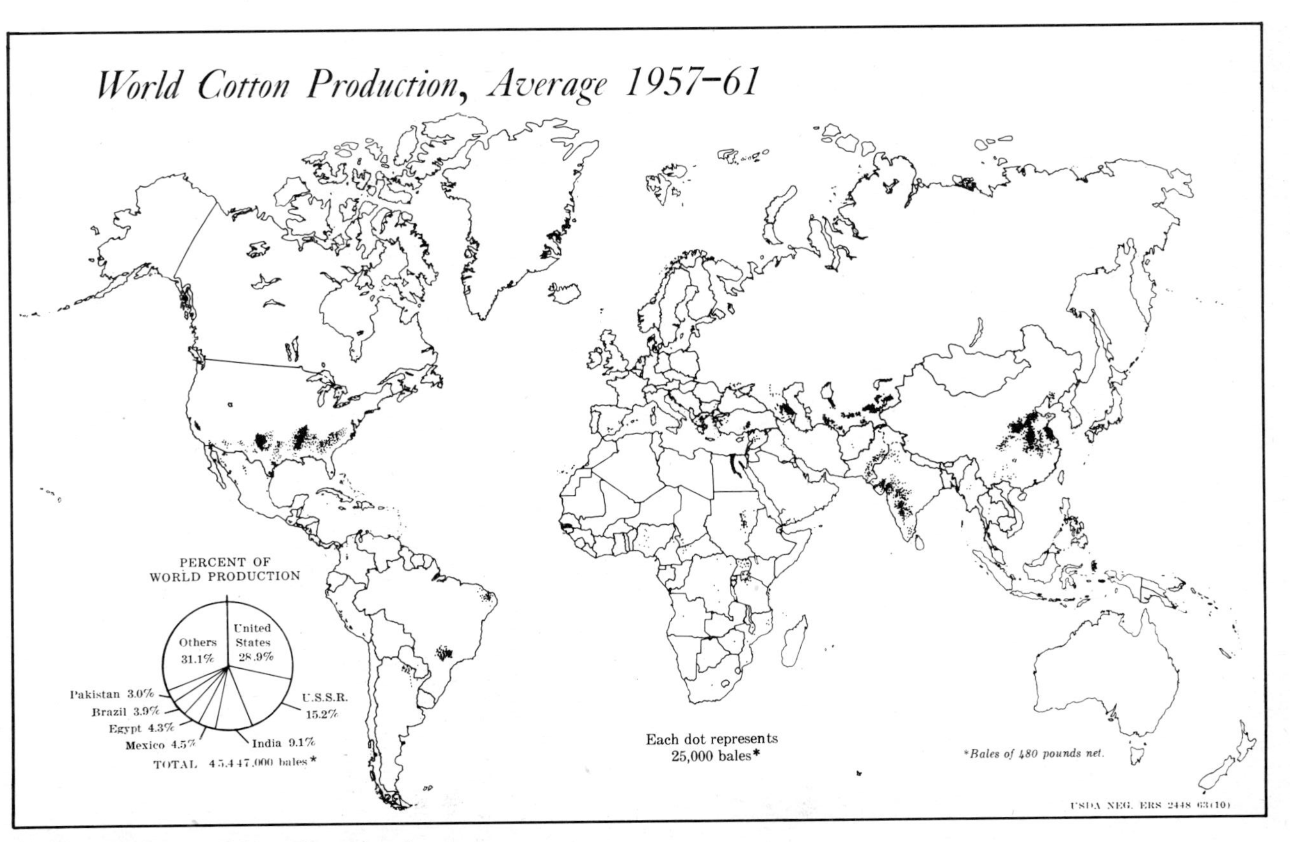

Figure 33–1. World production of cotton.

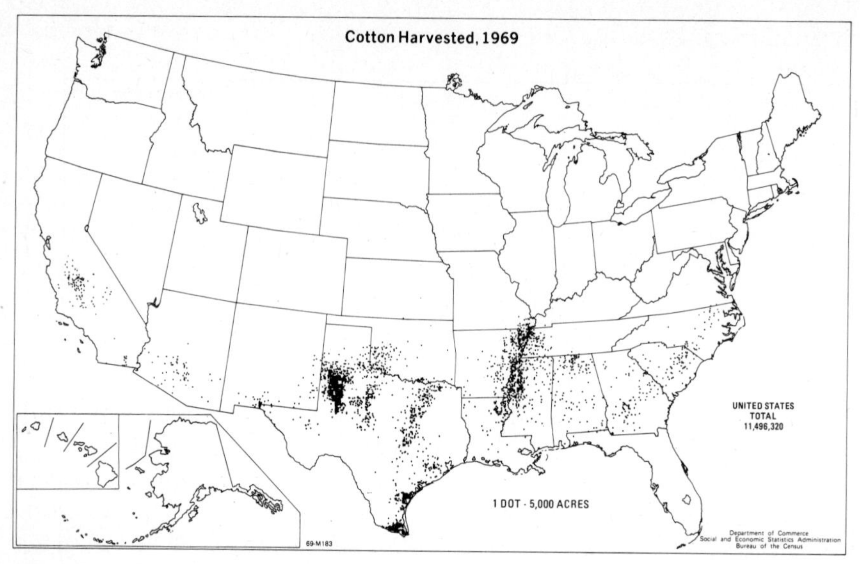

Figure 33–2. Acreage of cotton harvested in the United States in 1969.

considerable extent at that time. Furthermore, a cotton plant growing as a perennial in the tropics can attain the size of a small tree. Even today cotton is called *Baumwolle* (tree wool) by the Germans.

Columbus found cotton growing in the West Indies. Cotton fabrics probably 800 years old have been found in Indian ruins in Arizona. The crop was grown in the Virginia Colony in 1607.[12] Its culture soon spread throughout the south, but large-scale production began after the invention of the cotton gin in 1794.

There probably are two general centers of origin of the cotton plant: Indo-China and tropical Africa in the Old World, and South and Central America in the New World.[79] Separate origins are indicated by the fact that consistently fertile hybrids have never been obtained from crosses between the 26-chromosome American cottons and the 13-chromosome Asiatic cottons.[81] The three distinct types of cotton grown in the United States, i.e. Upland, Sea Island, and American-Pima are of American origin. Upland cotton is assumed to have descended from Mexican cotton or from natural crosses of Mexican and South American species.

ADAPTATION

Climatic conditions are favorable for cotton where the mean temperature of the summer months is not less than 77° F. The zone of cotton production lies between 37° north and 32° south latitude, except that in the Russian Ukraine, cotton is grown at up to 47° north latitude.

Three climatic essentials are freedom from frost for a minimum growing and ripening season, an adequate supply of moisture, and abundant sunshine.[17] General requirements are:

(1) a mean annual temperature of over 60° F., though where the distribution of rainfall, sunshine, and temperature is favorable, a mean of over 50° F. probably would be sufficient; (2) a frostless season of 180–200 days; (3) a minimum rainfall of 20 inches a year with suitable seasonal distribution—a maximum of 60 inches, or up to 75 inches, would not be excessive if distribution were favorable; (4) open sunny weather; areas recording "half cloudiness" annually have too little sunshine to be safe, and areas over three-fifths cloudy are unsuitable for cotton.[17]

The Sea Island and American-Pima types require about six months to mature, whereas the period for Upland varieties is about five months. In the United States, cotton is limited to sections that have 16 inches or more of rainfall without irrigation, but in parts of the irrigated southwestern cotton region the average annual rainfall is less than 6 inches. Cotton is grown on irrigated land in California, Arizona, and Nevada. Most of the crop in New Mexico, western Texas, and southwestern Oklahoma also is irrigated, and some irrigation is applied in other states.[80] All the cotton in Egypt and the Sudan is irrigated, as well as much of it in the Near East and Middle East.

The growing conditions most favorable for cotton are a mild spring with light frequent showers; a warm, moderately moist summer; and a dry, cool, prolonged autumn. Rainy weather when the bolls begin to open retards maturity, interrupts picking, and damages the exposed fiber. Irregular growth in cotton when irrigations are delayed causes lack of uniformity and lack of strength in the fiber.[35] Early killing frosts in the fall and high evaporation during the flowering period limit the yields of American-Pima cotton in Arizona.[24] American Upland cotton, being of indeterminate growth habit and very insensitive to length of day,[21] will produce flowers throughout the year under warm conditions.

The minimum, optimum, and maximum temperatures for the germination and early growth of cotton are about 60°, 93°, and 102° F., respectively. The most rapid growth and flowering of the plants occur at temperatures of 91° to 97° F.[73] The highest cotton yields in the United States are in areas where the mean July temperature averages 81° to 83° F.

Cotton grows well on moderately fertile soils. The soils in the cotton regions range from sands to very heavy clays with ranges in acidity from *p*H 5.2 to *p*H 8+. The best cotton lands are mixtures of clay and sandy loam that contains a fair amount of organic matter and a moderate amount of available nitrogen, phosphorus, and potash.[12] The heavier soils promote later maturity, larger vegetative growth, and greater boll-weevil damage. The best cotton regions, from the standpoint of both yield and quality, are perhaps the Mississippi Delta and the irrigated valleys of the Southwest. Cotton tolerates some salinity, but thrives best on nonsaline soils. A salinity of −5 bars of root osmotic potential may reduce cotton yields 50 per cent.

Environmental conditions that promote earliness in Mississippi are important because they help avoid excessive shedding due to midseason and late-season stress conditions, avoid boll-weevil infestation, and have the crop ready to pick before the onset of fall rains and cold weather.[49] Maturity is not affected by application of nitrates or by the stripping off of young bolls. Potash-deficient soils or fertilizers, low soil moisture, and other environmental contributions to earliness usually result in decreased yields.[42] Thick spacing, which rarely reduces yields, is the most practical way to obtain earliness. Thick spacing does not advance initial flowering appreciably but merely provides more plants to produce a proportionally larger number of early bolls, and restricts later branching and flowering.[20]

BOTANICAL DESCRIPTION

Cotton belongs to the family *Malvaceae*, or mallow family.

Upland cotton (*Gossypium hirsutum*) fibers range from $^3/_4$ to $1^1/_4$ inches or more in length and are of medium coarseness. The flowers are creamy white when they first open but they soon turn pink or red. The lint fibers adhere strongly to the seed. The bolls usually contain four or five locks (Figure 3–10).

Sea Island and American-Pima cottons (*G. barbadense*) have extra-long, fine fibers, $1^1/_2$ to 2 inches long, or even longer in some cases. The lint is readily detached from the seed. The petals are yellow with a purple spot at the base or claw. The bolls usually contain three locks. The term American-Egyptian was changed to American-Pima July 1, 1970.

The Asiatic cottons are classified as *G. arboreum* and *G. herbaceum*. The fibers are coarse and short, their length being mostly from $^1/_2$ to $^7/_8$ inches. Native cottons of the American Indians, such as the Hopis, likewise have short fibers. Upland and Asiatic cottons have shorter boll periods than do the Pima and Sea Island cottons.[5]

The cotton plant usually is considered an annual, although it is a long-lived perennial in the tropics where the mean temperature of the coldest months does not fall below 65° F. The plant is herbaceous, with a long taproot, and attains a height of 2 to 5 or more feet, with a main stem from which many branches arise. The taproot grows downward into moist soil at a rate of about 1 inch per day up to 5 weeks and nearly 0.38 inches daily for the entire 175-day growing period.[16] The leaves arise on the main stem in a regular spiral arrangement.

At the base of each cotton leaf petiole are two buds, the true axillary bud which continues to make a vegetative growth, and an extra-axillary bud which produces the fruiting branch.[29] A petioled, stipulate leaf with three, five or seven lobes arises at each node above the cotyledons (Figure 33–3). The leaves and stems are usually covered with fine hairs. The leaves are green except in a few red-leaf varieties.[60] Red-leafed varieties have been grown occasionally in the United States. Varieties with colored lint—green, brown, yellow, blue or red—are grown in the Soviet Union and other countries.

Figure 33–3. Cotton leaves, flowers, square (*upper right*), and unopened boll (*right*).

The flowers may appear arranged on alternate sides of the fruiting branch. Additional flowers appear on a branch at about 6-day intervals. There are three relatively large leaflike bracts at the base of the flower, above which is a true calyx that consists of five unequally lobed sepals. The corolla consists of five petals which range in color from white to yellow to purple in different types. The staminal column bears ten more or less double rows carrying 90 to 100 stamens, while the pistil is made up of from three to five carpels. The stigma tip extends up above the anthers. The fruit (Figure 3–10) is the enlarged ovary that develops into a three- to five-loculed capsule or boll. The bolls are $1\frac{1}{2}$ to 2 inches long among the common varieties; it requires about 60 to 80 bolls to produce a pound of seed cotton.[45] Other types have bolls not over 1 inch long. The earliest and latest bolls on a plant usually are smaller, with shorter fiber, than the intermediate bolls. The boll dehisces or splits open at maturity (Figure 33–4). Late unopened bolls are called bollies. The seeds are covered with lint hairs, i.e., the fibers[26] and usually also with short fuzz.

Flowering in the cotton plant begins from 8 to 11 weeks after planting. It continues until growth is stopped by frost, drought, insect attack, or other causes. In certain perennial types flowering awaits a suitable photoperiod. The fruit bud or square usually is discernible three to four weeks before the flower opens. Many of the buds, flowers, or young bolls drop off. As a result, only 35 to 45 per cent of the buds produce mature bolls under usual conditions. Most of the boll shedding occurs from 3 to 10 days after pollination.[27, 73] The period between flowering and the opening of the mature boll is about six to eight weeks, becoming longer as the season extends into cool autumn weather. The cotton plant shows a remarkable adjustment to its environment. Under severe drought conditions the plant may be

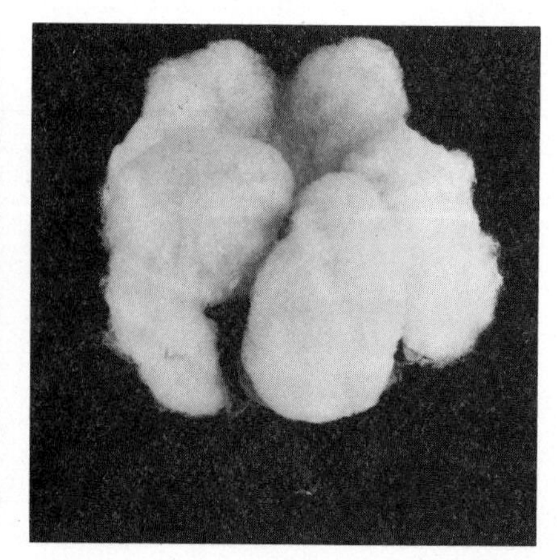

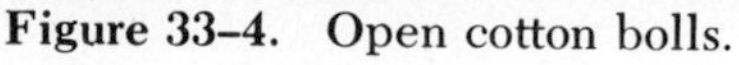

Figure 33–4. Open cotton bolls.

6 inches high and bear one boll. A plant of the same variety grown under irrigation may be 5 feet high and produce 40 bolls. The number of bolls that develop is kept in balance with the growth of the plant.[19, 73]

POLLINATION

Cotton is readily self-pollinated, while out-crossing is less than 5 per cent in some parts of Texas, but over 50 per cent in some states to the east. The extent of cross-pollination varies with the population of bees and other insects.[69]

Metaxenia, or the immediate effect of foreign pollen on the fertilized ovules, has reduced the lint length of Pima (long-staple) cotton when pollinated with Hopi (short-staple).[30]

LINT OR FIBER

Cotton fibers are slender single-cell hairs that grow out from certain epidermal cells of the cottonseed.[5] Their growth starts on the ovules about the time the flower opens. The fibers lengthen rapidly and attain full length in about 15 to 25 days when the seed has attained full length. For an additional 25 to 40 days, the cell walls of the fiber continue to thicken. Thickening occurs by the growth of two additional spiral rings each day on the inner surface of the cell wall. The fiber is cylindrical before the boll opens, but collapses and becomes more or less flattened and twisted (with convolutions) with the opening of the boll (Figure 33–5). The fiber is thin-walled, weak, and poorly developed when unfavorable conditions prevail during the time the fiber is thickening. A pound of lint may contain 100 million or more fibers. The fibers range in length from 6 mm. to over 50 mm. (Figure 33–6), and in thickness from 0.015 to 0.020 mm. Varieties with long fibers tend to have a low lint percentage in seed cotton, whereas those with short fibers have the highest percentage of lint.

The fiber length is enhanced by ample rainfall while the bolls are developing.[56] Thin-walled or immature fibers produce yarns that are unduly neppy or full of knots and snarls which cannot all be removed by the spinner. Dark-colored, plump, well-developed seeds yield a high percentage of mature fibers.[31]

Yarn strength is determined by the length, strength, and fineness of the fiber by an approximate respective ratio of 47:37:16.[5] Small-diameter fibers with thin walls contribute to yarn strength, but tend to cause a neppy or poor appearance in the yarn unless a combing process is used in spinning. Other things being equal, long fibers give a smoother and stronger yarn than do short fibers. Consequently, long-staple cotton is used in making the better grades of yarn. Short-

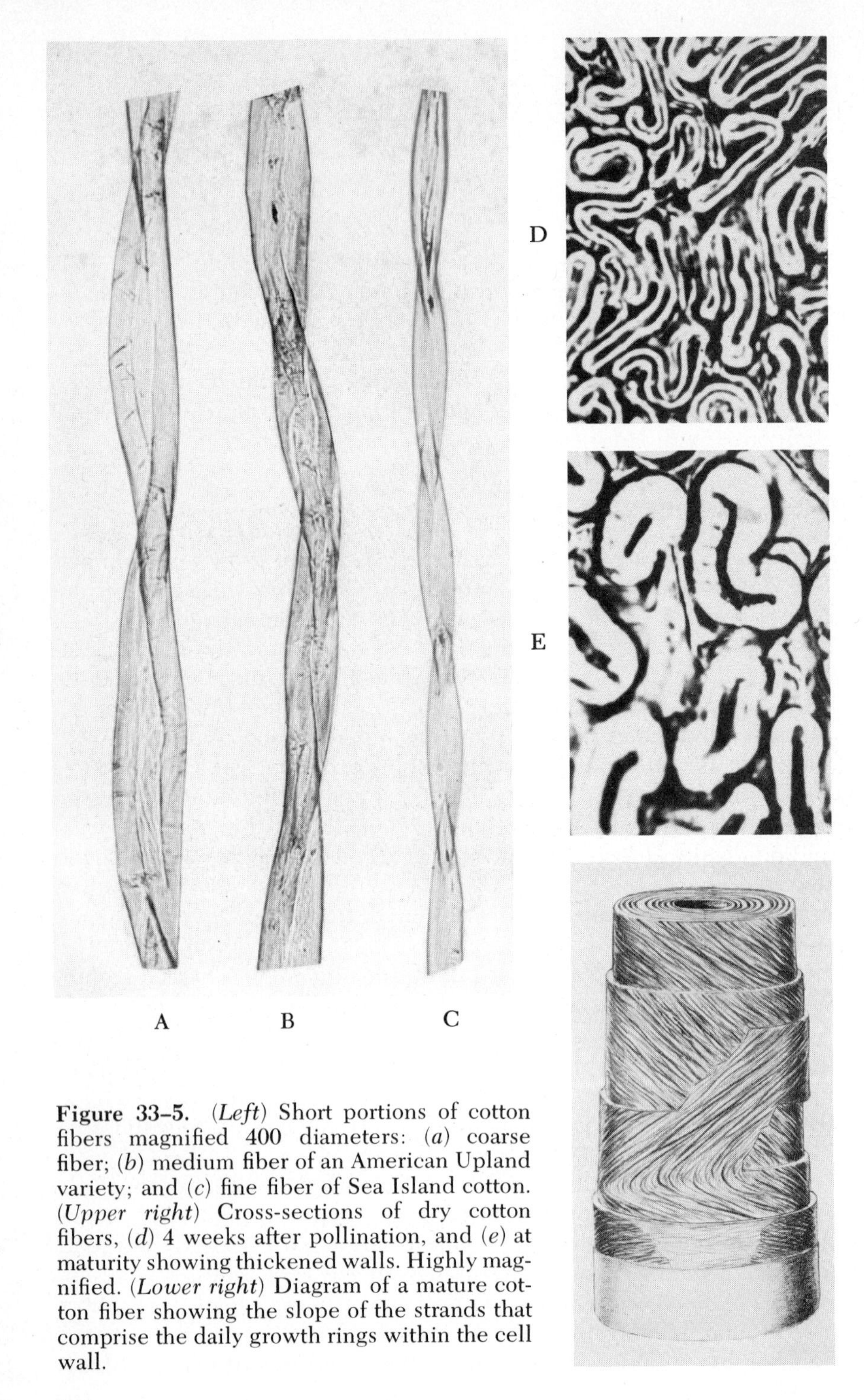

Figure 33–5. (*Left*) Short portions of cotton fibers magnified 400 diameters: (*a*) coarse fiber; (*b*) medium fiber of an American Upland variety; and (*c*) fine fiber of Sea Island cotton. (*Upper right*) Cross-sections of dry cotton fibers, (*d*) 4 weeks after pollination, and (*e*) at maturity showing thickened walls. Highly magnified. (*Lower right*) Diagram of a mature cotton fiber showing the slope of the strands that comprise the daily growth rings within the cell wall.

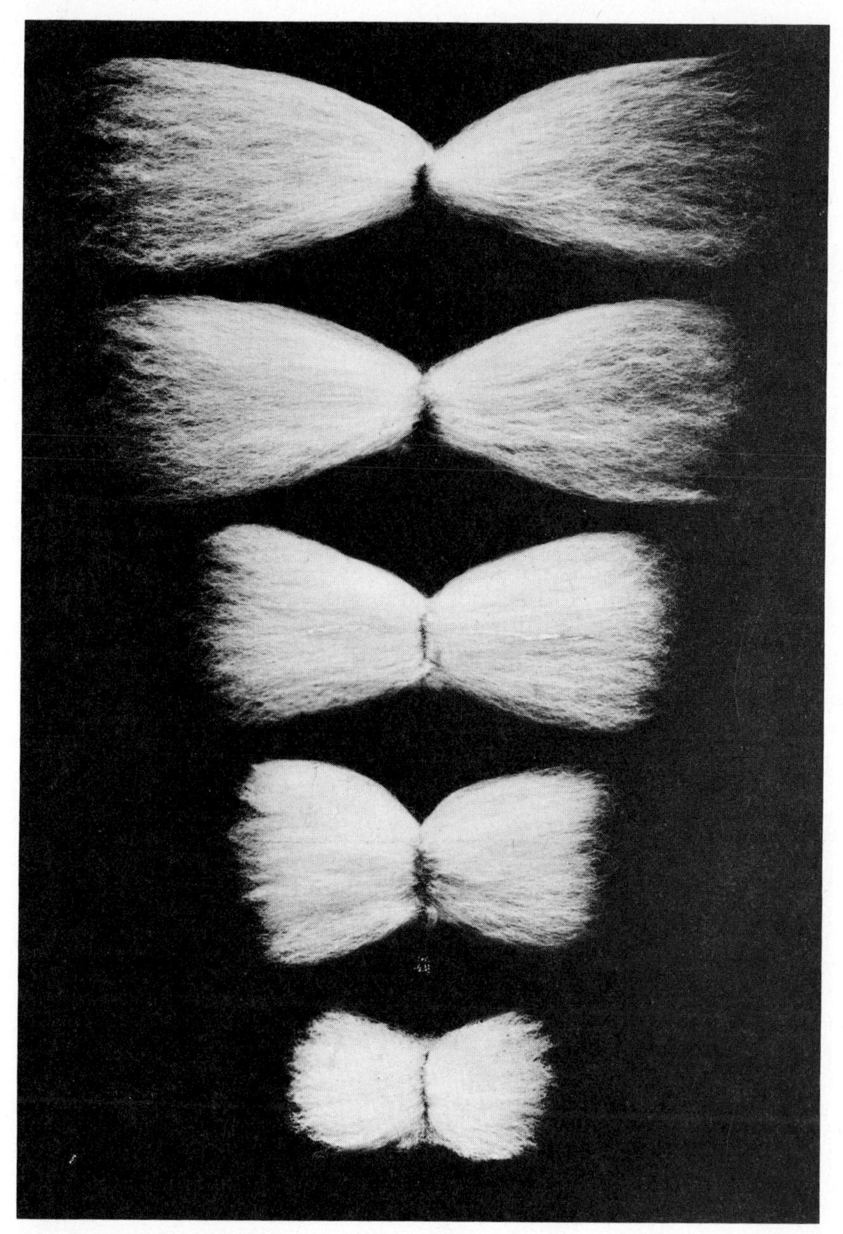

Figure 33–6. Combed fibers attached to the seeds of five types of cotton: (1) Sea Island, (2)American-Pima, (3) American Upland long staple, (4) American Upland short staple, and (5) Asiatic. (All natural size.)

staple varieties are grown in certain localities because they are more profitable. The stormproof varieties grown in western Texas are short-staple varieties.[25]

COTTONSEED

Normally, there should be nine cottonseeds in each lock, or 27 to 45 per boll. However, a lock usually contains one or two undeveloped (aborted) seeds called motes. The seeds are usually ovoid in shape (Figure 33–7). Most of the Upland cotton varieties have dark-brown seeds covered with fuzz, while Sea Island and American-Pima seeds are black and practically free from fuzz. A pound contains 3,000 to 5,000 seeds. Large-bolled varieties tend to have large seeds. The seed coat is a tough leathery hull that constitutes about 25 to 30 per cent of the weight of the seed. The oil content of the hulled kernel (or meat) varies from 32 to 37 per cent.[9]

Cottonseed stored in tight containers at a moisture below 8 per cent has retained its viability with only slight impairment for seven years.[61] Cottonseed, especially that intended for planting, may require artificial drying after ginning in order to avoid deterioration.[23]

Cottonseed is often delinted before planting to cause it to pass through the planter box more freely and germinate more rapidly. Cottonseed is delinted mechanically with delinting saw gins or by treatment with concentrated sulfuric and hydrochloric acids. Cotton-

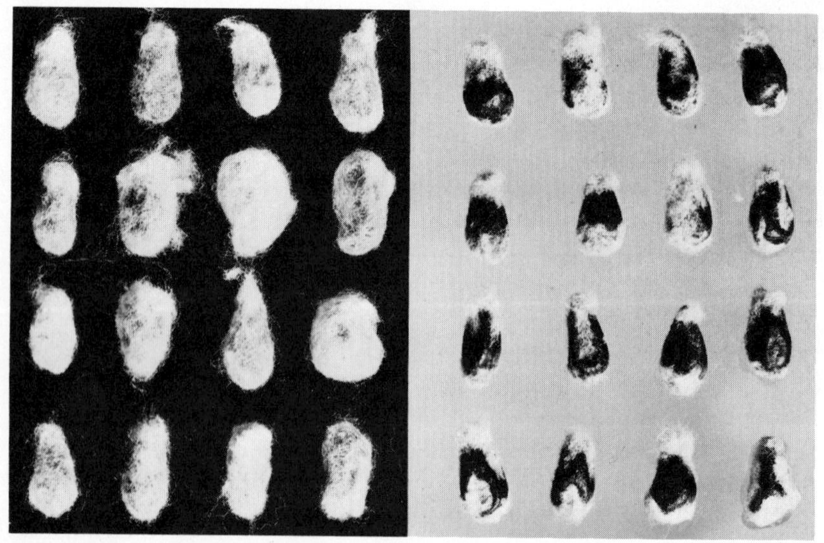

Figure 33–7. Cottonseed: (*left*) gin run; (*right*) delinted.

seed may be delinted very closely with not over 1 per cent of saw-cut injury, and 100 to 150 pounds of linters per ton of seed may be removed with safety.[23] Linters are a mixtures of soft or flaccid fibers and fuzz that escaped removal from the cottonseed in ginning. The sale value of the linters helps to defray the cost of mechanical delinting. Mechanical delinting is frequently done at oil mills. Often the seed is treated after delinting, or without delinting, with a fungicide to control seedling diseases.[23]

VARIETIES

More than 500 varieties of cotton were grown in the United States in the early years of the 20th century.[72] By 1970, strains of only 6 varieties occupied 67 per cent of the cotton acreage. The release of improved varieties only after wide regional tests, as well as the establishment of numerous one-variety communities, eliminated all mediocre varieties. The one-variety community system provides a more uniform market quality as well as high productivity within a district or locality. Some cotton varieties are widely adapted and are grown in several states in the humid cotton region. The average staple length of the American cotton crop increased 10 per cent from 1928 to 1962 by breeding for better staple quality.

Varieties differ in adaptation, staple length, fiber quality, boll size, disease resistance, and retention of the cotton in the locks. The extra long staple cottons (1¼ to 1⁹⁄₁₆ inches) now grown are the American-Pima varieties. Sea Island cotton with even longer staple, formerly was grown in the South Atlantic states, but its production practically ceased about 1944. Long-staple Upland cotton varieties, with fibers 1¹⁄₁₆ to 1³⁄₁₆ inches long, include selected strains of a Mexican variety, Acala, grown in California, Arizona, New Mexico, western Texas, and Oklahoma. Long-staple Upland varieties grown in the eastern and central states of the Cotton Belt[59] have a staple length ranging from 1¹⁄₁₆ to 1³⁄₁₆ inches. Medium-length staple is 1 to 1¹⁄₁₆ inches. The short-staple cottons are chiefly stormproof types grown in northwest Texas and adjacent areas, with staple lengths of ⅞ to 1 inch.[75]

American-Pima varieties have small bolls, i.e., 100 to 135 bolls per pound of seed cotton, most Upland varieties have 70 to 85 but large-boll varieties have 55 to 70 per pound.

American-Pima long-staple cotton is grown under irrigation on about 75,000 to 100,000 acres in Arizona, western Texas, New Mexico, and California.

Research in breeding with male-sterile cottons, or other procedures, may succeed in the development of commercial hybrid cottons with much higher yield potentials.

FERTILIZERS

Heavier fertilizer use is one of the improved practices that increased average cotton lint yields per acre from 180 pounds in 1930 to 495 pounds in 1972. From 95 to 100 per cent of the cotton fields in the southeastern states where the soils are low in fertility, received mineral fertilizers in 1970. Average rates in pounds per acre applied were 78–92 of N, 43–74 of P_2O_5, and 53–90 of K_2O. In the Coastal Plain the acre rates for maximum yields were 60, 26, and 50 pounds, respectively, of N–P–K elements.[76] Applications of nitrogen at 120 pounds per acre were profitable in an area in Louisiana.[80] About 96 pounds per acre of nitrogen, 60 pounds of P_2O_5, and 61 pounds of K_2O were applied to cotton in Mississippi in 1970. An 80-pound application of nitrogen, and 40 to 80 pounds of P_2O_5 where needed, can result in a yield of 2 bales per acre or more on irrigated land in the High Plains of Texas.

Nitrogen applied at rates of up to 63 pounds per acre were helpful to cotton yields in southeast Missouri but limestone applications were not beneficial even on soil of *p*H 5.3. The highest rates of nitrogen are applied to irrigated cotton in southern California and Arizona. Considerable phosphorus also is beneficial there but most soils contain ample potassium.[68]

Alluvial soils of the Mississippi Basin require mostly nitrogen applications of 80 to 120 pounds per acre and smaller amounts of P_2O_5 and K_2O. Irrigated cotton lands in the western half of the Cotton Belt usually require mostly nitrogen. However, growers in California and Arizona who fertilized applied about 130 pounds of N, 60 pounds of P_2O_5, and 16 to 24 pounds of K_2O per acre, in 1970. Nitrogen applications of 80 to 100 pounds for each prospective bale per acre of cotton are recommended.[51] Cotton grown without irrigation often is not fertilized in the drier parts of Oklahoma and Texas where small responses to fertilizer are common. Only 50 per cent of the cotton acreage in those states was fertilized with chemicals in 1970. In more than 150 experiments with cotton fertilizers the yield of lint was increased about 40 pounds for each 18 to 20 pounds of N, P_2O_5, and K_2O applied.[62] The best results were secured when one half the total nitrogen was applied in the mixed fertilizer, with the remainder applied later as a side dressing. About 12 per cent of the total seasonal intake of mineral nutrients occurs during the stage between seedling emergence and formation of squares. About 58 per cent of the mineral soil nutrients are taken up by the plant between the stages of square formation and boll formation. The remaining 30 per cent is taken up thereafter.[20] Fertilizer may be placed in bands 2 to 4 inches to each side and 2 to 4 inches below the level of the seed.[75] The material can be placed in such bands with a combination planter and fertilizer distributor. Supplementary nitrogen is applied as a side dressing in dry or liquid form.

Seed cotton equivalent to a bale of lint contains about 30 pounds of nitrogen, 14 pounds of phosphoric acid, 14 pounds of potash, 3 pounds of lime, and 4 pounds of magnesia. The lint portion contains about 1 pound of nitrogen and 6 pounds of mineral nutrients. About 98.5 per cent of the cotton fiber is carbon, hydrogen, and oxygen which the plant takes from air and water.[5] Thus, the cotton crop scarcely depletes the soil when the cottonseed meal is fed to livestock. The burs from the cotton gin can be returned to the field to supply organic matter and often increase the yield.

ROTATIONS

Winter legumes are often grown on cotton land as a means to reduce the amount of fertilizer needed for satisfactory cotton yields. Rotation is essential in some areas in order to reduce some cotton diseases such as root rot. The yields of seed cotton in six southern states were increased 200 pounds per acre by crop rotation as compared with continuous culture.[62] Growth of a winter legume preceding cotton increased the yield of seed cotton an average of 270 pounds per acre. Other crops in the rotation may be corn, small grains, and soybeans.[39] Any of these crops may be followed by a winter legume which is plowed under before cotton is planted. In irrigated areas, the cotton may follow alfalfa, sorghum, or an oil-seed crop. Cotton often is grown continuously on semiarid lands because it does not fit well in sequence with winter wheat or sorghum, the other important crops. Continuous cotton is possible also on clean land free from wilt infestation in the Yazoo-Mississippi delta.[66]

SEED-BED PREPARATION

The first step in seedbed preparation for cotton is stalk disposal, when cotton follows corn, sorghum, or cotton. A rotary stalk cutter often is used. These residues or a green manure are plowed under. Deep tillage with a chisel or subsoiler has increased cotton yields on soils with a hard pan. The land then may be formed into beds with a lister or middle buster.

Cotton is planted on ridges or beds in most of the humid Cotton Belt, while level or furrow plantings are more common in the western half of the Cotton Belt, except under irrigation. Low beds are preferable from the standpoint of weed control as well as for moisture conservation. Higher beds may be desirable in wet areas (Figure 33–9).

PLANTING PRACTICES

Cotton is best planted when the soil at a depth of 8 inches has warmed up to 60° F.,[61] and there is a high probability of an air temperature of 68° F. for 10 days following planting. Most of the crop is planted in March, April, or May, the later dates being applicable to the northern part of the Cotton Belt (Figure 33–8).

Planting on low W-type beds is helpful not only for obtaining good stands, but also for subsequent weed control, drainage, and irrigation. The W-type beds are alternate ridges and furrows. In irrigated areas the field often is watered before planting, but after the beds are prepared. Wings attached to the planter scrape off dry soil from the top of the beds and smooth them for planting. Rubber-tired wheels at the rear of the planter press the seeds into moist compacted soil. Most cotton is planted with multiple-row planters in the important areas of production (Figure 33–9). Knife or disk furrow openers are used.

A pound of cottonseed contains about 3,500 to 5,200 seeds. When the seeds are delinted, the number is about 18 per cent more. Cottonseed is planted 1 to 2 inches deep. Rates of 24 to 40 pounds of seed per acre have been prevalent in the humid regions. The average rate in the United States in 1970 was 26 pounds per acre. From 25 to 30 pounds of mechanically delinted seed—or even fewer pounds of acid-delinted, treated seed of medium size—will provide four to eight plants per foot of row when the field germination is 60 to 80 per cent. Such stands often are thinned with a mechanical chopper. The use of a precision planter and acid-delinted and treated seed reduces the planting rate and eliminates the thinning.[4] Most of the cotton grown in the semiarid regions is planted at a rate of about 20 to 30 pounds per acre but is not chopped. Delinted seed in semiarid regions is planted at a rate of only 7 to 10 pounds per acre because of better germination.

Although the cotton plant is sufficiently adaptable to produce satisfactory yields over a wide range of field stands, moderately close spacing has given the best yields. From 20,000 to 50,000 plants per acre are ample for maximum yields under humid or irrigated conditions.[51, 59] These stands are approximately one plant every 3 to 8 inches in the typical 38- to 40-inch rows.

Yields are similar with plants spaced 3 to 12 inches apart in the row. Crowded plants, which grow taller under favorable conditions, bear their first bolls higher on the stalk and somewhat later. A spacing of 12 to 18 inches gave the highest yields under semiarid conditions in southwestern Oklahoma.[50]

Planting 3 seeds per hill at a 9 inch hill spacing is feasible.[37] Narrow row planting with 60 to 80 thousand plants per acre matures cotton more uniformly and earlier and eliminates a second picking. The

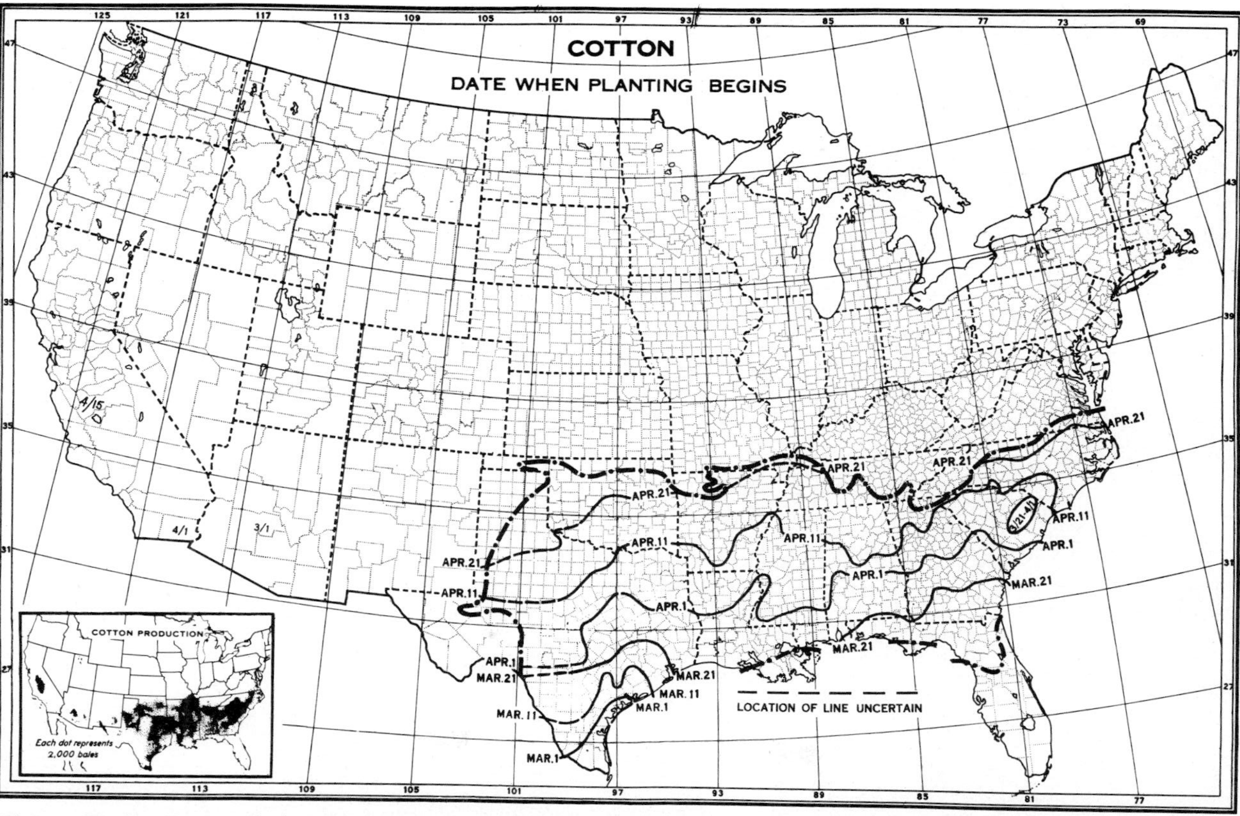

Figure 33–8. Dates of planting cotton in the United States.

Figure 33–9. Cultivating beds, planting cotton on beds, compacting the soil over the planted seed, and applying a herbicide in bands in one operation.

method is feasible in fields relatively free from weeds.[71] It also can be planted in paired rows 10 to 14 inches apart on beds spaced 38 to 40 inches between centers.

WEED CONTROL

Most grass seedlings and some annual broad-leaved weeds in cotton fields can be controlled by spraying herbicides on the field and then incorporating the material into the surface soil before planting the cotton.

Preemergence herbicides control early-season, small-seeded annual weeds when applied during planting or shortly thereafter. Applications in 12-inch bands over the planted rows during planting reduces chemical costs but the weeds between the rows must then be controlled by cultivation (Figure 33–9).

Most postemergence herbicides used in cotton fields are applied as basally directed sprays after the cotton plants are 3 to 6 inches tall or taller.

The flaming of weeds in the cotton rows, with propane or propane-butane as fuel, is effective after the cotton plants are 10 to 12 inches high, with their stems more than $^{3}/_{16}$ inch in diameter at the base (Figure 11–3). Flames directed at the proper height and angle kill the weeds without burning the cotton leaves.[70] The burners should be mounted individually to maintain a proper level at each row.

Cultivation, together with some cross-cultivation and hand hoeing, is the common method for weed control in cotton. Early cultivations often are made with a rotary hoe. Later, shallow cultivations with sweep or blade implements cut off the weeds with little injury to the cotton roots.

On some farms in New Mexico, grazing with geese has suppressed the weedy grasses in cottonfields.

IRRIGATION

All the cotton in California and Arizona is irrigated. The crop also is grown under irrigation in nearly all of New Mexico, much of Texas, and on appreciable acreages in Oklahoma, Arkansas, Louisiana, and other states.

When water is available, its daily consumption by transpiration and soil evaporation in the cottonfield averages about 1/5 inch per day. Maximum daily usage ranges from 1/4 to 2/5 inch at the peak of growth in midsummer. Thus, for high yields, from 24 to 42 inches of water in the soil from rainfall or irrigation is used in the growing season. The soil should contain available moisture throughout the root zone (4 to 6 feet deep) when the cotton is planted. Irrigation before planting time makes this possible.

Frequent irrigation to keep the soil moisture from dropping much below 50 per cent of its field capacity is desirable. In hot dry climates, this requires irrigation every 8 to 16 days. The soil moisture content can be determined with a tensiometer, or it may be estimated by an examination of a soil core taken with a probe or auger. The cotton plant needs water when the leaves turn to a dark or dull bluish-green color followed by wilting during the hot part of the day.[51] A need for water is evident when vegetative growth decreases to the point where flowers are opening near the top of the plants. Adequate, frequent irrigation keeps from 4 to 6 inches of growth above the upper flowers on the plant. Furrow or basin irrigation often produces higher yields than does sprinkler irrigation.

HARVESTING

About 98 per cent of the cotton was harvested by machines in the United States by 1970 (Figure 33–10), 70 per cent with pickers and 28 per cent with strippers or scrappers. Mechanical pickers are equipped with rotating steel spindles attached to revolving drums or moving bars. The spindles are fluted rods, smooth rods or barbed cones.[70] The moistened spindles pull the lint and seeds from the opened bolls, after which they are pushed from the spindle with a doffer, and then conveyed to a basket mounted on the picker. The harvested cotton contains fewer leaf fragments when smooth-leaf varieties are grown.

Figure 33–10. (*Top*) Four makes of cotton pickers. (*Bottom*) Cotton stripper. [Top photo by H. F. Miller; bottom photo courtesy J. I. Case Co.]

The stripper snaps the open and unopened bolls from the plants as the machine moves across the field. The strippers are equipped with inclined rollers which straddle the plant rows. The long-brush type of stripping roll is very effective.[33] Some of the strippers are equipped with a stick and bur remover so that less trash is hauled to the gin. The stripper was designed especially to harvest stormproof varieties of cotton, grown in northwest Texas and adjacent areas of New Mexico and Oklahoma where cotton formerly was hand-snapped.

A finger type stripper 13 feet wide is used to harvest narrow-row or broadcast cotton. The strippers are operated after frost, defoliants, or desiccants have removed most of the leaves from the plants. The burs on stripped cotton are removed at the gin, which also is equipped with boll breakers and extractors to remove the seed cotton from unopened bolls.

A cotton combine is a once-over picker or scrapper which cuts off the plants, strips out the seed cotton with spindles, and then chops the stalks and branches and returns them to the soil.

Hand picking of cotton is limited largely to rough terrain, or to small fields with low yields, where machine picking is not feasible or economical. Most of this is in the southeastern states. Hand picking one bale of cotton requires 45 to 70 man-hours. A two-row mechanical picker operator harvests it in less than an hour.

Some fields receive two pickings to avoid losses from weathering of the early cotton and save much that might drop to the ground before the late harvest. Mechanical gleaners that follow the picker and gather the cotton that has dropped to the ground are now in wide use.[41]

The lint and seed of cotton picked when the plants are wet from rain or dew will be damaged unless they are dried promptly.[27] Tramping or packing machine-picked cotton in the trailer likewise contributes to deterioration. Modern gins are equipped with driers as well as cleaning equipment. Picking by machine should be postponed until the relative humidity drops below 50 per cent each day,[59] unless the cotton is to be dried at the gin (Figure 33–11).

During the period from 1910 to 1914, about 276 hours of labor were expended in producing, harvesting, and handling a bale of cotton. This was accomplished in only 26 hours in 1970–71, following mechanized operations and higher yields.

DEFOLIATION, DESICCATION, AND TOPPING

Cotton to be harvested mechanically often is treated chemically to remove or dry the leaves.[74] These treatments eliminate much dampness, fiber staining, and trash from seed cotton. They also reduce difficulties from clogging of picker spindles with trash or leaf juices. Leaf

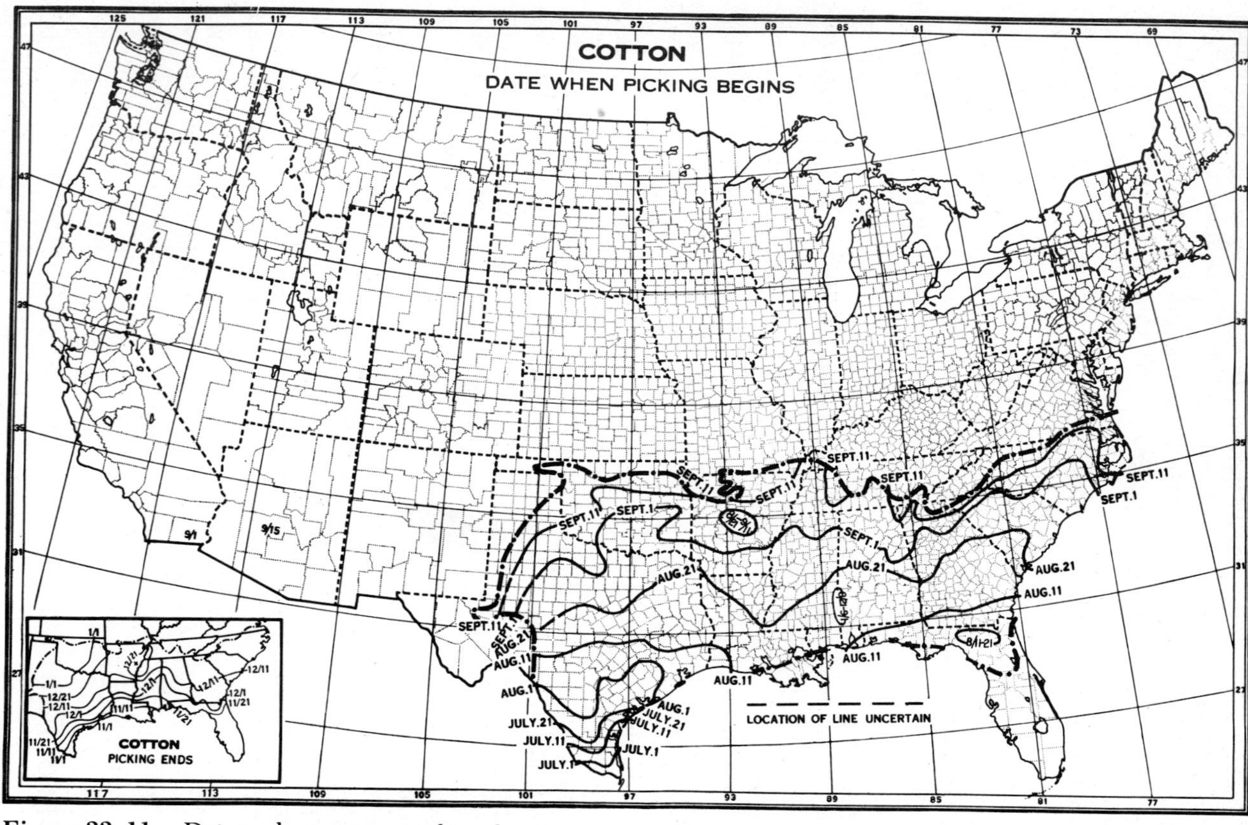

Figure 33–11. Dates when cotton picking begins.

removal provides better ventilation to reduce boll rotting on the lower branches of the plants. It also reduces lodging of the plants.

Defoliation induces abscission (separation) of the cells where the leaves are attached to the branches.[27] Most of the leaves drop off within a few days after proper treatment. Desiccation kills the leaves so that they dry and break off. At least 60 per cent of the bolls should be open when defoliants are applied because boll and fiber development are checked when the leaves are removed or killed.[6] Mature bolls not yet open are not affected. Either defoliation or desiccation hasten drying so that picking can be started about one hour earlier in the morning. Defoliants are most effective at temperatures between 90° and 100° F., but they are relatively ineffective below 60° F., or above 100° F.

Desiccants are more drastic than defoliants in their effect on plant development. Consequently, they should be applied 5 to 7 days later than for defoliants. Bottom defoliation, applied with a high-clearance field sprayer, permits early harvest of the bottom crop with a spindle picker. All except the lower spindles are removed for this operation.

Cotton plants that have ceased growth (cut out), after exhaustion of the soil moisture or mineral nutrients, or for other causes, usually drop their leaves without the need of defoliants.

The defoliants used include calcium cyanamide, a sodium chlorate–sodium metaborate mixture, magnesium chlorate-hexahydrate, and certain organic phosphorus compounds.

Pentachlorophenol is a desiccant that kills the plants promptly without defoliating them. It is not applied until 80 to 90 per cent of the bolls are open.[6]

Mechanical topping of cotton plants with high-clearance cutters is practiced in California to reduce lodging of the plants as well as to facilitate defoliation and the operation of spindle pickers. The plants are topped at a height of about 4 feet, or by cutting off the top 8 to 12 inches of the plants.[13, 18]

COTTON GINNING

After it has been picked, the seed cotton is hauled in bulk to the gin where it is unloaded with a pneumatic metal pipe that has telescoped sections at the outside end. Cotton ginned within a day after the harvest has more long fibers of improved spinning quality.[77]

Cotton ginning (Figure 33–12) involves separation of the fibers from the seed after removal of dirt, hulls, and other trash. Damp or wet cotton is put through a drier before ginning.[47] Dry lint contains 8 per cent moisture or less. The essential part of a saw gin is a set of circular saws that revolve rapidly with a portion projecting through a narrow slit between parallel ribs or bars of iron.[10] The teeth of the saws catch the lint, draw it between the ribs, and thus pull it from the

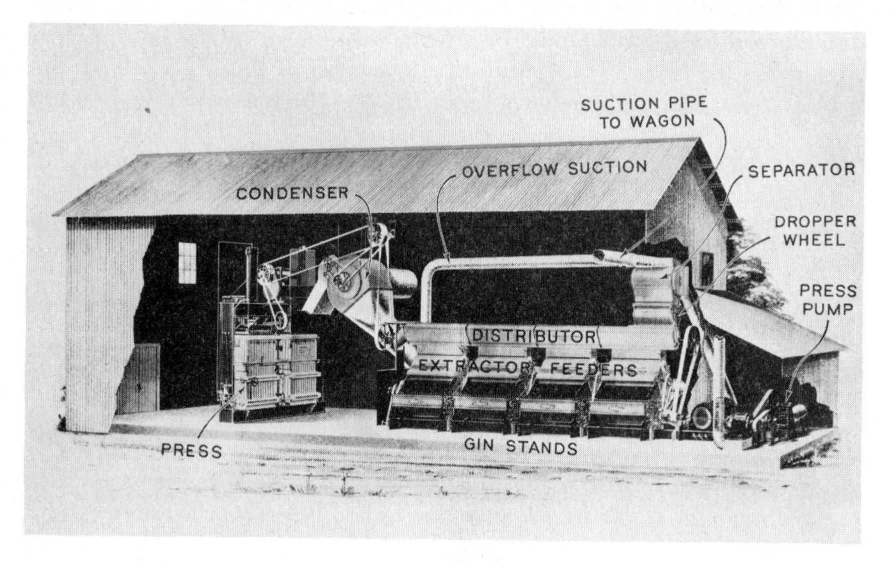

Figure 33–12. (*Top*) Sectional view of a gin plant. (*Bottom*) Section of a stand.

seeds (Figure 33–12). Since the seeds are too large to pass through, they are dropped into a conveyor trough or suction pipe. An air blast or a brush removes the lint from the saws. The lint is carried to the press by suction conveyers. American-Pima cotton is ginned on a roller gin.

At the press or baler the fluffy lint is pressed into bales of about 500 pounds (Figure 33–13). The bale is covered with about 6 yards of heavy (1¾ or 2 pounds per yard) jute bagging, and held by six asphalt-coated flat steel ties that weigh 9 pounds. The standard bale is about 54 × 27 × 45 to 48 inches in size,[1] or 38 to 40 cubic feet. It contains 480 pounds of lint, and 20 pounds of bagging and ties per 500-pound bale. For more economical export, transportation, and storage, the bales often are compressed.

In 1971 some 3754 active gins each processed an average of 2,674 bales of cotton produced on 2930 acres. Nearly all of them were equipped with driers and lint cleaners. Each equivalent bale of Upland cotton (480 pounds of lint plus 853 pounds of seed) received at the gin contained an average of 40 pounds of trash and dirt from hand-picked cotton, 139 pounds from machine-picked, 653 pounds from hand-scrapped, 708 pounds from machine-scrapped, and 836 pounds from machine-stripped.

Figure 33–13. Attaching the ties to a bale of cotton in a press.

UTILIZATION

Cotton lint is spun into thread to be woven into various fabrics (Figure 33–14). The finer threads are made from American-Pima cotton, while the coarser ones are made from the Upland varieties. About 73 per cent of the domestic cotton is used for clothing and household goods. The remainder is used in industry, mostly for awnings, bags, belts, hose, and twine. The short lint is utilized in carpets, batting, wadding, and low-grade yarns, as well as for stuffing material for pads and cushions.[18] Linters or fuzz are used mostly for stuffing and for making rayon and other cellulose products.

About 7 per cent of the cottonseed produced is planted each year, while small quantities are used for feed. The remainder is crushed for the oil, the meal being a by-product. A ton of cottonseed yields about 320 pounds (or 43 gallons) of oil, 900 pounds of meal, 135 pounds of linters, and 500 pounds of hulls, which leaves 145 pounds to account for trash, waste, and invisible losses. The production of linters averages more than 1½ million 500-pound bales annually. The seed is delinted before it is crushed for oil extraction.

From 20 to 45 per cent of the cotton produced each year, as well as some cotton fabrics, are exported. About one-third of the cottonseed oil, and some cottonseed meal, are exported. About 20 per cent of the total oil production is made into shortening, and 7 to 8 per cent into margarine, by the process of hydrogenation.

Cottonseed meal is a high-protein feed for livestock, but it contains a toxic substance called gossypol. This limits the proportion of cottonseed meal that can safely be added to poultry or livestock rations. Cottons with few glands on the leaves and in the seeds are low in gossypol. Production of glandless cotton is on the increase in the high plains of Texas. Fiber yield of glandless cotton is less than normal cotton. Use of glandless cotton however reduces the cost of oil processing and increases the protein meal quality of the seed meal. Some seed processors are paying a premium for the seed to offset the reduced fiber yield and insure a supply of glandless cotton.

Cotton is classed according to staple length, grade, and character. Staple length is estimated or measured from approximately a ¼-ounce sample pulled and straightened between the fingers. For the shorter staple lengths the lint is separated into classes: below ¾ inch, ¾ inch, 13/16, and ⅞. Above that length the classes differ by 1/32 inch. In any sample there can be found an array of different fiber lengths that range from very short to somewhat longer than the typical length. Upland cotton having fibers 1⅛ inches or longer is classed as long-staple.

Fiber fineness often is recorded as micronaire values[2] which approximate the weight of an inch of fiber in micrograms. It actually

Figure 33–14. A cotton mill converting yarn to thread.

measures the air flow at a standard pressure through a comparable weight and volume of fiber.

Quality factors that determine grade include color, leaf and other foreign matter, ginning preparation, and maturity. These values are estimated by visual comparison with standard samples. Character, including strength, fineness, pliability, and uniformity of length are determined by sight and touch. Instruments for determining length, strength, fineness, and structure are being used, mostly in research studies of the fiber. Color classes recognized are extra white, white, blue-stained, gray, spotted, yellow-tinged, light-stained, and yellow-stained. Within these classes are 32 grades for quality. For American Upland white cotton the numerical grades are as follows:

No. 1 or Middling Fair
No. 2 or Strict Good Middling
No. 3 or Good Middling
No. 4 or Strict Middling
No. 5 or Middling

No. 6 or Strict Low Middling
No. 7 or Low Middling
No. 8 or Strict Good Ordinary
No. 9 or Good Ordinary

Grades 1 to 7 are deliverable on future contracts.[1, 5] Grade 5, middling, is the basic grade for commercial transaction. Premiums are allowed for higher grades, while discounts are applied for grades lower than middling. Grades are established also for American-Pima and Sea Island cotton and for linters.

When a grower hauls his seed cotton to the gin he may sell it directly, but usually he has it custom-ginned and baled, and then stores it at a warehouse or hauls it home or to a buyer. The seed may be sold to the gin, or hauled home or to an oil mill. The identity of the cotton of a grower is maintained at the gin and when the bales are stored in a warehouse. Prices offered the grower are based upon the current future price for middling cotton of 1 inch staple at a terminal market, plus any premium for quality above that base (or less any discount for lower quality), less costs of handling, shipping, and the margin of the buyer.[7] A typical cotton warehouse is a large one-story open-sided building with the bales set on end in a single layer. Much of the cotton is not sold immediately, but is held under government loan.

DISEASES

Root Rot

Root rot, caused by the fungus *Phymatotrichum omnivorum*, is a destructive cotton disease, especially on calcareous soils of the Southwest as well as on the black waxy soils of Texas. Affected leaves and roots are yellow or bronze, and the plants wither and die. The fungus survives in the soil on dead roots of cotton or other taprooted plants or in a dormant sclerotial stage. Diseased plants occur in irregular areas over the field. The fungus attacks most dicotyledonous trees, shrubs, and herbs, but does not attack plants of the grass family.

The severity of the root-rot disease is reduced by rotation with grain or grass crops for two or three years, combined with deep tillage after these crops are harvested. Effective control was obtained in Arizona by turning under large amounts of organic matter in rows and then planting over the row.[36, 53]

Fusarium Wilt

Fusarium wilt, caused by the fungus *Fusarium oxysporium* F. *vasinfectum*, is serious on light sandy soils from the Gulf Coast plain to New Mexico. Infected plants are stunted early in the season, and the leaves turn yellow along the margins and between the veins. The cut stem of a wilted plant shows brown or black vascular tissues inside. The disease can be controlled by the use of resistant varieties and fertilizers with adequate amounts of potash. Several varieties are resistant to the fusarium wilt as well as to the nematode-wilt complex (Figure 33–15).

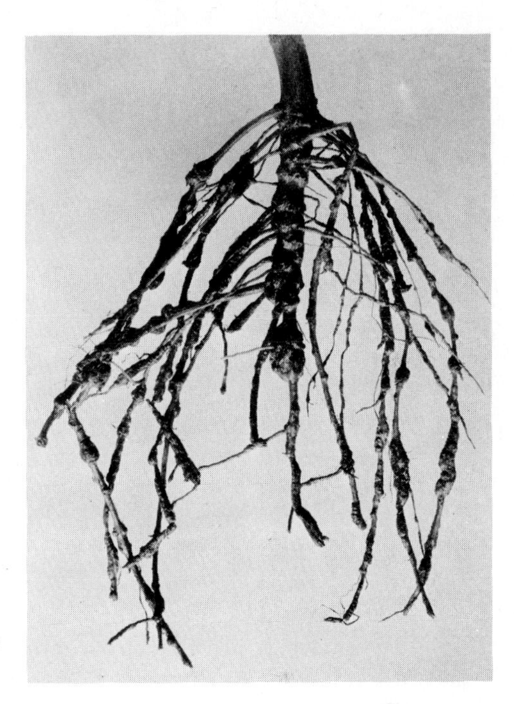

Figure 33–15. Root knot of the cotton plant.

Verticillium Wilt

Verticillium wilt is a disease caused by a soil-inhabiting fungus (*Verticillium alboatrum*) that attacks the roots of the cotton plant. It causes wilting, mottling, shedding of the leaves, and vascular discoloration of the roots and stems. Some varieties are resistant to the disease, while others are somewhat tolerant.

Other Diseases

Other diseases of cotton and their causal organisms are leaf blights, *Ascochyta gossypii;* and bacterial blight, or angular leaf spot or black arm, *Xanthomonas malvacearum* (Figure 33–16). These organisms are seed-borne, but they also live over winter on cotton residues. Crop rotation, trash disposal, and seed treatment reduce these and some other leaf diseases. Empire cottons are somewhat tolerant to *Ascochyta* blight, and some varieties are resistant to certain races of the bacterial blight organism. Anthracnose or sore shin is caused by the fungus, *Glomerella gossypii*. Seedling blights and damping-off

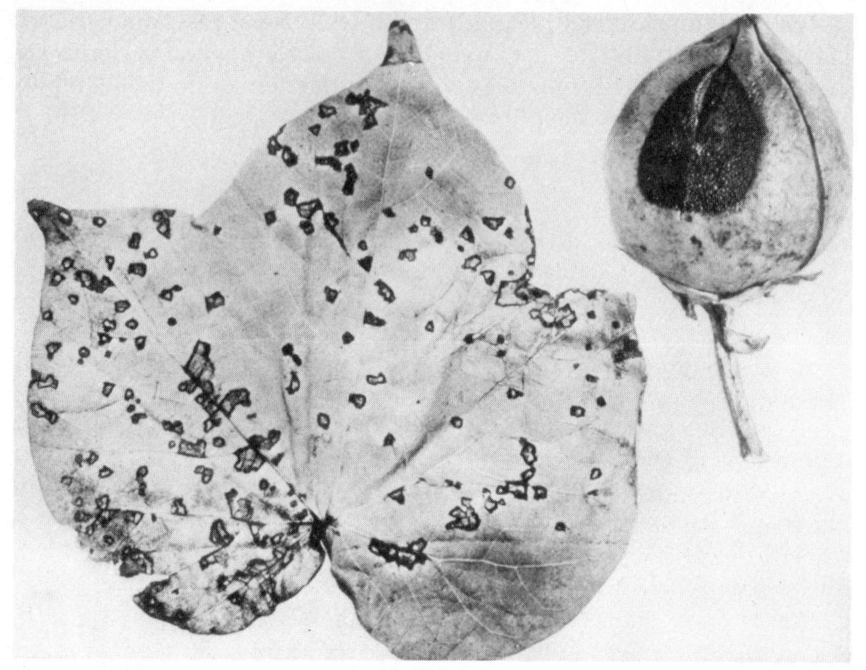

Figure 33–16. Angular leaf spot on leaf and boll.

diseases are caused by species of *Rhizoctonia*, *Fusarium*, and *Pythium*. They are partly controlled by seed treatment. Seed disinfectants should be mixed with the seed before planting. They may be put on the seed in the seed hoppers of the planter; in addition, a spray or dust may be mixed with the soil that covers the row during the planting operation.

Cotton-boll rots are caused by common soil-borne fungi that include species of *Aspergillus*, *Fusarium*, *Nigrospora*, and *Rhizopus*. Topping and defoliation keep the plants drier and reduce boll rotting. Types with "okra" leaves or "frego" boll bracts, also have less boll rots. Bolls that are attacked by bollworms are often infected with boll rots.

Several different fungi cause fiber deterioration in damp cotton. A rust fungus, *Puccinia stakmanii*, also attacks the crop.

Nonparasitic disorders of cotton include crazy top, caused by irregular irrigation; a potash deficiency sometimes called a rust; and crinkle leaf, which is a manganese toxicity that occurs on acid soils deficient in calcium.

The root-knot nematode, *Meloidogyne incognita;* reniform nematode, *Rotylenchulus reniformis;* and meadow nematode, *Pratylenchus* species, are very damaging to cotton. They may be partly controlled by fumigation of the soil during planting.[59] The fumigant is sprayed into furrows which are opened with a lister bottom and covered immediately. The cottonseed is planted in the covering soil. Crop rotation is helpful in nematode control. Certain varieties are somewhat resistant to the root-knot nematode.

INSECT PESTS

The most serious insect pests of cotton are the boll weevil (*Anthonomus grandis*), pink bollworm (*Pectinophora gossypiella*), cotton leafworm (*Alabama argillaceae*), bollworm (*Heliothis armigera*), cotton fleahopper (*Psallus seriatus*), cotton aphid (*Aphis gossypii*), red spider (*Tetranychus bimaculatus*), garden webworm (*Loxostege similalis*), Lygus bug or tarnished plant bug (*Lygus pratensis*), and cotton stainer (*Dysdercus suturellus*).[54, 59] Other insects that attack the cotton plant include the cotton leaf perforator (*Bucculatrix thurberiella*), cutworms, grasshoppers, and the fall armyworm (*Laphygma frugiperda*). At least 15 other insects are common pests of cotton in the United States. Cotton pests of the Eastern Hemisphere but not of the United States include the jassid (*Empoasca* species), cluster caterpillar (*Prodenia litura*), and the spiky bollworm (*Earias insulana*).

The boll weevil is probably the most destructive insect, but its depredations are confined to sections that have an annual rainfall of more than 25 inches. The adult female weevil punctures the squares or bolls, in which it deposits its eggs (Figure 33–17). This destroys the squares, and the larvae that develop within the boll destroy the lint in all or some of the locks. The bollworm and the pink bollworm also destroy the contents of the boll. The bollworm, also called the corn earworm, attacks corn, sorghum, and other crops.

The leaf worm and webworm devour the leaves and thus retard boll development. Aphids, red spiders, and fleahoppers, all sucking insects, check the development of the plants, bolls, and lint. Aphids

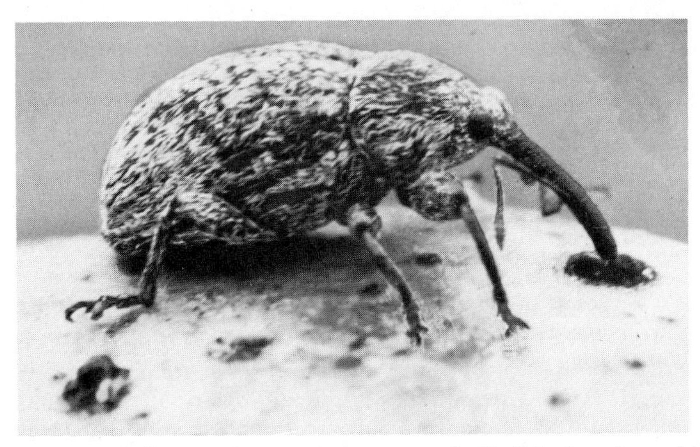

Figure 33–17. Boll weevil puncturing cotton boll. (Much enlarged.)

cause curling and shedding of leaves, and they also secrete honey dew which may contaminate the lint of opened bolls. Red spiders, which may have 17 generations in a year, feed on the under side of the leaves, causing them to redden and drop from the plant. The flea-hopper feeds on the juices of the terminal buds, small squares, and other young tender plant parts. The injured squares may shed while they are too small to be observed. The leaves become deformed.

The boll weevil and some of the other insects are checked by destruction of old cotton stalks and weeds in and near the cotton fields which harbor the pests. Plowing cotton stalks under more than 6 inches deep immediately after harvest is recommended as a control measure for the pink bollworm and boll weevil. The pink bollworm overwintering in the cottonseeds can be killed with heat.[8] The spread of the pink bollworm is being held in check by severe quarantine regulations.

More than 30 insecticides were recommended for the control of pests of cotton in the United States in 1963. All the important cotton pests can be kept under control by one or more of several insecticides. Repeated sprayings are necessary to control the pink bollworm. It is the most serious pest in the southwestern states.

REFERENCES

1. Anonymous. "The classification of cotton," *USDA Misc. Pub. 310*, pp. 1–54. 1938.
2. Anonymous. "Cotton fiber and spinning tests," Report of National Cotton Council, Memphis, Tenn. 1956.
3. Anonymous. "The impact of mechanization on cotton production," *USDA Agricultural Research Service Spec. Rept. ARS 22–34*. 1956.
4. Anonymous. "Planting in the mechanization of cotton production," *Southern Coop. Ser. Bul. 49*. 1957.
5. Anonymous. "Better cottons," USDA Progress Report. 1947.
6. Anonymous. "Harvest-aid chemicals for cotton—defoliants, desiccants, and second-growth inhibitors," *USDA Agricultural Research Service Spec. Rept. ARS 22–58*. 1960.
7. Anonymous. *Agricultural Statistics, 1962, USDA Ann. Publ.*, pp. 1–741. 1963.
8. Anonymous. "Controlling the pink bollworm on cotton," *USDA Farmers Bul. 2207*, pp. 1–12. 1965.
9. Bailey, A. E. *Cottonseed and Cottonseed Products,* Interscience, New York, pp. 1–960. 1948.
10. Bennett, C. A. "Ginning cotton," *USDA Farmers Bul. 1748* (rev.). 1956.
11. Brown, H. B. "Cotton spacing," *Miss. Agr. Exp. Sta. Bul. 212*. 1923.
12. Brown, H. B. "A brief discussion of the history of cotton, its culture, breeding, harvesting, and uses," *La. State Dept. Agr. and Immigration Bul.*, 5th ed. 1939.

13. Carter, L. M., Colwick, R. F., and Tavernetti, J. R. "Topping cotton to prevent lodging and improve mechanical harvesting," *USDA Agricultural Research Service Spec. Rept. ARS 42–67*. 1962.
14. Christidis, B. C., and Harrison, G. J. *Cotton Growing Problems*, McGraw-Hill, New York, pp. 1–633. 1955.
15. Cotton, J. R., and Brown, H. B. "Cotton spacing in southern Louisiana in relation to certain plant characters," *La. Agr. Exp. Sta. Bul. 246*. 1934.
16. Dennis, R. E., and Briggs, R. E. "Growth and development of the cotton plant in Arizona," *Ariz. Agr. Exp. Sta. Bul. A–64*, pp. 1–21. 1969.
17. Doyle, C. B. "Climate and cotton," in *Climate and Man*, USDA Yearbook, 1941, pp. 348–363.
18. Duggan, I. W., and Chapman, P. W. "Round the world with cotton," *USDA Adjustment Adm. Publ.*, pp. 1–148. 1941.
19. Eaton, F. M. "Physiology of the cotton plant," *Ann. Rev. Pl. Physiol.*, 6:299–328. 1955.
20. Eaton, F. M., and Ergle, D. R. "Mineral nutrition of the cotton plant," *Pl. Physiol.*, 32(3):169–175. 1957.
21. Eaton, F. M., and Rigler, N. E. "Effect of light intensity, nitrogen supply, and fruiting on carbohydrate utilization of the cotton plant," *Pl. Physiol.*, 20(3):380–411. 1945.
22. Elliot, F. C., and others. *Advances in Production and Utilization of Quality Cotton: Principle and Practice*, Iowa State Univ. Press, Ames, Iowa, pp. 1–532. 1968.
23. Franks, G. N., and Oglesbee, J. C., Jr. "Handling planting-seed at cotton gins," *USDA Production Res. Rpt. 7*. 1957.
24. Fulton, H. J. "Weather in relation to yield of American-Egyptian cotton in Arizona," *J. Am. Soc. Agron.*, 31:737–740. 1939.
25. Gerdes, F. L., Martin, W. J., and Bennett, C. A. "Cotton from boll to bale," *USDA Leaflet 211*. 1941.
26. Gore, U. R. "Morphogenetic studies on the influence of cotton," *Bot. Gaz.*, 97:118–138. 1935.
27. Hall, W. C. "Physiology and biochemistry of abscission in the cotton plant," *Tex. Agr. Exp. Sta. MP-285*. 1958.
28. Hamby, D. S., Ed. *The American Cotton Handbook*, Vol. I, pp. 1–518, 3rd Ed., 1965; Vol. II, pp. 519–1240, 3rd Ed., Interscience, New York. 1966.
29. Hancock, N. I. "Relative growth rate of the main stem of the cotton plant and its relationship to yield," *J. Am. Soc. Agron.*, 33:590–602. 1941.
30. Harrison, G. J. "Metaxenia in cotton," *J. Agr. Res.*, 42:521–544. 1931.
31. Hawkins, R. S. "Field experiments with cotton," *Ariz. Agr. Exp. Sta. Bul. 135*. 1930.
32. Hawkins, R. S. "Methods of estimating cotton fiber maturity," *J. Agr. Res.*, 43:733–742. 1931.
33. Humphries, R. T., Green, J. M., and Oswalt, E. S. "Mechanizing cotton for low cost production," *Okla. Agr. Exp. Sta. Bul. B-382*. 1952.
34. Kearney, T. H. "Self-fertilization and cross-fertilization in Pima cotton," *USDA Bul. 1134*, pp. 1–68. 1923.
35. King, C. J. "Effects of stress conditions on the cotton plant in Arizona," *USDA Tech. Bul. 392*, pp. 1–35. 1933.

36. King, C. J. "A method for the control of cotton root rot in the irrigated Southwest," *USDA Cir. 425*, pp. 1–9. 1937.
37. Larsen, W. E., and Cannon, M. D. "Planting cotton to a stand," *Ariz. Agr. Exp. Sta. Bul. A-46*, pp. 1–10. 1966.
38. Leding, A. R., and Lytton, L. R. "Cotton spacing experiments in the Mesilla Valley, N. Mexico," *N. Mex. Agr. Exp. Sta. Bul. 219*. 1934.
39. Leighty, C. E. "Crop rotation," in *Soils and Men*, USDA Yearbook, 1938, pp. 406–430.
40. Lewis, C. F., and Richmond, T. R. "Cotton as a crop," in *Advances in Production and Utilization of Quality Cotton* (F. C. Elliot and others) Iowa State Univ. Press, Ames, Iowa. 1968.
41. Lindsey, M. M., and others. "Mechanical gleaning and ginning of ground-loss cotton on the Yazoo-Mississippi delta," *Miss. Agr. Exp. Sta. Bul. 775*, pp. 1–20. 1969.
42. Ludwig, C. A. "Some factors concerning earliness in cotton," *J. Agr. Res.*, 43:637–657. 1931.
43. Mangialardi, G. J., Jr. "Multiple lint-cotton cleaning: Its effect on bale value fiber quality, and waste composition," *USDA Tech. Bul. 1456*, pp. 1–69. 1972.
44. Marsh, P. B., and Bollenbacher, K. "The fungi concerned in fiber deterioration. I. Their Occurrence," *Textile Res. J.*, 19(6):313–324. 1949.
45. McNamara, H. C., Hooton, D. R., and Porter, D. D. "Differential growth rates in cotton varieties and their response to seasonal conditions at Greenville, Tex.," *USDA Tech. Bul. 710*, pp. 1–44. 1940.
46. Meyer, J. R., and Meyer, V. G. "Origin and inheritance of nectarless cotton," *Crop Sci.*, 1:167–169. 1961.
47. Montgomery, R. A., and Wooten, O. B. "Lint quality and moisture relationships in cotton through harvesting and ginning," *USDA Agricultural Research Service Spec. Rept. ARS 42–14*. 1958.
48. Mooers, C. A. "The effect of spacing on the yield of cotton," *J. Am. Soc. Agron.*, 20:211–230. 1928.
49. Neely, J. W. "The effect of genetical factors, seasonal differences, and soil variations upon certain characteristics of Upland cotton in the Yazoo-Mississippi Delta," *Miss. Agr. Exp. Sta. Tech. Bul. 28*. 1940.
50. Osborn, W. M. "Cotton experiments at the Lawton (Oklahoma) Field Station 1916–1931," *Okla. Agr. Exp. Sta. Bul. 209*, pp. 1–31. 1933.
51. Peterson, G. D., Jr., Cowan, R. L., and Van Schaik, P. H. "Cotton production in the lower desert valleys of California," *Calif. Agr. Exp. Sta. Cir. 508*. 1962.
52. Pope, O. A., and Ware, J. O. "Effect of variety, location, and season on oil, protein and fuzz of cottonseed and on fiber properties of lint," *USDA Tech. Bul. 903*, p. 1–41. 1945.
53. Presley, J. T. "Cotton diseases and methods of control," *USDA Farmers Bul. 1745*. 1954.
54. Rainwater, C. F., and others. "Insects on cotton," in *Insects*, USDA Yearbook, 1952, pp. 497–514.
55. Rea, H. E. "The effect of tillage on eradication of cotton root rot," *J. Am. Soc. Agron.*, 25:764–771. 1933.

56. Reynolds, E. B., and Killough, D. T. "The effect of fertilizers and rainfall on the length of cotton fibers," *J. Am. Soc. Agron.*, 25:756–764. 1933.
57. Rhyne, C. L., Smith, F. H., and Miller, P. A. "The glandless leaf phenotype in cotton and its association with low gossypol content in the seed," *Agron. J.*, 51:148–152. 1959.
58. Roth, J. A., and Fisher, T. E. "Limestone and nitrogen application influence on cotton yields and soil tests," *Mo. Agr. Exp. Sta. Rsh. Bul. 988*, pp. 1–19. 1972.
59. Roussel, J. S., and others. *Louisiana Cotton Producers Handbook*, La. State U., Baton Rouge, La., pp. 1–44. 1963.
60. Shaw, C. S., and Franks, G. N. "Cottonseed drying and storage at cotton gins," *USDA Tech. Bul. 1262.* 1962.
61. Simpson, D. M. "Factors affecting the longevity of cottonseed," *J. Agr. Res.*, 64:407–419. 1942.
62. Skinner, J. J. "Use of commercial fertilizers in cotton production," *USDA Cir. 726*, pp. 1–26. 1945.
63. Smith, C. E., Jr. "The new world centers of origin of cultivated plants and the archaeological evidence," *Econ. Bot.*, 22(3):253–266. 1968.
64. Smith, H. P., and Byron, M. H. "Effects of planter attachments and seed treatment on stands of cotton," *Tex. Agr. Exp. Sta. Bul. 621*, pp. 1–16. 1942.
65. Smith, H. P., and others. "Mechanical harvesting of cotton as affected by varietal characteristics and other factors," *Tex. Agr. Exp. Sta. Bul. 580*, pp. 1–49. 1939.
66. Spurgeon, W. I., and Cooke, F. T., Jr. "Cost reduction research for cotton production systems in the Yazoo-Mississippi delta," *Miss. Agr. Exp. Sta. Bul. 783*, pp. 1–12. 1971.
67. Stansel, R. H. "The effect of spacing and time of thinning on the yield, growth, and fruiting characteristics of the cotton plant in 1925," *Tex. Agr. Exp. Sta. Bul. 360*, pp. 1–38. 1927.
68. Starbird, I. R., and French, B. L. "Costs of producing Upland cotton in the United States, 1969," *USDA Agr. Econ. Rpt. 227*, pp. 1–47. 1972.
69. Stephens, S. G., and Finkner, M. D. "Natural crossing in cotton," *Econ. Bot.*, 7(3):257–269. 1953.
70. Tavernetti, J. R., and Miller, H. F., Jr. "Studies on mechanization of cotton farming in California," *Calif. Agr. Exp. Sta. Bul. 747.* 1954.
71. Taylor, B. B. "Narrow row cotton gives better quality at lower cost," *Crops and Soils*, 24(3):7–9. 1971.
72. Tharp, W. H. "Cotton growing in the U.S.A.," *World Crops*, April:139–143. 1957.
73. Tharp, W. H. "The cotton plant—how it grows and why its growth varies," *USDA Handbook 178.* 1960.
74. Tharp, W. H., and others. "Effect of cotton defoliation on yield and quality," *USDA Prod. Res. Rpt. 46.* 1961.
75. Thurmond, R. V., Box, J., and Elliott, F. C. "Texas guide for growing irrigated cotton," *Tex. Agr. Exp. Sta. Bul. B-896.*
76. Walker, M. E., and others. "The effects of high fertilization on the production of cotton in the Coastal Plain," *Ga. Agr. Exp. Sta. Rsh. Rpt. 27*, pp. 1–13. 1968.

77. Ward, J. M., and Graves, J. W. "Effects of ginning of cotton on fiber quality as reflected in certain fiber properties," *Tex. Agr. Exp. Sta. Bul. 1039*, pp. 1–8. 1965.
78. Ware, J. O. "Cotton spacing. I. Studies on the effect of yield and earliness," *Ark. Agr. Exp. Sta. Bul. 230*, pp. 1–84. 1929.
79. Ware, J. O. "Plant breeding and the cotton industry," in *USDA Yearbook, 1936*, pp. 657–744.
80. Woolf, W. E., and others. "An economic analysis of irrigation, fertilization and seeding rates for cotton in the Macon Ridge area of Louisiana," *La. Agr. Exp. Sta. Bul. 620*, pp. 1–32. 1967.
81. Webber, J. M. "Cytogenetic notes on cotton and cotton relatives," *Science*, 80:268–269. 1934.

Chapter 34

Tobacco*

"By Hercules! I do hold it and will affirm it, before any prince in Europe, to be the most sovereign and precious weed that ever the earth tendered to the use of man."
"By Gad's me!" rejoins Cob, "I mar'l what pleasure or felicity they have in taking this roguish tobacco. It is good for nothing but to choke a man and fill him full of smoke and embers."
—Ben Jonson, *Every Man in His Humour*, 1598

ECONOMIC IMPORTANCE

Tobacco is grown on about 4 million hectares, with a production of about 4.7 million metric tons or 1.17 tons per hectare. The countries leading in tobacco production are United States, China, India, the U.S.S.R., Brazil, Japan and Turkey (Figure 34–1). Production in the United States from 1970 to 1972 was 1,787 billion pounds (0.81 million metric tons) with an average acre yield of 2,070 pounds (2.32 metric tons per hectare) on 860,000 acres (348,000 hectares). The leading states in tobacco production were North Carolina, Kentucky, South Carolina, Georgia, Virginia, and Tennessee. The farm value of the crop in the United States exceeds 1.4 billion dollars.

The per capita consumption of tobacco in the United States for adults (18 years old or older) in 1971 was about 9.5 pounds, approximatedly 7.7 pounds of which were smoked in cigarettes. Males over 18 years of age consumed over 4,000 cigarettes, 119 cigars, 1.1 pounds each of smoking and chewing tobacco and 0.2 pounds in snuff. In 1900 the average per adult was about 55 cigarettes, 110 cigars, 4 pounds of chewing tobacco, 1.5 pounds of smoking tobacco, and 0.25 pounds of snuff. This shift resulted in a large increase in bright flue-cured tobacco used primarily for cigarettes with a decrease in several other types. The American Indians usually smoked pipes, but occasionally devised cigars or cigarettes, chewed tobacco or used snuff. The wives of early Connecticut farmers learned to roll cigars, and a cigar factory opened in Connecticut in 1810. Today, all except the most expensive cigars and all commercial cigarettes are rolled by machines. Most of the cigarettes are attached to filters.

* For a complete discussion of tobacco, see Garner, W. W., *The Production of Tobacco*, rev. ed., McGraw-Hill—Blakiston, New York, 1951, pp. 1–520.

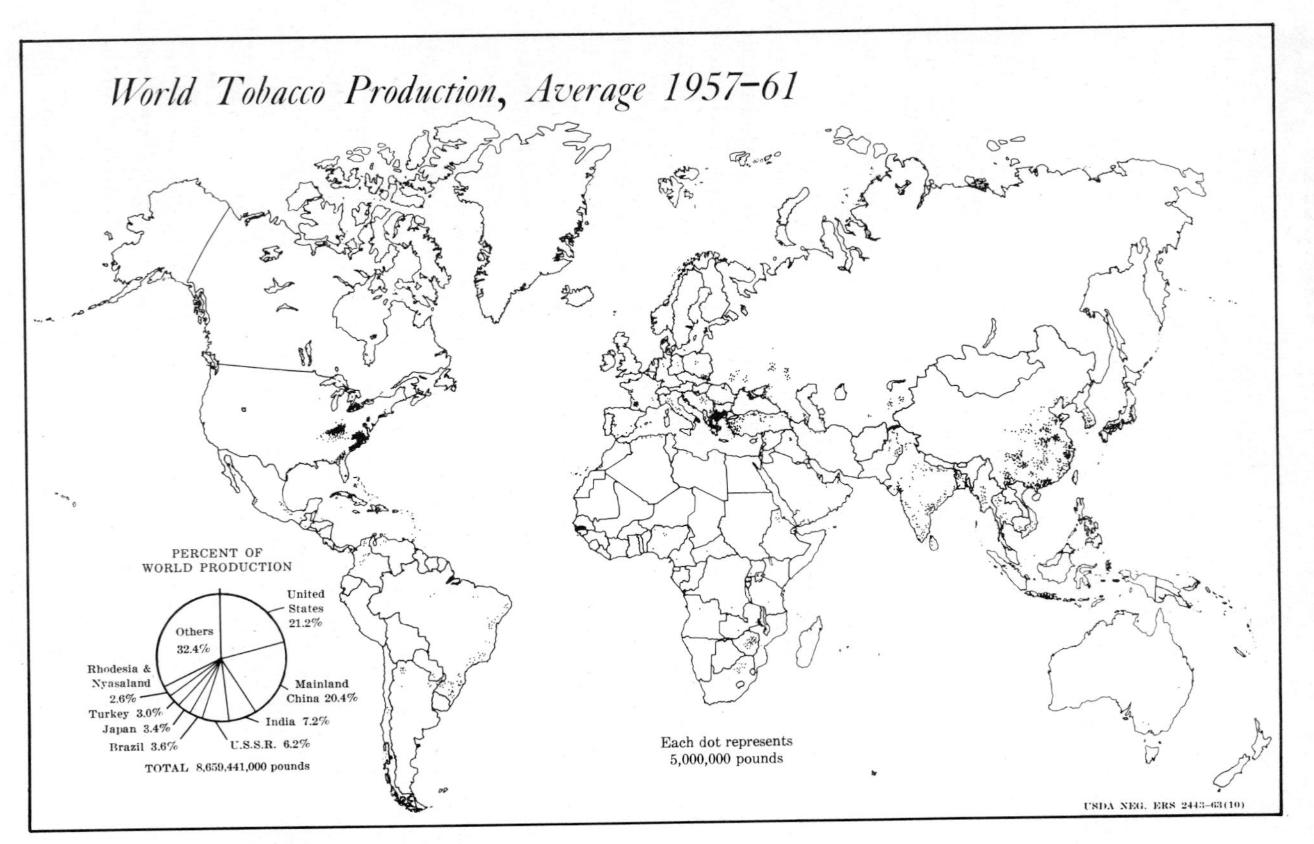

Figure 34–1. World tobacco production.

ORIGIN AND HISTORY

Tobacco (*Nicotiana tobacum*) is a native of the Western Hemisphere from Mexico southward. Two sailors saw an Indian smoking a cigarette with a corn husk or palm leaf wrapping, on San Salvador on November 6, 1492. The Indians of Yucatan were cultivating tobacco when the Spaniards first arrived there in 1519.[28, 29] It was grown in Spain and France by 1561 and in England in 1565, probably 20 years before Sir Walter Raleigh popularized it.[32]

Tobacco culture was begun by John Rolfe at Jamestown, Virginia, in 1612, and 15 years later 500,000 pounds of tobacco were shipped to England. Tobacco culture had begun in Southern Maryland in 1635, and tobacco continues to be the chief cash crop of that section.

In parts of Mexico and in the United States from the Mississippi River eastward and northward to Canada Indians were growing another species, *Nicotiana rustica*. The word *tobacco* comes from an Indian word for pipe. Three other tobacco species formerly grown by Indians in the United States[28] are *N. attenuata*, grown in the Southwest, in most of the plains area, and along the northern Pacific coast; *N. quadrivalvis*, grown in the Northwest; and *N. trigonophylla*, grown by the Yuma tribe of the Southwest.

ADAPTATION

Tobacco can be grown under a wide range of climatic and soil conditions, but the commercial value of the product depends largely on the environment under which it is produced. In the United States the culture of each important commercial type of tobacco is highly localized, due primarily to the influence of the climate and soil on the quality of the leaf (Figure 34–2).

Tobacco is grown from central Sweden at latitude 60° N southwest to Australia and New Zealand at latitude 40° S. The optimum temperature for germination of tobacco seed is about 88° F.[40] During the six to ten weeks that tobacco seedlings are in a cold frame or hotbed, the optimum temperature for growth is about 75° to 80° F., the minimum about 50° F., and the maximum 95° F. or higher. A very heavy growth of tobacco is undesirable because of its consequent poorer quality. Tobacco in the field grows most rapidly at about 80° F. It will mature at this temperature in about 70 to 80 days. In southern Wisconsin and the Connecticut Valley tobacco regions, where the mean summer temperature is about 70° F., a frost-free period of about 100 to 120 days from transplanting to maturity is required to mature the crop. In the central tobacco regions the average temperature during the growing season is about 75° F., while in northern Florida it

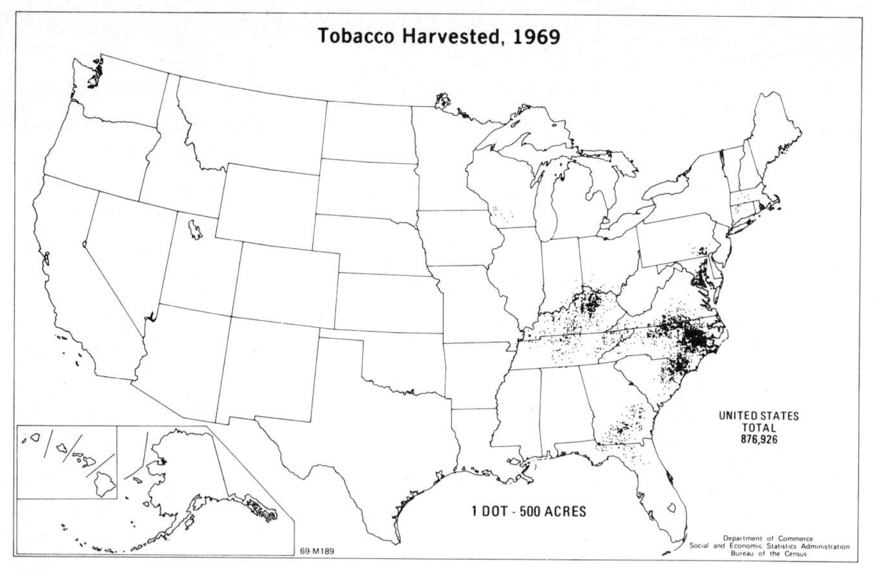

Figure 34–2. Acreage of tobacco in the United States in 1969.

is about 77° F.[27] Flue-cured tobacco thrives best where the daytime temperature is 70° to 90° F.[48]

A low rainfall results in tobacco high in nitrogen, nicotine, ether extract, acids, and calcium, but low in potash and soluble carbohydrates.[19] Most of the tobacco crop in America is grown where the annual precipitation is 40 to 45 inches. An autumn humidity that permits the harvested leaves to cure at the proper rate is important, yet the humidity should be sufficiently high at times during the year to make the leaves pliable enough to be handled without breakage. High winds are damaging to tobacco leaves in the field. Tobacco is unsuited to semiarid regions not only for this reason, but also because of the low humidity, low summer rainfall, and alkaline soils high in nitrogen.

The tobacco plant demands a well-drained soil. Each type of tobacco has its special soil requirements. Most of the tobacco sections have sandy loam or silt-loam soils, but in the Georgia[11] and Florida areas the soils are sandy, and in the Virginia section the soil is a clay, while some of the tobacco in the Miami Valley of Ohio is grown on a clay loam. Fire-cured tobaccos are grown on heavy loam soils not adapted to other types. In some tobacco areas, e.g., Lancaster County in Pennsylvania and the Miami Valley of Ohio, the soils are highly productive loams[15] of limestone origin, but in most of the tobacco sections of the piedmont and coastal plain of the South Atlantic states the soils are low in both organic matter and mineral nutrients. Heavy soils and fertile soils tend to produce tobacco high in nitrogen, nicotine, and calcium, but low in potash and carbohydrates. Light sandy soils have an opposite effect.[18]

Since the high returns per acre from good tobacco justify good, heavy fertilization, natural fertility usually is a secondary consideration. The proper relative proportions of calcium, magnesium, and potassium, which are essential to good burning qualities in tobacco—especially cigar tobacco[51]—usually are reached by fertilizer application. A somewhat acid soil (*p*H 5.5 to 6.5) is best for tobacco; a *p*H of 5.0 to 5.6 is optimum in Connecticut.[3] Excessive acidity (below *p*H 4.5 to 5) is often detrimental to quality or yield because it retards absorption of calcium, magnesium, and phosphorus, and leads to excessive absorption of manganese and aluminum. A soil near the neutral point favors development of the black root-rot disease. Tobacco is unsuited to soils in which the water table is less than 3 feet from the surface.

BOTANICAL DESCRIPTION

Tobacco belongs to the family *Solanaceae,* the nightshade family. *Nicotiana tobacum,* the common cultivated tobacco, is an amphidiploid, assumed to have originated from a natural cross between *N. sylvestris* and some other native species. The plants are 4 to 6 feet in height. They are terminated by a cluster (raceme) of up to 150 or more funnel-shaped flowers (Figure 3–12). The terminal flowers are first to open. The five petals are pink, but may be white or red in some varieties. The tobacco fruit is a capsule that splits into two or four valves at maturity. It may contain 4,000 to 8,000 seeds. A single tobacco plant sometimes produces 1 million seeds. The seeds are extremely small, about 1100 per gram or 5 million in 1 pound. The plants bear 12 to 25 leaves that range up to 2 feet in length and are covered with sticky hairs (Figure 34–5).

The flower of cultivated tobacco is normally self-fertilized although occasional outcrosses occur due to wind-borne or insect-borne pollen (Figure 4–3).

Of the two species of tobacco cultivated for their leaves, *Nicotiana tobacum,* which is native to Mexico or Central America, is grown in the United States as well as in most other tobacco-producing countries. The species native to the United States, *N. rustica,* is grown in India and certain other foreign countries. However, in the United States it is grown only sparingly or experimentally for making insecticides because of its high content of nicotine.[53] This latter species has pale yellowish flowers and a short thick corolla tube with rounded lobes. It is characterized further by thick, broadly ovate leaves covered with sticky hairs, and by a distinct naked petiole or leaf stalk. It grows to a height of only 2 to 4 feet.

More than 50 species of *Nicotiana* exist, chiefly in the Western Hemisphere or Australia. Of these species, *alata, glauca, sylvestris,* and *sanderae* are grown as ornamentals, while *rustica, longiflora, debneyi, glutinosa,* and *megalosiphon* have been used in breeding varieties of tobacco that resist diseases or nematodes.[12*]

The leaves of *N. tobacum* are arranged in a spiral on the stalk.[2] A leaf of flue-cured tobacco averages about 21 inches in length, 11 inches in width, and 147 square inches in area. The lamina (blade between veins) of the leaf is about 0.3 millimeters thick. The average weight of a green leaf is about 42 grams, while that of a cured leaf is about 6 grams, making 72 leaves per pound. Cell division ceases when a leaf reaches about one-sixth its normal size. All subsequent growth thereafter is due to cell enlargement and thickening of cell walls, together with increases in dry weight of the cell contents. During curing a leaf shrinks about 9 per cent in length, 15 per cent in width, and 24 per cent in area.[5, 74]

The upper leaves on the plant tend to be high in nitrogen, nicotine, and acid, but low in ash, calcium, and magnesium. The potash content shows a less definite decrease from top to bottom.[9, 33, 66]

CLASSIFICATION

The official grades for domestic tobacco divide the product into the 7 classes and 27 types shown in Table 34–1. The locations where most of the types are grown are shown in Figure 34–3.

Flue-cured tobaccos constituted about 61 per cent of the domestic

TABLE 34–1. Tobacco Class and Type in the United States

TYPE OF CURING	CLASS NO.	TYPE NO.	AVG. PRODUCTION 1970–72 (*1000 lbs.*)
Flue-cured	1	11–14	1,094,278
Fire-cured	2	21–23	41,727
Light air-cured	3A	31–32	568,357
Dark air-cured	3B	35–37	16,281
Cigar filler air-cured	4	41–46	30,302
Cigar binder air-cured	5	51–55	24,101
Cigar wrapper air-cured (shade grown)†	6	61–62	12,369
Miscellaneous (Louisiana perique)	7	72	160
Total			1,787,608

† Some fire-cured wrapper tobacco grown in Florida and Georgia (see Figure 34–4).

* See also Burk, S. G., and Heggestad, H. E., "The genus *Nicotiana:* A source of resistance to disease of cultivated tobacco," *Econ. Bot.*, 20:76–88. 1966.

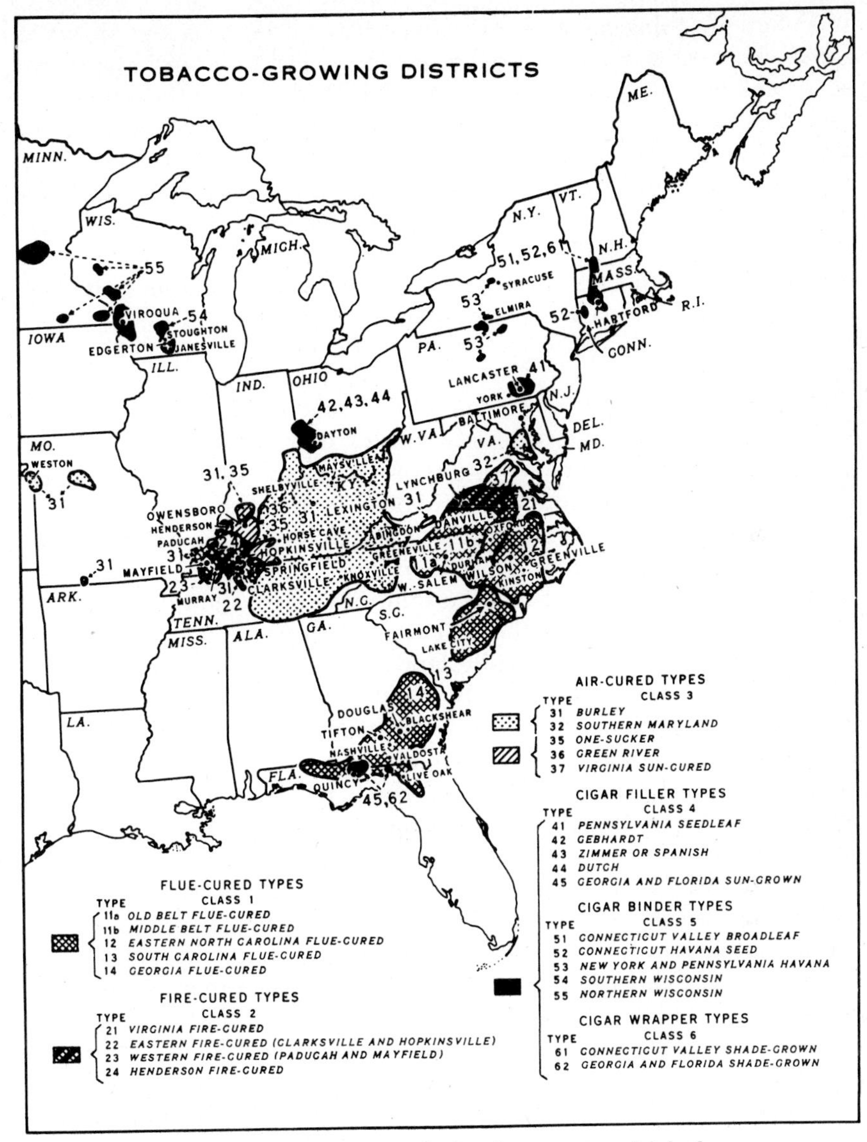

Figure 34–3. Types of tobacco, with the districts in which they are grown.

crop; air-cured, 33 per cent; cigar types (also air-cured), 4 per cent; and fire-cured, 2 per cent. Perique, a pungent type used in pipe-tobacco blends, is grown annually on about 200 acres only in St. James Parish in southern Louisiana. The aromatic "Turkish" types of tobacco are grown only occasionally in the southeastern states or in California.

Flue-cured tobacco, used chiefly for cigarettes, is also used for pipe and chewing tobaccos and for export. Fire-cured tobaccos are used chiefly for export, snuff, and plug wrappers. Dark air-cured types are used for chewing plug and export. Maryland (type 32) air-cured, is used for cigarettes and export. Burley (type 31), air-cured, is used for cigarette, pipe, and chewing tobaccos. The cigarette blend in 1966 was 52.7 per cent flue-cured tobacco, 35.6 per cent burley, 10.5 per cent Oriental (Turkish), and 1.2 per cent Maryland.[22]

The uses of a particular strain are not wholly restricted. For example, Connecticut Havana may be used for cigar wrapper as well as binder, and occasionally for filler leaf.

ROTATIONS

The chief consideration in determining a crop-rotation system for tobacco growing is the effect of the previous crop on the quality and yield of tobacco and on the incidence of tobacco diseases.[15, 25, 42] Most adapted crops succeed as well or better after tobacco than following other intertilled crops such as corn. Consequently, conditions that favor the high-priced tobacco crop obviously should govern the rotation. In Colonial times, the best tobacco crops often were obtained on cleared forest land or on idle land that had gone back to weeds, grass, and brush. In southern Maryland, tobacco following a natural two-year weed fallow gives better returns than tobacco following either bare fallow or legume crops in rotations.[10] In southern Maryland experiments, tobacco yielded well after red clover in a tobacco-wheat-red clover three-year rotation; it also yielded well when continuously cropped, with hairy vetch as a winter cover crop, but the tobacco quality was poorer than that of tobacco following ragweed and horseweed.

Flue-cured tobacco that follows corn, cotton, and small grain is of excellent leaf quality.[12] When grown after such legumes as cowpeas, crotalaria, lespedeza, peanuts, soybeans, and velvetbean, the tobacco ranges from poor to good. For tobacco following legumes, an appreciable increase in quality is obtained when nitrogen fertilization is reduced or heavy applications of potash are made to balance the deleterious effect of the excess nitrogen left over from legumes.

In Pennsylvania, cigar-binder tobacco is grown mostly by continuous cropping, whereas cigar-filler tobacco follows a legume in three-year or four-year rotations that include wheat or wheat and corn. A five-year rotation of tobacco, wheat, grass (two years), and corn also is practiced. In Kentucky, Ohio, and Virginia, tobacco usually follows a legume or grass.[6, 48, 61] In Connecticut much of the tobacco is grown in alternation with fall-sown grain, usually rye, as a cover crop, or after a grass meadow crop, because tobacco following a legume has a dark leaf.[3]

FERTILIZERS

Fertilizers are applied to all tobacco fields in the United States. An average 2,020 pound crop of flue-cured tobacco leaves contains about 32 pounds of nitrogen, 8 pounds of P_2O_5, 40 pounds of K_2O, 50 pounds of calcium, 52 pounds of magnesium, and 12 pounds of sulfur. A liberal amount of potash in tobacco leaves is essential to good burning quality. Nitrogen is necessary for good growth of tobacco, but an overabundance of nitrogen results in a low-quality[29] leaf. Excessive phosphorus shortens the leaf burn duration.[8] Magnesium is essential to prevent sand drown, a nutritional disorder caused by a deficiency of that element in some soils. Magnesium and calcium can be supplied in dolomitic limestone.[51] A 2000-pound crop of tobacco leaf requires 24 to 35 pounds of water-soluble magnesium per acre, with fully five times as much available calcium in the soil.

Flue-cured tobacco fields should be fertilized at rates in pounds per acre about as follows: N, 24 to 36; P_2O_5, 70 to 110; K_2O, 80 to 120; Cl, 18 to 24; CaO, 50 to 70; MgO, 16 to 24; SO_4, 65 to 100; or 10 per cent above these amounts in kilos per hectare. Maryland tobaccos require about 90 pounds per acre of P_2O_5, 120 of K_2O, 50 of CaO and 30 of MgO.[8] About 80 pounds of sulfur also is helpful. Fire-cured and perique tobaccos require about 750–1000 pounds per acre of a 4–12–8, 5–10–10 or 5–10–15 fertilizer or its equivalent. Dark air-cured tobaccos usually should receive the equivalent of 500 to 1000 pounds of a 5–10–10 fertilizer. Cigar-filler tobaccos may receive about 1000 to 1500 pounds of 8–12–4 fertilizer plus ten tons of barnyard manure. Cigar tobaccos in the Connecticut Valley may use about 2,500 pounds per acre of an 8–4–8 fertilizer, plus 2 to 4 per cent magnesia, when no manure is applied. About 1000 pounds of 3–9–18 fertilizer are used with heavy applications of manure.

Burley tobaccos require about 100 pounds of nitrogen and 120 to 250 pounds of potash per acre. Some of the soils in the bluegrass region contain ample phosphorus, but low-phosphorus soils in other regions may respond to 100 to 150 pounds of phosphoric acid. Only 50 pounds of nitrogen and a reduced amount of potash are applied when the field has received ten tons of manure. Side or top dressings of 8–0–24 fertilizer may be beneficial. The current higher-analyses fertilizers are applied at lower gross weights per acre than those listed above.

Some tobacco-transplanting machines are equipped to apply fertilizer in double bands 3 or 4 inches to each side of the row and below the level of the roots of the tobacco plants being planted. When the rate of application exceeds 750 pounds of fertilizer per acre, the remainder should be applied as a side dressing or the entire amount should be broadcast before planting.

SEEDING PRACTICES

Tobacco seed is planted in beds or cold frames 3 to 9 feet wide and as long as needed to provide sufficient plants for planting the field.[55] The number of square yards of bed area necessary to insure ample plants for an acre of tobacco land is about as follows: for cigar types, 33; for fire-cured, dark air-cured, and perique, 50; for flue-cured, 67; for Maryland and aromatic, 100; and for burley, 150.

Clean land for the beds is plowed or disked, raked, and then sterilized with wood fires, steam, or chemical disinfectants. Chemicals such as methyl bromide, calcium cyanamide, and urea are applied 60 to 90 days before the beds are seeded. Calcium cyanamide and urea are applied at the rate of 1 pound per square yard. Methyl bromide for weed and nematode control is sprayed on the soil from pressurized cans at the rate of 9 pounds per 100 square yards. The bed is covered immediately with a polyethylene cover. Insecticides also may be applied to the bed. Combinations of these soil treatments can kill soil-borne fungi, nematodes, weed seeds, and insects.

The beds should be fertilized at rates of ½ to 1½ pounds per square yard with a fertilizer that contains 4 to 8 per cent nitrogen, 4 to 12 per cent phosphoric acid, and 3 to 12 per cent potash. The fertilizer for flue-cured and Maryland tobaccos should also contain 1 per cent magnesia. The higher rates of fertilizer are desirable for the flue-cured tobacco; low rates, for fire-cured, burley, dark, air-cured, and shade-grown cigar-wrapper type; and medium rates (1 pound per square yard), for the other cigar types and aromatic tobacco.

An ounce of tobacco seed usually is mixed with a bushel of sand, sifted wood ashes, land plaster, or other inert materials to permit better distribution of the seed in the bed. An ounce of tobacco seed is sufficient for 1650 to 4950 square feet of seedbed to provide two to four plants per square inch of bed, which is enough plants for 1 to 6 acres. Seed is scattered over the bed three times to ensure more even distribution. The bed is then rolled to compact the seedbed and press the seeds into the soil. Then the bed is covered over with plastic, cloth, or glass.[55] A cloth is adequate in warm climates. Sometimes cloth also is laid on the soil until the seed germinates. The bed may require occasional watering.

TRANSPLANTING INTO THE FIELD

Land for tobacco usually is plowed in the spring or fall, at which time the cover crop or weed growth is turned under. The land is worked down into a good tilth before the tobacco is transplanted. Where rainfall is heavy the land may be thrown up into beds. Tobacco is transplanted six to ten weeks after the beds are sown, when the plants have six to eight leaves and are 6 to 8 inches (15 to 20 cm.) in height.[63] Transplanting is done in June in the northern districts, in May in the central districts as well as in the mountainous southern sections, in April in South Carolina, and from March 20 to April 15 in Florida and southern Georgia. The transplanting is done by machine or by hand. The plants are watered when they are being set in the field unless the soil is wet.

Distances between rows vary from 32 to 54 inches, depending on local district customs, while the spacing within the rows varies from 10 to 12 inches.[55] The space per plant in the field varies from 4 to 6 square feet in the northern tobacco districts including Kentucky[63] to 8 to 12 square feet in the southern districts, where the season is long and soils tend to be less fertile, and 15 square feet for perique tobacco.[60] Shade-grown tobacco is planted somewhat more thickly, i.e., with 3 to 4 square feet per plant, in order to keep shading costs per plant at a minimum.

The wide rows in the South permit use of beds and water furrows.

Tobacco is cultivated at frequent intervals, beginning a week to 10 days after transplanting. It usually is hoed soon after the first cultivation and sometimes again later in the season. The high acre value of the crop justifies intensive cultivation. Most annual grasses and some small broad-leaved weeds can be controlled by applying 4 pounds per acre of diphenamid immediately after transplanting the tobacco.

Figure 34–4. Shade tobacco in Florida. Three primings of the lower leaves have been harvested.

Cigar-wrapper tobacco growing under artificial shade (Figure 34–4) may cost $1,200 per acre. The object of shading is to produce smaller, thinner, and smoother leaves with smaller veins more suitable for fine cigar wrappers.[3] The effect of the shading is to increase the humidity and reduce the air currents, thereby retarding evaporation from the plants and the soil.[34] The shading material, a loosely woven cloth (eight to ten threads to the inch), is spread over and sewed to wires that are supported by stout posts extending 8 feet above the ground. In the Connecticut Valley the posts are placed 33 feet apart. Most of the cloth is used only one year.

Tobacco plants seem to grow very slowly at first because the seeds and consequently the seedlings are extremely small. At transplanting, the dry weight of the seedling is less than 0.05 per cent of the final dry weight of the plant.[71] Even 26 days after transplanting, less than 2 per cent of the total growth has been produced. Growth proceeds rapidly after that stage.

TOPPING

Topping consists in breaking off the top or crown of the plant at about the third branch below the flower head. Tobacco plants are topped to keep them from producing seed. This forces the synthesized carbohydrate and nitrogen materials to remain in the leaves for further growth and enrichment. The top leaves that are removed are highest in nicotine content.[33] Topping results in larger, thicker, and darker leaves that mature earlier and more uniformly than do those on untopped plants (Figure 34–5).

Topping causes a higher content of nicotine (especially in the upper leaves) and of sugars, but also a lower ash content. Topping begins in cigar and burley tobaccos when about half the plants show flower heads, in shade-grown wrapper and cigar-filler tobaccos before the blossoms open, in flue-cured tobaccos when the plants have 10 to 15 leaves, and in perique tobacco when the plants are two feet tall.[60] In the fire-cured tobaccos, the 3 or 4 bottom leaves on the plants often are removed also, with the result that only 8 to 12 leaves remain. In flue-cured tobacco, 10 to 14 leaves remain after topping in the piedmont section, while in the coastal plain 12 to 18 leaves are left. From 14 to 18 leaves remain on burley tobacco, 16 to 20 on Maryland tobacco, and nearly all on shade-grown tobacco. A field must be gone over two or three times in order for all plants to be topped at about the same stage of growth. After the topping, it is necessary to remove the suckers at intervals of every 7 to 10 days. Suckers are branches that develop in the leaf axils after the tops are removed.

Sucker growth can be suppressed by the application of chemicals such as maleic hydrazide (MH30), vegetable oils, or fatty acids to the

Figure 34–5. Flue-cured tobacco that has been topped.

plants immediately after topping. Maleic-hydrazide treatment increases the yield,[63] and the leaf weight, sugar content, and equilibrium moisture content, especially of flue-cured tobacco. It also reduces the specific volume or the filling power of the tobacco in the cigarette.[38] This reduced specific volume impairs the drawing quality of the cigarette, which is reflected in the value of the tobacco. Chemicals under trial promise to suppress sucker growth without a loss in quality.[70]

HARVESTING

The leaves of tobacco are ripe and ready to harvest when they turn a lighter shade of green[56] and have thickened so that when a section of the leaf is folded it creases or cracks on the line of folding. This stage is reached in the bottom leaves when the seed heads form, but in the middle leaves it is reached about two or three weeks later. Flue-cured and fire-cured types are harvested when fully ripe with the leaves showing light-yellow patches or flecks. Stalk cutting is done when the middle leaves are ripe.

Two general methods of harvesting are practiced: priming and stalk cutting. Priming is practiced on nearly all flue-cured tobacco, all the

Figure 34–6. Spearing tobacco stalks and hanging them on the stick.

shade-grown cigar-wrapper and Georgia-Florida sun-grown cigar-binder tobacco, and some of the Havana Seed cigar-binder tobacco. The other types are stalk-cut (Figure 34–6). Priming is the picking of the leaves when they are in prime condition, beginning with the lower three or four leaves of commercial size, followed by four or five successive pickings of two to four leaves each at intervals of about 5 to 10 days. The successive harvestings of the lower leaves cause gains in weight of the remaining upper leaves. This weight gain, together with the avoidance of translocation of materials from the leaf to the stalk during curing, results in primed leaves that weigh 20 to 25 per cent more than the leaves on cut stalks. A single priming of the lower leaves of burley tobacco increases the total yield, but the quality of the primed leaves is not typical of stalk-cured burley. After priming, the leaves are placed in baskets, mounted bins, or on burlap sheets and conveyed to the curing barn. The leaves are strung on sticks about 4½ feet long. A string is attached to one end of the stick, after which the leaves are strung either by a needle pushed through the base of the leaf midrib or by looping the string around the base of two to five leaves. When the string is full, it is tied to the other end of the stick (Figure 34–7): A stick holds 60 to 80 or more leaves. This system has been largely replaced by loose-leaf handling and bulk curing. Several men riding on a moving machine strip off the leaves and place them on a metal rack.

In stalk cutting, the stalks are cut off near the base and left on the ground until the leaves are wilted. The stalk is then hung upon a lath. This is done by piercing the stalk near the base with a removable metal spearhead placed on the end of the lath and sliding the stalks along the lath (Figure 34–6). A lath holds six to eight stalks. Some-

Figure 34–7. Tobacco leaves strung and hung on sticks.

times the stalks are hung on the lath with hooks or nails. Fire-cured and burley stalks are split down toward the base; after cutting they are placed astride the laths. Splitting of cut stalks hastens the drying of the stalks and stops translocation early. The laths, which bear about six plants, are placed upon a rack and hauled to the curing barn.

CHANGES IN TOBACCO PRODUCTION

Tobacco yields in the United States averaged about 800 pounds per acre from 1910 to 1914 and 2,080 pounds in 1970–71. The labor required per 100 pounds of tobacco was about 44 and 24 hours, respectively, or 352 and 499 hours per acre. About half of the labor is involved in harvesting, curing, and marketing the crop. Such high labor requirements limit the acreage of tobacco that a farm family can operate.

Motorized equipment for tillage, transplanting, topping, and spraying has reduced labor somewhat. Chemical sprays to suppress sucker growth can replace hand suckering. A crew riding on a mobile machine can strip and load tobacco leaves more rapidly than when walking. Much of the tobacco is now cured and marketed as loose-leaf without tying it into hands. Untied tobacco reduces labor by more than 20 per cent.[36] A few machines that strip the leaves mechanically are now in use. Tobacco plants can be topped with mechanical cutters, but with some loss of leaves. Heating the curing barns with fuel oil or gas requires less labor than the former practice of stoking a wood fire.[23]

CURING

Curing involves the processes of drying, decomposition of chlorophyll until the green color disappears from the leaf, changes in the nitrogen compounds including release of ammonia, hydrolysis of starch into sugars, and respiration or fermentation of the sugars.[16, 17, 50] Mineral salts also crystallize out, thus producing the grain of the leaf. Practically all the sugars are used up when tobacco is air cured, but considerable quantities remain in the leaves of fire-cured and flue-cured types. Losses of nicotine during curing range from 10 to 33 per cent. Too-rapid drying, like bruising the leaf, kills the tissue prematurely. This prevents complete decomposition of chlorophyll, which results in an undesirable green color. It also stops some of the other desirable chemical reactions. During curing there is a decrease in total weight of 84 to 88 per cent,[58] which includes a loss of dry matter of 12 to 20 per cent in primed leaves, with an additional 10 to 12 per cent of material translocated to the stalks when the leaves are not primed. Harvested tobacco contains about 75 to 80 per cent moisture. After the leaf is cured, it is allowed to regain moisture (the regained moisture being known as order or case) until the leaves containing about 24 to 32 per cent moisture are pliable enough to handle without breaking. Thus, the final net loss in total weight of the leaf during curing is about 75 per cent, i.e., two tons of fresh leaves will yield 1000 pounds of cured and cased product. Eight tons of cut stalks will yield about 1800 pounds of cured leaf.

Tobacco Barns

Barns for curing tobacco usually are high enough (16 to 20 feet to the plate) to hold three to five tiers of suspended leaves or stalks. The barns are built in bents with poles spaced at the proper distances for having sticks or laths hung over them. For primed tobacco on sticks 4½ feet long, the poles are spaced 50 inches apart horizontally and 22 to 30 inches apart vertically. For stalk curing on 4-foot laths the pole centers are 46 inches apart horizontally and 2½ to 5 feet apart vertically. Since the tobacco should hang at least 3 feet above the ground, the lower tier of poles is 6 to 9 feet high. A space of 1½ to 2½ cubic feet per plant is required for curing tobacco. A barn 17 feet high, 32 to 36 feet wide, and 140 to 160 feet long, may hold the crop from 3 or 4 acres. Barns for loose-leaf curing on metal racks are much lower.

Air Curing

Tobacco is air-cured in huge, tight barns sometimes 300 feet long, equipped with numerous ventilating doors on the walls that can be opened or closed to regulate the humidity and temperature (Figure 34–8). The relative humidity at the beginning should be about 85 per cent, but after the leaves begin to turn brown a lower humidity that permits rapid drying is advisable.[38] A condition called pole sweat or house burn occurs when the humidity exceeds 90 per cent for 24 to 48 hours with the temperature above 60° F. This condition causes injured, partly cured, and even fully cured leaves to begin to soften and decay. This damage is prevented by use of artificial heat to lower the relative humidity and promote drying. This may be provided by LP gas burners, coke stoves, or by several small charcoal fires built on the earth floor inside the barn. These fuels do not impart undesirable odors to the tobacco leaf. Artificial heat is required also when the temperature drops below 50° F. so that curing processes other than drying may continue. Air curing may require from four to eight weeks.[21]

Figure 34–8. Barn for air curing stalk-cut cigar-binder tobacco in Connecticut valley. Note the center driveways.

Flue Curing

The objective in flue curing is to hasten the early curing stages so that drying will be completed while the leaves are still a light-yellow color. Flue-cured or bright tobacco is riper and of a lighter green than other types when curing begins. The typical temperature and humidity requirements for the yellowing, drying, killing, and ordering stages of flue curing are illustrated in Figure 34–9. The yellow color of the leaf is fixed during the later period of the yellowing stage. The temperature is allowed to rise to 130° F. in about 50 hours, with a range in relative humidity of from 65 to 85 per cent in the yellowing stage. Drying is completed in about 40 or more hours as the temperature is raised to about 165° F. The leaves are killed during this stage, while

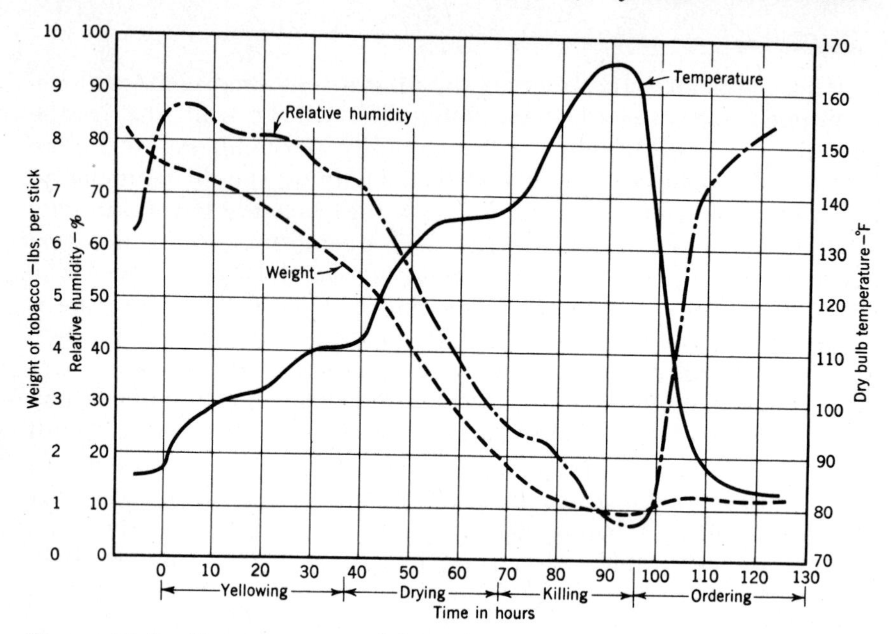

Figure 34–9. Temperature and humidity requirements for flue-cured tobacco. [After Moss and Teeter, N. Car. Agr. Exp. Sta.]

the relative humidity drops to less than 10 per cent. The heating is stopped after 4 or 5 days and the tobacco allowed to take up moisture. When drying is too rapid, the humidity in the barn can be increased by wetting the floor and walls, and when the humidity is too high the roof ventilators are opened. Improper curing may result in leaf discolorations such as black splotching, green veins, green leaf, dark sponging, scalding, and black stem.

For flue curing, a large sheet-iron pipe connected to the top of each furnace usually is extended to the far side of the barn, along it for a short distance, and then turned back and out through the wall on the furnace side.

Ventilation is provided for incoming air by openings near the bottom of the barn, and for the departing moist air by ventilating doors in the roof. The cold incoming air can be passed over hot flues inside the barn.[57] Most of the furnaces fired with wood, or coal,[22] have been replaced with types that burn propane gas, natural gas, stove oil, or furnace oil. The combustion gases pass through the tobacco by forced air ventilation or exhaust fans.[72] The temperatures for forced-air drying are 95–105° F. during yellowing, 105–125° and 125–135° during color fixing, and 135° to 145° to 165° F. during final drying.

Fire Curing

The usual system of fire curing is to allow the tobacco to yellow and wilt in the barn without fires for 3 to 5 days, then start slow fires to maintain the temperature at 90° to 95° F. until yellowing is completed, and finally raise the temperature to 125° to 130° F. until the leaves are dry. The fires are kept up for 3 to 5 days or more. The smoke from the open, hardwood fires in the barn imparts the characteristic odor and taste to the tobacco desired for chewing plug and snuff.

Perique tobacco is air-cured 8 to 14 days with the whole stalks hung on wire. Then the dry brown stripped leaves are moistened, stemmed, formed into twists, and packed in casks under heavy pressure with occasional loosening for a month until the leaves are black. Thereafter they are allowed to ferment in their own juice for 9 months.

HANDLING AND MARKETING

After the tobacco on the stalk is cured and has been cased or ordered enough to be pliable, it is ready to be stripped from the stalks by hand. In that condition, the leaf contains 24 to 32 per cent moisture. Often the tobacco is bulked in piles without being removed from the sticks when it is first taken down after curing. The bulking not only prevents further drying but helps to decompose any remaining chlorophyll. Both stripped and primed leaves are sorted into grades and sometimes tied into hands of 15 to 30 leaves, with a leaf used as a binder. Most of the flue-cured tobacco is now being sold as loose leaf without tying. A pound of cured wrapper tobacco contains 60 to 90 leaves.[39]

The hands of cigar tobaccos are packed in cases or in bundles (bales) which then are covered with paper. Other types, handled as loose-leaf tobacco, are delivered in baskets or in bundles wrapped temporarily in burlap or other cloth. Most of the tobacco is sold by auction[22] after it is graded and labeled (Figure 34–10). An auctioneer may sell tobacco at the rate of 300 to 400 individual lots per hour, but must alternate with another auctioneer from time to time to save his voice. Cigar tobacco is often sold at the farm soon after curing. The marketing season is July to December for flue-cured types, and December to February or March for burley fire-cured and air-cured types. However, Maryland tobacco is sold from May to July in the year after its production.[22]

The tobacco as marketed is moist enough to be pliable. This pliability prevents severe losses from breakage, but usually such tobacco is too moist for storage or export.[16] It is redried on racks in heated chambers or containers, or by a high-frequency current diathermic process. The latter process has been used for drying and heat fumigation

Figure 34-10. A tobacco auction. The leaves are tied into "hands." The auctioneer has folded his arms to indicate that a transaction has been completed.

of tobacco in hogsheads destined for export. Before redrying, tobacco sometimes is sweated or fermented in bulk piles to complete processes begun during curing. Export tobacco contains about 11 per cent moisture. Fermentation consists of allowing a pile or box of slightly moist tobacco to heat at temperatures of 85° to 120° F., the temperature depending upon the type of tobacco and method of bulking. Fermentation causes losses of 5 to 12 per cent in dry matter. Decreases in starch, sugar, nicotine, amino nitrogen, citric acid, and malic acid occur during fermentation.

Fermented tobacco is aged for one to three years before being used in manufacture. In aging, tobacco undergoes a decrease in nitrogen, nicotine, sugars, total acids, and *p*H, as well as in irritating and pungent properties. The aroma is improved by aging.[20]

The leaves with the midrib (stem) removed are known as strips. Stemming may be done just before the leaf is used in manufacturing or two years earlier when it is shipped and stored. The stem comprises about 25 per cent of the weight of the leaf, but varies from 17 to 30 per cent. Tobacco used for snuff is not stemmed. Stems, along with tobacco waste, are used in manufacturing nicotine sulfate for insecticides, although the stems contain only about 0.7 per cent nicotine as compared with the leaf, which usually contains 1 to 5 per cent. Some of the stems, which contain about 6 per cent potash, are used as fertilizer or for mulching. *Nicotiana rustica* leaves are very high (5 to 10 per cent) in nicotine content. When leaves of this species are grown, they are used for insecticide materials.[53] Most of the nicotine can be extracted by forcing the leaves through a series of rollers. Some nicotine is extracted from waste tobacco.

The shrinkage of tobacco purchased from the grower ranges from 30 to 40 per cent as a result of stemming, drying, sweating, and handling. A pound of tobacco that includes this waste will make about 40 cigars, or 350 cigarettes, or 1.06 pounds of snuff. One pound is sufficient to make 500 modern filter-type cigarettes. Tobacco stems, licorice, sugar, and other materials are added to unstemmed tobacco for making snuff. Chewing plug consists of about 60 per cent tobacco, while the remainder comprises sweetening and flavoring materials. A considerable quantity of cigar-filler tobacco is produced in Puerto Rico, but some is imported from other countries. Large quantities of aromatic tobacco for cigarette, and pipe tobacco blends, are received from Turkey, Greece, and countries near them.

TOBACCO QUALITY

Tobacco varies widely in quality, which depends upon the conditions of growing, curing, sorting, and handling.[73] During an auction observed by the senior author, prices for different lots of the same market type of leaf ranged from 4 cents to 36 cents per pound. The production of a good-quality leaf is essential to success, since low-grade tobacco brings less than its cost of production. Color, texture, and aroma are important characteristics. Dark fire-cured tobacco is expected to be strong, dark, and gummy. Flue-cured cigarette tobaccos are usually a light lemon-yellow color, while air-cured cigarette tobaccos are light brown. Cigar-wrapper leaf should be thin, uniform, and free from blemishes. The color of the wrapper, whether dark or light, is not a true measure of the mildness of a cigar, as is popularly assumed. A heavy sweating process that darkens the leaf also reduces the nicotine content. Furthermore, the color of the thin wrapper leaf does not reveal the color or mildness of the binder and filler which make up the greater part of a cigar. Most domestic cigars are now being made with a "reconstituted sheet" as a binder. The sheet is rolled out of ground cigar binder leaf, stems and scraps, mixed with a cohesive.[22] The highest prices are paid for shade-grown wrapper, followed by flue-cured and light air-cured types. The composition of flue-cured and burley tobacco is shown on page 868.

TYPICAL COMPOSITION OF TOBACCO IN PER CENTS (EXCEPT FOR *p*H)

	Flue-cured	*Burley*
Total nitrogen	2.0	3.4
Nicotine	2.3	3.0
Petroleum-ether extract	5.0	5.5
Total sugars	18.75	0.5
Acids	14.0	27.0
*p*H	5.1	6.0
Ash	12.0	19.0
Potash (K_2O)	3.25	–
Calcium oxide (CaO)	3.25	–
Magnesia (MgO)	7.0	–
Chlorine	0.70	–
Sulfur	0.60	–

Dark air-cured and fire-cured tobaccos contain 4 to 4.5 per cent nicotine.[55]

A high-quality tobacco usually is high in soluble carbohydrates and potash, but relatively low in crude fiber, nicotine, nitrogen, calcium, ash, and acids. An extremely low content (1 to 1.3 per cent) of nitrogen is not desirable. A *p*H not lower than 5.3 is preferred. The popular brands of cigarettes contain about 1.75 per cent nicotine. Cigars of all classes contain about 1.5 per cent nicotine. The nicotine contents of granulated pipe mixtures, scrap chewing tobacco, and chewing plug are about 2 per cent, 1 per cent, and 2 per cent, respectively.

A low nicotine content is desired in tobaccos grown for domestic use, despite the fact that nicotine is the characteristic active stimulative principle. Nicotine content varies greatly with climate, season, soil, maturity, and general cultural practices.[67]

Low-nicotine strains of tobacco are available, if a market for them is ever engendered. Nicotine is an alkaloid, an oily liquid, soluble in water, and has considerable volatility at ordinary temperatures. It boils at 250° C., and is extremely poisonous. The chemical formula is $C_{10}H_{14}N_2$. When oxidized it yields nicotinic acid, the amide of which is the familiar vitamin called niacin.

Cultivated tobacco contains traces of several other alkaloids. The nicotine content is higher in the leaf lamina (blade) than in the veins, but it is higher in the veins than in the midribs or stalks.[7] The alkaloid increases as the plant grows, with the result that the greatest amount is probably present when the plant is fairly mature. It is believed to decrease after this time.

Tobacco smoke contains carbon, carbon dioxide, carbon monoxide, oxygen, hydrogen sulfide, hydrocyanic acid, ammonia, nicotine, pyridine, methyl alcohol, resins, phenols, ethereal oils, nicotinic and glutamic acids, and other compounds.[27] Most of these compounds or elements, even oxygen, are toxic in excessive proportions. An extensive campaign to stop cigarette smoking because of its apparent tendency to increase the incidence of emphysema and lung cancer,[75] has reduced the per capita consumption of cigarettes only about 10 per cent. Pipe and cigar smokers are less likely to inhale the fumes into the lungs so that their cancer attacks usually are restricted to the mouth and throat.

DISEASES

The important tobacco diseases are black root rot, blackfire, wildfire, mosaic, blue mold, and frenching.[13, 14, 55]

Black Root Rot

Root rot, caused by *Thielaviopsis basicola,* is a fungus disease that produces black lesions and eventual decay of the roots. Diseased plants are small and stunted. The roots are small and begin decay at the tips. Frequently the plants become yellowish. The plants usually outgrow the disease in the field when the temperatures exceed 80° F. The fungus lives over in the soil. The most effective control measures are crop rotation, seedbed disinfection, and resistant varieties.

Mosaic

The mosaic disease (also called calico or walloon) is the most destructive virus disease of tobacco. It is present wherever tobacco is grown.[14] Plants infected with mosaic at transplanting time may be reduced 30 to 35 per cent in yield and 50 to 60 per cent in value.[49] The disease is characterized by a mosaic pattern of light and dark green or yellow areas. The different strains of the virus give various symptoms. Some strains cause mild mottling with no distortion of the leaves, while others cause prominent mottling, or distortion of new leaves. The virus that causes the common mosaic of tobacco is a crystalline nucleoprotein of high (40,000,000) molecular weight. The crystals are rod-shaped, about 270 millimicrons in length, and 15 millimicrons thick as revealed in an electron microscope photograph. These virus crystals possess many characteristics of living organisms such as reproduction and mutation. The leaves and leaf juices of plants containing mosaic virus are very infectious. The virus is spread to other plants by direct or indirect contact.

The tobacco mosaic virus remains infectious for long periods—as long as 30 years—in dead plant material.[41] It will overwinter in dry, compact, or water-logged soils. Perennial wild hosts and aphids that feed on them as well as on the tobacco plants are important factors in the dissemination of the disease. The disease also is spread by workers who use tobacco.

The disease can be partly controlled by (1) sterilizing the tobacco bed and tobacco cloths, (2) destroying near-by solanaceous weeds, (3) removing diseased plants, (4) destroying all tobacco refuse around the bed, (5) washing the hands with a trisodium phosphate solution after handling diseased plants, and (6) completely abstaining from use of natural leaf tobacco products during cultural operations.[46] Resistant varieties are grown.

Wildfire or Bacterial Leaf Spot

Wildfire, caused by *Pseudomonas tabaci,* spreads rapidly through the field. The symptoms are lemon-yellow spots on the leaves with small dead spots in the center of the yellow spots. The disease is best controlled by spraying the plants in the bed shortly after they come up and at weekly intervals thereafter until transplanted. The sprays used include streptomycin sulfate, streptomycin nitrate, Copper-A, and bordeaux mixture. Improved varieties are resistant. Ample potash should be applied for control in the field.

Blackfire or Angular Leaf Spot

Blackfire is a destructive bacterial leaf-spot disease in Virginia and Kentucky caused by *Pseudomonas angulata.* Control measures are similar to those recommended for the wildfire disease.

Bacterial Wilt

The organism *Pseudomonas solanacearum,* which causes bacterial wilt or Granville wilt, enters the roots, where it multiplies and clogs the water-conducting vessels, causing the plants to wilt. Control methods[64] are rotations that involve immune crops such as corn, cotton, sweetpotatoes, grass, clover, and small grains, in which tobacco is grown not oftener than once in four or five years. Since the wilt organism attacks other plants (tomatoes, peppers, peanuts, and various weeds), these must be excluded from the tobacco land. Resistant varieties are available.

Blue Mold

The blue mold (downy mildew) disease is caused by the fungus *Peronospora tabacina,* which attacks and destroys the leaves in the seedbed. It often kills large numbers of plants. The disease occurs in a mild form from Florida to Maryland every year. Serious damage occurs sporadically, depending upon weather conditions. The best remedy is regular spraying or dusting with fungicides. When the disease appears, the beds may receive a gas treatment with 2 to 3 pounds per 100 square yards of paradichlorobenzene crystals.[14, 50] Anthracnose, caused by *Rhizoctonia solani,* is controlled by the same seedbed treatments. The sore-shin stage of anthracnose, which occurs in the field, is not subject to control.

Root Diseases

At least six root diseases,[13, 14] which include the Granville wilt already described, damage the tobacco plant. Southern stem rot, caused by *Sclerotium rolfsii,* is controlled by crop rotation and soil fumigation. Fusarium wilt, caused by *Fusarium oxysporum* var. *nicotianae,* can be controlled by the growing of resistant varieties, by soil fumingation to control nematodes, and by crop rotation. Black shank, caused by *Phytophthora parasitica* var. *nicotianae,* can be controlled by growing resistant varieties as well as by crop rotation that avoids solanaceous plants.

Root knot is caused by a nematode or eelworm (*Meloidogyne* species) that produces galls on the roots. This disease as well as nematode root rot, caused by *Pratylenchus* species, is controlled by soil fumigation and crop rotation. Several varieties are resistant to the root-knot nematode.[23]

Other Diseases

Other diseases include two leaf spots, frog eye (or Wisconsin leaf spot), caused by the fungus *Cercospora nicotianae,* and brown root rot. Several varieties are somewhat resistant to some leaf spots. Two virus diseases, etch and vein banding, are checked by the control of aphids.[50]

Broom rape, a parasitic flowering plant, lives upon the tobacco plant in much the same manner as dodder lives on clover.

NUTRITIONAL DISORDERS

Frenching

Frenching[44] has been confused with mosaic. Newly frenched plants are nearly white at the growing point. The frenched leaves are narrowed and drawn, the tips sometimes bending sharply downward so as to form a cup of the underside of the leaf. The leaves may be reduced to narrow straps in severe cases. The disease is more prevalent in seasons of abundant rainfall. It appears to be associated with soil conditions, but usually only scattered spots or plants in the field are affected. No control method is recommended.

Sand Drown

Sand drown (magnesia hunger) is caused by deficiency of available magnesium. It occurs most frequently after heavy rains on sandy soils subject to the leaching of soluble minerals; hence its name. The leaves turn nearly white at the tips and along the margins, especially on the lower leaves. Sand drown is prevented when a fertilizer that contains 10 to 20 pounds of available magnesia has been applied. This can be supplied by the use of sulfate of potash-magnesia as the potash fertilizer. Dolomitic limestone high in magnesia content applied in the drill at the rate of 500 pounds per acre or broadcast at the rate of 1000 pounds per acre likewise prevents sand drown. Manure and other organic fertilizers also tend to prevent the disease.

Fleck

Fleck causes small gray or white spots on the upper surface of tobacco leaves.[35] It occurs following high concentrations, 4 or more parts per 100 million, of ozone in the atmosphere. Ozone apparently is formed by photochemical reactions that involve nitrogen dioxide, certain hydrocarbons, and other air-borne chemicals. Fleck occurs most frequently in fields that are in the proximity of metropolitan and manufacturing areas where industrial and automotive fumes are abundant.

Detection of Mineral Deficiencies

Tobacco serves as an excellent test plant for determination of mineral deficiencies in soils or culture solutions, because the deficiencies produce symptoms that can be recognized readily.[44, 51, 52]

INSECTS

The most serious insect pests of tobacco in the United States[45, 50] are the hornworms or greenworms, of which there are two important species: the tomato hornworm (*Manduca quinquemaculata*), and tobacco worm or southern tobacco worm (*P. sexta*) (Figure 34–11). Other serious insect pests are the tobacco fleabeetle (*Epitrix hirtipennis*) and the tobacco budworm (*Heliothis virescens*). The hornworms devour entire leaf blades, whereas the fleabeetle riddles the leaves with small holes. The fleabeetle causes the most damage on young plants. The budworm feeds in the top of the plants and cuts holes in the young leaves. By chewing into the bud it may puncture several unopened leaves at once. This insect is prevalent in the southern tobacco-growing areas. Cutworms attack the plants in the beds as well as after they are set in the field. All of these pests as well as aphids can be controlled with insecticides.[50] Destruction of tobacco-plant residues, unused plants in the beds, and weeds near the beds and fields helps in keeping down the insect populations.

The tobacco or cigarette beetle (*Lasioderma serricorne*) damages stored and manufactured tobacco. The tobacco moth (*Ephestia elutella*) attacks flue-cured and imported Turkish cigarette tobacco in storage.[4] The larvae devour the entire leaves except for the larger veins. They also foul other leaves with webs and excreta. They infest

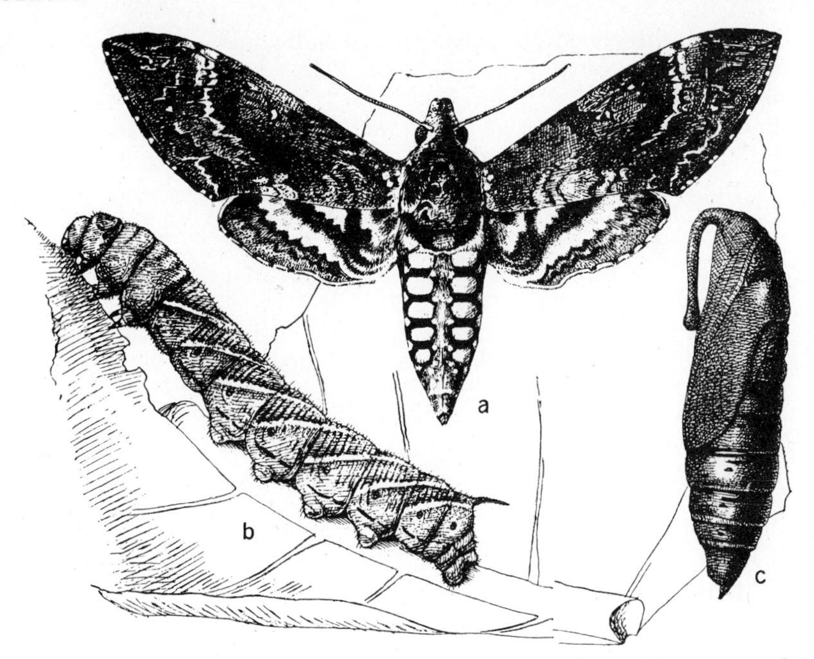

Figure 34–11. Southern tobacco hornworm: (*a*) adult, (*b*) larva, and (*c*) pupa.

tobacco stored in cases, hogsheads, or piles in warehouses or storage sheds. This pest is kept under control by periodic fumigation with hydrocyanic-acid gas in closed storages. In open storages pyrethrum sprays and dust are effective.

REFERENCES

1. Akehurst, B. C. *Tobacco*, Harlow (England), Longmans, pp. 1–551. 1968.
2. Allard, H. A. "Some aspects of the phyllotaxy of tobacco," *J. Agr. Res.*, 64:49–55. 1942.
3. Anderson, P. J. "Growing tobacco in Connecticut," *Conn. Agr. Exp. Sta. Bul. 564*. 1953.
4. Anonymous. "Stored tobacco insects, biology and control," *USDA Handbk. 233*, pp. 1–43. 1971 (Rev.).
5. Avery, G. S., Jr. "Structure and development of the tobacco leaf," *Am. J. Bot.*, 20:565–592. 1933.
6. Bachtell, M. A., Salter, R. M., and Wachter, H. L. "Tobacco cultural and fertility tests," *Ohio Agr. Exp. Sta. Bul. 590*. 1938.
7. Bacon, C. W. "Some factors affecting the nicotine content of tobacco," *J. Am. Soc. Agron.*, 21:159–167. 1929.
8. Bowling, J. D. "Phosphorus, potassium, calcium and magnesium requirements of Maryland tobacco grown on Monmouth soil," *Md. Agr. Exp. Sta. Bul. A-151*, pp. 1–48. 1967.
9. Bowman, D. R., Nichols, B. C., and Jeffrey, R. N. "Time of harvest—its

effect on the chemical composition of individual leaves of Burley tobacco," *Tenn. Agr. Exp. Sta. Bul. 291*. 1958.
10. Brown, D. E., and McMurtrey, J. E., Jr. "Value of natural weed fallow in the cropping system for tobacco," *Md. Agr. Exp. Sta. Bul. 363*. 1934.
11. Carr, J. M. "Bright tobacco in Georgia," *Ga. Coastal Plain Exp. Sta. Bul. 10*, pp. 1–31. 1928.
12. Clayton, E. E. "The genes that mean better tobacco," in *Plant Diseases*, USDA Yearbook, 1953, pp. 548–553.
13. Clayton, E. E., and others. "Control of flue-cured tobacco root diseases by crop rotation," *USDA Farmers Bul. 1952*, pp. 1–12. 1944.
14. Clayton, E. E., and McMurtrey, J. E., Jr. "Tobacco diseases and their control," *USDA Farmers Bul. 2033* (rev.). 1958.
15. Colwell, W. E. "Tobacco," in *Soil*, USDA Yearbook, 1957, pp. 655–658.
16. Cooper, A. H. "The curing and drying of tobacco," *Refrig. Engr.*, 38:207–9, 244–5. 1939.
17. Cooper, A. H., and others. "Drying and curing bright leaf tobacco with conditioned air," *J. Ind. Eng. Chem.*, 32:94–99. 1940.
18. Darkis, F. R., Dixon, L. F., and Gross, P. M. "Flue-cured tobacco: Factors determing type and seasonal differences," *J. Ind. Eng. Chem.*, 27:1152–1157. 1935.
19. Darkis, F. R., and others. "Flue-cured tobaccos: Correlation between chemical composition and stalk position of tobaccos produced under varying weather conditions," *J. Ind. Eng. Chem.*, 28:1214–1223. 1936.
20. Dixon, L. F., and others. "The natural aging of flue-cured cigarette tobacco," *J. Ind. Eng. Chem.*, 28:180–189. 1936.
21. Donaldson, R. W. "Quality tobacco in Massachusetts," *J. Am. Soc. Agron.*, 23:234–241. 1931.
22. Doub, Albert, Jr., and Wolfe, A. "Tobacco in the United States," *USDA Misc. Pub. 867*. 1961.
23. Elliot, J. M., and Marks, C. F. "Control of nematodes in flue-cured tobacco in Ontario," *Can. Dept. Agr. Pub. 1465*. 1972.
24. Fairholt, F. W. *Tobacco: Its History and Associations*, Singing Tree Press, Detroit, pp. 1–332. 1968.
25. Gaines, J. G., and Todd, F. A. "Crop rotations and tobacco," in *Plant Diseases*, USDA Yearbook, 1953, pp. 553–561.
26. Garner, W. W. "Some aspects of the physiology and nutrition of tobacco," *J. Am. Soc. Agron.*, 31:459–471. 1939.
27. Garner, W. W. "Climate and tobacco," in *Climate and Man*, USDA Yearbook, 1941, pp. 364–372.
28. Garner, W. W., Allard, H. A. and Clayton, E. E. "Superior germ plasm in tobacco," *USDA Yearbook, 1936*, pp. 785–830.
29. Garner, W. W., and others. "The nitrogen nutrition of tobacco," *USDA Tech. Bul. 414*. 1934.
30. Garner, W. W. *The Production of Tobacco*, McGraw-Hill-Blakiston, New York, pp. 1–520. 1951 (Rev.).
31. Grizzard, A. L., Davies, H. R., and Kangas, L. R. "The time and rate of nutrient absorption of flue-cured tobacco," *J. Am. Soc. Agron.*, 34:327–339. 1942.
32. Hahn, P. M. *Sold American*, The American Tobacco Company, Louisville, Kentucky. 1954.

33. Hamner, H. R., Street, O. E., and Anderson, P. J. "Variation in chemical composition of cured tobacco leaves according to their position on the stalk," *Conn. Agr. Exp. Sta. Bul. 433*, pp. 177–186. 1940.
34. Hasselbring, H. "The effect of shading on the transpiration and assimilation of the tobacco plant in Cuba," *Bot. Gaz.*, 57:257–286. 1914.
35. Heggestad, H. E., and Middleton, J. F. "Ozone in high concentrations as cause of tobacco leaf injury," *Science*, 129(3334):208–210. 1959.
36. Hayert, J. H. "Feasibility of Maryland tobacco leaf untied," *Md. Agr. Exp. Sta. M.P. 788*. 1971.
37. Jeffrey, R. N. "The effect of temperature and relative humidity during and after curing upon quality of White Burley tobacco," *Ky. Agr. Exp. Sta. Bul. 407*. 1940.
38. Jeffrey, R. N., and Cox, E. L. "Effects of maleic hydrazide on the suitability of tobacco for cigarette manufacture," *USDA Agricultural Research Service Spec. Rept. ARS 34–35*. 1962.
39. Jenkins, E. H. "Studies on the tobacco crop of Connecticut," *Conn. Agr. Exp. Sta. Bul. 180*, pp. 1–65. 1914.
40. Johnson, J., Murvin, H. F., and Ogden, W. B. "The germination of tobacco seed," *Wis. Agr. Exp. Sta. Res. Bul. 104*. 1930.
41. Johnson, J., and Ogden, W. B. "The overwintering of the tobacco mosaic virus," *Wis. Agr. Exp. Sta. Res. Bul. 95*. 1929.
42. Jones, J. P. "Influence of cropping systems on root-rots of tobacco," *J. Am. Soc. Agron.*, 20:679–685. 1928.
43. Karraker, P. E. "The comparative effect of muriate and sulfate of potash on the consumption and quality of White Burley tobacco," *Ky. Agr. Exp. Sta. Res. Bul. 341*. 1932.
44. Karraker, P. E., and Bortner, C. E. "Studies of frenching of tobacco," *Ky. Agr. Exp. Sta. Bul. 349*. 1934.
45. Lacroix, D. S. "Insect pests of growing tobacco in Connecticut," *Conn. Agr. Exp. Sta. Bul. 379*, pp. 84–130. 1935.
46. Lehman, S. G. "Practices relating to control of tobacco mosaic," *N. C. Agr. Exp. St. Bul. 297*. 1934.
47. Lunn, W. M., and others. "Tobacco following bare and natural weed fallow and pure stands of certain weeds," *J. Agr. Res.*, 59:829–845. 1939.
48. Matthews, E. M., and Hutcheson, T. B. "Experiments on flue-cured tobacco," *Va. Agr. Exp. Sta. Bul. 329*. 1941.
49. McMurtrey, J. E., Jr. "Effect of mosaic disease on yield and quality of tobacco," *J. Agr. Res.*, 38(5):257–267. 1929.
50. McMurtrey, J. E., Jr. "Tobacco production," *USDA Info. Bul. 245*. 1961.
51. McMurtrey, J. E., Jr. "Relation of calcium and magnesium to the growth and quality of tobacco," *J. Am. Soc. Agron.*, 24:707–716. 1932.
52. McMurtrey, J. E., Jr. "Nutrient deficiencies in tobacco," in *Hunger Signs in Crops*, 3rd ed., McKay, New York, pp. 99–141. 1964.
53. McMurtrey, J. E., Jr., Bacon, C. W., and Ready, D. "Growing tobacco as a source of nicotine," *USDA Tech. Bul. 820*. 1942.
54. McMurtrey, J. E., Jr., and Robinson, W. O. "Neglected soil constituents that affect plant and animal development," in *Soils and Men*, USDA Yearbook, 1938, pp. 807–829.
55. Miles, J. D. "Producing flue-cured tobacco plants under plastic cover," *Ga. Agr. Res.*, 3(2):8–10. 1961.

56. Moseley, J. M., and others. "The relationship of maturity of the leaf at harvest and certain properties of the cured leaf of flue-cured tobacco," *Tobacco Science*, 7:67–75. 1963.
57. Moss, E. G., and others. "Fertilizer tests with flue-cured tobacco," *USDA Tech. Bul. 12.* 1927.
58. Moss, E. G., and Teter, N. C. "Bright leaf tobacco curing," *N.C. Agr. Exp. Sta. Bul. 346*, pp. 1–25. 1944.
59. Reed, W. D., and Livingstone, E. M. "Biology of the tobacco moth and its control in closed storage," *U.S. Dept. Agr. Circ. 422*, pp. 1–39. 1937.
60. Rense, W. C. "The Perique tobacco industry of St. James Parish, Louisiana: A world monopoly," *Econ. Bot.*, 24(2):123–130. 1970.
61. Roberts, G., Kinney, E. J., and Freeman, J. F. "Soil management and fertilization for tobacco," *Ky. Agr. Exp. Sta. Bul. 379.* 1938.
62. Shear, G. M. "Factors affecting physiological breakdown of maturing tobacco," *Va. Agr. Exp. Sta. Tech. Bul. 74.* 1941.
63. Smiley, J. H. "Kentucky farm tobacco," *Rsh. Rpt. 195*, pp. 1–24. 1971.
64. Smith, T. E. "Control of bacterial wilt (*Bacterium solanacearum*) of tobacco as influenced by crop rotation and chemical treatment of the soil," *USDA Circ. 692*, pp. 1–16. 1944.
65. Smith, T. E., Clayton, E. E., and Moss, E. G. "Flue-cured tobacco resistant to bacterial (Granville) wilt," *USDA Circ. 727*, pp. 1–7. 1945.
66. Swanback, T. R. "Variation in chemical composition of leaves according to position on the stalk," *Conn. Agr. Exp. Sta. Bul. 422.* 1939.
67. Thatcher, R. W., Streeter, L. R., and Collison, R. C. "Factors which influence the nicotine content of tobacco grown for use as an insecticide," *J. Am. Soc. Agron.*, 16:459–466. 1924.
68. Thomas, R. P. "The relation of nitrate nitrogen and nitrification to the growth of tobacco following timothy," *Wis. Agr. Exp. Sta. Res. Bul. 105.* 1930.
69. Tso, T. C., and McMurtrey, J. E., Jr. "Preliminary observations on inhibition of tobacco suckers by vegetable oils and fatty acids," *Tobacco Science*, 7:101–104. 1963.
70. Tso, T. C. *Physiology and Biochemistry of Tobacco Plants*, Dowden, Hytchinson and Ross, Stroudsburg, Pa., pp. 1–393. 1972.
71. Vickery, H. B. "Chemical investigations of the tobacco plant. V. Chemical changes that occur during growth," *Conn. Agr. Exp. Sta. Bul. 375*, pp. 557–619. 1935.
72. Walker, E. K., and Vickery, L. S. "Curing flue-cured tobacco," *Can. Dept. Agr. Publ. 1312*, pp. 1–27. 1969.
73. Ward, G. M. "Physiological studies with the tobacco plant," *Dominion Dept. Agr. Tech. Bul. 37*, Ottawa, 1942.
74. Wolf, F. A., and Gross, P. M. "Flue-cured tobacco: A comparative study of structural responses induced by topping and suckering," *Bul. Torrey Bot. Club*, 64:117–131. 1937.
75. Wynder, E. L., and Hoffman, D. *Tobacco and Tobacco Smoke: Studies in Experimental Carcinogenesis*, Academic Press, New York, pp. 1–730. 1967.

Chapter 35

Sugarbeets

ECONOMIC IMPORTANCE

Sugarbeets are grown on about 7.7 million hectares, producing some 220 million metric tons or 28.6 metric tons per hectare. The leading countries in sugarbeet production are the U.S.S.R., the United States, France, Poland, West Germany, and Italy. The production in the United States from 1970 to 1972 was about 24.8 million metric tons on 550,000 hectares (1.37 million acres), yielding 45 metric tons per hectare, or 20 short tons per acre (Figure 35–1). In 1967, 18,300 growers produced sugarbeets on an average of 62 acres per farm. The average beet area per farm exceeds 100 acres in several states. Mechanized planting, thinning, harvesting, cultivation and the application of selective herbicides, together with some semiauto-

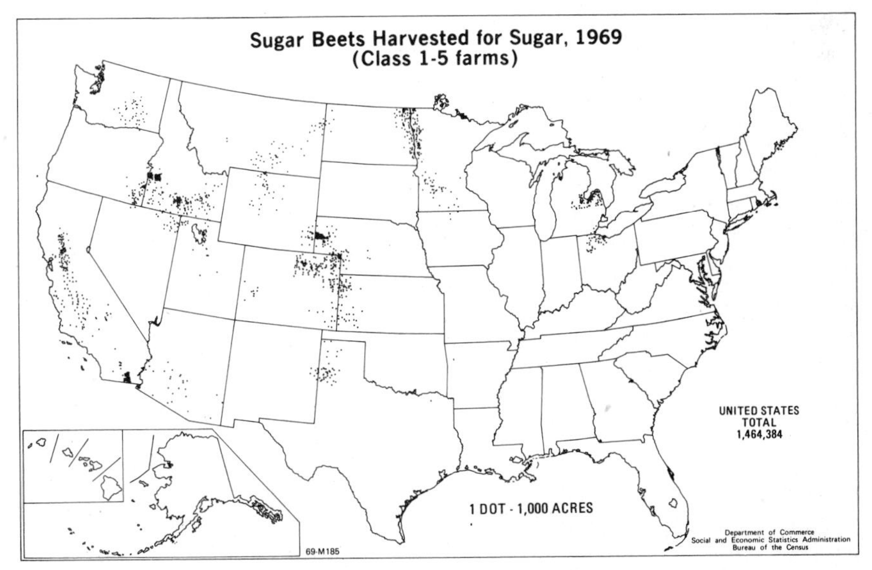

Figure 35–1. Acreage of sugarbeets in the United States in 1969.

matic sprinkler irrigation, have eliminated most of the hand labor formerly required for sugarbeet production in the United States.

Sugarbeets have been one of the most uniformly profitable cash crops in many irrigated valleys in the western United States. Sugarbeets not only facilitate diversification, but also provide an intertilled crop in rotations with hay and grain crops. Feeding the by-products of beet culture and manufacture forms the basis of an extensive sheep- and cattle-fattening industry around each beet-sugar factory.

HISTORY

The sugarbeet dates back over 200 years.[10] In 1747 a German chemist named Andrew S. Marggraf found that the kind of sugar in two cultivated species of beets was identical to that in cane. The first factory for the extraction of sugar from sugarbeets, built in 1802 in Silesia, was a failure. The percentage of sucrose in the sugarbeets used then was very low. Louis Vilmorin in France selected beets by progeny test methods and raised the sugar content from 7.5 per cent to 16 or 17 per cent. By 1880 sugarbeets had practically as high a sugar percentage as the varieties of today. The first successful commercial factory in America was erected at Alvarado, California, in 1870. General success of the industry in the United States dates from 1890.

About one-third of the world sugar supply is furnished by the sugarbeet. A popular prejudice against beet sugar, now almost forgotten, continued for many years despite the fact that beet sugar could not be distinguished from cane sugar. The pure sugar (sucrose) from the beet and the pure sugar from the cane are identical.

ADAPTATION

Sugarbeets are grown in favorable localized sections within feasible shipping distances from sugarbeet factories. A factory represents an investment of several million dollars, and usually has a capacity of 1000 to 5,500 tons of beets daily. Therefore, it is essential that several thousand (usually 10,000 or more) acres of beets be grown in order to keep the factory in operation during the season of 65 to 200 days after harvest begins.

Successful sugarbeet production is found only on fertile soils. They thrive best in soils of *p*H 6 to 7.5. A good percentage of soil organic matter supplied naturally, by manuring, or by legume residues, favors beet yields. Loam or sandy loam soils predominate in most of the sugarbeet areas, but heavy clay soils are used to some extent. These stickier soils usually are less favorable for high yields. They also increase the difficulties of lifting the beets and freeing them from adhering soil. The sugar content of beets is highest on soils that also produce the best tonnages.[47] However, nitrogen fertilizers applied in excess, or late in the season, lower the sugar content. Sugarbeets can endure large quantities (1 to 1.5 per cent) of saline salts (alkali) after they become established.[23]

Irrigation water is necessary for successful commercial sugarbeet production where the rainfall is less than about 18 inches, even in the cool northern boundary states of the United States. The seasonal water usage (evapotranspiration) for beet yields of 20 to 30 tons per acre ranges from 21 inches in the cool area in southern Canada to more than 40 inches in Arizona. Commercial sugarbeet production west of the 100th meridian is on irrigated land. Most of the sugarbeets grown in eastern North Dakota, Minnesota, north-central Iowa, and states to the east are not irrigated.

Sugarbeet seed germinates well when the soil temperature is about 16° C. but germination is most rapid at 28° C. The sugarbeet root has the highest sugar content when produced where the summer temperature averages about 67° to 72° F. The optimum temperature for plant growth is about 24° C. but 17–20° C. for root growth.[44] The plant is uninjured by cool nights. Cool autumn weather (15° C.) favors sugar storage in the roots. Temperatures of 30° C. or above retard sugar accumulation. Sugarbeets make a good vegetative growth in the South, but the roots are low in sugar content.[6] On the other hand, sugarbeets planted in the winter in California and harvested during very hot July weather have a high sugar content. The newly emerged seedling may be killed by a temperature of −4° C. Later the plants become more resistant to cold. The mature plant is able to withstand fall frosts, but a temperature as low as 26° F. usually causes foliage injury. After such injury, a period of warm weather may bring about a growth of new leaves as well as a sharp decline in the sugar content of the roots.[35]

Sunlight has a sanitary effect in disease control. The intensity of light is associated more closely with the utilization of nutrient elements than with their absorption. The relative sugar content is not influenced by light until growth is inhibited[47] by high or low intensities.

Initiation of seed stalks and flowers of the sugarbeet (a long-day plant) is brought about mainly by the cumulative effect of exposure to low temperatures followed or accompanied by the effects of long photoperiods. This combined effect of light and temperature has been termed photothermal induction.[42] Seed stalks are produced most abundantly and seed yields are highest when the weather is cool (less than 21° C. maximum temperature), wet, and cloudy (with less than 10.6 hours sunshine per day) for a period of about six weeks, when followed by two weeks of cool dry weather to stimulate seed production.[27] Early beets are able to flower at cooler temperatures than are those planted late.[36]

BOTANICAL CHARACTERISTICS

The sugarbeet (*Beta vulgaris*) is a herbaceous dicotyledon, a member of the family *Chenopodiaceae*, characterized by small, greenish, bracteolate flowers. The flowers are perfect, regular, and without petals. A large fleshy root develops. The species includes four groups: (1) sugarbeets, (2) mangelwurzels or mangels, (3) garden beets, and (4) leaf beets such as chard and ornamental beets. The sugarbeet sometimes is classed as a separate species, *B. saccharifera*, but there appears to be little justification for such a separation.

It is believed that a wild type of *Beta vulgaris*, often referred to as *Beta maritima*, was the progenitor of the sugarbeet. Several wild species of *Beta* native to Europe[10] have been crossed with *B. vulgaris.*

Vegetative Development

The sugarbeet, a biennial, normally completes its vegetative cycle in two years. During the first year it develops a large succulent root, in which much reserve food is stored. During the second year it produces flowers and seeds. Prolonged cool periods cause a seed stalk to be sent up the first year, a behavior known as bolting. Some strains of beets bolt more readily or at higher temperatures than do others.

The mature beet (Figure 35–2) is an elongated pear-shaped body composed of three regions: the crown, the neck, and the root.[1] The crown is the broadened, somewhat cone-shaped apex that bears a tuft of large succulent leaves and leaf bases. Just below it is the neck, a smooth thin zone, which is the broadest part of the beet. The root is cone-shaped and terminates in a slender taproot. The root is flattened on two sides and often is more or less grooved. The two depressions, which extend vertically downward or form a shallow spiral, contain the lateral rootlets indistinctly arranged in two double rows. The surface of the beet is covered by a thin cork layer that is yellowish white except on the aerial parts at places of injury.[4]

In cross section the beet is made up of a number of rings or zones of growth. The center of the beet has a more or less star-shaped core. Most cultivated beets have ivory-white flesh, but in poor strains the flesh is of a watery-green or yellowish hue.

The beet root is richest in sugar slightly above the middle with sugar quantity decreasing toward both ends. The tip of the root is lower in sugar than any other part except the center of the crown. There is little consistent relation between the internal structure, size, or shape of the root, and the sugar content.[3]

Figure 35–2. Sugarbeet top and root showing characteristic groove.

The beets grown in one season may have a higher average sugar percentage than those grown in another. Individual beets vary widely in sugar content, but usually range from 10 to 20 per cent.

The leaves are arranged on the crown in a close spiral.[1] Leaf growth after unfolding occurs by cell enlargement but root growth results from cell division.[44]

Flower and Seed Development

The sugarbeet normally sets seed as a result of wind or insect pollination. Either the bagging of branches or plant isolation is necessary for the production of pure strains.[7, 17]

During their second year the beets first produce a rosette of leaves like those of the first year, but after about six weeks of growth the newly formed leaves become progressively smaller and then the flower stalk develops.[2] The flower stalk grows rapidly and branches considerably. The mature inflorescence or *seed bush* is composed of large, paniculate, more or less open spikes that bear the flowers and later the seeds (Figure 35–3). The calyx and stigma adhere to the mature fruit.

The sugarbeet fruit is an aggregate, formed by the cohesion of two or more flowers that have grown together at their bases. They form a hard and irregular dry body, the so-called seed ball, which usually

Figure 35–3. Sugarbeet plant producing seed. A branch bearing monogerm seed balls is shown at lower left.

contains two to five seeds. The mature seed is a shiny, lentil-like structure about 3 millimeters long and $1\frac{1}{2}$ millimeters thick. The mature reddish-brown outer seed coat is very brittle and separates easily from the seed.[2]

A plant with monogerm seed borne in separated flowers was found in western Oregon in 1948.[12] The progeny of this plant was crossed with several multigerm varieties. Selection, back-crossing, and reselection transferred the monogerm character into improved disease-resistant varieties as well as into lines used for producing hybrid sugarbeet seed. Sugarbeet plants with cytoplasmic male-sterile flowers were reported in 1945. This character also was bred into improved strains of sugarbeets, which made hybrid sugarbeet-seed production a commercial success. The hybrid beets in the United States are diploid plants but polyploids are widely used in Europe.

HYBRIDS OR VARIETIES

Sugarbeets grown in the United States since 1967 are improved hybrids bred for disease resistance, monogerm seed, slow bolting, regional adaptation, high production, or other desired characters.[12, 32] Sugarbeet breeding for disease resistance was urgent because of former heavy losses from curly top, leaf spot, downy mildew, and blackroot diseases.

Beet-sugar companies in the United States have developed varieties or hybrids adapted to their particular areas. Many of these are resistant to curly top or leaf spot. Practically all sugarbeets are grown under contract with a sugar company which furnishes technical assistance and usually produces or procures the seed for the growers.

ROTATIONS

Sugarbeets are grown almost exclusively in rotations that involve legumes (alfalfa, red clover, and sweetclover), small grains, and, frequently, potatoes, corn, or beans.[45] The growing of sugarbeets in continuous culture soon results in depressed yields[34] and often encourages infestation of the land by the sugarbeet nematode or by disease-producing fungi and bacteria. Planting sugarbeets immediately after a legume usually is inadvisable. The early spring preparation essential to early planting of sugarbeets destroys the legume crop before it can make enough spring growth to provide much green manure. Consequently, such land is best left for later plowing in preparation for corn or potatoes. Furthermore, the frequent failure to kill all alfalfa and sweetclover plants by tillage is detrimental to sugarbeet yields. The growing of an intertilled crop, such as potatoes or corn, after a legume crop facilitates disease and weed control and also is favorable to beet stands and yields. Certain legume crops and weeds promote the occurrence of soil-borne organisms that cause damping off and black root of sugarbeets. Better stands of beets are obtained following corn, potatoes, small grain, or soybeans than after alfalfa, sweetclover, or red clover. Thus, sugarbeets usually follow the crop that follows the legume.

FERTILIZATION PRACTICES

Barnyard manure applied at the rate of 6 to 12 tons per acre is a common fertilizer used in sugarbeet production. Pasturing of legumes preceding beets adds some manure to the land. Applications of manure produce substantial increases (often five tons or more per acre) in yields of beets on irrigated land.[22] Sugarbeets require about 15 pounds of available or applied nitrogen per ton of roots. The potassium requirement is 200 to 400 pounds per acre. Most of the sugarbeets are grown on soils well supplied with potash. However, muck soils often require potash. The turning under of legume residues and application of branyard manure often supply most of the nitrogen required for beet production. However, nitrogen fertilizers applied to sugarbeet fields in increasing (and sometimes excessive) amounts may cause sprangled roots. Applications of more than 60 to 80 pounds of nitrogen per acre to fertile soils are likely to lower the sugar content of the beets. Suggested nitrogen applications for sugarbeets are 0 to 60 pounds per acre on highly fertile soils, 60 to 120 pounds on medium soils, and 120 to 180 pounds on soils of low fertility or where a heavy growth of residues is turned under.

Phosphorus is the dominant element in commercial fertilizer applied for beets. In the humid area, an application of some 200 pounds of superphosphate per acre usually suffices. Heavier applications are advisable in the irrigated sections of the Great Plains states on heavy, highly calcareous soils from which much of the available phosphorus has been removed by long cropping to alfalfa. There, 150 pounds of triple superphosphate (equivalent to 400 pounds of ordinary superphosphate) have given profitable returns.[35] On the more porous sandy soils of the Great Plains and intermountain region, from which calcium has been leached by irrigation water, the application of phosphorus has not been particularly beneficial.

Small applications of borax, not exceeding 15 pounds per acre, are beneficial on soils deficient in boron. On some muck soils beets respond to applications of 25 to 50 pounds of copper sulfate. Zinc or iron applications may be needed on heavily graded irrigated spots. Sodium is an essential element for sugarbeets but the soil supplies ample amounts.

CULTURAL METHODS

Fall plowing in preparation for sugarbeets usually is advisable except for very friable soils. Deep plowing (8 to 12 inches) usually is recommended,[29, 35] although experimental evidence justifying plowing friable soils deeper than 8 inches is inadequate. Chiseling or deep plowing annually or occasionally is practiced on very hard soils.

Sugarbeet roots penetrate below any plowing depth with the feeding roots going down 5 or 6 feet.[45] Tillage subsequent to plowing should provide a mellow seedbed for the small beet seedlings.

Seeding Practices

Planting usually is done with special 4- to 12-row (Figure 35–4) beet planters with rows 18 to 22 inches apart or in 34 cm. double rows spaced at one meter between centers. The planting is done in April or early May in the northern, in February and March or April in milder areas, and cool areas of the United States, and in the winter or autumn in parts of California and Arizona. Unprocessed monogerm seed or processed multigerm seed is planted at rates of 5 to 8 pounds per acre, or at least 10 to 12 seeds per foot. Monogerm seed, polished to remove some of the surrounding cork, usually is planted at the rate of 1.5 to 4 pounds per acre. One pound of seed per acre, or four to six seeds per foot of row, is sufficient to provide a stand of beets under favorable conditions. Such a rate with good seed placement may eliminate the need for thinning.

Thinning

Sugarbeets usually require thinning[40] because even precision planting may result in irregular plant stands. This often is done by cross-row cultivation with a flexible-toothed or knife weeder or cultivator.[5] Rotary (down-the-row) implements, which uproot the plants at fixed intervals, also are used (Figure 35–5). Electric-eye thinners

Figure 35–4. Six-row sugarbeet planter with fertilizer attachments. [Photo courtesy Great Western Sugar Co.]

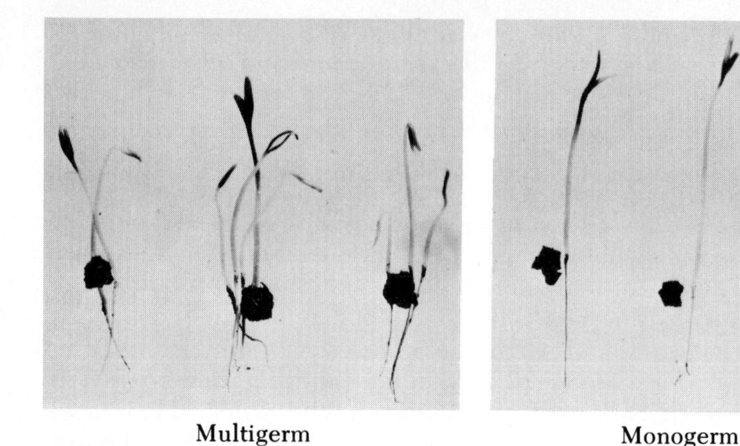

Multigerm Monogerm

Hand Thinned

Mechanized

Figure 35–5. The use of monogerm seed permits thinning, where necessary, by a down-the-row thinner. Unprocessed multigerm seed requires hand thinning and either hand or mechanical blocking.

can be used after precision planting of monogerm seeds on weed-free fields. Some growers thin the plants with long-handled hoes while chopping out the weeds in the row. The desired plant spacing is 8 to 12 inches apart in the row,[29] or 21,000 to 36,000 plants per acre.

When unprocessed multigerm seed is planted, the rows are blocked out by hand hoeing or by machine. Each clump that remains is thinned to a single plant by hand. A multigerm seed ball produces two or more plants in one place, but two adjacent plants do not develop normal beets. Hand blocking and thinning requires 20 to 40 man-hours per acre. Thinning is often done when the plants have six to eight leaves, but usually it begins about three weeks after planting when the seedlings have about four leaves. Thinning should be completed by the time the plants have eight to 10 leaves.

Weed Control

Special four-row or six-row beet cultivators are in common use for weed control. Those equipped with both knife-type and small sweep-type shovels are most satisfactory.

Weeds in the row are suppressed while small by cultivation with a harrow or flexible-shank weeder before the beets are up and shortly thereafter. Some hand hoeing prevents scattered weeds from going to seed. Herbicides are applied to much of the sugarbeet acreage.

Irrigation

In the West, beets are irrigated about every 10 to 14 days, or when water is available, or whenever the plants evidence the need of water by their dark-green color and the continuation of leaf wilting after sunset. Better yields result when the field is irrigated before the plants show signs of water need. The highest acre yields are obtained when the soil moisture at a depth of 1 foot is maintained at not less than 50 per cent of the total available water-holding capacity of the soil. The usual irrigation is 2 to 6 acre-inches at each application, or 12 to 36 inches for the season. The final irrigation should provide sufficient moisture for the sugarbeets to complete their growth. In addition, it should leave the soil moist enough for the roots to be dug. In some sections it is necessary to irrigate in the spring before or after planting or both in order to germinate the seed.

Harvesting

Sugarbeets should be left in the field until they reach a maximum sucrose content. Maturity is indicated by a browning in the lower

Figure 35–6. Three-row beet digger preceded by a six-row top harvester which windrows six rows of tops together. [Photo courtesy Great Western Sugar Co.]

leaves and a yellowing of the remaining foliage.[21] The contracting sugar company usually makes sugar analyses and instructs the growers as to when harvesting should begin.

The harvesting of sugarbeets in the United States was fully mechanized by 1958. The former hand operations of pulling the beets, knocking off the dirt, and piling, topping, and loading beets is now accomplished by mechanical harvesters that move along the row. Some machines cut off the tops of the standing plants and then lift, shake, and elevate the roots to a hopper.[5] Other harvesters lift the beets from the soil and convey them to rotating disk blades for topping (Figure 35–6). The beets are topped by the cutting off of the crown at the base of the lowest leaf scar. The crown contains little sugar, but it is high in mineral salts.

The topped beets are hauled in trucks to a railroad siding or directly to the sugar factory. At the factory, the beets are either processed immediately or are piled for processing within a few weeks.

SUGAR MANUFACTURE

In the manufacture of sugar the topped beets are first washed in a flume of rapidly flowing water.[43] They are then sliced mechanically with V-shaped knives into thin angular strips called cosettes, which are about the diameter of a lead pencil. The sugar is extracted from the slices by the diffusion process in large drums that contain warm juice followed by warm water at a temperature of 80° to 84° C. After separation of the juice from the pulp, milk of lime is added to the juice in large tanks to precipitate impurities and to neutralize oxalic and other organic acids. The acids combine with calcium to form less soluble salts which settle out of the solution. Excess calcium is precipitated as calcium carbonate by carbonation of the limed juice. The juice is filtered, further clarified, decolorized with sulfur dioxide, and again filtered. The juice is then concentrated to a sirup by being boiled under reduced pressure in steam-heated vacuum pans or evaporators called effects. The sirup is treated with sulfur dioxide, again filtered, and evaporation is continued until the sugar crystallizes. This mixture of sugar crystals and molasses is separated in centrifuges with perforated inner walls. The washed sugar crystals are then separated and dried in a granulator. The molasses or mother liquor is reworked several times to recover additional sugar, leaving a residue of final-discard molasses.

The average yield of refined sugar obtained in the United States is about 258 pounds per ton of beets, or over 5,000 pounds of sugar per acre in 1973. The average sugar content of the beets exceeds 15 per cent, but 16 per cent of the sugar is left in the molasses, pulp, or residues.

By-products of beet-sugar manufacture are pulp, molasses, and lime cake or waste lime from the filter presses. The beet pulp is the wet fibrous material left after the sugar is extracted from the sliced beets. The yield of wet pulp that contains 90 to 95 per cent water is about 1600 pounds per ton of beets. The final yield of wet pulp, after removal of part of the excess water by pressing or partial drying, is 400 to 600 pounds per ton of sliced roots. The pulp is used for stock feed either fresh, ensiled, or dried. Usually 2 to 3 gallons or more of final-discard molasses are obtained from each ton of beets. This molasses contains about 20 per cent water, 60 per cent carbohydrates (mostly sucrose, arabinose, and raffinose) and about 10 per cent ash or mineral matter in which potassium compounds predominate. The molasses often is added to the pulp and used as dried-molasses pulp. Considerable quantities are fed in mixture with other feeds. The remainder is used mostly in the manufacture of alcohol. The molasses also contains glutamic acid,[43] some of which is recovered for making monosodium glutamate. When added to foods, the monosodium glutamate stimulates human taste buds to a greater appreciation of food flavors.

The dry matter in the lime cake contains the equivalent of more than 80 per cent calcium carbonate; 10 per cent organic matter; and traces of potash, phosphorus, and nitrogen. It is suitable for liming soils, but most soils in the western sugarbeet areas are not in need of lime. For each ton of sliced beets, about 100 pounds of burned limestone are used, with water, to form milk of lime.[41]

By weight, beet tops consist of about one third crown and two thirds leaves. The green weight of tops ranges from 75 to 80 per cent of the weight of topped beets. Beet tops are fed mostly to sheep and cattle, either ensiled, fresh, or in dry form as cured in small piles in the field. Often the stock are turned into the field to eat the piled tops, but this is a wasteful practice. Beet tops are palatable and nutritious, but they are dangerous when fed in large quantities to horses and pigs because of their abundance of cathartic salts and oxalic acid. Ruminants are able to utilize large quantities without injury. Beet tops contain about two-thirds of the digestible nutrients found in corn silage.

SEED PRODUCTION

The discovery about 1925 that beets were able to overwinter in the field in the mild climates of the Southwest and Pacific Coast, started the American sugarbeet seed industry in 1932.[10, 12] The beet seed is planted in rows in August or September at a rate of 15 to 17 pounds per acre. The beets are left unthinned and given sufficient irrigation to keep them alive over winter. The next summer seed is produced on the overwintered plants.[42]

The sugarbeet seed crop is harvested with a windrower, which sometimes is equipped with vertical sickles to open the swath, as well as the usual horizontal sickle. A combine with a pick-up attachment threshes the crop. More than 13 million pounds of hybrid sugarbeet seed are produced annually in the United States. The yields are about 2,600 pounds per acre on more than 5,000 acres. The chief seed-producing areas are the Salt River Valley of Arizona, and the Willamette Valley of Oregon,[42] and in the Virgin River Valley of Utah.

The biennial sugarbeet plant becomes a winter-annual seed producer when grown in environments suitable for photothermal[42] induction of flowering. Thermal induction occurs at mean temperatures of about 45° to 55° F. and lasts for 90 to 110 days. The lengthening

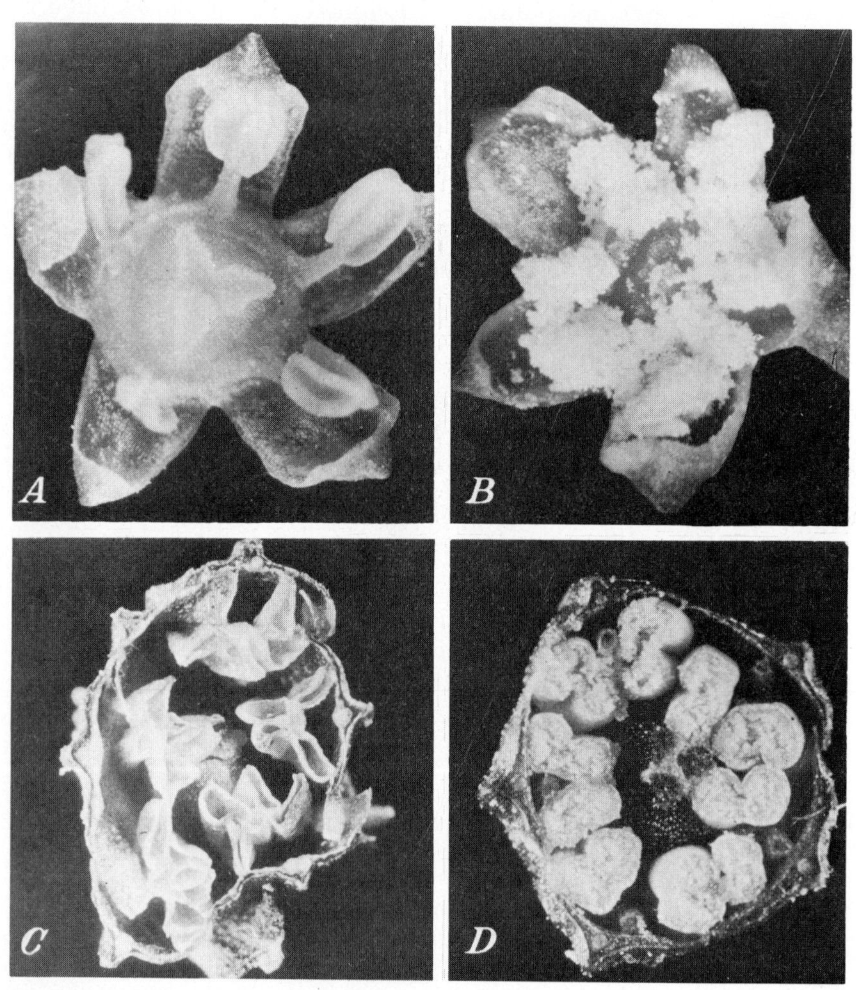

Figure 35–7. Sugarbeet flowers: (*A*) male-sterile, (*B*) male-fertile, (*C*) clipped empty male-sterile anthers, and (*D*) clipped male-fertile anthers full of pollen.

days of spring that follow cause a photoperiodic response, after which flowers and seeds are produced.

Production of sugarbeet seed formerly was largely a European enterprise. Nearly all of the seed planted in America was imported because of the labor involved in beet-seed production. By the old methods, the beets for seed were planted in the usual manner but were left closer together in the row at thinning time in order to obtain more but smaller beets, usually called stecklings. These were dug in the fall and stored in a pit or pile covered with straw or earth or both. In the spring, the stecklings were set out in the field for seed production.

Single cross or three-way cross hybrid sugarbeet seed is produced by interplanting a male-sterile strain with a suitable fertile pollinator parent (Figure 35–7). Seed of a male-sterile monogerm line usually is mixed with about 10 per cent seed of a multigerm pollinator before planting. The harvested seed is screened to remove the large multigerm self-pollinated seed balls from the mass of small monogerm hybrid seed. Hybrid seed also is produced by planting alternating strips of male-sterile and pollinator lines.

DISEASES

Cercospora Leaf Spot

Cercospora leaf spot, caused by the fungus *Cercospora beticola*, is one of the most prevalent of sugarbeet diseases.[11, 28] Circular sunken spots scattered over the leaf are about 1 to 2 millimeters in diameter. The center of a spot is ashy gray, but frequently the spot has a reddish-purple margin. When the spots are numerous, they coalesce, causing the leaf blades to become brown and dry. The outer leaves of the plant often show the blighting which is a result of infection that took place when they were unfolding. Under severe attack the affected leaves die, becoming brown or black, and the field looks brown or scorched.

The principal control measures for leaf spot are resistant varieties or hybrids and fungicidal sprays.[49]

Curly Top

Curly top is a virus disease that formerly caused heavy losses to sugarbeet growers in the intermountain region and in California. Production declined so much that several beet-sugar factories were abandoned. The insect vector of curly top is the beet leafhopper (*Circulifer tenellus*), often called the white fly.[11] This leafhopper breeds and feeds upon about 40 plant species, principally weeds of the goosefoot and mustard families.[9] Russian thistles and other weeds which spread on overgrazed range and abandoned fields caused the beet leafhopper and curly-top disease to increase. The leafhoppers move to the sugarbeet fields after the range plants dry up in the spring. A portion of the leafhoppers carry the curly-top virus after feeding upon weeds that are subject to curly top. The beet leafhoppers introduce and spread the virus throughout a sugarbeet field.

The typical curling is upward. It is usually accompanied, more or less, by a roughening and distortion of the leaf veins. These symptoms are accompanied by a shortening of the petiole as well as a general retardation in the growth of the entire plant. The stunted beets commonly develop large numbers of tiny rootlets. The most satisfactory control is the planting of resistant hybrids. Most of those grown west of the Rocky Mountains are at least fairly resistant to curly top.

Seedling Diseases

A general complex of damping-off diseases leads to the death of the sugarbeet plant at the time that sprouting takes place or as the seedling is emerging from the soil. This complex also includes later phases of attack which occur on plants that have partly recovered from damping off. This disease complex has been called black root by growers.[11] Death of the plant may occur at the time of sprouting or when the plants emerge from the soil. Some plants persist in spite of fungus attack, but these may remain stunted or eventually die.

A number of fungus pathogens are responsible for sugarbeet black root. Soil-inhabiting organisms such as species of *Pythium* and *Rhizoctonia* cause the death of sugarbeet plants. A seed-borne fungus, *Phoma betae,* also is serious. The most serious loss, however, apparently is caused by *Aphanomyces cochlioides,* a water mold. This fungus does not kill the plant outright but dwarfs it because of a persistent attack on the lateral or feeding roots.

The control of seedling diseases or black root involves (1) long rotations and crop sequences that keep soil infestation at a minimum, (2) proper drainage and fertilization so that the plants make vigorous growth, (3) seed treatment with fungicides and (4) the growing of resistant hybrids.

Root or Crown Rots

Sugarbeet roots may rot in midseason because of the attack of a number of fungus pathogens, including *Pythium* and *Phymatotrichum. Rhizoctonia* crown rot is probably the most serious.

In many cases the rotting of the mature root is a carry-over from an attack of the fungus in the seedling stage. This is followed by partial recovery and then a renewed spread of the fungus on the half-grown plants. Such plants usually show a cleft top, but sometimes the entire crown breaks away. Since the fungus tends to spread down the row, it is common to find a number of contiguous plants affected with crown rot. Crop rotation, good culture, and the control of seedling diseases serve to check the losses from root rots to a considerable extent.

Other Diseases

Downy mildew, caused by the fungus *Peronospora schachtii*, occurs in the coastal areas of California, but resistant sugarbeet hybrids are being grown there.[32] Virus yellows, a common disease of sugarbeets in Europe, is now present in several producing areas of the United States. Resistant varieties and hybrids are available.

Storage Disease

Sugarbeets must be harvested before severe freezing weather occurs. Consequently, deliveries are greatly in excess of processing capacity, with the result that the roots are piled at the factory where they may be stored for a month or more. The roots keep well as long as they are alive and reasonably cool. Frozen roots, wounded roots, roots topped excessively low, and those roots whose tails die because of excessive drying out are subject to attack by fungi. These fungi include not only such parasites as *Phoma betae, Rhizoctonia,* and *Phycomycetes* but also saprophytes such as *Fusarium, Penicillium, Aspergillus,* and the *Mucors.* Prevention of storage losses requires careful handling of the beets and avoidance of excessive drying.[46] Blowing cold night air into the beet piles lowers the temperature, reduces respiration, and checks storage rots. A controlled atmosphere of 6 per cent CO_2 and 5 per cent oxygen at 2° C. gives the best sugar retention.

DEFICIENCY DISORDERS

Sugarbeets show typical deficiency symptoms when the supply of any of the essential mineral elements is inadequate. Nitrogen starvation produces typical yellowing effects, as does sulfur deficiency also. Potash hunger manifests itself by a reddish coloration or bronzing. When phosphorus is deficient the plants show stunting and unbalanced proportions of roots to tops as well as characteristic necrotic blotches on the blades of the older leaves. In severe cases the leaves dry and shrivel, the leaf rolling in on the midrib from the tip to give a fiddle-neck effect. The severe aspects of phosphate deficiency have been called black heart,[30] a name that indicates the symptoms very poorly.

Boron deficiency causes blackening followed by death of the heart, leaves, blackening or black markings on the inner faces of the petioles, and cankers on the roots. The flesh beneath the dried necrotic spot or canker is brown or lead-colored. The sugarbeet requires a relatively large amount of boron to avoid the deficiency symptoms. Applications of about 10 pounds of boron per acre are adequate where this element has merely been leached out. In soils in which boron is bound as a result of high soil salinity, much heavier applications are necessary.[13] Boron deficiencies may occur in organic soils with a *p*H above 5.5 or in mineral soils with a *p*H above 6.6.

NEMATODES

The sugarbeet nematode (*Heterodera schachtii*), a minute whitish eelworm, attacks the roots of sugarbeets and kills or stunts the plants.[31] Stunted roots may be covered with short, hairy rootlets. Infestation is spread through soil (dump dirt) that is transferred at the beet dumps and then returned to the fields. The pest is controlled by long rotations with immune crops such as alfalfa, beans, potatoes, small grains, and corn, with sugarbeets grown at intervals not less than every four or five years. Fumigation of the soil with chemicals is restoring sugarbeet culture to fields once abandoned because of nematode infestation. Such fumigation is expensive. The land also must be kept free from susceptible weeds such as mustard, lambsquarters, knotweed, ladysthumb, purslane, curly dock, and black nightshade, as well as from susceptible crops such as mangels and turnips.

INSECTS

The beet leafhopper (*Circulifer tenellus*)[8, 21, 37] causes some damage by feeding on the leaf sap, but the chief-injury from this insect is the spread of curly-top virus. Losses are minimized by growing hybrids resistant to the virus. The beet webworm (*Loxastege sticticalis*) and the beet armyworm (*Spodoptera exigua*) devour the leaves. The latter insect may also attack the crown and roots of the plant. Fleabeetles puncture the leaves. Other insects that attack sugarbeets include grasshoppers, leaf miners, wireworms, white grubs, sugarbeet root aphid (*Pemphigus betae*), darkling beetles, salt marsh caterpillars, and spider mites.[16] Damage from these pests is reduced by applying suitable insecticides.

REFERENCES

1. Artschwager, E. "Anatomy of the vegetative organs of the sugar beet," *J. Agr. Res.*, 33:143–176. 1926.
2. Artschwager, E. "Development of flowers and seed on the sugar beet," *J. Agr. Res.*, 34:1–25. 1927.
3. Artschwager, E. "A study of the structure of sugar beets in relation to sugar content and type," *J. Agr. Res.*, 40:867–915. 1930.

4. Artschwager, E., and Starrett, R. C. "Suberization and wound-cork formation in the sugar beet as affected by temperature and relative humidity," *J. Agr. Res.*, 47:669–674. 1933.
5. Barmington, R. D., and McBirney, S. W. "Mechanizing of sugar beets," *Colo. Agr. Exp. Sta. Bul. 420-A*. 1952.
6. Brandes, E. W., and Coons, G. H. "Climatic relations of sugarcane and sugar beet," in *Climate and Man*, USDA Yearbook, 1941, pp. 421–438.
7. Brewbaker, H. E. "Self-fertilization in sugar beets as influenced by type of isolator and other factors," *J. Agr. Res.*, 48:323–337. 1934.
8. Cook, W. C. "The beet leafhopper," *USDA Farmers Bul. 1886*, pp. 1–21. 1941.
9. Cook. W. C. "Life history, host plants and migrations of the beet leafhopper in the Western United States," *USDA Tech. Bul. 1365*, pp. 1–122. 1967.
10. Coons, G. H. "Improvement of the sugar beet," *USDA Yearbook, 1936*, pp. 625–655.
11. Coons, G. H. "Some problems in growing sugar beets," in *Plant Diseases*, USDA Yearbook, 1953, pp. 509–524.
12. Coons, G. H., Owen, F. V., and Stewart, D. "Improvement of the sugar beet in the United States," *Adv. Agron.*, 7:89–139. 1955.
13. Cox, T. R. "Relation of boron to heart rot in the sugar beet," *J. Am. Soc. Agron.*, 32(5):354–370. 1940.
14. Creek, C. R. "Changes in sugar beet production in Colorado," *Colo. Exp. Sta. Bul. 530-S*, pp. 1–20. 1968.
15. Dawson, J. H. "Weed control in sugarbeets with cycloate," *USDA Tech. Bul. 1436*, pp. 1–24. 1971.
16. Dennis, R. E., and Nelson, J. M. "Sugar beets in Arizona," *Ariz. Agr. Exp. Sta. Bul. A-71*, pp. 1–6. 1971.
17. Down, E. E., and Lavis, C. A. "Studies on methods for control of pollination in sugar beets," *J. Am. Soc. Agron.*, 22:1–9. 1930.
18. Draycott, A. P. *Sugarbeet Nutrition*, Wiley, New York, pp. 1–250. 1972.
19. Frazier, R. D., and others. "Sugar beet nutrition studies in southern and west central Minnesota," *Minn. Agr. Exp. Sta. Misc. Rpt. 102*, pp. 1–15. 1970.
20. Hanson, E. W., Hansing, E. D., and Schroeder, W. T. "Seed treatments for control of disease," in *Seeds*, USDA Yearbook, 1961, pp. 272–280.
21. Harris, F. S. *The Sugar Beet in America*, Macmillan, Inc., New York, pp. 1–342. 1926.
22. Hastings, S. H., Nuckols, S. B., and Harris, L. "Influence of farm manure on yields and sucrose of sugar beets," *U.S. Dept. Agr. Tech. Bul. 614*, pp. 1–12. 1938.
23. Headden, W. P. "A soil study: The crop grown—sugar beets," *Colo. Agr. Exp. Sta. Bul. 46*, Part I. 1898.
24. Headden, W. P. "A soil study: The crop grown—sugar beets," *Colo. Agr. Exp. Sta. Bul. 58*, Part II. 1900.
25. Johnson, R. T., and others. *Advances in Sugarbeet Production—Principles and Practices*, Iowa State Univ. Press., Ames, Iowa, pp. 1–470. 1971.
26. Jones, D. P., and Jones, F. G. W. Wireworms and the sugarbeet crop: field trials and observations. *Ann. Appl. Biol. 34*. pp. 562–574. 1947.

27. Kohls, H. L. "The influence of some climatological factors on seedstalk development and seed yields of space-isolated mother beets," *J. Am. Soc. Agron.*, 29:280–285. 1937.
28. LeClerg, E. L. "Control of leaf spot of sugar beets," *Minn. Agr. Exten. Circ. 46.* 1934.
29. Lill, J. G. "Sugar beet culture in the North Central States," *USDA Farmers Bul. 2060.* 1964.
30. Maxson, A. C. "Manure and phosphate," *Through the Leaves*, 18(1):33–37. 1930.
31. Maxson, A. C. *Principal Insect Enemies of the Sugar Beet in the Territories Served by the Great Western Sugar Company*, Great Western Sugar Company, Denver, pp. 1–157. 1920.
32. McFarlane, J. S., Owen, F. V., and Murphy, A. M. "New hybrid sugar beet varieties for California," *J. Am. Soc. Sugar Beet Tech.*, 11(6):500–506. 1961.
33. McGinnis, R. A. *Beet-Sugar Technology*, Beet Sugar Dev. Foundation, Ft. Collins, Colo., pp. 1–835. 1971 (2nd Ed.).
34. Nuckols, S. B. "Yield and quality of sugar beets from various rotations at the Scotts Bluff (Nebr.) Field Station, 1930–35," *USDA Circ. 444*, pp. 1–14. 1937.
35. Nuckols, S. B. "Sugar beet culture under irrigation in the northern Great Plains," *USDA Farmers Bul. 1867*, pp. 1–52. 1941.
36. Palmer, T. B. *Sugar Beet Seed.* Wiley, New York, pp. 1–120. 1918.
37. Peay, W. E. "Sugarbeet insects—how to control them," *USDA Farmer's Bull. 2219.* 1968 (Rev.).
38. Raleigh, S. M. "Environmental factors affecting seed setting in sugar beets," *J. Am. Soc. Agron.*, 28:35–51. 1936.
39. Schneider, A. D., and Mathers, A. C. "Water use by irrigated sugar beets in the Texas High Plains," *Tex. Agr. Exp. Sta. M.P. 935*, pp. 1–7. 1969.
40. Skuderna, A. W., and others. "Agronomic evaluation tests on mechanical blocking and cross cultivation of sugar beets," *USDA Circ. 315*, pp. 1–23. 1934.
41. Skuderna, A. W., and Sheets, E. W. "Important sugar beet by-products and their utilization," *USDA Farmers Bul. 1718*, pp. 1–29. 1934.
42. Stewart, D. "New ways with seeds of sugar beets," in *Seeds*, USDA Yearbook, 1961, pp. 199–205.
43. Stout, M., McBirney, S. W., and Fort, C. A. "Developments in handling sugar beets," in *Crops in Peace and War*, USDA Yearbook, 1950–51, pp. 300–307.
44. Terry, N. "Developmental physiology of the sugar-beet," *J. Exp. Bot.*, 21(67):477–496. 1970.
45. Tolman, B., and Murphy, A. "Sugar beet culture in the intermountain area with curly top resistant varieties," *USDA Farmers Bul. 1903*, pp. 1–52. 1942.
46. Tompkins, C. M., and Nuckols, S. B. "The relation of type of topping to storage losses in sugar beets," *Phytopath.*, 20:621–635. 1930.
47. Tyson, J. "Influence of soil conditions, fertilizer treatments, and light intensity on growth, chemical composition, and enzymatic activities of sugar beets," *Mich. Agr. Exp. Sta. Tech. Bul. 108.* 1930.

48. Ulrich, A. "Influence of night temperature and nitrogen nutrition on the growth, sucrose accumulation and leaf minerals in sugar beet plants," *Plant Physiol.*, 39:250–257. 1955.
49. Weihang, J. L., and Finkner, R. E. "Fungicidal control of sugar beet leaf spot," *Neb. Agr. Exp. Sta. SB 502*, pp. 1–14. 1968.
50. Withers, R. V. "Economics of sugarbeet production in Idaho," *Ida. Agr. Exp. Sta. Bul. 517*, pp. 1–19. 1970.
51. Yaggie, R. A., and Loftsgard, L. D. "Sugar beet production costs and practices in the Red River Valley," *N. Dak. Exp. Sta. Bul. 466*, pp. 1–32. 1966.

Chapter 36

Potatoes

ECONOMIC IMPORTANCE

The world production of potatoes is nearly 300 million metric tons annually. Except for the cereals it is the world's most important food crop. They are grown on more than 22 million hectares with a yield of 13.3 metric tons per hectare. The countries leading in potato production are the Soviet Union, Poland, West Germany, the United States, East Germany, France, and China (Figure 36–1).

In the United States potatoes were harvested on an average of 1,357,000 acres (0.55 million hectares) from 1970 to 1972, with a production of more than 313 million hundred weight (14.2 million metric tons) and a yield of 231 cwt. per acre, or 25.9 metric tons per hectare. The leading states in potato production are Idaho, Maine, Washington, California, North Dakota, and New York.

ORIGIN AND HISTORY

The potato originated in the Andes highlands, probably in Peru or Bolivia where wild species of *Solanum* are found, several of which have been cultivated. It has been proposed that two diploid species intercrossed naturally to produce a tetraploid cultivated type. This was collected by European explorers, and later reselected to evolve the current typical cultivated species, *Solanum tuberosum*.[70] Potatoes apparently were cultivated well before 400 B.C.

Spanish explorers found the potato under cultivation in western South America in 1537, but not in Mexico. They had introduced it to Europe by 1565 or 1580. Potatoes reached England and Ireland somewhat later, but they did not become widely grown in any European country until after 1750. The potato was taken from England to Bermuda, and from there it was taken to Virginia in 1621. It was not generally grown in the United States until after it was introduced at Londonderry, New Hampshire, in 1719. The planting stock had been obtained from Ireland by Scotch-Irish immigrants, hence the name Irish potato.[30, 59]

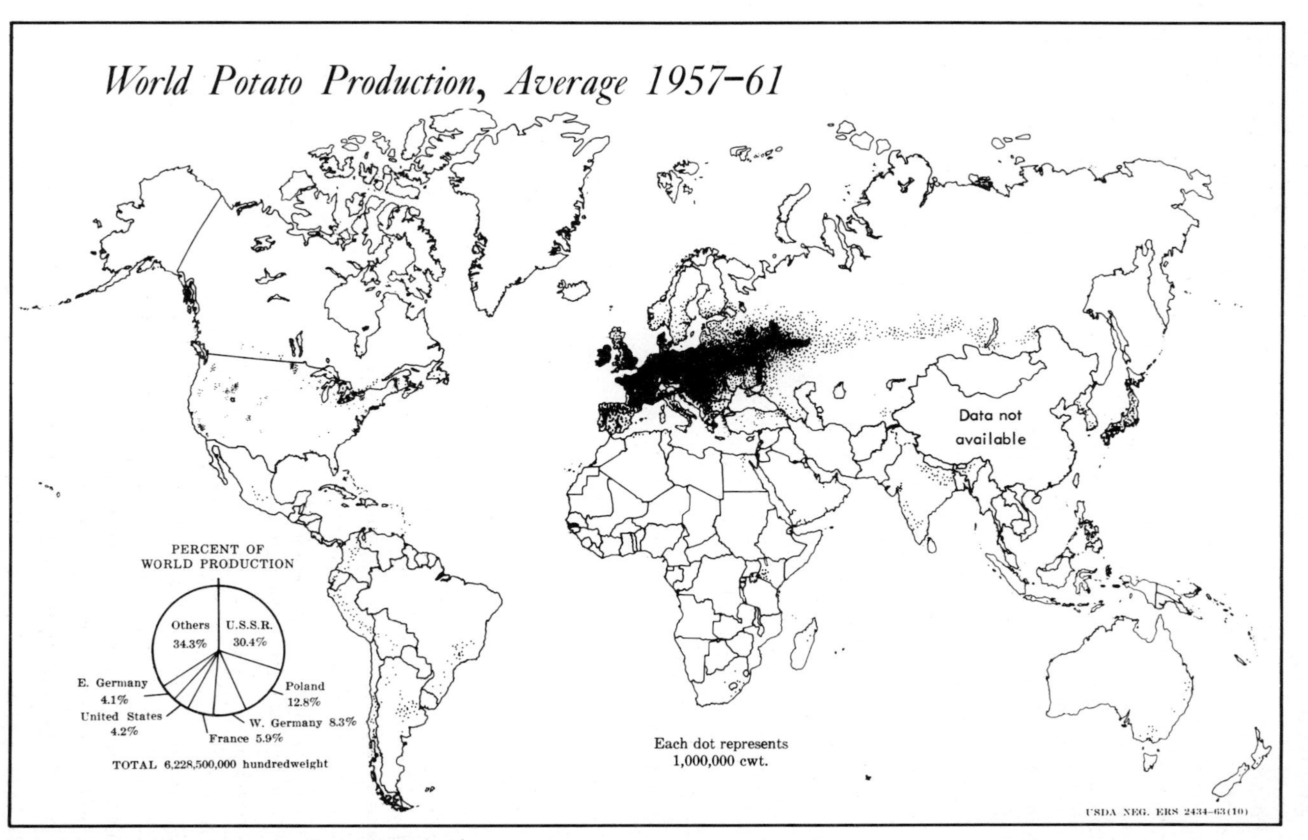

Figure 36–1. World potato production.

ADAPTATION

Potatoes are grown in every state in the Union, but most of the commercial crop is highly localized (Figures 36–2 and 36–3). The most important locality is Aroostook County, Maine. The concentrated areas either are especially suited to potato production, or are favorably situated to supply certain seasonal or regional markets. Winter and early spring potatoes are produced in southern areas with mild winters but with summers too hot for growing fall potatoes. Idaho, although distant from eastern markets, supplies most of the high-quality baking potatoes.

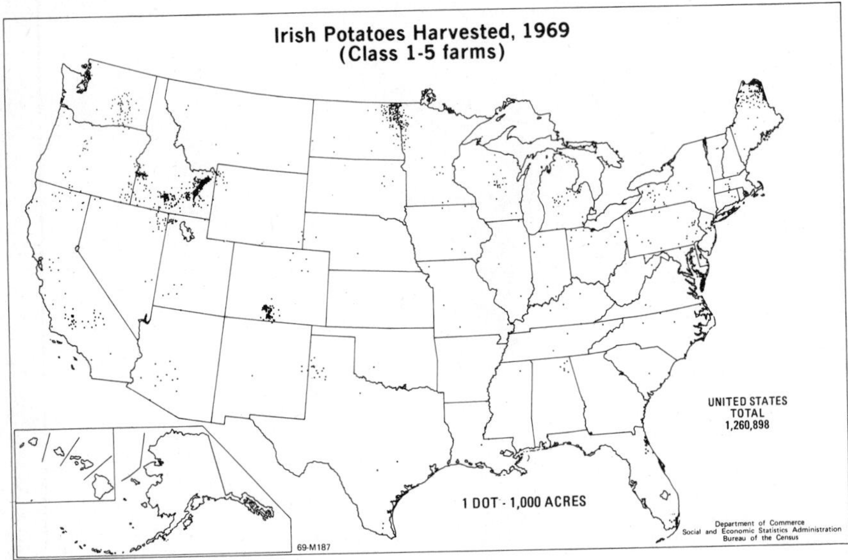

Figure 36–2. Acreage of potatoes in the United States in 1969.

The potato is a cool-weather plant that makes its best growth where the mean July temperature is about 70° F. or lower.[53] Main-crop potatoes are produced chiefly in cool climates such as are found in northern Europe and the northern United States. The young sprouts develop best at soil temperatures of about 75° F. but tuber growth is best at a temperature of 60 to 65° F. Tuber production is retarded at soil temperatures above 68° F., second growth may occur at 80° F, and growth is completely inhibited at 84° F., above which point the carbohydrates consumed by respiration exceed those produced by photosynthesis.[12] Potatoes grown in the South are planted in early spring or in the fall or winter so that growth takes place while the weather is cool. Where temperatures of 77° F. or higher prevail, the symptoms of mosaic are indistinct.[10] The potato plant withstands

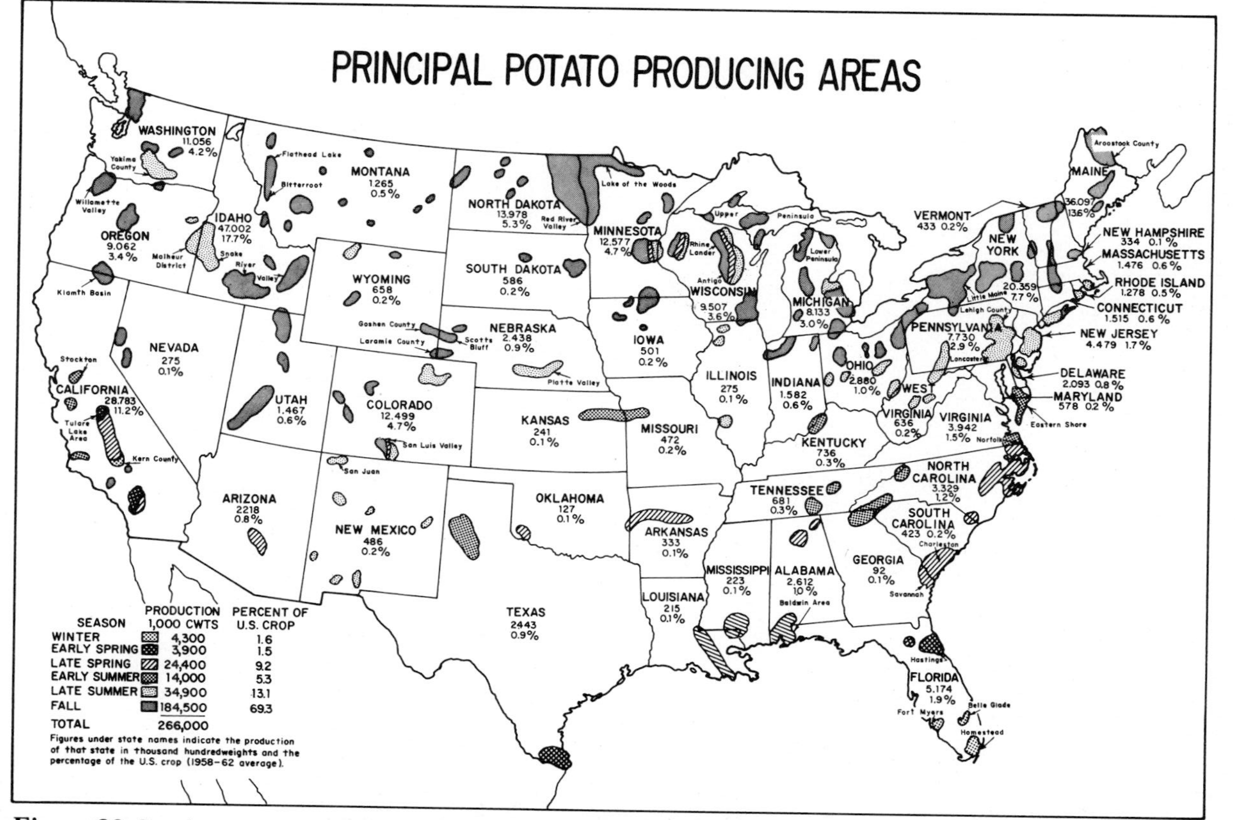

Figure 36–3. Areas in which potatoes are produced for the different seasonal markets.

light frosts, but frequently is injured by freezing in the fall, winter, or early spring. The freezing point of the tubers is 28.5° to 29.5° F., and when completely frozen the tissues disintegrate soon after thawing.

Long days, high temperatures, and high amounts of nitrogen favor a heavy growth of plants,[10] and prolong the growing season. Short days, cool temperatures, or a deficiency of nitrogen favor early tuberization. Days of intermediate length, cool temperatures, and ample nitrogen favor maximum tuberization. Although flower primordia (rudimentary flowers) can be formed in either long or short days and even in darkness,[29] flowering and seed formation are favored by long days and cool temperatures. Thus, potato plants commonly produce seeds in the more northern states but not in the middle latitudes or in the South. However, varieties differ in their fertility.[40] Unfruitfulness may be caused by abnormal chromosome behavior at the time of pollen formation. Considerable shedding of buds may occur even before pollen formation.

A uniform supply of soil moisture is essential to the production of good well-formed tubers. Interrupted growth followed by later favorable growing conditions, or excessive nitrogen or high temperatures may result in knobby or pear-shaped tubers. Potatoes are commercially successful in northern semiarid regions such as western Nebraska, which has produced the crop chiefly for the seed market.

Cool moist conditions favor development of the late blight disease. Warm weather favors reproduction of insects that transmit the mosaic viruses, which in turn accelerates the spread of mosaic diseases.

Commercial potato production is found on a wide range of soil types ranging through sandy loams, silt loams, loams, and peat.[9] Good yields of potatoes may be obtained in some fertile clay soils, but sticky soil adhering to the tubers interferes with digging and marketing. Well-drained porous soils are desirable. Either fertile soil or heavy fertilization is essential to high yields of potatoes. The optimum soil *p*H for potatoes is about 5.0 to 5.5 from the standpoint of both yield and scab retardation. The scab disease develops mostly on soils above *p*H 5. However, the soils in the north-central potato sections are nearly neutral in reaction, while they are alkaline in the irrigated intermountain and Great Plains potato regions. The highest recorded potato yields in the United States, exceeding 1,100 bushels per acre (74 metric tons per hectare) were secured on the subirrigated peat soils of the San Joaquin Delta in California.

BOTANICAL DESCRIPTION

The cultivated potato (*Solanum tuberosum*) belongs to the family *Solanaceae.* The plant is an annual having stout, erect, branched stems 1 to 2 feet long that are slightly hairy, and distinctly winged at the angles.[30] The slightly hairy leaves are 1 to 2 feet long and are comprised of one terminal leaflet, two to four pairs of oblong acute leaflets, and two or more short leaflets (Figure 36–4). Flowers are borne in compound, terminal cymes with long peduncles. The five petals are white, rose, lilac, or purple in color. The flower has five anthers and one pistil with a long style.[16, 30] The potato ball is a smooth, globose, green or brown berry (Figure 36–5) less than an inch in diameter. The rhizomes, which are 2 to 4 inches long, enlarge at the outer end to form a tuber. The tuber is a modified stem with lateral branches forming what are known as potato eyes.[7] Each eye contains at least three buds protected by scales. The eyes are arranged around the tuber in the form of a spiral.

The interior of the tuber shows a pithy central core with branches leading to each of the eyes. Surrounding the pith is the parenchyma, in which most of the starch is deposited. Toward the outer part of the tuber is the vascular ring that contains the cambium, as well as an outer cortex that contains the pink, red, or purple pigment of colored-skinned varieties (Figure 36–6). The potato skin (periderm) is a layer six to ten cells deep, composed largely of cork (or suberin) having a basic composition similar to that of fatty substances. Scales form on

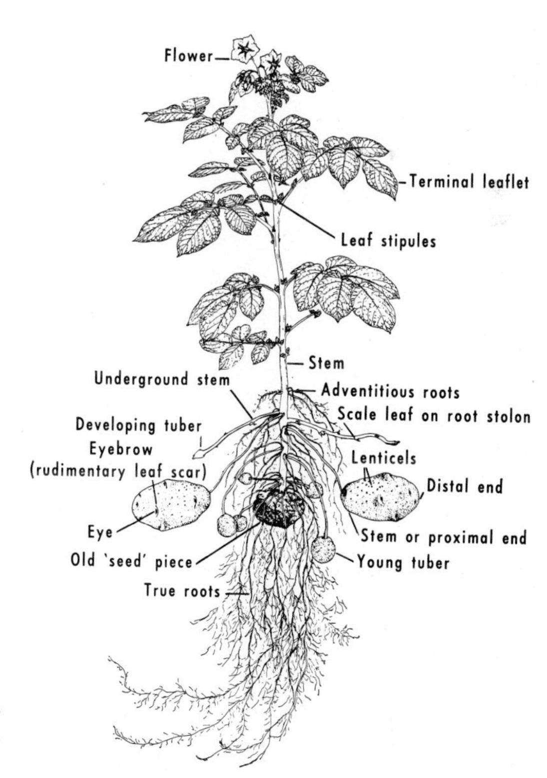

Figure 36–4. A potato plant. Stolons, which bear tubers, arise from stem tissue.

Figure 36–5. Potato "balls" that contain seeds.

the outer surface of the periderm. Openings in the periderm called lenticels become enlarged under moist conditions.

When a potato is cut and left in a suitable environment, the surface of the cut suberizes (corks over) to form a new skin—a wound periderm, which serves to protect the tuber from decay. When a tuber is exposed to sunlight for several days, either before or after digging, the skin of most varieties turns green as the result of development of chlorophyll. Some varieties turn purple. Along with this change, increased quantities of solanin are formed in the cortex. Solanin is an alkaloidal glucoside, bitter in taste, and poisonous when taken in sufficient quantities. The poisonous alkaloid in solution is called solanidine. The quantity of solanin in sunburned tubers may be more than 20 times that in normal tubers. Most of the solanin is discarded with the peelings.

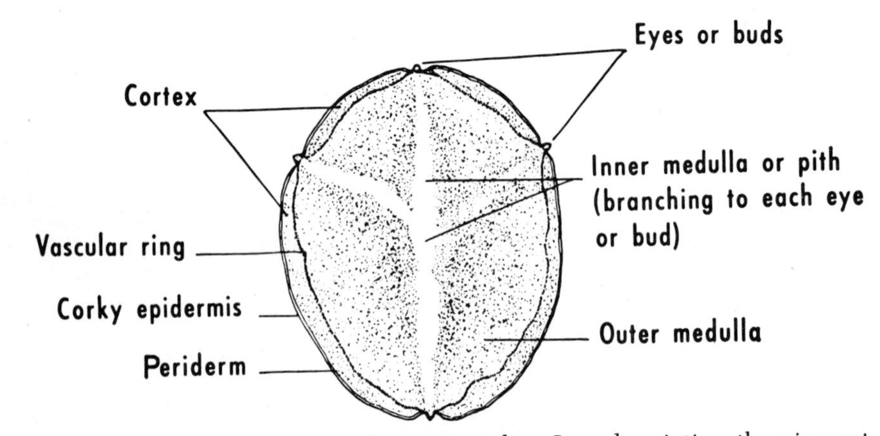

Figure 36–6. Cross section of a potato tuber. In red varieties, the pigment occurs in the periderm or outer cortex. The pith has a water-soaked appearance.

VARIETIES

A total of 89 varieties of potatoes were certified for growing in the United States and Canada in 1971.

Some of the distinctive characteristics of the varieties include the oblong tubers of Green Mountain (Figure 36–7), the red skin and short rounded shape of Triumph, and the long cylindrical tuber and russet skin of Russet Burbank, the Idaho baking potato.[16, 30] The ideal tuber seems to be short, wide, and flat with shallow eyes. Many of the newer varieties are of that type.

Two productive yellow-fleshed varieties distributed in the 1940s, were rejected by consumers of the United States. This yellow-flesh characteristic is popular in several European and Latin American countries because of its assumed high vitamin-A value as well as its rich appearance. Some of the potatoes of South America have purple flesh.

Development of new improved varieties and elimination of inferior or unadapted varieties have been important factors in the improvement of potato yields. Along with this have been the use of better planting stocks, increased fertilizer usage, better control of insects and diseases, better cultural methods, and concentration of the crop in the more favorable localities. All of these factors together are responsible for the consistent increases in potato yields from an average of 48 hundredweight around 1890 to 229 hundredweight per acre in 1971.

ROTATIONS

Potatoes usually succeed best after a green-manure crop or after a legume crop such as clover, alfalfa, sweetclover, vetch, or peas, with the residues turned under in the fall. In the northeastern states, the rotation usually is potatoes, small grain, clover. In the Corn Belt, the rotation may be corn, potatoes, small grains, clover. The potato field leaves a good seedbed that requires no plowing for small grains. Sometimes two consecutive crops of potatoes are grown. In the irrigated regions, potatoes often follow alfalfa that has been *crowned* in the fall and replowed in the spring. Sugarbeets follow the potato crop advantageously, and the beet field makes a good seedbed for sowing small grain and alfalfa.

In the South, potatoes planted in early spring often follow an early fall-sown green-manure crop such as crimson clover, vetch, winter peas, or lupines. Those potatoes planted in the fall or winter can follow a summer legume such as the velvetbean, lespedeza, cowpea, soybean, or crotalaria.

Potatoes respond well to green manures[67] except on peat and muck

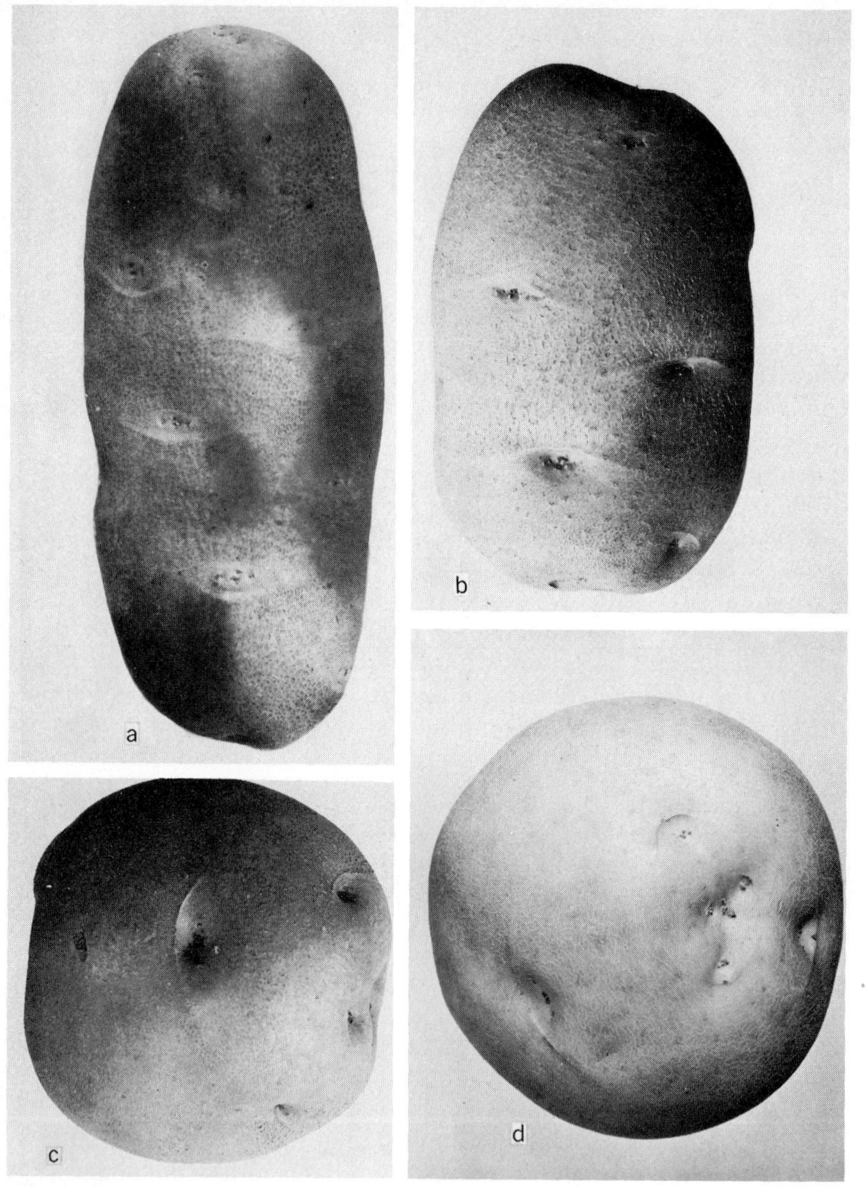

Figure 36–7. Tubers of four varieties of potatoes: (*a*) Russet Burbank (Netted Gem), (*b*) Green Mountain, (*c*) Triumph, and (*d*) Katahdin.

soils. Rotations in which potatoes are grown three or more years apart reduce losses from diseases caused by soil-borne organisms. Such diseases include scab, verticillium wilt, Fusarium wilt, and blackleg.

In the dryland regions, potatoes usually are planted on fallowed land, or after corn or beans. The potato crop is likely to fail on the dry lands unless the soil is moist to a depth of 3 feet or more at planting time. A spring grain crop usually follows the potatoes, the land being worked but not plowed before the grain is sown.

FERTILIZERS

A 300-cwt. crop of potatoes contains about 210 pounds of nitrogen, 26 pounds of phosphorus, and 236 pounds of potassium.

Potatoes respond well to applications of barnyard manure. Light applications of 6 to 8 tons per acre of manure that is well rotted or else turned under considerably in advance of planting time are recommended in order to avoid excessive scab injury. Also, the addition of 40 to 50 pounds of superphosphate to each ton of manure is advised in order to provide a better fertility balance.[30] Manure is used more generally for potatoes than for any other field crop except sugarbeets.

Nearly all of the commercial potato fields receive mineral fertilizers.

In Maine, potatoes give profitable responses to nitrogen applied at rates that range from 80 to 120 pounds per acre when the potatoes follow a nonlegume crop, or from 60 to 100 pounds per acre after a legume or applications of stable manure.[69] Excessive nitrogen rates may decrease the mealiness, increase skinning and bruising of the tubers, and increase the tyrosine content which causes browning of the cut flesh. Profitable rates of nitrogen application in other northeastern states range from 30 to 100 pounds per acre.[54]

A rate of 240 pounds of ammonium sulfate per acre is advisable for irrigated land in central California.

Under irrigated conditions in the Columbia Basin of Washington, rates up to 240 pounds of nitrogen per acre may be profitable for potatoes.[19] Increased yields of potatoes on irrigated land were obtained from nitrogen applications of up to 40 pounds per acre in south-central Colorado,[34] but very little benefit from fertilizers was evident on fertile land in northeastern Colorado.[31] Much of the nitrogen need in the intermountain region, as well as in many other areas, is often supplied with barnyard manure, or the plowing under of legume residues, or both.

Potatoes usually respond to phosphorus fertilizers, except on soils that are high in available phosphorus. Suitable rates of P_2O_5 range from 30 to 80 pounds per acre for moderate yields on fertile soils,[46] up to 160 pounds in southern Colorado.[34] Applications as high as 300 pounds per acre were beneficial in low-phosphorus soils in Washington, where potato yields of 500 to 600 bushels per acre were anticipated.[19] In Maine, a recommended rate has been 160 to 200 pounds per acre of P_2O_5, except on high-phosphorus soils where 80 to 100 pounds is ample.[69]

Potash is an important constituent in potato fertilizers. A good supply is essential for heavy starch production, and to increase the citric acid content of the flesh in order to avoid blackening of the cooked tuber. On peat and muck soils that are rich in nitrogen, abundant potash applications, and often some phosphorus also, are needed to balance the soil nutrients. Potash (K_2O) application rates of 100 to 120 pounds per acre usually are ample for potatoes in the East, except on soils low in potash where 140 to 200 pounds are desirable.[67] A rate of 100 pounds per acre was desirable on a low-potash soil in the Columbia River Basin of Washington.[19] From 20 to 60 pounds of potash often are ample on soils in the North Central states.[46] Potatoes grown on peat or muck soils may benefit from 100 to 300 pounds. The content of solids in the tubers often is somewhat higher where potassium sulfate rather than potassium chloride fertilizers are applied.[67] Ample potassium sulfate and ammonium sulfate reduces the percentage of small potatoes.

Heavier fertilizer applications prevail in the humid eastern areas than in central states where the soils are more fertile. The quantities used in Aroostook County in Maine may exceed the seasonal needs of the crop.

Consistently higher yields are secured when the fertilizer bands are 2 inches to the side and slightly below the potato seed piece rather than directly under the piece, or when mixed in the soil around the seed piece.[11, 47] The side placement of fertilizer bands is accomplished with a planter having a fertilizer attachment. Split applications of fertilizer usually are not beneficial to potatoes. Any side dressing with fertilizer should be applied soon after the potato plants have emerged. With extremely heavy fertilization (2,500 pounds per acre or more), part of the fertilizer may be broadcast and plowed under with the remainder applied in bands at planting. This avoids injury to the seed pieces.

Magnesium-deficiency symptoms, which have appeared in potatoes grown on acid soils along the Atlantic seaboard, are corrected by including magnesium in the fertilizer mixture. Some potato soils are deficient in calcium or are excessively acid, but heavy applications of lime increase the incidence of the scab disease. In general, liming of potato land where the soils have a reaction above *p*H 5.2 is not recommended. On more acid soils, finely ground limestone or hydrated lime may be applied at rates not exceeding 1,000 pounds per acre for the clover that precedes the potato crop. Liming prevents aluminum toxicity in soils below *p*H 5. About 100 pounds per acre of 65 per cent manganese sulfate is required for manganese-deficient potato soils in southern Florida.[30]

SEED POTATOES

The planting of certified, nearly disease-free seed has been rather generally adopted by successful commercial growers. This eliminates the problems of place effect and running out that result from infection by viruses causing mosaic, leaf roll, spindle tuber, and other diseases.[32] The former belief that potatoes grown on dry land are better for planting purposes than those grown under irrigation seems to be without foundation.[21]

Tuber Sprouting

Tubers left in the air or in dry soil will produce sprouts but no roots. In moist soil or some other wet medium, both sprouts and roots are formed. Large vigorous sprouts are an indication of quick emergence and relatively early maturity. In darkness or subdued light, the sprouts are long and lack green color. In the sunlight, the sprouts are short and soon turn green, and the tuber also becomes green. Occasionally the tubers are prepared for planting by being exposed to light, a process called greening.

Since each eye usually contains several buds, sprouts will be produced even after some of the sprouts are removed or damaged. The removal of one crop of sprouts is only slightly detrimental to the tuber, although growth and removal of sprouts reduce the vigor of subsequent sprouts.[5] However, all eyes do not sprout at once unless the tubers are cut up because of a phenomenon called apical dominance, i.e., the eyes nearest the blossom (outer) end of the tuber sprout first. Long storage destroys apical dominance.[5]

Sprouting of potato tubers in storage occurs at temperatures of 40° F. or higher, but only after completion of a rest period, which is caused by restricted oxidation.[4] The rest period may range from 4 to 16 weeks when the potatoes are stored at 70° F., or longer at cooler storage temperatures. Immature tubers may have a longer rest period than do those that are fully mature.

Size of Seed Piece

The planting of certified, nearly disease-free seed and the prevalent practice of closer planting have altered the former situation in which the planting of small tubers was hazardous because they often were diseased.[30] Small-size graded tubers from certified seed fields called No. 1 B (1½ to 2 inches in diameter) often are used for seed. Specialized seed-potato growers often plant thickly so the seed-crop tubers will not be too large. However, a frequent procedure is to plant larger tubers cut into pieces of 1¼ to 2 ounces in weight. It is desirable that each piece have one to three eyes. Block-like pieces 1½ to 2 inches in diameter are preferred. Tubers 5 to 8 ounces in weight can be cut into quarters of suitable planting size, while smaller tubers can be cut in half. Whole seed may be superior to cut pieces of the same size for seed purposes.[30, 37]

Preparation of Seed Tubers

Where conditions permit, potatoes are cut immediately before planting. The cut seed sometimes is dusted with lime or gypsum to keep the pieces from sticking together. Often labor is more readily available when fields are too wet for tillage operations, which makes it advisable to cut the seed before planting time. In the latter case, it is usually recommended that the tubers be suberized before planting. This consists of handling the tubers so as to promote development of cork tissue on the cut surface. The tubers are held in crates or sacks for a day or two after cutting and then emptied out to separate any pieces that have stuck together. Throughout the 10-day period of suberization, the temperature should be kept at about 60° F., with a high humidity (about 85 per cent) maintained by wetting the storage room or keeping wet sacks over the pile of cut seed.[71] Suberized potatoes usually can be stored up to 30 days after cutting without a reduction in stand or yield in the resultant crop. Shrinkage of the suberized tubers should not exceed 2 or 3 per cent in 30 days under good storage conditions.[39] Cutting may be done by special machines or by hand.[36]

Seed potatoes may be treated with disinfectants for the control of soft rots, common scab, rhizoctonia, and canker when disease-free certified seed is not used.[25]

Clean seed needs no treatment. Clean seed, or even treated seed, planted in scabby soil or in-soil heavily infested with the rhizoctonia organism will produce diseased tubers. Fungicides may be applied to the cut seed to prevent decay.

Certified Seed Potatoes

Planting certified seed has resulted in yields 31 to 219 bushels higher per acre than yields secured from uncertified seed supplies.[62] Certified seed production in the United States is ample for the total commercial domestic planting. Certified seed potatoes are exported in considerable quantities, but some are imported from Canada. The leading states in certified-seed production are Maine, Minnesota, Idaho, and North Dakota. Cool seasonal conditions in these states are very suitable for producing certified seed potatoes because mosaic-disease symptoms come to full expression in the field, and the diseased plants can be rogued or removed easily. Also, the cool temperature conditions in the North are suited to the growing of well-developed tubers. Moreover, the late harvest and cold winters facilitate the storage of seed supplies. Both field and bin inspections are required for certification. Certified seed potatoes grown in the North are planted in the southern states.

Seed growers may practice tuber selection as a means of eliminating undesirable or diseased plants from the foundation seed stocks.

Undesirable plants among the initial foundation stocks can be detected by planting a piece of each tuber early in a southern location to observe its behavior while keeping a record of and storing the remaining portion. Only those pieces that pass the preliminary inspection are planted later for increase. This procedure, called tuber index, is required of the producer of foundation stocks in the potato-certification rules of certain organizations. Other seed-potato growers test selected stock by the tuber-unit method. This consists of quartering the tubers and planting the four pieces in adjacent hills. The four progeny plants of any defective tubers can then be located and eliminated.

SEEDBED PREPARATION

Plowing for potatoes often is done in the fall to turn under legume residues, but spring plowing is satisfactory when it can be completed two to four weeks in advance of planting so that any vegetation has time to decompose. Deep plowing for potatoes usually is recommended. It is essential that the seedbed be prepared at least 2 inches deeper than the 3- to 5-inch depth at which potatoes are planted. Most of the potato roots are found within 1 foot of the surface of the soil, but since they can penetrate to a depth of 5½ feet, there is no object in plowing deep to facilitate root growth. Subsoiling in preparation for potatoes is of no advantage even in heavy Fargo clay underlaid by a clay subsoil.[52] However, heavy soils in California frequently are tilled with a chisel at a depth of 12 to 16 inches in preparation for potatoes. A very compact and fine seedbed such as is desired for small-seeded crops is not essential for potatoes. In the South, the land often is bedded up with a lister so that the potato rows on the beds will be well drained.

PLANTING

Most of the commercial potato crop is planted with two-row, three-row, or four-row machines. The popular automatic picker-planter selects the seed piece by jabbing it with a pointed spike, which may spread diseases to healthy seed pieces. Cup-type planters that require one or two operators are most suitable for planting whole tubers. The assisted feed planter, which requires one or two men to help distribute the seed pieces in the compartments so that one piece is dropped each time, gives the most uniform stands. Small-scale planting can be done by dropping the seeds into furrows by hand and covering them with a plow, cultivator, or harrow. Planting usually is 4 to 5 inches deep. It is deeper in sandy soil than in heavy soil, for dry soil than for wet soil, for late planting than for early planting, and for warm soil than for cold soil.

The average rate of planting potatoes in the United States in 1970 was 1700 pounds per acre, the rates ranging from about 600 to 2,200 pounds. Formerly a common rate of planting potatoes was 500 to 1000 pounds, a rate equivalent to a 1½-ounce seed piece every 15 to 30 inches in 42-inch rows. This low rate may still be a satisfactory rate in certain dryland sections. However, under irrigated or humid conditions where the soil is fertile or heavily fertilized, close planting results in better yields and fewer oversized tubers. About 2,200 pounds per acre are required for planting 1½-ounce seed pieces 8 inches apart in 34-inch rows. Varieties that produce few tubers per hill, require thick planting, often 6 inches apart in the row, so the tubers will not be too large for the best market demand.

Under irrigation in Colorado, planting 14 inches apart in 36-inch rows resulted in lower yields than was obtained from planting 8, 10, or 12 inches apart.[23] In general, the best yield and quality of tuber are obtained from relatively close spacing.[23] Thick planting and large seed pieces reduce the incidence of oversized tubers, hollow heart, and growth-cracked tubers.

In the northern part of the United States, the planting of potatoes often begins as early as the soil can be fitted after the land has thawed and becomes sufficiently dry for seedbed tillage. The minimum temperature for any sprout growth in the potato is 40° F. Planting in the late winter and spring generally is done when the mean air temperature has risen to about 50° to 55° F. In the spring the soil temperature at a 4-inch depth, where potatoes usually are planted, is about the same as the mean air temperature. In the northern portion of the country, this temperature is reached about 10 to 14 days before the average date of the last killing frost. In the central latitudes of the United States, planting is generally about four weeks before, but in the southern latitudes about six weeks before, the average last killing frost in the spring. The milder cold-spring periods in the South do not freeze the ground as deep as the potatoes are planted. This permits relatively earlier planting with little fear of spring frosts. Winter potatoes are planted in October or November in the southern parts of California and Florida (Figure 36–8). Spring planting begins about January 20 to February 1 in southern Georgia at latitude 31° N, and about May 10 to 20 in Aroostook County, Maine, at about latitude 46° N. Thus, in the low-altitude Atlantic coastal region, the planting date differs by 1 day for about every 8 miles, or for every ⅛ degree difference in latitude. For higher elevations at a given latitude, planting is delayed about one day for each additional 100 feet in altitude.

The period between planting and digging is about 3½ to 4 months for the early crop in all sections and for the main crop in the northern

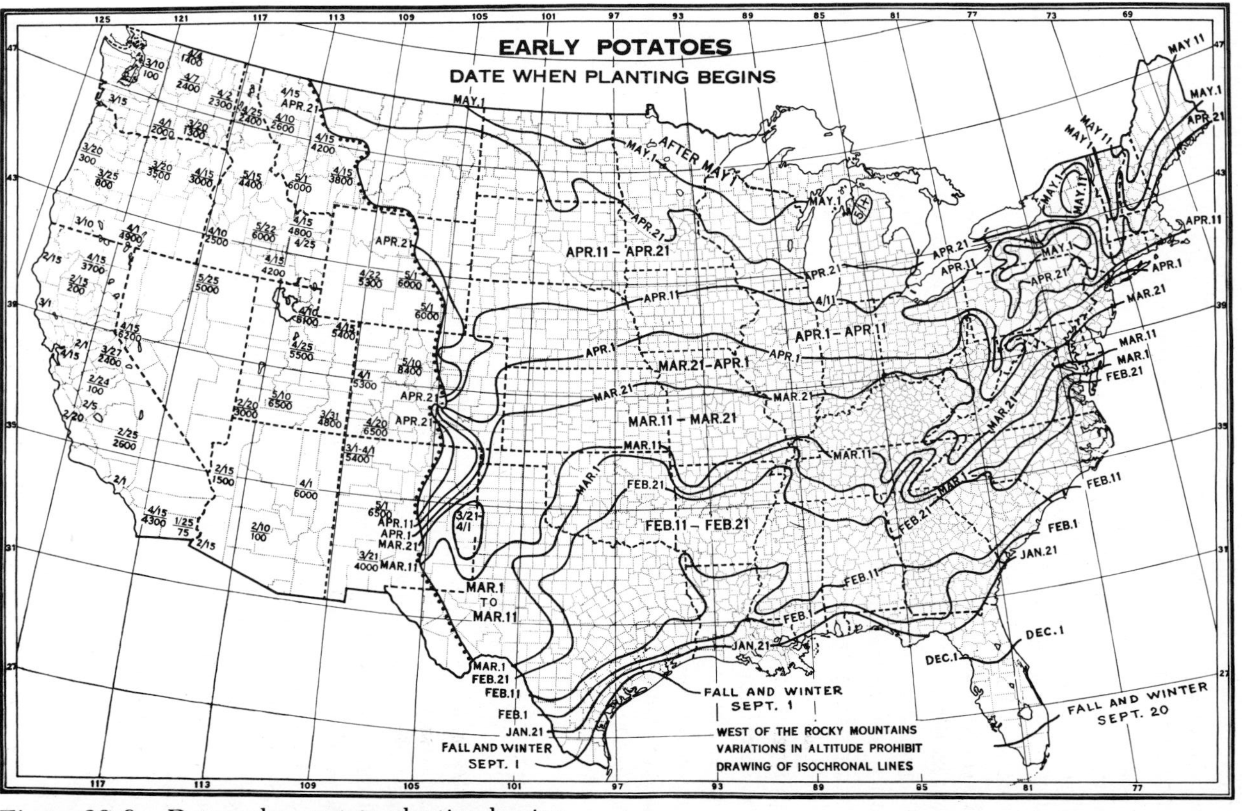

Figure 36–8. Dates when potato planting begins.

border states, and 3 to 3½ months for the late or fall crop where early growth is not retarded by cold weather. The late crop usually is planted sufficiently early for the tubers to be fully mature at or before the date of the average first killing frost. Thus, quick-maturing varieties can be planted two to four weeks later than can late (long-season) varieties. A late variety tends to produce the highest yields when planting is early and when the tubers are able to reach maturity before frost, whereas an early variety usually gives the highest yields when planting is late.[22] Nearly three-fourths of the crop consists of fall potatoes grown in the northern half of the United States.

Two crops of potatoes a year can be grown, sometimes in the same field, in the area from Maryland and Kentucky southward. The fall or late crop is planted from July to September in time for the tubers to mature before freezing weather. The seed for the late crop is obtained either from stored tubers, usually from the North, or from the early southern crop. Certain varieties of these early-crop tubers do not sprout promptly because the rest period has not been broken. Freshly harvested tubers of certain varieties will sprout promptly after chemical treatment. A fairly successful treatment is to soak freshly cut seed in a 1½ per cent solution of potassium or sodium thiocyanate for 1½ hours. The tubers should be cut through an eye or at the bud end. Whole tubers can be treated with ethylene chlorohydrin gas in an airtight container at 75° to 80° F. for 5 days.[30]

WEED CONTROL

Extra tillage beyond that necessary to kill weeds is of no benefit[38] because it merely injures the roots and dries out the soil. Preemergence tillage (or blind cultivation) to destroy small weeds is an accepted practice, a spike-tooth harrow or a light flexible-shank weeder being used once or twice before the potato plants are up. This usually is followed by a deep cultivation close to the row as soon as the plants are clearly visible, to loosen the soil for tuber development. Later, in some sections, the rows often are ridged to prevent the shallower tubers from being exposed to sunburn, or to protect them from freezing. Ridging, except to a slight degree, is not practiced in many sections, particularly not on the drylands where evaporation of moisture from the soil is increased by ridging.

Annual grasses as well as many broad-leaved weeds can be controlled by preplanting treatment with herbicides incorporated into the 2 to 4 inches of surface soil immediately after spreading. Preemergence sprays are also suitable herbicides for potato fields. For late-emerging weeds herbicides can be applied at the time of the last cultivation.

The growing of potatoes under a straw mulch lowers the soil temperature several degrees and increases the yields appreciably under hot conditions.[13] When 8 to 10 tons of straw per acre are applied, a mulch 4 inches deep is left after settling. In the cooler northern border states, mulching has lowered the yields.

Mulches should be applied after the soil begins to get warm. Weeds are able to penetrate the straw mulch when the rate of application is less than eight tons per acre, i.e., less than 8 to 10 inches deep when applied.

IRRIGATION

Under irrigated conditions, water should be applied often enough to keep the crop well supplied with soil moisture (above 50 to 60 per cent of field capacity), but water should always be applied before the plants show the need of moisture by a dark-green color in the leaves. Potatoes usually are irrigated in furrows. Sprinkler irrigation is often practiced, but this method tends to favor the increase of foliar diseases. The field should not be flooded. A seasonal total of 12 to 24 acre-inches of water in about six applications usually is ample for potatoes in the intermountain and Great Plains regions. On sandy soils in warmer regions, heavier irrigation may be necessary. The largest use of water is after the potatoes bloom, while the tubers are making rapid growth. Potato roots may absorb 57 per cent of the water used by the crop from the surface foot of soil, 23 per cent from the second foot, 13 per cent from the third foot, and 7 per cent from the fourth foot. The roots feed below the second foot only later in the season.

VINE KILLING

Killing the vines before harvest stops growth, hastens maturity, checks the formation of oversized and hollow-heart tubers, and reduces the spread of late blight infection from the vines to the tubers.[32] It may permit digging before heavy freezing occurs, and also limits the tuber size of seed potatoes. The tubers ripen quickly, the skins thicken, and digging is easier after the vines are killed. Vine killing is advisable only for the purposes listed above, since it limits yields and the specific gravity of the tubers. Various chemical dusts or sprays, including sodium meta-arsenite, DNBP, Diquat, and Paraquat have been used for vine killing.[42, 43] Vines also are killed with mechanical beaters and rotary mowers and by flaming. Vine destruction facilitates mechanical digging.

HARVESTING

Potatoes are harvested in all months of the year from January to December in some part of the United States. Early potatoes ordinarily

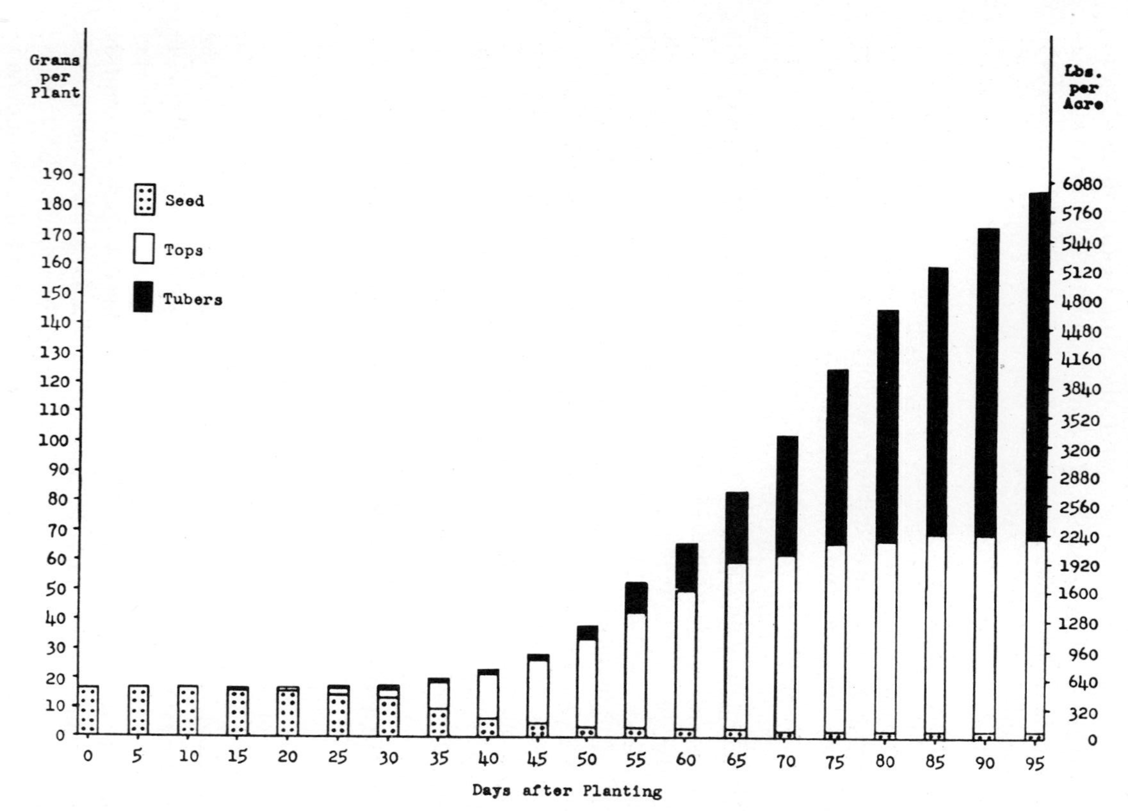

Figure 36–9. Dry weight of potato tops, tubers, and seed pieces at various stages of growth. [Courtesy Maine Agricultural Experiment Station.]

are dug as soon as they are large enough for the market. Such new potatoes, in addition to being small, have thin, poorly suberized skins that are easily damaged or rubbed off in handling, and they are watery and somewhat low in starch content. High seasonal prices justify their premature harvest. In potatoes not harvested early, the number of tubers per hill and the percentage of starch, protein, and ash continue to increase until most of the leaves are dead. The total weight of tubers may increase fivefold between blossoming and the stage at which all the leaves are dead.[5] (Figure 36–9.) In this same period, the sugar content may decrease from 1.0 down to 0.2 per cent, and the moisture content drops from 84 down to 80 per cent, but the crude-fiber content remains almost constant. The starch content of tubers may increase from 11.5 per cent to 15 per cent during the three weeks between the time when all the blooms have fallen and when 80 per cent of the leaves are dead. The total weight of tubers may double in the last 20 days of the growing period.[15]

Most commercial potatoes are dug with a mechanical digger (Figure 36–10). The small one-row and two-row diggers shake and screen out the soil through round bars and deposit the tubers on top of the rows. The tubers are then picked up by hand into baskets, crates, or bags and sometimes emptied into barrels in the field. They are passed over sorting machines to screen out small tubers, either at a warehouse or shed, or sorted by machine or by hand in the field before they are hauled. Mechanical harvesters that also convey the potatoes into trucks or bags are in wide use except on stony lands. Hand sorting and removal of stones and trash usually is done by workers who ride on the machine (Figure 36–10). Careful operation of the harvester is essential in order to keep mechanical bruising of the tubers at a minimum.[8]

Number One potatoes are $1\frac{7}{8}$ inch or more in diameter. Careful handling in all operations will avoid bruises or other damage to the tubers. They bruise less when the soil is warm. In hot dry weather, the tubers must be picked up within a few hours after digging in order to avoid sun scald. Most of the potatoes are washed with jets of water before they are sorted and graded.

Potatoes are marketed mostly in burlap bags of 100 pounds each.

STORAGE

For best keeping quality, potatoes going into storage should be held at about 60° F. with a humidity of 85 per cent for two weeks to permit injured tubers to heal. Thereafter, the temperature should be lowered to 36° to 40° F.,[45] when it is desired to keep them from sprouting for an indefinite period. Potatoes for the fresh market or processing should be stored at 45° F.[57] At 40° F., potatoes will remain dormant for three

Figure 36–10. Harvesting potatoes: (*Top*) two-row potato digger, followed by hand picking; (*Middle*) mechanical picking of dug potatoes with hand sorting; and (*Lower*) mechanical harvester.

to five months after harvest. Potatoes stored at 32° F. do not freeze, but they do not keep so well as when stored at 36° to 38° F. Potatoes are stored occasionally in refrigerated chambers, but usually, in cold climates, they are kept in pits, cellars, or warehouses, the latter often being mostly or partly underground. At low temperatures, the potatoes can be stored in large bins without danger of heating. At milder temperatures, they should be stored in sacks.[20] Ventilation of the storage place, often by forced air, is necessary to prevent spoilage, because the tubers are living organisms that continue respiration with evolution of heat, water, and carbon dioxide. The ventilating air flow should be 0.5–0.7 cubic feet per minute per hundred weight of potatoes.

Warm air admitted to cold potatoes causes condensation of moisture, which in turn encourages decay. Consequently, ventilators should be closed when the outside temperature is higher than that in the cellar. The relative humidity should be kept at 95 per cent to prevent shrinkage of the tubers.

Potatoes stored at temperatures below 50° F. undergo conversion of starch into sugar. At 36° F., 25 to 30 per cent of the starch is converted into sugar within six weeks,[30] and the tubers are sweet and soggy instead of mealy, and thus of poor cooking quality. When slices of such potatoes are French-fried or cooked into chips, they are of an unattractive dark-brown color because of caramelization of the sugar. Starch conversion is much less at 40° F., and scarcely perceptible at 50° F. The sugar reverts largely to starch when the storage temperature is raised to 50° to 70° F.[43] for a few weeks or sometimes for only a few days. For immediate or early use, potatoes should be stored at 50° to 70° F. At 60° to 70° F. nearly all the initial sugar present is eventually used up in respiration, and the tubers become more mealy. For home storage at moderate temperatures, sprouting can be checked by storing apples in the same compartment. The apples give off substances (probably ethylene) that check sprouting. When it is necessary to hold potatoes for some time at the higher storage temperatures or to ship them southward, the sprouting of tubers can be retarded by dusting, spraying, or storage house fume circulation with chemicals such as CIPC, TCNB, or maleic hydrazide.[43] Irradiation with gamma rays after 70 days of normal storage kept tubers from sprouting for 502 days, and they were firm and salable 300 days after treatment.[56]

USES

In the United States nearly 8 per cent of the potato crop is used for planting, and about 10 per cent is fed to livestock or lost by shrinkage and decay, chiefly the latter. The yearly consumption per capita was about 198 pounds in 1910 and declined to 103 pounds in 1952 but it increased to 119 pounds in 1971 when half of the potatoes were processed (Figure 36–11) in forms such as frozen (chiefly french fried), chips, shoestrings, and dehydrated, plus some canned alone or in mixtures.[64] The Incas of South America practiced freeze-drying of potatoes.[59] Dehydrated potatoes also are used in whisky manufacture. Some 8 to 12 million pounds are processed in starch and flour. In Germany many potatoes are processed into starch and alcohol, or are fed to livestock. Russians make vodka from potatoes.

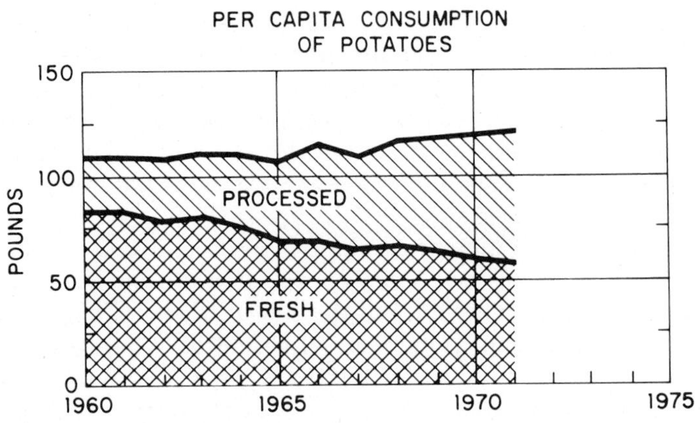

Figure 36–11. Trends in potato consumption patterns. [Courtesy USDA.]

For many years most of the new potatoes on the early market were Bliss Triumph, an early variety with a pink or red skin. Some improved early varieties with white skins that were adapted to the Southeast were harvested before maturity as new potatoes for northern markets. Most housewives refused to buy them unless the skins were dyed red, which soon became a standard practice. Applications of the herbicide 2,4-D to red-skinned varieties intensifies the red color.

The average whole potato consists of about 79 per cent water, 1 per cent ash, 2 per cent protein, 0.1 per cent fat, 0.6 per cent crude fiber, and 17 to 18 per cent nitrogen-free extract. About 20 per cent of a large sound tuber consists of peeling, i.e., periderm (or skin) plus some of the layers of the cortex.[14] The remainder is composed of about 78 per cent water, 2.2 per cent protein, 1 per cent ash, 0.1 per cent fat, 0.4 per cent crude fiber, and 18 per cent starch, sugar, and other carbohydrates. The dry matter consists of about 70 per cent starch, 15 per cent other carbohydrates, 10 per cent protein, 4.4 per cent ash, and 0.3 per cent fat. High starch content, high amount of solids, and high specific gravity are associated with mealiness and good cooking quality for chips, baking, boiling, mashing, or French frying. The potato solids and starch content can be measured by determination of the specific gravity of the whole tuber.

The most mealy baking potatoes have a specific gravity of 1.085 or higher, which indicates a solids content above 20 per cent as well as a starch content of about 15 per cent. Those potatoes most suitable for boiling, mashing, or French frying have specific gravities of 1.080 or higher, with at least 19.8 per cent solids and 14 per cent starch. Potatoes can be sorted mechanically for these differences by immersion in tanks that contain salt brine or other solutions of the desired specific gravity. The sorted potatoes later are washed and rinsed to remove the salts, which would otherwise damage the tubers.[30, 33]

Potatoes of high specific gravity may be produced by (1) early planting and delayed harvest to permit full maturity, (2) avoiding excessive applications of potassium and nitrogen, (3) providing adequate but not excessive amounts of irrigation water or fertilizers, (4) using slow-action chemicals for vine killing, and (5) when possible, growing the crop where sunny weather with temperatures below 70° F. prevails late in the season.[17, 30, 41]

DISEASES

Diseases Caused by Viruses

The virus diseases include mild mosaic, rugose mosaic, latent mosaic, leaf roll, spindle tuber, purple-top wilt, calico, and yellow dwarf. Tubers from infected plants carry the virus; when planted, they produce diseased plants. Such affected plants should be removed from fields producing seek stock, but it is not necessary to remove them from plantings for table use. The virus is spread considerably by insects, especially aphids. Use of certified seed reduces losses from virus diseases.[25, 26, 46]

MILD MOSAIC: Mild mosaic, caused by potato viruses A + X, is characterized by a definite mottling in which yellowish or light-colored areas alternate with similar areas of normal green in the leaf (Figure 36–12). A slight crinkling is usually present. Under conditions favorable for the disease, the margins of the leaves may be wavy or ruffled. Usually, affected plants will not produce more than three-fourths of the normal yield. Several varieties are resistant to this disease.

RUGOSE MOSAIC: Rugose mosaic, caused by potato viruses Y + X, is more destructive than mild mosaic. Plants from diseased tubers are dwarfed and the leaves are mottled, the mottled areas being smaller and more numerous than in mild mosaic. A distinct crinkling is always present. The undersides of lower leaves generally show blackening and death of veins. Tubers from plants with infection from the current

Figure 36–12. Potato leaflet showing mild mosaic.

season will produce plants with rugose mosaic the following season. Resistant varieties are available.

LATENT MOSAIC: This disease, caused by potato virus X, shows only a slight mottling of the leaflets. Virus-free seed should be planted.

LEAF ROLL: Leaf roll is caused by several strains of virus X. The symptoms are leathery leaflets that roll upward at the edge. These rolled leaves become yellow-green instead of dark green, and the plants are considerably dwarfed. Affected leaves are brittle to the touch. The yield of affected plants may be reduced to one-third or one-half of normal. Only certified seed stock known to have come from plantings that were free from leaf roll and resistant varieties should be used for planting.

SPINDLE TUBER: In most varieties, spindle tuber is well characterized by a pronounced elongation of the tubers. The tubers are small and may become pointed at one or both ends. The plants are more erect than normal and somewhat spindling in growth. There also is a narrowness of shoots, dwarfing, and a decidedly darker green color of the foliage. The disease may cause a marked reduction in yield. The affected tubers are often so poorly shaped that they are of low commercial grade. Certain insects, such as grasshoppers, fleabeetles, the tarnished plant bug, and the Colorado potato beetle, spread the disease. Cutting knives and picker planters also spread it. It can be controlled by use of good certified seed.

PURPLE-TOP WILT: This disease is caused by the aster-yellows virus that is transmitted by leafhoppers. It attacks both vines and tubers. The leaflets are discolored and rolled at the base. Control measures include control of the leafhoppers, destruction of weeds that carry the virus, and the planting of certified seed.

Common Scab

Common scab[1, 26, 48] is caused by a soil-inhabiting fungus (*Streptomyces scabies*) that also is carried on the tubers. The disease is characterized by round, corky pits on the skin of the potato (Figure 36–13).

Development of common scab is favored by an alkaline soil reaction, but it also occurs in slightly acid soils. An increase of the soil acidity will check development of the fungus. Addition of lime or barnyard manure to any soil, except a very acid one, tends to increase scab. The fungus develops best when the soil moisture is slightly below the optimum for development of the potato plant. In loose, well-aerated soils, it may develop readily even under very wet conditions. Application of 300 to 600 pounds of sulfur per acre on lighter types of soils to make the soil more acid usually reduces the severity of the attack somewhat.

Common scab can be practically eliminated on moderately infected tubers by treatment with hot formaldehyde or with acidulated mercuric chloride, if the field is clean. Several varieties are resistant to common scab.

Rhizoctonia Canker (Black Scurf)

Rhizoctonia canker, or black scurf, is caused by a widespread soil-inhabiting fungus (*Rhizoctonia solani*), that sometimes causes considerable reduction in stand and yield.

The most common symptom of the disease, irregular small black crusts on the skin, does little harm except to the appearance of the mature tubers (Figure 36–13). When infested tubers are planted, the fungus may attack and destroy the young sprouts. In this stage, brown decayed areas on the white underground stems often girdle the stems. Plants attacked in this manner may turn yellowish, the leaves may roll, and small greenish tubers may appear on the stems above the ground. The fungus thrives under moist, low-temperature conditions. It is most likely to attack potatoes planted in the cooler part of the year.

Rhizoctonia canker can be controlled in clean fields by treatment of infected seed stock with hot formaldehyde. Planting on a ridge tends to reduce the seriousness of this disease, since the soil on a ridge warms up more rapidly than on flat places, thus making conditions unfavorable for the fungus.

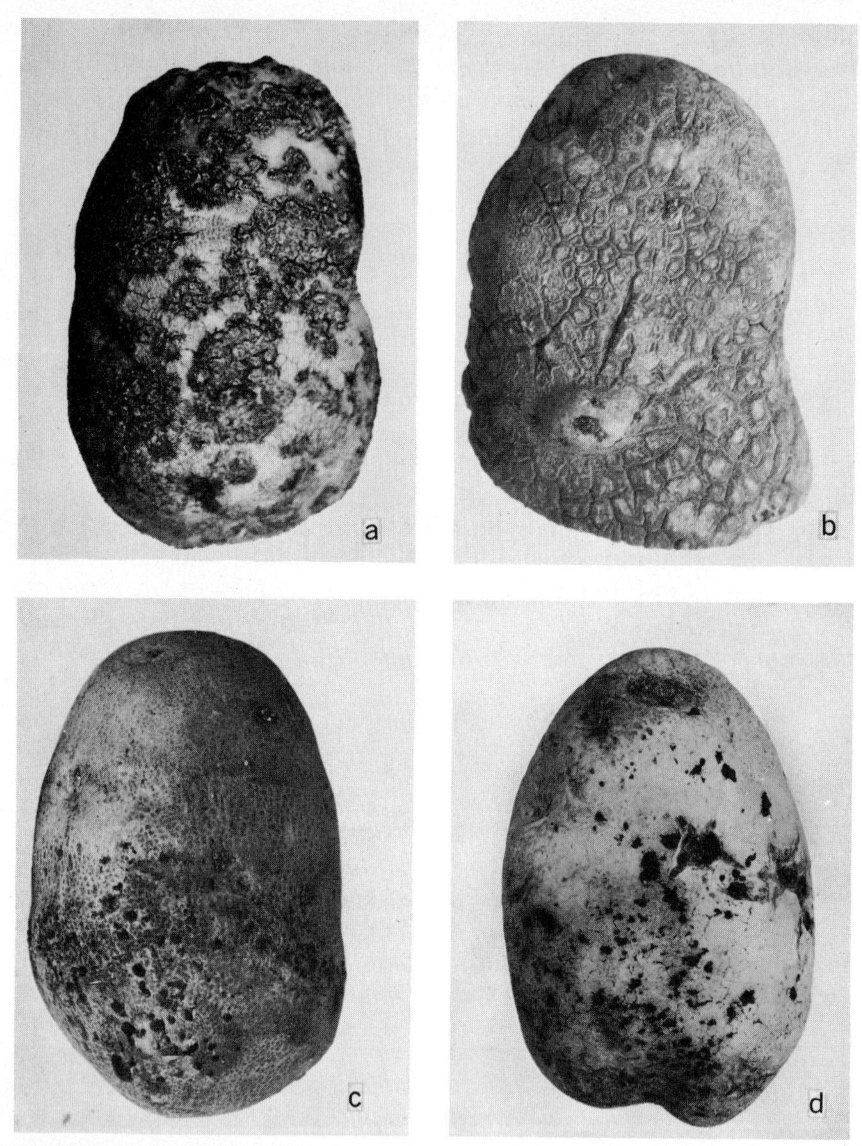

Figure 36–13. Tuber diseases: (*a*) scab, (*b*) russet scab, (*c*) black scurf, and (*d*) rhizoctonia.

Ring Rot

Ring rot, or bacterial ring rot, caused by *Corynebacterium sepedonica* is a very destructive seed-borne disease. The first symptom of the disease in the foliage is a wilting of the tips of leaves and branches. Later, the leaves become slightly rolled and mottled and fade to a pale-green followed by a pale-yellow color. Dead areas develop on the leaves, and the affected plants gradually die.[26]

When diseased hills are dug into, they usually contain tubers ranging from those that are apparently sound to those that are completely decayed. Decay of the tuber begins in a region immediately below the skin, thus causing a ring-rot appearance.[47]

Infected seed should not be planted. Only dependable certified seed should be used, to avoid possible loss from ring rot. Some varieties are resistant.

Late Blight

Late blight, caused by *Phytophthora infestans*, is usually seen first on the margins of the lower leaves. It works inward until entire leaves may be affected and killed. Irregular water-soaked spots develop at the margins of the leaves. Under conditions of high moisture, with warm day temperatures and cool nights, the spots enlarge rapidly and the infection spreads to other leaves. All the plants in a field may be killed in a few days. A noticeable odor is given off from the dead foliage in fields that are severely attacked. Late blight was the chief cause of the potato famine in Ireland in 1845–46.

The late-blight organism may also attack the tubers and produce in them slightly sunken brownish or purplish spots that enlarge until the entire tuber is affected. When such tubers are cut, the interior shows granular brick-red blotches, a distinctive characteristic of late blight. Infected tubers may rot in the field or in storage.[47]

A fixed-copper spray will control late blight, but it should be applied as soon as symptoms of the disease appear. The spraying should be repeated at weekly intervals so as to keep the foliage well covered.

The use of resistant varieties is the best means of controlling late blight.

Early Blight

Early blight, caused by the fungus *Alternaria solani*, is a very common disease of potatoes which causes noticeable damage in some years. It usually becomes serious several weeks before harvest. Small, scattered, dark circular spots are first produced on the lower leaves, which often become yellow. These spots enlarge, and the affected tissue dies. The enlarged spots develop a series of concentric rings that produce a target-board effect. Sprays give control of early blight.

INSECTS

Colorado Potato Beetle

About 120 different insects attack potatoes in the United States.[50, 51] The Colorado potato beetle (*Leptinotarsa decemlineata*) is one of the

Figure 36–14. Colorado potato beetle. (Much enlarged.)

most widespread, and formerly was one of the most destructive, insect pests of potatoes in the United States. It spread into Europe also. The adult, which appears in the spring, is a hard-shell beetle about ⅖ inch long, stout and roundish, and of a light yellowish color, with ten black stripes down its back (Figure 36–14). The female beetles lay orange-red eggs on the undersides of the potato leaves. The eggs hatch into small larvae, or soft-bodied slugs, that range in color from lemon to reddish brown marked with two rows of black spots on each side. The heads and legs are black. The slugs feed greedily and grow rapidly for a period of approximately two weeks, during which they devour large quantities of potato foliage. There may be one to two generations of this insect each year, depending upon climatic conditions.

The Colorado potato beetle can be controlled with insecticides.

Potato Fleabeetle

The adult potato fleabeetle (*Epitrix cucumeris*) reduces potato yields by eating the foliage, while the larvae may attack the tubers. The adult beetle is about 1/16 inch long, black in general appearance, with yellow legs. When disturbed, it jumps quickly and may readily disappear from sight. It eats holes in the leaves, which causes some leaves to dry and fall. The slender, white, wormlike larva, approximately ⅕ inch long, feeds on the roots and tubers, producing tiny tunnels near the surface of the tubers and pimple-like scars on the surface. The potato fleabeetle on the foliage may be controlled with the same insecticides used to control the Colorado potato beetle.

Potato and Tomato Psyllid

The potato and tomato psyllid (*Paratrioza cockerelli*) greatly reduces the yields of potatoes in Colorado, Nebraska, Wyoming, and New Mexico. The adult psyllids often migrate long distances from their winter hibernation quarters. The young psyllids or nymphs, which are flat, scalelike, and light yellow or green, feed on the underside of the leaves. The feeding causes a curling and yellowing of the leaves known as psyllid yellows, and the production of many small tubers in the hill. Several generations of the psyllid develop within a year. They are controlled by dusting or spraying, or by band applications of granulated insecticides at planting time, and with resistant varieties.

Potato Tuber Worm

The potato tuber worm (*Gnorimoschema operculella*) is the immature stage of a small gray moth that deposits its eggs on the foliage of potatoes and related plants. In stored potatoes, the moth lays eggs near the eyes of the tubers or in other depressions. The eggs soon hatch into small white tuber worms that develop a pinkish cast along the back as they mature. In the field the young tuber worms are leaf or stem miners. When the infested foliage dries, they find their way to the tubers through cracks in the soil. The tuber worms tunnel throughout the tubers and render them unmarketable.

The tuber worm is controlled in the field with suitable insecticides. All culls and infested tubers should be destroyed or fed to stock immediately after harvest. All volunteer potato plants should be destroyed. Tuber worms in stored potatoes may be killed by dusting the tubers with insecticides.

Potato Aphids

Several species of soft-bodied plant lice, or aphids, attack potato foliage and transmit leaf roll, mosaic, and other diseases. When abundant, these aphids reduce the yield of tubers by sucking the juices from the foliage. Potato aphids are controlled with dust or spray insecticides.

Seed-Corn Maggot

The seed-corn maggot (*Hylemya cilicrura*) is the immature, or maggot, stage of a small fly that lays its eggs on soil and decaying vegetable matter. Its feeding on potato seed pieces in the soil is accompanied by decay. The infested young potato plants are either killed or become so weakened that their growth and yield are reduced. The best control for the seed-corn maggot is to allow the potato seed pieces to heal, or suberize, before they are planted, or to mix suitable insecticides in the surface soil before planting.

Potato Leafhopper

The potato leafhopper (*Empoasca fabae*) causes a very destructive disease-like condition known as hopper burn. This condition begins with a yellowing of the leaf around the margin and tip, followed by a curling upward and rolling inward. The leaf changes in appearance from yellow to brown and then becomes dry and brittle. When the leafhopper infestation is heavy the entire plant may die prematurely. The potato leafhopper is a small green wedge-shaped insect, about 1/8 inch in length. Both young and adults feed on the underside of the leaves and suck the juices from the plants.

Leafhoppers and the resultant hopper burn can be controlled by spraying or dusting with suitable insecticides.

Wireworms

Potato tubers are often rendered unmarketable by small holes caused by wireworms. Wireworms bore clean tunnels that are usually perpendicular to the surface of the tuber and are lined with a new growth of plant tissue. Lands known to be infested with wireworms should be avoided for potato culture. Partial control is obtained by soil treatments with insecticides.

Nematodes

Root-knot nematodes (*Meloidogyne* species) that attack potato roots and tubers can be partly controlled by expensive soil treatments with Telone or DD before planting (Figure 36–15).

Figure 36–15. Applying DD, to control root knot nematode, and liquid ammonia, for nitrogen fertilization, to the soil in the same operation.

The golden nematode (*Heterodera rostochiensis*), a serious pest of potatoes in Europe, occurs in parts of Long Island in New York, with occasional small outbreaks elsewhere in the United States. The infested areas are under quarantine to restrict further spread of the pest. Partial control is obtained by soil fumigants such as DD; Vidden D, or Vorlex.[58] One or more varieties are resistant to the nematode.

REFERENCES

1. Afanasiev, M. M. "Potato diseases in Montana and their control," *Mont. Ext. Bul. 329*, pp. 1–46. 1970.
2. Akeley, R. V., and Stevenson, F. J. "Yield, specific gravity, and starch content of tubers in a potato breeding program," *Am. Potato J.*, 20(8):203–217. 1943.
3. Anon. "Potatoes and sweet potatoes," *USDA Handbk. 127*, pp. 1–16. 1967 (Rev.).
4. Appleman, C. O. "Study of rest period in potato tubers," *Md. Agr. Exp. Sta. Bul. 183*, pp. 181–226. 1914.
5. Appleman, C. O. "Potato sprouts as an index of seed value," *Md. Agr. Exp. Sta. Bul. 265*, pp. 239–258. 1924.
6. Appleman, C. O., and Miller, E. V. "A chemical and physiological study of maturity in potatoes," *J. Agr. Res.*, 33(6):569–577. 1926.
7. Artschwager, E. "Studies on the potato tuber," *J. Agr. Res.*, 27(11):809–835. 1924.
8. Bartlett, H. D., and Huntington, D. H. "Mechanical potato harvesting," *Maine Agr. Exp. Sta. Bul. 549*. 1956.
9. Boswell, V. R. "Vegetables," in *Soil*, USDA Yearbook, 1957, pp. 692–698.
10. Boswell, V. R., and Jones, H. A. "Climate and vegetable crops," in *Climate and Man*, USDA Yearbook, 1941, pp. 373–399.
11. Brown, B. E., and Cummings, G. A. "Fertilizer placement for potatoes," *Am. Potato J.*, 13:269–272. 1936.
12. Bushnell, J. "The relation of temperature to the growth and respiration of the potato plant," *Minn. Agr. Exp. Sta. Tech. Bul. 34*, pp. 1–29. 1925.
13. Bushnell, J., and Welton, F. A. "Some effects of straw mulch on yield of potatoes," *J. Agr. Res.*, 43(9):837–845. 1931.
14. Caldwell, J. S., Culpepper, C. W., and Stevenson, F. J. "Variety, place of production, and stage of maturity as factors in determining suitability for dehydration in white potatoes II," *The Canner*, June 3 and 10, 1944.
15. Carpenter, P. N. "Mineral accumulation in potato plants," *Maine Agr. Exp. Sta. Bul. 562*. 1957.
16. Clark, C. F., and Lombard, P. M. "Descriptions of and key to American potato varieties," *USDA Circ. 741* (rev.). 1951.
17. Cunningham, C. E., and others. "Yields, specific gravity and maturity of potatoes," *Maine Agr. Exp. Sta. Bul. 579*. 1959.
18. Cuthbert, F. P., and others. "The southern potato wireworm, *USDA Leaflet 534*, pp. 1–8. 1965.
19. Dow, A. I., and Kunkel, R. "Fertilizer tests with potatoes," *Wash. Agr. Ext. Cir. 335*. 1963.

20. Edgar, A. D. "Studies of potato storage houses in Maine," *USDA Tech. Bul. 615*, pp. 1–47. 1938.
21. Edmundson, W. C. "Effect of irrigation water on vigor and viability of seed potatoes," *USDA Tech. Bul. 216*, pp. 1–6. 1930.
22. Edmundson, W. C. "Time of planting as affecting yields of Rural New Yorker and Triumph potatoes in the Greeley, Colo., District," *USDA Circ. 191*, pp. 1–7. 1931.
23. Edmundson, W. C. "Distance of planting Rural New Yorker No. 2 and Triumph potatoes as affecting yield, hollow heart, growth cracks and second-growth tubers," *USDA Circ. 338*, pp. 1–18. 1935.
24. Findlen, H., and Glaves, A. H. "Vine killing in relation to maturity of Red River Valley potatoes," *USDA Tech. Bul. 1306*. 1964.
25. Folsom, D., Simpson, G. W., and Bonde, R. "Maine potato diseases, insects, and injuries," *Maine Agr. Exp. Sta. Bul. 469* (rev.). 1955.
26. Harris, M. R. "Diseases and pests of potatoes," *Wash. Agr. Ext. Bul. 553*. 1962.
27. Hawkes, J. G., and Hjerting, J. P. *The Potatoes of Argentina, Brazil, Paraguay and Uruguay*, Clarendon Press, Oxford, England, pp. 1–525. 1969.
28. Ivins, J. D., and Milthrope, F. S., Eds. *The Growth of the Potato*, Butterworths, London, pp. 1–328. 1963.
29. Jones, H. A., and Borthwick, H. A. "Influence of photoperiod and other factors on the formation of flower primordia in the potato," *Am. Potato J.*, 15(12):331–336. 1938.
30. Kehr, A. C., Akeley, R. V., and Houghland, G. V. C. "Commercial potato production," *USDA Handbook 267*. 1964.
31. Kunkel, R. "Factors affecting the yield and grade of Russet Burbank potatoes," *Colo. Agr. Exp. Sta. Tech. Bul. 62*. 1957.
32. Kunkel, R., Edmundson, W. C., and Binkley, A. M. "Results with potato vine killers in Colorado," *Colo. Agr. Exp. Sta. Tech. Bul. 46*. 1952.
33. Kunkel, R., and others. "The mechanical separation of potatoes into specific gravity groups," *Colo. Agr. Exp. Sta. Bul. 422-A*. 1952.
34. Kunkel, R., and others. "Fertilization of Red McClure potatoes in the San Luis Valley of Colorado," *Colo. Agr. Exp. Sta. Tech. Bul. 43*. 1951.
35. Lambeth, V. N. "Potato growing in Missouri," *Mo. Agr. Exp. Sta. Bul. 548*. 1951.
36. Leach, S. S., and others. "Mechanical potato cutters can produce high quality seed pieces," *Rsh. in Life Sci.*, 19(16):1–8. 1972.
37. LeClerg, E. L. "Effect of whole and cut seed on stand and yield of Irish potatoes," *La. Agr. Exp. Sta. Bul. 371*, pp. 1–8. 1943.
38. Lombard, P. M. "Comparative influence of different tillage practices on the yield of the Katahdin potato in Maine," *Am. Potato J.*, 13(9):252–255. 1936.
39. Lombard, P. M. "Suberization of potato sets in its relation to stand and yield," *Am. Potato J.*, 14(10):311–318. 1937.
40. Longley, A. C., and Clark, C. F. "Chromosome behavior and pollen production in the potato," *J. Agr. Res.*, 41(12):867–888. 1930.
41. Murphy, H. J., and Goven, M. J. "Factors affecting the specific gravity of the white potato in Maine," *Maine Agr. Exp. Sta. Bul. 583*. 1959.
42. Nelson, D. C., and Nylund, R. E. "Effect of chemicals on vine kill, yield

and quality of potatoes in the Red River Valley," *Am. Pot. J.*, 46(9):315–322. 1969.

43. Parent, R. C., and others. "Potato growing in the Atlantic Provinces," *Can. Dept. Agr. Pub. 1281*, pp. 1–28. 1967.
44. Ramsey, G. B., Wiant, J. S., and Smith, M. A. "Market diseases of fruits and vegetables; potatoes," *USDA Misc. Pub. 98*, 1949.
45. Rose, D. H., Wright, R. C., and Whiteman, T. M. "The commercial storage of fruits, vegetables and florists stocks," *USDA Circ. 278*, pp. 1–52. 1942.
46. Rost, C. O., Kramer, H. W., and McCall, T. M. "Fertilizers for potatoes in the Red River Valley," *Minn. Agr. Exp. Sta. Bul. 385*, pp. 1–16. 1945.
47. Rowberry, R. G., and others. "Potato production in Ontario," *Ont. Dept. Agr. & Food Publ. 534*, pp. 1–50. 1969.
48. Schultz, E. S. "Control of diseases of potatoes," in *Plant Diseases*, USDA Yearbook, 1953, pp. 435–443.
49. Shands, W. A., and Landis, B. J. "Potato insects—their biology and biological and cultural control," *USDA Handbook 264*. 1964.
50. Shands, W. A., Landis, B. J., and Reid, W. J., Jr. "Controlling potato insects," *USDA Farmers Bul. 2168* (rev.). 1965.
51. Shands, W. A. and Landis, B. L. "Controlling potato insects," *USDA F.B. 2168*, pp. 1–15. 1970 (Rev.)
52. Shepperd, J. H., and Churchill, O. O. "Potato culture," *N. Dak. Agr. Exp. Sta. Bul. 90*, pp. 81–126. 1911.
53. Smith, J. W. "The effect of weather on the yield of potatoes," *U.S. Monthly Weather Rev.*, 43:222–236. 1915.
54. Smith, O. "Potato production," in *Advances in Agronomy*, Vol. 1, Academic Press, New York, pp. 353–390. 1949.
55. Smith, Ora. *Potatoes, Production, Storing, Processing*, A.V.T. Pub. Co., Westport, Conn., pp. 1–632. 1968.
56. Sparks, W. C., and Iritani, W. M. "The effect of gamma rays from fusion product wastes on storage losses of Russet Burbank potatoes," *Ida. Agr. Exp. Sta. Rsh. Bul. 60*, pp. 1–26. 1964.
57. Sparks, W. C., and others. "Ventilating Idaho potato storage," *Ida. Agr. Exp. Sta. Bul. 500*, pp. 1–15. 1968.
58. Spears, J. F. "The golden nematode handbook," *USDA Handbk. 353*, pp. 1–81. 1968.
59. Stevenson, F. J. "The potato—its origin, cytogenetic relationships, production, uses, and food value," *Econ. Bot.*, 5:153–171. 1951.
60. Stevenson, F. J., and Akeley, R. V. "Newer potato varieties can be produced with less labor and expense than some of the old," *Am. Potato J.*, 19(8):153–161. 1942.
61. Stevenson, F. J., and Clark, C. F. "Breeding and genetics in potato improvement," in *USDA Yearbook, 1937*, pp. 405–444.
62. Stuart, W. "Seed potatoes and how to produce them," *USDA Farmers Bul. 1332*, (rev.). pp. 1–14, 1936.
63. Stuart, W. *The Potato, Its Culture, Uses, History, and Classification*, 4th ed., rev., Lippincott, Philadelphia, pp. 1–508. 1937.
64. Sullivan, G. H. "The potato industry," *Ind. Agr. Exp. Sta. Rsh. Bul. 862*, pp. 1–29. 1970.

65. Talburt, W. F., and Smith, O. *Potato Processing*, A.V.T., Westport, Conn., pp. 1–475. 1959.
66. Terman, G. L. "Green manure crops and rotations for Maine potato soils," *Maine Agr. Exp. Sta. Bul. 474.* 1949.
67. Terman, G. L. "Effect of rate and source of potash on yield and starch content of potatoes," *Maine Agr. Exp. Sta. Bul. 481.* 1950.
68. Terman, G. L., and others. "Rate, placement, and source of nitrogen for potatoes in Maine," *Maine Agr. Exp. Sta. Bul. 490.* 1951.
69. Terman, G. L., and others. "Rate, placement, and source of phosphorus fertilizers for potatoes in Maine," *Maine Agr. Exp. Sta. Bul. 506.* 1952.
70. Urgent, D. "The potato. What is the botanical origin of this important crop plant, and how did it first become domesticated?" *Science*, 170(3963):1161–1166. 1970.
71. Wright, R. C., and Peacock, W. M. "Influence of storage temperatures on the rest period and dormancy of potatoes," *USDA Tech. Bul. 424*, pp. 1–22. 1934.

Chapter 37

Sweetpotatoes and Yams

ECONOMIC IMPORTANCE

The sweetpotato (*Ipomoea batatas*) is grown in the warm parts of all continents. The total production, including that of true yams (*Dioscorea* species), exceeds 140 million metric tons grown on 16.6 million hectares with yields of 0.86 metric tons per hectare. The leading producing countries of both crops include Nigeria, Japan, China, Taiwan, Indonesia, Korea, Brazil, Uganda, and the Ivory Coast. The average production of sweetpotatoes in the United States from 1970 to 1972 was 12,522,000 cwt (600,000 metric tons) on 119,000 acres (48,000 hectares) or 106 cwt per acre (11.9 metric tons per hectare).

The states leading in sweetpotato production were North Carolina, Louisiana, Virginia, and Mississippi. The production of sweetpotatoes declined about 50 per cent from 1944 to 1964 but has been maintained since that time by higher yields despite a continued decline in acreage. The per capita consumption of sweetpotatoes in the United States declined from 31 pounds in 1920 to about 5 pounds in 1972. The production of sweetpotatoes in the United States declined partly because of heavy labor requirements for producing the crop.

Several features of sweetpotato culture make mechanization difficult. In 1959, about 72 minutes of human labor were required to produce 100 pounds of sweetpotatoes in the United States. The higher-yielding white-potato crop, by use of mechanical harvesters, was produced with only 11 minutes of labor per 100 pounds.

HISTORY

The original home of the sweetpotato apparently was Mexico, Central America, or South America, where a wild species, *Ipomoea trifida*, occurs.[32, 39]

The plant appears to have been introduced into the Pacific islands in early times. Its culture also may have spread to China and other Asiatic countries before 1492. Columbus found sweetpotatoes growing in the West Indies. The crop was grown in Virginia as early as 1648.

ADAPTATION

The sweetpotato thrives only in moderately warm climates. It requires a growing period of at least four months of warm weather. A slight frost kills the foliage, while prolonged exposure of the plant or roots to temperatures much below 50° F. causes damage.[13] However, no part of the United States is too hot or too moist for the sweetpotato to grow. Maximum yields of roots are obtained where the mean temperature for July and August exceeds 80° F. Where sweetpotatoes are grown commercially, the mean temperature for the summer months (June, July, and August) is above 70° F. It is higher than 75° F. in all regions except Maryland, Delaware, and New Jersey, as well as in the less-important growing areas of Iowa, southern Illinois, and southern Indiana.

Differences in day lengths at different latitudes in the United States do not greatly modify root production. Long days promote heavy vegetative growth. The plants make their growth during the long days of the summer. Relatively short days, high temperatures, and a long growing season are required for the sweetpotato to produce an abundance of seed or flowers. Consequently, the plants seldom bloom in the continental United States when grown in the field, but most varieties flower readily when grown in a warm greenhouse in the winter. An 11-hour day is favorable for flowering.[31] Sweetpotatoes produce seed readily in the West Indies or in other tropical regions.

Nearly all of the nonirrigated commercial sweetpotato crop is grown where the average rainfall exceeds 35 inches, but irrigation increased the yields in Missouri, Arkansas, and Louisiana.[24]

The sweetpotato grows best on sandy loam or loamy[10] fine sand soils with a pH of 5.6–6.5. Neutral soils induce the pox and scurf diseases.[41] A sandy soil with a clay subsoil is desirable, but good yields are obtained in very sandy soil types that are heavily fertilized.[9] Clay soils are undesirable because of harvest difficulties. The roots often have poor shapes when grown in clay or muck soils. Very fertile soils tend to be detrimental to root growth, but favorable to a heavy growth of vines. However, the crop succeeds well on fertile soils in the Southwest. A well-drained soil is desirable, but bedding or ridging of the field helps to overcome the effects of inadequate surface drainage. Sweetpotatoes thrive under subirrigated conditions in California, where the water table is only 4 or 5 feet below the surface.

BOTANICAL DESCRIPTION

The sweetpotato is a member of the *Convolvulaceae*, or morning-glory family. The sweetpotato (*Ipomoea batatas*) has funnel-shaped flowers $3/4$ to $1\frac{1}{2}$ inches wide to 1 to $1\frac{1}{2}$ inches long. The flowers usu-

Figure 37–1. Flowers and leaves of the sweetpotato.

ally have rose-violet or bluish petals, with a darker color in the throat. Each flower has five stamens and an undivided style (Figure 37–1). One to four small, black, flattened seeds are borne in each fertile capsule. Nearly all plants are self-sterile.[35]

The sweetpotato plant is a perennial, but in the United States it is killed by cold weather each winter. Thus, it is grown in this country only as an annual. The vines usually are 4 to 16 feet long, green or red to purple or both, and often somewhat hairy, especially at the nodes. The leaves are tinged with purple in some varieties. The characteristic leaves in many varieties are somewhat heart-shaped in outline, with the margins entire or toothed in the more popular varieties and deeply lobed in others. The roots have white, yellow, salmon, red, or purplish-red skin, while the flesh is white or various shades of yellow, orange, or salmon.[42] Roots with purple or partly purple flesh are encountered occasionally.

The sweetpotato root lacks the eyes or buds found in the potato tuber, but shoots arise from adventitious buds.[5] Starch is stored in the parenchyma cells of both the central core and the cortex.

VARIETIES

Sweetpotatoes are often considered as falling into two culinary classes: the firm (dry-fleshed) or Jersey type, and the soft (moist-fleshed) type or "yam." The industrial and feed varieties represent still other classes. The Jersey type formerly was favored in the markets of the North, while the soft-fleshed varieties were popular in the South. The difference between the two types lies in the structure and appearance of the flesh, because the Jersey type is higher in moisture content.[15] The name yam, as applied either to types or varieties of sweetpotatoes, is a misnomer. The true yam, a tropical plant that produces tubers instead of fleshy roots, belongs to the genus *Dioscoria*, family *Dioscoreaceae*. It is not grown in continental United States, except to a limited extent in gardens in southern Florida, but is widely grown in tropical Africa.

Most of the older sweetpotato varieties originated from root selections. The newer varieties were selected from controlled crosses or from bud mutations.[22, 40] The crossing is done in the greenhouse.

New varieties released to growers since 1940 have largely replaced the older varieties.[40] New varieties similar to the Jersey type, have orange-colored flesh, rather than the firm yellow flesh typical of the older Jersey types.

Varieties with a salmon or orange flesh have a higher carotene content than do the white or yellow varieties. Sweetpotatoes with white flesh, or with purple streaks in the flesh, are popular in some other countries but are unacceptable in the food markets of the United States. Soft-fleshed varieties in which more of the starch is converted to sugars and dextrin are most widely grown.

Varieties with a cream-colored flesh are suitable for stock feed or industrial use, but they also are preferred for human food in many countries.

ROTATIONS

Rotations of five years or longer with other crops such as corn, cotton, potatoes, small grains, and legume hays are helpful in sweetpotato production. Such rotations reduce the infestation by soil-borne organisms that cause sweetpotato diseases such as black rot and scurf.[25] Sweetpotatoes follow winter-cover crops advantageously because there usually is time to plow under the cover crop before the sweetpotato plants are set in the field.[9] Legume crops sown in the fall may save about one-half to two-thirds of the quantity of nitrogen fertilizer that would otherwise need to be applied. Excessive nitrogen merely increases vine growth.

FERTILIZERS

General fertilizer recommendations for high yields of sweetpotatoes are 20 to 60 pounds per acre of nitrogen, 80 to 120 pounds of phosphate, and 80 to 150 pounds of potash,[9] or 600 to 1500 pounds containing 20 to 30 units of N-P-K of which 33 to 50 per cent is potash.[41] About 1,200 pounds of a 3-8-8 mixture were recommended for New Jersey,[38] and 20-50-50 pounds of N-P-K is advisable in Kansas. There, more than 50 pounds of nitrogen per acre reduced yield.[21]

Experiments in New Jersey, Virginia, Maryland, Georgia, and Florida indicate that sweetpotatoes respond to quantities of potash in excess of 40 to 50 pounds per acre, whereas other experiments in North Carolina, South Carolina, Mississippi, and Louisiana show that quantity to be sufficient.[11] Ample potash sometimes produces thick roots,[38] but in other cases it has little influence on root shape.[11] Moderate liming may be helpful where the soil is below *p*H 5.0,[9] but a slightly acid soil prevents injury from scurf. Deep placement of fertilizer and lime may tend to develop long roots.

CULTURAL METHODS

Seedbed preparation for sweetpotatoes is about the same as that for other row crops in a particular region. Throwing up ridges or beds about 10 inches high in the field is believed to encourage growth of long roots.

Planting

Sweetpotato roots are bedded in warmed soil (preferably about 80° F.,[9] until the sprouts emerge, and at 70° F. thereafter. A month to six weeks after bedding and the danger of frost is past, the rooted sprouts (called slips or draws) are pulled and planted in moist warm soil in ridged cultivated rows. In the North, the roots are bedded in a greenhouse, an electrically heated bed, or a hotbed heated by an underlying layer of fresh fermenting horse manure. Sometimes the beds are mulched to retain the heat until the sprouts emerge. In the South, where the season is longer, a cold frame or an outdoor bed may be used, although heated beds permit earlier planting.

Sound small to medium-size roots are first treated with a recommended fungicide. Then they are laid in the bed, close together but not touching, and pressed into the surface of a 4- or 5-inch layer of sand. As the sprouts elongate, additional sand (2 or 3 inches) is added gradually to provide for root development along 2 or 3 inches at the base of the stem. Small roots usually are selected for bedding because more slips per bushel of roots are obtained.[2] About 150 to 240 square feet of bed planted with 10 to 14 bushels of roots provide sufficient slips for planting an acre, while 6 to 8 bushels or less are sufficient when two to three pullings are made. The plants are pulled or drawn by hand when they reach the desired height of 7 to 10 inches. (Figure 37–2), with 6 to 12 leaves.[8, 41] A new crop of sprouts develops in two or three weeks after the first sprouts are pulled. The lower yields

Figure 37–2. Sweetpotato sprouts or slips ready for planting.

usually obtained from the second and third pullings are merely a consequence of later planting. Prompt planting of the pulled sprouts is essential to the highest yields.[16]

The sprouts or draws are planted with a transplanting machine or by hand. One-row to 4-row machines are used.[21, 33, 41]

Sweetpotatoes for seed usually are grown from vine cuttings. These cuttings also are used for the late plantings[9] especially in the South. Cuttings are taken from plants as soon as the vines start to run; 8- to 12-inch sections of vine are pushed into the soil at desired intervals. Vine cuttings are cheaper than sprouts and less likely to spread diseases, but their use delays planting. One acre of plants will supply cuttings for eight acres of crop.

The plant spacing within the row may range from 10 to 24 inches without much effect upon total yield, yield of marketable roots, or starch content of the roots.[3] Typical spacing is 12 to 15 inches in 42-inch rows or 10,000 to 12,000 plants per acre. On highly productive soils, where high yields of market potatoes are expected, a closer spacing of 6 to 8 inches is advantageous. The thinner spacings tend to produce fewer small roots but more large (Jumbo) roots.[7] Uniform medium-size roots are desired for the food market, but small roots are desired for canning. The plants usually are set in ridges or beds 28 to 48 inches apart, wide rows being more common on infertile soils. Level planting is common in the drier regions of the United States.

Sweetpotatoes should be transplanted about one month after the average date of the last spring frost, or when the leaves of most oak trees have reached full size. Any delay beyond the optimum date not only results in marked decreases in yield,[3, 16] but also produces slender roots of low starch and carotene content.

Weed Control

The field is cultivated to control weeds until the vines spread between the rows. Early cultivation may be done with a modified rotary hoe.[9] For other cultivations the prevailing local implements, such as sweep cultivators, are satisfactory. Deep cultivation injures the roots. In laying by the crop, the rows are ridged up well by use of a disk cultivator, a middlebuster, or other convenient device.

HARVESTING

Sweetpotato harvest begins in August in the South, September in New Jersey, and October in California.

The maximum yield of sweetpotatoes is reached at the end of the growing season. The roots should be dug before, or immediately after, the first light frost. The number and, often, the size of roots continue to increase for a month after the maximum starch content is reached.[26] Since the sweetpotato vines interfere with the operation of digging machines, the vines usually are cut either during or before digging. The vines can be cut with a mower, shallow-running vertical knives, or, preferably, with shielded 8-inch colters attached to the digger.[33] Cutting vines prematurely decreases yields materially. Even pruning vines to obtain cuttings for planting reduces yields decidedly when more than one-half the runners are removed.[16]

As far as food use is concerned, the most important factor in digging is to avoid cutting or bruising the roots, which would promote rapid decay.[28] A wide-bottom (16-inch) plow run at the side of the row is preferable for digging sweetpotatoes that are to be stored. Usually the roots (Figure 37–3) are allowed to dry for a few hours before they are picked up by hand. A two-row sweetpotato harvester digs, tops, sorts, and loads the roots into containers. It saves labor but may injure more roots than by hand picking.

CURING

Uncured sweetpotato roots decay quickly at temperatures of 50° F. or lower.[29] About 20 to 40 per cent of the harvested sweetpotato crop may spoil before the roots are used. The storage room, as well as the containers, should be fumigated before the roots are harvested. The conditions for curing are a temperature of 85° to 90° F. and a relative

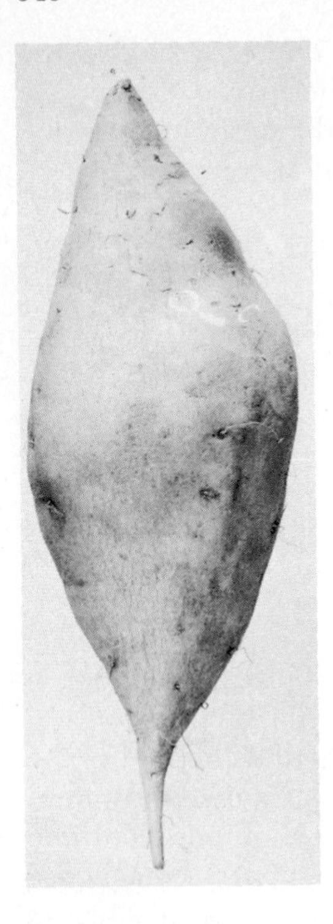

Figure 37–3. Sweetpotato root.

humidity of 85 to 90 per cent maintained for 4 to 7 days. Such conditions promote the healing of wounds in the roots, which check future fungus infection.[5, 28]

STORAGE

After curing, the sweetpotato roots should be stored at 55° to 60° F., with a relative humidity of 85 to 90 per cent. Storage of cured roots at 40° F. for two weeks or more tends to promote decay. Ventilation occasionally is necessary not only to regulate the temperature, but also to prevent condensation of moisture in the storage room or on the roots. The storage room should provide for artificial heat to keep the temperature from falling below 55° F.[29] Storage in baskets or crates causes less bruising and consequent shrinkage and decay than occurs when the roots are stored in slatted bins.

Curing causes a shrinkage of about 10 per cent due to moisture evaporation and loss of dry matter that results from respiratory oxidation of sugars. Subsequent shrinkage occurs during a five-month storage period,[36] but total shrinkage and spoilage under good conditions should not exceed 15 per cent. In storage, some of the starch and dextrin is converted to sugar. Under proper conditions the flesh softens somewhat, while the cooking quality, color, and flavor improve.[28, 41]

USES

Sweetpotatoes are used primarily for food, being prepared mostly from the fresh state. However, large quantities are frozen, canned, chipped or dehydrated. The culls are used mostly for stock feed. Sweetpotato vines make a satisfactory roughage or silage, but the best practice is to leave them on the field for soil improvement.[43]

Extraction of starch from sweetpotatoes[43] was carried on in Mississippi and Florida from 1934 to 1946. The starch was used chiefly for textile sizing or in laundering. The average yield was 10 to 11 pounds of starch per bushel of roots. Roots of high starch content (23 per cent) should yield 20 per cent starch or 12 pounds per bushel (60 pounds) with efficient operation.

Roots preferred by consumers are about 4 to 6 inches long, 1¾ to 3½ inches in diameter, and weigh not more than 1 to 1½ pounds.[17] Thick roots longer than 10 inches, called Jumbos, usually are not shipped. A standard bushel of sweetpotatoes varies in weight among different states from 50 to 60 pounds.

Peeled sweetpotatoes contain about 69 per cent water, 1.1 per cent ash, 1.8 per cent protein, 0.7 per cent fat, 1 per cent crude fiber, 28 per cent nitrogen-free extract (of which about 3 to 4.5 per cent is sugars and the remainder mostly starch), 0.2 per cent calcium, and 0.05 per cent phosphorus. The sweetpotato contains about 100 micrograms of carotene per gram as well as fair quantities of ascorbic acid and B vitamins. Compared with the Irish potato tuber, the sweetpotato root is higher in dry matter, starch, sugar, crude fiber, and fat, but lower in protein.

DISEASES

The diseases of sweetpotatoes[19, 25, 41] are of two main types: (1) those that attack the plants or roots in the field, and (2) storage diseases that attack the mature harvested roots. Some organisms attack the roots both in the field and in storage.

Stem Rot or Wilt

The stem rot or wilt disease is caused by two species of fungi, *Fusarium batatatis* and *F. hyperoxysporium*. The fungus is carried over in infected roots as well as in the soil, where it remains for several years. It causes a wilt in the sprouts in the bed and in the vines in the field, and also a rotting of the vines. It forms a blackened ring inside the root. Control measures consist of bedding disease-free roots that have been disinfected for 5 to 10 minutes in 1:1,000 bichloride of mercury, in clean or sterilized beds. The roots are then transplanted to noninfested fields. A good method for getting clean seed is selection of roots to be bedded from hills in which the stems, when split at harvest time, are somewhat resistant.

Foot Rot or Die-Off

Foot rot or die-off, caused by the fungus *Plenodomus destruens*, occurs in Virginia, Ohio, Iowa, Missouri, and other states. It stunts or kills the sprouts, rots or girdles the base of the stems, and spreads to the roots, where it develops brown spots. The black fruiting bodies of the fungus are visible on diseased stems. Infection usually spreads in the plant bed. Sanitary measures recommended for stem rot aid is control of foot rot.

Texas Root Rot

The Texas root-rot disease is caused by the fungus *Phymatotrichum omnivorum*. It occurs in irregular areas in many fields as well as in uncultivated areas, chiefly in Texas, Oklahoma, New Mexico, Arizona, and California. On the sweetpotato it causes a brown rot of the roots and a browning of the stems. Control for sweetpotatoes consists of avoiding areas in the field known to be heavily infested. Rotation with cereal or grass crops or green manuring with these crops is helpful.

Scurf

Scurf (called also soil-stain, rust, and Jersey mark) is caused by the fungus *Monilochaetes infuscans*. It affects the appearance of the roots by producing brown or blackish spots on the root surfaces. It is not a decaying organism, but it causes excessive shrinkage in storage. Control consists of crop rotation along with the use of clean planting stock. Seed treatment with bichloride of mercury (1:1,000) plus 3.7 ounces of wettable sulfur per gallon of solution is helpful. Roots produced from vine cuttings are more likely to be free from infection. Crop rotation is helpful but does not offer complete protection where the disease has occurred previously.[25] Several varieties are resistant or partly resistant.

Black Rot

Black rot, caused by the fungus *Ceratosomella fimbriata*, causes widespread heavy losses in the field and in storage.[24] It often destroys sprouts and yellows, wilts, and dries the foliage, after which black cankers develop in the underground parts of the stem. The infected roots show dark-brown or black spots that enlarge until most of the root is affected. Control measures are similar to those for stem rot.

Precautionary sanitary measures in and around the plant beds help prevent the spread of black rot. Resistant varieties or partly resistant varieties are available.

Soil Rot or Pox

The soil-rot or pox disease of sweetpotatoes is caused by the bacterialike fungus, *Actinomyces ipomoea,* that infests the soil. It is a serious widespread disease in certain important sweetpotato areas. Affected plants are stunted, have decayed roots, and often succumb to the disease. The surface of diseased mature roots shows pits or cavities with irregular jagged or roughened margins. These pits are filled with dead tissues at first, but this later sloughs off. Soil rot does not develop on soils in Louisiana that have a *p*H below 5.2.[25] The disease was largely checked there by applying sufficient sulfur to acidify the soil down to *p*H 5.2 or less. Several varieties are resistant to disease.

Virus Diseases

Yellow dwarf and ring spot are virus diseases that are evident on sweetpotato leaves. The planting of relatively virus-free seed stocks is desirable as a control measure.

Storage Rots

Soft rot, caused by the fungus *Rhizopus stolonifer,* is perhaps the most destructive storage disease of the sweetpotato. The mold enters the root through wounds or at the attachment end. It spreads quickly over the root, which causes complete decay. Control consists of good storage and proper curing of the roots to heal wounds.

Some varieties are resistant to the disease.

Dry rot, caused by the fungus *Diaporthe batatatis,* produces a firm brown rot. Java black-rot, caused by *Riplodia tubericola,* makes the roots dry and hard. Charcoal rot is caused by *Macrophomina phaseoli* (*Sclerotium bataticola*). It breaks down the interior of the root where it leaves black spore bodies.

Internal cork is a virus disease carried in the roots and spread in the field by aphids. Virus-free seed stocks should be planted.

INSECTS

Sweetpotato Weevil

The sweetpotato weevil (*Cylas formicarius* var. *elegantulus*)[1] deposits its eggs in small punctures in the stems or roots. They newly hatched larvae bore through the stems and roots. They leave a frass that is so unpalatable that hogs will not eat badly damaged roots. A complete weevil generation develops every few weeks throughout the year. It causes the most damage in the Gulf Coast region. The weevil is controlled by applying powdered insecticides in 7-inch bands along the row as well as on the plant bed. Lighter field applications may be desirable later. Other helpful control measures involve sanitation practices such as feeding or destruction of all plant residues including small roots in the field and plant beds; destruction of related wild hosts that belong to the genus *Ipomoea:* and the use of weevil-free planting roots, vine cuttings, or slips. Keeping sweetpotatoes off the field for a year is effective, provided all volunteer plants and weed hosts in the immediate vicinity are destroyed. The seed in storage may be dusted with insecticides.

Other Sweetpotato Insects

The adults of the sweetpotato leaf beetle (*Typophorus viridicyaneus*), a pest of much less importance than the weevil, emerge in late May to feed on the sweetpotato leaves. The larvae feed on the leaves and roots, enter the soil, and pupate in the fall. Some of the larvae, still in the roots at digging time, are transported to other fields.

Other insects that attack the sweetpotato include grasshoppers, fleabeetles, termites, wireworms, and white-fringed beetles.[18]

Insects harmful to the sweetpotato can be controlled with suitable insecticides.

YAMS

The true yams are grown in Puerto Rico, and formerly or occasionally in Florida, as well as in many tropical areas, especially Africa. The principal species are *Dioscorea alta* L., *D. rotunda* Poir., and *D. esculenta* Lowr., but several other species are grown occasionally. The plants are vines that climb on poles. The tubers are similar to potatoes in food value but of poorer keeping quality than sweetpotatoes. The plants require a 6- to 10-month warm growing season, ample soil moisture, good drainage, and fairly productive soils. The crop is propagated by planting cut, or whole small, tubers.[17, 30]

REFERENCES

1. Anonymous. "The sweetpotato weevil and how to control it," *USDA Leaflet 431.* 1958.
2. Anderson, W. S. "Sweetpotato plant production," *Miss. Agr. Exp. Sta. Bul. 349.* 1940.
3. Anderson, W. S., and others. "Regional studies of time of planting and hill spacing of sweetpotatoes," *USDA Circ. 725*, pp. 1–20. 1945.
4. Anderson, W. S., and Randolph, J. W. "Sweetpotato production. Time

of planting and hill-spacing studies," *Miss. Agr. Exp. Sta. Bul. 378*, pp. 1–22. 1943.
5. Artschwager, E. "On the anatomy of the sweetpotato root, with notes on internal breakdown," *J. Agr. Res.*, 27(3):157–166. 1924.
6. Artschwager, E., and Starrett, R. C. "Suberization and wound-periderm formation in sweetpotato and gladiolus as affected by temperature and humidity," *J. Agr. Res.*, 43:353–364. 1931.
7. Beattie, J. H., Boswell, V. R., and Hall, E. E. "Influence of spacing and time of planting on the yield and size of the Porto Rico sweetpotato," *USDA Circ. 327*, pp. 1–10. 1934.
8. Beattie, J. H., Boswell, V. R., and McCown, J. D. "Sweetpotato propagation and transplanting studies," *USDA Circ. 502*, pp. 1–15. 1938.
9. Boswell, V. R. "Commercial growing and harvesting of sweetpotatoes," *USDA Farmers Bul. 2020*. 1950.
10. Boswell, V. R. "Vegetables," in *Soil*, USDA Yearbook, 1957, pp. 692–698.
11. Boswell, V. R., and others. "Effect of potash on grade, shape, and yield of certain varieties of sweetpotatoes grown in South Carolina," *USDA Circ. 498*, pp. 1–23. 1938.
12. Boswell, V. R., and others. "Place and season effects on yields and starch content of 38 kinds of sweetpotatoes," *USDA Circ. 714*, pp. 1–15. 1944.
13. Boswell, V. R., and Jones, H. A. "Climate and vegetable crops," in *Climate and Man*, USDA Yearbook, 1941, pp. 373–399.
14. Brannon, L. W. "The sweetpotato leaf beetle," *USDA Circ. 495*, pp. 1–10. 1938.
15. Caldwell, J. S., and Magoon, C. A. "The relation of storage to the quality of sweetpotatoes for canning purposes," *J. Agr. Res.*, 33(7):627–643. 1926.
16. Clark, C. F. "Sweetpotato varieties and cultural practices," *Miss. Agr. Exp. Sta. Bul. 313*, pp. 1–27. 1938.
17. Coursey, D. G. *Yams, an Account of the Nature, Origins, Cultivation and Utilization of the Useful Members of the Dioscoreaceae*, Longmans, Longon, pp. 1–230. 1967.
18. Cuthbert, F. P., Jr. "Insects affecting sweetpotatoes," pp. 1–28. *USDA Agric. Handbook*. 1967.
19. Elmer, O. H. "Diseases of sweetpotatoes," *Kans. Agr. Exp. Sta. Bul. 495*, pp. 1–43. 1967.
20. Gatlen, G. O. "Marketing southern-grown sweetpotatoes," *USDA Bul. 1206*, pp. 1–46. 1924.
21. Greig, J. H. "Sweetpotato production in Kansas," *Kans. Agr. Exp. Sta. Bul. 498*, pp. 1–27. 1967.
22. Harter, L. L. "Bud sports in sweetpotatoes," *J. Agr. Res.*, 33(6):523–525. 1926.
23. Harter, L. L., and Weimer, J. L. "A monographic study of sweetpotato diseases and their control," *USDA Tech. Bul. 99*, pp. 1–118. 1929.
24. Hernandez, T. P., and others. "The value of irrigation in Louisiana," *La. Agr. Exp. Sta. Bul. 607*, pp. 1–15. 1965.
25. Hildebrand, E. M., and Cook, H. T. "Sweetpotato diseases," *USDA Farmers Bul. 1059* (rev.). 1959.

26. Kimbrough, W. D. "Studies of the production of sweetpotatoes for starch or feed purposes," *La. Agr. Exp. Sta. Bul. 348*, pp. 1–13, 1942.
27. Kushman, L. J. "Preparing sweetpotatoes for the market," *USDA Marketing Bul. 38*. 1967.
28. Kushman, L. J., and Wright, F. S. "Sweetpotato storage," *USDA Handbk. 358*, pp. 1–35. 1969.
29. Lutz, J. M., and Simons, J. W. "Storage of sweetpotatoes," *USDA Farmers Bul. 1442* (rev.). 1958.
30. Martin, F. W. "Yam production methods," *USDA Prod. Rsh. Rpt. 147–1–17*. 1972.
31. McClelland, T. B. "Studies of the photoperiodism of some economic plants," *J. Agr. Res.*, 37:603–628. 1928.
32. Nishiyama, I. "The origin of the sweet potato plant," *Contribution No. 291 from the Laboratory of Genetics*, Faculty of Agriculture, Kyoto University, Kyoto, Japan, 10 pp. 1961.
33. Park, J. K., Powers, M. R., and Garrison, O. B. "Machinery for growing and harvesting sweet potatoes," *S. Car. Agr, Exp. Sta. Bul. 404*. 1953.
34. Poole, C. F. "Improving the root vegetables," *USDA Yearbook, 1937*, pp. 300–325.
35. Poole, C. F. "Sweetpotato genetic studies," *Hawaii Agr. Exp. Sta. Tech. Bul. 27*. 1955.
36. Poole, W. D. "Harvesting sweetpotatoes in Louisiana," *La. Agr. Exp. Sta. Bul. 568*, pp. 1–24. 1963.
37. Randolph, J. W., and Anderson, W. S. "Sweetpotato production: Mechanical equipment studies," *Miss. Agr. Exp. Sta. Bul. 392*, pp. 1–96. 1943.
38. Schermerhorn, L. G. "Sweetpotato studies in New Jersey," *N.J. Agr. Exp. Sta. Bul. 398*, pp. 1–19. 1924.
39. Smith, C. E., Jr. "The new world centers of origin of cultivated plants and the archaeological evidence," *Econ. Bot.*, 22(3):253–266. 1968.
40. Steinbauer, C. E. "Principle sweetpotato varieties developed in the United States, 1940 to 1960," *Horticultural News*, 43(1). 1962.
41. Steinbauer, C. E. and L. J. Kushman, "Sweetpotato culture and diseases," *USDA Handbk. 388*, pp. 1–74. 1971.
42. Thompson, H. C. and Beattie, J. H. "Group classification and varietal descriptions of American varieties of sweetpotatoes," *USDA Bul. 1021*, pp. 1–30. 1922.
43. Thurber, F. H., Gastrock, E. A., and Guilbeau, W. F. "Production of sweetpotato starch," in *Crops in Peace and War*, USDA Yearbook, 1950–51, pp. 163–167.

Chapter 38

Miscellaneous Forage Crops

GENERAL TYPES

Several forage crops that are neither grasses nor legumes have been grown on limited acreages in the United States. Their production was nearly terminated by 1960 because of their limited value or high production costs.

Cassava (*Manihot esculenta* utilitissima) also called mandioca or yuca, is a perennial root crop native to Brazil, Venezuela, or Mexico.[24] It formerly was grown for forage in Florida and adjacent states. Cassava is an important food plant grown on nearly 10 million hectares chiefly in Brazil, Indonesia, Congo, Nigeria, and Uganda. The world production in 1970 was 90 million metric tons. The plant is a shrubby perennial sometimes more than 4 meters tall with various colored stems. It is propagated from stem cuttings and from sprouts. The flowers are monoecious and cross-pollinated mostly by insects. Cassava roots (Figure 38–1) are the source of tapioca starch which is used for food, adhesives or sizing.

The cortical layers of roots contain varying quantities of linamarin, a cyanogenetic glucoside, and an enzyme which releases the poisonous hydrocyanic acid. The roots are prepared for food often by grating and pressing to remove the juice which contains most of the glucoside. Boiling the roots destroys the enzyme and the linamarin is dissolved in the cooking water, which is discarded.

A plant called burnet (*Sanguisorba minor*), a deep-rooted perennial native to Europe, is utilized occasionally in pastures in California and Oregon. Sacaline (*Polygonum sachalinese*), a member of the *Polygonaceae* (buckwheat) family grown as an ornament, has been occasionally exploited as a forage crop, often under some other name such as Eureka clover.

Spineless cactus seldom was planted for forage because of its recognized slow growth and limited feed value. Nearly spineless forms of the prickly pear cactus (*Opuntia* species), native to Mexico, have long been observed. The wild spiny prickly pear is gathered, the spines usually singed off, and the pear is then fed to livestock in

Figure 38–1. Cassava roots.

emergencies, chiefly in southern Texas. The pears are watery as well as high in crude fiber, but they are satisfactory for feed.[11]

The chufa (*Cyperus esculentus*) is grown occasionally in the United States mostly in Florida. The chufa is a sedge-like plant with creeping rootstalks that produces small tubers or nuts. In some places it is a pernicious weed called yellow nutgrass. Chufas are planted from late spring to midsummer by dropping the nuts 6 to 12 inches apart in rows spaced 2½ to 3 feet apart. The planting rate is about 15 to 40 pounds per acre. The crop is dug with a plow or potato digger, the plants allowed to dry, and then the nuts knocked off with a flail. Most of the chufas are gathered by hogs.[19] The tops are used for hay. Chufas suffer little loss from diseases, but the nuts are infested by the Negro bug.

Comfrey has been exploited in the United States as a productive forage since about 1952, but the acreage is limited by high production costs and its unsuitability for mechanical harvesting. It is soon destroyed by grazing, and the thick leaves and stems with a moisture content of nearly 90 per cent make it difficult to preserve as hay or silage. Comfrey, a member of the *Boragineae* plant family, comprises three species, *Symphytum officinale*, *S. asperrium*, and *S. peregrinum*. It is a perennial that attains a height of 2 to 4 feet, with large hairy leaves and blue, white, or pink flowers. It is propagated by root divisions or root cuttings. Comfrey has a high protein content, but is low in digestibility. Economically, it cannot compete with alfalfa or clover as a forage.

Medical folklore credits comfrey with curative powers for more human ailments than is claimed for any other plant.[16]

JERUSALEM ARTICHOKE

Economic Importance

The Jerusalem artichoke has been grown only to a limited extent in the United States, although it has been advocated[2] as having great possibilities for food, feed, sugar, or alcohol. It was grown in the Pacific Coast states for many years, and later in Nebraska. The Jerusalem artichoke, native to America, was used for food by Indians who gathered the wild tubers or grew them in their clearings. It was found by Champlain on Cape Cod in Massachusetts and taken to France in 1605, where its culture has ranged from 80,000 to 130,000 acres annually. Small quantities of the tubers are produced and sold for food in the United States.

Adaptation

The highest yields of Jerusalem-artichoke tubers in the United States have been obtained in cool, mild, humid sections of the Pacific Northwest. The tops are killed by the first frost in the fall. The crop responds to a good supply of moisture.[18] The plants require a growing season of 125 days, but they do not develop flowers or tubers until the approach of shorter days in August. The Jerusalem artichoke is a short-day plant.[27] The crop is best adapted to rich light or medium loams. Heavy soils make digging of the numerous small tubers rather difficult.

Botanical Description

The name artichoke was early applied to the Jerusalem artichoke because its taste resembled that of the edible bracts of the true artichoke, the globe, French, or bur artichoke (*Cynara scolymus*). The Jerusalem part of the name has no reference to the Holy Land but arose from a corruption of the word *girasol*, the Italian name for sunflower. The Jerusalem artichoke (*Helianthus tuberosus*) grows to a height of 6 feet. It is similar to the wild sunflower except that the center of the head (disk) is light yellow instead of dark brown. The flower heads usually are about 2 to 3 inches across. The tubers (Figure 38–2), which resemble hand grenades, are ovoid but irregular due to knobs and to rings or sections of different diameter. The skin is much thinner than that of the potato and not corky. Well-developed tubers weigh 1 to 5 ounces. Assimilates stored in the stem stimulate growth in the tubers after the leaves cease photosynthesis.[13] Since the tubers

Figure 38–2. Tubers of Jerusalem artichoke.

remain alive in the ground over winter, they maintain the crop as a perennial. The Jerusalem artichoke easily becomes established as a weed because it is extremely difficult to gather all of the tubers.

Many varieties of the Jerusalem artichoke that have tubers with white, yellow, or red skin are grown in Europe.[3]

Cultural Methods

The Jerusalem artichoke cultural, fertilizer, and rotation practices are similar to those for potatoes.[4] Liming increased the yield of sugar per acre.[28] Either whole or cut tubers, preferably the former, may be used for planting. Recommended planting rates for 2- to 3-ounce tubers or cut pieces range from 300 to 1,300 pounds per acre in rows 3½ to 5 feet apart.[3, 15, 17, 18] Large seed pieces produce more stems but do not increase the size of tubers.

The tubers should be planted 4 inches deep as early in the spring as the land can be prepared. When some of the tubers are left during digging, a crop will develop the next year. The best method for eradication of the Jerusalem artichoke is to plow the land in late spring or early summer when the tops are about 1½ feet high and then immediately plant some smother crop.

Harvest

The digging of tubers starts after frost. It is done with a modified elevating potato digger, which also breaks down the stalks and permits the pickers to ride on the machine while sorting and bagging the tubers.[20]

When hogs gather the tubers, they leave enough seed tubers in the ground for a stand the following year. Tubers for planting often are not dug until spring because they keep better in the ground than in storage.[4]

Fresh green tops are suitable for feed, but when mature they are harsh, woody, unpalatable, low in digestible nutrients, and therefore are seldom harvested.

Average yields of tubers in the Midwest and East are about 5 to 6 tons per acre, but yields of 10 tons or more are obtained under favorable conditions.[4]

The tubers keep best in cold storage at a temperature of 31° to 32° F., with a relative humidity of 90 to 95 per cent.[14]

Uses

The Jerusalem-artichoke tuber is used chiefly for animal feed or human food. It is very low in digestibility and watery when cooked. In gross chemical composition the Jerusalem-artichoke tuber is similar to potatoes, but it contains no starch. The carbohydrates are mostly several polysaccharides, chiefly synanthrine,[1] which when hydrolyzed by acids or enzymes produce a simple very sweet sugar called levulose, fructose, or fruit sugar. Levulose often is prescribed for diabetics. Hydrolyzed cornstarch, or potato starch produces dextrose (glucose, or grape sugar). When cane sugar is hydrolyzed, or partly digested, it produces the sweet invert sugar that consists of equal quantities of dextrose and levulose. Honey is composed chiefly of invert sugar. Levulose is 3 to 73 per cent sweeter than cane sugar, but this depends upon who does the tasting.[27] Dextrose is 20 to 50 per cent less sweet than cane sugar. Equal mixtures of levulose and dextrose therefore should be sweet as honey. This is the basis for the hope that by producing and mixing levulose from artichokes and dextrose from grains or potatoes, the domestic sugar requirements of the United States could be supplied at home.[13]

ROOT AND LEAF CROPS

The chief root crops grown for forage include the mangel (or mangel-wurzel, mangold, or stock beet) (*Beta vulgaris*), turnip (*Brassica campestris* var. *rapa*), rutabaga or swede (*Brassica napus* var. *napobrassica*), and carrot (*Daucus carota*).[32] The chief leaf crops for forage are biennial rape (*Brassica napus* var. *biennis*) and thousand-headed kale (*Brassica oleracea* var. *acephala*). These root and leaf crops, except rape, were grown mostly in cool sections of the northern border states to supply succulent feed during the fall, winter, and spring, mostly on farms too small to afford a silo. In northern Europe and parts of Canada where corn does not succeed, the root crops have been important sources of winter feed.

The chief handicaps to the growing of root crops are the labor involved in growing, harvesting, and slicing them, and also the limited period of storage as compared with corn or grass silage.[9] Under favorable conditions, mangels, rutabagas, and turnips yield 20 to 30 tons per acre, whereas carrots yield about one-half as much, and sugarbeets two-thirds as much. Corn silage is higher in digestible nutrients.[8]

Adaptation

The root crops thrive where the mean summer temperature is 60° to 65° F. They do not succeed as summer crops where the mean temperature is much above 70° F. Mangels, turnips, rutabagas, kale, and rape will withstand light freezes. A moist climate, together with deep loam soils and soils high in organic matter are favorable for most root crops. Carrots thrive in rather light friable soils.

Origin

The beet, of which the mangel is a representative, like the turnip, rutabaga, and carrot, is native to Eurasia. The varieties grown in the United States are mostly of European origin or were developed from European varieties. Rape and kale are native to Europe. Turnips and rutabagas were grown for food and feed in the Roman Empire in the first century A.D.

Botanical Description

Mangels differ from sugarbeets (Chapter 35) chiefly in their larger roots, larger acre yields, and smaller sugar content, which is only about one-third or one-half as much. Mangel roots usually extend up above the surface of the soil for about two-fifths their length.

The turnip and rutabaga belong to *Cruciferae*, the mustard family, and the genus *Brassica*, which also includes rape and kale. The plants of this family are biennial or annual pungent-tasting herbs which bear flowers with four petals arranged in the form of a cross. The genus *Brassica* has yellow flowers, six stamens, and globose seeds borne in two-celled pods. The rutabaga can be distinguished from the turnip because it forms a neck on the top of the root to which the leaves are attached. Also, the older leaves of the rutabaga are smoother and of a lighter bluish color, and the roots of most varieties have yellow flesh in contrast with the white flesh of most turnip varieties (Figure 38–3).

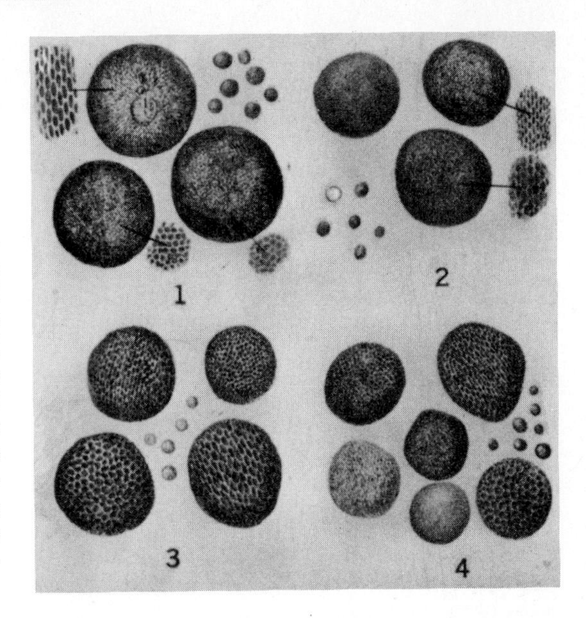

Figure 38–3. Rape seeds: (1) Dwarf Essex (winter) rape (*Brassica napus* var. *biennis*); (2) Oilseed (summer) rape (*B. napus* var. *annua*); (3) Annual turnip rape, or bird rape (*B. campestris*); and (4) Biennial turnip rape, or bird rape (*B. campestris* var. *autumnalis*). The turnip rapes, which also are processed for oil, are closely related to the turnips and the mustards.

Biennial rape plants resemble cabbage when young but later they grow to a height of 1½ to 2½ feet. They produce only leaves and branches without forming a head. Kale plants likewise are branched and leafy but are larger than those of rape.

The varieties of mangels[23] are either red-fleshed, yellow-fleshed, or white-fleshed. Types with long roots are not well suited to shallow soils. Carrot varieties for feed include white and yellow varieties.

Thousand-headed kale is a single variety. A similar crop, but with thick fleshy stems, is called marrow kale or marrow cabbage.[7]

FERTILIZER PRACTICES

For mangels, as well as other root crops, heavy applications of manure and 300 to 400 pounds of superphosphate alone or in addition to manure have been recommended.[8, 33] A complete fertilizer such as 400 to 700 pounds of 4-16-4 or 4-16-8 can be used where manure is not available. Acid soils may require lime.

Cultural Methods

The seedbed for root crops should be prepared as described for sugarbeets in Chapter 35. On wet lowlands planting the crops on ridges or beds is recommended.[8] A fine compact seedbed is essential to good stands.[11]

Rows can be spaced at the desired distance of 18 to 24 inches apart with special planters, sugarbeet drills or grain drills. The seeds should be planted in a seedbed of good tilth, about ⅓ to ¾ inch deep, except for carrots which should be 1/4 to 1/2 inch deep, the shallower depths applying to heavy soils. The usual rates of planting[26, 33] are 6 to 8 pounds per acre for mangels, 1 1/2 to 2 pounds for carrots, and 1 pound for rutabagas or turnips. When labor is available for thinning, the final desired spacing is for mangels, 12 inches; rutabagas, 10 to 12 inches; turnips, 10 inches; and carrots 6 to 8 inches. Broadcast turnips are planted at a rate of 4 to 5 pounds per acre.

Rape is planted at a rate of 2 to 3 pounds per acre in rows or 4 to 5 pounds when broadcast. Thousand-headed kale usually is sown in a bed or cold frame; later the plants are set out about 1 1/2 to 3 feet apart in rows about 3 1/2 feet apart. A pound of seed is sufficient for 4 or 5 acres.[29] When sown directly in the field to be thinned later, 8 to 12 ounces of seed per acre usually are ample.

Mangels and carrots should be planted in the spring as soon as the danger of heavy freezes is past, or a little before corn-planting time. Turnips and rutabagas may be planted in early spring for a summer crop, in the summer for a fall crop in cool climates only, and in early fall in the milder sections of the South and Pacific Coast when a winter crop is desired. Rape likewise may be sown in the spring, summer, or fall, depending upon the region as well as on the period in which the crop is to be pastured or fed. Early spring planting is recommended in Missouri.[12] In cool sections, rape often is sown in the summer as a catch crop or is sown with grain in the spring and then is pastured in the fall or late summer. Thousand-headed kale seed is planted in the spring; the plants are set out usually in June and are cut during the fall and winter as needed for feeding in a fresh condition.

Harvest and Storage

The root crops are usually harvested before a heavy freeze, or as soon as the leaves wither and turn yellow, an indication that growth has ceased. They are pulled by hand either directly or after they have been loosened with a beet lifter or plow. The tops can be twisted or cut off after pulling, or topped with a sharp hoe before pulling. The crown is not cut off as with sugarbeets. The harvested roots usually are stored in root cellars or outdoor pits. The best storage conditions are those described for potatoes, namely, a temperature of 36° to 40° F., ample ventilation, and humidity high enough to avoid excessive shrinkage. Turnips do not keep as long as mangels, rutabagas, or carrots. Consequently, they should be fed up first. When kept for three months in a root cellar in West Virginia without any special control of temperature, ventilation, or humidity, mangels lost considerable water and protein, while rutabagas lost water, carbohydrates, and some protein.[22]

Uses

Root crops replace silage as a succulent winter feed. They may be fed in quantities of 1 to 2 pounds daily per 100 pounds of live weight of the animal.[33] Large roots may be fed whole to poultry but should be sliced or chopped for livestock. Mangels, rutabagas, turnips, and carrots contain roughly 90 per cent water. In mixed rations for hogs, 100 pounds of grain are saved by each 440 to 780 pounds of roots fed.

Dwarf Essex (biennial) rape is grown in the United States and Canada largely as pasture for cattle, hogs, or sheep, being nearly equal to alfalfa for that purpose.

Pests

Root crops suffer relatively little damage from diseases in the regions where they are generally grown. The chief damage is from rots in storage.

While the root crops usually suffer relatively little damage from insects in the chief production areas, they occasionally are attacked by cutworms, grasshoppers, aphids, flea beetles, cabbage worms, and root maggots. These can be controlled with suitable insecticides.

PUMPKIN AND SQUASH

Economic Importance

The pumpkin and squash formerly were important crops for fall and winter feeding of livestock in the United States. By 1959, these crops were rarely grown for feed. The former practice of interplanting pumpkins with corn interferes with mechanized field operations.

Adaptation

The field or stock varieties of pumpkin and squash are adapted to humid conditions.[30, 31] They also thrive under a wide range of temperatures, but lack tolerance to frost or prolonged exposure near freezing. The large varieties require 110 to 120 frost-free days to reach maturity. They thrive in rich, well-drained, light soils. They prefer full sunlight but develop satisfactorily in a cornfield.

Botanical Description

The pumpkin and squash belong to the genus *Cucurbita*, of the family *Cucurbitacea* or the gourd family, which bears a fleshy fruit called a pepo. The large showy yellow flowers are monoecious.

The common pumpkin belongs to the species *Cucurbita pepo;* the Cushaw (or crookneck) pumpkin is *C. mixta*, the Kentucky Field pumpkin is *C. moschata*, and the large yellow varieties of field pumpkins (or squash) are *C. maxima*. The *C. pepo* species includes some small table varieties of pumpkins and summer squash. The *moschata* species include the Winter Crookneck and Butternut varieties of squash. The *maxima* species includes the Hubbard and Banana winter squashes. Thus the distinction between a pumpkin and a squash is rather confusing.

The peduncle of *C. pepo* is hard, angular, and ridged; the *C. moschata* peduncle is hard, smoothly angular, and flared; the *C. mixta* peduncle is hard, basically angular, and enlarged by hard cork; the *C. maxima* peduncle is soft, basically round, and enlarged by soft cork.[6, 25, 30, 34]

These species are normally cross-pollinated but do not show a loss of vigor when inbred. They usually are insect-pollinated. The different species do not intercross readily except with *C. moshata*. They do not cross with other cucurbits.

Both the pumpkin and squash originated in the Western Hemisphere and were cultivated by the Indians.[34]

Cultural Methods

Field varieties of pumpkin and squash are planted in hills 10 to 12 feet apart, 3 to 6 seeds per hill, and later thinned to 1 or 2 plants. On soils of low or moderate fertility, it is helpful to stir in a forkful of manure at each hill, leaving a cover of about 2 inches of unmixed soil in which the seed is planted. Planting is done after the danger from frost is past. The crop requires clean cultivation until the trailing vines start to run and the ground is partly shaded by the large leaves.

Harvest

For best keeping quality, field pumpkins and squashes should be harvested upon the approach of the first frost. Light frost kills the vines but does not damage the fruits appreciably. Field pumpkins and squash keep best when a short section of the stem is left on the fruit when it is cut from the vine. Careful handling at harvest is essential to prevent decay in storage. The best storage conditions for pumpkins and squash are a temperature of 50° to 55° F. with a relative humidity of 70 to 75 per cent. Such conditions often are available in a livestock barn or a house cellar. A preliminary curing for about two weeks at 80° to 85° F. will ripen immature specimens and aid in healing mechanical injuries. Field pumpkins and squash often do not keep more than two or three months.

Uses

Sliced or chopped pumpkins and squashes are fed as succulents during late fall and early winter. Yields of squash in the United States average about 9 tons per acre. Pumpkins grown among corn stalks yield about 1½ tons per acre.

REFERENCES

1. Bates, F. J. "Discussion," *Proceedings 2nd Dearborn Conference of Agriculture, Industry, and Science*, Farm Chemurgic Council and the Chemical Foundation, Inc., Dearborn, Mich., p. 159, 1936.
2. Boswell, V. R. "Growing the Jerusalem artichoke," *USDA Leaflet 116*, pp. 1–8. 1936.
3. Boswell, V. R., and others. "Studies of the culture and certain varieties of the Jerusalem artichoke," *USDA Tech. Bul. 514*, pp. 1–70. 1936.
4. Burlison, W. L. "Growing artichokes in America," *Proceedings 2nd Dearborn Conference of Agriculture, Industry, and Science*. Farm Chemurgic Council and the Chemical Foundation, Inc., Dearborn, Mich., pp. 111–120. 1936.
5. Carrier, L. *The Beginnings of Agriculture in America*, McGraw-Hill, New York, pp. 1–323. 1923.
6. Castetter, E. F., and Erwin, A. T. "A systematic study of squashes and pumpkins," *Iowa Agr. Exp. Sta. Bul. 244*, pp. 107–135. 1927.
7. Chapin, L. J. "Thousand-Headed kale and marrow cabbage," *Wash. Agr. Exp. Sta. Spec. Bul. 6*. 1912.
8. Delwiche, E. J. "Profitable root crops," *Wis. Agr. Exp. Sta. Bul. 330*, pp. 1–22. 1921.
9. Fraser, S., Gilmore, J. W., and Clark, C. F. "Root crops for stock feeding," *Cornell U. Agr. Exp. Sta. Bul. 243*, pp. 45–76. 1906.
10. Fraser, S., Gilmore, J. W., and Clark, C. F. "Culture and varieties of roots for stock feeding," *Cornell U. Agr. Exp. Sta. Bul. 244*, pp. 77–122. 1907.
11. Griffiths, D. "Prickly pear as stock feed," *USDA Farmers Bul. 1072*, pp. 1–24. 1920.
12. Hutcheson, C. B. "Growing rape for forage," *Mo. Agr. Ext. Serv. Circ. 3*, pp. 1–4. 1915.
13. Incoll, L. D., and Neales, T. F. "The stem as a temporary sink before tuberization in *Helianthus tuberosus* L.," *J. Exp. Bot.*, 21(67):469–476. 1970.
14. Jackson, R. F., Silsbee, C. G., and Proffitt, M. J. "A method for the manufacture of levulose," *J. Indus. and Eng. Chem.*, 16:1250–1251. 1924.
15. Johnson, H. W. "Storage rots of the Jerusalem artichoke," *J. Agr. Res.*, 43(4):337–352. 1931.
16. Kadans, J. M. *Modern Encyclopedia of Herbs*, Parker Publ. Co., West Nyack, N.Y., pp. 1–256. 1970.
17. Kiesselbach, T. A. "Should more Jerusalem artichokes be grown?" *U. Nebr. Agr. Ext. Circ. 108*, pp. 1–4 (mimeographed). 1937.
18. Kiesselbach, T. A., and Anderson, A. "Cultural tests with the Jerusalem artichoke," *Jour. Amer. Soc. Agron.*, 21(10):1001–1006. 1929.
19. Killinger, G. B., and Stokes, W. E. "Chufas in Florida," *Fla. Agr. Exp. Sta. Bul. 419*, pp. 1–16. 1946.

20. Lehmann, E. W., and Shawl, R. I. "New method is devised for harvesting artichokes," *47th Ann. Rpt. Ill. Agr. Exp. Sta.*, p. 200. 1934.
21. McRostie, G. P., and others. "Field roots in Canada. Classification, improvement and seed production," *Dom. Dept. Agr. Bul. 84,* Ottawa, pp. 1–51. 1927.
22. Morrow, K. S., Dustman, R. B., and Henderson, H. O. "Changes in the chemical composition of mangels and rutabagas during storage," *J. Agr. Res.*, 43(10):919–930. 1931.
23. Piper, C. V. *Forage Plants and their Culture,* rev. ed., Macmillan, New York, pp. 1–671. 1924.
24. Rogers, D. J. "Some botanical and ethnological considerations of *Manihot esculenta,*" *Econ. Bot.*, 19(4):369. 1965.
25. Russell, P. "Identification of the commonly cultivated species of *Cucurbita* by means of seed characters," *Wash. Acad. Sci.,* 14:265–269. 1924.
26. Schoth, H. A. "Root crop production for livestock," *USDA Leaflet 410.* 1957.
27. Shoemaker, D. N. "The Jerusalem artichoke as a crop plant," *USDA Tech. Bul. 33,* pp. 1–32. 1927.
28. Sprague, H. B., Farris, N. F., and Colby, W. G. "The effect of soil conditions and treatment on yields of tubers and sugar from the American artichoke," *J. Am. Soc. Agron.*, 27(5):392. 1935.
29. Stookey, E. B. "Kale and root crops," *Wash. Agr. Exp. Sta. Monthly Bul.*, 8(1):8–10. 1920.
30. Tapley, W. T., Enzie, W. D., and Van Eseltine, G. P. *The Vegetables of New York,* Vol. 1, Part IV, *The Cucurbits,* N.Y. Agr. Exp. Sta. Albany, pp. 1–131. 1937.
31. Thompson, R. C. "Production of pumpkins and squashes," *USDA Leaflet 141,* pp. 1–8. 1937.
32. Westover, H. L., and Schoth, H. A. "Experiments in growing roots as feed crops," *USDA Tech. Bul. 416,* pp. 1–15. 1934.
33. Westover, H. L., Schoth, H. A., and Semple, A. T. "Growing root crops for livestock," *USDA Farmers Bul. 1699,* pp. 1–12. 1933.
34. Whitaker, T. W., and Bohn, G. W. "The taxonomy, genetics, production, and uses of the cultivated species of *Cucurbita,*" *Econ. Bot.*, 4(1):52–81. 1950.

Chapter 39

Miscellaneous Industrial Crops

HOPS*

Economic Importance

Hops are harvested on about 63,000 hectares, with a production of nearly 96,000 metric tons or 1.5 metric tons per hectare. The leading countries in hop production are West Germany, the United States, the United Kingdom, Spain, the U.S.S.R., and Yugoslavia. From 1970 to 1972 the United States produced an average of nearly 49 million pounds (over 22,000 metric tons) on nearly 29,000 acres with a yield of 1,700 pounds per acre (1.9 metric tons per hectare). The crop was produced by some 225 growers in Washington, Oregon, Idaho, and California.

History

The hop is thought to be indigenous to the British Isles,[65] but cultivated hops were grown in Germany in A.D. 768, and in England late in the 15th century.[81] Hops were taken to America as early as 1629. The first commercial hop yard was established in New York in 1808. Cultivation of hops later extended into Wisconsin, but before 1869 they had become established in the Pacific Coast States.

Adaptation

Hops are grown under a wide range of climatic conditions, but mostly they are cultivated in mild regions with abundant early rainfall, or irrigation, followed by dry, warm weather. Such conditions are found in the Pacific Northwest, the valleys of the northern half of California accessible to cool ocean breezes, western New York, and northern and central Europe. The preferable soils are deep sandy or gravelly loams.[11]

* For more complete information on hops see Burgess, A. H., *Hops: Botany, Cultivation, and Utilization*, Interscience, New York, 1964, pp. 1–300.

Botanical Description

Hops belong to the tribe or subfamily *Cannabineae* of the family *Urticaceae*, the nettle family. The hop plant (*Humulus lupulus*) is a long-lived perennial.

VEGETATIVE GROWTH: Hop stems are herbaceous, roughened, angular, and hollow. Their color among different varieties is purplish-red, pale green, or green streaked with red. The stems above ground bear thin opposite lateral branches. The slender trailing or twining vines may grow to a length of 25 to 30 feet. The stems, as well as the leaf petioles, have several lines of strong hooked hairs which help them cling to their supports. After being cut back each year, new stems arise each spring from buds on the underground rhizomes.

INFLORESCENCES: The hop is a dioecious plant with sterile flowers in racemes or panicles, and fertile (pistillate) flowers in clusters or catkins. Occasional plants are monoecious. The commercial hops borne on the female plants are the spike-like pistillate inflorescences[48] that somewhat resemble fir cones in shape (Figure 39–1). The cones, botanically, are called strobiles, but colloquially they are referred to as beer blossoms. The strobiles, borne on lateral branches of the main stem, are 1 to 2½ inches in length.

The staminate flowers are about 6 millimeters in diameter. Since the male vines produce no hops, large numbers of them are undesirable. In a crop, 1 or 2 per cent male plants uniformly distributed

Figure 39–1. Machine-picked hops. [Courtesy E. Clemens Horst Co.]

throughout the hop yard will provide sufficient pollen for fertilization of the pistillate plants.[11]

FRUIT AND SEED: The fruit of the hop is purple, oval, and about the size of a mustard seed. The hop seed possesses a curved embryo and a very small amount of endosperm.[82] The crop is seldom reproduced from seed because of the great variability in the progeny. Seedless (unfertilized) hops weigh about 30 per cent less than those that develop seed. Their quality is better than that of seeded hops but the yield is smaller. Hop seeds contain a bitter, unpalatable oil.[54] Seedless hops are now being grown to a considerable extent because of their higher market value, by excluding all male plants.

The outer surface of the bracteoles, perianth, and bases of the bract-like stipules are covered with yellow pollen-like resinous grains. These latter are the lupulin or hop meal. The commercial value of well cured hops is based upon the amount and quality of the lupulin.

The principal active ingredients in the lupulin glands are:

1. Hop oil. A volatile essential oil with a characteristic aroma; hop oil constitutes about 0.2 to 0.8 per cent of the active lupulin-gland ingredients.
2. Nonresinous bitter principle.
3. Resins. Of these, the hard resin is practically tasteless and is devoid of brewing value. The soft alpha and beta resins are bitter. They impart this taste to the beer wort, and also have a preservative effect. Hops contain 10 to 18 per cent or more total resins.
4. Tannin, which is also bitter, constitutes 4 to 5 per cent of the hop.

The alpha-acids in the resins, which are the most important bitter principles, constitute from 4½ to 7 per cent of the weight of the dried hops in the important domestic varieties and 8 to 12 per cent in some English varieties.

Cultural Methods

Hops are propagated by cuttings from the numerous runners (rhizomes) sent out by the plant just below the surface of the ground. The excess shoots are pruned once a year, usually in the spring. These are cut into pieces 6 to 8 inches long that bear at least two sets of buds, which are planted promptly therafter.[11]

The hops are set in hills 6½ to 8 feet apart, with two to four cuttings in each hill. Very few strobiles are produced the first year. In fact, the vines do not reach full bearing until the second or third year.

Hop yards, once established, often are maintained for many years, missing hills being replaced at pruning time each spring. Winter cover crops such as vetch or small grain are grown between the hop rows to be plowed under in the spring. Fertilizers that are applied usually consist of 80 to 160 pounds of nitrogen per acre[8] and 50 to 100 pounds of P_2O_5. Most soils in the present domestic hop-growing areas contain ample potash, but sulfur, zinc, or boron may be needed.[11] These elements are supplied by applications of calcium sulfate, zinc sulfate, and borax.

Hop plants are trained on high wire trellises, usually about 18 feet above the ground (Figure 39–2). Heavy wires attached to upright poles set four hills apart are stretched across the field. Lighter trellis or "stringing" wires are strung across the heavy wires over each row, or over the middle between two rows. The trellis wires are hung under the cross wires with hooks where hops are picked by hand. Large end poles at the borders of the field are slanted outward and well anchored. Coir twine tied to the trellis wires and pegged to the ground at or near each hill provides strings for the vines to climb. The vines wind clockwise around the strings.[9]

Figure 39–2. Hops being harvested with a mobile machine. [Courtesy Oregon State University.]

Harvest

Ripe hops have the best quality. The strobiles or cones change from a bright green to a bright yellowish green. They become more resilient, sticky, crisp, and papery, while the pericarp on the hop fruits takes on a dark-purple color. The characteristic lupulin odor becomes very marked as hops approach ripeness.

By 1974 nearly all (domestic) hop vines were cut and loaded into trucks by mobile cutting machines that move down the rows, with men on elevated platforms cutting the vine tops free from the trellises (Figure 39–2). The cut vines are hauled to a stationary picker adjoining a drier. The picking machines strip the hops and most of the leaves from the vines with fingers attached to rotating drums. These then pass through a forced air stream to remove most of the leaves and stems and are conveyed onto a series of inclined "dribble belts" which carry the trash upward while the cleaned hops roll off the belt (Figure 39–3). The hops are then conveyed to the drier.

Curing

The drier reduces the 65 to 80 per cent moisture content of the hops down to 8 to 10 per cent by blowing heated air (140° to 150° F.) through layers of hops on a moving belt or in stationary chambers. The dried hops are then moved to a cooling chamber where they remain for 5 to 7 days or longer.[9] The hops are then pressed into bales.

Most of the hops are bleached and conditioned by the burning of 1 to 4 ounces of sulfur per 100 pounds of green hops, to produce the sulfur dioxide fumes which are blown up through the hops during drying.

Uses

The principal use of hops is in the manufacture of beer. The sweet beer wort is boiled with the hops. The flavor of the wort is improved by the extraction of the active ingredients of the hops. An appreciable portion of the crop is now sold as the extract from the dried hops. The essential oil of the lupulin glands imparts an aroma to the beer. The tannin probably serves to precipitate albuminous substances. The malic and critic acids in the hops tend to increase the acidity of the wort. About 8 to 13 ounces of hops are used for each barrel of beer.[82]

Diseases

Downy mildew is the most serious disease of hops in the world. This disease is caused by a fungus (*Pseudoperonospora humuli*). The disease attacks the shoots,[44] stems, and flower parts. It is spread in damp weather by wind, and from cuttings. Downy mildew may be

Figure 39–3. Hand-cut hop vines entering a stationary picker. [Courtesy Oregon State University.]

controlled by destroying the cut vines, escaped plants, and affected shoots, as well as by spraying or dusting. Stripping all leaves from the stems from the bottom up to within 4 feet of the trellis wire reduces the spread of downy mildew as well as of the red spider and the hop aphid. Sprays should be applied (1) when the vines are first trained, (2) when the vines get to the cross wires, (3) just before the female flower buds open, and (4) at other times when weather conditions are favorable to spread of the disease. Streptomycin sprays are effective against downy mildew.[45]

Sooty mold (*Fumago vagans*) blackens the cones. The infection is carried into the cones by aphids. Control of aphids largely prevents the trouble. Powdery mildew (*Sphaerotheca humuli*) is controlled by sulfur dusting. Plants affected by virus diseases, crown gall, and rots should be removed from the field and be replaced by healthy cuttings.

Insects

The chief insect that attacks hops is the hop aphid (*Phorodon humuli*). It is controlled by stripping the lower leaves from the vines and by insecticidal sprays and dusts. Other insects that attack hops include the cucumber beetle, thrips, leaf tier, hop-plant borer, hop butterfly, hop fleabeetle, and webworm.

The red spider is controlled by stripping the lower leaves or by dusting with malathion or other insecticides.

MINT

Economic Importance

Peppermint was harvested from an average of about 67,000 acres from 1970 to 1972, mostly in Oregon. Other producing states were Washington, Idaho, Indiana, Wisconsin, and Michigan. The average yield of oil was 58 pounds per acre with a production of nearly 3.9 million pounds. Spearmint was harvested on over 29,000 acres in Washington, Indiana, Idaho, Michigan, and Wisconsin. The oil production was nearly 1.9 million pounds, or 63 pounds per acre.

History

The distillation of mint oil for use in medicines was practiced in Egypt as early as A.D. 410. The Japanese were distilling mint oil for an eyewash as early as A.D. 984. Production began in England about 1696, and in the United States, at Ashfield, Massachusetts, in 1812. Production extended to Michigan in 1835, to Indiana about 1855, and to the Pacific Coast States over a half-century later.

Adaptation

Mint is grown mostly in humid sections of the northern states, where summer temperatures are moderate and the long summer days promote a high oil content. Soils peculiarly suited to mint production are deep, rich, well-drained soils, especially muck, or other soils that are high in organic matter. The muck soils usually are somewhat acid. The Houghton muck of northern Indiana has a reaction of *p*H 5.2 or less.[29] However, mint is grown in the alkaline, irrigated sandy loam soils of the Yakima Valley of Washington. The preferred *p*H range is 6.0 to 7.5. Peppermint thrives on alluvial sandy loam soils in western Oregon, and on irrigated lands in central Oregon.

Botanical Relationships

Mint belongs to the family *Labiatea*, the mint family, the plants of which have a four-sided or square stem. The three perennial species grown commercially are peppermint (*Mentha piperita*), spearmint (*M. spicata* or *M. viridis*), and Scotch spearmint (*M. cardiaca*). Japanese mint (*M. arvensis* var. *piperascens*) is grown to a limited extent in the United States. The inconspicuous flowers, which are borne in whorls on a spike about 2 inches long, produce very few seeds. The plants produce numerous rhizomes which are used for propagation or for field planting. Peppermint (Figure 39–4) plants usually are 3 feet or less in height.

Three types of peppermint are (1) black peppermint (black mint, English peppermint, or Mitcham mint), which has dark-green to purple stems and deep-green, broadly lanced, slightly toothed leaves, and, usually, light-purple flowers, (2) American peppermint (American mint or State mint) which has green stems and lighter green leaves, and is similar to black mint otherwise but not very satisfactory commercially, and (3) white peppermint, which is grown in England.[36]

Spearmint has light-green, pointed, spear-shaped leaves and pointed flower spikes that are longer than those of peppermint. Spearmint is the common home-garden plant grown for its leaves which give the mint flavor as well as the green decoration to drinks and foods (Figure 39–4).

Native species of mint, which include *M. canadensis*, contain oil of an inferior quality. They should be eradicated where they occur as weeds in commercial mint fields. Catmint, or catnip (*Nepeta cataria*), a member of the mint family, was grown on 40 acres in North

Figure 39–4. Flowers and leaves of peppermint (*left*); flowers, leaves, and rhizomes of spearmint (*right*).

Carolina in 1972. This native of Europe which escaped over the United States provides pleasure for cats.

Cultural Methods

Lands planted to mint should be comparatively free from weeds,[26] because many species of weeds affect the color or odor of the distilled oil. Ragweed is the most objectionable weed because it contains a light, volatile oil that distills over and contaminates the mint oils with a disagreeable flavor. A mixture of 1 per cent ragweed in spearmint discolors the oil and renders chewing gum unsalable.[29]

Other weeds such as smartweed, pigweed, horseweed, nettles, hemp, purslane, lambsquarter, buttermold, and Canada thistle impart disagreeable odors. Since their oils are heavier than water, they separate from mint oils after distillation. Clean land is best attained by planting the mint after fallow or after an intertilled crop such as potatoes or onions, which also are adapted to muck soils. Perennial grasses in mint fields are difficult to eradicate by cultivation after the mint plants spread between the rows, but they may be suppressed by earlier applications of herbicides. The herbicides should be applied and worked into the soil immediately after plowing the mint field in the spring.

Fertilizer application on muck soils usually consists of 250 to 500 pounds per acre of a 5–20–20 mixed fertilizer. Irrigated sandy loams in eastern Washington or eastern Oregon receive about 120 pounds of nitrogen per acre. The soils in western Oregon need about 200 pounds per acre of nitrogen from mineral fertilizers or well-rotted manure. Potassium, and sometimes phosphorus, also are beneficial.

The fields for mint should be prepared in time for early spring planting of the stolons. Sometimes strips of rye or other tall crops or permanent willow windbreaks are planted at intervals across the field to prevent drifting sand particles from damaging the young mint plants.

The fields that supply the rhizomes should have been cut early the previous year so the runners have time to build up adequate plant-food reserves. The rhizomes are plowed out, forked free from soil, and thrown into piles, and the piles covered lightly to prevent excessive drying. One acre produces enough rhizomes to plant ten acres or more.

The runners are planted with special machines or dropped into the furrows by hand, lengthwise and end to end. About 40 bushels of loosely packed rhizomes are required for planting an acre. Mint often is propagated by pulling up young rooted plants that arise from the rhizomes in established mint fields in the spring. These are set out with vegetable or tobacco transplanting machines.

A rotary hoe or a light weeder or harrow is used for cultivation until the mint plants are 5 to 6 inches high. Rubber-toothed harrows often are used to cultivate mint fields also. Row cultivation, after the first year, is precluded by the spreading of the plants between the rows. Spring tooth or other harrows or the rotary hoe may be used early in the spring, followed by hand weeding later in the season. After harvest, the fields are left until late fall or early spring when they are plowed. Fall plowing covers the rhizomes enough to protect them from winter cold. The plants come up from rhizomes each year. Fall plowing is not recommended in the coastal districts where the land is subject to overflow during the winter or early spring.

Harvest

The maximum oil yield of both peppermint and spearmint occurs about when the plants are in bloom. In Indiana the oil and menthol yield continue to increase up to the stage of 50 per cent bloom,[29] but a better quality of oil sometimes is obtained by cutting before blooming begins.[26] The crop must be cut before leaf shedding occurs because the oil is contained in small glands on the lower leaf surface. The crop usually is harvested after the plants are well in bloom in Oregon and Washington, except where cutting in early bloom permits the cutting of two crops in a season. Cutting is done in July, August, and September. The mint is cut with a mower equipped with lifter guards. It is raked a day or two later after it is partly dried. The crop is dry enough for distilling when the moisture content is down to 40 to 50 per cent, but it must be gathered and hauled to the still before the leaves are dry enough to shatter. The partly dried mint is gathered and chopped with a forage harvester and blown into a trailer or into portable distilling tubs,[36] or gathered from the swath with a hay loader and unloaded into the distilling tub with slings (Figure 39–5).

Figure 39–5. Gathering mint from windrow and loading into a distilling tub.

Distillation of the Oil

The oil is distilled from the mint with steam (Figure 39–6). Distillation is completed in 45 to 60 minutes. The vapors are condensed in water-cooled coils and passed into a tank receiver. About 6 gallons of water boil over with each gallon of oil. The oil, which is in the top layer, is drawn off into metal cans or drums.

Figure 39–6. Unloading mint into a distillation vat.

Uses

Peppermint oil should consist of 50 per cent or more menthol and 4 to 9 per cent esters. Peppermint oil is used in ointments, salves, and medicines as a flavoring, and also as a remedy for digestive disturbances. It has an important use in confections, tooth pastes, tooth powders, and perfumes. Spearmint oil contains at least 50 per cent carvone or carvol. The chief use of spearmint oil is in chewing gum, mint sauces, and mint flavors. A substantial portion of the mint oils is evaporated for use as a dry ingredient.

Crop Pests

The chief fungus diseases of mints, and their causal organisms, are Verticillium wilt (*Verticillium albo-atrum*), rust (*Puccinia menthae*), and anthracnose or leopard spot (*Sphaceloma menthae*). The wilt is controlled with resistant varieties and by postharvest flaming that kills the spores without harming the rootstocks. Rust and anthracnose are reduced by plowing under all mint residues or by applying fungicides. Rusted, early emerging mint shoots may be killed by herbicides or by flaming to reduce the amount of later infection. Powdery mildew, leaf spots, root rots, and nematodes also attack mints.[56]

The mint looper (*Rachiplusia ou*), other loopers or caterpillars, cutworms, and the mint fleabeetle (*Longtarsus waterhousi*) are the chief insect pests of mint.[26]

DILL

Dill is grown for seed in the United States on a few hundred acres, mostly in Oregon. Additional supplies are imported.

Dill (*Anethum graveolens*) belongs to the family *Umbelliferae*, the parsnip family. The dill plant is a perennial that develops flowers and seeds the year of seeding when the time of sowing is early enough so that the cold temperature requirements are satisfied. The plants reach a height of 3 to 4 feet.

Although a native of the Mediterranean region, dill is well adapted to the relatively cool climate of the northern United States, where it grows well on fertile soils. Dill is sown in cultivated rows 1½ to 3 feet apart either in early spring or so late in the fall that the plants do not emerge until early spring. The plants are later thinned so they are 6 to 15 inches apart in the row. The seed crop is cut with a windrower or mower when the first seeds are ripe, but while the plants are damp enough to avoid excessive losses from shattering. The crop is partly cured, after which it is threshed, all possible precautions being used to prevent shattering of the seed. The seed is then spread out in shallow piles, with occasional stirring, until dry.

When being harvested for oil production, the dill is cut when the most nearly ripe seed is turning brown. After partial field curing, the crop is hauled to the still where it is handled about as described for the distillation of mint oil. The yield of seed under good cultural conditions is reported to average about 500 to 700 pounds per acre. Oil yields are approximately 20 pounds per acre.

Dill is grown commercially chiefly for the fruits or seeds that are used to flavor foods, chiefly pickles, or for the volatile oil distilled from the harvested plants that is used for similar purposes, and in perfumes and medicines. The leaves and seed heads, often grown in herb gardens, also are used to season prepared foods.[104]

WORMSEED

The culture of wormseed constitutes a unique localized industry that has persisted for many years in Carroll County, Maryland. Wormseed is grown on a few acres for the production of wormseed (or Baltimore) oil.

Wormseed (*Chenopodium ambrosioides* var. *anthelminticum*) belongs to the *Chenopodiacea* or goosefoot family. The genus *Chenopodium* includes such plants as *C. album*, the ordinary lambsquarters weed, and *C. quinoa*.

Wormseed is known also as Chenoposium, American wormseed, and Jerusalem oak. The plant, a native of Europe, has become naturalized in waste places from New England to Florida and westward to California. It is a many-branched annual herb, 2 to 4 feet in height. The entire plant, including the fruit and seed, contains a volatile oil of disagreeagle odor that is used in worm remedies for man and beast. The seeds also are used in vermifuges.[79] The oil contains more than 65 per cent ascaridole, the active principle that kills several internal parasites.

Wormseed is sown in beds in the spring and transplanted to the field when the plants are about 4 inches high spaced about 18 inches apart in rows approximately 3 feet apart. The crop is harvested about the middle of September, usually with a modified binder. The rather loose bundles are shocked and then hauled to the still when partly dried. The oil is distilled out into a long trough type of condenser. Distillation is much like that described for mint. Yields of oil range from 30 to 90 pounds per acre.[80] A federal permit must be obtained to operate any still.

QUINOA

Quinoa (*Chenopodium quinoa*) was grown in the Andes region of Argentina, Peru, Bolivia, and Equador on some 35,000 hectares in 1970, with a seed production of about 25,000 metric tons. There it is called quinua (keen-wha). The seeds are consumed as a food substitute for cereals. Attempts to produce quinoa in the Rocky Mountain region of the United States have been disappointing.

WORMWOOD

Wormwood, absinth, or madderwort (*Artemisia absinthium*) formerly was produced in Michigan and Oregon. The plant is a hardy long-lived perennial, 2 to 4 feet in height, related botanically to sagebrush. The hairy, silvery shoots bear grayish-green leaflets and small yellow flower clusters. The seed is planted in a bed or directly in the field. The plants also can be propagated by division. Wormwood is harvested once a year with a mower or windrower.[79]

The oil distilled from the leaves and tops was used in tonics and external proprietary medicines. It is no longer used as a worm remedy (anthelmintic) because of its toxic properties when taken internally.

MUSTARD

Economic Importance

Nearly 7 million pounds of mustard seed was produced in 1964 in Montana, Minnesota, California, and other states. The yield was 421 pounds per acre on 16,363 acres.

Adaptation

Mustard seed is best adapted to sections with relatively cool summer temperatures, fair supplies of soil moisture during the growing season, and dry harvest periods. Such conditions are found in the "triangle" section of north-central Montana, and also in the low-rainfall but foggy coastal valley sections of southern California. In Montana, the soils where mustard is grown chiefly are medium loams that are slightly alkaline. In California, the brown mustard is grown mostly on sandy loam, while the yellow mustard is produced on either a heavy type of sandy loam or a light adobe soil.

Botanical Relationships

Mustard belongs to the family *Cruciferae* (or *Brassicaceae*), genus *Brassica*, described in Chapter 38. The species grown for their seeds are all annual herbaceous plants about 2 to 3 feet in height. They include (1) *Brassica alba*, the yellow or white mustard (Figure 39–7) (2) *B. juncea*, the brown mustard, and (3) *B. nigra*, the black mustard. In California, the brown mustard yields 500 to 1,500 pounds and the yellow mustard 250 to 1000 pounds per acre. The mustards that occur as weeds, the seeds of which are sometimes processed, include the white mustard and black mustard as well as the charlock or wild mustard, *B. arvensis;* ball mustard, *Neslia paniculata;* and, sometimes, other species.

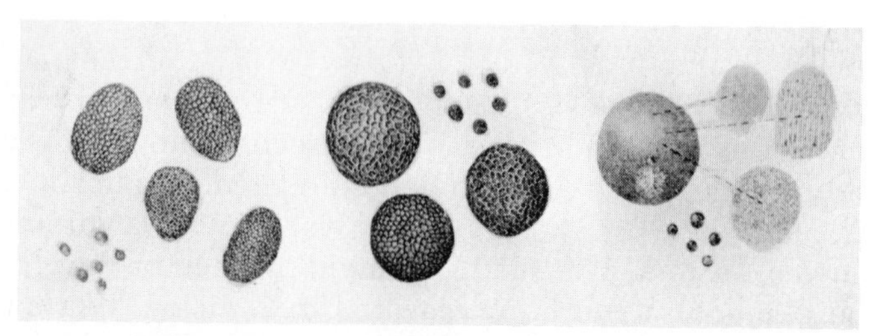

Figure 39–7. Seeds of black mustard (*left*), yellow mustard (*center*), and wild mustard or charlock (*right*).

Cultural Methods

The cropping system, seedbed preparation, and seeding date for mustard seed are all very similar to those for small grains in any particular region. Thus, in Montana, most of the mustard is sown in the spring on fallow land or as a catch crop where winter wheat has winterkilled. In California, the mustard is sown from January to March on fall-plowed land. It is sown as a companion crop with clover in Minnesota.

The seed is sown with a grain drill about 1 inch deep at a rate of 3 to 4 pounds per acre.

Mustard is harvested just before the pods open in order to avoid heavy losses from shattering of seed. It is cut with a swather and then threshed with a combine equipped with a pick-up attachment.

Uses

As a condiment, the chief use of mustard is the ground seed or paste spread on cooked meats and other foods. The counter-irritant properties of the nondrying mustard oil are the basis for the familiar mustard plaster as well as other medicinal applications. The brown mustard is best for medicinal uses. Whole seeds of yellow mustard are used for seasoning pickles.

Rape

Rape seed (Figure 39–8) is produced on an average of 9.8 million hectares, chiefly in India, China, Poland, Canada, and France. The production from 1970 to 1972 was about 7.3 million metric tons, or over 0.7 metric tons per hectare. The production in Canada on 1 to 2 million hectares is mostly in Saskatchewan, Alberta, and Manitoba. In the United States rape is harvested on about 1000 hectares, mostly in Montana and Minnesota, with a yield of one metric ton per hectare. The rape species belong to the genus *Brassica* (see Chapter 38). Of the species listed in the legend of Figure 38–3, the oilseed or summer rape, *B. napus* var. annua is the most important. The methods of producing seed rape are similar to those for mustard.

The oilseed (summer) rape is processed for oil (3a). The oil also is extracted from bird and turnip rape seeds that are not used for feeding caged birds. The world production of rapeseed oil in 1973 was about 2,480 thousand metric tons. It is used in the production of rubber products, lubricants, continuous casting of steel, brake fluids, and in tempering steel plates. It is an important edible vegetable oil in many countries other than those of North America, where it also is used as lamp fuel and for making soap.

The oil is extracted by volatile solvents or by cold pressing. The oil

Figure 39–8. Mature plants of seed rape (*Brassica napus* var. *annua*). It is the leading oilseed crop in Chile, Canada, and other countries.

content of the air-dry seed of less than 10.5 per cent moisture is about 40 per cent.

The erucic acid content of the oil ranges from 0.3 to 50.6 per cent. The new varieties now produced on most Canadian farms have a very low erucic acid content in the range of 0 to 7 per cent, which is more suitable as an edible oil but not for many industrial uses. The cessation of sperm oil imports is stimulating the production of oil from crambe, which is high in erucic acids. Another plant called meadow-foam (*Limnanthes* species) is being tested as an oil crop substitute for sperm whale oil.

The solvent processed rapeseed meal is similar to soybean meal in protein and amino acid content but lacks the vitamins. The meal is suitable for livestock feed but not in amounts exceeding 10 per cent of the ration for mature fowls or cattle, of 20 per cent for sheep or 4 per cent for young pigs. It is unsuitable for young fowls or pregnant or nursing sows because it contains glucosinolates (thioglucosides) which are antithyroid (goiter producing) substances. However, the toxins can be extracted from the meal with water.[62]

SESAME

Economic Importance

Sesame seed (or benne) was grown on about 5.9 million hectares from 1970 to 1972, chiefly in India, China, Sudan, Mexico, and Burma.

The production was 1.9 million metric tons or about 0.33 metric tons per hectare. The world production of sesame oil in 1973 was about 605,000 metric tons. Production in the United States in recent years has been limited to less than 1,000 hectares with a yield of 0.53 metric tons per hectare. Sesame has been grown in Texas, Arizona, New Mexico, Oklahoma, and Kansas.

Sesame may have originated in East Africa where several wild species are found. It was cultivated in the Near East by about 3000 B.C. and in Egypt by 1300 B.C.[60] It reached the United States in the 17th century. The crop was grown sporadically, but regular commercial production began about 1950. It usually is grown under contract.

Adaptation

Sesame requires a warm climate with a frost-free period of 150 days or more. Medium-textured soils are most favorable.[1]

Botanical Characteristics

Sesame (*Sesamum indicum*) belongs to the plant family *Pedaliaceae*, characterized by bell-shaped flowers and opposite leaves. It is an erect annual herb that reaches a height of 3 to 5 feet. The tubular, two-lipped flower is about ³⁄₄ inch long, with a pink or yellow corolla. The lower flowers begin blooming two or three months after seeding, but blooming continues for some time until the upper pods open.

The upright pods split open at the top at maturity which gave rise to the expression, "open sesame" (Figure 39–9). The seeds drop out when the plant is inverted, except with varieties with indehiscent pods. The seeds of different varieties are creamy white, dark red, brown, tan, or black.[1] The seeds somewhat resemble flaxseed in size, shape, and sometimes also in color. The yields and oil contents of nonshattering varieties usually are below those of the shattering varieties.

Cultural Methods

Sesame should be planted in late spring after the soil temperature is 70° F., or higher. It is planted in rows, usually with corn or cotton planters equipped with vegetable-seed planting plates. The rows usually are about 40 inches apart, although narrower rows 18 to 30 inches apart often give higher yields. Sound treated seeds, planted at the rate of 1 pound per acre, should provide stands of five to eight plants per foot of row.

Shattering varieties are harvested with a row binder or grain binder when the early pods are about to open. The bundles are cured in shocks, after which they are threshed with a combine drawn up to each shock. Special precautions are taken to avoid losses from shattered seeds when feeding the bundles into the combine. Non-

Figure 39–9. Spike and capsules of sesame.

shattering varieties are cut with a windrower, after which they are threshed with a pick-up combine.

Uses

The chief use of sesame is for the fixed (nonvolatile) oil the seed contains that is utilized mostly for edible purposes. The oil content may range from 50 to 56 per cent. It is suitable for a salad oil when combined with equal or larger proportions of other palatable oils. The oil has been used for lighting purposes in Asia. Decorticated seeds of sesame are sprinkled on the surface of certain types of bread, rolls, cookies, and cakes just before baking. This is the usage most familiar to Americans. Whole or ground sesame seeds are included in many food preparations.

Crop Pests

Some improved varieties of sesame are resistant to leaf-spot diseases, but the older varieties are susceptible. Wilts, blights, and charcoal rot also attack sesame.

Aphids, thrips, and other insect pests can be controlled with insecticides.

SAFFLOWER

Economic Importance

The world production of safflower oil was about 285,000 metric tons in 1973. Safflower probably originated in the Near East. It has been grown there and in India for centuries. It has been grown in Europe for more than 200 years, mainly for its reddish dye (carthamin).[64] Safflower was introduced into the United States over 100 years ago but its culture was not established until after 1940. It was grown on more than 300,000 acres in 1964, chiefly in California, Nebraska, Montana, and North Dakota with a seed production of over 553 million pounds, or 1840 pounds per acre.

Adaptation

Safflower is adapted to semiarid and irrigated regions of the western half of the United States where there is a frost-free period of 110 days or more.[3] The plants (Figure 39–10) are subject to diseases under humid conditions.[26] Young plants withstand temperatures of 20° to 25° F. or lower, but buds, flowers, and developing seeds are damaged by frost. The moisture requirements for safflower are similar to those for successful small-grain production. Warm weather and low humidity during the flowering period is helpful.

The plant tolerates high temperatures, considerable soil salinity, and rather high levels of sodium salts.[23]

Botanical Relationships

Safflower (*Carthamus tinctorius*) is a member of the family *Compositeae*. It is an annual, erect, glabrous, deep-rooting herb, 1 to 4 feet high, and branched at the top, with white or yellowish smooth pithy stems and branches. The flower heads are globular, ½ to 1½ inches in diameter, with white, yellow, orange, or red florets (Figure 39–10). It is largely self-pollinated except when many insects are present. The seed is smooth, obovoid, four-angled, white or cream-colored, and resembles a small sunflower seed. The leaves and outer floral bracts of the best adapted varieties bear short spines which make the plant a typical thistle in appearance (Figure 39–10). Varieties without spines are being tested. The plant is an annual, but it is grown as a winter annual in the southern parts of California and Arizona. The growing season is 110 to 150 days from spring sowing, but 200 or more days from fall sowing.

Figure 39–10. Safflower plant and seeds. [Courtesy Allis-Chalmers Manufacturing Co.]

Cultural Methods

Safflower usually replaces small-grain crops in rotations. It may follow an intertilled crop that has been kept free from weeds. It often is sown on fallowed land in dryland areas and is irrigated in Arizona and other arid areas. In northern California, safflower often follows rice. The rice lands usually contain ample moisture after the winter rains, along with that from flooding of the rice fields the previous year. Safflower fields in the semiarid regions often are unfertilized, but irrigated lands may receive from 50 to 120 pounds per acre of nitrogen and sometimes phosphate also.

Safflower seed is spring-sown from February 15 to April 20 in California, but from April 10 to May 20 in other areas.[3, 53] It is sown as a winter crop from early November to mid-January in the warm southern irrigated areas. It usually is sown with a grain drill at a rate of about 30 pounds per acre in dryland areas, but at somewhat higher rates under irrigation. On weedy land, safflower should be grown in cultivated rows 18 to 30 inches apart, with seeding rates of 15 to 25 pounds per acre.

Incorporation of suitable herbicide in sprays when the plants are 2 to 4 inches tall control wild oats and lambsquarter.

Harvest

The safflower crop is ready to harvest when the seeds are hard and dry. It may be allowed to stand until the stems also are dry enough for combine harvest because the crop does not lodge, the seeds seldom shatter, and the seeds are not eaten by birds. Thus, the crop is well adapted to harvesting with a combine.

Uses

The whole seed of improved varieties contains 32 to 40 per cent oil while the meats contain from 50 to 55 per cent. The meal from unhulled seed contains 18 to 24 per cent protein, but that from hulled seed has from 28 to 50 per cent protein.

Safflower oil is a drying oil with an iodine number of about 140 to 155. The oil is used for edible purposes as well as in soaps, paints, varnishes, and enamels. Although of lower drying value, it is superior to linseed oil for white inside paints because the painted surface remains white instead of turning yellow as does white linseed-oil paint when not exposed to sunlight. This is due to the absence of linolenic glycerides. The increased popularity of safflower oil for edible uses is due to its content of unsaturated fatty acids (75 per cent linoleic acid).

The flower heads of safflower yield a red dye, carthamin, formerly extracted commercially in India.

Crop Pests

Safflower is comparatively free from diseases in dry seasons, but it may be damaged seriously by leaf spot, rust, wilt, and root rot under moist conditions.[85] Rust, caused by *Puccinia carthami*, and leaf spot, caused by *Alternaria carthami*, infest the seeds. Partial control is effected by planting treated disease-free seed. Root rot, caused by *Phytophthora drechsleri*, is a severe disease on irrigated land. Some varieties are resistant. Verticillium wilt and Botrytis head rot attack safflower under humid or foggy conditions where the crop probably should not be grown.

The chief insects that attack safflower are lygus bugs, wireworms, aphids, leafhoppers, green peach aphids, and thrips.

CASTORBEAN

Economic Importance

Castorbeans are grown on all continents on about 1.3 million hectares annually with a production of 700,000 metric tons or 0.56 tons per hectare. The leading producing countries are Brazil, India, the U.S.S.R., and the United States. The world production of castor oil in 1973 was about 365,000 metric tons. Production in the United States was nearly 20,000 metric tons grown on about 20,000 acres with a yield of 0.95 tons of castorbeans per hectare. They are now grown in the central and southern Great Plains and the southwestern states. Formerly, castorbeans were imported for oil extraction in crushing plants along the Atlantic and Pacific coasts.[77]

History

The castorbean is generally believed to be a native of Africa. However, it may have originated in India, because wild forms occur in that country.

The castorbean was grown in Kentucky and New York as early as 1803. By 1850 some 23 oil-extraction mills were operating in the East and South and in the Mississippi Valley. After 1865, the chief area of production shifted from Missouri to Kansas and then to Oklahoma. The domestic crop declined to negligible quantities after 1900.

Production was revived by government supports during World War I, World War II, and the Korean conflict, but the growing of castorbeans practically ceased when peace was restored. Commercial mechanized production was achieved after 1957 when efficient harvesting machines and improved varieties became available.[100]

Adaptation

The castorbean is adapted to areas that have a frost-free period of 160 to 180 days, an average warm season (April to September), precipitation between 15 and 25 inches, an average July temperature above 76° F., and an average relative humidity at noon during July of less than 60 per cent.[19] The crop is well adapted to irrigated regions.

The gray-mold disease makes the castorbean a risky crop in the humid high-rainfall areas that comprise the Atlantic and Gulf coastal plains regions.

The soil for castorbeans should be fertile, well drained, and contain enough sand so that it warms up early in the spring. The crop should be planted on sites that are not subject to water erosion. Castorbeans are grown successfully in Egypt on irrigated coarse sandy soils of *p*H 8.7.[28]

Botanical Characteristics

The castorbean belongs to the family *Euphorbiaceae,* the spurge family. The castorbean (*Ricinus communis*) is a short-lived perennial in the tropics where it grows into a tree 30 to 40 feet in height. In the United States it is an annual, 3 to 12 feet in height, except in Florida and parts of Texas and California, where it can survive the winters (Figure 39–11).

The leaves are large, usually 4 to 12 inches wide or wider, alternate, and palmately divided into 5 to 11 lobes. The leaves of different varieties, which include numerous ornamental forms, may be green, purple, or red. The stems also may be green or red. The greenish-yellow flowers without petals are borne in racemes. The plant has an

Figure 39–11. Castorbean plant showing spike (*left*); castorbean seeds (*right*). The protuberance at the end of the seed is a strophiole or carnucle.

indeterminate blooming habit. It also is monoecious, with the pistillate flowers in the upper part of the raceme and the staminate flowers below. Consequently, it is frequently cross-pollinated. The fruit is a spiny or smooth, dehiscent or indehiscent, three-celled capsule. Each cell contains a hard-shelled seed that usually is mottled.

The pods of many varieties forcibly eject the seeds at maturity. Improved varieties for seed production retain most of the seeds for a time after maturity.

The general appearance of the seed is somewhat similar to that of a mottled bean, except that it is obovoid instead of kidney-shaped. It has a prominent hilum or caruncle at the end instead of a depressed hilum on the edge of the seed as in a bean.

Since the castorbean is not a legume, the seed is not truly a bean. Furthermore, the oil is not used to lubricate castors. The name castor apparently was coined by English traders who confused the oil with that from *Vitex agnus-castus* which the Spanish and Portuguese in Jamaica called *agno-casto.*

The castorbean also is called the castor-oil plant, Palma Christi, and mole bean.

The castorbeans grown in California are hybrids with normal internodes. The plants are about 7½ feet tall.[108] Hybrid seed is produced on pistillate plants that are pollinated by normal monoecious strains. Dwarf-internode varieties, and hybrids are grown in Texas.[7, 8] These dwarf varieties are 3 to 5 feet in height.

Cultural Methods

Irrigated land to be planted to castorbeans often is fertilized with 40 to 120 pounds per acre of nitrogen. From 20 to 60 pounds of phosphate are applied to soils deficient in phosphorus. Seedbed preparation for castorbeans is about the same as for cotton or corn.[2, 8, 103] Castorbeans grown under irrigation usually are planted on beds, either in shallow furrows or in moist soil that has been exposed by scraping off the ridge by a blade on the planter.

Treated seed is planted in rows 1 meter apart, at a depth of 1½ to 3 inches in moist soil on unirrigated fallow land, and in narrower rows on irrigated land.[9] Spring planting is deferred until the soil has warmed up to a temperature of 60° F. Desirable stands are 10,000 plants per acre on unirrigated land and 25 to 30,000 under irrigation. The seed size is 1,000 per pound.

Early cultivation is done with a rotary hoe, while later cultivation is performed with sweep or shovel cultivators.

Harvest

Castorbeans are gathered and hulled with special self-propelled machines or with special combine attachments. The machines have beaters that shake the capsules from the spikes. Augers and elevators convey the capsules to the huller, while the hulled and cleaned beans are carried to a hopper bin on the machine. Both two-row and four-row harvesters are used (Figure 39–12).[8, 75, 103]

Indehiscent (nonshattering) hybrids are grown so that harvest can be deferred until the castorbeans are ripe and dry. Much of the crop is harvested after the plants have been killed by freezes.[2, 8] Desiccants are applied when the crop is ripe before frost occurs.

From 5 to 10 per cent of the seeds may drop on the ground during or before harvest. These seeds produce volunteer plants the next spring, which must be eradicated before the planting of another crop. Stray castorbean plants or seeds in a forage or food crop could be fatal when consumed. Grain sorghum is planted late enough for castorbeans to be destroyed by cultivation. Stray or surviving plants can be killed with herbicides.

Castorbeans are now a fully mechanized crop in the United States.

Uses

The hulled castorbean contains 35 to 55 per cent oil, which averages about 50 per cent in well-developed seed. The oil usually is extracted by pressure without grinding or decorticating, and usually without heating the beans. Castor oil is a nondrying oil with an iodine number that ranges from 82 to 90. Recent processes of dehydrating castor oil give it drying-oil properties, so that it is now widely used in paints and varnishes. The oil also is used in producing sebacic acid which is

Figure 39–12. Two-row castorbean harvester.

used in making synthetic lubricants for jet aircraft and as a solvent or plasticizer in the manufacture of certain types of nitrocellulose plastics and fabrics. Castor oil also is used in the manufacture of linoleum, oil cloth, and printer's ink and artificial leather and the dyeing of textiles.[76] Its use as a laxative for ailing youths is all too well known to those of two generations ago. Now only one per cent of the oil is used as a purgative. Castor oil is used frequently in hydraulic cylinders, lubricants, cosmetics, soaps, and many other products.[8] Considerable castor oil is imported into the United States, chiefly from Brazil, but castorbeans are no longer imported.

The castorbean, as well as the press cake after oil extraction, are poisonous to humans, livestock, and poultry. The toxic principles are ricin, which is an albumen, and ricinine, which is an alkaloid. The press cake is used largely in mixed fertilizers under the name of castor pomace. The oil is nonpoisonous, but the leaves and stems of the plant contain a sufficient quantity of ricinine to be toxic. The eating of one castorbean seed is said to produce nausea, while ingestion of appreciable numbers often proves fatal.[2]

Crop Pests

Alternaria leaf spot, caused by *Alternaria ricini*, may defoliate castorbeans, while the same fungus also can cause the young capsule to mold.[2] Bacterial leaf spot, caused by *Xanthomonas ricinicola*, has been destructive in Texas.[8]

Some varieties are resistant to these leaf-spot diseases. Gray mold, caused by *Sclerotinia ricini*, attacks and destroys the racemes. It occurs in warm humid regions.[34] The castorbean is highly susceptible to the Texas root-rot disease caused by the fungus *Phymatotrichum omnivorum*. Consequently, the crop should not be planted in fields infested with that organism.

Insects that attack the castorbean include the false chinch bug, army-worms, corn earworm, stink bug, leafhoppers, leaf miners, caterpillars, grasshoppers and Japanese beetle, but usually the damage is not great. Nematodes also attack the castorbean to some extent.

PERILLA

Perilla is grown for its seeds, which contain a drying oil of excellent quality that in the United States has been used in paints, varnishes, and linoleum manufacture. It has not been grown in this country except in preliminary field trials. It is grown in Mainland China, Korea, Japan, and India.

Perilla requires a warm growing season of five to six months with an ample supply of moisture. It can be grown successfully in the

United States only in the humid Southeast and in the irrigated Southwest. The plants can withstand light frosts. Sandy loam soils of average fertility are suitable for perilla.

Perilla (*Perilla frutescens*, or *P. ocymoides*) belongs to the *Labiatae*, the mint family. A subspecies, *P. frutescens nankinensis*, also is grown. Perilla is a coarse annual plant indigenous to India, Japan, and Mainland China, where it is cultivated. The plants, which are 3 to 6 feet in height and much branched unless crowded, bear coarse-toothed leaves and have small inconspicuous white flowers borne in racemes 3 to 8 inches long. The seeds are round and brownish, and resemble mustard seeds. The average seed yield in Japan is about 560 pounds per acre.

The oil content of perilla seed ranges from 35 to 45 per cent, with an average of about 38 per cent. The extracted oil is yellow or greenish. It is similar to linseed oil in odor and taste, but it is higher in drying quality (iodine number about 200). Therefore, it is suitable for quick-drying paints and varnishes. In Japan the oil is used in the manufacture of paper umbrellas, oil papers, artificial leather, printer's ink, paint, varnish, and lacquer. Processed oils have largely replaced perilla oil in the United States.[1]

SUNFLOWER

Economic Importance

Sunflower seed was harvested on an average of about 8.5 million hectares from 1970 to 1972 with a production of about 10 million metric tons or 1.2 metric tons per hectare. The leading producing countries are the U.S.S.R., Argentina, Romania, Bulgaria, and Yugoslavia. The production of sunflower seed oil is nearly 4 million metric tons.[89] More than 160,000 hectares (400,000 acres) were harvested in the United States in 1972, chiefly in the Red River Valley of Minnesota and North Dakota. The yield of seed is about 1 metric ton per hectare or, 1,100 pounds per acre. Production extends northward in the Red River valley into Canada and southward to Texas.[88]

History

The cultivated sunflower is a native of America. It was taken to Spain from Central America before the middle of the 16th century. It was being grown by Indians for food at Roanoke Island in 1586, and in New England for hair oil in 1615.[102] Established culture of the sunflower in the United States followed introduction of improved varieties that had been developed in Europe before 1600.

The wild sunflower, the state flower of Kansas, flourishes in overgrazed pastures, and in fields, and waste places over western United States, particularly in the Great Plains States, strictly as a weed. The wild sunflower dramatically took possession of the wind-eroded fields and pastures of the southwestern dust bowl, where the efforts of man had failed, when the rains came in 1940 and 1941. Its rank-growing branched stems protected the lands from further wind injury while perennial vegetation became re-established.

Adaptation

The sunflower is adapted for seed production where corn is successful in the northern two-thirds of the United States. It formerly was grown for silage in cool northern and high-altitude regions where corn does not thrive.[91] The young plants will withstand considerable freezing until they reach the four to six-leaf stage. The ripening seeds likewise suffer little damage from frost. Between those stages the plants are more sensitive to frost. The plants show relatively little photoperiodic response.

Botanical Description

The sunflower (*Helianthus annuus*) belongs to the family *Compositae.* Plants of this family bear heads in which the fertile flowers are aggregated and are bordered by rays, the corollas of sterile flowers.

The cultivated sunflower is a stout erect annual, 5 to 20 feet in height,[102] with a rough hairy stem 1 to 3 inches in diameter which terminates in a head or disk 3 to 24 inches in diameter. The top of the disk is brown to nearly black. The rim of the disk is surrounded by pointed scales and 40 to 80 yellow rays. The seed is an elongated rhomboid achene. The stalk may produce several branch heads. The flowers are almost entirely cross-pollinated. Cytoplasmic male sterile types are now used for producing hybrid seed.[31]

The stalk of wild sunflower (also called *H. annuus*) usually is 1 inch or less in diameter, has many branches and bears numerous heads. The wild sunflower seeds are too small and chaffy for economic utilization.

A noticeable characteristic of the sunflower heads—the facing toward the sun throughout the day—accounts for both the common and botanical names for the plant. Their heliotropic movement results from a bending of the stem, a process called nutation, which tilts the head to the west in the afternoon.[74] After sunset the stem gradually straightens until it becomes erect about midnight. Thereafter, the stem gradually bends in the opposite direction up to as much as 90 degrees, so that the head faces east by sunrise. Soon afterward the stem starts to straighten until the head is erect again at noon. The leaves likewise face east in the morning, west in the evening, and upward at noon and midnight. However, at 10 P.M. the leaves are drooping downward. Stripping the leaves from the stalk stops all bending of the stem. Nutation ceases when anthesis (pollen shedding) begins, or shortly thereafter. Fully 90 per cent of the heads are facing east or northeast as they hang at maturity except where strong winds occur. Growers of tall varieties have taken advantage of this eastward nodding habit by planting the rows north and south. At harvest time they drive along the east side of each row and cut off the overhanging heads.[42]

Varieties

Early dwarf sunflower hybrid or varieties are grown in Canada, North Dakota, and Minnesota for harvesting with a combine. These types grow to a height of 3½ to 6 feet. Tall late hybrids or varieties are grown in California, in other warm parts of the country, and in home gardens.[73]

Cultural Methods

The seedbed for sunflowers should be prepared as for corn. Planting is done with a corn planter equipped with special sunflower plates, or with a grain drill with most of the feed cups closed off. Planting usually is done in May except in the central latitudes of the United States where April planting seems to be more favorable. Three to seven pounds per acre in drills give an ample stand. Plants of large varieties should be about 12 to 14 inches apart in rows. Small varieties require closer spacing (6 inches apart in 36-inch rows).[52]

Harvest

Sunflowers are mature when the backs of the heads are yellow and the outer bracts are beginning to turn brown. The seeds are ready to thresh and store when they contain not more than 8 to 9 per cent moisture. The large tall varieties of sunflowers are harvested by hand while the workers ride in a truck along the row. The cut heads can be threshed with a combine standing in the field. Dwarf sunflowers are cut with a combine. Extra slats or screening on the reel arms of combines are necessary to avoid catching and throwing out the crook-necked heads.

Uses

The head of the matured sunflower contains about 50 per cent of the dry matter of the whole plant.[102] Nearly one-half the weight of dried heads is seed. About 35 to 50 per cent of the seed consists of hull. The whole seed contains 24 to 45 per cent oil, whereas the hulled kernel contains 45 to 55 per cent or more. The expressed oil yield from whole seed ranges from 20 to 35 per cent. The residue is oil cake which contains about 35 per cent protein when made from whole seed.

Sunflowers are now grown primarily for the production of oil. This began in the United States in 1967 using high oil varieties developed in the Soviet Union. Hybrids or varieties with large, striped seeds are eaten as a confection, either roasted, raw, or in candies. The seeds also are fed to poultry, caged birds, or wild birds. The meal is fed to livestock but would be suitable for human feed in the form of chips or breads.[88] The meal is free from toxins. Discoloration of the white defatted meal into shades of beige, green or brown is due mainly to oxidation of chlorgenic acid at pH levels above 7.0.

Sunflower oil is mostly polyunsaturated. It is used primarily in shortening, cooking and salad oils and margarine. It smokes at about 150° C. (302° F.), which is too low for deep-fat frying at the usual temperature of about 200° C. (392° F.). Sunflower oil is used also for making soaps and paints. It is a semidrying oil with an Iodine Number of about 130. The oil cake is mostly fed to livestock, but some is used for fertilizer.

The growing of sunflowers for silage in the colder parts of the United States, ceased when forage harvesters became available. These machines facilitated the ensiling of adapted grasses and legumes that are more palatable and nutritious than sunflowers.

Diseases

Sunflowers are attacked by rust, Verticillium wilt, charcoal rot, downy mildew, powdery mildew, and black leaf.

Sunflower rust, caused by the fungus *Puccinia helianthi-mollis,* is the most common disease. It develops numerous brown (and later black) pustules on the leaves, which cause them to dry up. Destruction of so much leaf tissue checks the development of the stalk and head. Crop rotation and dusting with sulfur are the most effective remedies. Certain hybrids, European varieties, the ornamental sunflower (*Helianthus agyrophyllus*),[91] and the North Texas wild sunflower are resistant.

The charcoal-rot disease causes wilting and lodging following the decay of the pith in the stalk. Control consists of late planting, an ample soil moisture supply, and avoidance of severely infested soils. A wilt or stem rot is caused by the fungus *Sclerotina sclerotiorum.*

Insects

The chief insect enemies of the sunflower are the sunflower head moth (*Homoeosoma ellectellum*), cutworms, wireworms, white grubs, grasshoppers, aphids, thrips, sunflower beetles, seed weevils, carrot beetles, sunflower budworm, and webworms. The head moth is numerous in warm areas, but it can be controlled with insecticide sprays. At the present time no known method, cultural or chemical, is available for control of the carrot beetle which feeds on the root system of the plant. Some sunflower varieties or strains are resistant to the sunflower beetle.

CRAMBE

Crambe, or Abyssinian mustard (*Crambe abyssinica*), has been grown as an oilseed crop in Canada, Montana, Illinois, North Dakota, and Indiana. It was introduced into the United States from Ethiopia after 1940 and was first grown for one season only in Montana, in 1965. The plant, a member of the family *Cruciferae,* is an erect annual with large pinnately lobed leaves, white flowers and spherical seeds. Crambe is a cool-season crop adapted to northcentral and northwestern states and adjacent areas of Canada. Seed yields under semiarid conditions range from 500 to 1,000 pounds per acre, but may exceed 2,000 pounds under irrigation.[95, 96]

It is sown with a grain drill at rates of 8 to 15 pounds per acre, at a depth of 1 inch or less. The best time for sowing lies between the seeding of spring grain and the planting of corn. Fertilizer requirements are similar to those for small grains. Crambe does not compete well with some of the common weeds. Certain herbicides may be applied for weed control, but 2,4-D kills the crambe plants. The crop is mature when the seed pods are yellow or brown. It preferably is harvested by swathing and then combining. The threshed seeds are retained in the hulls, which comprise 20 to 30 per cent of the weight.

The unhulled seeds usually weigh 5 to 7 grams per 1,000 seeds and have a test weight of 24 to 29 pounds per bushel. Good quality seed contains about 32 per cent of oil and 20 to 30 per cent protein. The hulled seed may contain 44 per cent oil. The erucic acid content of the oil ranges from 50 to 60 per cent. Potential uses for the oil include plasticizers, rubber additives, waxes and a lubricant for molds in continuous casting of steel. The seed meal contains sulfur compounds (Thioglucosides) which are unpalatable to livestock.

GUAYULE

Economic Importance

Guayule (pronounced gway-oó-lay or wy-oó-lay) is the Indian name for a North American rubber crop. More than 33,000 acres of guayule were planted in California, Arizona, and Texas from 1942 to 1944. About 550 acres of old plantings were harvested in 1943. Larger quantities were harvested from natural stands in Mexico and Texas, while new plantings were harvested a few years later. Guayule production was discontinued in the United States by 1950 because synthetic rubber supplies were abundant.

History

Guayule is a native of northern Mexico or southwestern Texas. The plant was first collected for identification in 1852.[55] Indians long ago learned to extract the rubber by chewing the wood. The first experiments on the commercial extraction of rubber from guayule were begun about 1888. Several factories were established in Mexico in 1904 and shortly thereafter, and one at Marathon, Texas, in 1909. Attempts to cultivate guayule were begun in Mexico about 1907. A total of 8,000 acres was planted in California, Arizona, and elsewhere in the United States after 1912 but most of this had been plowed up by 1942.

Adaptation

Guayule is distributed naturally over an area of 130,000 square miles which extends southward from the vicinity of Fort Stockton, Texas, into Mexico.[55] This area has an annual rainfall of 10 to 15 inches with altitudes of 3,000 to 7,000 feet. Guayule might be cultivated in the United States in a belt about 150 miles wide north of the Mexican border, extending from the mouth of the Rio Grande River to southern California and northward to central California. In a dormant condition the plants may withstand temperatures of 5° to 10° F., but while growing actively they are injured somewhat by temperatures of 20° F. Guayule thrives under hot conditions. It may be grown either on irrigated or nonirrigated land.

Botanical Description

Guayule (*Parthenium argentatum*) is a slow-growing, widely branching, woody, long-lived perennial shrub that belongs to the family Compositae. Plants in the wild may live 40 to 50 years. It is seldom more than 30 inches tall at maturity. The plant develops a strong taproot, as well as a number of long shallow lateral roots from which new shoots may arise. The leaves are silver gray, as a result of a covering of thick, short white hairs. They are 1 to 2 inches in length. The fruit is an achene with ray and disk flowers attached after maturity. A pound of the fruit consists of about 600,000 achenes, of which about 90 per cent contain no seeds because of sterility that results from irregular chromosome numbers and other factors. Many seeds are produced without fertilization by apomixis. Fresh seed of guayule germinates very poorly, but after-ripening is completed in about a year, so that year-old seed can be planted without treatment with calcium hypochlorite.

Cultural Methods

Cultural methods[41] for guayule include planting treated, presprouted seed in beds with the seeds covered by $^{1}/_{10}$ inch of sand by

Figure 39–13. Harvester picking up and chopping guayule plants that have been undercut.

the planting machine. After eight months the seedlings are topped and then undercut by special machines. The seedlings are then transplanted 30 inches apart in each direction with a tractor-drawn six-row machine. Weeds are controlled largely by use of oil sprays.

The plants normally are suitable for harvest after four years of growth on irrigated land or five years on nonirrigated land. A special machine digs two rows at a time and leaves the plants in a single windrow (Figure 39–13). A special pick-up machine gathers, chops, and binds the material. It then conveys the bundles or bales into an accompanying truck. Mowing instead of digging the plants would recover about 65 to 75 per cent of the rubber, while it would also permit regrowth of guayule for another harvest in two or three years.

A special rotary cylinder machine with vacuum suction equipment is used to gather seed from mature plants in the field.

Rubber Extraction

The extraction of rubber from guayule involves the crushing of the material in corrugated rolls, after which it is macerated in water in a series of pebble mills to release the rubber. The particles of rubber aggregate into what are called worms that float and can be skimmed off. The worms are separated from adhering bark, scrubbed, dried, and pressed into blocks.

Uses

Good strains of guayule yield about 20 per cent rubber or about 1,600 pounds of rubber per acre every fourth or fifth year.[66]

Guayule rubber as recovered ordinarily contains 15 to 20 per cent resins. Removal of these resins (deresinating) leaves a product very similar to Hevea rubber, except that it does not make a satisfactory rubber cement. However, for many purposes, such as the fabric of tires, the presence of resin is beneficial in reducing friction. Some use may be found for the resins. Guayule plants make a hot fire when burned. Formerly they were used for smelting in Mexico. The native shrub is browsed by burros and goats.

Crop Pests

The principal diseases that attack guayule are Verticillium wilt caused by the fungus *Verticillium albo-atrum*, and Texas root rot, caused by *Phymatotrichum omnivorum*. There is no control for the former disease. The latter, which is not serious, can be evaded only by avoiding infected fields when planting. Several other root-rot diseases attack guayule.

Insects that attack guayule include grasshoppers, lygus bugs, lace-bugs, harvester ants, leaf-cutting ants, and termites.

The dodder weed sometimes is a parasite on guayule plants.

Other Rubber Sources

Hevea rubber trees are unable to survive in the continental United States, except in the southern tip of Florida. The roots of Russian dandelion or kok-saghyz (*Taraxacum koksaghyz*) yield an excellent quality of rubber, but yields of present strains cannot be expected to exceed 50 to 60 pounds of rubber per acre per year under average conditions. Goldenrod (*Solidago leavenworthii*) gives low yields of low-grade rubber by double-solvent extraction of the leaves that contain 4 to 7 per cent rubber. A vinelike rubber-bearing shrub, *Cryptostegia grandiflora*, grown in India and Madagascar, may yield 100 pounds per acre per year by excessive expenditure for hand labor.

Rabbitbrush (*Chrysothamus nauseosus*), a native shrub distributed over the drier regions of western United States, yields a good quality of rubber but the percentage is too small for commercial success.

HEMP

Economic Importance

Hemp (*Cannabis sativa*) is harvested for fiber on about 600,000 hectares, mostly in the U.S.S.R. and India, but also in southern and eastern Europe, Chile, and elsewhere. The average fiber yield is one-half metric ton per hectare. Also about 50,000 metric tons of hemp seed is harvested, mostly in the Soviet Union. Hemp was grown in the United States from early Colonial days but declined to about 200 acres by 1933. Production was chiefly in Kentucky, but also through the North Central states. Production resumed from 1940 to 1946 to replace imported abaca (Manila hemp) which was then unavailable. The fiber was needed for marine cordage.

History

Hemp probably is native to central Asia. It has been grown in China for many centuries. It was grown by the ancient Greeks.

Hemp escaped from cultivation in the United States many years ago. It now occupies large areas of uncultivated land, chiefly in the bottomlands along the lower Missouri River and its lower tributaries. Harvesting of the wild growth has not been feasible because of scattered and irregular stands and growth and because of mixtures with other types of vegetation. It is now being eradicated,[27] with herbicides, flaming, mowing, and tilling.

Adaptation

Hemp is adapted to sections that produce a good crop of corn. It requires a frost-free season of about four months for fiber production, or five months for seed production.[71] It will withstand light frosts. For good uniform growth of hemp, the rainfall or soil moisture should be ample throughout the season. A higher quality of fiber usually is obtained in the more humid climate east of longitude 95° because of better conditions for dew retting than commonly occur west of the meridian.

Well-drained, deep, fertile, medium-heavy loams have produced the best hemp. Barnyard manure, commercial fertilizers, or lime should be applied to soils where other crops respond to these supplements.

Botanical Description

Hemp (*Cannabis sativa*) belongs to the family *Urticaceae*, tribe *Cannabinaceae*. It is a stout, erect annual that grows to a height of 5 to 15 feet. It is dioecious; the male plants, which comprise one-half the crop, produce no seed.

Hemp is of considerable historical interest because it first revealed sex in plants to Camerarius in 1694. It has continued to serve as the subject for determining the nature of sex in plants. Under normal field conditions, hemp is strictly dimorphic (has two forms) as to flowers and vegetation parts. The carpellate and staminate plants show almost no sex reversal. However, individual bisexual plants occur under unfavorable conditions[59] such as reaction to old pollen,[7] mutilation,[67] and reduced light.[75]

The main stem of hemp is hollow and produces a few branches near the top. The bast fibers are not so fine as those of flax, even when the plants are grown close together. The staminate inflorescences are in axillary, narrow, loose panicles.[45] The pistillate flowers are in erect leafy spikes (Figure 39–14).

The hemp ovary matures into a hard ovoid achene.[63] The seeds mature on the lower part of the spikes first and on the upper part last.

Types of Varieties

The common fiber hemp has larger stalks than the drug-plant type, which is grown in the hot climates of India, Syria, and elsewhere, often being classified as a separate species, *Cannabis indica.* The dried leaves and flowers of this species are called hashish. The type grown in the United States, known as Kentucky or domestic hemp, is of Chinese origin.

Cultural Methods

Hemp should be planted on fall-plowed land just before the time for planting corn.[71] Hemp responds to fertilizers on some of the less

Figure 39–14. Pistillate (seed-bearing) hemp plant (*left*); staminate plant (*right*).

fertile soils of the Corn Belt States, but heavy nitrogen fertilization reduces the strength of the fibers.[47]

For seed production in Kentucky, the hemp usually is planted in hills 5 feet by 5 feet, with six to ten seeds per hill or about 1½ pounds of seed per acre. Later the hills are thinned to two or five plants per hill. Harvesting for seed occurs when the seeds in the middle branches are ripe. Harvesting and shocking when the plants are damp save much seed that otherwise would be shattered.

For fiber production, hemp should be sown with a grain drill ½ to 1 inch deep at the rate of 5 pecks (55 pounds) of seed per acre to produce fine uniform stems suitable for machine processing. Hemp is harvested for fiber when the male plants are in full flower and shedding pollen.[69] Earlier harvesting yields weak fiber. Harvesting is done with a modified 8-foot rice binder or a self-rake reaper that delivers the stalks in an even swath on the ground. The stalks are left in this position for dew retting. After retting is complete, the stalks are gathered up and bound by a pick-up binder. The bundles are shocked and later hauled to the processing plant any time after drying.

Preparation of Hemp for Market

In dew retting, the stalks lying on the ground are exposed to cool moist weather or, sometimes, to alternate freezing and thawing. The process is complete when the bark separates easily from the woody portion of the stem. The stalks are turned once or twice to provide uniform exposure. Water retting is practiced in Europe.

In the breaking process, the inner cylinder of brittle woody tissue is broken into short pieces called hurd, while the long flexible fibers are left largely intact. This permits the wood to be separated from the fibers.[46] Scutching is the beating or scraping off of the broken pieces of wood.

The rough fiber is combed out by drawing it over coarse hackles. Hemp tow consists of short, more or less tangled strands. A ton of dry retted hemp stalks processed by modern machine methods yields about 340 to 360 pounds of fiber, of which slightly more than half is line (long) fiber and the remainder is tow.

Hemp is used in commercial twine, small cordage, thread, hemp carpet twines, oakum, and marline.[46] It ranks next to Manila hemp for marine ropes.

Hemp Seed and Leaves

Hemp seed is often fed to caged birds after the germination of the seed has been killed in accordance with federal regulations (in order to prevent its use for planting). The seed contains 20 to 25 per cent of an oil that sometimes is extracted for use in making soft soaps or is used as a substitute for linseed or other oils. The glandular hairs secrete both a volatile oil and the strong narcotic resin (cannabin). The drug, called marijuana in North America and hashish or bhang in Asia, is obtained from the flowers and leaves of both *Cannabis sativa* and *C. indica.* Hemp is the source of the so-called "reefer" cigarettes that are sold illegally. Hemp can be grown only under a license issued by the Commissioner of Narcotics.

The presumed principal hallucinogen of marijuana is 9-tetrahydrocannabinol (THC).

Hemplike Plants

In addition to the true hemp (*Cannabis sativa*) several fiber plants sometimes called hemp are grown in other countries.[24] These include abaca or Manila hemp (*Musa textilis*), a relative of the banana plant grown in the Philippines and other Pacific islands; Sunn hemp (*Crotalaria juncea*), a legume cover crop also harvested for fiber in India and other parts of Asia; Mauritius hemp, the green aloe, or piteira (*Furcraea gigantea*) of Africa; New Zealand hemp (*Phormium tenax*);[68] jute or India hemp (*Corchorus capsularis*); and sisal hemp. The latter type includes sisal (*Agave sisalana*) grown in Africa and the West Indies, and henequen or Mexican or Cuban sisal (*Agave fourcroydes*). Henequen, sisal, and abaca are the chief fibers used in binder twine. Jute is used in making burlap. All the above fibers are used to make rope or twine. *Sansevieria* or bow-string hemp has been tested in Florida as a possible substitute for Manila hemp.

RAMIE

Ramie, also called China grass and rhea, has been grown in the Orient since the earliest records of agriculture and was introduced into the United States in 1855. It has been tested in many states, and small acreages were grown occasionally. Several hundred acres were planted in the Florida Everglades section in 1945 and 1946, but production ceased after 1961. Interest in the crop is stimulated not only by the fiber, which is the strongest of vegetable fibers, but also by the lustrous appearance of ramie cloth. It is grown in China, Japan, India, and the Soviet Union. China produces about 100,000 tons annually.

Adaptation

Ramie is a semitropical plant adapted to areas of the Southeast where the annual rainfall exceeds 40 inches. It also is adapted to interior irrigated valleys of California, where frost does not penetrate into the soil more than an average of 3 inches. The capacity of the plants to make a heavy vegetative growth necessitates a rich alluvial soil for a satisfactory crop. Excellent results have been obtained on the muck soils of the Florida Everglades.[61] Sandy soils often have proved unsatisfactory.

Botanical Relationships

Ramie (*Boehmeria nivea*) belongs to the family *Urticacae.* The plant is a perennial that sends up from the rootstocks new stalks which often reach a height of 8 feet. The stems are $\frac{1}{2}$ inch or less in diameter and covered with inconspicuous hairs (Figure 39–15). The flowers are small and monoecious, with the staminate flowers in the lower part of the cluster. Established plantings live ten or more years.[70]

Cultural Methods

About 500 pounds per acre of 2-4-36 commercial fertilizer that contains the essential minor elements (manganese, copper, and zinc) are recommended for the peat and muck soils of the Florida Everglades.[61] Ramie is propagated by dropping pieces of rootstocks about 6 inches long in an upright or slanting position in furrows with a transplanter. The furrows are then filled so that the upper ends are covered by about 2 inches of soil. The rootstock pieces are spaced 12 to 18 inches apart in rows $3\frac{1}{2}$ to 4 feet apart. The fields are cultivated not only to control weeds, but also to prevent the stand from becoming too thick. Young plants require six to eight cultivations before the plants are large enough (2 feet high) to shade the ground.

Figure 39–15. Leaves and inflorescence of ramie.

Harvest

Ramie is harvested from May to November in the Florida Everglades. Two to four crops are cut at intervals of two months or more.[13] The crop is ready to harvest when the growth of the stem has ceased and the staminate flowers are beginning to open. It is advisable to defoliate the plants by spraying the field with endothal.[14] Ramie is best gathered with a harvester-ribboner which cuts the stalks and separates the wood and bark from the ribbons of fiber. These ribbons contain cortical tissue and gums along with the fiber. They are hauled directly to a decorticator, which removes the cortical tissue. The ribbons are then squeezed through rollers to remove part of the water, brushed to untangle the fibers,[15] and then dried with artificial heat down to a moisture content of 10 to 12 per cent. The gums are removed later with chemical solvents. Annual yields of ungummed fiber in Florida have been about 2,000 pounds per acre.

Ramie fiber is similar to linen in appearance and use. The fabrics are called Swatow grass cloth, grass linen, Chinese linen, and Canton linen. The commercial fiber strands are 3 to 5 feet long and 0.002 to 0.003 inch in diameter. The ultimate fiber cells average 6 to 8 inches long.

KENAF

Kenaf (*Hibiscus cannabinus*) is grown for its soft fibers chiefly in Thailand, India, and the U.S.S.R., but also in Pakistan, other Asiatic countries, Cuba, and parts of Africa and the Central and South Americas. It has been grown in its native habitat of Thailand, India and Pakistan since ancient times. The annual production of kenaf fiber is nearly 800,000 metric tons. Kenaf has been grown in Florida on a small acreage since 1943, but it did not result in a firmly established fiber industry. In 1968 about 5,000 acres of kenaf stalks were cut in Florida for use as bean poles. Since 1956 the plant has been tested extensively as a possible source of pulp for paper making.[96]

Adaptation

Kenaf is adapted to the warm southeastern states of the United States where there is 20 to 25 inches of rainfall or reserve moisture during the four-to-five-month growing season. Well-drained soil with ample organic matter is desirable.

Botanical Description

Kenaf, sometimes called bimli-jute is a member of the plant family Malvaceae.[38] The plants grow to a height of up to 12 feet or more, with deeply lobed leaves. The yellow flowers with red centers open for 6 to 8 hours and then close. The petals drop off later. The flowers are largely self-pollinated. Kenaf is a short-day annual plant,[92] and most varieties bloom and produce seed only when the day length is less than 12½ hours. Some varieties are photoinsensitive (day neutral).

Cultural Methods

For pulp, kenaf may be planted at 5 to 8 pounds per acre in rows 20 to 36 inches apart to achieve stands of 75,000 to 100,000 plants per acre. It is best sown in the spring (March or April in South Texas) when the soil temperature reaches 55° F. or higher.[100] For fiber, the seed is drilled at 30 to 35 pounds per acre to space the plants 1 to 4 inches apart in 7-inch or 8-inch drill rows.[13, 106] The crop responds to heavy fertilizer applications on sandy soils.

Uses

Kenaf is harvested for fiber when the first flowers open, with a harvester-ribboner equipped with crushing rolls. The ribbons of long bast fiber are retted in tanks where bacterial action frees the fiber from adhering substances. The ribbons are then cleaned in a burnishing machine and dried.[15] Fiber yields as high as 2,000 pounds per acre have been obtained under good growing conditions.

For pulp the plants should be harvested when they cease growth or are killed by frost.[96, 99] The bast or outer bark fiber and the inner thick core of short, woody fibers would provide the pulp.[96, 97]

For seed production, kenaf is sown in southern latitudes in late summer so that the shortening days will induce flowering.[38, 92] The seed crop is harvested with a binder equipped with a high reel.

Diseases

Kenaf has suffered severe attacks by a fungus, *Colletotrichum hibisci*. Some selected strains of kenaf are resistant to certain races of this fungus. A rot caused by a species of *Botrytis* occurs under high humidities. It can be controlled by spraying with carbamate fungicides.

SANSEVIERIA

Sansevieria (bowstring hemp or snake plant) is grown for cordage fiber in tropical and subtropical countries. More than 50 species of the genus *Sansevieria* are native to Africa and Asia. Sansevieria is a popular ornamental garden plant in warm climates. It is a common potted house plant in the United States. Several species of *Sansevieria* have been grown experimentally in Florida as potential sources of marine cordage.[13, 15] The plants are perennials which are members of the lily family. The leaves are harvested for fiber extraction every three to five years. The leaves are 2 to 3 feet long. Sansevieria is propagated by planting leaf cuttings 6 to 8 inches long with the lower end of the cutting set in the soil in rows. Machines have been devised for making cuttings, for harvesting the leaves, and for separating the fiber. The cuttings are set with a transplanting machine.[41]

TARO

The taro or cocoyam (*Colocasia esculenta*) is grown on nearly 500 acres annually in Hawaii. The dasheen, a variety of taro of Chinese origin, is grown in Florida for the vegetable market. Taro, a native of Asia, spread to Africa and the Pacific islands many centuries ago. It is a popular food crop in tropical countries. It is grown for its starchy, edible corms and cormels.

Taro, a member of the plant family, *Araceae*, greatly resembles the ornamental elephant's-ear, *C. antiquorum*. The plants with their enormous leaves are usually 3 to 5 feet tall.

Taro culture is similar to that for growing potatoes, except that the plants need more space. Moreover, the corms are often planted in Hawaii in mud flats.[37] Top corms from taro plants are planted in the spring. The crop is dug after the tops are dead, perhaps eight months later.

Nearly 10 million pounds of taros are processed into flour or bread annually in Hawaii. Fresh corms also are peeled, mashed, and fermented to produce the dish called poi. Taro also is baked or boiled like potatoes. Most varieties of raw taro are inedible because of numerous irritating needle-like oxylate crystals in the flesh.

TEASEL

The teasel was introduced from Europe about 1840. Its culture was concentrated largely on the shores of Skaneateles Lake in New York.

The teasel, a native of southern Europe, belongs to the Dipsaceae family. It escaped from cultivation in England, New York, Oregon, and other states. The fuller's teasel plant (*Dipsacus fullonum*) is an erect herbaceous, branched, monoecious biennial, 2 to 6 feet in height, with opposite or whorled leaves. The upper pairs of leaves are united to form a cup that holds water around the stem. The blue or lilac flowers are in heads or whorls, surrounded by a many-leaved involucre. A low leafy plant is produced in the first season, but flower stalks shoot up the second year.

The teasel crop is propagated by seeds planted in cultivated rows in the spring. It is ready for harvest in the summer of the second year. In harvesting, the stems are cut a few inches below the heads with a knife as soon as the blossoms have fallen. The cut heads are then hauled to a drying yard.

The commercial product of teasel is the dried dense ovoid flower head that bears numerous pointed bracts which end in hooked spines. These heads are used to raise the nap on woolen cloth (Figure 39–16). The heads of the fuller's teasel or clothier's teasel were formerly attached to revolving cylinders that engaged fibers of a passing bolt of cloth and raised the nap to make the goods feel softer and warmer.[5] Hooks made of plastics have replaced teasels, and its culture in the United States has ceased.

Figure 39–16. Teasel head.

CHICORY

Chicory is grown in European countries as well as in South Africa. The chicory plant, a native of Europe, was introduced into the United States for production many years ago. Chicory culture began in Michigan in 1890 and continued until about 1960. It is now naturalized over a large part of the North, where it is a common weed pest in meadows and waste places.

Chicory thrives best where the average temperature during the growing season does not exceed 70° F.[20] It will stand considerable drought. The soil requirements are similar to those for sugarbeets in Michigan.

Chicory (*Cichorum intybus*) is a member of the family *Compositae*. It is also called succory, blue sailors, blue daisy, coffee weed, and bunk. Chicory is a perennial, erect herb 1½ to 6 feet in height, with long slender fleshy roots that resemble parsnips. The ray flowers are bright blue and fertile.[49]

Sugarbeet machinery was used for planting and cultivating chicory in Michigan. The roots are dug and topped in the same way as for sugarbeets. Yields of fresh chicory roots range from five to ten tons per acre.

The chicory roots are shipped to a drying plant where they are cut up into 1-inch cubes and then dried. The dried product is shipped to factories where it is roasted and ground as needed.

Dried chicory root is mixed with coffee or used as a coffee substitute. The chicory imparts a special flavor and bitterness to the beverage brewed from the coffee mixture familiar to those who have tasted the French drip coffee of the deep South. The leaves of seedlings, as well as shoots that sprout from roots planted in a dark cellar, are eaten as fresh vegetables.

PYRETHRUM

Pyrethrum is grown for its flowers, which contain two active principles, pyrethrin I and pyrethrin II, that have insecticidal properties. It has been grown at times in California, Colorado, Pennsylvania, and other states, but most of the supply now is imported. From 750 to 2,800 tons of pyrethrum flowers and extract are imported annually. The crop has been grown extensively in Yugoslavia, Japan, East Africa, and Brazil, where labor is available for picking the flowers by hand. Synthetic pyrethrins are now available also.

Botanical Characteristics

Pyrethrum (*Chrysanthemum cinerariaefolium*) is a herbaceous perennial that belongs to the family *Compositae*, the flower being a typical white daisy in appearance. The plant forms a clump 6 to 12 inches in diameter at the base from which the stems flare upward and outward (Figure 39–17). The height ranges from 15 to 40 inches. The plants do not flower during their first year, but thereafter they bloom each season. When grown in the southern states, pyrethrum often lives only as an annual or biennial.[22]

Cultural Methods

The methods of establishing pyrethrum fields are very similar to the early phases of tobacco culture (Chapter 34). About 9 ounces of good seed planted in 300 square feet of bed space should produce 15,000 to 30,000 plants. The plants from spring sowing reach a height

Figure 39–17. Pyrethrum plant in bloom (*left*); leaves and pods of pyrethrum (*right*).

of 3 to 5 inches about three months after sowing. They are then ready for transplanting. Pyrethrum also can be planted in late summer or in the fall. The plants are then set out the next spring. Methods of transplanting are the same as for tobacco. The plants are set in rows at a rate of about 10,000 plants per acre. About 300 to 500 pounds per acre of a 4-12-4 fertilizer has been recommended.[22] It is applied between the rows and worked into the soil either in early spring or after the crop is harvested.

Harvest

The pyrethrum flowers are ready for gathering when one-half to three-fourths of the florets on most of the disks are opened. A higher content of pyrethrins as well as higher total yields are obtained when harvesting is early than when delayed until the plants are three-fourths in bloom.[22, 80] An efficient hand worker can gather in a day only about 100 pounds of fresh flowers, equivalent to 25 pounds of dried flowers. Strippers operated by hand or by machine gather too much trash along with the flowers.

Soon after picking, the flowers must be dried on trays in the sun, on floors of large drying sheds, or in artificial driers.

Uses

Yields of dried pyrethrum flowers average about 700 to 800 pounds per acre where the crop is adapted. No crop is obtained in the year the plants are set out. Fair yields are obtained in the second year, but better yields are produced in the third and fourth years. Thereafter yields tend to decline. Dried flowers contain about 1 per cent total pyrethrins, the stems about 0.1 per cent, and the leaves practically none. The pyrethrin content is affected only slightly by soil type and fertilization. The powdered dried flowers of pyrethrum constitute the well-known insect powder used by our grandparents to control fleas, lice, bedbugs, and other household pests. Pyrethrins dissolved in kerosene or other petroleum derivatives, when supplemented by other chemicals constituted the common fly sprays. An important use of pyrethrum is for dusting vegetables nearly ready for harvest, when other insecticides would leave a residue of poisons on edible portions of the plants. Synthetic pyrethins have since replaced much of the pyrethrum.

Crop Pests

Several diseases attack pyrethrum, causing the death of plants which then require annual replacements. These diseases are caused by the fungi *Septoria chrysanthemella, Diplodia chrysanthemella, Sclerotium rolfsii, S. delphinii, Rhizoctonia,* and various species of *Fusarium, Sclerotinia, Alternaria,* and *Phytophthora.*

Insects have not been especially damaging to pyrethrum. However, the fact that the plant contains an insecticide is no assurance that certain insects might not attack pyrethrum severely.

BELLADONNA

Belladonna was grown as an emergency crop on 275 acres in 1918, and again on about 700 acres in 1942 when imports of this essential drug plant were cut off by war. The drug, which is obtained from the belladona, is used to dilate the pupils of the eyes and relieve muscle spasms. Atropine is an example of this drug; an antidote for organic phosphate poisoning. Annual imports of the herb and root, mostly from Yugoslavia, Italy, and the Soviet Union, average 190,000 pounds.

Belladonna is best adapted to the Northeastern, North Central, and Pacific Coast states.[57]

The name belladonna[80] means beautiful lady. The belladonna plant (*Atropa belladonna*) is a poisonous perennial herb of the *Solanaceae* (nightshade) family. It grows 2 to 6 feet in height, with ovate, entire leaves and purple bell-shaped flowers.

The sowing, transplanting, culture, harvest, and curing of belladonna are very similar to the methods used for tobacco. The 1942 crop was produced by tobacco growers in Kentucky, Tennessee, Virginia, Ohio, Pennsylvania, and Wisconsin. It can be cut two to four or more times in a season.

The leaves, stems, and roots of belladonna contain the alkaloid, atropine, used to dilate the pupils of the eyes. Extracts of the plant are used internally or externally as a sedative.

Belladonna plants are attacked not only by the Rhizoctonia disease, but also by the same insects that attack potatoes and other solanaceous crops. These insects include the Colorado potato beetle, cutworms, fleabeetles, and tobacco worms.[79]

HENBANE

The leaves, flowers, and stems of henbane (*Hyoscyamus niger*) contain several poisonous alkaloids that are used in the preparation of medicines. Henbane was grown on a small acreage in Montana, Michigan, and other states during World War II. The crop from about 200 acres will supply our ordinary domestic requirements. Supplies usually are imported. Henbane was introduced from Europe, but it escaped from cultivation.

Henbane is adapted to the relatively cool, long, summer-day conditions of the northern border and Pacific Coast states.

The plant is an annual or biennial that belongs to the family *Solanaceae*. The biennial type, which is the type usually grown, produces a crop the first year, but it may not live over winter. It can be harvested the second year when it survives. The plant has a nauseous odor. It is covered with hairs. The flowers have a five-toothed calyx and an irregular funnel-shaped corolla. The seeds are borne in a capsule.[79]

Henbane usually is sown in pots under glass in midwinter, the seedlings being transplanted to 3-inch pots and then transplanted to the field in May. The plants are set about 15 inches apart in rows 30 inches or more apart. The leaves and stems are harvested by hand when the plants are in full bloom, which usually is in August. The drying process and facilities required are about the same as for tobacco.

Extracts of henbane are used as sedatives for the relief of coughs, spasmodic asthma, spasms, and other ailments.

The henbane plant is extremely susceptible to tobacco-mosaic virus. It often is destroyed by the disease.

GINSENG

About 160,000 pounds of dried ginseng roots are produced on 100 to 200 acres annually in the United States.[105]

Ginseng (*Panax quinquefolium*), also called American ginseng, sang, red berry, and five-fingers, occurs sparingly in shady sloping locations in rich, moist soil in the hardwood forests from Maine to Minnesota and southward to the mountains of northern Georgia and Arkansas. The gathering of wild ginseng for export began in the 18th century, but natural supplies eventually were greatly depleted. Ginseng culture was started after 1870 by Abraham Whisman at Boones Path, Virginia.[103]

Ginseng might truly be regarded as a glamour crop. In the United States it has been the cause of visions of great wealth because of its high price per pound, although the profits hoped for by its promotors seldom have been realized. In eastern Asia it has been regarded as a medicine that possesses extraordinary virtues as a cure-all, an aphrodisiac, a heart tonic, and a remedy for exhaustion of body and mind. Older branched roots that resemble the human form are especially prized.[79] The chief market for ginseng is in the Orient although it is sold in food stores in the United States.

Ginseng can be produced by dividing the roots of old plants, but this entails a risk of transmitting diseases.[94] Ginseng usually is propagated by seeds which are stratified in moist sand or sawdust. They are kept in a cool damp place until they are ready to germinate, which usually is in the second spring after they are gathered. The seeds are then planted under dense natural or artificial shade in a bed of deep leaf mold or compost. The plants are spaced about 6 or 8 inches apart. The roots are ready to dig and dry for the market not earlier than the sixth year after seeding, when they have reached a length of about 4 inches.

GOLDENSEAL

Goldenseal (*Hydrastis canadensis*) was formerly grown in the United States.

Goldenseal is native to the United States, being found in patches in high, open woods, and usually on hillsides or bluffs, from western New England to Minnesota and south to Georgia and Missouri.

The goldenseal plant is an erect perennial with a hairy stem, about 1 foot in height, with two branches at the top. One branch bears a leaf, while the other bears a leaf and a flower. The greenish-white flower is followed by a fleshy, berry-like head that contains 10 to 20 small hard black shiny seeds. The rootstock is about 2 inches long, 3/4 inch thick, and bright yellow when fresh, with yellow flesh. It bears numerous fibrous rootlets. Goldenseal, which belongs to the *Ranunculaceae* or crowfoot family, also is called yellowroot, yellow puccoon, orange-root, yellow Indian paint, turmeric root, Indian turmeric, Ohio curcuma, ground raspberry, eyeroot, eyebalm, yellow-eye, jaundice root, and Indian dye.

Goldenseal was used by Indians as well as by the early settlers of eastern North America as an external remedy for sore mouth and inflamed eyes. It was also used internally as a bitter tonic for stomach and liver troubles.[90] Commercial demand began about 1860. The collection of goldenseal from wild areas for domestic and export use had so largely exhausted the supply by 1904 that its cultivation was begun by ginseng growers.

Goldenseal is propagated from seeds, rootstock division, or buds or sprouts on the fibrous roots. Rootstock division is the most popular method. The seeds are partly separated from the pulp of the fresh fruit and stratified in moist sand until October, and then planted in beds. They are transplanted a year or two later.

The soil and shade conditions for goldenseal are about the same as for ginseng, except that about 600 pounds per acre of sulfate of potash or an equivalent quantity of potash from some other source should be applied. The rootstocks are ready for digging three or four years from the time of rootstock or bud planting or five years after seed planting. They are dug in the autumn after the tops are dead, and then sorted, washed, and dried in trays. Yields of 2,000 pounds or less of dried roots may be obtained.

Goldenseal is subject to certain diseases which preclude its culture in the southeastern piedmont states.

POPPY

Seed of the opium poppy (*Papaver somniferum*) that has been rendered unviable is imported into the United States, chiefly for use in the bakery trade. About 8 million pounds are imported annually. The seeds are sprinkled on certain types of bread and rolls before baking. The seeds produce an edible oil that has an iodine number of about 135. White-seeded types are fed to birds under the name of maw seed. Considerable quantities of opium for medicinal purposes are imported from Asia Minor, India, and Mainland China. Federal narcotic laws prohibit the growing of the opium poppy in the United States except under permit. There is no commercial production here because of the risk of facilitating illicit drug traffic.

The opium poppy is adapted to sections where certain other species of the Oriental poppy are known to be adapted as ornamentals. *P. somniferum* is an annual plant that belongs to the family *Papaveraceae*. It is a native of Greece and the Orient. The plants grow to a height of 4 feet and have large white, purple, reddish-purple, or red flowers.

Opium is the dried milky juice obtained by lancing the fertile but unripe seed capsules. Opium contains about 19 distinct alkaloids which include morphine and codeine. Paregoric and laudanum are manufactured from opium. Morphine can be extracted from the dried plant.[33]

REFERENCES

1. Anonymous. "Sesame production," *USDA Farmers Bul. 2119*. 1958.
2. Anonymous. "Castorbean production," *USDA Farmers Bul. 2041* (rev.). 1960.
3. Anonymous. "Growing safflower—an oilseed crop," *USDA Farmers Bul. 2133* (rev.). 1961.

3a. Applequist, L. A., and Ohlson, B., Ed. Rapeseed Cultivation, Processing and Utilization. Elsenier, New York, pp. 1–391. 1973.

4. Artschwager, E. "Growth studies on guayule (*Parthenium argentatum*)," *USDA Tech. Bul. 885*, pp. 1–19. 1945.
5. Bailey, L. H. *Cyclopedia of American Agriculture*, Vol. II, *Crops*, Macmillan, Inc., New York, 1907.
6. Bell, J. M., and others. "Oil and meal from Canadian rapeseed," *Can. Dept. Agr. Pub. 1183*. 1967 (Rev.).
7. Bessey, E. A. "Effect of age of pollen upon the sex of hemp," *Am. J. Bot.*, 15:405–411. 1928.
8. Brigham, R. D., and Spears, B. R. "Castorbeans in Texas," *Texas Agr. Exp. Sta. Bul. B-954*. 1961.
9. Brigham, R. D. "Performance of dwarf internode castor (*Ricinus communis* L.) under two row spacings," *Tex. Agr. Exp. Sta. MP-1070*, pp. 1–14. 1972.
10. Brooks, S. N., and Keller, K. R. "Effect of time of applying fertilizer on yield of hops," *Agron. J.*, 52:516–518. 1960.
11. Brooks, S. N., Horner, C. E., and Likens, S. T. "Hop production," *USDA Inf. Bul. 240*. 1961.
12. Burgess, A. H. *Hops: Botany, Culture and Utilization*, Interscience, New York, pp. 1–300. 1964.
13. Byrom, M. H. "Progress with long vegetable fibers," in *Crops in Peace and War*, USDA Yearbook, 1950–51, pp. 472–476.
14. Byrom, M. H. "Ramie production machinery," *USDA Info. Bul. 156*. 1956.
15. Byrom, M. H., and Whittemore, H. D. "Long fiber burnishing, ribboning, and cleaning machine," *USDA Agricultural Research Service Spec. Rept. ARS 42–49*. 1961.
16. Coffey, D. L., and Warren, G. F. Effect of activated carbon and other adsorbents on the biological activity of certain herbicides. *Abstr. Meetings Weed Sci. Soc. Amer.* 52. 1967.
17. Clarkson, R. E. *Herbs—Their Culture and Uses*, Macmillan, Inc., New York, pp. 1–226. 1944.

18. Clarkson, V. A. "Effect of black polyethylene mulch on soil and micro-climate temperature and nitrate level." *Agron. J.* 52:307–309. 1960.
19. Classen, C. E., and Kiesselbach, T. A. "Experiments with safflower in western Nebraska," *Nebr. Agr. Exp. Sta. Bul. 376,* pp. 1–28. 1945.
20. Cormany, C. E. "Chicory growing in Michigan," *Mich. Agr. Exp. Sta. Spec. Bul. 167,* pp. 1–11. 1927.
21. Crane, J. C. "Kenaf–fiber plant rival of jute," *Econ. Bot.,* 1(3):334–350. 1947.
22. Culbertson, R. E. "Pyrethrum–A new crop," *Proceedings 2nd Dearborn Conference of Agriculture, Industry, and Science,* Farm Chemurgic Council and the Chemical Foundation, Inc., Dearborn, Mich., pp. 292–297. 1936.
23. Dennis, R. E., and Rubis, D. D. "Safflower production in Arizona," *Ariz. Agr. Exp. Sta. Bul. A-47,* pp. 1–24. 1966.
24. Dewey, L. H. "Hemp," *USDA Yearbook, 1913,* pp. 283–346.
25. Domingo, W. E., and Crooks, D. M. "Investigations with the castorbean plant," *J. Am. Soc. Agron.,* 37(9):750–762; (10):812–820; (11):910–915. 1945.
26. Dunlap, A. A. "Septoria leaf spot on safflower in north Texas," *Plant Disease Reporter,* 25(14):389. 1941.
27. Eaton, B. J. "Identifying and controlling wild hemp," *Kans. Agr. Exp. Sta. Bul. 555,* pp. 1–12. 1972.
28. El-Hamidi, A., and others. "Effects of nitrogen and spacing on castor-bean in sandy soils in Egypt," *Expl. Agric.* 4:61–64. 1968.
29. Ellis, N. K., and others. "A study of some factors affecting the yield and market value of peppermint oil," *Purdue U. Agr. Exp. Sta. Bul. 461,* pp. 127. 1941.
30. Feustel, I. C., and Clark, F. E. "Opportunities to grow our own rubber," in *Crops in Peace and War,* USDA Yearbook, 1950–51, pp. 367–374.
31. Fick, G. N., and Swallers, C. M. "Higher yields and greater uniformity with hybrid sunflowers," *N. Dak. Farm. Rsh.,* 29(6):7–9. 1972.
32. French, H. E., and Humphrey, H. O. "Experiments on sunflower oil," *U. Mo. Bul.* 27(7); *Mo. Agr. Exp. Sta. Series* 25. 1926.
33. Fulton, C. C. *The Opium Poppy and Other Poppies.* U.S. Treas., Dept. Bureau of Narcotics, pp. 1–85. 1944.
34. Godfrey, G. H. "Gray mold of castor bean," *J. Agr. Res.,* 23:679–715. 1923.
35. Gould, G. E. "Insect pests of mint," *Purdue U. Agr. Exp. Sta. Circ. 211,* pp. 1–8. 1935.
36. Green, R. J., Jr. "Mint farming," *U.S. Dept. Agr. Info. Bul. 212* (rev.). 1963.
37. Greenwell, A. H. B. "Taro–with special reference to its culture and uses in Hawaii," *Econ. Bot.,* 1(3):276–289. 1947.
38. Haarer, A. E. *Jute Substitute Fibers,* Wheatland Journals, Ltd., London, pp. 1–210. 1952.
39. Hannay, A. H. "The sunflower, its cultivation and uses," *USDA Library List 20.* 1941.
40. Hauser, E. A., and Le Beau, D. C. "Studies in compounding guayule rubber," *India Rubber World,* 106(5):447–449. 1942.

41. Hellwig, R. E., and Byrom, M. H. "Sansevieria planting machinery," *USDA Agr. Res. Serv. ARS 42–34.* 1959.
42. Hensley, H. C. "Production of sunflower seed in Missouri," *Mo. Coll. Agr. Ext. Cir. 140.* 1924.
43. Hoag, B. K., and others. "Safflower in North Dakota," *N. Dak. Agr. Exp. Sta. Bul. 447,* pp. 1–16. 1969.
44. Hoerner, G. R. "Downy mildew of hops," *Oreg. Ext. Bul. 440,* pp. 1–11. 1932.
45. Horner, C. E. "Chemotherapeutic effects of streptomycin on establishment and progression of systemic downy mildew infection in hops," *Phytopath.,* 53(4):472–474. 1963.
46. Humphrey, J. R. "Marketing hemp," *Ky. Agr. Exp. Sta. Bul. 221,* pp. 1–43. 1919.
47. Jordan, H. V., Lang, A. L., and Enfield, G. H. "Effects of fertilizers on yields and breaking strengths of American hemp, *Cannabis sativa,*" *J. Am. Soc. Agron.,* 38(6):551–562. 1946.
48. Kadans, J. M. *Modern Encyclopedia of Herbs.* Parker Publ. Co., West Nyack, N.Y., pp. 1–256. 1970.
49. Kains, M. G. "Chicory growing," *USDA Div. Bot. Bul. 19,* pp. 1–52. 1898.
50. Kirby, R. H. *Vegetable Fibers—Botany, Cultivation and Utilization,* Interscience, New York, pp. 1–464. 1963.
51. Kittock, D. L., and Williams, J. H. "Castorbean production as related to length of growing season: II. Date of planting tests," *Agron. J.,* 59:456–458. 1967.
52. Kittock, D. L., and Williams, J. H. "Effects of plant population on castorbean yield," *Agron. J.,* 62(4):527–529. 1970.
53. Knowles, P. F., and Miller, M. D. "Safflower in California," *Calif. Agr. Exp. Sta. Manual 27.* 1960.
54. Kuhlman, G. W., and Fore, R. E. "Cost and efficiency in producing hops in Oregon," *Oreg. Agr. Exp. Sta. Bul. 364,* pp. 1–57. 1938.
55. Lloyd, F. E. *Guayule, a Rubber Plant of the Chihuahuan Desert.* Carnegie Inst. Wash., Washington, pp. 1–213. 1911.
56. Maloy, O. C., and Skotland, C. B. "Diseases of mint," *Wash. Ext. Circ. 357,* pp. 1–4. 1969.
57. McCollum, W. B. "The cultivation of guayule," *India Rubber World,* 105(1):33–36. 1941.
58. McGregor, W. G. "Safflowers in Canada," *Mimeographed Rpt. 62,* Central Experimental Farm, Ottawa, pp. 1–8. 1943.
59. McPhee, H. C. "The genetics of sex in hemp," *J. Agr. Res.,* 31:935–943. 1925.
60. Nayar, N. M., and Mehra, K. L. "Sesame: Its uses, botany, cytogenetics and origin," *Econ. Bot.,* 24(1):20–31. 1970.
61. Neller, J. F. "Culture, fertilizer requirements and fiber yields of ramie in the Florida Everglades," *Fla. Agr. Exp. Sta. Bul. 412,* pp. 1–40. 1945.
62. Owen, D. F., and others. "A process for producing nontoxic rapeseed protein isolate and an acceptable feed by-product," *Cer. Chem.,* 48(2): 91–96. 1971.

63. Percival, J. *Agricultural Botany* (8th ed.), Duckworth, London, pp. 1–839. 1936.
64. Peterson, W. F. "Safflower culture in the west-central plains," *USDA Inf. Bul. 300,* pp. 1–22. 1965.
65. Polhamus, L. G. "Rubber from guayule," *Agriculture in Americas,* 5(2):1–4. 1945.
66. Polhamus, L. G. *Rubber–Botany, Production and Utilization,* Interscience, New York, pp. 1–449. 1962.
67. Pritchard, F. J. "Change of sex in hemp," *J. Hered.,* 7:325–329. 1916.
68. Puri, Y. P., and others. "Anatomical and agronomic studies of phormium in western Oregon," *USDA Prod. Rsh. Rpt. 93,* pp. 1–43. 1966.
69. Robbins, W. W. *Botany of Crop Plants,* Blakiston, Philadelphia, pp. 1–639. 1931.
70. Robinson, B. B. "Ramie fiber production," *USDA Circ. 585,* pp. 1–15. 1940.
71. Robinson, B. B. "Hemp," *USDA Farmers Bul. 1935,* pp. 1–16. 1943.
72. Robinson, R. G. "Mustard and rape, oilseed crops for Minnesota," *Minn. Ext. Bul. 311,* pp. 1–12. 1964.
73. Robinson, R. G., and others. "The sunflower crop in Minnesota," *Minn. Ext. Bul. 299,* pp. 1–30. 1967.
74. Schaffner, J. H. "The nutation of *Helianthus,*" *Bot. Gaz.,* 29:197–200. 1900.
75. Schaffner, J. H. "The fluctuation curve of sex reversal in staminate hemp plants induced by photoperiodicity," *Am. J. Bot.,* 18:424–430. 1931.
76. Schoenleber, L. G. "Mechanization of castor bean harvesting," *Okla. Agr. Exp. Sta. Bul. B-591.* 1961.
77. Shrader, J. H. "The castor-oil industry," *USDA Bul. 867,* pp. 1–40. 1920.
78. Sievers, A. F. "Methods of extracting volatile oils from plant material and the production of such oils in the United States," *USDA Tech. Bul. 16,* pp. 1–36. 1928.
79. Sievers, A. F. "Production of drug and condiment plants," *USDA Farmers Bul. 1999,* pp. 1–99. 1948.
80. Sievers, A. F. "Belladonna in the United States in 1942," *Drug and Cosmetic Industry,* p. 7. May 1943.
81. Sievers, A. F., Lowman, M. S., and Hurst, W. M. "Harvesting pyrethrum," *USDA Circ. 581,* pp. 1–18. 1941.
82. Smith, D. C. "Varietal improvement in hops," in *USDA Yearbook, 1937,* pp. 1215–1241.
83. Smith, D. C. "Influence of moisture and low temperature on the germination of hop seeds," *J. Agr. Res.,* 58:369–381. 1939.
84. Sulerud, G. I. "An economic study of the hop industry in Oregon," *Oreg. Agr. Exp. Sta. Bul. 288,* pp. 1–77. 1931.
85. Thomas, C. A., Klisewicz, J., and Zimmer, D. "Safflower diseases," *USDA Agricultural Research Service Spec. Rept. ARS 34–52.* 1963.
86. Thompson, F. C. "Hop growing," *Agriculture,* 77:157–160. 1970.
87. Talley, L. J., and others. "Sunflower food products, *Tex. Agr. Exp. Sta. MP 1026.* 1972.
88. Talley, L. J., and Burns, E. E. "Sunflower, a survey of the literature,

June 1967 to January, 1971," *Texas Agr. Exp. Sta. MP 992*, pp. 1–21. 1972.

89. Trotter, W. K. "Potential for oilseed sunflowers in the United States," *USDA AER 237*, pp. 1–55. 1973.
90. Van Fleet, W. "Goldenseal under cultivation," *USDA Farmers Bul. 613*, pp. 1–15. 1914.
91. Vinall, H. N. "The sunflower as a silage crop," *USDA Bul. 1045*. 1922.
92. Walker, J. E., and Sierra, M. "Some cultural experiments with kenaf in Cuba," *USDA Cir. 854*. 1950.
93. Weiss, E. A. *Castor, Sesame and Safflower*, Barnes and Noble, New York, pp. 1–901. 1971.
94. Whetzel, H. H., and others. "Ginseng diseases and their control," *USDA Farmers Bul. 736*, pp. 1–23. 1916.
95. White, G. A., and Higgins, J. J. "Culture of crambe—A new industrial oilseed crop," *USDA Prod. Rsh. Rpt. 95*, pp. 1–20. 1966.
96. White, G. A., and others. "Culture and harvesting methods for kenaf—an annual crop source of pulp in the southeast," *USDA Prod. Rsh. Rpt. 113*, pp. 1–38. 1970.
97. White, G. A. "New crops on the horizon," *Seed World*, 110(4):20–22. 1972.
98. White, W. J., and Putt, E. D. "Sunflower production for grain," *Spec. Bul. 69*, Agr. Supplies Board, Ottawa, 1943.
99. Whiteley, E. L. "Influence of date of planting on the yield of kenaf (*Hisbiscus cannabinus* L.)," *Agron. J.*, 63(1):135–136. 1971.
100. Whiteley, E. L. Influence of date of harvest on the yield of kenaf (*Hisbiscus cannabinus* L.)," *Agron. J.*, 63(3):509–510. 1971.
101. Whittemore, H. D., and Hellwig, R. E. "Kenaf," *Soil and Crop Sci. Soc. Fla.*, 18:338–340. 1958.
102. Wiley, H. W. "The sunflower plant: Its cultivation, composition, and uses," *USDA Div. Chem. Bul. 60*. 1901.
103. Williams, L. O. "Ginseng," *Econ. Bot.*, 11(2):344–348. 1957.
104. Williams, L. O. "Drug and condiment plants," *USDA Handbk. 172*, pp. 1–37. 1960.
105. Williams, L. O. "Growing ginseng," *USDA Farmers Bul. 2201*. 1964.
106. Wilson, F. D., and Joyner, J. F. "Effects of age, plant spacing and other variables on growth yield and fiber quality of kenaf (*Hibiscus cannabinus* L.)," *USDA Tech. Bul. 1404*, pp. 1–19. 1969.
107. Zimmerman, L. H., Miller, M. D., and Knowles, P. F. "Castorbeans in California," *Calif. Agr. Exp. Sta. Cir. 468*. 1958.
108. Zimmerman, L. "Castorbeans: A new oil crop for mechanical production," *Adv. Agron.*, 10:257–288. 1958.

PART FIVE

APPENDIX

Table A–1. Seeding; Seed and Plant Characteristics

CROP	BOTANICAL NAME	SEEDING RATE PER ACRE		SEEDS PER		WEIGHT PER BUSHEL	GERMINATION TIME	TEMPERATURE TYPE	GROWTH HABIT	CHROMOSOME NUMBER	PHOTOPERIODIC REACTION
		Close Drills (lb.)	*Rows (lb.)*	*Pound (thousands)*	*Gram (no.)*	*(lb.)*	*(days)*				
Alfalfa	*Medicago sativa* L.			220	500	60	7	C	P	16	L
Humid areas		10–20									
Irrigation		10–15									
Semiarid		8–10									
Alyceclover	*Alysicarpus vaginalis* DC.	10–20		275	660	60	21	W	A	8	
Bahiagrass	*Paspalum notatum* Flügge	10–12		150	336		21	W	P	20	
Barley	*Hordeum vulgare* L.	72–96		13	30	48	7	C	A, WA	7	L
Bean											
Adzuki	*Phaseolus angularis* Wight		20–25	5	11	60	10	W	A	11	
Field	*Phaseolus vulgaris* L.		40–75	1–2	4	60	8	W	A	11	S; N
(dryland)			5–20								

Temperature type: C = cool-weather growth; W = warm-weather growth.
Growth habit: A = annual; W.A. = winter annual; B = biennial; P = perennial; P(A) = perennial but grown as annual.
Chromosome number: reduced (gametic) number (N).
Photoperiodic reaction: L = long day; S = short day; N = day neutral or indeterminate; I = intermediate.

TABLE A–1. Seeding; Seed and Plant Characteristics (*continued*)

CROP	BOTANICAL NAME	SEEDING RATE PER ACRE		SEEDS PER		WEIGHT PER BUSHEL	GERMINATION TIME	TEMPERATURE TYPE	GROWTH HABIT	CHROMOSOME NUMBER	PHOTOPERIODIC REACTION
		Close Drills (lb.)	*Rows (lb.)*	*Pound (thousands)*	*Gram (no.)*	*(lb.)*	*(days)*				
Lima	*Phaseolus limensis* Macf.		120	0.4	2	56	9	W	P(A)	11	S
Lima (baby)	*Phaseolus lunatus* L.		60	0.5				W	A	11	S; N
Mung	*Phaseolus aureus* Roxb.		10–15	11	24		7	W	A	11	S
Tepary	*Phaseolus acutifolius latifolius* Freem		10–12	2	5			W	A	11	S
Beet (see mangel and sugarbeet)											
Beggarweed (Florida) unhulled	*Desmodium purpureum* DC.	10 30–40		194	400	60	28	W	P(A)	11	S
Bentgrass											
Astoria	*Agrostis tenuis* Sibth.	40–60		5,500	12,000		28	C	P	14	

Colonial	*Agrostis tenuis*	40–60	8,000	19,000	20–40	28	C	P	14	
Creeping	*Agrostis palustris* Huds.	40–60	7,700	17,000		28	C	P	14	L
Highland	*Agrostis tenuis*	40–60	9,100	20,000		28	C	P	14	
Velvet	*Agrostis canina* L.	40–60	11,000	24,000		21	C	P	7; 14	
Belladonna	*Atropa belladonna* L.		480	1,000				P	36	
Bermudagrass	*Cynodon dactylon* (L)	6–8	1,800	3,900	40(14)	21	W	P	15; 18	L
(unhulled)	Pers.	10–15	1,300	2,900						
Berseem (see clover, Egyptian)										
Big trefoil	*Lotus ulginosus* Schkuhr.	4–6	1,000	1,900	60	7	C	P	7	L
Birdsfoot trefoil	*Lotus corniculatus* L.	8–12	375	800	60	7	C	P	12	
Black medic	*Medicago lupulina* L.	10–15	300	600	60 (hulled)	7	C	A	8	L(?)
Bluegrass										
Annual	*Poa annua* L.	15–25	1,200	2,600		21	C	A	7; 14	N
Bulbous	*Poa bulbosa* L.	20–25	450	1,000		35	C	P	14	L
Canada	*Poa compressa* L.	15–25	2,500	5,500	14	28	C	P	21; 28	L
Kentucky	*Poa pratensis* L.	15–25	2,200	4,800	14	28	C	P	14; 28; 35	N

TABLE A–1. Seeding; Seed and Plant Characteristics (*continued*)

CROP	BOTANICAL NAME	SEEDING RATE PER ACRE		SEEDS PER		WEIGHT PER BUSHEL	GERMINATION TIME	TEMPERATURE TYPE	GROWTH HABIT	CHROMOSOME NUMBER	PHOTOPERIODIC REACTION
		Close Drills (lb.)	*Rows (lb.)*	*Pound (thousands)*	*Gram (no.)*	*(lb.)*	*(days)*				
Nevada	*Poa nevadensis* Vasey	6–10		1,000	2,300		21	C	P	31; 32; 33	
Rough	*Poa trivialis* L.	15–25		2,500	5,600		21	C	P	7	
Texas	*Poa arachnifera* Torr.	6–10		1,100	2,500		28	C	P	21	
Wood	*Poa nemoralis* L.			3,200	7,000		28	C	P	14; 21	
Bluestem											
Big	*Andropogon furcatus* Muhl.	15–20		150	340		28	W	P	35; 20	S
Little	*Andropogon scoparius* Michx.	12–20		260	560		28	W	P	20	L
Sand	*Andropogon hallii* Hack.	10–15		100	230		28	W	P	30; 35	
Bonavist	*Dolichos lablab* L.		20–25	1	2	60		W	P(A)	11; 12	N

Bromegrass											
Awnless	*Bromus inermis* Leyss.	15–20		137	300	14	14	C	P	21; 28; 35	L
Mountain	*Bromus marginatus* Nees.	10–20		64	140		14	C	P	21; 28; 35	
Broomcorn	*Sorghum vulgare* Pers.		2–4	25	60	(44–50)	10	W	A	10	S
Buckwheat (common)	*Fagopyrum saggitatum (esculentum)* Grlib	36–60		20	45	48	6	W	A	8	L; N
Buckwheat (tartary)	*Fagopyrum tataricum* Gaertn.	24		26	60	48			A	8	
Buffalograss (burs)	*Buchloë dactyloides* Engelm.	15–20		50	110		28	W	P	28; 30	
Buffalograss (caryopses)		5		330	738						
Buffelgrass	*Cenchrus ciliaris* L.				?	?		W	P	18	S
Burclover											
California (out of bur)	*Medicago hispida* Gaertn.	15–20		209	303	50	14	C	WA	7	
Spotted (in bur)	*Medicago arabica* Huds.	50–75		22	49	8–12	14	C	WA	8	
Buttonclover	*Medicago orbicularia*	15–20		150				C	A	8	

TABLE A–1. Seeding; Seed and Plant Characteristics (*continued*)

CROP	BOTANICAL NAME	SEEDING RATE PER ACRE		SEEDS PER		WEIGHT PER BUSHEL	GERMINATION TIME	TEMPERATURE TYPE	GROWTH HABIT	CHROMOSOME NUMBER	PHOTOPERIODIC REACTION
		Close Drills (lb.)	*Rows (lb.)*	*Pound (thousands)*	*Gram (no.)*	*(lb.)*	*(days)*				
Burnet	*Sanguisorbe minor* Scop.							W	P	14	
Canarygrass	*Phalaris canariensis* L.	20–25		68	150	50			A	6	L
Carpetgrass	*Axonopus compressus* (Swertz) Beauv.	8–12		1,350	2,500	18–36	21	W	A	6	N
Carrot (stock)	*Daucus carota* L.		2	405	800	50	28	C	B	9	L
Cassava (roots)	*Manihot utilissima (esculenta)* Pohl.		900					W	P	18; 36	N
Castorbean	*Ricinus communis* L.		7–10	1	2	50		W	P(A)	10	L
Cheat	*Bromus secalinus* L.	30–50		71	160			C	WA	7; 14	
Chicory	*Cichorium intybus* L.		1–2	420	940		14	C	P	Ca 9	L

Chickpea	*Cicer arietinum* L.	60		1	2		7	C	A	7; 8	N; L
Chufa	*Cyperus esculentus* L		15–40			44		W	P	54	
Clover											
Alsike (alone)	*Trifolium hybridum* L.	6–8		680	1,500	60	7	C	P	8	L
(with timothy)		2–4									
Alyce (see alyceclover)											
Bur (see burclover)											
Cluster	*Trifolium glomeratum* L.	3–4		1,300	2,900	60	10	C	WA	7; 8	L
Crimson	*Trifolium incarnatum* L.	15–25		150	330	60	7	C	WA	7; 8	L
(unhulled)		40–60									
Egyptian (berseem)	*Trifolium alexandrinum* L.	15–20		210	460	60	7	C	WA	8	L
Hop	*Trifolium agrarium* L.	8–12		830	1,800	60			WA	7	L
Ladino	*Trifolium repens* L.	5–7		860	1,900	60	10	C	P	8; 12; 14; 16	L
Lappa	*Trifolium lappaceum* L.	4–5		680	1,500	60	7	C	WA	8	L
Large Hop	*Trifolium procumbens* L.	3–4		2,500	5,400	60	14	C	WA	7	L

TABLE A–1. Seeding; Seed and Plant Characteristics (*continued*)

CROP	BOTANICAL NAME	SEEDING RATE PER ACRE		SEEDS PER		WEIGHT PER BUSHEL	GERMINATION TIME	TEMPERATURE TYPE	GROWTH HABIT	CHROMOSOME NUMBER	PHOTOPERIODIC REACTION
		Close Drills (lb.)	*Rows (lb.)*	*Pound (thousands)*	*Gram (no.)*	*(lb.)*	*(days)*				
Low Hop	*Trifolium dubium* (minus) L.	4–5		860	1,900	60	14	C	WA	14; 16	L
Persian	*Trifolium resupinatum* L.	4–8		640	1,400	60	7	C	WA	8	L
Red	*Trifolium pratense* L.	8–12		260	600	60	7	C	P	7; 14	L
(with timothy)		4–6									
Rose	*Trifolium hirtum* All.	15–20		160	360						
Sour (see sourclover)											
Strawberry	*Trifolium fragiferum* L.	4–6		290	640	60	7	C	P	8	L
Sub (subterranean)	*Trifolium subterraneum* L.	20–25		55	120	60	14	C	WA	8	L
Sweet (see sweetclover)											
White	*Trifolium repens* L.	5–7		700	1,500	60	10	C	P	8; 12; 14; 16	L

Zigzag	*Trifolium medium* L.								P	40; 48; 49	L
Cloudgrass	*Agrostis nebulosa* Bois. and Reut										
Corn											
Field—for grain	*Zea mays* L.		6–18	1.2	3	56	7	W	A	10	S
—for silage			8–18								
Pop	*Zea mays* L.		3–6	3	7	62–68	7	W	A	10	S
Sweet	*Zea mays* L.		12–18	2	4	50	7	W	A	10	S
Cotton						32					
Upland	*Gossypium*										N
	hirsutum L.			4	8	28–33	12	W	P(A)	26	
Eastern humid			24–40								
Mississippi Valley			24–48								
Drier areas			16–32								
Irrigated			25–30								N
American Pima (S.W. irrigated)	*Gossypium barbadense* L.		35–40	4	9		12	W	P(A)	26	
Cowpea	*Vigna sinensis* Endl.	75–120	30–45	2–6	8	60	8	W	A	12	S
Crambe	*Crambe abyssinica* Hochst ex R. E. Tries	8–5		86	190	27		C	A	45	
Crested dogtail	*Cynosurus cristatus* L.	15–25		860	1,900		21	C	P	7	L
Crested wheatgrass (see wheatgrass)											

TABLE A–1. Seeding; Seed and Plant Characteristics (*continued*)

CROP	BOTANICAL NAME	SEEDING RATE PER ACRE		SEEDS PER		WEIGHT PER BUSHEL	GERMINATION TIME	TEMPERATURE TYPE	GROWTH HABIT	CHROMOSOME NUMBER	PHOTOPERIODIC REACTION
		Close Drills (lb.)	*Rows (lb.)*	*Pound (thousands)*	*Gram (no.)*	*(lb.)*	*(days)*				
Crotalaria											
Showy	*Crotalaria spectabilis* Roth.	15–25	4–6	30	80	60	10	W	A		S
Striped	*Crotalaria mucronata* Desv. (*striata* DC.)	10–20	3–4	75	215	60	10	W	A	8	S
"Intermedia"	*Crotalaria intermedia* Kotschy.	10–15	2–4	96	210	60	10	W	A		S
"Juncea"	*Crotalaria juncea* L.	20–40	5–7	16	35	60	10	W	A	8	S
"Lanceolata"	*Crotalaria lanceolata*	8–12	2–3	170	375	60	10	W	A	8	S
Crownvetch	*Coronilla varia* L.	5–10		140	300				P	12	
Dallisgrass	*Paspalum dilatatum* Poir.	8–25		340	485	12–15	21	W	P	20; 25	S
Digitalis (foxglove)	*Digitalis purpurea* L.								B	28	L

Dill	*Anethum gravolens* L.		5	410	900		21		A; P	11	L
Fenugreek	*Trigonella foenum-graecum* L.	15–30				60			A	8	L
Dropseed, Sand	*Sporobolus cryptandrus* (Torr.) A. Gray	3–5		5,450	12,000		42	W	P	9	
Fescue											
Chewings	*Festuca rubra* var. *commutata* Gaud.	15–40		615	1,400	14–30	21	C	P	21	
Hair	*Festuca capillata* Lam.			1,500	3,200		28	C	P	7	
Meadow	*Festuca elatior* L.	10–25		230	500	14–24	14	C	P	7; 14; 21; 35	
Red	*Festuca rubra* L.	15–40		400	900		21	C	P	7; 21; 28; 35	
Sheeps	*Festuca ovina* L.	25		530	1,167	10–30	21	C	P	7; 14; 21; 35	
Tall	*Festuca arundinacea*	10–25		227	500			C	P	21	
Feterita (see sorghum)											

TABLE A–1. Seeding; Seed and Plant Characteristics (*continued*)

CROP	BOTANICAL NAME	SEEDING RATE PER ACRE		SEEDS PER		WEIGHT PER BUSHEL	GERMINATION TIME	TEMPERATURE TYPE	GROWTH HABIT	CHROMOSOME NUMBER	PHOTOPERIODIC REACTION
		Close Drills (lb.)	*Rows (lb.)*	*Pound (thousands)*	*Gram (no.)*	*(lb.)*	*(days)*				
Field pea											
(large-seeded)	*Pisum arvense* L.	120–180		4	8	60	8	C	A	7	L
(small-seeded)		90–120									
Austrian winter	*Pisum arvenu* L.	30–90		5	11		8	C	WA		
Flax (for seed)	*Linum usitatissimum* L.	28–42		82	180	56	7	C	A	15	L
(for fiber)		84		113	250						
Gamagrass	*Tripsacum dactyloides* L.	Veg.		7	15			W	P	35; 18; 36	S
Giant panic grass	*Panicum antidotale*	2–6		610	1,450		28		A	9	
Goldenseal	*Hydrastis canadensis* L.								P	13	
Gramagrass											
Black	*Bouteloua eripoda* Torr.	7–9		560	1,200			W	P	21	

Blue	*Bouteloua gracilis* (H.B.K.) Lag.	8–12		900	1,980		28	W	P	20; 21; 14	N; S
Hairy	*Bouteloua hirsuta* Lag.	10–15		980	2,200					21	
Side oats	*Bouteloua curtipendula*							W	P		
	Torr.	15–20		200	442		28	W	P	21; 28; 35	S; I
(caryopses)				730	1,600						
Grasspea	*Lathyrus sativus* L.	70–80		5			60		A	7	L
Guar	*Cyamopsis psoralides* DC.	40–60	10	5	11	60		W	A	7	
Guayule	*Parthenium argentatum* Gray			600	1,300				P	36	
Guineagrass	*Panicum maximum* Jacq.	Veg.		1,000	2,200		28	W	P	18	
Hairy indigo	*Indigofera hirsuta*	3–10		200		55		W	A	16	S
Hardinggrass	*Phalaris tuberosa* var. *stenoptera* Hitchc.	25–30		340	750		28		P	14	
Hegari (see sorghum)											

TABLE A–1. Seeding; Seed and Plant Characteristics (*continued*)

CROP	BOTANICAL NAME	SEEDING RATE PER ACRE		SEEDS PER		WEIGHT PER BUSHEL	GERMINATION TIME	TEMPERATURE TYPE	GROWTH HABIT	CHROMOSOME NUMBER	PHOTOPERIODIC REACTION
		Close Drills (lb.)	*Rows (lb.)*	*Pound (thousands)*	*Gram (no.)*	*(lb.)*	*(days)*				
Hemp	*Cannabis sativa* L.	44		27	45	44	7	C	A	10	S
Henbane	*Hyoscyamus niger* L.								A; B	17	L
Hop	*Humulus lupulus* L.		Veg.						P	10	
Horsebean	*Vicia faba* L.	120–200		1–2	3	47	10	C	A	6; 7	L
Hungariangrass (see millet–foxtail)											
Indiangrass	*Sorghastrum nutans* (L). Nash			170	365		21	W	P	20	I
Jerusalem artichoke (tubers)	*Helianthus tuberosus* L.	300–800						C	P	51	S; N
Job's tears	*Coix lachyryma-jobi* L.								A	10	N
Johnsongrass	*Sorghum halepense* (L.) Pers.	20–30		130	290	28	35	W	P	20	S
Kafir (see sorghum)											

Kale (Thousand-headed)	*Brassica oleraceae* DC.	$\frac{1}{2}$–$\frac{3}{4}$		140	315		10	C	B	9	N; L
Kenaf	*Hibiscus cannabinua* L.	30–35	2–4	14			7	W	A	18	S
Kidney vetch	*Anthyllis vulneraria* L.	20		150	320	60				6	
Koa Haole	*Leucaena leucocephala*										
Kudzu	*Pueraria thunbergiana* Benth.	Veg.	6–10	37	81	54	14	W	P	22 or 24	
Lentil	*Lens esculenta* Moench. J.		5–8	9	20	60		W	A	7	L
Lespedeza											
Chinese (Sericea)	*Lespedeza cuneata* (Dum. de Cours) G. Don.	30–40		372	820	35	28	W	P	10	S
(scarified)		15–20		335	820	60					
Common and Tenn. 76	*Lespedeza striata* Hook & Arn	25–30		343	750	25	14	W	A	10	S
Kobe	*Lespedeza striata*	30–35		185	750	30	14	W	A	10	S
Korean	*Lespedeza stipulaceae* Maxim	20–25		240	525	45	14	W	A	10	S

TABLE A–1. Seeding; Seed and Plant Characteristics (*continued*)

CROP	BOTANICAL NAME	SEEDING RATE PER ACRE		SEEDS PER		WEIGHT PER BUSHEL	GERMINATION TIME	TEMPERATURE TYPE	GROWTH HABIT	CHROMOSOME NUMBER	PHOTOPERIODIC REACTION
		Close Drills (lb.)	*Rows (lb.)*	*Pound (thousands)*	*Gram (no.)*	*(lb.)*	*(days)*				
Siberian	*Lespedeza hedysaroides* (Pallas) Kitagawa	10–15		370	820	60	21	W			
Lovegrass, Weeping	*Eragrostis curvula* (Schrad. Nees	1/4		1,500	3,300		14	W	P	20	
Lehmann	*Eragrostis lehmanniana* Nees	1–3		4,245	9,400			W	P	20	
Sand	*Eragrostis trichoides* (Nutt.) Nash	1–2	1	1,800	4,000			W	P		
Lupine											
Blue	*Lupinus angustifolius* L.	70–90		3	7	60	10	C	A	20; 24	N; S
White	*Lupinus albus* L.	120–160		3	7	60	7	C	A	Ca 20	N
Yellow	*Lupinus luteus* L.	45–60		4	9	60	21	C	A	Ca 23	N; L
Mangel	*Beta vul-*										

Meadow fescue (see fescue)											
	Alopecurus pratensis L.	15–25		540	1,200	6–12	14	C	P	14	L
Meadow foxtail											
Meadowfoam	*Limnonthes spp*	10–15									
Melilotus indica (see sourclover)											
Milk vetch	*Astragalus cicer* L.										
Millet											
Browntop	*Panicum ramosum* L.	10–20	4–10	140	300		14	W	A	18	
Foxtail	*Setaria italica* (L.) Beauv.	20–30		220	470	50	10	W	A	9	S
Japanese (barnyard)	*Echinochloa crusgalli frumentacea* W. F. Wight	20–25		155	320	35	10	W	A	18; 24; 28	
Pearl (cattail)	*Pennisetum glaucum* L.	16–20	4–6	85	190		7	W	A	7	S
Proso	*Panicum miliaceum* L.	15–35		80	180	56	7	W	A	18; 21; 36	S
Ragi (finger)	*Eleusine coracana* (L.) Gaertn							W	A	18	S
Milo (see sorghum)											

TABLE A–1. Seeding; Seed and Plant Characteristics (*continued*)

CROP	BOTANICAL NAME	SEEDING RATE PER ACRE		SEEDS PER		WEIGHT PER BUSHEL	GERMINATION TIME	TEMPERATURE TYPE	GROWTH HABIT	CHROMOSOME NUMBER	PHOTOPERIODIC REACTION
		Close Drills (lb.)	*Rows (lb.)*	*Pound (thousands)*	*Gram (no.)*	*(lb.)*	*(days)*				
Mint											
Peppermint	*Mentha piperita* L.	Veg.							P	18	
Spearmint	*Mentha spicata* L.	Veg.							P	18	
Molassesgrass	*Melinis minutiflora* Beauv.			6,800	15,000		21	W	P	18	
Mustard			3–4			58					
Black	*Brassica nigra* Koch			570	1,250		7	C	A	8	L
Brown	*Brassica juncea* Coss.	3		280	624		7	C	A	18	
White	*Brassica alba* Rabenh.	4		73	160		5	C	A	12	
Napiergrass	*Pennisetum purpureum* Schumach.	Veg.		1,402	3,100		10	C	P	14	
Narrowleaf trefoil	*Lotus tenuis*	8–12		375	800	60	7	C	P	6	L

Oats											
Common	*Avena sativa* L.	48–128		14	30	32	10	C	A; WA	21	L
Red	*Avena byzantina* C. Koch	48–128		14	30	32	10	C	A; WA	21	L
Oatgrass, tall meadow	*Arrhenatherum elatius* (L.) Mert & Koch	30–40		150	330	11–14	14	C	P	14	
Orchardgrass	*Dactylis glomerata* L.	20–25		590	1,440	14	18	C	P	14	N; L
Paragrass	*Panicum purpurescens* (var. *barbinode*) Raddi	Veg.						W	P	18	
Patridgepea	*Cassia fasiculata* michx.										
Pea (see field pea)											
Peanut	*Arachis hypogeae* L.		20–40	1	1–3	20–30	10	W	A	20	S
Pepper	*Capsicum frutescens* L.		½	75	165		14	W	A	12	N; S
Perilla	*Perilla frutescens* Britt. (*ocymoides*)		4–5			37			A	19; 20	S

TABLE A–1. Seeding; Seed and Plant Characteristics (*continued*)

CROP	BOTANICAL NAME	SEEDING RATE PER ACRE		SEEDS PER		WEIGHT PER BUSHEL	GERMINATION TIME	TEMPERATURE TYPE	GROWTH HABIT	CHROMOSOME NUMBER	PHOTOPERIODIC REACTION
		Close Drills (lb.)	*Rows (lb.)*	*Pound (thousands)*	*Gram (no.)*	*(lb.)*	*(days)*				
Pigeonpea	*Cajanus cajan* (*indicus* Spreng)		8–10	8	18	60			P	11	
Popcorn (see corn)											
Poppy (opium)	*Papaver somniferum* L.					46			A	11	L
Potato (tubers)	*Solanum tuberosum* L.		1,000			60		C	P(A)	12; 24	L; N
Pumpkin (hills)	*Cucurbita pepo* L.		3–4	2	4		7	W	A	12	
Cushaw	*Cucurbita moshata* Duchesne		3–4	2	4				A	12; 20; 24	
Pyrethrum	*Chrysanthemum cineriaefolium* Vis.		½						P	9	L
Ramie (Chinagrass)	*Boehmeria nivea* Gaud.	Veg.							P	14	

Rape						50					
Oilseed (summer)	*Brassica napus* var. *annua* Koch	3–4		160	345		7	C	A	19	L
Biennial turnip	*Brassica campestris* var. *autummalis* DC.		2	240	535		7	C	B	10	
Bird Annual turnip	*Brassica campestris* L.	3–4		190	425		10	C	B	10	
Winter	*Brissica napus* var. *biennis*	2–3		104	230	50	7	C	B	19	L
(broadcast)	(Schubl. and Mart.) Reichb.	4–6									
Redtop	*Agrostis alba* L.	10–12		5,100	11,000	14	10	C	P	14; 21	
Reed canarygrass	*Phalaris arundinacea* L.	8–12		550	1,200	44–48	21	C	P	14	L
Rescuegrass	*Bromus catharticus* Vahl.	25–30		70	145	8–12	35	W		14; 21	
Rhodesgrass	*Chloris gayana* Kunth	10–12		1700	4,700	8–12	14	W	P	10; 20	
Rice	*Oryza sativa* L.	67–160		15	65	45	14	W	A	12	S
Ricegrass, Indian	*Oryzopsis hymenoides*	8–10		140	310		42	C	P	24	

TABLE A–1. Seeding; Seed and Plant Characteristics (*continued*)

CROP	BOTANICAL NAME	SEEDING RATE PER ACRE		SEEDS PER		WEIGHT PER BUSHEL	GERMINATION TIME	TEMPERATURE TYPE	GROWTH HABIT	CHROMOSOME NUMBER	PHOTOPERIODIC REACTION
		Close Drills (lb.)	*Rows (lb.)*	*Pound (thousands)*	*Gram (no.)*	*(lb.)*	*(days)*				
Roughpea	*Lathyrus hirsutus* L.	20		14	40	55	14	C	WA	7	
Rough-stalked meadow grass (see bluegrass, rough)											
Rutabaga	*Brassica napo-brassica* Mill.		1–2	200	430	60	14	C	B	18	L
Rye	*Secale cereale* L.	28–112		18	40	56	7	C	A; WA	7	L
Ryegrass											
Italian	*Lolium multiflorum* Lam.	25–30		227	500	24	14	C	WA	7	L
Perennial	*Lolium perenne* L.	25–30		330	500	24	14	C	P	7	L
Sacaline	*Polygonum sachalinense* F. Schmidt								P	22	
Safflower	*Carthamus tinctorius* L.	20–100		8–13	22	45			A	12	L; N

Sainfoin	*Onobrychis viciaefolia* Scop.	30–35		23	50	55	10		P	11	
Sand dropseed	*Sporobolus cryptandrus* (Torr.) A. Gray	½		3–5	5,000			W	P	9	
Serradella	*Ornithopus sativus* Brot.	15–20				36			A	8	
Sesame (benné)	*Sesamum indicum* L.		5	100	220	46	10	W	A	13; 26	S
Sesbania	*Sesbania macrocarpa* Muhl. (*exaltata*)	25–30		40	105	60	7	W	A	12	
Siratro	*Phaseolus atropurpureus* DC.										
Smilograss	*Oryzopsis miliacea* (L.) Benth. and Hook.			990	2,000		42		P	12	
Sorghum	*Sorghum bicolor* L. Moench						10	W	A	10	S
feterita			3–6	13	33	56	10	W			
hegari			3–6	20	44	56	10	W			
kafir		15–45	3–6	20	55	56	10	W			
milo			2–5	15	33	56	10	W			
sorgo		15–75	4–8	28	50	50	10	W			
sorgo (Sumac)		15–75	3–6	40	88	50	10	W			

TABLE A–1. Seeding; Seed and Plant Characteristics (*continued*)

CROP	BOTANICAL NAME	SEEDING RATE PER ACRE		SEEDS PER		WEIGHT PER BUSHEL	GERMINATION TIME	TEMPERATURE TYPE	GROWTH HABIT	CHROMOSOME NUMBER	PHOTOPERIODIC REACTION
		Close Drills (lb.)	*Rows (lb.)*	*Pound (thousands)*	*Gram (no.)*	*(lb.)*	*(days)*				
Sourclover	*Melilotus indica* All.	15–20		300	660	60	14	C	WA	8	
Soybean (small-seeded)	*Glycine max* Merrill (*Soja max* Piper)	60	15–20	8	18	60	8	W		20	S; N
(medium-seeded)		90	20–30	2–3	6–13	60					
(large-seeded)		120	30–45	1	2	60					
Spelt (see wheat)											
Squash	*Cucurbita maxima* Duchesne		2	6	14		7	W	A	20; 24	
Sudangrass	*Sorghum vulgare sudanense* (Piper) Hitchc.	20–35	4–6	55	120	40	10	W	A	10	S
Sugarbeet	*Beta vulgaris* (*saccharifera*) L.		2–8	22	48	15		C	B	9	L

Sugarcane	*Saccharum officinarum* L.	Veg.						W	P	40	S
Sulla (sweetvetch)	*Hedysarium coronarium* L.	20–25		100	220	60			P(A)	8	S
Sunflower (seed)	*Helianthus annus* L.		3–7	3–9	13	24	7		A	17; 34	N
(silage)			5–10								
Sweetclover											
White	*Melilotus alba* Med. (hulled)	12–15		250	570	60	7	C	B; A	8	L
Yellow	*Melilotus officinalis* Lam. (unhulled)	30–45		250	570	60	7	C	B	8	L
Sweetpotato	*Ipomoea batatas* Lam.		Veg.			55–60		W	P(A)	45	S
Sweet vernalgrass	*Anthoxanthum odoratum* L.	15–25		730	1,600		14	C	P	5	
Switchgrass	*Panicum virgatum* L.			370	815		28	W	P	18; 36; 35; 28	S
Tall meadow oatgrass (see oatgrass)											
Teasel	*Dipsacus fullonum* L.								B	10	
Teosinte (annual)	*Euchlaena mexicana* Schrad.	3–5		7	15				A	10	

TABLE A–1. Seeding; Seed and Plant Characteristics (*continued*)

CROP	BOTANICAL NAME	SEEDING RATE PER ACRE		SEEDS PER		WEIGHT PER BUSHEL	GERMINATION TIME	TEMPERATURE TYPE	GROWTH HABIT	CHROMOSOME NUMBER	PHOTOPERIODIC REACTION
		Close Drills (lb.)	*Rows (lb.)*	*Pound (thousands)*	*Gram (no.)*	*(lb.)*	*(days)*				
(perennial)	*Euchlaena perennis* Hitchc.								P	20	S
Timothy											
(alone, spring)	*Phleum pratense* L.	8–12		1,230	2,500	45	10	C	P	7; 21	L
(alone, fall)		3–4									
(with clover)		4–6									
Tobacco	*Nicotiana tobacum* L.		1/250	5,000	11,000		14	W	A	24	N
Townsville "lucerne"	*Stylasanthes humilis* HBK.										
Trefoil (see big trefoil, narrow-leaf trefoil, and birdsfoot trefoil)											
Turnip	*Brassica rapa* L.		1–2	154	535	55	7	C	B	10	L; N
(broadcast)		4–5									
Vaseygrass	*Paspalum urvillei* Steud.	8–15		440	970		21	W	P	20; 30	S

Velvetbean (with corn)	*Stizolobium utile (deeringianum* Bort)		30–45	1	2	60	14	W	A	11	
Velvetgrass	*Holcus lanatus* L.	10–25		1,200			14	C	P	7	
Vetch											
Common (alone)	*Vicia sativa* L.	40–80		7	19	60	10	C	A; WA	6; 7	L
Hairy (alone)	*Vicia villosa* Roth	20–40		21	36	60	14	C	WA; B	7	
(with grain)		20									
Hungarian	*Vicia pannomica* Crantz	40–80		11	24	60	10	C	A; WA	6; 7	L
Monantha	*Vicia monantha* Desf.	30–70		10		60	10	C	A; WA	7	L
Narrowleaf (wild)	*Vicia angustifolia* L.	20–30		27	60	60	14	C	A	6	
Purple	*Vicia atropurpurea* Desf.	40–70		9	20	60	10	C	WA; A	7	
Woolypod	*Vicia dasycarpa* Ten.	25–50		11	25	60	14	C			
Wheat											
Club	*Triticum compactum Host*	60–90		20–24	48	60	7	C	A; WA	21	L
Common	*Triticum vulgare* Vill. *(aestivum* L.)	30–120		12–20	35	60	7	C	A; WA	21	L

TABLE A–1. **Seeding; Seed and Plant Characteristics** *(continued)*

CROP	BOTANICAL NAME	SEEDING RATE PER ACRE		SEEDS PER		WEIGHT PER BUSHEL	GERMINATION TIME	TEMPERATURE TYPE	GROWTH HABIT	CHROMOSOME NUMBER	PHOTOPERIODIC REACTION
		Close Drills (lb.)	*Rows (lb.)*	*Pound (thousands)*	*Gram (no.)*	*(lb.)*	*(days)*				
Durum	*Triticum durum* Desf.	60–90		8–16	26	60	10	C	A	14	L
Emmer	*Triticum diccoccum* Schrank	60–90				40	7	C	A; WA	14	L
Spelt	*Triticum spelta* L.	60–90				40	7	C	A; WA	21	L
Wheatgrass											
Crested (Fairway)	*Agropyron cristatum* (L.) Gaertn.	12–20	4–6	320	714		14	C	P	7	
Crested (Standard)	*Agropyron desertorum* Fisch.	12–20	4–6	190	425	20–24	14	C	P	14	
Slender	*Agropyron trachycaulum* Malte (*pauciflorum*)	12–20		150	320		14	C	P	14	
Western	*Agropyron smithii* Rydb.	12–20		110	235		35	C	P	21; 28	L

Wild rice	*Zizania aquatica* L.	50–110					A	15	S
Wild rye, Canada	*Elymus canadensis* L.	10–12	120	261	21	C	P	14; 21	
Wormseed	*Chenopodium ambrosioides anthelminticum* Gray						A	16+	
Wormwood	*Artemisia absinthium* L.						P	9	

Temperature type: C = cool-weather growth; W = warm-weather growth.
Growth habit: A = annual; W.A. = winter annual; B = biennial; P = perennial; P(A) = perennial but grown as annual.
Chromosome number: reduced (gametic) number (N).
Photoperiodic reaction: L = long day; S = short day; N = day neutral or indeterminate; I = intermediate.

Average Seeding Rates Per Acre in the United States in 1970

CROP	BUSHELS	CROP	POUNDS
	(no.)		*(no.)*
Winter wheat	1.12	Corn	12.7
Durum wheat	1.37	Sorghum	7.1
Other spring wheat	1.22	Rice	137.0
Oats	2.57	Peanuts	119.0
Barley	1.63	Dry beans	51.0
Rye	1.41	Field peas	160.0
Buckwheat	1.16	Sweetpotatoes	570.0
Flaxseed	0.77	Cotton	26.0
Soybeans	1.13	Potatoes	1690.0
Cowpeas	0.68*	–	–

* 1962 data.

Seeds per gram, calculated seeds per pound, and germination time are from Federal Seed Act Regulations.

Weights per bushel listed are legal United States weights, most widely adopted, state legal weight, or customary weight when no legal weight has been established. Federal legal weights have been established for barley, beans, castorbeans, buckwheat, corn, flaxseed, potatoes, rye, and wheat.

TABLE A–2. The Percentage Composition of Crop Products

GRAINS, SEEDS, AND MILL CONCENTRATES

Product	*Moisture* (%)	*Ash* (%)	*Crude Protein* (%)	*Ether Extract*[a] (%)	*Crude Fiber* (%)	*Nitrogen-Free Extract*[b] (%)	*Calcium* (%)	*Phosphorus* (%)
Barley	9.6	2.9	12.8	2.3	5.5	66.9	0.07	0.32
Barley feed	7.9	4.9	15.0	4.0	13.7	54.5	0.03	0.41
Bread, kiln-dried	10.5	2.1	12.5	1.6	0.4	72.9	0.03	0.12
Brewers' dried grains:								
18–23 per cent protein	7.9	4.1	20.7	7.2	17.6	42.5	0.16	0.47
23–28 per cent protein	7.7	4.3	25.4	6.3	16.0	40.3	0.16	0.47
Brewers' rice	11.6	.7	7.0	0.8	0.6	79.3	0.03	0.25
Buckwheat	12.6	2.0	10.0	2.2	8.7	64.5		
Buckwheat middlings	12.4	4.6	28.0	6.6	5.3	43.1		
Corn, shelled	12.9	1.3	9.3	4.3	1.9	70.3	0.01	0.26
Corn bran	10.0	2.1	10.0	6.6	8.8	62.5	0.03	0.14
Corn chop	11.3	1.4	9.8	4.1	2.1	71.3	0.01	0.26
Corn (ear) chop	10.7	2.0	8.2	3.4	9.2	66.5		
Corn-feed meal	10.8	1.9	10.5	5.3	2.9	68.6	0.04	0.38
Corn-germ meal	7.0	3.8	20.8	9.6	7.3	51.5	0.05	0.59
Corn-gluten feed	9.5	6.0	27.6	3.0	7.5	46.4	0.11	0.78
Corn-gluten meal	8.0	2.2	43.0	2.7	3.7	40.4	0.10	0.47
Corn-oil meal	8.7	2.2	22.1	6.8	10.8	49.4	0.06	0.62
Cottonseed, whole pressed	6.5	4.3	29.6	5.8	25.1	28.7		
Cottonseed cake	7.5	5.9	44.1	6.4	10.3	25.8		
Cottonseed feed, 32 per cent protein	8.3	4.8	32.1	6.4	15.3	33.1	0.20	0.73
Cottonseed hulls	8.7	2.6	3.5	1.0	46.2	38.0		

[a] Fat.
[b] Carbohydrates except fiber.

Cottonseed meal:								
33–38 per cent protein	7.4	5.2	36.6	5.6	15.3	29.9	0.28	1.30
38–43 per cent protein	7.3	6.1	41.0	6.5	11.9	27.2	0.19	1.11
Over 43 per cent protein	7.2	5.8	43.7	6.5	11.1	25.7	0.18	1.15
Distillers' (corn) dried grain	7.0	2.4	28.3	9.4	14.6	38.3	0.04	0.29
Distillers' (rye) dried grain	6.1	2.4	17.9	6.3	15.9	51.4	0.13	0.43
Feterita	9.1	1.7	14.2	2.9	1.4	70.7		
Hemp cake	10.8	18.0	30.8	10.2	22.6	7.6		
Hempseed, European	8.8	18.8	21.5	30.4	15.9	4.6		
Hominy feed	9.5	2.9	11.2	8.3	6.3	61.8	0.03	0.44
Kafir	11.9	1.7	11.1	3.0	2.3	70.0	0.01	0.25
Kafir-head chops	10.4	3.9	10.9	2.5	6.0	66.3	0.09	0.20
Linseed meal:								
33–38 per cent protein	8.5	5.6	35.3	5.4	8.3	36.9	0.36	0.84
38–43 per cent protein	8.5	5.3	40.4	5.8	7.5	32.5	0.33	0.74
Malt	7.7	2.9	12.4	2.1	6.0	68.9		
Malt sprouts	7.3	6.1	28.1	1.8	13.3	43.4	0.26	0.68
Mesquite beans and pods	6.6	4.5	13.0	2.7	22.8	50.4		
Millet, foxtail	10.1	3.3	12.6	4.3	8.4	61.3		
Millet, proso or hog millet	9.8	3.4	12.0	3.4	7.9	63.5		
Milo	9.3	1.6	12.5	3.2	1.5	71.9		
Milo-head chops	10.4	4.3	10.7	2.6	7.1	64.9		
Molasses, cane	24.0	6.8	3.1			66.1	0.35	0.06
Oats, grain	7.7	3.5	12.5	4.4	11.2	60.7	0.10	0.40
Oat chops	8.9	3.9	12.8	5.0	11.8	57.6	0.10	0.36
Oat clips	9.0	9.3	11.8	4.5	22.7	42.7		
Oat groats, ground rolled	10.4	2.6	17.3	6.6	1.8	61.3	0.08	0.43
Oat hulls	5.8	6.5	4.3	1.9	30.8	50.7	0.09	0.12
Oatmeal	8.9	2.3	16.5	4.8	3.6	63.9	0.08	0.43
Oat millfeed	6.9	6.0	6.3	2.2	27.9	50.7	0.20	0.22
Peanuts, kernels	5.5	2.3	30.2	47.6	2.8	11.6	0.06	0.38

TABLE A–2. The Percentage Composition of Crop Products (*continued*)

GRAINS, SEEDS, AND MILL CONCENTRATES

Product	*Moisture* (%)	*Ash* (%)	*Crude Protein* (%)	*Ether Extract*[a] (%)	*Crude Fiber* (%)	*Nitrogen-Free Extract*[b] (%)	*Calcium* (%)	*Phosphorus* (%)
Peanuts, shells on	6.0	2.8	24.7	33.1	18.0	15.4		
Peanut meal:								
38–43 per cent protein	6.4	4.4	41.6	7.2	16.0	24.4	0.10	0.50
43–48 per cent protein	6.7	4.6	45.1	7.2	14.2	22.2	0.17	0.55
Over 48 per cent protein	7.0	5.0	51.4	4.8	9.2	22.6		
Rapeseed, brown Indian	5.7	6.4	21.0	41.2	12.5	13.2		
Rapeseed, common	7.3	4.2	19.5	45.0	6.0	18.0		
Rice, rough	9.7	5.4	7.3	2.0	8.6	67.0	0.10	0.10
Rice, bran	8.8	12.2	12.8	13.8	12.2	40.2	0.10	1.84
Rice hulls	6.5	21.9	2.1	0.4	44.8	24.3	0.08	0.06
Rice polish	10.0	7.6	12.4	13.2	2.8	54.0	0.03	1.52
Rice-stone bran	8.4	11.9	12.5	13.0	11.1	43.1		
Rye	9.5	1.9	11.1	1.7	2.1	73.7	0.04	0.37
Rye feed	10.2	4.0	15.6	3.2	4.3	62.7		0.59
Rye middlings	9.5	4.4	16.7	3.7	5.5	60.2		
Sesame seed	5.5	6.5	20.3	45.6	7.1	15.0		
Sesame-seed cake	9.8	10.7	37.5	14.0	6.3	21.7		
Sorgo	12.8	2.1	9.1	3.6	2.6	69.8		
Soybeans	8.0	4.8	38.9	18.0	4.8	25.5	0.22	0.67
Soybean meal:								
38–43 per cent protein	7.8	5.8	41.7	5.8	6.2	32.7	0.29	0.67
43–48 per cent protein	8.2	6.0	44.7	4.6	5.8	30.7	0.34	0.71
Sunflower seed	6.9	3.2	15.2	28.8	28.5	17.4		
Sunflower hulls	10.5	2.6	4.4	3.4	57.0	22.1		
Sunflower kernels	6.9	4.2	29.4	43.9	2.6	13.0		
Velvetbeans	9.8	3.1	26.2	4.8	6.0	50.1		

Wheat	10.6	1.8	12.0	2.0	2.0	71.6	0.05	0.38
Wheat bran	9.4	6.4	16.4	4.4	9.9	53.5	0.10	1.14
Wheat, brown shorts	10.8	4.0	17.8	4.8	5.8	56.8		
Wheat-flour middlings	10.4	3.3	18.8	4.0	4.2	59.3	0.09	0.80
Wheat, gray shorts	11.0	4.1	17.5	4.4	5.4	57.6	0.08	0.86
Wheat, mixed-feed	9.9	4.4	18.2	4.4	6.9	56.1	0.11	0.96
Wheat, red-dog	11.1	2.2	18.3	3.4	2.3	62.7	0.12	0.83
Wheat, standard middlings	10.4	3.9	17.0	4.3	5.4	59.0	0.09	0.90
Wheat, white shorts	10.9	2.2	15.6	3.7	2.4	65.2		
Wheat waste, shredded	8.0	1.6	12.4	1.6	2.6	73.8		
DRIED FORAGES								
Alfalfa hay	7.2	8.0	15.4	1.6	30.3	37.5	1.51	0.21
Alfalfa-leaf meal	8.5	14.4	20.9	2.6	15.7	37.9	1.42	0.25
Alfalfa meal	8.2	10.0	15.2	2.2	27.5	36.9	1.56	0.22
Alfalfa meal, dehydrated	6.6	10.0	16.9	2.6	25.4	38.5		
Alfalfa-stem meal	9.1	7.7	11.4	1.3	36.1	34.4		
Alsike clover hay	10.5	8.8	14.4	2.5	24.7	39.1	0.78	0.20
Australian saltbush hay	6.7	16.9	16.1	1.8	21.5	37.0		
Barley hay	15.0	6.4	6.7	1.6	21.4	48.9	0.17	0.25
Barley straw	14.2	5.7	3.5	1.5	36.0	39.1		
Bermudagrass hay	8.9	7.9	7.2	1.7	24.9	49.4	0.60	0.16
Black gramma hay	5.5	7.0	4.3	1.3	31.4	50.5	0.22	0.09
Blue gramma hay	10.9	8.5	6.7	1.8	27.9	44.2		
Bluegrass hay, immature	7.3	7.9	15.2	3.0	23.7	42.9	0.45	0.35
Bluegrass hay, bloom	11.9	7.0	9.3	3.4	27.9	40.5	0.30	0.21
Bluejoint grass hay	7.5	6.9	6.7	3.0	34.2	41.7		
Bromegrass hay	14.0	9.7	9.3	1.8	26.6	38.6		
Buckwheat straw	9.9	5.5	5.2	1.3	43.0	35.1		
Buffalograss hay	6.2	10.8	5.6	1.7	26.1	49.6		
Burclover hay	8.7	12.3	15.7	3.0	25.5	34.8	1.11	0.15
Corncobs	10.7	1.4	2.4	0.5	30.1	54.9		
Corn fodder	11.8	5.8	7.4	2.4	23.0	49.6		
Corn husks	9.8	2.9	2.9	0.7	30.7	53.0		
Corn leaves	11.8	8.5	8.1	2.2	24.4	45.0		

TABLE A–2. The Percentage Composition of Crop Products (*continued*)

DRIED FORAGES

Product	*Moisture* (%)	*Ash* (%)	*Crude Protein* (%)	*Ether Extract*[a] (%)	*Crude Fiber* (%)	*Nitro-gen-Free Extract*[b] (%)	*Calcium* (%)	*Phos-phorus* (%)
Corn stalks	11.7	4.6	4.8	1.8	32.7	44.4		
Corn stover	10.7	6.1	5.7	1.5	30.3	45.7	0.45	0.10
Cowpea hay	9.7	12.9	17.5	2.8	20.5	36.6	1.84	0.25
Cowpea straw	9.7	5.3	7.4	1.3	41.5	34.8		
Crabgrass hay	9.0	7.9	6.5	2.2	32.1	42.3	0.33	0.17
Crimson clover hay	9.6	8.6	15.2	2.8	27.2	36.6	1.18	0.13
Feterita fodder	13.3	6.4	8.7	1.9	21.5	48.2	0.27	0.19
Field-pea hay	10.6	8.3	16.1	2.7	24.8	37.5		
Flax straw	6.2	3.8	7.8	2.1	46.9	33.2		
Hegari fodder	13.5	8.2	6.2	1.7	16.7	53.7	0.17	0.18
Hegari stover	15.1	9.7	4.5	1.9	26.6	42.2	0.38	0.09
Johnsongrass hay	7.2	7.2	8.1	2.8	30.4	44.3	0.55	0.40
Kafir fodder	9.1	7.8	6.6	2.1	28.4	46.0	0.31	0.05
Kafir stover	12.6	9.0	5.8	1.7	27.5	43.4		
Lespedeza hay	7.9	6.2	11.9	2.8	28.5	42.7	0.80	0.25
Little bluestem hay	8.6	4.9	4.0	1.6	35.4	45.5		
Meadow fescue hay	11.6	7.0	6.6	2.0	31.6	41.2		
Millet hay, foxtail	7.0	8.2	9.2	2.8	28.0	44.8		
Millet hay, pearl or cattail	10.1	9.7	9.0	1.8	32.3	37.1		
Natalgrass hay	7.5	4.8	3.7	1.4	39.5	43.1	0.49	0.32
Oatgrass, tall, hay	8.1	6.4	9.4	2.7	29.8	43.6		
Oat hay	11.8	5.7	6.1	2.4	27.1	46.9	0.27	0.22
Oat straw	8.1	7.6	4.4	2.5	36.2	41.2	0.23	0.20

Orchardgrass hay, immature	9.9	6.0	8.1	2.6	32.4	41.0	0.31	0.18
Orchardgrass hay, mature	9.9	7.0	6.9	3.0	32.7	40.5		
Prairie hay (Colorado, Wyoming)	5.5	7.2	7.0	2.4	31.3	46.6		
Prairie hay (Kansas, Oklahoma)	9.5	7.5	4.4	2.3	30.4	45.9	0.55	0.07
Prairie hay (Minnesota, South Dakota)	11.6	7.2	6.0	2.4	30.3	42.5	0.44	0.11
Red-clover hay	7.0	10.0	16.1	2.6	23.6	40.7	1.01	0.14
Red clover, mammoth, hay	12.2	7.5	12.8	3.3	27.1	37.1		
Redtop hay	8.9	5.2	7.9	1.9	28.6	47.5	0.35	0.18
Rhodesgrass hay	8.6	8.4	5.3	1.2	33.4	43.1		
Rice straw	8.9	13.5	4.5	1.6	34.0	37.5	0.18	0.05
Rye hay	6.4	4.7	5.9	2.0	37.4	43.6	0.27	0.22
Rye straw	7.1	3.2	3.0	1.2	38.9	46.6		
Ryegrass, perennial, hay	10.2	8.6	8.6	4.1	24.5	44.0	0.17	0.11
Ryegrass, Italian, hay	8.5	6.9	7.5	1.7	30.5	44.9		
Ryegrass hay	8.3	8.5	6.3	2.0	33.0	41.9		
Sedge, western species	5.4	6.7	11.6	2.4	27.4	46.5		
Slender wheatgrass	7.5	6.6	7.8	2.1	30.8	45.2		
Sorgo fodder	11.6	6.0	5.3	2.4	26.0	48.7	0.27	0.15
Sorgo hay	5.8	9.5	9.5	1.9	26.8	46.5	0.31	0.09
Soybean hay	8.4	8.9	15.8	3.8	24.3	38.8	1.26	0.22
Soybean straw	8.7	7.4	5.7	2.5	34.6	41.1		
Sudangrass hay	5.3	8.1	9.7	1.7	27.9	47.3	0.47	0.24
Sweetclover hay	8.1	7.5	16.2	2.8	25.9	39.5	0.74	0.08
Sweetclover straw	5.1	3.4	6.7	1.2	49.6	34.0		
Timothy hay	7.1	5.8	7.5	2.9	30.2	46.5	0.31	0.13
Vetch, hairy, hay	13.1	8.4	20.9	2.7	24.2	30.7	0.25	0.30
Western needlegrass hay	9.9	6.2	5.5	2.7	33.2	42.5		
Western wheatgrass hay	8.6	8.7	8.4	2.3	31.9	40.1		
Wheat hay	9.6	4.2	3.4	1.3	38.1	43.4	0.14	0.15
Wheat straw	6.8	5.4	4.3	3.4	36.8	43.3		
White clover hay	7.2	9.4	15.6	2.2	22.7	42.9	1.31	0.28
Wiregrass hay	8.5	7.3	6.6	1.3	34.6	41.7		

TABLE A–2. The Percentage Composition of Crop Products (*continued*)

GREEN FORAGES

Product	*Moisture* (%)	*Ash* (%)	*Crude Protein* (%)	*Ether Extract*[a] (%)	*Crude Fiber* (%)	*Nitrogen-Free Extract*[b] (%)	*Calcium* (%)	*Phosphorus* (%)
Alfalfa, immature	79.4	2.9	5.2	0.7	3.8	8.0	0.28	0.09
Alfalfa, in bloom	77.2	1.8	3.2	0.6	7.8	9.4	0.39	0.07
Alsike clover, immature	81.2	2.4	4.9	0.6	3.1	7.8	0.26	0.09
Alsike clover, in bloom	74.8	2.0	3.9	0.9	7.4	11.0	0.21	0.06
Barley, immature	83.4	1.5	2.8	0.7	3.6	8.0	0.06	0.07
Barley, mature	77.1	1.6	2.2	0.5	6.4	12.2	0.05	0.07
Bluegrass, Kentucky, immature	70.5	2.5	5.0	1.2	7.5	13.3	0.15	0.13
Bromegrass, immature	77.5	2.9	4.3	0.9	5.2	9.2	0.14	0.10
Cabbage	90.5	0.9	2.4	0.3	1.2	4.7	0.06	0.02
Canada bluegrass, immature	74.1	2.5	4.3	1.3	6.8	11.0	0.11	0.12
Corn fodder:								
Dent, immature	79.0	1.2	1.7	0.5	5.6	12.0		
Dent, mature	73.4	1.5	2.0	0.9	6.7	15.5		
Flint, immature	79.8	1.1	2.0	0.7	4.3	12.1		
Flint, mature	77.1	1.1	2.1	0.8	4.3	14.6		
Cowpeas	82.5	2.5	3.4	0.5	4.0	7.1	0.18	0.05
Crimson clover	80.9	1.7	3.1	0.7	5.2	8.4	0.28	0.04
Kafir	73.0	2.0	2.3	0.7	6.9	15.1		
Lespedeza, Korean, immature	74.1	2.4	4.6	0.8	5.8	12.3	0.34	0.11
Meadow fescue, immature	78.8	2.6	4.0	0.9	4.7	9.0	0.15	0.11
Meadow foxtail, immature	73.9	2.8	4.5	1.2	5.6	12.0	0.15	0.12
Millet, foxtail	71.1	1.7	3.1	0.7	9.2	14.2	0.09	0.05
Oatgrass, tall, immature	78.4	3.0	4.3	1.0	4.6	8.7	0.11	0.13
Oats, immature	82.6	1.7	2.9	0.7	3.3	8.8	0.07	0.07

Oats, mature	72.0	2.1	2.7	0.9	7.4	14.9	0.08	0.08
Orchardgrass, immature	78.3	2.8	3.4	1.0	5.3	9.2	0.14	0.13
Orchardgrass, in bloom	73.0	2.0	2.6	0.9	8.2	13.3		
Pricklypear	78.9	4.3	.7	0.4	2.6	13.1		
Rape	85.7	2.0	2.4	0.6	2.2	7.1		
Red clover, immature	81.2	2.7	5.0	0.8	3.0	7.3	0.27	0.10
Red clover, in bloom	70.8	2.1	4.4	1.1	8.1	13.5	0.44	0.07
Red fescue, immature	70.5	2.8	4.1	0.9	8.2	13.5	0.16	0.13
Redtop, immature	76.8	2.8	4.1	0.9	5.4	10.0	0.15	0.10
Reed canarygrass, immature	80.7	2.4	3.5	0.7	4.3	8.4	0.13	0.10
Rye, immature	80.8	2.3	4.5	1.1	3.4	7.9	0.10	0.10
Rye, mature	76.6	1.8	2.6	0.6	11.6	6.8	0.08	0.06
Ryegrass, Italian, immature	77.3	2.5	3.5	1.0	5.2	10.5	0.13	0.12
Ryegrass, perennial, immature	75.9	3.0	3.8	0.9	5.4	11.0	0.15	0.12
Sorgo	77.3	1.3	1.5	1.0	6.2	12.7		
Soybeans	73.9	2.9	4.0	1.1	7.6	10.5	0.28	0.05
Sweetclover, immature	75.3	2.2	5.3	0.7	6.7	9.8	0.26	0.07
Sweetcorn	79.1	1.3	1.9	0.5	4.4	12.8		
Timothy, immature	74.9	2.3	4.1	0.9	5.4	12.4	0.12	0.11
Timothy, in bloom	61.6	2.1	3.1	1.2	11.8	20.2	0.13	0.05
Wheat, immature	82.3	2.1	3.8	0.9	3.0	7.9	0.07	0.10
Wheat, mature	68.7	2.6	2.4	0.7	8.6	17.0	0.06	0.08
White clover, immature	82.0	2.1	4.9	0.6	3.1	7.3	0.23	0.09
White clover, wild, immature	81.2	2.2	5.2	0.6	2.9	7.9	0.25	0.10
SILAGES, ROOTS, TUBERS, AND BY-PRODUCTS								
Alfalfa silage	68.9	2.7	5.7	1.0	8.8	12.9		
Alfalfa-molasses silage	68.6	3.4	5.8	1.0	8.4	12.8		
Beet pulp, dried	9.2	3.2	9.3	0.8	20.0	57.5	0.66	0.06
Beet pulp, molasses, dried	8.0	5.2	11.6	0.7	16.4	58.1	0.59	0.09
Carrots	88.6	1.0	1.1	0.4	1.3	7.6		
Cassava	63.8	1.4	1.0	0.3	0.8	32.7		

TABLE A–2. The Percentage Composition of Crop Products (*continued*)

SILAGES, ROOTS, TUBERS, AND BY-PRODUCTS

Product	*Moisture* (%)	*Ash* (%)	*Crude Protein* (%)	*Ether Extract*[a] (%)	*Crude Fiber* (%)	*Nitrogen-Free Extract*[b] (%)	*Calcium* (%)	*Phosphorus* (%)
Corn silage	73.8	1.7	2.1	0.8	6.3	15.3	0.08	0.08
Corn silage, immature	79.1	1.4	1.7	0.8	6.0	11.0		
Corn silage, mature	70.9	1.4	2.4	0.9	6.9	17.5		
Corn stover silage	80.7	1.8	1.8	0.6	5.6	9.5		
Cowpea silage	77.8	2.1	3.2	0.9	6.5	9.5		
Hegari silage	66.3	3.4	2.3	0.8	6.7	20.5		
Jerusalem artichokes	78.7	1.1	2.5	0.2	0.8	16.7		
Mangel-wurzel	90.8	1.0	1.4	0.2	0.9	5.7	0.02	0.02
Napiergrass silage	67.5	1.8	1.2	0.7	14.4	14.4	0.10	0.10
Parsnips	80.0	1.3	2.2	0.4	1.3	14.8		
Pea-vine silage	75.1	1.7	3.0	0.9	8.1	11.2		
Potatoes	78.9	1.0	2.1	0.1	0.6	17.3	0.01	0.06
Red-clover silage	72.0	2.6	4.2	1.2	8.4	11.6		
Rutabagas	88.6	1.2	1.2	0.2	1.3	7.5	0.05	0.04
Sorgo silage	74.7	1.4	1.6	1.0	6.9	14.4	0.09	0.04
Soybean silage	75.6	2.6	2.4	0.8	9.6	9.0	0.29	0.10
Sugarbeets	78.0	1.0	1.5	0.1	2.9	16.5	0.05	0.06
Sugarbeet pulp	90.5	0.4	.9	0.2	2.2	5.8		
Sunflower silage	77.9	2.1	1.8	1.6	6.5	10.1		
Sweetclover silage	70.2	2.9	6.1	1.0	9.7	10.1		
Sweetpotatoes	71.1	1.0	1.5	0.4	1.3	24.7	0.02	0.05
Turnips	90.6	0.8	1.3	0.2	1.2	5.9	0.05	0.05

From USDA *Yearbook*, 1939. For further information on the composition and digestibility of crops, crop products, and range plants see: "Atlas of nutritional data on United States and Canadian feeds." Natl. Acad. Sci. Washington, D.C. pp 1–772. 1971

TABLE A-3. Conversion Tables

MASS

1 microgram (μgm.) = 0.000001 gm.
1 milligram (mgm.) = 0.001 gm.
1 centrigram (cgm.) = 10 mg. = 0.01 gm.
1 decigram (dgm.) = 100 mg. = 0.1 gm.
1 gram (gm.) = 1,000 mg. = 15.432356 grains
1 dekagram (dkgm.) = 10 gm.
1 hectogram (hgm.) = 100 gm.
1 kilogram (kgm.) = 1,000 gm. = 2.204622341 lb.
1 metric quintal = 100 kg. = 220.46 lb.
1 metric ton = 1,000 kg. = 2,204.62 lb.

AVOIRDUPOIS WEIGHTS

1 grain (gr.) = 64.798918 mgm.
1 ounce (oz.) = 437.5 gr. = 28.349527 gm.
1 pound (lb.) = 16 oz. = 7,000 gr. = 453.5924 gm.
1 short hundredweight = 100 lb.
1 long hundredweight = 112 lb.
1 short ton = 2,000 lb.
1 long.ton = 2,240 lb.

LENGTH

1 kilometer (km.) = 1,000 m. = 0.62137 mile
1 hectometer (hm.) = 100 m. = 328 ft. 1 in.
1 dekameter (dkm.) = 10 m. = 393.7 in.
1 meter (m.) = 1 m. = 39.37 in. = 3.28 ft.
1 decimeter (dm.) = $^1/_{10}$ m. = 3.937 in.
1 centimeter (cm.) = $^1/_{100}$ m. = 0.3937 in.
1 millimeter (mm.) = $^1/_{1000}$ m. = 0.03937 in.
1 micron (μ) = $^1/_{1000}$ mm.
1 millimicron (mμ) = $^1/_{1000}$ μ
1 Ångström (Å) = $^1/_{10}$ mμ
1 mile = 5,280 ft. = 1,760 yd. = 320 rd. = 80 chains
1 chain = 66 ft. = 22 yd. = 4 rd. = 100 links
1 rod = $16^1/_2$ ft. = $5^1/_2$ yd. = 25 links
1 yard = 3 ft. = 36 in. = 0.9144 m.
1 foot = 12 in. = 30.48006 cm.
1 link = 7.92 in.
1 inch = 2.540005 cm. = 25.4 mm.

AREA

1 centare (ca.) = 1 square meter
1 are (a.) = 100 sq. m.
1 hectare (ha.) = 10,000 sq. m. = 2.47104 A.
1 labor (lah-bore') = $177^1/_7$ A.
1 square league = 25 labors = 4,409 A.
1 Old Spanish league = 2.63 miles
(Parts of Texas are surveyed in leagues and labors)
1 sq. in. = 6.4511626 sq. cm.
1 sq. ft. = 144 sq. in.
1 sq. yd. = 9 sq. ft.
1 sq. rd. = 1 perch = $30^1/_4$ sq. yd. = $272^1/_4$ sq. ft.

TABLE A–3. Conversion Tables (*continued*)

1 acre = 160 sq. rd. = 4,839 sq. yd. = 43,560 sq. ft. = 0.404687 ha.
1 sq. mi. = 1 section = 640 A.
1 township = 36 sections = 23,040 A.

VOLUME

1 cubic centimeter (cc.) = 0.06102338 cu. in.
1 cubic decimeter = 1,000 cc. = 61.02338 cu. in.
1 cubic meter = 1,000,000 cc. = 35.31445 cu. ft.
1 cu. in. = 16.387162 cc.
1 cu. ft. = 1,728 cu. in. = 28,317.016 cc.
1 cu. yd. = 27 cu. ft. = 46,656 cu. in.

CAPACITY (liquid measure)

1 gill = 118.292 ml.
1 pt. = 4 gi. = 473.167 ml. = 0.473167 1.
1 qt. = 2 pt. = 8 gi. = 0.946333 1.
1 gal. = 4 qt. = 8 pt. = 3.785332 1. = 231 cu. in.
1 liter (1.) = 1,000 milliliters (ml. or cc.) = 2.11342 pt. = 1.05671 qt.
1 teaspoon = 4.93 ml.
3 teaspoons = 1 tablespoon = 14.79 ml.
2 tablespoons = 1 fluid oz. = 29.578 ml.
16 tablespoons = 8 fluid oz. = 1 cup = 236.583 ml.
2 cups = 16 fluid oz. = 1 pt. = 473.167 ml.

CAPACITY (dry measure)

1 liter = 61.025 cu. in. = 0.908102 dry qt.
1 dekaliter (dkl.) = 10.1. = 0.28378 bu.
1 hectoliter (hl.) = 100 1. = 2.8378 bu.
1 U.S. bu. per A. = 0.8708 hl. per ha.
1 hl. per ha. = 1.1483 bu. per A.
1 dry pt. = 33.6003125 cu. in. = 0.550599 1.
1 dry qt. = 2 dry pt. = 1.101198 1.
1 peck = 8 dry qt. = 8.80958 1.
1 bushel = 4 pk. = 32 dry qt. = 35.2383 1.
1 Winchester (U.S.) bu. = 1.244 cu. ft. = 2,150.42 cu. in.
1 Imperial (British) bu. = 2,219.36 cu. in. = 1.0305 Winchester bu.

CONVERSION CENTIGRADE TO FAHRENHEIT TEMPERATURE

1° C. = 1.8° F. 1° F. = 5/9° C.
° C. = (° F. − 32) × 5/9 ° F. = 9/5° C. + 32

° C.	° F.	° C.	° F.	° C.	° F.
100	212	50	122	0	32
90	194	40	104	−10	14
80	176	30	86	−20	− 4
70	158	20	68	−30	−22
60	140	10	50	−40	−40

WATER

1 gal. = 8.355 lb.
1 cu. ft. = 7.48 gal. = 62.42 lb.
1 acre-inch = 113 tons (approximately)
1 acre-foot = 43,560 cu. ft. = 323,136 gal.

TABLE A–3. Conversion Tables (*continued*)

1 second foot = 1 cu. ft. per sec. flow past a given point
1 second foot = 7½ gal. per sec. = 450 gal. per min.
1 second foot in 24 hours = 1.983 acre-feet
1 second foot in 1 hour = 1 acre-inch (approximately)
1 second foot = 40 miner's inches (in some states by statute)
1 second foot = 50 miner's inches (in other states)

DRY SOIL

1 cu. ft. muck = 25 to 30 lb.
1 cu. ft. clay and silt = 68 to 80 lb.
1 cu. ft. sand = 100 to 110 lb.
1 cu. ft. loam = 80 to 95 lb.
1 cu. ft. average soil = 80 to 90 lb.
1 acre foot (43,560 cu. ft.) = 3,500,000 to 4,000,000 lb. (2,000 tons)
The soil surface plow depth (6⅔ in.) is usually calculated as 2 million lb. (1,000 tons) per acre.
The volume of compact soil increases about 20 per cent when it is excavated or tilled.

GRAIN STORAGE

To compute the approximate capacity of a grain bin in bushels from the measurements of the bin in feet:
Rectangular bins – Length × width × height × 0.8 = bu.
Round bins – Diameter squared × height × ⅝ = bu.
Ear corn in crib:
Length × average width × average depth in feet × 0.4 = bu.

WEIGHTS, MEASURES AND CONVERSION FACTORS USED BY THE U.S. DEPARTMENT OF AGRICULTURE

Barley
Flour 1 bbl. (196 lb.) ≎ 9 bu. barley
Malt 1.1 bu. (34 lb.) ≎ 1 bu. (48 lb.) barley
Beans, dry 1 sack = 100 lb.
Beer 1 bbl. = 31 gal.
Broomcorn 1 bale = 333 lb.; 6 bales = 1 ton
Buckwheat flour 1 bbl. (196 lb.) ≎ 7 bu. buckwheat
Corn
Meal (degermed) 1 bbl. (196 lb.) ≎ 6 bu. corn
Meal (nondegermed) 1 bbl. (196 lb.) ≎ 4 bu. corn
Ear corn (dry) 2 level bu. or 3 heaping half-bushels weighing 70 lb. ≎ 1 bu. (56 lb.) shelled corn
Cotton 1 lb. ginned ≎ 2.86 lb. unginned
Lint 1 bale (gross) = 500 lb.
1 bale (net) = 480 lb.
Flour 1 bbl. = 196 lb. 1 sack = 100 lb.
Hops 1 gross bale = 200 lb.
Jerusalem artichoke 1 bu. = 50 lb.
Linseed oil 2½ gal. (19 lb.) ≎ 1 bu. flaxseed
Oatmeal 1 bbl. (196 lb.) ≎ 10⁸⁄₉ bu. oats
Oil (corn, cottonseed, soybean, peanut, linseed, etc.) 1 gal. = 7.7 lb. (in trade), castor oil 1 gal. = 8 lb.
Potatoes 1 bu. = 60 lb; 1 bbl. = 165 lb; 1 sack = 100 lb.

TABLE A–3. Conversion Tables (*continued*)

Rice
  Rough 1 bag = 100 lb.; 1 bbl. = 162 lb.
  Milled 1 pocket (100 lb.) $\approx$ 154 lb. (3.42 bu.) rough (unhulled)
Rye flour 1 bbl. (196 lb.) $\approx$ 6 bu. rye
Sorgo sirup 1 gal. = 11.55 lb.
Sugar 1 ton domestic 96° raw sugar $\approx$ 0.9346 tons refined sugar
Sugarcane sirup 1 gal. = 11.35 lb.
Sweetpotatoes 1 bu. = 55 lb.
Tobacco 1 lb. stemmed $\approx$ 1.33 lb. unstemmed
  Maryland (hogshead) = 600–800 lb.
  Flue-cured (hogshead) = 900–1,100 lb.
  Burley (hogshead) = 1,000–1,200 lb.
  Dark air-cured (hogshead) = 1,000–1,250 lb.
  Virginia fire-cured (hogshead) = 1,050–1,350 lb.
  Kentucky and Tennessee fire-cured (hogshead) = 1,350–1,650 lb.
  Cigar leaf (case) = 250–365 lb.
        (bale) = 150–175 lb.
Turnips (without tops) 1 bu. = 54 lb.
Vegetables and other dry commodities 1 bbl. = 7,056 cu. in. = 105 dry qt.
Wheat flour 1 bbl. (196 lb.) $\approx$ 4.7 bu. wheat
        100 lb. $\approx$ 2.33 bu. wheat
(Bushel weights of grains and seeds are shown in Appendix Table A–1.)

Glossary of Agronomic Terms*

A HORIZON: The surface and subsurface soil that contains most of the organic matter and is subject to leaching.

ABORTIVE: Imperfectly developed.

ABSCISSION: The natural separation of leaves, flowers, and fruits or buds from the stems or other plant parts by the formation of a special layer of thin-walled cells.

ACHENE (AKENE): A dry, hard, one-celled, one-seeded, indehiscent fruit to which the testa and pericarp are not firmly attached.

ACID SOIL: A soil with a *p*H reaction of less than 7.0 (usually less than 6.6). An acid soil has a preponderance of hydrogen ions over hydroxyl ions. Litmus paper turns red in contact with moist acid soil.

ACUMINATE: Gradually tapering to a sharp point, as in a leaf of grass.

ACUTE: Sharp-pointed but less tapering than acuminate.

ADVENTITIOUS: Arising from an unusual position on a stem (or at the crown of a grass plant), e.g., roots and buds.

AERIAL ROOTS: Roots that arise from the stem above the ground.

AFTERMATH (ROWEN): The second or shorter growth of meadow plants in the same season after a hay or seed crop has been cut.

AGGREGATE: A mass or cluster of soil particles or other small objects.

AGEOTROPIC: Lacking a geotropic response, as in stolons, rhizomes, and lateral roots which grow neither erect nor downward.

AGROBIOLOGY: A phase of the study of agronomy dealing with the relation of yield to the quantity of an added or available fertilizer element.

AGROECOLOGY: The study of the relation of crop varietal characteristics to their adaptation to environmental conditions.

AGROLOGY: The study of applied phases of soil science and soil management.

AGRONOMY: The science of crop production and soil management. The name is derived from the Greek words *agros* (field) and *nomos* (to manage).

ALEURONE: The outer layer of the cells of the endosperm of the seed. It sometimes contains pigment.

ALKALI SOIL: A soil, usually above *p*H 8.5, containing alkali salts in quantities that usually are deleterious to crop production.

ALKALINE SOIL: A soil with a *p*H above 7.0, usually above 7.3.

ALTERNATE: Placed singly rather than opposite, e.g., one leaf at a node, or flower parts of one whorl opposite to intervals of the next.

AMMONIFICATION: *Formation of*

* Adapted from (1) reports of committees on terminology of the American Society of Agronomy, (2) glossaries in USDA yearbooks for 1936, 1957, and 1961, and (3) various botanical glossaries.

ammonia or ammonium compounds, in soils.

AMYLOPECTIN: The branched-chain (waxy or glutinous) starch.

AMYLOSE: The straight-chain fraction of normal starch.

ANGIOSPERMS: The higher seed plants.

ANNUAL: A plant that completes its life cycle from seed in one year.

ANNULAR: Forming a ring; circular.

ANTHER: The part of the stamen that contain the pollen.

ANTHESIS: The period during which the flower is open and, in grasses, the period when the anthers are extended from the glumes.

ANTHOCYANIN: A water-soluble plant pigment that produces many of the red, blue, and purple colors in plants.

APETALOUS: Without petals.

APICULATE: Having a minute pointed tip.

APOGEOTROPIC: Turning upward in response to a stimulus opposed to the force of gravity, e.g., seedling stems growing erect.

APOMIXIS: A type of asexual production of seeds, as in Kentucky bluegrass.

APPRESSED: Pressed close to the stem.

AQUATIC PLANT: A plant that lives in water.

ARID CLIMATE: A dry climate with an annual precipitation usually less than 10 inches and not suitable for crop production without irrigation.

ARISTATE: Awned.

ARTICULATE: Jointed.

ARVICULTURE: Crop science.

ASEXUAL REPRODUCTION: Reproduction without involving the germ or sexual cells.

ASH: The nonvolatile residue resulting from complete burning of organic matter.

ASSOCIATION: A biologically balanced group of plants.

ATTENUATE: Gradually narrowing to a slender apex or base.

AURICLES: Ear-shaped appendages, as at the base of barley and wheat leaves.

AUXINS: Organic substances that cause stem elongation.

AWN: The beard or bristle extending from the tip or back of the lemma of a grass flower.

AXIL: The angle between the leaf and stem or between a branch or pedicel and its main stem.

AXILLARY: In the axil.

AXIS: The main stem of a flower or panicle.

B HORIZON: The subsoil layer in which certain leached substances (e.g., iron) are deposited.

BACKCROSS: The cross of a hybrid with one of the parental types.

BANNER: The upper or posterior, usually broad, petal of a papilionaceous legume flower; the standard.

BARBED: Furnished with rigid points or hooks.

BEAK: A point or projection, as on the glume of the wheat flower.

BEARD: The awn of grasses.

BED: (1) A narrow flat-topped ridge on which crops are grown with a furrow on each side for drainage of excess water. (2) An area in which seedlings or sprouts are grown before transplanting.

BED UP: To build up beds or ridges with a tillage implement.

BIENNIAL: Of two years' duration; a plant germinating one season and producing seed the next.

BINDER: A machine for cutting a crop and tying it into bundles with twine.

BINE: A twining vine stem, as in the hop.

BINOMIAL: The two Latin names for a plant species.

BIOMETRY: The application of statistical methods to biological problems.

BLADE: The part of the leaf above the sheath.

BLIND CULTIVATION: Cultivating with a harrow, weeder, rotary hoe, or other implement to kill weeds before a seeded or planted crop has come up.

BOLL: The subspherical or ovoid fruit of flax or cotton.

BOLT: The formation of an elongated stem or seed stalk, as in the sugar-beet.

BOOT: (1) The upper leaf sheath of a grass. (2) The stage at which the inflorescence expands the boot.

BRACE ROOT: An aerial root that functions to brace the plant, as in corn.

BRACT: A modified leaf subtending a flower or flower branch.

BRAN: The coat of a caryopsis of a cereal removed in milling, consisting of the pericarp, testa, and, usually, the aleurone layer.

BRANCH: A lateral stem.

BRISTLE: A short stiff hair.

BROADCAST: To sow or scatter seed on the surface of the land by hand or by machinery.

BROWN SOILS: A group of soils formed in semiarid climates having a brown surface soil and a zone of carbonate accumulation at a depth, usually of 1 to 2 feet.

BROWSE: (1) The leaves and twigs of woody plants eaten by animals. (2) To bite off and eat leaves and twigs of woody plants.

BUD: An unexpanded flower or a rudimentary leaf, stem, or branch.

BULB: A leaf bud with fleshy scales, usually subterranean.

BUR: A prickly or spiny fruit envelope.

BUSH-AND-BOG: A heavy cut-away disk-tillage implement used on rough or brushy pasture land.

C HORIZON: The layer of weathered parent rock material below the B horizon of the soil but above the unweathered rock.

CALCAREOUS SOIL: An alkaline soil containing sufficient calcium and magnesium carbonate to cause visible effervescence when treated with hydrochloric acid.

CALICHE: A more or less cemented deposit of calcium carbonate often mixed with magnesium carbonate at various depths, characteristic of many of the semiarid and arid soils of the Southwest.

CALLUS: A hard protuberance at the base of the lemma in certain grasses.

CALYX: The outer part of the perianth, composed of sepals.

CAMBIUM: The growing layer of the stem.

CAPILLARY: (1) Hair-like. (2) A very small tube.

CAPITATE: With a globose head.

CAPSULE (POD): Any dry dehiscent fruit composed of more than one carpel.

CARBOHYDRATES: The chief constituents of plants, including sugars, starches, and cellulose, in which the ratio of hydrogen molecules to oxygen molecules is 2:1.

CARBONATE ACCUMULATION ZONE: A visible zone of accumulated calcium and magnesium carbonate at the depth to which moisture usually penetrated in semiarid soils before they were brought under cultivation.

CARINATE: Keeled.

CAROTENE: A yellow pigment in green leaves and other plant parts (the precursor of vitamin A). Beta carotene has the formula $C_{40}H_{56}$.

CARPEL: A simple pistil or one element of a compound pistil.

CARUNCLE: A strophiole.

CARYOPSIS: The grain or fruit of grasses.

CATKIN: A spike with scale-like bracts.

CELL: The unit of structure in plants. A living cell contains protoplasm, which includes a nucleus and cytoplasm within the cell walls.

CELLULOSE: (1) Primary cell-wall substance. (2) A carbohydrate having the general formula $(C_6H_{10}O_5)n$.

CEREAL: A grass cultivated for its edible seeds or grains.

CHAFF: (1) A thin scale, bract, or glume. (2) The glumes and light plant-tissue fragments broken in threshing.

CHALAZA: That part of the ovule where all parts grow together.

CHARTACEOUS: Having the texture of paper.

CHECK ROW: A row of seeds or plants spaced equally in both directions.

CHERNOZEM SOIL: A dark to nearly black grassland soil high in organic matter developed in a subhumid climate.

CHESTNUT SOIL: A soil having a dark-brown surface developed under mixed tall and short grasses in a subhumid to semiarid climate.

CHISEL: A tillage implement with points about a foot apart that stir the soil to a depth of 10 to 18 inches.

CHLOROPHYLL: The green coloring matter of plants that takes part in the process of photosynthesis. It occurs in the chloroplasts of the plant cell.

CHLOROSIS: The yellowing or blanching of leaves and other chlorophyll-bearing plant parts.

CHROMOSOME: That which carries genes (or factors), the units of heredity. Chromosomes are dark-staining bodies visible under the microscope in the nucleus of the cell at the time of cell division.

CILIATE: Fringed with hairs on the margin.

CLAVATE: Club-shaped.

CLAY: Small mineral soil particles less than 0.002 millimeters in diameter (formerly less than 0.005 millimeters).

CLEISTOGAMOUS: Fertilized without opening of the flowers.

CLIMATE: The total long-time characteristic weather of any region.

CLONE: A group of organisms composed of individuals propagated vegetatively from a single original individual.

COLEOPTILE: The sheath covering the first leaf of a grass seedling as it emerges from the soil.

COLEORHIZA: A sheath covering the tip of the first root from a seed.

COLLAR: The area on the outer side of the leaf at the junction of the sheath and blade.

COLLOID: A fine particle usually 10^{-6} to 10^{-4} millimeters in diameter which carries an electric charge. A wet mass of colloidal particles is glue-like in consistency.

COLUMELLA: The axis of a capsule which persists after dehiscence.

COMBINE: A machine for harvesting and threshing in one operation.

COMPANION CROP: A crop sown with another crop, as a small grain with forage crops. Formerly called a nurse crop.

CONSUMPTIVE USE: The use of water in growing a crop, including water used in transpiration and evaporation.

CONTINENTAL CLIMATE: Climate typical of the interior of large land masses having wide extremes of temperature.

CONTOUR FURROWS: Furrows plowed at right angles to the slope, at the same level or grade, to intercept and retain runoff water.

CONVOLUTE: Rolled longitudinally, with one part inside, one outside.

CORDATE: Heart-shaped with point upward.

CORIACEOUS: Leathery.

CORM: The hard, swollen base of the stem.

COROLLA: The inner part of the perianth, composed of petals.

CORRELATION: A relation between two variable quantities such that an increase or decrease of one is associated (in general) with an increase or decrease in the other.

CORRELATION COEFFICIENT (r): The degree of correlation, which ranges from +1 to −1. A correlation coefficient of zero means that the two variables are not interrelated. An r value of +1 or −1 indicates complete association. A positive (plus) correlation means that high values of one variable are associated with high values of the other. A negative (minus) correlation means that as one variable increases the other tends to decrease.

CORYMB: A flat-topped or convex flower cluster in which the pedicels are unequal and the outer flowers blossom earliest.

COTYLEDONS: (1) The first leaves of a plant as found in the embroyo. (2) The major portion of the two halves of a pea or bean (legume) seed.

COVER CROP: A crop grown between orchard trees or on fields between cropping seasons to protect the land from leaching and erosion.

CRENATE: Dentate (toothed) with rounded teeth.

CROSS: (1) A hybrid. (2) To produce a hybrid.

CROSS DRILL: To drill seed in two directions, usually at right angles to each other.

CROSS-FERTILIZATION OR CROSS-POLLINATION: Fertilization secured by pollen from another plant.

CROWN: The base of the stem where roots arise.

CULM: The jointed stem of grasses.

CULTIVAR: See variety.

CURE: To prepare for preservation by drying or other processes.

CUTICLE: The outer corky or waxy covering of the plant.

CUTIN: A complex fatty or waxy substance in cell walls, particularly in the epidermal layers.

CUTTING: A part of a plant to be rooted for vegetative propagation.

CYME: A somewhat flat-topped determinate inflorescence.

CYTOLOGY: The study of the structure, function, and life history of the cell.

CYTOPLASM: The contents of a cell outside the nucleus.

CYTOPLASMIC MALE STERILITY: Male sterility caused by the cytoplasm rather than by nuclear genes; transmitted only through the female parent.

DECIDUOUS: Plants or trees that shed leaves or awns at a particular season or stage.

DECUMBENT: Curved upward from a horizontal or inclined position.

DEFLOCCULATE: To separate or break down soil aggregates of clay into their individual particles.

DEHISCENCE: The opening of valves or anthers, or separation of parts of plants.

DENITRIFICATION: The reduction of nitrates to nitrites, ammonia, and free nitrogen in the soil.

DENTATE: Toothed.

DETASSEL: To remove the tassel.

DETERMINATE INFLORESCENCE: Flowers that arise from the terminal bud and check the growth of the axis.

DIADELPHOUS (STAMENS): Collected in two sets.

DICOTYLEDONOUS PLANTS: Plants producing two cotyledons in each fruit.

DIFFERENTIATION: The process whereby cells and tissues become structurally unlike during the process of growth and development.

DIGITATE: Fingered; compound with parts radiating from the apex of support.

DIOECIOUS: Having stamens and pistils in separate flowers upon different plants.

DIPLOID: Having two sets of chromosomes. Body tissues of plants are ordinarily diploid.

DISARTICULATING: Separating at maturity.

DOMINANT: Possessing a character which is manifested in the hybrid to the apparent exclusion of the contrasted character from the other (the recessive) parent.

DORMANCY: An internal condition of a seed or bud that prevents its prompt gemination or sprouting under normal growth conditions.

DORSAL: Relating to the back or outer surface.

DOUBLE CROSS: The result of mating two single crosses, each of which had been produced by crossing two distinct inbred lines.

DRILL: (1) A machine for sowing seeds in furrows. (2) To sow in furrows.

DRILL ROW: A row of seeds or plants sown with a drill.

DUCK-FOOT CULTIVATOR: A field cultivator equipped with small sweep shovels.

ECHINATE: Armed with prickles.

ECOLOGY: The study of the mutual relations between organisms and their environment.

ECOTYPE: A variety or strain adapted to a particular environment.

EDAPHOLOGY: Soil science, particularly the influence of soil upon vegetation.

EGG: The female reproductive cell.

ELLUVIATION: Removal of material by solution or suspension.

EMARGINATE: Notched at the end.

EMBRYO: The rudimentary plantlet within a seed; the germ.

EMBRYO SAC: The sac in the embryo containing the egg cell.

EMERGENCE: Coming out of a place, as a seedling from the soil or a flower from a bud.

ENDEMIC: Indigenous or native to a restricted locality.

ENDOCARP: Inner layer of pericarp.

ENDOSPERM: The starchy interior of a grain.

ENSILAGE: Silage.

ENSILE: To make into silage.

EPICARP (EXOCARP): The outer layer of pericarp formed from the ovary wall.

EPICOTYL: The stem of the embryo or young seedling above the cotyledons.

EPIDERMIS: External layer of cells.

EPIGYNOUS FLOWER: Flower with a corolla that seems to rise from the top of the ovary.

EROSION: The wearing away of the land surface by water, wind, or other forces.

ETHER EXTRACTS: Fats, oils, waxes, and similar products extracted with warm ether in chemical analysis.

EXOTIC PLANT: An introduced plant not fully naturalized or acclimated.

EXSERTED: Protruding beyond a covering.

F_1: The first filial generation; the first-generation offspring of a given mating.

F_2: The second filial generation; the first hybrid generation in which segregation occurs.

FACTOR (GENETIC OR HEREDITARY): The gene or unit of heredity.

FALLOW: Cropland left idle, usually for one growing season, while the soil is being cultivated to control weeds and conserve moisture.

FASICLE: A compact bundle or cluster of flowers or leaves.

FERTILE PLANT: A plant capable of producing fruit.

FERTILITY (PLANT): The ability to reproduce sexually.

FERTILITY (SOIL): The ability to provide the proper compounds in the proper amounts and in the proper balance for the growth of specified plants under the suitable environment.

FERTILIZATION (PLANT): The union of the male (pollen) nucleus with the female (egg) cell.

FERTILIZATION (SOIL): The application to the soil of elements or compounds that aid in the nutrition of plants.

FIBROUS ROOT: A slender thread-like root, as in grasses.

FILAMENT: The stalk of the stamen which bears the anther.

FILIFORM: Thread-shaped.

FLESHY ROOT: A thickened root containing abundant food reserves, e.g., carrot, sweetpotato, and sugarbeet.

FLOAT: (1) A land leveller. (2) A plank clod masher.

FLOCCULATE: To aggregate individual particles into small groups or granules.

FLORET: Lemma and palea with included flower (stamens, pistil, and lodicules).

FODDER: Maize, sorghum, or other coarse grasses harvested whole and cured in an erect position.

FOLIATE: Leaved.

FORAGE: Vegetable matter, fresh or preserved, gathered and fed to animals.

FRIABLE: Easily crumbled in the fingers; nonplastic.

FRUIT: The ripened pistil.

FUNGICIDE: A chemical substance used as a spray, dust, or disinfectant to kill fungi infesting plants or seeds.

FUNGUS: A group of the lower plants that causes most plant diseases. The group includes the molds and belongs to the division *Thallophyta*. Fungi reproduce by spores instead of seeds, contain no chlorophyll, and thus live on dead or living organic matter.

FUSIFORM: Feather-shaped; enlarged in the middle and tapering at both ends.

GENE: The unit of inheritance, which is transmitted in the germ cells.

GENETICS: The science of heredity, variation, sex determination, and related phenomena.

GENICULATE: Bent abruptly or kneelike, as the awn of wild oats.

GENOTYPE: The hereditary make up of characteristics of a plant or a pure line or variety.

GEOTROPIC: Turning downward in response to a stimulus caused by the force of gravity, e.g., the roots of a seedling growing downward.

GERM CELL: A cell capable of reproduction or of sharing in reproduction.

GLABROUS: Smooth.

GLANDULAR: Containing or bearing glands or gland cells.

GLAUCOUS: Covered with a whitish, waxy bloom, as on the sorghum stem.

GLUMES: The pair of bracts at the base of a spikelet.

GLUTEN: The protein in wheat flour which enables the dough to rise.

GLUTINOUS: Glue-like.

GOPHER: (1) A crop-damaging burrowing rodent. (2) A type of plow.

GRAIN: (1) A caryopsis. (2) A collective term for the cereals. (3) Cereal seeds in bulk.

GRASS: A plant of the family *Gramineae*.

GREEN MANURE: Any crop or plant grown and plowed under to improve the soil, especially by addition of organic matter.

GULLY EROSION: Erosion which produces channels in the soil.

GYNOPHORE: A pedicel bearing the ovary, e.g., the peg which penetrates the soil and bears the peanut pod.

HABIT: Aspect or manner of growth.

HALOPHYTE: A plant tolerant of soils high in soluble salts.

HAPLOID: Single; containing a reduced number of chromosomes in the mature germ cells of bisexual organisms.

HARDPAN: A hardened or cemented soil horizon.

HAULM: A stem.

HAY: The herbage of grasses or comparatively fine-stemmed plants cut and cured for forage.

HEAD: A dense roundish cluster of sessile or nearly sessile flowers on a very short axis or receptable, as in red clover and sunflowers.

HELIOTROPIC: Turning towards the sun.

HERB: A plant that contains but little wood and that goes down to the ground each year.

HERBACEOUS: Having the characteristics of an herb.

HERBAGE: Herbs, collectively, especially the aerial portion.

HERBICIDE: A weed killer or any chemical substance used to kill herbaceous plants.

HERMAPHRODITE: Perfect, having both stamens and pistils in the same flower.

HETEROSIS: Increased vigor, growth, or fruitfulness that results from crosses between genetically different plants, lines, or varieties.

HETEROZYGOUS: Containing two unlike genes of an allelomorphic pair in the corresponding loci (positions) of a pair of chromosomes. The progeny of a heterozygous plant does not breed true.

HEXAPLOID: A plant or crop that

carries three genomes (sets of paired haploid chromosomes).

HILUM: The scar of the seed; its place of attachment.

HIRSUTE: Hairy.

HOMOGAMY: The maturing of anthers and stigmas at the same time.

HOMONYM: The same name for two different plants.

HOMOZYGOUS: Containing like germ cells. Homozygous plants produce like progeny for the character under observation.

HORIZON, SOIL: A layer of soil approximately parallel to the land surface with more or less well-defined characteristics. See also A horizon, B horizon, and C horizon.

HORMONE: A chemical growth-regulating substance that can be or is produced by a living organism.

HULL: (1) A glume, lemma, palea, pod, or other organ enclosing a seed or fruit. (2) To remove hulls from a seed.

HUMID CLIMATE: A climate with sufficient precipitation to usually support forest vegetation. In the agricultural sections of the United States the precipitation in humid regions usually exceeds 30 to 40 inches annually.

HUMUS: The well decomposed, more or less stable part of the organic matter of the soil.

HUSK: (1) The coarse outer envelope of a fruit, as the glumes of an ear of maize. (2) To remove the husks.

HYALINE: Thin and translucent or transparent.

HYBRID: The offspring of two parents unlike in one or more heritable characters.

HYBRIDIZATION: The process of crossing organisms of unlike heredity.

HYDROPHYTE: A plant adapted to wet or submerged conditions.

HYDROPONICS: The growing of plants in aqueous chemical solutions.

HYPOCOTYL: The stem of the embryo or young seedling below the cotyledons.

HYPOGYNOUS: With parts under the pistil.

IMPERFECT FLOWER: Flower lacking either stamens or pistils.

INBRED: Resulting from successive self-fertilization.

INDEHISCENT: Not opening by valves or slits.

INDETERMINATE INFLORESCENCE: An inflorescence in which the flowers arise laterally and successively as the floral axis elongates.

INDURATE: Hard.

INFLORESCENCE: The flowering part of the plant.

INSECTICIDE: A chemical used to kill insects.

INTEGUMENTS: Coats or walls of an ovule.

INTERNODE: The part of the stem or branch between two nodes.

INTERTILLED CROP: A crop planted in rows, followed by cultivation between the rows.

INVOLUCRE: A circle of bracts below a flower or flower cluster.

ION: An electrically charged element, group of elements, or particle.

JOINT: (1) A node. (2) The internode of an articulate rachis. (3) To develop distinct nodes and internodes in a grass culm.

KEEL: (1) A ridge on a plant part resembling the keel of a boat, as the glume of durum wheat. (2) The pair of united petals in a legume flower.

KERNEL: The matured body of an ovule.

LAMINA: The blade of a leaf, particularly those portions between the veins.

LAMINATED: In layers or plates.

LANCEOLATE: Lance-shaped, as a grass leaf.

LAND PLANE: A large, wheeled machine used to level the land.

LATERALLY COMPRESSED: Flattened from the sides, as the spike of emmer.

LAX: Loose.

LEACH: To remove materials by solution.

LEAF: The lateral organ of a stem.

LEGUME: (1) Any plant of the family *Leguminoseae.* (2) The pod of a leguminous plant.

LEMMA: The outer (lower) bract of a grass spikelet enclosing the caryopsis.

LETHAL SUBSTANCE: A substance or hereditary factor causing death.

LEY: See pasture, rotation.

LIGNIN: A constituent of the woody portion of the fibrovascular bundles in plant tissues, made up of modified phenyl propane units.

LIGULE: (1) A membranous projection on the inner side of a leaf at the top of the sheath of wheat and many other grasses. (2) A strap-shaped corolla as in the ray flower of the sunflower.

LIME: Calcium oxide or quicklime (CaO); often, also, calcium carbonate ($CaCO_3$), or calcium hydroxide, hydrated or slaked lime ($Ca[OH]_2$).

LIMESTONE: Rock composed essentially of calcium carbonate.

LINEAR: Long and narrow, with nearly parallel margins.

LISTER: An implement for furrowing land, often having a planting attachment.

LOAM: A soil composed of a mixture of clay, silt, and less than 52 per cent sand.

LOBE: A rounded protion or segment of any organ.

LODICULES: The pair of organs at the base of the ovary of a grass floret that swell and force open the lemma and palea during anthesis.

LOESS: Geological deposit of relatively uniform fine material, mostly silt, presumably transported by wind.

LONGEVITY: Length of life, usually of seeds or plants of longer than average life.

MARL: A crumbling deposit of calcium carbonate mixed with clay or other impurities.

MASLIN (MESLIN): Grains grown in mixture or the milled product thereof.

MEADOW: An area covered with fine-stemmed forage plants, wholly or mainly perennial, and used to produce hay.

MEIOSIS: The division of the sexual cells in which the number of chromosomes is halved.

MELLOW SOIL: A soil that is easily worked or penetrated.

MEMBRANE: A thin soft tissue.

MESOCARP: Middle layer of pericarp.

MESOCOTYL: The subcrown internode of a grass seedling.

MESOPHYLL: Tissues between two epidermal layers, as in the interior of a leaf.

MESOPHYTE: A plant that thrives under medium conditions of moisture and salt content of the soil.

METABOLISM: The life processes of plants.

METAXENIA: Direct effect of the pollen on the parts of a seed or fruit other than the embryo and endosperm.

MICROCLIMATE: (1) A local climatic condition that differs from surrounding areas because of differences in relief, exposure, or cover. (2) Experimental, controlled plant-growth chambers.

MICRONUTRIENT: A mineral nutrient element that plants need only in trace or minute amounts.

MICROPYLE: Opening through which the pollen tube passes.

MIDDLEBUSTER: A double shovel plow or lister.

MITOSIS: Cell division involving the formation and longitudinal splitting of the chromosomes.

MOISTURE TENSION: The force at which water is held by the soil. One atmosphere equivalent tension equals a unit column of water of 1,000 centimeters.

MONADELPHOUS: Uniting stamens into one set.

MONOCOTYLEDON: A plant having one cotyledon, as in the grasses.

MONOECIOUS: Having stamens and

pistils in separate flowers on the same plant.

MORPHOLOGICAL: Referring to structure or texture.

MOW (mō): To cut with a mower or scythe.

MOW (mou): (1) A place for indoor hay storage. (2) To place hay in a mow.

MUCK: Fairly well decomposed organic soil material relatively high in mineral content, dark in color, and accumulated under conditions of imperfect drainage.

MULCH: A layer of plant residues or other materials on the surface of the soil. Formerly, a loose layer of cultivated surface soil.

MUTATION: A sudden variation that is later passed on through inheritance.

NATURALIZED PLANTS: Introduced species that have become established in a region.

NECROSIS: Discoloration, dehydration, and death of plant parts.

NECTARY: Any gland or organ that secretes sugar.

NEMATOCIDE: A substance that can be used to kill nematodes.

NERVE: A vein on a leaf, glume, or lemma.

NEUTRAL SOIL: A soil neither acid nor alkaline, with a *p*H of about 7.0 or between 6.6 and 7.3.

NITRIFICATION: Formation of nitrates from ammonia.

NITROGEN FIXATION: The conversion of atmospheric (free) nitrogen to nitrogen compounds brought about chemically, or by soil organisms, or by organisms living in the roots of legumes.

NITROGEN-FREE EXTRACT: The unanalyzed substance of a plant (consisting largely of carbohydrates) remaining after the protein, ash, crude fiber, ether extract, and moisture have been determined.

NODE: The joint of a culm where a leaf is attached.

NODULE: A tubercle formed on legume roots by symbiotic nitrogen-fixing bacteria of the genus *Rhizobium.*

NUCELLUS: The ovule tissue within the integuments.

NUCLEOLUS: A small spherical body within a cell nucleus.

NUCLEUS: A body of specialized protoplasm containing the chromosomes within a cell.

NURSE CROP: See companion crop.

NUTRIENT: A chemical element taken into a plant that is essential to the growth, development, or reproduction of the plant.

OBOVATE: Egg-shaped with larger end above.

OBTUSE: Rounded at the apex.

OCEANIC CLIMATE: A climate modified by the tempering effect of ocean water.

ONE-WAY: A tillage implement having a gang of disks that throw the soil in one direction.

ORGANIC FARMING: Growing crops without applying pesticides and mineral fertilizers in an inorganic form.

OSMOSIS: Diffusion of substances through a cell wall or other membrane.

OUTCROSS: A cross to an individual not closely related.

OVARY: Seed case of the pistil.

OVULES: Unripe seeds in the ovary.

OXIDATION: A chemical change involving addition of oxygen or its chemical equivalent, or involving an increase in positive or decrease in negative valence. Burning is oxidation.

PALEA (PALET): Inner (upper) bract of a floret lying next to the caryopsis in grasses. It usually is thin and papery.

PALISADE CELLS: Elongated cells perpendicular to the epidermis on the upper side of a leaf.

PALMATE: Radiately lobed or divided.

PANICLE: An inflorescence with a main axis and subdivided branches, as in oats and sorghum.

PAPILIONACEOUS: Butterfly-shaped, as the flower of legumes.

PAPPUS: The teeth, bristles, awns, and so on, surmounting the achene of the sunflower and other *Compositae.*

PARASITIC: Living in or on another living organism.

PARENCHYMA: Soft cellular tissue.

PARTHENOGENESIS: The development of a new individual from a germ cell without fertilization.

PASTURE: An area of land covered with grass or other herbaceous forage plants, used for grazing animals.

- **NATIVE PASTURE:** A pasture covered with native plants or naturalized exotic plants.
- **PERMANENT PASTURE:** A pasture of perennial or self-seeding annual plants kept for grazing indefinitely.
- **ROTATION PASTURE (LEY):** A pasture used for a few seasons and then plowed for other crops.
- **TAME PASTURE:** A pasture covered with cultivated plants and used for grazing.
- **TEMPORARY PASTURE:** A pasture grazed during a short period only, not more than one crop season.

PASTURE RENOVATION: Improvement of a pasture by tillage, seeding, fertilization, and, sometimes, liming.

PASTURE SUCCESSION: A series of crops for grazing in succession.

PEAT: Slightly decomposed organic matter accumulated under conditions of excessive moisture.

PEDALFER: The soil of a humid region in which iron and aluminum have leached into the B horizon, but which has no carbonate zone.

PEDICEL: (1) A branch of an inflorescence supporting one or more flowers. (2) The stalk of a spikelet.

PEDOCAL: A soil of a semiarid or arid region having a distinct carbonate zone.

PEDOLOGY: Soil science.

PEDUNCLE: The top section of the stalk that supports a head or panicle.

PEPO: The fruit of cucurbitaceous plants, as the pumpkin, squash, and gourd.

PERENNIAL: Living more than one year but, in some cases, producing seed the first year.

PERFECT FLOWER: A flower having both pistil and stamens.

PERIANTH: The floral envelope including the calyx or calyx and corolla.

PERICARP: The modified and matured ovary wall, as the bran layers of a grain.

PERIGYNOUS: Located around the pistil.

PERISPERM: Nucellus.

PETAL: A division of the corolla.

PETIOLE: The stalk of a leaf.

***p*H:** The designation for degree of acidity or hydrogen-ion activity. pH value = logarithm 1/CH, in which CH is the concentration of active hydrogen ions expressed in gram ions per liter of a solution.

PHENOTYPE: The organism as exemplified by its expressed characters but not necessarily as all its progeny will appear.

PHLOEM: Portion of a vascular bundle containing the sieve tubes through which are transported the food materials manufactured in the plant leaves.

PHOTOPERIODISM: The response of plants to different day lengths or light periods.

PHOTOTHERMAL INDUCTION: The initiation of flowering in a plant resulting from the combined effect of light and temperature.

PHOTOTROPISM: The growing or turning toward the light.

PHYLLOTAXY: The arrangement of leaves upon the stem.

PHYTOBEZOAR: A ball in the intestinal tract of an animal, formed from plant hairs.

PHYTOTOXIC: Injurious to plant life or life processes.

PICK-UP: An attachment for a combine or other implement to gather cut crops from a windrow and convey them to the machine.

PILOSE: Covered with soft slender hairs.

PINNATE LEAF: A compound leaf with leaflets arranged on each side of a common petiole, as in the pea leaf.

PISTIL: The seed-bearing organ of a flower consisting of the ovary, style, and stigma.

PISTILLATE: Provided with pistils but without stamens.

PLACENTA: The part of the ovary to which the ovules are attached, as in the pea pod.

PLANT: (1) Any organism belonging to the plant or vegetable kingdom. (2) To set plants or sow seeds.

PLANTER: A machine for opening the soil and dropping tubers, cuttings, seedlings, or seeds at intervals.

PLASTID: A protoplasmic body in a cell which may or may not contain chlorophyll.

PLUMOSE: Feather-like.

POD: Any dry dehiscent fruit.

PODZOL SOIL: A soil having an organic mat and a very thin organic-mineral layer above a gray leached layer, formed under a forest in a cool humid climate.

POLLEN: The male germ cells produced in the anthers.

POLLINATION: The transfer of pollen from the anther to the stigma.

POLYADELPHOUS (STAMENS): Separate, or in more than two groups.

POLYCROSS: A selection, clone, or line naturally out-crossed through random pollination by all other strains in the same isolated block.

POLYPLOID: A plant having three or more basic sets of chromosomes.

PRAIRIE SOIL: A soil having a very dark-brown or grayish-brown surface horizon ranging from brown to lighter-colored material, formed under tall grasses in a temperate climate under relatively humid conditions and without a carbonate zone below.

PRIMARY ROOT: (1) A seminal root. (2) A main root.

PROCAMBIUM: Fibrovascular tissue formed before the differentiation into xylem and phloem.

PRODUCTIVITY (OF SOIL): The capability of a soil to produce a specified plant or sequence of plants under a specified system of management.

PROFILE (OF SOIL): A vertical section of the soil through all its horizons and extending into the parental material.

PROTANDROUS: Having anthers that shed their pollen before the stigmas are receptive.

PROTEIN: Nitrogenous compounds formed in plants, composed of 50 to 55 per cent carbon, 19 to 24 per cent oxygen, 6.5 to 7.3 per cent hydrogen, 15 to 17.6 per cent nitrogen, usually 0.5 to 2.2 per cent sulfur, and 0.4 to 0.9 per cent phosphorus. Carbohydrates, the chief constituents of carbonaceous feeds, contain only 40 to 44 per cent carbon.

PROTOGYNOUS: Having pistils ready for fertilization before anthers are matured.

PROTOPLASM: The contents of a living cell.

PUBERULENT: Minutely pubescent.

PUBESCENT: Covered with fine, soft, short hairs.

PULSE: Leguminous plants or their seeds, chiefly those plants with large seeds used for food.

PULVINUS: The swelling at the base of the branches of some panicles which causes them to spread.

PUNCTATE: Dotted.

PURE LINE: A strain of organisms that is genetically pure (homozygous) because of continued inbreeding, self-fertilization, or other means.

RACE: A group of individuals having certain characteristics in common because of common ancestry —generally a subdivision of a species.

RACEME: An inflorescence in which the pediceled flowers are arranged on a rachis or axis.

RACHILLA (LITTLE RACHIS): The axis of a spikelet in grasses.

RACHIS: The axis of a spike or raceme.

RADICLE: That part of the seed which upon vegetating becomes the root.

RANGE: An extensive area of natural pastureland. If unfenced, it is an open range.

RAY: The branch of an umbel; marginal ligulate flowers of a composite head.

REACTION (OF SOIL): The degree of acidity or alkalinity of the soil expressed as *p*H.

RECEPTACLE: The swollen summit of flower stalk.

RECESSIVE: Possessing a transmissible character which does not appear in the first-generation hybrid because it is masked due to the presence of the dominant character coming from the other parent.

RECIPROCAL CROSS: A cross between the same two strains but with the pollen and pistillate parents reversed.

REPLICATION: Multiple repetition of an experiment.

RESPIRATION: The process of absorption of oxygen and giving out of carbon dioxide.

RETICULATE: In a network.

RHIZOME: A subterranean stem, usually rooting at the nodes and rising at the apex; a rootstock.

RILL EROSION: Erosion producing small channels that can be obliterated by tillage.

ROGUE: (1) A variation from the type of a variety or standard, usually inferior. (2) To eliminate such inferior individuals.

ROOT: The part of the plant (usually subterranean) which lacks nodes.

ROOT CAP: A mass of cells protecting the tip of a root.

ROOT HAIR: A single-celled protrusion of an epidermal cell of a young root.

ROOTLET: A small root.

ROOTSTOCK: A rhizome.

ROSETTE: A cluster of spreading or radiating basal leaves.

RUDIMENTARY: Underdeveloped.

RUGOSE: Wrinkled; rough with wrinkles.

RUNNER: A creeping branch or stolon.

RUSH: A plant of the family *Juncaceae*.

SAC: A pouch.

SAGITTATE: Shaped like an arrowhead.

SALINE SOIL: A soil containing an excess of soluble salts, more than approximately 0.2 per cent, but with a *p*H of less than 8.5.

SALT: The product, other than water, of the reaction of a base with an acid.

SAND: Small rock or mineral fragments having diameters ranging from 0.05 to 2.0 millimeters.

SAPROPHYTIC: Living on dead organic material.

SCABROUS: Having a rough surface.

SCALE: Any thin appendage; morphologically, a modified degenerate leaf.

SCLERENCHYMA: Lignified tissue in referring to thick-walled fibers.

SCRAPER: (1) A machine for beating the spikelets from broomcorn fibers. (2) A blade for removing soil from revolving disks or wheels on field implements.

SCUTELLUM: A shield-shaped organ surrounding the embryo of a grass seed, morphologically a cotyledon.

SECOND BOTTOM: The first terrace level of a stream valley above the flood plain.

SECONDARY ROOT: A branch or division of a main root.

SEDGE: A plant of the family *Cyperaceae*, especially of the genus *Carex*.

SEED: (1) The ripened ovule enclosing a rudimentary plant and the food necessary for its germination. (2) To produce seed. (3) To sow.

SEEDLING: (1) The juvenile stage of a plant grown from seed. (2) A plant derived from seed (in plant breeding).

SEGREGATION: Separation of hybrid

progenies into the different hereditary types representing the combination of characters of the two parents.

SELECTION: The choosing of plants having certain characteristics for propagation. Natural selection occurs under competitive conditions or adverse environment.

SELFED (SELF-POLLINATED): Pollinated by pollen from the same plant.

SEMIARID CLIMATE: A climate which, in the United States, usually has an annual precipitation of between 10 and 20 inches or more. Prevailing plants in a semiarid region are short grass, bunchgrass, or shrubs.

SEMINAL: Belonging to the seed.

SEMINAL ROOT: A root arising from the base of the hypocotyl.

SEPAL: A division of the calyx.

SEPTUM: A partition or dividing wall.

SERRATE: Having sharp teeth pointing forward.

SESSILE: Without a pedicel or stalk.

SETACEOUS: Bristle-like.

SHEATH (BOOT): The lower part of the leaf that encloses the stem.

SHEET EROSION: Erosion by removal of a more or less uniform layer of material from the land surface.

SHOCK: (1) An assemblage or pile of crop sheaves or cut stalks set together in the field to dry. (2) To set into shocks.

SHOOT: (1) A stem with its attached members. (2) To produce shoots. (3) To put forth.

SIBLINGS (SIBS): Offspring of the same parental plants.

SILAGE: Forage preserved in a succulent condition by partial fermentation in a tight container.

SILO: A tight-walled structure for making and preserving silage.

SILT: Small mineral soil particles of a diameter of 0.002 to 0.05 millimeters.

SIMPLE: (1) In botany—without subdivisions, as opposed to compound; e.g., a simple leaf. (2) In heredity—inheritance of a character controlled by not more than three factor-pair differences.

SINGLE CROSS: The first generation hybrid between two inbred lines.

SINUOUS: Wavy.

SLIP: A cutting, shoot, or leaf to be rooted for vegetative propagation.

SOD: (1) Turf. (2) Plowed meadow or pasture.

SOIL: The natural medium for the growth of land plaats on the surface of the earth, composed of organic and mineral materials.

SOLUM: The upper part of the soil profile; the A and B horizons.

SOW: To place seeds in a position for growing.

SPATULATE: Shaped like the tip of a spatula.

SPECIFIC COMBINING ABILITY: The response in vigor, growth, or fruitfulness from crossing with other particular inbred lines.

SPICATE: Arranged in a spike.

SPIKE: An unbranched inflorescence in which the spikelets are sessile on the rachis, as in wheat and barley.

SPIKELET: The unit of inflorescence in grasses, consisting of two glumes and one or more florets.

SPORE: Single-celled reproductive bodies produced by fungi.

SPORT: (1) abrupt deviation from type of a plant.

SPROUT: (1) A young shoot. (2) To produce sprouts. (3) To put forth sprouts from seeds. (4) To remove sprouts, as from potato tubers.

SQUARE: An unopened flower bud of cotton with its subtending involucre bracts.

STALK: A stem.

STAMEN: The pollen-bearing organ of a flower.

STAMINATE: Having stamens but no pistils.

STAND: The density of plant population per unit area.

STANDARD: (See banner.)

STELLATE: Star-shaped.

STERILE: Incapable of sexual reproduction.

STIGMA: The part of the pistil that receives the pollen.

STIPULE: One of two leaf-like appendages arranged in a pair at the base of the petiole.

STOCK: A supply of seed of a crop or variety.

STOLON: A modified propagating, creeping stem above ground that produces roots.

STOLONIFEROUS: Bearing stolons.

STOMA: An opening in the epidermis for the passage of gases and water vapor.

STOMATA: Plural of stoma.

STOOL: The aggregate of a stem and its attached tillers, i.e., a clump of young stems arising from a single plant.

STRAIN: A group of plants derived from a variety.

STRAW: The dried remnants of fine-stemmed plants from which the seed has been removed.

STRIATE: Marked with parallel lines or ridges.

STRIP CROPPING: The growing of dense crops alternating with intertilled crops or fallow in long narrow strips across a slope approximately on a line of contour or across the direction of prevailing winds.

STROBILE: An inflorescence or fruit with conspicuous imbricated bracts, as in the hop.

STROPHIOLE (CARUNCLE): A swollen appendage near the micropyle of a seed.

STRUCTURE: The morphological aggregates in which the individual soil particles are arranged.

STUBBLE: The basal portion of the stems of plants left standing after cutting.

STYLE: The slender part of the pistil supporting the stigma.

SUBERIZE: To form a protective corky layer on a cut surface of plant tissue.

SUBHUMID CLIMATE: A climate with sufficient precipitation to support a moderate to dense growth of tall and short grasses but usually unable to support a dense deciduous forest. In the United States this usually means a precipitation of 20 to 30 inches or more.

SUBSOIL: That part of the solum below plow depth or below the A horizon.

SUBSPECIES: A taxonomic rank immediately below that of a species.

SUCCULENT: Juicy, fleshy.

SUCKER: (1) A tiller. A shoot produced from a crown or rhizome, or, in tobacco, from axillary buds. (2) To produce suckers. (3) To remove suckers.

SURFACE SOIL: The upper 5 to 8 inches of the soil, or, in arable soils, the depth commonly stirred by the plow.

SUTURE: The line of junction between contiguous parts.

SWARD: The grassy surface of a pasture.

SWATH: A strip of cut herbage lying on the stubble.

SWEAT: To emit moisture as does damp hay or grain, usually with some heating taking place at the same time.

SWEEP: A double-bladed V-shaped knife on a cultivating implement.

SYMBIOTIC NITROGEN FIXATION: The fixation of nitrogen by bacteria infesting the roots of legumes while benefiting the legume crop.

SYNONYM: A different name for the same species or variety.

TAPROOT: A single central root.

TASSEL: (1) The staminate inflorescence of maize composed of panicled spikes. (2) To produce tassels.

TAXONOMY: The science of classification.

TEDDER: An implement for stirring hay in the swath or windrow.

TENDRIL: A leaflet or stem modified for climbing or anchorage, as in the pea.

TERETE: Cylindrical and slender, as in grass culms.

TERMINAL: Situated at or forming

the extremity or upper bud, flower, or leaf.

TERRACE: A channel or embankment across a slope approximately on a contour to intercept runoff water.

TESTA: The seed coat.

TETRAPLOID: Having four times the primary chromosome number.

TILL: To plow or cultivate soil.

TILLER: (1) An erect shoot arising from the crown of a grass. (2) To produce tillers.

TILTH: The physical condition of the soil with respect to its fitness for the planting or growth of a crop.

TOPSOIL: The surface soil, usually the plow depth or the A horizon.

TRANSPIRATION: The evaporation of moisture through the leaves.

TRICHOME: A plant hair.

TRUNCATE: Ending abruptly as if cut off across the top.

TUBER: A short thickened subterranean branch.

TURF: The upper stratum of soil filled with the roots and stems of low-growing living plants, especially grasses.

TURGID: Distended with water or swollen.

UMBEL: An indeterminate type of inflorescence in which the peduncles or pedicles of a cluster seem to rise from the same point, as in a carrot-flower cluster.

UNDULATE: Having a wavy surface.

UNISEXUAL: Containing either stamens or pistils, but not both.

UNIT CHARACTER: A hereditary trait that is transmitted by a single gene.

VALVE: One of the parts of a dehiscent fruit or a piece into which a capsule splits.

VARIATION: The occurrence of differences among individuals of a species or variety.

VARIETY (CULTIVAR): A group of individuals within a species that differ from the rest of the species.

VEIN: A bundle of threads of fibrovascular tissue in a leaf or other organ.

VENATION: The arrangement of veins.

VENTRAL: On the lower or front side.

VERRUCOSE: Warty.

VERTICILLATE: Whorled.

VILLOUS: Bearing long, soft straight hairs.

VIRUS: Ultramicroscopic protein bodies, the presence of certain types of which cause mosaic and other diseases in plant tissues. Viruses reproduce in plant tissues.

VISCID: Sticky.

WEED: A plant that in its location is more harmful than beneficial.

WHORL: An arrangement of organs in a circle around the stem.

WINDROW: (1) Curing herbage dropped or raked into a row. (2) To cut or rake into windrows.

WING: The lateral petal of the papilionaceous flower of a legume.

WINTER ANNUAL: A plant that germinates in the fall and blooms in the following spring or summer.

XEROPHYTE: A plant adapted to arid conditions.

XYLEM: The woody part of a fibrovascular bundle containing vessels, the water-conducting tissue.

ZYGOTE: Product of united gametes.

Index

Index

Boldface numbers refer to illustrations.